ISLPED'06

AA001024

Proceedings of the

2006 International Symposium on Low Power Electronics and Design

Tegernsee, Germany • October 4-6, 2006

Sponsored by

ACM SIGDA and IEEE Circuits and Systems Society

with technical support from

the IEEE Solid-State Circuits Society

and

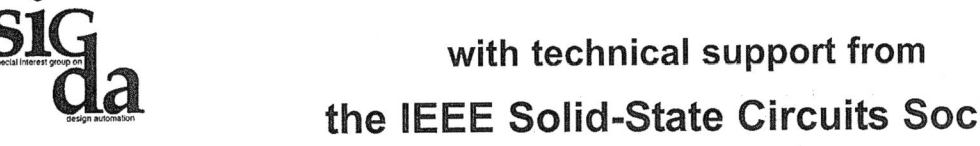

the IEEE Electron Devices Society

The Association for Computing Machinery
1515 Broadway
New York, New York 10036

Copyright © 2006 by the Association for Computing Machinery, Inc. (ACM). Permission to make digital or hard copies of portions of this work for personal or classroom use is granted without fee provided that copies are not made or distributed for profit or commercial advantage and that copies bear this notice and the full citation on the first page. Copyright for components of this work owned by others than ACM must be honored. Abstracting with credit is permitted. To copy otherwise, to republish, to post on servers or to redistribute to lists, requires prior specific permission and/or a fee. Request permission to republish from: Publications Dept., ACM, Inc. Fax +1 (212) 869-0481 or <permissions@acm.org>.

For other copying of articles that carry a code at the bottom of the first or last page, copying is permitted provided that the per-copy fee indicated in the code is paid through the Copyright Clearance Center, 222 Rosewood Drive, Danvers, MA 01923.

Notice to Past Authors of ACM-Published Articles

ACM intends to create a complete electronic archive of all articles and/or other material previously published by ACM. If you have written a work that has been previously published by ACM in any journal or conference proceedings prior to 1978, or any SIG Newsletter at any time, and you do NOT want this work to appear in the ACM Digital Library, please inform permissions@acm.org, stating the title of the work, the author(s), and where and when published.

ISBN: 1-59593-462-6

Additional copies may be ordered prepaid from:

ACM Order Department
PO Box 11405
New York, NY 10286-1405

Phone: 1-800-342-6626
(US and Canada)
+1-212-626-0500
(all other countries)
Fax: +1-212-944-1318
E-mail: acmhelp@acm.org

ACM Order Number 477064
IEEE Catalog No. 06TH8925
Printed in the USA

2006 International Symposium on Low Power Electronics and Design

Tegernsee, Germany
4-6 October 2006

IEEE Catalog Number: CFP06LOW-POD
ISBN: 978-1-59593-462-8

Message from the Chairs

Welcome to the 11th ACM/IEEE International Symposium on Low Power Electronics and Design!

The 2006 edition of the ACM/IEEE International Symposium on Low Power Electronics and Design (ISLPED) welcomes you to the beautiful, on-the-lake setting of the Tegernsee area in Germany, within a one hour drive from Munich. On a par with this great location, a strong and diverse technical program features two keynote talks from leaders in the field, a technical panel, four embedded tutorials, and, new for this year, three half-day tutorials scheduled for the last day of the symposium.

Dr. Christoph Kutter, Senior VP Products and Systems, Infineon Technologies, will give a keynote on design challenges for mobile communication devices. The Thursday keynote is given by Barry Dennington, Senior Vice President, NXP Semiconductors/CTO/SoC Design Technology, who will present an industrial view on the demands of a convergent communications, consumer and automotive market, the technology challenges a semiconductor company faces and the solutions that lead to actual products with a few examples. The Friday panel is titled "Flexibility and Low Power; a Contradiction in Terms?". Moderated by Peter Wintermayr, Editor in Chief of Markt & Technik, panelists Reiner Hartenstein, Univ. Kaiserslautern; Heinrich Meyr, RWTH Aachen Univ. and Chief Scientific Officer CoWare; and Steve Leibson, Tensilica, will discuss whether configurable or re-configurable computing offer solutions for low power design.

The four embedded tutorials during the first two days of the conference are organized around two major themes: (1) low power circuit design and technologies; and (2) low power systems and energy management. Two embedded tutorials on sub-threshold design and variability/low power design from experts from MIT, University of Michigan, and IBM Research are paired with two others on energy harvesting for battery limited systems and low power portable applications from researchers from NEC Labs and University of California. New for this year, ISLPED will offer three half-day tutorials on hot topics such as leakage aware design, micropower analog design, and addressing fault-tolerance/variability and power issues from system to circuit levels. The tutorials feature presenters from OFFIS, Intel, University of North Carolina, Carnegie Mellon University, and IBM Research.

Out of 214 submissions received in early March, only 75 strong technical papers were accepted for presentation in paper or poster sessions, yielding an acceptance rate of 26% for regular and short papers (56 papers), or 35% including 19 posters. Topics range from low power analog and RF design, power aware circuit design and tools, and architecture techniques for low power, to system and application level power management. The program is organized in twelve technical sessions featuring long (30 min) and short (20 min) paper presentations, as well as two interactive poster sessions that will provide an additional venue for authors and symposium attendees to interact in an informal setting. Following the tradition of ISLPED, this year's event also includes exhibits featuring tools and methodologies from leading vendors of low power or power-aware design tools. New for this year, the program includes an industry session that will highlight contributions showcased by companies participating in the ISLPED exhibits. Winning entries to the annual Low Power Design Contest will also be featured in a separate technical session.

Such a rich and strong program would have been impossible without the help of an outstanding Technical Program Committee who has worked for more than six weeks reviewing and selecting the best papers. Our many thanks go to the 2006 ISLPED officers who have made everything work like clockwork: Anand Raghunathan, Vice-General Chair; Ulf Schlichtmann, Local Arrangements; Vijay Narayanan, Treasurer; Qing Wu, Exhibits; Barry Pangrle, Design Contest; Tony Givargis, Website and Submission system; Farzan Fallah, Jihong Kim, Reiner Hartenstein, Publicity; and Yung-Hsiang Lu, Registration.

ISLPED is grateful for the support it has received from ACM Special Interest Group on Design Automation and the IEEE Circuits and Systems Society. ISLPED also receives technical co-sponsorship from the IEEE Solid State Circuits and the IEEE Electron Devices Societies and is thankful for the generous financial support from Bosch, BullDAST s.r.l., Cadence, ChipVision, IBM, Infineon, Intel, Magma Design Automation, Nokia, OFFIS, Philips, ST Microelectronics, Synopsys, and Texas Instruments.

We hope you will find ISLPED an exciting venue for interacting with fellow researchers in the field!

Wolfgang Nebel and Mircea Stan
General Co-chairs

Diana Marculescu and Joerg Henkel
Technical Program Co-chairs

Table of Contents

ISLPED'06 Executive Committee & Symposium Officers xi

Technical Program Committee xii

ISLPED'06 Sponsors & Supporters xv

Keynote
Session Chair: Wolfgang Nebel (*University of Oldenburg and OFFIS*)

- **Design Challenges for Mobile Communication Devices** 1
  C. Kutter (*Infineon Technologies AG*)

Session 1: Emerging Technologies and Designs for Low Power
Session Chair: Chris Kim (*University of Minnesota*), Co-Chair: Ali Keshavarzi (*Intel Corporation*)

- **Analysis of Super Cut-off Transistors for Ultralow Power Digital Logic Circuits** 2
  A. Raychowdhury, X. Fong, Q. Chen, K. Roy (*Purdue University*)

- **Variation-Driven Device Sizing for Minimum Energy Sub-threshold Circuits** 8
  J. Kwong, A. P. Chandrakasan (*Massachusetts Institute of Technology*)

- **Robust Level Converter Design for Sub-threshold Logic** 14
  I. J. Chang (*Purdue University*), J.-J. Kim (*IBM T.J. Watson Research Center*), K. Roy (*Purdue University*)

- **Integrated Solar Energy Harvesting and Storage** 20
  N. Guilar, A. Chen, T. Kleeburg, R. Amirtharajah (*University of California at Davis*)

- **Energy/Power Breakdown of Pipelined Nanometer Caches (90nm/65nm/45nm/32nm)** 25
  S. Rodriguez, B. Jacob (*University of Maryland*)

Session 2: Microarchitectural Techniques for Low Power
Session Chair: Russ Joseph (*Northwestern University*), Co-Chair: Qing Wu (*SUNY Binghamton*)

- **Stall Cycle Redistribution in a Transparent Fetch Pipeline** 31
  E. L. Hill, M. H. Lipasti (*University of Wisconsin*)

- **Selective Writeback: Exploiting Transient Values for Energy-Efficiency and Performance** 37
  D. Balkan, J. Sharkey, D. Ponomarev, K. Ghose (*State University of New York*)

- **Energy-Efficient Dynamic Instruction Scheduling Logic through Instruction Grouping** 43
  H. Sasaki, M. Kondo, H. Nakamura (*The University of Tokyo*)

- **Independent Front-end and Back-end Dynamic Voltage Scaling for a GALS Microarchitecture** 49
  G. Magklis, P. Chaparro, J. González, A. González (*Intel Barcelona Research Center*)

- **Synergistic Temperature and Energy Management in GALS Processor Architectures** 55
  YK. Zhu (*University of Rochester*), D. H. Albonesi (*Cornell University*)

Session 3: Circuit Techniques for Scaled Technologies
Session Chair: Dongsheng Brian.Ma (*University of Arizona*), Co-chair: Matthew Ziegler (*IBM*)

- **A Two-Port SRAM for Real-Time Video Processor Saving 53% of Bitline Power with Majority Logic and Data-Bit Reordering** 61
  H. Fujiwara, K. Nii, J. Miyakoshi, Y. Murachi (*Kobe University*), Y. Morita (*Kanazawa University*),
  H. Kawaguchi, M. Yoshimoto (*Kobe University*)

- **A High-Speed Variation-Tolerant Interconnect Technique for Sub-Threshold Circuits Using Capacitive Boosting** 67
  J. Kil (*Intel Corporation*), J. Gu, C. H. Kim (*University of Minnesota*)

- **A Dual-V_{DD} Boosted Pulsed Bus Technique for Low Power and Low Leakage Operation** 73
  H. S. Deogun, R. Senger, D. Sylvester (*University of Michigan*), R. Brown (*University of Utah*), K. Nowka (*IBM Corporation*)

- **Time-Borrowing Multi-Cycle On-Chip Interconnects for Delay Variation Tolerance** 79
  K. Bowman, J. Tschanz, M. Khellah (*Intel Corporation*), M. Ghoneima, Y. Ismail (*Northwestern University*), V. De (*Intel Corporation*)

- **A Pulsed Low-voltage Swing Latch for Reduced Power Dissipation in High-Frequency Microprocessors** 85
  P.-F. Lu, N. Cao, L. Sigal, P. Woltgens, R. Robertazzi, D. Heidel (*IBM T.J. Watson Research Center*)

Session 4: Power Management and Application Specific Architectures
Session Chair: Massoud Pedram (*University of Southern California*), Co-chair: Vivek Tiwari (*Intel Corporation*)

- **Temporal Vision-Guided Energy Minimization for Portable Displays** 89
  W.-C. Cheng, C.-F. Hsu, C.-F. Chao (*National Chiao Tung University*)

- **Dynamic Current Modeling at the Instruction Level** 95
  J. Rizo-Morente, M. Casas-Sanchez, C. J. Bleakley (*UCD School of Computer Science and Informatics*)

- **Reducing Idle Mode Power in Software Defined Radio Terminals** 101
  H. Lee, T. Mudge (*University of Michigan*), C. Chakrabarti (*Arizona State University*)

- **Power Reduction in an H.264 Encoder Through Algorithmic and Logic Transformations** 107
  M. G. Koziri, G. I. Stamoulis (*University of Thessaly*), I. X. Katsavounidis (*InterVideo, Inc.*)

- **Energy-efficient Motion Estimation using Error-Tolerance** 113
  G. V. Varatkar, N. R. Shanbhag (*University of Illinois at Urbana-Champaign*)

Poster Session 1: Low Power Circuits and Microarchitectures
Session Chair: Tanay Karnik (*Intel Corporation*)

- **Variability-Aware Device Optimization under I_{ON} and Leakage Current Constraints** 119
  J. Jaffari, M. Anis (*University of Waterloo*)

- **A 0.5-V FD-SOI Twin-Cell DRAM with Offset-Free Dynamic-V_T Sense Amplifiers** 123
  R. Takemura, K. Itoh, T. Sekiguchi (*Hitachi Ltd.*)

- **Utilizing Reverse Short Channel Effect for Optimal Subthreshold Circuit Design** 127
  T.-H. Kim, H. Eom, J. Keane, C. Kim (*University of Minnesota*)

- **Logic Circuits Operating in Subthreshold Voltages** 131
  J. Nyathi, B. Bero (*Washington State University*)

- **A New Mismatch-Dependent Low Power Technique with Shadow Match-Line Voltage-Detecting Scheme for CAMs** 135
  J. Zhang, Y. Ye (*Harbin Institute of Technology*), B. Liu (*National Cheng Kung University*)

- **Thread-Associative Memory for Multicore and Multithreaded Computing** 139
  S. Wang, L. Wang (*University of Connecticut*)

- **Hierarchical Value Cache Encoding for Off-Chip Data Bus** 143
  C.-H. Lin, C.-L. Yang (*National Taiwan University*), K.-J. King (*IBM STG xSeries Development*)

- **Reducing Cache Traffic and Energy with Macro Data Load** 147
  L. Jin, S. Cho (*University of Pittsburgh*)

- **Modelling Macromodules for High-Level Dynamic Power Estimation of FPGA-based Digital Designs** 151
  A. Reimer, A. Schulz (*University of Oldenburg*), W. Nebel (*University of Oldenburg & OFFIS Research Institute*)

Industry Session: Highlights of Industrial Low Power Tools and Flows.........155
Session Chair: Qing Wu (*SUNY Binghamton*)

Session 5: Thermal and Energy Aware Design
Session Chair: Domenik. Helms (*OFFIS*), Co-chair: Ulf. Schlichtmann (*Technical University Munich*)

- **Thermal Via Allocation for 3D ICs Considering Temporally and Spatially Variant Thermal Power** 156
  H. Yu, Y. Shi, L. He (*University of California at Los Angeles*), T. Karnik (*Intel Labs.*)

- **Dynamic Thermal Clock Skew Compensation Using Tunable Delay Buffers** 162
  A. Chakraborty, K. Duraisami, A. Sathanur, P. Sithambaram (*Politecnico di Torino*),
  L. Benini (*Università di Bologna*), A. Macii, E. Macii, M. Poncino (*Politecnico di Torino*)

- **An Efficient Chip-level Time Slack Allocation Algorithm for Dual-Vdd FPGA Power Reduction** 168
  Y. Lin, Y. Hu, L. He (*University of California at Los Angeles*), V. Raghunathan (*Purdue University*)

- **A Novel Approach for Variation Aware Power Minimization during Gate Sizing**.........174
  V. Mahalingam, N. Ranganathan, J. E. Harlow III (*University of South Florida*)

Session 6: Energy Management for Sensor and Memory Systems
Session Chair: Hiroshi Nakamura (*University of Tokyo*), Co-chair: Vijay Raghunathan (*NEC Labs America*)

- **Adaptive Duty Cycling for Energy Harvesting Systems**.........180
  J. Hsu, S. Zahedi, A. Kansal, M. Srivastava (*University of California at Los Angeles*),
  V. Raghunathan (*NEC Labs America*)

- **Power Reduction of Multiple Disks Using Dynamic Cache Resizing and Speed Control**.........186
  L. Cai (*Purdue University*), Y.-H. Lu (*Purdue University*)

- **Lifetime Aware Resource Management for Sensor Network Using Distributed Genetic Algorithm** 191
  Q. Qiu, Q. Wu (*Binghamton University*), D. Burns, D. Holzhauer (*Air Force Research Laboratory*)

- **Everlast: Long-life, Supercapacitor-operated Wireless Sensor Node**.........197
  F. Simjee, P. H. Chou (*University of California at Irvine*)

Embedded Tutorial 1
Session Chair: Vivek De (*Intel Corporation*)

- **Model to Hardware Matching: For nano-meter Scale Technologies**.........203
  S. R. Nassif (*IBM Research*)

Embedded Tutorial 2
Session Chair: Pai Chou (*University of California at Irvine*)

- **Low Power Light-weight Embedded Systems** 207
  M. Sarrafzadeh, F. Dabiri (*University of California at Los Angeles*), R. Jafari (*University of Texas at Dallas*),
  T. Massey, A. Nahapetan (*University of California at Los Angeles*)

Keynote
Session Chair: Mircea Stan (*University of Virginia*)

- **Low Power Design from Technology Challenge to Great Products** 213
  B. Dennington (*NXP Semiconductors/CTO/SoC Design Technology*)

Session 7: Leakage Control and Dynamic Power Optimization
Session Chair: Alberto Macii (*Politecnico di Torino*), Co-chair: Massimo Poncino (*Politecnico di Torino*)

- **A Novel Dynamic Power Cutoff Technique (DPCT) for Active Leakage Reduction in Deep Submicron CMOS Circuits** 214
  B. Yu, M. L. Bushnell (*Rutgers University*)

- **Analysis and Modeling of Subthreshold Leakage of RT-Components under PTV and State Variation** 220
  D. Helms, G. Ehmen (*OFFIS Research Institute*), W. Nebel (*University of Oldenburg*)

- **Power Optimization in a Repeater-Inserted Interconnect Via Geometric Programming** 226
  W. T. Cheung, N. Wong (*The University of Hong Kong*)

- **Input-specific Dynamic Power Optimization for VLSI Circuits** 232
  F. Hu (*Intel Corporation*), V. D. Agrawal (*Auburn University*)

- **Two-phase Fine-grain Sleep Transistor Insertion Technique in Leakage Critical Circuits** 238
  Y. Wang, Y. Liu, R. Luo, H. Yang, H. Wang (*Tsinghua University*)

Session 8: Memory Hierarchy and Caches
Session Chair: Bruce Jacob (*University of Maryland*), Co-chair: Trevor Mudge (*University of Michigan*)

- **Register File Caching for Energy Efficiency** 244
  H. Zeng, K. Ghose (*State University of New York at Binghamton*)

- **L-CBF: A Low-Power, Fast Counting Bloom Filter Architecture** 250
  E. Safi, A. Moshovos, A. Veneris (*University of Toronto*)

- **A Low Power SRAM Architecture Based on Segmented Virtual Grounding** 256
  M. Sharifkhani, M. Sachdev (*University of Waterloo*)

- **Process Variation Aware Cache Leakage Management** 262
  K. Meng, R. Joseph (*Northwestern University*)

- **Substituting Associative Load Queue with Simple Hash Tables in Out-of-Order Microprocessors** 268
  A. Garg (*University of Rochester*), F. Castro (*Universidad Complutense Madrid*),
  M. Huang (*University of Rochester*), D. Chaver, L. Piñuel, M. Prieto (*Universidad Complutense Madrid*)

Session 9: RF CMOS Building Blocks
Session Chair: David Binkley (*University of North Carolina at Charlotte*),
Co-chair: Domine Leenaerts (*Philips Semiconductors*)

- **A Novel Power Optimization Technique for Ultra-Low Power RFICs** 274
  A. Shameli, P. Heydari (*University of California at Irvine*)

- **A CMOS Analog Frontend for a Passive UHF RFID Tag** 280
  A. Facen, A. Boni (*University of Parma*)

- **High-Speed Low-Power Frequency Divider with Intrinsic Phase Rotator** 286
  S. Henzler, S. Koeppe (*Infineon Technologies AG*)

Session 10: Temperature-Aware Design and Microarchitectures
Session Chair: Peter Feldmann (*IBM*), Co-chair: Kanishka Lahiri (*NEC*)

- **An Optimal Analytical Solution for Processor Speed Control with Thermal Constraints** 292
  R. Rao, S. Vrudhula, C. Chakrabarti (*Arizona State University*), N. Chang (*Seoul National University*)

- **Temperature-Aware Floorplanning of Microarchitecture Blocks with IPC-Power Dependence Modeling and Transient Analysis** 298
  V. Nookala, D. J. Lilja, S. S. Sapatnekar (*University of Minnesota*)

- **Power Efficiency for Variation-Tolerant Multicore Processors** 304
  J. Donald, M. Martonosi (*Princeton University*)

- **Power-Conscious Configuration Cache Structure and Code Mapping for Coarse-Grained Reconfigurable Architecture**..........310
  Y. Kim, I. Park, K. Choi, Y. Paek (*Seoul National University*)

- **Dynamic Thermal Management for MPEG-2 Decoding**..........316
  W. Lee, K. Patel, M. Pedram (*University of Southern California*)

Poster Session 2: Low Power Mixed-Signal and Digital Systems
Session Chair: Qinru Qiu (*State University of New York Binghamton*)

- **A Low-Power Active Substrate-Noise Decoupling Circuit with Feedforward Compensation for Mixed-Signal SoCs**..........322
  S. Guo, H. Lee (*The University of Texas at Dallas*)

- **Power-Efficient Pulse Width Modulation DC/DC Converters with Zero Voltage Switching Control**..........326
  C. Long, S. Reddy, S. Pamarti, L. He (*University of California at Los Angeles*), T. Karnik (*Intel*)

- **Behavioral Modeling of Opamp Gain and Dynamic Effects for Power Optimization of Delta-Sigma Modulators and Pipelined ADCs**..........330
  A. A. Hamoui, T. Alhajj, M. Taherzadeh-Sani (*McGill University*)

- **Low-power Fanout Optimization Using MTCMOS and Multi-Vt Techniques**..........334
  B. Amelifard (*University of Southern California*), F. Fallah (*Fujitsu Laboratories of America*), M. Pedarm (*University of Southern California*)

- **A New Technique for Jointly Optimizing Gate Sizing and Supply Voltage in Ultra-Low Energy Circuits**..........338
  S. Hanson, D. Sylvester D. Blaauw (*University of Michigan*)

- **Considering Process Variations During System-Level Power Analysis**..........342
  S. Chandra (*University of California at San Diego & NEC Laboratories America*), K. Lahiri, A. Raghunathan (*NEC Laboratories America*), S. Dey (*University of California at San Diego*)

- **Synchronization-Driven Dynamic Speed Scaling for MPSoCs**..........346
  M. Loghi, M. Poncino (*Politecnico di Torino*), L. Benini (*Università di Bologna*)

- **Power Phase Variation in a Commercial Server Workload**..........350
  W. L. Bircher, L. K. John (*The University of Texas at Austin*)

- **Reducing Power through Compiler-Directed Barrier Synchronization Elimination**..........354
  M. Kandemir, S. W. Son (*The Pennsylvania State University*)

- **Minimizing Energy Consumption of Banked Memories Using Data Recomputation**..........358
  H. Koc (*Syracuse University*), O. Ozturk, M. Kandemir, S. H. K. Narayanan (*The Pennsylvania State University*), E. Ercanli (*Syracuse University*)

Design Contest Presentations
Session Chair: Barry Pangrle (*ArchPro Design Automation*)

Embedded Tutorial 3
Session Chair: Sachin Sapatnekar (*University of Minnesota*)

- **Energy Optimality and Variability in Subthreshold Design**..........363
  S. Hanson, B. Zhai, D. Blaauw, D. Sylvester (*University of Michigan*), A. Bryant, X. Wang (*IBM T.J. Watson Research Center*)

- **Sub-Threshold Design: The Challenges of Minimizing Circuit Energy**..........366
  B. H. Calhoun (*University of Virginia*), A. Wang (*Texas Instruments*), N. Verma, A. Chandrakasan (*Massachusetts Institute of Technology*)

Embedded Tutorial 4
Session Chair: Philippe Royannez (*Texas Instruments*)

- **Design and Power Management of Energy Harvesting Embedded Systems**..........369
  V. Raghunathan (*NEC Labs America*), P. H. Chou (*University of California at Irvine*)

Panel

- **Flexibility and Low Power; A Contradiction in Terms?**
  Can Configurable or Re-Configurable Computing Offer Solutions? 375
  Moderator: P. Wintermayr (*Markt und Technik*), R. Hartenstein (*University of Kaiserslautern*),
  H. Meyr (*RWTH Aachen University & CoWare*), S. Leibson (*Tensilica*)

Session 11: Low Power, Low Voltage Circuits and DC/DC Converters

Session Chair: Ben Calhoun (*University of Virginia*), Co-chair: Seonghwan Cho (*KAIST*)

- **Efficient Scan-Based BIST Scheme for Low Power Testing of VLSI Chips** 376
  M. Shah (*Dhirubhai Ambani Institute of Information and Communication Technology*)

- **Modeling and Analysis of Leakage Induced Damping Effect in Low Voltage LSIs** 382
  J. Gu, J. Keane, C. Kim (*University of Minnesota*)

- **Dithering Skip Modulator with a Novel Load Sensor**
  for Ultra-wide-load High-Efficiency DC-DC Converters 388
  H.-W. Huang (*National Taiwan University*), H.-H. Ho, K.-H. Chen (*National Chiao Tung University*),
  S.-Y. Kuo (*National Taiwan University*)

- **Adaptive On-Chip Power Supply with Robust One-Cycle Control Technique** 394
  D. Ma, J. Wang, P. Vozqua (*The University of Arizona*)

- **Robust Multiple-Phase Switched-Capacitor DC-DC Converter**
  with Digital Interleaving Regulation Scheme 400
  D. Ma (*The University of Arizona*)

Session 12: Low Power Architectures and Systems

Session Chair: Yung-Hsiang Lu (*Purdue University*),
Co-chair: Radu Marculescu (*Carnegie Mellon University*)

- **A Low Power Viterbi Decoder Implementation Using Scarce State Transition**
  and Path Pruning Scheme for High Throughput Wireless Applications 406
  J. Jin, C.-Y. Tsui (*The Hong Kong University of Science and Technology*)

- **SmartSaver: Turning Flash Drive into a Disk Energy Saver for Mobile Computers** 412
  F. Chen, S. Jiang, X. Zhang (*The Ohio State University*)

- **An Energy-Efficient Virtual Memory System with Flash Memory**
  as the Secondary Storage 418
  H.-W. Tseng, H.-L. Li, C.-L. Yang (*National Taiwan University*)

- **Maximizing the Lifetime of Embedded Systems Powered**
  by Fuel Cell-Battery Hybrids 424
  J. Zhuo, C. Chakrabarti (*Arizona State University*), N. Chang (*Seoul National University*),
  S. Vrudhula (*Arizona State University*)

Author Index 430

ISLPED 2006 Executive Committee & Symposium Officers

General Chairs: Wolfgang Nebel, *University of Oldenburg and OFFIS*
Mircea Stan, *University of Virginia*

Vice General Chair Anand Raghunathan, *NEC Laboratories America*

Program Chairs: Joerg Henkel, *University of Karlsruhe*
Diana Marculescu, *Carnegie Mellon University*

Treasurer: Vijaykrishnan Narayanan, *Penn State University*

Exhibits Chair: Qing Wu, *State University of New York at Binghamton*

Design Contest Chair: Barry Pangrle, *ArchPro Design Automation*

Publicity Chairs: Farzan Fallah (US), *Fujitsu Labs of America*
Reiner Hartenstein (Europe), *Technical University Kaiserslautern*
Jihong Kim (Asia), *Seoul National University*

Local Arrangements Chair: Ulf Schlichtmann, *Technische Universität München*

Web Chair: Tony Givargis, *University of California, Irvine*

Registration Chair: Yung-Hsiang Lu, *Purdue University*

Executive Committee: Brock Barton, *Texas Instruments*
David Blaauw, *University of Michigan*
Robert Brodersen, *University of California, Berkeley*
Anantha Chandrakasan, *Massachusetts Institute of Technology*
Kiyoung Choi, *Seoul National University*
Jason Cong, *University of California, Los Angeles*
Vivek De, *Intel Corporation*
Giovanni DeMicheli, *EPF Lausanne*
Christian Enz, *EPF Lausanne*
Mary-Jane Irwin, *Penn State University*
Rajiv Joshi, *IBM*
Enrico Macii, *Politecnico di Torino*
Farid Najm, *University of Toronto*
Massoud Pedram, *University of Southern California*
Christian Piguet, *CSEM*
Jan Rabaey, *University of California, Berkeley*
Kaushik Roy, *Purdue University*
Takayasu Sakurai, *University of Tokyo*
Christer Svensson, *Linkoping University*
Vivek Tiwari, *Intel Corporation*
Ingrid Verbauwhede, *K.U. Leuven*
M. Balakrishnan, *Indian Institute of Technology, Delhi*
Juergen Becker, *University of Karlsruhe*
Frank Bellosa, *University of Karlsruhe*

Program Committee: Luca Benini, *University of Bologna*
Lucien Breems, *Philips Research*
David Brooks, *Harvard University*
Alper Buyuktosunoglu, *IBM T. J. Watson Research Center*
Ben Calhoun, *University of Virginia*
Kevin Cao, *Arizona State University*
Chaitali Chakrabarti, *Arizona State University*
Srimat Chakradhar, *NEC Laboratories America*
Anantha Chandrakasan, *Massachusetts Institute of Technology*
Naehyuck Chang, *Seoul National University*
Seonghwan Cho, *Korea Advanced Institute of Science and Technology*
Pai Chou, *University of California, Irvine*
Jason Cong, *University of California, Los Angeles*
Nikil Dutt, *University of California, Irvine*
Farzan Fallah, *Fujitsu Labs of America*
Joan Figueras, *University of Politecnica de Catalunya*
Vassilios Gerousis, *Cadence Design Systems*
Patrick Girard, *LIRMM, France*
Helmut Graeb, *Technical University of Munich*
Rajesh Gupta, *University of California, San Diego*
Lei He, *University of California, Los Angeles*
Payam Heydari, *University of California, Irvine*
Ed Huijbregts, *Magma Design Automation*
Wei Hwang, *National Chiao Tung University*
Ali Keshavarzi, *Intel Corporation*
Chris Kim, *University of Minnesota*
Jihong Kim, *Seoul National University*
Suhwan Kim, *Seoul National University*
Volkan Kursun, *University of Wisconsin*
Kanishka Lahiri, *NEC Laboratories America*
Domine Leenaerts, *Philips Research*
Yung-Hsiang Lu, *Purdue University*
Enrico Macii, *Politecnico di Torino*
Radu Marculescu, *Carnegie Mellon University*
Vasily Moshnyaga, *Fukuoka University*
Trevor Mudge, *University of Michigan, Ann Arbor*
Satyen Mukherjee, *Philips Research*
Vijaykrishnan Narayanan, *Penn State University*
Chandra Narayanaswami, *IBM T. J. Watson Research Center*
Sreedhar Natarajan, *Emerging Memory Technologies*
Wolfgang Nebel, *University of Oldenburg and OFFIS*
Vojin Oklobdzija, *University of California, Davis*
Barry Pangrle, *ArchPro DA*
Christian Piguet, *CSEM, Switzerland*
Massimo Poncino, *Università di Verona*
Anand Raghunathan, *NEC Laboratories America*
Vijay Raghunathan, *NEC Laboratories America*
Nagarajan Ranganathan, *University of South Florida*

Program Committee (continued):

Philippe Royannez, *Texas Instruments*
Takayasu Sakurai, *University of Tokyo*
Ulf Schlichtmann, *Technische Universität München*
Eike Schmidt, *ChipVision Design Systems*
Naresh Shanbhag, *University of Illinois, Urbana-Champaign*
Youngsoo Shin, *Korea Advanced Institute of Science and Technology*
Dinesh Somasekhar, *Intel Corporation*
George Stamoulis, *University of Thessaly*
Vivek Tiwari, *Intel Corporation*
Nestoras Tzartzanis, *Fujitsu Labs of America*
Kimiyoshi Usami, *Shibaura Institute of Technology*
Kees Veelenturf, *Philips Research*
Johan Vounckx, *Interuniversity MicroElectronics Center*
Alice Wang, *Texas Instruments*
Gu-Yeon Wei, *Harvard University*
Chris Wilkerson, *Intel Corporation*
Qing Wu, *Binghamton University*
Futao Yamaguchi, *Sony Corporation*
Patrick Yue, *Carnegie Mellon University*
Matthew Ziegler, *IBM T. J. Watson Research Center*

Additional reviewers:

Yuvraj Agarwal
Remus Albu
Saif Ali
Geoff Balke
Sudarshan Banerjee
Lichun(Luke) Bao
Jens Becker
Mainak Biswas
Keith Bowman
Jeff Brateman
Saif Ali Butt
Le Cai
Ashutosh Chakraborty
Saumya Chandra
Youngjin Choi
Radu Cornea
Suleyman Demirsoy
Bnoit Dufort
Karthik Duraisami
Matthew Exon
Jeff Furlong
Christophe Giacomotto
Neeraj Goel
Olga Golubeva
Per Gunar
Aseem Gupta

Soonhoi Ha
Narender Hanchate
Domenik Helms
Douglas Herbert
M. Hillers
Chang-Tao Hsieh
Xuejue Huang
Michael Hübner
Ilya Issenin
Hans Jacobson
Vipul Jain
Wei Jiang
Taotao Jin
Nicole Kaczoreck
Jung-Chun Kao
Himanshu Kaul
Taeho Kgil
Muhammad Khellah
Minyoung Kim
NamSung Kim
Ralf Koenig
Ranjith Kumar
Sarath Kumar
Benjamin Lee
Hyunsek Lee
Seongsoo Lee

Additional reviewers (continued):

Bing Li
Xin Li
Yuan Lin
Zhiyu Liu
Mirko Loghi
Shih-Lien Lu
Guojie Luo
Zhen Ma
Alberto Macii
V. Mahalingam
Bingfeng Mei
Yongguo Mei
Maurice Meijer
Daniel Messaguer
Kurt Metzger
Kirill Minkovich
Saraju Mohanty
Carlos Morra
Nikola Nedovic
Umit Ogras
Young-Hwan Park
Marcel Pelgrom
Eddie Pettis
Deyi Pi
Ravishankar Rao
Vikram Rao
Srivaths Ravi
David Roberts
Aminghasem Safarian
Amir Safarian

Oliver Sander
Ashoka Sathanur
Stefan Schmermbeck
Manuel Schmidt
Walter Schneider
Milan Schulte
Arne Schulz
Amin Shameli
Aviral Shrivastava
Dirk Siemer
Prassanna Sithambaram
Krishnan Srinivasan
Chiawei Su
Sriiram Sundararaman
Sherif Tawfik
Clark Taylor
Michael Ullmann
Paul van de Wiel
Harry Veendrick
Milena Vratonjic
C.K. Wang
C.S. Wang
Mark Woh
Changjiu Xian
Nick Zamora
Bart Zeydel
Yan Zhang
Ying Zhang
Zhiru Zhang
Jianli Zhuo

Sponsors:

Technical Support from:

and corporate support from:

Keynote

Design Challenges for Mobile Communication Devices

Dr. Christoph Kutter
Business Group Communication Solutions, Infineon Technologies AG
Munich, Germany
christoph.kutter@infineon.com

ABSTRACT

System on Chips (SoC) for mobile devices, such as GSM/EDGE/UMTS, have strongly conflicting requirements. On one hand the demand for processing performance is steadily increasing with every new standard and on the other hand extremely low power dissipation is demanded. The performance demands vary strongly, depending on the phone modes and activities e.g. stand-by mode vs. talk mode vs. high performance application modes such as video processing and gaming.

Next to the development of new telecommunication standards the silicon technologies develop according to the shrink path. Scaling of physical structures, especially the gate thickness, induces larger leakage in DSM technologies. The shrink is accompanied with a further reduction of supply voltage that helps to reduce the dynamic power dissipation but also reduces the leverage of performance improvement. To reduce the leakage and to reach the targets of design projects new low power measures have to be defined and implemented by integration into technology, libraries, design tools, and the design flow.

In recent years several low power features have been developed to address both the static leakage power consumption and the dynamic active power consumption. These features, or a combination of them, can be tailored to dynamically varying performance needs of the SoC in different modes meaning different use cases.

Categories and Subject Descriptors: B.7 INTEGRATED CIRCUITS, C.3 SPECIAL-PURPOSE AND APPLICATION-BASED SYSTEMS

General Terms: Management, Design

Keywords: SoC, Design for Low Power, Leakage

Copyright is held by the author/owner(s).
ISLPED'06, October 4–6, 2006, Tegernsee, Germany.
ACM 1-59593-462-6/06/0010.

Analysis of Super Cut-off Transistors for Ultralow Power Digital Logic Circuits

Arijit Raychowdhury, Xuanyao Fong, Qikai Chen, Kaushik Roy

Department of Electrical and Computer Engineering, Purdue University, IN, USA

Abstract

Super cut-off devices with sub-60mV/decade subthreshold swings have recently been demonstrated and being extensively studied. This paper presents a feasibility analysis of such tunneling devices for ultralow power subthreshold logic. Analysis shows that this device can deliver 800X higher performance (@iso-I_{OFF}) compared to a MOSFET. The possible use of this device as a sleep transistor in conjunction with the regular Si MOSFET shows 2000X average improvement in leakage power compared to Si MOSFETs.

Categories and Subject Descriptors: B.7

General Terms: Performance

Keywords: Carbon nanotube FETs, Tunneling transistors.

I. INTRODUCTION

Over the last couple of decades considerable attention has been given to the design of high performance and high power microprocessors. As technology scaling continues despite the many challenges it faces, the transistor feature size is decreased by ~30% thereby leading to ~40% improvement in performance every generation. This, however, has adversely affected the transistor's operation as a switch. Larger short channel effects, increased leakage and rising power densities, stand in the way to further scaling. In recent years, to mitigate some of these issues, the demand for power sensitive designs has grown significantly. Coupled with this is the tremendous growth of battery-operated hand-held gadgets, like personal digital assistants and cell phones. This has accelerated the need for ultralow power digital systems. Circuits, as well as architectural (such as voltage scaling, switching activity reduction, pipelining and parallelism) and CAD techniques (of device sizing, interconnect and logic optimization) have been incorporated in most of the digital systems. Further, energy recovery and adiabatic logic [1] have been investigated for reducing the dynamic power consumption, but this design style requires use of high quality on-chip inductors.

On the technology front, increasing the I_{ON}/I_{OFF} ratio has always been a daunting challenge. Decreased gate control in nanometer transistor geometries, has resulted in an exponentially increasing OFF current. Subthreshold leakages, gate tunneling leakages, as well as junction leakages, have all become significant. In recent years, multi-gate devices (e.g., FinFETs, Tri-gates) as well as the use of compound semiconductor materials have been demonstrated and they exhibit, both in theory and experiments, better short channel immunity and higher mobilities. Although improving the I_{ON}/I_{OFF} ratio by decreasing the subthreshold slope (S), has been the ultimate challenge for device designers, the fundamental limit of S=60mV/decade has always been the barrier. Sub-100nm bulk MOSFETs have approximately 80-90mV/decade of subthreshold swing (S), whereas, FinFETs offer a value close to 65mV/decade. The reason for this limit (S=2.3k_BT/q), as would be discussed shortly, comes from the thermal emission of carriers over a channel barrier.

Permission to make digital or hard copies of all or part of this work for personal or classroom use is granted without fee provided that copies are not made or distributed for profit or commercial advantage and that copies bear this notice and the full citation on the first page. To copy otherwise, or republish, to post on servers or to redistribute to lists, requires prior specific permission and/or a fee.

ISLPED '06, October 4–6, 2006, Tegernsee, Germany.

Copyright 2006 ACM 1-59593-462-6/06/0010...$5.00.

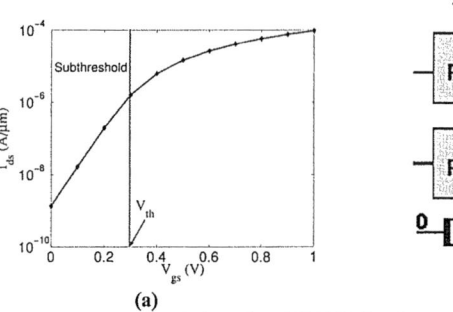

(a)   (b)

Fig 1: (a) I_d-V_g characteristics of an NMOS showing the subthreshold region of operation (b) Schematic diagram of an NMOS Sleep transistor

Very recently, work has been done in designing a new genre of Field Effect Transistors that rely on band-to-band tunneling instead of thermal emission. Remarkably low subthreshold swings (<40mV/decade) [2-4] have been observed in such transistors and rigorous theoretical analysis proves, most indubitably, the opportunity for obtaining even lower values of S. Lateral Si [4] as well as vertical Si/SiGe MOSFETs [3] have been fabricated with S as low as 15mV/decade. Pioneering work in the field of carbon nanotube transistors has also revealed the possibility of achieving sub-60mV/decade subthreshold swings [2]. This class of devices, with very sharp subthreshold current roll-off is referred to as ***super cut-off devices*** in the rest of this paper.

In this paper, we investigate the feasibility of using super cut-off devices for ultralow power logic design. Since carbon nanotube transistors (CNFETs) with their near-ballistic channels, excellent electrostatics and improved thermal properties, have been the centre of attention among all the novel device technologies in a post Si era (2015 timeframe), this paper will discuss the possibilities of using super cut-off tunneling carbon nanotube field effect transistors (T-CNFETs) in digital logic design. More specifically, we will explore the use of lower subthreshold swing of T-CNFETs in digital subthreshold logic design and as sleep transistors in regular CMOS design.

Design of digital sub-threshold logic has been investigated with transistors operating in the sub-threshold region where the supply voltage, (V_{dd}, corresponding to Logic 1) is less than the threshold voltage *(V_{th})* of the transistor [5] (Fig. 1a). In such a technique, the sub-threshold leakage current of the device is used for necessary computation. *It results in high transconductance gain of the devices (thereby providing near ideal voltage transfer characteristics of the logic gates) and reduced gate input capacitance.* Its impact on system design is an exponential reduction of power at the cost of reduced performance. The target frequency of operation of subthreshold logic circuits is a few KHz to MHz. One of the chief advantages of Si MOSFET based subthreshold logic comes from the lower energy per switching. In subthreshold logic, the device capacitance decreases from its value of C_{ox}, because the device depletion capacitance becomes comparable to the oxide capacitance. The switching power (CV^2) is significantly lower than its superthreshold value. Hence for the same frequency of operation, the dynamic power dissipated in a subthreshold transistor (CV^2f) is lower than its superthreshold value not only due to lower V_{DD} but also due to lower C. In this paper, we will study the feasibility of

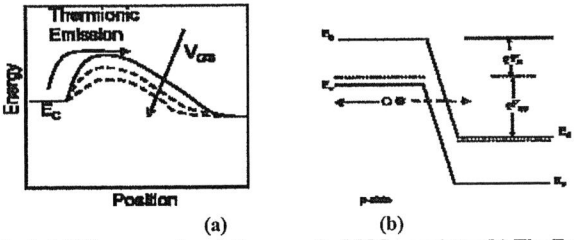

(a)   (b)

Fig 2: (a) The energy band diagram of a MOS transistor (b) The Energy Band diagram at the source-channel junction of a tunneling transistor.

using super cut-off transistors in subthreshold operation. With S<60mV/decade, these transistors offer higher g_m than their MOSFET counterparts. This results in higher performance at iso-I_{OFF} and equal V_{DD}. The intricacies of circuit design, the design challenges and some of the solutions are also discussed.

In the second half of the paper, we will investigate the use of super cut-off devices as sleep transistors in a regular CMOS fabric. Sleep transistors are pull down or pull-up devices that are turned off during the standby mode, in order to realize a stacking effect and hence lower the leakage of the circuit [6]. A pull-down sleep transistor insertion methodology has been shown in Fig. 1b. The ultralow leakage super cut-off devices can be potentially integrated with regular high performance Si MOSFETs (that provide high performance) and used as sleep transistors. In this paper we will present results of rigorous numerical mixed-mode simulations (both at the device as well as the circuit levels).

The rest of the paper is divided as follows. Section II will discuss the super cut-off device physics vis-à-vis standard MOSFETs. Carbon nanotube transistors and their different varieties would be introduced in Section III and the simulation methodology will be presented in IV. Section V and VI will present feasibility studies of using these devices in low-power subthreshold operations and as sleep transistors to reduce stand-by leakage. Finally conclusions are drawn in Section VII.

II. THE ESSENTIALS OF SUPER CUT-OFF DEVICES

Before going into circuit design, it is worthwhile to mention the basic physics of operation of these devices and contrast it to traditional MOSFETs. A model for tunneling FETs has been presented along with a formulation for its threshold voltage.

(a) MOSFET operation: MOSFETs work on the principle of carrier injection over a potential barrier at the source-channel junction (Fig. 2a). In the subthreshold state of a device ($V_{gs} < V_{th}$), the current flowing from the drain to the source of a transistor is known as the subthreshold current. The subthreshold current flowing through a transistor is given by [5],

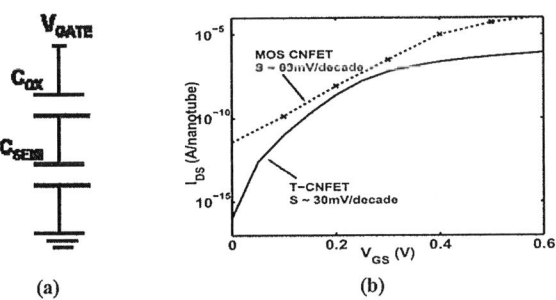

(a)   (b)

Fig 3: (a) The equivalent capacitor network for a CNFET (b) The I_{DS}-V_{GS} characteristics of a MOS CNFET and a T-CNFET illustrating s sub-60mV/decade subthreshold slope.

$$I_{sub} = \frac{w_{eff}}{L_{eff}} \mu \sqrt{\frac{q \varepsilon_{si} N_{cheff}}{2\Phi_s}} v_T^2 exp\left(\frac{V_{gs} - V_{th}}{m v_T}\right)\left(1 - exp\left(\frac{-V_{ds}}{v_T}\right)\right) \quad (1)$$

where, N_{cheff} is the effective channel doping, Φ_s is the surface potential, m is the body effect coefficient related to the subthreshold swing and v_T is the thermal voltage given by kT/q. Using the charge sharing model and following the procedure given in [17], the threshold voltage can be expressed as :

$$V_{th} = V_{FB} + (\Phi_{s0} - \Delta\Phi_s) + \gamma\sqrt{\Phi_{s0} - V_{bs}}\left(1 - \lambda\frac{X_d}{L_{eff}}\right) + \Delta V_{NWE} \quad (2)$$

where, V_{FB} is the flat-band voltage, Φ_{s0} is zero bias surface potential, γ is the body factor, X_d is the depletion layer thickness, λ is a fitting parameter (~1) and ΔV_{NWE} is the narrow-width correction factor as given in [7].

The subthreshold slope (S) (often referred to as the inverse subthreshold slope) for the short channel device, considering the penetration of the drain induced electric fields in the centre of the channel is given by [7]:

$$S \approx 2.3\frac{mkT}{q}\left(1 + \frac{11t_{ox}}{X_d} e^{-\pi L/(X_d + 3t_{ox})}\right) \quad (3a)$$

where the body effect coefficient, m is given by:

$$m = 1 + \frac{3t_{ox}}{X_d} \quad (3b)$$

For an ideal MOSFET, with *no short-channel effects* and $t_{ox} << X_d$ (i.e., m=1):

$$S \geq 2.3\frac{kT}{q} \Rightarrow S \geq 60mV / decade @ 300K \quad (4)$$

Thus, even with excellent electrostatics and enhanced transport in the device, MOSFETs are limited to a subthreshold swing of 60mV/decade.

(b) Tunneling transistors: The origin of tunneling devices dates back to tunneling diodes with p^+-n^+ junctions. Consider the band structure of a p^+-i tunnel junction, as shown in Fig. 2b. A gate bias lowers the potential in the channel region (i-region) and splits the Fermi level. This introduces a very high electric field at the junction which causes significant current to flow through the junction due to tunneling of electrons from the valence band of the p-region to the conduction band of the n-region (Fig. 2b) [8]. Tunneling occurs when the total voltage drop across the junction (effective gate bias (V_{app}) + built-in voltage (ψ_{bi})) is more than the band-gap (Σ_g). Here the applied voltage V_{app} is the effective gate voltage V_{eff}. The tunneling current per unit width through the junction is given by [8]:

$$I_{b-b} = A \frac{E V_{app}}{\Sigma_g^{1/2}} exp\left(-B \frac{\Sigma_g^{3/2}}{E}\right)$$

$$A = \frac{\sqrt{2m^*}q^3}{4\pi^3\hbar^2}, \text{ and } B = \frac{4\sqrt{2m^*}}{3q\hbar} \quad (5)$$

where, m^* is the effective mass of electron, E is the electric field at the junction, q is the electronic charge and $\hbar$ is the reduced Plank's constant. For a p^+-i-n^+ transistor ψ_{bi}~$\Sigma_g/2$ and hence, conduction due to tunneling commences when:

$$V_{eff} = \Sigma_g/2 \quad (6)$$

Again, V_{eff} is related to the gate bias by (Fig. 3a):

$$V_{eff} = V_{GATE} \frac{C_{ox}}{C_{ox} + C_{semi}} \quad (7)$$

where, C_{semi} is the semiconductor capacitance.

From (6) and (7), the threshold voltage of the transistor can be approximated to:

3

From (5) the subthreshold slope can be expressed as:

$$S = \left[\frac{d(\log_{10} I)}{d(V_{GATE})}\right]^{-1} = \left[\left(\frac{1}{V_{eff}}.\frac{dV_{eff}}{dV_{GATE}} + \frac{E + B\Sigma_g^{3/2}}{E}\frac{dE}{dV_{GATE}}\right).\log_e 10\right]^{-1} \quad (9)$$

(9) clearly shows that in tunneling devices, contrary to MOSFETs, there is no fundamental limit which bounds the subthreshold slope. Simplifying further, for S< 60mV/decade, we require:

$$\left(\frac{1}{V_{eff}}.\frac{dV_{eff}}{dV_{GATE}} + \frac{E + B\Sigma_g^{3/2}}{E}\frac{dE}{dV_{GATE}}\right). > 38.461V^{-1} \quad (10)$$

Hence, with the proper choice of material properties, and the right amount of electric field, it is indeed possible to obtain S<60mV/decade in these devices. Note however, that the expression for S in (9) is derived from the tunneling current, once the conduction band in the channel goes below the valence band in the source. Hence (9) describes the slope just before the threshold voltage is reached. To estimate the drain current before the E_c (channel) and E_V (source) cross, a quantum description of the device potential is required and numerical simulation, as has been discussed in Section IV needs to be performed.

Fig. 3b illustrates the I_{DS}-V_{GS} characteristics of a MOSFET with a good slope (63mV/decade), and a T-CNTFET (obtained through device simulations) (The simulation methodology will be discussed in Section IV). Note, however, that the description of tunneling transistors, presented above, ignores higher order effects present in nanoscaled transistors, like quantization of states inside the channel itself and the simultaneous tunneling from the drain to the channel. More detailed models will be discussed in Section IV. Suffice it is to mention that the model described above can faithfully represent the essentials of the device.

Different kinds of tunneling transistors have already been demonstrated. The Si-SiGe vertical FET [3], the lateral channel Si FET [4] and the T-CNFET [5] are a few examples. In this paper, however, we will focus on T-CNFETs. In a post Si era, CNFETs promise extremely high integration densities, better scalability, and higher performance. Hence, in the rest of the paper focus will be on T-CNFETs and comparisons will be made with ideal MOS-CNFETs.

III. TYPES OF CARBON NANOTUBE TRANSISTORS

Over the last six years, different flavors of carbon nanotube transistors (CNFETs) have been demonstrated and studied. Device physicists propose three basic types of transport in these transistors.

The first is a MOSFET-like device [9] in which the ungated portion (source and drain regions) is heavily doped (Fig. 4a). In this case, the on-current is limited by the amount of charge that can be induced in the channel by the gate and not by the doping in the source.

The second is that the device operates like a Schottky barrier FET [10], in which the gate field modulates the width of a tunnel barrier at the source (Fig. 4b). The source drain metal for these CNFETs can be mid-gap or band-edge. It is obvious that the first device will give a higher on current and hence would define the upper limit of performance.

More recently, experimentalists have demonstrated a band to band tunneling CNFET with a subthreshold slope of 40mV/decade [2]. Rigorous numerical simulations done by Koswatta et.al. [11] have shown that it is indeed possible to achieve sub-60mV/decade subthreshold swings in these devices. The device under study in [11] is a p-i-n device where tunneling occurs from the valence band of the source to the conduction band of the channel. Such p-i-n devices have been extensively studied in both silicon [4] and III-V materials [12]. One important difference between Si and carbon, is that Si is an indirect band-gap material. Hence, the tunneling needs to be phonon assisted whereas, in a direct bandgap material, like T-CNFETs (Fig.

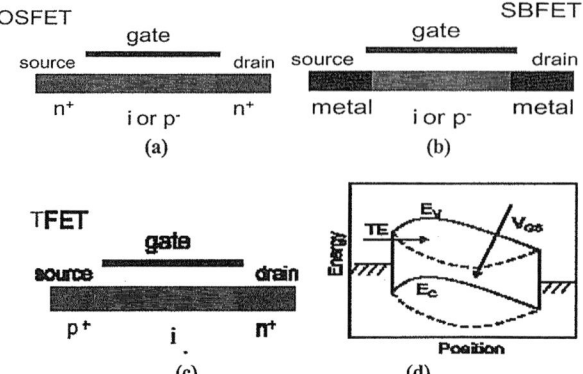

Fig 4: (a,b,c) Schematic diagrams of MOS-FET, SB-FET and T-FET using carbon nanotubes (d) Energy band diagram showing the principle of operation in a SB-CNFET

4c), tunneling can occur without the loss or absorption of phonons. This leads to the possibility of ballistic transport in these devices. Further, the extremely light effective mass of electrons (~0.01m_o) in carbon nanotubes helps in the process. Hence, T-CNFETs are an extremely promising device structure for future technologies.

It is worthwhile to mention, in this context, the difference between the SBCNFET and the T-CNFET, since both the devices rely on electron tunneling. While in SB-CNFETs, tunneling occurs from the conduction band of the source metal to the conduction band of the intrinsic channel (Fig. 4d), in T-CNFETs, the tunneling is from the valence band of the p-type source to the conduction band of the channel. Although SB-CNFETs, are tunneling devices, a closer inspection of the band structure reveals that at low bias conditions (i.e., in subthreshold) the current is dominated by the source/channel barrier and the thermal injection of carriers over the barrier. It is only at high bias conditions, that the ON current is given by the tunneling of carriers from source to channel.

Furthermore, since MOS-CNFETs have better I-V characteristics (in terms of S and higher ON currents) than SB-CNFETs, we will use MOS-CNFETs for comparisons with T-CNFETs in the subsequent discussions.

IV. SIMULATION METHODOLOGY

The device simulation methodology used in this paper has been extensively discussed in [13] and [11] and will be briefly explained for the convenience of the readers. The p^+-i-n^+ device under consideration has a 20nm long channel with a (10,0) chirality and a coaxial gate with 2nm thick HfO$_2$ dielectric [11]. The tube diameter is ~1nm with a band-gap voltage of ~0.84V. CNFETs were simulated by solving the Schrödinger equation using the non-equilibrium Green's function (NEGF) formalism [13], self-consistently with the Poisson equation. An atomistic description of the nanotube using a tight binding Hamiltonian with an atomistic (p_z orbital) basis was employed. The charge density was computed by integrating the local density-of-states (LDOS) over energy,

Here e is the electron charge, sgn(E) is the sign function, $E_{FS,D}$ is the

$$Q(z) = (-e)\int_{-\infty}^{+\infty} dE \cdot \text{sgn}[E - E_N(z)] \\ \{D_S(E,z)f(\text{sgn}[E - E_N(z)](E - E_{FS})) \\ + D_D(E,z)f(\text{sgn}[E - E_N(z)](E - E_{FD}))\} \quad (11)$$

source (drain) Fermi level, $E_N(z)$ is the charge neutrality level and $D_{S,D}$ is the LDOS due to the source (drain) contact. Along with the NEGF transport equation, 3D Poisson equation is solved iteratively, to obtain the self consistent electrostatic potential [13]. Once the self-consistent potential profile is obtained, the source-drain current is computed by Landauer's equation:

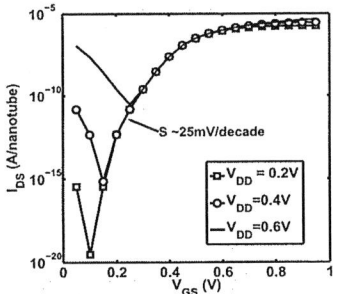

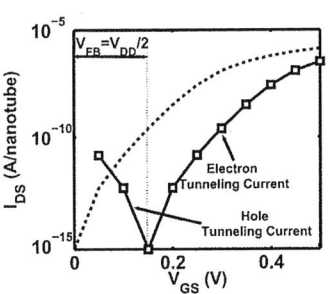

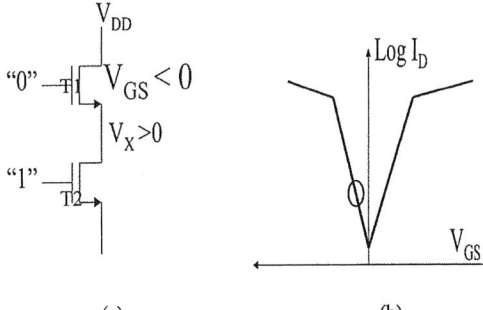

Fig 5: I_{DS}-V_{GS} characteristics of the simulated T-CNFET for different values of V_{DD}. Note that the current minima for this symmetric device occurs at $V_{DD}/2$.

Fig 6: Changing the flat band voltage of the p-i-n device to obtain an NMOS. Note that for the PMOS the change in $V_{FB} = -V_{DD}/2$

Fig 7: A 2-transistor stack (a) the circuit schematic. Note that $V_{GS} < 0$ for T1. (b) A carton showing the $I_D - V_{GS}$ characteristic of a single NMOS. The operating point of T1 has been circled.

$$I = \frac{4e}{h} \int dE \cdot T(E)[f(E - E_{FS}) - f(E - E_{FD})] \quad (12)$$

$$\text{where } T(E) = trace(\Gamma_S G \Gamma_D G^+) \quad (13)$$

It may be noted that a ballistic transport has been assumed in the channel. Since we are assuming ballistic transport (where the carrier momentum is conserved), the source contribution and the drain contribution of the channel charge can be easily obtained. This enables us to compute the device capacitance as:

$$N_{CHANNEL} = N_{DRAIN} + N_{SOURCE}$$

$$\Rightarrow C_{GS} = \frac{\partial N_{SOURCE}}{\partial V_{GS}}; C_{GD} = \frac{\partial N_{DRAIN}}{\partial V_{GS}} \quad (14)$$

Once the I-V and C-V characteristics were obtained, a discrete time numerical solver was used to simulate logic gates. Also note that, the valence band of carbon nanotubes is the mirror image of the conduction band. Consequently, the I-V characteristics of the PMOS are the identical to that of the NMOS.

Fig. 5 illustrates the I_{DS}-V_{GS} characteristic of the T-CNFET for different values of V_{DS}. In this device, when the E_c in the channel of the T-CNFET is below the E_V in the source, the tunneling current diminishes abruptly. Further, as V_{GS} is increased the bands bend more and more resulting in an increasingly thinner barrier at the source-channel junction. This results in super cut-off characteristics in these devices. When a negative V_{GS} is applied, the bands bend upwards, resulting in the tunneling of holes from the drain to the channel and giving rise to ambipolar conduction. Fig. 5 shows how the T-CNFET has both electron conduction when the V_{GS} is high and hole tunneling conduction when V_{GS} is low or negative. For a symmetric p-i-n device (as in our case), the point of minimum current occurs at $V_{GS} = V_{DD}/2$.

It is interesting to note that the diameter plays a vital role in determining the current characteristics of the T-CNFET. The diameter of a carbon nanotube is related to its band-gap as:

$$\Sigma_g = \frac{0.84eV}{d} \quad (15)$$

where d is the diameter (in nm). A smaller diameter increases the band-gap, thereby reducing the OFF current further. However, this also reduces the ON current (or operating current of the device). Hence, a proper trade-off of the ON current and the OFF current can be obtained by selecting the correct diameter. Based on discussions and results published in [14] and [11] a diameter of 1nm has been chosen for this work.

In the same simulation set-up a MOS-CNFET has also been simulated. This device has the same geometry as the p^+-i-n^+ device and has a n-i-n doping profile. The transport in the MOS-CNFET is

determined by the amount of charge induced in the channel due to the gate voltage.

V. USE OF SUPER CUT-OFF DEVICES IN SUBTHRESHOLD LOGIC

1. Choice of Flat band voltage

Before studying the use of these devices in subthreshold circuits, a few words of caveat are in order. Firstly, as mentioned earlier, the p-i-n device exhibits ambipolar conduction, with the current minima at $V_{GS}=V_{DD}/2$. Hence, to use it successfully in CMOS Logic, there needs to be a flat-band shift (ΔV_{FB}) of $-V_{DD}/2$ in the case of the NMOS (Fig. 6) and $V_{DD}/2$ in the case of the PMOS. This problem, in connection to SB-CNFETs has also been discussed in [14].

The shift of the flat-band potential, as has been discussed above, makes it possible to use the SB CNFET in an inverter. However, in complementary digital circuits, the transistor stack is often employed (N stack in NAND gates and P stack in NOR gates). The ambipolar nature of conduction poses a possible threat to the successful use of these devices in transistor stacks. This is illustrated in Fig. 7. Consider an input vector condition "01" where the top transistor (T1) at zero-gate voltage and the bottom transistor (T2) is held at a high gate voltage. For normal operations of the device, this should be the off-state of the stack. Now the intermediate potential (V_x) is slightly positive due to a finite amount of current flowing through the stack. The bottom transistor is "on" with $V_{GS}=V_{DD}$. For the top transistor (T1) $V_{GS} = -V_x$. In such a condition, as T_1 has a negative V_{GS} it will start conducting. Depending on the value of V_x and the subthreshold slope of these devices, the net current flowing through the stack can be significantly high causing very high subthreshold leakage and even functional failure (when the total current is comparable to the operating current of the transistors). It is thus important to design the flat-band voltage for the devices to obtain minimum leakage in the transistor stack.

Fig. 8 shows how the total OFF current of a two and a three transistor stack varies with varying flat-band voltage. The input vector applied to the stack is "01". It can be noted that at $V_{DD}=300mV$, a shift of flat band potential (ΔV_{FB}) of 150mV gives minimum leakage current. Simulations have been carried out for other values of V_{DD} and at each case the minimum leakage current is obtained when the stack transistors have a flat-band potential of $V_{DD}/2$. Thus we may conclude that $\Delta V_{FB} = -V_{DD}/2$ can be used for correct digital operation in ambipolar T-CNFETs.

B. Power Performance trade-off in subthreshold T-CNFETs

The operating current in subthreshold circuits is the leakage current flowing through the device when the transistor is not fully turned on. For T-CNFETs, it is extremely difficult to define a traditional threshold voltage. Hence, for subthreshold operation, we will define

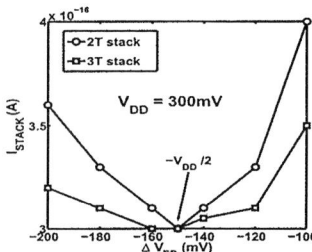
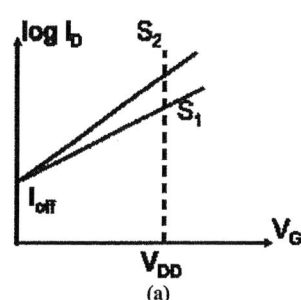

(a)

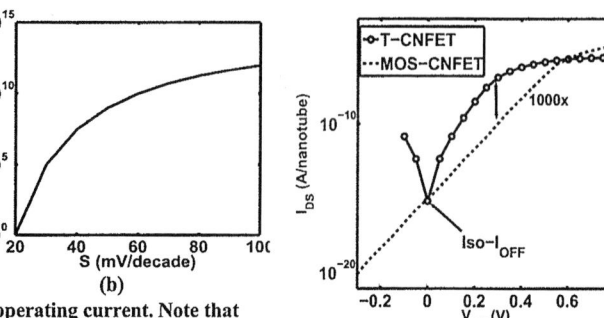

Fig 8: Leakage current vs. the shift in flat band voltage in a 2T and 3T stack. Note that in both cases, the current minima is obtained at ΔV_{FB} = -V_{DD}/2.

Fig 9: (a) The role of S in determining the operating current. Note that the smaller the S, higher the operating current at iso-I_{OFF} (b) Variation of normalized delay with the subthreshold slope for iso-I_{OFF} conditions.

Fig 10: Shifting V_{FB} of MOS-CNFET to match I_{OFF} of T-CNFET

the exponential part of the I-V characteristics as subthreshold conduction and we will use V_{DD} <= 300mV.

To understand the role of subthreshold swing on subthreshold circuits, consider an inverter driving another inverter. The subthreshold-ON current of a device is given by:

$$I_{ON}^{SUBTH} = I_{OFF} 10^{\frac{V_{DD}}{S}} \quad (16)$$

Let us consider that both the devices have the same I_{OFF} and are operated at iso-V_{DD} (Fig. 9a) resulting in identical switching energy CV^2. The switching delay is given by:

$$T = \frac{CV_{DD}}{I_{ON}^{SUBTH}} = \frac{CV_{DD}}{I_{OFF}} 10^{-\frac{V_{DD}}{S}} \quad (17)$$

This shows that at iso-switching energy, a better subthreshold swing leads to an exponentially faster operation. The relation between T and S has been illustrated in Fig. 9b. This proves that the T-CNFETs would fare better than MOS-CNFETs in the subthreshold region. Let us consider the feasibility of using T-CNFETs in subthreshold region in the following cases:

(i) Iso-I_{OFF} Condition: First, let us consider a fixed I_{OFF} for both the T-CNFET and the MOS-CNFET. We will consider a shift in the V_{FB} (by 175mV in our case) of MOS-CNFETs, such that the OFF current coincides with that of the T-CNFET (Fig. 10). This results in identical standby currents in both the devices and gives us the opportunity to compare and contrast the dynamic power and the switching delay of both devices.

Fig. 11 shows the delay of an FO4 inverter built with T-CNFET and the MOS-CNFET for different values of V_{DD}. It may be noted that at iso-I_{OFF} conditions, the T-CNFET delivers more than 1000X higher current than the MOS-CNFET at V_{DD} = 300mV. This results in ~800X improvement in performance. Since the switching capacitance for both the devices are comparable, the switching energy CV_{DD}^2 is also identical.

Hence, in conclusion, at iso-I_{OFF} conditions, the T-CNFET delivers more than 800X performance improvement compared to MOS-CNFETs.

(ii) Iso-Delay Condition:

Now, let us consider the scenario, where an inverter is driving a fixed capacitive load under a specified delay constraint. Consider that the V_{th} of the MOS-CNFET has been adjusted to 300mV (Fig. 12) (identical to that of the T-CNFET). This gives us the opportunity of studying the effect of the higher current densities in MOS-CNFETs. To obtain higher driving currents, it is prudent to use carbon nanotube FETs in parallel in an array configuration. The centre-to-centre distance of the CNFETs (the pitch) is taken to be 2nm (equal to twice the diameter). Fig. 13a illustrates the ratio of the widths of T-CNFET vs. MOS-CNFET for identical switching delay. Note that, for low V_{DD} of operation, the difference of the current between the T-CNFET and the MOS-CNFET is larger than at high

V_{DD}. Hence the ratio of the sizes (proportional to the number of parallel tubes) of MOS-CNFET and T-CNFET is large (~10^4X) for V_{DD}~100mV and ~70X for V_{DD} = 300mV.

Having fixed the delay and the size required to meet the delay target, it will be essential to look at the leakage power dissipated. The high gain in leakage obtained by using T-CNFETs, is reduced at lower V_{DD} because a large number of transistors have to be used to meet the delay target. Fig. 13b shows the leakage currents in the inverters for iso-switching delay. Note that the leakage power in a T-CNFET is 7X lower than the MOS-CNFET at V_{DD} = 200mV and 10^4X lower at V_{DD}=300mV. In conclusion it can be noted that T-CNFETs with their super cut-off characteristics are favorable for subthreshold circuits. These devices have low operating currents, requiring wider area compared to their MOS counterpart. Nevertheless, the extremely low leakage can cause significant improvement in battery lifetime of handheld devices.

VI. SUPER CUT-OFF DEVICES AS SLEEP TRANSISTORS

Sleep transistor insertion has emerged as a promising method for

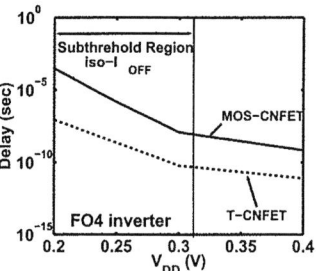
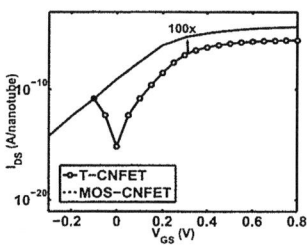

Fig 11: Variation of delay with V_{DD} for a MOS-CNFET and a T-CNFET for iso-I_{OFF} conditions.

Fig 12: Shifting the V_{FB} for MOS-CNFET to obtain iso-V_{th} condition

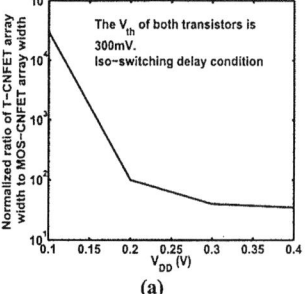

(a)

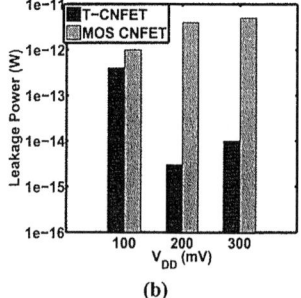

(b)

Fig 13: (a) Ratio of T-CNFET array width to MOS-CNFET array width for identical switching speed. (b) Leakage power consumption for iso-delay. Note that at lower V_{DD} a large number of T-CNFETs need to be used, thereby reducing leakage power savings.

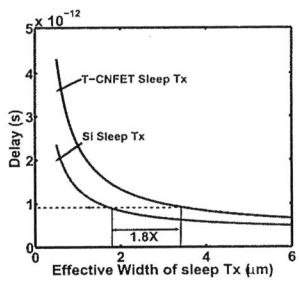

Fig 14: Variation of delay of an FO4 inverter with the size of the sleep transistor.

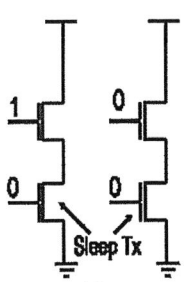

(a)

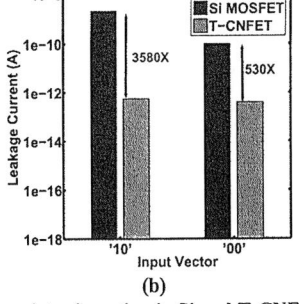

(b)

Fig 15: Leakage after sleep transistor insertion in Si and T-CNFET technologies.

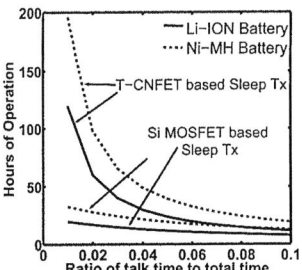

Fig 16: Increase in battery life-time of a typical cell phone application with T-CNFET and Si MOSFET based sleep Tx.

standby power reduction in high performance circuits [6,15]. In the standby mode, the sleep transistor is turned off, thereby creating stacking effect and reducing the overall leakage of the cell.

In this section, let us investigate the use of super cut-off T-CNFETs for sleep transistor applications in a regular Si fabric. It has often been argued that CNFETs would complement traditional Si MOSFETs in future technology nodes. Use of T-CNFETs as sleep transistors in a Si based logic circuit would take advantage of both worlds. An integrated circuit with Si MOSFETs in one layer and carbon nanotube transistors in another has been envisioned. As an add-on to the existing Silicon technology, carbon nanotube transistors hold a lot of promise.

The super cut-off T-CNFETs, can be effectively used as near ideal sleep transistors. When the T-CNFET is OFF, the leakage through it will be orders of magnitude lower than its Si MOSFET counterpart. This would lead to significant savings in terms of leakage.

For delay estimations, consider an FO4 inverter with $W_P=3um$ and $W_n = 1um$. The delay of the inverter increases as the size of the sleep transistor is decreased. First let us consider traditional bulk Si MOSFET technology. In this paper, we have used the 32nm Predictive Technology Models [16] and an operating V_{DD} of 700mV. Fig. 14 shows how the delay of the inverter varies with the width of the Silicon sleep transistor. Note that an optimal size of the sleep transistor is ~1.8um. Now let us consider a situation where the sleep transistor is fabricated with T-CNFETs. To compensate for their low ON currents, it is prudent to use T-CNFET arrays. Fig. 14 also illustrates the variation of delay with the width of the T-CNFET. By width is meant the effective width of the T-CNFET array when the carbon nanotubes are laid down in parallel with a pitch of 2nm.It may be noted that the optimal width for the T-CNFET array is ~3.3um. Hence for identical delay, the effective width of the T-CNFET array is ~1.8X wider than the Si MOSFET.

Now let us consider the standby mode when the sleep transistor is turned OFF. The two possible configurations corresponding to input vectors "00" and "10" have been shown in Fig. 15a. Fig. 15b shows the leakage current through the stack for different input vector conditions. Note that the use of super cut-off devices reduces the total leakage current by ~2000X. A few important observations are:

• Silicon sleep transistors show distinct stacking effect. From Fig. 15b it can be seen that, the '00' vector shows more than an order of magnitude improvement in leakage current compared to the '10' vector. This is due to the negative V_{GS} of the top transistor. However, if a super cut-off sleep transistor (as in T-CNFET) is used, the stacking effect does not exist. This is because a super cut-off device has extremely low leakage current and determines the total leakage in the stack, independent of the input vector condition. This is critical because it reduces the input vector dependence of leakage and simplifies the sleep transistor insertion methodology.

• Compared to Si transistors, the size of the T-CNFET array is about 1.8X for iso-delay. This may apparently seem to contradict the

observation in Fig. 13a where the T-CNFET array was shown to have more than an order of magnitude area penalty. However, it should be remembered, that the comparison in Fig. 13a was with MOS-CNFETs which have 10X higher current densities than SI MOSFETs. Hence, in a mixed technology, T-CNFETs can be effectively used in conjunction with Si transistors to provide ultralow leakage at a slight increase in the area. Then again, in such a technology, the CNTs could possibly be grown in a different active layer altogether, and the Si area would not be affected.

Now let's consider a cell-phone as a typical example of sleep transistor insertion. Fig. 16 shows the increase in battery life-time of a typical cell phone for different talk times with a 1.2Wh Li-ION and a 1.96Wh Ni-MH battery. A typical cell phone consumes ~1W of active power out of which ~30% is in the digital signal processor (DSP). The rest of the power is lost in leakage. A cell phone is a typical example where the device is more often in standby mode and wastes a considerable portion of the battery energy in leakage (~30-35%). Hence, it is essential to save leakage energy to improve the battery life-time. Most of the leakage energy is wasted in the DSP. From Fig. 16, it can be noted that decreasing the standby power by T-CNFET sleep transistor insertion, the total battery lifetime can be increased by more than 4X for low activities. Thus T-CNFETs can be effectively used in battery operated devices for increasing the battery life-time.

VII. CONCLUSIONS

This paper presents a through feasibility analysis of super cut-off transistors for ultralow power digital design. Tunneling CNFETs have been studied for subthreshold operation. Simulation results show 800X improvement in performance (@iso-I_{OFF}) when T-CNFETs are used instead of MOS-CNFETs. T-CNFETs have also been shown to be an excellent choice for sleep transistor applications in a regular Si CMOS process. T-CNFETs, with ultralow leakage currents can deliver more than 2000X improvement in leakage power (@iso-Delay) when compared to Si MOSFET based sleep transistors.

References:

1. J. Lim et al., Electronics Letters, Vol. 34, No. 4, pp. 344 -346, 1998.
2. Appenzellar et. al. PRL, Vol. 93, pp. 196805, 2004
3. P. F. Wang, Solid State Electronics, Vol. 48, pp. 2281, 2004
4. K. K. Yong, Trans on Electron Devices, Vol. 36, pp. 399, 1989.
5. Raychowdhury et. al., TVLSI, Vol 13, 2005, pp 1213-1224
6. M. C. Johnson, et.al.,ACM/IEEEDAC, 1999.
7. Taur & Ning, *Fund. Of Mod. VLSI Devices*, Cambridge Univ. Press, 1998
8. S.M. Sze, *Phy. Of Semicond. Devices*, John-Wiley, 1969
9. Ali. Javey et. al., Nanoletters, Vol 5, No. 2, 2005, Page(s): 345-348
10. Ali. Javey et. al., Nanoletters, Vol. 4, 2004, Pages(s). 1319-1322.
11. S. O. Koswatta at. al. IEDM Tech. Digest, pp. 525-528, 2005.
12. T. Baba, Jpn. J. Appl. Phy., vol 31, no. 4B, pp: L455-L457, 1992
13. J. Guo et. al., The Int J.on Multiscale Comp Engg, Vol. 2, pp: 257, 2004.
14. Raychowdhury et. al., Proc. Of IEEE Nano 2004, pp. TH-2-2-1
15. S. Narendra et. al., Proc. of ISLPED, 2001
16. www.eas.asu.edu/~ptm/

7

Variation-Driven Device Sizing for Minimum Energy Sub-threshold Circuits

Joyce Kwong
jyskwong@mtl.mit.edu

Anantha P. Chandrakasan
anantha@mtl.mit.edu

Massachusetts Institute of Technology
50 Vassar St., Room 38-107
Cambridge MA 02139, USA

ABSTRACT

Sub-threshold operation is a compelling approach for energy-constrained applications, but increased sensitivity to variation must be mitigated. We explore variability metrics and the variation sensitivity of stacked device topologies. We show that upsizing is necessary to achieve robustness at reduced voltages and propose a design methodology to meet yield constraints. The need for upsizing imposes an energy overhead, influencing the optimal supply voltage to minimize energy. Finally, we characterize performance variability by summing delay distributions of each stage in an arbitrary critical path and achieve results accurate to within 10% of Monte Carlo simulation.

Categories and Subject Descriptors: B.8.1 [Reliability, Testing, and Fault-Tolerance]

General Terms: Performance, Design, Reliability

Keywords: Sub-threshold circuits, Minimum energy point, Delay model

1. INTRODUCTION

In sub-threshold circuits, the power supply is set below the transistor threshold voltage V_T to obtain energy savings when speed is not the primary constraint [1]. Authors of [2][3] derived analytical expressions for the optimum V_{DD} to minimize energy in sub-threshold and showed its dependence on major circuit parameters. Sub-threshold circuits rely on leakage currents that are exponentially dependent on V_T and are therefore more sensitive to process variation than traditional above-threshold designs.

It was suggested in [4] that minimum size devices are theoretically optimal for minimizing energy in sub-threshold. However, minimum size devices have increased sensitivity to V_T variation because σ_{V_T} is roughly proportional to $(WL)^{-\frac{1}{2}}$. If a minimum size circuit does not function at the optimum V_{DD} due to degraded logic output swing, it is necessary to

upsize devices to improve robustness at the expense of increased energy consumption. Therefore, variability must be considered when analyzing the minimum energy operating point.

Previous work in [5] addresses intra-die variation by providing statistical models for energy and delay of an inverter chain in sub-threshold. An empirical expression for the optimum voltage is shown as a function of logic depth, assuming complete functionality at V_{min}. Work in [6] presents a unified delay variability expression for strong- and weak-inversion and applies it to a NAND gate. Researchers have also proposed various approaches to optimize delay yield by tuning V_{DD}/V_T or choosing gates of different drive strengths, for example in [7]. However, functional yield was not considered until [8][9], which address unsatisfactory V_{OH} and V_{OL} in sub-threshold inverters whose output levels are degraded by leaking devices, such as in a register file. Body biasing is another option for mitigating variation in sub-threshold [10] when a triple-well process is available.

We address inter- and intra-die variation and show that functionality in sub-threshold circuits may be compromised without proper design for variations. We first explore variability metrics for the inverter and logic gates with stacked devices, and propose a metric to size logic gates for a fixed failure rate under process variation. We then examine the energy versus V_{DD} profile given the failure rate constraint and find the optimum sizing and supply voltage. We present an efficient methodology to model delay variability of a chain of logic gates and characterize the effect of yield-based sizing constraints on performance variability.

2. VARIABILITY METRICS AND DEVICE SIZING

A commonly used expression for sub-threshold current is given by [11]

$$I_{sub} = I_o e^{\frac{V_{GS} - V_T + \eta V_{DS}}{n V_{th}}} (1 - e^{\frac{-V_{DS}}{V_{th}}}) \qquad (1)$$

$$I_o = \mu_o C_{ox} \frac{W}{L} (n-1) V_{th}^2 \qquad (2)$$

where n is the sub-threshold swing factor, V_{th} the thermal voltage, and η the DIBL coefficient. The nominal current scales linearly with W/L, while standard deviation of V_T distribution reduces with $(WL)^{-\frac{1}{2}}$, thus lowering sub-threshold current variation. This section explores how sizing affects

---

Permission to make digital or hard copies of all or part of this work for personal or classroom use is granted without fee provided that copies are not made or distributed for profit or commercial advantage and that copies bear this notice and the full citation on the first page. To copy otherwise, to republish, to post on servers or to redistribute to lists, requires prior specific permission and/or a fee.
ISLPED'06, October 4–6 2006, Tegernsee, Germany
Copyright 2006 ACM 1-59593-462-6/06/0010 ...$5.00.

variability in output swing and active current in the inverter and stacked device topologies.

2.1 Logic Gate Output Swing

In the sub-threshold regime, the ratio of active to idle currents in a logic gate is much lower than in strong inversion. If, for example, process variation strengthens NMOS relative to PMOS, a pull-up network will not be able to drive the logic gate output fully to V_{DD} because of idle leakage in the pull-down network. This degradation in gate output swing is illustrated in Figure 1(a). The solid line shows the voltage transfer characteristic (VTC) of a minimum size inverter in a 65nm technology at skewed global process corner. Dashed lines plot the VTCs when random local V_T mismatch is applied to the inverter. One case shows a severely degraded V_{OL}, which can cause functional error if it is above the input low threshold (V_{IL}) of the succeeding gate. Therefore, V_T variation significantly impacts circuit functionality in deeply scaled technologies.

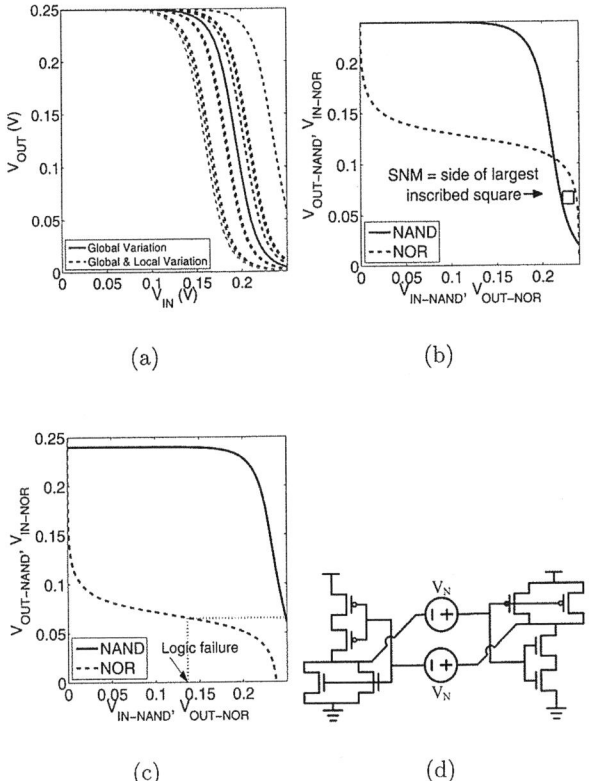

(a) (b)

(c) (d)

Figure 1: (a) Inverter VTCs at skewed process corner with random V_T mismatch. (b) Butterfly plot of NAND/NOR gates with functional output levels. (c) Butterfly plot of NAND with failing V_{OL}. (d) Example circuit for verifying logic gate output levels.

A consistent metric is necessary to determine whether a logic gate has sufficient V_{OL} and V_{OH} levels. Arbitrary limits, such as 10% and 90% of V_{DD}, do not scale well across global process corners. For example, at the strong-PMOS weak-NMOS corner, strong leakage through PMOS raises V_{OL} of all gates above ground. This also shifts VTCs to the right, and thus logic gates can tolerate higher V_{OL} in the preceding gate. Instead of arbitrary limits, we propose using butterfly plots to verify output voltage levels, specifically in the context of standard cell design.

2.1.1 Use of the Butterfly Plot

To verify V_{OL} of a given gate, we superimpose its VTC with the mirrored VTC of NOR, since the latter has the most stringent V_{IL} requirement from stacked devices in the pull-up network and parallel devices in the pull-down. Similarly, we verify V_{OH} using the NAND VTC, which has the worst case V_{IH}.

In Figure 1(b), a NAND gate has sufficient output swing such that $V_{OL-NAND}$ produces a logic high output in a succeeding NOR gate. In contrast, the NAND gate in Figure 1(c) exhibits $V_{OL-NAND}$=65mV and produces a NOR output of 136mV, close to mid-rail and thus causing logic failure.

A gate with failing output levels is analogous to a 6T SRAM cell displaying negative static noise margin (SNM), in that the butterfly plots for both cases do not contain an inscribed square. Therefore, we can also apply [12] to find the side of the largest inscribed square, illustrated in Figure 1(b). Figure 1(d) shows an equivalent circuit for this measurement on two back-to-back logic gates. Because the VTC is input-dependent, all inputs are varied simultaneously to obtain the worst case V_{IH} and V_{IL}.

It was shown in [13] that the SNM of two back-to-back gates G1 and G2 is equal to the maximum noise that can be applied to all gates in an infinitely long chain of alternating G1 and G2, before logic failure occurs. Thus when verifying a standard cell G using the butterfly plot, we essentially assume that all logic paths in a synthesized circuit are composed of alternating G and NAND3 gates with the same two skewed VTCs. To accurately model the failure rate of a custom-designed logic path, we would plot VTCs of all gates and trace the signal propagation through the path. Exact modeling is not possible for standard cell design where the target circuit is unknown. Therefore, although the butterfly plot does not reflect the exact mismatch conditions in a circuit, it does provide a guideline for sizing standard cells consistently to account for local variation.

2.1.2 Failure Rate From Insufficient Output Swing

We now define logic failure as having no inscribed square in the butterfly plot and measure how the failure rate varies with V_{DD} and device sizing. To consider logic gates with up to three stacked devices, we verify the INV, NAND2, and NOR2 gates against NAND3 and NOR3, which give the most stringent V_{IH} and V_{IL} requirements respectively. Sizing of NAND3 and NOR3 are fixed to provide a starting point for designing the remaining gates.

The failure rate is estimated from a 5k-point Monte Carlo simulation at worst case temperature. V_T of transistors in the gate under test and global (inter-die) process conditions are randomized such that the Monte Carlo runs are analogous to sampling logic gates across multiple dies. Figure 2(a) shows the failure rate versus V_{DD} of an inverter at various widths normalized to minimum size. Simulated values in markers are fitted to an exponential function ae^{bx}, drawn as a solid line. Note that the failure rate decays more quickly when W=1.66 compared to W=1. Furthermore, zero samples failed in the 5-k point run at higher voltages, as indicated by arrows on the graph.

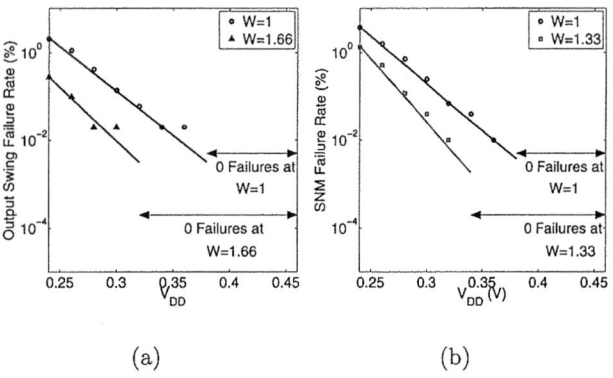

(a)   (b)

Figure 2: Failure rate of (a) inverter and (b) static register vs. V_{DD}, plotted for various NMOS and PMOS widths (normalized to minimum size).

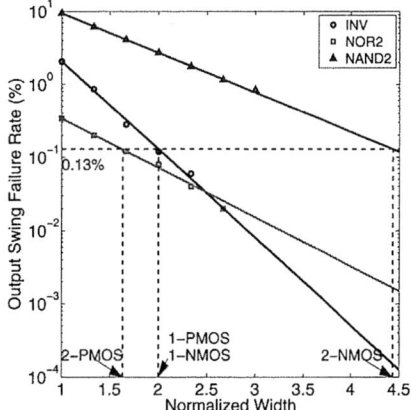

Figure 3: Output swing failure rate of the inverter, NAND2, and NOR2, plotted against device width (normalized to minimum size). V_{DD} is set at 240mV for demonstration.

Figure 3 plots the failure rate versus normalized device width of INV, NAND2, and NOR2. In the inverter, both device sizes are varied simultaneously. In NAND2 and NOR2, the critical two-transistor stack is changed while the two parallel devices are kept constant. The failure rates also decay exponentially with widths. By increasing the device width or V_{DD}, the failure rate can be made to approach 0.

2.2 Noise Margin in Registers

The concept of noise margin is also relevant in sub-threshold register design, where data retention is a particular challenge. Dynamic registers suffer from charge leakage, which worsens in sub-threshold due to slow circuit speeds. Therefore, we consider the static transmission-gate based register. Similar to SRAM cells, the data retention capability of the register is reflected in the hold static noise margin of its cross-coupled inverters. Figure 4 shows the equivalent circuit for measuring the register SNM, accounting for the

voltage drop across T2 and the worst case leakage across T1. This circuit is used in a Monte Carlo simulation while varying the V_T of each transistor and inter-die process conditions. Figure 2(b) plots the resulting failure rate in the cross-coupled inverters. Similar to the case of logic gates, the failure rate decreases exponentially to zero when either width or V_{DD} is increased.

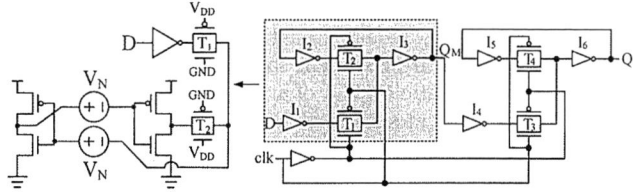

Figure 4: Static register schematic and equivalent circuit for measuring SNM.

2.3 Current Variability

In addition to output swing, active current variability is another metric of interest since it relates directly to variation in propagation delay. With the common assumption that V_T is normally distributed, sub-threshold current can be modeled as a lognormal random variable. From the property of lognormal distributions, the coefficient of variation of active current is given by

$$\sigma_{I_{sub}}/\mu_{I_{sub}} = \sqrt{e^{(\frac{\sigma_{V_T}}{nV_{th}})^2} - 1} \qquad (3)$$

It was observed in [5] that as V_{DD} reduces, the sub-threshold swing factor n decreases. This leads to higher uncertainty in the sub-threshold current through a single device. To examine the impact of topology, Figure 5 plots simulated $\sigma_{I_{sub}}/\mu_{I_{sub}}$ versus device width for static CMOS primitives consisting of one to three devices in series. Variability decreases with larger widths as expected. Stacked device topologies clearly display lower spread in active currents.

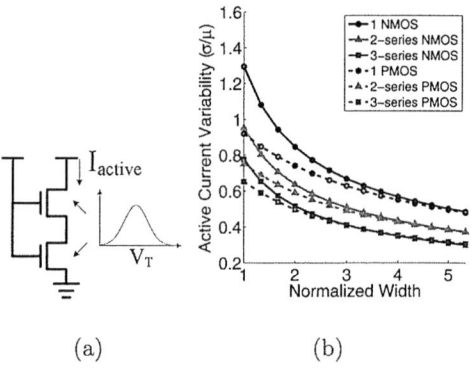

(a)   (b)

Figure 5: (a) Monte Carlo setup for current variability measurement. (b) Active current variability of different CMOS primitives vs. device width (normalized to minimum size) at V_{DD}=300mV.

10

2.4 Constant Yield Device Sizing

We now address the issue of device sizing for single and stacked device topologies, given the metrics of output swing and current variability. In above-threshold design, series devices are sized to give equivalent resistance as the inverter. However, in sub-threshold design when the objective is to minimize energy, device sizes should be kept as small as possible while satisfying variability constraints.

Compared to a single device, stacked devices display lower current spread but higher uncertainty in output levels, which may lead to functional errors. Reducing the error rate clearly takes precedence, so output swing rather than current variability should be considered first in sizing decisions.

The output swing failure rate versus width plot of Figure 3 illustrates a sizing methodology for single and stacked devices. Suppose we constrain all topologies to have the same failure rate, or interchangeably, a constant yield. We obtain the required device sizes by drawing a horizontal line at the desired failure rate, then finding where this line intersects the failure curve and the corresponding x-axis value. In Figure 3, a target failure rate of 0.13% requires a single and 2-stack NMOS to be sized at 2 and 4.43 times minimum width respectively. 1-PMOS is sized the same as 1-NMOS as both devices are varied together in simulation. The 2-stack sizing here can be used for any static CMOS gate with two series NMOS, since it was derived from NAND2 where two leaking parallel PMOS give the worst case V_{OL}.

Because the failure rate reduces at higher V_{DD}, the required size for a given yield constraint also decreases. The resulting energy trade-off will be analyzed in Section 3.1. Table 1 lists device widths for a constant failure rate of 0.13% while V_{DD} is varied at 20mV intervals. 0.13% represents the 3σ tail of a normal distribution and is chosen for demonstration. It should be noted that such a target allows sizing logic gates consistently, but does not relate in a straightforward way to the failure rate of a circuit built from these gates. As mentioned previously, this value is a pessimistic estimate because it assumes that every second gate in the circuit is NAND3 or NOR3. Furthermore, failing logic gates tend to cluster on die at process corners.

Table 1: Required widths (normalized to minimum size) vs. V_{DD} for constant failure rate=0.13%

$V_{DD}(V)$	0.24	0.26	0.28	0.30	0.32	0.34
1-NMOS	2	1.67	1.33	1	1	1
2-NMOS	4.43	2.93	2.3	2.27	1.3	1
1-PMOS	2	1.67	1.33	1	1	1
2-PMOS	1.63	1	1	1	1	1

3. MINIMUM ENERGY OPERATION

The total energy per operation consumed by an arbitrary circuit is modeled in [2] as

$$E_T = E_{DYN} + E_L = C_{eff}V_{DD}^2 + W_{eff}I_{leak}V_{DD}t_dL_{DP} \quad (4)$$

E_{DYN} and E_L model the dynamic switching and leakage energy per cycle respectively. C_{eff} and W_{eff} denote the average total switched capacitance and normalized width contributing to leakage current. t_d and I_{leak} represent the delay and leakage current of a characteristic inverter, while L_{DP} is the logic depth in terms of the inverter delay. As V_{DD} decreases, E_{DYN} is lowered quadratically. The leakage

current reduces because of DIBL, but t_d goes up exponentially at sub-threshold voltages and causes a similar increase in leakage energy. The two opposing trends give rise to an optimal supply voltage V_{DDopt} at which total energy is minimized, assuming the circuit is functional.

Section 2 has shown that functionality is no longer guaranteed at low supply voltages when V_T variation is significant. Reducing the probability of logic failure requires either upsizing devices or increasing V_{DD}, which must be considered when finding V_{DDopt}. This can be accounted for within the framework of [2] by treating C_{eff} and W_{eff} as a function of V_{DD}. The resulting energy versus V_{DD} characteristic of an inverter chain and 32-bit Kogge-Stone adder are simulated in a 65nm process and presented as examples.

3.1 Minimum Energy Point with Yield Constraint

Figure 6 plots C_{eff} and W_{eff} versus V_{DD} for the Kogge-Stone adder under two sizing schemes. The solid line plots energy of designs satisfying an upper bound on the output swing failure rate, derived from constant yield sizing of Table 1. The dashed line indicates an adder with only minimum size devices. Note that W_{eff} is obtained by normalizing the adder leakage current to that of a characteristic inverter [2]. DIBL affects leakage through the two circuits differently as V_{DD} decreases, causing a slight increase in W_{eff} in this case. V_{DDcrit} denotes the critical operating voltage at which minimum size devices can be used to satisfy the yield constraint. When $V_{DD} \geq V_{DDcrit}$, the circuit under both schemes are identical.

It should be noted that once the yield constraint is set, V_{DDcrit} can be found immediately from Table 1 and the topology of a given circuit. For example, a circuit without stacked devices does not require upsizing when $V_{DD} \geq V_{DDcrit} = 300$mV. In contrast, a circuit with stacks of two NMOS has $V_{DDcrit} = 340$mV.

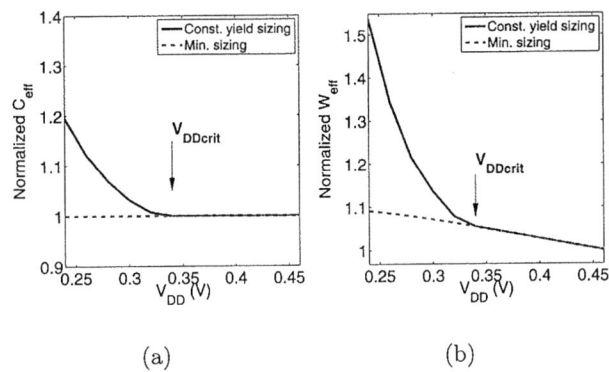

Figure 6: (a) C_{eff} and (b) W_{eff} for adder with constant yield (CY) and minimum sizing (MS).

The switching, leakage, and total energy of the inverter chain and adder are then calculated according to Equation 4. Figure 7(a) plots the energy versus V_{DD} characteristic of the inverter chain at nominal process and temperature. Total energy in both constant yield and minimum sized chains are dominated by the dynamic component. Therefore, the optimum supply voltage of the minimum size chain (dashed

11

line) is the lowest V_{DD} at which yield constraints are met. By definition, this is equal to V_{DDcrit}. In the constant yield sizing scheme (solid line), reducing the supply below V_{DDcrit} necessitates an increase in device widths. The resulting rise in C_{eff} dominates total energy. In this situation, there is no benefit from upsizing in order to operate at lower V_{DD}. The optimum operating point is with minimum sizing at the lowest V_{DD} permitted by the failure rate constraint.

When the minimum size circuit does have a local minimum in its energy characteristic, three scenarios exist depending on the relationship between V_{DDcrit} and the optimum V_{DD} of the constant yield ($V_{DDopt-CY}$) and minimum sizing ($V_{DDopt-MS}$) schemes.

Case 1) $V_{DDopt-MS} > V_{DDcrit}$: No upsizing is required to operate at the minimum energy point, therefore a minimum sized circuit at $V_{DDopt-MS}$ yields optimum energy.

Case 2) $V_{DDopt-MS} < V_{DDopt-CY} < V_{DDcrit}$: A minimum size circuit cannot operate at $V_{DDopt-MS}$ without violating failure rate constraints. A circuit suitably upsized to operate at $V_{DDopt-CY}$ yields optimum energy while satisfying yield requirements.

Case 3) $V_{DDopt-MS} < V_{DDopt-CY} = V_{DDcrit}$: At V_{DDcrit}, the circuit under both sizing schemes are identical. Therefore a minimum size circuit operating at V_{DDcrit} provides minimum energy.

An example of case 2 is seen in Figure 7(b) for a synthesized 32-bit Kogge-Stone adder with interconnect parasitics extracted from layout. Ignoring failure rate constraints, the minimum size adder (dashed line) has an optimum supply voltage of $V_{DDopt-MS} = 280$mV. When we account for failure rate constraints, the effect of constant yield sizing (solid line) is to add energy overhead when $V_{DD} < V_{DDcrit}$. This shifts the local minimum to the right, hence $V_{DDopt-CY} > V_{DDopt-MS}$. Here $V_{DDopt-CY}$ is also $< V_{DDcrit}$, therefore the adder with constant yield sizing at $V_{DDopt-CY} = 300$mV consumes 10.1% less energy than a minimum size adder at $V_{DDcrit} = 340$mV. In this example, constant yield sizing results in a small reduction in energy due to the shallow minimum of the energy versus V_{DD} curve.

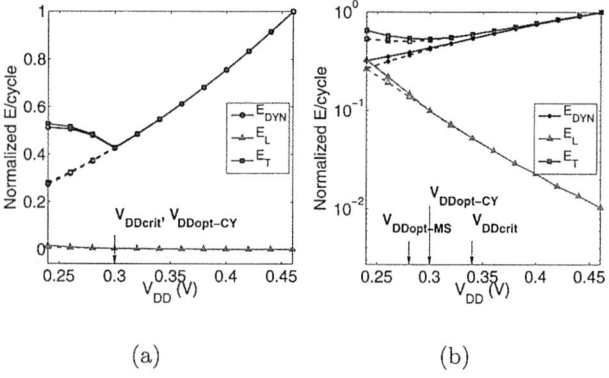

(a)   (b)

Figure 7: Energy vs. V_{DD} of (a) 11-stage inverter chain and (b) 32-bit adder. Solid and dashed lines indicate CY and MS sizing respectively.

4. PERFORMANCE VARIABILITY

4.1 Delay Variability Modeling

Circuits in sub-threshold display significantly higher delay variability than in above-threshold, therefore proper modeling is essential for timing verification. This section presents a methodology to efficiently model the delay distribution of a chain of logic gates. Using this model, we characterize the delay variability of the Kogge-Stone adders of Section 3.1.

From [2], the delay of a sub-threshold logic gate can be modeled as

$$t_d = \frac{KC_gV_{DD}}{I_o e^{\frac{V_{GS}-V_T}{nV_{th}}}} \tag{5}$$

where K is a delay fitting parameter, C_g is the output capacitance, and the denominator models the gate active current. Both the active current and t_d are lognormally distributed with the same σ parameter. Therefore, delay variability is also given by Equation 3. It depends on σ_{V_T}, which decreases as $(WL)^{-\frac{1}{2}}$, and the sub-threshold swing n, which decreases with V_{DS}. To the first order, σ/μ does not depend on input slew or load capacitance.

The critical path delay in sub-threshold is a sum of lognormal random variables (RVs), typically approximated as another lognormal RV. Authors of [5] derived an expression for the propagation delay of a chain of identical inverters using the Wilkinson approximation. Here we employ the Schwartz-Yeh method [14] to model the sum of non-identically distributed lognormal RVs. The delay of an arbitrary critical path can then be obtained by summing the pre-characterized distributions of each logic gate in the path.

The Schwartz-Yeh method is an iterative algorithm for calculating the sum of lognormal RVs, but requiring much less computation time than Monte Carlo simulation. The modeling methodology using this algorithm is described as follows:

1) Characterize mean delay and standard deviation (μ_{gate}, σ_{gate}) of each logic gate in a cell library, under one input slew and output load condition.

2) Simulate the (N-stage) critical path of interest at nominal process corner and without V_T variation. The delay of the jth stage in the critical path gives μ_{j-path}, for j=1 to N.

3) For each gate j in the critical path, let $\sigma_{j-path} = \sigma_{j-gate} \times \mu_{j-path}/\mu_{j-gate}$, where σ_{j-gate} and μ_{j-gate} are characterized in 1). Since the delay variability σ_j/μ_j is approximately constant across input slew and load conditions, this scales the pre-characterized standard deviation of each gate to the input slew and load conditions in the actual critical path.

4) μ_{j-path} and σ_{j-path} characterize the distribution of each stage, and are input to the Schwartz-Yeh algorithm to generate the delay distribution of the entire critical path.

The above methodology is applied to a three-stage chain consisting of INV-NAND-NOR and to the critical path of a 32-bit Kogge Stone adder at 300mV. Table 2 compares statistical model results with a 1-k point Monte Carlo simulation randomizing V_T of all transistors. The model estimates the mean and standard deviation of the path delay to within a few percent of the Monte Carlo results. This shows that keeping σ/μ constant provides a good approximation.

Table 2: Delay distribution parameters from statistical model and Monte Carlo simulation at 300mV. Values are normalized to FO4 delay.

	Model	Monte Carlo	% Difference
INV-NAND-NOR Chain			
μ	4.957	4.692	5.65%
σ	1.561	1.493	4.51%
Kogge-Stone Critical Path			
μ	36.52	37.13	1.65%
σ	7.038	7.262	3.09%

This method is used to characterize the delay distribution of 1) 32-bit adder with constant yield sizing at $V_{DDopt-CY} = 300$mV, and 2) adder with minimum size devices at $V_{DDcrit} = 340$mV. Table 3 shows that the first adder exhibits larger mean and 3σ delay, since $V_{DDopt-CY} < V_{DDcrit}$. However, the delay variability of both adders are comparable, indicating that upsized devices in the first adder offset increased variability from operating at a lower supply voltage.

Table 3: Delay distribution comparison of two adders from Section 3.1. Values are normalized to FO4 inverter delay at $V_{DDopt-CY}$.

	Const. Yield Sizing	Min. Sizing
μ	90.88	44.92
σ	17.46	8.857
$\mu + 3\sigma$	143.3	71.49
σ/μ	0.1921	0.1972

4.2 Energy Variability

From a 1k-point Monte Carlo simulation, we characterize the energy distribution of the adder with constant yield sizing at $V_{DDopt-CY}$ and the other with minimum size devices at V_{DDcrit}. As suggested in [5], the switched capacitance is verified to vary negligibly with V_T mismatch and is treated as deterministic. Figure 8(a) shows that even though the former adder employs larger devices, it displays lower mean leakage current due to DIBL, and lower variability as an additional benefit. The first adder exhibits lower mean total energy but higher variability in Figure 8(b). The latter effect results from the delay term in leakage energy having larger mean and standard deviation at 300mV compared to 340mV. Note that the leakage component is a product of two dependent lognormal RVs, so E_T is not strictly lognormally distributed.

5. CONCLUSION

In this paper, we have examined the effect of variation and sizing on single and stacked device topologies in subthreshold circuits. Compared to a single device, stacked devices exhibit lower current variability but a higher probability of logic failure from insufficient output swing. We introduced the use of butterfly plots to verify logic gates as well as registers against process variation, and showed that upsizing is necessary to mitigate degraded output levels. The need for upsizing to meet a given yield constraint imposes an energy overhead and impacts the optimum sizing and supply voltage at which energy is minimized. We presented a methodology to model delay variation in an arbitrary critical path using the delay distribution of each stage. Finally, we compared the delay and energy variability of the

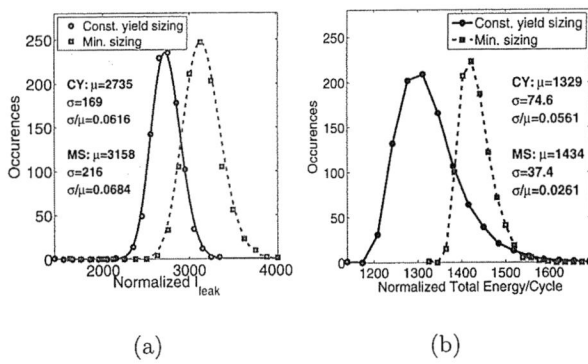

(a)   (b)

Figure 8: (a) Leakage current and (b) total energy for two adders of Section 3.1, normalized to those of characteristic inverter at $V_{DDopt-CY}$.

proposed sizing scheme with a minimum size circuit, and showed that energy reduction is possible without compromising yield or performance variability.

6. ACKNOWLEDGEMENTS

This research is supported by DARPA and Texas Instruments. The authors are grateful to N. Verma, B. Ginsburg, and D. Finchelstein for helpful discussions.

7. REFERENCES

[1] A. Wang and A. Chandrakasan, "A 180mV FFT Processor Using Sub-threshold Circuit Techniques," in *ISSCC*, 2004, pp. 292–293.

[2] B. H. Calhoun and A. Chandrakasan, "Characterizing and Modeling Minimum Energy Operation for Subthreshold Circuits," in *ISLPED*, 2004, pp. 90–95.

[3] B. Zhai, *et al.*, "Theoretical and Practical Limits of Dynamic Voltage Scaling," in *DAC*, 2004, pp. 868–873.

[4] B. H. Calhoun, *et al.*, "Device Sizing for Minimum Energy Operation in Subthreshold Circuits," in *CICC*, Oct. 2004, pp. 95–98.

[5] B. Zhai, *et al.*, "Analysis and Mitigation of Variability in Subthreshold Design," in *ISLPED*, 2005, pp. 20–25.

[6] Y. Cao and L. T. Clark, "Mapping Statistical Process Variations Toward Circuit Performance Variability: An Analytical Modeling Approach," in *DAC*, June 2005, pp. 658–663.

[7] S. H. Choi, *et al.*, "Novel Sizing Algorithm for Yield Improvement under Process Variation in Nanometer Technology," in *DAC*, 2004, pp. 454–459.

[8] J. Chen, *et al.*, "Robust Design of High Fan-In/Out Subthreshold Circuits," in *IEEE Int. Conf. on Computer Design (ICCD)*, Oct. 2005, pp. 405–410.

[9] ———, "Maximum-Ultra-Low Voltage Circuit Design in the Presence of Variations," in *IEEE Circuits and Devices Magazine*, Jan.-Feb. 2006, pp. 12–20.

[10] N. Jayakumar and S. P. Khatri, "A Variation-tolerant Sub-threshold Design Approach," in *DAC*, 2005, pp. 716–719.

[11] V. De, *et al.*, "Techniques for Leakage Power Reduction," in *Design of High-Performance Microprocessor Circuits*, A. Chandrakasan, *et al.*, Eds. IEEE Press, 2001, ch. 3, pp. 46–62.

[12] E. Seevinck, *et al.*, "Static Noise Margin Analysis of MOS SRAM Cells," *IEEE J. Solid-State Circuits*, vol. SC-22, no. 5, pp. 748–754, Oct. 1987.

[13] J. Lohstroh, *et al.*, "Worst-case Static Noise Margin Criteria for Logic Circuits and Their Mathematical Equivalence," *IEEE J. Solid-State Circuits*, vol. SC-18, no. 6, pp. 803–807, June 1983.

[14] S. Schwartz and Y. Yeh, "On the Distribution Function and Moments of Power Sums with Log-Normal Components," *Bell Sys. Tech. Journal*, vol. 61, no. 7, pp. 1441–1462, Sept. 1982.

Robust Level Converter Design for Sub-threshold Logic

Ik Joon Chang
School of ECE, Purdue University,
West Lafayette, IN 47906 ,USA
1-765-426-5707
ichang@purdue.edu

Jae-Joon Kim
IBM T. J. Watson Research
Center,
Yorktown Heights, NY 10598,USA
1-914-945-3873
jjkim2@us.ibm.com

Kaushik Roy
School of ECE, Purdue University
West Lafayette, IN 47906 ,USA
1-765-494-2361
kaushik@ecn.purdue.edu

Abstract

The large supply voltage difference between sub-threshlold core logic and I/O makes it extremely challenging to convert signals from core circuit to I/O circuit. In this paper, we propose two novel circuits, *Clock Synchronizer* and *Reduced Swing Inverter* to design dynamic and static level converters for sub-threshold logic. Circuit simulations shows that our level converters work at frequency > 500Khz between 20°C and 40°C with a supply voltage of 0.25V.

Categories and Subject Descriptors

B.6.1 [Logic Design]

General Terms

Design

Keywords

Sub-threshold logic, level converter, low power circuit design

1. Introduction

The demand of low power circuit design has increased significantly due to the explosive market growth of battery-operated portable applications like laptop computers and cellular phones. In order to reduce energy consumption, many techniques such as voltage scaling [1], switching activity reduction [2] are widely used. However, these methods are not sufficient in some portable applications and medical electronics in which ultra-low power consumption has higher priority than performance and area. To cope with this, digital sub-threshold logic design have been explored [3-4].

Since extremely low supply voltage (VDD) is used in sub-threshold logic, energy consumption can be improved significantly. The ultra-low core VDD, however, can be vastly different from high I/O supply voltage (VDDH), and hence it is important to convert signals from core logic to I/O circuits. Lowering I/O supply voltage may not be a solution for this problem since there is a limit to scaling of I/O supply voltage due to the large impedance load and high noise immunity requirement for I/O circuit. In addition, conventional level converters which were designed to convert low super-threshold input to high super-threshold output are not effective to convert sub-threshold input to super-threshold voltage output. Therefore, the design of level converter which can convert sub-threshold input signal to I/O signal is critical for the success of sub-threshold logic.

Permission to make digital or hard copies of all or part of this work for personal or classroom use is granted without fee provided that copies are not made or distributed for profit or commercial advantage and that copies bear this notice and the full citation on the first page. To copy otherwise, or republish, to post on servers or to redistribute to lists, requires prior specific permission and/or a fee.
ISLPED '06, October 4–6, 2006, Tegemsee, Germany
Copyright 2006 ACM 1-59593-462-6/06/0010...$5.00.

In this paper, we propose two level converter designs for sub-threshold logic. The first level converter design is based on dynamic circuit style. In this design, we propose a new clock synchronization scheme and a novel keeper structure to overcome the limitation of conventional dynamic level converting schemes. In the second level converter design, we propose a DCVSL static level converter with the *Reduced Swing Inverter*. In this scheme, the pull-up PMOS transistors are weakened by the *Reduced Swing Inverter*s so that the on-current of sub-threshold pull-down NMOS transistors can overcome initial PMOS current. This is not possible in conventional static level converter design.

2. Conventional Level Converters

In this section, we discuss conventional level converters. Several different types of level converters have been proposed in literature [5]. Fig. 1 shows two representative static level converters and a dynamic level converter. As shown in Fig. 1 (c), a static level converter is embedded for clock shifting inside a dynamic level converter. If a dynamic level converter uses a different clock path from core logic, synchronization problem may result from mismatch between delays of both clock paths. Therefore, the clock converted from low voltage is employed for dynamic level converters, which means that an extra static level converter is required for a dynamic level converter. Since a dynamic level converter can be embedded in a flip-flop, it can be employed as a level converting flip flop for multiple VDD logic. However, an extra flip flop is not required for the I/O part. Thus, static level converters are more efficient than dynamic level converters for I/O circuit.

The conventional static level converters, however, have critical problems when converting sub-threshold input voltage to higher super-threshold output voltage. Fig. 2 shows the reason why the conventional static level converter does not function correctly for ultra low voltage input. Level conversion process of DCVSL level converters takes 3 steps as shown in Fig. 2 (a). When MN1 turns on, MN1 pulls down 'OUTB' and turns on MP2. Since MP2 is on and MN2 is off, 'OUT' is charged to '1' and MP1 turns off. Note that MP1 and MN2 are on simultaneously in step 2. In order to discharge the node 'OUTB', MN1 should be much stronger than MP1. By controlling the size of each transistor, this problem can be solved at normal VDD. However, on-current of transistors decreases exponentially as gate voltage is reduced in sub-threshold region. Since only pull down transistors such as MN1 and MN2 are in sub-threshold region, we cannot obtain appropriate pull down strength to overcome the current from MP1 and MP2 only by sizing as shown Fig. 2 (b). Other static level converter designs experience similar problems.

Furthermore, thick oxide device should be employed for I/O circuit since maximum possible voltage between gate and drain of

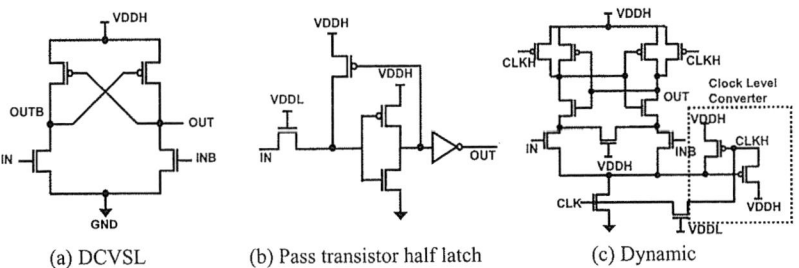

(a) DCVSL   (b) Pass transistor half latch   (c) Dynamic

Fig. 1 Conventional level converters

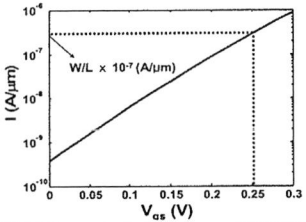

Fig. 4 I/V Characteristic Curve of NMOS
(W/L = 1µm / 0.3µm)

normal oxide device is much less than I/O voltage. Thick oxide device has higher threshold voltage (V_t) than normal oxide device, which makes it more challenging to design static level converters for ultra low voltage.

In order to solve these problems, several techniques have been proposed [6-7]. Fig. 3 shows a static level converter based on these techniques. By doubling input voltage, we can increase the strength of pull down transistors. In addition, a normal oxide NMOS transistor stacked with a zero V_t NMOS transistor is employed instead of thick oxide devices. Due to a zero V_t NMOS transistor, we can avoid large voltage difference between gate and drain of normal oxide NMOS devices. Since V_t of normal oxide transistors is lower than thick oxide transistors, the use of stacked NMOS transistors makes pull down stronger and results in lowering input voltage significantly. In our simulation, however, these techniques still failed to convert sub-threshold voltage to I/O

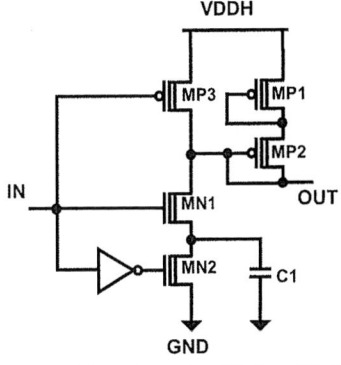

Fig. 5 The schematic of *Reduced Swing Inverter*

voltage due to large difference of on-current between super-threshold and sub-threshold regions. Therefore, it is difficult to make static level converter for sub-threshold logic using these schemes.

3. Level Converter Design for Sub-threshold Logic

As discussed in the previous section, conventional static level converters cannot be applied for communication between sub-threshold logic and I/O circuit. In this section, we present a static and a dynamic converter design that can convert sub-threshold signal to I/O signal. For both the designs, a novel circuit named *Reduced Swing Inverter* has been employed. The *Reduced Swing Inverter* is discussed in section 3.1.

Our dynamic level converter is introduced in section 3.2. Conventional dynamic level converters use clock signals boosted by static level converters for clock synchronization. However, we cannot apply the same method for sub-threshold logic since static level converters do not work correctly at such low voltages. Therefore, in our dynamic converter design, we assumed that a high voltage clock would be delivered through an extra clock path. In such a scenario, level converters have different clock paths from core logic, which can lead to synchronization problems. To solve this issue, we have developed a novel circuit called *Clock Synchronizer*. This *Clock Synchronizer* generates a new high voltage clock synchronized with low voltage clock of core logic. In section 3.3, we propose static level converter based on DCVSL logic using the *Reduced Swing Inverter*. In our static level converter design, both pull up and pull down transistors work in sub-threshold domain, and hence the static level converter may be sensitive to process and temperature variation. We discuss design of these issues and introduce ways to make the design robust to these variations in the section 3.3.

The proposed techniques were implemented using IBM 130nm technology. The V_t of normal oxide NMOS in IBM 130nm is 0.19V. However, V_t for ultra low VDD is higher due to reduced DIBL [8]. For example, the I/V characteristic curve of Fig. 1

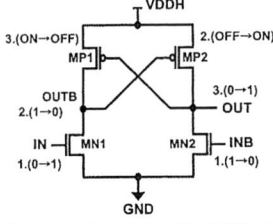

(a) Level conversion steps of DCVSL level converters

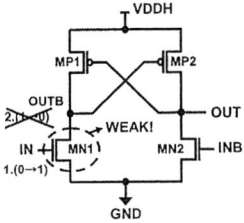

(b) Level conversion fails in ultra low voltage input due to weak pull down

Fig. 2 Failure mechanism of conventional level converter when applied to sub-threshold logic

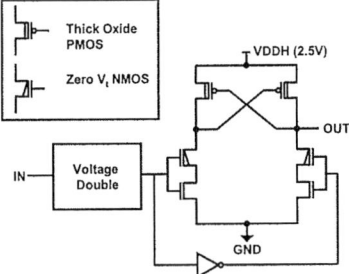

Fig. 3 Level converter based on voltage doubler and stacked NMOS transistors

15

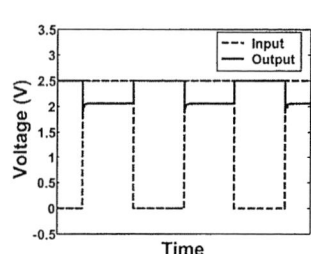

Fig. 6 Simulation result of *Reduced Swing Inverter*

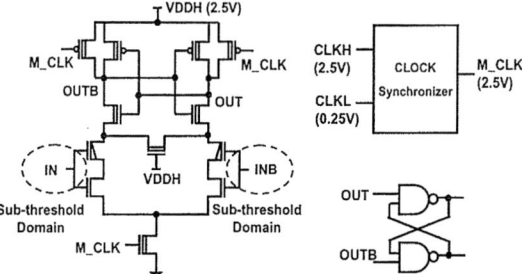

Fig. 7 Dynamic level converter for sub-threshold logic

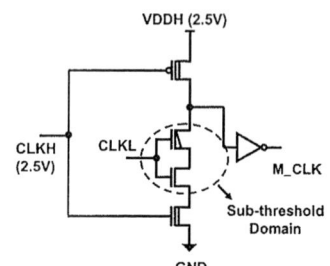

Fig. 8 The schematic of *Clock Synchronizer*

which is simulated for 0.3V VDD shows that V_t is around 0.25V. Our designs are optimized for 0.25V VDD.

3.1 Reduced Swing Inverter

The fundamental problem of conventional level converters is that pull down transistors are too weak compared to pull up transistors when the converters are driven by sub-threshold input signals. In super-threshold logic, we can make pull down transistors strong by sizing them suitably. However, we should note that for level converters driven by sub-threshold input signals, only the pull down transistors are in sub-threshold domain. As a result, it is difficult to obtain strong pull down network required for level conversion with device upsizing only. As discussed in section 2, the gate signal of low VDD domain is boosted to solve this problem in the previous research [6]. The boosting efficiency, however, decreases as voltage decreases. At ultra-low voltage such as 0.2V, signals are not boosted high enough for level conversion. Therefore, in this design, we try to make the NMOS transistors relatively strong compared to the PMOS transistors by reducing the PMOS V_{gs} values. For this purpose, we propose a *Reduced Swing Inverter* design as shown in Fig. 5.

The operating principle of the *Reduced Swing inverter* is as follows: When the input signal is '0', MP3 turns on and 'OUT' is charged to 'VDDH'. If the input signal is changed to '1', both MN2 and MP3 are turned off. Since 'OUT' is disconnected from GND, this node is not completely discharged. However, the voltage of 'OUT' drops by a certain value due to charge redistribution between the output node and the capacitance C1. This results in reduced output voltage swing. Since there is no direct on-current path like in a conventional inverter, static power consumption can be saved. In this scheme, sub-threshold current may destroy the voltage value of 'OUT'. In order to keep 'OUT' from losing its charge, we add two diodes, MP1 and MP2. If 'OUT' goes down below a certain voltage, the diodes turn on and 'OUT' is recharged. Due to diode voltage drop, the 'OUT' voltage value is not raised beyond a certain point, which means that the voltage of 'OUT' can be preserved constantly when the input is '1'. The voltage drop of 'OUT' can be controlled by changing the number and sizes of diodes, which is very critical for level converters. In our design, two diodes are cascaded as shown in the figure. We employed PMOS as a diode to compensate the effect of inter-die variation in our level converters. The details are discussed in section 3.3.

Fig. 6 shows the simulation result of a *Reduced Swing Inverter*. An input pulse of 2.5V results in output swing between 2.05V and 2.5V. Since the swing range of the output is limited around high VDD, we can obtain weak pull up transistors by driving PMOS with the proposed *Reduced Swing Inverter*.

3.2 Dynamic Level Converter Design for Sub-threshold Logic

Similar to other dynamic logic styles, clock signal is required for pre-charge of dynamic level converter. In order to synchronize level converters with core logic, dynamic level converters should use the same clock as core logic. Since low voltage clock cannot be used for pre-charge signal of dynamic level converters, clock level shifting is required. In conventional technologies, a static level converter is embedded inside a dynamic level converter for shifting the clock. As discussed in section 3.2, static level converters do not work correctly in the sub-threshold region. This limits the application of dynamic level converters as well. Hence, there is a great need to devise new synchronization schemes between core clock and I/O clock since the conventional dynamic level converter can easily convert sub-threshold input signal to high output signal if there is no clock synchronization problem.

In this work, we assume that PLL delivers high voltage clock signal for level converters through different paths from core logic. Since it is difficult to design a PLL that works with sub-threshold VDD, we use high voltage VDD for the power supply of the PLL. Therefore, the PLL generates a high voltage clock, which we level down for applying to the core logic. If PLL provides high voltage clock for level converters through an extra clock path, we can receive high voltage clock which has the same period as the leveled down clock for the core logic. However, different clock paths for level converters and core logic may result in synchronization problem. Furthermore, low voltage clock buffers (for core logic clock) have large delay compared to high voltage clock buffers (for level converter clock), which results in large clock skew. In our design, we propose a novel circuit called *Clock Synchronizer* in order to solve this clock skew problem. The *Clock Synchronizer* generates modified clock (M_CLK) from high voltage clock (CLKH) and sub-threshold voltage clock (CLKL). Since M_CLK is synchronized with CLKL, we can use M_CLK as clock signal of dynamic level converters as shown in Fig. 7.

In order to make M_CLK synchronized with CLKL, we use domino logic as drawn in Fig. 8. CLKH is employed for pre-charge and evaluation signal of the *Clock Synchronizer*. Since PMOS is off during evaluation of domino logic, the *Clock Synchronizer* works well even for ultra low voltage input. Fig. 9 shows timing diagram of the *Clock Synchronizer*'s input and output. As discussed above, the clock buffers of sub-threshold logic have larger delay than clock buffers of super-threshold logic. Though the lengths of clock paths are different, CLKH would arrive faster than CLKL due to large difference between I/O voltage and sub-threshold VDD. Since CLKH arrives earlier, the footer transistor of the *Clock Synchronizer* turns on. However, evaluation does not start since the top evaluation transistors are not yet turned on. Evaluation occurs only when CLKL signal is high

16

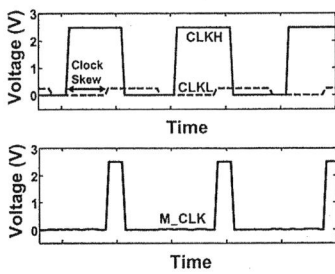

Fig. 9 Timing diagram
of *Clock Synchronizer* input and output

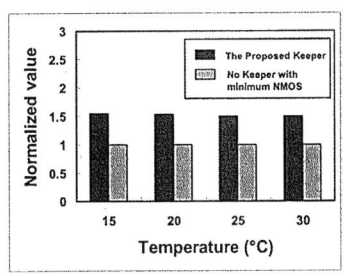

Fig. 11 Normalized noise margin comparison
between two designs

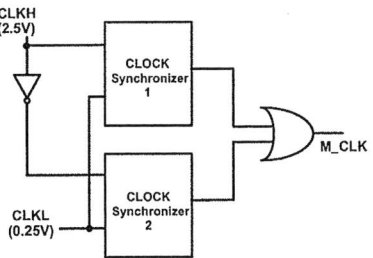

Fig. 12 Duplicated *Clock Synchronizer*

and the top two evaluation transistors are turned on. M_CLK can therefore be synchronized with CLKL as shown in Fig. 9.

However, this *Clock Synchronizer* design has some inherent problems. First, the noise margin of *Clock Synchronizer* is an issue. Since the *Clock Synchronizer* is based on domino logic style, it may be susceptible to noise. In conventional domino logic, keepers are used for increasing the noise margin. However, weak pull down transistors limit the application of conventional keepers in this design. Hence, there is a necessity to implement a proper keeper for this circuit. The second problem is that the *Clock Synchronizer* does not work correctly when the clock skew is larger than half clock period. Since sub-threshold logic is very sensitive to process and temperature variations, low voltage clock may suffer from clock jitter. Therefore, we may not guarantee that low voltage clock follows high voltage clock within half clock cycle period. In the next subsections, we propose the solutions to these two problems.

3.2.1 Keeper Design for Clock Synchronizer

As discussed above, the proposed *Clock Synchronizer* suffers from small noise margin similar to other dynamic circuits. In conventional domino logic, keepers are employed to raise the noise margin. However, if conventional keepers are employed here, the *Clock Synchronizer* may have problems in evaluation due to weaker pull down strength of the NMOS transistors operating in sub-threshold region than keeper PMOS operating in super-threshold region.

In this work, a new keeper is proposed using a *Reduced Swing Inverter*. The *Reduced Swing Inverter* drives the keeper as shown in Fig. 10. In section 3.1, we verified that the output range of a *Reduced Swing Inverter* is limited between 2V and 2.5V. As a result, before the evaluation phase, MP1 turns on weakly since the voltage of 'Keep' node stays around 2V. During evaluation, MN1 should discharge the 'PRE_OUT' node. Since MP1 is weakened

by using a *Reduced Swing Inverter*, we can make MN1 relatively strong compared to MP1 with conventional sizing techniques.

In order to verify the usefulness of the proposed keeper, we compare two designs. For the first design, we do not use any keepers and employ a minimum size NMOS as evaluation device, in order to maximize noise margin. For the second one, the proposed keeper is used and NMOS is sized properly. Fig. 11 shows normalized noise margins of both designs which are measured at 500 Khz and operated at nominal process corner. The figure shows that the noise margin is improved by about 50% with the 'weak keeper' design. We also compared the delay distribution of the two designs (measured by 1000 Monte Carlo simulations), and found that the mean and worst delay of the second design are smaller than the first. If the NMOS width is increased in the first design, we can obtain the same performance as the second one. However, it results in much smaller noise margin. Thus, we conclude that noise margin is significantly improved with the proposed keeper.

3.2.2 Modified Clock Synchronizer

If clock skew is larger than half clock cycle, the *Clock Synchronizer* is in pre-charge mode when CLKL signal starts to rise. Since pre-charge device is on and footer NMOS is off in such as case, the *Clock Synchronizer* does not work properly. In order to solve this problem, we propose duplicated *Clock Synchronizer* as shown Fig. 12. One *Clock Synchronizer* uses CLKH as the pre-charge signal. The other *Clock Synchronizer* receives the inverted CLKH as its pre-charge signal. Since complementary signaling is employed for pre-charging the two *Clock Synchronizer*s, one of the two paths is evaluated when CLKL rises. The OR gate at the output generates the M_CLK signal that rises only when CLKL becomes high. Therefore, we can obtain high voltage signal synchronized with CLKL.

However, there still exists one problem in this design. Fig. 13 shows the simulation result of the design proposed in Fig. 12.

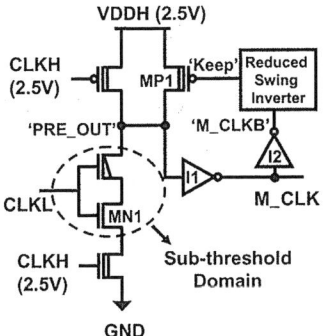

Fig. 10 Proposed *Clock Synchronizer*. Weak keeper is designed by employing *Reduced Swing Inverter*.

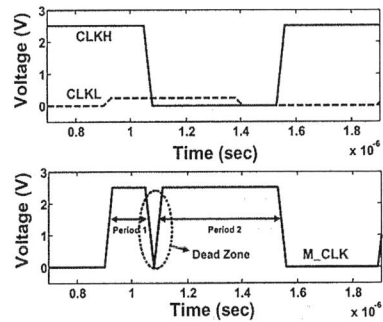

Fig. 13 The simulation result of design proposed in Fig. 12

17

During 'period 1', *Clock Synchronizer* 1 is evaluated. It is pre-charged and the output node of *Clock Synchronizer* 1 becomes 0V, when the CLKH signal goes low. Meanwhile, *Clock Synchronizer* 2 starts evaluation at the falling edge of CLKH. Note that V_{gs} of NMOS is in sub-threshold region during evaluation of *Clock Synchronizers*, which means that the evaluation speed of *Clock Synchronizers* is slow compared to the pre-charge speed. Therefore, the fall delay at the output node of Clock Synchronizer 1 is much smaller than the rise delay of Clock Synchronizer 2. This results in a 'Dead Zone' as shown in Fig. 13. When CLKH falls, one input to the OR gate falls before the other input rises. Thus, both inputs to the OR gate is 0 between 'period 1' and 'period 2'. During this period, M_CLK will be 0V.

If we use M_CLK as pre-charge signal of our dynamic level converters, unexpected pre-charging occurs during 'Dead Zone'. In other words, dynamic level converter has two evaluation phases during one clock cycle. Since input signals of level converters may be changed during 'period 1', the first evaluation output may be different from the second evaluation. To ensure correct operation, the hold time of dynamic level converter inputs should be longer than 'period 1'. However, the width of 'period 1' is unpredictable since the clock paths of low voltage clock are different from high voltage clock. Therefore, we need to eliminate the 'Dead Zone'.

If the rise delay of *Clock Synchronizer* output node is shorter the fall delay, one input of the OR gate is changed to '1' before other input becomes '0' and the output is not changed between 'period 1' and 'period 2'. Thus, the 'Dead Zone' can be removed if the fall delay of *Clock Synchronizer* is larger than its rise delay. However, it is difficult to obtain larger fall delay than rise delay with *Clock Synchronizer* in Fig. 10 due to large difference between V_{gs}'s of evaluation NMOS and pre-charge PMOS. In order to obtain long falling delay, we modify the design of the *Clock Synchronizer* as shown in Fig. 14. The operating principle of the modified Clock Synchronizer is as follows:

When 'V_1' is discharged in the evaluation mode, 'V_2' is charged. Conversely, 'V_2' discharges during the pre-charge of '$V1$'. Since both the evaluation path of 'V_1' and the discharge path of 'V_2' are in sub-threshold domain, the *Clock Synchronizer* includes a sub-threshold operation for both pre-charge and evaluation phases.

Therefore, we can obtain a longer falling delay of the Clock Synchronizer compared to its rising delay by suitable sizing MN1 and MN2. We verified this technique from Circuit simulations, which show that the 'Dead Zone' is removed in the modified *Clock Synchronizer* design.

3.3 Static Level Converter Design for Sub-threshold Logic

In section 3.2, we present a dynamic level converter which can be applied for sub-threshold logic. For the design, we assumed that high voltage clock was supplied from PLL. Though the proposed dynamic level converter works properly for sub-threshold logic, it consumes large clock power due to large capacitance of clock paths. Therefore, a static level converter is more efficient for I/O circuit since it does not require any clock signal.

Conventional static level converters have problem in converting sub-threshold input signal due to the weak pull down of NMOS operating in sub-threshold region. However, we can obtain relatively strong NMOS compared to PMOS by using *Reduced Swing Inverters*. Fig. 15 shows our static level converter design. Since the output range of a *Reduced Swing Inverter* is limited around 2.5V, V_{gs} of MP1 and MP2 is reduced. In other words, both PMOS transistors are operated in near-sub-threshold region, which means that pull up transistor is weakened by the *Reduced Swing Inverters*. Therefore, the strength of NMOS required for level conversion is reduced. By sizing transistors suitably, we can design a static level converter which can convert sub-threshold voltage inputs to I/O voltage. Simulation results show that our static level converter can shift 0.15V signal to 2.5V with nominal corner variation at room temperature (Fig. 16).

It is worthwhile to note that this level converter is sensitive to process and temperature variation since V_{gs} of both NMOS and PMOS is very low. Although we can compensate both variations with adaptive body biasing techniques [9], these techniques would reduce the extent of variation. Since even small variations have large effect on sub-threshold logic, the level converter should be designed carefully to deal with variations, which raises several critical issues in our static level converter design.

The most important issue is the output swing variation of *Reduced Swing Inverter*. When the input of a Reduced Swing Inverter is '1', MN2 and MP3 are turned off (Fig. 5). Diodes are also turned off when output node rises beyond a certain point, which leads to high impedance state of output node. Though there is no ON current during this time, sub-threshold current flows. Thus, the output is determined by voltage division between diodes and MN2 (Fig. 5). Since both are in sub-threshold region, they are sensitive to process and temperature variations. Therefore, output swing of a Reduced Swing Inverter may be susceptible to variations as well. In the proposed static level converter, the output

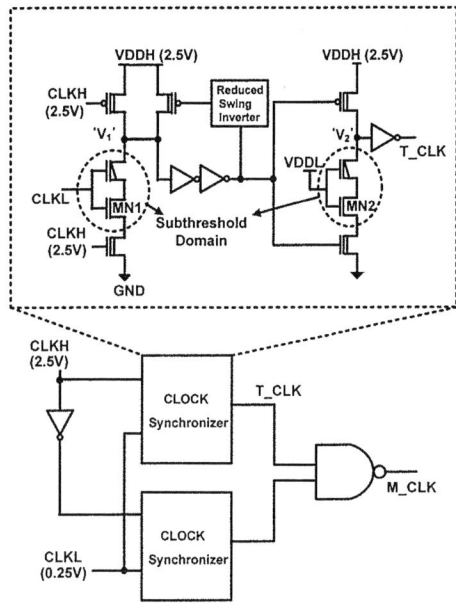

Fig. 14 Modified *Clock Synchronizer*

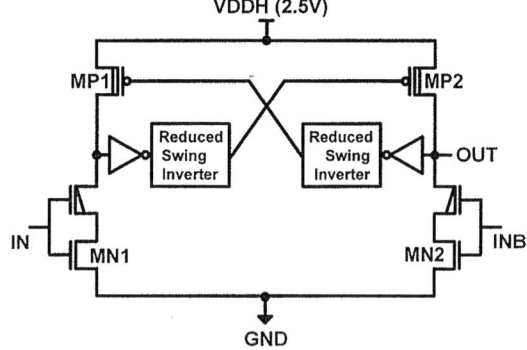

Fig. 15 A static level converter for sub-threshold logic

swing is V_{gs} of pull up transistors. As a result, small variation of output swing may have high impact on PMOS strength, which makes suitable sizing of our static level converter very difficult.

In order to solve this problem, we implemented diodes inside the *Reduced Swing Inverter*s with PMOS. Device parameters move to the same direction due to die to die variations [10]. Though process variations exist within a die, the impact is less than die to die variations [10]. Thus, all PMOS transistors of the static level converter have similar strength. If PMOS diodes become stronger than nominal value, the output swing of the *Reduced Swing Inverter*s decreases due to low resistance of diodes. Conversely, weak PMOS diodes result in larger output swing. As a result, V_{gs} of MP1 and MP2 goes up as PMOS becomes weaker. Since large V_{gs} of PMOS transistors leads to strong pull up, the effect of die to die variation can be negated by employing PMOS transistors for diodes.

In addition, for more robust design, we increased output swing of the *Reduced Swing inverter*s as much as possible. The impact of process and temperature variations is reduced as V_{gs} becomes larger. Therefore, we can make the pull up transistors of the level converter less sensitive to these variations by increasing the output swing of the *Reduced Swing inverter*s. In order to verify robustness of the static level converters, we simulated the circuit in HSPICE (1000 Monte Carlo simulations). The results show that our static level converter work at >500Khz between 20°C and 40°C. Though only DCVSL based level converter is proposed in this work, other static level converters can be modified for sub-threshold logic by employing the Reduced Swing Inverter.

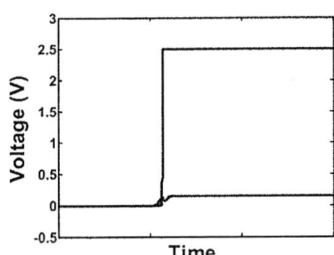

Fig. 16 Waveform for level conversion from 0.15V input to 2.5V output signal at typical corner variation and 25°C

4. Simulation Results

The level converters are designed in IBM 130nm technology. The noise margin of the *Clock Synchronizer* is very critical for the design of the dynamic level converter since the performance of the dynamic level converter is dependent on this. From Monte Carlo simulations, 30mV is guaranteed as the noise margin at frequency >500kHz and a temperature of 40°C.

In order to compare the effect of process variation between static and dynamic designs, we tried corner simulations at room temperature (25°C) with each corner variation, which are fast fast (FF), fast slow (FS), slow fast (SF), and slow slow (SS). Oxide thickness was fixed to nominal value. Then, we inspected the influence of temperature variation. From the first simulation, we can obtain the worst corner (SS corner). At this corner, we changed only temperature from 20°C to 30°C.

Fig. 17 summarizes our simulation results. As shown in Fig. 17 (a), the delay of the static level converter is around 20% of the dynamic level converter's delay. Due to performance loss for clock generation of the *Clock Synchronizer*, the dynamic level converter is slower than the static one. For both designs, the delay

of the worst corner increases by 14% compared to the best corner. Therefore, the sensitivity to process variations of both designs can be regarded as almost equal. Fig. 17 (b) shows that two level converter have similar sensitivity to temperature variation as well. Therefore, we can conclude that the static level converter is as robust to process and temperature variations as the dynamic level converter. Furthermore, the static level converter requires less clock power consumption than dynamic level converter. Though the design process for robust design of the static level converter is very complex as discussed in section 3.3, it is more efficient for sub-threshold logic.

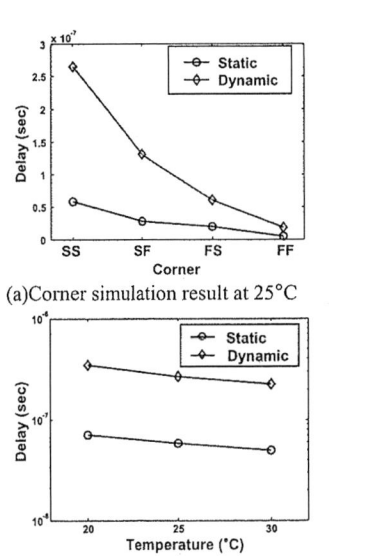

(a)Corner simulation result at 25°C

(b) Delay of static and dynamic level converters at SS corner

Fig. 17 Simulation result (SS= slow slow, SF = slow fast, FS=fast slow, FF=fast fast, VDD=0.25V, I/O VDD=2.5V)

5. Conclusion

We proposed new dynamic and static level converters for sub-threshold logic which overcome the limitations of conventional level converters. We developed a new circuit style to solve the clock synchronization problem between sub-threshold and super-threshold voltage domain in dynamic level converter. We also propose a novel *Reduced Swing Inverter* circuit for correct functioning of static level converter and for the keeper design in the dynamic level converter. Extensive Circuit simulation results show that the static level converter is more efficient than dynamic level converter although the design of static level converter requires more careful device sizing to cope with large process variation in sub-threshold region.

REFERENCES

[1] C. Chen, et al, IEEE *Trans. Circuits Systems. II*, vol. 49, Jun, 2002.

[2] G. Palumbo, et al, IEEE *Int. Symp. Circuit and Systems*, vol. 4. 2002.

[3] C. Kim, et al, IEEE *Trans. VLSI Syst.*, vol. 11, no. 6, Dec. 2003.

[4] A. Wang, et al, IEEE *JSSC*, vol. 40, no. 1, pp. 310-319, Jan. 2005.

[5] F. Ishihara, et al, IEEE *Trans. VLSI Syst*, vol. 12, no. 2, Feb. 2004

[6] Y. Kanno, et al, Symposium on VLSI Circuits Digest of Technical Papers, 2000.

[7] Y. Kanno, et al, VLSI Technology, Systems, and Applications, 2001. Proceedings of Technical Papers. 2001 International Symposium.

[8] Y. Taur and T. H. Ning, *Fundamentals of Modern VLSI Devices*, New York: Cambridge Univ. Press, 1998.

[9] J.W.Tschanz, et al, , IEEE *JSSC*, vol. 37, no. 1, pp1396-1402, 2002

[10] S. R. Nassif, et al, CICC 2001

Integrated Solar Energy Harvesting and Storage

Nathaniel Guilar*, Albert Chen, Travis Kleeburg, Rajeevan Amirtharajah
Micropower Circuits and Systems Group, Department of Electrical and Computer Engineering
University of California, Davis, CA 95616

ABSTRACT

To explore integrated solar energy harvesting as a power source for low power systems such as wireless sensor nodes, an array of energy scavenging photodiodes based on a passive-pixel architecture for imagers and have been fabricated together with storage capacitors implemented using on-chip interconnect in a 0.35 μm CMOS logic process. Integrated vertical plate capacitors enable dense energy storage without limiting optical efficiency. Measurements show 225 μW/mm^2 output power generated by a light intensity of 20k LUX.

Categories & Subject Descriptors

B.7 [Integrated Circuits]: General.

General Terms

Design, Experimentation, Measurement.

Keywords

Energy Harvesting, Low Power, Photodiode.

1. INTRODUCTION

Solar energy harvesting has been proposed to extend the lifetime of wireless sensor networks beyond limitations imposed by batteries [1]. To reduce system cost and volume it is desirable to integrate energy harvesting and storage with data processing circuits. Recent advances in very low power signal processing architectures for sensors [2] has created the opportunity to use CMOS photodiodes, similar to those used in digital cameras, for solar energy harvesting. Moreover, the increase in interconnect capacitance as CMOS processes scale provides an opportunity to store the harvested energy without requiring battery materials to be integrated. This paper describes a test chip incorporating an array of photodiodes and storage capacitors developed to explore the maximum energy per area that can be gathered from a solar source and stored in a standard CMOS process.

* N. Guilar is supported by a GAANN Fellowship

Permission to make digital or hard copies of all or part of this work for personal or classroom use is granted without fee provided that copies are not made or distributed for profit or commercial advantage and that copies bear this notice and the full citation on the first page. To copy otherwise, or republish, to post on servers or to redistribute to lists, requires prior specific permission and/or a fee.
ISLPED'06, October 4–6, 2006, Tegernsee, Germany.
Copyright 2006 ACM 1-59593-462-6/06/0010...$5.00.

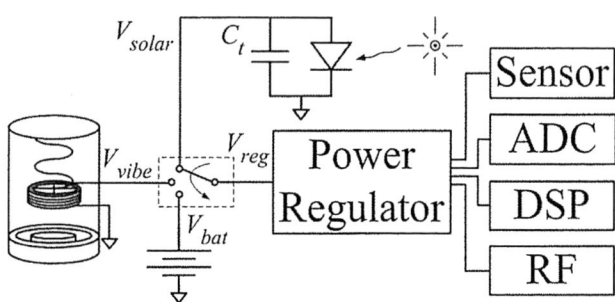

Figure 1: A low power wireless sensor node system powered from energy scavengers and a battery. Energy sources are labeled *Vsolar*, *Vvibe* and *Vbat* for the solar, mechanical vibration and battery, respectively. A mux switches between the unregulated energy sources.

2. ENERGY SCAVENGING PHOTODIODES

Figure 1 shows the block diagram of a typical wireless sensor node powered by light, mechanical vibration, and a battery. Light energy and vibrations are converted to electrical energy by photodiodes [1] and electromechanical transducers [2], respectively. The system's energy gathering ability will depend on environmental conditions which change over time and the scavenged energy needs to be regulated before being used. The sensor node consists of an analog-to-digital converter (ADC) that samples sensor data, a DSP core, and an RF transceiver. For low duty cycles, average power for this system (estimated from the literature [2,3,4]) can be under 5 μW.

The layout and design of an energy scavenging photodiode must balance several competing factors. The charge generated in the depletion region of the photodiode is stored in on-chip capacitors; therefore the physical layout of the diodes should facilitate both the solar energy harvesting and capacitive energy storage. Light reaching the photodiode depletion region must first pass through the passivation layers and around the storage capacitance, which is constructed on top of the diode to reduce area. Optical efficiency (OE), defined as the fraction of incident light onto the chip's surface which reaches the photodiode [5], is influenced by three loss factors: reflection losses, absorption losses, and critical angle losses. Once photons reach the photodiode, the quantum efficiency (QE) determines how many photons will generate electron-hole pairs. The product of OE and QE should be maximized by the geometry of the energy scavenging photodiode and storage capacitances to maximize output power.

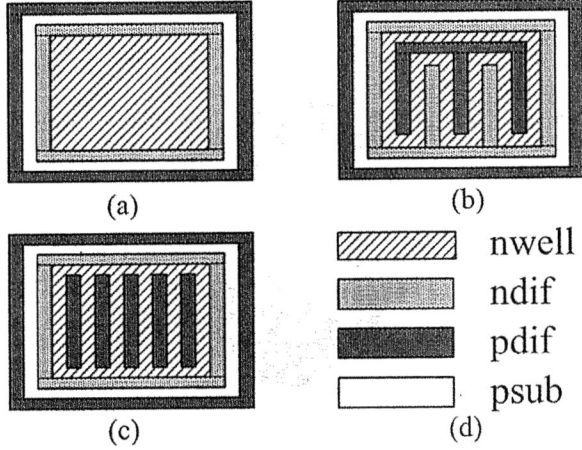

Figure 2: Top view of photodiodes. (a) D1, (b) D2, (c) D3, (d) Layer key.

To explore the photodiode design tradeoffs experimentally, three different geometries were fabricated and tested. The first design (D1) is shown in Figure 2(a) and is similar to a passive pixel structure used for a CMOS imager [6]. The p-substrate and n-well form the diode. The second design (D2) is shown in Figure 2(b). This structure uses fingers of p-diffusion inside the n-well in addition to the n-well to p-substrate junction to form more diodes connected in parallel with the well-substrate diode. The final design (D3) is shown in Figure 2(c). The D2 layout is similar to the D3 layout except every other p-diffusion finger is replaced with an n-diffusion.

A goal of this work is to determine the maximum energy per area that can be gathered from solar energy and stored in a standard CMOS logic process. To determine the energy we must first calculate the capacitance per area. The capacitance analysis will start with the energy relationship, which can then be divided up as the sum of two components: the metal capacitance and the junction capacitance. The total capacitance is given as

$$Ct = \frac{2 \cdot E_v}{\Delta V_{solar}^2} = Cm + Cd$$

(1)

, where E_v is the energy stored in the capacitor, ΔV_{solar} is the voltage difference between the metal plates, and Ct is the total capacitance, which is made up of the metal capacitance Cm and the diode capacitance Cd.

When the mux in Figure 1 switches to the solar cell the total charge accumulated in the storage capacitance is shared with the capacitance at the input of the regulator. This charge sharing will yield a potential at the input of the regulator which is less then the initial stored potential before the switch. With ideal switching the voltage at the input of the regulator can be written as

$$V_{reg} = \frac{Ct}{Ct + Cr} V_{solar}$$

(2)

, where Cr is the input capacitance of the regulator present at the right of the rotating switch in Figure 1. It is seen here that to limit this attenuation factor the storage capacitance should be as large as possible. A capacitor structure that simultaneously enables a high OE and a high capacitance density is the vertical parallel plate structure shown in Figure 3, which passes light through the plates vertically. The fingers forming the n-well p-diffusion junctions of D2 and D3 can align with the vertical plates allowing for easy routing.

Capacitance simulations were carried out using Momentum [7] assuming four metal layers, a minimum vertical parallel plate separation of 0.6 µm, a silicon dioxide dielectric of thickness 0.64 µm, and an average metal thickness of 0.71 µm. D3 has less metal capacitance than the other two designs due to the need to connect the p-diffusion fingers together. For a diode area of 338 µm² in the given technology, the theoretical limit on the maximum obtainable capacitance (TL1) is close to 1pF while the semi-empirical upper bound (SEUB) is 0.616pF, determined as in [8]. D2 has a capacitance density close to 41% of SEUB (neglecting diffusion capacitance C_d).

The total pn junction capacitance is the sum of the capacitances from the diffusion-well diode, Dpdif-nw, and the substrate-well diode, Dpsub-nw. This capacitance can be modeled by a particular capacitance per unit area, Cj and a particular side wall capacitance per unit length, Cjw. These capacitances are in general nonlinear (voltage-dependent). The width of the depletion region will shrink with an increase in applied forward bias voltage across the junction. This decrease in depletion width will lead to an increase in the depletion capacitance, which can be modeled using a square root dependence on applied voltage. The pn junction capacitances can be written as

$$Cd = \frac{C_{do}}{\sqrt{1 - \dfrac{V_{Solar}}{\varphi_o}}} = \sum_{i=1}^{N} \left(C_{sw} P_i + C_j A_i \right)$$

(3)

, where φ_o is the built in potential of the junction, Ai is the area and Pi is the perimeter of the ith diode. Table 1 summarizes the simulated capacitive characterization of the three diodes shown in Figure 2. The Cd given here is calculated with a junction voltage of 0.55 V, which is close to the open circuit voltage of the photodiodes under normal indoor lighting conditions.

Table 1: Capacitive Characterization (25°C, Area = 338µm²).

	D1	D2	D3	TL1	SEUB
Cm (pF)	0.254	0.254	0.216	1.004	0.616
Cdo (pf)	0.070	0.178	0.285	–	–
Cd (pF)	0.113	0.286	0.460	–	–

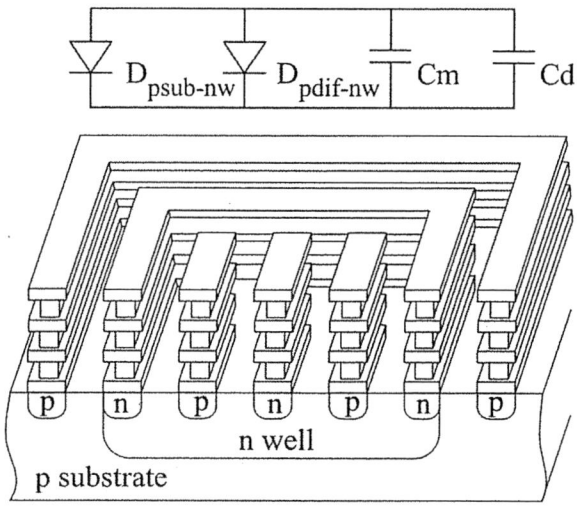

Figure 3: Side view cutaway of D2. Metal connected to p- and n- diffusions correspond to top and bottom capacitor plates, respectively.

3. EXPERIMENTAL RESULTS

Figure 4(a-c) shows the measured output power plotted versus light intensity and load resistance for the three photodiodes. The experimental light source was a typical tungsten filament incandescent bulb chosen to model indoor environmental conditions. At high light intensities, D3 generates more power per area than the other designs. However, D3 shows the highest sensitivity of output power to varying load resistance. For all three diodes, the optimal load resistance is a function of light intensity, with high intensity yielding lower optimal load resistance. D1 is least sensitive to light intensity and load resistance. No significant power loss was observed from the blocking of light by the storage capacitance.

To verify the operation of the photodiodes, a ring oscillator was constructed on-chip that uses the scavenged energy. Figure 5 shows a schematic of the prototype. It consists of a light source, integrated energy scavenging photodiodes, storage capacitance (Ct), a ring oscillator and buffers to drive the signal off chip. The nine stage ring oscillator employs current starving techniques with both pmos and nmos transistors to enable frequency tuning of the oscillator. A level shifting output buffer was used so that the oscillator's operation could be observed with little loading. Bias voltages V_p, V_n and the power for the output buffer was generated off-chip for testing purposes. In a single-well p-type substrate process, photodiodes working under forward bias drive the n-well below ground, potentially causing unwanted substrate currents to flow. Therefore, two test chips are required to demonstrate the oscillator. In a twin- or triple-well process this limitation can be eliminated. Figure 6 plots the ring oscillator frequency and the open circuit voltage (V_{OC}) versus light intensity. With one series diode, $V_{Solar} = 0.55$ V and the oscillator dissipates 67.3 pW (0.48 fJ per cycle) at a maximum frequency of 140 kHz.

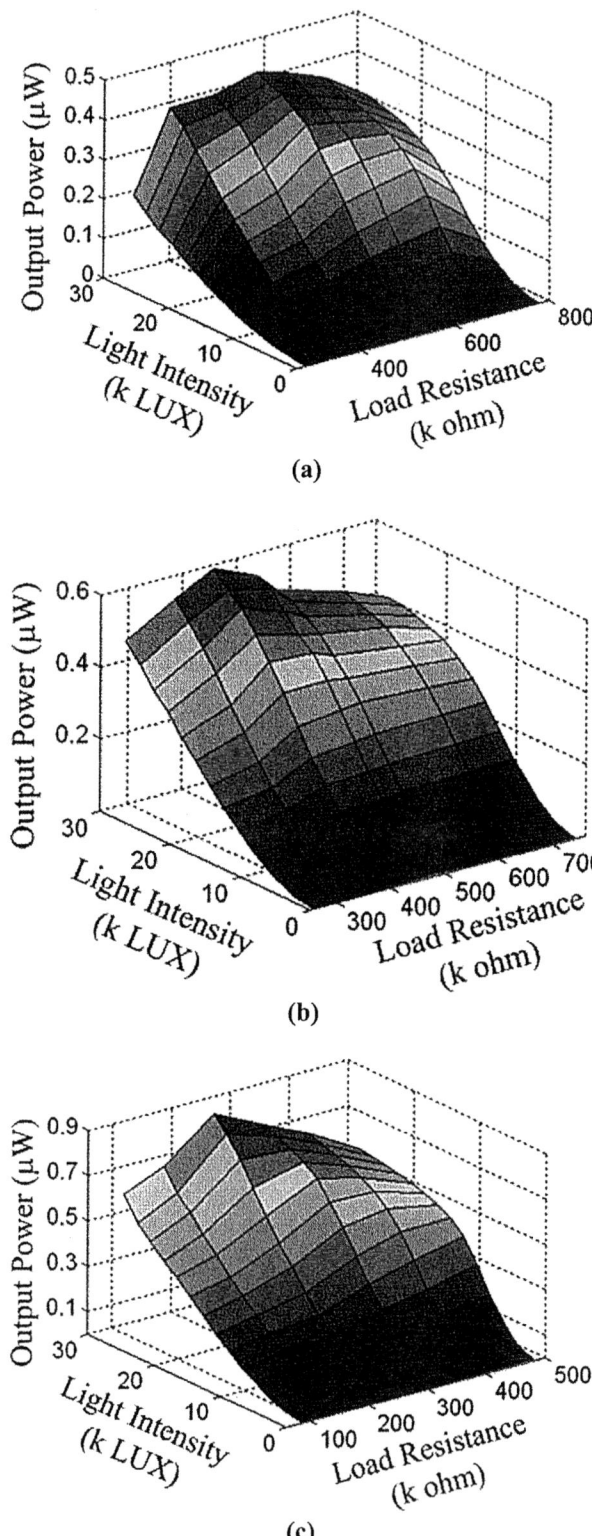

Figure 4: Output power vs. incident light intensity and load resistance, active area = 3000 μm², (a) D1, (b) D2, (c) D3.

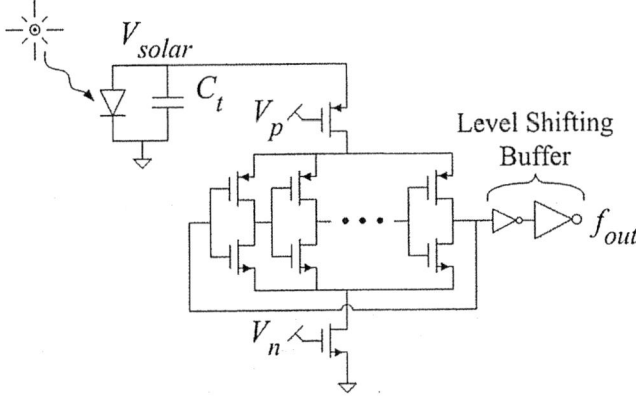

Figure 5: Test setup showing photodiode, current starved ring oscillator and level shifting output buffer.

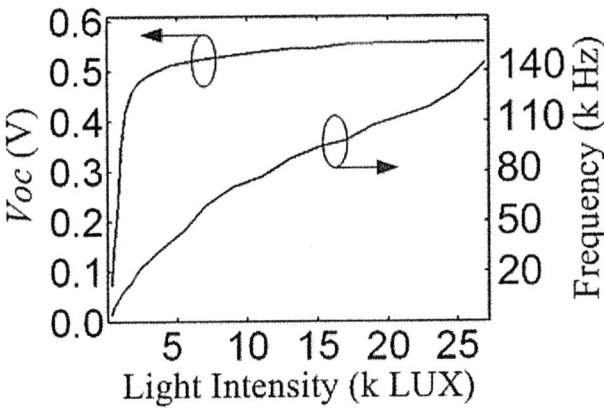

Figure 6: Measured plots of D3 showing V_{OC} and ring oscillator frequency.

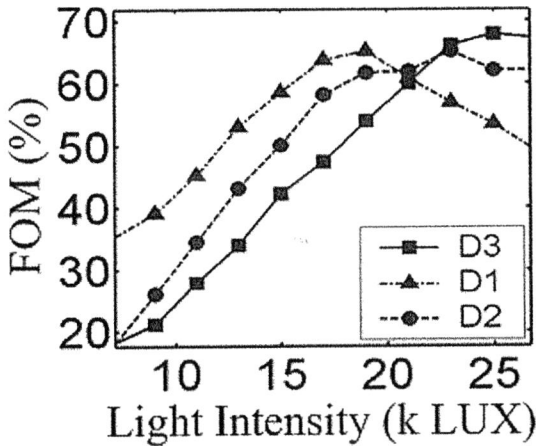

Figure 7: Measured figure of merit (FOM) plotted vs. input light intensity.

4. CONCLUSION

Figure 7 plots a figure-of-merit (FOM) for each photodiode which is defined as the maximum delivered power for a given level of illumination divided by the product of the open circuit voltage (V_{OC}) and the short circuit current (I_{SC}) [5]. Design D1 has the best FOM for low to moderate light intensities, but falls off significantly at high intensity. Design D3 has the largest FOM with high light conditions. For comparisons in terms of scaling each diode was laid out in three different sizes: 36 μm^2, 108 μm^2 and 338 μm^2. Larger photodiodes showed slightly better metrics which are largely attributed to lateral photocurrents [9,10]. In more advanced technologies the capacitance per unit area will increase as the minimum distance between plates shrinks. However, for the case where vertical parallel plate storage capacitors are constructed above the photodiodes this decrease in distance will negatively impact the OE due to a smaller aperture size. Figure 8 shows a die photograph of two D3 diodes connected in parallel. Table 2 summarizes the results with an incident light intensity of 20k LUX, similar to being outside on a sunny day. Based on these results, design D3 with area 150 μm x 150 μm can deliver 5 μW to power the system described above. D1 needs 184 μm x 184 μm and D2 needs 164 μm x 164 μm to deliver 5 μW. Without illumination, the system energy must be supplied by the integrated storage capacitors. For a 25 mm^2 total diode area consisting of 3 diodes in series with the metal storage capacitances for each diode connected in parallel, D1, D2, and D3 can supply enough energy for the DSP in [2] to produce 687, 745, and 903 output samples respectively. Future work will involve testing the photodiodes' response with incident light from a green laser (532 nm wavelength) to determine OE and QE as well as fabricating the diodes in polysilicon on top of the die to exploit more area for energy scavenging.

Table 2: Measured Performance (25°C, Area=338μm^2).

Parameters	D1	D2	D3
Power (nW)	50	63	76
Energy Stored (fJ)	26	35	31
FOM (%)	65	66	62
Capacitance, C_m (pF)	0.245	0.254	0.216
V_{OC} (mV)	465	525	533
I_{SC} (nA)	165	182	230

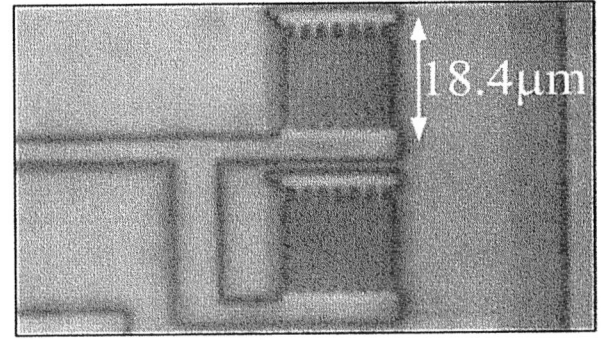

Figure 8: Die photograph of two D3 diodes connected in parallel.

ACKNOWLEDGMENTS

The authors are grateful to S. Bruss, P. Hurst, A. Knoesen, S. Lewis, R. Perry, and D. Yankelevich for their help with designing and testing the prototype.

5. REFERENCES

[1] A. Kansal and M. Srivastava, "An environmental energy harvesting framework for sensor networks," *Intl. Symp. Low Power Elec. and Design (ISLPED) '03*, pp. 481-6, Aug. 2003.

[2] R. Amirtharajah and A. Chandrakasan, "A micropower programmable DSP using approximate signal processing based on distributed arithmetic," *IEEE JSSC*, Vol. 39, No. 2, p. 337-47, Feb. 2004.

[3] M. Scott, B. Boser, and K. Pister, "An ultralow-energy ADC for smart dust," *IEEE JSSC*, Vol. 38, No. 7, p. 1123-9, July 2003.

[4] B. Otis, Y. Chee, and J. Rabey, "A 400 μW-Rx, 1.6mW-Tx super-regenerative transceiver for wireless sensor networks, *ISSCC 05*, pp. 396-7, 606, Feb. 2005.

[5] P. Catrysse, P. and B. Wandell, "Optical efficiency of image sensor pixels," J. Opt. Soc. Am., Vol. 19, No. 8, August 2002.

[6] I. Fujimori, C. Wang, and C. Sodini, "A 256 × 256 CMOS differential passive pixel imager with FPN reduction techniques," IEEE JSSC, Vol. 35, pp. 2031-7, December 2000.

[7] Advanced Design Systems software package 2005, Agilent Inc.

[8] R. Aparicio and A. Hajimiri, "Capacity Limits and Matching Properties of Integrated Capacitors," *IEEE JSSC*, vol. 37, no. 3, pp. 384-93, March 2002.

[9] J. Soo Lee, R. I. Hornsey, and D. Renshaw, "Analysis of CMOS Photodiodes-Part I: Quantum Efficiency," *IEEE Trans. on Electron Devices*, Vol. 50, No. 5, May 2003.

[10] J. Soo Lee, R. I. Hornsey, and D. Renshaw, "Analysis of CMOS Photodiodes-Part II: Lateral Photoresponse," *IEEE Trans. on Electron Devices*, Vol. 50, No. 5, May 2003.

Energy/Power Breakdown of Pipelined Nanometer Caches (90nm/65nm/45nm/32nm)

Samuel Rodriguez and Bruce Jacob

Electrical and Computer Engineering Department
University of Maryland, College Park
{samvr,blj}@eng.umd.edu

ABSTRACT

As transistors continue to scale down into the nanometer regime, device leakage currents are becoming the dominant cause of power dissipation in nanometer caches, making it essential to model these leakage effects properly. Moreover, typical microprocessor caches are pipelined to keep up with the speed of the processor, and the effects of pipelining overhead need to be properly accounted for.

In this paper, we present a detailed study of pipelined nanometer caches with detailed energy/power dissipation breakdowns showing where and how the power is dissipated within a nanometer cache. We explore a three-dimensional pipelined cache design space that includes cache size (16kB to 512kB), cache associativity (direct-mapped to 16-way) and process technology (90nm, 65nm, 45nm and 32nm).

Among our findings, we show that cache bitline leakage is increasingly becoming the dominant cause of power dissipation in nanometer technology nodes. We show that subthreshold leakage is the main cause of static power dissipation, and that gate leakage is, surprisingly, not a significant contributor to total cache power, even for 32nm caches. We also show that accounting for cache pipelining overhead is necessary, as power dissipated by the pipeline elements is a significant part of cache power.

Categories and Subject Descriptors

B.3.2 [Memory Structures] : Cache memories.

General Terms

Design, Performance.

Keywords

Cache design, nanometer design, pipelined caches.

1. INTRODUCTION

Power dissipation has become a top priority for today's microprocessors. What was previously a concern mainly for mobile devices has also become of paramount importance for general-purpose and even high-performance microprocessors, especially with the recent industry emphasis on processor "Performance-per-Watt."

As transistors become steadily smaller with improving fabrication process technologies, device leakage currents are expected to significantly contribute to processor power dissipation [7]. Cache power dissipation has typically been significant, but increasing static power will have a greater impact on the cache since most of its transistors are inactive (dissipating no dynamic power, only static) during any given access. It is therefore essential to properly account for these leakage effects during the cache design process.

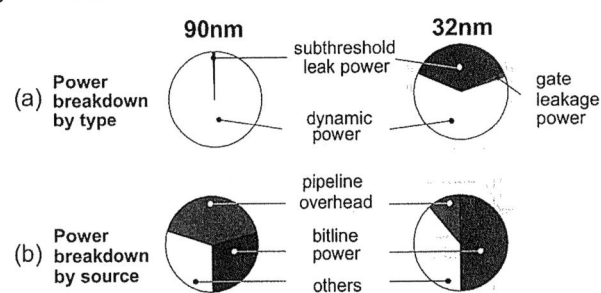

Figure 1: Power breakdowns of a 64kB-4way cache. (a) Dynamic and static power dissipation components for a 64kB-4W cache in 90nm and 32nm, (b) Major components of power dissipation for a 64kB-4way 90nm and 32nm pipelined cache

Moreover, microprocessor caches are obviously not designed to exist in a vacuum – they exist to complement the processor by hiding the relatively long latencies of the lower level memory hierarchy. As such, typical caches (specifically level-1 caches and often level-2) are pipelined and clocked at the same frequency as the core. Explicit pipelining of the cache will involve a non-trivial increase in access-time (because of added flop delays) and power dissipation (from the latch elements and the resulting additional clock power), and these effects must be accounted for properly, something which is not currently done by existing publicly available cache-design tools.

In this paper, we use analytical modeling of cache operation combined with nanometer BSIM3v3/BSIM4 SPICE models [5,13] to analyze the behavior of various cache configurations. We break down cache energy/power dissipation to show how much energy/power each individual part of a cache consumes, and what fraction can be attributed to dynamic (switching) and static (subthreshold leakage and gate tunneling) currents. We explore a three-dimensional cache design space by studying caches with different sizes (16kB to 512kB), associativities (direct-mapped to 16-way) and process technologies (90nm, 65nm, 45nm and 32nm).

Among our findings, we show that cache bitline leakage is increasingly becoming the dominant cause of power dissipation in nanometer technology nodes. We show that subthreshold leakage is the main cause of static power dissipation, but surprisingly, gate leakage tunneling currents do not become a significant contributor to total cache power even for the deep nanometer nodes. We also show that accounting for cache pipelining overhead is necessary, as power dissipated by the pipeline elements are a significant part of cache power.

2. BACKGROUND

2.1 Leakage

Dynamic power is dissipated by a circuit whenever transistors switch to change the voltage in a particular node. In the process, energy is consumed by charging (or discharging) the node's parasitic capacitive loads and, to a lesser degree, by possible short-circuit currents that flow during the small but finite time window when the transistor pull-up and pulldown networks are fully or partially turned "on," providing a low-impedance path from supply to ground.

On the other hand, static power is dissipated by leakage currents that flow even when the device is inactive. Different leakage mecha-

---

Permission to make digital or hard copies of all or part of this work for personal or classroom use is granted without fee provided that copies are not made or distributed for profit or commercial advantage and that copies bear this notice and the full citation on the first page. To copy otherwise, or republish, to post on servers or to redistribute to lists, requires prior specific permission and/or a fee.
ISLPED'06, October 4–6, 2006, Tegernsee, Germany.
Copyright 2006 ACM 1-59593-462-6/06/0010...$5.00.

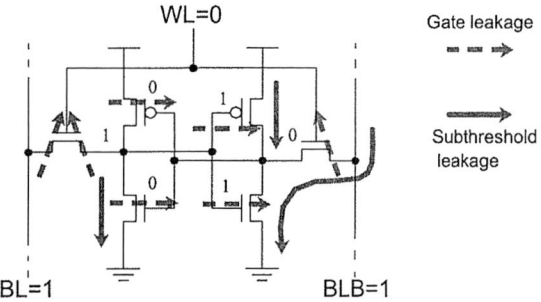

Figure 2: Memory cell leakage currents. An inactive six-transistor memory cell (6TMC) showing the subthreshold leakage and gate leakage currents flowing across the devices

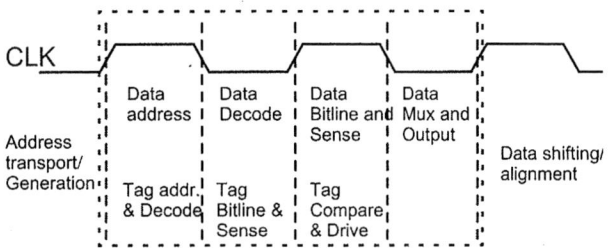

Figure 3: Cache pipeline diagram. The shaded region shows the part of the pipeline that we model

nisms exist for MOS transistors [2], but the two most important ones are lumped into the subthreshold leakage current and the gate leakage current. The subthreshold leakage current has been extensively studied [2,3,6,18,22]. It is mainly caused by the generational reduction in the transistor threshold voltage to compensate for device speed loss when scaling down the supply voltage, with the consequence of exponentially increasing subthreshold leakage current. The gate leakage current has been less extensively studied because of its relatively smaller value compared to subthreshold leakage for older technologies, but is increasingly receiving more attention [10,17,18,23] as it is expected to be comparable to the subthreshold leakage current in the deep nanometer nodes. Gate leakage currents flow whenever voltages are applied to transistor terminals, producing an electric field across gate oxides that are getting thinner (20 to less than 10 angstroms) resulting in significant leakage currents due to quantum tunneling effects.

Figure 2 shows a typical 6-transistor memory cell (6TMC) and the typical leakage currents involved for the memory cell idle state (i.e. wordline is off, one storage node is "0" and the other is "1"). Although a sizeable number of transistors in a cache are active for any given access, the vast majority of memory cells are in this inactive state, dissipating static power. Cache designers currently account for subthreshold leakage current (most problematic of which is the bitline leakage since this not only affects power, but also circuit timing and hence functionality), but not much attention is given to gate leakage in the publicly available cache design tools like CACTI [25,19,20] and eCACTI [14]. For better accuracy, this paper also considers the gate leakage currents as shown in the 6TMC diagram.

2.2. Pipelined Caches

To keep up with the speed of a fast microprocessor core while providing sufficiently large storage capacities, caches are pipelined to subdivide the various delays in the cache into different stages, allowing each individual stage to fit into the core's small clock period. Figure 3 shows a typical pipeline diagram for a cache. The given timing diagram shows operations being performed in both phases of the clock. Figure 4 shows a possible implementation of a pipeline latch that easily facilitates this phase-based operation.

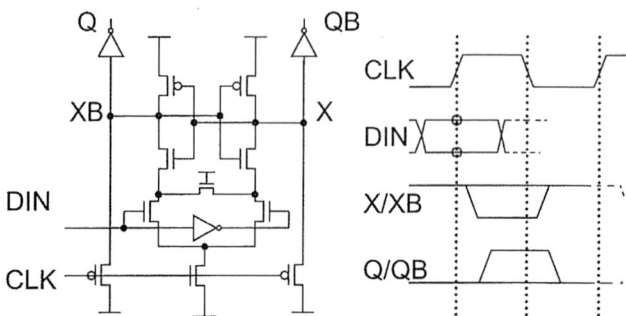

Figure 4: Pipeline latch. An example of a pipeline latch that can be used to implement the phase-based operations in the cache [8][15]. Also shown is the latch's timing diagram

The major publicly-available cache analysis tools CACTI and eCACTI use implicit pipelining through wave pipelining, relying on regularity of the delay of the different cache stages to separate signals continuously being shoved through the cache instead of using explicit pipeline state elements. Unfortunately, this is not representative of modern designs, since cache wave-pipelining is not being used by contemporary microprocessors [21,24]. Although wave pipelining has been shown to work in silicon prototypes [4], it is not ideally suited for high-speed microprocessor caches targeted for volume production which have to operate with significant process-voltage-temperature (PVT) variations. PVT variations in a wave-pipelined cache cause delay imbalances which, in the worst case, lead to signal races that are not possible to fix by lowering the clock frequency. Hence, the risk for non-functional silicon is increased, resulting in unattractive yields. In addition, wave-pipelining does not inherently support latch-based design-for-test (DFT) techniques that are critical in the debug and test of a microprocessor, reducing yields even further. On the contrary, it is easy to integrate DFT "scan" elements inside pipeline latches that allow their state to be either observed or controlled (preferably both). This ability facilitates debugging of microprocessor circuitry resulting in reduced test times that directly translate to significant cost savings. Although not immediately obvious at first, these reasons make it virtually necessary to implement explicit pipelining for high-volume microprocessors caches.

3. EXPERIMENTAL METHODOLOGY

Although initially based on CACTI and eCACTI tools, the analysis tool we have developed bears little resemblance to its predecesors. In its current form, it is, to the best of our knowledge, the most detailed and most realistic (i.e. similar to realistically implementable high-performance caches) cache design space explorer publicly available.

The main improvements of our tool compared to CACTI/eCACTI are the following:

- More optimal decode topologies [1] and circuits [16] along with more realistic device sizing ensures that cache inputs present a reasonable load to the preceeding pipeline stage (at most four times the load of a 1X strength inverter).[1]
- Accurate modeling of explicit cache pipelining to account for delay and energy overhead in pipelined caches.
- Use of BSIM3v3/BSIM4 SPICE models and equations to perform calculation of transistor drive strengths, transistor RC parasitics, subthreshold leakage and gate leakage. Simulation characteristics for each tech node are now more accurate.[2]
- More accurate RC interconnect parasitics by using local, intermediate and global interconnects (as opposed to using the

---

1. Both CACTI and eCACTI produce designs with impractically large first-stage inverters at the cache inputs. This shifts some of the burden of driving the cache decode hierarchy to the circuits preceding the cache, resulting in overly optimistic delay and power numbers.

26

same wire characteristics for all interconnects, as is being done by CACTI and eCACTI), and accurate analytical modeling of these structures. In addition, a realistic BEOL-stack[3] was used for each technology node.

An important note to make when discussing dynamic and static power and/or energy is that it is only possible to combine the two into a single measurement by assuming a specific frequency. Dynamic power inherently describes how much energy is consumed in a single switching event, and an assumption of how often that event happens is necessary to convert the energy value into power dissipation (hence the activity factor and frequency components in standard power equations). On the other hand, static leakage inherently describes the amount of current flow, and hence the instantaneous power, at any given time. Converting this into energy requires an assumption of the amount of time the current is flowing. Table 1 shows the values for

---

2. For delay and dynamic power computations, CACTI and eCACTI use hardcoded numbers based on 0.80um technology and use linear scaling to translate power and delay numbers to the desired technology.

3. The Back-End-Of-Line stack refers to the fabrication steps performed after the creation of the active components. Use of a realistic BEOL-stack results in more accurate modeling of the interconnect. In CACTI and eCACTI, a single interconnect characteristic was assumed.

frequency and supply voltages that were used for this study, where the values are chosen from historical [26] and projected [11,12] data.

Table I. VDD and frequency used for each tech node (T=50 C)

	250nm	180nm	130nm	90nm	65nm	45nm	32nm
VDD	2.0 V	1.8 V	1.6 V	1.4 V	1.3 V	1.2 V	1.1 V
Freq. (GHz)	0.8 GHz		1.5GHz	1.8GHz	2.4GHz	2.6GHz	2.8GHz	3.0GHz

4. RESULTS AND DISCUSSIONS

4.1. Dynamic and static power

Figure 5 shows the power dissipation of the different cache configurations as a function of process technology. Each column of plots represents a specific process technology, while each row represents a specific cache size. Each plot shows the dynamic, subthreshold leakage, and gate leakage, and total power as a function of associativity for the given cache size and technology node.

The most basic observation here is that total power is dominated by the dynamic power in the larger technology nodes but is dominated by static power in the deep nanometer nodes (with the exception of very highly-associative small to medium caches).

This can be seen from the plots for the 90nm and 65nm nodes, where the dynamic power comprises the majority of the total power. In the 45nm node, the subthreshold leakage power is significant

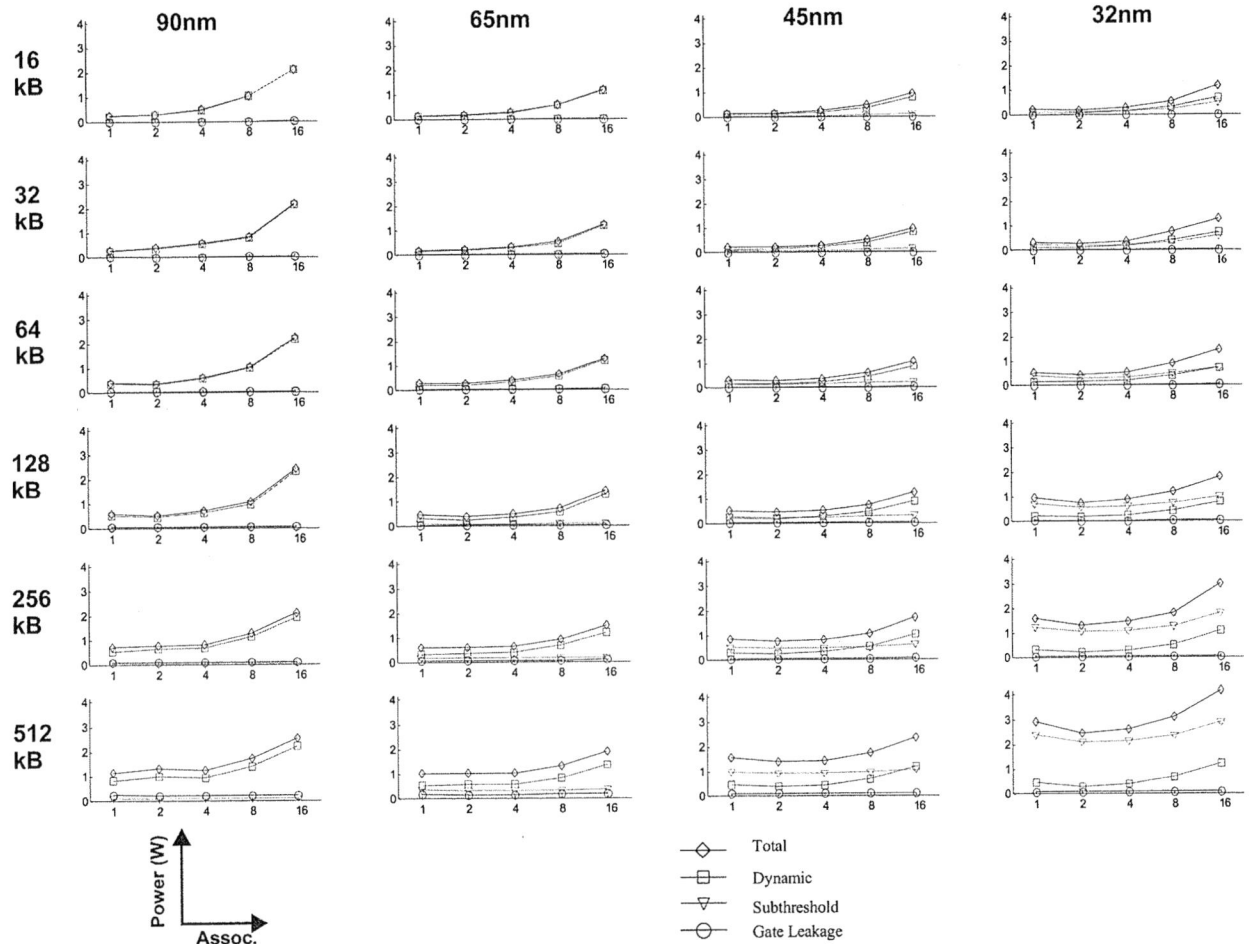

Figure 5: Power consumption vs. technology node of different cache configurations. The plots show how total power consumption is broken down into dynamic power and static power (due to subthreshold leakage and gate leakage). Each column of plots represents a single tech node, while a single row represents a specific cache size. Within each plot, the power dissipation is shown in the y-axis, with increasing associativites represented in the x-axis.

27

enough that it becomes the dominant component for some configurations. For small caches of any associativity, dynamic power typically dominates because there are fewer leaking devices that contribute to subthreshold leakage power. But as cache sizes increase, the number of idle transistors that dissipate subthreshold leakage power also increases, making the subthreshold leakage power the dominant component of total power except for the configurations with high associativity (which requires more operations to be done in parallel, resulting in dissipation of more dynamic power). In the 32nm node, the subthreshold leakage power is already comparable to the dynamic power even for small caches (of any associativity), and starts to become really dominant as cache sizes increase (even at high associativities where we expect the cache to burn more dynamic power).

A surprising result that can be seen across all the plots is the relatively small contribution of gate leakage power to the total power. This is surprising given all the attention that gate leakage current is receiving and how it is expected to be one of the dominant leakage mechanisms. It turns out that this is a result of many factors, including the less aggressive gate oxide thickness scaling in the more recent ITRS roadmaps [11,12], and that Vdd scaling and device size scaling tend to decrease the value of the gate tunneling current. The net effect of a slightly thinner gate oxide (increasing tunneling), slightly lower supply voltage (decreasing tunneling), and smaller devices (decreasing tunneling) from one generation to the next might actually result in an effective decrease in gate leakage.

Some of the plots in Figure 5 (e.g. the 256k and 512k caches for the 45nm and 32nm node) exhibit a sort of "saddle" shape, which shows that increasing cache associativity from direct-mapped to 2-way or 4-way does not automatically cause an increase in power dissipation, as the internal organization may allow a more power optimal implementation of set-associative caches compared to direct-mapped caches, especially for medium to large-sized caches.

A final observation for Figure 5 is that technology scaling is capable both of increasing or decreasing the total power of a cache. Which direction the power goes depends on which power component is dominant for a particular cache organization. Caches with dominant dynamic power (e.g. small caches/highly associative caches) will enjoy a decrease in cache power as dynamic power decreases because of the net decrease in the CV^2f function, while caches with dominant static power (e.g. large caches/medium-sized low-associativy caches) will suffer from an increase in cache power as the static power increases.

4.2. Detailed power breakdown

Figure 6 shows a detailed breakdown of cache power for three different cache configurations (16kB-4W, 32kB-4W, 64kB-4W). Each plot represents a single configuration implemented in the four nanometer nodes. For each of these nodes, total cache power is shown, along with its detailed breakdown. Power dissipation for each component, as shown by each vertical bar in the plot, is also broken down into its dynamic, subthreshold leakage, and gate leakage components.

The first notable point here is that for these cache configurations, going from 90nm to 65nm results in power decrease; going from 65nm to 45nm results in a smaller power decrease; finally, going from 45nm to 32nm causes a significant power increase. Again, these observations are caused by the intertwined decrease and increase of dynamic and subthreshold leakage power, respectively, as we go from one technology node to the next.

A second point is that for most of the plots, it can be seen that the two main contributors to cache power are the bitlines and the pipeline overhead (with the data_dataout component, which accounts for the power in driving the output data buses, also significant for some configurations). Pipeline overhead, shown in the plots as data_pipe_ovhd and tag_pipe_ovhd, accounts for the power dissipated in the pipeline latches along with the associated clock tree circuitry.

This is an important observation as it shows that the overhead associated with pipelining the cache is often a very significant component of total power, and that most of this power is typically dynamic. This can be easily seen in Figure 7, which shows the fraction of total power dissipated by pipeline overhead for all the configurations studied. This is especially noteworthy since we model aggressive clock gating where only latches that need to be activated actually see a clock signal. Consequently, we expect the pipeline overhead to be even more significant for circuits that use less aggressive clock-gating mechanisms. Lastly, it should be noted that since most of the power in the pipeline overhead is dynamic, it should decrease for smaller technology nodes, as seen in Figure 5.

Another point of importance is that although dynamic power in general tends to decrease because of the net effect of increased frequency but decreased supply voltage and device capacitances, this is the case only for circuits with device-dominated loading (i.e. gate and diffusion capacitances). Circuits with wire-dominated loading may actually see dynamic power go up as we go from one generation to the next, since a wire capacitance decrease due to shorter lengths and slightly smaller coupling area will typically be offset by the shorter distances between wires that cause significantly increased capacitances. This can be seen in Figure 6, where the dynamic power for the data_dataout component increases for all three configurations, mostly since the load for data_dataout is wire-dominated, as it involves only a few high-impedance drivers driving a long wire across the entire cache.

The power breakdown of representative cache configurations as shown in Figure 6 shows that most of the power in a pipelined, nanometer cache is dissipated in the bitlines, the pipeline overhead, and possibly the data output drivers. To get a better view of the design space, we can lump together similar components in order to show the entire space. In Figure 8, we subdivide total power into five categories -- bitline power, pipeline overhead power, decoder power, data processing power (data/tag sensing, tag comparison, data output muxing and driving), and lump all remaining blocks into a single component labeled "others." Results for the 65nm and 32nm nodes for all the configurations are then plotted together.

It is interesting to see that while the power dissipation due to the bitlines monotonically increases as cache size increases for both technology nodes, the power due to the other components does not necessarily do so. For instance, the pipeline power of the 65nm 128k-16 way and 256kB-16 way caches, where the pipeline power is significantly reduced from one configuration to the next. The reason here is that the bitline power mainly depends on the cache size (assuming static power is the dominant cause of bitline power) and not the cache implementation. The other power components, on the other hand, greatly depend on the particular implementation, so their values may noticeably fluctuate from one implementation to the next, especially if there exist implementations that cause an increase in the power of one component but results in a significant power improvement in

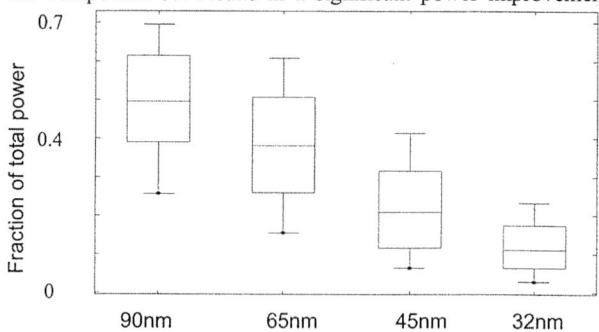

Figure 7: Pipeline overhead contribution to total cache power.
Distribution of data showing the fraction of total power attributed to pipeline overhead for all cache configurations, where each plot shows min, max, median, and quartile information

28

some other part of the cache. It is important to realize that the optimization scheme that we used optimized total power, and not component power. As long as a specific implementation has a better power characteristic compared to another, relative values of specific cache components were not considered important.

It is easily seen from the different cache configurations in Figure 8 that that pipeline power (as represented by pipe_ovhd) will typically be a non-trivial contributor to the total cache power. Any analysis that fails to account for this pipelining overhead and instead assumes that caches can be trivially interfaced to a microprocessor core will be inaccurate. Although the figure also shows that the contribution of the pipeline overhead to total power is reduced from 65nm to 32nm compared to the contribution of the "dataproc" component (mainly because of the different effect of technology to the dynamic power of

the two components), it should be noted that our study modeled conservative scaling of the dielectric constants of the interlevel dielectrics (i.e. we did not model aggressively scaled low-K interconnects). If wire delay becomes especially problematic in future generations and more aggressive measures are taken to improve the interconnect performance, we can expect the relative contribution of the pipeline overhead to the total power to increase in significance.

A final observation that we want to point out from Figure 8 is that increasing cache associativity may not necessarily result in a significant increase in total power, especially for the larger caches at 65nm and below. The power for these configurations are typically dominated by the bitline leakage, reducing the effect of any increase in power due to higher associativity. In some cases, the total power dissipation can actually decrease with an increase in associativity, as a

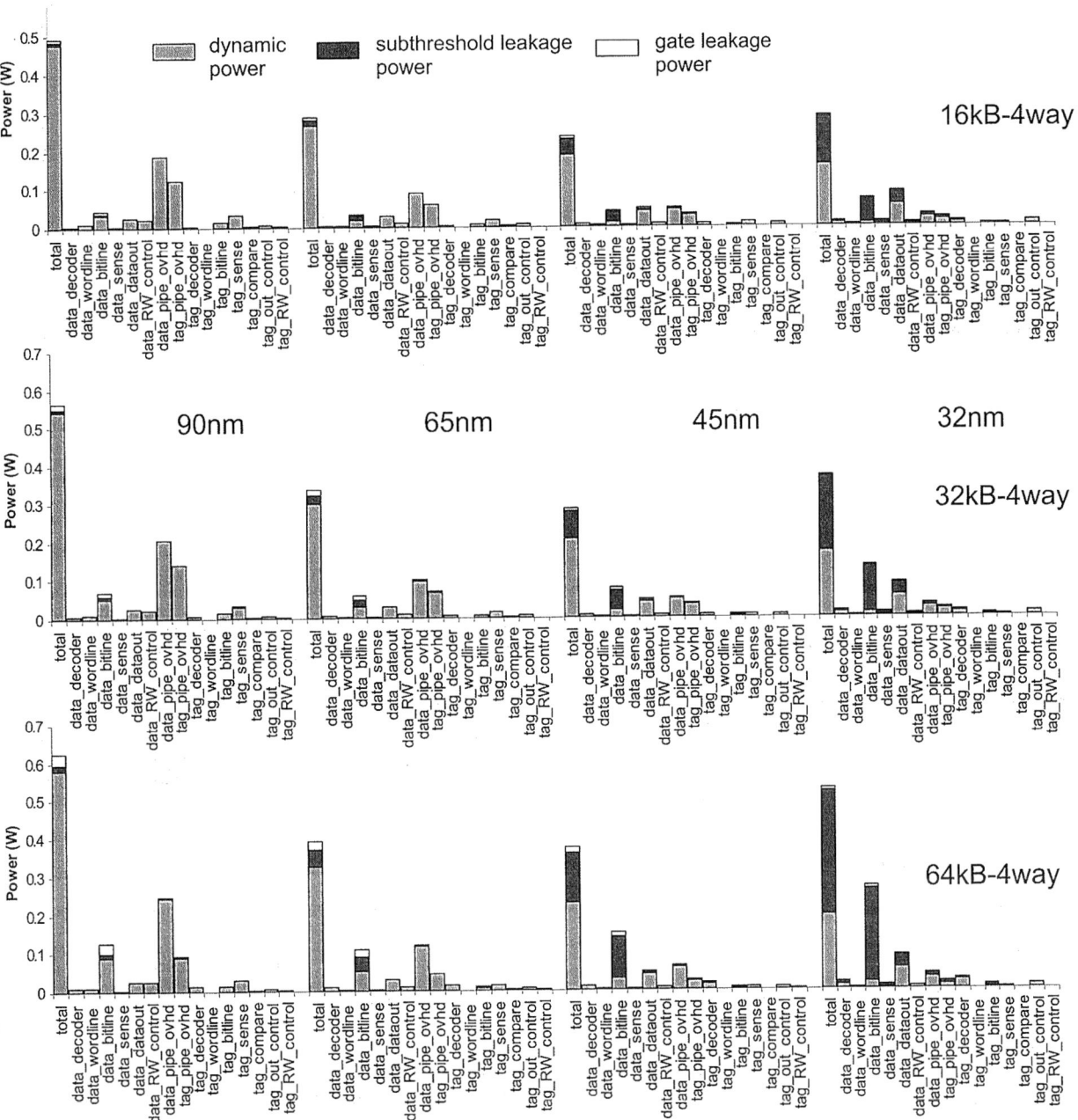

Figure 6: Detailed cache power breakdown. Detailed power breakdown for three different cache configurations showing the amount of dynamic, subthreshold leakage and gate leakage power being consumed by different parts of the cache. Each plot shows four sets of valuescorresponding to different technology nodes. (Note that all three plots use the same scale)

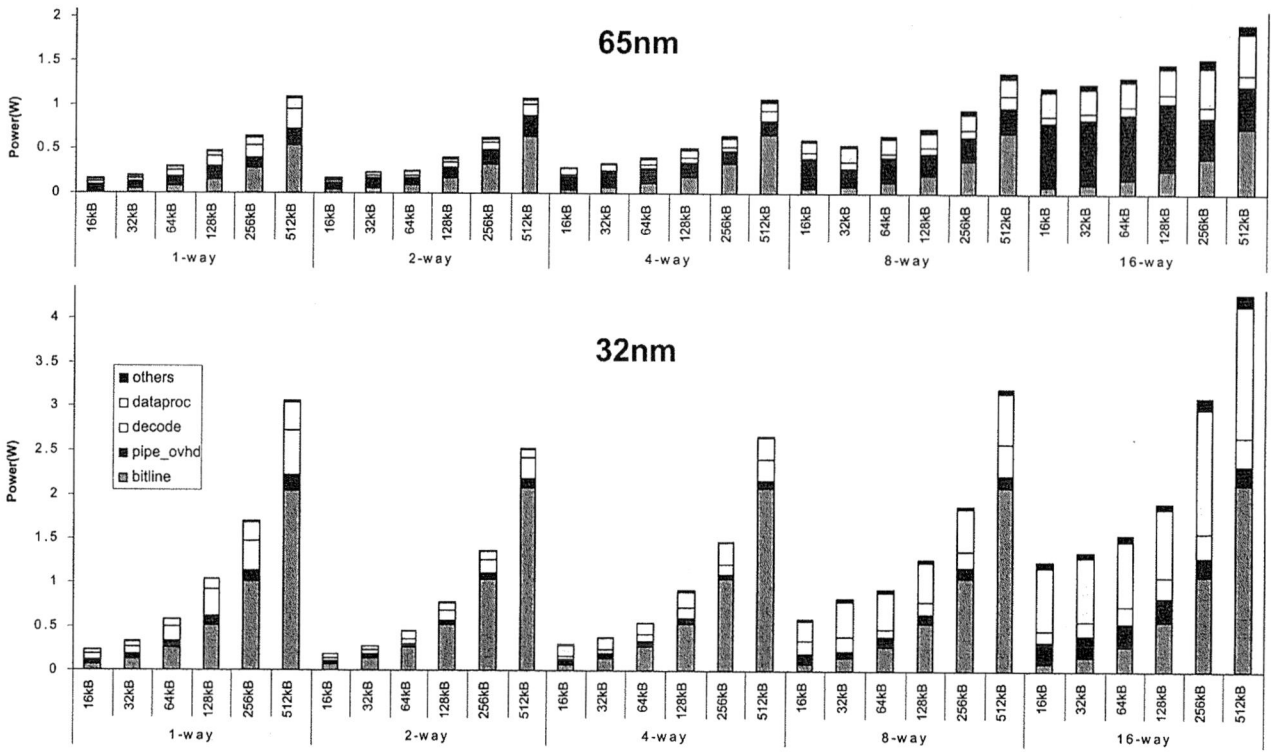

Figure 8: Power contributions of different cache parts. Power breakdown showing major cache power contributors for the 65nm and 32nm technology nodes for different cache sizes and associativities. (Note that the y-axis for both plots use the same scale)

particular configuration is able to balance the overall static and dynamic power for all the cache components. This knowledge will be useful in enabling large, high-associativity caches suitable for use in L2 caches (and higher).

5. CONCLUSION

We have presented a detailed power breakdown of the different nanometer pipelined cache configurations in a three-dimensional design space consisting of different cache sizes, associativities and process technologies.

We have shown that the power of nanometer caches will continue to increase, and that this increase will be primarily driven by static power due to subthreshold leakage currents. In addition, we saw that static power due to gate leakage tunneling currents does not contribute significantly to the cache power, even in the deep nanometer technology nodes.

Lastly, we have argued that practical microprocessor caches virtually require the use of pipeline latches to facilitate designing PVT variation-tolerant caches and debug and test-friendly designs. But in using explicit pipelining, accurate cache analysis requires accounting for the overhead introduced by these pipeline latches, as these represent a significant portion of the cache's power dissipated (in some corner cases being the dominant cause of power).

6. REFERENCES

[1] B.Amrutur and M.Horowitz, "Fast low-power decoders for RAMs," IEEE JSSC, vol.36, no.10, Oct 2001.
[2] Mohab Anis, "Subthreshold leakage current: challenges and solutions," ICM 2003.
[3] S.Borkar, "Circuit techniques for subthreshold leakage avoidance, control and tolerance," IEDM 2004.
[4] W.P.Burleson et al., "Wave-pipelining: A tutorial and research survey," IEEE Trans. VLSI Systems, vol.6, no.5, Sep 1998.
[5] Y.Cao et al., "New paradigm of predictive MOSFET and interconnect modeling for early circuit design," Proc. of CICC, pp.201-204, 2000.
[6] Z.Chen, M.Johnson, L.Wei and K.Roy, "Estimation of standby leakage power in CMOS circuits considering accurate modeling of transistor stacks," in Proc. Int.Symp.Low Power Electronics and Design, 1998.

[7] B.Doyle, et al., "Transistor elements for 30nm physical gate lengths and beyond," Intel Technology Journal, Vol.6, Page(s): 42-54, May 2002.
[8] P.Gronowski et al., "A 433-MHz 64-b quad-ussue RISC microprocessor, "JSSC, vol.31, no.11, Nov.1996, pp.1687-1696.
[9] J.P.Halter and F.Najm, ,"A gate-level leakage power reduction method for ultra-low-power CMOS circuits," in Proc. IEEE Custom Integrated Circuits Conf., 1997.
[10] F.Hamzaoglu and M.R.Stan, "Circuit-level techniques to control gate-leakage for sub-100nm CMOS," ISLPED August 2002.
[11] "International Technology Roadmap for Semiconductors 2005 Edition," Semiconductor Industry Association, http://public.itrs.net.
[12] "International Technology Roadmap for Semiconductors 2003 Edition," Semiconductor Industry Association, http://public.itrs.net.
[13] Predictive technology model. http://www.eas.asu.edu/~ptm
[14] M.Mamidipaka and N.Dutt, "eCACTI: An enchanced power estimation model for on-chip caches," Center for Embedded Computer Systems, Technical Report TR 04-28, Oct. 2004.
[15] J.Montanaro et al., "A 160-MHz, 32-b, 0.5-W CMOS RISC microprocessor, JSSC, vol.31 no.11, Nov. 1996, pp.1703-1714.
[16] H.Nambu, et.al., "A 1.8-ns access, 550MHz, 4.5-Mb CMOS SRAM," IEEE J. Solid-State Circuits, vol.33, no.11, pp.1650-1658, Nov. 1998.
[17] R.Rao, J.L. Burns and R.B.Brown, "Circuit techniques for gate and subthreshold leakage minimization in future CMOS technologies," , 2003. ESSCIRC '03..
[18] R.M.Rao, J.L.Burns and R.B.Brown, "Circuit techniques for gate and subthreshold leakage minimization in future CMOS technologies," ESSCIRC '03. Proceedings of the 29th European Solid-State Circuits Conference, 2003.
[19] G.Reinman and N.Jouppi, "CACTI 2.0: An integrated cache timing and power model," WRL Research Report 2000/7, Feb.2000.
[20] P.Shivakumar and N.Jouppi , "CACTI 3.0: An integrated cache timing, power and area model," WRL Research Report 2001/2, Aug.2001.
[21] Riedlinger, R., Grutkowski, T., "The high-bandwidth 256 kB 2nd level cache on an Itanium microprocessor," ISSCC 2002.
[22] Y.Ye, S.Borkar and V.De, "A new technique for standby leakage reduction in high-performance circuits," in Symp.VLSI Circuits Dig.Tech.Papers, 1998.
[23] Y.C.Yeo et al., "Direct tunneling gate leakage current in transistorswith ultrathin nsilicon nitride gate dielectric," IEEE Electron Device Letters, Nov. 2000.
[24] Weiss, D., Wuu, J.J., Chin, V., "An on-chip 3MB subarray-based 3rd levl cache on an Itanium microprocessor," ISSCC 2002.
[25] S.J.Wilton and N.P.Jouppi, "CACTI: An enhanced cache access and cycle time model," IEEE JSSC, vol.31, no.5, May 1996.
[26] The Tech Report. http://techreport.com/cpu/.

30

Stall Cycle Redistribution in a Transparent Fetch Pipeline

Eric L. Hill and Mikko H. Lipasti

Electrical and Computer Engineering
University of Wisconsin - Madison

{elhill, mikko}@ece.wisc.edu

ABSTRACT

Power and power density are now primary design constraints for modern high performance microprocessors. Up to 70% of the dynamic power consumed can be attributed to the clocking system. A consequence of this trend is that clock gating has emerged as both a necessary and efficient method to significantly reduce dynamic power.

Transparent pipelining, a recently proposed fine-grain clock gating technique, has the potential to significantly reduce clock power above and beyond conventional pipestage-level clock gating. Previous studies of transparent pipelining have focused on the circuit and implementation-related issues of this approach, while neglecting the broader microarchitectural implications. This paper aims to quantify the microarchitectural opportunities that are afforded by the use of transparent pipelining in a processor's fetch pipeline. We develop a technique, based on stall cycle redistribution, designed to improve the performance of transparent pipelining on fetch and other high utilization pipelines. We show that stall cycle redistribution can dramatically reduce the clocking overhead of an aggressively pipelined Cell-like microprocessor.

Categories and Subject Descriptors

C.1.3 [**Processor Architectures**]: Other Architecture Styles – Pipeline processors.

General Terms

Performance, Algorithms.

Keywords

pipeline gating, dynamic power, instruction fetch, microarchitecture.

1. INTRODUCTION

Modern processors retire instructions at an average rate much lower than their peak rate; this discrepancy creates an opportunity for reducing needless switching activity and reducing dynamic power consumption by gating clock signals to unused circuits or stalled portions of the instruction processing pipeline. For conventional unit- and pipestage-level gating techniques, savings in dynamic power consumption are inversely proportional to the average utilization of the unit or pipeline stage [3]. In other words, units with low average utilization are attractive candidates for

clock gating, since a substantial portion of their dynamic switching activity can be cut off and converted into power savings. Hence, functional blocks such as floating-point units, which are hardly used in many workloads, become prime candidates for clock gating. In such units, dynamic power can be directly estimated by counting the number of units of work (e.g. floating-point instructions) that exercise the unit, and assuming that the unit would consume no dynamic power the rest of the time, whenever no such units of work were flowing through it. This straightforward approach extends itself to pipestage-level gating, where individual pipestages can be gated, since each unit of work that traverses a pipeline still requires latching activity at each stage. As a result, microarchitectural techniques that seek maximum benefit from clock gating simply need to minimize the total units of work traversing a pipeline, with no consideration for their arrival rate distribution.

However, the introduction of transparent pipeline gating [10][11] fundamentally changes this microarchitectural tenet of clock gating. In transparent pipeline gating, pipeline latches remain transparent, dynamically converting a pipelined implementation into a multicycle combinational path whenever low pipeline utilization will allow. Stated simply, a transparent pipeline only imposes opaque latches when two or more units of work are in the pipeline and are in danger of racing with each other, and must be separated by an opaque latch to maintain correctness and prevent the race. Of course, opaque latches require clock edges in order to capture new values, and hence consume dynamic clock power, while transparent latches simply propagate their inputs to their outputs, requiring much less power; herein lies the opportunity for reducing dynamic power with transparent latches. However, since the need for opaque latches in a pipeline is determined not solely by the presence of a unit of work traversing that pipeline, but rather by the distance from that unit of work to the preceding and following units of work (i.e. the distribution of the work unit arrival rate), we can no longer estimate power consumption simply by measuring the utilization of a pipeline, and assuming that dynamic clock power was consumed by a latch when and only when a unit of work was present in the corresponding pipestage. Instead, to estimate power consumption in a transparent pipeline, we must establish the number of stages separating each unit of work from its predecessor, and use that distance to determine how many of the latches in the pipeline must be opaque to prevent races between these two concurrent entries in the pipeline.

The simple fact that power consumption in a transparent pipeline varies as a function of the distribution of work unit arrival rate-- rather than being directly proportional to the arrival rate--creates a microarchitectural optimization opportunity that does not exist for prior approaches to clock gating. As we will describe in Section 2, inter-unit gaps ranging from a single cycle between subsequent arrivals all the way up to gaps as long as the entire transparent pipeline will provide progressively greater clock power savings in

---

Permission to make digital or hard copies of all or part of this work for personal or classroom use is granted without fee provided that copies are not made or distributed for profit or commercial advantage and that copies bear this notice and the full citation on the first page. To copy otherwise, or republish, to post on servers or to redistribute to lists, requires prior specific permission and/or a fee.

ISLPED'06, October 4–6, 2006, Tegernsee, Germany.

Copyright 2006 ACM 1-59593-462-6/06/0010...$5.00.

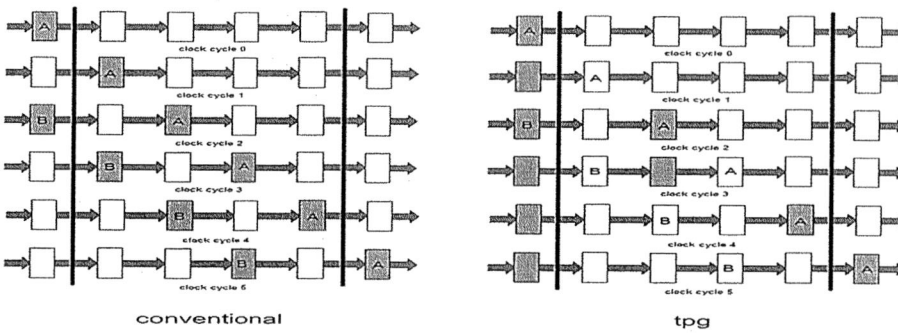

Figure 1. Conventional vs. Transparent Pipeline Clocking

a transparent pipeline. However, gaps larger than the length of the pipeline provide no further savings, since all clock edges have already been inhibited when the gap matches the length of the pipeline. In this work, we seek to alter the distribution of the inter-unit gaps to best take advantage of transparent clock gating, without actually reducing the arrival rate. Since gaps between units in an instruction processing pipeline are caused by stall conditions (e.g. cache misses, branch mispredictions, and structural or data hazards), we call such an approach stall cycle redistribution.

This paper studies opportunities for stall cycle redistribution in the context of an aggressively pipelined processor front end modeled after the IBM/Sony Cell PPU. We chose to study machines similar to the Cell processor due to the fact that it's deep pipeline and in-order execution core imply that pipeline latches will constitute a significant portion of the overall power consumption. Because of these design attributes, we feel that transparent pipeline gating has the potential to be a valuable power saving technique, and thus complementary techniques to increase its effectiveness are interesting to study. In our pipeline, stalls are caused by instruction cache misses, branch mispredictions, and issue stalls. We find that transparent pipeline gating of the front end provides some power reduction over conventional opaque pipeline gating, but that the benefit is much less than one might expect from the relatively low utilization of the front end for our applications (the Cell front end has a peak rate of 2 instruction per cycle, but we retire only 0.5 instructions per cycle for a typical program, leading to a 25% front-end utilization). The main reason for this is that the stall conditions that provide benefit by causing gaps in the pipeline are not uniformly distributed. Rather, every time a stall condition (e.g. a cache miss) occurs, the stall condition persists for many cycles, creating a large inter arrival gap in the pipeline. However, once the stall resolves, the pipeline quickly fills up, and all latches must become opaque.

To combat this poor distribution of stall cycles, we develop a heuristic for redistributing stall cycles; this is described in Section 3. We show that the use of our heuristic dramatically decreases the amount of clocking overhead in a microprocessor core. Section 4 evaluates these techniques in terms of clock power savings, Section 5 discusses related work, and Section 6 concludes the paper.

2. TRANSPARENT PIPELINE GATING

This section briefly reviews the correctness requirements for pipeline clocking and clock gating, and describes the mechanics of transparent pipeline gating.

2.1 Stateholders

Jacobson described the concept of stateholders in order to reason about the correctness requirements of a pipeline. A stateholder is defined as a latch in opaque mode, which protects a downstream data packet from being overwritten by an upstream packet [10]. Essentially, the stateholder of a data packet is the latch that is currently preventing a data race from occurring between that data packet and the next packet following it in the pipeline. Using the concept of stateholders, the differences between transparent and conventional pipelining techniques can be concisely illustrated. Figure 1 shows the progression of state holder allocation as two items A and B move down a pipeline clocked using the both the conventional and transparent pipeline clocking models. Each rectangle represents a pipeline stage (a latch and associated combinational logic). The shaded and unshaded rectangles represent stages with closed and open latches, respectively. Items residing in the two endpoint pipeline stages are always clocked for safety reasons. Under the traditional model, as each piece of data propagates into a new stage, the pipeline latches in that stage are clocked, allocating a new state holder for that data packet. In Figure 1 it is clear that stateholders are allocated somewhat inefficiently. In cycle 1, A is unnecessarily allocated a new stateholder. Because there is no valid data in the pipeline stage directly behind A, simply keeping the latch in stage 0 opaque during cycle 1 is a sufficient condition for preventing a data race. Transparent pipelines derive their primary benefit by detecting when data races are about to occur (rather than always assuming they can potentially happen), and only allocating new stateholders in those cases. Figure 1 also illustrates how stateholders would be allocated for a transparent pipeline. The key difference here is that a new stateholder for A is only allocated during cycles 0, 2, 4, and 5, which correspond to when A enters the pipeline, when B overwrites A's stateholder, and when B leaves the pipeline. Because Item B has no following data packet, only one stateholder is allocated in cycle 2. Jacobson's seminal work on transparent pipelining made the observation that latches only need to be clocked to separate rather than to propagate data [10]. He also makes the observation that only floor(N/k) clock pulses (or stateholder updates) are needed to propagate a data item to the end of an N stage pipeline, when there are k pipeline stages separating it from the next valid data item. The implicit assumption of conventional pipelining clocking theory is that there is never any separation between valid pieces of data (k = 1), so pipeline latches should always be clocked to avoid data races. In reality, this assumption is a sufficient but not necessary condition for correct pipeline operation. Transparent pipelines only enforce the necessary conditions for correct pipeline operation, and thus have the potential to significantly reduce the number of clock pulses needed during workload execution to ensure race-free operation.

2.2 Implementation Assumptions
The details of implementing a transparent pipeline are discussed thoroughly in [10]. This clock gating technique is intended for static CMOS circuits, which is the logic style used by most current low-power designs, as domino circuits tend to be power-hungry. Feedback and bypass paths also need to be registered with flip-flops. Because these paths are generally narrow (e.g. a single status bit), their associated clocking overhead is small compared to the forward pipeline paths, which are many bits wide. The cost of control logic can also be amortized over the wide latches that exist in the frontend. Additionally, if the cost of the control logic for a deeper pipeline is prohibitive, a pipeline composed of multiple transparently pipelined segments can be constructed. A pipeline built in this fashion would still be able to capture the majority of the benefit of a longer transparent pipeline, but at a lower cost.

2.3 Microarchitectural Opportunity
Prior work on this topic has shown that the maximum clock power reduction is attained when transparent pipelining is applied to pipelines with low utilization [10]. Examples of pipelines with low utilization include floating-point or media processing functional units that are underutilized when running conventional integer workloads, since there are relatively few instructions of those types of instructions being executed. In this work, we apply transparent pipelining to fetch pipelines, which typically have high utilization rates. Intuitively, pipelines with very high utilization rates should see little to no benefit from transparent pipelining (or clock gating in general) because the k term in floor(N/k) converges to 1. Given this, an opportunity exists to decrease the power consumption of a transparent fetch pipeline by controlling the arrival rate of data items entering the pipeline. This is equivalent to increasing the distance (in terms of pipeline stages) separating each valid data item that propagates through the pipeline. One important property of transparent pipelines that we take advantage of is the fact that coarse-grained gaps between units of work are less useful than gaps of a finer granularity. Once a gap is longer than the length of the transparent pipeline (k>N), no additional power benefits can be obtained (because floor(N/k) is already 0). As two items become separated by an increasing number of pipeline stages, the clock power that is saved by each additional stage of separation decreases. For example, if an item X is followed by a succeeding item Y in the immediately preceding pipestage, there is no benefit from transparent pipelining (because floor(N/k)=N). If instead X and Y are separated by 2 pipestages, transparent pipelining has half the cost (floor(N/k)=N/2). Continuing with this example, separating the data items by 3 stages has one-third of the cost, and larger separations have cost benefits that can be derived in a similar fashion. The key point of this example is to illustrate that the first stage of separation is the most important. This key observation is the primary motivation behind the concept of stall cycle redistribution. The goal of this work is to leverage the presence of these coarse-grained stalls as an opportunity to inject additional inter-unit gaps of a finer granularity into a front-end transparent pipeline, effectively redistributing these stall cycles to enhance the benefits provided by a transparent pipeline. A more in depth discussion of stall cycle redistribution is provided in Section 3.

2.4 Front End Pipelines
High performance microprocessors generally have very aggressive front-end pipelines in order to give the back end of the machine a sufficient supply of instructions to find ILP. Because of this, a great deal of architecture research has been aimed at minimizing any and all factors that would contribute to a front-end pipeline being underutilized. Because of this focused research effort, frontend pipelines generally enjoy low instruction cache miss rates and high branch prediction accuracy rates. As a consequence of these trends, the immediate clock power reduction benefits of applying transparent pipelining to a microprocessor are much less than one might expect (this intuition is validated by our results in Section 4). The deep front end pipeline we model is based on the fetch pipeline of the IBM/Sony Cell microprocessor [16]. In addition to the aggressive pipelining (23 stages, 12 in the front end), the Cell processing units lack an instruction window (which typically consumes a large amount of power). Both of these design attributes indicate that the overhead of pipeline latches constitute a significant fraction of the total core power.

3. STALL CYCLE REDISTRIBUTION

3.1 General Concept
A more general way to frame this problem is to view the microprocessor frontend as a queueing system. The front end of the machine can be defined as a first come, first serve (FCFS) queue with 1 server (instructions move through the front end in groups). The arrival rate distribution of this system can be described as deterministic, as instructions are generally fetched as aggressively as possible for performance reasons discussed previously. While this default arrival rate distribution ensures that the frontend will almost never be a performance bottleneck, it precludes any power gains transparent pipelining may have over valid based clock gating. In contrast, the service time distribution of the queueing system is somewhat more varied. Service times can vary based on whether or not instructions in a particular fetch group encounter cache misses, are multicycle operations, or have data dependencies on previous instructions. In Figure 2 we use the time diagram notation suggested by [14] to visually illustrate the front end as a queueing system. The lower line in this figure represents the queue insertion point (the point in time when a group of instructions is fetched). The upper line represents the queue exit point (the point in time when the first instruction from the group is issued). The variables C_n and C_{n+1} represent consecutive groups of instructions that were fetched during program execution. The time T_{n+1} represents the inter arrival time between C_n and C_{n+1} entering the front end queue. This term is relatively small, representing the fact that instructions are generally fetched very aggressively for maximum performance. The term S_{n+1} represents the total time the instruction group C_{n+1} spends in the queueing system. This term has two components, the

service time X_{n+1}, and the wait time W_{n+1}. This wait time represents the *fetch slack* of instruction group C_{n+1}. Given this precise definition, the fetch slack of a group of instructions can be precisely measured by tracking groups of instructions moving through the fetch pipeline, recording for each group the amount of time the group is the oldest group in the front end that has not issued an instruction. This is equivalent to knowing the wait time for the instruction group. The ideal way to exploit this fetch slack would be to delay the arrival time of C_{n+1} as much as possible (essentially increasing T_{n+1} by W_{n+1}). This would have the effect of creating pipeline bubbles between C_n and C_{n+1}, allowing the benefits of transparent pipelining to be visible, without delaying the service time of C_{n+1}. It is also important to note that the fetch slack or wait time (W_{n+1}) of C_{n+1} is very closely related to the service time of the previous instruction group C_n (represented by term X_n).

In this work we argue that certain portions of the service time distribution are predictable, and because the fetch slack of an instruction group is closely related to the service time of the previous instruction group, that fetch slack is also predictable in many cases. Fetch slack can either come from data dependences that arise between instructions or can be due to cache misses. Data dependences should be extremely easy to predict, as the parents of an instruction are usually the same for each dynamic instance of that instruction (any differences in parents would be due to control flow, which is also predictable). While cache misses are much harder to predict, on average they account for less than 5% of the observed slack across all benchmarks. We do not explicitly present this characterization data due to space reasons. The goal of the predictive flow control techniques presented in this section is to leverage the more predictable components of the service time distribution in order to dynamically change the arrival rate distribution to a format more amenable to a transparent pipeline. The predictability of the service time distribution is essential because it allows us to be confident that the changes made to the inter arrival distribution will not result in performance loss.

3.2 Slack Predictor Design

We propose a low overhead scheme for slack prediction. At the beginning of the pipeline, before the instruction cache is accessed, the slack predictor is read using the fetch address as an index. The prediction information for each fetch group is represented by a few extra bits in its corresponding BTB entry. These bits represent the predicted slack, along with the confidence in the prediction. This means that there are also BTB entries allocated to fetch groups that do not contain a taken branch, and there will be additional BTB activity when our slack prediction scheme is enabled. If there is a slack s predicted for a particular fetch group (along with sufficient confidence), the fetching of that particular group will be deferred for s cycles, effectively inserting an inter-unit gap of s stages into the transparent pipeline. It should be noted that since inter-unit gaps of longer than the length of the pipeline yield no additional power benefits, fetch groups with larger slacks are only delayed by the number of cycles equal to the pipeline length. This also minimizes any performance loss that may be incurred by injecting extra inter-unit gaps into the pipeline in the event of an incorrect prediction. At commit time, when the branch predictor is updated, the slack predictor is also trained.

In addition to a prediction scheme, a means by which to verify predictions made is also needed. This is a non-trivial problem, because if the fetch of a group of instructions is delayed for n cycles, and the observed slack of that fetch group is 0, it can either mean that the prediction was correct and the group of instructions

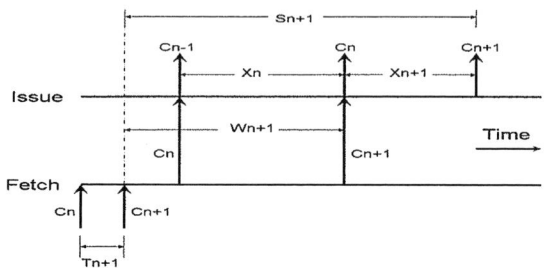

Figure 2. Time-diagram notation for frontend queue.

was fetched just in time, or that the group of instructions was fetched too late and performance was actually hurt. Our solution to this problem is to randomly sample fetch groups during instruction fetch that have predicted slack and not delay them. If the prediction that was made was really correct, this undelayed instance should have slack greater than or equal to the slack that we predicted at fetch time. In some respects this is an inverted version of the delay-and-observe method used in [8] used to verify whether or not instructions had slack. Fetch groups that have non-zero slack, but low confidence are simply trained by making sure that the observed slack was greater than or equal to the predicted slack.

4. EVALUATION

All experiments performed in this study were conducted using a detailed near-register-transfer-level full system simulator [5]. The performance simulator was augmented with a detailed activity counter model in order to estimate power consumption, based on Wattch [4]. The per-structure estimates of power per access correspond to 100 nm process technology and a clock frequency of 2 GHz. In order to model pipestage-level clock gating, our simulator deviates from Wattch's approach to modeling clock power in that we model the power consumed by pipeline latches and precharge logic separately. The additional space overhead due to slack and confidence bits is included in the power per access estimate for the BTB. Similar to [21], we do not model power for distributing the clock, as Wattch's approach appears quite specific to the design style of the Alpha 21264 and is not representative of more recent designs like the Cell PPU. Frontend latch counts were calculated by assuming a latch growth factor of 1.1 for each additional pipeline stage. Also, because we are only focusing on techniques that reduce dynamic power, no leakage power consumption is modeled. The machine configuration parameters used for our experiments are shown in Table 1. For our experiments we use the SPECINT2000 benchmark suite, along with two commercial workloads (specjbb and tpc-h). All simulations done were for a period of 250 million instructions.

Figure 3 shows the relative stage-level clock pulse savings for a transparent fetch pipeline over a traditionally clocked pipeline. The vertical axis represents the number of clock pulses that are needed to propagate groups of instructions safely through the frontend pipeline, normalized to the baseline case of a pipeline that is clocked in the conventional manner. The total height of the stacked bars represents the latch activity factor of a pipeline that

uses stall cycle redistribution and conventional pipestage-level gating. In other words, this height represents the amount of latch activity that is eliminated by removing wrong-path traffic. The height of the lower bars represents the relative latch activity factor when both stall cycle redistribution and transparent pipeline gating are used. The case where transparent pipeline gating is used alone is not shown because the reduction in latch activity factor is negligible across all benchmarks. From Figure 3, it is clear that

Table 1. Machine Parameters

InOrder Execution	4-wide fetch, 2-wide issue/commit
Pipeline Depth	10 stage transparent front end pipeline (with 2 conventional latches at endpoints)
Branch Prediction	Gshare (8k entry), 32 RAS, 1024 set 4-way BTB, 8 cycle misprediction latency
Memory System (latency)	32KB 4-way 64B IL1/DL1 (1/4), 512KB 4-way 64B unified L2 (12)
Slack Prediction	4 confidence bits / >4 high conf threshold / predictions checked randomly 10% of the time

the use of stall cycle redistribution in conjunction with transparent pipelining reduces the overhead of the frontend pipeline latches significantly more than either technique in isolation.

Figure 4 shows a comparison of the frontend energy-delay product for the baseline, stall cycle redistribution with opaque pipeline gating, and stall cycle redistribution with transparent pipeline gating configurations. The frontend energy delay product is simply the energy delay product only considering structures that reside in the fetch pipeline of the microprocessor (e.g. the instruction cache, branch predictors, and front end latches). We chose to use the frontend energy delay product as an evaluation metric because that is the portion of processor where our technique primarily reduces power. Our technique lowers the overall energy delay product by 6% on average, but we do not explicitly include the results due to space constraints. The vertical axis is normalized to the baseline processor configuration, and the stacked bars represent the contribution of each front end component to the total energy delay. It is important to note that like any fetch throttling technique, a processor using stall cycle redistribution will fetch less aggressively down wrong paths relative to a processor without throttling. This means that there will be a reduction in the utilization of both the instruction cache and the branch predictor. The addition of transparent pipeline gating further reduces the total frontend energy-delay product across all benchmarks.

We do not show an explicit performance comparison because the observed performance variation for all benchmarks is negligible. In all of our experiments, we observed a performance loss of less than 2% in the worst case. This lack of performance loss implies that our throttling technique is accurate. This is not surprising as stall cycle redistribution injects pipeline bubbles by taking advantage of the program order dataflow dependence relationships

between instructions, which are very predictable. Looking at the energy-delay results, there are 4 benchmarks for which stall cycle-redistribution does not perform as one would like. These benchmarks are crafty, specjbb, tpc-h, and vortex. The reason for this is that the extra predictor related BTB updates are

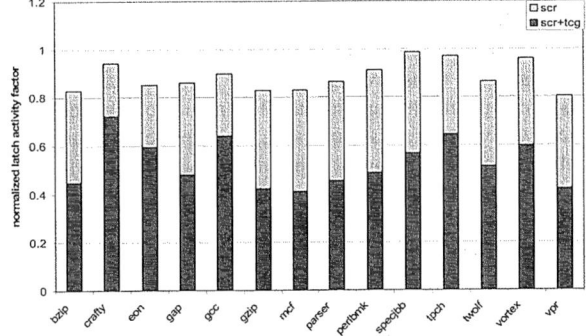

Figure 3. Latch Activity Factor Comparison.

compensating for the energy that is saved by the reduced activity in the latches and the instruction cache.

5. RELATED WORK

There has been a significant body of prior research focused on reducing dynamic power consumption in conventional high performance microprocessors, particularly in the instruction fetch and decode logic. These techniques primarily focus on exploiting low confidence branches to remove wrong path traffic [17][2][13]. While these works primarily derive their benefit from eliminating traffic (or units of work) moving through the pipeline, our work is not only concerned with the magnitude of work that propagates through the pipeline, but also the number and size of **inter-unit gaps** that exist between those units of work. The distribution of inter-unit gaps in the stream of pipeline traffic is uniquely interesting in the context of transparent pipelining and irrelevant in the context of these earlier studies.

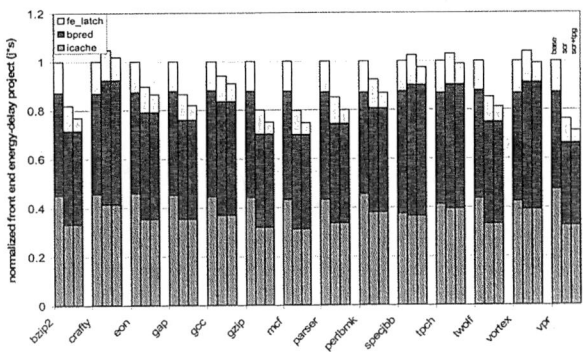

Figure 4. Frontend Energy-Delay Product.

Prior work on dynamic pipeline scaling, which dynamically varies the length of a pipeline by making latches transparent [6][15][19], was elegantly generalized by Jacobson in his seminal work on transparent pipeline gating[10]. Transparent pipeline gating was also studied in the context of a modern out-of-order processor in [11].

Fields et al. develop several notions of slack in [7], which we use as theoretical underpinnings for our study. Additionally, Muthler et. al proposed an out of order fetch architecture that used the notion of static slack to fetch and execute critical instructions earlier [18]. Our work is distinguished from these studies because we evaluate how slack can be used to enhance the performance of a transparent pipeline.

6. CONCLUSION

In this paper, we developed and evaluated a method to enhance the performance of transparent pipelining. While previous work on transparent pipelining applied the technique to pipelines which typically have low utilization rates, we explored a transparently pipelined front end, a part of the microprocessor pipeline which typically has very bursty utilization. Predictably, we found that for most benchmarks, the number of inter-unit gaps created as a result of branch mispredictions and instruction cache misses were not sufficient to make transparent pipelining an attractive design choice. In order to increase the attractiveness of transparent pipelining for microprocessor front ends, we developed a novel predictor based strategy that exploits existing instruction dependences in order to inject additional inter-unit gaps into the fetched stream of instructions, dramatically enhancing the effectiveness of a transparent pipeline in a high-utilization environment with a negligible impact on performance.

This work was supported in part by donations from IBM and Intel and by NSF grants CCR-0133437 and CCF-0429854.

7. REFERENCES

[1] C. J. Anderson, J. Petrovick, J. M. Keaty, J. Warnock, G. Nussbaum, J. M. Tendler, C. Carter, S. Chu, J. Clabes, J. DiLullo, P. Dudley, P. Harvey, B. Krauter, J. LeBlanc, Lu Pong-Fei, B. McCredie, G. Plum, P. J. Restle, S. Runyon, M. Scheuermann, S. Schmidt, J. Wagoner, R. Weiss, S. Weitzel, B. Zoric. Physical Design of a Fourth-generation POWER GHz Microprocessor. In Proc. of the 2001 International Solid-State Circuits Conference, Februrary 2001.

[2] A. Baniasadi, A. Moshovos. Instruction Flow-based Front-end Throttling for Power-aware High-performance processors. In Proc. of the 2001 International Symposium on Low Power Electronics and Design, August 2001.

[3] D. Brooks, P. Bose, S. Schuster, H. M. Jacobson, P. Kudva, A. Buyuktosunoglu, J. Wellman, V. V. Zyuban, M. Gupta, P. W. Cook. Power-Aware Microarchitecture: Design and Modeling Challenges for Next-Generation Microprocessors. IEEE Micro, December 2000.

[4] D. Brooks, V. Tiwari, M. Martonosi. Wattch: A Framework for Architectural-Level Power Analysis and Optimizations. In Proc. of the 25th Annual International Symposium of Computer Architecture, May 2000.

[5] H. Cain, K. Lepak, B. Schwarz, and M. H. Lipasti. Precise and Accurate Processor Simulation. In Workshop on Computer Architecture Evaluation using Commercial Workloads.

[6] Aristides Efthymiou and Jim D. Garside. Adaptive pipeline depth control for processor power-management. In ICCD, pages 454–457. IEEE Computer Society, 2002.

[7] B. A. Fields, R. Bodík, M. D. Hill. Slack: Maximizing Performance Under Technological Constraints. In Proc. of the

29th Annual International Symposium of Computer Architecture, May 2002.

[8] B. A. Fields, S. Rubin, R. Bodík. Focusing Processor Policies via Critical-path Prediction. In Proc. of the 28th Annual Internation Symposium of Computer Architecture, July 2001.

[9] H. P. Hofstee. Power Efficient Processor Architecture and The Cell Processor. In Proc. of the 11th Annual International Symposium on High Performance Computer Architecture, February 2005.

[10] H. M. Jacobson. Improved clock-gating through transparent pipelining. In Proc. of the 2004 International Symposium of Low Power Electronics and Design. August 2004.

[11] H. M. Jacobson, P. Bose, Z. Hu, A. Buyuktosunoglu, V. V. Zyuban, R. Eickemeyer, L. Eisen, J. Griswell, D. Logan, B. Sinharoy, J. M. Tendler. Stretching the Limits of Clock-Gating Efficiency in Server-Class Processors. In Proc of the 11th Annual Internation Symposium on High Performance Computer Architecture, February 2005.

[12] J. A. Kahle, M. N. Day, H. P. Hofstee, C. R. Johns, T. R. Maeurer, D. Shippy. Introduction to the Cell Multiprocessor. IBM Journal of Research and Development. July/September 2005.

[13] T. Karkhanis, J. E. Smith, P. Bose. Saving Energy with Just In Time Instruction Delivery. In Proc. of the 2002 International Symposium of Low Power Electronics and Design, August 2002.

[14] L. Kleinrock. Queueing Systems, Volume I: Theory. Wiley Interscience, New York, 1972.

[15] Jinson Koppanalil, Prakash Ramrakhyani, Sameer Desai, Anu Vaidyanathan, and Eric Rotenberg. A case for dynamic pipeline scaling. In CASES '02: Proceedings of the 2002 international conference on Compilers, architecture, and synthesis for embedded systems, pages 1–8, New York, NY, USA, 2002. ACM Press.

[16] K. Krewell. Cell Moves into the Limelight. In Microprocessor Report, February 2005.

[17] S. Manne, A. Klauser, D. Grunwald. Pipeline Gating: Speculation Control for Energy Reduction. In Proc. of the 25th Annual International Symposium of Computer Architecture, June 1998.

[18] G. A. Muthler, D. Crowe, S. J. Patel, S. Lumetta. Instruction Fetch Deferral using Static Slack. In Proc. of the 35th Annual International Symposium of Microarchitecture, November 2002.

[19] Hajime Shimada, Hideki Ando, and Toshio Shimada. Pipeline stage unification: a low-energy consumption technique for future mobile processors. In Ingrid Verbauwhede and Hyung Roh, editors, ISLPED, pages 326–329. ACM, 2003.

[20] J. E. Smith. An analysis of pipeline clocking. Technical report, University of Wisconsin, March 1990.

[21] W. Ye, N. Vijaykrishnan, M. Kandemir, M. J. Irwin. The Design and Use of SimplePower: A Cycle-Accurate Energy Estimation Tool. In Proc. of the 36th Annual Design Automation Conference, June 1999.

Selective Writeback: Exploiting Transient Values for Energy-Efficiency and Performance

Deniz Balkan   Joseph Sharkey   Dmitry Ponomarev   Kanad Ghose

Department of Computer Science,
State University of New York, Binghamton, NY 13902-6000
{dbalkan, jsharke, dima, ghose}@cs.binghamton.edu

ABSTRACT

Today's superscalar microprocessors use large, heavily-ported physical register files (RFs) to increase the instruction throughput. The high complexity and power dissipation of such RFs mainly stem from the need to maintain each and every result for a large number of cycles after the result generation. We observed that a significant fraction (about 45%) of the result values are delivered to their consumers via the bypass network (consumed "on-the-fly") and are never read out from the destination registers. In this paper, we first formulate conditions for identifying such transient values and describe their micro-architectural implementation; then we propose a technique to avoid the writeback of such transient values into the RF. With 64-entry integer and floating point register files, our technique achieves an 11% performance improvement and 29% reduction in the RF energy consumption compared to the baseline machine with the same number of registers. Furthermore, for the same performance target, the Selective Writeback scheme results in a 38% reduction in the energy consumption of the RF compared to the baseline machine.

Categories and Subject Descriptors

C.1 [**Processor Architectures**]: Other Architecture Styles – *Pipeline processors.*

General Terms: Performance, Design

Keywords: Register Files, Energy-Efficiency

1. INTRODUCTION

The physical register file (RF) is one of the key datapath components of modern high-performance microprocessors. To support large instruction windows for effective exploitation of available ILP, large RFs are required in order to buffer speculative state of the program. However, as the number of entries in the RF increases, the access to the RF (which can potentially limit the cycle time [4], [17], [15], [8], [22]) is likely to require multiple cycles, especially in the era of higher frequencies and dominating wire delays. At the same time, read access to the RF typically lies on the critical schedule-to-execute path, which makes it desirable to perform this access within a single cycle. If the RF access is

pipelined over several cycles, then the load-hit and branch misspeculation penalties increase and additional bypass stages are needed, negatively impacting both performance and complexity. Finally, large RFs also dissipate significant amount of power. The energy dissipated in the register file has been reported to account for 10% to 25% of the total chip energy [1], [4]. The situation is further exacerbated in the SMT processors, where the pressure on the register file is increased as larger physical register files are needed to support multiple thread contexts.

Traditional register management mechanisms are designed to easily support precise interrupts, but result in an inefficient use of registers. Specifically, a physical register allocated for the destination of an instruction is deallocated only when the next instruction writing to the same architectural (logical) register commits. Such a conservative register management guarantees that until all instructions between the two consecutive definitions of the same architectural register commit, the earlier definition is available and can be resurrected should the later definition be squashed as a result of a branch misspeculation, exception or interrupt. However, in this scheme registers remain allocated for a significant number of cycles, requiring large RFs to be used. An alternative to using large RFs is to use smaller number of registers, but manage them more effectively. To this end, a number of solutions have been proposed, including techniques for late register allocation, early register deallocation and register sharing. One set of solutions [17], [20], [21] targets early register deallocation. In all of these schemes, however, each and every result value is still written into the RF and the validity of the physical register is used as one of the conditions for its early deallocation. In this paper, we overcome this restriction and propose a scheme that in some cases deallocates physical registers right after instruction execution and *avoids the writebacks* of the produced values into the RF.

The specific contributions and the key results of this paper are:

• We introduce the *Selective Writeback (SWB)* mechanism - an aggressive scheme for early deallocation of physical registers whose values (which we call transient) are obtained by all of their consumers through the bypass network. We show that such early deallocations are possible in 45% of the cases across the SPEC 2000 benchmarks. Compared to the previously proposed early deallocation techniques, our solution is more proactive in nature, as we deallocate a register immediately after the instruction that targets this register completes the execution, *without even writing the produced transient value into the RF*. Such values are simply dropped from the datapath.

• We evaluate the impact of this mechanism on the performance as well as the energy dissipations within the RF. The RF energy is

---

Permission to make digital or hard copies of all or part of this work for personal or classroom use is granted without fee provided that copies are not made or distributed for profit or commercial advantage and that copies bear this notice and the full citation on the first page. To copy otherwise, or republish, to post on servers or to redistribute to lists, requires prior specific permission and/or a fee.
ISLPED'06, October 4–6, 2006, Tegernsee, Germany.
Copyright 2006 ACM 1-59593-462-6/06/0010...$5.00.

reduced because the writes of *transient* values are avoided. For 64-entry RFs, *selective writeback* results in 11% IPC improvement and 29% dynamic energy savings within the RF compared to the baseline machine with the same number of registers. Furthermore, for the same performance target, our scheme achieves a 38% reduction in the energy consumption of the RF compared to the baseline machine (accounting for the dissipations in the additional logic).

• We compare the performance of the SWB scheme against some previously proposed techniques for register file optimizations and show that our mechanism outperforms the previous solutions for the majority of the benchmarks as well as on the average.

2. MOTIVATION AND DEFINITIONS

It is well-known that most of the register instances in a datapath are consumed within a very few cycles following their generation [18], [23]. Following the work of [23], we define a value targeting a register to be ***short-lived*** if the architectural register allocated as a destination of the instruction X has been redefined (used as a destination by another instruction) before the value generated by X is written back. We call the instruction that redefines a register allocated to hold a short-lived value as the ***redefiner***. In our simulations of the SPEC 2000 benchmarks, about 85% of all generated register values were identified as short-lived.

We define a produced result value as ***transient*** if the following conditions are true:

C1) The value must be short-lived.

C2) There must be at most one instruction that consumes the value.

C3) The only consuming instruction must be selected for issue before the value is produced – this ensures that the value is obtained for consumption off of the bypass network.

C4) There must be no branch instructions between the value-producing instruction and its redefiner.

C5) The sole consumer of the value must not be subject to a replay caused by a load latency misprediction or a memory dependence misprediction

Figure 1 shows the percentage of transient values across the execution of SPEC 2000 benchmarks. The details of our simulation framework are given in Section 4. The bars from left to right correspond to the cumulative percentages of the various conditions (C1 through C5) described above. For example, the leftmost bars show the percentage of all generated values when condition C1 is valid, the next set of bars shows the percentage of cases when both conditions C1 *and* C2 are satisfied and so on. The rightmost set of bars depicts the percentage of transient values, as all 5 conditions are satisfied. On the average across all benchmarks, about 45% of the produced results are transient.

The results of Figure 1 imply that almost half of the generated results are not read from the RF, are not needed to recover from a branch misprediction, and the consumers of these results are not subject to any memory replay traps. The only reason to store these values in the register file is to allow for the reconstruction of the precise state after exceptions or interrupts. It is precisely the goal of this paper to introduce mechanisms that avoid writing such

values into the RF and instead rely on periodic RF checkpointing to correctly handle exceptions and interrupts. The transient values are simply "dropped" from the datapath right after their generation and the corresponding physical registers are immediately released.

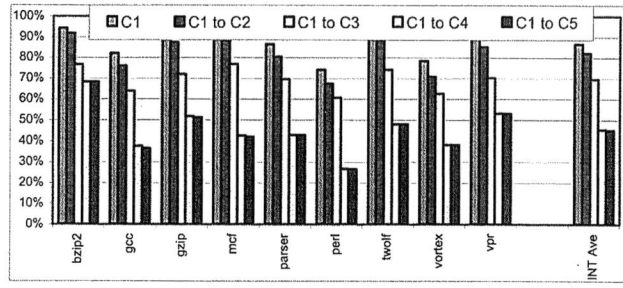

(a)

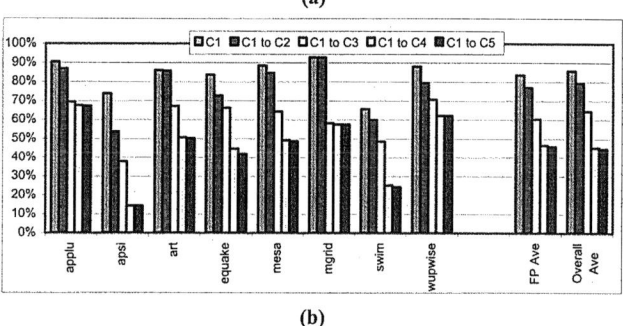

(b)

Figure 1. Various Statistics about Short-lived Values for integer (a) and floating-point (b) benchmarks.

3. DETECTING TRANSIENT VALUES

It is conceivable that some of the conditions for transient values (specifically, C2 and C4) can be determined by the compiler to simplify the hardware; however in the following discussions we describe an all-hardware implementation. Surprisingly, this mechanism is quite simple and relies on the use of two marker bits per rename table entry and two status bits per physical register. The two marker bits associated with each rename table entry are as follows:

a) The FC ("first consumer renamed") bit: this bit is set when the first consumer of a physical register representing the current instance of an architectural register is renamed.

b) The MCB ("multiple consumers or following branch") bit: this bit is set directly when a second or subsequent consumer of a physical register representing the current instance of an architectural register is renamed. This bit is also gang set for all entries in the rename table when a branch instruction is encountered.

When the rename table entry for mapping an architectural register is set up, the FC and MCB bits in the newly created entry *are both cleared*. The FC bit is set, as described above – on renaming the first consumer instruction of the associated physical register. When a consumer of a physical register is renamed, the FC bit of the rename table entry for that physical register is checked – if that bit is found to be already set, the MCB bit is set to indicate that more than one consumer exists for that register. A trivial extension of the logic also handles the case when more than one initial consumer of a physical register are renamed in the same cycle. The MCB bit is also set, irrespective of the number of consumers, as described above on encountering a branch. Thus if the MCB bit in the rename table entry for a physical register is set, the implication is that either condition C2 or C4 are not valid, that is, the associated value is not transient.

The two additional bits – *both set to a one at the time of allocating a physical register* - are needed for detecting transient values are each associated with a physical register and are as follows:

c) The NT ("not transient") bit. This bit indicates if the associated register value is transient (NT = 0) or not (NT = 1). The initialization of NT marks the associated value as transient by default.

d) The CI ("consumer issued") bit, which is cleared when any consumer of a physical register issues. Recall that the physical register has to be read by the issued instruction, so this bit can be easily cleared as the issued instruction has initiated a register file read – either from the RF (or off the bypass network).

At the time of creating the new instance of an architectural register R (that is, at the time of renaming the renamer of the instruction that generates/generated the previous instance of R), the rename table entry to be overwritten is read out. This entry is for the previous instance of R, say physical register P. The MCB bit for P is then copied to the NT bit of P. If the MCB bit is a zero, the NT bit then indicates that conditions C2 and C4 are both valid, as well as the fact that a renamer was also encountered. Ignoring condition C5 for the time being, if both the NT and CI bits of a physical register are both zero at the time of initiating a write to the register, the writing of the register can be abandoned. Put in other words, the value targeting the register is transient, as it has no branch instructions preceding its renamer, nor multiple consumers (NT = 0) and its only consumer has issued (CI = 0) Note that the physical register's address is typically available at least one cycle preceding the actual write in modern datapaths for waking up dependent instructions.

In processors that use speculative scheduling based on load latency prediction, condition C5 is implemented by resetting the NT and CI bits of the affected physical registers. In addition, no value is dropped (even if it is determined as transient) during the time when such recovery takes place. To support SWB in processors that use memory dependence prediction [7], we use a technique similar to what is described in [20]. Specifically, we ensure that the consumer of a transient value is older than the oldest store instruction whose address has not yet been computed. As shown in Figure 1, the percentage of values identified as transient is reduced only slightly by imposing this additional condition. In our evaluations, we model both types of memory-related speculations. The energy savings and potential performance gains from avoiding writebacks into the register file come at the cost of maintaining and managing just three bits per physical register, as described earlier.

4. MAINTAINING THE PRECISE STATE

When some register values are discarded right after their generation, traditional ROB-based mechanisms for restoring the precise state can no longer be used. Therefore, to maintain the precise state in the event of exceptions or interrupts, we rely on the use of checkpointing (We note here that branch mispredictions are handled just as in traditional designs due to the condition C4 as described in Section 2). Specifically, we create periodic checkpoints of the RF in the following manner. Assume that the instruction I1 is the youngest dispatched instruction in the ROB at the instant when the decision to create a checkpoint is made. At this moment, the most recent instance of every architectural register is either the value in the RF or it is defined by some in-flight instruction, whose result has not been generated. For example, if the value produced by the instruction writing to architectural register X has been dropped, then there is another in-flight instruction down the stream that writes into the same architectural register. Consequently, if we allow all the instructions between the *ROB_head* (which points to the commitment end of the ROB) and instruction I1 to unconditionally write their results to the register file, then at the time of I1's commitment a precise state of the register file will be available. At this point, the full checkpoint of the register file can be created. Basically, the selective dropping of values is avoided during the checkpoint creation period, i.e. all values are written into the register file. The specific circuit implementation of checkpointing is not central to this paper. One can either use a separate register file for this purpose (as in [20]), or embed the checkpoint within the register file itself, by backing up each bitcell with a shadow copy [9]. For our evaluations, we assumed the Checkpointed Register File (CRF) design proposed in [9], where each bitcell is backed up by a shadow cell.

Table 1. Configuration of the Simulated Processor

Parameter	Configuration
Machine width	4-wide fetch, 4-wide issue, 4 wide commit
Window size	64-entry issue queue, 64 entry LSQ, 128–entry ROB
Registers	Various sizes studied, as indicated in the text
Function Units and Latency (total/issue)	4 Int Add (1/1), 1 Int Mult (3/1) / Div (20/19), 2 Load/Store (2/1), 4 FP Add (2), 1FP Mult (4/1) / Div (12/12) / Sqrt (24/24)
L1 I–cache	32 KB, 2–way, 32 byte line, 1 cycle hit time
L1 D–cache	32 KB, 4–way, 32 byte line, 2 cycles hit time
L2 Cache unified	512 KB, 4–way, 128 byte line, 8 cycles hit time
BTB	1024 entry, 4–way set–associative. Minimum branch misprediction penalty – 10 cycles
Branch Predictor	Combined with 1K entry Gshare, 10 bit global history, 4K entry bimodal, 1K entry selector
Memory	128 bit wide, 120 cycles
TLB	64 entry (I), 128 entry (D), fully associative

When an exception occurs, the processor state is recovered to a previous checkpoint. However, if upon resumption of the execution from the checkpointed state we continue to drop transient values, the same exception can potentially reoccur again, thus inhibiting any forward progress. To prevent this scenario, we avoid dropping the values until the exception reoccurs, at which point the exception can be handled precisely. However, some exceptions may not reoccur in the course of re-execution, so we need some mechanism to identify when to resume selective dropping of the register values. Our solution is as follows. After the checkpoint is created, we maintain a counter of the number of instructions committed since the creation of the checkpoint. When an exception occurs for the first time (an exception is recognized when the instruction commits), the value of this counter is saved in a register. In the course of re-execution, we avoid dropping the values until the number of committed instructions reaches the saved value of the counter. At this point, the exception would have either reoccurred or is guaranteed not to be encountered again. Since our checkpointing period is rather small and exceptions generally occur infrequently, the overall impact on the performance is negligible, as we detail in Section 5.

To buffer a large number of store instructions between two consecutive checkpoints, we use the approach described in [20] and also used in a few others works. The values are stored within the local cache hierarchy, but their propagation to the main memory is avoided until it is safe to do so. Each cache line updated in this manner is marked *Volatile*, using one extra bit for each cache line. When a processor needs to rollback to a checkpoint, all cache lines marked *Volatile* are invalidated using the gang-invalidate signal. When the precise state is created (as described in the previous section), the *Volatile* bits are cleared.

5. SIMULATION METHODOLOGY

For estimating the energy savings and the performance gains achieved by using the SWB scheme, we used a significantly modified version of the Simplescalar simulator [6] that explicitly models the issue queue, the reorder buffer, the load/store queue, the register renaming logic and other out-of-order execution mechanisms associated with a datapath where a unified register file is used. Register file is read after the instructions are issued. The studied processor configuration is shown in Table **1**. For load-latency prediction, we used the load hit/miss predictor which was used for the Alpha 21264 processor. A 5-bit saturating counter is used for each entry where the counter is incremented by 1 in case of hit and decremented by 2 in case of miss. Load instructions are predicted to hit in the cache if the most significant bit of the counter is 1. We also used the store set predictor described in [7] for speculating on memory dependencies.

We used 9 integer SPEC 2000 benchmarks (*gcc, gzip, parser, perlbmk, twolf, vortex, mcf, bzip* and *vpr*) and 8 floating point SPEC 2000 benchmarks (*applu, art, mesa, mgrid, swim, apsi, equake* and *wupwise*). We had difficulties compiling the other benchmarks (mostly those written in Fortran) in our simulation framework. Benchmarks were compiled using the Simplescalar GCC compiler (with –O4 optimizations) that generates code in the portable ISA (PISA) format. Reference inputs were used for all the simulated benchmarks. The results from the simulation of the first 1 billion instructions were discarded and the results from the execution of the following 500 million instructions were used.

For estimating the energy dissipated in the course of accessing register files, the event counts gleaned from the simulator were used, along with the energy dissipations, as measured from the actual hand–crafted VLSI layouts using industry-standard Cadence® design tools. CMOS layouts for the register files and the bit–vectors in a 0.18 micron 6 metal layer process (TSMC) were used to get an accurate idea of the energy dissipations for each type of transition.

6. RESULTS AND DISCUSSIONS

We now describe the energy implications of the SWB scheme and then allude to the performance aspects of our mechanism, comparing the results against some previously proposed techniques.

6.1 Energy Considerations

We first present the energy implications of SWB on the RF, ignoring the dissipations in the auxiliary structures, such as the additional bit-vectors. We then take the dissipations of those components into consideration. First, we consider the reduction in the overall energy dissipations (or energy per task). Figure 2 presents the energy reduction achievable within the register file if the SWB scheme is used. The comparison is given for a datapath with 64 integer and 64 floating point registers. There are two bars depicted in the figure, the bar on the left presents the energy savings within the register file itself when the dissipations in the additional logic required by SWB are ignored. On the average, SWB reduces the RF's energy consumption by 36% - The energy reduction on individual benchmarks ranges from 50% for *bzip* to 13% for *apsi*. In the presented results, we assumed that the energy reduction is only a consequence of the smaller number of writebacks. The bar on the right depicts the energy reduction in the RF, if the dissipation in all additional bit-vectors and arrays required by the SWB scheme are taken into consideration. Even with the additional energy dissipation, there is still a reduction of 29% in the RF energy. The overall processor energy savings depends on the percentage of energy attributed to the RF. The range of these savings will be between 3% and 7% (considering that RFs contribute between 10% and 25% to the overall processor energy, as mentioned in the introduction).

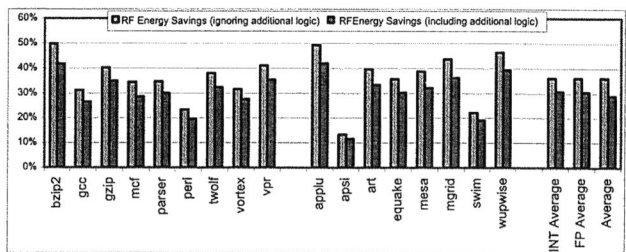

Figure 2. Percentage of Energy Savings within the RF.

Finally, we also evaluated the energy savings within the RF, if the machine employing the SWB scheme is configured to provide the same IPC performance as the baseline processor. As detailed later in the section, SWB with 80-entry RFs provides the same IPC performance on the average as the baseline machine with 96-entry RFs. However, higher energy savings can be achieved comparing these configurations, as in addition to having fewer writebacks, a smaller RF size also further decreases the energy per access. On the average, the energy savings (as well as the power savings since the performance is the same) are 45% in this case if the additional logic is ignored and 38% if the additional logic is accounted for.

6.2 Performance Considerations

Next, we evaluate the performance of the SWB scheme described in Section 3 for 64-entry register files, (64-entry integer RF + 64-entry floating point RF). Figure **3** depicts the commit IPC of the SWB scheme for different checkpointing frequencies. The figure has three curves, showing the average commit IPC of the integer benchmarks, the average commit IPC of the floating point benchmarks and the overall average IPC. For this experiment, we assumed that no exceptions or interrupts occurred, but we still avoided the dropping of the transient values in the course of constructing a checkpoint. Thus, performance difference among the various points on these curves only comes from the reduced pressure on the RF. It can be seen from the figure that the performance improvement achieved by the SWB scheme is small when checkpoints are created very frequently. For the checkpoining periods greater than 500 cycles, the performance gains are almost equal to those with no checkpointing in place – i.e. when all transient values identified as such are dropped. Based on the results presented in Figure 3, we conclude that the optimal checkpointing interval is 500 cycles, and we use this number for the rest of our experiments. We also simulated the effects of exceptional events, on the performance of the SWB scheme. Our experiments indicated that unless exceptions occur exceedingly often, such as a page fault occurring once every 1000 memory operations (which is, of course, an unrealistically high rate), the performance is very close to that of the system with no exceptions. We observed similar results for other RF sizes as well. For a checkpointing frequency of 500 cycles, the average performance gains of the SWB scheme compared to the baseline machine are almost 11%, ranging from 0.8% for *equake* to 39% for *art*. The average IPC gain for integer benchmarks is 6.2% and the average IPC gain for floating point benchmarks is 16.2%.

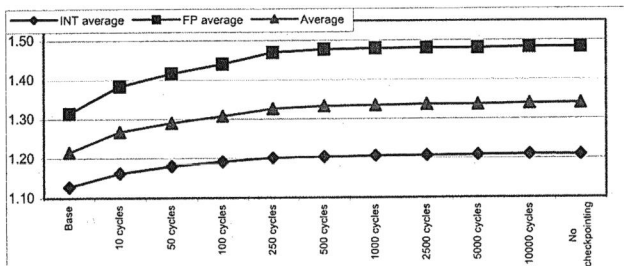

Figure 3. Commit IPCs of the SWB Scheme for different checkpointing periods.

6.3 Performance Comparison with Previous Techniques

Figure 4 compares the performance of SWB with some previously proposed techniques for optimizing register files. These alternative schemes are summarized in Section 6, we avoid repeating these discussions here for the sake of space.

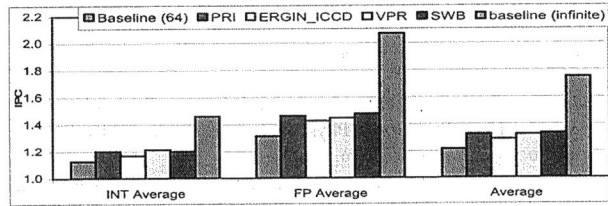

Figure 4. IPC comparison of the SWB scheme with PRI, VPR and Ergin_ICCD schemes.

In Figure 4, the first set of bars shows the performance of the baseline case. The second set of bars shows the IPCs of the scheme of [9], which deallocates the physical registers at the time of commitment or at the time of renaming the renamer instruction. We refer to this scheme as Ergin_ICCD to reflect the first author name and the publication venue. The third set of bars depicts the IPCs of the Virtual Physical Registers scheme of [11]. We refer to it as VPR in the rest of the paper. For the VPR scheme, we assume that 32 reserved registers are in use – the best configuration suggested in [11]. The next set of bars shows the performance of Physical Register Inlining [17], where narrow-width register values are stored directly within the rename table. We have implemented all these techniques in our simulation framework and therefore can compare the results on an even footing.

On average, SWB outperforms the baseline by 10.9%, compared to 5.35% for Ergin_ICCD scheme, 7.7% for VPR and 10.2% for PRI. It is interesting to notice that SWB outperforms the previous schemes on most of the benchmarks, but there are some exceptions. For example, PRI performs better than SWB for *gzip, swim, art* and *parser*. This is because a larger percentage of narrow-width values can be stored in the rename table for these benchmarks, than the percentage of values that can be detected and predicted as transient. For *bzip2, gzip, applu, vortex,* and *wupwise* the performance of SWB falls short of the VPR scheme and for *swim and wupwise*, Ergin_ICCD scheme outperforms SWB. In all other cases, SWB shows superior performance compared to other techniques. It is also interesting to observe either VPR or PRI perform better for various benchmarks with respect to each other.

6.4 Sensitivity Analysis to the RF Size

Figure 5 depicts the commit IPC values for the baseline machine, the SWB scheme for register files ranging from 48 to 128 entries. The commit IPC values are presented for the overall averages. Similar trends can be observed in all three graphs. For the same number of physical registers, the SWB scheme results in the following IPC improvements over the baseline machine on the average across all simulated benchmarks: 10.8% for 48 registers, 10.9% for 64 registers, 10.9% for 72 registers, 8.1% for 80 registers, and 7.7% for 96 registers. Note that with 128 registers only 1.57% IPC improvement is possible since the datapath with 128 registers already performs almost as well as the one with infinite number of registers, i.e. with 128 registers there are very few stalls due to the lack of physical registers.

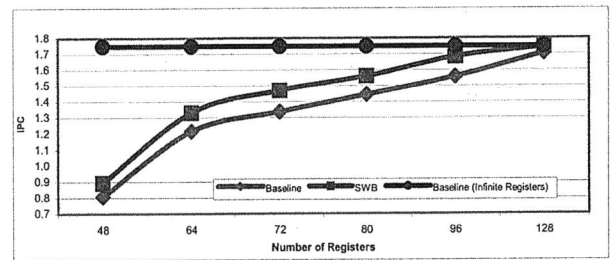

Figure 5. IPC values for the baseline machine, the SWB scheme with various register file sizes

It is also interesting to compare the number of registers needed to provide the same level of performance for the various schemes. As shown in Figure 5, the performance of the baseline machine with 96-entry RFs is matched by the SWB scheme with 80-entry RFs and. Consequently, for the same IPC, the use of fewer registers with SWB can reduce the wire delays and possibly decrease the cycle time, if the register file access lies on the critical timing path. The quantification of these additional potential advantages of SWB is beyond the scope of this paper.

We do not directly compare the energy advantages of SWB with the previously proposed schemes for optimizing register files, because energy reduction was not the goal of those techniques. With those schemes, it is possible that the additional structures and/or data movements required to support the complicated register management will outweigh the energy advantages of having somewhat smaller register files. The energy advantages of SWB come from avoiding a significant fraction of writebacks to the register file, which is a unique feature of our solution. Alternative register file optimizations for energy, such as banking

[4][15][22][24] can be used in conjunction with our technique to provide additional energy reduction.

In summary, SWB achieves higher performance AND lower energy consumption per instruction (or per task) at the same time. It is possible that the power consumption may increase on some configurations due to much faster execution time (because more work is done per cycle). However, the desired power / performance targets can always be achieved using DVFS techniques.

7. RELATED WORK

Several techniques have been proposed in the recent literature to reduce the register file pressure by using the early deallocation of physical registers [9], [17], [20], [21]. These techniques are close in spirit to our selective writeback proposal, but there are fundamental differences. In all of these works, each and every generated result is still written into the register file and validity of the register value is one of the conditions for the earlier register deallocation. Energy reduction was not the goal of the previous proposals for the early deallocation of registers. The second set of solutions delays the actual allocation of physical registers until the time that the result is written back [11]. The third set of solutions reduces the number of registers through the use of register sharing [3], [14].

There is a large body of work that targets the energy reduction in the register files through reducing the number of register ports and register file banking [4], [15], [22], [24]. The common feature of these techniques is that they all encounter a small performance loss due to various reasons. In contrast, our techniques improve energy-efficiency and performance at the same time. Also, all proposed register file banking schemes can be used in conjunction with our technique. Alternative register file organizations (mainly using various forms of caching) have also been explored for reducing the access time (which goes up with the number of ports and registers), particularly in wire-delay dominated circuits [5], [8]. In [19], register file usage was optimized using compiler support to exploit dead value information. A compiler-assisted early register release is explored in [13]. In [10], a technique to pack multiple narrow-width results into the same physical register is proposed to reduce register file pressure. Narrow-width operands were also exploited to reduce the area, access time and energy consumption [16] as well as port complexity [1] of the RFs. The concept of partial value locality was exploited for reducing the register file power, area and delay in [12].

8. CONCLUDING REMARKS

We introduced the notion of transient values and showed that 45% of all the results produced in a typical superscalar datapath are transient in nature. We presented Selective WriteBack (SWB) - a microarchitectural technique that eliminates the writes of transient values to the register file which results in the following power/performance advantages compared to the traditional superscalar datapath:

For the same register file size, SWB improves the performance by 10.8% for 48 registers, 10.9% for 64 registers, 10.9% for 72 registers, 8.1% for 80 registers, and 7.7% for 96 registers. over the baseline machine on the average across the simulated SPEC 2000 benchmarks.

For a processor with 64-entry RFs, SWB also reduces energy by 29%, when the energy dissipated by the additional logic needed by the scheme is taken into account. For the same performance target, the SWB scheme results in 38% reduction in the RF energy compared to the baseline machine.

Finally, we also showed that SWB outperforms several recently proposed schemes for register file optimizations.

9. ACKNOWLEDGMENTS

We thank Oguz Ergin for his contributions during the early stages of this work. This research was supported in part by the National Science Foundation, award numbers CNS 0454298 and EIA 9911099, and by the Integrated Electronics Engineering Center at SUNY Binghamton.

10. REFERENCES

[1] Aggarwal, A., Franklin, M., "Asymmetrically-ported Register Files", *in Proc. of ICCD 2003*.

[2] Azevedo, A., et.al., "Profile-based Dynamic Voltage Scheduling using Program Checkpoints in COPPER Framework", *in Proc. of DATE, 2002*.

[3] Balakrishnan, S., Sohi, G., "Exploiting Value Locality in Physical Register Files", *in Proc. of MICRO-36, 2003*.

[4] Balasubramonian, R.,et. al., "Reducing the Complexity of the Register File in Dynamic Superscalar Processor", *in Proc. of MICRO-34, 2001*.

[5] Borch, E. et al, "Loose Loops Sink Chips", *in Proc. of HPCA-8, 2002*.

[6] Burger, D. and Austin, T. M., "The SimpleScalar tool set: Version 2.0", *Tech. Report, Dept. of CS, Univ. of Wisconsin-Madison, June 1997*.

[7] Chrysos G., J.Emer, "Memory Dependence Prediction using Store Sets", *in Proc. of ISCA-25, 1998*.

[8] Cruz, J-L. et. al., "Multiple-Banked Register File Architecture", *in Proc. of ISCA-27, 2000*.

[9] Ergin O., et.al., "Increasing Processor Performance through Early Register Release", *in Proc. of ICCD, 2004*.

[10] Ergin O., et.al., "Register Packing: Exploiting Narrow-Width Operands for Reducing Register File Pressure", *in Proc. of .MICRO-37, 2004*.

[11] Gonzalez, A., Gonzalez, J., Valero, M., "Virtual-Physical Registers", *in Proc. of HPCA-4, 1998*.

[12] Gonzalez, R. et. al. "A content Aware Register File Organization", *in Proc. of ISCA-31, 2004*.

[13] Jones T. et al., "Compiler Directed Early Register Release", *in Proc. of PACT 2005*.

[14] Jourdan, S.,et. al. "A Novel Renaming Scheme to Exploit Value Temporal Locality through Physical Register Reuse and Unification", *in Proc. of MICRO-31, 1998*.

[15] Kim, N., Mudge, T., "Reducing Register Ports Using Delayed Write-Back Queues and Operand Pre-Fetch", *in Proc. of ICS-17, 2003*.

[16] Kondo M. and Nakamura H. "A Small, Fast and Low-Power Register File by Bit-Partitioning", *in Proc.of HPCA-11, 2005*.

[17] Lipasti, M., et.al., "Physical Register Inlining", *in Proc. of ISCA-31, 2004*.

[18] Lozano, G. and Gao, G., "Exploiting Short-Lived Variables in Superscalar Processors", *in Proc. MICRO-28, 1995*.

[19] Martin, M., Roth, A., Fischer, C., "Exploiting Dead Value Information", *in Proc. of MICRO-30, 1997*.

[20] Martinez, J., et. al., "Cherry: Checkpointed Early Resource Recycling in Out-of-order Microprocessors", *in Proc. of MICRO-35, 2002*.

[21] Monreal, T., Vinals, V., Gonzalez, A., Valero, M. "Hardware Schemes for Early Register Release", *in Proc. of ICPP-02, 2002*.

[22] Park, I., Powell, M., Vijaykumar, T., "Reducing Register Ports for Higher Speed and Lower Energy", *in Proc. of MICRO-35, 2002*.

[23] Ponomarev, D., et. al."Reducing Datapath Energy Through the Isolation of Short-Lived Operands", *in Proc. of PACT-12, 2003*.

[24] Tseng, J., Asanovic, K., "Banked Multiported Register Files for High Frequency Superscalar Microprocessors", *in Proc. of ISCA-30, 2003*.

Energy-Efficient Dynamic Instruction Scheduling Logic through Instruction Grouping

Hiroshi Sasaki, Masaaki Kondo, and Hiroshi Nakamura

Research Center for Advanced Science and Technology,
The University of Tokyo

{sasaki, kondo, nakamura}@hal.rcast.u-tokyo.ac.jp

ABSTRACT

Dynamic instruction scheduling logic is quite complex and dissipates significant energy in microprocessors that support superscalar and out-of-order execution. We propose a novel microarchitectural technique to reduce the complexity and energy consumption of the dynamic instruction scheduling logic. The proposed method groups several instructions as a single issue unit and reduces the required number of ports and the size of the structure for dispatch, wakeup, select, and issue. The present paper describes the microarchitecture mechanisms and shows evaluation results for energy savings and performance. These results reveal that the proposed technique can greatly reduce energy with almost no performance degradation, compared to the conventional dynamic instruction scheduling logic.

Categories and Subject Descriptors:
C.1.1 [**Computer Systems Organization**]: PROCESSOR ARCHITECTURES – *RISC/CISC, VLIW architectures*

General Terms: Design, Performance, Measurement

Keywords: Dynamic Instruction Scheduling, Instruction Grouping, Issue Queue

1. INTRODUCTION

General purpose microprocessors usually apply superscalar and out-of-order execution. These microprocessors should dynamically extract instruction level parallelism (ILP) for high performance because they are required to execute various types of programs efficiently and must also have to run a number of legacy binary codes. In addition, embedded processors are required to adapt to such techniques in order to dynamically extract ILP. This is due to upgrading of facilities and continuing diversification of portable devices and embedded systems, such as cellular phones and PDAs.

However, dynamic instruction scheduling logic for superscalar and out-of-order execution has a serious problem in that it consumes a significant amount of energy due to the complicated nature of its hardware logic [6][16]. Increased energy consumption is extremely disadvantageous in both portable devices, which have a limited battery life, and high-end systems, in which increased heat generation associated with increased energy consumption becomes a serious problem. As a result, it is becoming increasingly difficult to implement such complicated dynamic instruction scheduling logic on all types of processors. Therefore, reducing both the

energy consumption and the complexity of the dynamic instruction scheduling logic is essential in modern microprocessors.

The central structure, which enables aggressive out-of-order execution, is the dynamic instruction scheduling logic, which is composed of the (i) instruction queue (instruction window), which holds the renamed instructions until their source operands become available, and the (ii) instruction scheduling logic, which selects the instructions to issue from among the ready instructions in the instruction queue. A large instruction queue and a wider machine width are essential in order to achieve high performance by extracting ILP to the greatest extent possible from the target program. However, enlarging the queue size and the number of ports generally increase its complexity and energy consumption. Therefore, numerous studies have been conducted in order to reduce the energy consumption or shorten the cycle time by reducing the complexity of the instruction queue and instruction scheduling logic.

We propose a novel microarchitectural technique and hardware implementation to reduce the complexity and energy consumption of dynamic instruction scheduling logic by grouping instructions together in the instruction queue. The fundamental concept of the proposed method is to group several instructions together and let the dynamic instruction scheduling logic treat them as a single instruction. Thus, grouping should be performed in the dispatch stage when instructions are written into the instruction queue. In the present paper, we propose the grouping of two instructions by using dependence information. If issuing one instruction is the only requirement for waking up the other instruction, this pair will be grouped together. By treating the grouped instructions as a single instruction, dynamic instruction scheduling logic becomes capable of holding and issuing a greater number of instructions without increasing the size or number of ports of the instruction queue or the selection logic. This is because the amount of hardware required to treat the grouped instruction is only that which is required for a single instruction, although two instructions can be handled. Compared to the conventional dynamic instruction scheduling logic, the proposed technique can achieve approximately the same or higher performance and can significantly reduce the energy consumption by requiring a smaller amount of hardware.

The remainder of this paper is organized as follows. The conventional dynamic instruction scheduling logic is briefly described in the next section, followed in Section 3 by a description of the proposed technique and microarchitectural implementation. Section 4 describes the evaluation environment and assumptions. The results and discussions are presented in Section 5, and Section 6 describes related studies. Finally, conclusions are presented in Section 7.

2. INSTRUCTION SCHEDULING LOGIC

Dynamic instruction scheduling logic is central part of the processor that supports superscalar and out-of-order execution. Instructions are dispatched to the instruction queue after the register renaming stage, where all of the unnecessary serialization of

---

Permission to make digital or hard copies of all or part of this work for personal or classroom use is granted without fee provided that copies are not made or distributed for profit or commercial advantage and that copies bear this notice and the full citation on the first page. To copy otherwise, to republish, to post on servers or to redistribute to lists, requires prior specific permission and/or a fee.
ISLPED'06, October 4–6, 2006, Tegernsee, Germany.
Copyright 2006 ACM 1-59593-462-6/06/0010 ...$5.00.

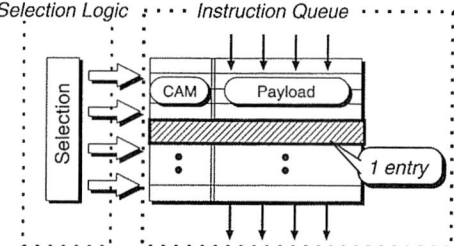

Figure 1: Conventional Dynamic Instruction Scheduling Logic

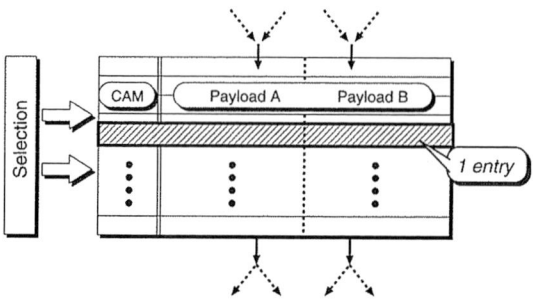

Figure 2: Image of Grouping Instructions

program operations imposed by the reuse of registers are avoided. The instruction scheduling is composed of two phases: instruction wake-up and instruction selection. The wake-up logic is responsible for waking up instructions in the instruction queue, which *sleep* until their source operands become ready (available). The selection logic selects instructions from the ready instructions in the instruction queue to be issued regardless of the program order. Selected instructions then typically broadcast their destination tags through all of the entries in the instruction queue, and each instruction associatively compares its own source register tags with the broadcasted tags. When the tag matches, the corresponding tag is set as ready. After all of the source operands become ready, the instruction becomes ready to be selected for execution.

The dynamic instruction scheduling logic is fully set associative and can dispatch, wake-up, and select instructions for the purpose of issuing every cycle to/from any of its entries. Therefore, access to the dynamic instruction scheduling logic occurs at dispatch (write to the instruction queue), wake-up (associatively matching the broadcasted tags to the local tags in the instruction queue), select (selection logic selects from the ready instructions in the instruction queue), and issue (read out from the instruction queue) for every cycle. Due to such complication, its performance is critical to the overall processor frequency and energy dissipation. In [6][16], it is reported that the energy of the instruction queue account is approximately 25% of the total chip power.

Figure 1 shows the conventional dynamic instruction scheduling logic of a four-way out-of-order superscalar processor. The right-hand side of the figure shows the instruction queue, and the shaded line indicates a single entry, which is composed of the CAM logic for tag matching and the payload RAM (op codes, physical register number, offsets, etc.). The left-hand side shows the selection logic. The instruction queue has four read and write ports for dispatch and issue, respectively, and the selection logic has the ability to select four instructions per cycle.

3. PROPOSED TECHNIQUE

The objective of the present study is to reduce the complexity and energy consumption of dynamic instruction scheduling logic. We propose the grouping of several instructions together in order to reduce the required hardware of dynamic instruction scheduling logic. In this section, we describe the outline and microarchitectural implementation of the proposed technique.

3.1 Outline

The proposed technique groups several instructions together and treats the groups as single instructions in order to reduce the complexity and energy consumption of the dynamic instruction scheduling logic. Grouping is performed at the dispatch stage and dynamic instruction scheduling logic will treat the grouped instructions as a single instructions until being issued. Figure 2 indicates an image of grouping two instructions and dispatching the group to a single entry of the instruction queue of a four-way superscalar proces-

sor. Two instructions will be a grouping candidate if one instruction can definitely be issued in the cycle following the first instruction. The payload area can hold up to two instructions, and the grouped instructions are stored to a single entry. As two instructions are grouped together, wake-up, select, and issue need only be performed once in order to handle both instructions. Thereby, the number of ports for wake-up, select, and issue required per instruction can be reduced by half (the width will be doubled, but the area increases roughly in proportion to the square of the number of ports) and the complexity of the dynamic instruction scheduling logic can be greatly reduced. As a result, the energy consumption can be significantly reduced because reducing the complexity (number of ports) of a structure is comparable to reducing the energy consumption.

However, in order to achieve sufficient throughput, compared to the conventional instruction scheduling logic, instructions must be grouped to the greatest possible extent. For example, if no instructions are grouped in a four-way superscalar, then the processor can issue only up to two instructions. The methodology of grouping instructions and the microarchitectural implementation to support the proposed technique will be discussed in the following.

3.2 Grouping Instructions

In this section, we will discuss what sort of instructions can be grouped together, as well as how the grouping of two instructions can be achieved.

3.2.1 Grouping Condition

By grouping two instructions, waking up, selecting, and issuing only the first instruction leads to the issuing of both instructions because the second instruction should be issued in the cycle following the first instruction. We decided to group a single latency operation for the first instruction because this can simplify the logic and the hardware used to support the proposed technique, which is described in Section 3.3. The candidates of grouping, the latency of which is single cycle, is an integer ALU operation. We are required to implement the proposed technique in the instruction queue, in which the integer instructions are dispatched. The details of hardware implementation of the proposed dynamic instruction scheduling logic are described in Section 3.3. Instruction pairs, which are candidates of the proposed grouping technique, can be classified into the following two types.

[I] *The issuing of one instruction is the only requirement for issuing the other instruction.*

[II] *Both instructions can be issued in the same cycle.*

We first discuss about Type [I], which consists of two different types of pairs, each of which will be explained using an example.

$$\left(\begin{array}{l} \text{Instruction A} : \; add \; r5 \leftarrow r3, \, r2 \\ \text{Instruction B} : \; add \; r4 \leftarrow r5, \, R \end{array} \right. \tag{1}$$

We can see that right operand of Instruction B is ready, and the left operand r5 is the destination of Instruction A. Therefore, only the condition of issuing Instruction A enables Instruction B to be issued in the next cycle.

The pair of instructions below also belong to Type [I].

$$\left(\begin{array}{l} \text{Instruction A}: add\ r5 \leftarrow r3,\ r2 \\ \text{Instruction B}: add\ r4 \leftarrow r5,\ r3 \end{array} \right. \tag{2}$$

The left operand of Instruction B is the destination of Instruction A and is the same as Example (1). Furthermore, the right operand r3 of Instruction B is the left operand of Instruction A, so if Instruction A is ready for issue, r3 is also ready. For this reason, the above instructions can be grouped together, as in the case of Example (1).

Next, we describe Type [II]. The instructions of Example (3), given below, are both ready for issue, because both the left and right operands are ready. Since both instructions are ready, they can be grouped, although there are no data dependencies between these two instructions. In this case, the previous instruction in program order is issued first. When grouping is not performed, these two instructions might be able to be executed in the same cycle. Therefore, the second instruction will be issued one cycle later and may cause a performance degradation. However, in the proposed technique, it is important to obtain a high throughput by grouping as many instructions as possible. Moreover, ready instructions usually stay in the instruction queue for only a short time, and most of these instructions are not critical to the performance. Thus, even if the second instruction is issued one cycle later, there is almost no performance degradation. Therefore, we group ready instructions.

Although more patterns that belong to the above two types exist, we decided to group only three types of pairs described above.

$$\left(\begin{array}{l} \text{Instruction A}: add\ r1 \leftarrow R,\ R \\ \text{Instruction B}: add\ r2 \leftarrow R,\ R \end{array} \right. \tag{3}$$

In the next section, we will discuss how to find the groupable instructions at the dispatch stage.

3.2.2 Finding Grouping Candidates

In order to achieve high throughput, finding and grouping as many instructions as possible at the dispatch stage is required. Thus, we first search the instructions in the same stage to see if there are any instructions that can be grouped (i.e., we try to detect a group from the scope of the machine bandwidth).

Out-of-order superscalar processors perform register renaming to avoid unnecessary serialization of program operations imposed by the reuse of registers. At this time, dependency between instructions dispatched at the same cycle is resolved. This structure is implemented by a simple combinational circuit. With slight changes to this structure, searching the groupable instructions described in Section 3.2.1 becomes possible. Searching these three types is performed in parallel, and when a pair of groupable instructions are found, they will be dispatched into the same entry of the instruction queue.

Furthermore, in an effort to aggressively group more instructions, after searching in the same stage, the proposed technique will search to see if the dispatching instruction can be grouped with an instruction that is already in the instruction queue. In this case, we will only search and group the pair that agrees with the pattern of Example (1) of Section 3.2.1 as a target. That is, when Instruction B of (1) is at dispatch, we will search the Instruction A from among instructions in the instruction queue. Because the processor should know the condition of the physical register at the dispatch stage, the processor accesses to the physical register state-table during the renaming stage. We will make slight change to this state-table for the purpose of searching the groupable instruction in the instruction queue. There are generally four states for a physical register entry

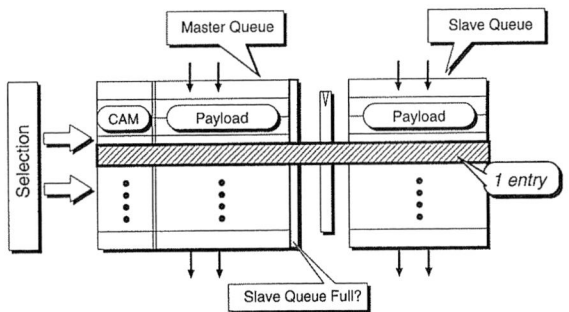

Figure 3: Basic Implementation

(two bits). We add a new *queue field* to each entry, the bit width of which is $log_2(queue\_size)$. The entry number of the instruction queue in which the producer instruction of the physical register is held is written in this field. When dispatching an instruction that may be grouped with an instruction in the instruction queue (a single-cycle latency operation), the state of the source register is checked by reading the state-table. If one of the source registers is ready and a valid value is written in the *queue field* of the other source register, the dispatching instruction can be grouped with the instruction in the instruction queue. When dispatching a single-cycle latency operation, the entry number that the instruction will be dispatched to should be written in the *queue field*.

3.3 Proposed Instruction Scheduling Logic

3.3.1 Basic Implementation

Figure 3 shows the proposed implementation of the proposed dynamic instruction scheduling logic of a four-way superscalar processor. Among two grouped instructions, the instruction that will be issued first will be dispatched to the Master Queue, and the other will be dispatched to the Slave Queue. The two grouped instructions are dispatched to the same entry and will be treated as a single unit. The Master Queue has a single-bit flag that indicates whether an instruction is dispatched to the Slave Queue side of its entry. This flag will be set when an instruction has been dispatched into the Slave Queue. Up to two instructions can be dispatched and issued to/from the Master and Slave Queues, respectively. In addition, the selection logic is able to select two instructions from the Master Queue only.

In addition, there is a latch between the Master and Slave Queues. When an instruction in the Master Queue is selected for issue and the flag is set, its access is latched and the instruction in the Slave Queue will be issued in the next cycle. If the flag is not set, the access will stop and prevent energy consumption. Therefore, only the Master Queue has the CAM logic for associative searching because the instructions in the Slave Queue are automatically issued after the instruction in the Master Queue.

As discussed above, the implementation of dynamic instruction scheduling logic shown in Figure 3 enables the proposed technique. However, there is a limit at dispatch because the number of write ports of the Master and Slave Queues is two for each. For example, writing three instructions to the Master Queue and one instruction to the Slave Queue at dispatch is impossible due to a port conflict. Thus, the throughput of the front domain becomes lower than the conventional implementation. The front domain is a structure that is critical to performance, so it is thought that the performance will be degraded by the proposed implementation, shown in Figure 3. Therefore, in order to tackle the above problem, we propose the following improvement of the dynamic instruction scheduling logic shown in Figure 3.

45

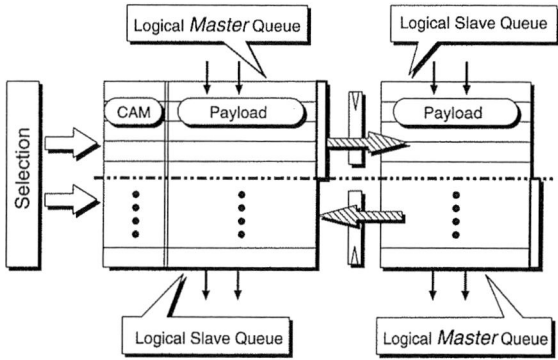

Figure 4: Improved Implementation

3.3.2 Improved Implementation

We resolve the low flexibility of the dispatch stage, which is a fundamental problem of the dynamic instruction scheduling logic of Figure 3, by physical improvement.

Figure 4 shows the improved dynamic instruction scheduling logic. The dotted horizontal line in the center of the figure composes a border. The left side is the Master Queue, and the right side is the Slave Queue in the upper half, as shown in Figure 3. In contrast, in the lower half, the left side is the Slave Queue and the right side is the Master Queue. By such improvement, the dispatching flexibility increases, and it becomes possible to write three instructions to the Master Queue and one instruction to the Slave Queue by writing two instructions to the upper left Master Queue, one instruction to the lower right Master Queue, and one instruction to the upper right Slave Queue. Nevertheless, we must keep in mind that the requirement of dispatching three instructions to the upper right Slave Queue may occur (all instructions are grouped with the instructions in the upper left Master Queue), although such cases are very rare. The implementation shown in Figure 4 supports the theoretical behavior of grouping two instructions at the dispatch stage and treat the group as a single issue unit with almost no restriction on the number of ports.

Finally, it is expected that the proposed technique will achieve higher performance when the instruction queue size is small. This is because, in such cases, the performance bottleneck to extract ILP is the size of the instruction queue, and the proposed technique can hold more instructions than the conventional implementation because it can hold up to two instructions in a single entry. When the instruction queue size is sufficiently large, almost no performance degradation is expected. In addition, regardless of the instruction queue size, the energy consumption can be significantly reduced because the complexity is greatly reduced.

4. EXPERIMENTAL SETUP

4.1 Evaluation Environment

We evaluated the performance and energy consumption of the proposed technique by the implementation described in Section 3.3.1 and Section 3.3.2, compared to a conventional RISC-like dynamically scheduled superscalar processor. We used the SimpleScalar Tool Set [1] as the base simulation environment. We extended SimpleScalar so that the proposed microarchitecture, shown in Figure 3 and Figure 4, was evaluated. For estimating the energy consumption, we used the Wattch [3] extension.

We used all of the programs of the SPEC CPU2000 integer benchmark suite, along with the *ref* input set and mpeg2 encode program from the MediaBench [10]. The programs were compiled by the DEC C compiler for Alpha AXP instruction set architecture (ISA).

Table 1: Processor Configuration

Fetch & Decode width	4
Branch prediction	Combined bimodal (4K-entry) gshare (4K-entry), selector(4K-entry)
BTB	1,024 sets, four-way
Mis-prediction penalty	three cycles
Instruction queue size - floating-point	32
Issue width - integer - load/store - floating-point	4 2 2
Reorder buffer size	96
Commit width	4
L1 I-cache	32 KB, 32 B line, two-way one-cycle latency
L1 D-cache	32 KB, 32 B line, two-way two-cycle latency
L2 unified cache	512 KB, 64 B line, eight-way 10-cycle latency
Memory latency	100 cycles
Bus width	16 B
Bus clock	1/4 of processor the core

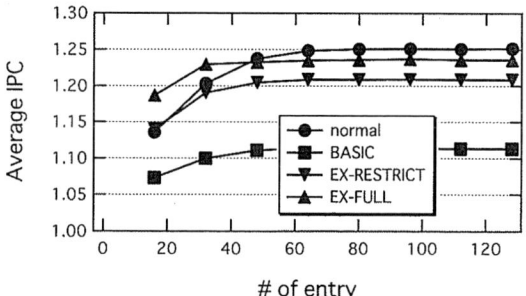

Figure 5: Average IPC

We fast-forwarded two billion instructions and simulated 200 million instructions for all of the evaluated programs.

4.2 Assumption

Table 1 shows the processor configuration assumptions used for the evaluation. In order to group several instructions, we assumed an integer-load/store queue. Both the integer and load/store instructions are dispatched to this queue. By this assumption, we can group not only two integer ALU operation, but also the integer ALU operation for the first instruction and a load instruction for the second instruction. The size of the integer-load/store queue was varied throughout the evaluation from 16 to 128. We ignored the energy consumption of the additional hardware as negligible, as described in the previous section.

5. RESULTS

5.1 Performance

We first evaluate the relationship between performance and the instruction queue size. Figure 5 shows the number of committed instructions per cycle (IPC) for the conventional processor (denoted as normal) and the proposed technique for instruction queues of various size.

BASIC represents the result of basic implementation of Figure 3 in Section 3.3.1. EX (Extension) -RESTRICT and EX-FULL both use the implementation of Figure 4. EX-RESTRICT and EX-FULL differ in that EX-RESTRICT searches only for grouping candidates among the instructions at the same stage, whereas EX-FULL also searches for the Example (1) of Section 3.2.1 from the instruction queue. In addition, BASIC searches for Example (1) of Section 3.2.1 from the instruction queue. We must pay an attention to

46

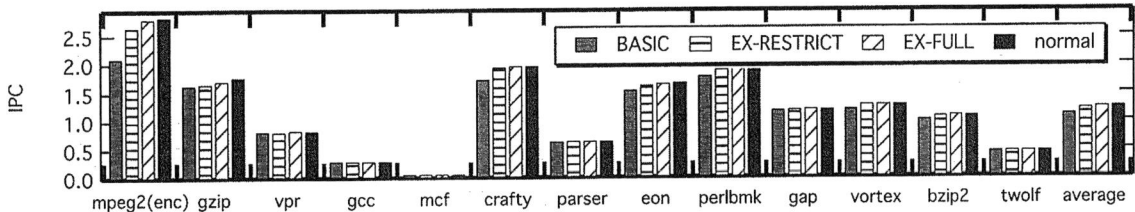

Figure 6: IPC for all benchmarks (48-entry instruction queue)

the fact that the size (number of entries) of the instruction queue for the proposed techniques is different from the conventional processor in that a single entry can hold up to two instructions, as shown in Figure 3. Figure 5 illustrates the results and shows the average IPC numbers for all of the evaluated programs.

As seen from all of the curves in this figure, the IPC degrades gradually as the instruction queue size decreases. This is because the number of instructions to execute in parallel in order to exploit ILP is limited by the lack of free entry of the instruction queue. The IPC degradation values of BASIC, EX-RESTRICT and EX-FULL are 11.1%, 3.5%, and 1.1%, respectively, if the number of entries is larger than 64. However, the performance of EX-FULL is superior to that of normal processor for a smaller number of entries.

In the following, we will examine in detail of the case of the 48-entry instruction queue. This configuration was selected because it is an actual and reasonable size for the four-way superscalar processor. Figure 5 shows that the performance of EX-FULL and the normal processor are approximately the same. Figure 6 presents the commit IPCs for all of the evaluated programs, and Figure 7 shows the percentage of grouped instructions (sum of all types). For example, benchmarks such as *mpeg2*, *gzip*, and *crafty* show a performance degradation in BASIC and EX-RESTRICT, but there is almost no performance loss for EX-FULL. The reason for this can be explained with the help of Figure 7, which shows that EX-FULL grouped a sufficient number of instructions and achieved a high throughput, compared to BASIC and EX-RESTRICT.

Figure 8 shows the breakdown of the percentage of grouped instructions of EX-FULL. In the figure, (1), (2), and (3) indicate the cases described in Section 3.2.1. The figure shows that more than 30% of the ready instructions was grouped in *gzip*, *gcc*, *crafty*, *parser*, *eon*, *vortex*, and *bzip2*. On the other hand, a pair of instructions belonging to (1) and (2) was grouped more than 30% in *mpeg2*, *vpr*, *crafty*, *gap*, *bzip2*, and *twolf*. For all applications, the average of instructions grouped for types (1), (2), and (3), were 20.6%, 6.3%, and 32.7%, respectively. This figure indicates that grouping these three types is compensated for covering all kinds of characteristics of benchmarks and prevented performance degradation.

5.2 Energy Consumption

The proposed technique saves energy due to the reduced number of ports of the instruction queue. In addition, the number of instructions to select and issue is reduced, which simplifies the instruction scheduling logic. Figure 9 presents the average energy savings of the instruction scheduling logic of EX-FULL. The bars represent the normalized energy of the dispatch, wake-up, select and issue. In the figure, the bars indicated by "48 vs. 48" on the horizontal axis represent energy consumption of 48-entry EX-FULL normalized to that of 48-entry normal. In 48-entry EX-FULL, the energy saving was 56%, 7%, 52%, and 58% for dispatch, wake-up, select, and issue, respectively, compared to the conventional dynamic instruction scheduling logic. Except for the wake-up stage, the re-

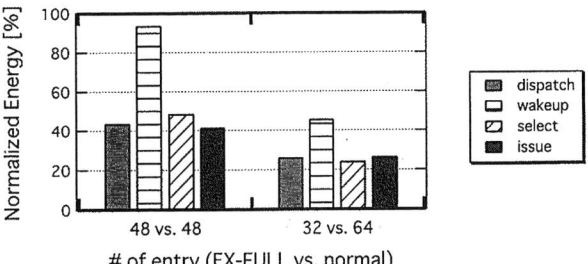

Figure 9: Average Energy of EX-FULL

duction in energy consumption is large. In the case that energy consumption of the wake-up stage is dominant, we should consider halving the number of entries of the proposed implementation in order to reduce the energy consumption of the wake-up logic. When comparing the performance of EX-FULL with 32 entries (which can hold up to 64 instructions) and normal with 64 entries, the performance degradation of EX-FULL is approximately 1.5%. In addition, the reduction in energy consumption was approximately 74%, 55%, 76%, and 73% for dispatch, wake-up, select, and issue, respectively. The results show that the large amount of energy consumption of the dynamic instructions scheduling logic can be reduced with almost no performance degradation.

6. RELATED WORK

Palacharla et al. [12] first analyzed the tradeoff between complexity and clock speed on the superscalar processor. They indicated that instruction wake-up and selection logics to be the most critical factors in the processor pipeline and proposed the dependency-based instruction queue. Canal and González [5] discussed reordering the instructions before they entered the instruction queue using predicted operation latencies. Michaud and Seznec [11] also proposed a data-flow pre-scheduling technique. Lebeck et al. [9] presented an instruction queue design, in which long latency operation (such as cache misses) are moved into a large waiting instruction buffer (WIB). Raasch et al. [13] divided the instruction queue into small segments and formed a pipeline. They used dynamic data dependencies to control instructions from segment to segment.

Dynamic adaptation techniques are another approach to reduce the energy consumption of the instruction queue. Folegnani and González [6] proposed a technique to dynamically disable the wake-up for empty entries and ready operands. In addition, they reduced the size of the instruction queue dynamically. Buyuktosunoglu et al. [4] introduced the combination of fetch gating and dynamic instruction queue adaptation. They broke the instruction queue into chunks and disabled them on the fly in run time using the information of the number of valid entries in the instruction queue.

In addition, there are approaches based on the observation that several instructions are dispatched with at least one source operand ready. Kim and Lipasti [7] considered the problem whereby current microprocessors are designed to process instructions with one and

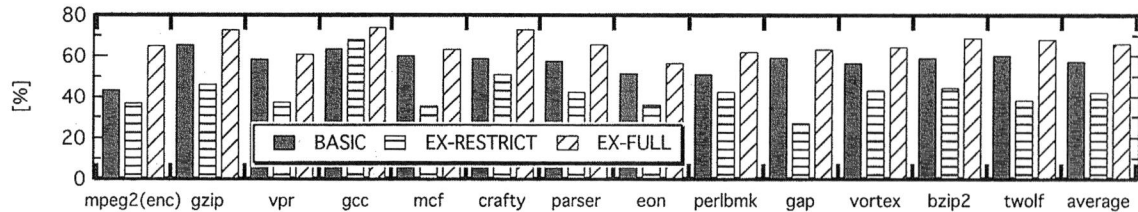

Figure 7: Percentage of Grouped Instructions (48-entry instruction queue)

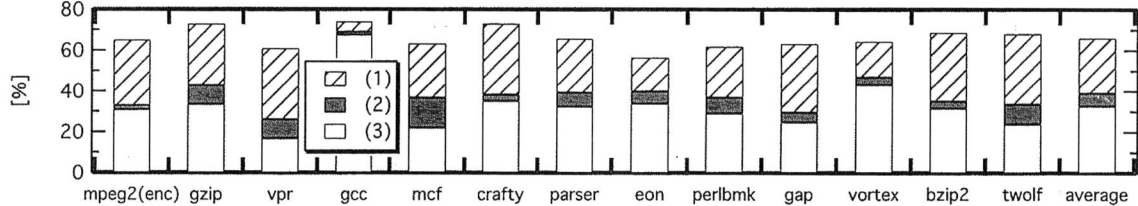

Figure 8: Breakdown of Grouped Instructions of EX-FULL (48-entry instruction queue)

two source operands at the same cost. They proposed the half-price architecture, which removes this overdesign by decoupling half of the tag match logic from the wake-up bus. Sharkey and Ponomarev [15] introduced instruction packing. They packed two instructions having at least one ready source operand at the time of dispatch into one entry of the instruction queue. Although packing two instructions into the same instruction queue entry is similar, this procedure can be used in conjunction with the present approach. They also applied this technique to SMT processors and reported as being more effective for SMT processors [14].

Kim and Lipasti [8] reduced scheduling complexity by pipelining wake-up and select stage by grouping several instructions into Macro-op (MOP). Grouping instructions is similar to the present concept, although their objective was to ensure back-to-back execution of dependent instructions by eliminating single-latency operations. Bracy et al. [2] discussed the use of mini-graph, which is a dataflow graph that has an arbitrary internal size and shape but the interface of a single instruction. The concept of dataflow mini-graphs is similar to that of the present instruction grouping, but dataflow mini-graphs were applied in order to achieve high performance.

7. CONCLUSION

In the present work, we introduced a novel dynamic instruction scheduling logic that groups several instructions. The proposed technique enables the dynamic instruction scheduling logic to hold and issue more instructions without increasing the size or number of ports. In addition, in order to support the proposed technique, we introduced a hardware implementation that minimizes performance degradation. The results showed energy reduction of 56%, 7%, 52%, and 58% for dispatch, wake-up, select, and issue, respectively. In addition, we achieved higher IPC performance in smaller instruction queues, and even when the queue size was large the performance degradation was only 1.3%. Future research will focus on grouping instructions not only at the dynamic instruction scheduling logic but also in the whole pipeline stage in order to reduce the complexity of the processor.

Acknowledgment

This work was supported in part by STARC (Semiconductor Technology Academic Research Center).

REFERENCES

[1] T. M. Austin, E. Larson, and D. Ernst. Simplescalar: An infrastructure for computer system modeling. *IEEE Computer*, 35(2):59–67, 2002.

[2] A. Bracy, P. Prahlad, and A. Roth. Dataflow mini-graphs: Amplifying superscalar capacity and bandwidth. In *MICRO*, pp. 18–29, 2004.

[3] D. Brooks, V. Tiwari, and M. Martonosi. Wattch: a framework for architectural-level power analysis and optimizations. In *ISCA*, pp. 83–94, 2000.

[4] A. Buyuktosunoglu, T. Karkhanis, D. H. Albonesi, and P. Bose. Energy efficient co-adaptive instruction fetch and issue. In *ISCA*, pp. 147–156, 2003.

[5] R. Canal and A. González. Reducing the complexity of the issue logic. In *ICS*, pp. 312–320, 2001.

[6] D. Folegnani and A. González. Energy-effective issue logic. In *ISCA*, pp. 230–239, 2001.

[7] I. Kim and M. H. Lipasti. Half-price architecture. In *ISCA*, pp. 28–38, 2003.

[8] I. Kim and M. H. Lipasti. Macro-op scheduling: Relaxing scheduling loop constraints. In *MICRO*, pp. 277–290, 2003.

[9] A. R. Lebeck, T. Li, E. Rotenberg, J. Koppanalil, and J. Patwardhan. A large, fast instruction window for tolerating cache misses. In *ISCA*, pp. 59–70, 2002.

[10] C. Lee, M. Potkonjak, and W. H. Mangione-Smith. Mediabench: A tool for evaluating and synthesizing multimedia and communications systems. In *MICRO*, pp. 330–335, 1997.

[11] P. Michaud and A. Seznec. Data-flow prescheduling for large instruction windows in out-of-order processors. In *HPCA*, pp. 27–36, 2001.

[12] S. Palacharla, N. P. Jouppi, and J. E. Smith. Complexity-effective superscalar processors. In *ISCA*, pp. 206–218, 1997.

[13] S. E. Raasch, N. L. Binkert, and S. K. Reinhardt. A scalable instruction queue design using dependence chains. In *ISCA*, pp. 318–, 2002.

[14] J. J. Sharkey and D. V. Ponomarev. Efficient instruction schedulers for smt processors. In *HPCA*, 2006.

[15] J. J. Sharkey, D. V. Ponomarev, K. Ghose, and O. Ergin. Instruction packing: reducing power and delay of the dynamic scheduling logic. In *ISLPED*, pp. 30–35, 2005.

[16] S. A. Taylor, M. Quinn, D. Brown, N. Dohm, S. Hildebrandt, J. Huggins, and C. Ramey. Functional verification of a multiple-issue, out-of-order, superscalar alpha processor - the dec alpha 21264 microprocessor. In *DAC*, pp. 638–643, 1998.

Independent Front-end and Back-end Dynamic Voltage Scaling for a GALS Microarchitecture

Grigorios Magklis, Pedro Chaparro, José González, Antonio González

Intel Barcelona Research Center, Intel Labs – UPC

{grigorios.magklis, pedro.chaparro.monferrer, pepe.gonzalez, antonio.gonzalez}@intel.com

ABSTRACT

In recent years, Globally Asynchronous Locally Synchronous (GALS) designs and dynamic voltage scaling (DVS) have emerged as some of the most popular approaches to address the ever increasing microprocessor energy consumption. In this work, we propose two on-line algorithms for adjusting dynamically, and independently, the voltage and frequency of the front-end and back-end domains of a novel two-domain microprocessor. We evaluate our mechanisms for both internal and external voltage regulators, and we present optimal dynamic voltage scaling results for the proposed microarchitecture. Our schemes achieve average improvement of 12% of the energy-delay^2 metric, when using internal voltage regulators.

Categories and Subject Descriptors

C.1.0 [**Processor Architectures**]: General

General Terms

Algorithms, Design.

Keywords

DVS, GALS, MCD, energy efficiency, microarchitecture

1. INTRODUCTION

Microprocessor power consumption has increased significantly in recent years, so much so that *energy efficiency* (or power efficiency) has become one of the main targets of microprocessor architects. Energy efficiency can be addressed statically by changing the microarchitecture design, or dynamically by employing run-time mechanisms to adapt the hardware to applications. One static technique is employing a Globally Asynchronous Locally Synchronous design [4]. GALS achieves better energy efficiency by reducing the complexity and power dissipation of the clock distribution, which constitutes a large part of the total processor power [10][11]. One of the most successful run-time techniques for improving energy efficiency is dynamic voltage and frequency scaling (DVS for short) [1]. GALS systems have the unique ability to operate each domain at different frequency and voltage, which allows applying DVS independently to different parts of the processor. It has been shown that per-domain adaptation is significantly more energy efficient compared to global adaptation [20].

Permission to make digital or hard copies of all or part of this work for personal or classroom use is granted without fee provided that copies are not made or distributed for profit or commercial advantage and that copies bear this notice and the full citation on the first page. To copy otherwise, or republish, to post on servers or to redistribute to lists, requires prior specific permission and/or a fee.
ISLPED '06, October 4–6, 2006, Tegernsee, Germany.
Copyright 2006 ACM 1-59593-462-6/06/0010...$5.00.

In this work, we study the effect of fine-grain DVS on a clustered GALS microprocessor. Our microarchitecture utilizes clustering on both the front-end and back-end of the microprocessor, resulting in a very efficient baseline [3]. We propose to further increase the efficiency of the system by separating the core into two clock domains (front-end and back-end) and by allowing independent DVS for each domain. Our work is the first to propose a front-end/back-end split of the microprocessor and to propose DVS control for *all* domains of a GALS microprocessor. Moreover, we perform a more complete study than previous works, by including the energy costs of internal voltage regulators, and leakage energy in our results. Third, we use a significantly different microarchitecture than previous MCD-like studies.

We have evaluated a previous state-of-the-art work based on queue utilization [20] for our microarchitecture and it achieves near zero energy-delay^2 improvement. The reason is the difference in the microarchitectures studied. Queue utilization is not a sufficiently good indicator of performance for our microarchitecture. Firstly, because our pipeline is much wider and has a far greater degree of out-of-order execution compared to previous proposals. Secondly, we have more queues in the microarchitecture than MCD-like proposals (due to synchronization FIFOs and clustering), which requires a significantly more complex mapping of queue utilization to performance.

2. GALS MICROARCHITECTURE

Figure 1 shows our GALS microarchitecture. The processor consists of three clock domains, shaded grey: front-end (FE), back-end (BE) and memory (MEM). We consider the memory domain external to the core of the microprocessor (but still on-chip). The processor follows a clustered design.

The FE contains a branch predictor, a trace cache, and an IA32 decoder, the ROB, the dispatch and the commit logic. The FE is divided into two clusters [3]. Instruction fetch, decode and steering is centralized. After the steering logic decides the destination BE cluster [8], it directs the instruction to the corresponding FE cluster. Each FE cluster has its own rename table and ROB, which are simpler and smaller than in a monolithic design. Renaming and allocation proceeds independently and in parallel. The commit logic must maintain order between the two ROBs, which increases its latency, but it is not on the critical path of the execution [3].

The BE is also divided into two clusters. Each cluster consists of an integer and a floating-point out-of-order execution engines. The load-store queue and the first-level data cache are centralized, and shared among the clusters. Addresses are calculated at the execution clusters and then are passed to the LSQ for resolution and issuing to the cache. Special *copy* instructions explicitly communicate register values between the clusters using point-to-point links [14].

Table 1. Voltage-frequency levels.

Level	mV	GHz	Level	mV	GHz
0	700	3400	7	875	5000
1	725	3700	8	900	5200
2	750	3900	9	925	5400
3	775	4200	10	950	5600
4	800	4400	11	975	5800
5	825	4600	12	1000	6000
6	850	4800			

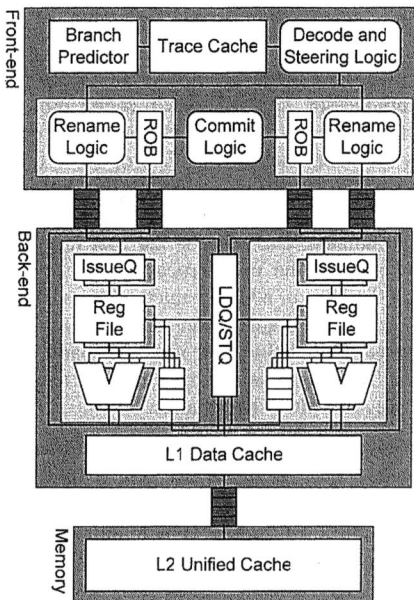

Figure 1. Clustered GALS microarchitecture.

Each domain has its own local clock network that distributes a reference clock signal to the domain. We assume that the skew between the domain reference clocks can be arbitrary. This allows to run each domain at a different frequency, and to apply DVS to each domain independently of the others but inter-domain communication must be synchronized correctly to avoid meta-stability We assume synchronizing FIFOs for inter-domain communication, similar to previous studies [5][16][17]. We chose this domain separation so as to minimize the performance loss due to synchronization delays. We keep the schedule–execute–bypass loop inside a domain so we can perform back to back scheduling of dependent instructions. Moreover, we do not put each cluster in a different domain because that would not allow back-to-back execution of copy- and load-dependent instructions.

3. DYNAMIC VOLTAGE SCALING

Domains can execute through voltage changes, similar to previous studies [11][12][16][20]. Each domain includes an on-chip digital clock multiplier connected to a single, shared, external PLL [6][13]. This limits our DVS choices (Table 1), but allows frequency changes without stopping the domain. Each domain includes an internal voltage regulator [9]. This allows for extremely fast voltage switching (in just a few nsec). Hazucha *et al.* report 94% efficiency (output power divided by input power) for their design. We are more conservative and assume only 90% efficiency. Our power model accounts for power losses due to the regulator inefficiency (independently for each regulator).

The goal of our proposed mechanisms is to improve energy efficiency of the baseline microarchitecture by adapting the voltage

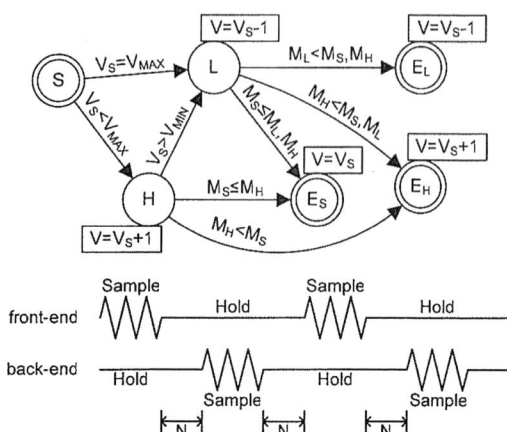

Figure 2. *Best-neighbor* state machine and meta-control.

and frequency of the front-end and the back-end domains of the processor. There are several metrics for energy efficiency. In this study, we adopt the *energy-delay2* (ED^2) metric proposed by Brooks *et al.* [1]. Thus, we need a mechanism—at the microarchitecture level—to estimate ED^2. The ED^2 metric, for an interval, is calculated as $(E_{dynamic}+E_{static}) \cdot N^2$, where N is the value of the cycle counter for the interval and $E_{dynamic}$ and E_{static} are the dynamic and static energy consumption for this interval.

We propose to use the mechanism devised in [7] to measure dynamic energy. The microprocessor uses performance counters to measure the activity on the various units. The total activity, during a time interval, is multiplied by the energy-per-access register associated with the unit (fixed at design time), to calculate the total energy of the unit. In this work, we estimate the static energy of a circuit, using the formula by Zhang *et al.* [21]; we assume that there exists a mechanism to estimate static energy consumption at run-time.

For a specific execution interval, if we plot the ED^2 metric over all voltage and frequency (V-F) settings, we will observe a roughly U-shaped graph. The lowest point in the graph is the optimal ED^2 for this execution interval. All points to the left of the optimal (*i.e.*, those with lower V-F) have higher ED^2 because the application loses too much performance compared to the energy reduction achieved. Correspondingly, to the right of the optimal, ED^2 is higher because the application consumes too much energy for the performance benefit realized.

3.1. Best-Neighbor Adaptation

For *best-neighbor* adaptation, the controller operates in two distinct phases: *sample* and *hold*. In the *sample* phase, the system calculates the ED^2 for the different V-F settings in the neighborhood of the current V-F level by trying each configuration for an interval. At the end of the *sample* phase, the controller decides the V-F setting with the minimum ED^2. This setting is then applied and the controller moves to the *hold* phase. In this phase, the controller maintains the chosen V-F setting for several intervals.

Figure 2 shows the state machine of the domain controller, for the *sample* phase. The controller starts at state S, with voltage V_S. After one interval, it moves to state H, to sample the ED^2 at a higher V-F setting (if not already running at maximum V-F). Next interval, it moves to state L, to sample ED^2 at a lower V-F setting. In the figure, M_S, M_L, and M_H denote the ED^2 of the corresponding states. Then, the controller chooses the best V-F setting of the three for the *hold* phase. States E_S, E_L and E_H correspond to the

50

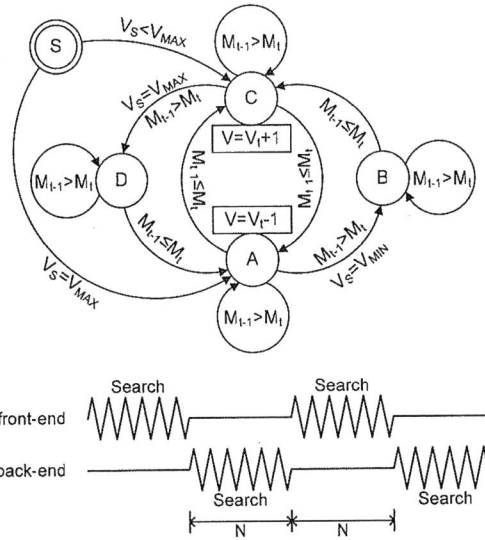

Figure 3. *Greedy-search* state machine and meta-control.

terminal states of the controller. Upon entering one of the E states, the controller sets the V-F to the appropriate values, resets to the S state, and waits until the *hold* phase ends.

Imagine the situation where we are trying to make a decision on the front-end V-F setting. If during the *sample* phase, we do not hold the back-end V-F constant then we do not know if the benefit in ED^2 that we see during an interval is because of the front-end or the back-end V-F changes. To avoid this situation there is a meta-controller to force the domain controllers to operate in an inter-leaved fashion. The meta-controller ensures that when one of the domains is in the *sample* phase the other one will be in the *hold* phase and vice versa. Figure 2 shows the meta-controller operation graphically. The parameter N defines how many intervals should pass between the *sample* phases of the two domains.

3.2. Greedy-Search Adaptation

The goal of *greedy-search* is to approximate the optimal ED^2 by following an imaginary U-shaped curve. Figure 3 shows the state machine of the domain controller. The controller starts at state S and after one interval it moves to either state A or C, depending on the starting V-F (the V-F changes accordingly). After this point, the controller uses the history of the last two intervals t and t-1 to make decisions for the upcoming interval t+1 (M_t and M_{t-1} denote the ED^2 of the corresponding intervals). If the ED^2 was reduced, then the last V-F change was correct and we continue changing in the same direction. Otherwise, if the ED^2 increased, then we reverse direction.

The assumption is that if the last V-F change was beneficial then we are in the right direction to reach the valley of the ED^2 U-curve. If the last change was not beneficial then we are going in the wrong direction (either we started wrong or we reached the bottom of the U-curve and started climbing the other way) and we should change. Ideally, this mechanism will force the system to stay close to the lowest point of the U-curve.

There is also a meta-controller to interleave the two domain controllers. The meta-controller periodically (every N intervals) disables the currently active controller and enables the other one (Figure 3). When a controller is de-activated, the voltage and frequency for the corresponding domain remain at the last setting

Table 2. Microarchitectural parameters.

Front-end	
Fetch	24K inst. trace cache, 6 inst./cycle, 5 cycle fetch-to-dispatch
Decode, rename and steer	3+3 inst./cycle, 1 cycle latency, plus 1 cycle wire delay to synch, FIFOs
ROB	256+256 entries, commit 3+3 inst./cycle
Back-end (configuration shown per cluster)	
Synch. FIFOs	1 FIFO per issue queue, 24 entries each
Issue queues	48-entry INT, 2 inst./cycle and 48-entry FP, 2 inst./cycle and 24-entry COPY, 1 inst./cycle
Register file	256-entry INT, 256-entry FP
Inter-cluster	1 cycle latency, 1 copy/cycle
L1 data	32KB, 4-way, 3 cycle hit, 2 read ports, 1 write port, 256-entry LSQ
Memory	
L2 unified	2MB, 16-way, 13 cycle hit, ≥ 500 cycle miss, 1 read port, 1 write port

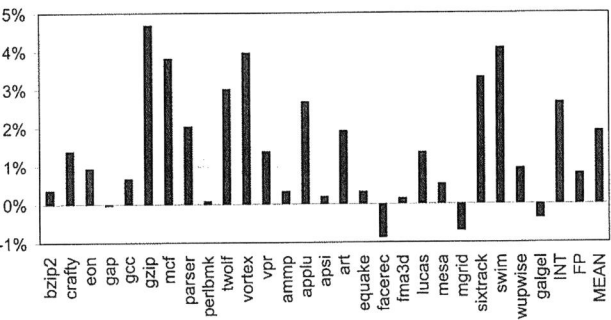

Figure 4. GALS performance loss due to inter-domain synchronization.

chosen until the controller is re-activated. Every time a controller is re-activated, it starts from the S state.

4. PERFORMANCE RESULTS

We use a cycle accurate simulator that executes traces of IA32 binaries, including OS code. The simulator also includes a power estimation module, based on an enhanced version of CACTI [18], utilizing activity counters (similar to Wattch [2]) and a leakage module utilizing the formula by Zhang *et al.* [21]. The assumed technology is 45nm. We initialize domain clocks at random values at the beginning of simulation time to account for clock skew. Table 2 shows the main microarchitectural parameters of the processor. For our experiments, we use the twenty-six applications of the SPEC CPU2000 benchmark suite, compiled with the latest Intel® compiler with full optimizations, and run with the reference input set. We simulate 200M IA32 instructions from the middle of the application.

We also employ off-line simulation to calculate the optimal V-F scheduling for a given execution. For the off-line analysis and all simulations that compare with the off-line, we simulate only 20M instructions, due to simulation time constraints. First, we simulate an application for all combinations of FE and BE V-F settings, and we collect statistics at fixed instruction intervals. Then we construct a timeline, *i.e.*, a sequence of intervals, for each application. To emulate a control mechanism C, we choose a V-F combination A, according to the control policy of C, for an interval of execution. We then add the statistics of A for this interval to the off-line total statistics, and repeat this process for the next interval. We stop when we reach the end of the timeline. The average error of an off-line vs. a dynamic run is about 0.45%.

51

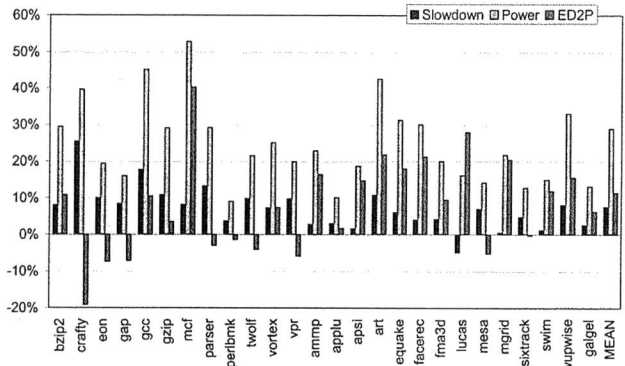

Figure 5. *Best-neighbor* slowdown, power reduction, and ED^2 improvement.

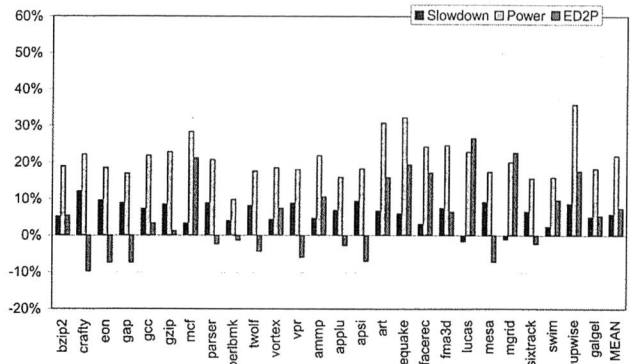

Figure 6. *Greedy-search* slowdown, power reduction, and ED^2 improvement.

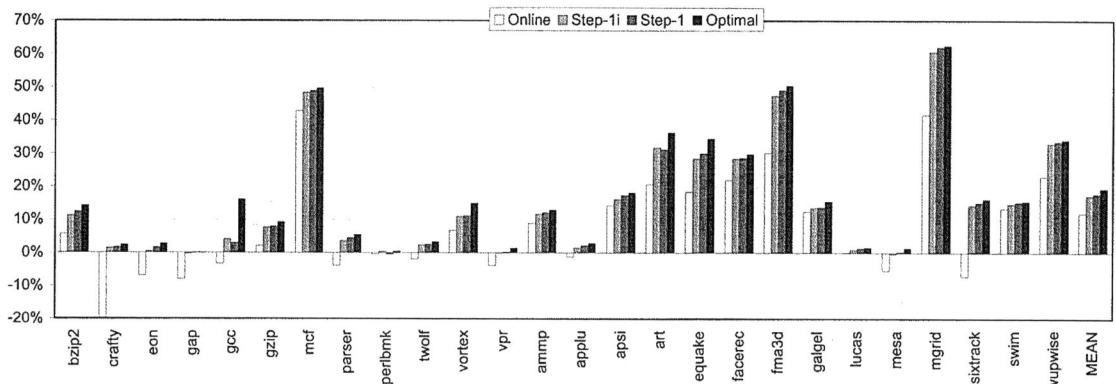

Figure 7. ED^2 improvement of *best-neighbor* (*Online*) compared to various off-line mechanisms.

4.1. GALS Performance

In this section, we evaluate the performance loss due to inter-domain synchronization whereas in following sections we evaluate the proposed DVS schemes. Figure 4 shows the slowdown of our GALS microprocessor running at maximum frequency, compared to an identical, fully synchronous one. All of the performance loss shown in this figure is due to inter-domain synchronization penalties. The worst-case slowdown is 4.7% for *gzip* with an average of less than 2% over all benchmarks.

There are two sources of performance loss in the microprocessor: the branch misprediction loop and the cache miss loop. Branch misprediction involves two synchronizations: (a) from the FE to the BE, when we decode a branch and before the branch is resolved, and (b) from the BE to the FE, to communicate the misprediction after the branch is resolved. L1 misses also involve two synchronizations: (a) from the BE to the L2, to notify for the miss and request the data, and (b) from the L2 to the BE, when the request is serviced. Integer benchmarks exhibit higher slowdown on average compared to floating-point ones (2.6% vs. 0.8%), mainly due to higher branch misprediction rate.

4.2. Fine-grain DVS

Figure 5 shows the performance loss, power reduction and ED^2 improvement of *best-neighbor*. The comparison is performed against the clustered GALS processor running all domains at the highest V-F. The interval is 10K instructions and $N=1$. This translates to 50K instructions for the *hold* phase on average. Figure 6 shows the same results for *greedy-search*. The interval length is similarly set, and the meta-controller interleaves the domains every 90K instructions ($N=9$). For both mechanisms, N was cho-

sen after an extensive search process, utilizing the off-line simulator. For many values of N close to the ones chosen, the results do not vary significantly.

Summarizing the two figures, the average ED^2 improvement of *best-neighbor* is about 11.5% while *greedy-search* achieves roughly 7.5%. The average slowdown of *greedy-search* is lower than *best-neighbor*, mainly due to its ability to perform large frequency changes faster: *best-neighbor* performs only a single-step change per turn, while *greedy-search* is allowed up to nine changes ($N=9$). This is beneficial when the application suffers abrupt changes in demand. *Greedy-search* reacts faster and reaches the new stable state earlier. *Best-neighbor* has the benefit of being more stable. If the domain is already running at the best V-F then the *sample* phase will choose not to change; *greedy-search* will constantly move around the best setting.

4.3. Optimal Fine-grain DVS

In order to evaluate the effectiveness of our schemes we calculate the best ED^2 for each application, utilizing off-line simulation. Due to space restrictions, we show results only for *best-neighbor*. Since the *hold* phase is roughly 50K instructions, we use this number as the interval length for all off-line mechanisms. We compare our schemes against three different upper bounds.

The first scheme is called *Optimal*. At each interval, it searches all combinations of V-F for the two domains and chooses the best one. This is the best we can do with any kind of dynamic mechanism. Our DVS mechanisms are restricted to single-step V-F changes. The second upper bound scheme, called *Step-1*, is used to discover if this restriction has an inherent inefficiency. At each

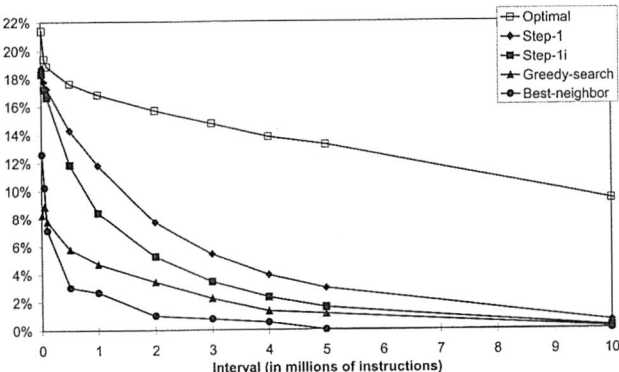

Figure 8. ED^2 **improvement for various interval sizes.**

interval, the V-F of each domain is changed by one level, towards the direction of the optimal V-F. Our third off-line mechanism, called *Step-1i*, accounts for the meta-controller by interleaving the two domains. This mechanism also accounts for single-step V-F changes, similarly to *Step-1*. This represents a tighter upper bound for our DVS mechanisms.

Figure 7 shows the ED^2 achieved with the four schemes. The results for *best-neighbor*, labeled *Online*, are not exactly the same as those in Figure 5 but very similar, because now we simulate only 20M instructions, while in Section 4.2 we simulated 200M instructions. *Online* achieves an average ED^2 improvement of 12% over all benchmarks, while *Step-1i*, *Step-1* and *Optimal* achieve 17.2%, 17.8% and 19.4% respectively. This means that we achieve about 70% of the efficiency of *Step-1i*, which is the best we could do given the two restrictions of our mechanisms: single-step V-F changes and domain interleaving.

Comparing *Step-1i* and *Step-1*, we can see that interleaving has minimal impact on the efficiency of the mechanisms (*Step-1i* achieves 97% of *Step-1*). This means that we can develop simpler control systems without loosing efficiency, which is exactly what we propose: simple per-domain control that ignores interactions and side effects from the rest of the domains, and simple meta-control to co-ordinate the domain controllers. Comparing *Step-1* and *Optimal*, we can see that single-step V-F changes have bigger impact than interleaving (*Step-1* achieves 92% of *Optimal*) due to the slow change rate in the face of sharp changes in program behavior. When the control interval is short, as in this case, the time it takes to stabilize to the new V-F is short. When the interval is long, the stabilization delay may incur significant inefficiency.

Figure 8 shows the sensitivity to the interval size. As expected, power efficiency gains are smaller when moving to longer intervals. This is because the mechanisms—including *Optimal*—cannot take advantage of short application phases; they must adjust V-F to account for the average behavior over long periods. The efficiency loss due to single-step V-F changes becomes more pronounced as the interval length increases. Interleaving the domains also affects the efficiency of the control, as can be seen by comparing *Step-1i* to *Step-1*, but not so much as single-level V-F changes. We can also see that for intervals of 500K instructions and above, *greedy-search* outperforms *best-neighbor*. When the interval length becomes significant compared to the application phase length, the faster V-F changes of *greedy-search* pay off. For short intervals, the stability of *best-neighbor* wins.

The determining factor of the interval length is the speed of the voltage regulator (VR). The left side of Figure 8 corresponds to

fast internal VRs (voltage changes in nsec), while the right side corresponds to slow external VRs (voltage changes in msec). The middle could be either internal or external VRs (voltage changes in μsec). We conclude that *best-neighbor* is best for fast VRs, where *greedy-search* is best otherwise.

We have also evaluated the best known previous work, based on queue utilization [20]. In that work, the utilization of the queues that connect the fetch and the execution domains is tried to be maintained inside some pre-defined bounds.. The ED^2 improvement for this mechanism was close to zero, for all benchmarks (we did not put it in our figures, because it does not show).

This is because of the great differences in the MCD-like microarchitecture of [20] and our proposal. In [20] the back-end is split into domains according to operation type, and most importantly, the instruction queues are the synchronizers. We have special synchronization FIFOs before the instruction queues, and we follow a clustered design. A simple queue model (for mixed-clock FIFO, instruction queue or combination of both), like the one assumed in [20] does not work for our microarchitecture.

Moreover, in our simulations we include the energy consumption of the internal VRs, and leakage energy. Both of these factors influence the improvement we see with DVS, and have been ignored from all previous studies (including [20]), as far as we know. Finally, we claim that our baseline microarchitecture is highly energy efficient, much more than previous proposals. This is due to extensive use of clustering both at the front-end and at the back-end. MCD-like microprocessors may be energy efficient, but we think are very hard to design, due to the non-deterministic latencies of domain crossings.

5. RELATED WORK

Chapiro [4] was the first to introduce the idea of GALS systems. Since then there have been several published works on GALS and DVS. Iyer and Marculescu [11] propose a microprocessor with five domains: fetch, decode and rename, integer pipeline, floating-point pipeline, and memory pipeline (includes first level cache). They conclude that the performance loss due to domain synchronization was significant but that GALS could be more efficient than fully synchronous designs using fine-grain adaptation mechanisms.

Semeraro *et al.* [16][17] propose a Multiple Clock Domain (MCD) processor, with four domains: front-end (fetch and dispatch), integer, floating-point, and memory (with first and second level cache). This separation results in minimal slowdown compared to a globally synchronous design. Moreover, they describe an off-line mechanism to obtain almost optimal DVS. Semeraro *et al.* [15] propose an interval-based hardware control mechanism for the domains of the MCD (all but the front-end), called the *Attack/Decay* [17]. The *Attack/Decay* uses the rate of change of the occupancy of the issue queues of the MCD to decide if the frequency should increase or decrease for the next interval. They report energy-delay product improvement of about 85% of what was achieved with their off-line mechanism.

The authors in [12] combine clustering with MCD into a Clustered Multiple Clock Domain (CMCD) design. The CMCD consists of four back-end clusters (each with a local first level cache), a shared front-end, and a shared second level cache each in a separate domain. They also propose a mathematical model that relates the fetch queue utilization, the branch prediction accuracy, the front-end frequency and the application performance. They use the model in a control mechanism to adapt the voltage and frequency of only the front-end domain, achieving close to optimal results.

Wu *et al.* [19] model the MCD domains as queue systems and propose a feedback control DVS system based on a Proportional-Integral (PI) controller. The controller uses the occupancy of the domain input queue over some interval of time and responds with a frequency for the upcoming interval. The goal is to maintain occupancy close to a pre-defined nominal value. Wu *et al.* [20] propose a DVS mechanism for the MCD that reacts to workload changes instead of making decisions at fixed time intervals. The controller utilizes both the queue occupancy and the rate of change of the occupancy. When a metric consistently exceeds a predefined threshold for a consecutive number of cycles an action is taken.

6. CONCLUSIONS

We have presented and evaluated a novel GALS microarchitecture with minimal IPC degradation (less than 2%). We have also described how to enable independent voltage and frequency control for the domains of the microprocessor, and how to measure energy consumption, including leakage, at run-time. Finally, we have proposed and evaluated two mechanisms for dynamically adapting the frequency and voltage of the domains.

To our knowledge, our proposed dynamic adaptation mechanisms are the first ones to control all domains of a GALS microprocessor. We perform an extensive evaluation of our mechanisms, comparing with optimal control and under different voltage regulator speeds. We conclude that our *best-neighbor* mechanism performs best for internal (fast) voltage regulators while our *greedy-search* mechanism performs best in all other cases. When using internal voltage regulators *best-neighbor* achieves about 12% ED^2 improvement, or 70% of an optimal algorithm with similar characteristics.

7. REFERENCES

[1] D. M. Brooks *et al.* Power-Aware Microarchitecture: Design and Modeling Challenges for Next-Generation Microprocessors. *IEEE Micro*, 20(6), Nov./Dec. 2000.

[2] D. Brooks, V. Tiwari and M. Martonosi. Wattch: A Framework for Architectural-Level Power Analysis and Optimization. In *International Symposium on Computer Architecture*, June 2000.

[3] P. Chaparro, G. Magklis, J. González and A. González. Distributing the Frontend for Temperature Reduction. In *International Symposium on High-Performance Computer Architecture*, Feb. 2005.

[4] D. M. Chapiro. *Globally Asynchronous Locally Synchronous Systems*. PhD thesis, Stanford University, 1984.

[5] T. Chelcea and S. M. Nowick. Robust Interfaces for Mixed-Timing Systems with Application to Latency-Insensitive Protocols. In *Design Automation Conference*, June 2001.

[6] T. Fischer *et al.* A 90-nm Variable Frequency Clock System for a Power-Managed Itanium Architecture Processor. *IEEE Journal of Solid State Circuits*, 41(1), Jan. 2006.

[7] J. González and A. González. Dynamic Cluster Resizing. In *International Conference on Computer Design*, Oct. 2003.

[8] J. González, F. Latorre and A. González. Cache Organizations for Clustered Microarchitectures. In *Workshop on Memory Performance Issues*, June 2004.

[9] P. Hazucha *et al.* Area-Efficient Linear Regulator with Ultra-Fast Load Regulation. *IEEE Journal of Solid-State Circuits*, 40(4), April 2005.

[10] A. Hemani *et al.* Lowering Power Consumption in Clock by Using Globally Synchronous Locally Synchronous Design Style. In *Conference on Design Automation*, June 1999.

[11] A. Iyer and D. Marculescu. Power and Performance Evaluation of Globally Asynchronous Locally Synchronous Processors. In *International Symposium on Computer Architecture*, May 2002.

[12] G. Magklis, J. González and A. González. Frontend Frequency-Voltage Adaptation for Optimal Energy-Delay^2. In *International Conference on Computer Design*, Oct. 2004.

[13] T. Olsson *et al.* A Digitally Controlled Low-Power Clock Multiplier for Globally Asynchronous Locally Synchronous Designs. In *International Symposium on Circuits and Systems*, May 2000.

[14] J. M. Parcerisa, J. Sahuquillo, A. González and J. Duato. Efficient Interconnects for Clustered Microarchitectures. In *International Conference on Parallel Architectures and Compilation Techniques*, Sept. 2002.

[15] G. Semeraro *et al.* Dynamic Frequency and Voltage Control for a Multiple Clock Domain Microarchitecture. In *International Symposium on Microarchitecture*, Nov. 2002.

[16] G. Semeraro *et al.* Hiding Synchronization Delays in a GALS Processor Microarchitecture. In *International Symposium on Asynchronous Circuits and Systems*, April 2004.

[17] G. Semeraro *et al.* Energy Efficient Processor Design Using Multiple Clock Domains with Dynamic Voltage and Frequency Scaling. In *International Symposium on High-Performance Computer Architecture*, Feb. 2002.

[18] P. Shivakumar and N. P. Jouppi. CACTI 3.0: An Integrated Cache Timing, Power, and Area Model. WRL Research Report 2001/2, Aug. 2001.

[19] Q. Wu, P. Juang, M. Martonosi and D. W. Clark. Formal Online Methods for Voltage/Frequency Control in Multiple Clock Domain Microprocessors. In *International Conference on Architectural Support for Programming Languages and Operating Systems*, Oct. 2004.

[20] Q. Wu, P. Juang, M. Martonosi and D. W. Clark. Voltage and Frequency Control with Adaptive Reaction Time in Multiple-Clock-Domain Processors. In *International Symposium on High-Performance Computer Architecture*, Feb. 2005.

[21] Y. Zhang *et al.* HotLeakage: A Temperature-Aware Model of Subthreshold and Gate Leakage for Architects. Technical Report CS-2003-05, Dept. of Computer Science, University of Virginia, Mar. 2003.

Synergistic Temperature and Energy Management in GALS Processor Architectures *

YongKang Zhu
Department of Electrical and Computer Engineering
University of Rochester
Rochester, NY 14627

David H. Albonesi
Computer Systems Laboratory
Cornell University
Ithaca, NY 14853

ABSTRACT

We propose a synergistic temperature and energy management scheme for GALS processors. Localized DVS is applied in domains that contain hotspots, permitting other critical domains to run unabated, thereby reducing performance cost relative to global DVS, and also creating execution slack in peripheral cooler domains that can be exploited to save energy. The reduction in energy in turn creates a steeper temperature gradient between the domains, permitting heat to flow more easily out of the hotspot domain. This symbiotic cyclical relationship between temperature and energy management leads to both significantly better performance, *and* lower energy, than the use of DTM alone.

Categories and Subject Descriptors: C.1.0 Processor Architectures: General

General Terms: Reliability, Performance

Keywords: Dynamic Voltage Scaling (DVS), Dynamic Temperature Management (DTM)

1. INTRODUCTION

The relentless scaling of transistor dimensions, coupled with a slowdown in supply voltage scaling and rapidly increasing leakage power, has led to unprecedentedly high on-chip power density levels. In response, microarchitectural techniques for Dynamic Temperature Management (DTM) have been proposed for maintaining suitable operating temperatures with reduced packaging and cooling costs [1, 18].

Global DTM techniques, such as Dynamic Voltage Scaling (DVS) and global clock throttling, though effective in reducing chip temperatures, have the disadvantage of impacting global microprocessor performance due to the global reduction in clock frequency, even in cases in which the thermal emergency is isolated to a small region of the die.

The differences in the logic composition and logic density among chip units, and in the utilization of these units as

*This work was supported in part by NSF grants CCR-0304574 and CCF-0541321, and by an IBM Faculty Partnership Award.

Permission to make digital or hard copies of all or part of this work for personal or classroom use is granted without fee provided that copies are not made or distributed for profit or commercial advantage and that copies bear this notice and the full citation on the first page. To copy otherwise, to republish, to post on servers or to redistribute to lists, requires prior specific permission and/or a fee.
ISLPED'06, October 4–6, 2006, Tegernsee, Germany.
Copyright 2006 ACM 1-59593-462-6/06/0010 ...$5.00.

Table 1: Thermal characteristics of a fully synchronous microprocessor without any DTM control for SPEC2000 programs.

	Max Temp (degrees C)	Three Hottest Units
crafty	92.4	iExec, IntQ, iReg
eon	98.0	fAdd, fReg, LSQ
gzip	90.2	iExec, IntQ, iReg
mesa	92.8	fAdd, fReg, IntQ
equake	96.5	iExec, IntQ, iReg
facerec	89.6	fAdd, fReg, fMul
fma3d	91.7	iExec, iReg, IntQ
galgel	125.6	fMul, fAdd, fReg

applications execute, means that the thermal hotspots on the die may be isolated to a small subset of all the chip units for any given application. Table 1 shows the thermal characteristics of the SPEC2000 programs used in this paper. One observation is that the units in the front-end are never among the hottest; therefore, there is little need to ever throttle performance in that domain for temperature purposes. Note also that for any given application, the hottest units are located within at most two of the regions of the die (integer, floating point, or load-store).

These results indicate that a *localized* response to temperature emergencies may be effective in maintaining acceptable temperature levels while maintaining global performance. One such approach, localized throttling of the clock within the region of interest, was previously proposed [17]. However, this approach only impacts frequency, and therefore is often too gentle in addressing serious thermal emergencies [17]. On the other hand, *localized DVS*, which can be realized by dividing the processor into several clock/voltage domains [10, 15], has the advantage of being a localized *and* vigorous response to thermal emergencies.

In this paper, *localized, DVS-based, DTM* is proposed via a Globally Asynchronous, Locally Synchronous (GALS) microprocessor called MCD (Multiple Clock Domain) [15]. In MCD, the major microprocessor functions are located in separate clock/voltage domains. The advantage of this approach, in terms of DTM, is that a localized *and* strong response can be made to the particular unit which is overheating at any given point of execution. This effectively reduces the thermal problem at the local level, permitting other domains to maintain full speed operation, resulting in less performance overhead compared to a fully synchronous processor with DVS-based DTM. The added performance cost of MCD is, of course, inter-domain synchronization. This cost is shown to be offset by the lower performance overhead afforded by localized DVS-based DTM control.

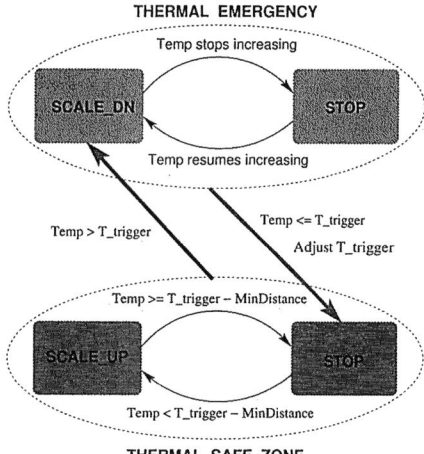

Figure 1: Adaptive DTM algorithm with dynamic setting of control parameters.

An interesting benefit of applying DVS to a subset of the chip is that this may create *execution slack* in the cooler areas of the die. This arises if one or more of the hotspot domains are on the critical path of execution at the time that it is slowed down (or becomes the case due to the slowdown). In this case, the peripheral domains need not run at full speed, and in fact, *Dynamic Energy Management (DEM)* techniques can be used to slow these domains to save energy without compromising performance. This slowing down of the peripheral domains, in turn, permits more lateral heat transfer from the hotspot domain, which reduces the degree that DVS has to be applied to reach acceptable temperature levels. This creates a *symbiotic relationship* between these two techniques (DTM within the hotspot domains and DEM within the others), in which each technique benefits from the use of the other. The application of localized DVS to achieve synergistic temperature and energy management is an interesting area for research that is explored for the first time in this paper.

The rest of this paper is organized as follows. The next section discusses related work, while Section 3 describes a new DTM algorithm that adapts its parameters to fit varying temperature characteristics. A combined DTM and DEM algorithm for MCD processors is presented in Section 4, followed by a description of the evaluation methodology in Section 5. Results are presented in Section 6, and the paper concludes and discusses areas for future work in Section 7.

2. RELATED WORK

A wealth of research has been conducted on architecture level dynamic thermal management. One approach restricts instructions from entering the processor core [1, 14, 17], while another uses dynamic frequency scaling, perhaps coupled with DVS [1, 5, 18]. Other DTM schemes include global clock gating (as used in Pentium 4 processors [6]), dual pipeline activity migration [7], instruction steering to a low power pipeline [12], and some floor planning techniques [18].

A promising approach is to combine several schemes, since different techniques may be well suited for different levels of thermal stress [16]. The framework proposed by Huang et al. [8] selects different schemes to control the temperature of different thermal phases.

Li et al. study how DTM affects the performance and power consumption of SMT and CMP architectures [11]. Furthermore, Powell et al. propose SMT thread assignment and CMP thread migration schemes to control power density [13]. The DVS-based DTM algorithm described in the next section can be extended and applied to both SMT and CMP, and the ability to tune individual domains within the processor core can potentially complement SMT and CMP approaches. This combination is left as future work.

3. ADAPTIVE DVS-BASED DTM ALGORITHM

Figure 1 shows the proposed DTM algorithm, which does not specify any target voltage. Rather, a trigger temperature is specified for engaging DVS, and temperature is sampled at a fine enough granularity to catch small changes. For a fully synchronous processor, one trigger temperature is used for all units; for an MCD machine, each domain has its own trigger temperature. In both cases, thermal sensors are located at each hotspot unit.

Each domain is initially in a thermal safe zone (Figure 1). When any monitored unit temperature exceeds the trigger temperature, the domain enters the thermal emergency zone and the voltage and frequency are reduced at a constant speed. This continues until the maximum temperature is observed to stop increasing and is below the hard limit. As in Intel's XScale processor [4], all circuits operate during the period of voltage and frequency transition. After the scaling down process is terminated, the temperature either stays stable between the trigger temperature and the hard limit; resumes increasing sometime later, in which case DVS resumes; or gradually decreases below the trigger temperature and enters the safe zone.

A fixed trigger temperature would not fit optimally in all thermal phases, so the algorithm sets it to an initial value and then adjusts it everytime the thermal safe zone is entered after an emergency. Increasing the trigger temperature may be risky in terms of safe temperature control, and we handle this by requiring a minimum separation of 0.5 degrees between the trigger temperature and the hard limit, and forcing the upper limit of the trigger temperature to be lowered dynamically; in our algorithm, the limit is lowered by 0.5 degrees for every three thermal violations.

If after returning to the safe zone the temperature still decreases, the voltage is increased until the temperature is close to the trigger temperature. This minimum temperature distance (called *MinDistance*) must be carefully chosen. If the value is too small, the temperature may oscillate around the trigger thus incurring an alternating scaling down and up process. Too large a value increases stability, but may degrade performance. While dynamic adjustment of the minimum distance is possible, a static value is used in the algorithm.

The last parameter is the voltage transition speed. In the proposed algorithm, the scaling up process occurs at the fastest practical speed, while scaling down occurs at a slower rate. This reduces the chances that a scaling down process results in a subsequent need to scale up, thereby incurring less oscillation and potentially benefiting performance.

4. SYNERGISTIC LOCALIZED DTM AND DEM

Since localized heating occurs much faster than chip wide heating due to slow lateral heat propagation [18], localized

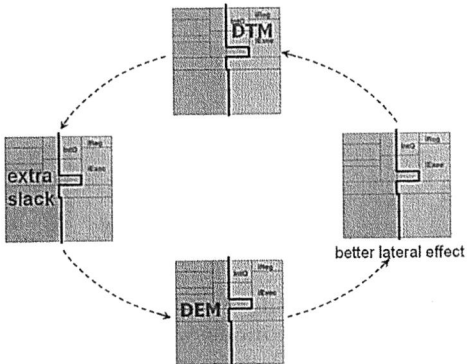

Figure 2: Synergy between DTM and DEM. Hotspot domains are colored with *pink* (lighter shade, right) and non-hotspot domains are colored with *blue* (darker, left).

DTM techniques, such as DVS within a given MCD domain, can be effective in controlling overall chip temperature. To circumvent a potential thermal violation, the throughput of these hot domains is inevitably reduced. This has the salient advantage of introducing execution slack in other domains which can be exploited by energy saving techniques.

On the other hand, effective energy saving techniques often reduce temperature in a local region as a by-product of reducing power consumption. If the local units happen to be the hot spots, then these techniques may avert a thermal emergency that would have otherwise occurred.

If the units being addressed by the energy management technique are peripheral to the hot spot region, the cooling of these neighboring regions permits additional heat transfer from the hot spots due to the better lateral effect resulting from a steeper temperature gradient. This, in turn, requires a less severe response in the hot spot domain, reducing the performance overhead associated with temperature control. This symbiotic cycle of mutually beneficial operation of the two techniques is shown in Figure 2.

The DEM algorithm adopted in this paper is similar to the improved *Attack/Decay* DVS algorithm proposed in [21] except that there are two queue occupancy change thresholds, one for each change direction. Having two thresholds permits more flexible control, and by setting different threshold values, the algorithm can be made more performance oriented or more energy aware.

The combined DTM and DEM approach operates as follows. Temperature sensors located at the hot spots independently trigger DTM control within each domain (using the algorithm described in Section 3). Hardware monitors embedded within each domain track statistics for the DEM algorithm. Within each domain, the DTM algorithm always has priority; the DEM algorithm operates only when DTM is not engaged. Whenever DTM is triggered, the DEM algorithm is disabled; and only when the thermal safe zone is reached again can the DEM algorithm be re-engaged. Note that since the DEM algorithm can only lower voltage and frequency below their nominal operating points that it cannot aggravate the ability of the DTM algorithm to control temperature. Clearly, if there is slack that is being exploited within a domain by the DEM algorithm then either DTM triggering may be avoided altogether, or when DTM is triggered, the severity of the DTM response is lower, reducing the performance overhead.

Table 2: Microarchitectural parameters.

Configuration Parameter	Value
Branch predictor:	
Level 1	1024 entries, history 10
Level 2	1024 entries
Bimodal predictor size	1024
Combining predictor size	4096
BTB	4096 sets, 2–way
Branch Mispredict Penalty	7
Decode/Issue/Retire Width	4/6/11
L1 Data Cache	64KB, 2–way set associative
L1 Instruction Cache	64KB, 2–way set associative
L2 Unified Cache	1MB, direct mapped
L1 cache latency	2 cycles
L2 cache latency	12 cycles
Integer ALUs	4 + 1 mult/div unit
Floating–Point ALUs	2 + 1 mult/div/sqrt unit
INT Issue Queue Size	20 entries
FP Issue Queue Size	15 entries
Load/Store Queue Size	64
Physical Register File Size	72 integer, 72 floating–point
Reorder Buffer Size	80

Table 3: Temperature modeling and thermal management parameters.

Temperature sampling interval	10000 cycles of a 3GHz clock
Thermal threshold	85 Degrees (Celsius)
Nominal frequency	3.0 GHz
Nominal voltage	1.4 Volt
Ambient air temperature	45 Degrees (Celsius)
Convection thermal resistance	0.8 K/W
Convection thermal capacitance	140.4 J/K
Die	0.5 mm thick
Heat spreader	1.0 mm thick, 3 cm × 3 cm
Heat sink	6.9 mm thick, 6 cm × 6 cm
Max temp sensor reading error	0.5 Degrees (Celsius)
Temp sensor resolution	0.5 Degrees (Celsius)

The algorithm used in this paper is a slight modification of that described above, as its first priority is to minimize the performance cost of DTM, with energy efficiency a secondary concern. The algorithm attempts to minimize the performance effects of the simultaneous engagement of DTM and DEM in the same domain. Therefore, once a domain enters the thermal emergency zone for a given application, the modified *Attack/Decay* DEM algorithm is disabled within that domain for the rest of the application run. (In practice, the DEM algorithm could be periodically re-enabled to account for phase behavior.) Thus, geographically, each of the algorithms controls different parts of the chip, with the DTM scheme operating in the domains that contain hot spots, and the DEM algorithm in those that do not. (However, as demonstrated in the next section, the triggering of DEM early in application execution may prevent DTM from ever needing to be engaged.) This non-overlapping of controlled domains avoids complex interactions between these two algorithms and yet achieves both good temperature control and energy efficiency.

To summarize, there are three effects that make localized DVS within an MCD processor efficient: a. it provides a localized, but vigorous, response to the particular area of the die that is undergoing a thermal emergency, permitting unaffected areas to continue to operate at full speed; b. the use of DEM within a particular domain may reduce the number of thermal emergencies in that domain, and the severity of the response that is required by the DTM algorithm; and

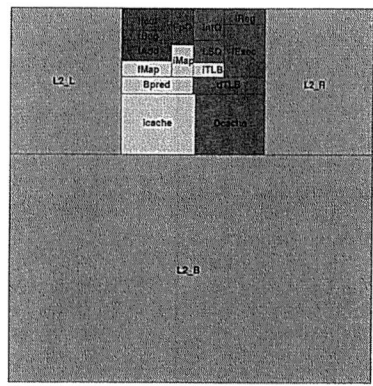

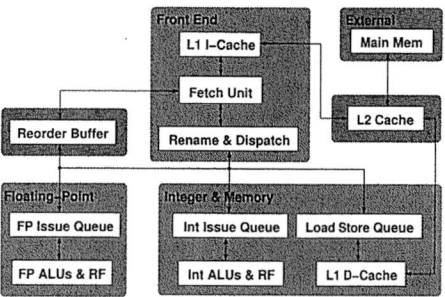

Figure 3: Floorplan (top) and logical domain partitioning (bottom) of the MCD processor.

c. a temperature response in one or two domains may create execution slack in adjacent domains. This permits DVS in these peripheral domains to be engaged, creating lateral heat flow away from the hotspot domains. This in turn permits a gentler response within the hotspot domains, leading to less performance loss.

5. EVALUATION METHODOLOGY

The evaluation methodology uses the MCD simulation framework [15], which is based on the SimpleScalar and Wattch toolkits [2, 3] and the HotSpot temperature modeling tool [9]. The microarchitecture and temperature modeling parameters are shown in Tables 2 and 3. Temperature sensors are placed at all relevant units and assumed to have a maximum reading error of 0.5 degrees Celsius, and a resolution of 0.5 degrees Celsius as well. The maximum voltage is 1.4V and there are 11 frequency levels, ranging from 3GHz to 1GHz. The fastest voltage transition speed is 16.7 mV per μs. The chip floorplan, and the logical domain partitioning of the MCD processor (proposed in [21]), are shown in Figure 3.

Of the SPEC2000 benchmark programs, the eight with the most severe thermal problems were chosen (refer to Table 1 for their characteristics); the remaining benchmarks generated no or very few thermal emergencies. For each DTM result with each benchmark, three simulation runs were conducted, with each run taking the steady state temperatures from the previous run as the initial temperatures, except for the first run which sets the initial temperature at 80 degrees Celsius and operated without any DTM control. For runs with DTM control, initial temperatures were clipped based on the pre-specified hard limit, which is set at 85 degrees Celsius.

Each benchmark was fast-forwarded 2 billion instructions, followed by the two-phase warm up for 300 million instruc-

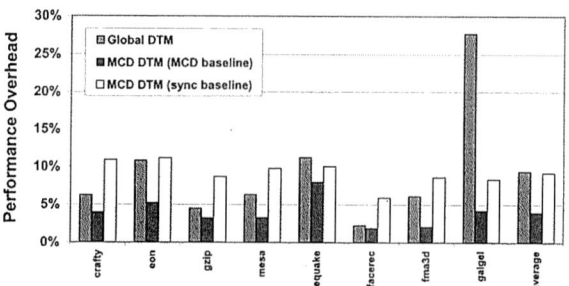

Figure 4: Performance degradation of Global DTM relative to fully synchronous without DTM, MCD DTM relative to MCD without DTM, and MCD DTM relative to fully synchronous without DTM.

tions total as suggested in [18], the warm up of various structures (like branch predictor and caches) for 100 million instructions, and then the warm up of different chip units to reach representative temperature values, for another 200 million instructions. The statistics used to generate the bar graphs in the next section were then gathered for the next 300 million instructions. However, Figures 5 and 7 include the data from the last 500 million instructions of execution.

6. RESULTS

In this section, localized DTM within an MCD processor is first compared with global DTM in a fully synchronous machine without considering DEM. Then, the combined DTM and DEM algorithm is evaluated in Section 6.2.

6.1 Localized Versus Global DVS-based DTM

The DVS-based DTM algorithm described in Section 3 was applied to both fully synchronous and MCD processors (each domain independently implementing the algorithm). Figure 4 shows the corresponding performance degradation. Comparing the bars on the left (fully synchronous microprocessor with DVS-based DTM) with the maximum temperatures in Table 1 shows that for high temperature programs like *galgel*, *eon* and *equake*, the performance cost is high as well, more than 10%. The worst performance cost is 27% for *galgel*, which is the program that has the highest temperature. For lower temperature programs such as *facerec* and *gzip*, the performance cost is also lower, as expected.

The performance cost of using localized DVS within MCD (relative to the baseline MCD machine) is on average 2.5 times less than that of global DVS (relative to the performance of a fully synchronous machine). Since domains in an MCD processor are independent, the performance impact of DVS is largely confined within the domain where the hot spots are located. Maintaining full speed operation in other domains is especially important when one or all of the other domains are very performance critical. If there happens to be slack in the hot spot domain at the time DVS is applied, then the performance loss is further reduced. Due to the lower performance cost of targeted, localized DTM on an MCD machine, even when the inter-domain synchronization performance cost of MCD is accounted for (over 5% on average – see Figure 6), its performance overall is competitive with that of the fully synchronous machine with DTM.

While for all benchmarks the performance cost of localized DTM within MCD is less than global DTM, the difference is particularly striking for *galgel*. Since *galgel* has the most severe thermal problem among all the programs (Table 1), it also requires the largest voltage and frequency reduction

58

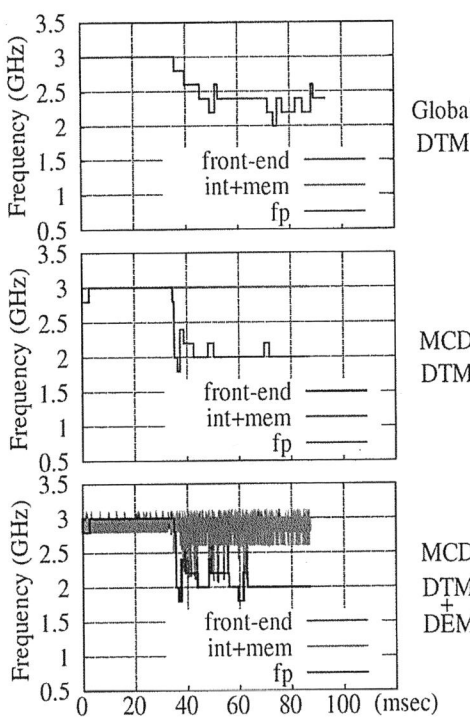

Figure 5: Frequency profiles for *galgel* when running the DTM algorithm on a fully synchronous machine (top, where all three curves overlap), an MCD machine (middle), and an MCD machine with DEM also applied in the non-hotspot domains (bottom).

to maintain acceptable temperatures. As shown in Figure 5, for the MCD DTM case, only the floating point domain frequency is reduced, as the other two domains do not contain hot spots. For Global DTM, all three domain frequencies must be reduced by the same amount. This has a significant performance cost for *galgel* since for this benchmark, all three domains are performance critical, each containing critical paths through the execution dataflow graph. Therefore, the ability to maintain full speed operation in two of the three domains through localized DTM yields a large performance advantage.

One advantage of global DVS is the cooling of neighboring units results in better heat removal from the hotspot due to a steeper temperature gradient; therefore, the voltage does not have to be lowered as much in the affected domain as in the MCD case, as seen for the floating point domain in *galgel* (Figure 5). However, this factor did not have nearly as large a performance impact as the ability to keep the front-end and integer/memory domains running at full frequency.

6.2 Combined DTM and DEM Algorithms

Figure 6 shows the performance overhead and energy savings, relative to the baseline fully synchronous machine without DTM or DEM, of the Global and MCD machines with DTM, and the MCD machine with both DTM and DEM. The last bar in each set shows the performance degradation (due to synchronization) of the baseline MCD machine. (The other MCD bars include this baseline degradation.) For MCD, the combined DTM and DEM approach achieves over a factor of two greater energy savings *and* better performance compared to the use of DTM alone. The energy benefit is a result of the overall lower voltage in all domains:

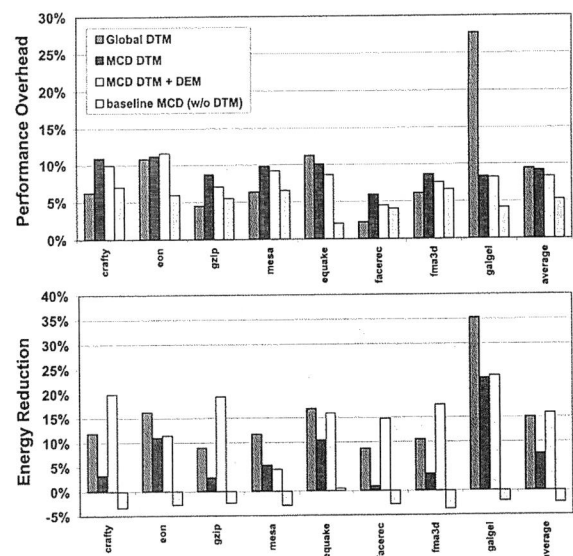

Figure 6: Performance degradation and energy savings for different schemes. The baseline is a fully synchronous machine without any DTM control or energy saving techniques.

in the hotspot domains due to DTM and/or DEM, and in the other domains due to the additional execution slack that can be exploited. The performance benefit comes from the DEM algorithm exploiting the extra slack in non-hotspot domains through DVS, creating a better lateral temperature effect, or from DEM being engaged in the hotspot domains preventing them from heating to the same degree; thus, the DTM algorithm operating in the hotspot domains does not need to reduce voltage and frequency as much to maintain acceptable temperature levels. The only exception is *eon*, which has a slightly higher performance overhead with the combined approach. This is due to the fact that *eon* is the only program among the eight where the DEM algorithm fails to save energy, and therefore the lateral effect is actually slightly degraded. Although the effect is minor, a DEM algorithm based on a more formal control theoretic approach [19, 20] may yield more consistent results.

There are two reasons why DEM lessens the performance impact of DTM. The first is the lateral effect as mentioned previously. Figure 5 shows this effect for *galgel*. In the MCD DTM+DEM case, DEM is activated for both the front-end and integer+memory domains, while DTM operates in the floating point domain. In comparing the floating point frequency curves, once DTM is activated, the floating point frequency for the DTM+DEM case is not as aggressively scaled as compared to DTM alone. From the period in which the floating point frequency drops to 2GHz, to the point where both the DTM and DTM+DEM floating point frequency curves remain flat (at roughly 72ms), the DTM+DEM floating point frequency is about 40MHz higher on average than for DTM alone.

On the other hand, Figure 7 shows the frequency curves for *facerec* for both DTM and DTM+DEM. For the latter, DEM is aggressively engaged immediately at execution, long before the DTM algorithm is engaged. This results in a significant energy savings over the DTM case, thereby obviating the need for DTM to be triggered. This case, in which DEM alone maintains acceptable temperatures (although DTM is of course available should this situation change),

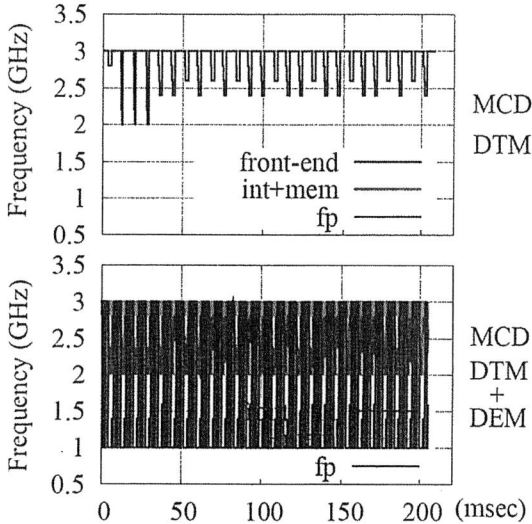

Figure 7: Frequency profiles with DTM alone (top) and with both DTM and DEM (bottom) for *facerec*.

happens for several of the benchmarks, while others benefit more from the lateral effect as in *galgel*.

Compared to DTM on a fully synchronous machine, MCD with DTM and DEM achieves a lower performance overhead and greater energy savings, even with the performance cost of inter-domain synchronization taken into account. While much of the performance benefit comes from *galgel*, most of the benchmarks demonstrate a comparable performance and energy tradeoff with MCD compared to Global. The ability to perform vigorous localized temperature management in MCD, and to exploit the synergy with DEM, provides an advantage in a thermally constrained environment that serves to offset the synchronization cost.

7. CONCLUSIONS AND FUTURE WORK

In this paper, a DVS-based algorithm for localized temperature control in MCD processors was proposed and compared with its use in a conventional fully synchronous design. The ability to provide a focused and vigorous response was shown to have a significantly lower performance cost. Furthermore, the symbiotic relationships between localized DTM and DEM within an MCD processor was examined. Due to several complementary effects, the addition of localized DEM to a localized DTM approach yields a significant performance benefit *and* improves energy efficiency. The use of these complementary techniques permits an MCD processor to be performance competitive with a fully synchronous design in a temperature constrained environment, even when accounting for the synchronization costs.

An interesting area for future work is to explore the integration of localized intra-core DTM and DEM policies with higher level CMP DTM approaches. Furthermore, the robustness of the new DTM algorithm will be compared against other DVS-based DTM approaches.

Acknowledgments

The authors would like to thank Alper Buyuktosunoglu for his guidance and feedback.

8. REFERENCES

[1] D. Brooks and M. Martonosi. Dynamic Thermal Management for High Performance Microprocessors. In *Proc., of the 7th Intl. Symp. on High Performance Computer Architecture*, Jan. 2001.

[2] D. Brooks, V. Tiwari, and M. Martonosi. Wattch: A Framework for Architectural-Level Analysis and Optimization. In *Proc., of the 27th Intl. Symp. on Computer Architecture*, Jun. 2000.

[3] D. Burger and T. Austin. The SimpleScalar Tool Set, Version 2.0. Technical Report 1342, Dept. of Computer Science, Univ. of Wisconsin, Jun. 1997.

[4] L. T. Clark. Circuit Design of XScale Microprocessors. In *2001 Symp. on VLSI Circuits, Short Course on Physical Design for Low Power and High Performance Microprocessors*, Jun. 2001.

[5] M. Fleischmann. Crusoe Power Management – Reducing the Operating Power with LongRun. In *Proc., of the HOT CHIPS Symp. XII*, Aug. 2000.

[6] S. Gunther, F. Binns, D. M. Carmean, and J. C. Hall. Managing the Impact of Increasing Microprocessor Power Consumption. In *Intel Technology Journal*, 2001.

[7] S. Heo, K. Barr, and K. Asanovic. Reducing Power Density through Activity Migration. In *Proc., of the 2003 Intl. Symp. on Low-Power Electronics and Design*, Aug. 2003.

[8] M. Huang, J. Renau, S. M. Yoo, and J. Torrellas. A Framework for Dynamic Energy Efficiency and Temperature Management. In *Proc., of the 33rd Intl. Symp. on Microarchitecture*, Dec. 2003.

[9] W. Huang, S. Ghosh, K. Sankaranarayanan, K. Skadron, and M. R. Stan. Compact Thermal Modeling for Temperature-Aware Design. In *Proc., of the 41st Design Automation Conf.*, Jun. 2004.

[10] A. Iyer and D. Marculescu. Power and Performance Evaluation of Globally Asynchronous Locally Synchronous Processors. In *Proc., of the 29th Intl. Symp. on Computer Architecture*, May 2002.

[11] Y. Li, K. Skadron, Z. Hu, and D. Brooks. Performance, Energy and Thermal Considerations for SMT and CMP Architectures. In *Proc., of the 11th Intl. Symp. on High Performance Computer Architecture*, Feb. 2005.

[12] C. H. Lim, W. Daasch, and G. Cai. A Thermal-Aware Superscalar Microprocessor. In *Proc., of the 3rd IEEE Intl. Symp. on Quality Electronic Design*, Mar. 2002.

[13] M. Powell, M. Gomaa, and T. N. Vijaykumar. Heat-and-run: Leveraging SMT and CMP to Manage Power Density Through the Operating System. In *Proc., of the 11th Intl. Conf. on Architectural Support for Programming Languages and Operating Systems*, Oct. 2004.

[14] H. Sanchez, B. Kuttanna, T. Olson, M. Alexander, G. Gerosa, R. Philip, and J. Alvarez. Thermal Management System for High Performance PowerPC Microprocessors. In *Proc., of the 42nd IEEE Intl. Computer Conf.*, 1997.

[15] G. Semeraro, G. Magklis, R. Balasubramanian, D. H. Albonesi, S. Dwarkadas, and M. L. Scott. Energy-Efficient Processor Design Using Multiple Clock Domains with Dynamic Voltage and Frequency Scaling. In *Proc., of the 8th Intl. Symp. on High Performance Computer Architecture*, Feb. 2002.

[16] K. Skadron. Hybrid Architectural Dynamic Thermal Management. In *Proc., of the 2004 Conf. on Design, Automation and Test in Europe*, Feb. 2004.

[17] K. Skadron, T. Abdelzaher, and M. Stan. Control-Theoretic Techniques and Thermal-RC Modeling for Accurate and Localized Dynamic Thermal Management. In *Proc., of the 8th Intl. Symp. on High Performance Computer Architecture*, Feb. 2002.

[18] K. Skadron, M. R. Stan, W. Huang, S. Velusamy, K. Sankaranarayanan, and D. Tarjan. Temperature-Aware Microarchitecture: Extended Discussion and Results. Technical Report CS-2003-08, Dept. of Computer Science, Univ. of Virginia, Aug. 2003.

[19] Q. Wu, P. Juang, M. Martonosi, and D. W. Clark. Formal Online Methods for Voltage/Frequency Control in Multiple Clock Domain Microprocessors. In *Proc., of the 11th Intl. Conf. on Architectural Support for Programming Languages and Operating Systems*, Oct. 2004.

[20] Q. Wu, P. Juang, M. Martonosi, and D. W. Clark. Voltage and Frequency Control with Adaptive Reaction Time in Multiple Clock Domain Processors. In *Proc., of the 11th Intl. Symp. on High Performance Computer Architecture*, Feb. 2005.

[21] Y. Zhu, D. H. Albonesi, and A. Buyuktosunoglu. A High Performance, Energy Efficient, GALS Processor Microarchitecture with Reduced Implementation Complexity. In *Proc., of the 2005 IEEE Intl. Symp. on Performance Analysis of Systems and Software*, Mar. 2005.

A Two-Port SRAM for Real-Time Video Processor Saving 53% of Bitline Power with Majority Logic and Data-Bit Reordering

Hidehiro Fujiwara[1] Koji Nii[1] Junichi Miyakoshi[1] Yuichiro Murachi[1] Yasuhiro Morita[2]

Hiroshi Kawaguchi[3] Masahiko Yoshimoto[3]

[1]Graduate School of Science and Technology, Kobe University
1-1 Rokkodai-Cho, Nada-ku, Kobe 657-8501, Japan

[2]Graduate School of Science and Technology, Kanazawa University
Kakuma-Machi, Kanazawa, Ishikawa 920-1192, Japan

[3]Department of Computer and Systems Engineering, Kobe University
1-1 Rokkodai-Cho, Nada-ku, Kobe 657-8501, Japan

Phone: +81-78-803-6234   fujiwara@cs28.cs.kobe-u.ac.jp

ABSTRACT

We propose a low-power two-port SRAM suitable for real-time video processing. In order to minimize discharge power on a read bitline, a majority-logic decides if input data are inverted in a write cycle, so that "1"s are in the majority. In video data, since more significant bits of adjacent pixel data are fortunately lopsided to either "0" or "1" with higher probability, the data bits in the pixels are reordered in each digit group to exploit the majority logic. The speed and area overheads are 4% and 11% in a 90-nm process technology, respectively. The proposed SRAM achieves 53% power reduction on the bitlines, and saves 43% of a total power when considered as an H.264 reconstructed-image memory.

Categories and Subject Descriptors

B.3.1 **[Semiconductor Memories]**: Static memory (SRAM) design

General Terms

Design

Keywords

Low power SRAM, two-port SRAM, real-time image processing, majority logic, data-bit reordering.

1. INTRODUCTION

As the ITRS predicts, memory area is becoming larger, and will occupy 90% of an SoC in 2013 [1]. In a real-time video SoC, this trend is going on, and an H.264 encoder for a high-definition television requires, at least, a 500-k bit memory as a search-window buffer, which consumes 40% of a total power [2]. As a process technology is scaled down, a large-capacity SRAM will

Permission to make digital or hard copies of all or part of this work for personal or classroom use is granted without fee provided that copies are not made or distributed for profit or commercial advantage and that copies bear this notice and the full citation on the first page. To copy otherwise, or republish, to post on servers or to redistribute to lists, requires prior specific permission and/or a fee.
ISLPED'06, October 4–6, 2006, Tegernsee, Germany.
Copyright 2006 ACM 1-59593-462-6/06/0010 ...$5.00.

be implemented on a chip as a frame buffer or restructured-image memory, which could consume a larger portion of power. To save an SRAM power in real-time video applications, we propose a low-power two-port SRAM in this paper.

A two-port SRAM is suitable for real-time video processing since it can make one read and one write at the same time in a clock cycle [2-5]. In general, a read port has a single read bitline for area efficiency, and the proposed SRAM also has the same structure in Fig. 1 (a). Two nMOS transistors for a read wordline (RWL) and read bitline (RBL) are added to the conventional single-port 6-T SRAM, which frees a static noise margin (SNM) in a read operation [6]. Therefore, a large β ratio is not required, and two nMOS driver transistors can be minimized.

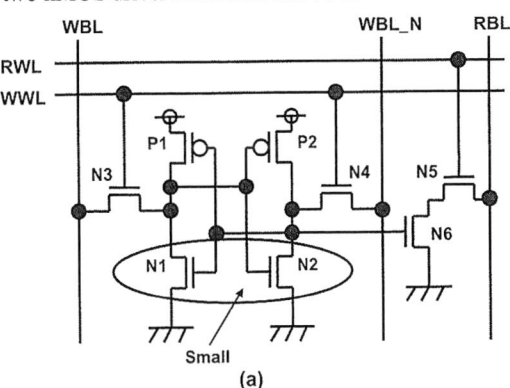

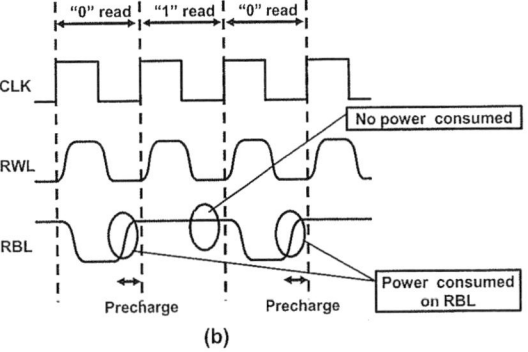

Figure 1. A 8-T two-port SRAM cell.
(a) Circuit, and (b) operation waveforms in read cycles.

Figure 1 (b) illustrates simplified operation waveforms in read cycles. Since a precharge scheme is adopted and an RBL is precharged to V_{dd} before the beginning of a clock cycle, charge and discharge power is consumed on the RBL when "0" is read. In contrast, no power is consumed on the RBL when "1" is read. For low-power operation, hence, it is good to write "1"s as much as possible. The possibility that "1" is read is increased, which in turn reduces the RBL power.

For this purpose, we append majority logic to a two-port SRAM. Although majority logic has been used on transmission lines to save an I/O bus power [7], it has not been utilized in a memory bus as far as we know. In the next section the concept of an SRAM with majority logic is introduced.

Other than the majority logic, we also exploit statistical similarity in video data for further power reduction. Adjacent pixels have correlation one another, which implies more significant bits of the adjacent pixel data are lopsided to either "0" or "1" with higher probability. We reorder the data bits of the adjacent data in each digit group to effectively boost the majority-logic function, which is discussed in Section 3 considering H.264 codec.

In Section 4, we describe design and evaluation of a 72-k bit SRAM in a 90-nm process technology, with the proposed features. The final section summarizes this paper.

2. SRAM WITH MAJORITY LOGIC

Figure 2 (a) illustrates the concept of an SRAM with majority logic. In order to maximize the number of "1"s, the majority-logic circuit counts the number of "1"s, and decides if the input data are inverted in a write cycle, so that "1"s are in the majority. The inversion information ("1" means inversion) is stored in an additional flag bit as depicted in Fig. 2 (b). In a read cycle, the process is reversed. Output data is inverted if a flag bit is true, so that the original data can be read.

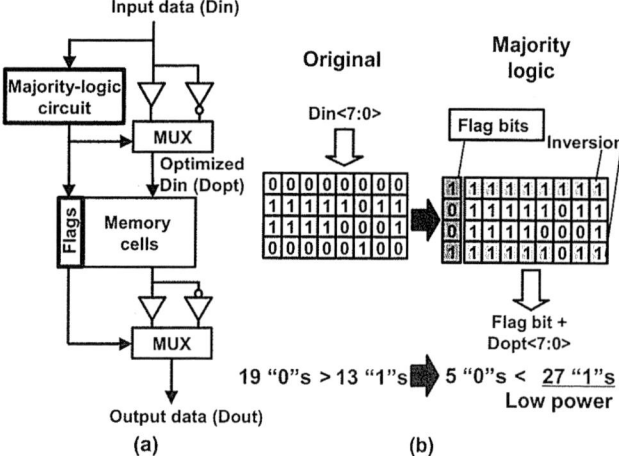

Figure 2. Concept of SRAM with majority logic.
(a) Schematic, and (b) flag bit.

The power-reduction factor on the RBL is briefly discussed in Fig. 3. The bit width of data is assumed to be eight here. If the number of "1"s in input data is eight, the data is not inverted and thus, one "0" is stored only in a flag bit and there are no "0"s in the data themselves. This means one charge/discharge is made on RBLs, which is a power overhead. If the number of "1"s are four

or less, the input data is inverted by the majority logic to maximize the number of "1"s, and the RBL power is reduces.

When input data have a random pattern, the number of charges/discharges is four out of eight RBLs in the conventional two-port SRAM. However, the majority logic reduces this value to 3.27 although the number of the RBLs is increased to nine. This indicates that the majority logic statistically saves 18% of an RBL power even if the data is random.

Note that it is important to consider which the inversion information in the flag bit should be "0" or "1", because the RBL power even on the flag bit depends on the value as well. If the number of "0"s in whole data is more than that of "1"s, the inversion information should be "1" to maximize the number of "1"s. As previously mentioned, we chose "1" as an inversion flag based on statistical analysis of HDTV test sequences, which is described in detail in the following section.

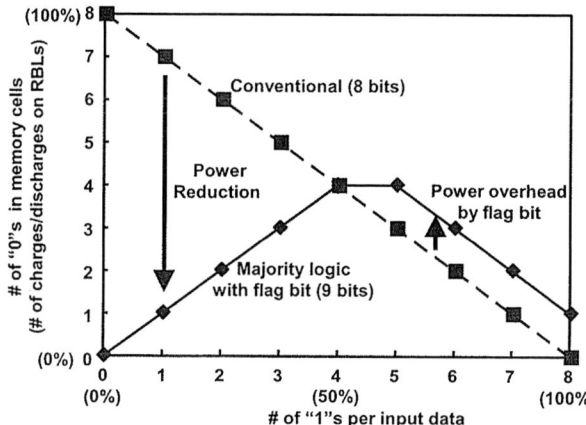

Figure 3. A comparison of RBL powers between the conventional and proposed two-port SRAM with majority logic.

3. DATA-BIT REORDERING

3.1 Statistical Characteristics of Video Image

In H.264 codec, the YUV format is adopted as pixel data. An example is illustrated in Fig. 4. One pixel is comprised of 8-bit luma (Y signal) and 4-bit chroma (U and V signals). In this study, only luma data are considered. In an image, adjacent pixels have strong correlation one another, and the correlation becomes stronger in a more significant bit. Namely, the most significant bits (MSBs) in contiguous data tend to be lopsided to either "0" or "1" with high probability, while in the least significant bits (LSBs), the values of the bits are random. Thus, correlations in each digit are somewhat different one another.

The distributions of the number of "1"s in different digit groups are represented in Fig. 5 when eight-pixel data (8×8 bits) are rearranged in each digit group as shown in Fig. 4. It can be seen that the MSB group tends to have "0", which was pointed out in the previous section. The distribution in the LSB group is normally-distributed (strictly, it is binominal distribution), and the same tendencies can be observed even in the 2nd- and 3rd-digit groups.

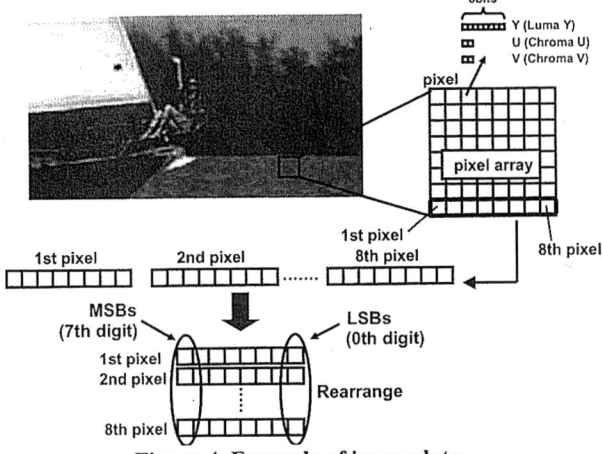

Figure 4. Example of image data.

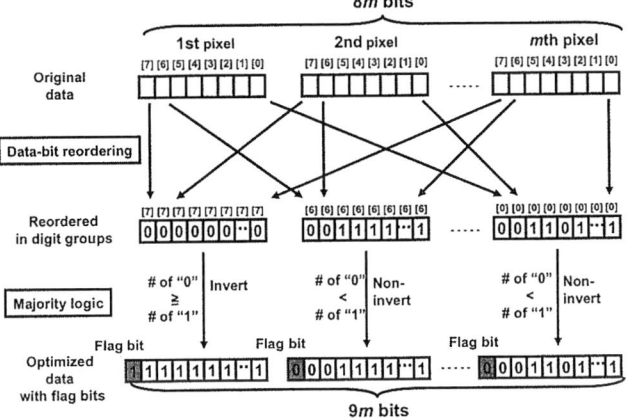

Figure 5. Distributions of the number of "1"s in different digit groups, extracted from the HDTV test sequences, "Market" and "Church".

Figure 6. Data-bit reordering and majority logic.

As discussed in the previous section, the power reduction on the RBLs is theoretically expected thanks to the majority logic even if input data is normally-distributed. Besides, further power reduction is promising because the image data is lopsided to "0"s in more significant digit groups as indicated in Fig. 5. We exploit these characteristics to reduce the RBL power.

The rearrange of digits is called data-bit reordering in this paper. Again, we explain data-bit reordering illustrated in Fig. 6. In a write cycle, data comprised of m pixels ($8m$ bits) are reordered in

each digit group. The optimum value of m is discussed in the next subsection. If the number of "0"s in a digit group is equal to or larger than that of "1", that is in other words, if the number of "0" is equal to or larger than $m/2$, bits in the digit group are inverted. Alternatively, if the number of "0"s is smaller than that of "1"s, they are not inverted. The majority logic and data-bit reordering maximize the number of "1"s in image data, and optimize the RBL power.

In a read cycle, the optimized data are either inverted or non-inverted according to a flag bit in a digit group. If a flag bit is "1", the data are inverted, and then the reordered bits are put back to the original pixel data.

3.2 Optimum Value of m

In order to obtain the optimum value of m, statistic analysis is carried out with the original image and reconstructed image, extracted from ten HDTV test sequences, "Bronze with Credit" (Bronze), "Building along the Canal" (Canal), "Church" (Church), "Intersections" (Inters), "Japanese Room" (Jpnroom), "European Market" (Market), "Yachting" (Sail), "Street Car" (Stcar), "Whale Show" (Whale), and "Yacht Harbor" (Yacht). The original image is encoded, and then a reconstructed image is generated in a local decoding loop and utilized for motion estimation. The encoding configuration is shown in Fig. 7.

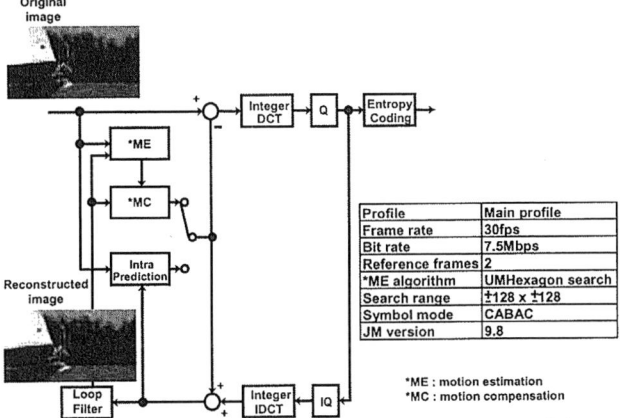

Profile	Main profile
Frame rate	30fps
Bit rate	7.5Mbps
Reference frames	2
*ME algorithm	UMHexagon search
Search range	±128 x ±128
Symbol mode	CABAC
JM version	9.8

*ME : motion estimation
*MC : motion compensation

Figure 7. H.264 encoding process and simulation conditions.

Although pixel data are segmented every m pixels, data-bit reordering does not cause addressing problem since consecutive pixels are simultaneously processed in a video memory.

Figure 8 illustrates the normalized RBL powers in comparison with the conventional two-port SRAM. Figure 8 (a) shows a case where the original image is utilized as input data and only the majority logic is applied to a set of pixels. The number of pixels to which the majority logic is applied, is varied (one, two, and four pixels). The figure demonstrates that 20% of the RBL power is saved on average only with the majority logic and without data-bit reordering, even though the flag bit is appended.

As illustrated in Fig. 8 (b), the saving factor is further extended to 45% with both the majority logic and data-bit reordering, which indicates that the statistical characteristic of image data is well exploited. Moreover, the maximum power save is achieved when the proposed two-port SRAM is utilized as a reconstructed image that has stronger correlation than the original image, as shown in Fig. 8 (c). 53% reduction of the RBL power is possible. The proposed SRAM is suitable for real-time video codec such as

63

MPEG2, MPEG4, and H.264 which require a large-capacity reconstructed-image memory for motion estimation.

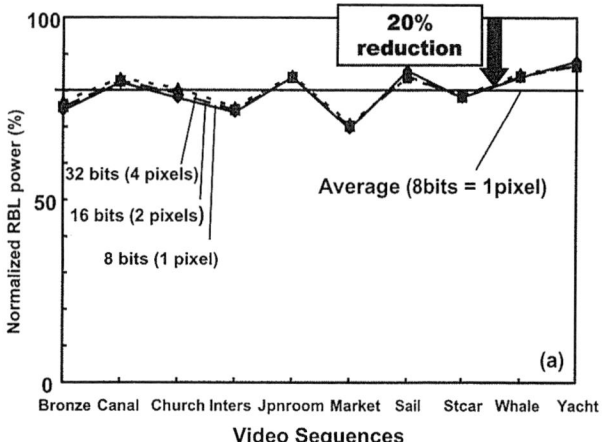

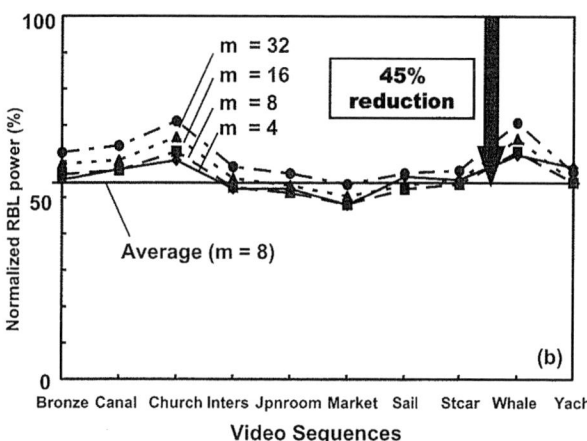

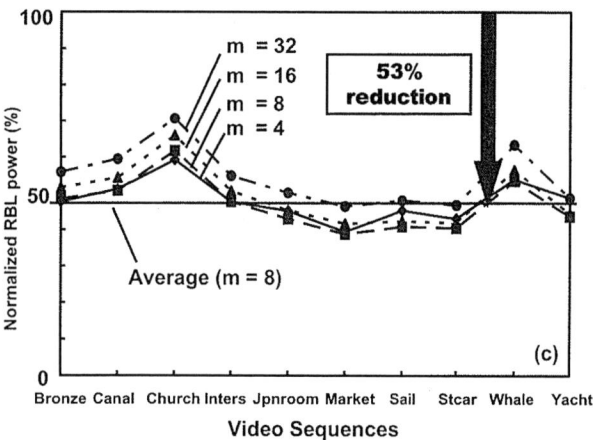

Figure 8. Normalized RBL powers in the (a) original image only with majority logic, (b) original image with both majority logic and data-bit reordering, and (c) reconstructed image with both majority logic and data-bit reordering. 100% signifies a case of the conventional two-port SRAM.

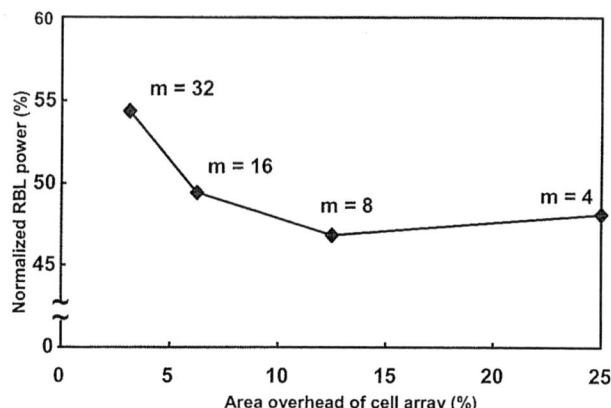

Figure 9. Normalized RBL power vs. area overhead in the proposed SRAM used as a reconstructed-image memory.

In the both cases of the original image and reconstructed image, $m=4$ or $m=8$ is optimum in terms of power reduction. However, if m is small, area overhead becomes large. Figure 9 shows the relationship between power and area overhead when m is changed. As m is increased, the area overhead becomes decreased, but the RBL power is raised. This is because correlation in a digit group becomes weaker as m is larger. The area overhead is 12.5% when $m=8$. $m=4$ is not a design choice due to its large area overhead.

4. DESIGN IN 90-nm TECHNOLOGY

4.1 Memory Cell

Figure 10 is a layout of the proposed two-port SRAM cell in a 90-nm process technology. The cell area is $3.15 \times 0.76 \ \mu m^2$. The schematic have been already shown in Fig. 1 (a), and the transistor sizes are shown in Fig. 10. Figure 11 is a block diagram, where a capacity of memory cells is 72-k bits. $m=8$ is chosen in this design, and thus 64-k bits are for data themselves, and the other 8-k are flag bits. A hierarchical RBL structure is applied to void a speed overhead of a single-bitline scheme [6]. The write bitlines (WBLs and WBL_Ns) do not have precharge transistors since they are dedicated for data write-in.

4.2 Write/Read Circuitry with Majority Logic

In a write circuit, there should be a majority-logic circuit. However, it would require 29 cell gates for eight-bit majority logic if designed as a digital circuit [7, 8], which might result in a large area overhead. Figure 12 is the proposed write circuit with majority logic. The majority-logic circuit is achieved with a precharge logic, and thus the majority logic is evaluated by pull-down networks in flip-flops. A sense amplifier amplifies a voltage difference between majority signals, JL and JL_N.

Figures 13 (a) and (b) show the operation waveforms of the majority logic in both cases that "1"s and "0"s are in the majority. When the number of "0"s is eight, eight pull-downs and one dummy rapidly sink a JL node, while the falling time of the JL node takes slower when the number of "0" is four. In this case that the numbers of "1" and "0" are same, the voltage difference between JL and JL_N would be theoretically zero, which might cause a meta stable in the sense amplifier. However in reality, there is some voltage difference between them thanks to the dummy circuit. J in the dummy circuit always sinks the JL node so as not to make the differential voltages same, even if the number of "1"s is as same as that of "0"s. However, as m is increased, the voltage difference could be closer. From this point of view, it can be said that $m=8$ is a good design choice.

Figure 14 is a read circuit resuming the original data, which inverts data bits depending on a flag bit. The conditional inversion is implemented with EX-ORs.

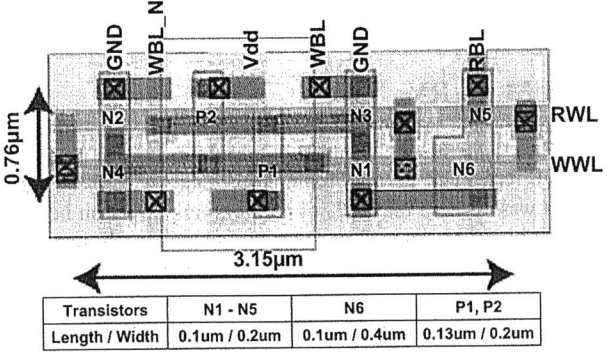

Figure 10. Memory cell layout

Transistors	N1 - N5	N6	P1, P2
Length / Width	0.1um / 0.2um	0.1um / 0.4um	0.13um / 0.2um

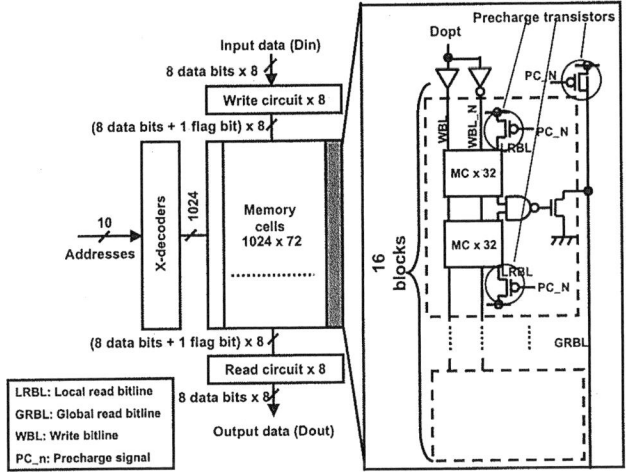

Figure 11. Block diagram of the proposed 72-k bit SRAM.

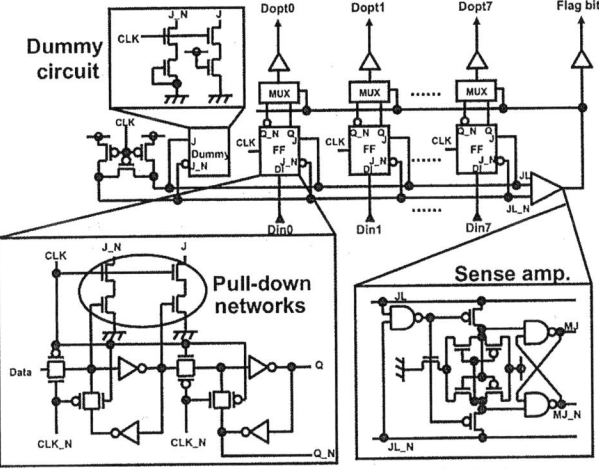

Figure 12. Write circuit with majority logic.

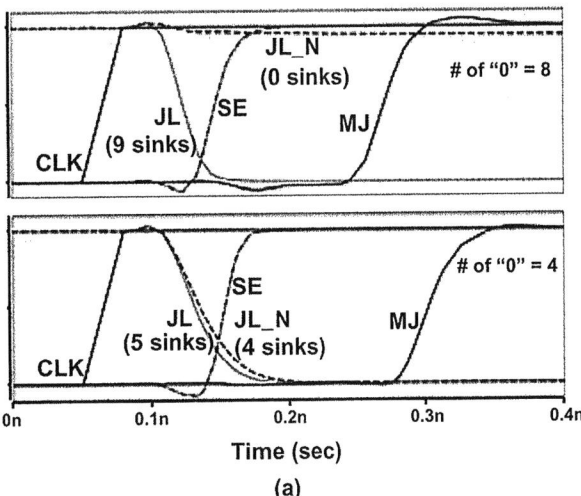

(a)

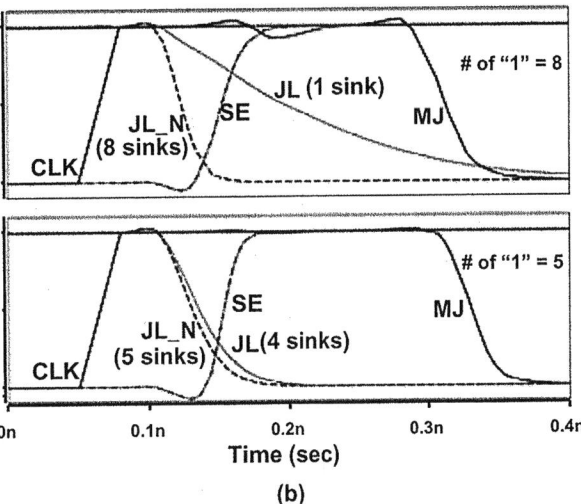

(b)

Figure 13. Operation Waveforms of majority logic when (a) "0"s and (b) "1"s are in the majority.

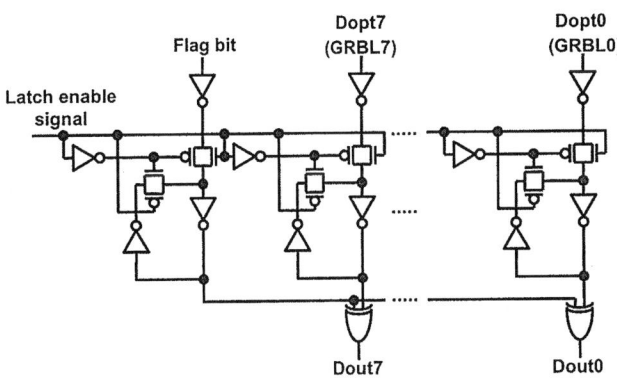

Figure 14. Read circuit resuming the original data.

Although some additional circuits are implemented to the write and read circuits for the majority logic, the speed overhead in a read cycle is 4% in a simulation. Alternatively, there is no speed overhead in a write cycle since it is canceled out by a wordline delay. The area overhead of the majority logic itself is less than 1%, and overall area overhead including the flag bits becomes 11%. The power caused by the majority logic is only 21 μW, which is negligible compared with a bitline power.

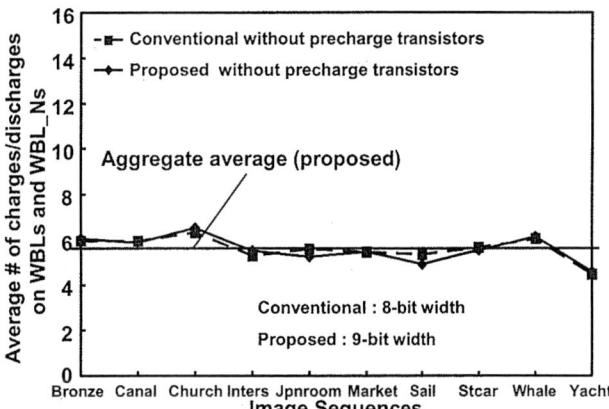

Figure 15. The number of toggles in a write cycle.

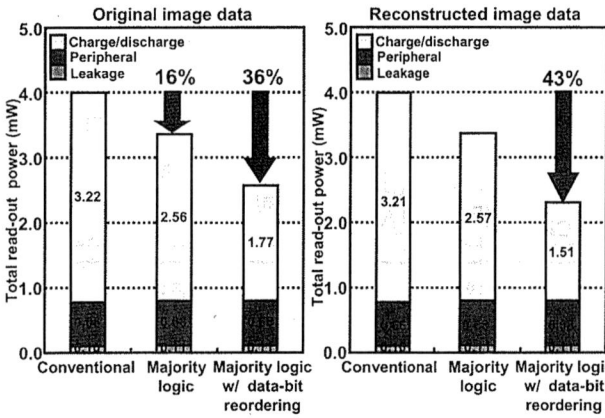

Figure 16. Readout power reduction (T=25°C).

4.3 Power Estimation

Since there are no precharge transistors on the write bitlines, it does not consume any power as far as same data are successively written. From this point of view, the statistical characteristic of image data helps to save the write-bitline power. Figure 15 shows the average number of charges/discharges in a write cycle. Even though the bit width is increased to nine by a flag bit, the reduction factor in the proposed SRAM is as same as that of the conventional one thanks to the majority logic. Consequently, the proposed SRAM has no power overhead on write bitlines.

In Fig. 16, the total read-out powers are compared between the conventional and proposed SRAMs. In a video memory, power reduction in a read operation is technically important since readout is made more frequently than write-in. In the conventional 64-k bit SRAM, a total read-out power is estimated at 4.0 mW, at a supply voltage of 1.0 V and frequency of 300MHz. The RBL

power occupies 80% of the power in the conventional SRAM, and thus the proposed SRAM reduces this part. Only with the majority logic, 16% of the total read-out power can be saved in the proposed 72-k bit SRAM when the original H.264 image is considered. In addition, 36% reduction can be achieved with both the majority logic and data-bit reordering. As a reconstructed-image memory, the proposed SRAM further saves 43% of the total read-out power.

5. SUMMARY

We proposed a two-port SRAM with majority logic and data-bit reordering. This SRAM is suitable for real-time image processing in which data have statistical similarity. Only with the majority logic, it was verified that 20% of read-bitline power can be saved in a 72-k bit SRAM in a 90-nm process technology. Moreover, the saving factor is extended to 45% thanks to data-bit reordering. The proposed SRAM achieves 53% power reduction on the bitlines when considered as an H.264 reconstructed image, and saves 43% of the total read-out power. The speed and area overheads are 4% and 11%, respectively.

6. REFERENCES

[1] International Technology Roadmap for Semiconductors 2005, http://www.itrs.net/Common/2005ITRS/Home2005.htm.

[2] J. Miyakoshi, Y. Murachi, K. Hamano, T. Matsuno, M. Miyama, and M. Yoshimoto, "A Low-Power Systolic Array Architecture for Block-Matching Motion Estimation," IEICE Trans. Electronics, Vol.E88-C, No.4, pp.559-569, Apr. 2005.

[3] Y. Murachi, K. Hamano, T. Matsuno, J. Miyakoshi, M. Miyama, and M. Yoshimoto, "A 95 mW MPEG2 MP@HL Motion Estimation Processor Core for Portable High-Resolution Video Application," IEICE Trans. Fundamentals, Vol.E88-A, No.12, pp.3492-3499, Dec. 2005.

[4] S. Ishiwata, T. Yamakage, Y. Tsuboi, T. Shimazawa, T. Kitazawa, S. Michinaka, K. Yahagi, A. Oue, T. Kodama, N. Matsumoto, T. Kamei, M. Saito, T. Miyamori, G. Ootomo, and M. Matsui, "A Single-Chip MPEG-2 Codec Based on Customizable Media Embedded Processor," IEEE J. Solid-State Circuits, Vol.38, No.3, pp.530-540, Mar. 2003.

[5] Y-W. Huang, T-C. Chen, C-H.Tsai, C-Y. Chen, T-W. Chen, C-S. Chen, C-F. Shen, S-Y. Ma, T-C. Wang, B-Y. Hsieh, H-C. Fang, and L-G. Chen, "A 1.3TOPS H.264/AVC Single-Chip Encoder for HDTV Applications," IEEE Int. Solid-State Circuits Conf., pp.128-129, Jan. 2005.

[6] K. Takeda, Y. Hagihara, Y. Aimoto, M. Nomura, Y. Nakazawa, T. Ishii, and H. Kobatake, "A Read-Static-Noise-Margin-Free SRAM Cell for Low-Vdd and High-Speed Applications," IEEE J. Solid-State Circuits, Vol.41, No.1, pp.113-121, Jan. 2006.

[7] M. R. Stan, and W. P. Burleson, "Bus-Invert Coding for Low Power I/O," IEEE Trans. VLSI Systems, Vol.3, No.1, pp.49-58, Mar. 1995.

[8] Y. Shin, and K. Choi, "Narrow Bus Encoding for Low Power Systems," Asia and South Pacific Design Automation Conf., pp.217-220, Jan. 2000.

A High-Speed Variation-Tolerant Interconnect Technique for Sub-Threshold Circuits Using Capacitive Boosting

Jonggab Kil	Jie Gu	Chris H. Kim
Intel Corporation	University of Minnesota	University of Minnesota
1900 Prairie City Road	200 Union Street SE	200 Union Street SE
Folsom, CA 95630	Minneapolis, MN 55455	Minneapolis, MN 55455
+1-916-356-9968	+1-612-625-1029	+1-612-625-2346
jonggab.kil@intel.com	jiegu@umn.edu	chriskim@umn.edu

ABSTRACT

This paper describes an interconnect technique for sub-threshold circuits to improve global wire delay and reduce the delay variation due to PVT fluctuations. By internally boosting the gate voltage of the driver transistors, operating region is shifted from sub-threshold region to super-threshold region enhancing performance and improving tolerance to PVT variations. A clock distribution network using the proposed drivers shows an 89% reduction in 3σ clock skew value. A 0.4V test chip has been fabricated in a 0.18μm 6-metal CMOS process to demonstrate the effectiveness of the proposed scheme. Measurement results show 2.6X faster switching speed and 2.4X less delay sensitivity under temperature variations.

Categories and Subject Descriptors

B.7.1 **[Integrated Circuits]**: Types and Design Styles - *Microprocessors and microcomputers; VLSI (very large scale integration)*

General Terms

Performance, Design, Measurement

Keywords

Global interconnect, sub-threshold circuit, capacitive boosting, clock skew, variation tolerance

1. INTRODUCTION

With aggressive CMOS scaling, on-chip global interconnects have become the bottleneck for high-speed operation due to the increase in RC per length of minimum wires and near-constant die size. To mitigate the global interconnect delay problem, metal wires have been scaled in a selective fashion. The upper layer metals are remained thick and wide to reduce the wire resistance. As such, the wire pitch is not scaled as aggressively in these

Permission to make digital or hard copies of all or part of this work for personal or classroom use is granted without fee provided that copies are not made or distributed for profit or commercial advantage and that copies bear this notice and the full citation on the first page. To copy otherwise, or republish, to post on servers or to redistribute to lists, requires prior specific permission and/or a fee.
ISLPED'06, October 4–6, 2006, Tegernsee, Germany.
Copyright 2006 ACM 1-59593-462-6/06/0010...$5.00.

layers to maintain a low interwire capacitance. This ensures a low RC value for global signals and power networks. The lower layer metals on the other hand, are scaled at approximately the same rate as the devices for the local interconnects. Low-k inter-dielectric materials and copper wires have been deployed for a one time improvement in RC delay. The wire delay can be made proportional (instead of quadratic) to wire length using tapered wires and efficient buffer insertion techniques [1]-[3]. Despite the various process and circuit techniques for wire RC reduction, global wire delay will continue to become the performance limiter as the delay of logic and short interconnects continue to scale faster than that of global interconnects. Fig. 1 illustrates this trend where the proportion of global interconnect delay with respect to total system delay rapidly increases with technology scaling for a constant die size.

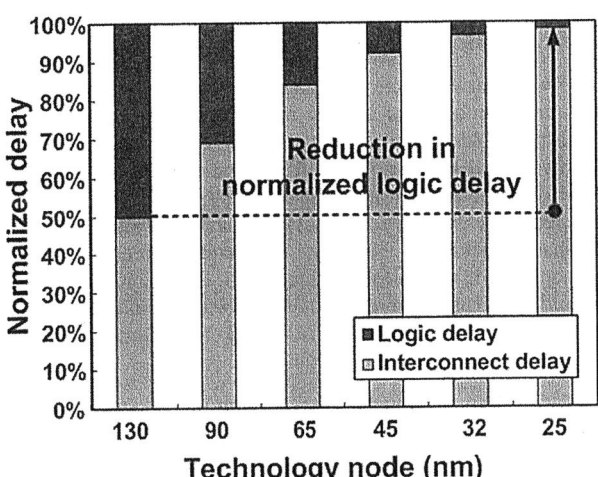

Figure 1. Scaling trend of logic delay and interconnect delay (source: International Technology Roadmap for Semiconductors [9]).

Interconnect delay becomes even more problematic in VLSI systems operating at low supply voltages. The extreme case for this is sub-threshold circuits where the supply voltage is lower than the threshold voltage. Sub-threshold operation can achieve orders of magnitude lower power consumption compared to conventional super-threshold operation and can be used in applications such as medical devices, portable electronics, and sensor networks where performance is of secondary importance [4]-[7]. By simply scaling down the supply voltage, undesirable characteristics of scaled CMOS, such as Drain Induced Barrier

Lowering (DIBL), quantum mechanical gate tunneling, and punch through can also be alleviated. In the sub-threshold region, the MOS gate capacitance is significantly less than that in the super-threshold region due to the channel depletion capacitance that appears in series with the oxide capacitance [8]. Unlike MOS gate capacitance, the wire capacitance value is independent of the supply voltage. As a result, the CV/I delay of wires increase more steeply than that of logic gates in sub-threshold circuits, exacerbating the global interconnect delay problem. Fig. 2 shows the logic delay in proportion to interconnect delay as the supply voltage is reduced from 1V to 0.2V. Evidently, the interconnect delay will dominate the overall system delay as the relative logic delay decreases at sub-threshold voltages due to the reduced MOS gate capacitance.

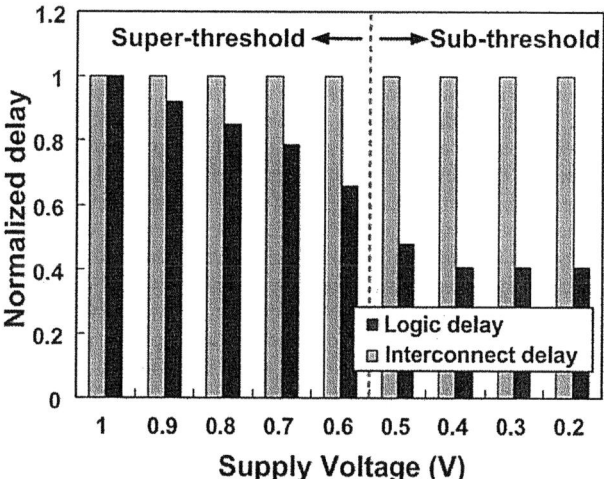

Figure 2. Simulation results on logic delay and interconnect delay at different supply voltages (0.18μm CMOS process).

With global wire delay dictating the overall system performance, its variation also has a larger impact on system performance in sub-threshold circuits. Due to the exponential relationship between weak-inversion current and the PVT parameters, performance and power dissipation of sub-threshold circuits are exceedingly sensitive to PVT fluctuations. Note that random dopant fluctuation becomes the main source of parameter fluctuation in sub-threshold since DIBL is much reduced [10]. To build efficient sub-threshold circuits with operating frequencies up to the MHz range, it is crucial to minimize the global interconnect delay and its variation.

In this paper, we propose a high-speed variation-tolerant interconnect technique for sub-threshold circuits based on capacitive boosting. The central idea is to shift the operating region of the final interconnect drivers from the sub-threshold to super-threshold region via bootstrapping techniques. Boosting techniques such as the one proposed in this work are extremely effective for improving interconnect performance in sub-threshold circuits since current is an exponential function of gate voltage. For example, a 100mV boost in gate voltage can offer a 10X increase in drive current. Circuit techniques to achieve high boosting efficiency, minimum leakage, and no start-up time constraints were deployed. The proposed scheme was applied to a clock distribution network to verify the effectiveness in reducing the clock skew. A 0.4V, 0.18μm test chip was successfully fabricated and tested. Measurement results show 2.6X higher switching speed and 2.4X reduced delay sensitivity with a 12% leakage power overhead at 0.4V.

2. PROPOSED SUB-THRESHOLD INTERCONNECT TECHNIQUE

2.1. Conceptual idea

Fig. 3 shows the principle of the proposed sub-threshold interconnect technique based on capacitive boosting. The 0.4V input signal is boosted to 0.8V for the NMOS driver and to -0.4V for the PMOS driver using internal gate capacitors. The supply voltage is 0.4V while the threshold voltages of the PMOS and NMOS are 0.51V and -0.51V, respectively. Owing to the exponential behavior of current in the sub-threshold region, 100X higher operating current can be achieved with a 0.4V boost in driver voltage as shown in Fig. 3. Note that for the same amount of voltage boost, the increase in drive current will be significantly less in super-threshold circuits. Hence voltage boosting offers greater speed benefits in the sub-threshold region where current is an exponential function of the gate-to-source voltage. The improvement in drive current comes at the expense of leakage current since the gate input to N5 (P5) must be preset to 0.4V (0V) before the boosting occurs. However, this has minimal impact on total chip power dissipation due to the limited number of speed-critical interconnects (buses and clock buffers). The leakage power can be further reduced by using minimum sized transistors P5 and N5 in Fig. 3 since the operating current is significantly increased via the boosting technique.

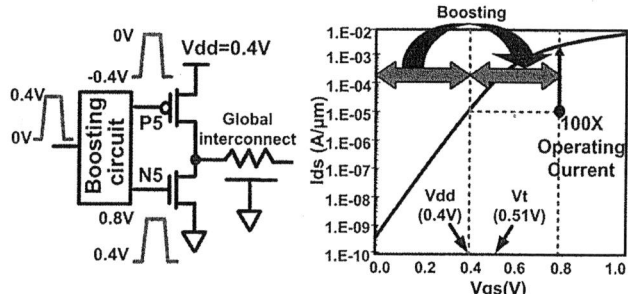

Figure 3. Concept of proposed sub-threshold interconnect driver using boosting technique (left). Effectiveness of boosting technique in sub-threshold region (right).

2.2. Circuit design

Circuit implementation and operating waveforms of the proposed interconnect driver operating at 0.4V are shown in Fig. 4. $V_{BOOST-N}$ and $V_{BOOST-P}$ are boosted by capacitors C_1 and C_2 to increase the operating currents of N5 and P5 which drive the long RC wire. Circuit operation for boosting the gate voltage of N5 is as follows (boosting operation of P5 is similar to that of N5). In order for V_{IN-BAR} to offset the voltage level of V_{BOOST_N} using C_1, V_{BOOST_N} is preset to 0.4V before V_{IN-BAR} makes the low-to-high transition. This is realized by P4 which connects $V_{BOOST-N}$ to 0.4V while V_{IN-BAR}=0V. After the low-to-high transition of V_{IN-BAR}, P4 is cut off so that the boosted voltage V_{BOOST_N}, stays at 0.7V while N5 is driving the RC interconnect. Due to the parasitic capacitance on the node V_{BOOST_N}, the boosting voltage does not reach the ideal 0.8V value. The Preset Signal Generator (PSG) circuit generates a $V_{PRESET-N}$ of -0.25V using C_4 during the preset of $V_{BOOST-N}$. This enables a fast preset by overdriving P4, minimizing the start-up time despite the low drive current. On the other hand, a V_{PRESET_N} of 0.7V is generated by the PSG circuit during the boosting operation by connecting V_{BOOST_N} to V_{PRESET_N} via P3. This eliminates the reverse current through P4 which can adversely discharge the boosted voltage.

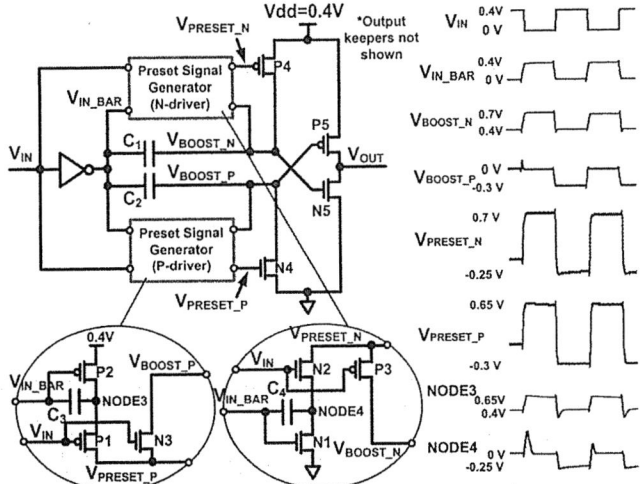

Figure 4. Circuit implementation and operation waveforms of the proposed interconnect driver.

2.3. Boosting efficiency in sub-threshold region

The boosting efficiency is defined in Fig. 5 as the ratio between the boosting capacitance (C_{boost}) and the total capacitance ($C_{boost}+C_{node}$) which consists of the boosting capacitance and the node capacitance. The boosting capacitance is implemented using a MOS capacitor and the node capacitance consists of the gate capacitances of P5, N5, P4, and N4, as well as the junction capacitances of all the other devices attached to the boosted node. To obtain a high drive current via efficient boosting, the boosting capacitance implemented using a MOS capacitor must be significantly larger than the node capacitance. Unfortunately, the boosting MOS capacitance reduces in the sub-threshold region since the depletion capacitance appears in series with the oxide capacitance in a weak-inversion device. Fig. 6 (left) shows the reduction in boost capacitance in the sub-threshold region. To achieve a high boosting efficiency in our design, 40% of the total driver area is dedicated to boosting capacitors. N5 and P5 which can make up 50% of the total node capacitance are also minimized. Fig. 6 (right) verifies the boosting efficiency of the proposed circuit at different voltages. The boosting efficiency is maximized at 0.6V and reduces to 59% at 0.3V mainly due to the reduction in boosting capacitance in the weak-inversion region. At 0.4V operation, boosting efficiency of 70% was achieved which is sufficient for a significant drive current boost.

	Boosting cap.	Node cap.
Pull-up	C_2	P5, N4, N3, P1
Pull-down	C_1	N5, P4, P3, N2

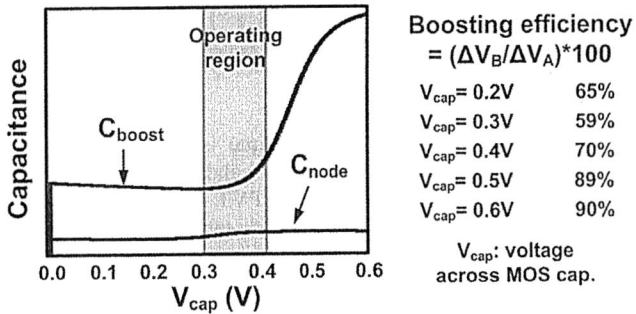

Figure 6. Boosting capacitance, node capacitance, and boosting efficiency at different voltages across MOS capacitors.

2.4. Sensitivity to PVT variation

In addition to the speed benefit, the proposed interconnect technique reduces the impact of PVT variation on global interconnect performance. Fig. 7 shows the rise delay and fall delay under supply (0.4V±5%) and temperature (0~40°C) variations. The conventional driver in the sub-threshold region is highly sensitive to voltage and temperature variation since the drive current is an exponential function of the PVT parameters. This results in a 167.5ns (161.2ns) variation in rise (fall) delay. The switching speed of the proposed interconnect varies significantly less even for the worst case corner conditions because the driver transistors are no longer in the sub-threshold region. Delay variation is reduced from 167.5ns to 50.2ns for the rise delay, and from 161.2ns to 45.1ns for the fall delay.

$$\Delta V_B = \frac{C_{boost}}{C_{boost}+C_{node}} * \Delta V_A$$

Boosting efficiency

Figure 5. Definition of boosting efficiency.

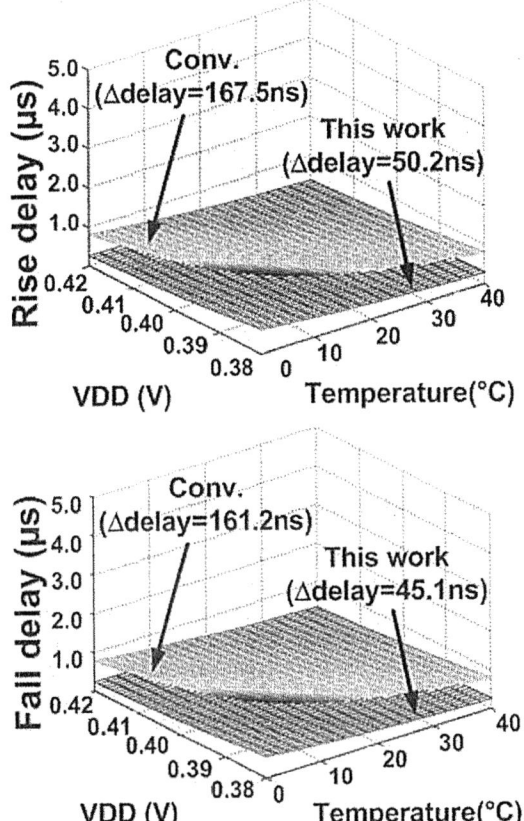

Figure 7. Rise and fall delay with respect to voltage and temperature variation.

69

3. CLOCK SKEW REDUCTION USING PROPOSED INTERCONNECT TECHNIQUE

The proposed interconnect technique is applied to a realistic clock distribution network [11] to validate the effectiveness in reducing the clock skew caused by PVT variations. Fig. 8 shows the simplified H-tree clock network topology where the clock signal paths are symmetrically routed across the chip for identical delays from clock source to the final load. The hierarchical topology combines four clock buffer stages with each buffer driving four clock buffers as well as the long interconnects. Clock skew is the signal arrival time difference between two different locations in a die. An ideal H-tree should be perfectly matched and will have zero clock skew without any within-die PVT variations. The worst case clock skew occurs when two clock paths are experiencing the extreme opposite PVT conditions. However, this would lead to unrealistically pessimistic estimates on clock skew [12, 13]. Hence in this work, we follow the clock skew distribution analysis based on Monte Carlo simulations assuming that the local supply voltage and threshold voltages are Gaussian random variables [11].

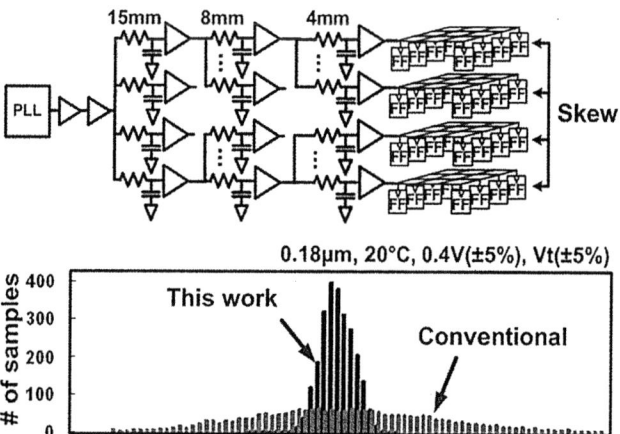

Figure 8. Sub-threshold clock network and clock skew distribution comparison.

Simulation results in Fig. 8 indicate a 9X reduction in 3σ clock skew value using the proposed driver. The 3σ value for VDD and Vt was assumed to be 5% of their nominal values. The average clock tree delay reduced from 251ns to 51ns and the standard deviation reduced from 30.7ns to 5.1ns.

4. TEST CHIP MEASUREMENTS

A test chip was fabricated in a 0.18μm 6-metal triple-well CMOS process to demonstrate the effectiveness of the proposed sub-threshold interconnect scheme. Threshold voltage of the NMOS and PMOS were 0.51V and -0.51V, respectively. The die photo of the test chip and the layout comparisons are shown in Fig. 9. Although the device count for the proposed driver has gone up to 18 (device count for conventional repeater is 4) the increase in layout area was only 32% since the transistors can be minimized thanks to the higher operating current using the proposed boosting technique. The 4 MOS capacitors for the boosting operation occupy 40% of the total driver layout area. Eight stages of conventional drivers and proposed drivers were implemented with each stage driving a 10mm long on-chip wire (Fig. 10). To improve the accuracy of the delay measurement, a differential

method was used to obtain the delay difference between the two paths and a bypass path. The peripheral and I/O circuit delay is cancelled out by subtracting the delay of the bypass path (t_{bypass}) from the delay of the other two paths ($t_{conv}+t_{bypass}$, $t_{proposed}+t_{bypass}$). The core, peripheral, and I/O circuits operate at 0.4V, 1.0V, and 1.8V respectively. Level converters were employed for the interface between sub-threshold and super-threshold circuits. The sub-threshold level-down converter contains a pull-up NMOS N6 to speed up the low-to-high transition in the internal node, A1. A dual-rail level-up converter was designed with an extra PMOS switch P7 (or P8) to reduce the contention current between P9 (or P10) and N7 (or N8).

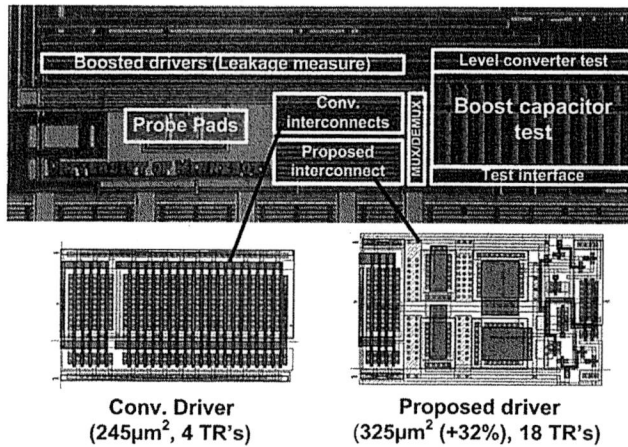

Figure 9. Chip microphotograph and driver layout comparison. The test chip was fabricated in a 0.18μm 6-metal CMOS process.

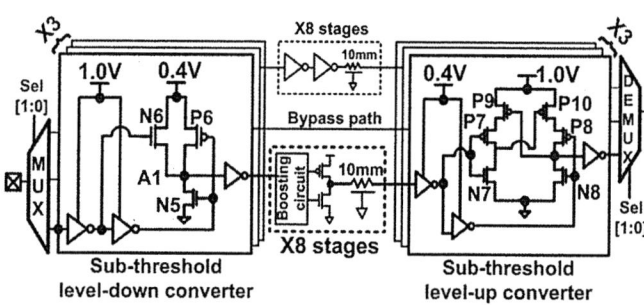

Figure 10. Organization of the test chip for differential delay measurements. Level converters are designed for the interface between sub-threshold and super-threshold circuits.

Fig. 11 shows the measured waveforms from the three different paths together with the input trigger signal. Delay of the proposed driver was 0.18μs which is 2.6X shorter than that of a conventional driver. Delay of the bypass path was 0.13μs which is comparable to the core interconnect delay. The delay improvement is less than the ideal amount of boost in operating current shown in Fig. 3 due to the following reasons: (i) minimum size driver transistors for reducing leakage power, (ii) extra number of logic stages required in the boosting circuit and (iii) reduced boosting effect because of the node capacitance. Fig. 12 shows measured delay improvements of the proposed interconnect technique at different core voltages. The delay improvement is 1.7-1.8X at 0.5V and 2.6-2.9X at 0.4V. This confirms that the proposed boosting technique becomes more efficient at low supply voltages due to the exponential current behavior in the sub-threshold regime.

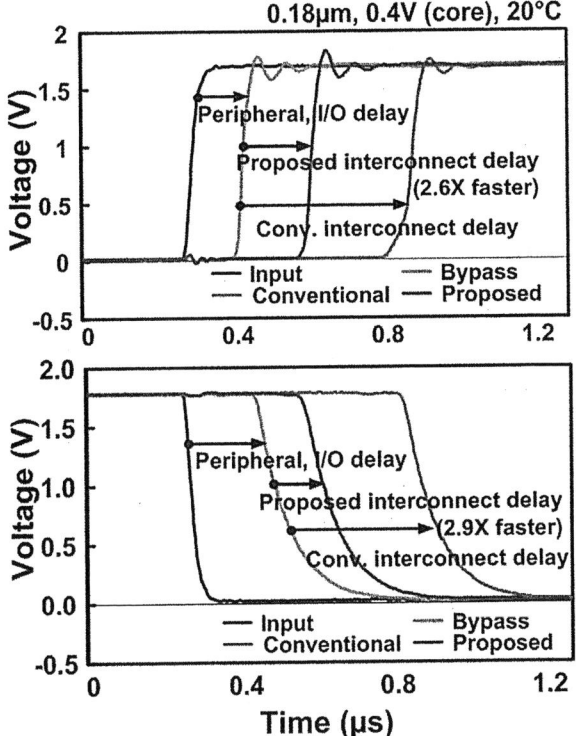

Figure 11. Measured waveforms (input trigger signal and output waveforms from the 3 interconnect paths) from the test chip.

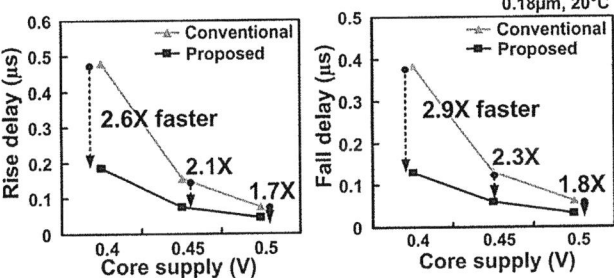

Figure 12. Rise and fall delay measurement data for different core voltages indicating 2.6-2.9X improvement in interconnect speed.

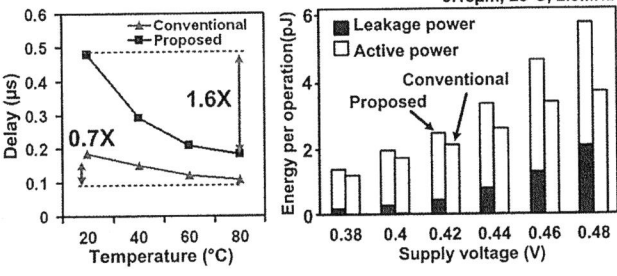

Figure 13. Measurement data from test chip. Delay variation with respect to temperature (left). Energy per operation comparison between conventional and proposed interconnect driver (right).

Fig. 13 shows the measured delay variation and power dissipation of the proposed and conventional interconnect drivers. The delay

variation of the conventional and proposed driver were 1.6X and 0.7X, respectively for a temperature range of 20-80°C. Delay of the proposed driver is less sensitive to PVT variations since the driver transistors are no longer operating in the sub-threshold region due to the boosted gate voltages. Measured energy per operation is shown in Fig. 13 (right) for the conventional and proposed drivers. Due to the leakage power during the preset of the gate voltages, energy per operation of the proposed driver increases by 12% when operated at 0.4V. The leakage current of the proposed boosting technique reduces exponentially in the deep sub-threshold region as shown in Fig. 13. Power dissipation comparison between conventional and proposed buffer is shown in Fig. 14 for different operating frequencies. Measurement results show good agreement with the simulation data. The proposed driver consumes 41% less power (or 49%+ higher performance) compared to the conventional driver for 4MHz (or 5.1μW) operation since the boosting technique allows the proposed driver to run at a lower supply voltage for the same frequency of operation.

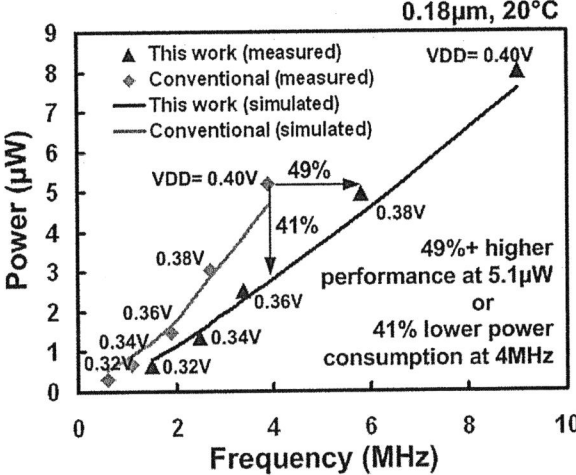

Figure 14. Power consumption versus frequency. (Measurement data is shown in dots and simulation data is shown in lines)

5. CONCLUSION

Digital sub-threshold logics are becoming increasingly popular for ultra-low power applications where performance is of secondary importance. Global interconnect drivers used for on-chip buses and clock distribution networks significantly suffer from performance degradation. This is because unlike gate capacitance, wire capacitance does not scale as the supply voltage is lowered. Another issue with sub-threshold interconnects is the large variability in performance under PVT variations; drive current in the sub-threshold region is an exponential function of threshold voltage, supply voltage, and temperature. In this paper, we proposed a capacitive boosting technique that can mitigate the performance and variability issues in sub-threshold interconnects. Owing to the exponential relationship between current and gate-to-source voltage in the sub-threshold region, 100mV boost in gate voltage can offer 10X improvement in drive current. This makes boosting techniques extremely effective for global interconnect drivers. A test chip was fabricated in a 0.18μm 6-metal CMOS process to demonstrate the proposed ideas. Measurement results show 41% less power consumption for same performance (or 49%+ higher performance for same power consumption) with 2.4X reduced delay sensitivity under temperature variations.

6. REFERENCES

[1] K. Yamashita and S. Odanaka, "Interconnect scaling scenario using a chip level interconnect model," *IEEE Trans. Electron devices,* vol. 47, no.1, pp. 90-96, Jan. 2000.

[2] D. Sylvester, C. Hu, O. S. Nakagawa, and S.-Y. Oh, "Interconnect scaling: Signal integrity and performance in future high-speed CMOS designs," *Symposium on VLSI Technology Dig. Tech. Papers*, pp. 42-43. Jun. 1998.

[3] S. Dhar and M. A. Franklin, "Optimum buffer circuits for driving long uniform lines," *IEEE Journal of Solid-State Circuits*, vol. 26, pp. 32-40, Jan. 1991.

[4] C.H. Kim, and K. Roy, "Ultra-low power DLMS adaptive filter for hearing aid applications," *IEEE Transactions on Very Large Scale Integration (VLSI) Systems*, vol.11, no.6, pp. 352-357, Dec. 2003.

[5] A. Wang, and A.P. Chandrakasan, "A 180mV FFT processor using subthreshold circuit techniques," *International Solid-State Circuits Conference*, pp. 292-293, Feb. 2004.

[6] B.H. Calhoun, and A.P. Chandrakasan, "Ultra-dynamic voltage scaling using sub-threshold operation and local voltage dithering in 90nm CMOS," *International Solid-State Circuits Conference*, pp. 300-302, Feb. 2005.

[7] B.H. Calhoun, A.Wang, and A.P. Chandrakasan, "Modeling and sizing for minimum energy operation in subthreshold circuits," *IEEE Journal of Solid-State Circuits,* vol. 40, no. 9, pp. 1778-1786, Sept. 2005.

[8] Y. Taur, and T. Ning, *Fundamentals of Modern VLSI Devices*, Cambridge University Press, 1998.

[9] International Technology Roadmap for Semiconductors, http://public.itrs.net/

[10] B. Zhai, S. Hanson, D. Blaauw, and D. Sylvester, "Analysis and mitigation of variability in subthreshold design," *International Symposium on Low Power Electronics and Design*, pp. 20-25, Aug. 2005.

[11] D. Harris and S. Naffziger, "Statistical Clock Skew Modeling With Data Delay Variations," *IEEE Transactions on Very Large Scale Integration (VLSI) Systems,* vol.9, no.6, pp. 888-898, Dec. 2001.

[12] M. Shoji, "Elimination of process-dependent clock skew in CMOS VLSI," *IEEE Journal of Solid-State Circuits,* vol.21, no.5, pp. 875-880, Oct. 1986.

[13] G. Geannopoulos, and X. Dai "An adaptive digital deskewing circuit for clock distribution networks," *International Solid-State Circuits Conference*, pp. 400-401, Feb. 1998.

A Dual-V$_{DD}$ Boosted Pulsed Bus Technique for Low Power and Low Leakage Operation

Harmander S. Deogun, Robert Senger, Dennis Sylvester, Richard Brown[†] and Kevin Nowka[‡]

Department of EECS, University of Michigan, Ann Arbor, MI, US 48109

[†]College of Engineering, University of Utah, Salt Lake City, UT, US 84112

[‡]Austin Research Laboratories, IBM Corporation, Austin, TX, US 78758

{hdeogun,rsenger,dmcs@umich.edu}, [†]{brown@coe.utah.edu}, [‡]{nowka@us.ibm.com}

ABSTRACT

In this paper, we propose a new dual-VDD bus technique that is well suited for low power operation. This technique adapts a static pulsed bus architecture to use dual-VDD power supplies. During quiescent periods, the bus system idles at the lower of the two VDD supplies, thereby lowering static power dissipation. When actively transitioning, the inverters in the bus system are temporarily boosted to the higher VDD supply to provide the needed drive strength for performance. Since the VDD boosting is done in a pulsed manner, the bus system is in a high VDD state only when required, ensuring lower power operation without sacrificing performance. This technique yields up to a 50% reduction in total power over traditional static buses and up to a 35% reduction in total power over standard static pulsed buses, with a 12-15% delay improvement.

Categories and Subject Descriptors

B.7.1 [**Hardware**]: Types and Design Styles.

General Terms

Performance and Design.

Keywords

Dual-VDD, repeaters, pulsed bus, leakage.

1. INTRODUCTION

The International Technology Roadmap for Semiconductors lists the management of overall power consumption as one of the major challenges facing the semiconductor industry [1]. Minimizing power consumption is a problem that has been, and will continue to be, attacked on a variety of design fronts. Of particular interest is the disproportionate amount of power required by communication buses that propagate signals across

This work was partially supported by the NSF and SRC.

Permission to make digital or hard copies of all or part of this work for personal or classroom use is granted without fee provided that copies are not made or distributed for profit or commercial advantage and that copies bear this notice and the full citation on the first page. To copy otherwise, or republish, to post on servers or to redistribute to lists, requires prior specific permission and/or a fee.
ISLPED '06, October 4–6, 2006, Tegernsee, Germany.
Copyright 2006 ACM 1-59593-462-6/06/0010...$5.00.

increasingly large and complex microprocessor chips. These buses have to propagate signals over longer distances and at higher frequencies for each new generation. A first-order analysis in [2] projects that the fraction of total cells in a functional logic block used for repeaters will grow from 6% at the 90nm node to 70% at the 32nm node. This rapid increase in repeaters, which are generally sized aggressively to maintain delay and signal skew, dramatically increases the total power of the integrated circuit. Bus design techniques that reduce both dynamic and static power will be a critical design requirement as technology scaling continues and on-chip communication consumes a larger percentage of the total system power.

Recently, there has been significant work focusing on reducing leakage power. In our work, in addition to leakage power, we aim to reduce dynamic switching power, which still comprises the majority of the total power dissipation. An effective method for reducing both static and dynamic power has been the use of dual voltage supplies (dual-VDD). In general, a lower supply voltage can be used for gates on non-timing critical paths while the higher VDD is used on timing critical paths.

Several dual-VDD approaches have been proposed and implemented for power reduction in generic circuit blocks [3-8]. In these block-level approaches, the critical path portion of the circuit typically runs on the nominal VDD (which we will refer to as VDD_H) and the non-critical portion runs on the reduced VDD (VDD_L) to conserve dynamic power. There have been a variety of enhancements [9, 10] to the original clustered voltage scaling (CVS) algorithm [3] that improve upon the assignment of VDD_H and VDD_L gates in an effort to save additional power. All of these approaches are limited by the constraint that gates powered by VDD_L can never directly drive gates powered by VDD_H unless a level converter is inserted. The level converter is required to raise the VDD_L output signal up to VDD_H to avoid the large static current that would be drawn through the PMOS transistors of downstream VDD_H gates if they were to be driven by a VDD_L input. However, inserting level converters costs power, delay and area which limits the effectiveness of dual-VDD in general.

A dual-VDD bus buffer was introduced in [11] that reduces delay and power compared to standard static bus buffers powered by VDD_H. The primary idea behind this approach was to power the two inverters comprising the buffer with VDD_L and use the VDD_H supply temporarily to speed up high going output transitions. This dual-VDD bus buffer yielded up to a 17%

energy improvement and a 22% performance improvement over a standard bus operating at VDD_L. Our focus in this work is on providing much larger dynamic and static power improvements.

In this work, we implement a dual-VDD static pulsed bus design that employs the second power supply to reduce total power dissipation (both static and dynamic) without sacrificing performance. The static pulsed bus technique, first proposed in [12], eliminates worst-case coupling capacitance in buses by preventing opposite phase transitions on adjacent bus lines. The additional second power supply incurs cost, area, and circuit complexity overhead that will be justified by the significant improvements in power and performance shown later.

The remainder of this paper is organized as follows. Section 2 reviews static pulsed bus design and introduces our dual-VDD boosted pulsed bus technique as well as analyzing its design constraints. Section 3 presents detailed simulation results in a 65nm SOI technology and Section 4 concludes the paper.

2. DUAL-VDD BOOSTED PULSED BUS

2.1 Static Pulsed Bus Design

A static pulsed bus [12] generates a pulse that is propagated down the bus line every time a transition is detected on the bus input. The pulses on all of the bus lines are of the same polarity (e.g. low-high-low), thereby eliminating the worst-case crosstalk by preventing adjacent lines from switching in opposite directions. In a static pulsed bus, bus lines only pulse in one direction regardless of the input transition and the pulses are decoded at the receiving end. Pulsed operation improves the total power and performance of the bus system by reducing capacitive coupling. Furthermore, by capitalizing on the fact that only the leading edge of the pulse is performance critical, inverter beta ratios can be skewed to favor single transitions. This enables more power savings via smaller devices. These two benefits outweigh the added switching activities on the lines to yield overall lower power.

2.2 Dual-VDD Boosted Pulsed Bus Design

Our dual-VDD pulsed bus design modifies the static pulsed bus technique to improve total power consumption with the addition of the second power supply rail. Similar to the static pulsed bus, the dual-VDD boosted pulsed bus operates by sensing a transition at the bus input and subsequently sending a pulse

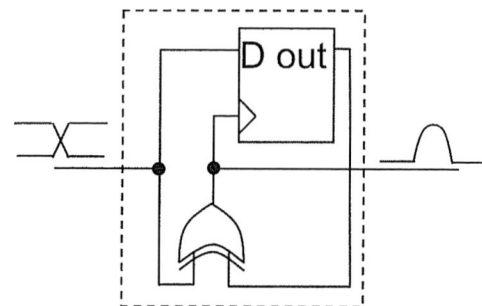

Figure 2: Pulse generation logic.

down the wire instead of the actual transition itself. The pulse at the input end of the bus line is generated by a pulse generation (PG) module that sends a low-high-low pulse for every high-low or low-high input transition. This PG module improves upon the pulse generation module from [12] by using the inherent delays in the latch and XOR gate to create an appropriately sized pulse, rather than relying on a long string of inverters. Many inverters need to be strung together in heavily scaled technologies (sub-90nm) to create a suitable pulse width, which in turn increases power consumption due to leakage and dynamic switching in the inverter chains.

Alternating stages (even stages) of the bus line have two supply voltages that are switched by the pulse. This pulse is propagated down the bus line until it is received and decoded by a toggle flip-flop (TFF) as was shown in [12]. The bus line connects to the CLK input of the toggle flip-flop whereby every pulse causes a switch in the output of the toggle flip-flop to capture the actual value of the input state. As mentioned above, since switching activity has increased, we leverage reduced effective coupling capacitance and downsized non-critical devices to yield power savings. In our circuit, the lower VDD becomes a third (very powerful) knob to compensate for the higher activity levels.

Figure 1 is a block diagram of the dual-VDD pulsed bus setup. Alternating inverter stages, starting with the second stage, have a multiplexed local VDD supply that powers the bus repeater with VDD_L during quiescent states and boosts the repeater to VDD_H during active transitions. The boosting is triggered by the pulsed data on the bus line itself such that no extra signaling or bus wiring is required. The power supply boost occurs only for the even half of the bus line repeaters since the odd stages drive an inverted pulse from VDD_L to 0 V to VDD_L and thus

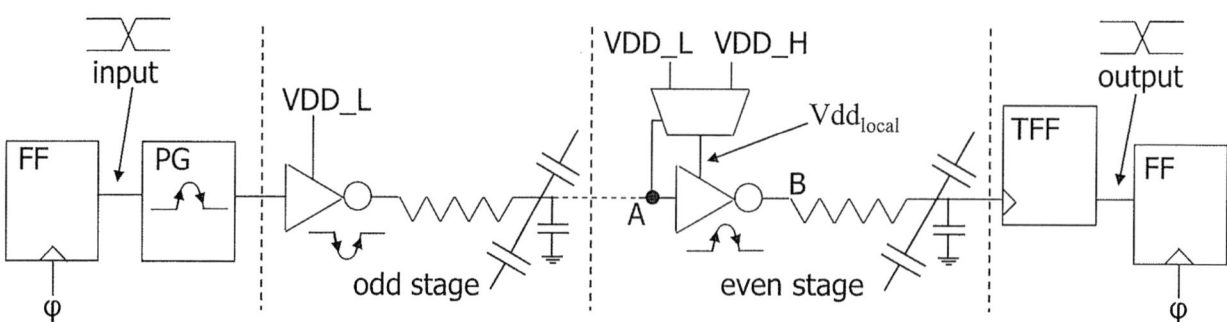

Figure 1: Block diagram of dual-VDD boosted pulsed bus.

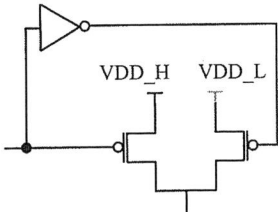

Figure 3: Power MUX logic.

do not need the VDD_H boost. To save power, the odd stage repeaters are statically powered by VDD_L. The pulse generation logic (PG) is shown in Figure 2.

The MUX that boosts the even stage power supply for drive strength is controlled by the data input of the boosted inverter. Figure 3 illustrates the circuit schematic of the power MUX used to switch between VDD_H and VDD_L. It consists of two PMOS transistors and a single inverter. During boosted operation, the input to the MUX is low, turning ON the PMOS transistor connected to VDD_H and turning OFF the PMOS connected to VDD_L. During quiescent operation, the reverse is true. Since we do not incorporate a level converter in the MUX (to reduce area, power and to increase MUX switching speed), there will be some static current draw from the weakly OFF PMOS transistor connected to VDD_H. This static current supplements the current drawn through the ON PMOS transistor connected to VDD_L. As detailed in Section 4, our results show a total power savings with this power MUX design.

The node connecting the input of the even stage inverter (Node A in Figure 1) with the boosting MUX select line pulses from VDD_L to 0V and then back to VDD_L. The output node of this repeater (Node B in Figure 1) pulses from 0V to VDD_H (for drive strength) and then back to 0V. This behavior of a boosted even-stage in the middle of the bus line is shown in Figure 4 (the nodes referred to are those labeled in Figure 1).

Since alternating stages of the bus line always pulse data in the same direction, the repeaters can be skewed to speed up the leading edge of the pulse, similar to the original static pulsed bus implementation. This also allows us to reduce the total width of the repeaters per stage. Another effect of skewing the repeaters is that the pulse will widen as it propagates down the bus line. In this case, the repeaters need to be sized such that the pulse does not widen to the point where it causes a logical error. This constraint is further discussed in Section 2.4.

2.3 Second Power Supply Voltage

The value of a second, lower power supply voltage is typically 0.7×VDD_H [6,8]. For our boosted pulsed bus system, the nominal VDD_H was held constant at 1V while the second supply, VDD_L, was swept from 0.9V to 0.5V. At each VDD_L voltage point, the power, delay, and pulse width of the system was measured and plotted in Figure 5 (normalized to the 0.9V case). Note that the inverters were sized for a VDD_L supply voltage of 0.7V and were not resized as we swept VDD_L. The number above each pair of VDD_L points is the pulse width of a typical pulse at that supply voltage in picoseconds. We set the clock to 2 GHz and used a switching activity of 0.1 for the input data on the bus line. The results in Figure 5 are somewhat

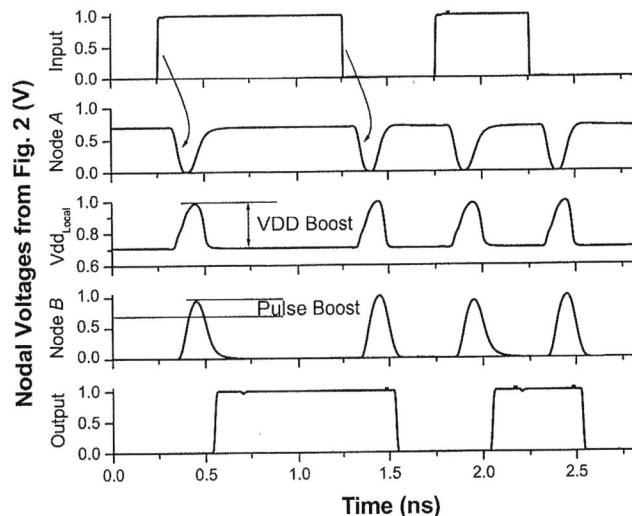

Figure 4: Waveform depicting local V$_{DD}$ boosting at a typical even stage.

counterintuitive since with lowering supply voltage the delay is reduced. Additionally, the 0.5V data point does not follow the trend of lower power as expected, instead, it increases in power.

We first discuss the delay trend. In general, we would expect to see increasing delay with decreasing supply voltage. However, in the case where the delay is dominated by capacitive coupling, the trend is reversed. For a static pulsed bus, there can never be a situation where adjacent lines to a bus line switch in opposite directions. This means that there can never be an effective doubling of coupling capacitance. Therefore worst-case performance is significantly improved over a traditional bus.

Furthermore, the amount of charge stored in the wire capacitance is also related to the voltage at which the wire is held. With a dual-VDD bus system, the bus wire can be maintained at the lower supply voltage for the majority of the time. Even stages still switch at VDD_H to provide the same current drive to the odd stages as in the traditional and static pulsed bus cases. Since the odd stages are powered by VDD_L, but driven by a VDD_H pulse, the PMOS is super cut-off (it has a source voltage of VDD_L and a gate voltage of VDD_H) and the NMOS is ON with less contention from a super cut-off PMOS. Thus performance is improved because the NMOS can pull down more quickly.

Additionally, from Equation 1 below we see that a reduction in V$_{Swing}$ leads to a linear reduction in delay (t_D). Since V$_{Swing}$ in the odd stages is reduced by using a smaller second supply voltage, we see a near linear reduction in the delay (Figure 5), which correlates well with Equation 1. Reduced voltage swing, coupled with constant drive current, leads to a further decrease in delay compared to a single VDD static pulsed bus.

$$t_D = \frac{C_L V_{Swing}}{I_d} \qquad (1)$$

As VDD_L is lowered, it is generally expected that power should be reduced. This trend is confirmed in Figure 5 for a VDD_L of 0.9V to 0.6V. The reduced power is primarily due to two mechanisms. First, the effective coupling capacitance is reduced as discussed previously. Second, the odd stages have a reduced voltage swing of VDD_L to 0 V. In the case of the 0.5V second power supply, the power is dramatically increased. This can be explained by noticing that the pulse width increases as VDD_L is reduced since VDD_L has a harder time pulling up the RC dominated bus line. As the pulse width increases, even stages of the bus system spend more time at the VDD_H level and thus there is more power consumption.

Considering Figure 5, we select the VDD_L power supply to be at 0.7V when reporting results later. Although 0.6V yields better power and performance, it also represents a less stable design point since it is near a very sharp inflection and will be sensitive to noise and/or process variation. Choosing VDD_L to be 0.7×VDD_H is also consistent with previously published reports [6, 8, 11].

2.4 Design Constraints

In static pulsed bus design, repeaters are skewed to favor the leading edge of the pulse in order to speed propagation. Since this widens the pulse as it travels down the bus line, inverter skewing and width reduction cannot be performed arbitrarily. If the pulse grows too wide, then the trailing edge of the pulse might collide with the leading edge of the next pulse, causing erroneous data transmission. Furthermore, the pulse cannot be too narrow or else it will either attenuate completely or fail to meet the setup time (T_{Setup}) or hold time (T_{Hold}) of the receiving toggle flip-flop.

To handle this, we must impose constraints on the minimum and maximum pulse width through proper sizing of the repeaters. These timing constraints are described by Equations 2 and 3, where T_T represents the arrival time at the toggle flip-flop for the pulse's trailing edge and T_L represents the arrival time (at the toggle flip-flop) for the pulse's leading edge.

Figure 6 is the timing diagram detailing the leading and trailing edges for the pulse output at the final stage of the bus. The pulse is widest at this point due to having propagated through all of the skewed inverters. It is at this final stage where Equations 2 and 3 need to be satisfied.

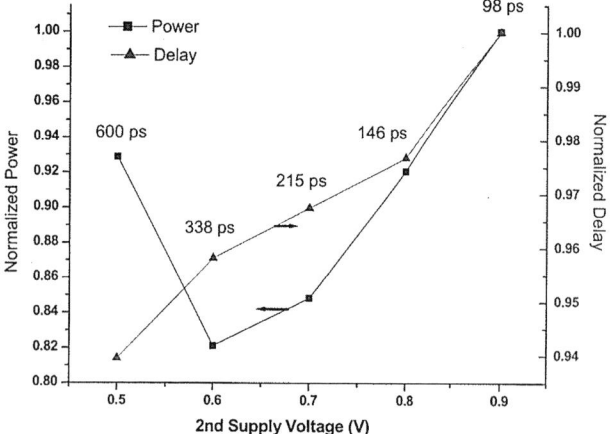

Figure 5: Power, delay and typical pulse width (number above data point) as a function of second supply voltage, normalized to 0.9V.

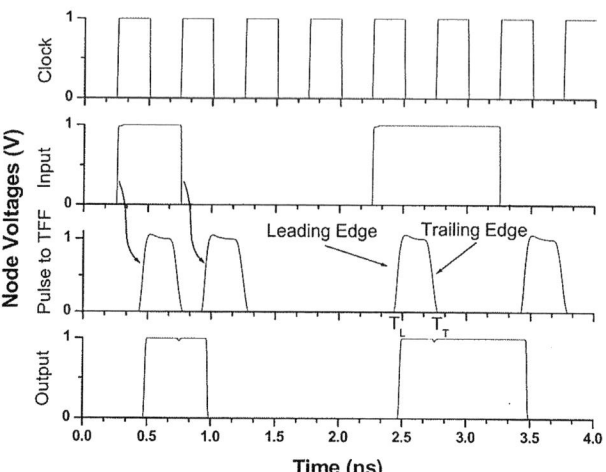

Figure 6: Timing waveform depicting leading and trailing edges.

$$T_T - T_L > T_{Setup} + T_{Hold} \qquad (2)$$

$$T_T - T_L < T_{CLK} - T_{Setup} - T_{Hold} \qquad (3)$$

3. RESULTS

The dual-VDD boosted pulsed bus was set up and simulated in an industrial 65nm SOI technology at 85°C. Since an SOI technology was used, the body was left floating to minimize area and increase performance. In the case of a bulk technology, the body of the PMOS devices in even stages would be tied to VDD_H since dynamically raising or lowering the body bias would be a slow and power consuming process. The PMOS body for odd stage inverters would be statically tied to VDD_L.

This new technique was compared to both the static pulsed bus technique and traditional static bus design. The bus setup consisted of eight inverters placed every 1mm along an 8mm bus line to emulate on-chip bus structures. The traditional bus was sized by sweeping the repeater widths and finding the best power-delay point. Repeater sizes for our dual-VDD boosted pulsed bus and the original static pulsed bus were obtained by manually skewing the optimized traditional bus widths to minimize delay. We simulated a 3-bit bus and tested delay in the middle wire to capture coupling effects.

Results in Table 1 show that the new dual-VDD boosted pulsed bus technique reduces total power by up to 50% over a traditional static bus and by up to 35% over a static pulsed bus. The dynamic power consumption is much greater than static power consumption since the aggregate number of gates is small. The large bus inverters dissipate significant dynamic power during active switching as a result of having to drive a large RC load (i.e., the bus wire). The aforementioned power savings come *in addition to a 12-15% performance improvement*. The large gains in both power and delay point to significant benefits from exploiting multiple supply voltages on-chip in a range of circuit fabrics (in this case communication structures) rather than solely in combinational logic blocks as have been previously explored. The cost here lies in design complexity and must be carefully weighed against substantially lower power and improved delays in buses.

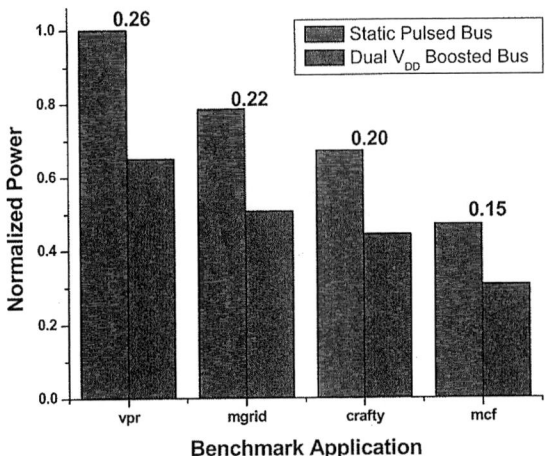

Figure 7: Total power comparison for dual-VDD boosted pulsed bus and a static pulsed bus for various benchmark applications.

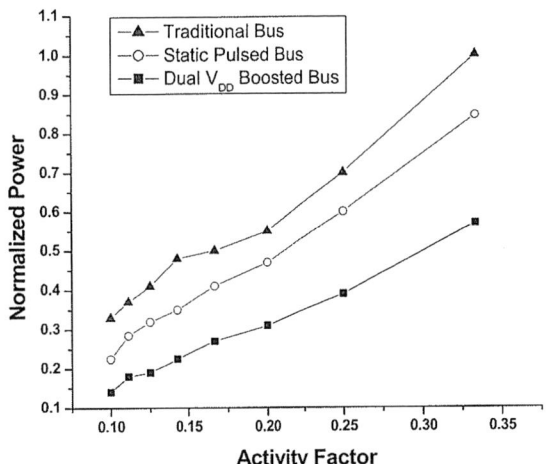

Figure 8: Activity factor versus normalized total power for three bus design techniques.

We obtained representative bus activity factors by simulating 100,000 cycles of different benchmark applications and applied these activity factors to each of the bus designs. The varying activities were derived from bus traces of a 64-bit ALPHA architecture processor executing a subset of benchmark applications. The results show that with varying activities between 0.15 and 0.26, our dual-VDD boosted pulsed bus achieves an average total power savings of 35% over a static pulsed bus. Figure 7 summarizes this data, normalized to the static pulsed bus. In this figure, the number on top of each column is the average activity factor for each of the respective benchmark applications.

To further investigate how the power of each bus design technique varies with switching activity, we plot the normalized total power against the activity factor. This plot is presented in Figure 8 where it can be seen that for varying activity factors, our dual-VDD boosted pulsed bus reduces total power between 32% and 38% over a static pulsed bus, and between 38% and 51% over a traditional static bus.

We also examined the maximum instantaneous current during switching for each of the three bus design techniques. We found that our dual-VDD boosted pulsed bus method reduced the maximum instantaneous current during active switching by up to 64% over a static pulsed bus and by up to 60% over a traditional static bus. A reduction in instantaneous current eases power grid design since the grid does not need to support large spikes of

current. Moreover, smaller power supply currents will lead to less circuit timing impact. Figure 9 is a waveform showing the currents of each of these bus design techniques during a period of active switching.

We compare these results to first-order theoretical projections by calculating the total power dissipation using Equation 4,

$$P = C_L V_{DD}^2 \, \alpha f + t_{SC} V_{DD} I_{switch} \, \alpha f + V_{DD} I_{leakage} \qquad (4)$$

where C_L is the load capacitance, f is the clock frequency, α is the switching activity and t_{SC} is the time the circuit spends in short circuit (switching time). We normalize all values to the static pulsed bus case and assume that f, α, and t_{SC} are constant for both designs and that C_L increases for the dual-VDD boosted design (due to the addition of the MUX hardware). We break Equation 4 into two parts for the dual-VDD boosted pulsed bus to model the different powers dissipated in the even and odd stages and then sum them for total power (Equation 5). We substitute in the appropriate values of the variables in Equation 4 for the dual-VDD boosted pulsed bus (P_{dual}) and compare it to the ordinary static pulsed bus (P_{spb}). The result from Equation 6 shows a theoretical 39% improvement in total power, very close to our simulated result of 35%.

$$P_{dual} = P_{odd} + P_{even} \qquad (5)$$

$$1 - \frac{P_{dual}}{P_{spb}} = 0.39 \qquad (6)$$

Table 1: Summary of delay, total dynamic and static power consumption and ratio of dynamic to static power for each of the various bus configurations in an industrial 65nm SOI technology with a 1V nominal supply voltage, (0.7V lower supply for dual-VDD), 85°C operating temperature and typical process corner. Results are normalized to the traditional static bus case.

Bus Design Technique	Dynamic Power	Static Power	Delay	Dynamic/Static Power Ratio
Traditional Static Bus	1.00	1.00	1.00	11.1 / 1
Static Pulsed Bus	0.76	0.78	0.97	10.9 / 1
Dual-VDD Boosted Pulsed Bus	0.51	0.28	0.85	20.4 / 1

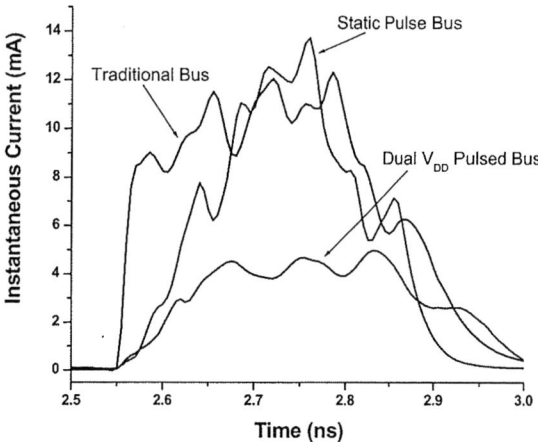

Figure 9: Maximum instantaneous current through VDD_H during active switching for each of the discussed bus design techniques.

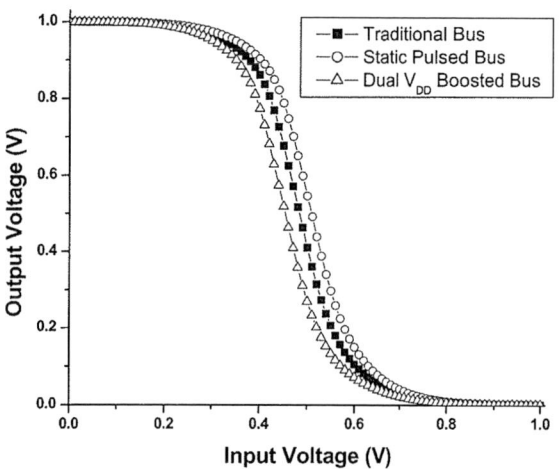

Figure 10: Bus inverter transfer curve for each bus design technique.

Finally, we plot the voltage transfer characteristics of the repeaters for the traditional static bus and compare them to those of a static pulsed bus and our dual-VDD boosted pulsed bus. Figure 10 is a plot of output voltage versus input voltage for all three of the bus designs (even stage for the dual-VDD boosted pulsed bus). The noise margins (defined using a unity gain model) of the dual-VDD bus are within 20-30 mV of the traditional static bus margins.

4. CONCLUSIONS

We have introduced a novel dual-VDD boosted pulsed bus technique for total power reduction. Using this technique, total bus power can be reduced by up to 50% over a traditional bus and by up to 35% over a static pulsed bus. The power reductions come with a performance improvement of over 12% due primarily to reduction in the effective coupling capacitance and reduced signal swing in the odd stages. We have shown that this technique maintains noise margins very close to those found in a traditional bus design. Benchmark applications were utilized to justify savings at realistic activity factors. We showed that the achievable savings are consistent through a wide range of data

switching rates. Finally, the simulation results were verified against a first order theoretical power model of our bus. The dual-VDD boosted pulsed bus can be a useful alternative to other bus design techniques for total power reduction while improving performance.

5. REFERENCES

[1] International Technology Roadmap for Semiconductors, *http://public.itrs.net,* 2005.

[2] P. Saxena, "The scaling challenge: can correct-by-construction design help?" *International Symposium on Physical Design,* 2002.

[3] K. Usami and M. Horowitz, "Clustered Voltage Scaling Technique for Low-Power Design," *Proc. International Symposium on Low Power Electronics and Design,* pp. 3-8, 1995.

[4] C. Chen et. al., "On Gate Level Power Optimization using Dual Supply Voltages," *IEEE Transactions on VLSI,* vol. 9, pp. 616-629, Oct. 2001.

[5] K. Usami et. al., "Automated Low-Power Technique Exploiting Multiple Supply Voltages Applied to a Media Processor," *IEEE Journal of Solid Static Circuits,* pp. 463-472, Mar. 1998.

[6] M. Hamada, Y. Ootaguro and T. Kuroda, "Utilizing Surplus Timing for Power Reduction," *Proc. Custom Integrated Circuits Conference,* pp. 89-92, 2001.

[7] M. Takahashi et. al., "A 60-mW MPEG4 Video Codec Using Clustered Voltage Scaling with Variable Supply-Voltage Scheme," *IEEE Journal of Solid State Circuits,* pp. 1772-1780, Nov. 1998.

[8] T. Kuroda and M. Hamada, "Low-Power CMOS Digital Design with Dual Embedded Adaptive Power Supplies," *IEEE Journal of Solid State Circuits,* pp. 652-655, Apr. 2000.

[9] S. Kulkarni, A. Srivastava and D. Sylvester, "A New Algorithm for Improved VDD Assignment in Low Power Dual VDD Systems," *Proc. International Symposium on Low Power Electronics and Design,* pp. 200-205, 2004.

[10] D. Chinnery and K. Keutzer, "Linear Programming for Sizing, Vth and Vdd Assignment," *International Symposium on Low Power Electronic Design,* pp 149-154, 2005.

[11] H. Kaul and D. Sylvester, "A Novel Buffer Circuit for Energy Efficient Signaling in Dual-VDD Systems," *Great Lakes Symposium on VLSI,* pp. 462-467, 2005.

[12] M. Khellah et al., "Static Pulsed Bus for On-Chip Interconnects," *Symposium on VLSI Circuits Digest. of Technical Papers,* pp. 78-79, 2002.

Time-Borrowing Multi-Cycle On-Chip Interconnects for Delay Variation Tolerance

Keith Bowman, James Tschanz, Muhammad Khellah, Maged Ghoneima[†], Yehea Ismail[†], and Vivek De

Circuit Research Lab, Intel Corporation, Hillsboro, OR
[†]Electrical Engineering Department, Northwestern University, Evanston, IL
keith.a.bowman@intel.com

Abstract

Insertion of time-borrowing (TB) flip-flops in multi-cycle repeater-based on-chip interconnects enables significant improvements in mean performance and energy by averaging systematic and random within-die (WID) delay variations across multiple interconnect segments. A statistically-based analytical model is derived to design a TB N-cycle interconnect with optimal delay variation tolerance. The model elucidates the dependency of the transparency window required to achieve data delay averaging on the delay variation mismatch between interconnect segments. Statistical circuit simulations and analyses in a 65nm process technology demonstrate that TB multi-cycle interconnects enable a 4-6% mean maximum clock frequency (FMAX) improvement and a corresponding 10% average energy savings over optimally designed multi-cycle interconnects with conventional master-slave flip-flops. The maximum mean FMAX benefit ranges from 4.0-7.5%, corresponding to approximately a bin-split shift in the FMAX distribution. For 1.41X larger WID delay variations, the maximum mean FMAX gain rises to 5-10%.

Categories & Subject Description

B.7.1 [Integrated Circuits]: Types and Design Styles
B.8.0 [Performance and Reliability]: General

General Terms: Algorithm, design, performance, reliability

Keywords: Parameter fluctuations, parameter variations, within-die variations, intra-die variations, variation tolerant, time borrowing, interconnect, multi-cycle interconnect

1. Introduction

Within-die (WID) parameter variations adversely impact the mean of the maximum clock frequency (FMAX) distribution for microprocessors [1]. As process technology continues scaling toward smaller dimensions, WID parameter variations are expected to worsen [2]-[3], and consequently, further degrade the relative mean FMAX, thus shifting product FMAX distributions toward relatively slower clock frequencies.

In microprocessor designs, numerous signals are routed over long repeater-based interconnects, requiring these signals to be propagated across multiple clock cycles. For today's microprocessors, a large number of signals require 2 to 6 clock cycles. As technology scaling continues, transistor switching delay reduces while interconnect latency increases [4], [5]. Furthermore, the clock frequency continues to increase as technology scales

Permission to make digital or hard copies of all or part of this work for personal or classroom use is granted without fee provided that copies are not made or distributed for profit or commercial advantage and that copies bear this notice and the full citation on the first page. To copy otherwise, or republish, to post on servers or to redistribute to lists, requires prior specific permission and/or a fee.
ISLPED'06, October 4–6, 2006, Tegernsee, Germany.
Copyright 2006 ACM 1-59593-462-6/06/0010...$5.00.

while the number of long-distance interconnects remains relatively unchanged [6]. The combination of these technology scaling trends reduces the maximum distance that a binary signal can propagate within a clock cycle, and consequently, increases the number of clock cycles required to transmit a binary signal at a fixed length [7].

In designing multi-cycle interconnects that span long distances, flip-flops are generally desired over a latch-based design with non-overlapping clock phases due to the difficulty in generating a consistent 50% duty cycle for latches as compared to just maintaining a stable clock edge for flip-flops. In addition, interconnect segments require repartitioning for a latch-based design [8], which negatively impacts the interconnect segment optimization.

Traditionally, time-borrowing (TB) flip-flops have been used to reduce the effect of clock skew and jitter on nominal FMAX [8]-[11]. In this paper, the transparency window of TB flip-flops in multi-cycle on-chip interconnects is expanded to alleviate the mean FMAX degradation caused by WID parameter variations. Since the differences in max-delay and min-delay for repeater-based on-chip interconnects are typically much smaller than those in logic blocks, the transparency window can be a relatively large fraction of the cycle time without creating min-delay failure problems. This allows effective usage of TB flip-flops in multi-cycle interconnects to average delay variations across multiple segments, thus mitigating the impact of systematic and random WID variations on mean FMAX as dictated by interconnect delays. Moreover, the absorption of clock skew and jitter by the transparency windows provides additional slack in the nominal cycle time in comparison to a multi-cycle interconnect with conventional master-slave flip-flops, where TB flip-flop and repeater transistors can be downsized to reduce energy. The performance and energy advantages of TB interconnects over conventional non-time borrowing interconnects will become larger with technology scaling as: (1) Cycle time reduces faster than interconnect segment delay, (2) Number of cycles required to propagate a signal increases, and (3) WID variations escalate.

2. TB N-Cycle Interconnect

Conventional master-slave flip-flops, which are referred throughout this paper as non-time borrowing (NTB) flip-flops, are commonly applied in multi-cycle interconnects. In Fig. 1, an NTB flip-flop circuit schematic is presented, and the application of this flip-flop into an N-cycle interconnect is described in Fig. 2. The N-cycle interconnect consists of N repeater-based interconnect segments separated by NTB flip-flops, where $t_{DATA(i)}$ denotes the delay of the ith interconnect segment, and $t_{CK(1)}$ represents the single-cycle clock skew and jitter. Each interconnect segment is composed of a driving flip-flop, a number of repeater segments, a protection inverter, and a receiving flip-flop. As depicted in Fig. 2, NTB flip-flops enforce a hard clock edge between adjacent

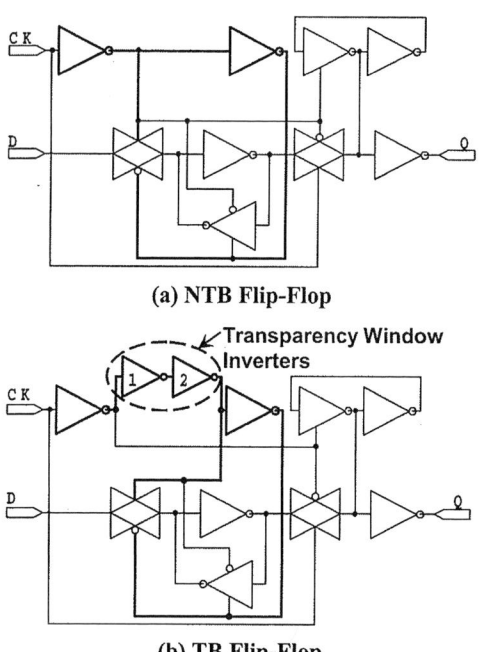

(a) NTB Flip-Flop

(b) TB Flip-Flop

Fig 1. (a) Non-time borrowing (NTB) and (b) time-borrowing (TB) flip-flops.

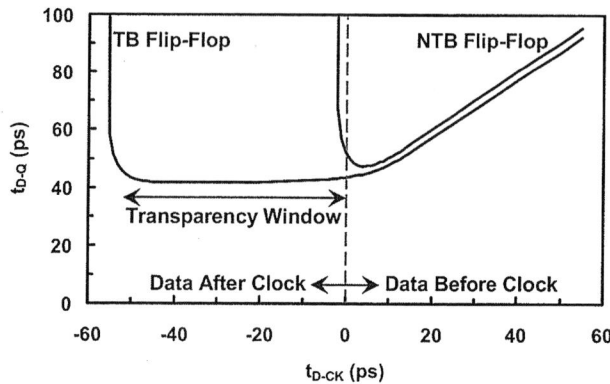

Fig. 3. Data-to-output delay (t_{D-Q}) versus data-to-clock delay (t_{D-CK}) for NTB and TB flip-flops.

interconnect segments, where data is allowed to pass through the flip-flop only when the data arrives before the setup time. For an NTB interconnect segment, $t_{DATA(i)}$ equals the sum of the flip-flop clock-to-output delay, the delay through the repeater segments and the protection inverter, and the setup time for the receiving flip-flop. The cycle time of one NTB interconnect segment is the sum of $t_{DATA(i)}$ and $t_{CK(1)}$. For an N-cycle interconnect, the slowest interconnect segment delay limits the cycle time ($t_{CYCLE}=1/FMAX$), resulting in more severe degradation in mean FMAX as N increases [1].

As illustrated in Fig. 1, a TB flip-flop can be designed from an NTB flip-flop. For terminology clarification, note that the TB flip-flop is sometimes referred as a pulsed latch [8]-[9], a hybrid-latch flip-flop [10], or a time-window based flip-flop [11]. Observing Fig. 1, the TB flip-flop includes extra inverter delays in the clock path of the master latch, which creates a transparency window (TW) for the flip-flop that is approximately equal to the delay of the extra inverters. For an odd number of TW inverters, the

polarity of the connections in the master latch pass gate and tri-state feedback inverter is reversed. Note that the TW is achieved at a cost of increasing flip-flop clocking energy. Adding or removing the extra inverters changes the TW delay and flip-flop clocking energy in discrete time steps. In Fig. 3, the data-to-output delay (t_{D-Q}) is plotted versus the data-to-clock delay (t_{D-CK}) for an NTB flip-flop and a TB flip-flop with four TW inverters. The TW delay is measured as the delay from the clock edge ($t_{D-CK}=0$ps) until the value of t_{D-CK} that corresponds to the onset of t_{D-Q} exponentially increasing. From Fig. 3, the TW delay equals ~53ps. In Fig. 2, the TB multi-cycle interconnect is constructed by inserting NTB flip-flops at the driver and receiver of the first and last interconnect segments, respectively, and TB flip-flops for the intermediate interconnect segments, which provides a TW at each adjacent interconnect segment to allow data to flow through the flip-flop until the late edge of the TW. Applying the TW data from Fig. 3 into the TB N-cycle interconnect in Fig. 2, data from a preceding interconnect segment can arrive ~53ps after the clock edge and still flow through the TB flip-flop and onto the next interconnect segment. For a TB N-cycle interconnect, the clock skew and jitter are amortized over the N interconnect segments, where the transparency windows for the previous N-1 interconnect segments absorb clock uncertainties. In addition, a setup time delay is only considered during the last interconnect segment.

3. Optimal Delay Variation Tolerant Design

A derivation of statistically-based analytical equations is presented for designing a TB N-cycle interconnect for optimal delay variation tolerance. Initially, clock skew and jitter are omitted to

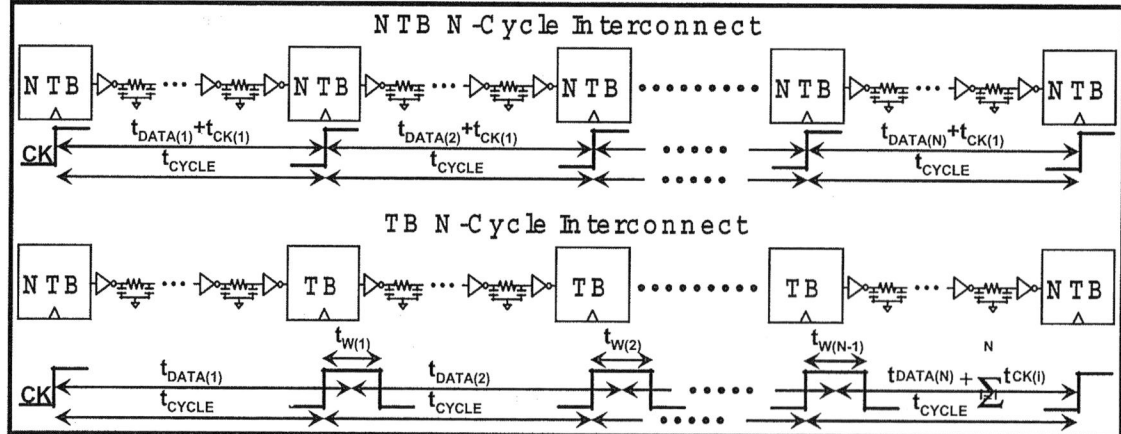

Fig. 2. NTB and TB N-cycle interconnects, where t_{CYCLE} is cycle time, $t_{DATA(i)}$ is i^{th} interconnect segment delay, $t_{W(i)}$ is i^{th} transparency window, $t_{CK(1)}$ is single-cycle clock skew and jitter, and $t_{CK(i)}$ is extra long-term jitter for i^{th} cycle ($i>1$).

80

simplify the model derivation, and then the model is later extended to account for these effects. The dependency of data delay averaging on the TW is illustrated in Fig. 4, where an example nominal timing diagram for a TB 2-cycle interconnect with zero clock skew and jitter is described. Ideal data delay averaging occurs in a multi-cycle interconnect when t_{CYCLE} equals the average of the interconnect segment delays, which is the best-case scenario for delay variation tolerance. For a TB 2-cycle interconnect, ideal averaging occurs when:

$$t_{CYCLE} = \frac{t_{DATA(1)} + t_{DATA(2)}}{2} \quad (1)$$

Observing the timing diagram in Fig. 4, two timing constraints must be satisfied to achieve ideal data delay averaging. First, if $t_{DATA(1)}$ is sufficiently slower than $t_{DATA(2)}$ such that $t_{DATA(1)}-t_{W(1)}>t_{W(1)}+t_{DATA(2)}$, then t_{CYCLE} is limited by $t_{DATA(1)}$, where $t_{CYCLE}=t_{DATA(1)}-t_{W(1)}$. Thus, the first constraint is defined as:

$$t_{DATA(1)} - t_{W(1)} \le t_{W(1)} + t_{DATA(2)} \quad (2)$$

Second, if $t_{DATA(1)}$ is faster than $t_{DATA(2)}$, then $t_{CYCLE}=t_{DATA(2)}$. The second constraint for ideal data delay averaging is given as:

$$t_{DATA(1)} \ge t_{DATA(2)} \quad (3)$$

Rearranging terms in (2) and (3), the constraints are simplified as:

$$0 \le t_{DATA(1)} - t_{DATA(2)} \le 2t_{W(1)} \quad (4)$$

The "delay variation mismatch" between $t_{DATA(1)}$ and $t_{DATA(2)}$ is defined as $\delta_{DATA(1)}-\delta_{DATA(2)}$, where $\delta_{DATA(i)}$ denotes the delay deviation from nominal conditions for the ith interconnect segment. The distribution of $\delta_{DATA(1)}-\delta_{DATA(2)}$ is assumed to have a median of zero, implying that positive and negative values for $\delta_{DATA(1)}-\delta_{DATA(2)}$ are equally probable. From (4), $t_{DATA(1)}-t_{DATA(2)}$ can range from 0 to $2t_{W(1)}$ while achieving ideal data delay averaging. Maximizing the percentage of ideal data delay averaging in (4), $t_{DATA(1)}$ and $t_{DATA(2)}$ are optimally designed for delay variation tolerance when the nominal delays equal the midpoint of the allowed variation range as:

$$t_{DATA(1),NOM} - t_{DATA(2),NOM} = t_{W(1)} \quad (5)$$

Parameter $t_{DATA(i),NOM}$ represents the ith interconnect segment delay under nominal conditions ($\delta_{DATA(i)}=0$). Applying (1) such that $t_{CYCLE,NOM}=0.5(t_{DATA(1),NOM}+t_{DATA(2),NOM})$, where $t_{CYCLE,NOM}$ is the nominal cycle time, the segment delays are derived as:

$$t_{DATA(1)} = t_{CYCLE,NOM} + 0.5t_{W(1)} + \delta_{DATA(1)} \quad (6)$$

$$t_{DATA(2)} = t_{CYCLE,NOM} - 0.5t_{W(1)} + \delta_{DATA(2)} \quad (7)$$

Reviewing Fig. 4, (6) and (7) correspond to the nominal data arriving in the middle of the TW, which enables equal averaging of both positive and negative values for $\delta_{DATA(1)}-\delta_{DATA(2)}$. Substituting (6) and (7) into (4), the relationship between interconnect segment delay variation mismatch and the TW required for ideal data delay averaging is derived as:

$$|\delta_{DATA(1)} - \delta_{DATA(2)}| \le t_{W(1)} \quad (8)$$

From (8), ideal data delay averaging between $t_{DATA(1)}$ and $t_{DATA(2)}$ is achieved if $t_{W(1)}$ is greater than the absolute value of $\delta_{DATA(1)}-\delta_{DATA(2)}$. The standard deviation of $\delta_{DATA(1)}-\delta_{DATA(2)}$ is given as:

$$\sigma_{DATA(1)-DATA(2)} = \sqrt{\sigma_{DATA(1)}^2 + \sigma_{DATA(2)}^2 - 2\rho_{(1)}\sigma_{DATA(1)}\sigma_{DATA(2)}} \quad (9)$$

$\sigma_{DATA(i)}$ is the standard deviation in the ith interconnect segment delay, and $\rho_{(k)}$ is the correlation between two segment delays separated by k segments. From (8), the TW required to achieve a desired percentage of ideal data delay averaging is calculated as:

$$t_{W(1)} = m\sigma_{DATA(1)-DATA(2)} \quad (10)$$

The parameter "m" is the number of standard deviations that maps to a desired percentage of ideal data delay averaging. For a normal Gaussian delay distribution, "m" of 1 ensures a 68% probability $(+/-\sigma_{DATA(1)-DATA(2)})$ of ideal data delay averaging as described in Fig. 4. As "m" increases, the TB interconnect becomes more delay variation tolerant at a cost of increased flip-flop clocking energy. Since the standard deviation in delay from the TW is significantly smaller than $\sigma_{DATA(1)-DATA(2)}$, the statistical impact of TW delay variability on the TB interconnect design is negligible.

Note if the delay variation mismatch between $t_{DATA(1)}$ and $t_{DATA(2)}$ is greater than $t_{W(1)}$, then ideal data delay averaging does not occur. Nevertheless, partial data delay averaging is achieved for any finite TW with optimally designed interconnect segments (6)-(7). As an example of partial data delay averaging for a TB 2-cycle interconnect (6)-(7) with $t_{CYCLE,NOM}=250ps$ and $t_{W(1)}=20ps$, if variations induce $\delta_{DATA(1)}=20ps$ and $\delta_{DATA(2)}=-20ps$ ($t_{DATA(1)}=280ps$, $t_{DATA(2)}=220ps$), then $t_{CYCLE}=t_{DATA(1)}-t_{W(1)}=260ps$, which provides a 10ps advantage over an NTB interconnect in a similar condition. If $t_{W(1)}=40ps$ for this example, then ideal data delay averaging occurs, where $t_{CYCLE}=250ps$.

In analyzing a TB N-cycle interconnect, ideal data delay averaging occurs when:

$$t_{CYCLE} = \frac{t_{DATA(1)} + \dots + t_{DATA(N)}}{N} = \frac{1}{N}\sum_{i=1}^{N} t_{DATA(i)} \quad (11)$$

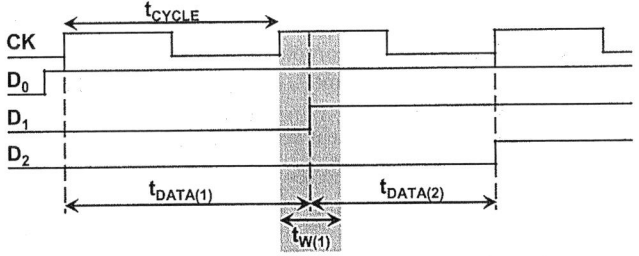

(a)

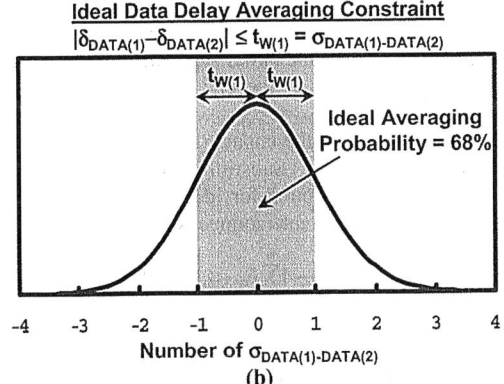

(b)

Fig. 4. (a) Nominal timing for a TB 2-cycle interconnect with zero clock skew and jitter, where D_0, D_1, and D_2 are input data for driving, intermediate, and receiving flip-flops, respectively. (b) Probability density versus number of standard deviations in $\delta_{DATA(1)}-\delta_{DATA(2)}$ ($\sigma_{DATA(1)-DATA(2)}$), where $\delta_{DATA(i)}$ is the delay deviation for the ith interconnect segment.

Reviewing the timing constraint for ideal averaging in a 2-cycle interconnect as given in (2), a TB N-cycle interconnect contains N-1 similar constraints given as:

$$\frac{1}{k}\left(\sum_{i=1}^{k} t_{DATA(i)} - t_{W(k)}\right) \leq \frac{1}{N-k}\left(t_{W(k)} + \sum_{i=k+1}^{N} t_{DATA(i)}\right) \ \forall \, k:1 \leq k \leq N-1 \quad (12)$$

Reviewing the timing constraint in (3), an additional set of N-1 similar constraints are defined as:

$$\frac{1}{k}\sum_{i=1}^{k} t_{DATA(i)} \geq \frac{1}{N-k}\sum_{i=k+1}^{N} t_{DATA(i)} \ \forall \, k:1 \leq k \leq N-1 \quad (13)$$

From (12) and (13), the relationships between the interconnect segment delays and the transparency windows are derived as:

$$0 \leq \left(\frac{1}{k}\right)\sum_{i=1}^{k} t_{DATA(i)} - \left(\frac{1}{N-k}\right)\sum_{i=k+1}^{N} t_{DATA(i)} \leq \frac{N}{k(N-k)} t_{W(k)} \quad (14)$$
$$\forall \, k:1 \leq k \leq N-1$$

From (14), there are N-1 separate delay ranges for unique combinations of interconnect segment delay variation mismatches that must be satisfied to achieve ideal data delay averaging. Similar to the steps in deriving (5) from (4), the percentage of ideal data delay averaging is maximized when the nominal interconnect segment delays equal the midpoint of the allowed variation range as:

$$\left(\frac{1}{k}\right)\sum_{i=1}^{k} t_{DATA(i),NOM} - \left(\frac{1}{N-k}\right)\sum_{i=k+1}^{N} t_{DATA(i),NOM} = \frac{1}{2}\left(\frac{N}{k(N-k)} t_{W(k)}\right) \quad (15)$$
$$\forall \, k:1 \leq k \leq N-1$$

Solving the N-1 equations in (15) along with (11) for nominal conditions, the TB N-cycle interconnect with zero clock skew and jitter is optimal for delay variation tolerance when a uniform TW (t_W) is applied at each interconnect segment, and the first, intermediate, and last interconnect segment delays equal $t_{CYCLE,NOM}+\frac{1}{2}t_W$, $t_{CYCLE,NOM}$, and $t_{CYCLE,NOM}-\frac{1}{2}t_W$, respectively, under nominal conditions. The TB N-cycle interconnect model with zero clock skew and jitter is derived as:

$$t_{W(1)} = t_{W(2)} = \dots = t_{W(N-1)} = t_W \quad (16)$$

$$t_{DATA(1)} = t_{CYCLE,NOM} + 0.5t_W + \delta_{DATA(1)} \quad (17)$$

$$t_{DATA(i)} = t_{CYCLE,NOM} + \delta_{DATA(i)} \ \forall \, i:2 \leq i \leq N-1 \quad (18)$$

$$t_{DATA(N)} = t_{CYCLE,NOM} - 0.5t_W + \delta_{DATA(N)} \quad (19)$$

The model in (16)-(19) corresponds to the nominal data arriving in the middle of the TW, which ensures equal averaging of both fast and slow segment delays. As a comparison, if the nominal data delay arrives at the beginning of the TW ($t_{DATA(1),NOM}=t_{DATA(N),NOM}=t_{CYCLE,NOM}$), then the TB interconnect cannot take advantage of fast delays in the initial segments. Conversely, if the nominal data delay arrives at the end of the TW ($t_{DATA(1),NOM}=t_{CYCLE,NOM}+t_W$, $t_{DATA(N),NOM}=t_{CYCLE,NOM}-t_W$), then the TB interconnect cannot reduce the impact of slow delays in the initial segments through averaging. Substituting (16)-(19) into (15), the dependency of the TW required for ideal averaging on the delay variation mismatch between interconnect segments is derived as:

$$\frac{2k(N-k)}{N}\left|\left(\frac{1}{k}\right)\sum_{i=1}^{k} \delta_{DATA(i)} - \left(\frac{1}{N-k}\right)\sum_{i=k+1}^{N} \delta_{DATA(i)}\right| \leq t_W \ \forall \, k:1 \leq k \leq N-1 \quad (20)$$

The TB N-cycle interconnect model with clock skew and jitter omitted (16)-(19) is now extended to include these effects. Single-cycle jitter is defined as the clock edge variation in consecutive clock cycles. Long-term jitter is the accumulated clock edge variation across multiple cycles. The dependency of long-term

clock jitter on the number of clock cycles is presented through measured data in Fig. 5. For a small number of clock cycles, jitter primarily results from voltage controlled oscillator (VCO) variation, where jitter is proportional to the square root of the number of cycles [12]-[13]. For a large number of clock cycles, the phase-locked loop (PLL) phase detector and loop filter sense the VCO error and adjust the VCO input accordingly, leading to a saturation of the long-term jitter. The sum of clock skew and single-cycle clock jitter is defined by $t_{CK(1)}$. The parameter $t_{CK(i)}$ represents the extra long-term jitter for the ith cycle (i>1). As an example, the clock skew and long-term jitter for a TB 3-cycle interconnect is calculated as $t_{CK(1)}+t_{CK(2)}+t_{CK(3)}$. Note that $t_{CK(i)}$ is a monotonically decreasing function as "i" increases.

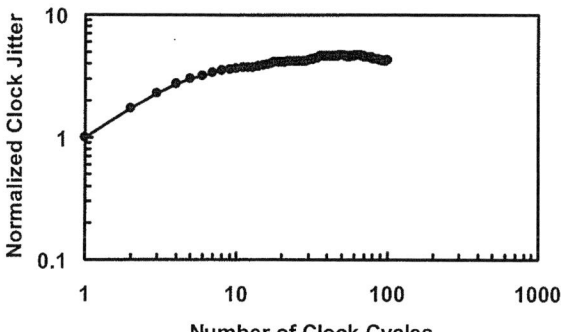

Fig. 5. Measured clock jitter versus number of clock cycles.

The TW in (16) can be divided into two regions, a left-side delay ($t_{W,L}=\frac{1}{2}t_W$) and a right-side delay ($t_{W,R}=\frac{1}{2}t_W$), where $t_{W,L}$ and $t_{W,R}$ enable data delay averaging of fast and slow delays, respectively, in the initial interconnect segments. Since clock jitter is a dynamic variation versus time, the worst-case delay range from jitter is assumed to eventually occur for each interconnect segment, which expands the middle of $t_{W(k)}$ by $t_{W(k),M}$ as illustrated in Fig. 6, where $t_{W(k)}$ and $t_{W(k),M}$ are the TW and the clock skew and jitter delay range for the kth segment, respectively. Using a TB 2-cycle interconnect as an example, two worst-case conditions for clock skew and jitter exist. First, a clock skew and jitter of $t_{CK(1)}$ and $t_{CK(2)}$ could occur in the 1st and 2nd interconnect segment delays, respectively, or second, $t_{CK(2)}$ and $t_{CK(1)}$ could occur in the 1st and 2nd interconnect segment delays, respectively. For the first scenario, $t_{DATA(1)}+t_{CK(1)}$ would place the nominal data arriving at the boundary of $t_{W(k),M}$ and $t_{W,R}$. For the second condition, $t_{DATA(1)}+t_{CK(2)}$ would place the nominal data arriving at the boundary of $t_{W,L}$ and $t_{W(k),M}$. For a TB 2-cycle interconnect, the delay range from clock skew and jitter for $t_{W(1)}$ equals $t_{CK(1)}-t_{CK(2)}$. Thus, $t_{W(1)}$ in a TB 2-cycle interconnect is expanded from (10) as:

$$t_{W(1)} = t_W + t_{CK(1)} - t_{CK(2)} \quad (21)$$

Parameter t_W represents the portion of the TW dedicated to data delay averaging such that $t_W=m\sigma_{DATA(1)-DATA(2)}$ in (21). From (5), (21), and $t_{CYCLE,NOM}=0.5(t_{DATA(1),NOM}+t_{DATA(2),NOM}+t_{CK(1)}+t_{CK(2)})$, the nominal interconnect segment delays are derived as:

$$t_{DATA(1),NOM} = t_{CYCLE,NOM} + 0.5t_W - t_{CK(2)} \quad (22)$$

$$t_{DATA(2),NOM} = t_{CYCLE,NOM} - 0.5t_W - t_{CK(1)} \quad (23)$$

For a TB 3-cycle interconnect, the worst-case delay range in clock skew and jitter for $t_{W(1)}$ is $t_{CK(1)}-t_{CK(3)}$ and for $t_{W(2)}$ is $(t_{CK(2)}+t_{CK(3)})-(t_{CK(2)}+t_{CK(3)})=t_{CK(1)}-t_{CK(3)}$. For this design, the first, intermediate, and last interconnect segment delays would equal $t_{CYCLE,NOM}+\frac{1}{2}t_W-t_{CK(3)}$, $t_{CYCLE,NOM}-t_{CK(2)}$, and $t_{CYCLE,NOM}-\frac{1}{2}t_W-t_{CK(1)}$, respectively, under nominal conditions. For practical design considerations,

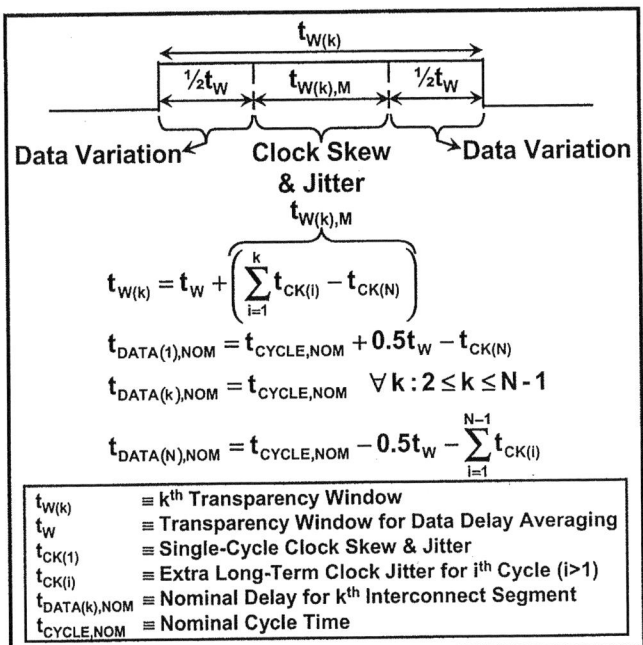

$$t_{W(k)} = t_W + \left(\sum_{i=1}^{k} t_{CK(i)} - t_{CK(N)} \right)$$

$$t_{DATA(1),NOM} = t_{CYCLE,NOM} + 0.5t_W - t_{CK(N)}$$

$$t_{DATA(k),NOM} = t_{CYCLE,NOM} \quad \forall k : 2 \le k \le N\text{-}1$$

$$t_{DATA(N),NOM} = t_{CYCLE,NOM} - 0.5t_W - \sum_{i=1}^{N-1} t_{CK(i)}$$

$t_{W(k)}$	$\equiv$ kth Transparency Window
t_W	$\equiv$ Transparency Window for Data Delay Averaging
$t_{CK(1)}$	$\equiv$ Single-Cycle Clock Skew & Jitter
$t_{CK(i)}$	$\equiv$ Extra Long-Term Clock Jitter for ith Cycle (i>1)
$t_{DATA(k),NOM}$	$\equiv$ Nominal Delay for kth Interconnect Segment
$t_{CYCLE,NOM}$	$\equiv$ Nominal Cycle Time

Fig. 6. TB N-cycle interconnect model for optimal delay variation tolerance.

however, intermediate interconnect segment delays ($t_{DATA(2)} = \ldots = t_{DATA(N-1)}$) are nominally designed equal to the nominal cycle time ($t_{CYCLE,NOM}$). Applying this constraint, $t_{w(2)}$ must increase by $t_{CK(2)}$ to satisfy the $t_{w,R}$ requirement in $t_{w(2)}$ to average slow delays in the first two interconnect segments. In addition, the last interconnect segment delay would account for $t_{CK(2)}$ as $t_{DATA(3),NOM} = t_{CYCLE,NOM} - \frac{1}{2}t_W - t_{CK(1)} - t_{CK(2)}$. Since intermediate interconnect segment delays are nominally designed equal to $t_{CYCLE,NOM}$, the worst-case delay range in clock skew and jitter for $t_{w(1)}$ and $t_{w(2)}$ are $t_{CK(1)} - t_{CK(3)}$ and $t_{CK(1)} + t_{CK(2)} - t_{CK(3)}$, respectively. For a TB N-cycle interconnect, the clock skew and jitter delay range for the kth TW is calculated as:

$$t_{W(k),M} = \sum_{i=1}^{k} t_{CK(i)} - t_{CK(N)} \qquad (24)$$

The statistically-based model for optimal delay variation tolerance in a TB N-cycle interconnect is provided in Fig. 6, where $t_{w(k)}$ consists of three distinct regions. The left and right portions of the window are equally sized to average data delay variations, and the middle portion is designed for clock skew and jitter.

4. Mean FMAX & Energy Comparison

The design of an NTB interconnect segment in M4 is optimized to provide the minimum energy for a nominal cycle time target of 250ps in a 65nm CMOS technology [14] with a V_{DD} of 1.2V and a temperature of 110°C, resulting in 6 repeaters, a protection inverter, and a flop-to-flop interconnect segment length of 1500µm. Starting with the optimal NTB interconnect segment, NTB flip-flops are replaced with TB flip-flops for the intermediate segments of a multi-cycle TB interconnect. *Thus, the comparison between TB and NTB multi-cycle interconnects is based on an optimal NTB interconnect design.* As listed in Fig. 7, the flip-flop clocking energy for the intermediate TB interconnect segments (TB-to-TB) increases relative to an NTB interconnect segment (NTB-to-NTB) as the TW increases from inserting extra inverters in the clock path of the master latch. Since clock skew and jitter are absorbed by the TW, additional slack is available in the

Energy/Cycle	NTB-to-NTB	TB-to-TB (TW=27ps)	TB-to-TB (TW=53ps)
Flip-Flop Clock	14.4 fJ	20.9 fJ	26.4 fJ
Data-Path (Activity Factor=0.1)	41.9 fJ	33.7 fJ	33.7 fJ
Total Active	56.3 fJ	54.6 fJ	60.1 fJ
Idle	8.0 fJ	5.5 fJ	5.5 fJ
Total Average (Clock Gating Factor=0.9)	12.8 fJ	10.4 fJ	11.0 fJ

Fig. 7. Energy per cycle components for a 250ps $t_{CYCLE,NOM}$.

nominal cycle time. TB flip-flop and repeater transistors are downsized until the nominal segment delay equals 250ps, which reduces active and average energies for the interconnect data-path as described in Fig. 7. Total active energy is the sum of the flip-flop clock and data-path energies, where the data-path activity factor is assumed equal to 0.1. Average energy is calculated from both the active energy and the idle clock gating energy. The clock gating factor is assumed equal to 0.9, which is based on clock gating statistics for a recent microprocessor.

Monte Carlo simulations are performed using a statistical SPICE-like circuit simulator in a 65nm process technology [14] to generate individual segment delay distributions ($\sigma_{DATA(i)}$) and correlations ($\rho_{(k)}$), where WID transistor, interconnect, and voltage variations are considered as given in Fig. 8. The WID parameter variations are categorized into systematic-WID (correlated) and random-WID (uncorrelated) components [3]. Since the systematic-WID variations contain a device-to-device correlation based on the distance between the devices, the device locations are required in the simulation. Although the sources of metal variation are mostly deterministic (e.g., metal thickness depends on metal density), the simulations are performed by assuming a random correlated variation model. These interconnect segment delay statistics are applied into an additional Monte Carlo analysis to determine the FMAX distribution as dictated by delay variations in a multi-cycle interconnect, including the effects of transparency windows, clock skew, and clock jitter.

WID-Variation Sources	Systematic (Correlated)	Random (Uncorrelated)
Channel Length	∏	∏
Channel Width	∏	∏
Overlap Capacitance	∏	∏
Threshold Voltage		∏
Metal Thickness	∏	∏
Metal Width/Spacing	∏	∏
ILD Thickness	∏	∏
Supply Voltage	∏	

Fig. 8: WID variation sources applied in simulation.

In Fig. 9, mean FMAX, active energy, and average energy of multi-cycle NTB and TB interconnects are compared for a range of cycles, representing present (N=3&6) and future (N=9&12) interconnect designs. For TB interconnects, the TW of each segment is optimally calculated through the model in Fig. 6, where the portion of the window dedicated to data delay averaging (t_W) is swept as a variable. In Fig. 9(a), the mean FMAX of TB interconnects is ~3.5% higher than NTB interconnects at equal active energy. Additional mean FMAX gains can be obtained by expanding the transparency window through a larger t_W at the expense of higher clocking energy until averaging of data delay variations approaches ideality. Since the impact of additional clocking energy in TB flip-flops is mitigated by clock gating, TB interconnects enable 4-6% mean FMAX gain with a corresponding 10% average energy savings as described in Fig. 9(b). In Fig. 10, the maximum mean FMAX benefit ranges from 4.0-7.5% for WID variations typical of a 65nm process technology. This range of FMAX improvement corresponds to approximately a bin-split increase in the FMAX distribution. As a comparison, an equivalent increase in FMAX from process technology results in ~2X larger average energy. For 2X larger WID delay variance (1.41X larger standard deviation), the maximum mean FMAX gain rises to 5-10%.

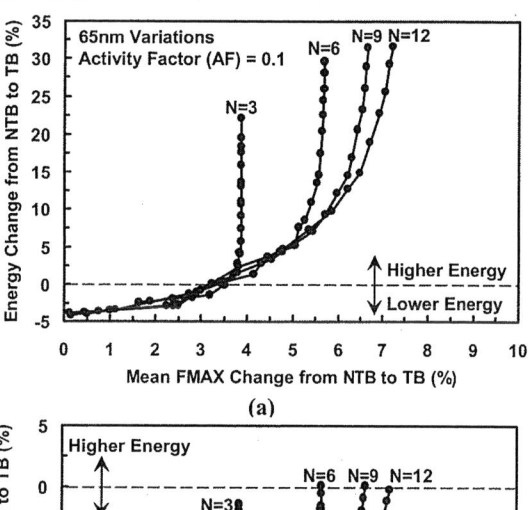

(a)

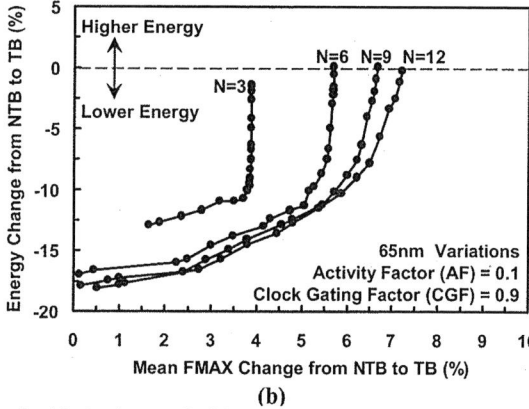

(b)

Fig. 9. (a) Active and (b) average energies per cycle versus mean FMAX trade-off from NTB to TB N-cycle interconnect.

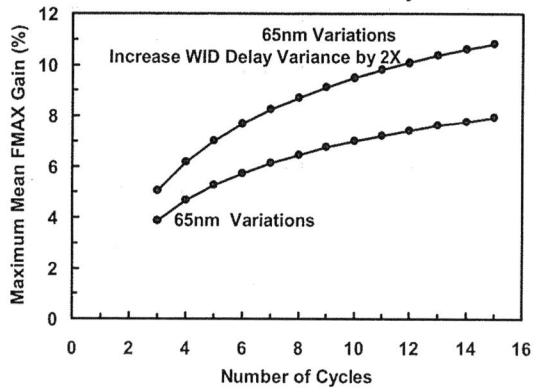

Fig. 10. Maximum mean FMAX gain from NTB to TB N-cycle interconnect versus N.

5. Conclusion

The transparency window of a TB flip-flop in a multi-cycle interconnect provides an opportunity to average systematic and random WID delay variations across multiple interconnect segments, thus mitigating the adverse impact of WID parameter variations on the mean FMAX as dictated by interconnect delays. Transparency windows as well as the first and last interconnect segment delays are optimized for delay variation tolerance through a newly developed statistically-based analytical model, which enables insight into the dependency of the transparency window required for data delay averaging on the delay variation mismatch between interconnect segments. Mean FMAX gains of 4-6% and a corresponding average energy savings of 10% are achieved in comparison to optimally designed multi-cycle interconnects with conventional master-slave flip-flops in a 65nm CMOS technology. The maximum mean FMAX benefit ranges from 4.0-7.5%, corresponding to approximately a bin-split shift in the FMAX distribution. As a comparison, ~2X larger average energy cost is required to achieve an equivalent FMAX improvement through process technology. For 1.41X larger WID delay variations, the maximum mean FMAX gain increases to 5-10%.

6. Acknowledgements

The authors express sincere appreciation to Tao Chen, Mozhgan Mansuri, and Nasser Kurd of Intel for their kind support.

7. References

[1] K. A. Bowman, Steven G. Duvall, and J. D. Meindl, "Impact of Die-to-Die and Within-Die Parameter Fluctuations on the Maximum Clock Frequency Distribution for Gigascale Integration," *IEEE JSSC*, pp. 183-190, Feb. 2002.

[2] S. Borkar, et al., "Parameter Variations and Impact on Circuits and Microarchitecture," in *40th DAC Proc.*, June, 2003, pp. 338-342.

[3] H. Masuda, S. Ohkawa, A. Kurokawa, and M. Aoki, "Challenge: Variability Characterization and Modeling for 65- to 90-nm Processes," in *IEEE CICC*, Sept. 2005, pp. 593-600.

[4] J. A. Davis, et al., "Interconnect Limits on Gigascale Integration (GSI) in the 21st Century," *Proc. IEEE*, pp. 306-324, Mar. 2001.

[5] M. T. Bohr, "Interconnect Scaling – The Real Limiter to High Performance ULSI," in *IEEE IEDM*, Dec. 1995, pp. 241-244.

[6] V. De and S. Borkar, "Technology and Design Challenges for Low Power and High Performance," in *Proc. ISLPED*, Aug. 1999, pp. 163-168.

[7] M. Khellah, et al., "A Skewed Repeater Interconnect Architecture for On-Chip Energy Reduction in Microprocessors," in *IEEE Proc. 2005 ICCD*, Oct. 2005, pp. 253-257.

[8] V. G. Oklobdzija, *The Computer Engineering Handbook*, CRC Press, 2002.

[9] D. Harris and M. A. Horowitz, "Skew-Tolerant Domino Circuits," *IEEE JSSC*, pp. 1702-1711, Nov. 1997.

[10] H. Partovi, et al., "Flow-Through Latch and Edge-Triggered Flip-Flop Hybrid Elements," in *IEEE ISSCC*, Feb. 1996, pp. 138-139.

[11] V. G. Oklobdzija, "Clocking and Clocked Storage Elements in a Multi-Gigahertz Environment," *IBM J. Res. & Dev.*, pp. 567-583, Sept. 2003.

[12] J. A. McNeill, "Jitter in Ring Oscillators," *IEEE JSSC*, pp. 870-879, June 1997.

[13] M. Mansuri and C. K. Yang, "Jitter Optimization Based on Phase-Locked Loop Design Parameters," *IEEE JSSC*, pp. 1375-1382, Nov. 2002.

[14] P. Bai, et al., "A 65nm Logic Technology Featuring 35nm Gate Lengths, Enhanced Channel Strain, 8 Cu Interconnect Layers, Low-k ILD and 0.57 μm² SRAM Cell," in *IEEE IEDM*, Dec. 2004, pp. 657-660.

A Pulsed Low-Voltage Swing Latch for Reduced Power Dissipation in High-Frequency Microprocessors

Pong-Fei Lu, Nianzheng Cao, Leon Sigal, Pieter Woltgens, R. Robertazzi, D. Heidel

IBM T. J. Watson Research Center
Yorktown Heights, NY 10598

ABSTRACT

We have reported previously [1] a low-swing latch (LSL) with superior performance-power tradeoff compared to the conventional pass-gate master-slave latch. In this paper, hardware results are presented for the proposed LSL with pulsed clock waveforms. The motivation is to combine low-voltage swing with pulsed signals to further reduce overall system power in high-frequency microprocessors. We have designed a 65-bit accumulator loop experiment to mimic a microprocessor pipeline stage. The local clock buffer design features a mode switch to toggle between two-phase (c1/c2) master-slave clocking and one-phase pulsed (c2 only) clocking. Our data show that 15-25% system power saving can be achieved in pulsed mode compared to non-pulsed mode. Power contribution from individual components is also presented.

Categories and Subject Descriptors
B.6 [**Hardware**]: logic design
General Terms: Design
Keywords: Latch, low-power, pulse latch

1. INTRODUCTION

Modern microprocessors leverage deep pipelined design for high frequencies, which has led to explosive growth of the latch count [2]. To date, up to 50% of the overall chip AC power is consumed by the latches and the clock distribution [3,4]. Majority of the clock power is spent in the local clock buffers (LCB) that drive the local clock load of the latch (16-32 latches per LCB in a typical dataflow stack). To mitigate the power consumption, there has been increased attention to power-efficient latch circuit and clock gating design techniques. One way to reduce the clocking power is to lower the voltage swing of LCBs which drive high capacitive loads. To facilitate the low-swing clock however, the latch needs to be operational with clocks below the nominal supply voltage other circuitry operates on. In an earlier study [1], we have reported such latch circuit, and demonstrated in a 90nm technology, the new scheme could reduce the AC power dissipation by as much as 12%. In our latch design (Fig. 1), a gated-inverter clock input stage replaces the pass-gate of a conventional master-slave (MS) latch to avoid problems with leakage in the clock splitter and the weakened drive of the pass-gate P-FET. Similar circuit topology has been used in the 'split latch' configuration in [5,6]. The LSL design is pin compatible

Permission to make digital or hard copies of all or part of this work for personal or classroom use is granted without fee provided that copies are not made or distributed for profit or commercial advantage and that copies bear this notice and the full citation on the first page. To copy otherwise, or republish, to post on servers or to redistribute to lists, requires prior specific permission and/or a fee.
ISLPED'06, October 4–6, 2006, Tegernsee, Germany.
Copyright 2006 ACM 1-59593-462-6/06/0010...$5.00.

with the conventional pass-gate MS latch, and has roughly equal total device width. In this paper, we report further power reduction opportunities using pulsed mode operation. In the pulsed mode, the clocking power is reduced since only one clock, instead of two in the master-slave mode, is distributed. Hardware data showed roughly 25% and 15% reduction in AC power in pulsed mode for the conventional MS latch and LSL, respectively. Furthermore, LSL hardware demonstrated 7% less power consumption over conventional MS implementation when both are running in pulsed mode.

Low-swing clocking scheme reported in [7,8] either used reduced voltage pervasively ('G+L': global + local) or for global clock distribution only ('G-only'). In our experiment, the global clock is distributed full-swing, and the reduced V_{dd} is for local clock only. This is to mitigate the edge rate sensitivity and the concern of clock skew/jitter at reduced voltages, so that low-swing clocking can be applied strategically without impacting the quality of global clock.

2. HARDWARE DESIGN AND CLOCKING SCHEME

The experiment is a latch-bound accumulator loop consisting of two 64-bit adders in series, with the second adder configured as a decrementer (Fig. 2). The experiment mimics a logic dominated pipeline stage in a microprocessor design. Key components include a clock generator, control logic to start/stop the clock, LCB, and two duplicate accumulator circuits with the control MS latch and LSL (Fig. 2), respectively. For the control experiment using the conventional MS latches, all circuits run on the same voltage, V_{dd}, whereas for the LSL experiment, a different power supply, V_{LCB}, was provided for the local clock buffers (LCB) to independently adjust the clock voltage swing. In order to separate out power measurements in different parts of the circuits, the latches, the LCB, and the adders are in different power supply domains accessible through individual external test pins. The global clock is supplied by an external clock generator and the power due to its distribution is not included in the experiments.

The operational modes of the LCBs are programmable through the LCB controller. The LCB controller can turn the clocks into pulsed mode through a mode switch (Fig. 3). In the non-pulsed mode, both c1 and c2 clock signals are active (2-phase with 50% duty cycle) and the latch is operated in master-slave style. In pulsed mode, the c1 clock is kept high, making the L1 part of the latch flushed, and only c2 clock is actively switching (Fig.3). It effectively transforms the MS latch into a single phase pulsed latch [9,10]. The nominal clock pulse width in pulsed mode is about 5 FO4 and is adjustable. The de-activation of c1 switching in pulsed mode contributes to the LCB power saving by shutting down the c1 clock driver (Fig. 3). In the meantime, the single-phase pulsed operation reduces latch latency as well as allows time borrowing to enhance frequency performance. The advantage and tradeoff of pulse latches are well documented in [11,12]. While increasing the pulse width is beneficial for the speed, it also increases the minimum bound of the fast paths which requires rigorous padding to avoid early mode failure. Since in our accumulator loop design the minimum delay is through two 2-way XOR gates (~5.4 FO4), which is longer than the pulse width, we did not encounter problems with fast paths. On the other hand, the minimum pulse width has to be wide enough for data to be written into the latch under the worst condition. In an optimal operation where data arrives during the transparency, the clock period can be made as small as the sum of the latch latency and the longest critical path delay in the accumulator. Since the accumulator portion is identical, the latch delay difference translates directly into the cycle time difference between the two sets of experiments.

3. MEASUREMENT RESULTS

In order to separate out the AC power from the total power consumption, we first established the DC leakage power by turning the clock off while keeping all power supplies on and all inputs quiet. In subsequent power measurements the AC power was derived by subtracting the total measured power by the DC leakage power. The AC measurements were performed by first determining the maximum frequency, F_{max}, at the targeted supply voltage. The functionality of the accumulator was verified at low frequencies using a complex add pattern over 50,000 cycles. The clock frequency was then gradually increased until the experiment failed to produce the expected result. Special care is taken to eliminate transient power spikes by inserting 'warm up' cycles between the scan and AC testing. For power measurements, the clock frequency was fixed at F_{max} and the experiment was run for an extended period of time so that average currents could be measured. The total system power consists of power dissipation from LCBs, latches and adders, which were measured separately then summed up.

Fig.4 plots the system power vs. F_{max} in non-pulsed and pulsed mode for the control MS and the LSL experiments. For the control MS experiment, V_{dd} on each curve is varied from 0.9 V to 1.2 V at 0.1 V intervals. Note that all circuits are operated at the same voltage when V_{dd} is changed. For LSL experiment however, the V_{dd} is fixed at 1.2 V (for latches and adder) and only V_{LCB} for clock voltage supply is swept (from 0.8 V to 1.2 V at 0.1 V intervals). The power saving in pulsed mode is evident in Fig. 4 for both control MS and LSL experiment. For example, at F_{max} around 2.25 GHz, the power saving in pulsed mode for control MS compared to the non-pulsed mode is about 27%. The substantial performance gain (~50 ps at 1.1 V in Fig. 4) in pulsed latch relative to the MS latch is attributed to the time borrowing [12] as follows. In the MS mode, the data is launched off the c2 rising edge and captured by c1 falling edge; i.e. all paths have exactly one-cycle delay time allowance. In the pulsed mode, the data is launched off the c2 rising edge and captured by the c2 falling edge. The pulse width defines the transparency window which benefits the late mode timing if there is a dominant long path more than a cycle's worth. The latter is corroborated by the data shown in Fig. 5 which plots the F_{max} with two different pulse

widths. The wider pulse width results in higher frequency across the entire V_{dd} range measured.

For the LSL experiments, the pulsed mode improves the overall frequency and power performance so that the power-frequency curves shift significantly to the lower right. A typical point from the overlap region where the comparison can be made shows 15% power saving at 2.35 GHz. Fig. 4 also illustrates the power reductions of LSL over control MS experiment when both are in non-pulsed or pulsed mode. In non-pulsed mode, the power saving of 12% is measured at 2.25 GHz for LSL compared to control MS experiment. Similar results were reported in our earlier paper [1]. In pulsed mode, LSL achieves 7% power saving over control MS at around 2.5 GHz. The power saving by LSL in pulsed mode is about half of that in non-pulsed mode, consistent with the fact that only one local clock (c2) is switching. Fig.4 indicates that the power saving of LSL over control MS implementation is not uniform across the power supply voltage range. The lower clock swing in LSL needs to be weighted against the performance loss in latch delay due to the reduced clock overdrive ('N0/N1' in Fig. 1(b)). For example, in Fig. 4 as V_{LCB} drops below about 0.9V (at V_{dd}=1.2 V) the pulsed LSL actually consumes more power than conventional MS latch at 1.1 V. This suggests that careful decisions need to be made when choosing (V_{dd}, V_{LCB}) parameters for LSL. In the latter case it would be better to lower V_{dd} and raise V_{LCB} to achieve overall power saving. However, we do not have data covering the full range of (V_{dd}, V_{LCB}) pair.

The hardware power measurement of individual voltage supply domain makes it possible for us to study the power-frequency characteristics for each component of the total system power, namely LCBs, latches and adders. Fig. 6 presents the power components for control MS experiment in non-pulsed and pulsed mode. For all the components, pulsed mode shifts the power-frequency curves to the lower right direction, indicating better performance and power efficiency. The total system power saving is from the power difference at same frequency for all components. It should be no surprise that the largest power difference between the non-pulsed and pulsed mode is from LCB, since half of the local clock driver is shut down in pulsed mode. Similarly, Fig.7 plots the power components for LSL experiment in non-pulsed and pulsed mode. The power-frequency characteristics for LSL are distinctively different from control MS experiment. For adders and latches, the power-frequency relationships collapse into the same linear curve because the voltage supply is maintained at 1.2 V for their V_{dd} domains. For fixed V_{dd}, circuit power is linearly proportional to the operational frequency. In other words, they do not contribute to any power savings. The LCB power-frequency curves however, are quite different between the two modes. Although following the same trend, the power in pulsed mode is much lower, roughly half of the non-pulsed mode, consistent with the expectation of power reduction by reduced local clock switching load.

4. CONCLUSION

We have shown successful operations of pulsed low-voltage swing latch in our hardware experiment using IBM 90 nm SOI technology. Substantial power saving measured in testing hardware illustrates the potential advantages in microprocessor designs. Our previous report compared the low-swing latch and the conventional transmission gate MS latch and demonstrated 12% overall system power saving. In this paper we examined the pulsed mode latch operation and showed that the total system power can be further reduced. The combined advantage of low clock swing and pulsed operation can provide significant opportunity in a power constrained design. Also, the unique design of LCB control to switch between non-pulsed and pulsed mode facilitates the mixing of latch types for robustness or performance.

ACKNOWLEDGEMENT: We would like to thank SRDC Technology Group in Fishkill, NY for processing the testchip.

REFERENCES

[1]. Pong-Fei Lu et al., IEEE Int SOI Conference, pp. 165-167, 2004.
[2]. V. Srinivasan et al., 35th IEEE/ACM Int. Symp. On Microarchitecture 2002
[3]. J. Clabes et al., Techical Digest of ISSCC 2004, p. 56.
[4]. B. Curran, Technical Digest of ISSCC 2001, p. 238.
[5]. D. W. Dobberpuhl et al, IEEE JSSC, Vol. 27, No. 11, pp. 1555-1567 (1992)
[6]. N. Nedovic et al., Integrated Circuits and System Design, pp. 211-215 (2000)
[7]. R. Krishnamurphy et al., 2002 Symposium on VLSI Circuits, p. 128.
[8]. M. Tokumasu et al., Proceedings of CICC, pp. 129-132, (2002)
[9]. J. D. Warnock & D. Wendle, U.S. Patent 6,822,500 B1 (Nov. 23, 2004)
[10]. B. Curran et al., Technical Digest of ISSCC 2006, p. 436
[11]. S. H. Unger and C.-J. Tan, IEEE Transactions on Computers, C-35(10), p.880, Oct. 1986.
[12]. V.G. Oklobdzija, IBM J. Res & Dev. 47(5/6) Sept./Nov. 2003.

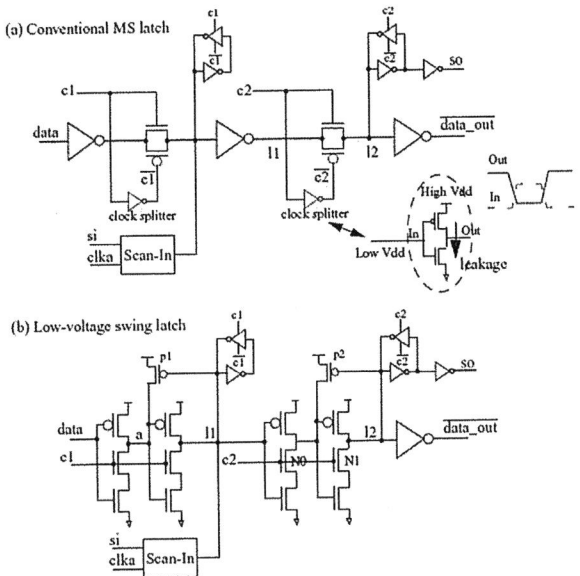

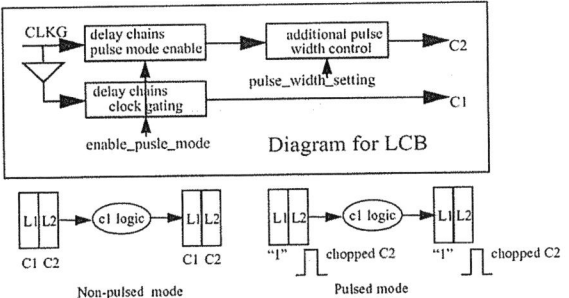

Fig 1: (a) Conventional transmission gate master-slave latch, and (b) the low-voltage swing master-slave latch. The two latches are pin compatible and LSSD compliant. A low-swing clock in (a) would result in incomplete turn-off of the PFET in the splitter when the clock is high, which in turn causes leakage as well as weakened PFET in the transmission gate.

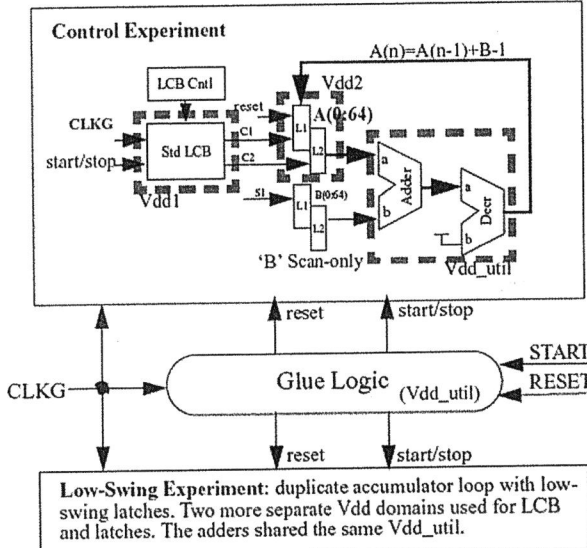

Fig. 2: Experimental setup: A 'glue logic' block controls the side-by-side experiments. There are two adders in series in the accumulator loop. The second adder is configured as a decrementer by tying the 'b' input to a '1.' The loop performs a function of $A(n)=A(n-1)+B-1$ for cycle n, where constant B is programmable through scan-only latches. Multiple V_{dd} domains (dashed line) are used to facilitate power measurements for different components. The pulse mode switch is inside LCB control.

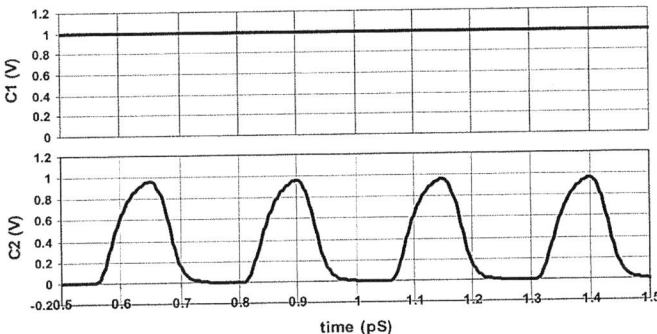

Fig. 3: Functional diagram of LCB and the simulated waveforms in pulsed mode (right). Notice that in non-pulsed mode both c1 and c2 are switching to synchronize the operation. In pulsed mode, only c2 clock switches while c1 clock is held high.

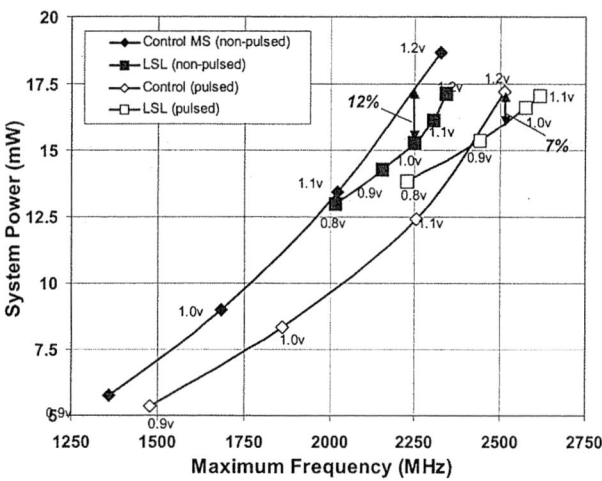

Fig. 4: AC Power vs. F_{max} for control and LSL experiment in non-pulsed and pulsed mode. In the MS experiment, all voltages are changed together; while in the LSL experiment, the V_{dd} is fixed at 1.2 V and only LCB supply is changed.

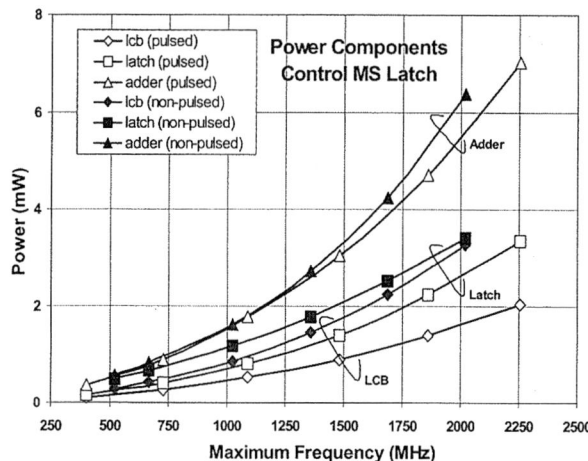

Fig. 6: AC power for individual components vs. F_{max} for control MS latch experiment. The same voltage V_{dd} applied to all components (LCBs, latches and adders) is swept from 0.6V to 1.1V at 0.1V intervals.

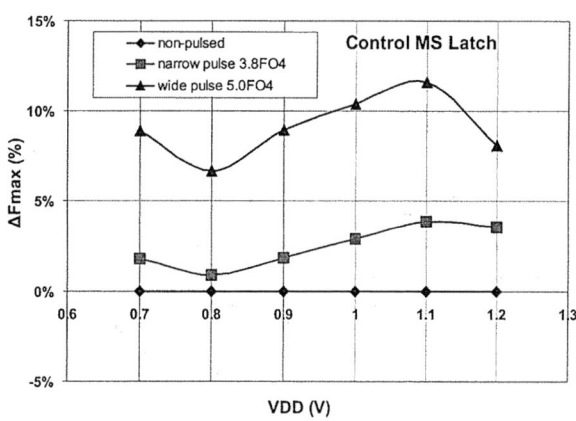

Fig. 5: Comparison of F_{max} for non-pulsed and pulsed MS latch. The y-axis is the percentage change in F_{max} relative to the referenced non-pulsed case. The wider pulse width results in higher operational frequency.

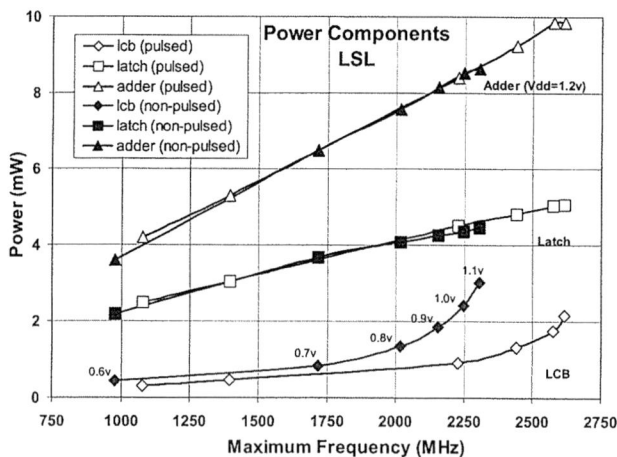

Fig. 7: AC power for individual component vs. F_{max} for LSL experiment. V_{LCB} is swept from 0.6V to 1.1V at 0.1V intervals. V_{dd} is fixed at 1.2V for the adder and the latches.

Temporal Vision-Guided Energy Minimization for Portable Displays

Wei-Chung Cheng
Department of Photonics and
Display Institute
National Chiao Tung University
Hsinchu 30010, Taiwan
1(800)409-9811#31207

waynecheng@mail.nctu.edu.tw

Chih-Fu Hsu
Department of Photonics and
Display Institute
National Chiao Tung University
Hsinchu 30010, Taiwan
1(800)409-9811#59319

cfhsu.iod93g@nctu.edu.tw

Chain-Fu Chao
Department of Photonics and
Display Institute
National Chiao Tung University
Hsinchu 30010, Taiwan
1(800)409-9811#59319

coreychao.iod93g@nctu.edu.tw

ABSTRACT

This paper presents a novel backlight driving technique for liquid crystal displays. By scaling the intensity, frequency, and duty cycle of the backlight, this technique not only increases the perceived brightness but also prolongs the service time of rechargeable batteries. The increased brightness comes from a perceptual effect of temporal vision – a brief flash appears brighter than a steady light of the same intensity, called Brücke brightness enhancement effect. The prolonged service time comes from the relaxation phenomenon – a lithium-ion battery lasts longer by pulsed discharge. Combining these two effects, a great amount of service time can be obtained at the cost of flickering. We performed visual experiments to parameterize the Brücke effect and derived an optimization algorithm accordingly. To demonstrate the potential energy savings of this technique, we profiled the power consumption of an Apple iPod and fabricated an LED driving module. Based on experimental data, 75% of energy consumption can be saved and the service time can be extended to 300%.

Categories and Subject Descriptors

I.4.3 [**Image Processing and Computer Vision**]: Enhancement
– *grayscale manipulation.*

General Terms

Algorithms, Measurement, Design, Human Factors.

Keywords

Power minimization, TFT-LCD, temporal vision, backlight management.

1. Introduction

Power consumption has become the most critical issue for battery-powered electronics. In literature, researchers have found that the display consumes a major portion of the total power consumption in portable devices. The liquid crystal display (LCD) has been widely used in battery-powered portable

Permission to make digital or hard copies of all or part of this work for personal or classroom use is granted without fee provided that copies are not made or distributed for profit or commercial advantage and that copies bear this notice and the full citation on the first page. To copy otherwise, or republish, to post on servers or to redistribute to lists, requires prior specific permission and/or a fee.
ISLPED'06, October 4–6, 2006, Tegernsee, Germany.
Copyright 2006 ACM 1-59593-462-6/06/0010...$5.00.

electronics such as laptop computers, personal digital assistants, cell phones, and global positioning systems. In these portable applications that require small-sized displays, the thin-film-transistor LCD (TFT-LCD) technology is favored thanks to its superb image quality, low manufacturing cost, compact size, and low-voltage power source compared with the other display technologies including CRT (tube), plasma, projection, organic/inorganic LED, etc. The low optical efficiency of the TFT-LCD panel is the major cause of its high power consumption.

A TFT-LCD monitor, as shown in Figure 1, consists of two major components: TFT-LCD panel and backlight module. Each sub-pixel on the panel can be considered as a voltage-controlled light valve. The light valve modulates the amount of light emitted from the backlight to the red, green, or blue color filter. The TFT-LCD panel transmits light for a bright sub-pixel and blocks light for a dark sub-pixel. In other words, in a transmissive display, the desired luminance is obtained by absorbing unwanted light, and energy is wasted in the process. Generally, only less than 5% of light can be delivered to the viewer, while the rest 95% is wasted in the monitor. The nominal transmittance rate of each layer (cf. Figure 1) is listed as follows.

Layer	Transmittance
Front polarizer	90%
Color filter	30%
TFT – aperture ratio	80% - 95%
Liquid crystals	95%
Rear polarizer	50%

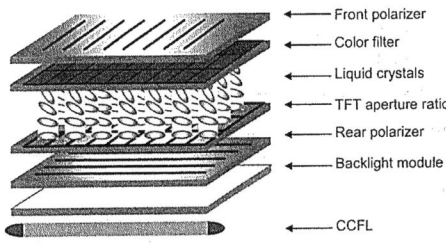

Figure 1. Structure of a CCFL-backlit TFT-LCD panel. From top to bottom: front polarizer, color filters, LC, TFT, rear polarizer, and backlight module.

More than 50% of light is blocked by the rear polarizer. The *aperture ratio*, representing the area percentage not occupied by the TFTs and wires, depends on the pixel circuit design and can be as high as 95%. About 5% of light is absorbed by the liquid crystals. Each of the red, green, and blue color filters transmits roughly 1/3 of visible wavelength. The front polarizer blocks another 10% right before the light exits the panel. The power consumption of the LCD panel is almost constant so it is independent of the panel transmittance (i.e. pixel value). On the contrary, the power consumption of the backlight is a strong function of its output luminance.

Terminologies

Luminance is a physical measure defined as cd/m^2. The luminance of an object can be measured by a luminance meter. *Brightness* is the attribute of a visual sensation according to which an area appears to emit more or less light. *Lightness* is the brightness of an area judged related to the brightness of a similarly illuminated area that appears to be white or highly transmitting. Brightness and lightness are psychophysical terms and cannot be measured by instruments. Intuitively speaking, brightness represents the perceived luminance when there is only one single color in sight, while lightness represents the relative brightness of the color when the reference white is also present [1].

Backlight Luminance Scaling

The luminance of an LCD, L, is the product of the backlight luminous intensity b and the panel transmittance t: $L \cong b \cdot t$. One can decrease the backlight luminous intensity to save the power consumption. The panel transmittance should be increased accordingly such that the luminance remains the same. In addition, for LCDs, higher transmittance can reduce the light leakage problem of liquid crystals and increase the image quality in terms of color saturation and viewing angles.

Consider a pixel consisting of red, green, and blue sub-pixel. Its color is determined by the product of the backlight luminous intensity (b_w) and the transmittance of each sub-pixel (t_R, t_G, t_B):

$$\begin{bmatrix} L_R \\ L_G \\ L_B \end{bmatrix} = b_W \cdot \begin{bmatrix} t_R \\ t_G \\ t_B \end{bmatrix} \qquad (1)$$

$L_R:L_G:L_B$ is the luminance ratio of red, green, and blue. For example, white is obtained at the ratio of 0.27:0.67:0.06. An LCD generates different colors by changing the transmittance ratio of sub-pixels $t_R:t_G:t_B$. By increasing (t_R, t_G, t_B), one can lower b_W to save power and preserve (L_R, L_G, L_B). This class of techniques is called *backlight scaling*. Note that (t_R, t_G, t_B) are bounded by [0,1]. When (t_R, t_G, t_B) need to be greater than 1, the original luminance can not be recovered and image distortion in terms of brightness/contrast occurs.

Backlight Scaling Algorithms

Backlight scaling is by far the most effective technique for reducing power consumption in a transmissive display. To compensate for the visual quality loss due to reduced luminance, proper image enhancement is necessary. Choi *et al.* proposed a technique that increases the pixel values (t) to recover the original luminance (L) [3].

$$\begin{bmatrix} L_R \\ L_G \\ L_B \end{bmatrix} = (\alpha \cdot b_W) \cdot \begin{bmatrix} t_R / \alpha \\ t_G / \alpha \\ t_B / \alpha \end{bmatrix} \qquad (2)$$

Choi's algorithm can preserve the luminance of the dark regions, but the bright regions will be over-saturated. In their work, the number of over-saturated pixels was chosen to evaluate the image quality loss.

Since preserving the original luminance is not always possible, finding a proper alternative transformation of luminance, $L^*=f(L)$, is the key of backlight scaling algorithms. Cheng *et al.* proposed an algorithm to compensate for the luminance loss by increasing the contrast [4]. The following linear transformation was used:

$$L^* = \begin{cases} 0, & L < gl \\ c(L - gl), & gl \leq L \leq gu \\ \alpha \cdot b_W, & gu < L \end{cases} \qquad (3)$$

where c, gl, and gu are constants generated by the optimization algorithm. Although Cheng's algorithm is a compromise between preserving the brightness and preserving the contrast, it does preserve the original *tonality*, i.e., the proportional difference between bright and dark regions. The relationship between brightness and contrast, however, was employed without substantial support.

Iranli *et al.* proposed using *histogram equalization*, an image processing algorithm that balances the number of pixels on each graylevel, to perform the image enhancement [5]:

$$L^* = h'(L), \qquad (4)$$

where h' is the derivative of the cumulative distribution function of the histogram. Histogram equalization can reproduce each gray level distinctly without over-saturation or under-saturation. However, tonality will be distorted when the original histogram tends to be irregular.

Backlight Blinking

Conventional requirements of backlight design are spatial, temporal, and chromatic uniformity. Recently, the modern LCD technologies call for different backlight driving methods. For example, *backlight blinking* is adopted by LCD-TVs to deal with the *motion blur* problem. Unlike CRT monitors, because of the longer response time of liquid crystals, when steady backlight is used in an LCD, the fast moving edges appear to be blurred and degrade the sharpness of motion pictures. One solution is to pulse-drive (or "blink") the backlight in order to generate CRT-like pulses. According to temporal vision study, a pulse-type display (e.g. CRT) is immune from motion blur than a hold-type display (e.g. LCD).

For a blinking backlight driven by square wave of frequency f, duty cycle d, and intensity b_w, equation 1 can be rewritten as

$$\begin{bmatrix} L_R \\ L_G \\ L_B \end{bmatrix} = (d \cdot b_W) \cdot \begin{bmatrix} t_R \\ t_G \\ t_B \end{bmatrix}. \qquad (5)$$

The duration of each blink is d/f. Although the time-average luminance can be simply determined to be $d \cdot b_w$, the perceived brightness of blinking backlight, however, is not a simple function of luminance.

2. Temporal Vision

A number of temporal visual effects are involved in our study.

Brightness of Flashing Light

The *Talbot-Plateau* law states that the brightness of a temporally modulated stimulus, when fused, is equal to the brightness of a steady light with the same time-averaged luminance. The concept is similar to pulse-width-modulation (PWM). Consider a series of flashes in a square waveform, which can be qualified by its frequency *f*, duty cycle *d*, and intensity (magnitude) *m*. The Talbot-Plateau law recognizes that two flashing lights have the same brightness if $m_1d_1 = m_2d_2$. Note that when $d=100\%$ the light is steady. In other words, if two flashing lights have the same time-average luminance, they have the same brightness.

The *Broca-Sulzer* effect states that the suprathreshold stimuli, whose duration is on the order of 50-100 milliseconds, appear brighter than stimuli of either shorter or longer durations [6]. This effect was discovered in 1902, when the flashing pattern of lighthouses was of great interest. The findings still inspire the design of electronics nowadays. For example, how to design the flashing pattern of the LED warning indicators of a cell phone such that it can efficiently draw the user's attention with minimum energy. Figure 2 depicts the time course of flashing lights of different intensity.

The Broca-Sulzer effect cannot be intuitively comprehended based on daily visual experience: A longer flash looks dimmer than a shorter flash. For example, "a 50ms light looks dimmer than a 40ms light" means that our visual sensation of a 40ms flash will be subtracted 10ms later. The phenomenon has been revisited by modern techniques and found to be caused by the neural mechanism.

The *Brücke brightness enhancement effect* is that when a light is flickered on and off, its brightness varies according to the frequency of flicker, reaching a maximum over a narrow range of frequencies at approximately 5 to 20 Hz, depending on the intensity of the flickering light [6].

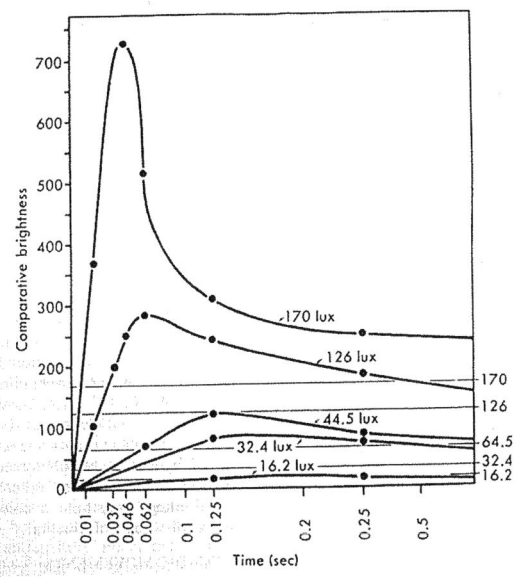

Figure 2. Broca-Sulzer effect: Brightness of flashes having various luminances, as functions of flash duration [6].

Flickering

The *critical flicker fusion frequency* (CFFF) indicates the transition from the perception of flicker to that of fusion occurs over a range of temporal frequencies. The *Ferry-Porter law* predicts that the CFFF increases as the luminance of the flashing stimulus increases:

$$f_{CFF}(L) = a \cdot \log(L) + b. \qquad (6)$$

The CFFF is not only a function of luminance, but also the stimulus size (*Franit-Harper law*), wavelength (*Hecht-Shlaer law*), etc. In this paper, only luminance is considered, because the display size is fixed and the colors of displayed image have to be preserved.

3. Experimental Platform

Power/Energy Characterization

To demonstrate our concept, we measured and characterized the power and energy consumption of an Apple iPod®, a globally available portable device which is capable of playing mp3 music and mpeg4-compressed video clips. Its major components include a hard disk drive, an LED-backlit LCD, a lithium-ion battery, a button/wheel interface, and a video processor (cf. Figure 3).

Figure 3. iPod's backlight consists of four white LEDs in series.

Figure 4 shows the power profile of playing a video clip. The spikes in the very beginning occurred when a video clip started to play, where access to the hard disk drive consumed a significant amount of power. The following table lists the power consumed in different states of the 1.8", dual-disk, 4,200 rpm Toshiba MK6008GAH hard disk drive.

State	Power (mW)
Start	1800
Reading, Writing, Seeking	1000 - 1100
Idle	400
Standby, Sleep	7 - 12

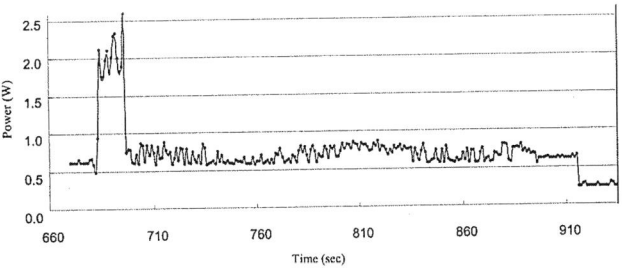

Figure 4. Power profiling of iPod playing a 323-second video clip.

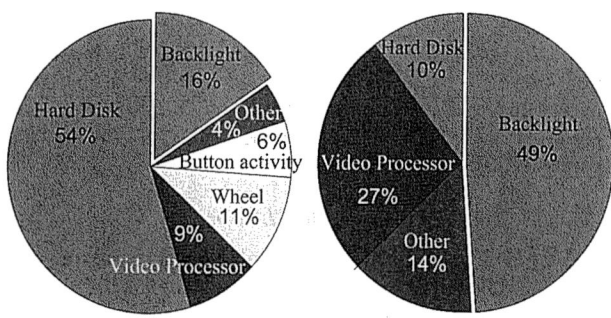

Figure 5. Power (left) and energy (right) breakdown of iPod playing a 323-second video clip. The backlight consumes 49% of total energy.

Although the hard disk drive consumes up to 54% of power spontaneously, compared with 16% by the backlight, it entered the idle state right after the video data was fetched. The right pie chart shows the breakdown of energy consumption. The backlight consumes as much as 49% of the total energy.

We performed luminance and power characterization of the LED backlight by a Konica-Minolta CS-200 chroma meter. The results are shown in Figure 6.

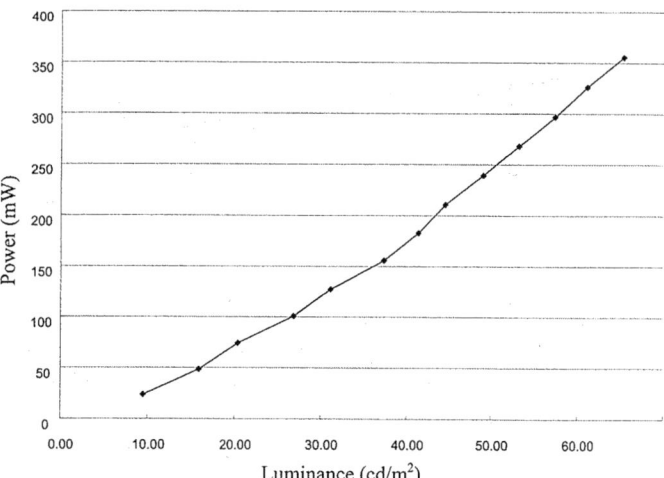

Figure 6. Power vs. luminance of the iPod backlight.

Based on the measurement data, the luminance vs. power relationship can be modeled by the following quadratic function:

$$P = 0.0374b^2 + 3.2194b - 10.5760 \quad (\text{mW}) \quad (7)$$

The *relaxation phenomenon* of lithium-ion batteries is that: after draining an impulse of current, if the cell is allowed to relax long enough, then the concentration gradient decreases and a charge recovery takes place at the electrode [7]. We used the *B#* battery simulator [8] to simulate the relaxation phenomena over a range of different duty cycles. The following parameters were used: V_{init}=4.5V, $V_{cut-off}$=3.7V, I=0.8A, and $V_{cut-off}$=3.7V. The simulation results are shown in Figure 7. The battery service time can be modeled by an exponential function:

$$S = 64 \cdot (9.5256 e^{-2.2885d}) \quad (\text{second}) \quad (8)$$

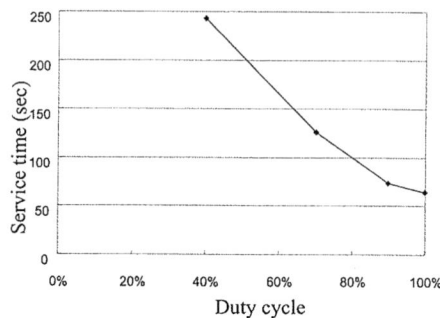

Figure 7. Relaxation phenomenon [7] simulated by the B# battery simulator [6]. Under the given conditions, as duty cycle decreases, the service time increases exponentially.

LED Driver and Illuminance Sensor

We fabricated an LED driving module, which is capable of driving six LED backlights with controllable intensity, frequency, and duty cycle. We also crafted a simple illuminance sensor by using a photoresistor and an analog-to-digital converter IC to detect the ambient light.

Figure 8. Fabricated LED driving module (left) and illuminance sensor (right).

4. Visual Experiments

We conducted visual experiments to parameterize the Brücke brightness enhancement effect, Ferry-Porter law and the relationship between favorite display luminance and ambient light.

Ferry-Porter Law

The conventional psychophysical *method of adjustment* was used to find the CFFF of three observers in a dark laboratory [2]. The iPod backlight was driven by 50% square waveforms at frequency 20, 25, 30, 35, and 40 Hz. The 4cm*5.5cm backlight was placed 70 cm away from the observer. Each observer was asked to find the CFFF by adjusting the intensity. The results are shown in Figure 9.

By linearly fitting the data to equation 6, the Ferry-Porter law can be modeled as

$$f_{CFF}(L) = 9\log(L) + 28. \quad (\text{Hz}) \quad (9)$$

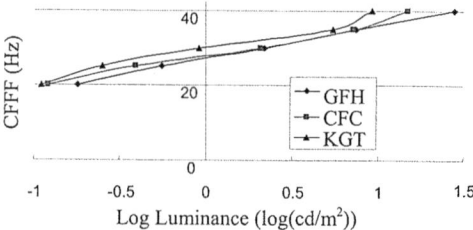

Figure 9. Experimental data of the Ferry-Porter law from three observers as CFFF vs. log(luminance).

Brücke Brightness Enhancement Effect

In the experiment of Brücke effect, two 1-degree white LED lights were placed side by side for the observer to match. One was driven by adjustable constant intensity, while the other was driven by 50% square waveforms over a range of 30, 40, and 50 cd/m^2. Each observer was asked to match the brightness by adjusting the steady light. The results are shown in Figure 10.

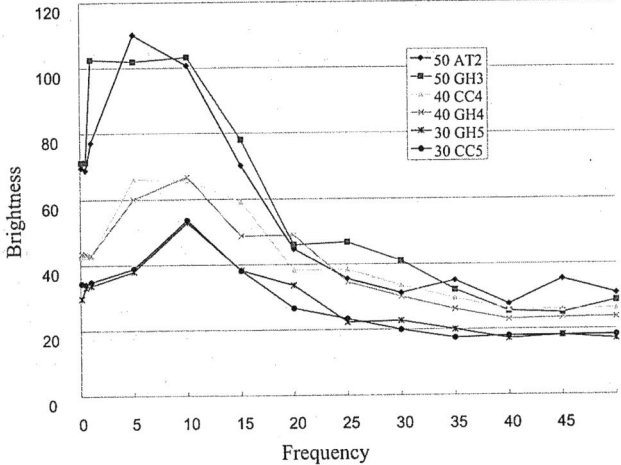

Figure 10. Experimental data of the Brücke effect: Brightness vs. frequency from 3 observers at luminance 30, 40, and 50 cd/m^2.

In Figure 10, flashes with higher intensity have higher brightness. According to equation 9, the CFFFs are 43, 42, and 41 Hz for 50, 40, and 30 cd/m^2, respectively. Beyond 43 Hz, the brightness was about half of the intensity because of the 50% duty cycle. When the frequency approached to zero (DC), the brightness reached about the full intensity. At these low frequencies, the observers could distinct the on-cycles from the off-cycles and chose the high brightness of on-cycles to match. The brightness reached the maximum of about twice intensity around 10 Hz. The frequency range between CFFF and 10 Hz of Figure 10 is redrawn as brightness vs. period in Figure 11.

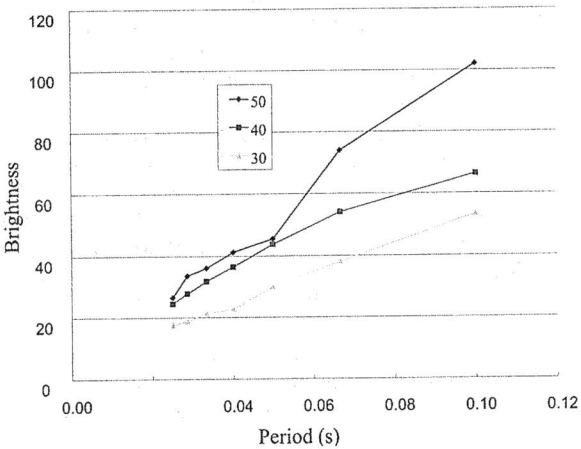

Figure 11. In the range between CFFF ($\cong$ 42 Hz) and 10 Hz, the relationship between brightness and 1/f can be linearly approximated.

Summarizing the above observations, we approximate the Brücke effect by:

$$L'_{50\%}(L,f) = \begin{cases} L/2, & f_{CFF}(L) < f \\ \dfrac{L}{2} + (\dfrac{1}{f} - \dfrac{1}{f_{CFF}(L)})(\dfrac{3L/2}{0.1 - \dfrac{1}{f_{CFF}(L)}}), & 10 < f < f_{CFF}(L). \end{cases} \quad (10)$$

Surrounding Effects

The surround luminance is one of the most important factors in visual sensation. When the user adapts to a dark surround, he/she may dim the backlight to the lowest level and still has the full range of lightness and chroma. In the mean time, a considerable amount of power savings is achieved without any side effect. In literature of display ergonomics such as TCO'03, the luminance ratio of display to surround is recommended to be set between 10:1 and 100:1. We conducted visual experiments to find the relationship between favorite display luminance vs. surround luminance. We visited three users in different offices and measured the surround illuminance. The users were asked to perform different tasks at different levels of surround illuminance with their favorite display luminance. If we assume the reflectance of the surround is similar to middle gray (i.e. 18% reflectance, Munsell N5), then the reflected luminance can be estimated as a linear function of illuminance. The results are shown in Figure 12.

The tasks included movie watching, web surfing, and text editing. Generally the favorite display luminance increases linearly as the surround illuminance. The movie watching task had much lower display luminance because the display was driven in the *direct draw* mode. For the same user, text editing had lower display luminance than web surfing in order to reduce eye strain. The curves have different trends in the bright portion (>100 lux) and dark portion (<100 lux). The reason may be the users switching between the photopic mode (light adapted, cone-dominating) and scotopic mode (dark adapted, rod-dominating).

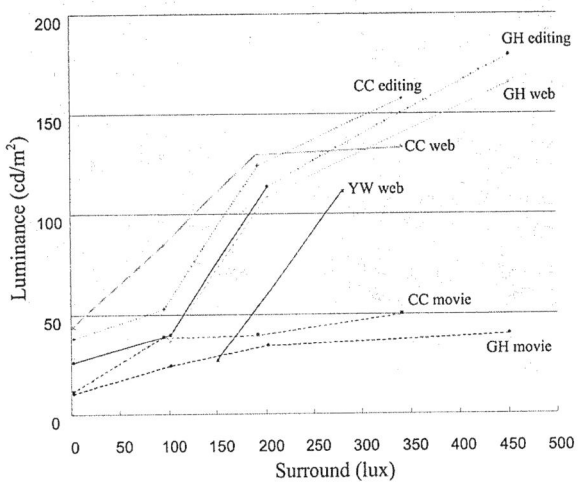

Figure 12. Favorite display luminance vs. surround illuminance of three users on movie watching, web surfing, and text editing.

5. Proposed Algorithm

Based on equation 10, we can reduce the magnitude and frequency of the backlight and still obtain the same brightness. Assume the following system consisting the abovementioned blocks.

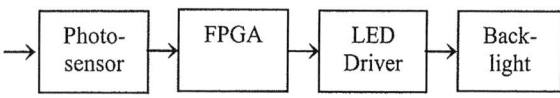

For the brightness of a steady light of luminance L, we blink the backlight with the following duty cycle, magnitude and frequency instead:

$$d^* = 50\% \tag{11}$$

$$m^* = L/2 \tag{12}$$

$$f^* = \frac{2}{3}f_{CFF}(L) + \frac{10}{3}. \tag{13}$$

By equation 8 and 10, we can estimate the power consumption and battery service time as follows.

	Baseline	Backlight Scaling	Proposed
Power consumption	100%	50%	25%
Battery service time	100%	235%	314%

When the backlight intensity is reduced to 50%, the existing backlight scaling techniques [3-5] can cut power consumption by 50% and extend battery service time to more than 235%. The proposed technique, on the other hand, can cut power consumption by 75% and extend battery service time to more than 300% while preserving the same brightness at the cost of flickering. The visual effects of different techniques are shown as follows.

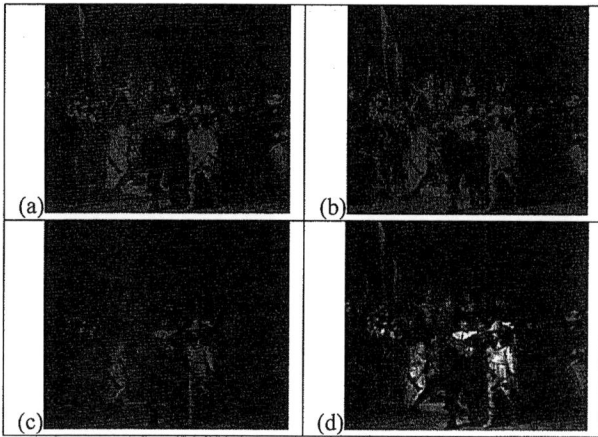

Figure 13. Simulated visual effects of (a) Choi, (b) Iranli, (c) Cheng, and (d) Proposed algorithm without showing the flickering.

Since flickering cannot be reproduced on paper, we suggest the following simple experiment to experience the flickering. Use a CRT monitor driven by analog signals from a computer, in which the refresh rate is adjustable. Note that an LCD, which re-samples the refresh rate, will not do. First, dial the luminance of the CRT (commonly labeled as "brightness" mistakenly) to the maximum, where the flickering is more pronouncing. Shortening the viewing distance can also enhance the perceived flickering. Then adjust the refreshing rate to the point where flickering starts to appear. Let this refreshing rate, say 75 Hz, be the critical fused flickering frequency. To experience a 20% loss in flickering, adjust the refresh rate from 75 to 60 Hz and observe the flickering.

6. Conclusions

We have presented a novel backlight driving technique for liquid crystal displays. By scaling the intensity, frequency, and duty cycle of the backlight, this technique not only increases the perceived brightness but also prolongs the service time of rechargeable batteries. A great amount of energy can be saved at the cost of flickering. We have employed the Brücke brightness enhancement effect from temporal vision and the relaxation phenomenon of lithium-ion batteries. Although the preliminary results are encouraging, this study is still in its infancy. Our future works include developing a metric for measuring the flickering, refining the visual experiments and collecting more data, and implementing the proposed algorithm in FPGA.

7. ACKNOWLEDGMENTS

This work was partially supported by MOEA Technology Development for Academia Project #94-EC-17-A-07-S1-046, MOE ATU Program "Aim for the Top University" #95W803, and Chunghwa Picture Tubes Ltd. #94C148.

8. REFERENCES

[1] G. Wyszecki and W. S. Stiles, *Color Science*, John Wiley and Sons, 1982.

[2] R. Sekuler and R. Blake, *Perception, 3rd Ed.*, NY: McGraw-Hill, 1994.

[3] N. Chang, I. Choi, H. Shim, "DLS: Dynamic backlight luminance scaling of liquid crystal display," *IEEE Tran. VLSI*, Vol. 12, No. 8, Aug. 2004, pp. 837-846.

[4] W-C. Cheng and M. Pedram, "Power minimization in a backlit TFT-LCD display by concurrent brightness and contrast scaling," *Proc. of Design Automation and Test in Europe*, Feb. 2004, pp. 252-259.

[5] A. Iranli, H. Fatemi, M. Pedram, "HEBS: Histogram equalization for backlight scaling," *Proc. of Design Automation and Test in Europe*, Mar. 2005, pp. 346-351.

[6] R. A. Moses and W. M. Hart, Jr., *Adler's Physiology of the eye*, 8th Ed., Washington, DC: The C. B. Mosby Company, 1987, pp. 435-436.

[7] C. F. Chiasserini, R. R. Roy, "Pulsed battery discharge in communication devices," *Mobicom 1999*, pp. 88-95.

[8] C. Park, J. Liu, P. Chou, "B#: A battery emulator and power-profiling instrument," *IEEE Design and Test of Computers*, Vol. 22, No. 2, Mar/Apr, 2005, pp. 150-159.

Dynamic Current Modeling at the Instruction Level

Jose Rizo-Morente
UCD School of Computer
Science and Informatics,
Belfield Campus
Dublin, Ireland
jose.rizo-morente
@ucd.ie

Miguel Casas-Sanchez
UCD School of Computer
Science and Informatics,
Belfield Campus
Dublin, Ireland
miguel.casassanchez
@ucd.ie

C.J. Bleakley
UCD School of Computer
Science and Informatics,
Belfield Campus
Dublin, Ireland
chris.bleakley
@ucd.ie

ABSTRACT

Estimation of processor current consumption is important for the design of low power systems. This paper proposes a novel method for estimating the dynamic current consumption of a processor. The method models dynamic current as the output of a linear system excited by a signal comprised of the total current due to each instruction. System identification is performed by cross-correlation of a pseudo-random stimulus with the measured current. The method was applied to the Texas Instruments TMS320VC5510 DSP and was found to provide an average correlation of 93% between estimated and measured dynamic current across a range of benchmarks.

Categories and Subject Descriptors

C.4 [**Computer System Organization**]: Performance of Systems—*Modelling techniques measurement techniques*; C.3 [**Computer System Organization**]: Special-Purpose and Application-Based Systems—*Signal processing systems real-time and embedded*; C.1.4 [**Processor Architectures**]: Other Architectures Styles—*Pipeline processors*; I.6.4 [**Computing Methodologies**]: Simulation and modelling —*Model Validation and Analysis*

General Terms

Measurements, Experimentation, Design, Verification

Keywords

Dynamic Instruction-level current model, current and power measurement in a processor

1. INTRODUCTION

Power consumption has become an important issue in embedded system design. As a consequence power aware techniques have been applied at every stage in the design cycle: technology, circuit, logic, architecture, system and software. Traditionally, all efforts were focused on hardware design while software was ignored from the power perspective. While the underlying hardware is the source of power dissipation, the program executing on the architecture needs to be taken into account since the machine is under its control.

The impact of software on power consumption is particularly evident for complex architectures, such as DSP processors. Modern DSP processors offer the programmer a high degree of parallelism. This leads to a high variation in power consumption between instructions which exercise the maximum number of parallel functional units and those which exercise only one. For these processors, power cannot be accuracy predicted from cycle count alone.

Instruction-level power models estimate the current consumption of a processor by analysis of the instructions to be executed. This is considerably more time and cost efficient that using electronic equipment to directly measure power consumption. However, at present, almost all instruction-level models estimate mean power consumption over a long time windows, typically hundreds or thousands of instructions in duration. Although useful for some applications, this is not sufficient for others for which current would ideally be estimated on a cycle-by-cycle basis.

Estimation of dynamic current is important for several applications. Knowledge of the dynamic current consumption of an application can assist programmers in optimizing software for power consumption. Recent work on iterative compilation and compile-space exploration has shown the utility of accurate power estimation within an automated design flow [3]. Battery life can be extended by smoothing the current peaks which occur in mobile devices [8]. It has been shown that encryption systems may be cracked by current analysis attacks. These attacks could be prevented by re-ordering instructions such that the current consumption of the processor is constant [6].

In this paper, a new instruction-level model for dynamic current consumption estimation is presented. A method for identification of the model is described. The model and method are applied to a commercial DSP processor. The results of the estimation technique are presented and compared with measurements.

The rest of the paper is organized as follows: Section 2 reviews related work already published in the literature; Section 3 introduces the dynamic current model. Section 4 explains the identification method. Section 5 describes the application of the model and method to the target processor. Results are given in Section 6. Conclusions and future work are provided in Section 7.

Permission to make digital or hard copies of all or part of this work for personal or classroom use is granted without fee provided that copies are not made or distributed for profit or commercial advantage and that copies bear this notice and the full citation on the first page. To copy otherwise, to republish, to post on servers or to redistribute to lists, requires prior specific permission and/or a fee.
ISLPED'06, October 4–6, 2006, Tegernsee, Germany.
Copyright 2006 ACM 1-59593-462-6/06/0010 ...$5.00.

2. RELATED WORK

The power consumption of processors has been modelled at the hardware and instruction levels. Hardware-level power models calculate power and energy consumption from detailed descriptions of the hardware, such as circuit, gate, register transfer and system models. Hardware-level models are slow to simulate and so are impractical for estimation of the power consumption of entire processors. These models often cannot be applied due to the lack of circuit and gate level information [16]. Instruction-level power models deal only with instructions and functional units from the software point of view without detailed knowledge of the underlying circuit architecture. In this context, the term 'instruction' corresponds to the combination of factors such as operation code, addressing mode, operand formats and processor resources which determine execution. Several Instrucion Level Power Analysis (ILPA) models have been proposed.

The ILPA model, introduced by Tiwari et al. [16, 14, 15], states that the energy consumption of a program can be computed by summing up the energy cost associated with each instruction plus the incremental inter-instruction cost associated with switching between consecutive pairs of instructions. In order to populate the model, the current consumption of each instruction, and of each instruction combination, must be measured for the target processor. This method has a small margin of error, typically 2 to 4 percent for simple processors. Tiwari's methodology has been succesfully applied to numerous processors [2].

ILPA models are not intended to model the dynamic variation of the processor current consumption. The models attribute a total current consumption to each instruction or instruction parameter and do not model when the current is actually drawn. As such, ILPA are effective in deriving the long term mean current but do not attempt to estimate the cycle-to-cycle current variation.

In [7] the problem of modeling the dynamic current trace is addressed. The model is based on a gamma function. This function is used to approximate the dynamic current consumption due to execution of an individual instruction. Results for the SC140 DSP processor core show an average error of less than 2.2%, and an average correlation coefficient of 98% between estimated and measured current traces. Although good results are reported, the gamma function is not general enought to describe instruction current consumption profiles for all processors. The basic current shape for the SC140 architecture was well described by gamma functions but other processors have different dynamic current consumption profiles. The gamma function is described and compared to the instruction current profile of our target processor in Section 3.

3. DYNAMIC MODEL

The Tiwari instruction-level model estimates the mean current consumed by a segment of code by summing the current consumption due to each instruction and due to inter-instruction effects [16]. Typically, the Tiwari model estimates the mean current consumption over several thousand clock cycles. The instruction and inter-instruction current consumptions are determined experimentally by putting the instructions in infinite loops and measuring the current. Based on this, the mean energy consumption of a program is expressed in the form:

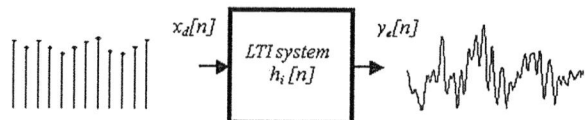

Figure 1: System model.

$$E_p = \sum_i (B_i N_i) + \sum_{i,j} (O_{i,j} N_{i,j}) + \sum_k E_k \qquad (1)$$

where B_i is the base cost of instruction i, N_i is the number of occurrences of instruction i, $O_{i,j}$ is the circuit state overhead for the instruction pair i, j, $N_{i,j}$ accounts for number of successive appearances of instruction i and j and E_k is the energy costs dissipated to other effects.

The Tiwari model does not capture the dynamic current consumption of the processor. This paper proposes a new dynamic model that estimates the current drawn by the processor in every clock cycle.

The dynamic current behaviour is determined by the power supply system including the processor, regulator, decoupling components and printed board circuit [9]. We model the dynamic current consumption of a processor power supply system as the output of a linear system excited by an input signal consisting of the total current consumption due to each instruction including the inter-instruction effect. Thus the estimated instantaneous current $y_e[n]$ can be calculated as the convolution of the discrete input signal, $x_d[n]$, with the system impulse reponse $h_i[n]$:

$$y_e[n] = \sum_{k=0}^{N} h_i[k] x_d[n-k] \qquad (2)$$

The input signal $x_d[n]$ can be derived by applying the conventional static Tiwari model to each individual instruction in the execution trace. For this work, a previously published static model [2] for the TMS320VC5510 DSP processor was employed. As will be shown in the next section, the system impulse response $h_i[n]$ can be determined using system identification techniques. The model is depicted in Fig. 1.

Our model assumes that:

- The system is Linear-Time Invariant (LTI). For the current and voltage ranges in question, the passive components in the processor power supply system behave in a linear fashion. The only non-linear effect arises from the DC-DC voltage regulator. Voltage regulators have a reference voltage and a feedback loop [4] which introduce some non-linear behaviour. However, as will be shown, this non-linear effect does not significantly reduce the accuracy of the model.

- The dynamic current profile for all instructions is the same. Clearly, different instructions draw different proportions of their total current consumption in different stages of the processor pipeline. For example, some instructions require memory accesses while others do not. Since the impulse response of the system is significantly longer than the depth of the processor pipeline, the error introduced by ignoring pipeline effects is small.

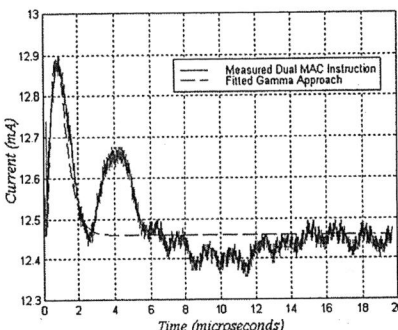

Figure 2: Difference between the gamma model and the actual dual MAC current consumption.

- The static model is accurate. For complex instruction set architectures, such as the one consider herein, it is not feasible to characterize the entire instruction set. In this work, uncharacterized instructions are matched to functionally similar characterized instructions and the total current consumption is assumed to be equal. This is a source of error and is an item for future work. The static model does not consider the data dependency of current consumption. Experiments were conducted across a range of instructions operating on data at varying Hamming distances. For practical DSP routines, the mean data dependency error was found to be 1.5% of total current per instruction. The static model does not cover accesses to different memories banks. Current estimation was conducted on the basis that source code and data is placed in SARAM and DARAM respectively. Different memory configuration may give rise to an error of up to 12% in total current estimation for a single instruction.

The model presented in this paper differs from that used by Muresan [7] in the basic function employed to describe the instantaneous current consumption due to instruction execution. The Muresan model uses a mathematical function, gamma, to approximate the current trace due to execution of a single instruction. In contrast, our model uses a generalized impulse response. The gamma function is defined as:

$$g(t; n, \lambda) = \frac{\lambda(\lambda t)^n}{n!} e^{-\lambda t} \qquad (3)$$

where t is the time variable of the instruction current trace, and parameters λ and n are determined by fitting the gamma function to the measured current trace. Unfortunately, the gamma function is not sufficient to accurately represent all possible current traces. For example, the current trace obtained for our target processor is not well modelled by a gamma function. Fig. 2 represents both the fitted gamma function, and the captured current trace resulting from execution of a dual MAC instruction. It can be seen that the gamma function describes just the first part (50.1%) of the measured trace. The remainder of the trace is not modelled (49.1%). This leads to a significant reduction in the accuracy of dynamic current estimation.

4. SYSTEM IDENTIFICATION

In order to apply our dynamic model, it is necessary to estimate the impule response of the system. In general, system identification is divided into parametric and non-parametric methods [5]. In parametric methods, a system model is assumed and identification is oriented towards extraction of the model parameters. In non-parametric methods, no assumption is made about the system model and identification is used to directly compute the system response. Non-parametric methods include: correlation analysis, transient-response analysis, and frequency response analysis [5], [1].

In this paper, system identification is carried out by means of cross-correlation analysis. Consider the cross-correlation, $R_{xy}[l]$, of the system input and output signals, $x[n]$ and $y[n]$ respectively:

$$R_{xy}[l] = \sum_{n=-M}^{+M} x[n]y[n-l] =$$
$$\sum_{n=-M}^{+M} h[n]R_{xx}[l-n]$$

where $R_{xx}[l]$ is the auto-correlation of the input signal.

In the case that the auto-correlation of the input signal tends to a Dirac impulse function $\delta[l]$, then the cross-correlation of the input and output signals tends to the impulse response of the system $h[l]$:

$$R_{xx}[l] = \delta[l] \longrightarrow R_{xy}[l] = h[l] \qquad (4)$$

During system identification, the input signal is controlled by modifying the program to be executed. The static current model is utilized to determine the input signal. The corresponding system output is measured using a current probe. The program to be executed is specially designed so that the auto-correlation of its static current is a Dirac impulse. Maximal length sequences are used for this purpose. Pseudo-random binary sequences may be generated using Linear Feedback Shift Registers (LFSRs). A maximal length LFSR sequence (m-sequence) has the property that its period is equal to $2^m - 1$ where m is the number of bits in the LFSR [5]. Its circular auto-correction $R_{mm}(l)$ is given by:

$$R_{mm}[l] = \begin{cases} 1, & \text{if } l = k(2^{m-1}), \\ \frac{-1}{2^m-1}, & \text{if } l \neq k(2^{m-1}). \end{cases}$$

As well as having an impulsive circular auto-correlation, m-sequences also have the advantage that they are two valued. Thus the desired input can be generated by executing a stimulus program utilizing just two statically characterized instructions.

A further advantage of this system identification method is that no electrical parameters of the processor chip or printed circuit board need to be determined. This allows dynamic models to be derived by third-parties, not just the chip manufacturer.

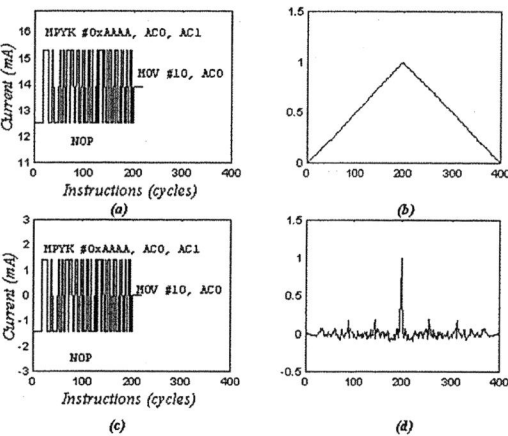

Figure 3: Input Generation: (a) Actual input signal and (b) its auto-correlation. (c) Zero mean input signal and (d) its auto-correlation

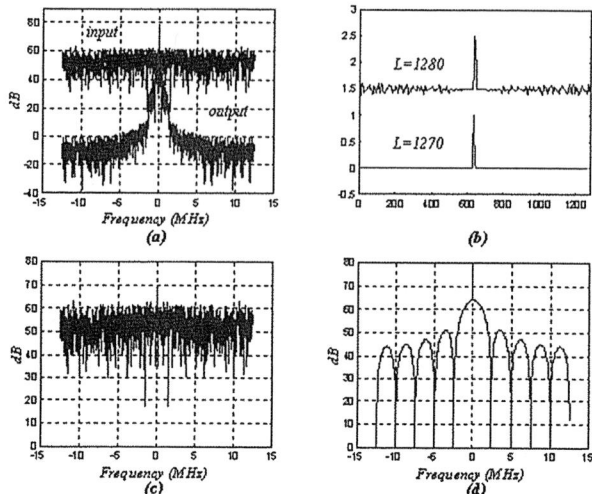

Figure 4: Input Generation: (a) Input and output spectrums with $X_c=1$; (b) Auto-correlation for LFSR signals of 1280 and 1270 length; (c) Input spectrum with $X_c=1$; (d) input spectrum with $X_c=10$ and 1270 length.

5. METHOD

5.1 Measurement framework

The target DSP processor used for the study is a Texas Instruments TMS320VC5510, with variable core voltage (0.9-1.6 V) and operating frequency of up to 200 MHz. It incorporates several special architectural features pertinant to the study. Among them, there are low power capabilities, parallel features, on-chip Single Access RAM (SARAM) and Double Access RAM (DARAM), two independent 40 bit MAC units and one 40 bit ALU [11].

The physical measurement methodology was applied using the C5510 Development Software Kit [10], with a 1.6 V core voltage and 24 MHz operating frequency connected to a PC running TI Code Composer Studio (CCS) version 2.56. CCS was used to download and run the test programs. External software routines were used to trigger measurements using a digital storage scope. The current drawn was measured with a non intrusive 0.1 mA resolution current probe. The probe bandwidth is 50 MHz, enough for these experiments. In order to avoid noise, each measurement was averaged over 128 instances. The measurements are completely repeatable.

5.2 Identification procedure

The cross-correlation method requires that the input signal meets certain requirements. Firstly, the inputs must be discrete and have a finite number of levels as they must be generated from instructions. Secondly, the auto-correlation of the input signal must be close to an ideal impulse. Thirdly, the inputs must be zero mean in order for the auto-correlation to be delta shaped. Fourthly inputs must concentrate as much energy as possible in the bandwidth of the system to be identified.

The first three requirements were met by using maximal length LFSRs to generate the stimulus. The binary LFSR outputs were mapped to a low consumption NOP instruction and a high consumption MPY instruction. The mean was subtracted to shape the auto-correlation as an impulse. These signals along with their auto-correlation functions are illustrated in Fig. 3.

The fourth requirement is related to the system bandwidth. Efficient input spectrums for identification have most

energy concentrated within the bandwidth of the system. The bandwith of an LFSR derived input signal can be controlled by altering the duration of the LFSR period. In our experiments, this was achieved by adjusting X_c, the number of processor clock cycles per LFSR output. This leads to an input signal length of L cycles where:

$$L = (2^m - 1)X_c \qquad (5)$$

For a clock frequency of F_s, this time scaling increases energy in the frequency range from 0 to F_s/X_c.

Several experiments were conducted to estimate the bandwidth of the system. These experiments were performed at a processor clock frequency of 24 MHz. Input signals were applied to the system with X_c factors varying from 1 to 20. For each experiment, both the input and output spectrum were analysed along with the system identification results. 95% of the energy system was found to be within the range 0-1.5 MHz. X_c equal to 10 increases the energy in the range of 0-2 MHz and was found to provide the most accurate identification.

Fig. 4 (a) shows the input and output magnitude spectrums for $X_c=1$. The output spectrum magnitude is -20 dB for frequencies greater than 2 MHz. Fig. 4 (d) shows the spectrum of the input signal generated with $X_c=10$. This signal has more energy within the system bandwidth, providing more accurate system identification. Due to the time scaling process, certain harmonic frequency components are not present in this input signal. Since these components are outside the system bandwith, this does not signifcantly reduce the accuracy of the identification process.

The period of the input signal must be longer than the impulse response of the system and less than the storage capacity of the digital oscilloscope. In our case, X_c and m were chosen to be 10 and 7 respectively, resulting in $L=1270$. Fig. 4 (b) plots the circular auto-correlations of length 1280 and 1270 input signals, illustrating the low noise property of the latter.

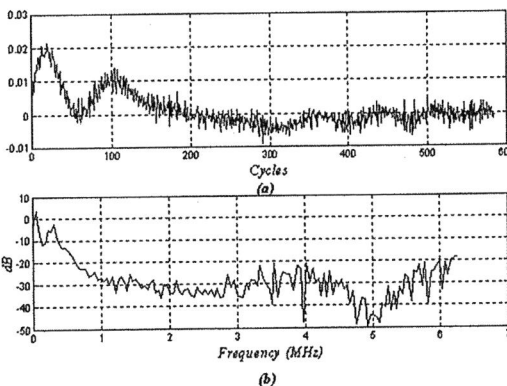

Figure 5: (a) Estimated impulse response along with (b) its spectrum

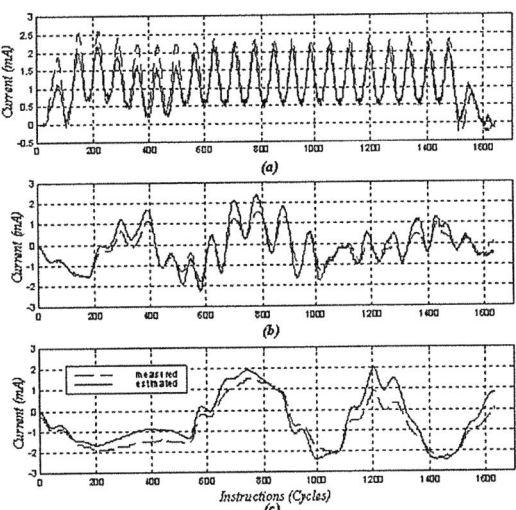

Figure 6: Estimated and measured current consumption for: (a) periodic pseudo-random sequence with $X_c=10$; (b) aperiodic pseudo-random sequence with $X_c=10$; (c) aperiodic pseudo-random sequence with $X_c=30$.

6. RESULTS

The impulse response and transfer function obtained for the TMS320VC5510 are plotted in Fig. 5. The impulse response is truncated to 600 cycles since later samples are close to zero. This also ensures that the filter has unit DC gain. As noted in Fig. 5 (b), the bandwidth of the system is less than 1 Mhz. This bandwidth does not depend on the processor clock frequency but on the electrical properties of the power supply system.

Additional identification experiments were performed at processor clock frequencies of 48 Mhz and 200 Mhz. The experiments resulted in impulse responses that showed squared correlation coeffients ($CC^2[\%]$) of 98% and 91% when compared with the impulse response obtained at 24 Mhz. The effect of changing the instructions that form the stimulus was also assessed. The experiment was repeated with inputs comprised of various instructions including MPY, MOV, MAC, ADD and Dual MAC. The worst $CC^2[\%]$ value between the estimated impulses was 96%.

The dynamic current consumption of the processor was estimated and measured for ten pseudo-random instruction sequences. The pseudo-random sequences were comprised solely of characterized NOP and MPY instructions. Fig. 6 shows the estimated and measured traces for three of the sequences. The accuracy of the results was assessed by calculation of the square of the correlation coefficient ($CC^2[\%]$), the relative error of the mean current ($RelMError[\%]$) and the root mean square error ($RMSError$) between the estimated and measured vectors. An average CC^2 of 91.11% was obtained while the average $RelMError$ and the average $RMSError$ were 1.2% and 0.338 mA (6.3%) respectively.

The dynamic current consumption was also estimated for five benchmarks from the TI Signal and Image Libraries [12], [13]. These benchmarks cover a range of DSP algorithms such as complex Fast Fourier Transform (FFT), filtering, convolution, correlation and math functions. The benchmarks exercise most the features of the target DSP: software loops, zero-overhead loops, parallel and single instructions. The benchmarks are briefly described in the following list:

- Iir4. Computes a cascaded IIR filter of 16 biquad sections using 32-bit coefficients and 32-bit delay buffers. Each biquad section is implemented using Direct-form II with 4 coefficients.

- Cfft. Computes a complex 512-points IFFT.

- jpeg_qize. Performs the quantization step in image/video compression.

- Log10. Computes log base 10 of a vector.

- Dct_idct. Comprises 2-D forward Discrete Cosine Transform (FDCT), format conversion and Inverse Discrete Cosine Transform (IDCT) for a 8x8 block.

For each benchmark, the estimated and measured current traces were obtained and analyzed. Fig. 7 shows the current consumption traces for the benchmarks and Table 1 gives the corresponding $RelMError$, $RMSError$, $WCError$ (Worst Case Error), CC_g^2 (Squared Correlation Coefficient for the fitted gamma approach), and CC^2 figures.

Almost all current estimates provide a CC^2 in excess of 93.12%. The proposed method clearly outperforms the gamma technique which only provides a mean CC^2 of 65%. However, some variations in the current consumption are not well captured. The Worst Case single Error was 15.5%. As was noted in Section 2, the static model does not contain characterized current values for every instruction of the instruction set. When an instruction is not characterized, errors may be introduced in the signal $x_d[k]$. Other potential sources of errors are discussed in Section 4. The $RelMError$ results indicate that the model also accurately estimates the mean current drawn by the processor.

7. CONCLUSIONS AND FUTURE WORK

This paper presented a novel instruction-level dynamic power model. An identification method was described which allows the model to be applied to third-party processors. The model and identification method were applied to the Texas Instrument TMS320VC5510 DSP. Results were presented and the accuracy of the model in estimating dynamic current was assessed by comparison with measurements across a range of DSP benchmarks. The results showed

Table 1: Benchmarks results.

Benchmark	$RelMError[\%]$	$RMSError[mA](\%)$	$WCError[\%]$	$CC_g^2[\%]$	$CC^2[\%]$
cfft	0.15	0.034 (2.4)	6.42	38.00	87.25
iir4	5.06	0.111 (8.3)	15.5	72.74	97.67
log10	3.07	0.080 (5.13)	8.44	47.93	94.93
jpeg_qize	0.33	0.044 (4.18)	10.41	81.33	93.12
Dct_idct	-1.68	0.073 (5.5)	10.46	86.35	93.31

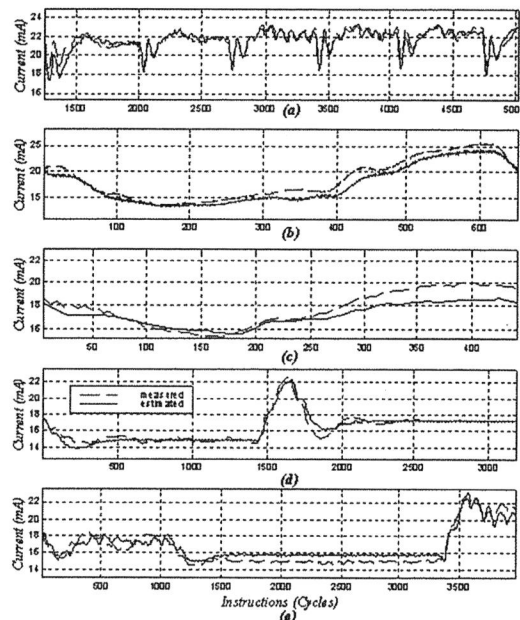

Figure 7: Measured and estimated current consumption for benchmarks: (a) cfft Kernel; (b) iir4; (c) log10; (d) jpeg_qize; (e) Dct_idct

that the model describes the dynamic behavior of the processor within an average correlation of 93% and estimates mean current with an average error of 2.05%.

For the processor under investigation, the method is more accurate than previously published methods. The new method is more general and so is applicable to a greater range of processors. The system identification method is also more robust and easier to apply than that described previously.

Future work includes analysis of the effect of the different components within the power supply system. Extension of the static model to complex instruction set architectures and modeling of dynamic frequency and voltage scaling is also planned.

8. REFERENCES

[1] M. L. Bengt Johansson. Possibilities of obtaining small-signal models of DC-to-DC power converters by means of system identification. In *Proc. Telec. Energy Conf*, pages 65–75, 2000.

[2] M. Casas-Sanchez, J. Rizo-Morente, and C. Bleakley. Power consumption characterisation of the texas instruments TMS320VC5510 DSP. In *Proc. of Power and Timing Modelling, Optimization and Simulation (PATMOS)*, pages 561–570, 2005.

[3] G. Fursin, M. O'Boyle, and P. Knijnenburg. Evaluating iterative compilation. In *Proc. Languages and Compilers for Parallel Computers (LCPC)*, pages 305–315, 2002.

[4] T. P. L.D. Smith, R.E. Anderson and T. Roy. Power distribution system design methodology and capacitor selection for modern CMOS technology. *IEEE Trans. on advanced packaging*, 22(3):284–291, 1999.

[5] L. Ljung. *System Identification: theory for the user.* Prentice-Hall, Inc., Upper Saddle River, NJ, USA, 1986.

[6] R. Muresan and C. Gebotys. Current flattening in software and hardware for security applications. In *CODES+ISSS '04: Proc. of the 2nd IEEE/ACM/IFIP Int. Conf. on Hardware/software codesign and system synthesis*, pages 218–223, 2004.

[7] R. Muresan and C. Gebotys. Instantaneus current modeling in a complex VLIW procesor core. *ACM Trans. on Embedded Computing Systems (TECS)*, 4(2):415–451, 2005.

[8] D. Rakhmatov, S. Vrudhula, and C. Chakrabarti. Battery-conscious task sequencing for portable devices including voltage/clock scaling. In *DAC '02: Proc. of the 39th Conf. on Design automation*, pages 189–194, 2002.

[9] L. Smith, R. Anderson, D. Forehand, T. Pelc, and T. Roy. Power distribution system design methodology and capacitor selection for modern CMOS technology. *IEEE Trans. on advanced packaging*, 22(3):284–291, 1999.

[10] Spectrum Digital Inc. *TMS320VC5510 DSK Technical Reference*, 2002.

[11] Texas Instruments Inc. *TMS320C55x DSP CPU Reference Guide*, 2004.

[12] Texas Instruments Inc. *TMS320C55x DSP Library Programmer Reference*, 2004.

[13] Texas Instruments Inc. *TMS320C55x Image/Video Processing Library Programmer Reference Guide*, 2004.

[14] V. Tiwari, M. Lee, S. Malik, and M. Fujita. Power analysis and low-power scheduling techniques for embedded DSP software. *Fujitsu Scientific and Technical Journal*, 5:215–229, 1995.

[15] V. Tiwari, M. Lee, S. Malik, and M. Fujita. Power analysis and minimization techniques for embedded DSP software. *IEEE Trans. on VLSI Systems*, 5(1):1–14, 1997.

[16] V. Tiwari, S. Malik, and A. Wolfe. Power analysis of embedded software: A first step towards software power minimization. *IEEE Transactions on VLSI Systems*, 2(4):437–445, 1994.

Reducing Idle Mode Power in Software Defined Radio Terminals

Hyunseok Lee, Trevor Mudge
Dept. of EECS
University of Michigan
{leehzz,tnm}@eecs.umich.edu

Chaitali Chakrabarti
Dept. of EE
Arizona State University
chaitali@asu.edu

ABSTRACT

In this paper, we propose a processor which is optimized for idle mode operation of a software defined radio (SDR) terminal. Since a SDR terminal spends most of its time in the idle mode, reducing the power consumption in this mode directly translates to longer terminal standby time. Workload analysis of idle mode operations of contemporary standards showed that these are dominated by FIR filtering, which can be easily parallelized. This analysis was used in the design of the idle mode processor. The key architectural components are an SIMD unit for the parallel computations that dominate the workload, a conventional scalar unit for the sequential computations, and a control unit which supports efficient data memory access and loop control. The idle mode processor was modeled with Verilog and synthesized using standard cells in 0.13 micron technology. It consumes about 9mW at 1.08V.

Categories and Subject Descriptors

C.14 [**Processor Architecture**]: Other Architecture Styles – Mobile processors

General Terms

Design, Performance

Keywords

Software defined radio, SDR, Wireless terminal, Baseband processor, Idle mode, Low power, SIMD

1. INTRODUCTION

Software defined radio (SDR) is a wireless communication system whose function blocks are implemented by flexible software routines instead of fixed hardware, so that various wireless protocols can be easily supported on the same platform and future changes can be smoothly accommodated. In

this paper we consider the baseband processor, which is digital hardware that performs complex signal processing algorithms for communicating over unreliable wireless channels. The baseband processor deals with the most computationally demanding part of the wireless communication system and, as a result, is usually realized as an ASIC. Although the baseband processor can be used both at the basestation and the wireless terminal, we focus on the baseband processor in the wireless terminal. This makes power reduction in the baseband processor particularly important.

The operation of the wireless terminal can be classified into two modes: *active mode* when a wireless terminal actively transmits and receives user traffic, and *idle mode* when a wireless terminal waits passively to respond to communication requests from other terminals. Because the workload of the active mode is substantial, most previous research on SDR platforms has focused on the efficient support of the active mode operations. However, for the end user, one of the most desirable features of a wireless terminal is long battery life [1]. A significant portion of battery power is used in the idle mode. Although the energy dissipated by a single idle mode operation is very small, the aggregated effect is substantial, because a wireless terminal spends most of its time in the idle mode. In the case of W-CDMA terminal, the idle time can be as much as 99% [2][3].

The main topic of this paper is to minimize the idle mode power consumption of the baseband processor for SDR terminals without limiting programmability too much. We will show that this can be achieved, because the number of distinct signal processing kernels of the idle mode is limited.

The first step in our work is to identify the signal processing algorithms used in the idle mode and to characterize their computation patterns. We analyze the idle mode operation scenario of several contemporary standards, including GSM, GPRS, EDGE, IS-95, cdma2000, W-CDMA, IEEE 802.11/a/b/g, WiMax, and WiBro. In the workload analysis described in Section 2 we find that, although wireless terminals perform many operations in the idle mode, the majority falls into the class of finite impulse response (FIR) filtering which can be represented by $\sum_{i=0}^{L-1} c_i \cdot x[i+n]$. The absence of data dependency between the terms means that the multiplications can easily be parallelized. There still exist some sequential computations in the idle mode operation; however, their impact is not as significant.

Based on these observations, we propose an idle mode processor which consists of three sub-units: a single instruction multiple data (SIMD) unit, a scalar unit, and a control unit. The SIMD unit handles the parallel workload, namely

---

Permission to make digital or hard copies of all or part of this work for personal or classroom use is granted without fee provided that copies are not made or distributed for profit or commercial advantage and that copies bear this notice and the full citation on the first page. To copy otherwise, to republish, to post on servers or to redistribute to lists, requires prior specific permission and/or a fee.

ISLPED'06, October 4–6, 2006, Tegernsee, Germany.
Copyright 2006 ACM 1-59593-462-6/06/0010 ...$5.00.

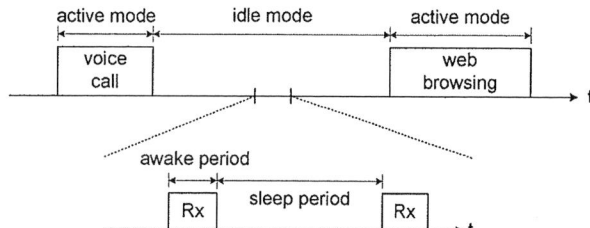

Figure 1: The active and idle operation modes of wireless terminal, and the awake and sleep periods in the idle mode

the FIR filter; the scalar unit handles the sequential workload; and the control unit assists the SIMD and scalar units with automatic data memory address generation and loop control. The operations in the FIR filter can be efficiently mapped onto the SIMD unit allowing for significant power savings – parallelization makes possible lower operation frequency and slower but more power efficient circuits. The proposed idle mode processor can also be used in the active mode, because the key idle mode computation kernels are also among the computationally-intensive kernels used in the active mode.

To validate our idea, we implement a hardware model of the proposed architecture using Verilog and synthesize it using 0.13 micron technology. This model is used as an input to Synopsys' PrimePower, a commercial power evaluation tool. The experiments show that the proposed idle mode processor consumes about 9mW at 1.08V when performing the idle mode operations. This is more than a 80% dynamic power reduction compared to state-of-the-art SDR processors optimized for the active mode [4]. Such significant power reduction is due to the lower operation frequency made possible by the simplified SIMD architecture. Further power reduction in low activity periods can be achieved by using more advanced silicon process technology and techniques such as clock gating.

This paper is not the first attempt to optimize the power consumption of hand held devices when they are in a low activity mode. There are several low power general purpose processors targeted for hand held devices. Examples are personal data assistants based on Intel's Xscale [5] and IBM's PowerPC 405 [6]. However, the application domain of these processors is quite different from that of the baseband processor. While there have been a number of projects that explore SDR platforms [7][8][9][10][11], these efforts are primarily focussed on the active mode, and ignore the idle mode. The contribution of this paper is to characterize the operation of a SDR terminal in the idle mode, and to propose an architecture that is optimized for idle mode operation and can be used as part of the baseband processor of an SDR terminal.

2. WORKLOAD ANALYSIS

2.1 Operation Modes of Wireless Terminal

Wireless terminals change their operation state according to use activity as depicted in Figure 1. When there are active user applications, wireless terminals employ all of their functionality to support high data rate communications (*active mode*). However, even when there exist no

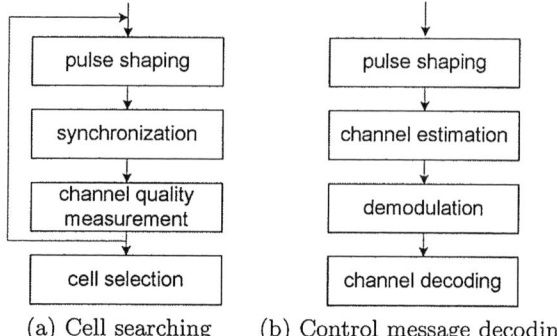

(a) Cell searching   (b) Control message decoding

Figure 2: Generalized idle mode operation flow of wireless terminals in basestation based networks

active user applications, wireless terminals can not turn off completely, because requests from other terminals may be received at unpredictable times. So, wireless terminals activate a subset of their communication components to receive the requests (*idle mode*). In the idle mode, the operation is intermittent. Instead of continuously monitoring for signals, terminals completely power off their receiver (*sleep period*) and activate it only for short pre-scheduled periods (*awake period*).

2.2 Operations in Idle Mode

In this section, we discuss in detail the idle mode operation of wireless terminals. We consider idle mode operation for two cases: a wireless network with basestations and a wireless network without basestations (ad hoc network).

2.2.1 Operation of Network with Basestation

Most wireless networks such as GSM, GPRS, EDGE, IS-95, cdma2000, and W-CDMA use basestation. In such networks, idle mode operations consist of two steps, cell searching and control message decoding, as shown in Figure 2. In the cell searching step, the terminal selects basestations that provide best quality channel among candidates. In the control message decoding step, the terminals receive and decode control messages from selected basestations.

Cell searching: As shown in Figure 2(a), the cell searching step consists of four procedures: pulse shape filtering, synchronization, channel quality measurement, and cell selection. Pulse shape filtering is a common low pass filtering operation used in all wireless terminals. It is typically implemented with an FIR filter. The synchronization procedures of most wireless networks have considerable similarity: a basestation broadcasts a reference signal containing a synchronization code, the terminals extract timing and frequency information from this signal by matched filtering of received signal with the synchronization code. After the synchronization, terminals measure the quality of channel provided by the synchronized basestation. Channel quality is measured by the fidelity of the decoded reference signal. As shown in Figure 2(a), the terminals iteratively perform the pulse shaping, synchronization and channel quality measurement procedures against all candidate basestations. Cell selection is to select a set of basestations that provide best quality channel.

Control message decoding: After the cell searching, the pulse shaped signal is passed to the channel estimator which

102

estimates the wireless channel characteristic. The estimated characteristics are used to demodulate the received signal. Demodulation is to reconstruct the most probable information sequence from the received signal. Terminals of TDMA networks typically use the FIR filter and a Viterbi decoder for the channel estimation and demodulation. Terminals of CDMA networks perform matched filtering for the channel estimation. The demodulated signal is processed by the channel decoder which performs forward error correction. Turbo decoders and Viterbi's decoders are typical examples of channel decoder. The decoded result is forwarded to upper layers. Although the channel estimation and channel decoding procedures are the ones with heaviest workloads in the active mode, their workload is not substantial in the idle mode because the control message decoding is performed only once while the synchronization and channel quality measurement procedures are performed multiple times against all candidate basestations. Thus, the idle mode workload of basestation based networks is dominated by the synchronization and channel quality measurement procedures.

2.2.2 Operation of Ad Hoc Network

The operation flow shown in Figure 2 is not valid for ad hoc networks. To avoid severe power penalty, a wireless terminal in an ad hoc network does not continuously broadcast the reference signal to synchronize all terminals within a network. Instead, terminals attach a preamble sequence at the header of all transmitted frames. In order to detect the transmission of a new frame, terminals compute the correlation between the received signal and its delayed signal. This type of computation is called a '**sliding window**' and its detailed characteristics will be discussed in the followed subsection. Wireless local area networks (WLAN) such as IEEE 802.11/a/b/g sometimes operate in the ad hoc mode.

2.3 Key Computation Patterns in Idle Mode

There are two key kernels that dominate the idle mode operations: FIR filtering and sliding window. For instance, the FIR filter operation occupies about 80% of the idle mode workload of W-CDMA terminal [12]. It follows that the power efficiency of the idle mode processor is dominated by the power efficiency of these kernel computations.

2.3.1 FIR Filter

The computation pattern of the FIR filter which is used for the pulse shaping and the synchronization can be represented by following equation:

$$y[n] = \sum_{i=0}^{L-1} c_i \cdot x[i+n] \qquad (1)$$

where $x[n]$ is the input signal, c_i is the filter coefficient, and L is the filter length. The coefficient of the FIR filter is typically an arbitrary number. However, in matched filtering with the synchronization code sequence which consists of 1 or -1, the multiplications shown in Equation (1) can be simplified into conditional complements as follows:

$$c_i \cdot x[i+n] = \begin{cases} -x[i+n], & \text{if } c_i = -1 \\ x[i+n], & \text{otherwise} \end{cases} \qquad (2)$$

As shown in Figure 3, there exist two methods to implement the computation pattern shown in Equation (1):

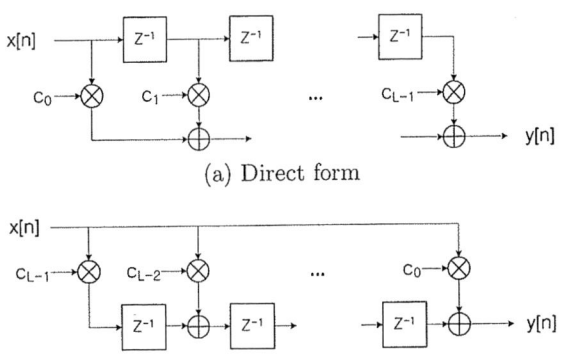

Figure 3: Two implementation methods for FIR filter

direct form and transpose form. Although they are functionally identical, these two implementation methods have different power costs.

There exist a lot of parallelism in the FIR filter. In Equation (1), the L multiplications for one output $y[n]$ can be performed in parallel. Furthermore, the multiplications for other outputs $y[n+1], \ldots, y[n+k]$ also can be done in parallel if resources are available. Despite such high parallelism, the FIR filter requires only one new input data, $x[n]$, to produce a new output, $y[n]$. The characteristics of the FIR filter vary according to the type of wireless network. The range of the filter order L is 30~300 taps. These can be broken into 32-wide parallel chunks for SIMD processing. The input data, $x[n]$, and filter coefficient, c_i, are represented by 1~16 bits.

2.3.2 Sliding Window

The computation pattern of the sliding window operation which is used for the frame detection can be represented as follows:

$$y[n] = \sum_{i=0}^{L-1} P_{i,n} \qquad (3)$$

where $P_{i,n} = x[i+n] \cdot x[i+n-D]$, $x[n]$ is the input signal at time n, L is the correlation length, and D is the delay between input signals. It is equivalent to auto-correlation of $x[n]$ with delay D and can be implemented with L multiplications and $L-1$ additions per output. However, the computation shown in Equation (3) can be further simplified due to the following relation between outputs:

$$y[n] = y[n-1] - P_{0,n-1} + P_{L-1,n} \qquad (4)$$

Because we can reuse the previously computed $P_{0,n-1}$, the operation shown in Equation (4) additionally requires only one multiplication, addition, and subtraction for the generation of the next output. Therefore, the sliding window operation can be classified as a scalar workload and implemented by the scalar unit.

2.4 Processing Time of Idle Mode

In addition to the limited number of computation patterns that have to be supported, the idle mode operation also has more relaxed processing time requirements compared to that of the active mode. To avoid buffer overflow, a baseband processor must finish the operations on the current frame

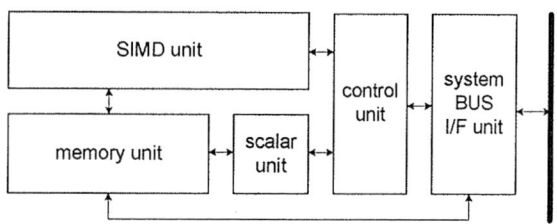

Figure 4: Architecture of the idle mode processor

```
for(i=0; i<Max_Sample; i++) {
    for(j=0; j<Filter_Len; j++) y[i] += x[j+i]*c[j];
}
```

(a) C routine for the FIR filter with direct form

```
Loop:   shift.v      VR1  ← VR1, (AR1)
        mul.v        VACC ← VR2, VR1
        reduction.v  (AR2) ← VACC
        inc          AR1, #1
        inc          AR2, #1
        dec          SR1, #1
        jnz          SR1, Loop
```

(b) Pseudo assembly routine for SIMD implementation

```
Loop:  shift.v      VR1 ← VR1, (AR1++)
       mul.v        VACC ← VR2, VR1
       reduction.v  (AR2++) ← VACC | djnz SR1, Loop
```

(c) Pseudo assembly routine after applying automatic address generation and loop control

```
Loop:  reduction.v (AR2++)
    ←{mul.v VR2,{shift.v VR1, (AR1++)}} | djnz SR1, Loop
```

(d) Pseudo assembly routine after cascading arithmetic units

Figure 5: FIR filter routine optimized for the SIMD implementation, automatic address generation and loop control, and cascading arithmetic units

before the next frame arrives. In the idle mode, the inter-frame arrival time is longer than that of the active mode, by at least an order of magnitude. Thus, it is important for the idle mode processor to exploit the longer processing time for power reduction.

3. PROCESSOR DESIGN

3.1 Overall Processor Architecture

At the top level, the idle mode processor consists of five units as shown in Figure 4: SIMD unit, scalar unit, control unit, memory unit, and system BUS interface unit. The SIMD unit implements the FIR filter. The scalar unit implements unparallelizable workloads, including the sliding window. The memory unit stores operation data and program. The system BUS interface unit is used to communicate with other entities in the system, such as the analog front-end. The control unit manages all the units; it includes hardware address generators and hardware loop counter in addition to conventional program counter, instruction decoder, and control logic.

In order to explain the following high level design decisions, we represent the FIR filter operation in 'C' and pseudo assembly routines. As shown in Figure 5(a), the operation

of FIR filter consists of two loops: the inner loop consists of operations for one output and the outer loop represents the multiple iterations of the filter operation. Through the use of the proposed hardware features, the 'C' routine shown in Figure 5(a) can be simplified into a single line of assembly routine as shown in Figure 5(d).

SIMD unit: The SIMD architecture is appropriate for the inner loop operation shown in Figure 5(a). The inner loop operation consists of identical computations with regular operand access patterns. The SIMD architecture is power efficient; the parallel execution of the computations of the inner loop helps reduce operations frequency and this directly translates to dynamic power reduction. The hardware complexity needed to provide operands to the SIMD datapath is low. This is because the inner loop operation is a single input and single output system, which can be implemented by a low power single port memory. Figure 5(b) describes the corresponding pseudo assembly routine that can be mapped to the SIMD architecture.

Control unit: The hardware address generators and loop controller in the control unit maximize the utilization of SIMD unit. As shown in Figure 5(b), only three out of the seven instructions are SIMD instructions. The other scalar instructions are related to address generation and loop control. Fortunately, the FIR filter operation has regular data memory access pattern. So, it is possible to automate the address generation without significant overhead. Because the FIR filter is single input – single output systems, two address generation units are sufficient, one for input data and the other for output data. The two loop control related instructions can also be automated easily by using decrementing counter and zero detection logic. The result of deploying automatic address generation and loop control is shown in Figure 5(c). In this routine, 'djnz' stands for decrease and then jump if the decrement result is not zero.

Scalar unit: To support the sequential workload, the idle mode processor consists of a scalar unit as well. Because the workload assigned to the scalar unit is control intensive and also includes computations such as multiplication, division, and logical operations, a conventional low power general purpose processor is a suitable choice. A multiplication and accumulation (MAC) unit is useful for minimizing the operation cycle of the scalar unit.

Memory unit: The memory unit consists of data and instruction memories. We estimated that 100 Kbytes data memory and 100 Kbytes instruction memory are sufficient for idle mode operation. To reduce power consumption, memories are divided into smaller sub-banks. Such a structure significantly reduces the dynamic power, because only one sub-bank is activated during read or write. Although sub-banking demands more silicon area, we use this scheme because power reduction is crucial in the idle mode processor. In addition, for further power reduction, each sub-bank is implemented with a single port memory. Data is distributed among the sub-banks in a way such that data read and write can be done in parallel.

System BUS interface unit: Because the idle mode processor is part of the wireless terminal, an interface with the system BUS is essential. According to our previous work [12], a low speed BUS is enough for the baseband processor, because communication rate between kernel algorithms is quite low. A client logic of advanced micropro-

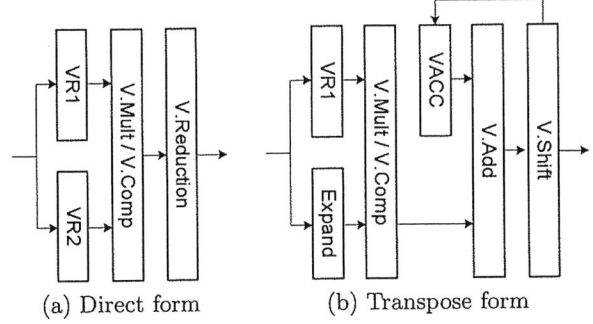

(a) Direct form (b) Transpose form

Figure 6: Two implementation methods of SIMD datapath

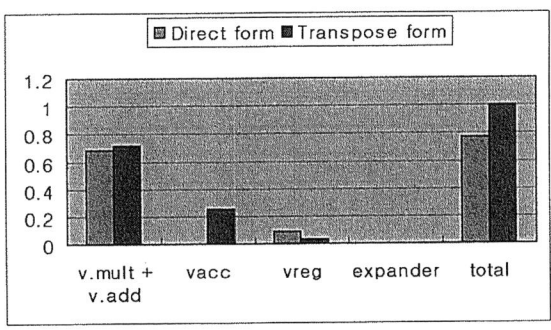

Figure 7: Normalized average dynamic power of SIMD datpaths: direct form vs. transpose form

cessor bus architecture (AMBA) bus is a typical example of a system BUS interface unit.

3.2 Detailed Design of SIMD Datapath

Direct form SIMD datapath: As discussed before, the FIR filter can be implemented in two ways. Figure 6 depicts the detailed structures of the SIMD datapath based on both the direct form and the transpose form. According to the dynamic power analysis in Section 4, the direct form SIMD datapath has about 20% better power performance. Thus, the SIMD datapath is implemented with the direct form.

Cascading arithmetic units: As shown in Figure 6(a), the second design decision on the SIMD datapath is to cascade arithmetic units such as the adder and multiplier so that the temporary results do not have to be stored in a register file. This is motivated by the fact that accessing the register file is power expensive. Generally cascading of the arithmetic units limits system flexibility, a feature which is important for the SDR system. However, the limited flexibility is not a problem in the idle mode processor, because the SIMD datapath with cascaded function units is flexible enough to cope with many varieties of FIR filter operations, which can appear at the idle mode operation of wireless protocols.

Pipelining: The SIMD datapath has three pipeline stages: read, execution-1, and execution-2. The idle mode workload is suitable for pipelining, because it can fully utilize the pipeline by repeating identical computations many times. For instance, in Figure 5(a), the inner loop operation is repeated for each input sample (Max_Sample times). Due to the absence of data dependencies, no data forwarding path is required. So, the SIMD datapath can be pipelined by simply inserting flip flops between pipeline stages. Although these flip flops contribute additional power, this can not be more than the power reduction achieved from operating at a lower frequency and by the preventing of glitch propagation.

4. EXPERIMENT AND ANALYSIS

Methodology: We validated the functional correctness of the proposed idle mode processor architecture with a Verilog model. For power evaluation, we synthesized it using the TSMC13 library based on 0.13 micron technology. We estimated the dynamic power of the proposed processor at the gate level with the help of the commercial power evalu-

ation tool, Synopsys' PrimePower. For memory generation, we used Artisan's memory compiler.

System configuration for experiment: We assumed that the SIMD width is 32. We used this number for compatibility with our SDR processor for active mode [4]. In general, the optimal SIMD width depends on design constraints such as silicon area and the type of silicon processing technology. Also, the W-CDMA was chosen as the application for the experiment since it has the most complex idle mode operation scenario. The implemented idle mode operation of the W-CDMA terminal was based on [13].

Minimum operation frequency estimation: In the measurement of dynamic power, the allowed minimum operation frequency is an important parameter. Although the length of the awake period is one of the design parameters, we assume it to be 30 msec: 27 msec for the cell searching and 3 msec for the control message decoding. From the behavioral model, we extracted the cycle count for idle mode operation. We used the allowed operation time and the cycle count to calculate the minimal operation frequency of the idle mode processor. For our design, the minimum frequency was about 36Mhz but we assumed it to be 50Mhz (adding about a 30% margin) for the dynamic power calculations.

Direct form vs. transpose form: Figure 7 shows the normalized average dynamic power of the SIMD datapaths based on the direct form and transpose form. It shows that the direct form implementation is about 20% more power efficient. This power efficiency comes from the absence of a vector accumulator (VACC) which stores the partial operation results in the direct form implementation. As show in Figure 3, the direct form stores input data having lower data precision whereas the transpose form stores partial computation results with higher data precision. Because a storage element requires more power compared to other logic elements, the VACC results in the transpose form SIMD datapath consuming higher power.

System dynamic power: Table 1 shows the dynamic power consumption of the idle mode processor. The most power consumer is the data memory. Although we applied memory sub-banking to reduce power, the data memory still dissipates about 43% of the total power. This is because the data memory is heavily accessed in most operations of the idle mode processor. The power consumption of the instruction memory is about one fifth the amount of the data memory, because it is only accessed for instruction reads by the control unit. The SIMD unit still consumes significant

Units	Subblocks	Peak Power (mW)	Average Power (mW)	Area (%)
SIMD	V.Mult	13.68	1.80	10.49
	V.Comp	2.24	0.24	1.96
	V.Reduction	3.80	1.21	2.31
	V.Regs	0.90	0.38	1.96
Scalar	ALU	1.21	0.70	0.83
	Regs	0.53	0.31	0.07
Control	Address gen.	0.16	0.13	0.24
	Loop control	0.12	0.12	0.09
	Ctrl. logic	0.22	0.13	0.52
Memory	I-mem	0.85	0.49	37.88
	D-mem	3.93	3.32	43.29
BUS I/F	BUS I/F	0.16	0.04	0.02
Total	(rounded)	——	9mW	$22mm^2$

Table 1: Idle mode processor: Dynamic power @(50Mhz, 1.08V) of each block during idle mode operation for the W-CDMA terminal, and area of each block

power (41%) even though we designed it with several low power techniques which are optimal for the idle mode operations. Thus, if we made the SIMD unit more programmable to support the algorithms in the active mode, the energy efficiency of the idle mode processor will be degraded substantially. Although the peak power of the vector multiplier is very high, its average is only 20% of total power. This is because the activation time of the vector multiplier is 10% of total operation time. Note that the vector multiplier only participates in the pulse shaping operation. The power overhead of the vector reduction logic is almost equivalent to that of the vector multiplier in average, because it is active throughout entire vector operation including the pulse shaping and the synchronization. The dynamic power of the scalar, control, and system BUS interface units is not substantial, approximately 16% of total power.

Table 1 also describes the area of each component. The scalar unit, the control unit, and the system BUS interface unit occupy negligible area (2.2%) compared to the SIMD unit (16.7%). However, the majority of the area (81%) is occupied by memories. Note that their sizes are likely to be determined by the overall SDR operation and not just the idle mode operation.

Further power reduction: The dynamic control of operand precision is an effective way for reducing power. This is because the operand precision varies across different wireless protocols even for the same operation. The dynamic control of operand precision directly affects the power consumption of dominating blocks. Second, use of low power implementation would result in additional power reduction. The current implementation only relies on a synthesis tool and memory compiler. Finally, synthesis of the processor in 90nm technology will help in achieving further power reduction.

5. CONCLUSION

In this paper we proposed a hardware platform for reducing the power consumption during idle mode operation of a SDR terminal. Our workload analysis showed that the idle mode operation consists of a substantial amount of computations with simple SIMD parallelism, and a much smaller amount of control intensive sequential computation. Furthermore, although a wide variety of operations are performed in the idle mode, two computation kernels, namely, FIR filtering and sliding window, account for 80% of the computations in the idle mode. Based on this analysis, we proposed an architecture for the baseband processor that consists of three sub-units: 1) an SIMD unit for the parallel workload; 2) a scalar unit for the sequential workload; and 3) a control unit having hardware address generation and loop control. The use of a specialized SIMD unit for the dominant computations, and a fully programmable scalar unit for control intensive sequential operations resulted in a power efficient architecture for the idle mode processor. We estimated that the architecture, when synthesized in 0.13 micron technology, could consume about 9mW at 1.08V.

6. ACKNOWLEDGEMENT

This work was supported by Samsung Electronics and NSF–ITR CCR–0325898 and CCR–0325761.

7. REFERENCES
[1] John Walko. Mobile phone users demand decent batteries. *EETimes*, Sep. 2005.

[2] UMTS Forum. UMTS/IMT-2000 spectrum. *Technical Report* 6, 1998.

[3] UMTS Forum. 3G offered traffic characteristics. *Technical Report* 33, 2003.

[4] Y. Lin et al. SODA: A low-power architecture for software radio. In *International Symposium on Computer Architecture*, pages 89–101, June 2006.

[5] L. Clark et al. An embedded 32-b microprocessor core for low-power and high-performance applications. *IEEE Journal of Solid-State Circuits*, 36(11):1599–1608, Nov. 2001.

[6] K. Nowka et al. A 32-bit PowerPC system-on-a-chip with support for dynamic voltage scalaing and dynamic frequency scaling. *IEEE Journal of Solid-State Circuits*, 37(11):1441–1447, Nov. 2002.

[7] J. Glossner et al. A software defined communications baseband design. *IEEE Communication Magazine*, 41(1):120–128, Jan. 2003.

[8] Simon Knowles. The SoC future is soft. In *IEE Cambridge Branch Seminar*, Dec. 2005.

[9] A. Duller et al. Parallel processing — the picoChip way! In *Communicating Processing Architectures*, pages 125–138, Sep. 2003.

[10] B. Mei et al. Architecture exploration for a reconfigurable architecture template. *IEEE Design and Test of Computers*, 22(2):90–101, March/April 2005.

[11] C. Berkel et al. Vector processing as an enabler for software-defined radio in handsets from 3G+WLAN onwards. In *Software Defined Radio Technical Conference*, Nov. 2004.

[12] H. Lee et al. Software defined radio — A high performance embedded challenge. In *High Performance Embedded Architectures and Compilers*, pages 6–26, Nov. 2005.

[13] Y. Wang et al. Cell search in W-CDMA. *IEEE Journal on Selected Areas in Communications*, 18(8):1470–1482, Aug. 2000.

Power Reduction in an H.264 Encoder Through Algorithmic and Logic Transformations

Maria G. Koziri
Department of Computer and
Communication Engineering,
University of Thessaly
37 Glavani St.,
382 21 Volos, Greece
+30 2421074979
mkoziri@uth.gr

George I. Stamoulis
Department of Computer and
Communication Engineering,
University of Thessaly
37 Glavani St.,
382 21 Volos, Greece
+30 2421074979
georges@uth.gr

Ioannis X. Katsavounidis
InterVideo, Inc.
46430 Fremont Blvd.,
Fremont, CA 94538, USA
+16 504921567
ioannis@intervideo.com

ABSTRACT

The H.264 video coding standard can achieve considerably higher coding efficiency than previous video coding standards. The keys to this high coding efficiency are the two prediction modes (Intra & Inter) provided by H.264. Unfortunately, these result in a considerably higher encoder complexity that adversely affects speed and power, which are both significant for the mobile multimedia applications targeted by the standard. Therefore, it is of high importance to design architectures that minimize the speed and power overhead of the prediction modes. In this paper we present a new algorithm, and the logic transformations that enable it, that can replace the standard Sum of Absolute Differences (SAD) approach in the two main prediction modes, and provide a power efficient hardware implementation without perceivable degradation in coding efficiency or video quality.

Categories and Subject Descriptors

B.2.4 **[Arithmetic and Logic Structures]**: High-speed Arithmetic – *algorithms, cost/performance*.

General Terms

Algorithms, Performance, Design.

Keywords

H.264 encoder, intra prediction, low-power implementation, motion estimation.

1. INTRODUCTION

Video has always been the backbone of multimedia technology. In the last two decades, the field of video coding has been revolutionized by the advent of various standards like MPEG-1 to MPEG-4 and H.261 to H.263, each addressing different aspects of multimedia. H.264[1] is a new standard which adds one more step

---

Permission to make digital or hard copies of all or part of this work for personal or classroom use is granted without fee provided that copies are not made or distributed for profit or commercial advantage and that copies bear this notice and the full citation on the first page. To copy otherwise, or republish, to post on servers or to redistribute to lists, requires prior specific permission and/or a fee.
ISLPED'06, October 4–6, 2006, Tegernsee, Germany.
Copyright 2006 ACM 1-59593-462-6/06/0010...$5.00

---

in the endeavor towards video coding excellence and provides one-stop solution for wide range of applications. The standard has been developed by the Joint Video Team (JVT) comprised of both ISO/IEC and ITU-T. The primary goal of H.264 is to achieve higher compression while preserving video quality. The motivation for compression is to compensate for the ever-present constraints of limited channel capacity.

This increased compression efficiency of the new ITU-T H.264/MPEG-4 Advanced Video Coding (AVC) standard will lead to new application areas. Applications concerning broadcasting over cable, satellite, cable modem, terrestrial, etc., will benefit from the new standard. As for the field of mobile communication, H.264 will play an important role because the compression efficiency will be doubled in comparison to the coding schemes previously specified by Third-Generation Mobile (3GPP and 3GPP2) for streaming. The video coding technique used in H.264 follows a flexibility to be used in low-delay real-time applications.

Each picture of a video, which can either be a frame or a field, is partitioned into fixed-size macroblocks that cover a rectangular picture area of 16×16 samples of the luma component and 8×8 samples of each of the two chroma components (for the case of Y:U:V 420 color format). All luma and chroma samples of a macroblock are either spatially or temporally predicted, and the resulting prediction residual is transmitted using transform coding. For this purpose, each color component of the prediction residual is subdivided into blocks of various sizes. Macroblocks are grouped in slices, which generally represent subsets of a given picture that can be decoded independently. H.264/AVC supports five different slice-coding types. The simplest one is the I slice (where "I" stands for intra). In I slices, all macroblocks are coded without referring to other pictures within the video sequence (i.e. Intra Prediction). On the other hand, prior-coded images can be used to form a prediction signal for macroblocks of the predictive-coded P and B slices (where "P" stands for predictive and "B" stands for bi-predictive) (i.e. Inter Prediction). The remaining two slice types are SP (switching P) and SI (switching I), which are specified for efficient switching between bitstreams coded at various bit-rates.

One computational element that is met in both, Inter and Intra Prediction modes, is that of the Sum of Absolute Differences (SAD). In this paper we present a new algorithm that can replace SAD in the two main prediction modes, and the circuit that implements the proposed algorithm, which results in better speed and significantly lower power. The rest of this paper is organized as follows: Section 2 reviews Intra and Inter Prediction, as well as the basic concepts of SAD. In Section 3 a description of the proposed algorithm is introduced. Section 4 presents the coding efficiency and video quality results achieved from the software implementation of the new algorithm, while Section 5 presents the delay and power improvements achieved by the hardware implementation of the proposed algorithm.

2. INTRA AND INTER PREDICTION

2.1 Intra Prediction

In H.264, intra-prediction with block sizes of 4×4, 8×8 and 16×16 is used to compress I-Macroblocks. Intra coding refers to the case where only spatial redundancies within a video picture are exploited. The resulting frame is referred to as an I-picture. I-pictures are typically encoded by directly applying the transform to the different macroblocks in a frame. As a consequence, encoded I-pictures are large in size since no temporal information is used as part of the encoding process. In order to increase the efficiency of the intra coding process in H.264, spatial correlation between adjacent macroblocks in a given frame is exploited. The idea is based on the observation that adjacent macroblocks tend to have similar properties. Therefore, as a first step in the encoding process for a given macroblock, one may predict the macroblock of interest from the surrounding macroblocks, typically the ones located on top and to the left of the macroblock of interest, since those macroblocks would have already been encoded. After a prediction block, P, is formed based on previously encoded and reconstructed blocks, it is subtracted from the current block prior to encoding. For the luma samples, P is formed for each 4×4, 8×8 block or for a 16×16 macroblock. There are a total of nine optional prediction modes for each 4×4 luma block, nine modes for each 8x8 block, four modes for a 16×16 luma block and four modes for the chroma components [2].

2.2 Inter Prediction

For all other types of Macroblocks, H.264 uses inter-prediction to compress them. Inter coding uses the temporal model. In this case the predicted frame is created from one or more past or future frames ('reference frames'). The process of finding the best predicted frame is known as Motion Estimation. The accuracy of the prediction can usually be improved by compensating for motion between the reference frame(s) and the current frame (Motion Compensation). A practical and widely-used method of motion compensation is to compensate for movement of rectangular sections or 'blocks' of the current frame. H.264 is a block-based motion-compensated hybrid transform codec, which supports a variety of block sizes (denoted as modes), varying from 16×16, 8×16, 16×8, 8×8, 8×4, 4×8 to 4×4 pixels. Usually the criterion to find the matching block is the energy in the residual formed by subtracting the candidate block from the current M×N block, and the candidate region that minimizes the residual energy is chosen as the best match. However, in order to reduce the computational complexity, most real world application, among them H.264, uses the sum of absolute differences (SAD) [3].

2.3 Sum of Absolute Differences (SAD)

The encoder typically selects the prediction mode, both in Intra and Inter Prediction, for each block that minimises the difference between the predicted block P and the block to be encoded, C. The selection is done by using SAD, also known as L_1 distance between the vectorized form of the two blocks, P and C.

L_1-distance is a positive-definite metric defined over vectors in k-dimensional vector spaces by the corresponding L_1 norm of the difference vector, as follows:

Let $\underline{x}, \underline{y} \in R^k$. Then,

$$L_1(\underline{x},\underline{y}) = \sum_{i=1}^{k} | x_i - y_i | \qquad (1)$$

It is easy to show the following properties of the L_1 norm

1. $L_1(x,y) = L_1(y,x)$ (symmetric)
2. $L_1(x,y) \geq 0$, with $L_1(x,y) = 0 \Leftrightarrow x=y$ (positive-definite)
3. $L_1(x,y) + L_1(y,z) \geq L_1(x,z)$ (Triangle inequality)

In addition to the L1-norm, there are other metrics used in vector spaces. The well-known Euclidean distance is perhaps the most popular, also known as L2-norm, defined as

$$L_2(\underline{x},\underline{y}) = \left(\sum_{i=1}^{k} | x_i - y_i |^2 \right)^{1/2}$$

It is easy to show that L_2 has the same properties (1)-(3) as L_1.

We can consider the nth-norm of two vectors as:

$$L_n(\underline{x},\underline{y}) = \left(\sum_{i=1}^{k} | x_i - y_i |^n \right)^{1/n}$$

and show that it has the same properties (1)-(3) above.

The limit case of n→∞, known as L_∞ can be shown to be

$$L_\infty(\underline{x},\underline{y}) = \lim_{n \to \infty} L_n(\underline{x},\underline{y}) = \lim_{n \to \infty} \left(\sum_{i=1}^{k} | x_i - y_i |^n \right)^{1/n} \Rightarrow$$

$$L_\infty(\underline{x}, \underline{y}) = \max_{i \in \{1, \cdots k\}} | x_i - y_i |$$

Clearly, all L_n-norms try to quantify in a single number the amount of difference between two vectors. There is great debate over which norm is the best to use in order to express error in signal processing, and especially audio, image and video [4]. Traditionally, researchers use the L_2-norm as the minimization criterion for improving signal processing and compression. That is the main reason the peak signal-to-noise ratio (PSNR, a logarithmic representation of the L_2-norm) has been used throughout the signal processing literature to express signal quality. On the other hand, SAD, which is the L_1-norm, has been used as the basic computational block to find block matches in video compression, since it does not require the additional complexity of the multiplier needed for L_2-norms. This is a necessary compromise – one of many one needs to make – in order to have a practical implementation of a video encoder.

The spirit of our work is the same: we offer an alternative compromise to the L_1-norm, an alternative that is similar in spirit (but yet distinct) to the L_∞-norm that yields very good results at a significant reduction in computations and power, especially from a hardware point of view.

3. PROPOSED ALGORITHM

In this section we introduce a new technique for approaching the problem of both, Intra and Inter mode decision. The base of this technique is to avoid the stage of addition, which increases significantly the power and delay cost at the hardware level.

For a given 4×4 block, according to equation (1), a total of 16 subtractions and 15 additions are needed in order to produce the SAD for one mode. Therefore, for the nine modes used in the Intra Prediction Mode, for example, we need 144 subtractions and 135 additions. After computing all modes, a comparison between the results decides which mode will be used. In the stage of Intra Prediction we need to find the prediction mode, thus, a qualitative approach may give the same results as a quantitative one, thus, removing the aforementioned computational cost.

Based on the above observation, we propose to compare the differences for the available modes, after calculating the differences among the predicted and the original pixels, instead of adding them. This comparison will conclude to the mode with the most minimum differences. Thus, the addition stage is completely bypassed. It must be noted here that the new algorithm would not have achieved better delay and power results if we had used the standard comparator [5], which has inferior power and delay characteristics when compared to adders. It was through a synergy with the novel implementation described in Section 5 that we were able to achieve these goals with the proposed algorithm.

In order to implement the new approach, we first calculate the absolute difference between the corresponding pixels for each mode. This can be written as:

$$M_{k_{ij}} = |C_{ij} - P_{k_{ij}}| \qquad (2)$$

where M, C, P are 4×4 arrays, k indicates the mode (in the case of the Intra Prediction k is in the range of $0 \leq k \leq 8$), and i, j are the indices specifying the single pixel within the 4×4 array.

Two successive M_k arrays comprise of a pair and a comparison among them is performed. The array with the largest number of minimum values is chosen. In the next step the array chosen from each pair forms a new pair with the array chosen from its successive pair and the same procedure is repeated until we end up with just one pair of arrays. The array chosen by the last comparison is the one which corresponds to the best mode.

The comparison among two arrays indicates the array with the largest number of minimum values in the following way. Let's assume that we have the following function:

$$f_{k_i} = \begin{cases} 1, M_{k_i} \leq M_{k_i+1} \\ \\ 0, M_{k_i} > M_{k_i+1} \end{cases} \qquad (3)$$

$$F_k = \sum_{i=0}^{15} f_{k_i} \qquad (4)$$

where M_k is the array with the differences for mode k and i (with $0 \leq i \leq 15$) is the number of differences . According to equation (4) we chose M_k if $F_k < F_{k+1}$, otherwise we chose M_{k+1}.

4. SOFTWARE MEASUREMENTS

The proposed mode decision scheme has been integrated with the H.264 JM10.2 [6] codec for performance evaluation. It is compared with the original codec (which uses SAD) of H.264 in terms of the average bits/frame and the average PSNR. The test sequences are the Akiyo, foreman, coastguard and hall QCIF video sequences. All four QCIF video sequences are 176×144 and have framerate 15 fps. The total number of frames used in the simulation is 300 for each sequence, with an I-Frame rate of 15 frames. The motion estimation scheme used is the "Full search" scheme with the search range set to 16 and number of reference frames set to 1.

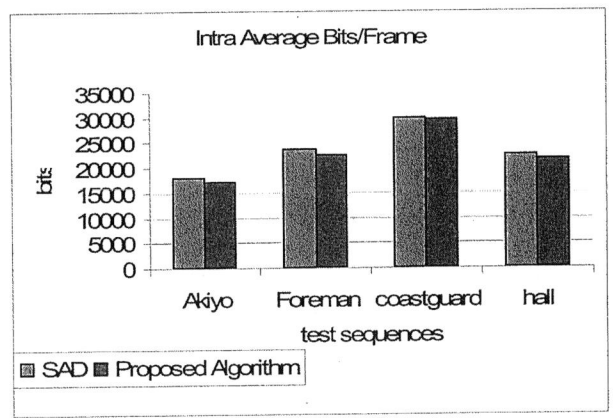

Figure 1. Average bits/frame for Intra Frames

In the simulation process the RD-optimization was set off, which does not allow any rate control. The quantization parameter (QP) was set to QP=28, which is the most common case. Figure 1 shows the average bits/frame of the intra frames in the video sequences produced by the encoder for the four test sequences. The respective average for the inter frames is shown in Figure 2. We see that the proposed scheme produces a small overhead in the average of the bits/frame.

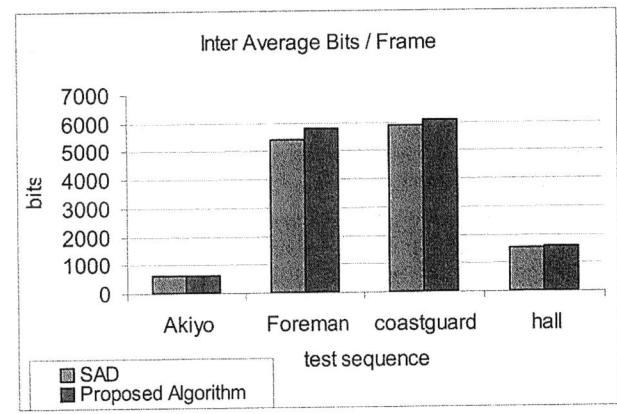

Figure 2. Average bits/frame for Inter Frames

The measurements used in the graphs are shown, with more details, in Table 1. In this Table the percentage difference for each sequence is also presented.

Table 1. Average bits/frame

Test Seq.	Average bits/frame					
	Intra			Inter		
	SAD	Proposed Arch.	% dif.	SAD	Proposed Arch.	% dif.
akiyo	18107.35	17276.40	-4.6	626.11	637.01	1.7
Foreman	23569.20	22572.80	-4.2	5403.14	5781.42	7.0
coastguard	29989.90	29593.55	-1.3	5888.91	6097.69	3.5
Hall	22439.75	21724.90	-3.2	1521.99	1575.49	3.5

Results show that for the intra frames we have an average decrease of 3.3% in the average bits/frame, and a respective average increase of 3.9% for the inter frames. The performance evaluation process also showed that the average PSNR remains practically the same, as the measurements gave us an average difference of 0.2% which is among the limits of statistical error. The aforementioned results confirm that the proposed algorithm achieves the same efficiency as that of SAD, with a significant reduction in the complexity of the encoder.

5. IMPLEMENTATION

The architecture described in Section 4 is shown in Figure 3, in which the "mode i" blocks refer to the absolute differences for each prediction mode and the "comparator" blocks implement the functions described by (3) and (4) by which the best mode, according to the proposed algorithm, is chosen.

This design is based on the Intra Prediction Mode for 4×4 blocks. That means that the circuit can compare 9 4×4 blocks at a time, as shown in Figure 3. Nevertheless, the circuit can be used as is in the Inter Prediction. For example, with a search range of 16×16 we have a total of 1089 available positions. Using the appropriate number of instances (i.e. 176) of the proposed circuit we can predict the position which gives us the smallest motion vector. The input word-lengths used were 128 bits (for an entire 4x4 block of 8-bit numbers), which is consistent with common video standards.

The comparison is not done by a standard comparator [7] but by specialized bit-wise circuits developed for this application. The comparison between absolute difference i of mode j and absolute difference i of mode k are done by the circuit, the block diagram of which is shown in Figure 4, while the circuit, the block diagram of which is shown in Figure 5 determined which of the two modes has the most minimum values. In case of a tie, the lower order mode is preferred.

The absolute difference selection circuit comprises of 3 stages, as shown in Figure 4. In the first stage we determine which of the 8 bits of the two inputs are equal. As output of this stage we have two 8-bit vectors, the vector *equal* which indicates the bits that are equal among the two input vector and the vector *comp* which is a copy of the second input.

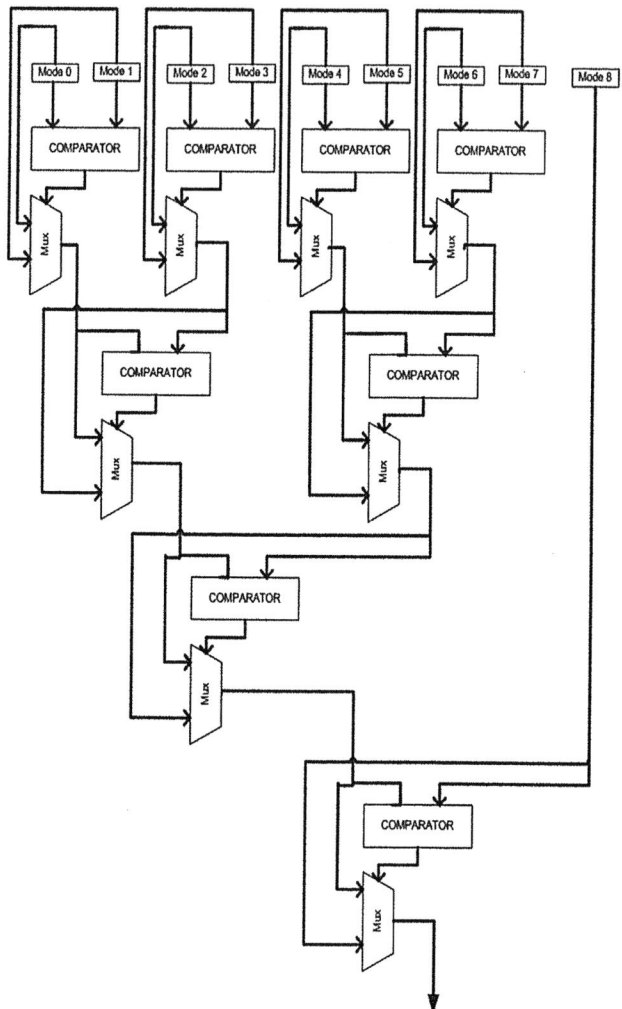

Figure 3. Block diagram of the implemented circuit

The other two stages are those that determine which one of the initial input vectors represents an 8-bit number with greater magnitude. In the second stage we have as inputs the vectors *equal* and *comp*. We part the 8-bit vectors in two triples and one dyad. For each one of these parts we perform a comparison that determines if the respective bits of *comp* are greater or not. For the triples we use a three-bit comparator and for the dyad we use a two-bit comparator.

The three-bit comparator works as follows. If the third bit of *comp* is set to 1 and, at the same time, the third bit of *equal* is set to 0, then the number represented by *comp* is greater than the other. The same applies when the third bit of *equal* and the second bit of *comp* are set to 1 and, at the same time the second bit of *equal* is set to 0, or when the third and second bit of *equal* and the first bit of *comp* are set to 1 and, at the same time the first bit of *equal* is set to 0. In the same sense works the two-bit comparator.

At the end of this stage we have as output 3 bits that show if we have equality and three other bits that show, for each part of the initial comp, if it is greater or not. In the last stage the output of the second stage is input in a three-bit comparator that gives a bit *comp*, that is 1 if the second of the initial vectors is greater and 0 if it is smaller, and a bit *equal* the two initial vectors are equal.

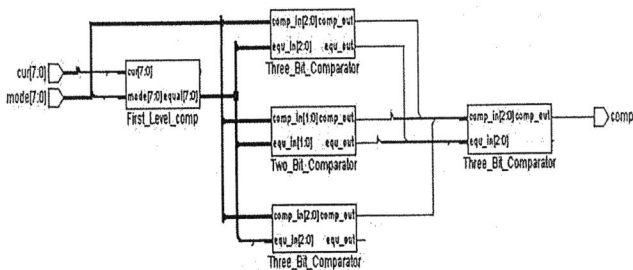

Figure 4. Absolute Difference Selection Circuit – Block Diagram

Each mode has a total of 16 absolute differences of pixel values i.e. 8-bit vectors. The circuit just described gives us a bit for each of these 16 differences. To be accurate it gives us 1 if the difference of the first mode is smaller than or equal to the difference of the second mode. So, at the end of this stage we end up with a 16-bit vector. The next thing to do is to determine which of the two modes has the most minimum values. This is done by the circuit, which's block diagram is shown in Figure 5 This circuit, also, consists of three stages.

In the first stage we part the 16-bit vector in four quadruplets. For each of the quadruplet we count the number of "1". This is done with the circuit named FirstStage in Figure 5. This circuit takes as input a 4-bit vector and has output a 5-bit vector named *sum*. The maximum number of "1" that can be found in the input is four. In this case the bit *sum(4)* is set to one. If we have three "1" in the input vector the bit *sum(3)* is set to 1, if we have 2 the bit *sum(2)* is set to 1, if we have 1 the bit *sum(1)* is set to 1 and finally if there are no "1" in the input vector the bit *sum(0)* is set to 1.

The bit-vectors produced in the first stage are the input to the second stage. In this stage we have two circuits named SecondStage. Each one of them takes as input two 5-bit vectors that represent a value between 4 and 0, and has as output an 8-bit vector. As in the first stage, each bit of the output represents a number of "1". So, if the 8th bit of the output is set to 1, that means that we a have a total of 8 "1", if the 7th bit is set to 1 we have 7 "1" and so on.

Finally we have the third stage, were the inputs are the two 8-bit vectors produced by the previous stage. Here we have the circuit named ThirdStage which has one bit as output. This bit is set to 1 if the total number of "1" (represented by the two 8-bit vectors as described) is greater than or equal to 8.

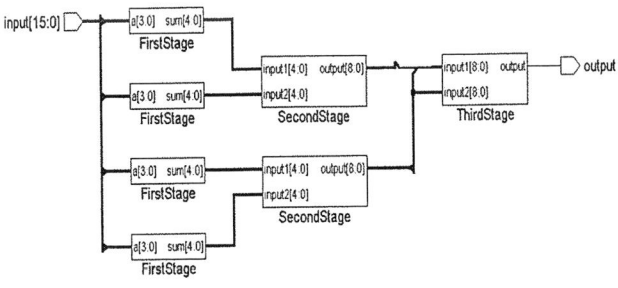

Figure 5. Mode Selection Circuit – Block Diagram

The implementation using SAD and the one using the new algorithm have been captured using VHDL and synthesized to allow comparisons of delay and power on a common reference. They have both been implemented on a 130nm CMOS technology (1.08V) using the UMC standard cell library. The architecture has been simulated using the Mentor Graphics ModelSim® SE 6.0 and synthesized using the Synopsys Design Compiler®. The power estimates for the two circuits were obtained by Synopsys PrimePower® using the four test sequences as inputs.

In order to assess the power-performance characteristics of both implementations we proceeded to form their respective power-delay curves, which are shown in Figure 6. The data points for the power-delay curves were determined as follows:

i. A specific delay target was selected for each implementation, starting from the best delay possible, and relaxing the timing requirement by 1ns at a time.

ii. The circuit was synthesized with Synopsys Design Compiler®, which also provided the delay for each case.

iii. The average power dissipation was calculated by Synopsys PrimePower®.

The operating frequency was set according to the maximum delay of the synthesized circuit, as reported in step (ii).

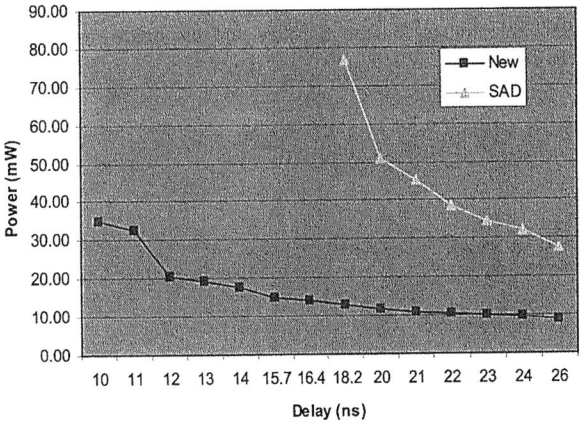

Figure 6. Power – Delay curves for both implementations

From Figure 6 it can be readily concluded that the new algorithm and its implementation outperforms the SAD approach in both power and performance, as its power-delay curve lies below and to the left of the one for SAD.

Table 2. Performance

Architecture	Performance		
	Critical Path (ns)	*Max Clk Frequency (MHz)*	*Cell area*
SAD	18.15	55	132356
Proposed Algorithm	10	100	58100

A more detailed analysis reveals that the design created contains less than 59k gates and can operate at frequencies of up to 100 MHz. The best delay for SAD that Design Compiler® could achieve was 18.15ns vs. 10ns for the proposed implementation, a 45% reduction, whereas the best area for SAD was 132k gates vs. 58k gates for the proposed algorithm, a 56% reduction.

The power reduction that was achieved ranges from 6X at 18.15ns to 3X at 26ns. The observed reduction is due to the reduced complexity of both the prediction algorithm and its circuit implementation. The power reduction at the best delay for both approaches is over 2X in favor of the proposed approach over SAD.

6. CONCLUSIONS

In this paper we presented a new mode selection algorithm for both Intra and Inter prediction modes in H.264 along with logic transformations for its hardware implementation. The proposed replacement of the SAD algorithm with a simple comparison was integrated with the H.264 JM10.1 codec for performance evaluation and was implemented using standard circuit synthesis tools. The resulting circuit has a clear advantage in power (3X-6X), delay (2X), and area (2X) over the one implementing the SAD algorithm, while software measurements show that the proposed algorithm produces an insignificant overhead in the bit rate, and the quality of the produced video sequence is almost the same as that of the one produced by the original H.264 codec.

Therefore, the new algorithm, by reducing the complexity of the computation, can achieve significantly faster execution time and significant reduction in power consumption without having any perceivable impact on the file size and the quality of the final video sequence.

7. REFERENCES

[1] ITU-T Rec. H.264 / ISO/IEC 11496-10, *Advanced Video Coding*, Final Committee Draft, Document JVT-J010d1, December 2003.

[2] Wiegand, T., Sullivan, G., Bjontegaard, G., and Luthra, A. , *Overview of the H.264/AVC video coding standard*, IEEE Transactions on Circuits and Systems For Video Technology, vol. 13, no. 7, pp.560-576, July 2003.

[3] Richardson, I., *H.264 and MPEG-4 Video Compression – Video Coding for Next-generation Multimedia*, John Wiley & Sons, 2003.

[4] Toivonen, T., Heikkila, J., *A new rate – minimizing matching criterion and a fast algorithm for block motion estimation*, International Conference on Image Processing (ICIP 2003).

[5] Mano, M., *Digital Design*, 3rd edition.

[6] Joint Video Team (JVT) Reference Software JM10.2, at http://iphome.hhi.de/suehring/tml/download/

[7] Wang, C., Lee, P., Wu, C., and Wu, H., *High Fan-In Dynamic CMOS Comparators with Low Transistor Count*, IEEE Transactions on Circuits and Systems – I:Fundamental theory and applications, vol. 50, no. 9, pp. 1216-1220, September 2003.

Energy-efficient Motion Estimation using Error-Tolerance

Girish V. Varatkar and Naresh R. Shanbhag
Coordinated Science Laboratory, University of Illinois at Urbana-Champaign
1308 W Main St., Urbana, IL 61801
[varatkar, shanbhag]@uiuc.edu

ABSTRACT

Presented is an energy-efficient motion estimation architecture using error-tolerance. The technique employs overscaling of the supply voltage (voltage overscaling (VOS)) to reduce power at the expense of timing errors, which are then corrected using algorithmic noise-tolerance (ANT) techniques. Referred to as input subsampled replica ANT (ISR-ANT), the proposed technique incorporates an input subsampled replica of the main sum of absolute difference (MSAD) block for obtaining the motion vectors in the presence of errors induced by VOS. Simulations show that the proposed technique can save up to 60% power over an optimal error-free present day system in a $130nm$ CMOS technology. Power savings increase to 79% in a $45nm$ predictive process technology.

Categories and Subject Descriptors: B.8.2 [Performance and reliability]: Performance Analysis and Design Aids

General Terms: Algorithms, Design, Reliability

Keywords: Low-power, noise-tolerance

1. INTRODUCTION

Next generation wireless multimedia communications standards such as digital video broadcast (DVB) [1], fourth generation (4G) mobile systems [2] need to provide services such as video transmission on hand-held units. These units need to be energy-efficient while providing a high quality of service. The MPEG-4 encoder is the most computationally intensive block in a video processor. The motion estimation (ME) kernel consumes 66%-94% of the encoder cycles [3]. Therefore, low-power motion-estimation architectures and implementations are of great interest.

Low-power motion estimation is a well-studied subject [4]-[6]. However, most, if not all, approaches assume error-free computation. The proposed work relaxes this assumption in order to push the boundaries of achievable energy-efficiency. In particular, most algorithmic low-power approaches focus on heuristics to reduce the number of macro-

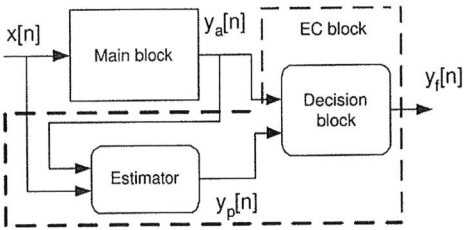

Figure 1: An ANT-based system.

blocks processed per motion vector [3]. These include employing a simpler distance criterion [8], fixed or adaptive search area [9] and temporal or spatial prediction [10]. Several VLSI architectures are proposed for ME with various trade-offs between gate-count, I/O bandwidth and throughput [3][11][12][13]. A typical motion estimation accelerator consists of a RAM for search area and current block, an address generation unit, a datapath consisting of a processor element and a control unit. The datapath power consumption is found to be 75% of the total ME power consumption for full search motion estimation algorithm and 50% of the total ME power consumption for three step search algorithm [7]. Therefore, it is important to reduce the datapath power consumption. Scaling of supply (V_{dd}) and threshold voltage (V_t), has been commonly employed to reduce the total datapath power consumption [14]-[15]. The benefits of conventional voltage scaling are limited by the (V_{dd}, V_t) combination at which the worst case critical path delay is equal to the clock period [15].

In this paper, we present a novel low-power ME architecture that is based on the concept of algorithmic noise-tolerance (ANT) [16]. In ANT (see Fig. 1), a main block is assumed to make intermittent errors which are then corrected by an error-control block (**EC**). The **EC** block includes an estimator and a decision block. The error mechanism can vary but if power reduction is being targeted then the errors are tailored to arise from voltage overscaling (VOS) [16]. In VOS, the supply voltage is reduced beyond $V_{dd-crit}$, i.e.,

$$V_{dd} = K_{vos}V_{dd-crit}, \qquad (1)$$

where $0 < K_{vos} \leq 1$ and $V_{dd-crit}$ is the supply voltage below which timing violations occur. These violations are referred to as VOS errors which are then corrected by employing ANT techniques. Thus, the combination of VOS and ANT can reduce power beyond that achievable by conventional voltage scaling alone.

Permission to make digital or hard copies of all or part of this work for personal or classroom use is granted without fee provided that copies are not made or distributed for profit or commercial advantage and that copies bear this notice and the full citation on the first page. To copy otherwise, to republish, to post on servers or to redistribute to lists, requires prior specific permission and/or a fee.
ISLPED'06, October 4–6, 2006, Tegernsee, Germany.
Copyright 2006 ACM 1-59593-462-6/06/0010 ...$5.00.

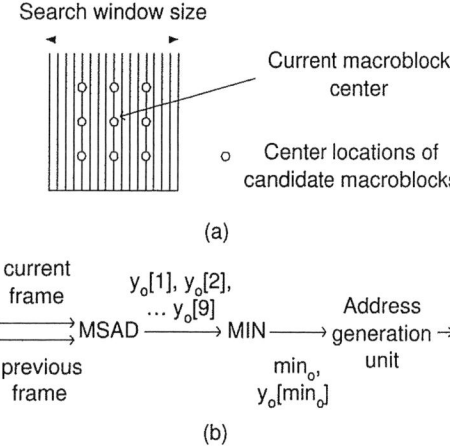

(a)

(b)

Figure 2: The three step search (TSS) algorithm: (a) search window, and (b) block level implementation.

Many ANT techniques exist. These include, the reduced precision replica (RPR) ANT [17] technique where the estimator is a reduced precision replica of the main block. In prediction-based ANT [16], the estimator is a predictor that exploits the correlation in the output of the main block. As VOS errors are input-dependent, the adaptive error-cancellation technique [18] employs an error estimator to estimate and cancel VOS errors at the main block output.

In this paper, we present a novel ANT technique for correcting VOS errors in ME architectures and study the achievable power savings. Simulations using an IBM $130nm$ CMOS process show that up to 60% power savings can be achieved over an optimal error-free architecture. Power savings increase to 79% in a $45nm$ predictive process technology [19][20].

Section 2 of the paper describes the ME algorithm and a straightforward application of ANT to ME referred to as motion-vector replica ANT (MVR-ANT). In section 3, we present our main contribution: a new technique referred to as input subsampled replica ANT (ISR-ANT) for energy-efficient motion estimation. In section 4, we present simulation results for ISR-ANT datapath designed using IBM $130nm$ process technology and using $45nm$ predictive technology models.

2. PRELIMINARIES

In this section, we present preliminaries of ME and ANT. This is done by first introducing the three-step search ME algorithm and then demonstrating a straightforward but ineffective application of ANT resulting in the MVR-ANT. The latter will then be modified to form the proposed ISR-ANT in section 3.

2.1 The Three Step Search (TSS) Algorithm

A motion estimation algorithm reduces temporal redundancy between consecutive video frames. In block matching motion estimation algorithms, the current video frame is partitioned into non-overlapping macroblocks of size sixteen pixels by sixteen pixels. For each macroblock in the current frame, the motion estimation algorithm searches for the best matching macroblock in the previous frame. There exist hundreds of algorithms for efficient search [3]-[13] since

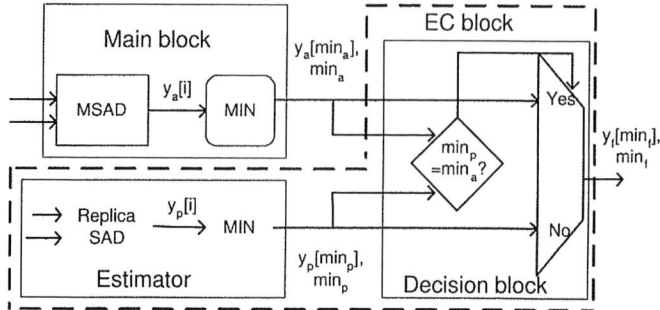

Figure 3: The MVR ANT architecture.

the ME algorithm is not standardized. For energy-efficiency purpose, we select an algorithm that is suitable for VLSI implementation. The full-search block matching algorithm is the most optimal and the most suitable algorithm for VLSI implementation due to regularity in computation. However, it demands a huge amount of computation.

The TSS algorithm is a commonly employed sub-optimal block matching algorithm [21] because of its simplicity, reduced amount of computation compared to full search, and near optimal performance. In this paper, we choose TSS algorithm to demonstrate effectiveness of ANT technique. Our proposed ANT technique does not depend upon the choice of block matching algorithm used and it can be extended to other block matching algorithms. In TSS algorithm, an initial step size Δ, typically equal to half of the search window size is chosen. The location of the center of the current macroblock inside the search window is shown in Fig. 2(a). Next, nine candidate macroblocks $M[1:9]$ with their center locations as shown in Fig. 2(a), are chosen from the previous frame for comparison. Eight of these candidate macroblocks have their centers at a distance of $\pm\Delta$ in the x and y direction from the current macroblock. The ninth macroblock is at the same location as the current macroblock.

The sum of absolute differences (SAD) for each of the nine macroblocks are calculated by the **MSAD** block by summing up the absolute difference between the corresponding pixels in the candidate macroblocks and the current macroblock. Let **MSAD** block have input pixel streams denoted by $a[k]$ and $b[k]$ from the current macroblock and the candidate macroblock, respectively. The output of **MSAD** are the nine candidate SAD values denoted by $y_o[i]$ for $1 \leq i \leq 9$, where,

$$y_o[i] = \sum_{k=1}^{256} |a[k] - b[k]| \qquad (2)$$

for $1 \leq i \leq 9$. The index corresponding to the best match is obtained as,

$$y_o[min_o] = min\{y_o[1], y_o[2], ..., y_o[9]\}$$
$$min_o = argmin\{y_o[1], y_o[2], ..., y_o[9]\} \qquad (3)$$

The motion vector is the vector difference between $M[min_o]$ and the current block. Next, $\pm\Delta$ is halved and the center of the search window is moved to coincide with that of $M[min_o]$. Previous steps are repeated till the Δ becomes less than 1. A block level implementation of TSS is shown in Fig. 2(b). The **MSAD** block calculates the SAD in Eq. 2 while the **MIN** block determines min_o using Eq. 3.

114

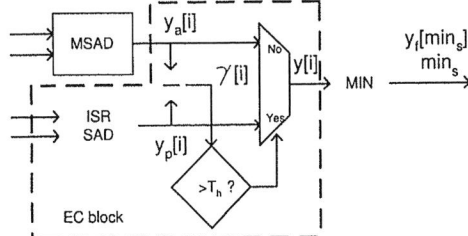

Figure 4: The ISR-ANT architecture.

2.2 Motion Vector Replica (MVR) ANT

A straightforward application of the ANT framework (see Fig. 1) results in the motion vector replica (MVR) ANT. A MVR ANT-based ME (see Fig. 3) has a main block and an error control block (**EC**). The address generation block is not shown. The main block is a complete ME engine that includes the **MSAD** and **MIN** blocks. The main block is made energy-efficient via VOS but makes intermittent errors. These VOS errors degrade the output peak signal-to-noise ratio ($PSNR$) if left uncorrected. Let the error-free **MSAD** block outputs be denoted as $y_o[i]$ for $i = 1, .., 9$. Under VOS, the **MSAD** output denoted as $y_a[i]$ is given by

$$y_a[i] = y_o[i] + \eta[i] \qquad (4)$$

where $\eta[i]$ is the VOS error. Next, we define the main block output min_a as follows.

$$y_a[min_a] = min\{y_a[1], y_a[2], ..., y_a[9]\}$$
$$min_a = argmin\{y_a[1], y_a[2], ..., y_a[9]\} \qquad (5)$$

The **EC** block has an estimator and a decision block. The estimator estimates the correct motion vector and is designed to have low complexity and hence error-free operation, compared to the main block. This means that the estimator output will not be as accurate as the correct main block output. Therefore, the estimator block operates in an error-free albeit inaccurate manner. If the main block and the estimator outputs differ, then the decision block employs the estimator output as the final corrected output $y_f[min_f]$ as shown in Fig. 3.

A simple **EC** block is one whose estimator is a reduced precision replica of the main block [17]. For example, the **MSAD** block can have 8-bit pixels as input while the replica **SAD** block can employ reduced input bit precision. In fact, it is known [22] that the average $PSNR$ degrades by less than $0.5dB$ if 3-bit inputs are employed for computing SAD. If the **MSAD** block employs 3-bit inputs, then the replica **SAD** must operate with less than 3-bit input pixel values resulting in very inaccurate estimates of SAD. Thus, the MVR-ANT is not very effective in power reduction.

3. INPUT SUBSAMPLED REPLICA (ISR) ANT ARCHITECTURE

In this section, we describe the main contribution of this paper referred to as the input subsampled replica (ISR) ANT-based ME architecture. We make the following modifications to the MVR ANT **EC** block to generate ISR-ANT as shown in Fig. 4.

1. We employ an estimator based on input subsampling, where an estimate of the **MSAD** output is calculated

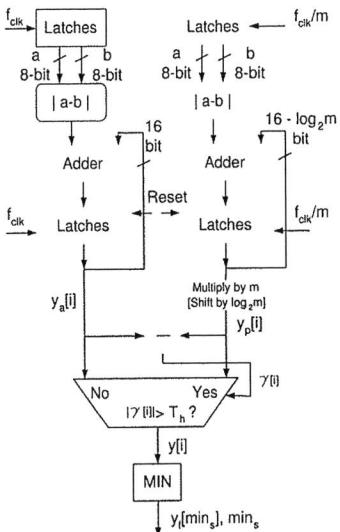

Figure 5: The ISR-ANT datapath.

by employing an **ISR-SAD** block which subsamples the input streams $a[k]$ and $b[k]$ by a factor of m.

$$y_p[i] = m \times \sum_{k=1}^{\lfloor 256/m \rfloor} |a[mk] - b[mk]| \qquad (6)$$

Let $e_p[i]$ denote the SAD estimation error.

$$e_p[i] = y_p[i] - y_o[i] \qquad (7)$$

2. We modify the decision block as follows. We detect and correct VOS errors at the output of the **MSAD** block instead of the **MIN** block.

Note, the **ISR-SAD** output $y_p[i]$ is an estimate of the error-free sum $y_o[i]$ for $1 \leq i \leq 9$. Hence, a threshold $T_h = \max|e_p[i]|$ can be chosen in such a way that $|e_p[i]| < T_h$. Let $\gamma[i]$ denote the difference between the **MSAD** output $y_a[i]$ and **ISR-SAD** output $y_p[i]$, i.e.,

$$\gamma[i] = y_a[i] - y_p[i] \qquad (8)$$

An error is declared if $|\gamma[i]| > T_h$. The decision block employs the **ISR-SAD** output $y_p[i]$ as input to the **MIN** block if error is detected. If there is no error, the **MSAD** output $y_a[i]$ is employed as input to the **MIN** block.

The ISR-ANT algorithm is described by the following steps.

1. An initial step size Δ is chosen. Eight blocks at a distance of $\pm\Delta$ from the center (around the center block) are picked for SAD computation and comparison. The **MSAD** and the **ISR-SAD** blocks, each calculate the nine candidate SADs.

2. An error is declared if $|\gamma[i]| > T_h$ where $T_h = \max|e_p[i]|$.

3. If an error is declared then $y[i] = y_p[i]$ else $y[i] = y_a[i]$.

4. The direction in which the block distortion $y[i]$ is minimum is chosen.

$$y_f[min_s] = min\{y[1], y[2], ..., y[9]\}$$
$$min_s = argmin\{y[1], y[2], ..., y[9]\} \qquad (9)$$

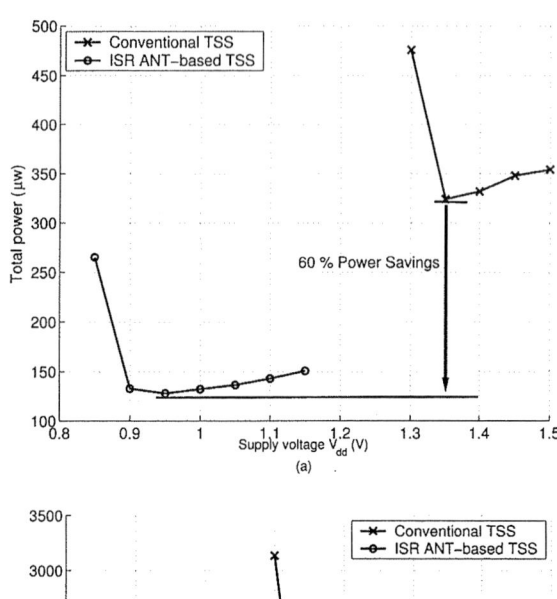

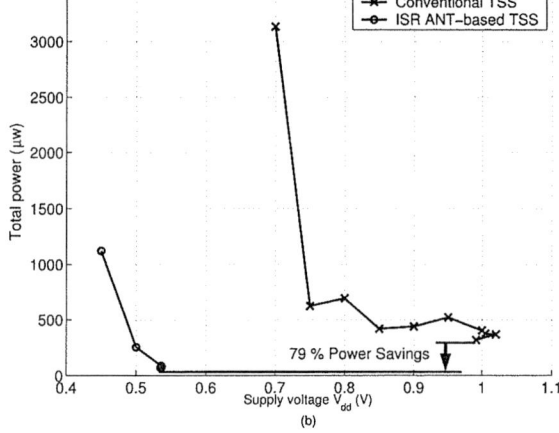

Figure 6: Supply voltage and body bias for full adder delay $= 150ps$ and $225ps$ for: (a)$130nm$, and (b) $45nm$ process technology.

Figure 7: Power consumption for conventional TSS and ISR-ANT-based TSS for: (a)$130nm$, and (b) $45nm$ process technology.

5. The step size Δ is halved. The center of the search window is moved to the point with the minimum distortion. Previous steps are repeated till the step size becomes less than 1.

ISR-ANT works well under the following assumptions:

1. The magnitude of VOS error in **MSAD** block output is large so that it is easy to detect errors in the output.

2. The **ISR-SAD** and the decision blocks are error-free.

The errors due to voltage-overscaling occur in the most significant bits (MSBs) due to least-significant bit (LSB) first nature of computation in **MSAD** . As a result, the magnitude of the VOS error in **MSAD** block output is large. The **ISR-SAD** block has only N/m inputs to process as compared to N inputs for the **MSAD** block. Hence it is able to operate in an error-free manner.

In ISR-ANT architecture as shown in Fig. 5, the **MSAD** block consists of a modulus block computing the absolute difference between 8-bit luminance values. These absolute differences are accumulated by an 16-bit adder whose outputs are latched at a frequency f_{clk}. The **ISR-SAD** block

uses pixel luminance values which are subsampled by a factor of m. Hence, the adder in **ISR-SAD** has $16 - log_2 m$ bits and its outputs are latched at f_{clk}/m. After 256 cycles of the frequency f_{clk} clock, the absolute value of $\gamma[i]$ is compared with the threshold T_h and the multiplexer output $y[i]$ is fed to the **MIN** block.

4. SIMULATION RESULTS AND DISCUSSION

In this section, we compare the power vs. performance trade-offs between conventional error-free architecture and ISR-ANT in an IBM $130nm$ CMOS process and $45nm$ predictive technology models [19][20]. Three different video clips are evaluated: flower garden (low motion), mobile calendar (medium motion) and football (high motion).

4.1 Simulation Set-up

We first determined the system level throughput requirements for motion estimation in real time encoding of MPEG-II main profile at main level [24] to be a CIF frame size of 288 by 352 pixels at the rate of 30 frames/s. Next, we simulated both architectures with ripple carry adders using an HDL simulator. This was done to determine the maximum delay

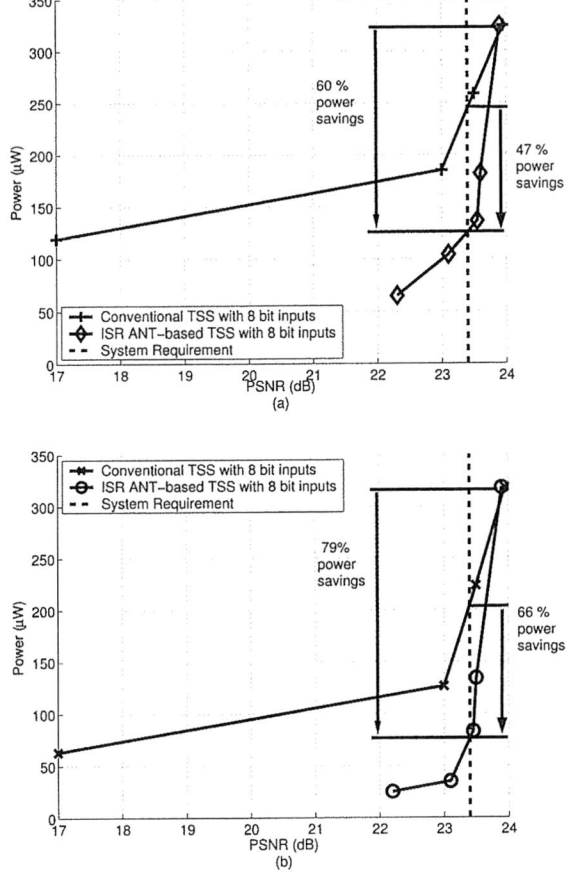

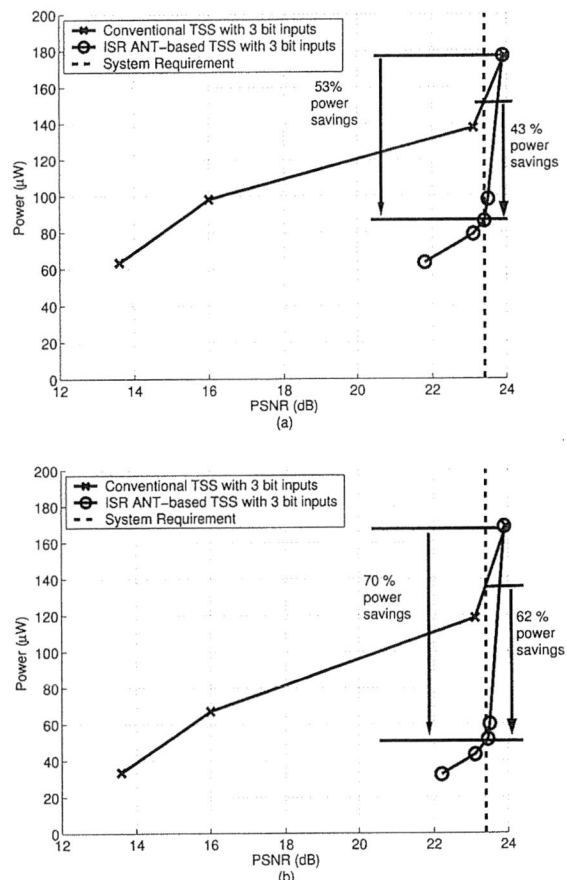

Figure 8: Power vs. *PSNR* plot for 8-bit input in: (a)130*nm*,** and (b) **45*nm* **process technology.**

Figure 9: Power vs. *PSNR* plot for 3-bit input in: (a)130*nm*,** and (b) **45*nm* **process technology.**

of a 1-bit adder (T_{FA}) necessary to support this system level throughput. The conventional architecture was found to require a $T_{FA} \leq 150ps$ for error-free operation. We simulated ISR-ANT with various subsampling ratios and determined that the system performance requirement as defined in 4.3 is met for $m \leq 4$ for the ISR-ANT. Therefore, the subsampling ratio is set to $m = 4$ for all simulations. With $m = 4$, we determined that T_{FA} needs to be less than 225ps in order for the **EC** block to correct the resulting errors effectively. For $T_{FA} \geq 225ps$, the VOS errors in **MSAD** degrade the ISR-ANT performance significantly.

Next, we characterized a full adder with the mirror structure [23] in terms of delay, dynamic power and leakage power consumption. Isodelay curves were obtained by varying the supply and body bias voltage combinations (V_{dd},V_b) via HSPICE in 130*nm* IBM process technology and 45*nm* predictive technology. The adder outputs were loaded with identical mirror full adders to determine the worst case output delay.

4.2   Power vs. Supply and Body Bias Voltage

As mentioned earlier, the error-free conventional architecture and ISR-ANT architecture needs to operate with $T_{FA} \leq 150ps$ and $T_{FA} \leq 225ps$ respectively. Fig. 6(a) shows the (V_{dd},V_b) combinations that result in a constant full-adder delay $T_{FA} = 150ps$ and $T_{FA} = 225ps$ in 130*nm* IBM technology. Similarly, Fig. 6(b) shows the isodelay plots in

the 45*nm* process technology. Similar isodelay curves are derived for intermediate delay values. These plots are useful in determining the power-optimum (V_{dd},V_b) combination for the conventional and ISR-ANT architectures.

We simulate the conventional architecture and the ISR-ANT architecture using HSPICE to obtain the power consumption for both the architectures operating at the (V_{dd},V_b) combinations obtained from the isodelay curves similar to Fig. 6. Thus we determine the power-optimum (V_{dd},V_b) combination for the conventional and ISR-ANT architectures. We plot the power dissipation of the two architectures in Fig. 7.

In the 130*nm* IBM process technology (see Fig. 7(a)), the conventional architecture achieves a minimum power consumption of $324\mu W$ at ($V_{dd-crit} = 1.35V$, $V_b = 0.5V$), while the ISR-ANT architecture achieves a minimum power consumption of $128\mu W$ at ($V_{dd} = 0.95V$, $V_b = 0.3V$). The **EC** block operates at lower voltage, $V_{dd-EC} = 0.6V$. Thus, ISR-ANT achieves power-reduction of 60% over the conventional system.

The results for 45*nm* technology are shown in Fig. 7(b). We see that the conventional architecture achieves a minimum power consumption of $318\mu W$ at ($V_{dd-crit} = 1.0V$, $V_b = -0.6V$), whereas ISR-ANT achieves a minimum power consumption of $69\mu W$ at ($V_{dd} = 0.53V$, $V_b = -0.6V$). Here, the **EC** block operates at a voltage of $V_{dd-EC} = 0.4V$. Power reduction of 79% is achieved via ISR-ANT.

4.3 Power vs. Performance Trade-off

In this subsection, we compare the power savings for the conventional and ISR-ANT architectures as a function of the $PSNR$. We simulate both architectures using the HDL simulator in order to determine the output motion vectors. We reconstruct the current frame from these motion vectors and the previous frame. Reconstruction error is calculated as the difference between the reconstructed frame and the actual current frame luminance values. The $PSNR$ is calculated as

$$PSNR(dB) = 20log_{10}\frac{255}{\sigma_r} \qquad (10)$$

where σ_r^2 is the reconstruction noise power. As mentioned earlier, the $PSNR$ calculations were done for three different clips. The flower garden clip has slowly moving garden background. The mobile calendar clip shows objects moving with medium speed. The football clip has fast player movement. We show a plot of power dissipation vs. $PSNR$ for mobile calendar clip in Fig. 8 for 8-bit inputs. The plots for other clips are similar and hence are not shown. We set the desired $PSNR$ requirement to be $0.5dB$ less than the $PSNR$ obtained using the error-free conventional architecture. Plots in Fig. 8(a) and (b) for $130nm$ and $45nm$ process nodes, respectively, are obtained by reducing the supply voltage from $V_{dd-crit}$ where the full-adder delay is $T_{FA} = 150ps$. A power-optimum value for V_b is obtained for each value of V_{dd} from the isodelay curves similar to those in Fig. 6. First, note that $PSNR$ of the conventional architecture drops severely as the supply voltage is reduced from $V_{dd-crit}$. The ISR-ANT architecture is seen to be robust to VOS. At a $PSNR = 23.9dB$, we see that 60% and 79% total power savings are achieved in $130nm$ and $45nm$, respectively, with the error-free conventional architecture as a reference. The power savings become 47% and 66% in $130nm$ and $45nm$, respectively, when compared at the desired $PSNR = 23.4dB$.

We can reduce the complexity of the **MSAD** block by reducing its precision from 8-bits to 3-bits. Figure 9 shows the power savings when the input is quantized to 3-bits. At a $PSNR = 23.9dB$, we see that 53% and 70% total power savings are achieved in $130nm$ and $45nm$, respectively, with the error-free conventional architecture as a reference. The power savings become 43% and 62% in $130nm$ and $45nm$, respectively, when compared at the desired $PSNR = 23.4dB$.

We note that if we subsample the inputs of **MSAD** in the conventional architecture by $m = 4$ in order to reduce its complexity, then the $PSNR$ degrades from $PSNR = 23.9dB$ to $PSNR = 22.8dB$. This $PSNR$ loss is unacceptable and it indicates that ISR-ANT is a unique approach to power reduction.

5. CONCLUSIONS

In this paper we have applied error-tolerance for energy-efficient motion estimation. We minimized the total power by jointly optimizing supply voltage and body bias in $130nm$ and $45nm$ process technologies. The proposed ISR-ANT technique was shown to be robust to VOS errors. The ISR-ANT technique is agnostic to the actual source of errors and thus it would be interesting to study its performance in the presence of random errors such as soft-errors due to particle hits as well as errors due to process variations.

6. REFERENCES

[1] http://www.dvb.org

[2] http://www.4gmf.org

[3] P. Kuhn, "Algorithms, complexity analysis and VLSI architectures for MPEG-4 motion estimation," Kluwer Academic Publishers, Boston 1999.

[4] M. A. Elgamel, et. al., "A comparative analysis for low power motion estimation VLSI architectures," in *IEEE Workshop on Signal Processing* , pp. 149–158, 2000.

[5] F. Dufaux, et. al., "Motion estimation techniques for digital TV: A review and a new contribution," in *Proc. of the IEEE* , vol. 83, No. 6, pp. 858–876, June 1995.

[6] Po-Chih Tseng, et. al., "Advances in hardware architectures for image and video coding- a survey," in *Proc. of the IEEE*, vol. 93, No. 1, pp. 184–197, Jan. 2005.

[7] R. Steven Richmond II, et. al., "A low-power motion estimation block for low bit-rate wireless video," in *Proc. of ISLPED*, August 2001.

[8] M. Ghanbari, "The cross-search algorithm for motion estimation," in *IEEE Trans. on Comm.*, vol. 38, No. 7, pp. 950–953, July 1990.

[9] J. Minocha, et. al., "A low power data-adaptive motion estimation algorithm," in *IEEE Workshop on Multimedia Signal Processing*, pp. 685–690, Sept. 1999.

[10] B. Zeng, et. al., "Optimization of fast block motion estimation algorithms," in *IEEE Trans. on Circuits and Systems for Video Technology*, vol. 7, No. 6, pp. 833–844, Dec. 1997.

[11] S. Kim et. al., "A fast motion estimator for real-time systems," in *IEEE Trans. on Consumer Electronics*, vol. 43, No. 1, pp. 24–33, Feb. 1997.

[12] S. Dutta et. al., "A flexible parallel architecture adapted to block matching motion estimation algorithms," in *IEEE Trans. on Circuits and Systems for Video Technology*, vol. 6, No. 1, pp. 74–86, Feb. 1996.

[13] P. Lakamsani, "An architecture for enhanced three step search generalized for hierarchical motion estimation algorithms," in *IEEE Trans. On Consumer Electronics*, vol. 43 no. 2, pp. 221-227, May 1997.

[14] A. P. Chandrakasan, et. al., "Minimizing power consumption in digital CMOS circuits," in *Proc. of IEEE*, Vol. 83, April 1995.

[15] R. Gonzalez, et. al., "Supply and threshold voltage scaling for low-power CMOS," in *IEEE Journal of Solid-state Circuits*, Vol. 31, No. 3, pp. 395-400, March 1999.

[16] R. Hegde, et. al., "Soft digital signal processing," in *IEEE Trans. on VLSI*, vol. 9 pp. 813-823, December 2001.

[17] B. Shim, et. al., "Low-power digital signal processing via reduced-precision redundancy," in *IEEE Trans. on VLSI Systems*, vol. 12 pp. 497-510, May 2004.

[18] L. Wang, et. al., "Low-power filtering via adaptive error-cancellation," in *IEEE Trans. on Signal Processing*, vol. 51 pp. 575-583, February 2003.

[19] http://www.eas.asu.edu/ ptm

[20] Y. Cao et. al., "New paradigm of predictive MOSFET and interconnect modeling for early circuit design," in *Proc. of CICC*, pp. 201-204, 2000.

[21] T. Koga, "Motion compensated interframe coding for video conferencing," in *Proc. NTC*, 1981, Ch. 9.6.1-9.6.5.

[22] Y. Baek, et. al., "An efficient block matching criterion for motion estimation and its VLSI implementation," in *IEEE Trans. on Consumer Electronics*, vol. 42 no. 4, pp. 885-892, November 1996.

[23] J. M. Rabaey, et. al., "Digital integrated circuits A design perspective,"' Pearson Education, NJ: Prentice-Hall, 2003.

[24] S. C. Hsia, "VLSI implementation for low-complexity full-search Motion estimation," in *IEEE Trans. on Circuits and Systems for Video Technology*, vol. 12, No. 7, pp. 613–619, July. 2002.

[25] B. Liu, et. al., "New fast algorithms for the estimation of block motion vectors," in *IEEE Trans. on Circuits and Systems for Video Technology*, vol. 3 no. 2, pp. 148-157, April 1993.

Variability-Aware Device Optimization under I_{ON} and Leakage Current Constraints

Javid Jaffari
ECE Department, University of Waterloo
Waterloo, ON N2L 3G1, Canada
jjaffari@vlsi.uwaterloo.ca

Mohab Anis
ECE Department, University of Waterloo
Waterloo, ON N2L 3G1, Canada
manis@vlsi.uwaterloo.ca

ABSTRACT

In this paper, a novel device optimization methodology is presented that is constrained by the total leakage and the ON current of the device. The devised technique locates a maximum yield rectangular cube in a three-dimensional feasible space composed by oxide thickness, halo peak doping, and halo characteristic length parameters. The center of this cube is considered as the maximum yield design point with the highest immunity against variations. Monte-Carlo simulations show that the optimized Bulk-MOS device for 45 nm gate length satisfies the on current and leakage constraints under a variability of up to 30% in the three parameters.

Categories and Subject Descriptors: B.7.1 [Types and Design Styles]: Advanced technologies

General Terms: Algorithms, Design, Reliability.

Keywords: Device Design, Process Variation, Optimization, Leakage Current, Performance.

1. INTRODUCTION

Scaling of CMOS technology brings up enormous challenges that must be resolved by designers. As the silicon industry moves toward nanometer designs, the two most important design challenges cited are the growing leakage power dissipation [1] and the increasing variability in device characteristics [2] which threaten the life time of silicon technology, and why CMOS scaling is coming to an end. In fact, the leakage power problem is further compounded by its strong dependence on the process parameters and hence on their variations [3]. As a result, circuits experiencing variability, now exhibit very high leakage power consumption, pushing them over the power budget. Analytical models have been developed in [4] to estimate the mean and standard deviation for the three major components of the leakage current (subthreshold, gate direct tunneling, and Band-To-Band-Tunneling) under variability. Three devices have been designed, each having a dominant leakage current component. Based on the developed analytical models, it has been

shown that although these different devices have relatively equal total leakage current (TL), the variance of TL is considerably different. This re-emphasizes the importance of including process variability in the design of devices.

Motivated by the above challenges, the design of CMOS devices must be revised to include the variability of leakage current in computing the I_{ON}/I_{OFF} ratio. The objective of this work is to re-design the CMOS device to increase its reliability by enhancing its immunity against process variations. The work proposes a methodology for the design of CMOS devices that not only reduces TL but also its variance while accounting for the drive-in current (I_{ON}) of the device. With the aid of the proposed methodology, the designer would define two bounds on I_{ON} and TL, and can now exploit the allowable design space for variability to maximize the device's yield. Oxide thickness, halo peak doping, and halo characteristic length are considered as the main design variables.

First, a feasible three-dimensional space composed by the above mentioned process parameters is formed. Any point in this space would satisfy the defined constraints on I_{ON} and TL. Then, analytical models will be used to represent feasible space boundaries and process parameter distributions. Finally, the yield maximizing step attempts to place a rectangular cube in the feasible space such that the device lies in the center of the cube has more immunity against variabilities.

2. PROBLEM DEFINITION

In this paper, as shown in Figure 1, a symmetric Bulk-MOS device structure with source/drain extension (SDE) and Gaussian halo doping profile is selected. The device's parameters are: (Gate length (L_g), SDE length (L_{SDE}), oxide thickness (T_{ox}), junction depth (Y_j), transistor Width (W)) and the doping profile (channel doping (N_{dep}), halo peak (N_{halo}), and halo characteristic length (l_c)) [5].

Permission to make digital or hard copies of all or part of this work for personal or classroom use is granted without fee provided that copies are not made or distributed for profit or commercial advantage and that copies bear this notice and the full citation on the first page. To copy otherwise, to republish, to post on servers or to redistribute to lists, requires prior specific permission and/or a fee.
ISLPED'06, October 4–6, 2006, Tegernsee, Germany.
Copyright 2006 ACM 1-59593-462-6/06/0010 ...$5.00.

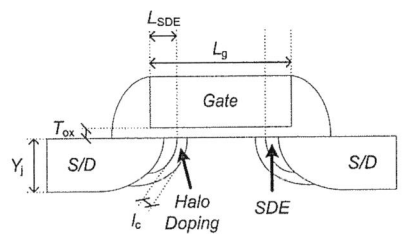

Figure 1: Symmetrical Bulk–MOSFET structure

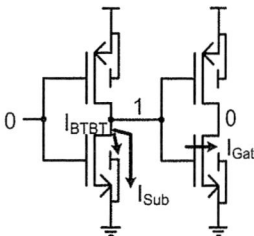

Figure 2: Total Leakage (TL) estimation scheme

In the literature, a number of techniques have proposed guidelines to find the optimum parameter values for the device to maximize the I_{ON}/I_{OFF} ratio. For example, the short channel effects have been found to be independent of the junction depth Y_j and simulation results showed that I_{ON} increases significantly with the increase in Y_j [6]. Additionally, a detailed study of the impact of lateral doping abruptness of SDE on device's I_{ON} and I_{OFF} was presented in [7] and proposed some design guidelines for the SDE. Also, leakage currents have been shown to be more sensitive to the variations in T_{ox}, Halo doping profile (N_{halo}, l_c), and L_g [4].

In this work, L_g which is typically varied to design different versions of a device, has been set to the technology's minimum feature size. As a result, the following three device parameters which are the dominant sources of variability will be used in the proposed yield optimization problem, while other parameters get values based on the ITRS [8].

i) T_{ox}: Oxide thickness

ii) N_{halo}: Halo peak doping concentration

iii) l_c: Characteristic lateral decay length of Gaussian halo

Considering a three-dimensional space composed by the T_{ox}, N_{halo}, and l_c, the yield optimization problem can be represented as follow:

$$\begin{array}{c} \underset{x=(T_{ox},N_{halo},l_c)}{\text{Maximize}} \quad Yield = P_x\{C(x)=1\} \\ \text{Subject to}: \quad \text{Variation in } x \text{ elements} \end{array} \quad (1)$$

where C, denotes a binary random variable defined based on desired bounds on I_{ON} and total leakage, and $P_x\{C(x)=1\}$ represents the probability that a device with parameters (x) satisfies constraints in presence of variation. Therefore, C is formulated by (2).

$$C(x) = (I_{ON}(x) \geq I_{ON-Min}) \, \& \, (TL(x) \leq TL_{Max}) \quad (2)$$

where I_{ON-Min} and TL_{Max} are minimum and maximum desirable current for drive-in and total leakage, respectively. It should be noted that by assigning different values to I_{ON-Min} and TL_{Max}, various devices for high-performance, low-power, or general-purpose applications can be designed.

TL consists of three components. Each of which is a function of one or more of three process parameters under consideration. I_{gate}, I_{BTBT}, and I_{sub} are exponentially depends on T_{ox}, N_{halo}, and V_{th}, respectively [1], while V_{th} is a function of all above process parameters [4]. To have a more realistic indication of the total leakage in digital circuits, as shown in Figure 2, all of the worst case leakage components are added together [8]:

$$\begin{array}{l} TL(T_{ox}, N_{halo}, l_c) = \\ I_{sub}(T_{ox}, N_{halo}, l_c) + I_{BTBT}(N_{halo}) + I_{gate}(T_{ox}) \end{array} \quad (3)$$

3. FEASIBLE SPACE AND DESIGN VARIABLES DISTRIBUTIONS

To solve the yield optimization problem formulated in (1), at first, a feasible solution space needs to be identified. This feasible space is a set of x points where the yield condition C is satisfied. To find such region, the equation set (4) is used to create surfaces which determine the desired space.

$$\begin{array}{l} I_{ON}(T_{ox}, N_{halo}, l_c) = I_{ON-Min} \\ TL(T_{ox}, N_{halo}, l_c) = TL_{Max} \end{array} \quad (4)$$

To find the surface(s), two ranges of T_{ox} and l_c are swept and the nearest N_{halo} value which satisfy each (4) is found. An iterative-based Gauss-Newton algorithm [11] is used to search the appropriate N_{halo} by running MEDICI 2-D device simulator in each iteration [10].

Figure 3 shows the surfaces extracted for $I_{ON-min} = 500\mu A/\mu m$ and $TL_{max} = 1.5\mu A/\mu m$. Figure 3-(a) shows that the total leakage constraint generate two surfaces. The upper surface is where the I_{BTBT} dominates other leakage components due to high value of N_{halo}, and the lower surface illustrates the region where I_{sub} contributes the most to TL due to low N_{halo} causing threshold voltage reduction [12]. Moreover, it can be seen that a lower T_{ox} value causes a tighter feasible region due to intense increase in gate direct tunneling and subthreshold leakage currents. In addition, small values of l_c need higher N_{halo} to compensate the SCE. As a result, the feasible space where $TL_{max} \leq 1.5\mu A/\mu m$ lies between these two surfaces. The feasible space corresponding to $I_{ON-min} \geq 500\mu A/\mu m$ is located below the I_{ON} surface since less N_{halo} causes lower threshold voltage which increases I_{ON}. Figure 3-(b) shows how these surfaces together can form the feasible solution space.

The next step is to fit the extracted surfaces with analytical equations. To find appropriate fitting parameter values, a least square error method has been applied which shows negligible (in this case: less than 0.2%) error. Fitting sur-

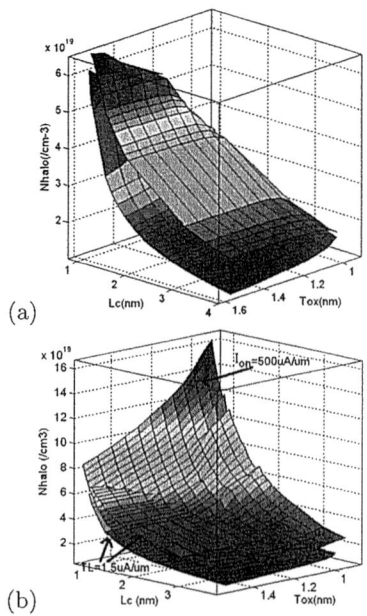

Figure 3: Surfaces obtained for constraints

faces to analytical equations will reduce the run-time of the yield maximizing step. The following formats have been chosen for I_{ON} and TL surfaces.

$$I_{ON} : N_{halo} = S_1\left(T_{ox}, l_c\right) = \alpha \cdot T_{ox}^{-\beta} \cdot l_c^{-\gamma} \qquad (5)$$

$$TL : \begin{cases} N_{halo} = S_2\left(T_{ox}, l_c\right) = \\ -\lambda_1 \exp\left(-\phi_1 \cdot T_{ox} - \varphi_1 \cdot l_c^{-\eta_1}\right) + \varsigma_1 \cdot l_c^{-\upsilon_1} \\ \\ N_{halo} = S_3\left(T_{ox}, l_c\right) = \\ \lambda_2 \exp\left(-\phi_2 \cdot T_{ox} - \varphi_2 \cdot l_c^{-\eta_2}\right) + \varsigma_2 \cdot l_c^{-\upsilon_2} \end{cases} \qquad (6)$$

where α, β, γ, λ_1, ϕ_1, φ_1, η_1, ς_1, υ_1, ϕ_2, φ_2, η_2, ς_2, and υ_2 are fitting parameters. The process parameters' units are as follows $N_{halo}(cm^{-3})$, $l_c(\mu m)$, and $T_{ox}(\mu m)$.

By using the analytical format of surfaces, the feasible space can be defined as F_c. F_c denotes a subset of the three-dimensional space points where the yield condition (C) is satisfied. Also, for each technology T_{ox} and l_c should be selected between two minimum and maximum bounds.

$$F_c = \left\{ x \, \middle| \, \begin{array}{l} T_{ox} \in \left[T_{ox}^{\min}, T_{ox}^{\max}\right], l_c \in \left[l_c^{\min}, l_c^{\max}\right], \\ N_{halo} \leq S_1\left(T_{ox}, l_c\right), \\ N_{halo} \in \left[S_3\left(T_{ox}, l_c\right), S_2\left(T_{ox}, l_c\right)\right] \end{array} \right\} \qquad (7)$$

One remainder step before solving the yield optimization problem of (1) is to model the probability distribution function (PDF) of the design parameters. Variability of each design parameter (x elements) is considered to be independent and the distribution is assumed to be Gaussian [3]. Gaussian distribution does not have a closed form cumulative distribution function (CDF) which is needed for the yield evaluation, so we utilized the Kumaraswamy's distribution model [13], a double-bounded probability density function (DB-PDF). The probability distribution function (PDF) $f(z)$ of this model is in the form of:

$$f(z) = abz^{a-1}(1-z^a)^{b-1}$$
$$z = \frac{x - x^{lb}}{x^{ub} - x^{lb}}, \qquad x^{lb} \leq x \leq x^{ub} \qquad (8)$$

where x^{ub} and x^{lb} represent upper and lower bounds of random variable x. Depending on the values chosen for parameters a and b, DB-PDF can take various shapes. In this work, a truncated Gaussian shape with range $t = x^{ub} - x^{lb}$ has been used, and x^{ub} and x^{lb} are set to $x + 3\sigma_x$ and $x - 3\sigma_x$, respectively. The closed-form CDF of this model $F(z)$ called DB-CDF is easily available from its PDF integral:

$$F(z) = 1 - (1 - z^a)^b \qquad (9)$$

4. YIELD MAXIMIZATION

As mentioned earlier, to find the optimum process parameters which maximize immunity of the design against variability, a tolerance rectangular cube should be placed in the feasible space. For illustrative purposes, the problem is shown in the two dimensional space in Figure 4. It is assumed that l_c is constant and equal to 2nm. Therefore, the problem is reduced to inscribe a rectangle in a two-dimensional feasible region. The center of the rectangle is the optimum yield point (x^c), and other points show emerging devices due to variability.

However, the actual problem is three-dimensional, so the yield maximization objective is to find a rectangular cube such that the portion of points lies in the feasible region be maximized. This rectangular cube is represented as below:

$$Cube(x^l, x^u) = \left\{ x \in \Re^3 | x^l \leq x \leq x^u \right\} \qquad (10)$$

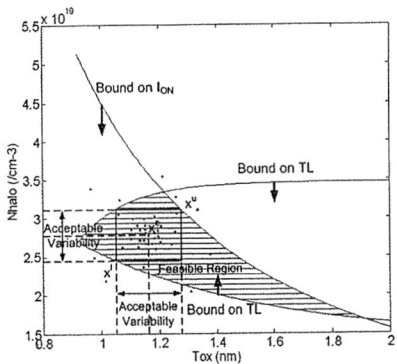

Figure 4: Problem in two-dimensional space

where x^l and x^u are the coordinates of its two extreme corners. This cube should be contained in the feasible space $Cube\left(x^l, x^u\right) \subseteq F_c$, which means every 8 devices, obtained by combinations of design corner elements $(T_{ox}^l, N_{halo}^l, l_c^l)$ and $(T_{ox}^u, N_{halo}^u, l_c^u)$, should be within the feasible region.

Furthermore, due to the symmetrical nature of design variables PDF, the final device is placed in the center of the obtained maximum yield cube ($x^c = \frac{x^l + x^u}{2}$).

Since the design variables do not have equal variances, the maximum volume cube does not necessarily correspond to the maximum yield and the yield function must be maximized directly [14]. The final step is to introduce an analytical yield function. The probabilistic $Yield$ of (1) is substituted by the analytical function. The optimization problem (1) can be re-written as:

$$\begin{array}{ll} \underset{x^l, x^u}{\text{Maximize}} & Yield\left(x^l, x^u\right) \\ \text{Subject to :} & \\ Cube\left(x^l, x^u\right) \subseteq F_c & \\ x^u \geq x^l & \end{array} \qquad (14)$$

where $Yield(x^l, x^u)$, the analytical function of yield, is the portion of transistors lie in the rectangular cube with two extreme corner coordinates x^l and x^u. Equation (13) shows how the idea behind DB-CDF has been used to estimate $Yield$.

In (13), each t indicates the distribution range of the corresponding design variable. It should be noted that, here, a designer is free to assign different standard deviations (σ) and ranges to design variables. For example, low $\sigma_{T_{ox}}$ can be assigned to T_{ox} standard deviation percentage in comparison to others since the oxide thickness is more controllable during VLSI fabrication process. The Sequential Quadratic Programming (SQP) [11] has been used to effectively solve (14).

5. RESULTS AND DISCUSSION

To verify the method, MEDICI template files which contain structural information of a Bulk-NMOS transistor were developed. The value of T_{ox}, N_{halo}, and l_c can be dynamically changed during the runtime of surface extraction procedure. V_{DD} was set to 0.9v, and the simulations were done in $100°C$. $I_{ON-min} \geq 500\mu A/\mu m$ and $TL_{max} \leq 1.5\mu A/\mu m$ were selected as the design constraints. Then, two ranges of 7 to 16 and 7 to 40 Angstrom were selected for T_{ox} and l_c and assigned to T_{ox}^{min}, T_{ox}^{max}, l_c^{min}, and l_c^{max}, respectively.

$$Yield\left(x^l, x^u\right) = P\left\{T_{ox}^l \leq T_{ox} \leq T_{ox}^u\right\} \times P\left\{l_c^l \leq l_c \leq l_c^u\right\} \times P\left\{N_{halo}^l \leq N_{halo} \leq N_{halo}^u\right\}$$

$$= \left(F\left(\frac{T_{ox}^u - T_{ox}^{lb}}{t_{T_{ox}}}\right) - F\left(\frac{T_{ox}^l - T_{ox}^{lb}}{t_{T_{ox}}}\right)\right) \times \left(F\left(\frac{l_c^u - l_c^{lb}}{t_{l_c}}\right) - F\left(\frac{l_c^l - l_c^{lb}}{t_{l_c}}\right)\right) \times \left(F\left(\frac{N_{halo}^u - N_{halo}^{lb}}{t_{N_{halo}}}\right) - F\left(\frac{N_{halo}^l - N_{halo}^{lb}}{t_{N_{halo}}}\right)\right)$$

$$= \left(F\left(\frac{T_{ox}^u - \left(T_{ox}^c - 3\cdot\sigma_{T_{ox}}\cdot T_{ox}^c\right)}{6\cdot\sigma_{T_{ox}}\cdot T_{ox}^c}\right) - F\left(\frac{T_{ox}^l - \left(T_{ox}^c - 3\cdot\sigma_{T_{ox}}\cdot T_{ox}^c\right)}{6\cdot\sigma_{T_{ox}}\cdot T_{ox}^c}\right)\right) \times \left(F\left(\frac{l_c^u - \left(l_c^c - 3\cdot\sigma_{l_c}\cdot l_c^c\right)}{6\cdot\sigma_{l_c}\cdot l_c^c}\right) - F\left(\frac{l_c^l - \left(l_c^c - 3\cdot\sigma_{l_c}\cdot l_c^c\right)}{6\cdot\sigma_{l_c}\cdot l_c^c}\right)\right) \quad (13)$$

$$\times \left(F\left(\frac{N_{halo}^u - \left(N_{halo}^c - 3\cdot\sigma_{N_{halo}}\cdot N_{halo}^c\right)}{6\cdot\sigma_{N_{halo}}\cdot N_{halo}^c}\right) - F\left(\frac{N_{halo}^l - \left(N_{halo}^c - 3\cdot\sigma_{N_{halo}}\cdot N_{halo}^c\right)}{6\cdot\sigma_{N_{halo}}\cdot N_{halo}^c}\right)\right)$$

Table 1: Results for different parameters' standard deviations

3σ	T_{ox} (%)	10	15	6
	N_{halo} (%)	10	15	20
	l_c (%)	10	15	20
x^c	T_{ox} (nm)	1.161	1.161	1.165
	N_{halo} (cm^{-3})	2.89E19	2.89E19	2.72E19
	l_c (nm)	1.911	1.908	2.042
$Yield$	(%)	100	100	98
I_{ON}	Mean ($\mu A/\mu m$)	599	604	597
	S.D. ($\mu A/\mu m$)	27	40	38
I_{Sub}	Mean ($nA/\mu m$)	478	497	493
	S.D. ($nA/\mu m$)	103	151	226
I_{BTBT}	Mean ($nA/\mu m$)	178	175	178
	S.D. ($nA/\mu m$)	94	125	24
I_{Gate}	Mean ($nA/\mu m$)	113	131	105
	S.D. ($nA/\mu m$)	45	84	24
TL	Mean ($nA/\mu m$)	769	803	777
	S.D. ($nA/\mu m$)	108	170	227

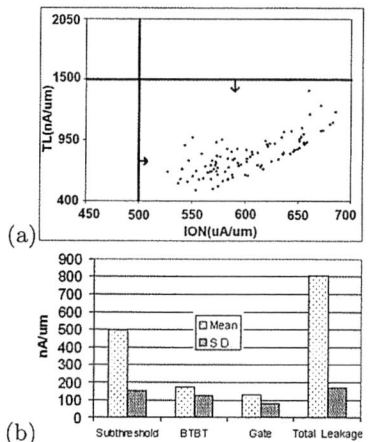

Figure 5: Monte Carlo Simulation for $3\sigma = 15\%$

The surfaces satisfying mentioned design constraints were extracted by MEDICI. Then, the (5) and (6) equations were fitted to the obtained surfaces. Finally, the center point coordinates, the yields obtained by Monte Carlo simulations, means, and standard deviations of currents are given in Table 1 for different parameters' standard deviations. It can be seen that using the extracted process parameters causes the emerging devices to satisfy all defined constraints in presence of up to $6\sigma = 30\%$ variations for the three parameters. The devices' total leakage and drive-in currents are shown in Figure 5(a). The mean and standard deviation of three major leakage current components and TL of the mentioned device are given in Figure 5(b).

6. CONCLUSIONS

In this research, a new device design approach has been proposed. This method tries to find appropriate values for oxide thickness, halo peak doping, and halo characteristic length for a known device structure such that the extracted design leads the transistor to maximally satisfy two constraints on saturation and total leakage currents in presence of variability. The algorithm tries to find a maximized yield rectangular cube such that it is inscribed in the analytically modeled feasible space. The center of this cube is considered as the maximum yield design point. This method takes into account different possible variances on process parameters and desired saturation and leakage current for a particular application. To the best of our knowledge, this is the first design methodology that maximizes the variability-driven MOS device yield while satisfying the bounds on leakage power and drive-in currents.

Adding gate length and other advanced channel doping profiles (e.g. Retrograde Well) to design variables and solving the problem for various applications (High-Performance and Low-Power) can be applied for future works.

7. REFERENCES

[1] K. Roy et al, "Leakage current mechanisms and leakage reduction technique in deep-submicrometer CMOS circuits," Proc. IEEE, vol. 91, no. 2, pp. 305–327, Feb. 2003.

[2] S. R. Nassif, "Design for Variability in DSM Technologies," in Proc. of IEEE ISQED, 2000, pages 451–454.

[3] A. Srivastava et al, "Modeling and analysis of leakage power considering within-die process variations," in Proc. of ACM ISLPED, 2002, pages 64–67.

[4] S. Mukhopadhyay and K. Roy, "Modeling and Estimation of Total Leakage Current in Nano-scaled CMOS Devices Considering the Effect of parameter Variation," in Proc. of ACM ISLPED, 2003, pages 172–175.

[5] S. Mukhopadhyay et al, "Accurate Estimation of Total Leakage Current in Scaled CMOS Logic Circuits Based on Compact Current Modeling," in Proc. of ACM DAC, 2003, pages 169–174.

[6] R. Gwoziecki et al, "Junction Design Guideline for 0.18m CMOS," in Proc. of IEEE ESSDERC, 1997, pages 388–319.

[7] M. Y. Kwong et al, "Impact of Lateral Source/Drain Abruptness on Device Performance," IEEE Trans. Electron Devices, vol. 49, no. 11, Nov. 2002, pp. 1882–1890.

[8] International Technology Roadmap for Semiconductors, 2003.

[9] T. Sakurai and A. R. Newton, "Alpha-power law MOSFET model and its application to CMOS inverter and other formulas," IEEE J. of Solid-State Circuits, vol. 25, no. 2, pp. 584–593, Apr. 1990.

[10] Synopsys Taurus MEDICI Version V-2003.12, December 2003.

[11] T. F. Colemen and Y. Zhang, Optimization Toolbox for use with Matlab, The Mathworks Inc., Natick MA, 2005.

[12] Y. Taur and T. H. Ning, Fundamental of Modern VLSI devices, Cambridge University Press, 1998.

[13] P. Kumaraswamy, "A generalized probability density function for double-bounded random processes," J. of Hydrology, (46) 1980, pp. 79–88.

[14] K. Ponnambalam et al, "Probabilistic design of systems with general distributions of parameters," Intl. J. of Circuit Theory and Appl., vol. 29, no. 6, 2001, pp. 527–536.

A 0.5-V FD-SOI Twin-Cell DRAM with Offset-Free Dynamic-V_T Sense Amplifiers

Riichiro Takemura, Kiyoo Itoh, and Tomonori Sekiguchi

Hitachi Ltd., Central Research Laboratory

1-280 Higashi-koigakubo Kokubunji-shi, Tokyo 185-8601 JAPAN

+81-42-323-1111

{riichiro.takemura.vp, kiyoo.itoh.pt, tomonori.sekiguchi.ns}@hitachi.com

ABSTRACT

Three DRAM technologies, which are a leakage- and soft-error-free planar-capacitor SOI cell, a data-line shielded twin (2-T) cell array, and an offset-free dynamic-V_T sense amplifier suitable for low-voltage mid-point sensing, are presented and evaluated. New noise-generation mechanisms are also shown. Using the experimental data of an ultrathin BOX double-gate fully-depleted SOI MOST, a 1.5-ns cycle-time 65-nm 2-kb subarray was found to be feasible for embedded applications, even at 0.5 V.

Categories and Subject Descriptors

B.7.1 [**Integrated Circuits**]: Types and Design Styles – *Memory technologies.*

General Terms: Performance, Design

Keywords: FD-SOI, twin-cell DRAM, low-voltage RAM, and dynamic-V_T sense amplifier.

1. INTRODUCTION

Major challenges in developing deep-sub volt embedded (e-) DRAMs are to achieve a high signal-to-noise-ratio (S/N) cell array with a logic-compatible planar cell and high-speed midpoint sensing (i.e., a half-V_{DD} data-line precharge), and to reduce the speed variation of peripheral circuits caused by the threshold (V_T) variation of MOST. In particular, noise in the array must be significantly reduced. However, the details for doing this remain unknown.

In this paper, a leakage-and soft-error-free SOI cell and array, a new data-line-shielded twin cell (2-T cell), and an offset-free dynamic-V_T sense amplifier suitable for low-voltage high-speed mid-point sensing are presented and evaluated. New noise generation mechanisms are also clarified.

2. PROPOSED FD-SOI DRAM
2.1 FD-SOI MOSTs

The challenges are overcome by using an ultrathin BOX (buried oxide) double-gate fully depleted (FD) SOI MOST [1, 2] with an ultrathin and lightly doped channel, as shown in Fig. 1. If the NMOS is used in the one-transistor one-capacitor cell, the

Permission to make digital or hard copies of all or part of this work for personal or classroom use is granted without fee provided that copies are not made or distributed for profit or commercial advantage and that copies bear this notice and the full citation on the first page. To copy otherwise, or republish, to post on servers or to redistribute to lists, requires prior specific permission and/or a fee.

ISLPED'06, October 4–6, 2006, Tegernsee, Germany.

Copyright 2006 ACM 1-59593-462-6/06/0010...$5.00.

junction leakage (Fig. 2) and soft-error rate of the cell become much smaller than those of the conventional bulk-MOS cell due to the small junction area. Furthermore, the offset voltage (i.e., V_T mismatch between paired MOSTs, δV_T) of sense amplifiers, which is derived from the intrinsic V_T variation (σ_{int}) and is reportedly as large as 30 – 40 mV, even in the 100-nm bulk MOST [3], becomes negligible, as seen in Fig. 3. The small offset increases the S/N and enables low-voltage high-speed midpoint sensing. The speed variation of peripheral circuits is significantly reduced by the reduced intrinsic and extrinsic V_T variations. In addition, V_T can be controlled by the back-gate voltage, even in the forward-bias range, without junction leakage, as shown in Fig. 4.

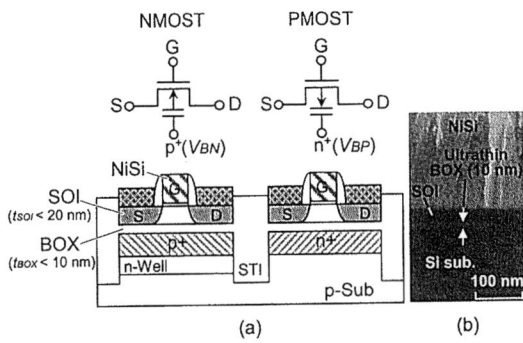

Figure 1 Double-gate FD-SOI MOST [1]. Schematic cross sectional view (a) and SEM image of fabricated device (b).

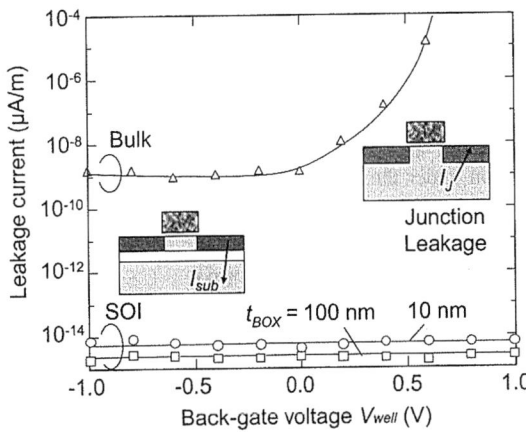

Figure 2 Experimental leakage current characteristics of double-gate FD-SOI MOST [1].

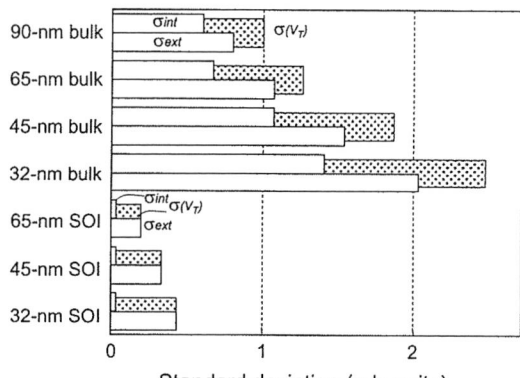

Figure 3 Predicted V_T-variation of FD-SOI MOSTs [2].

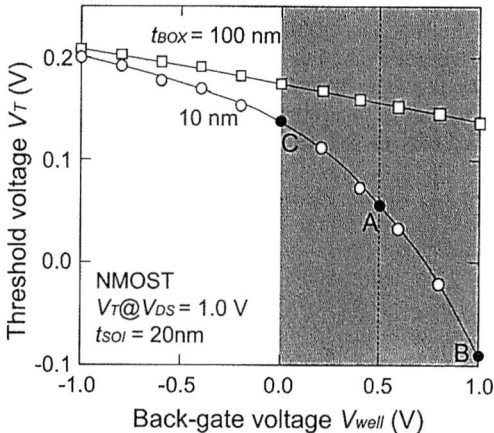

Figure 4 Experimental V_T-dependence on back-gate voltage of FD-SOI MOSTs [1].

2.2 Structures of Array and Peripheral Circuits

A proposed double-gate SOI DRAM structure, which maximizes the advantages of the FD-SOI MOST, is shown in Fig. 5. A MOS planar capacitor is used; the plate (PLT) of which is at the half-V_{DD}, as usual. To maintain extremely low junction-leakage characteristics in a MOST, no implant is added to the stored node. Thus, the work function of the capacitor plate material is different from that of MOST gate materials, so the MOS capacitor is depleted even when a full V_{DD} is written from the data line. The V_Ts of the cell and peripheral MOSTs are adjusted using ion implantation under the BOX and/or well-bias voltages V_{BA}, V_{BN}, and V_{BP}. Thus, multi-V_T peripheral circuits are achieved, which is indispensable for low-power DRAMs.

2.3 Shielded-Twin Cell Array

Due to the above-described advantages of the array structure using the FD-SOI MOST, the S/N design of the cell is confined to only the signal and noise in an array. A proposed 2-T layout and its array architecture are shown in Fig. 6. To avoid coupling noise from adjacent data lines, each data line is shielded by interspersed quiescent data lines, as shown in the figure. A conventional one-transistor cell (1-T cell) is shown for reference. For the same cell size, the noise of the 1-T cell is also negligible due to well-

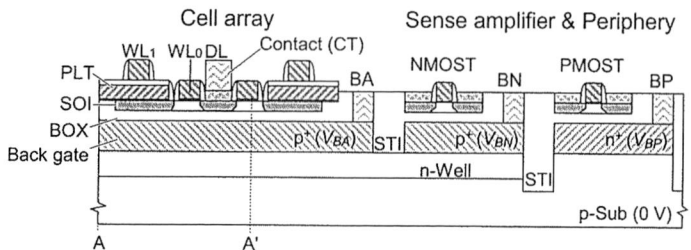

Figure 5 Cross section of proposed SOI DRAM.

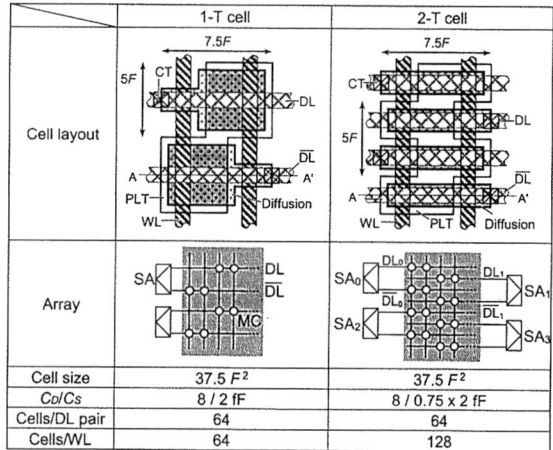

Figure 6 Memory cell, array architectures, and 65-nm design parameters.

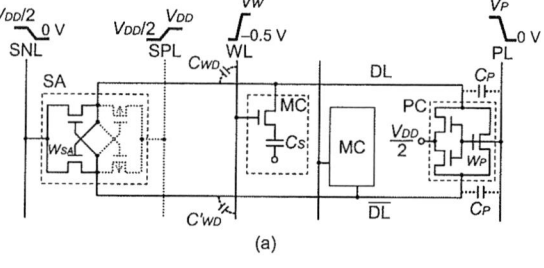

(a)

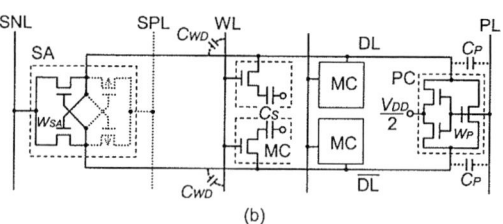

(b)

Figure 7 Noise sources in 1-T cell (a) and 2-T cell (b).

separated data lines. The following simulation using experimental data of the MOST demonstrated that the 2-T cell array is ideal for low-voltage operation due to improved S/N and sensing stability.

Noise-generation circuits and their operating mechanisms before sensing are shown in Figs. 7 and 8, respectively. Major noise sources are the coupling capacitances of the precharger/equalizer (PC) to the data line (DL), CP, and the word line (WL) to the data lines, C_{WD} and C'_{WD}. A common-mode noise, $\Delta(C_P)$, coupled to a pair of data lines when a precharge line (PL) is turned off, is given as $\Delta(C_P) \cong (V_P - V_{DD}/2 - V_{TP})C_{GP}/(C_D + C_{GP})$ where V_P,

	1-T cell		2-T cell
	'H' read	'L' read	
DL Voltage V(DL)	$V_{DD}/2 - \Delta(C_P)$ $+ \Delta(C_{WD}) + v_{S1}$	$V_{DD}/2 - \Delta(C_P)$ $+ \Delta(C_{WD}) - v_{S1}$	$V_{DD}/2 - \Delta(C_P)$ $+ \Delta(C_{WD}) + v_{S2}$
$\overline{DL}$ Voltage V($\overline{DL}$)	$V_{DD}/2 - \Delta(C_P)$	$V_{DD}/2 - \Delta(C_P)$	$V_{DD}/2 - \Delta(C_P)$ $+ \Delta(C_{WD}) - v_{S2}$
Signal component	$V(DL) - V(\overline{DL})$ $= v_{S1} + \Delta(C_{WD})$	$V(\overline{DL}) - V(DL)$ $= v_{S1} - \Delta(C_{WD})$	$V(DL) - V(\overline{DL})$ $= 2 v_{S2}$
V_G of on-NMOST in SA	$V_{DD}/2 - \Delta(C_P)$ $+ \Delta(C_{WD}) + v_{S1}$	$V_{DD}/2 - \Delta(C_P)$	$V_{DD}/2 - \Delta(C_P) +$ $\Delta(C_{WD}) + v_{S2}$

Figure 8 Signal component and gate voltage (V_G) of turned-on NMOST in sense amp. v_{S1}/v_{S2}: signal voltage of 1-T and 2-T cells.

V_{TP}, C_{GP}, and C_D are the amplitude of the precharge-line pulse, the V_T and gate capacitance of the precharger/equalizer MOST, and the data-line capacitance, respectively. A voltage, $\Delta(C_{WD})$, coupled to the data line when the word line is activated, is given as $\Delta(C_{WD}) \cong (V_W - V_{DD}/2 - V_{TM})C_{GM}/(C_D + C_S + C_{GM})$ where V_W, V_{TM}, C_{GM}, and C_S are the amplitude of the word pulse, the V_T and gate capacitance of the memory-cell MOST, and the cell capacitance, respectively. Here, the small $\Delta(C'_{WD})$ is neglected. Note that for the 1-T cell, the signal component decreases by $\Delta(C_{WD})$ when a 0-V ('L') stored cell is read, while it increases by $\Delta(C_{WD})$ when a V_{DD} ('H') stored cell is read. For the 2-T cell, $\Delta(C_{WD})$ and $\Delta(C_P)$ are coupled to both data lines, so the signal component always remains constant. $\Delta(C_P)$ and $\Delta(C_{WD})$ become significant when C_S and C_D are sufficiently small, which may occur in high-speed e-DRAMs with a planar capacitor and a short data line. This is because the signal voltage ratios, $v_S = (V_{DD}/2)C_S/(C_D + C_S)$, are given as $\Delta(C_P)/v_S = (C_{GP}/C_S)(C_D + C_S)/(C_D + C_{GP})$ and $\Delta(C_{WD})/v_S = (C_{GM}/C_S)(C_D + C_S)/(C_D + C_S + C_{GM})$ on the assumptions that $V_P = V_{DD} + V_{TP}$ and $V_W = V_{DD} + V_{TM}$. Thus, their detrimental effects increase as C_S approaches C_{GP} and C_{GM}.

Simulated signal waveforms at V_{DD} of 0.5 V for different channel widths (W_Ps) of a precharge MOST for a fixed channel width (97 nm) of a cell MOST are shown in Fig. 9. Obviously, the data-line voltages in the steady state decrease with a large W_P. For the 1-T cell, the signal component depends on the stored data, as expected. For example, for $WP = 50$ nm, the signal voltage is as small as 38 mV for an 'H' read despite being 54 mV for a 'L' read. In contrast, for the 2-T cell, the signal voltage is a fixed value of 42 mV, which is larger than that of the 1-T cell. Note that in the 1-T cell, a C_D-imbalance caused by C_S between a pair of data lines during signal amplification, which is not shown in Fig. 8, generates additional noise for a 'L' signal that is about 4 – 8 mV, as shown in Fig. 10. The 2-T cell completely cancels the noise.

For the 1-T cell array, the gate voltage, V_G, of a turned-on NMOST in the sense amplifier is the highest for the 'H' read, while it is the lowest for the 'L' read (Figs. 8 and 9). This implies a speed difference in amplification as discussed below. In contrast, for the 2-T cell, the V_G is independent of the stored data, implying no speed difference.

3. OFFSET-FREE DYNAMIC-V_T SENSE AMPLIFIER AND ARRAY SPEED

The sensing speed is improved by the FD-MOST enabling an offset-free sense amplifier, as mentioned above. The sensing

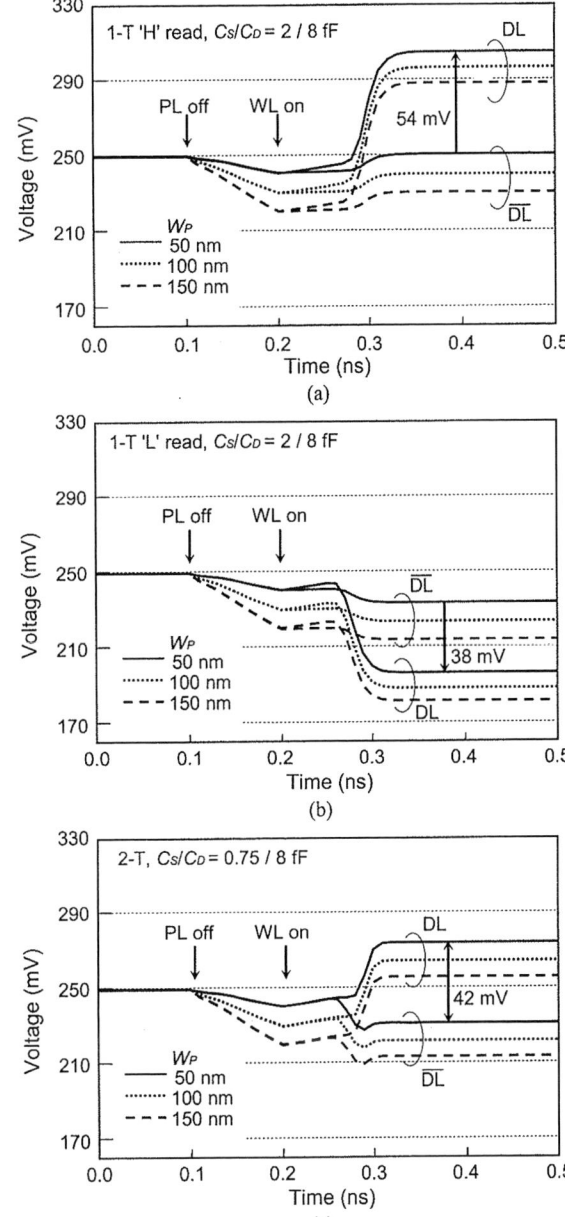

Figure 9 Signal waveforms of 1-T 'H' read (a), 1-T 'L' read (b), and 2-T read (c). $V_P/V_{TP} = 1.2/0.1$ V, $V_W/V_{TM} = 1.2/0.2$ V, W/L(cell) = 97/65 nm, and $t_{OX} = 2.2$ nm.

speed of the amplifier is further enhanced by the proposed gate-well (body) connection (Fig. 11) with no restrictions on body bias due to the body being surrounded by an insulator. For example, with a V_{DD} of 1 V and a BOX thickness, t_{BOX}, of 10 nm, as shown in Fig. 4, the V_T of an NMOST is 0.06 V at point A before starting amplification when the gate and well are at a bias of $V_{DD}/2$ (0.5 V). However, V_T decreases or increases depending on the well biasing during amplification. At the final stage of amplification, the V_T is either negative (−0.1 V at point B) with a V_{DD} well or positive (0.14 V at point C) with a 0-V well. This is also the case for PMOSTs. This new dynamic enhancement/depletion (E/D) MOS circuit accelerates

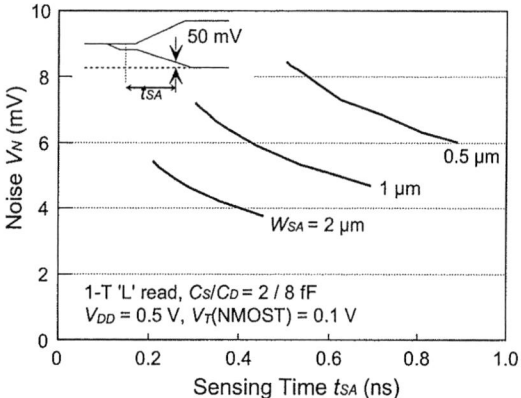

Figure 10. C_D-imbalance noise.

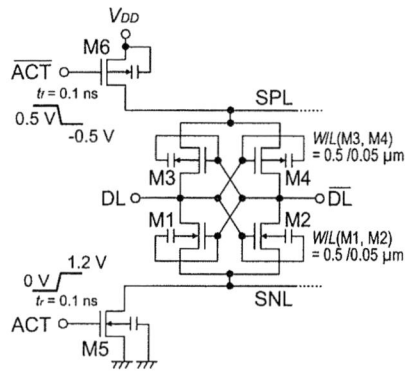

Figure 11 Gate-well connected sense amplifier.

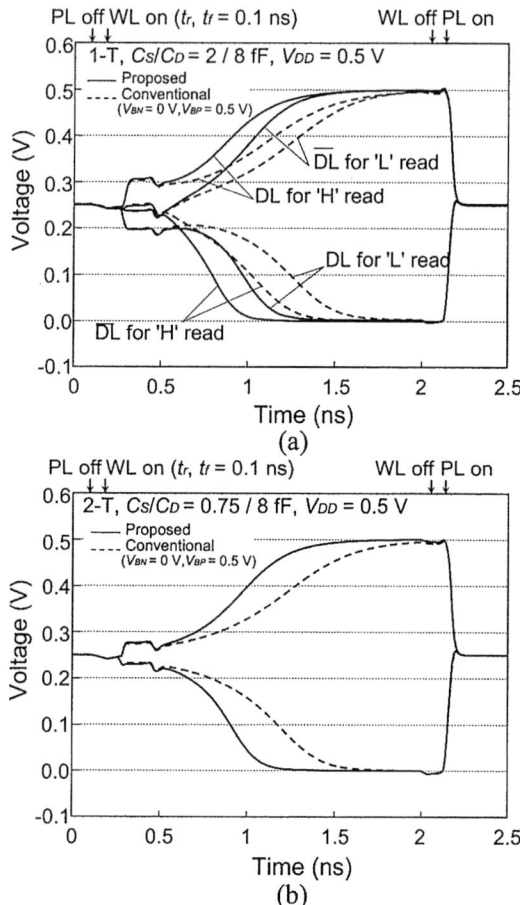

Figure 12 Waveforms of conventional SA and gate-well connected SA in 1-T cell (a) and 2-T cell (b). $V_T(SA) = 0.2$ V@V_{DS}, $V_{BS} = 0.5$, 0 V, $\delta V_T = 0$ V, $V_W = 1.2$ V, $V_P = 1.2$ V, $W/L(SA) = 0.5/0.05$ μm, $W/L(cell) = 97/65$ nm, and $W_P = 50$ nm.

amplification. This dynamic circuit also reduces the subthreshold current of the sense amplifiers after amplification with a resultant high V_T. Even when $V_{DD} = 0.5$ V, such an E/D circuit is achieved with a thinner BOX layer (e.g., $t_{BOX} = 2$ nm), as expected from observing Fig. 4.

Simulated waveforms for a 65-nm 0.5-V 2-kb subarray (Fig. 6) with $t_{BOX} = 10$ nm are shown in Fig. 12. In the 1-T cell, a difference in amplification speed develops as expected due to a difference in the above-mentioned V_G, unlike the 2-T cell. The difference is a source of malfunctions. For example, when all pairs of data lines in an array receive the 'H' read signal, except the pair that receive the 'L' read signal, an increased voltage at the sense-amplifier drive line (SNL in the figure) caused by the earlier start of the amplification of the 'H' read signals delays the start of the amplification of the 'L' read signal [4, 5]. In addition to the C_D-imbalance noise, eventually, the V_G difference is an obstacle to low-voltage operation of the 1-T cell. Thus, the 2-T cell is ideal for ultralow-voltage high-speed DRAMs. The gate-well connected sense amplifier obviously increases the speed of sensing. The expected cycle time is as fast as 1.5 ns even at 0.5 V, which could be further improved by a dynamic E/D sense amplifier with a thinner BOX.

4. CONCLUSION

Evaluations indicated that the proposed SOI cell, data-line shielded 2-T cell array, and an offset-free dynamic-V_T sense amplifier took full advantage of the FD-SOI structure. Through clarifying new noise components essential for ultralow-voltage midpoint sensing, a 1.5-ns cycle-time 65-nm 2-kb subarray was found to be feasible for embedded applications, even at 0.5 V.

5. ACKNOWLEDGEMENTS

The authors are grateful to Drs. T. Kawahara and R. Tsuchiya for their discussion.

6. REFERENCES

[1] R. Tsuchiya, et al., *IEDM Tech. Dig.*, Dec. 2004, pp. 631-634.

[2] M. Yamaoka, et al., *Symp. VLSI Circuits Dig.*, Jun. 2004, pp. 28-291.

[3] J. Y. Sim, et al., *ISSCC Dig., Tech. Papers*, Feb. 2003, pp. 310-311.

[4] M. Aoki, et al., *IEEE J. Solid-State Circuits*, vol. 23, no. 5, pp. 1113-1119, Oct. 1988.

[5] J. Okamura, et al., *IEEE J. Solid-State Circuits*, vol. 25, no. 1, pp. 18-23, Jan. 1990.

Utilizing Reverse Short Channel Effect for Optimal Subthreshold Circuit Design

Tae-Hyoung Kim, Hanyong Eom, John Keane, and Chris Kim
Department of Electrical and Computer Engineering, University of Minnesota, MN, USA
{kimxx692, eomxx001, jkeane, chriskim}@umn.edu

ABSTRACT

The impact of the Reverse Short Channel Effect (RSCE) on device current is stronger in the subthreshold region due to the reduced Drain-Induced-Barrier-Lowering (DIBL) and the exponential dependency of current on threshold voltage. This paper describes a device size optimization method for subthreshold circuits utilizing RSCE to achieve high drive current, low device capacitance, less sensitivity to random dopant fluctuations, and better subthreshold swing. Simulation results using ISCAS benchmark circuits show that the critical path delay and power consumption can be improved by up to 10.4% and 34.4%, respectively.

Categories & Subject Descriptors

B.7.1 [**Integrated Circuits**] : Types and Design Styles

General Terms: Design, Performance

Keywords: Subthreshold operation, Subthreshold Circuits, Reverse short channel effect, digital circuits, PVT variations, Optimization.

1. INTRODUCTION

Emerging ultra-low power applications such as portable devices, medical instruments, and wireless sensor networks have extremely stringent power budgets. Subthreshold circuits, which operate at supply voltages lower than the threshold voltage (V_{th}), are considered to be promising candidates for ultra low power systems. Recently, a significant amount of research has been done dealing with subthreshold circuits. Soeleman et al. analyzed various logic styles for subthreshold operation [1]. The impact of PVT variations on subthreshold circuits was investigated in [2] and [3]. Circuits such as analog voltage references, subthreshold SRAMs, tiny-XOR circuits, and adaptive filters for hearing aid applications have been demonstrated [4][5][6][7][8]. New transistor scaling trends dedicated for subthreshold circuits have been suggested in [9].

Short channel devices have been optimized for regular superthreshold circuits to meet various device objectives such as high mobility, reduced Drain-Induced-Barrier-Lowering (DIBL), low leakage current, and minimal V_{th} roll-off. However, a transistor that is optimized for superthreshold logics may not be

Permission to make digital or hard copies of all or part of this work for personal or classroom use is granted without fee provided that copies are not made or distributed for profit or commercial advantage and that copies bear this notice and the full citation on the first page. To copy otherwise, or republish, to post on servers or to redistribute to lists, requires prior specific permission and/or a fee.
ISLPED'06, October 4–6, 2006, Tegernsee, Germany.
Copyright 2006 ACM 1-59593-462-6/06/0010...$5.00.

optimal for achieving low performance and low power in the subthreshold region. Although it would be ideal to have a dedicated process technology optimized for subthreshold circuits, mainstream CMOS technology will continue to scale aiming at optimal performance in conventional superthreshold circuits. In order to design optimal subthreshold circuits using CMOS devices that are targeted for superthreshold operation, it is crucial to develop design techniques that can utilize the side effects that appear in this new regime. SCE (or V_{th} roll-off) is an undesirable phenomenon in short channel devices where V_{th} decreases as the channel length is reduced. Variation in device critical dimensions translates into a larger variation in the threshold voltage as SCE worsens with increasing DIBL [10]. Traditionally, non-uniform HALO doping was used to mitigate this problem by making the depletion widths narrow and hence reducing the DIBL effect. As a byproduct of HALO, a short channel device shows RSCE behavior where the V_{th} decreases as the channel length is increased [11][12]. In subthreshold circuits, the SCE mechanism is not as strong as in superthreshold circuits. On the other hand, RSCE is still significant enough to affect the subthreshold performance.

This paper describes a sizing method that utilizes RSCE to improve performance and power consumption in subthreshold circuits. In section II, we will illustrate general transistor sizing considerations for subthreshold circuits. Section III describes the proposed method utilizing the RSCE to improve drive current, capacitance, process variation, and subthreshold swing. Experimental results are presented in section IV. Finally, we draw conclusions in section V.

2. GATE SIZING CONSIDERATIONS FOR SUBTHRESHOLD CIRCUITS

Conventional superthreshold logics require special modifications in order to achieve optimal performance and power in subthreshold operation. For example, PMOS to NMOS width ratio (PN ratio) and stacked device sizing must be reevaluated for the subthreshold operating voltage. The optimal PN ratio changes in the subthreshold region because the weak-inversion current is also a strong function of the subthreshold slope and is affected significantly by other secondary effects such as narrow width effect, SCE, and RSCE. The significant reduction in the optimal PN ratio with a lower supply voltage can be attributed to the difference in V_{th} and subthreshold slope. Selection of the proper effective width of stacked transistors is also crucial for achieving optimal performance. The effective width of a transistor in a stack of n devices is roughly $1/n$ in the strong-inversion region. This means that in order for an n-stack to conduct the same amount of current as a single transistor, the devices in the stack must each be sized up by a factor of n. Previous sizing methods for subthreshold logics were based on the traditional assumption that the minimum channel length is optimal for speed and power. This was true in the superthreshold region, but it does not hold true in subthreshold logic because of RSCE. Therefore, a new sizing method suitable for subthreshold circuit which considers the impact of RSCE on drive current, device capacitance, and subthreshold slope is indispensable.

3. PROPOSED TRANSISTOR SIZING METHOD UTILIZING RSCE

3.1 Reverse Short Channel Effect

Fig. 1 (top) shows the threshold voltages as a function of channel length at VDD=1.2V and VDD=0.2V. In the superthreshold region (1.2V), a strong V_{th} roll-off behavior is observed at the minimum channel length due to the high DIBL effect (point A in Fig. 1). To compensate the worsening V_{th} roll-off caused by DIBL in small dimensions, non-uniform p+ doping in the source-body and drain-body boundaries are used to reduce the amount of control the drain has over the channel by making the depletion layer width narrow. These highly doped regions, called HALO implants, can also suppress the body punchthrough [13]. However, as a byproduct of using HALO implants, the threshold voltage decreases as the channel length increases. This phenomenon is known as the RSCE. Fig. 2 (top) illustrates this trend by showing the effective surface doping in the longitudinal direction. RSCE becomes more significant with process scaling due to the higher HALO doping required to negate the aggravating V_{th} roll-off as shown in Fig. 2 (bottom). RSCE is not a major concern in conventional superthreshold designs since SCE is dominant in minimum channel length devices in that region. However, in the subthreshold region, only the RSCE effect is present due to the significantly reduced DIBL. This causes the V_{th} to decrease monotonically, and operating current to increase exponentially, with longer channel length.

3.2 Optimal Channel Length for Maximum Current-Per-Width

As the V_{th} behavior changes significantly in the subthreshold region, the optimal channel length yielding maximum current-per-width changes accordingly. This is illustrated in Fig. 3, where V_{th} and current-per-width are plotted versus channel length. Maximum current-per-width is obtained at the minimum channel length (0.12μm) for VDD=1.2V because W/L is maximized. However, the optimal channel length increases to 0.55μm at VDD=0.2V since the lower V_{th} caused by RSCE provides an exponential increase in current. Current is also proportional to W/L which makes it eventually decrease at channel lengths longer than the optimal. The optimal channel length for maximum current-per-width can be derived theoretically by taking the derivative of the current equation. The RSCE-affected threshold voltage can be expressed as

$$V_{th} = V_{th0} + K_1 \left(\sqrt{1 + \frac{K_2}{L_{eff}}} - 1 \right) \sqrt{\Phi_S} \qquad (1)$$

where V_{th0} is the zero-bias threshold voltage of a long channel device, K_1 and K_2 are technology parameters, L_{eff} is the effective channel length, and Φ_s is the surface potential. DIBL is omitted because its effect is negligible in the subthreshold region, and the

body effect is ignored for simplicity. The optimal channel length obtained is

$$L_{eff} = \frac{-K_2 + \sqrt{K_2{}^2 - 4K_3}}{2} \quad , \quad K_3 = -\frac{K_1{}^2 \Phi_S}{m^2 V_t{}^2} K_2 \qquad (2)$$

The optimal channel length calculated using the analytical expression in (2) is 0.58μm which is very close to 0.55μm from simulation. We can also compare the current at the optimal channel length given by (2) with that at minimum channel length. The maximum current-per-width is 2.5X larger than that at the minimum channel length. However, using a longer channel length can have a negative impact on device capacitance which can affect the CV/I delay. In the following section, we derive the optimal channel length for maximum performance.

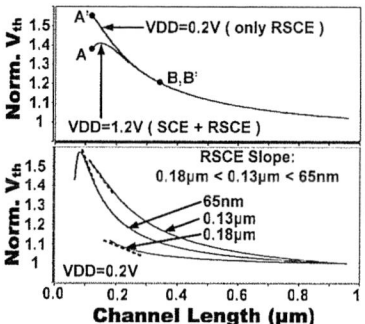

Fig. 1. Dependency of normalized V_{th} on channel length for VDD=1.2V and 0.2V

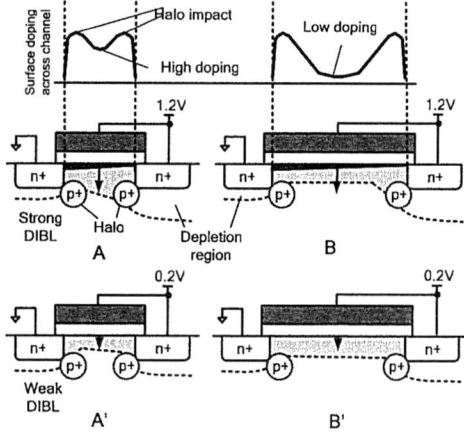

Fig. 2. Device cross sections corresponding to A, A', B, and B' in Fig. 1. Surface doping across channel is shown to illustrate the RSCE.

3.3 Optimal Channel Length for Maximum Performance

Another factor to consider when increasing the channel length for optimal subthreshold sizing is the increase in device capacitance. Delay and power consumption increases linearly with capacitance. Fig. 4 shows the different components of device capacitance in the subthreshold region. C_{DEP} is the depletion width, $C_{GS,GD}$ is the overlap capacitance per width, C_{OX} is the oxide capacitance, and C_J is the junction capacitance per width. To show the effectiveness of increasing the channel length, the capacitances of a transistor having a constant current is plotted versus channel length in Fig. 5. Note that the device width can be reduced as the channel length is increased since RSCE lowers the V_{th} and exponentially increases the device current. This was not the case for superthreshold circuits where the decrease in W/L had a larger impact on current than the reduction in V_{th} due to RSCE. Increasing the channel length alone has no effect on junction capacitance (C_J). However, since the device width is reduced simultaneously for constant current, the junction capacitance also goes down with a longer channel length as shown in Fig. 5. The increase in gate capacitance (C_G) is moderate between channel lengths of 0.12µm and 0.36µm for two reasons. First, the reduction in width makes the increase in gate area smaller. Second, the RSCE associated with longer channel length makes the depletion capacitance (C_{DEP}) smaller since the depletion layer width increases as shown in Fig. 2. At channel lengths longer than 0.36µm however, C_G increases rapidly since the RSCE is significantly weaker, and gate area must be increased to drive the same current. As a result, there exists a minimum point in total capacitance for iso-current at a channel length of 0.36µm. By using this optimal channel length, we can reduce delay and power consumption in subthreshold circuits.

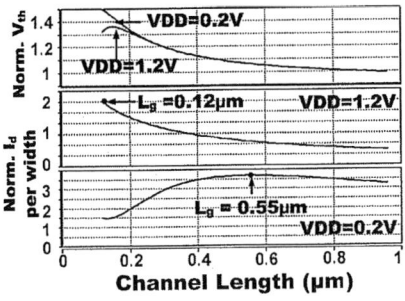

Fig. 3. Dependency of normalized V_{th} and current-per-width on channel length

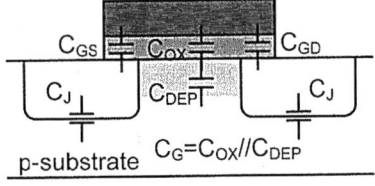

Fig. 4. Capacitance in subthreshold MOS device

3.4 Impact of Process Variation

Random Dopant Fluctuation (RDF) causes random parameter mismatches even between devices with identical layout in close proximity. The standard deviation (σ) of the threshold voltage distribution caused by RDF is proportional to $(WL)^{-1/2}$. Using the proposed sizing method, the gate area for optimal performance increases from 0.24µm² (=2µm×0.12µm) to 0.35µm² (=0.98µm×0.36µm) as shown in Fig. 5. This interesting characteristic leads to less threshold voltage variations for the proposed sizing scheme. Fig. 6 and Fig. 7 show the Monte Carlo simulation results of the delay and power distribution using a static inverter designed by the proposed and conventional scheme. σ/μ of the delay and power distribution is reduced by 37.5% and 70%, respectively, resulting in a squeezed distribution for the proposed sizing scheme. Simulation results show a 13%

improvement in average delay while simultaneously achieving a 31% reduction in average power dissipation.

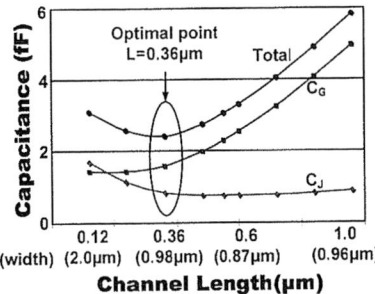

Fig. 5. Capacitance vs. channel length for constant current

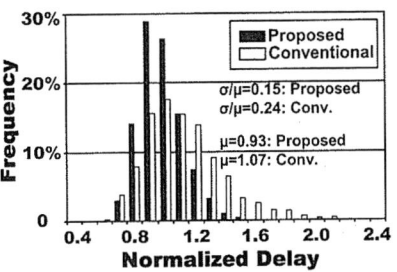

Fig. 6. Delay distribution comparison for a static inverter.

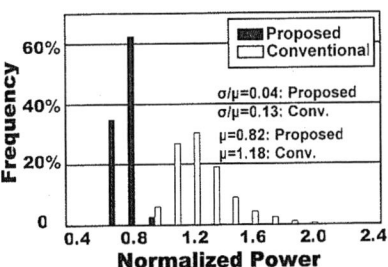

Fig. 7. Power distribution comparison for a static inverter.

3.5 Subthreshold Swing and Ion-to-Ioff Ratio

Subthreshold swing (S) is a critical parameter that determines the relationship between subthreshold current and the gate voltage. It is defined as the amount of V_{GS} required to change the subthreshold current by an order of magnitude. S has generally been considered a process-dependent parameter. A small S is preferred for higher on-current for a given off-current value. Our proposed sizing scheme utilizes a longer channel length which reduces S and improves the Ion-to-Ioff ratio. The subthreshold swing can be represented as

$$S = m\frac{kT}{q}\ln 10 \quad (mV/dec) \tag{11}$$

where $\quad m = 1 + \dfrac{C_{DEP}}{C_{OX}}, \quad C_{OX} = \dfrac{\varepsilon_{ox}}{t_{ox}}, \quad C_{DEP} = \dfrac{\varepsilon_{si}}{W_{DEP}} \tag{12}$

and kT/q is the thermal voltage.

As explained in section III-C, RSCE increases the depletion width underneath the channel and lowers the depletion capacitance, C_{DEP}. This alters the value of m in (12) and reduces S. In this specific process technology, the subthreshold slope of the proposed method is 71mV/dec which is 16mV lower than that of the conventional minimum channel device. The improved subthreshold slope reduces the off-current by 30% for the same on-current and increases the Ion-to-Ioff by a 2.5X, from 190 to 483 at 0.2V.

4. EXPERIMENTAL RESULTS

A delay chain composed of inverters, 2-input NANDs and 2-input NORs was used for the experiments to verify the effectiveness of the proposed sizing scheme. Layout of the sample delay chain is shown in Fig. 8. The conventionally sized gates have a taller layout than the gates sized using the proposed scheme. This is due to the fat devices in the proposed scheme having longer channel lengths and narrower widths. Minimum channel length is used for the PMOS devices since the RSCE was significantly weaker for PMOS transistors in this particular technology. In future technology nodes where RSCE is severe in both PMOS and NMOS devices, our proposed sizing scheme can be applied in general. The layout area of the proposed scheme is 18% smaller compared to that using conventional sizing. The delay variation of the proposed scheme is 38.7% smaller than that of the conventional method for each process corner. The power savings range from 10% to 39%, mainly depending on the current of the conventional scheme which is sensitive to process variations. We tested our sizing method in more general logic paths by synthesizing a number of ISCAS benchmark circuits, as well as different component circuits used in that suite. Two cell libraries were created, each containing inverters, two-input NANDs, and two-input NORs. Critical path delays and power consumption are compared in Table 1. Improvements in delay range from 7.8% to 10.4% depending on the type of logics used in the critical path. In addition, a simultaneous power reduction of 8.4% to 34.4% is achieved with the proposed scheme.

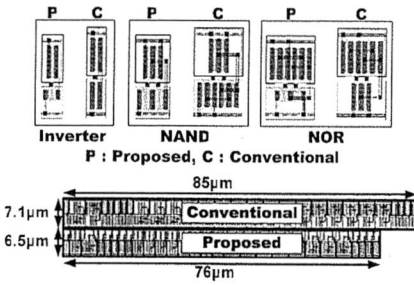

Fig. 8. Layout comparison for basic logic gates and sample delay chain.

5. CONCLUSIONS

RSCE was not a major concern in superthreshold designs. Rather, DIBL and V_{th} roll-off were the main considerations. However, in the subthreshold region, RSCE must be considered for optimal device sizing. We proposed a novel device size optimization scheme which can achieve high drive current, low device capacitance, and high Ion-to-Ioff ratio by utilizing the RSCE. Circuits using the proposed sizing scheme are more robust against RDF because of the increased gate area at the optimal performance point. The average delay was improved by 13% while average power dissipation was reduced by 31%. The proposed scheme also offers a tighter delay and power

distribution by improving the σ/μ by 37.5% and 70%, respectively.

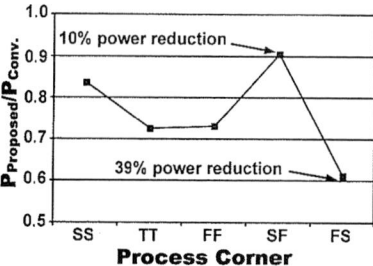

Fig. 9. Comparison of average power for corner parameters.

Table 1. Delay and power comparison for ISCAS.

Circuit	0.2V, Temp=27° (Delay / Power)		
	Conv. (ns)	Proposed (ns)	Improvement
C6288	119 / 2.24	107 / 1.50	10.1 / 33.0 %
C1355	443 / 0.64	397 / 0.42	10.4 / 34.4 %
74283	231 / 1.60	207 / 0.14	10.4 / 13.8 %
74L85	121 / 0.38	110 / 0.26	9.1 / 31.6 %
74182	103 / 0.40	95 / 0.38	7.8 / 8.4 %

6. REFERENCES

[1] H. Soeleman, et al., "Robust subthreshold logic for ultra-low power operation", IEEE Trans. VLSI Systems, Volume 9, pp. 90-99, Feb. 2001.

[2] A. Bryant, et al., "Low-power CMOS at Vdd=4kT/q", in Proc. Device Research Conference, pp. 22-23, 2001.

[3] B. Zhai, et al., "Analysis and mitigation of variability in subthreshold design", in Proc. ISLPED, pp. 20-25, Aug. 2005.

[4] E. Vittoz, et al., "CMOS analog integrated circuits based on weak inversion operations", IEEE J. of Solid-State Circuits, Volume 12, pp. 224-231, June 1977.

[5] A. Wang, et al., "A 180-mV subthreshold FFT processor using a minimum energy design methodology", IEEE J. of Solid-State Circuits, Volume 40, pp. 310-319, Jan. 2005.

[6] B.H. Calhoun, et al., "A 256k Sub-threshold SRAM using 65nm CMOS," in Proc. ISSCC, pp. 628-629, Feb. 2006.

[7] B.H. Calhoun, et al., "Ultra-dynamic voltage scaling using subthreshold operation and local voltage dithering in 90nm CMOS", in Proc. ISSCC, pp. 300-301, Feb. 2005.

[8] C.H. Kim, et al., "Ultra-low-power DLMS adaptive filter for hearing aid applications", IEEE Trans. VLSI Systems, Volume 11, pp. 1058-1067, Dec. 2003.

[9] J.J. Kim, et al., "Double gate-MOSFET subthreshold circuit for ultra-low power applications", IEEE Trans. Electron Devices, Volume 51, pp. 1468-1474, Sept. 2004.

[10] R. R. Troutman, "VLSI Limitations from drain-induced barrier lowering," IEEE Trans. Electron Devices", Volume 26, pp. 461-469, Apr. 1979.

[11] C.Y. Lu, et al., "Reverse short-channel effects on threshold voltage in submicrometer salicide devices", IEEE Electron Device Letters, Volume 10, pp. 446-448, Oct. 1989

[12] C. Subramanian, et al., "Reverse short channel effect and channel length dependence of boron penetration in PMOSFETs", in Proc. IEDM, pp. 423-426, Dec. 1995.

[13] Y. Taur, et al., "25nm CMOS Design Considerations", in Proc. IEDM, pp. 789-792, Dec. 1998.

Logic Circuits Operating in Subthreshold Voltages

Jabulani Nyathi and Brent Bero
School of EECS
Washington State University
102 Spokane St, Pullman, WA 99164-2752
Phone: 509-335-1157

{jabu,bbero}@eecs.wsu.edu

ABSTRACT

In this paper different logic circuit families operating in the subthreshold region are analyzed. Their performance in terms of power and speed are of particular interest. The study complements existing work that has reported static CMOS circuit performance under different body biasing schemes in the subthreshold region. Further it offers assurances on noise margins with scaling going beyond the 100 nm technology node. Simulations have been performed at the 180 nm technology node using a 6 metal layer TSMC process. A tunable body biasing scheme that allows bulk CMOS circuits to operate efficiently at subthreshold as well as above threshold voltages is introduced. The scheme improves a five-stage NAND ring oscillator switching speed 6X better than the static CMOS configuration while dissipating 18 % less power.

Categories and Subject Descriptors

D.3.3 [**Low Power High-speed Subthreshold CMOS Design**]: Suitable Logic families and body biasing techniques — swapped body biasing, tunable body biasing and dynamic threshold CMOS

General Terms

Performance, Design, and Theory.

Keywords

Subthreshold, ultra-low power, medium-to-high speed, logic styles, noise margins, body biasing, off current.

1. INTRODUCTION

Logic gates are the fundamental building block for high performance data path circuits and they continue to be a topic of interest especially as technologies are scaled to the nanometer regime. However increased leakage currents have led to the design of subthreshold circuits that use these currents to drive the logic [1]. Considerable work on limiting leakage currents has been performed for above threshold systems [2], [3] and there is a distinct difference between designing for ultra-low power and designing for high speed.

Much of the work on subthreshold has focused only on the ultra-

Permission to make digital or hard copies of all or part of this work for personal or classroom use is granted without fee provided that copies are not made or distributed for profit or commercial advantage and that copies bear this notice and the full citation on the first page. To copy otherwise, or republish, to post on servers or to redistribute to lists, requires prior specific permission and/or a fee.
ISLPED'06, October 4-6, 2006, Tegernsee, Germany.
Copyright 2006 ACM 1-59593-462-6/06/0010…$5.00.

low power aspect of this regime treating speed as a very secondary metric. There have not been significant efforts toward establishing what logic styles are ideal for subthreshold, what optimizations need to be made for both power and speed, in addition noise margin issues have not been analyzed. This study is a first step at addressing these issues. In [1] two logic families are discussed and methods for keeping the subthreshold currents stable under temperature and process variations are presented. Section 2 of this paper presents the transistor (pMOS/nMOS) configuration that provides more current in subthreshold. In Section 3 a comparison of body biasing schemes in different logic styles is presented, offering designers a choice of improved speed, ultra-low power or a good speed-power trade-off. Section 4 discusses noise margins while in Section 5 the impacts of fan-in and fan-out on subthreshold logic circuits are evaluated and some concluding remarks appear in Section 6.

2. MOS OPERATION IN SUBTHRESHOLD

In this study the SPICE device parameters used were extracted from MOSIS and these measured values represent an average from a fabrication lot. These parameters are dependable and previous work has shown a good correlation between systems' simulated results and their measured values.

Device leakage currents have been extensively studied primarily with a view of identifying leakage current sources and being able to keep these currents at a minimum [2], [3]. The focus on leakage currents is beginning to shift from this view of minimization to actually using these currents to drive logic [1], [4]. Digital circuits tend to simplify transistor operation, allowing devices to be viewed as switches. When the devices are operated at power supply voltages exceeding the threshold voltages they already exhibit some ill-effects. They do not operate as ideal switches and these effects cannot be viewed differently in subthreshold. In fact the problems are more pronounced in the subthreshold region since these leakage currents are relied upon to drive logic. The leakage currents that are expected to drive logic are present in the devices' OFF state and must be minimized in order to maintain the ultra-low power benefits.

There are a number of variables that influence device response and in this study the bulk terminal voltage is the variable of choice. Varying the bulk terminals' potentials provides further insights into device response to changes at the gate and drain/source terminals. The standard configuration has the bulk of the nMOS tied to the ground terminal, while that of the pMOS is tied to the power supply voltage (V_{DD}) for an inverter. This prevents forward biasing the source/drain-to-bulk p/n^{+} junctions. If the bulk terminal of the nMOS device is raised above ground and the power supply voltage is below threshold, there is a noticeable increase of the drain current. Similarly lowering the bulk voltage of the pMOS device leads to increased drain current. The bulk voltage (V_B) allows the p/n^+ junction to be forward biased thus allowing current flow from the bulk into the source/drain regions. Figure 1 compares the drain currents of a device (nMOS) whose bulk is raised to that of a device whose bulk terminal is at logic 0. With increasing gate-to-source voltage and the bulk voltage at 600 mV the drain current increases by at least an order of magnitude particularly at voltages below the threshold. Above the threshold voltage the drain current is dominated by the saturation current and is thus comparable to the drain current when the bulk is at ground for nMOS and V_{DD} for pMOS. The increase in current in the subthreshold region could lead to increased switching speeds while potentially dissipating less power. Simulations performed at 350, 250 and 180 nm with the bulk terminal of the nMOS at a voltage above 0 V show an increase in the drain-to-source current (I_{DS}) as shown in Figure 1. The pMOS yields similar increases in current when its bulk is tied to ground instead of the standard configuration (bulk tied to V_{DD}). The ability to increase subthreshold currents in this manner calls for the examination of the OFF current (I_{OFF}) with the modified bulk potential. The devices do not effectively turn OFF and would thus dissipate power even when there is no useful work being performed. A configuration that increases the subthreshold currents when the devices are turned ON and reduces these currents when the devices are in their OFF states requires that the bulk of each transistor be tied to its gate. The approach is termed subthreshold dynamic threshold voltage (Sub-DTMOS) and is presented in detail [1].

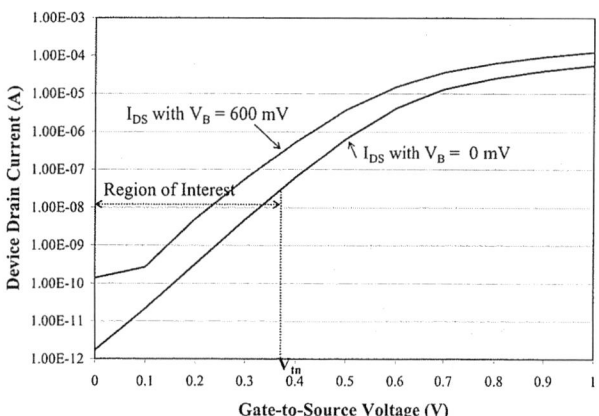

Figure 1. 180 nm nMOS operation with standard body biasing compared to a configuration with V_B=600 mV.

Both the nMOS and pMOS devices show increased OFF current when their bulk terminals are tied to V_{DD} and ground respectively. These observations have led to several body biasing schemes some of which include (i) swapped body biasing (SBB), (ii) dynamic threshold CMOS (DTMOS), (iii) adaptive body biasing (ABB) and (iv) multi-threshold CMOS (MTCMOS) just to name a few. The SBB scheme allows the pMOS devices' well to be connected to ground while the bulks of the nMOS devices are connected to the power supply voltage (V_{DD}). Simulations show that swapped body biasing has diminishing returns past the threshold voltage as a result the technique has been used at low power supply voltages for energy efficient designs. The dynamic

threshold CMOS circuit has the bulk of the transistor tied to its gate allowing the threshold voltage of the device to change dynamically with the gate input voltage. Increasing the power supply voltage to above threshold with this configuration in effect leads to the device turning OFF.

In this study an adaptive body biasing scheme that selects between operation in the subthreshold or above threshold regions and offers significant performance improvements over a range of power supply voltages $(V_{DD}=0.5*V_{th0}$ through $V_{DD}=1.8$ V in a 180 nm TSMC process) has been experimented with. Control of the bulk terminal in this manner offers an option of having tunable circuits that can operate in the ultra-low power range (at voltages below threshold) or at high speed (at voltages above threshold). The approach is termed tunable body biasing (TBB). No simulations have been performed with MTCMOS configuration in this study since this approach is intended for designs that need to limit leakages for above threshold operations. The following section of this paper presents different logic families operating in subthreshold under different body biasing schemes and details the gates' performance evaluations.

3. LOGIC FAMILIES IN SUBTHRESHOLD

Analysis of different logic styles operating in the subthreshold region is essential. This could provide a designer with a meaningful choice depending on what the design calls for: speed, power, reliability (based on noise margins), or a good compromise between high speed and ultra-low power. The study provides some insights on different logic style performance in the subthreshold region. In this study static CMOS with standard body biasing (nMOS devices' bulk at ground and pMOS devices' bulk terminals at V_{DD}) is considered as reference. Admittedly there are more logic families than discussed here, but these are the most representative ones and results obtained can easily be extended to other logic families. It would also be important to ensure that the work covers the effects of parametric variations (environmental as well as process variations) at these low voltages and an interested reader is referred to [1].

3.1 Performance Evaluation for Differing Body Biasing Schemes with Static CMOS

Before examining the different logic style performance attributes under subthreshold voltages, it is instructive to present performance evaluations of static CMOS under different body biasing schemes. The propagation delay simulation results appear in the bar chart displayed in Figure 2. The results show that swapped body biasing, dynamic threshold CMOS and the tunable body biasing schemes all lead to shorter propagation delays.

Propagation delays of Figure 2 are those of static CMOS transmission gate, inverter, NAND2, NOR2 and a two input XOR under the different body biasing conditions. Figure 3 shows the transistor configurations for each of the body biasing schemes that have been simulated to produce the results displayed in Figure 2. In subthreshold voltages DTMOS, SBB and TBB outperform the traditional body biasing technique by approximately a factor of six-to-ten in terms of propagation delay. The tunable body biasing (TBB) scheme is a result of the observation that in subthreshold swapping the bulk connections yields improved delays while above threshold swapping the bulk terminal connections degrades the delays significantly. In order to successfully bridge the speed-

132

power gap the bulk terminals have to be controlled such that the $V_{control}$ signals of Figure 3(c) reflect these findings. It is an accepted fact that circuits operating in the subthreshold region run at significantly low frequencies (e.g. 200 kHz to 500 kHz at 50 % V_{th0}) since the target metric is ultra-low power. It is the intention in this work to explore the possibilities of bridging the power-speed gap between ultra-low power circuits/systems and that of high speed systems.

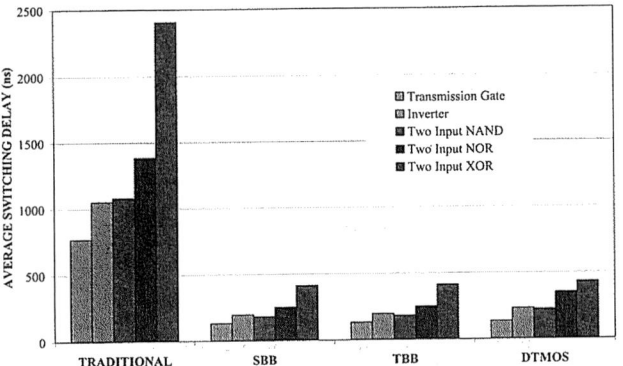

Figure 2. Static CMOS delays under different biasing conditions (standard, SBB, TBB and DTMOS).

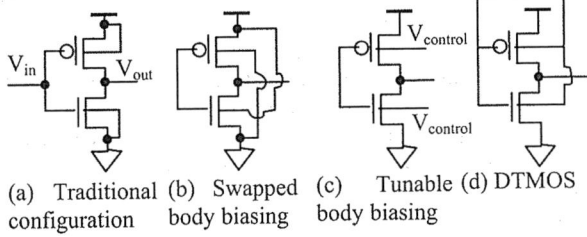

(a) Traditional configuration (b) Swapped body biasing (c) Tunable body biasing (d) DTMOS

Figure 3. Different body biasing schemes' configurations.

A more direct way of achieving such performance gains is to increase the power supply voltage to as close to the threshold voltage as possible. At 75 % of the threshold voltage the speed is improved significantly (achieving 8X the frequencies recorded at a V_{DD} that is $0.5*V_{th0}$)

DTMOS, SBB and TBB all have comparable delays however DTMOS and SBB are only good for power supply voltages below threshold. If operation has to span the entire power supply voltage scale from 0V to V_{DD}, it is better to use the TBB approach. The additional circuits required to provide the control voltages ($V_{control}$) leads to a 9 % increase in power dissipation for a 32-bit linear feedback shift register and an area overhead of 4 %. In the following delay characteristics of different logic styles configured in the traditional manner are presented and the findings compared to the results of the same logic gates using the swapped body biasing scheme, dynamic threshold CMOS and tunable body biasing schemes.

3.2 Logic Family Gate Delays Under the Various Biasing Schemes

Research on subthreshold circuit operation has focused on device sizing, the ultra-low power gains and most importantly the best configuration (i.e. how to bias the bulk terminals for best results).

Very little attention has been paid to the influence of logic style on performance in this regime. Extensive simulations at the 180 nm technology node have been performed for static CMOS, pseudo-nMOS, Domino and pass transistor logic styles. Of interest are the signal propagation delays and Figure 4 shows a comparison of a two input NAND gate's propagation delays under the different logic styles and biasing schemes. The simulations have been performed only at subthreshold voltages with power supply voltages ranging from $0.5*V_{th0}$ to V_{th0}. Circuit behavior is such that the propagation delays improve with increasing power supply voltage irrespective of biasing scheme nor logic style. The simulation results depicted in Figure 4 are those obtained with a power supply voltage of $0.75*V_{th0}$.

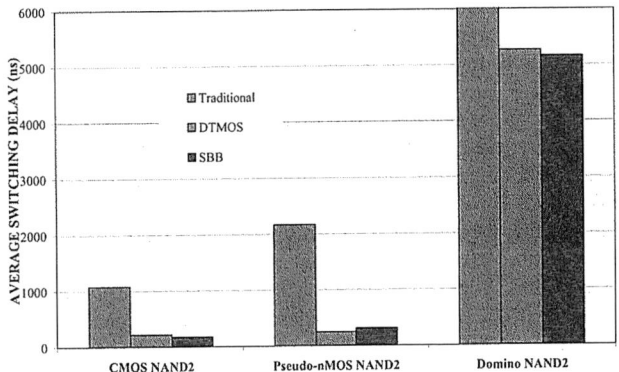

Figure 4. Comparison of propagation delays of a two input NAND gate in different logic styles under different biasing techniques (standard, SBB, TBB and DTMOS).

Simulations on inverters, NAND, NOR, XOR and XNOR gates have been performed but only the propagation delays of the NAND gate are shown since all the other gates show similar performance patterns. It is apparent from the graph that Domino logic has significantly longer delays and static CMOS is the better of the three logic styles under any of the bias conditions. Domino logic has longer delays owing to the additional transistor of the pull-down network driven by a clock input. Pseudo-nMOS on the other hand suffers longer delays due to the need to overcome the pull-up device in the event there is a path from the output node to the ground terminal. Pass Transistor Logic will always be fast however this logic style's delays exhibits heavy dependence on the input pattern. A two-input AND gate experiencing switching activity on both inputs simultaneously will see a slow output response compared to a static CMOS AND gate that has the majority of its devices closer to the power supply rails. These reasons therefore leave static CMOS being the logic style to offer better performance. It must be noted that as is the case at above threshold operation the other logic styles can still find wide spread usage in subthreshold operation.

4. NOISE MARGINS

A subject of importance as power supply voltages are scaled further into subthreshold is that of noise margins. At such small voltages it is possible that a slight change at the input might be misinterpreted by the circuit resulting in an incorrect output. The voltage levels that define the three regions of a digital circuit namely logic 0, the undefined region and logic 1 are examined. Simulations of an inverter under the three body biasing schemes that have been studied with static CMOS used as a reference point once again are recorded. This exercise has been extended to examine a NAND gate's noise margins.

Figure 5 represents the inverter input signal and the corresponding outputs for both the traditional and TBB inverters. The stair step representation of the input allows for better observation of the circuit response to each defined voltage level (the stair). The circuit response to the input can be analyzed using a ramp as an input, but the start of the transition region becomes less distinct with such an input. The power supply voltage is at 376.2 mV and from Figure 5 it can be determined the regions at which the output voltage reaches steady state and the transition region is clear. It has been determined that the logic 0 is valid for an input range of 0-175 mV while a logic 1 is valid for the 200-376.2 mV range with the uncertain region occupying the 175-200 mV range. The nMOS' bulk terminal is at the ground terminal and that of the pMOS at V_{DD} for these results. When the bulk terminals' connections are swapped (SBB or TBB) the ranges are as follows: logic 0 is defined in the range 0-200 mV, logic 1 is in the 225-376.2 mV range and the uncertain region occupies the 200-225 mV range. Tunable body biasing has skewed noise margins due to the fact that the inverter is forward biased, causing it to identify a logic 1 more easily. The analysis has been performed only on the static CMOS logic style since it is a foregone conclusion that the pseudo-nMOS logic style degrades noise margins considerably. The static CMOS configuration definitely offers better noise margins. The swapped body biasing scheme gives improved propagation delays but has poor noise margins. The analysis shows that scaling to 90 nm and beyond would still guarantee proper logic levels with full signal swing.

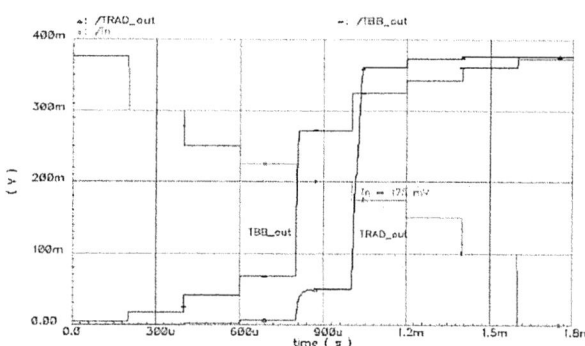

Figure 5. Traditional and TBB inverter input and output signals showing how noise margins are determined.

5. FAN-IN—FAN-OUT CONSIDERATIONS

At such low operating voltages it is highly likely that stacking devices could lead to erroneous results. It is thus instructive to evaluate circuit behavior under different fan-in conditions. All the gates analyzed in this study have a fan-out of four (FO4) achieved by having each gate drive four unit sized inverters. The simulation results displayed in Figure 6 are those of the NAND and NOR gates simulated in static CMOS. The swapped body biasing scheme is compared to the standard configuration as the fan-in increases while each gate maintains a constant fan-out (FO4). The NOR gate has longer propagation delays owing to the series p-type devices. The swapped body biasing scheme allows for fast switching hence improved/shorter delays. Increasing the fan-in to

five and above is impractical since the RC delay of the series devices significantly increases the propagation delays. These results do not reveal anything new but the exercise remains valuable in that it eliminates the need for speculative evaluation regarding gate performance in subthreshold.

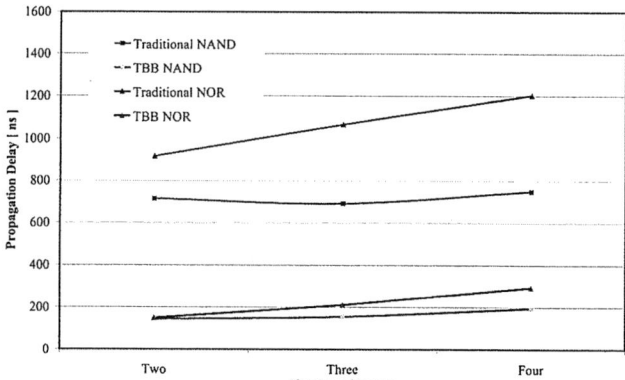

Figure 6. Effects of fan-in on propagation delay.

6. CONCLUDING REMARKS

This paper has explored the relevance of different logic styles operating in the subthreshold region, concluding that static CMOS outperforms both Domino and pseudo-nMOS in all aspects (except area). The pass transistor logic style though better than static CMOS is limited by the heavy dependence of delays on the input patterns. It has been shown that a newly introduced TBB NAND gate improves propagation delays 6 times better than the standard biased CMOS NAND. Noise margins for devices operating in subthreshold have also been analyzed and the conclusion drawn is that even with scaling beyond 90 nm gate lengths valid logic levels are attainable. The TBB configured circuits have a smaller range of the undefined region and could thus be more sensitive to small fluctuations of the gate signals.

7. REFERENCES

[1] Soeleman, H., Roy, K. and Paul, B. C. "Robust Subthreshold Logic for Ultra-Low Power Operation," *IEEE Transactions on Very Large Scale Integration (VLSI) Systems,* Vol. 9, No 1, February 2001, pp. 90-99.

[2] Kao, J., Narenda, S. and Chandrakasan, A., "Subthreshold Leakage Modeling and Reduction Techniques," *IEEE/ACM International Conference on Computer Aided Design (ICCAD),* November 10-14, 2002, pp.141-148.

[3] S. Yang, W. Wolf, N. Vijaykrishnan, T. Xie and W. Wang, "Accurate Stacking Effect Macro-Modeling of Leakage Power in Sub-100 nm Circuits," *18th International Conference on VLSI Design,* 2005, pp. 165-170.

[4] C. Hyung-II Kim, H. Soeleman and K. Roy, "Ultra-Low-Power DLMS Adaptive Filter for Hearing Aid Applications," *IEEE Transactions on Very Large Scale Integration (VLSI) Systems,* Vol. 11, No. 6, December 2003, pp. 1058-1067.

A New Mismatch-Dependent Low Power Technique with Shadow Match-Line Voltage-Detecting Scheme for CAMs

Jianwei Zhang
Microelectronics Center
Harbin Institute of Technology
Harbin, China

zjw@hit.edu.cn

Yizheng Ye
Microelectronics Center
Harbin Institute of Technology
Harbin, China

yeyizhen@public.hr.hl.cn

Binda Liu
Department of Electrical Engineering
National Cheng Kung University
Taiwan, China

bdliu1@spic.ee.ncku.edu.tw

ABSTRACT

A new mismatch-dependent low-power technique is presented for content-addressable memories (CAMs). With a novel shadow match-line voltage-detecting scheme, the word circuits realize fast self-disable of the charging paths in case of mismatches. Since the majority of CAMs words are mismatched, a significant power is reduced with a high search speed. Simulation results show the proposed 256-word×144-bit ternary CAM, using 0.13-μm 1.2-V CMOS process, achieves 0.51 fJ/bit/search for the word circuit with less than 900 ps search time. The achievement illustrates a 77% energy-delay-product (EDP) reduction as compared to the speed-optimized current-saving scheme.

Categories and Subject Descriptors

B.3.2 [**Memory Structures**]: Design Styles—*Associative memories*; B.7.1 [**Integrated Circuits**]: Types and Design Styles—*Memory Technologies*

General Terms: Design, Performance

Keywords: CAM, low power, high speed, mismatch-dependent, voltage detecting

1. INTRODUCTION

Content addressable memories (CAMs) compare input search data against a table of stored data, and return the address of the matching data. CAMs, especially fully parallel ones, have a single clock cycle throughput that makes them faster than other hardware- and software-based search systems. Thus CAMs are extensively used in many high speed data searching applications, such as asynchronous transfer mode, communication networks, LAN bridges/switches, database, and lookup table. Still, there is an increasing demand for CAMs with lower power and higher search speed in these applications. So low energy per search and high search speed are two major design goals of CAMs.

Figure.1 shows the basic block diagram of a CAM [8], which consists of an array of storage elements (M), a column of sense amplifiers (SA) and search-lines (SL) drivers. Each row of array stores one word (144-bit in this paper) and has an associated match-

Permission to make digital or hard copies of all or part of this work for personal or classroom use is granted without fee provided that copies are not made or distributed for profit or commercial advantage and that copies bear this notice and the full citation on the first page. To copy otherwise, or republish, to post on servers or to redistribute to lists, requires prior specific permission and/or a fee.
ISLPED'06, October 4-6, 2006, Tegernsee, Germany.
Copyright 2006 ACM 1-59593-462-6/06/0010...$5.00.

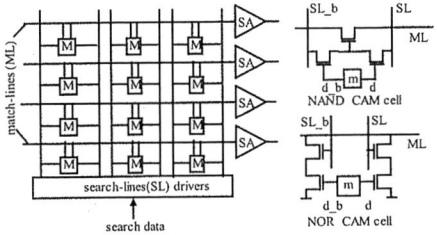

Figure 1. Simplified CAM architecture with two types of cells

line (ML) to indicate matches/mismatches. Two types of CAM cell are also shown. The search result is output by SA.

One way to reduce the power is to decrease the switching capacity on the MLs by using the NAND architecture [2]. Due to its terrible search speed, the NOR architecture is more preferred. In a conventional precharge-high NOR architecture [3], all the MLs are precharged high in the reset phase, and then discharged to GND in case of mismatches. Due to the highly capacitive MLs and that the majority of CAMs words are mismatched, the word circuits contribute greatly to the total CAM power [1]

To reduce power while at a low price of the search speed, many techniques have been developed around the NOR architecture. Several focused on reducing word circuit power by directly [4][5] or indirectly [6] decreasing the voltage swing of MLs. While in these approaches, equal power is consumed independent of the number of mismatched bits. Based on [6], a mismatch-dependent method [7][8], named current saving, is developed, which dynamically allocates less power to MLs with more mismatches through a current saving control (CSC) block. As a result, a large amount of power is saved. The main drawback of the current saving approach is that it cannot realize low power and high search speed simultaneously.

In this work, a more effective mismatch-dependent technique with shadow match-line voltage-detecting (SMLVD) scheme is presented, which achieves low power and high search speed simultaneously. It also keeps the low-power feature of non-resetting search-lines. Simulation results show the proposed 256-word×144-bit ternary CAM, using 0.13-μm 1.2-V CMOS process, achieves 0.51 fJ/bit/search for the word circuit with less than 900 ps search time. The Energy-Delay-Product accounts for only 19.9% that of the conventional precharge-high architecture [3], 23.0% and 20.9% that of the speed- and power-optimized current-saving approaches [7][8].

2. NEW MISMATCH-DEPEDENT LOW POWER TECHNIQUE WITH SMLVD SCHEME

A general architecture of the proposed SMLVD scheme is shown in Figure 2. It is composed of a CAM array, level shifters and minimum energy-overhead voltage detectors (MEVDs). Also included is a dummy word that mimics the case of one-bit mismatch and sends out ENCODER_EN to a priority encoder (not shown in this figure) as a timing control signal. Some detailed circuits are also shown in Figure 2.

The CAM array consists of 256-word memory rows. Each word is composed of 144-bit NOR type CAM cells as shown in Figure 3 (a). Note that the lower part of the four bit-compare circuits (TN1~TN4 in Figure 3) in the CAM cell is connected to a wire named Shadow Match-Line (SML) instead of GND. This forms a major difference as compared to the ordinary NOR cells, which enables adoption of the SMLVD scheme through which both low power and high search speed can be obtained as explained later in this paper.

In case of a match, there will be no paths created from ML to SML and the SML voltage will stays at zero if it is initially precharged to GND. The circuit model of this case is shown on the left of Figure 3 (b). Here C_{ML}/C_{SML} denotes the stray capacity on ML/SML. While in case of mismatches, at least one path will be created from ML to SML and the SML voltage will rise steadily if ML is being charged by current flow. The circuit model of this case is shown on the right of Figure 3 (b). Here R_{ML} denotes the conducting resistor of one path from ML to SML, while n is the mismatched bits or the number of conducting paths.

So the idea of the design can be easily formed. By using a voltage detector/comparator, non-zero voltage of SML can be detected that corresponding to the cases of mismatches. The lower sense threshold voltage the voltage comparator has, the faster the mismatch is detected. Since the majority of CAMs words are mismatched, a large amount of power will be saved if the charging paths to MLs are broken as soon as mismatches are detected.

Usually low sense-threshold voltage detector operates with large working current. If this working current can be reused to charge up the MLs in order to determine a match/mismatch, a further power reduction can be achieved. Based on this idea, a minimum energy-overhead voltage detector (MEVD) is devised with detailed circuit shown in Figure 2 (c). MEVD also has self-disable function to break the charging path whenever a mismatch is detected on the corresponding word circuit. In order to provide an appropriate

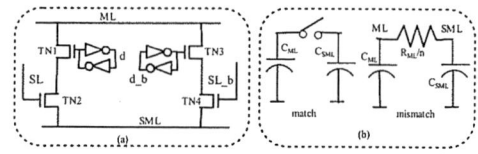

Figure 3. (a) Ternary NOR cell in this work. (b) ML/SML circuit model for match/mismatch state

common-mode level to MEVD, a level shifter is introduced that features low power consumption. The details of the level shifter are shown in Figure 2 (b).

Then the overall circuit operation of Figure 2 (a) can easily be understood. In the reset phase, all MLs and SMLs are precharged to GND. In the evaluate phase, MEVD begins working with its working current charging up MLs. Whenever a mismatch of the word circuit is detected, the output signal SMLS toggles its polarity and the corresponding charging path to the ML is disabled.

Further details of related circuits will be illustrated in the following sections. Throughout the paper, we use M0 to represent full-match, M1 1-bit-mismatch and Mn n-bit-mismatch.

2.1 Level Shifter

The proposed level shifter does not consume static power. Its operation is as follows. See Figure 2 (b). In the reset phase with SEARCH_EN_b (an inverted signal of SEARCH_EN) being logic "1", the three nMOS switches MN6~MN8 are on, and both the outputs, V_IN and V_REF, equal V_{bias}. Then in the evaluation phase, V_IN follows the changes of SML while V_REF keeps constant. It is easy to prove that the linear slope that V_IN changes with SML equals $C_L/(C_S+C_L)$, where C_S is the stray capacitance on V_IN, and C_L is the capacitance of selected capacitor as shown in Figure 2 (b).

In order to eliminate the impact of V_{bias} variation during evaluation phase, V_REF is driven by MN8 instead of directly by V_{bias}. MN6 and MN8 are of the same sizing .V_{bias} can be generated on-chip or supplied off-chip.

2.2 MEVD

As shown in Figure 2 (c), MEVD mainly consists of three parts: the first stage amplifier (MP1~MP2 and MN1~MN2), the second stage amplifier (INV) and a positive feedback path (MN4~MN5). MP3 and MN3 are MOS switches allowing enable or disable of the first stage amplifier. While switch MP4, accompanied by MN4, is used to pull up node PT in the reset phase. Note that the lower part of the first stage amplifier is connected to ML instead of GND. Thus the major part of the MEVD's working current is reused to charge up ML to determine a match/mismatch. Such architecture leads to the

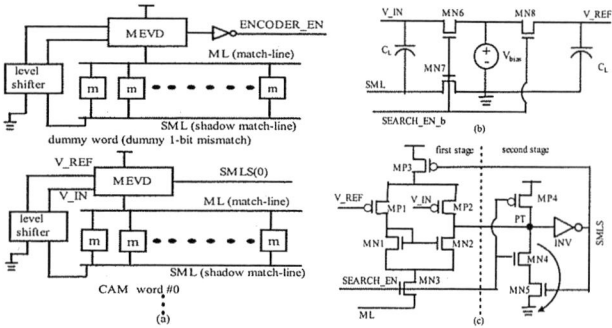

Figure 2. (a) General architecture of the proposed SMLVD scheme. Detailed circuits of (b) Level shifter and (c) MEVD

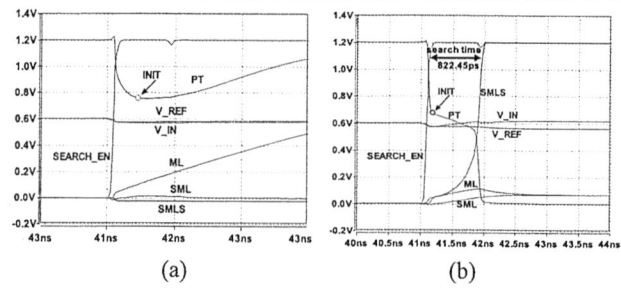

Figure 4. Simulation results of MEVD in case of (a) M0 (b) M1

minimum energy-overhead introduced by the first stage amplifier that usually has a large working current.

2.2.1 Operation Principle

MEVD of Figure 2 (c) works as follows. In the reset phase with SEARCH_EN equals logic "0" (or SEARCH_EN=0), node PT is charged high with MP4 on and MN4 off. Also all MLs and SMLs are reset to GND and both the input pairs (V_REF and V_IN) to the MEVD equal V_{bias}. In the evaluation phase with SEARCH_EN equals logic "1", MEVD first setups its initial working state by pulling down node PT to a certain level (point INIT in Figure 4 (a) and (b)) which is still above the sense threshold voltage of INV. Simultaneously, a large current flows through the first stage amplifier and also the voltage of ML begins to rise. In case of M0, SML still stays at GND for there are no charging paths created from ML to SML. While the voltage of node PT will eventually goes up as the voltage of ML increases. So in this case, V_{PT} always stays higher than the sense threshold of INV and the output of MEVD keeps un-toggled. The current continues to flow through MEVD. While there is no need to worry about the power dissipation of M0 for most MLs are statically mismatched. We need a timing signal to tell that the un-toggled result SMLS is the final result and the case is a full match, so a dummy word circuit is needed to generate such timing signal. As explained later in this section, an M1 has the longest time to tell the final result, so the dummy word circuit mimics an M1 to indicate the priority encoder (not shown in this paper) the validity of the input data.

In case of mismatches, after a high SEARCH_EN, SML eventually goes up as there are paths created from ML to SML. As can be seen from Figure 4 (b), in less than 900 ps, the voltage difference of V_IN and V_REF reaches the sense threshold of MEVD (about 40 mV), and the corresponding output SMLS goes up signaling a mismatch. Meanwhile, the high SMLS turns off the switch MP3, so the charging path to ML is broken and MEVD is disabled. This function is called self-disable. During the high transition of SMLS, the nMOS switch MN5 turns on and the positive feedback loop is formed (note that MN4 is "on" in the evaluate phase). As a result of the feedback loop, V_{PT} falls faster and reaches GND finally. Also it should be pointed out that the search result is latched by this loop until next reset phase.

2.2.2 Mismatch-Dependent Power Dissipation

Due to the self-disable function, when a mismatch is detected, the current through MEVD to ML is cut off and thus a further power dissipation is prevented. This subsection will show that when the mismatches bits increase, the power of the word circuit decreases.

Firstly, we concentrate on the circuit's search time. The search time defined here is the time period from the rising edge of SEARCH_EN to the rising edge of SMLS (both referred to 90% of VDD) as shown

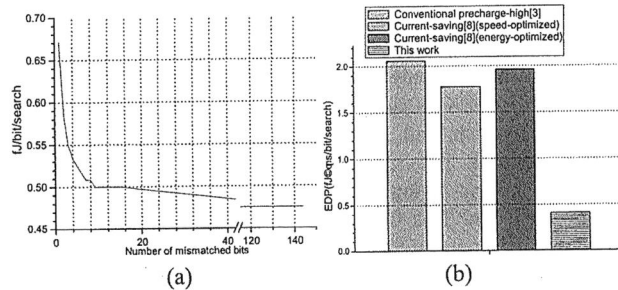

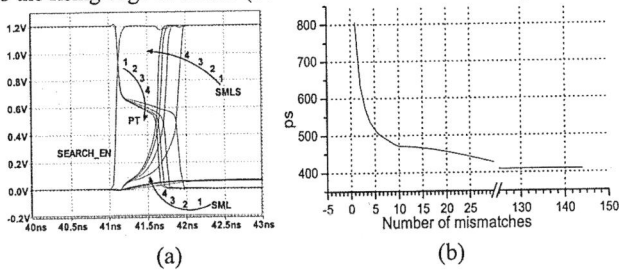

Figure 5. (a) Simulation result of four cases. 1~4 represent M1~M4. (b) Search time as a function of mismatched bits.

Figure 6. (a) Energy consumption of the word circuit as a function of the mismatched bits. (b) EDP comparison of four types of approaches

in Figure 4 (b). Figure 5 (a) shows the simulation results of four cases M1~M4. As can be seen that, when the mismatched bits increase from 1 to 4, V_{SML} grows faster resulting in a faster falling of V_{PT}, The search time is also decreased. Figure 5 (b) plots the curve of search time as a function of the mismatched bits in the full range (M1 to M144). As can be seen that the search time of word circuit rapidly falls from about 800 to 470 ps when the mismatched bits increases from 1 to 10; after that, the descendent trend smoothes and finally reaches 408 ps. We take search time of M1 as the proposed scheme's search time for it is the slowed one among various mismatches.

As explained before, shorter search time may lead to less energy consumption due to the self-disable function of MEVD. Figure 6 (a) validates this by the energy consumption of the word circuit as a function of the number of mismatched bits. As can be seen that the power dissipation rapidly falls from 0.67 to 0.50 fJ/bit/search when the mismatched bits increases from 1 to 10; after that, the descendent trend smoothes and finally reaches 0.48 fJ/bit/search. Since the majority of CAMs words are mismatched, the proposed word circuit consumes 0.5 fJ/bit/search in most cases.

Table 1 gives a comparison of power (word circuits only) and speed for four types of circuit. The proposed scheme in this paper was implemented under the same condition (0.13-μm 1.2-V CMOS process) as that of [7][8]. We also re-realized the precharge-high NOR architecture [3] as a reference for comparison that consumes 2.06 fJ/bit/search for the word circuit. The value is almost the same as that (2.1 fJ/bit/search) mentioned in [7][8]. This indicates that the stray capacitances on the MLs are almost the same for the same NOR memory array, and that the comparison is carried out on a fair foundation. The search time of the precharge-high circuit is referred to as the time period from the enable signal (90% VDD) of a search to the 50% fall of V_{ML}. The power figure of the proposed scheme is

Table 1. Power and speed comparison of four types of circuit

	Conventional precharge-high [3]	Current-saving [8] (speed optimized)	Current-saving [8] (energy optimized)	This work
Energy dissipation (fJ/bit/search)	2.06	0.71	0.56	0.51
Search time (ns)	1.0	2.5	3.5	0.8
EDP(fJ. ns/bit/ search)	2.06	1.78	1.96	0.41

derived in case of 7-bit mismatches, which is the same as that in [8]. As can be seen from Table 1, the word circuits of our proposed scheme consumes only 0.51 fJ/bit/search, which achieves a reduction of 75.2% as compared with the conventional precharge-high [3] approach, and 28.2% with the speed-optimized current-saving approach [8]. Even when compared with the energy-optimized current-saving approach [8], our approach gets 8.9% energy reduction. This figure may seem not remarkable, but note that the search speed of the proposed scheme is rather faster than all the other three approaches. It seems fairer to carry out the comparison from view of energy-delay-product (EDP). Figure 6 (b) gives such EDP comparison. As can be seen that the EDP of the proposed scheme accounts for only 19.9% that of the conventional precharge-high architecture, 23.0% and 20.9% that of the speed- and power-optimized current-saving approaches.

2.2.3 Simulation with Process Variation Consideration

When dealing with a voltage detector, an obvious concern is the sensitivity to process variation, for transistor mismatch will occur due to local variation during fabrication. We emulate this imbalance by changing the transistor sizing and adding additional capacitance as stray capacitance, which is the same approach as in [7][8]. This section shows the robustness of the proposed scheme.

When considering process variation, two special cases, M1 and M0, are considered. This is because that M1 is the hardest-to-detect, while M0 may give wrong result in case of large transistor mismatch. Figure 7 (a) and (b) give the worst-case models of M0 and M1 given 20% transistor mismatch. We focus on the two input transistors (MP1 and MP2) of the MEVD because they are the most susceptible to transistor mismatch. For M0 where both the input signals stay at the same level, it is most prone to be wrong if MP1 get a larger sizing than MP2. This is because that V_{PT} will be lower than ordinary that may result in a high SMLS, signaling a false mismatch. For M1 where the voltage of V_IN is slightly higher than that of V_REF, it is most prone to be wrong if C_{SML} gets larger and MP2 gets a larger sizing than MP1. Due to the limited power supply rejection ration (PSRR) of MEVD, a high SMLS may never occur as V_{ML} rises in such situation.

Figure 7 (c) and (d) gives the normal-case versus worst-case simulation result of M0 and M1. As it can be seen that, for worst-case M0, V_{PT} is lower than ordinary resulting in a small rise of SMLS (about 60mV) that it can be negligible. As for worst-case M1, the circuit still gives proper operation except that the search time is lengthened to about 1.6 ns. So we can see robustness of the proposed scheme given 20% transistor mismatch.

3. CONCLUSIONS

This paper presents a new mismatch-dependent low-power technique for CAMs that adopts a novel shadow match-line voltage-detecting scheme. A voltage detector/comparator with minimum energy-overhead is introduced that not only realizes fast search operation but also effectively reduces power dissipation of the word circuit. Nevertheless, the proposed scheme shows remarkable robustness considering process variation. Simulation results show that for a 900 ps search operation, under 0.13-μm 1.2-V CMOS process, the proposed word circuit of the 256×144-bit TCAM consumes only 0.51 fJ/bit/search, which achieves a reduction of 75.2% as compared with the conventional precharge-high [3] approach, and 28.2% with the speed-optimized current-saving

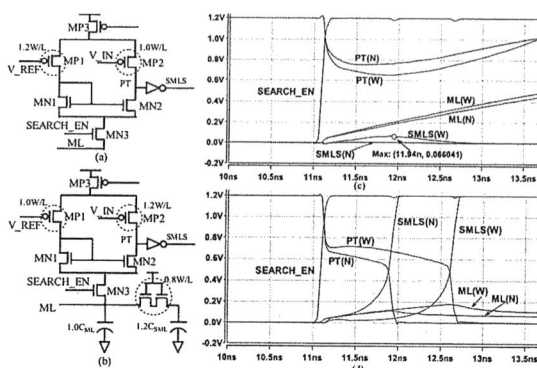

Figure 7. The worst-case models of MEVD for (a) M0 (b) M1. Normal-case .VS. worst-case simulation for (c) M0 (d) M1 (N denotes normal case, W denotes worst case).

approach [8]. Even when compared with the energy-optimized current-saving approach [8], the proposed approach gets 8.9% energy reduction. Including delay factor, the achievement will be more remarkable. In view of energy-delay-product, the proposed scheme accounts for only 19.9% that of the conventional precharge-high architecture, 23.0% and 20.9% that of the speed- and power-optimized current-saving approaches.

4. REFERENCES

[1] Y. L. Hsiao, D. H. Wang, and C. W. Jen, Power modeling and low-power design of content addressable memories, in *Proc. IEEE Int. Symp. Circuits and Systems*, vol. 4, 2001, pp. 926-929.

[2] F. Shafai, K. J. Schultz, G. F. R. Gibson, A. G. Bluschke, and D. E. Sompppi, Fully parallel 30-MHz 2.5-Mb CAM, *IEEE J. Solid-State Circuits*, vol. 33, pp. 666-676, Apr. 1998.

[3] P. Lin and J. Kuo, A 1-V 128-kb four-way set-associative CMOS cache memory using wordline-oriented tag compare (WLOTC) structure with the content-addressable-memory (CAM) 10-transistor tag cell, *IEEE J. Solid-State Circuits*, vol. 36, pp. 666-675, Apr. 2001.

[4] H. Miyatake, M. Tanaka, and Y. Mori, A design for high-speed low-power CMOS fully parallel content-addressable memory macros, *IEEE J. Solid-State Circuits*, vol. 36, pp. 956-958, June 2001.

[5] G. Kasai, Y. Takarabe, K. Furumi, and M. Yoneda, 2300MHz/200MSPS 3.2W at 1.5V Vdd, 9.4Mbits ternary CAM with new charge injection match detect circuits and bank selection scheme, in *Proc. IEEE Custom Integrated Circuits Conference*, 2003, pp. 387-390.

[6] I. Arsovski, T. Chandler, and A. Sheikholeslami, A ternary content-addressable memory (TCAM) based on 4T static storage and including a current-race sensing scheme, *IEEE J. Solid-State Circuits*, vol. 38, pp. 155-158, Jan. 2003.

[7] I. Arsovski and A. Sheikholeslami, A current-saving match-line sensing scheme for content-addressable memories, in *IEEE Int. Solid-State Circuits Conf. Dig. Tech. Papers*, 2003, pp. 304-305.

[8] I. Arsovski and A. Sheikholeslami, A mismatch- dependent power allocation technique for match- line sensing in content-addressable memories, *IEEE J. Solid -State Circuits*, vol. 38, no. 11, pp. 1958-1966, Nov 2003.

Thread-Associative Memory for Multicore and Multithreaded Computing

Shuo Wang and Lei Wang
Department of Electrical and Computer Engineering
University of Connecticut
371 Fairfield Road, U-2157
Storrs, CT 06269
Email:{shuo.wang, leiwang}@engr.uconn.edu

ABSTRACT

Presented in this paper is the thread-associative memory microarchitecture for multicore and multithreaded processor design. Memory contention among concurrent threads in chip multithreaded processing has become a limiting factor for performance improvement. The proposed thread-associative memory addresses this challenge by incorporating thread-specific information explicitly into on-chip memory hardware. The proposed technique can be utilized at different levels of memory hierarchy. Furthermore, it is not just a technique for performance enhancement but also a solution for energy efficiency. Trace-driven simulations on a $32KB$ L1 data cache demonstrate 36.6% maximum performance improvement and up to 15.1% total energy reduction, with 20.3% dynamic energy reduction and 9.9% leakage energy reduction.

Categories and Subject Descriptors:
B.3.0 [Memory Structures]: General;
C.5.0 [Computer System Implementation]: General

General Terms: Design, Performance

Keywords: Cache Mapping, Memory System, Multicore, Multithreading

1. INTRODUCTION

The relentless scaling of semiconductor process combined with the advances in computer architecture has been the primary force that drives an explosive growth in computational performance. Research in computer architecture aims at exploiting higher levels of parallelism in instruction processing. Simultaneous multithreading (SMT) [1], [2] and chip multiprocessors (CMP) [3], [4] exploit instruction-level and thread-level parallelism jointly to improve instruction throughput. The SMT-CMP based architecture, however, is challenged by the incompatible memory hierarchy that is developed originally for single-threaded computing. As each processing core in SMT-CMP systems is extended with the capability of multithreaded execution, competition for memory resources is exacerbated. This not only hurts the overall performance but also affects the energy efficiency. Therefore, thread-aware memory microarchitecture needs to be developed to effectively address the thread interference and memory contention.

Existing research on inter-thread conflicts has been focusing on modeling [5] and software optimization [6], [7]. It was shown that these techniques might be only affordable in large and less active L2 or L3 caches in high-performance processors. For embedded multicore systems-on-chip with limited memory resources, the effectiveness of these techniques becomes marginal. In this paper, we propose a *thread-associative memory microarchitecture* to address the emerging challenge of memory contention in multicore and multithreaded computing systems. The proposed technique incorporates thread-specific information explicitly into on-chip memory hardware to deal with inter-thread conflicts. Simulation results demonstrate this technique is not just a technique for performance enhancement but also a solution for energy efficiency.

We discuss the microarchitectural aspect of the proposed thread-associative memory in section 2. A detailed analysis of energy efficiency is presented in section 3. Section 4 evaluates the proposed technique, and section 5 concludes the paper.

2. THREAD-ASSOCIATIVE MEMORY MICROARCHITECTURE

In traditional computing platforms, cache memory deals with cache replacement through different memory mapping methods such as direct mapping, fully-associative mapping, and set-associative mapping. Although inducing extra cost over direct mapping, set-associative mapping can reduce miss rate caused by intra-thread conflicts and hence improve the cache performance.

The same idea can be extended to managing memory access from concurrent threads in multicore and multithreaded environments. To illustrate the thread-associative memory microarchitecture, we use a two-rail thread-associative cache as an example. Here, the rails define the memory mapping for concurrent threads. The concept of *rail* is quite different from the conventional definition of *way* in set-associative mapping. In set-associative cache, cache blocks are grouped into sets. A set with n blocks is called n-way set-associative. In a thread-associative cache, each way within a set is fur-

---

Permission to make digital or hard copies of all or part of this work for personal or classroom use is granted without fee provided that copies are not made or distributed for profit or commercial advantage and that copies bear this notice and the full citation on the first page. To copy otherwise, to republish, to post on servers or to redistribute to lists, requires prior specific permission and/or a fee.
ISLPED'06, October 4–6, 2006, Tegernsee, Germany.
Copyright 2006 ACM 1-59593-462-6/06/0010 ...$5.00.

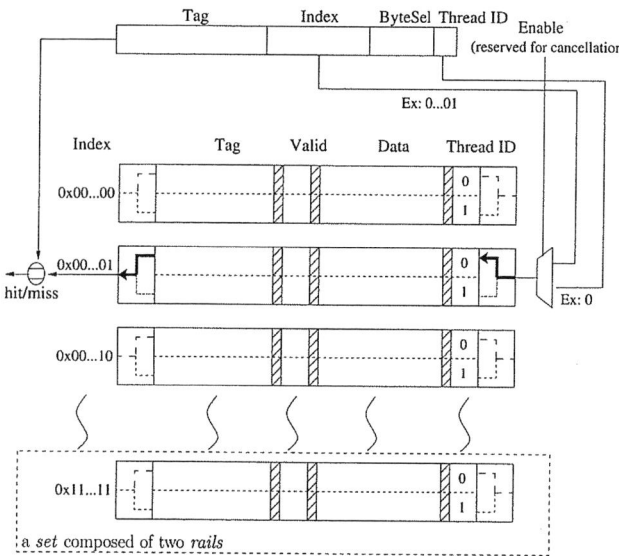

Figure 1: Thread-associative cache microarchitecture.

ther grouped into rails. If there are m rails in a set, then it is called m-rail thread-associative. The number of rails within a set is determined by the number of concurrent threads. Figure 1 shows the microarchitecture of a two-rail thread-associative cache. For the sake of simplicity, each rail employs direct mapping for access management. Note that fully-associative mapping and set-associative mapping can also be applied to manage the rail access in thread-associative cache.

As shown in Figure 1, each index contains two identical physical memory rails, each of which is assigned to a specific thread. The cache address is composed of four parts: tag, index, byte selection and thread ID. A memory address can be mapped into either of the two rails, which share the same index but belong to different threads. The thread ID determines which rail to actually access. The tag comparison will be performed only for the rails belonging to the thread that initiates memory access. Note that the thread-associative cache can also be assessed using set-associative mapping for each thread-specific rail. For example, if thread-associative memory uses n-way set-associative mapping for each rail, a memory access requested by a specific thread will be directed to the corresponding rails in n different cache ways. The tags of these cache rails are then compared to determine a cache hit or miss. Thus, the memory mapping will be determined jointly by tag, index, byte selection as well as the thread ID. An additional "Enable" bit is reserved for deactivating thread-associative in case that performance degradation is observed. Both thread ID and "Enable" bit can be generated by the operating system.

In comparison with traditional cache design, the proposed thread-associative memory introduces new microarchitectural features to handle inter-thread conflicts. Traditional cache memory does not distinguish access requests from concurrent threads but rather treats them as from a single thread. This leads to a significant increase in inter-thread conflicts, particularly in memory-intensive applications such as multimedia processing and database operations. Employing the thread-associative memory, the access requests from

concurrent threads are directed into separate memory rails. Cache replacements for concurrent threads are decoupled, thereby avoiding the inter-thread conflicts. When applied in conjunction with set-associative mapping for each thread, intra-thread conflicts could be reduced as well. As a result, we can expect improvements in both memory access and energy efficiency. In the next section, we present an analytical model to quantify the performance of the proposed thread-associative memory.

3. ANALYSIS OF ENERGY EFFICIENCY

Energy efficiency is a key design issue in memory circuits. The proposed thread-associative memory microarchitecture reduces inter-thread conflicts, which in turn improves energy efficiency by minimizing the energy overheads related to memory misses.

In memory circuits, both dynamic power and leakage power impose significant constraints on chip-level power optimization. The dynamic component of energy dissipation, denoted by $E_{dynamic}$, is a function of memory access activities, as summarized by [9]

$$E_{dynamic} = AC \cdot E_{hit} + MC \cdot E_{miss}, \qquad (1)$$

where AC is the access count, MC is the miss count. A memory access (a hit or a miss) will involve a dynamic energy of E_{hit}. If a miss occurs, an extra energy of E_{miss} is induced. This extra energy includes the energy of accessing the lower level of memory, the stall energy and the energy of cache line filling. It can be approximated as proportional to the hit energy, i.e.,

$$E_{miss} = K \cdot E_{hit}, \qquad (2)$$

where the factor K is determined by the process technology and the level of memory hierarchy being considered. In [10], the value of K was selected to be 50 and 200.

To compare the energy efficiency, we normalize the dynamic energy over the number the instructions being executed. We denote the Ed_{shared} and $Ed_{threaded}$ as the dynamic energy per instruction in the conventional cache and thread-associative cache, respectively. The reduction in dynamic energy achieved by the proposed thread-associative cache can be expressed as

$$\frac{Ed_{threaded}}{Ed_{shared}} = \frac{AC + K \cdot MC_{threaded}}{AC + K \cdot MC_{shared}}$$
$$= \frac{AC + K \cdot MC_{threaded\_intra}}{AC + K \cdot (MC_{shared\_intra} + MC_{shared\_inter})}. \qquad (3)$$

Thus, we find that dynamic energy reduction is a function of miss count reduction achieved by removing inter-thread conflicts among concurrent threads. Note that according to (3), dynamic energy reduction is also a function of the ratio K between E_{hit} and E_{miss}. When K increases, dynamic energy reduction approaches miss count reduction.

In addition to dynamic energy, leakage energy is also a serious problem in deep submicron memory circuits. We estimate the leakage energy using the following method

$$E_{leak} = E_{leak}^{ave} \cdot C_{exe}, \qquad (4)$$

where E_{leak} is the total leakage energy consumed during the execution of a program, E_{leak}^{ave} is the average leakage energy per cycle, and C_{exe} is the number of execution cycles. The value of C_{exe} is a sum of the effective cycles for processing the instructions and the cycles of memory stalls in cache misses, such as

$$C_{exe} = IC \cdot CPI + MC \cdot PPM, \qquad (5)$$

where IC is the total instruction count, CPI is the cycles per instruction excluding memory stalls, and PPM is the average penalty (in terms of cycles) per miss.

We normalize the E_{leak} over the number of instructions to obtain the average leakage energy per instruction. The reduction in memory leakage energy achieved by the proposed thread-associative memory can be expressed as

$$\frac{El_{threaded}}{El_{shared}} = \frac{IC \cdot CPI + PPM \cdot MC_{threaded}}{IC \cdot CPI + PPM \cdot MC_{shared}}$$
$$= \frac{IC \cdot CPI + PPM \cdot MC_{threaded\_intra}}{IC \cdot CPI + PPM \cdot (MC_{shared\_intra} + MC_{shared\_inter})}, \qquad (6)$$

where El_{shared} and $El_{threaded}$ are the leakage energy per instruction of the conventional cache and the thread-associative cache, respectively.

Similar to dynamic energy reduction, leakage energy reduction is achieved by the reduction of miss count through removing inter-thread conflicts in thread-associative memory. Combining (3) and (6), we derive the total energy reduction per instruction EPI achieved by the proposed thread-associative memory, expressed as

$$\frac{EPI_{threaded}}{EPI_{shared}} = \frac{Ed_{threaded} + Ed_{threaded}}{Ed_{shared} + El_{shared}}$$
$$= \frac{Ed_{threaded}}{Ed_{shared}} \cdot \alpha_d + \frac{El_{threaded}}{El_{shared}} \cdot \alpha_l, \qquad (7)$$

where α_d and α_l are the ratios of dynamic energy and leakage energy, respectively, to the total energy. From (7), the total energy reduction is the weighted average of dynamic energy reduction and leakage energy reduction.

4. EVALUATION AND DISCUSSION

In this section, we study the access performance and energy efficiency of the proposed thread-associative memory in a multithreaded computing environment.

Our simulations were obtained from a dual-threaded simulator based on the trace-driven simulator Dinero IV [11]. As this simulator does not provide the energy results directly, we collect the statistical data from the simulator and then derive the cache performance and energy efficiency based on the analytical model presented in section 3. Some parameters in these models are given as: $CPI = 2$, $PPM = 24$, $K = 50$, $\alpha_d = 50\%$, and $\alpha_l = 50\%$. We also assume each main memory access takes 100 cycles. Table 1 lists the configuration of the dual-threaded simulator. While these parameters assume specific architecture and energy profile, the trends observed are likely to be reproduced in different computing environments. In these simulations, we focus on two-rail thread-associative L1 data cache as inter-thread conflicts are more pronounced at this level of memory [8]. Note that

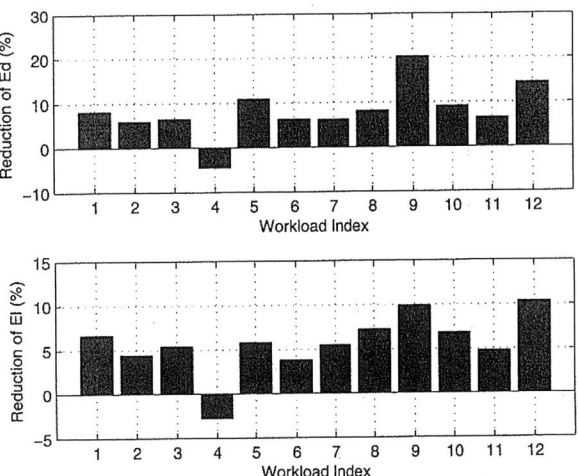

Figure 2: Energy reduction in the proposed thread-associative cache. (a) Dynamic energy reduction. (b) Leakage energy reduction.

Table 1: Configuration of the simulation environment.

parameters	L1-instr.	L1-data	L2	L3
size	32KB	32KB	256KB	4MB
block size	32B	32B	32B	32B
associativity (conventional)	DM	1-rail&DM	DM	DM
associativity (proposed)	DM	2-rail&DM	DM	DM
fetch	demand	demand	demand	demand
write alloc	N/A	always	always	always
write back	N/A	always	always	always
hit cycles	1	1	10	30

thread-associative mapping can be applied to other levels of memory hierarchy as well.

All the simulations employ the SPEC CPU2000 [12] trace files collected from the Stream-Based Trace Compression (SBC) [13]. In total, twelve pairs of workloads are randomly selected. Table 2 shows the workloads, the results of misses per instruction and total energy reduction per instruction. The reductions of dynamic energy and leakage energy per instruction are shown in Figures 2(a) and 2(b), respectively. From these results, we observe that the proposed thread-associative cache achieves both performance improvement (in terms of the reduction in misses per instruction) and energy reduction in nearly all the cases. The maximum performance improvement is 36.6%, and the maximum energy reduction is 15.1%, with 20.3% dynamic energy reduction and 9.9% leakage energy reduction.

The improvement in performance and energy efficiency is achieved by eliminating inter-thread conflicts while keeping the intra-thread conflicts under control. In the proposed thread-associative memory, each thread will access a specific memory space that shares the same address with other

Table 2: Simulation results of memory access performance and total energy reduction.

index	workloads	MPI reduction	EPI reduction
1	applu&mesa	9.58%	7.43%
2	apsi&lucas	6.81%	5.20%
3	art&mesa	7.06%	5.94%
4	crafty&wupwise	-6.25%	-3.61%
5	equake&perlbmk	29.40%	8.37%
6	gap&parser	9.68%	5.04%
7	mcf&mesa	6.55%	5.86%
8	mcf&sixtrack	8.56%	7.71%
9	mesa&sixtrack	36.57%	15.10%
10	mgrid&vortex	11.30%	8.00%
11	perlbmk&mgrid	8.57%	5.65%
12	swim&sixtrack	16.67%	12.41%

concurrent threads, but does not interfere with each other due to the thread-associative mapping. This effectively removes the inter-thread conflicts, which account for about 26%−60% of the total memory misses. Due to the real-time thread dynamic and scaled accessible space for each thread, the intra-thread conflicts are increased around 6%−90%. Nevertheless, we observe an overall improvement in performance and energy efficiency, as the reduction in inter-thread conflicts plays the dominant effect.

There is only one exception among the twelve workloads. In the crafty&wupwise case, the thread-associative cache results in slight degradation in performance and energy efficiency. This problem occurs when the increase in intra-thread conflicts is large enough to offset the reduction in inter-thread conflicts. This is partly due to the direct mapping in each thread-specific rail that we use in this paper for demonstration purpose. In our future work, we expect to solve this problem by applying set-associative mapping for each thread. In addition, we reserve a bit of "Enable" in the thread-associative memory (see Figure 1). In the extreme case where there is only one thread being processed or significant performance degradation is observed, the "Enable" bit can deactivate the thread-associative mapping, thereby avoiding the degradation of performance and energy efficiency.

Indeed, the thread-associative memory microarchitecture exploits the memory reference locality of each thread and hence leads to overall performance optimization. Given the trends towards massive parallelism in multicore and multithreaded systems, interference between simultaneously executed threads is expected to be a serious problem. The proposed thread-associative memory microarchitecture is a promising and scalable solution to ensure the performance and energy efficiency for future multicore and multithreaded computing.

5. CONCLUSIONS

In this paper, we propose a thread-associative memory microarchitecture to address the emerging challenge of memory contention in multicore and multithreaded computing systems. The proposed technique incorporates thread-specific information explicitly into on-chip memory hardware to deal with inter-thread conflicts. Simulation results on a $32KB$ L1 data cache demonstrate 36.6% maximum performance improvement and up to 15.1% total energy reduction, with 20.3% dynamic energy reduction and 9.9% leakage energy reduction of. Thread-associative memory can be applied to a wide range of multicore and multithreaded computing systems from high-end processors to low-power embedded systems-on-chip. Future work is being directed towards integration of the proposed technique and other low power techniques, and joint exploration of thread-associativity and set-associativity for cache design.

Acknowledgment

This research was supported in part by the University of Connecticut Faculty Research Grant 446751.

6. REFERENCES

[1] D. M. Tullsen, S. J. Eggers, and H. M. Levy, "Simultaneous multithreading: Maximizing on-chip parallelism," *Proc. Intl. Symp. Computer Architecture*, pp. 392-403, Jun. 1995.

[2] J. L. Lo, et al., "Converting Thread-Level Parallelism Into Instruction-Level Parallelism via Simultaneous Multithreading," *ACM Trans. on Computer Systems*, pp. 322-354, Aug. 1997.

[3] K. Olukotun *et al.*, "The Case for a Single Chip Multiprocessor," *Proc. 7th Intl. Conf. Architectural Support for Programming Languages and Operating Systems*, ACM Press, New York, pp. 2-11, 1996.

[4] L. Hammond, B. Nayfeh, and K. Olukotun, "A Single-Chip Multiprocessor," *Computer*, vol. 30, no. 9, pp. 79-85, Sept. 1997.

[5] D. Chandra, et al., "Predicting inter-thread cache contention on a chip multi-processor architecture," *Proc. Intl. Symp. on High Performance Computer Architecture*, pp. 340-351, Feb. 2005.

[6] G. E. Suh, L. Rudolph, and S. Devadas, "Dynamic partitioning of shared cache memory," *The Journal of Supercomputing*, pp. 7-26, Apr. 2004.

[7] A. Settle, et al., "A Dynamically Reconfigurable Cache for Multithreaded Processors," *The Journal of Embedded Computing (JEC)*, special issue, Dec. 2005.

[8] S. J. Eggers, et al., "Simultaneous multithreading: A platform for next-generation processors," *IEEE Micro*, pp. 12-19, 1997.

[9] C. Zhang, and F. Vahid, "A self-tuning cache architecture for embedded systems," *Design, Automation and Test Conference in Europe*, pp. 142-147, 2004.

[10] C. Zhang, et al., "A way-halting cache for low-energy high-performance systems," *Low Power Electronics and Design*, pp. 126-131, Aug. 2004.

[11] J. Edler and M. D. Hill, "Dinero IV Trace-Driven Uniprocessor Cache Simulator," at *http://www.cs.wisc.edu/∼markhill/DineroIV/*, 2006.

[12] SPEC CPU2000 at *http://www.spec.org/cpu/*.

[13] A. Milenkovic and M. Milenkovic, "Exploiting Streams in Instruction and Data Address Trace Compression," *Proceedings of the IEEE 6th Annual Workshop on Workload Characterization*, pp. 99-107, Oct. 2003.

Hierarchical Value Cache Encoding for Off-Chip Data Bus

Chung-Hsiang Lin, Chia-Lin Yang
Department of Computer Science and
Information Engineering
National Taiwan University
Taipei, Taiwan

{r94040,yangc}@csie.ntu.edu.tw

Ku-Jei King
IBM STG xSeries Development
kujei@tw.ibm.com

ABSTRACT

Off-chip data bus consumes a significant part of system power. Recent works use small caches (Value Cache) at each side of the off-chip data bus, and transmit cache indexes instead of data values to reduce bus switching activity. A larger VC has a higher VC hit rate, but it also incurs more switching activity on a VC hit. In this paper, we propose the hierarchical VC design concept that provides a good tradeoff between VC capacity and bus switching activity. Our experimental results show that the proposed hierarchical VC design reduces the off-chip data bus energy by 60.2%.

Categories and Subject Descriptors: C.4 [Performance of systems]: Design studies - data bus encoding, low power design.

General Terms: Algorithms, Measurement, Performance, Design, Experimentation.

Keywords: Data bus encoding, Hierarchical value cache, Energy.

1. INTRODUCTION

Power consumption is becoming a critical design criterion for embedded systems and mobile computing devices [1]. Since the off-chip bus wires are associated with very large capacitances, they have been shown to be a major contributor to system power consumption, between 9.8% and 23.2% according to the study [2].

Recent works on bus power reduction use small caches (called value cache, or VC for short) at each side of the off-chip data bus [3][4][2][5][6]. These VCs keep track of the data values that have recently been transmitted over the bus. If a data value hits in the sender's VC, its VC index, rather than the data value, is transmitted over the bus. This method is called FVE (Frequent Value Encoding). To transmit the data in the VC using only 1 bit switching activity, the size of VC is limited to the width of the data bus. That is, with a 32-bit bus, the VC could have only 32 en-

tries. The effectiveness of FVE depends on the VC hit rate. A larger VC results in a higher VC hit rate at the cost of more switching activity on a VC hit. Suresh et al. [6] proposed the TUBE architecture which observes the switching patterns of both contiguous and non-contiguous bits, and identifies hot (silent) bits that switch more (less) than other bits. Hot and silent bits are encoded separately. In order to partition bits, this approach requires off-line profiling. In this paper, we propose a hierarchical value cache architecture that provides a good tradeoff between the VC capacity and the bus switching activity without resorting to off-line profiling.

The concept of hierarchical value cache is to organize the value caches in a hierarchy where lower-level VCs have higher hit rates but incur more switching activity on VC hits. The idea is similar to the memory hierarchy design where lower-level caches have higher hit rates but longer hit latencies. A naive hierarchical value cache architecture is to employ larger value caches in the lower levels of the hierarchy, and all value caches store full data values. This is referred to as the Hierarchical Unified Value Cache (HUVC) in this paper. This architecture reduces the bus switching activity at the cost of high area overhead. Therefore, we propose another hierarchical value cache design: the Hierarchical Combinational Value Cache (HCVC). Instead of storing full data values in one VC as in the HUVC, the level i of the HCVC contains 2^{i-1} VCs that store only partial values. Our experimental results show that the HCVC reduces the switching activity of the off-chip data bus by 67.2%, and reduces the bus energy by 60.2%. Compared to the TUBE [6], the HCVC reduces 17.9% more bus energy with less than 1/3 of the VC capacity.

The rest of this paper is organized as follows. In Section 2, we describe the design of hierarchical value caches. In Section 3 and Section 4, we show our experimental setup and results, respectively. Finally, we conclude in Section 5.

2. HIERARCHICAL VALUE CACHE

The concept of the hierarchical value cache is to organize the value caches in a hierarchy where lower-level VCs have higher hit rates but incur more switching activity to transmit encoded data. In this paper, we propose two types of hierarchical value cache architectures. The first one is the Hierarchical Unified Value Cache (HUVC) which stores full data values in each value cache as shown in Figure 1. This architecture reduces the bus switching activity at the cost of high area overhead. To minimize the area overhead, we propose the Hierarchical Combinational Value Cache (HCVC)

Permission to make digital or hard copies of all or part of this work for personal or classroom use is granted without fee provided that copies are not made or distributed for profit or commercial advantage and that copies bear this notice and the full citation on the first page. To copy otherwise, to republish, to post on servers or to redistribute to lists, requires prior specific permission and/or a fee.
ISLPED'06, October 4–6, 2006, Tegernsee, Germany.
Copyright 2006 ACM 1-59593-462-6/06/0010 ...$5.00.

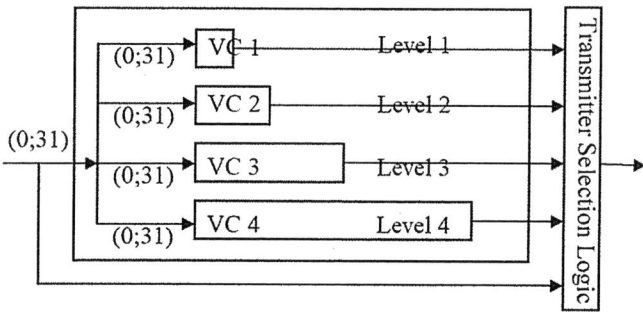

Figure 1: 4-Level Hierarchical Unified Value Cache Architecture

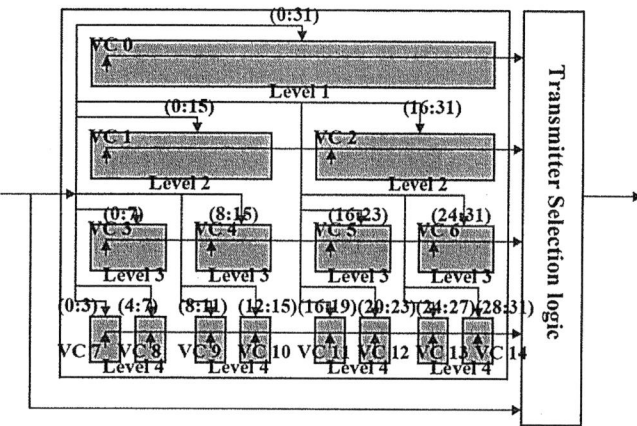

Figure 2: 4-Level Hierarchical Combinational Value Cache Architecture

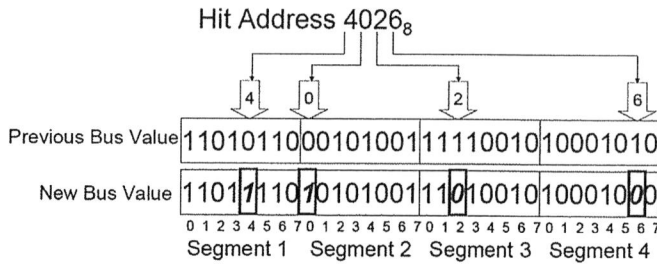

Figure 3: HUVC VC index encoding

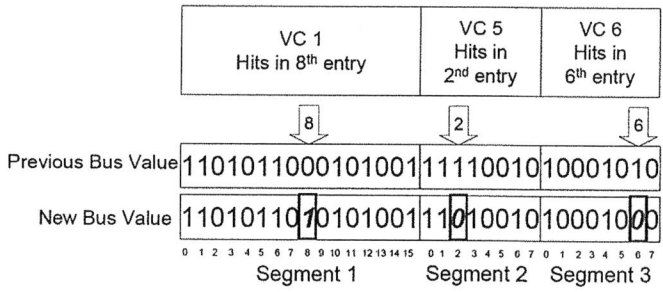

Figure 4: HCVC VC index encoding

of the ith level is determined by the following formula:

$$\left(\lfloor \tfrac{32}{i} \rfloor + 1\right)^{(32 \bmod i)} \times \left(\lfloor \tfrac{32}{i} \rfloor\right)^{i - (32 \bmod i)}$$

For a 4-level HUVC, and 32-bit data bus, the total VC size is 22.4KB. The size of the VC is too large to be feasible in practice.

2.2 Hierarchical Combinational Value Cache

Figure 2 shows a 4-level Hierarchical Combinational Value Cache (HCVC). Except for the first level, all VCs store partial data values. The level i of the HCVC contains 2^{i-1} VCs. A data value is partitioned into 2^{i-1} segments, and each VC stores one data segment. Similar to the HUVC, an incoming data is simultaneously checked with all VCs of every level of the HCVC. A data value could hit in different levels of the hierarchy. The uppermost VC hit index is encoded. Figure 4 shows the process of converting the VC hit indexes to the bus value. We assume that the upper 16 bits of the data value hit in the 8th entry of VC1, the next 8 bits hit in the 2nd entry of VC5, and the last 8 bits hit in the 6th entry of VC6. So we partition the data bus into three segments, and each segment represents the hit index of its corresponding VC. In the example shown in Figure 4, we switch the 8th, 2nd, and 6th bit of the first, second, and third segments. The HCVC scheme requires n control signals, where n is the number of VCs. The ith control signal is switched to indicate the ith VC hit. For a 4-level HCVC, we need 15 control signals. The total VC size of the ith level is $\frac{32}{2^{i-1}}$ words. For a 4-level HCVC with 32-bit data bus, the total VC size is only 240 bytes.

One design complexity of the HCVC is in the control signals. One way to implement this is to add extra wires for control signals. The alternative is to utilize the address bus to transmit control signals. Since a cache block often contains multiple words, the off-chip address bus has much lower utilization than the data bus. For a 4-level HCVC, we could use 15 of 32 address lines to transmit control signals.

architecture as shown in Figure 2. The level i of the hierarchy contains 2^{i-1} VCs that store only partial values. Below we describe these two architectures in details.

2.1 Hierarchical Unified Value Cache

The architecture of Hierarchical Unified Value Cache (HUVC) is shown in Figure 1. Each VC in the HUVC stores 32-bit values, and is managed in the LRU policy. An incoming data is simultaneously checked with the VC on every level, and the uppermost VC hit index is encoded. Each hit in the ith level of the HUVC incurs i bits switching activity. By switching any i bits of a 32-bit data bus, we can get $\frac{32!}{(32-i)!i!}$ numbers. That is, we could have $\frac{32!}{(32-i)!i!}$ entries. However, it would require complicated logics to map VC indexes to bus values. For easy VC index encoding, we partition the data bus into i segments and switch one bit in each segment. Figure 3 shows an example of the VC index encoding method in the 4th level VC. We partition the 32-bit data bus into 4 segments. Each segment represents one digit of an 8-based number which represents a VC hit index. In the example shown in Figure 3, we assume the VC hit index to be 4026_8, so we switch the 4th bit in segment 1, the 0th bit in segment 2, the 2nd bit in segment 3 and the 6th bit in segment 4. The HUVC scheme requires n control signals, where n is the depth of the VC hierarchy. The ith control signal is switched to indicate that the VC of level i hits.

According to our VC index encoding method, the VC size

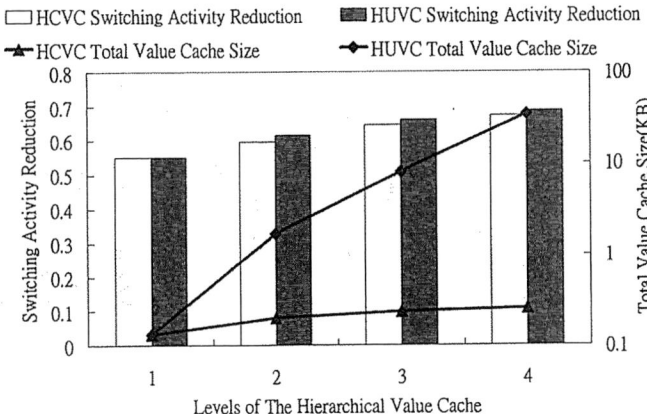

Figure 5: Switching Activity Reduction vs. Total Value Cache Size

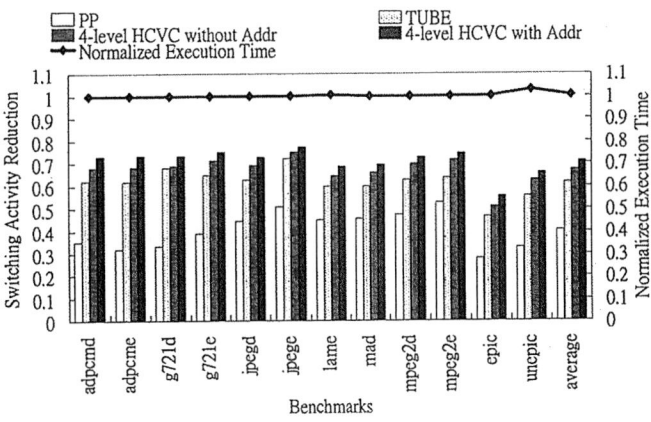

Figure 6: Switching Activity Reduction of HCVC, PP and TUBE.

The remaining 17 lines can be used to transmit encoded data. This design would cause performance degradation, but it would also allow larger VCs. In Section 4.1, we will evaluate its impact on both performance and energy aspects.

3. EXPERIMENTAL SETUP

We modified the SimpleScalar ([7] and [8]) tool set to simulate the proposed hierarchical value cache. To accurately evaluate the performance impact of transmitting encoded data over the address bus, we model lockup-free caches which can have up to 8 outstanding misses. The target processor is an ARM-like processor which has a 32KB, 2-way L1 data cache, and 32KB, direct-mapped L1 instruction cache with 100MHz off-chip buses. We evaluate the proposed scheme on 12 applications from the MediaBench suite [9]. We compare our scheme with two previous works, the Power Protocol (PP) [2] and the TUnable Bus Encoder (TUBE) [6]. The VC sizes of the PP and the TUBE are 1K bytes, and 896 bytes, respectively.

To estimate the implementation complexity of the hierarchical value cache, we consider the transmission latency, energy and area overheads. To evaluate the overheads of value caches as content addressable memory (CAM) cells, we scale the result in [6]. To evaluate the overheads of selection logic, we model the functionality of selection logic by doing a layout-level description using $0.18\mu m$ technology by the *Cadence* and the Synopsys' *Design Compiler*. The bus energy model adopted in this paper is similar to [6]. The load capacitance and supply voltage is set to 20pF and 3.3 Volts for experimental results presented in the next section.

4. EXPERIMENTAL RESULTS

In this section, we first evaluate the bus switching activity reduction achieved with the proposed hierarchical value cache. We then analyze the performance, energy and area overheads. Finally, we evaluate the bus energy reduction.

4.1 Bus Switching Activity Reduction

We first compare the average reduction of switching activity of the HUVC and the HCVC in 1, 2, 3, and 4 levels in Figure 5. We also show the required value cache size of the HUVC and the HCVC. In this set of experiment, we do not utilize the address bus to transmit encoded data. Note

that we also include the switching activity in the control signals. We can see that deeper VC hierarchy leads to more bus switching activity reduction for both the HUVC and HCVC. However, for the HUVC, the VC capacity also increases significantly. The VC size of a 4-level HUVC is more than 10 times of a 4-level HCVC. In contrast, the VC size of the HCVC only increases slightly with deeper pipeline. The results show that the HCVC performs comparable to the HUVC with much smaller VCs. Therefore, for the results presented below, we all assume a 4-level HCVC for our scheme.

Figure 6 shows the reduction rates of the bus switching activity for our schemes (with/without utilizing the address bus to transmit encoded data), the PP [2] and the TUBE [6]. Without utilizing the address bus, the HCVC can reduce 67.2% switching activity on the average and outperform the PP and the TUBE by 26.7% and 5.5%, respectively. Note that our scheme requires smaller VCs than both the PP and the TUBE. The total VC capacity of the PP and the TUBE are 1024 bytes and 896 bytes, while the HCVC only requires 240 bytes. Note that since the HCVC requires only less than 1/3 of the VC capacity of the TUBE, the HCVC shows more advantage over the TUBE in terms of bus energy reduction. We will compare the bus energy reduction of the HCVC and the TUBE in Section 4.3.

Utilizing the address bus to transmit encoded data could further reduce the bus switching activity by 3.7% because it allows larger VCs. The downside of this scheme is performance degradation. In Figure 6, we also show the execution time of the HCVC utilizing the address bus normalized to that of the baseline HCVC. We can see that utilizing address bus only degrades performance by 0.3% on the average. The unepic shows the largest performance impact, about 2.9%. Previous results show that the unepic is the most memory intensive among the applications tested in this paper [9]. As unepic has many contiguous memory requests, the address bus receives burst requests. When the address bus is occupied with encoded data, the memory requests sent back to back are blocked. If the number of blocked memory requests exceeds the maximal number of the outstanding requests, the CPU will stall. This is the reason why we see more performance degradation in unepic than other applications.

145

4.2 Analysis on the Performance/energy/area Overhead

A. Performance Overhead Analysis

The size of the largest value cache of a 4-level HCVC is 196 bytes. The estimated access latency is 5ns, and the delay of the selection logic is 1.2ns. The total latency of HCVC is 6.2ns (161MHz). Due to complicated selection logic, the latency of the HCVC is slightly larger than the latency of the TUBE (5.71ns).

The off-chip bus traffic is mainly due to cache misses. The HCVC introduces an additional cycle delay at both the encoder and decoder sides. Since the data transfers to/from memory are in the granularity of cache blocks, encoding/decoding consecutive intra-block entries can be pipelined. We would encounter an additional 2-cycle penalty on every cache block transfer. Assuming 100-cycle off-chip latency, the HCVC incurs a penalty of 0.3% in terms of the total number of execution cycles on the average.

B. Energy Overhead Analysis

To estimate the energy overhead of the HCVC, we consider three components: VC access energy, selection logic energy and pipeline registers energy. The VC access energy of the HCVC is the sum of the energy of all VCs. The estimated energy overhead of VCs access is 40.2pJ. The estimated power of the selection logic is 7.2mW. With the 1.2ns delay described above, the estimated energy of the selection logic is 8.6pJ. The power of registers is 11.8mW and the maximum delay of the registers is 0.12ns. The energy of pipeline registers is 1.4pJ. The total energy overhead is 50.2pJ.

The total energy overhead of the TUBE is 103.1pJ as indicated in their paper [6] so our scheme incurs less energy overhead than the TUBE. Although our scheme requires more complicated controls than the TUBE, our value caches are much smaller than those in the TUBE.

C. Area Overhead Analysis

We consider the area overhead in three components: VC area, selection logic area and pipeline registers area. We find the total area of the CAM cells in the HCVC to be $0.019mm^2$. The area overhead of the pipeline registers and selection logic are $0.014mm^2$ and $0.007mm^2$, respectively. The total area overhead of the HCVC is $0.039mm^2$.

The area overhead of the TUBE is $0.052mm^2$ as indicated in their paper [6]. The HCVC can reduce more switching activity with less area overhead than the TUBE.

4.3 Bus Energy Reduction

Figure 7 shows the bus energy reduction of the HCVC. We also show the TUBE for comparison. The results show that the HCVC reduce the bus energy by 17.9% compared to the TUBE. Utilizing the address bus to transmit encoded data can further reduce the bus energy by 1.3%.

5. CONCLUSION

In this paper, we propose a hierarchical value cache concept that organizes value caches in a hierarchy where lower-level VCs have higher hit rates but incur more switching activities on VC hits. We present a hierarchical value cache design called HCVC which uses multiple VCs in each level to store partial data. The HCVC provides a good trade-off between the VC capacity and the bus switching activity. Simulation results show that on the average the HCVC re-

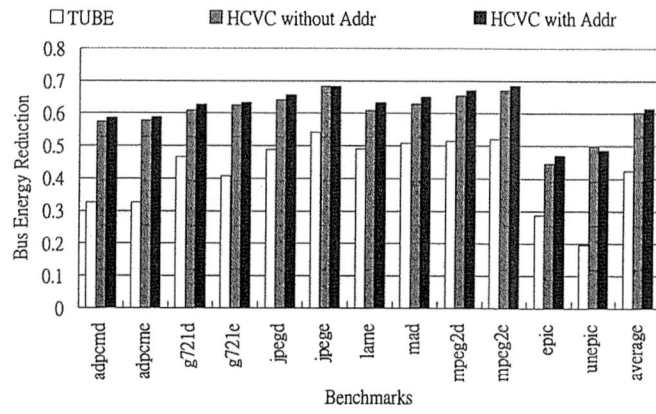

Figure 7: Bus Energy Reduction of HCVC and TUBE

duces bus switching activities by 67.2%, and bus energy by 60.2%. The HCVC also achieves more energy savings with lower energy/area overheads than the state-of-art bus encoding scheme TUBE.

6. ACKNOWLEDGEMENT

This work is supported in part by research grants from ROC Industrial Technology Research Institute 95-S-B43, National Science Council NSC 95-2752-E-002-008-PAE and IBM faculty award.

7. REFERENCES

[1] "National technology roadmap for semiconductors," *Semiconductor Industry Association*, 2001.

[2] K. Basu, A. Choudhary, J. Pisharath and M. Kandemir, "Power Protocol: Reducing Power Dissipation on Off-Chip Data Buses," *MICRO*, 2002.

[3] Jun Yang and Rajiv Gupta, "FV Encoding for Low-Power Data I/O," *ACM/IEEE International Symposium on Low Power Electronic Design*, 2001.

[4] Dinesh C Suresh, Banit Agrawal, Jun Yang, Walid Najjar and Laxmi Bhuyan, "Power Efficient Encoding Techniques for Off-Chip Data Buses." *CASES*, 2003.

[5] Dinesh C. Suresh, Banit Agrawal, Walid A. Najjar and Jun Yang, "VALVE: Variable Length Value Encoder for Off-Chip Data Buses," *ICCD*, 2005.

[6] Dinesh C Suresh, Banit Agrawal, Walid Najjar and Jun Yang, "A Tunable Bus Encoder for Off-Chip Data Buses," *ISLPED*, 2005.

[7] T. M. Austin., "The SimpleScalar/ARM Toolset. http://www.eecs.umich.edu/ taustin/simplescalar."

[8] T. M. Austin and D. Burger., "The SimpleScalar Architectural Research Tool Set. http://www.cs.wisc.edu/ mscalar/simplescalar.html."

[9] Chunho Lee, Miodrag Potkonjak and William H. Mangione-Smith, "MediaBench: A Tool for Evaluating and Synthesizing Multimedia and Communications Systems," *Intl. Symp. on Microarchitecture*, 1997.

Reducing Cache Traffic and Energy with Macro Data Load

Lei Jin   Sangyeun Cho
Department of Computer Science
University of Pittsburgh
{jinlei,cho}@cs.pitt.edu

ABSTRACT

This paper presents a study on *macro data load*, an efficient mechanism to enhance *loaded value reuse*. A macro data load brings into the processor a maximum-width data value the cache port allows, saves it in an internal structure, and facilitates reuse by later loads. A comprehensive limit study using a generalized *memory value reuse table* (MVRT) shows the significantly increased reuse opportunities provided by macro data load. We also describe a modified load store queue design as an implementation of the proposed concept. Our quantitative study shows that over 35% of L1 cache accesses in the SPEC2k integer and MiBench programs can be eliminated, resulting in a related energy reduction of 24% and 35% on average, respectively.

Categories and Subject Descriptors

C.4 [**Performance of Systems**]: Design studies; C.1.1 [**Processor Architectures**]: Single Data Stream Architectures

General Terms

Design, performance

Keywords

Memory hierarchy, LSQ design, low power

1. INTRODUCTION

Just as caches filter memory accesses so that main memory sees much less traffic, cache read traffic can be tackled within a processor by store-to-load and load-to-load forwarding techniques [8, 10]. Cache traffic reduction can lead to significantly decreased energy consumption in the cache.

In this work, we propose and study *macro data load*, an efficient mechanism to uncover additional opportunities for load-to-load forwarding by utilizing the spatial locality that

Permission to make digital or hard copies of all or part of this work for personal or classroom use is granted without fee provided that copies are not made or distributed for profit or commercial advantage and that copies bear this notice and the full citation on the first page. To copy otherwise, to republish, to post on servers or to redistribute to lists, requires prior specific permission and/or a fee.
ISLPED'06, October 4–6, 2006, Tegernsee, Germany.
Copyright 2006 ACM 1-59593-462-6/06/0010 ...$5.00.

exists in cache port wide *macro data*. When this mechanism is in effect, memory loads always bring a macro data from the processor cache. For example, a byte load would trigger a 64-bit macro data transfer in a 64-bit processor[1]. The processor then provides the necessary data portion to the load instruction, while saving the loaded macro data in a separate data storage. The saved macro data can be retrieved by a later load targeting the whole data or a smaller part of it. Most previous works focused on reusing the exact data described by its address and size [8, 10]. The proposed macro data load mechanism capitalizes on the largely underutilized data cache port bandwidth which is mainly due to abundant narrow data accesses in programs.

As an implementation of the proposed concept, we present a load store queue design and the necessary microarchitectural changes. The implementation cost and complexity are shown to be very modest. We perform a detailed performance study using a realistic superscalar processor model incorporating the microarchitectural changes. Our results show that the proposed approach eliminates over 30% of L1 cache traffic leading to a commensurate energy reduction, compared with a conventional processor design.

The rest of this paper is organized as follows. First, we give a limit study on how macro data loads generate more opportunities for loaded value reuse in Section 2. Then Section 3 describes a set of microarchitectural techniques to support macro data loads efficiently. Section 4 presents our quantitative evaluation using a realistic processor model. Related works are summarized and contrasted with our work in Section 5. Lastly, conclusions are drawn in Section 6.

2. VALUE REUSE IN MEMORY ACCESSES

2.1 Evaluation model

To analyze the degree of data reuse among memory instructions, we constructed a 64-bit machine model with *memory value reuse table* (MVRT), a conceptual hardware structure that tracks the address, the value, and the type of memory instructions. Allocation of a new entry and replacement of an old entry is done in a FIFO fashion. MVRT is parameterized and can have a varied number of entries.

The memory value reuse algorithm works as follows. Whenever a new memory instruction is executed, it is recorded in MVRT. If it is a store, all the MVRT entries with a previous

---

[1]This does not usually incur a change in cache designs. A high-performance cache provides a full port wide value on loads (*e.g.*, [2]) and the processor core selects the necessary portion using its internal data alignment logic.

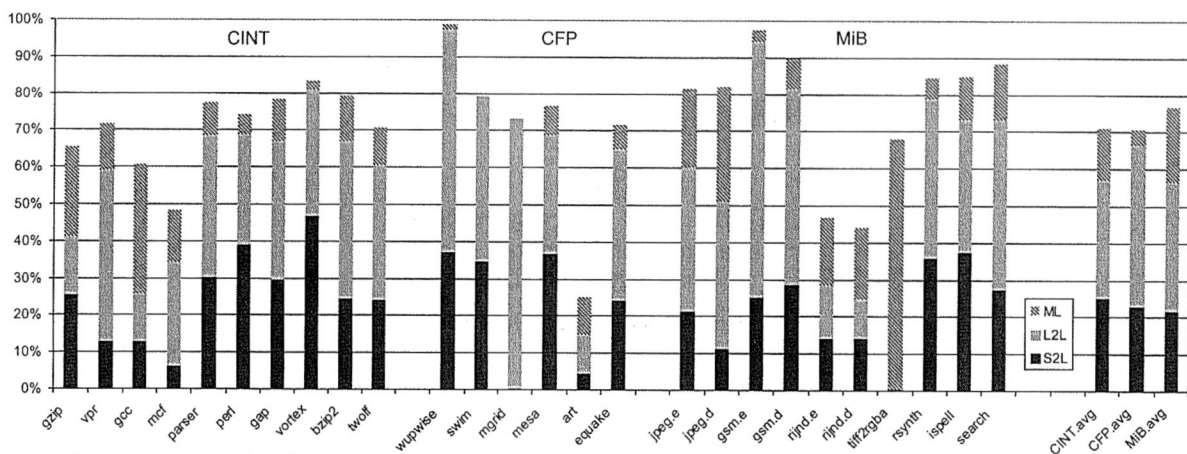

Figure 1: Percentage of loads reusing memory values. Two segments from bottom stand for loads finding their values from prior stores ("S2L" – store-to-load) and from prior loads("L2L" – load-to-load). Each top segment shows the extra opportunities offered by macro data loads ("ML").

memory instruction that overlaps in the address space with the store are invalidated. If the new instruction is a load, MVRT is searched to find a valid entry with a matching address, in which case, the load becomes redundant since the valid data can be provided from a previous memory instruction (either store or load).

For all experiments, we use a set of SPEC2k integer programs (dubbed "CINT" hereafter), SPEC2k floating-point programs ("CFP") [13], and MiBench programs ("MiB") [5]. After skipping the initialization phase, we collect analysis data from two billion instructions or until the end of execution if it comes first. Programs were compiled with gcc 2.7.2 targeting PISA [3] at the -O3 optimization level.

2.2 Results

Although we performed a comprehensive limit study and analysis per data size, per memory region, and using different ISA, we only report a small set of results here due to space limitations. More complete results can be found in [6].

2.2.1 Maximum memory value reuse

In this subsection, we use a 256-entry MVRT to study how many loads find their reuse value from (1) previous stores only, (2) previous stores and loads without macro data loads, and (3) previous stores and loads with macro data loads. Figure 1 shows that on average 70% or more loads find their values within MVRT. Roughly, 20–25% of loads get a reuse value from stores and 30–40% of loads from previous loads if macro data loads are not used. Macro data loads consistently boost the number of loads that reuse a previously loaded data value. 13.6% (CINT) and 20.1% (MiB) more loads reuse memory values. In CFP programs the conventional load-to-load forwarding performs well and allows nearly 44% of loads to find their data in MVRT. Macro data loads provide a small benefit of only 4.3% additional loads. Considering only load-to-load value reuse, macro data loads provide 42.3% (CINT), 9.8% (CFP), and 57.2% (MiB) more reuse opportunities, relative to a conventional load-to-load forwarding technique.

2.2.2 Sensitivity to MVRT size

We changed the MVRT size from 16 to 256 and repeated

our experiments. Several important observations are made from the results, although the graph is not shown here due to limited space.

First, a larger MVRT captures more value reuse opportunities. The number of covered loads increases almost linearly as we keep doubling the MVRT size, although the slope gradually dwindles after 64 entries in CINT and CFP.

Second, macro data loads expose significantly more opportunities for load-to-load forwarding in all the studied MVRT configurations, especially when MVRT is small. With a 32-entry MVRT, for example, there are 105% (CINT), 46% (CFP), and 188% (MiB) more loaded value reuse.

Third, as a result of our second observation, the area-effectiveness of MVRT, from the viewpoint of memory value reuse, is substantially improved. In the case of CINT and MiB, the total achievable degree of reuse with a 32-entry MVRT with macro data loads is comparable to that with a 256-entry MVRT without macro data loads.

3. MICROARCHITECTURE FOR LOADED VALUE REUSE

3.1 A modified LSQ design

Many recent high-performance processors implement LSQ to allow out-of-order execution of memory instructions, as well as to bypass correct store data to dependent loads [1, 4, 7, 15]. Adding the load-to-load forwarding feature in LSQ then becomes a natural extension since it already provides the necessary storage to save a loaded data and the functionality to identify it [8]. We present an example LSQ design that supports both store-to-load and load-to-load forwarding with the proposed macro data load mechanism. Figure 2 shows the LSQ, data cache, and how data from them are circulated via various buses.

A typical LSQ design is composed of two memory structures: *address-matching tag* and *data storage*. The data storage, not used for a load entry in a conventional design, can be used to hold a loaded macro data. We provide the necessary datapath to guide the data as observed at the cache port to the desired LSQ entry (Figure 2(d)).

The address tag portion is often implemented with an

148

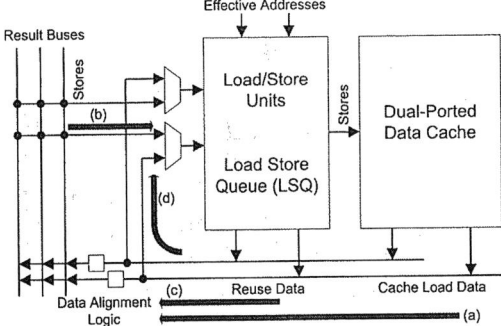

Figure 2: The LSQ structure with the macro data reuse mechanism implemented. (a) Normal load data path. (b) Store data path. (c) Reuse data path for both store-to-load and load-to-load forwarding. (d) LSQ update path. The depicted LSQ design is based on the published AMD K7 design [4].

associative memory logic such as *content addressable memory* (CAM), which helps resolve memory ordering conflicts quickly and locate a previous store for forwarding. To detect a load-to-load forwarding instance, however, LSQ should perform a *partial-match searching* since the source macro data can be larger than and inclusive of the dependent load, potentially leading to different address bits in several LSB positions. The tag has a few related bits in addition to the address, including V (valid), P (data present), and SL (store or load) bits. In a conventional design, an LSQ entry is invalidated once the corresponding instruction retires. We keep the V bit on as long as the data in the entry is up-to-date.

3.2 Pipeline design issues

To reduce L1 cache traffic, cache access should be suppressed whenever the target data can be found in LSQ. This requires that cache access be delayed and serialized with LSQ lookup. This arrangement however is likely to incur increased cache access latency for loads that do not reuse values and can potentially lead to decreased performance thereof. The performance degradation will highly depend on how many loads find their values in LSQ in this case. It is noted that the serialization delay may be avoided if we can (1) predict accurately whether or not a load will find its value in LSQ; and (2) selectively bypass the waiting time for loads that are likely to find their values in cache. This idea has been explored in similar, yet different contexts [11].

4. QUANTITATIVE EVALUATION

4.1 Experimental setup

We perform experiments using a detailed execution-driven simulator derived from sim-outorder in the SimpleScalar tool set [3]. We model a modest 4-issue processor, which has 32kB 2-way set associative L1 I/D cache, with 64B block size and 2-cycle latency. L2 cache is 2MB 4-way associative, with 128B block size and 10-cycle latency. Main memory has 120-cycle latency. The numbers of RUU entries is 128 while the LSQ size is set to 64, similar to recent high performance processors [4, 7, 15]. The data reuse latency in LSQ is set to one cycle. Section 2.1 describes the setup for our

benchmarks.

We consider three configurations to reduce cache traffic: S2L, L2L and ML. They save load traffic on store-to-load forwarding, additionally on load-to-load forwarding (similar to [8]), and additionally on macro data load hits. The baseline processor, to which we compare these three configurations, accesses LSQ (for store-to-load forwarding) and cache at the same time to minimize latency and thus will have the highest number of cache accesses.

4.2 Evaluation results

4.2.1 L1 cache traffic

Figure 3 confirms the observations made in Section 2. With only store-to-load forwarding (*i.e.*, S2L), the cache traffic reduction is limited: 10% (CFP) or less (CINT and MiB). Only two programs among all the studied programs, namely wupwise and rsynth, show a traffic reduction of 15% or more. With ML, however, there is a significant reduction in cache accesses, 35% (CINT), 33% (CFP), and 38% (MiB). Four programs, namely bzip2, mgrid, jpeg.e, and gsm.d, had over 50% of cache traffic reduction. With L2L, the cache traffic reduction was 27% (CINT), 30% (CFP), and 23% (MiB), considerably lower than that of ML.

Although the results are in accordance with our limit study results presented in Figure 1, the actual traffic reduction is less than the maximum potential due to three factors: (1) speculative memory references occupy available LSQ entries and cause frequent pipeline flushing, not allowing memory value reuse between distant references; (2) speculative loads execute and generate cache traffic; and (3) memory reordering results in later loads accessing cache prior to or simultaneously with earlier loads, losing the reuse opportunities.

4.2.2 Energy consumption

We used the CACTI 100nm model [12] to consider the energy consumption related with both LSQ and cache. Since a memory reuse scheme can increase LSQ activities (*e.g.*, loads update the data array in LSQ), it is important to consider not only cache but also LSQ as we evaluate the energy consumption impact of different reuse schemes.

Experimental results show that S2L achieves a limited energy reduction of less than 10%. L2L performs better and results in additional savings totalling up to around 20%. ML performs consistently better than S2L and L2L, with a notable energy reduction of over 35% for MiB.

4.2.3 Performance impact

When the latency of memory value reuse is shorter than the cache access latency, increase in the number of loads finding their values from LSQ can lead to improved average memory latency. Our simulation configuration captures this case by setting the reuse latency to be one cycle and the cache access latency to be two cycles.

Simulation results show that the ML configuration is performance competitive with the baseline machine, in which, LSQ and cache are accessed simultaneously. For CFP and MiB, ML even performs slightly better. This is due to (1) many loads find their values from LSQ; and (2) the increased latency seen by the remaining loads is well tolerated by the out-of-order processor model.

Among the three traffic-optimized configurations, only the

149

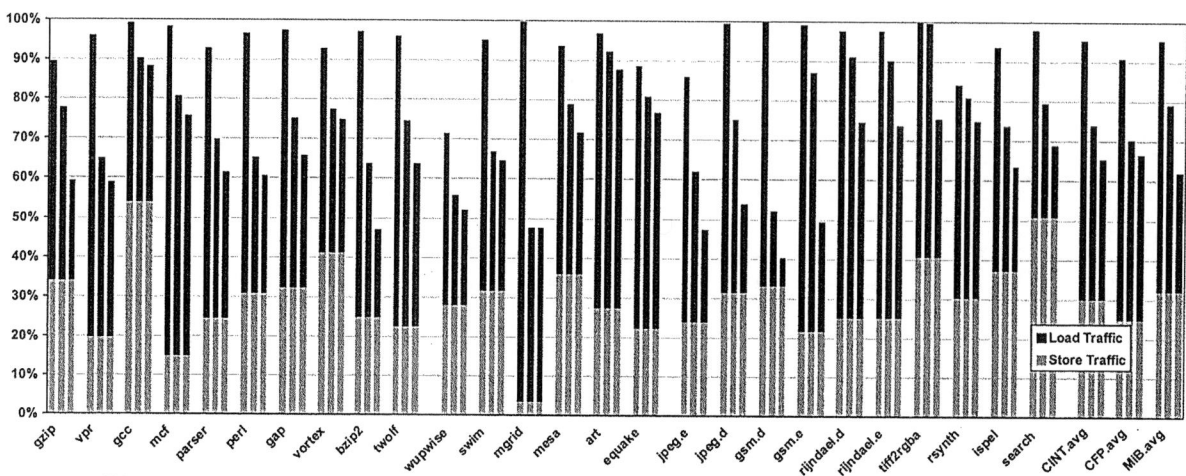

Figure 3: Cache traffic of S2L, L2L, and ML (from left) relative to the baseline.

ML configuration achieves a competitive performance level with the baseline design. It is noted that increase in execution time can have a detrimental effect on overall energy consumption.

5. RELATED WORK

Önder and Gupta proposed *value address association structure* (VAAS) to eliminate redundant loads and silent stores [10]. Nicolaescu *et al.* [8] proposed *cached load store queue* (CLSQ) to detect redundant loads and provide reuse data. In their design, each data entry in LSQ is allowed to cache a loaded value as well as to keep store data. Both VAAS and LSQ manage memory accesses with a FIFO policy and therefore our limit study with MVRT accurately models and predicts their performance. Our study showed that macro data loads give significantly boosted loaded value reuse compared with these techniques, given the same storage space to keep data.

More recently, Nicolaescu *et al.* [9] proposed *wide cached load store queue* (WCLSQ) to take advantage of spatial locality by having each LSQ entry keep multiple words or by increasing the LSQ data width to accomodate a large 16-byte or even 32-byte memory block. To fill WCLSQ, however, the cache should be accessed multiple times or the cache port should be widened to match the WCLSQ width. This approach potentially maximizes the loaded data reuse opportunities, but its effectiveness is offset by increased LSQ size (and thus energy consumption per access), increased initial overhead to fill WCLSQ, and much decreased area efficiency due to short stores occupying large data entries. Compared to WCLSQ, our proposal exploits only the freely available cache port bandwidth and requires no change to the cache design, promoting reuse of existing designs. It is also compatible with popular *cache subbanking* techniques [14].

6. CONCLUSIONS

This paper introduced and studied cache port wide macro data loads to enhance loaded value reuse in high-performance processors. As future microprocessors will be critically constrained by power consumption, performance-competitive, power-efficient design techniques will become even more important. Our work shows that the proposed macro data

load scheme is practical and at the same time effective in reducing L1 cache traffic and energy.

7. REFERENCES

[1] D. Boggs, A. Baktha, J. Hawkins, D. T. Marr, J. A. Miller, P. Roussel, R. Singhal, B. Toll, and K. S. Venkatraman. "The Microarchitecture of the Intel Pentium 4 Processor on 90nm Technology," *Intel Technology Journal*, 8(1), Feb. 2004.

[2] D. Bradley, P. Mahoney, and B. Stackhouse. "The 16kB single-cycle read access cache on a next-generation 64b Itanium microprocessor," *Proc. Int'l Solid State Circuits Conf.*, pp. 110 – 111. Feb. 2002.

[3] D. Burger and T. M. Austin. "The SimpleScalar Tool Set, Version 2.0," *Computer Sciences Dept. TR*, No. 1342, Univ. of Wisconsin, June 1997.

[4] K. Diefendorff. "K7 Challenges Intel," *Microprocessor Report*, Vol. 12, No. 14, pp. 1 – 7, Oct. 1998.

[5] M. R. Guthaus, J. S. Ringenberg, D. Ernst, T. M. Austin, T. Mudge, and R. B. Brown. "MiBench: A Free, Commercially Representative Embedded Benchmark Suite," *Proc. Annual Workshop Workload Characterization*, Dec. 2001.

[6] L. Jin and S. Cho. "A Characterization Study on Memory Value Reuse," *Proc. Workshop Memory Performance Issues, during Int'l Symp. High-Performance Computer Architecture*, Feb. 2006.

[7] R. E. Kessler. "The Alpha 21264 Microprocessor," *IEEE Micro*, 19(2):24 – 36, March/April 1999.

[8] D. Nicolaescu, A. Veidenbaum, and A. Nicolau. "Reducing Data Cache Energy Consumption via Cached Load/Store Queue," *Proc. Int'l Symp. Low-Power Electronics and Design*, pp. 252, Aug. 2003.

[9] D. Nicolaescu, A. Veidenbaum, and A. Nicolau. "Caching Values in the Load Store Queue," *Proc. Int'l Symp. Modeling, Analysis, and Simulation of Computer and Telecomm. Systems*, pp. 580 – 587, Oct. 2004.

[10] S. Önder and R. Gupta. "Load and Store Reuse Using Register File Contents," *Proc. Int'l Conf. Supercomputing*, pp. 289 – 302, June 2001.

[11] T. Sha, M. Martin, and A. Roth. "Scalable Store-Load Forwarding via Store Queue Index Prediction," *Proc. Int'l Symp. Microarchitecture*, Nov. 2005.

[12] P. Shivakumar and N. P. Jouppi. "CACTI 3.0: An Integrated Cache Timing, Power, and Area Model," *HP WRL Research Report 2001/2*, Aug. 2001.

[13] Standard Performance Evaluation Corporation. http://www.specbench.org.

[14] C.-L. Su and A. M. Despain. "Cache Designs for Energy Efficiency," *Proc. Hawaii Int'l Conf. System Sciences*, pp. 306 – 315, Jan. 1995.

[15] J. M. Tendler, J. S. Dodson, J. S. Fields, H. Le, and B. Sinharoy. "POWER4 System Microarchitecture," *IBM J. Research & Development*, 46(1), Jan. 2002.

Modelling Macromodules
for High-Level Dynamic Power Estimation
of FPGA-based Digital Designs*

Axel Reimer	Arne Schulz	Wolfgang Nebel
University of Oldenburg	University of Oldenburg	University of Oldenburg and
Germany	Germany	OFFIS Research Institute
		Germany
axel.reimer@uni-oldenburg.de	arne.schulz@uni-oldenburg.de	nebel@o.s.de

ABSTRACT

We present our approach for a new macromodule power model library which can be used in high-level dynamic power estimation for FPGA technologies. The approach adapts a previously published high-level estimation flow for ASIC technologies. Due to the different architectures (ASIC vs. FPGA) the presented approach builds on an iterative optimization step during the model generation phase.

Categories and Subject Descriptors:
B.8.2: Performance Analysis and Design Aids.

General Terms:
Measurement, Reliability.

Keywords:
High-level power estimation, FPGA power estimation, RT-level power modeling.

1. INTRODUCTION

State-of-the-art FPGAs have the capacity to implement millions of gates [3] and their application has migrated from being primarily a prototyping platform to their use in low to medium volume production designs. As in ASIC designs the biggest impact on the overall power consumption reduction can be achieved at high-level steps of the design flow. Here a fast and wide design space exploration is possible and an optimal design can be implemented right the first time. This leads to the need of high-level power estimation.

To our knowledge only few approaches are available for high-level power estimation of FPGAs, mostly concerning the power-efficient implementation of designs with efficient ressource allocation, typically the efficient use of SRAM-based look-up-tables (LUT). The logic blocks in modern

---
*This work was partly funded by the German DFG.

Permission to make digital or hard copies of all or part of this work for personal or classroom use is granted without fee provided that copies are not made or distributed for profit or commercial advantage and that copies bear this notice and the full citation on the first page. To copy otherwise, to republish, to post on servers or to redistribute to lists, requires prior specific permission and/or a fee.
ISLPED'06, October 4–6, 2006, Tegernsee, Germany.
Copyright 2006 ACM 1-59593-462-6/06/0010 ...$5.00.

FPGAs are built up of LUTs and registers. In the design flow, the high-level designs are synthesized and mapped on specific devices, transforming a circuit description from a functional form into a network of LUTs. LUT-based technology mapping has been studied extensively in recent years, under the focus of area and / or the logic depth as well as the power consumption [6].

The known high-level estimation approaches use on the one hand a neural network as presented by the Fraunhofer Institute [4] and an estimation based on a HCDFG (hierarchical control data flow graph) [7] on the other. The first approach aims at fast estimations of digital filter structures but has the disadvantage of the need of being trained for the specific application field. The second approach mainly focuses on a quick performance estimation with the option of synthesis of the given high-level description [7]. In our approach the FPGA-design consists of characterized register-transfer-level components, called macromodules, which can be estimated with a high-level estimation tool.

The paper is organized as follows. First we describe the background of the modelling process in the context of high level power estimation. In the consecutive section we describe our modelling approach which is generally divided into two parts. On the one hand we figure out which parameters have to be modelled and on the other we present our characterization flow. The fourth section depicts the evaluation results. Finally, the paper ends with a conclusion.

2. HIGH-LEVEL POWER ESTIMATION FOR FPGA-TECHNOLOGY

Before looking at the power estimation we have to understand, where the power dissipation in FPGAs takes place and what has to be done to perform a fast high-level power estimation. Then we motivate the proposed approach and address the interconnect power estimation.

Several studies of FPGA power consumption have appeared in the literature [6][7][8][9]. These contributions have shown that power dissipation in FPGA devices is predominantly in the programmable interconnection network, which is used to build up the functionality with the LUTs. This fact differs substantially from custom ASIC technologies. The difference in the sources of power dissipation between these two technologies is based in the composition of their interconnect structures: FPGA interconnect consists of prefabricated wire segments of various lengths, with used and unused routing switches attached to each wire segment [6].

High-level power estimation itself consists typically of different steps. First the high-level description of the algorithm has to be analyzed and in a second step a quick high-level synthesis from the specification layer down to RT layer has to be performed. In this step the scheduling, allocation and binding has to be set in an optimal way. During the binding the functionality has to be mapped on arithmetic functions, which have been allocated from a technology library. Additionally during simulation – often the execution of the compiled source code – the activity has to be tracked. Applying this internal switching activity enables a quick power estimation for the assumed circuit.

Modelling the interconnect is, due to the fact that this contributes the largest fraction of the dynamic power dissipation, an important aspect in the modelling process. In this paper we divide the interconnect in a so called "local interconnect", which provides the connections inside a macromodule, and the "global interconnect", which respects the connections between the macromodules, the (on-chip) memories and the pins. The estimation of the power dissipation of interconnect is described in [11].

The high-level power estimation process itself can be divided into four parts. The first part is the estimation of an initial architecture of a given high-level design description. Within our flow this is done by the commercial tool ORINOCO which provides an interface to C/C++ and SystemC and builds up an internal graph. These informations are then used to perform a quick architecture prediction step consisting of allocation, binding and scheduling. The second part especially needed for dynamic power estimation is the collection of activity data of the design. This is done also automatically by ORINOCO using a concept of instrumented code which stores the processed data in a file during execution of the design description. The third necessary part is a technology library which provides power information (and additionally information like timing and area) for specific macromodules. The information is used by the estimation tool to map the internal representation of the design on the available functional components, memories etc. This macromodule power library currently only exists for ASIC technologies. The last part is the integration of the above mentioned parts into a power estimator.

The next section describes in detail, our approach to cope with the special needs and constraints for FPGA-technology to build a macromodel power library.

3. MODELLING APPROACH

In this section we describe the modelling process for the FPGA power estimation. We separate this into two steps. In section 3.1 the identification of suitable modelling variables is described. In section 3.2 we present the characterization process of the selected components.

3.1 RT Models for High-Level Power Estimation

The first step is the definition of modelling variables suitable for high-level power estimation. The goal is to obtain an adequate abstraction of the FPGA-Design.

We identified five model variables for our high-level power estimation approach: the FPGA-device, basic-component, Hamming distance (hd), signal distance (sd) and bitwidth.

According to our experiments these variables provide an adequate compromise between modelling and characterization on the one hand and accuracy on the other.

No one will argue that the concrete FPGA-device family has a big influence on power consumption. Because we want to use the Xilinx Virtex XCV800 for our evaluation this model variable is fixed.

Further we need the components which are our primitives for the power estimation process. The Xilinx ISE provides a tool named Coregen to create register-transfer-level components which are mapped on the LUT-structure. These components can be instantiated directly in the Sourcecode (e.g. Verilog). This makes it possible to use these components for characterizing the power. The Coregen components represent the second model variable. This has the advantage that every macromodule consists of LUTs and routing resources. By using macromodules, the connections and LUTs in one macromodule, the local interconnect, are modelled in a single modelvariable.

Since we want focus on dynamic power consumption which depends on the input data stream, it is important to have another modelvariable regarding the transitions of bitvectors. In our approach we chose the variable set mentioned in [2] to get an abstraction from the transitions. In [2] the following variables are used:

1. Hamming distance (hd), the 1-normalized percentage of bits changing during a transition.

2. Signal distance (sd), the 1-normalized percentage of bits remaining the value "1" during a transition.

3. 0 distance ($0d$), the 1-normalized percentage of bits remaining the value "0" during a transition.

Because the data is propagated from the inputs to the outputs of combinatoric designs we can consider the distances at the inputs only without restricting the generality.. Only two of these three distances are linearly independent. They meet the following equation:

$$hd + sd + 0d = 1 \qquad (1)$$

Finally, we chose the components bitwidth as a model variable to make our model scalable.

The modelvariables do not consider the power of the connections between components which we call the "global interconnect". Since these are connections without complex logic and the data-dependent components are normally placed nearby, we currently neglect them and use the available ASIC-approach at this point, without limiting the generality. Our experiments validate this assumption.

In the next section the characterization process is described.

3.2 Characterization for FPGA-Technology

So far, we have presented the modelling variables for our power library. The next challenge is to determine the power values for our components. If we implement a 16-bit multiplication on the FPGA-device and estimate the power with XPower we get the power result for the whole FPGA including pads and static power consumption. In order to get only the relevant dynamic power consumption, two improvements are necessary: subtraction of the static power and the subtraction of the power used for pads and their connections.

The first improvement can be done easily. Since the static power is spearately reported by XPower and is nearly constant, it simply can be subtracted from the total power results XPower provides. The second refinement is more sophisticated. Although XPower provides a detailed breakdown of the dynamic power consumption, an unambiguous separation of the macromodule and the pads including their connections is not directly possible. The power of pads and their connections depends on the input data activity and output data activity. So every pad uses a different amount of power. We will describe the handling of the pads after the characterization process of the components.

The process of characterizing the components is completely automated. First, Coregen is used to create a template of the component. This component is instantiated in an FPGA-Design which only contains the component itself and its pad connections. This design is the basis for the characterization process. In order to obtain the most exact power results, the design passes through the whole flow including the mapping and place & route phase. After this, Netgen is used to create Verilog sourcecode and a file in standard-delay format (sdf) for the most accurate simulation. Additionally, a design-fitting testbench is created. Since we abstracted transitions to their Hamming- and signal distance we use these variables to classify the different input-data streams. We generate every possible combination of input-streams for characterization with the sets $hd = \{0.0, 0.25, 0.5, 0.75, 1.0\}$ and $sd = \{0.0, 0.25, 0.5, 0.75, 1.0\}$.

The combinations are limited by equation (1). With every combination a simulation with MentorGraphics ModelSim is done. ModelSim creates a vcd-file which is used by XPower to estimate the power consumption. This characterization process is done for every component for the bitwidth set $bw = \{8, 16, 24, 32\}$. The power values are saved in a tabular format. The file contains the power of the component itself including the pad connections. Before we describe how the pad power is subtracted, we explain the remaining flow.

Since only a grid of Hamming and signal distance is calculated a regression has to be done. The regression is done using MacAnova [12] while building the power library from our characterization files. In a technology file it is possible to set the capacitance value for the pads. We also define which component can be used for which operation (e.g. the component "adder" is used for the operation "+").

The pad connection have not been considered so far. This implies that estimating designs on high-level with this characterization data will lead to an overestimation of power consumption because every component is supposed to be connected to the pads. The solution to this challenge is to detach the pad power from our characterization data. This implies that the high-level estimation needs to know these pad capacitances in order to estimate designs correctly.

In order to estimate pad-values, we use ORINOCO as a "pad estimator". We create a small C-procedure only containing the operation mapped to the characterized component (e.g. "+" for adder). For the first estimation a pad capacitance value is needed. Because we do not know the value by now, we guess an initial positive pad capacitance value C_pad (Figure 1).

Now the pad power dissipation for all characterized bitwidths and input-streams is calculated by ORINOCO. This estimation is based on C_{pad}, the number of pad connections n_{pad} and their activity α_{pad}. The pad power values are writ-

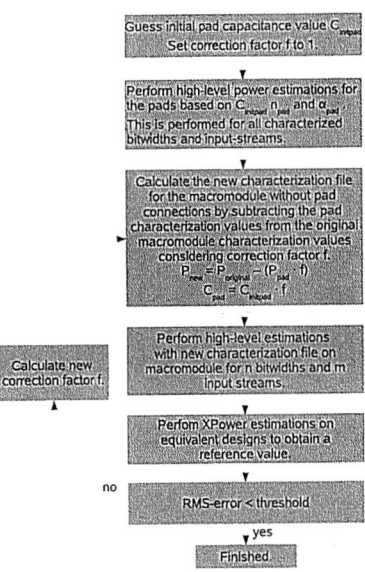

Figure 1: Iterative estimation of pad capacitances

ten in a text-file and can then be automatically subtracted from the component characterization text file.

Since we have chosen a possible pin-value we evaluate the components by doing many automated estimations of the components and compare them to the XPower results. In order to fit the results we multiply the pad-estimation by a correction factor f to find the best value. This fitting-step has to be iterated a few times until the pad capacities provide good results. At the moment it is only possible to do this fitting-step empirically.

At this point the modelling process is done. We are capable of creating an FPGA power library for our components which can be used to do high-level power estimations of FPGA designs. The advantage of our modelling process is that every component has to be characterized only once for the target FPGA-device.

We take into account the overhead for the creation of the power library because it has to be done only once and it can then be used for estimating the power of many designs. The experimental validation of our model is described in the next section.

4. EXPERIMENTAL RESULTS

In this section we provide the results of our high-level estimation approach. We evaluated it for components and small designs without registers. Since registers have to be modelled in a different way we concentrated on the modelling of combinatorical components.

As mentioned before, the input-data has a big influence on the dynamic power consumption. The characterization was done with a stepwidth of 0.25 of Hamming and signal distance. For the evaluation we try to use streams, which mostly do not fit this grid.

We created nine random input-data streams with defined signal-distance probabilities of $0.1 - 0.9$ for every bitwidth. This leads to bitstreams, which have a high variation of different Hamming and signal distances and show whether the regression is valid for FPGA-components. The comparison value is generated by XPower. In Figure 2 the eval-

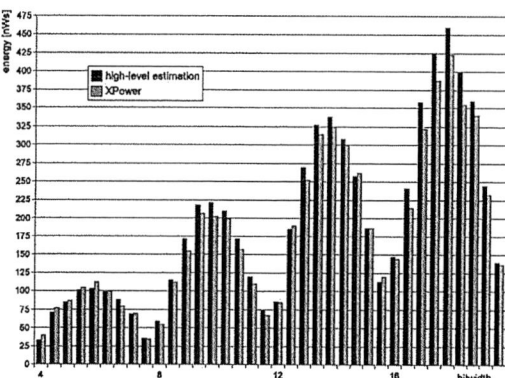

Figure 2: Adder (RMS-error = 6.68 %)

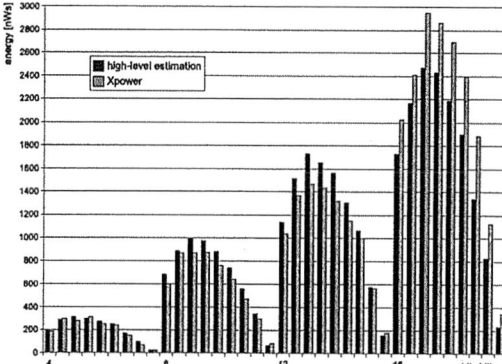

Figure 3: Multiplier (RMS-error = 15.79 %)

uation results of the adder component for the bitwidths 4, 8, 12 and 16 are shown. The results look the same for the whole bitwidth range of evaluations. In Figure 2 all the nine streams are listed. From left to right the signal distance probability decreases from 0.9 to 0.1. This leads to the conclusion, that the results in the middle of every bitwidth have the highest transition occurence.

The RMS-error of the results for the adder component is 6,68 %. Since the estimation works for the adder, we evaluate a larger component: a multiplier. The results are presented in the same way in Figure 3. It can be seen that the estimation is not as good as for the adder. The RMS-error is 15.79 % which can be explained by the more complex logic of the multiplier. We further evaluated our estimation approach on a small design consisting of some additions, subtractions and multiplications. We built a small design containing the well-known three binomic equations. At the same time, we constructed a C-algorithm which calculates the same equations and estimated the power with our high-level estimation approach. The estimation results are presented in Figure 4. In this figure we decided to use a different visualization method to present all tested input-streams and bitwidths. This diagram is based on the equations:

$$\text{XPower\%} = \frac{\text{XPower}}{\text{Xpower} + \text{estimation}} \quad (2)$$

$$\text{estimation\%} = \frac{\text{estimation}}{\text{Xpower} + \text{estimation}} \quad (3)$$

This means that a horizontal line at 50 % would be a perfect

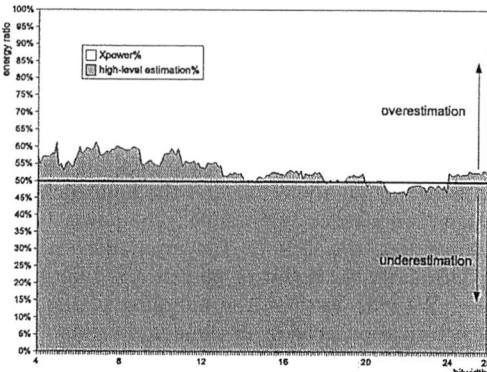

Figure 4: Binomic equations (RMS-error = 12.61 %)

result because then the estimation exactly fits the XPower result.

With an RMS-error of 12.61 % the results are nearly the same as for single components and the estimation approach seems to be applicable on FPGAs.

5. CONCLUSION

We presented an approach for the creation of macromodules for power estimation of FPGA designs. The experimental results show the capabilities of this approach. With an RMS-error of 6.68 % , 15.79 % and 12.61 % repectively the high-level estimated power values nearly fit the XPower results after placement & routing.

Since combinatorical ciruits were evaluated the next step will be the modelling of registers and memories. Additionally we aim at an automation of the empirical characterization of the pad capacitances and at the evaluation of our approach for other FPGA vendors and device families.

6. REFERENCES
[1] G. De Micheli, *Synthesis and Optimization of Digital Circuits,* McGraw-Hill, 1994.
[2] G. Jochens, L. Kruse, E. Schmidt, and W. Nebel, "A New Parameterizable Power Macro-Model for Datapath Components", in *Proc. of DATE*, 1999.
[3] Xilinx, *Virtex-4 Family Overview* http://direct.xilinx.com/bvdocs/publications/ds112.pdf
[4] A. Monostori, H. H. Frühauf, G. Kokai, *Quick Estimation of Resources of FPGAs and ASICs Using Neural Networks* in Proceedings of IWA05, 2005.
[5] S. Bilavarn, G. Gogniat, J. L. Philippe, L. Bossuet, *Fast prototyping of reconfigurable architectures from a C program* in Proceedings of ISCAS, vol. 5, 2003.
[6] J. H. Anderson, F. N. Najm, *Power-Aware Technology Mapping for LUT-Based FPGAs* in Proceedings of ICFPT02, 2002.
[7] S. Bilavarn, G. Gogniat, J. L. Philippe, *FPGA Area Time Power Estimation for DSP Applications* in Proceedings of ICSPAT00, 2000.
[8] L. Shang, A. Kaviani, K. Bathala, *Dynamic power consumption in the VIRTEX II FPGA family* in Proceedings of ACM ISFPGA, 2002.
[9] K. W. Poon, A. Yan, S. J. E. Wilton, *A Detailed Power Model For Field-Programmable Gate Arrays* in ACM TODAES, 2005
[10] V. George, J. M. Rabaey *Low-Energy FPGAs - Architecture and Design* Kluwer Academic Publishers, ISBN 0-7923-7428-2, 2001
[11] J. H. Anderson, F. N. Najm *Interconnect Capacitance Estimation for FPGAs* in Proceedings of ASP-DAC04, 2004.
[12] C. Bingham, G. Oehlert *An Introduction to MacAnova* University of Minnesota, School of Statistics, Technical Report Number 60, 1995

Highlights of Industrial Low Power Tools and Flows

Chair: Qing Wu
SUNY Binghamton

I1.1 **Cadence Design Systems: Low Power Everywhere**
Michael O'Sullivan, Cadence Design Systems

I1.2 **ChipVision: Low Power Design and Tools at ChipVision**
Laila Kabous, ChipVision

I1.3 **BullDAST: PowerChecker: A Framework for RTL-to-Layout Dynamic and Leakage Power Estimation and Optimization**
Alberto Bonanno, BullDAST, s.r.l.

I1.4 **OFFIS: System Level Power Analysis, Optimization, and Tools from OFFIS**
Milan Schulte, OFFIS

I1.5 **Infineon: Controlling Leakage Power in nanoCMOS SOC's**
Domenik Helms, OFFIS (on behalf of Infineon)

Thermal Via Allocation for 3D ICs Considering Temporally and Spatially Variant Thermal Power *

Hao Yu, Yiyu Shi, Lei He
Electrical Engineering Dept.
University of California
Los Angeles, CA 90095
{hy255,yshi,lhe}@ee.ucla.edu

Tanay Karnik
Circuit Research
Intel Labs
Hillsboro, OR 97124
tanay.karnik@intel.com

ABSTRACT

All existing methods for thermal-via allocation are based on a steady-state thermal analysis and may lead to excessive number of thermal vias. This paper develops an accurate and efficient thermal-via allocation considering temporally and spatially variant thermal-power. The transient temperature is calculated using macromodel by a structured and parameterized model reduction, which generates temperature sensitivity with respect to thermal-via density. By defining a thermal-violation integral based on the transient temperature, a nonlinear optimization problem is formulated to allocate thermal-vias and minimize thermal violation integral. This optimization problem is transformed into a sequence of subproblems by Lagrangian relaxation, and each subproblem is solved by quadratic programming using sensitives from the macromodel. Experiments show that compared to the existing method using steady-state thermal analysis, our method is 126X faster to obtain the temperature profile, and reduces the number of thermal vias by 2.04X under the same temperature bound.

Categories and Subject Descriptors: B.7.2[Hardware]: Integrated circuits – Design aids

General Terms: Algorithms, Design

Keywords: Thermal Management and Simulation, Model Order Reduction, SQP Optimization

1. INTRODUCTION

3D integration [1, 2] to stack multiple active layered ICs is effective to improve the deep-submicron interconnect performance and increase the transistor packing density. However, due to the increased power density, the heat dissipation is extremely important in 3D-ICs [1]. It is well known that excessively high temperature can significantly degrade interconnect/device reliability and performance [3–5]. One effective heat-removal approach is to use thermal vias to improve the thermal conductivity. Fig. 1 shows the topology of typical 3D-IC designs including the active device layers, thermal-vias, and the substrate.

Because of different workloads and dynamic power management techniques such as clock gating technique extensively used in the modern VLSI design, power has both temporal and spatial variations. A transient thermal-power is the running average of the cycle-accurate power over the scale of the thermal constant [6]. A cycle-accurate micro-architecture level thermal simulation

---

* This paper is partially supported by NSF CAREER award CCR-0093273/0401682 and Intel. Address comments to lhe@ee.ucla.edu.

Permission to make digital or hard copies of all or part of this work for personal or classroom use is granted without fee provided that copies are not made or distributed for profit or commercial advantage and that copies bear this notice and the full citation on the first page. To copy otherwise, to republish, to post on servers or to redistribute to lists, requires prior specific permission and/or a fee.
ISLPED'06, October 4–6, 2006, Tegernsee, Germany.
Copyright 2006 ACM 1-59593-462-6/06/0010 ...$5.00.

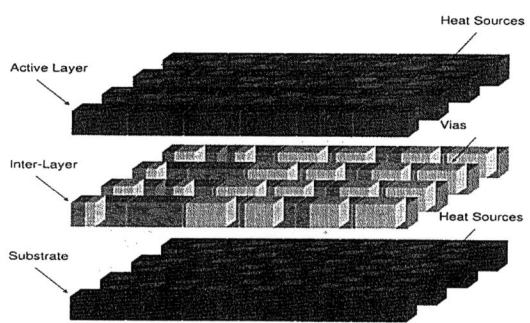

Figure 1: 3D-IC topology including: active device layers, inter-layer dielectrics, vias, and the substrate.

Hotspot [7] has been developed based on a thermal *RC* model to calculate the transient temperature. Assuming steady-state thermal analysis (based on thermal resistance model), thermal-via allocation has been studied during the placement [8] and routing [9]. Because the steady-state analysis ignores the temporal and spatial variations of the transient thermal-power, to obtain a solution without thermal violation, the methods in [8, 9] have to assume a maximum thermal-power *simultaneously* for all regions. Because it is rare for different regions to simultaneously reach their maximum thermal-power, the methods in [8, 9] may lead to excessive number of thermal vias. In addition, [7–9] directly solve the matrix-formed state equation. It can not efficiently calculate the nominal temperature and its sensitivity with respect to the thermal-via density for large sized circuits. The design procedure is either based on iterations [8], or based on an approximated square-root relation [9] between temperature and thermal-vias. It may not converge or may lead to inaccurate results. Therefore, accurate and efficient solutions to calculate temperature and temperature sensitivity should be developed.

In this paper, an accurate yet efficient thermal-via allocation is proposed that considers the temporal and spatial variations of the thermal-power. The transient temperature is calculated using macromodel by a *structured and parameterized model reduction*, which also generates the temperature sensitivity with respect to the thermal-via density. By defining a *thermal-violation integral* based on the transient temperature, a nonlinear optimization problem is formulated to allocate thermal-vias and minimize thermal violation integral. This optimization problem is transformed into a sequence of subproblems using Lagrangian relaxation, and each subproblem is solved by quadratic programming with the sensitives provided by the macromodel. Experiments show that compared to the steady-state thermal analysis, our method is 126X faster to obtain the temperature profile, and reduces the number of thermal vias by 2.04X under the same temperature bound.

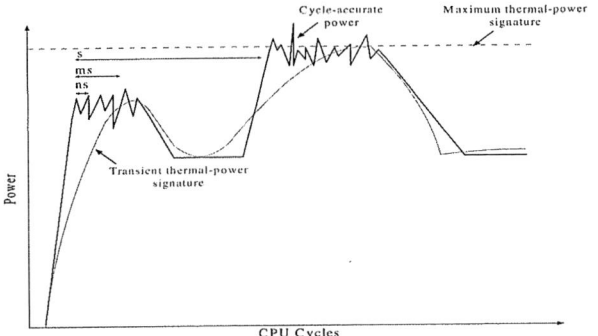

Figure 2: The definitions of cycle-accurate power, transient thermal-power signature, and maximum thermal-power signature at the different scale of time constant.

The rest of the paper is organized below. In Section 2, we first present the preliminary for 3D thermal model and analysis. In Section 3, we discuss a structured and parameterized reduction to generate the macromodel. In Section 4, we formulate a nonlinear optimization to accurately allocate the thermal-via driven by the thermal-violation integral. In Section 5, we present experimental results and conclude the paper in Section 6.

2. PRELIMINARY

2.1 Thermal Model

There is a well-known duality between electrical and thermal systems (See Table 1). As temperature is analogous to voltage, the heat flow can be modeled by a current passing though a pair of thermal resistance and capacitance driven by the current source, modeling the power dissipation.

The 3D layout can be uniformly discretized into N tiles by the finite difference method. Our design variable here is the thermal-via density. The larger the thermal-via density in one tile, the more heat that can be convected away through layers to the heat sink. In this paper, K critical tiles are assumed to be specified by users. An ith tile has a thermal-via area A_i. Because A_i is related to the thermal-via density ρ_i by $\rho_i = A_i/a$, A_i is used to represent the thermal-via density at ith tile in the sequel. Note that a is the unit area of thermal-via determined by the process.

The equivalent thermal circuit by nodal analysis (NA) in the frequency(s) domain is

$$[G_0 + sC_0 + \sum_{i=1}^{K} A_i(g_i + sc_i)]x(\mathbf{A}, s) = Bu(s)$$

$$y(\mathbf{A}, s) = L^T x(\mathbf{A}, s), \qquad (1)$$

where $\mathbf{A} = [A_1, ..., A_K]$ is a parameter-vector of thermal-via density. Note that G_0 and C_0 ($\in R^{N \times N}$) are conductive and capacitive matrices of discretized thermal networks, and $\sum_{i=1}^{K} A_i g_i$ and $\sum_{i=1}^{K} A_i c_i$ are conductive and capacitive matrices of thermal vias, respectively. In addition, $x(\mathbf{A}, s)$ ($\in R^N$) is the state variable of node temperatures, B ($\in R^{N \times p}$) is the adjacent matrix to select input u, and L ($\in R^{N \times p}$) is the adjacent matrix to select output y. The notations are summarized in Table 2.

The thermal-via is inserted as follows. An insertion (incident) matrix X ($\in R^{N \times N}$) is used to record the location and the number of added vias. If a via is added between two nodes m and n at two between two vertical-adjacent layers, its insertion matrix is

$$X(k, l) = X(l, k) = \begin{cases} -1 & \text{if } k = m,\, l = n \\ \sum_l |X(k, l)| & \text{if } k = l \\ 0 & \text{else} \end{cases} \quad . \qquad (2)$$

Accordingly, we have $g_i = (k_1/t)X_i$ and $c_i = (k_2 t)X_i$, where k_1 and k_2 are thermal conductive/capacitive constants of the

Temperature	Voltage state variables ($x(t)$)
Input Thermal-Power	Input Current sources ($u(t)$)
Thermal conductance	Electrical conductance (G)
Thermal capacitance	Electrical capacitance (C)

Table 1: Thermal and electrical duality

$N(K)$	number of tiles (critical tiles)
p	number of input/output ports
q	order of reduced models
G_0, C_0	nominal thermal RC state matrices
A_i	via density of ith tile
$x(y)$	state variable of temperature (at output)
$x^{(0)}(y^{(0)})$	nominal temperature (at output)
$x^{(1)}(y^{(1)})$	$1st$-order sensitivity (at output)
$x^{(2)}(y^{(2)})$	$2nd$-order sensitivity (at output)

Table 2: Notation list

thermal-via, w and t are the width and thickness of the thermal-via.

Moreover, note that u ($\in R^{p \times 1}$) is the current source to model the thermal-power input. There are several types of thermal-power as defined in [6]. A *thermal power* is defined by the running average of the cycle-accurate (often in the range of ns) power over several thermal time constants (often in the range of ms). When the set of architectural model/constraints and the particular instruction sets and working loads driving the chip are available, a *transient thermal-power signature* can be further defined as the thermal power with a worst-case trace input [6]. In addition, a constant *maximum thermal-power signature* is defined as the maximum of the transient thermal-power signature. Fig. 2 illustrates differences of these power definitions.

2.2 Thermal Analysis

The direct solution in [7–9] is not efficient to solve (1) for large sized circuits. Similar to the macromodeling for the electrical RC network, moment matching based model order reduction can be used to obtain a compact thermal RC model, which not only has a smaller matrix size but also preserves the dominant system response. The existing macromodeling approach from electrical analysis is mainly based on the subspace projection [10] by expanding the system equation (1) at some frequency points. After projection, an order reduced state equation can be obtained with preserved low-order moments to represent the dominant response of the original system.

To further obtain the sensitivity information, the parametrized moments [11] can be obtained by expanding (1) at selected parameter points. However, because the parameterized moments have coupled frequency and parameter variables, its dimension grows exponentially, preventing practical use. This is improved in [12] by separately expanding moments of parameters from the frequency. It results in an augmented state matrix containing the nominal state and the expanded states, i.e., sensitivities with respect to parameters. Nevertheless, all these approaches [10–12] apply a flat projection during the reduction. The reduced state matrices and state variables have coupled nominal values and sensitivities. It is unknown how to separate parametrized sensitivities from the reduced macromodel, and apply those sensitivities in the optimization.

3. STRUCTURED AND PARAMETERIZED MACROMODEL

In this Section, we will show that the separated nominal temperature and its sensitivities can be obtained by a structured and parameterized reduction, and apply this technique to obtain a structured and parameterized macromodel for the thermal RC network. Here the parameter to be expanded is the thermal-via density A_i.

Because the output sensitivity is large with respect to the frequency but small with respect to the geometric parameter, the temperature state variable $x(A_1, ..., A_K, s)$ can be approximated by the Taylor expansion:

$$x(\mathbf{A}, s) = \sum_{i_1}^{\infty} \cdots \sum_{i_K}^{\infty} x_{1,...,K}^{(i_1+...+i_K)}(s)(\delta A_1)^{i_1} \cdots (\delta A_K)^{i_K}. \quad (3)$$

This is similar to the method in [12] modeling variations for the electrical system. Substituting (3) in (1), explicitly matching the moment for each A_i up to the second-order, we can reformulate (1) into an augmented parameterized state equation:

$$(G_{ap} + sC_{ap})x_{ap} = B_{ap}u(t), \quad y_{ap} = L_{ap}^T x_{ap}, \quad (4)$$

with

$$G_{ap} = \begin{bmatrix} G_0 & 0 & \cdots & 0 & 0 & 0 & \cdots & 0 \\ A_1 g_1 & G_0 & \cdots & 0 & 0 & 0 & \cdots & 0 \\ \vdots & \vdots & \ddots & \vdots & \vdots & \vdots & \ddots & \vdots \\ A_K g_K & 0 & \cdots & G_0 & 0 & 0 & \cdots & 0 \\ 0 & A_1 g_1 & 0 & \cdots & G_0 & 0 & \cdots & 0 \\ 0 & A_2 g_2 & A_1 g_1 & 0 & \cdots & G_0 & \cdots & 0 \\ \vdots & \vdots & \vdots & \vdots & \ddots & \vdots & \ddots & \vdots \\ 0 & 0 & \cdots & A_k g_K & \cdots & 0 & \cdots & G_0 \end{bmatrix} \quad (5)$$

and

$$\begin{aligned} x_{ap} &= [x_0^{(0)}, x_1^{(1)}, ..., x_K^{(1)}, x_{1,1}^{(2)}, ..., x_{K,K}^{(2)}]^T \\ B_{ap} &= [B, 0, ..., 0, 0, ..., 0]^T \\ L_{ap} &= [L, \delta A_1 L, ..., \delta A_K L, \delta A_1 \delta A_1 L, ..., \delta A_K \delta A_K L]^T. \end{aligned}$$

Note that C_{ap} has the same lower-triangular structure as G_{ap} does. In addition, the system state variable y_{ap} at output for those critical tiles can be also divided into three parts: nominal value $y^{(0)} = y_0^{(0)}$ ($\in R^1$), first-order sensitivity $y^{(1)} = \{y_1^{(1)}, ..., y_K^{(1)}\}$ ($\in R^K$), and second-order sensitivity $y^{(2)} = \{y_{1,1}^{(2)}, ..., y_{K,K}^{(2)}\}$ ($\in R^{K \times K}$). As a result, solving (4) results in the nominal value of temperature $y^{(0)}$, and its according first-order sensitivity $y^{(1)}$ and second-order sensitivity $y^{(2)}$ with respect to each parameter A_i.

Because the dimension of the system equation (4) is large, its order needs to be reduced using projection with preserved moments (of s) up to q-th order. A flat projection matrix V can be constructed recursively using Arnoldi method [12]. However, directly projecting (4) by V leads to a reduced macromodel losing the lower-triangular block structure of G_{ap} and C_{ap}. As a result, $y^{(0)}$, $y^{(1)}$ and $y^{(2)}$ are coupled with each other.

Instead of using the flat projection matrix V, we introduce a structured projection matrix

$$\mathcal{V} = diag[V_0, \underbrace{V_1, ..., V_K}_{K}, \underbrace{V_{K+1}, ..., V_{K^2}}_{K^2}], \quad (6)$$

by partitioning V according to the dimension of $x^{(0)}$, $x^{(1)}$ and $x^{(2)}$. As a result, the order-reduced state matrices

$$\widetilde{G}_{ap} = \mathcal{V}^T G_{ap} \mathcal{V}, \quad \widetilde{C}_{ap} = \mathcal{V}^T C_{ap} \mathcal{V}, \quad \widetilde{B}_{ap} = \mathcal{V}^T B_{ap}, \quad \widetilde{L}_{ap} = \mathcal{V}^T L_{ap}.$$

Because $V \subseteq \mathcal{V}$, a q-th ordered projection by $\mathcal{V}$ still preserves at least q moments according to [13].

The time-domain transient response of the reduced model can be solved by Backward-Euler method. The reduced system equation at time instant t with time step h is

$$\widetilde{G}_{ap} + \frac{1}{h}\widetilde{C}_{ap}\widetilde{x}_{ap}(t) = \frac{1}{h}\widetilde{C}_{ap}\widetilde{x}_{ap}(t-h) + \widetilde{B}_{ap}u(t)$$
$$\widetilde{y}_{ap}(t) = \widetilde{L}_{ap}^T \widetilde{x}_{ap}(t). \quad (7)$$

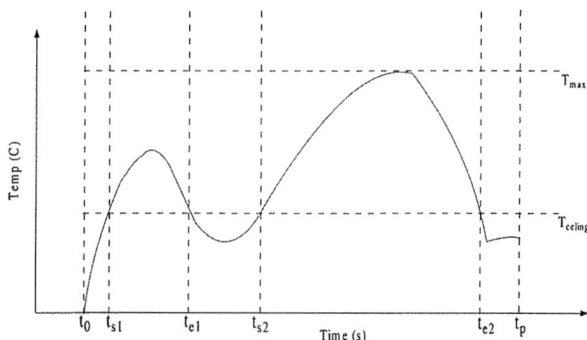

Figure 3: Figure of merit using thermal-violation integral with defined ceiling temperature under an input of transient thermal-power signature.

where

$$\widetilde{G}_{ap} = \begin{bmatrix} \widetilde{G}_0 & 0 & \cdots & 0 & 0 & 0 & \cdots & 0 \\ A_1 \widetilde{g}_1 & \widetilde{G}_0 & \cdots & 0 & 0 & 0 & \cdots & 0 \\ \vdots & \vdots & \ddots & \vdots & \vdots & \vdots & \ddots & \vdots \\ A_K g_K & 0 & \cdots & \widetilde{G}_0 & 0 & 0 & \cdots & 0 \\ 0 & A_1 \widetilde{g}_1 & 0 & \cdots & \widetilde{G}_0 & 0 & \cdots & 0 \\ 0 & A_2 g_2 & A_1 \widetilde{g}_1 & 0 & \cdots & \widetilde{G}_0 & \cdots & 0 \\ \vdots & \vdots & \vdots & \vdots & \ddots & \vdots & \ddots & \vdots \\ 0 & 0 & \cdots & A_k \widetilde{g}_K & \cdots & 0 & \cdots & \widetilde{G}_0 \end{bmatrix} \quad (8)$$

and

$$\widetilde{y}_{ap} = [\widetilde{y}^{(0)}, \widetilde{y}^{(1)}, \widetilde{y}^{(2)}]^T = [\widetilde{y}_0^{(0)}, \widetilde{y}_1^{(1)}, ..., \widetilde{y}_K^{(1)}, \widetilde{y}_{1,1}^{(2)}, ..., \widetilde{y}_{K,K}^{(2)}]^T.$$

Note that the reduced $\widetilde{C}_{ap}$ has the same structure as $\widetilde{G}_{ap}$.

Because the reduction preserves the block structure, the reduced nominal value $\widetilde{y}^{(0)}$, first-order sensitivity $\widetilde{y}^{(1)}$ and second-order sensitivity $\widetilde{y}^{(2)}$ at output (critical tiles) can be solved independently. The temperature profile at those critical tiles perturbed by the parameter is

$$\widetilde{y}(\mathbf{A}, t) = \widetilde{y}^{(0)}(\mathbf{A}, t) + \widetilde{y}^{(1)}(\mathbf{A}, t) + \widetilde{y}^{(2)}(\mathbf{A}, t), \quad (9)$$

A thermal-via planning based on the accurate yet efficient transient simulation with $\widetilde{y}(\mathbf{A}, t)$ can be consequently design. Note that as the reduced system still has the lower-triangular structure, (7) can be efficiently solved using block back substitution, where there is only one factorization cost from the diagonal block, i.e., the reduced block of nominal state matrix.

4. THERMAL-VIA ALLOCATION

In this Section, an accurate figure of merit, thermal-violation integral is first defined to consider the transient temperature profile. A thermal-via allocation can consequently be formulated as a nonlinear optimization problem, which is relaxed and solved by a sequence of quadratic programmings with use of sensitivities provided from the structured and parameterized macromodel.

4.1 Thermal-Violation Integral

A *thermal-violation integral* is defined by the integral of the transient temperature above a user-specified ceiling temperature $T_{ceiling}$:

$$\begin{aligned} f_i(\mathbf{A}) &= \int_{t_0}^{t_p} max[\widetilde{y}(\mathbf{A}, t), T_{ceiling}]dt \\ &= \int_{t_s}^{t_e} [\widetilde{y}(\mathbf{A}, t) - T_{ceiling}]dt, \quad (10) \end{aligned}$$

158

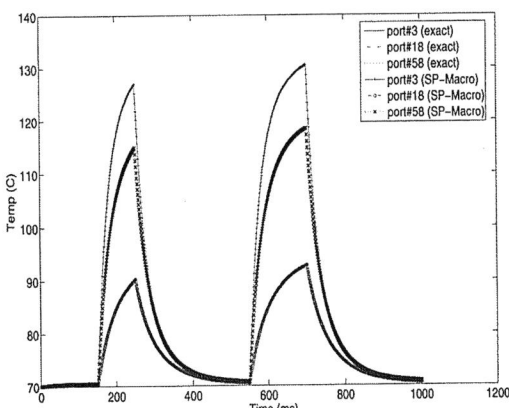

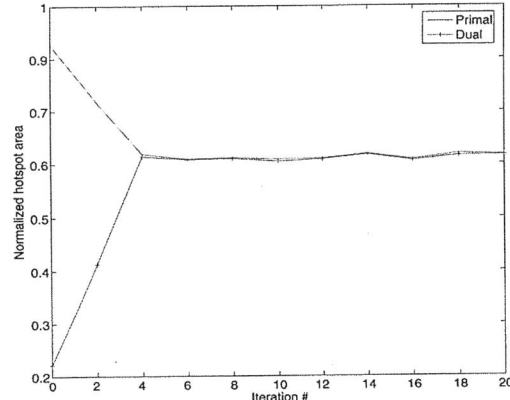

Figure 4: Transient temperature responses of exact and structured and parameterized macro (SP-Macro) models at port 3, 18, and 58 of layer-1 with step-response input. The macromodels are visually identical to those exact models.

Figure 5: Convergence of subgradient optimization of primal and dual problems. The hotspot is represented by violation integral normalized to the maximum. α_0 here is set to 0.7.

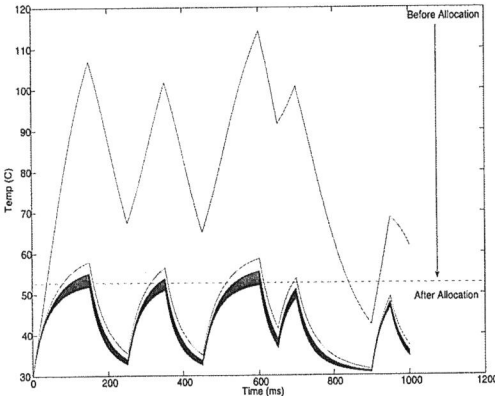

where $\mathbf{A} = [A_1, ..., A_K]$ is parameter vector of thermal-via density, t_0 and t_p define time-period, and the interval $[t_s, t_e]$ is determined by comparing

$$max[\widehat{y}(\mathbf{A}, t), \quad T_{ceiling}],$$

which can contain multiple intervals. As shown in Fig. 3, the integral is actually the area above the $T_{ceiling}$. This definition captures the fact that a thermal violation occurs only when the temperature is above the temperature bound for a long enough period. A similar merit is used for noise estimation in [14].

Moreover, the figure of merit for a group of critical tiles in the entire circuit is

$$f(\mathbf{A}) = \sum_{i=1}^{K} f_i(\mathbf{A}). \qquad (11)$$

It is called *total thermal-violation integral*. The total thermal-violation integral is used as an accurate objective function in the sequel to be minimized by allocating thermal vias.

Note that for the steady-state analysis, the input of the maximum thermal-power signature results in a constant maximum temperature T_{max}. Hence the hotspot reduction by the steady-state solution is equivalent to reduce a rectangular area defined between T_{max} and $T_{ceiling}$, obviously an over-estimated violation integral (See Fig. 3). It becomes even worse for the total violation integral. The reason is that each critical tile has a different transient thermal-power signature, and hence their maximum usually does not happen at the same time. As a result, the thermal-violation integral from a transient solution is more accurate to guide the thermal-via allocation than from a steady-state one.

4.2 Problem Formulation

To minimize the total violation integral, thermal vias are allocated at each pair of adjacent layers. With consideration of the congestion from vertical signal vias, A_{max} and $(A_i)_{max}$ are the *total* available space and *local-tile* available space for inserting thermal vias, which are assumed to be provided by the user. Accordingly, an optimization problem is formulated as

Problem 1

$$min \; f(\mathbf{A})$$

$$s.t. \; \sum_{i=1}^{K} A_i \leq A_{max}, \qquad (12)$$

$$0 \leq A_i \leq (A_i)_{max}, (i = 1, ..., K). \qquad (13)$$

where the constraint (12) is a *global constraint* implying that the total thermal-via density is limited by the A_{max}, and the con-

Figure 6: Iterative optimizations showing the hotspot reduction by thermal-via allocation under the input of transient thermal-power signature at port 32 of layer-1. The ceiling temperature is $52°C$.

straint (13) is a *local constraint* implying that the local thermal-via density at ith tile is limited by $(A_i)_{max}$. Moreover, to compute $f(\mathbf{A})$, t is discretized into finite intervals and Problem 1 becomes semi-definite [14], which can be further solved using Lagrangian relaxation.

Using matrix $\mathbf{U}$ ($\in R^{(K+1) \times (K)}$)

$$\mathbf{U} = \begin{bmatrix} 1 & 0 & \cdots & 0 \\ 0 & 1 & \cdots & 0 \\ \vdots & \vdots & \ddots & \vdots \\ 0 & \cdots & \cdots & 1 \\ 1 & 1 & \cdots & 1 \end{bmatrix}, \qquad (14)$$

the constraints (12) and (13) become

$$\mathbf{UA} \leq \mathbf{A}_{max}, \qquad (15)$$

where $\mathbf{A}_{max} = [(A_1)_{max}, (A_2)_{max}, ..., (A_K)_{max}, A_{max}]^T$. To efficiently solve Problem 1, the below Lagrangian relaxation is used to transform the original problem into a sequence of subproblems.

The constraint function can be added to the objective function using a vector of Lagrangian multiplier $\lambda = [\lambda_1, ..., \lambda_K]$. As

159

are first-order sensitivities, and

$$
H = \begin{bmatrix} \int_0^{t_p} \widetilde{y}_{1,1}^{(2)} dt & \cdots & \int_0^{t_p} \widetilde{y}_{1,K}^{(2)} dt \\ \vdots & \ddots & \vdots \\ \int_0^{t_p} \widetilde{y}_{K,1}^{(2)} dt & \cdots & \int_0^{t_p} \widetilde{y}_{K,K}^{(2)} dt \end{bmatrix}
$$

is the Hessian matrix composed by the second-order sensitivities. Both the first and second order sensitivities can be efficiently solved by (7) independently.

The sequential subgradient optimization procedure is outlined in Algorithm 1, where α_k is the step size usually determined through a geometric regression [15]. Note that because the projec-

---

Algorithm 1 Subgradient Optimization using Structured Parameterized Macromodel

---

Initialize: $(\mathbf{A}_0, \alpha_0, \lambda_0, H_0, k)$;
Solve: $\widetilde{y}_0$ using (7);
Solve: $\delta \mathbf{A}_0 = quadprog(\lambda_0, \widetilde{y}_0)$;
Set: $\mathbf{s}_0 = \frac{\mathbf{UA}_0 - \mathbf{A}_{max}}{\|\mathbf{UA}_0 - \mathbf{A}_{max}\|}$;
Set: $\lambda_1 = \lambda_0 + \alpha_0 \cdot \mathbf{s}_0$;
while $|L(\lambda_{k+1}) - L(\lambda_k)| > TOL$ **do**
    $\mathbf{s}_k = \frac{\mathbf{UA}_k - \mathbf{A}_{max}}{\|\mathbf{UA}_k - \mathbf{A}_{max}\|}$;
    $\lambda_{k+1} = \lambda_k + \alpha_k \cdot \mathbf{s}_k$;
    $\delta \mathbf{A}_k = quadprog(\lambda_k, \widetilde{y}_k)$;
    $\mathbf{A}_{k+1} = \mathbf{A}_k + \delta \mathbf{A}_k$;
    Update $(G_{ap})_{k+1}$ and $(C_{ap})_{k+1}$ with $\mathbf{A}_{k+1}$;
    Solve $\widetilde{y}_{k+1}$ using (7) with updated macromodel;
    $k = k + 1$;
end while

---

tion (6) preserves the block structure, the reduced state matrices can be repeatedly used when updating the new parameter vector $\mathbf{A}$. Therefore, there is only one reduction needed. In addition, since the reduced model is much smaller than the original one, and the factorization cost only comes from the nominal blocks in diagonal, its nominal value and sensitivities can be efficiently solved by the back-substitution of (7). Therefore, the optimization procedure in Algorithm 1 is computationally efficient.

5. EXPERIMENTS

Our structured and parametrized macromodeling (SP-Macro) and thermal-via allocation are both implemented in MATLAB, and run on Linux workstation with Intel Pentium IV 2.66G CPU and 2G RAM. The examples have following settings. k_1 (thermal conductive constant) is $100W/m \cdot K$ for silicon and $400W/m \cdot K$ for copper, and k_2 (thermal capacitive constant) is $1.75 \times 10^6 J/m^3 \cdot K$ for silicon and $3.55 \times 10^6 J/m^3 \cdot K$ for copper. The substrate is 500um thick, the device layer is 6um thick and interlayer thickness is 1um thick. 4 silicon layers are used and the thermal-via is assumed to be copper. The unit via area is $2 \times 2um^2$. The overall chip size is $2 \times 2cm^2$, and the number of individual modules and its according size are from MCNC benchmarks. A random power distribution at each node is used. 90% of tiles have power densities from 0 to $2 \times 10^6 W/m^2$, and their clock gating pattern has a period of 500ms, where the power in the standby mode is 5% of the running mode. The other 10% of tiles having power densities from $3 \times 10^6 W/m^2$ to $9 \times 10^6 W/m^2$, and their clock gating pattern has a period of 250ms where the power in the standby mode is 20% of the running mode.

A detailed 3D thermal RC circuit is used to verify the proposed algorithm. It has 4 layers and each layer contains about 10K tiles. 64 tiles of each layer are selected as critical tiles. The total thermal-via density constraint is 3000, and the local via number constraint is randomly generated from 10 to 100. Structured and parameterized model reduction is first applied to generate SP-Macro for the thermal-via allocation considering the transient effect. Then the entire circuit is used to generate the steady-state map of the temperature profile.

For SP-macro and original models, Fig. 4 compares the time-domain transient temperature at selected three critical tiles (3,

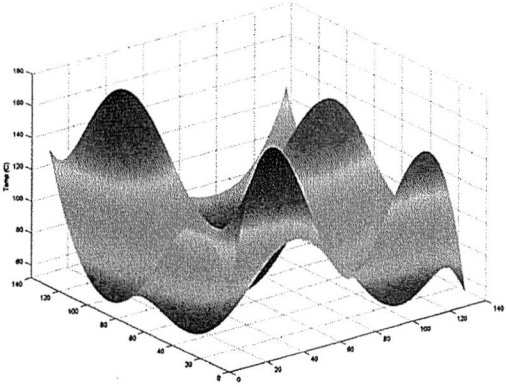

Figure 7: Steady-state temperature map of top layer (layer-1) before thermal-via allocation.

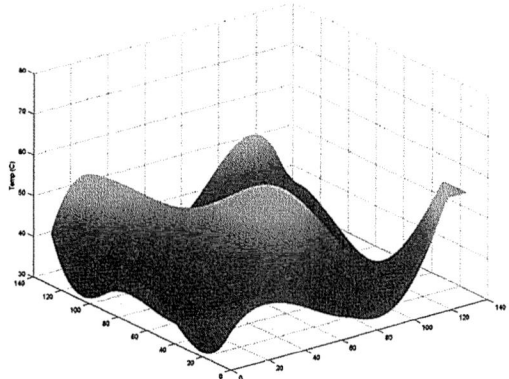

Figure 8: Steady-state temperature map of top layer (layer-1) after thermal-via allocation using transient temperature profile.

a result, the primal problem (Problem 1) has a following dual problem:

$$L(\mathbf{A}, \lambda) = f(\mathbf{A}) + \lambda \cdot h(\mathbf{A}) \tag{16}$$

where

$$h(\mathbf{A}) = \mathbf{UA} - \mathbf{A}_{max}. \tag{17}$$

This relaxed problem can be transformed into a sequential sub-problems by subgradient optimization [15]. At each iteration, each subproblem is constructed from a quadratic approximation of the nonlinear objective function, and a linearization of the constraints about the solutions from previous iteration. The optimization terminates when the convergence criterion is achieved. This called as *sequential quadratic programming* (SQP) [15].

Expanding $f(\mathbf{A})$ and $h(\mathbf{A})$ with respect to $\mathbf{A}$ up to the second-order, an approximated equivalent subproblem is

$$min \quad \nabla f(\mathbf{A})^T \delta \mathbf{A} + \frac{1}{2} \delta \mathbf{A}^T H \delta \mathbf{A} \tag{18}$$

$$s.t. \quad \nabla h(\mathbf{A}) \cdot \delta \mathbf{A} \leq h(\mathbf{A}). \tag{19}$$

(19) can be solved by the standard quadratic programming, where

$$\nabla f = \int_0^{t_p} \widetilde{y}^{(1)} dt, \quad \nabla h = const.$$

total/critical tile#	total via #	original/ceiling temp ($^\circ C$)	direct			SP-macro			
			solve dc (s)	solve tran (s)	allo-via	redu ckt (s)	solve sens (s)	qp-prog plan (s)	allo-via
256/30	704	120/40	1.64	10.27	440	0.12	0.19	0.15	360
1024/60	2818	120/40	12.62	130.12	2281	1.08	0.96	0.42	1609
4096/80	5980	140/50	341.13	3872.98	5620	12.92	6.28	1.92	3217
8192/100	8218	140/50	7809.12	NA	8021	46.27	16.92	8.98	4382
16384/120	18000	160/60	NA	NA	17600	120.89	101.23	23.65	9280
32768/200	24000	160/60	NA	NA	23800	262.12	257.21	42.78	11660

Table 3: Experiment setting and results of thermal-via planning time and number. The allocated thermal-via of steady-state analysis is based on the reduced macromodel with the use of thermal-violation integral defined by the maximum temperature.

18, 58) using (9). 16 moments are used for the moment matching. The reduced models are visually identical to original ones. Fig. 5 shows the subgradient optimization procedure after few iterations, where the dual problem converges with the primal problem. The ceiling temperature is $52^\circ C$ and, the transient temperature at one port is cooled down to the ceiling point as shown in Fig. 6. Clearly, the gradient approach greedily minimizes the thermal-violation integral. Fig. 7 and 8 further show the steady-state temperature map across the top layer (layer-1). The initial chip temperature at the top layer is around $150^\circ C$, and its temperature profile at steady-state is shown in Fig. 7. In contrast, the allocation results in a cooled temperature profile that closely approaches the ceiling temperature as shown in Fig. 8.

Table 3 further analyzes the runtime scalability and allocated thermal-via density by the proposed method and the direct steady-state analysis. Because directly solving steady-state equation needs to handle large sized matrix, it has a long runtime and uses a lot of memory. In contrast, the macromodel can efficiently match the transient response using around 20 moments. For a circuit with 8192 tiles, our model reduces runtime by 126X (62s versus 7809s) compared to the steady-state analysis. More importantly, due to the use of our accurate figure of merit: the thermal-violation integral, which considers the transient effect, our allocated thermal-via density is much smaller than the one by steady-state analysis under the same targeted ceiling temperature. Because directly solving steady-state equation can not generate the sensitivity for the optimization, the allocated thermal-via of steady-state analysis is based on the reduced macromodel, where the thermal-violation integral is defined by the maximum temperature (See Fig. 3). For a circuit with 32768 tiles, our design reduces 2.04X (11660 versus 23800) thermal vias compared to the steady-state analysis.

6. CONCLUSIONS

An accurate yet efficient thermal-via allocation is proposed for the thermal-aware design of 3D ICs. The previous thermal-via allocations [8,9] use the direct steady-state analysis and ignore the temporal and spatial variations of the thermal-power. They are inefficient to generate the nominal temperature and its sensitives for large sized circuits. More importantly, they result in a design with excessive number of thermal vias.

In this paper, to consider the temporally and spatially variant thermal-power input, a structured and parameterized model order reduction is used to obtain a macromodel, which can efficiently provide the transient nominal temperature and its sensitives to thermal-via densities. A thermal-violation integral of the transient temperature is then defined to accurately capture the thermal violation, and a nonlinear optimization is formulated to minimize the thermal-violation integral. In addition, using parameterized sensitivities provided from the macromodel, the relaxed subproblems of the formulated problem are efficiently solved by a sequence of quadratic programming, where the reduced macromodel can be repeatedly used during the gradient search. Clearly, the proposed structured and parameterized macromodel can be used for a number of integrity-driven physical synthesis.

7. REFERENCES

[1] K. Banerjee, S. J. Souri, P. Kapur, and K. C. Saraswat, "3D ICs: A novel chip design for improving deep submicron interconnect performance and systems-on-chip integration," *Proc. IEEE*, pp. 602–633, 2001.

[2] W. Davis and et al., "Demystifying 3D ICs: the pros and cons of going vertical," *IEEE Design and Test of Computers*, pp. 498–510, 2005.

[3] C. C. Teng, Y. K. Cheng, E. Rosenbaum, and S. M. Kang, "iTEM: A temperature-dependent electromigration reliability diagnosis tool," *IEEE Trans. on CAD*, pp. 882–893, 1997.

[4] K.Banerjee, A.Mehrotra, A.Sangiovanni-Vincentelli, and C. Hu, "On thermal effects in deep sub-micron VLSI interconnects," in *ACM/IEEE DAC*, 1999.

[5] W. Huang, E. Humenay, K. Skadron, and M. R. Stan, "The need for a full chip and package thermal model for thermally optimized IC designs," in *ACM/IEEE ISLPED*, 2005.

[6] V. Tiwari, D. Singh, S. Rajgopal, G. Mehta, R. Patel, and F. Baez, "Reducing power in high-performance microprocessors," in *ACM/IEEE DAC*, 1998.

[7] M. R. Stan, K. Skadron, M. Barcella, W. Huang, K. Sankaranarayanan, and S. Velusamy, "Hotspot: a dynamic compact thermal model at the processor-architecture level," *Microelectronics Journal*, pp. 1153–1165, 2003.

[8] B. Goplen and S. Sapatnekar, "Thermal via placement in 3D ICs," in *ACM ISPD*, 2005.

[9] J. Cong and Y. Zhang., "Thermal via planning for 3D ICs," in *IEEE/ACM ICCAD*, 2005.

[10] A. Odabasioglu, M. Celik, and L. Pileggi, "PRIMA: Passive reduced-order interconnect macro-modeling algorithm," *IEEE Trans. on CAD*, pp. 645–654, 1998.

[11] L. Daniel, O. C. Siong, L. S. Chay, K. H. Lee, and J. White, "A multiparameter moment matching model reduction approach for generating geometrically parameterized interconnect performance models," *IEEE Trans. on CAD*, pp. 678–693, 2004.

[12] X. Li, P. Li, and L. Pileggi, "Parameterized interconnect order reduction with explicit-and-implicit multi-parameter moment matching for inter/intra-die variations," in *IEEE/ACM ICCAD*, 2005.

[13] E.J.Grimme, *Krylov projection methods for model reduction (Ph. D Thesis)*. Univ. of Illinois at Urbana-Champaign, 1997.

[14] C. Visweswariah, R. A. Haring, and A. R. Conn, "Noise considerations in circuit optimization," *IEEE Trans. on CAD*, pp. 679–690, 2000.

[15] M. S. Bazaraa, H. D. Sherali, and C. M. Shetty, *Nonlinear Programming: Theory and Algorithms*. John Wiley and Sons, 1993.

Dynamic Thermal Clock Skew Compensation using Tunable Delay Buffers

A. Chakraborty[†]   K. Duraisami[†]   A. Sathanur[†]   P. Sithambaram[†]

L. Benini[◦]   A. Macii[†]   E. Macii[†]   M. Poncino[†]

[†] Politecnico di Torino
10129 Torino, ITALY

[◦] Università di Bologna
40136 Bologna, ITALY

ABSTRACT

The thermal gradients existing in high-performance circuits may significantly affect their timing behavior, in particular by increasing the skew of the clock net and/or altering hold/setup constraints, possibly causing the circuit to operate incorrectly. The knowledge of the spatial distribution of temperature can be used to properly design a clock network that is able to compensate such thermal non-uniformities. However, re-design of the clock network is effective only if temperature distribution is stationary, i.e., does not change over time. In this work, we specifically address the problem of dynamically modifying the clock tree in such a way that it can compensate for temporal *variations of temperature. This is achieved by exploiting the buffers that are inserted during the clock network generation, by transforming them into* tunable *delay elements. Temperature-induced delay variations are then compensated by applying the proper tuning to the tunable buffers, which is computed off-line and stored in a tuning table inserted in the design. We propose an algorithm to minimize the number of inserted tunable buffers, as well as their* tunable range (which directly relates to complexity). *Results show that clock skew is kept within original bounds with minimum area and power penalty. The maximum increase in power is 23.2% with most benchmarks exhibiting less than 5% increase in power.*

Categories and Subject Descriptors: B.8.2 Performance Analysis and Design Aids: Miscellaneous

General Terms: Algorithms, design, performance.

Keywords: Clock tree, clock skew, tunable delay buffers, temperature aware design methodology.

1. INTRODUCTION

Integration densities in nanometer CMOS devices have dramatically increased over the last thirty years, resulting in a continuous increase in power density. Microprocessors are a striking example of this trend: Their power density has

doubled every three years [1]. Many techniques have been developed to decrease power consumption and power density by exploiting local circuit idleness (e.g., dynamic power management, clock-gating). An unfortunate side effect of these techniques is the increase of the variance of power dissipation over the chip area and over time, which generally translates into many areas with different power densities, as well as a non-stationary power dissipation profile.

These time and space variations in power dissipation causes temperature gradients in the substrate and, as a consequence, non-uniform substrate temperature: Gradients of 50° C have been measured across the substrate in high-performance ICs [2]. Temperature variations over time are even more insidious, as it is much harder to characterize them off-line at design time. In fact, these variations depend on a number of environmental conditions and on thermal transients that have time constants much longer than the typical simulation times used for design-time power estimation. For this reason a number of closed-loop techniques (e.g. clock throttling, thermal shutdown) have been developed for run-time temperature sensing and temperature-aware power management [3, 4, 5].

Thermal runaway is not the only risk connected with temperature variations. Temperature has also a significant impact on transistors and on global interconnects [6]. Temperature effects on MOS transistors are well-known. However, higher temperatures also increase interconnect delays, further degrading circuit performance. This is mainly due to the linear dependency that exists between temperature and resistivity of metal interconnect. Temperature-induced delay variations in interconnects are extremely critical for clock distribution networks whose delays must be carefully tuned to avoid synchronization errors. Hence, clock distribution networks and clock analysis and synthesis tools cannot neglect the impact of spatial and temporal temperature variations in today's nanometer technologies.

In [7] a technique for the design of a temperature-aware zero-skew clock distribution network was proposed that addresses the drawbacks of traditional clock-tree routing that assume a uniform thermal profile. This method was extended in [8] to deal with non-zero skew, thus achieving clock trees with smaller total wirelength. Even though these approaches take into account temperature variations across various regions of the chip, they do not account for *variations over time*. In other words, they optimize clock distribution for a given (non uniform) temperature profile, but they cannot account for the fact that such a profile can

---

Permission to make digital or hard copies of all or part of this work for personal or classroom use is granted without fee provided that copies are not made or distributed for profit or commercial advantage and that copies bear this notice and the full citation on the first page. To copy otherwise, to republish, to post on servers or to redistribute to lists, requires prior specific permission and/or a fee.
ISLPED'06, October 4–6, 2006, Tegernsee, Germany.
Copyright 2006 ACM 1-59593-462-6/06/0010 ...$5.00.

greatly change during system operation, depending on environmental conditions and workloads.

Dynamically tunable clock trees have been proposed by several authors in the literature, to boost yield under increasing process variations [9, 10]. In this paper we propose the use of dynamically tunable clock trees, in conjunction with distributed on-chip temperature sensing, to dynamically adapt the delays in the clock tree distribution network. To support this adaptive approach, we have developed novel algorithms for (i) insertion of variable-delay buffers in the clock tree, (ii) minimization of the range of the adjustable delays, and (iii) generation of the appropriate delay tuning commands in response to measured temperature variations.

Experimental results over a set of realistic, large-scale benchmarks, and based on clock distribution networks obtained from industry-strength physical design tools, show that clock skew compensation is achieved with minimum increase in area and power.

This paper is organized as follows: Section 2 surveys the state-of-the-art. In Section 3 an implementation of a tunable delay buffer is discussed. Section 4 describes the core of our contribution, while Section 5 shows the effectiveness of the proposed methodology. Section 6 discusses the results in detail. Finally, Section 7 closes the paper.

2. PREVIOUS WORK

Traditional approaches for the design the clock network historically focused on generating a clock network (typically a tree) with minimum wire-length, with zero [11, 12] or bounded skew [13, 14], and possibly combined with wire sizing and/or with buffer insertion [15, 16].

All these approaches assume a constant temperature along the clock network. As discussed in Section 1, this assumption is not anymore realistic, and the effects of substrate temperature variations over time and space must be taken into account to guarantee proper operation of a circuit.

The work of [7] was the first one proposing an adaptation of the basic Deferred-Merge-Embedding algorithm [12] that accounts for skew variations due to *spatial* temperature gradients along a clock network. That work was improved in [8], in which a more general approach is proposed that allows non-zero skew bound, and incremental clock tree generation thus minimizing the changes on the original clock tree structure. None of these approaches, however, deals with *temporal* temperature variations: they optimize the clock network starting from a given (non-uniform) temperature profile, without considering that the thermal profiles may change over time in response to workloads and environmental changes.

A widely used solution for online, post-silicon tuning of delays is the use of *tunable delay buffers* (TDBs). In [9] TDBs are used to implement a post-silicon clock optimization technique for reducing timing violations induced by process variations, through a statistical timing driven clock scheduling algorithm. A similar solution was proposed in [17, 18], where TDBs are inserted into the clock network for redistributing slacks and correcting timing failures through post-silicon clock tuning. These approaches are generalized into a methodology in [10], which describes a clock-tree synthesis flow driven by statistical timing analysis.

3. TUNABLE DELAY BUFFERS

Tunable Delay Buffers (TDBs) are typically used to perform on-line skew compensation to counter the effect of increased delay in regular buffers due to temperature or other type of variations. They normally consist of a number of taps providing fixed delays when activated.

Various implementations of TDBs exist in literature. In [19] a chain of inverters feeding a multiplexer was used. A similar implementation with transmission gates was used in [20] for clock tuning on System on Chips (SoCs). The disadvantage of these approaches was that the inverters in the path are continuously switching leading to a huge power overhead. [9] overcame this drawback by using a pair of inverters with capacitive loads in between them. These loads were achieved using transmission gates and NMOS transistors connected to them, which can be activated using the appropriate control signals to achieve variable delays. The conceptual architecture of such a TDB is shown in Figure 1. Our imple-

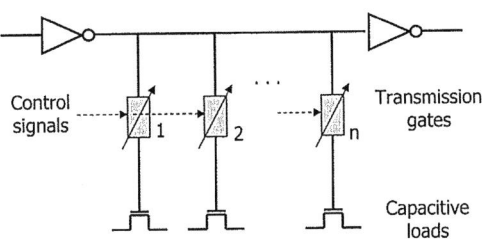

Figure 1: Tunable Delay Buffer

mentation of the TDB is very similar to the one used in [9], except that the inverter pair and taps have been appropriately sized to meet our delay requirements. This sizing is necessary to keep the delay of our TDB almost equal to the regular buffer from the library it is replacing when all taps are off. Propagation delay and normalized power values for various taps are as shown in Figure 2, and refer to a 90nm technology library by STMicroelectronics.

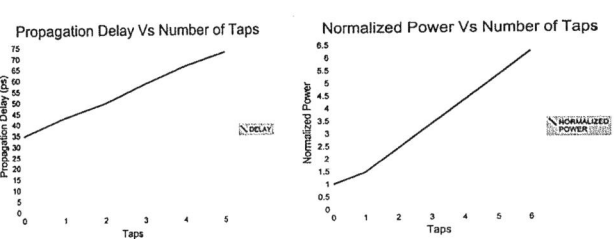

Figure 2: Propagation Delay (Left) and Normalized Power (Right) Vs Number of Taps.

Taps	0	1	2	3	4	5
Area	1.14l	1.82l	2.51l	3.19l	3.88l	4.56l

Table 1: Area Values vs. Number of Taps.

The area information for various taps are reported in Table 1: values represent the total width of the inverter pair multiplied by the minimum length l. The power values show a substantial increase per tap when seen in isolation. This is not the case, however, in the actual clock tree where other parasitic components are present. Since we replace only a small number of buffers in the clock network, the overall contribution of TDBs to the total clock tree power is within acceptable bounds.

4. DYNAMIC THERMAL COMPENSATION METHODOLOGY

Our objective is to devise an on-line (i.e., dynamic and without external control) hardware mechanism that allows selecting the appropriate tuning of clock buffers so as to thermally compensate the skew variations induced by a thermal gradient. The proposed methodology consists of two main phases, and is summarized by the high-level flow of Figure 3.

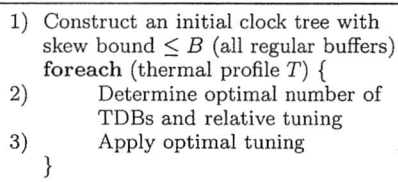

1) Construct an initial clock tree with skew bound $\leq B$ (all regular buffers)
   foreach (thermal profile T) {
2)   Determine optimal number of TDBs and relative tuning
3)   Apply optimal tuning
   }

Figure 3: Thermal Compensation Flow.

Initially (Phase 1) an initial buffered clock tree with a given skew bound is built, assuming a given, fixed thermal profile (either uniform or non-uniform). Phase 2 concerns the *calculation of the optimal number of TDBs and their relative tuning*, for a given change of the temperature profile; this step has to be done for every thermal variation occurring during the lifetime of the design. Phase 3 is then simply the application of this tuning configuration to the TDBs.

It is important to emphasize that our methodology assumes that we initially generate a regular (i.e., non-tunable) buffered clock tree, for which any standard algorithm can be used [11, 12]; the insertion of TDBs and their tuning is carried out as a post-processing phase. This issue is an important added-value because it allows us to apply the methodology to clock distribution network generated with industry-strength commercial clock routing tools (and thus not limited to purely binary clock trees).

A direct implementation of the above flow would require implementing in hardware the "algorithm" that solves the optimal tuning problem, thus allowing a true on-line, dynamic tuning. Such a solution, however, is clearly infeasible due to potentially high hardware complexity (e.g., the determination of the tuning may require multiple clock cycles). Therefore, we resort to an alternative scheme, which, provides very good results for the typical thermal gradients experienced by a circuit. Specifically, we envision a *partially off-line* scheme, in which the solutions to the optimal tuning are pre-calculated (e.g., based on some typical or expected thermal gradients) during a *characterization step*, and stored into a lookup table inside the system. The tuning values for the set of thermal configurations applied is properly adapted in such a way that, during normal operation of the system, the actually measured thermal map will cause to select the appropriate tuning configuration. This scenario is conceptually depicted in Figure 4.

The figure shows an example clock tree with p TDBs and the tuning table that stores, for each thermal transition, the tuning of the p TDBs. The routing of the signals from the sensors and to the TDBs is not shown for ease of readability. In the following, we will discuss the three main issues involved with the proposed methodology: (i) the algorithm for optimizing the tunable range under a given thermal profile, (ii) the intersection of the various tuning solutions to derive a good solution on average, and (iii) the implementation issues, in particular how the solution is affected by the number and type of sensors.

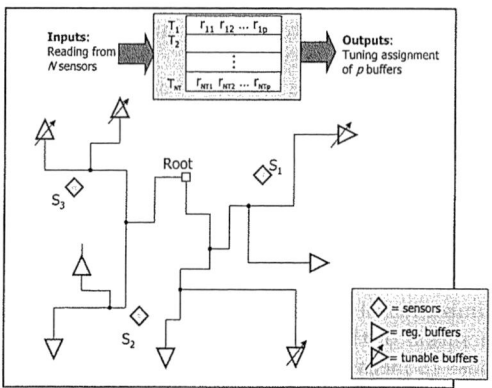

Figure 4: Partially Adaptive Tuning Scenario.

4.1 Terminology and Definitions

Let TR be the original clock tree, with root r, containing m buffers, and whose skew is bounded by B. Let d_i be the signal (Elmore) delay from clock source r to sink i. The skew $Skew(TR)$ of a tree TR is defined as the maximum difference between the delay of any two sinks, i.e., $Skew(TR) = \max_{\text{sinks } i,j} |d_i - d_j|$

Let T be a generic temperature profile, consisting of a temperature value for each grid point (x, y). We account for temperature-induced delay variations of a wire by defining ED_T, the *temperature dependent Elmore delay*, as defined by Equation 1, for a generic wire segment of length L [6]:

$$ED_T = ED + (c_0 \cdot L + C_L)\rho_0\beta \int_0^L T(x)dx - c_0\rho_0\beta \int_0^L x \cdot T(x)dx \tag{1}$$

where $ED = \frac{\rho_0 c_0 L^2}{2} + \rho_0 L C_L$ is the nominal Elmore Delay, β is the temperature coefficient of resistance and L is length of the wire. Notice that the definition of the temperature-dependent delay of the wire only uses the distribution of temperature along the dimension of the wire length.

The delay d_i along a root-to-sink path of a tree is actually the sum of a set of Elmore delays (for the wires) plus the delay of the buffers on that path; therefore, its dependency on temperature is directly accounted when computing the delays for the individual wire segments through Equation 1. Buffer delay, however, is also affected by temperature variations, in terms of an increase in the driver resistance [6]. This is accounted directly by deriving a parametric formula for the nominal buffer delay, as shown in Section 3.

We assume that a tunable buffer can have a set of M delay values $[\delta_1, \ldots, \delta_M]$, δ_1 being the minimum delay value. Delay values are equally spaced, i.e., $\delta_i - \delta_{i-1}$ is constant, $\forall i$. The nominal delay δ_{nom} of a buffer (i.e., before any tuning) is assumed to coincide with the smallest value of the range, i.e., $\delta_{nom} = \delta_1$.

The delay value of a TDB is important for calculating the root-to-sink delays; however, in our scenario, it is even more relevant to define the *relative tuning* of a buffer, i.e., its change in delay; this quantity can be expressed as a positive integer $r_i \in [0, +r_{max}]$. Obviously, the cardinality M of the set of delays is identical to the number of tuning values, i.e., $r_{max} + 1 \equiv M$.

Notice that since $r_i \geq 0$, we assume that it is only possible to increase the delay of a TDB, and never decrease it. In principle, this can increase excessively the delay of a path.

However, for the clock signal it is the skew that matters (thus, difference of path delays) rather than absolute path delays; allowing delays to increase only simplifies the implementation of the buffer. $r_i = 0$ indicates no variation with respect to the previous delay value. We represent the overall tuning assignment by an m-element array $\mathbf{R}$, with initial value $\mathbf{R} = [0, ..., 0]$.

Based on these quantities, we parameterize the definition of skew as $Skew(TR, \mathbf{R}, T)$ as the skew of the tree TR, with the m TDBs tuned according to the tuning vector $\mathbf{R}$, and under temperature profile T.

4.2 Optimization of the Tunable Range

The problem we are solving can be formulated as follows: *Given a tree TR with skew bound B, with m buffers - and their positions -, and a temperature profile T, find the optimal assignments of the tuning $\mathbf{R} = \{r_i\}, i = 1, ..., m$ such that* $Skew(TR, \mathbf{R}, T) \leq B$.

Since the overhead (power and area cost) of a TDB is proportional to the tunable range, as shown in Section 3, The optimal assignment of r_i implies minimizing the overall magnitude of the r_i's. Buffers with a non-zero tuning value after the optimization will correspond to TDBs.

This problem can be cast into the following ILP formulation:

$$\begin{cases} \text{Minimize} \quad r_{tot} = \sum_{i=1}^{m} r_i, \text{such that} \\ \\ Skew(TR, \mathbf{R}, T) \leq B \\ 0 \leq r_i \leq r_{max}, \forall i \end{cases} \quad (2)$$

This apparently simple formulation hides the actual complexity of the problem behind the constraint on the skew, which is in fact defined in terms of MAX/MIN operators, and cannot be transformed directly into a *linear* constraint. Therefore, we expand the skew constraint into a set of inequalities according to its definition, i.e., in terms of path delays. The latter can be expressed as a linear function of the r_i's as follows.

Let d_{nom} be the delay of a path when all its k TDBs are tuned at their default position (i.e., all $r_j \equiv 0, \forall$ buffer $j = 1, ..., k$). The delay for a generic tuning assignment $\mathbf{R} = [r_1, ..., r_k]$ is thus simply $d(\mathbf{R}) = \delta_{nom} + \sum_{j=1,...,k} r_j$

EXAMPLE 1. *Consider a path a containing two TDBs b_1 and b_2. Let the tunable range of both buffers be $[0, 1, 2]$. The delay of the path can be expressed as $d = d_{nom} + r_1 + r_2$. Therefore, the maximum delay of the path corresponds to the assignment $r_1 = 1, r_2 = 1$, while the minimum delay corresponds to the $[0, 0]$ assignment (i.e., identical to the nominal path delay).*

Using path delays the ILP can thus be re-formulated as follows:

$$\begin{cases} \text{Minimize} \quad r_{tot} = \sum_{i=1}^{m} r_i, \text{such that} \\ \\ d_a(\mathbf{R}) - d_b(\mathbf{R}) \leq B \quad \forall \text{ pairs of paths } (a, b) \\ 0 \leq r_i \leq r_{max}, \forall i \end{cases} \quad (3)$$

In the above formulation, the number of constraints is quadratic in the number of clock sinks (since there is one path per leaf in a tree). Although this is a potentially very large number, it can be drastically reduced by observing that not all path needs to be considered. More specifically, we can avoid considering paths *that share the same set of buffers*. This is

because the skew between paths that have the same set of TDBs are not affected by their tuning.

This criterion partitions the set of paths into a set of m classes, each one having the same set of buffers. Therefore, there at most $m(m - 1)$ of these constraints; considering that clock tree generation tools tend to insert as few buffers as possible, this number is much smaller the number of path pairs.

4.3 Impact of Sensor Distribution

The above formulation is in principle independent of the number and position of the temperature sensors, in the sense that it assumes the availability of a temperature value at each grid point. In practice however, only a limited number of sensors are available on the chip.

This affects the methodology in two ways. First, the point-by-point temperature profile assumed by the ILP formulation can only be estimated by assuming some spatial distribution of the temperature sensed at each sensor (e.g., a linear decrease as we get farther away from the sensor). Therefore, although typical on-chip CMOS sensors [21, 22, 23] are able to report arbitrarily high precision in terms of absolute temperature reading, for our purposes such an accuracy is not required; the availability of a set of temperature ranges is more than enough.

Second, and more important the number of sensors affects the practical implementation of the algorithm. In our scheme we store the optimal tuning solution for a given temperature profile in a table; the number of entries of this table corresponds thus to the number of possible transition between thermal "states" (i.e., a set of temperature readings) of the system. If we assume there are N sensors, each with a reading that can be quantized into a set of q ranges, there are $N_S = q^N$ possible thermal states and exactly $N_T = N_S(N_S - 1)$ table entries.

These considerations also affects the "characterization" phase of the methodology. Even if we could in principle characterize the system behavior by applying a potentially huge number of thermal profiles, this is in practice useless, since the system will not be able to distinguish more than N_S distinct profiles.

Notice that this is not an issue if the thermal profile is stationary over time, as in [7, 8]. In our case, conversely, adaptation is dynamic and thus it has to be measured on-line. Therefore, the number of sensors and their resolution does matter.

4.4 Overlapping The Tunable Ranges

Each transition from one thermal state to another defines one tuning configuration of the m buffers. Since the tunable clock tree must be designed for every possible thermal state transition, we must be able to determine the number and tuning range of the TDBs that satisfies all possible configurations. There are two main options for the definition of a common tuning:

- An *optimal* scheme that chooses for each TDB the *worst-case* tuning requirement; since tuning values are relative, this is achieved by selecting, for the i-th buffer (the i-th column of the table) the maximum and minimum value. Their difference, divided by the tuning step determines the number of tuning intervals.

$$r_i = \frac{(\max_{j=1,\ldots,N_T} r_{ij} - \min_{j=1,\ldots,N_T} r_{ij})}{Tuning\ Step}$$

- An *approximate* scheme in which we limited the number of TDBs to be used. After applying of the above optimal scheme to determine the ideal tuning requirements, we select as tunable only a subset of the buffers. Various options are possible such as selecting only those with the largest tuning intervals.

Notice that both schemes encompass the determination of the number of TDBs. A buffer does not require tunability if a generic r_i is 0 after the merging.

5. EXPERIMENTAL SETUP

The overall flow of the methodology is depicted in Figure 5. Given an RTL, we perform its placement, clock tree generation and global routing. The clock-tree geometry is extracted from the DEF format. In the next step, a series of thermal profiles are read in. For each thermal profile, we calculate the insertion delays to each sink and generate skew constraints for the ILP Solver.

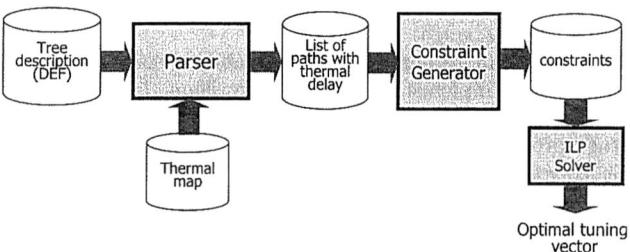

Figure 5: Flow of the Proposed Methodology.

5.1 Benchmarks

For our experiments, we used various open-source designs from opencores.org. Our benchmarks include: 1) I8051 - Intel 8051 Microcontroller, 2) Vga - A VGA controller, 3) Ethernet - A 10G Ethernet MAC core, and 4) WishBone - A portable SOC IP core. The main features of these benchmarks are presented in Table 2.

Values are relative to a $130nm$ low-leakage library and a $90nm$ low-power standard Vt library from STMicroelectronics.

Benchmark	Area		# Buffers		# Sinks
	130nm	90nm	130nm	90nm	
I8051	0.11	0.07	9	9	325
Vga	0.16	0.09	33	67	1523
Ethernet	1.33	0.46	265	133	9968
WishBone	0.58	0.32	33	35	690

Table 2: Description of our Benchmarks

5.2 Generation of Thermal Profile

For a given RTL, we calculated the power dissipation in each module using Synopsys Power Compiler. This data was fed into *HotSpot* [24] along with physical location of the modules. The thermal map so obtained was then perturbed [-20, +20] Kelvin to take into account the hotspot migration over time. A set of 5 thermal profiles thus generated are used as a series of thermal profiles the chip is expected to work in.

5.3 Physical Design

We used relative placement feature of Silicon Ensemble tool to place cells belonging to a module in a fixed region. To reduce impact of this constraint on interconnect lengths, we first placed the design without region constraint. The placement thus obtained was used to infer the natural region where the cells can be constrained to be placed. The design was then placed again under region constraint. For all our experiments an operating frequency of 500MHz was used. The clock tree was generated using the CTGen tool. We constrained the transition time of clock signal at the sinks to be less than 10 times the transition time of the clock input signal. The skew bound was kept at a value of 200ps which is 10% of clock period.

5.4 Constraint Generation

The geometric information of clock tree was extracted from the DEF output of physical design. We applied the series of thermal profiles obtained for the design and calculated insertion delay to each sink for each profile. The nominal (with uniform thermal profile) delay was extracted from the SDF parasitics file generated during physical design. The values of insertion delays were used to generate the skew constraints to be solved by ILP solver. In this paper we have used a freely available Mixed Integer Programming Solver *lp_solve* [25]. We have interfaced this solver with Matlab and all the experiments were run on Matlab.

6. RESULTS

In Table 3, we report the results for the flow in Section 5 for both the 130nm and 90nm technologies. We see from the table that our method is able to compensate large skew values with very small number of TDBs, in most cases, the only exception being benchmark VGA_130, which required 31 TDBs (out of 33 buffers) with a significant amount of total tuning (Column *Taps*).

The complexity of the solution seems to be strongly affected by by the relative increase of skew due to the applied thermal maps. VGA and Wishbone have a worst-case skew increase of about 90% (from 18 to 34ps at 130nm) and about 80% (from 39 to 61ps), respectively. However, at 90nm, comparable increases of skew do require fewer TDBs to be compensated. Conversely designs such as Ethernet exhibit smaller relative skew increase, which can be compensated with fewer TDBs. The main evaluation metric for our methodology is the total clock tree power. The latter is calculated by obtaining the total capacitance seen by each net in clock tree obtained by parasitic extraction after detailed routing. The final power values for the clock tree with TDBs is calculated by adding the power component due to extra capacitance of the taps and due to the extra sizing of the buffers. We see from the results that increase in clock power is directly dependent on the number of TDBs and the total number of taps. While in some cases the increase is sizable (between 10 and 25%), this occurs for the 130nm implementations. In the 90nm implementation, conversely, we measured a marginal increases in clock power (less than 3.5% in all cases).

Designs	Orig Skew ps	Worst Skew ps	New Skew ps	Buf	TDB	Taps	Max Tap	Clk Pow mW	Clk Pow New mW	Rise %
Ethernet_130	114.0	135.1	≤114.0	265	8	21	3	16.96	17.30	2.0
VGA_130	18.1	34.0	≤18.1	33	31	49	2	2.61	3.10	18.8
Wishbone_130	39.2	61.2	≤39.2	33	17	28	3	1.30	1.60	23.1
I8051_130	14.3	4.5	≤14.3	9	8	10	2	0.20	0.22	11.2
Ethernet_90	171.3	217.3	≤171.3	133	5	13	6	31.51	31.62	0.3
VGA_90	45.2	71.5	≤45.2	67	5	8	3	4.10	4.20	2.4
Wishbone_90	50.4	84.3	≤50.4	35	1	5	5	2.70	2.79	3.3
I8051_90	18.1	27.7	≤18.1	9	3	4	2	0.31	0.32	3.2

Table 3: Skew Results.

7. CONCLUSIONS

Tunable delay buffers (TDBs) are an effective option to design variable-delay interconnections, which is commonly adopted in many high-performance circuits to compensate undesired timing variations due to various sources.

In this work we have investigated how these elements can be used to match temperature-induced delay *in a dynamic way*, that is, by selectively slowing down or speeding up portions of the clock tree, according to non-stationary temperature distributions, in order to keep the clock skew within the originally desired bounds.

We proposed a partially off-line methodology for clock tuning that requires the characterization of typical thermal profiles to pre-calculate the tuning configurations of the TDB, as well as an algorithm for the calculation of the minimum required number of TDBs and their tuning range.

Results, evaluated on a industry-strength CAD flow with realistic benchmark circuits, show that by replacing a limited subset of the buffers in the clock distribution network with TDBs, we are able to compensate for worst-case temperature gradients.

8. REFERENCES

[1] K. Skadron et al., "Temperature-Aware Computer Systems: Opportunities and Challenges," IEEE Micro, Vol. 23, No.6, Nov-Dec 2003, pp. 52-61.

[2] S. Borkar et al., "Parameter Variation and Impact on Circuits and Microarchitectures," DAC'03: Design Automation Conference, 2003, Jun. 2003, pp. 338-342.

[3] D. Brooks, M. Martonosi, "Dynamic thermal management for high-performance microprocessors" HPCA'01: High-Performance Computer Architecture, Jan. 2001, pp. 171-182.

[4] J. Srinivasan, S. V. Adve, "Predictive Dynamic Thermal Management for Multimedia Applications," ICS'03: International Conference on Supercomputing, June 2003.

[5] K. Skadron, "Hybrid architectural dynamic thermal management," DATE'04: Design Automation and Test in Europe Conference, Feb. 2004, pp. 10–15.

[6] Amir H. Ajami, Kaustav Banerjee, Massoud Pedram, "Modeling and Analysis of Nonuniform Substrate Temperature Effects on Global ULSI interconnects.", IEEE Transactions on CAD, Vol. 24, No. 6, June 2005, pp. 849–861.

[7] M. Cho, S. Ahmed, D. Z. Pan, "TACO: Temperature Aware Clock-tree Optimization," ICCAD'05: International Conference on Computer-Aided Design, Nov. 2005, pp. 582-587.

[8] A. Chakraborty et al.,"Thermal Resilient Bounded-Skew Clock Tree Optimization Methodology," DATE'06: Design Automation and Test in Europe, March 2006, pp. 832–837

[9] J.-L. Tsai, D. Baik, C. C.-P. Chen, K. K. Saluja, "A Yield Improvement Methodology Using Pre- and Post-silicon Statistical Clock Scheduling," ICCAD'04: International Conference on Computer-Aided Design, Nov. 2004, pp. 611–618.

[10] J.-L. Tsai, L. Zhang, C. Chen, "Statistical Timing Analysis Driven Post-Silicon-Tunable Clock-Tree Synthesis," ICCAD'05: International Conference on Computer-Aided Design, Nov. 2005, pp. 575-581.

[11] R. S. Tsay, "Exact Zero Skew Clock Routing" ICCAD'91: International Conference on Computer-Aided Design, Nov. 1991, pp. 336-339.

[12] T. H. Chao, Y-C.Hsu, J-M.Ho, K.D.Boese, A.B. Kahng, "Zero Skew Clock Routing with Minimum Wirelength," IEEE Transactions on Circuits and Systems II, Vol. 39, No. 11, Nov. 1992, pp. 799 - 814.

[13] J. Cong, C. K. Koh, "Minimum-Cost Bounded-Skew Clock Routing," ISCAS'95: International Symposium on Circuits and Systems, May 1995, pp. 215-218.

[14] D. J. H. Huang, A. B. Kahng, C.-H. A. Tsao, "On the Bounded-Skew Clock and Steiner Routing Problems," DAC'95: Design Automation Conference, Jun 1995, pp. 508–513.

[15] Y.P. Chen, D.F. Wang, "An algorithm for Zero-Skew Clock Tree Routing with Buffer Insertion," EDTC'96: European Design and Test Conference, Mar. 1996, pp. 230–236.

[16] Jeng-Liang Tsai, Tsung-Hao Chen, Chen, C.C.-P, "Zero Skew Clock Tree Optimization with Buffer insertion/sizing and wire sizing.", IEEE Transactions on CAD, Vol. 23, No. 4, Jun 2004, pp. 565-572.

[17] S. Tam et al., "Clock generation and distribution for the first IA-64 microprocessor," IEEE Journal of Solid-State Circuits, Vol. 35, No. 11, Jun 2000, pp. 1545–1552.

[18] E. Takahashi, Y. Kasai, M. Murakawa, T. Higuchi, "A post-silicon clock timing adjustment using genetic algorithms," VLSI'03: Symposium on VLSI circuits, pp. 13–16.

[19] A.DeHon, "In-System Timing Extraction and Control through Scan-Based, Test-Access Ports," ITC'94: International Test Conference 1994, Oct. 2004, pp. 350-359.

[20] Y.Elboim, A.Kolodny, R.Ginosar, "A Clock Tuning Circuit for System-on-Chip," IEEE transactions on VLSI, Vol. 11, Issue 4, Aug. 2003, pp. 616-626.

[21] S. Kaxiras, P. Xekalakis "4T-Decay Sensors: A New Class of Small, Fast, Robust, and Low-Power, Temperature/Leakage Sensors," ISLPED'04: International Symposium on Low Power Electronics and Design, Aug. 2004, pp. 108–113.

[22] P. Chen, C.-C. Chen, C.-C. Tsai, W.-F. Lu "A Time-to-Digital-Converter-Based CMOS Smart Temperature Sensor," IEEE Journal of Solid-State Circuits, Vol. 40, No. 8, August 2005, pp. 1642–1648.

[23] M. A. P. Pertijs "A CMOS Smart Temperature Sensor With a 3σ Inaccuracy of 0.5°C From 50°C to 120°C" IEEE Journal of Solid-State Circuits, Vol. 40, No. 2, February 2005, pp. 454–461.

[24] K.Skadron et.al "Temperature-Aware Microarchitecture," ISCA'03: International Symposium on Computer Architecture, Jun 2003, pp. 2-13.

[25] ftp://ftp.es.ele.tue.nl/pub/lp_solve

An Efficient Chip-level Time Slack Allocation Algorithm for Dual-Vdd FPGA Power Reduction [*]

Yan Lin[1], Yu Hu[1], Lei He[1] and Vijay Raghunat[2]
Electrical Engineering Dept., UCLA, Los Angeles, CA[1]
Purdue University, La Fayette, IN[2]

ABSTRACT

To reduce FPGA power, a linear programming (LP) based time slack allocation algorithm, EdTLC-LP, has been proposed recently for Vdd-programmable interconnects without using Vdd-level converters for mixed wire lengths. However, it takes a long time to solve the LP problem for time slack allocation. In this paper, we develop EdTLC-NW, a slack allocation algorithm based on min-cost network flow to reduce runtime. Compared to single Vdd FPGA with power-gating, EdTLC-LP and EdTLC-NW reduce interconnect power by 52.71% and 52.52%, respectively. EdTLC-NW achieves as good results as EdTLC-LP but runs $8X$ faster on average. Furthermore, the speedup increases for larger circuits and EdTLC-NW is $20X$ faster for the largest circuit.

Categories and Subject Descriptors: B.7.2 [Integrated Circuits]: Design aids

General Terms: Algorithms, Design

Keywords: Low power, time slack, FPGA

1. INTRODUCTION

FPGA power modeling and reduction has become an active research area recently. [1, 2] present power evaluation frameworks for generic parameterized FPGA architectures, and show that both interconnect and leakage power are significant for nanometer FPGAs. [3] studies the interaction of a suite of power-aware FPGA CAD algorithms without changing the existing FPGAs. [4] proposes a configuration inversion method to reduce leakage power of multiplexers. In addition, dual-Vdd and Vdd programmability have been applied to FPGA to reduce power. [5, 6] are the first work introducing dual-Vdd and field programmability of Vdd to FPGA. Vdd programmability has been applied to both FPGA logic blocks [5, 6] and interconnects [7, 8, 9].

A Vdd-level converter is needed when a low-Vdd (VddL) circuit elements drives a high-Vdd (VddH) circuit element to avoid excessive leakage. [8] inserts a level converter in front of each interconnect switch to provide the fine-grained Vdd programmability for interconnects. However, it has

---

[*]This paper is partially supported by NSF grant CCR-0306682. Address comments to lhe@ee.ucla.edu.

Permission to make digital or hard copies of all or part of this work for personal or classroom use is granted without fee provided that copies are not made or distributed for profit or commercial advantage and that copies bear this notice and the full citation on the first page. To copy otherwise, to republish, to post on servers or to redistribute to lists, requires prior specific permission and/or a fee.
ISLPED'06, October 4–6, 2006, Tegernsee, Germany.
Copyright 2006 ACM 1-59593-462-6/06/0010 ...$5.00.

been shown in [10] that this fine-grained Vdd-level converter insertion may introduce large leakage and area overhead. Recently, a few approaches have been presented without directly using level converters in Vdd-programmable interconnects. [9] uses the positive feedback PMOS transistor in the level-restore buffer as an alternative level converter with much reduced area and power overhead. [7] enforces that all the routing trees driven by (driving) a logic block have the same Vdd-level as the source (sink) logic block when level converters are inserted at CLB inputs (outputs). [11] uses a smaller granularity, a routing tree, as the unit in Vdd-level assignment. [10] further allows a mix of Vdd-levels within a routing tree, but only VddH switches can drive VddL switches. [12] extends the algorithms in [10] for mixed interconnect wire lengths.

In this paper, we use the same circuit design from [10, 12] and aim to improve the linear programming (LP) based time slack allocation algorithm, *EdTLC-LP*. In EdTLC-LP, time slack is allocated to each routing tree by formulating the problem as an LP problem to minimize power. However, it takes unacceptable runtime for EdTLC-LP to solve the LP problem for time slack allocation (more than 10 hours for the largest circuit *clma* on a 1.9GHz Xeon machine). Our contribution is to formulate the time slack allocation problem as a min-cost network flow problem and present a new algorithm, *EdTLC-NW*, which significantly reduces the run-time. Using single-Vdd FPGA with power-gating as the baseline, EdTLC-LP and EdTLC-NW reduce interconnect power by 52.71% and 52.52%, respectively. EdTLC-NW achieves as good results as EdTLC-LP but runs $8X$ faster on average. Furthermore, the speedup increases for larger circuits and EdTLC-NW achieves up to $20X$ speedup in overall runtime.

The rest of the paper is organized as follows. Section 2 introduces background and modeling. Section 3 reviews the LP based budgeting. Section 4 describes netflow based budgeting for interconnects with mixed wire lengths. Section 5 discusses the experimental results. Section 6 concludes this paper.

2. PRELIMINARIES
2.1 Delay and Power Modeling with Dual-Vdd

To make the presentation simple, we summarize the notations frequently used in this paper in Table 1. They will be explained in detail when first used.

A directed acyclic timing graph $\mathcal{G}(\mathcal{V}, \mathcal{E})$ [13] is constructed to model the circuit for timing analysis. The Elmore delay model is used to calculate the routing delay. We define the

$\mathcal{G}(\mathcal{V}, \mathcal{E})$	timing graph
$\mathcal{PI}$	set of all primary inputs and register outputs
$\mathcal{PO}$	set of all primary outputs and register inputs
$\mathcal{FO}_v$	set of all fanout vertices of vertex v in $\mathcal{G}$
$\mathcal{SRC}$	set of vertices corresponding to routing tree sources
$\mathcal{R}_i$	i^{th} routing tree in FPGA
$\mathcal{FO}_{ij}$	set of fanout switches of jth switch in $\mathcal{R}_i$
$\mathcal{SL}_{ij}$	set of sinks in the fanout cone of jth switch in $\mathcal{R}_i$
$a(v)$	arrival time of vertex v in $\mathcal{G}$
$d(u,v)$	delay from vertex u to vertex v in $\mathcal{G}$
N_r	total number (#) of routing trees in FPGA
c_{ij}	load capacitance of jth switch in $\mathcal{R}_i$
l_{ik}	# of switches in the path from source to k^{th} sink in $\mathcal{R}_i$
S_{ik}	allocated slack for k^{th} sink in $\mathcal{R}_i$
p_{i0}	vertex in $\mathcal{G}$ corresponding to the source of $\mathcal{R}_i$
p_{ik}	vertex in $\mathcal{G}$ corresponding to k^{th} sink of $\mathcal{R}_i$
$f_s(i,j)$	transition density of jth switch in $\mathcal{R}_i$
$N_k(i)$	# of sinks in $\mathcal{R}_i$
$N_s(i)$	total # of switches in $\mathcal{R}_i$
$N_l(i)$	# of VddL switches in $\mathcal{R}_i$
$F_n(i)$	estimated # of VddL switches in $\mathcal{R}_i$
W_{ik}	power weight associated with k^{th} sink in $\mathcal{R}_i$
w_{ik}	power weight associated with $e(i,j)$ in $\mathcal{G}$

Table 1: Notations frequently used in this paper

fanout cone of a switch as the sub-tree of the routing tree rooted at the switch. Dynamic power occurs when a signal transition happens at the gate output. Although timing change may change the transition density, we assume that the transition density for an interconnect switch will not change when VddL is used. Let v_{ij} indicate Vdd-level of jth switch in $\mathcal{R}_i$ as follows

$$v_{ij} = \begin{cases} 1 & \text{if Vdd-level of } j^{th} \text{ switch in } \mathcal{R}_i \text{ is VddH} \\ 0 & \text{if Vdd-level of } j^{th} \text{ switch in } \mathcal{R}_i \text{ is VddL} \end{cases}$$

The interconnect power reduction P_r using programmable dual-Vdd can be expressed as

$$P_r = \sum_{i=0}^{N_r-1} \sum_{j=0}^{N_s(i)-1} (1-v_{ij})(0.5 f_{clk} f_s(i,j) c_{ij} \Delta Vdd^2 + \Delta P_s(i,j)) \tag{1}$$

which is the sum of dynamic and leakage power reduction. N_r is the total number of routing trees, $f_s(i,j)$ is the transition density of jth switch in i^{th} routing tree $\mathcal{R}_i$, $N_s(i)$ is the number of switches in $\mathcal{R}_i$, and $\Delta P_s(i,j)$ and c_{ij} are the leakage power reduction and load capacitance of each switch, respectively.

Dual-Vdd tree based level converter insertion [10, 12] is used in this paper. A mix of Vdd-levels within one routing tree is allowed, and the Vdd-level constraints are

$$v_{ik} \leq v_{ij} \qquad 0 \leq i < N_r \wedge 0 \leq j < N_s(i) \wedge k \in \mathcal{FO}_{ij} \tag{2}$$

i.e., no VddL switch should drive VddH switches. $\mathcal{FO}_{ij}$ gives the set of fanout switches of jth switch in $\mathcal{R}_i$.

3. LINEAR PROGRAMMING BASED BUDGETING

In this section, we review the LP based time slack allocation algorithm, *EdTLC-LP*, for mixed interconnect wire lengths [12]. Time slack is first allocated to each routing tree by formulating the problem as an LP problem considering the load capacitance of each switch explicitly. A bottom-up assignment algorithm is then performed to achieve the optimal solution within each routing tree for the allocated time slack. A refinement step is finally performed to leverage surplus time slack.

3.1 Estimation of Interconnect Power Reduction

Estimating power reduction given the allocated slack is the key for the LP and netflow based algorithms. There is an upper bound for slack, which is the delay increase when VddL is assigned to all the switches in a tree. Clearly, slack more than the upper bound cannot lead to more VddL switches. The slack upper bound constraints can be expressed as

$$0 \leq S_{ik} \leq D_{ik} \qquad 0 \leq i < N_r \wedge 1 \leq k \leq N_k(i) \tag{3}$$

where $N_k(i)$ is the number of sinks in $\mathcal{R}_i$ and D_{ik} is the delay increase of the path from the source to k^{th} sink in $\mathcal{R}_i$ when VddL is assigned to all the switches in that path.

Let l_{ik} represent the number of switches in the path from the source to the kth sink in $\mathcal{R}_i$. Slack S_{ik} is first transformed into s_{ik}, which is expressed in number of switches as follows,

$$s_{ik} = \frac{S_{ik}}{D_{ik}} \cdot l_{ik} \tag{4}$$

Let c_{ij} represent the load capacitance of the jth switch in $\mathcal{R}_i$. Let C_{ik} represent the total load capacitance of the switches in the path from the source to the kth sink in $\mathcal{R}_i$. *Sink list $\mathcal{SL}_{ij}$ is defined as the set of sinks in the fanout cone of the jth switch in $\mathcal{R}_i$.* The number of VddL switches given the allocated slack is then estimated as

$$F_n(i) = \sum_{j=0}^{N_s(i)-1} min(\frac{s_{ik}}{C_{ik}} \cdot c_{ij} : \forall k \in \mathcal{SL}_{ij}) \tag{5}$$

The rationale is that we consider k^{th} sink with minimum $s_{ik} c_{ij}/C_{ik}$ in sink list $\mathcal{SL}_{ij}$ as the most critical sink to j^{th} switch in $\mathcal{R}_i$.

The dynamic/leakage power reduction of the tree $\mathcal{R}_i$ is estimated as the sum of the dynamic/leakage power reduction of each switch in $\mathcal{R}_i$ and can be expressed as,

$$P_{dr}(i) = 0.5 f_{clk} \cdot \Delta Vdd^2 \sum_{j=0}^{N_s(i)-1} [min(\frac{s_{ik}}{C_{ik}} \cdot c_{ij} : \forall k \in \mathcal{SL}_{ij}) \cdot f_s(i,j) \cdot c_{ij}] \tag{6}$$

$$P_{lr}(i) = \sum_{j=0}^{N_s(i)-1} [min(\frac{s_{ik}}{C_{ik}} \cdot c_{ij} : \forall k \in \mathcal{SL}_{ij}) \cdot \Delta P_s(i,j)] \tag{7}$$

where $\Delta P_s(i,j)$ is the leakage power difference of jth switch in $\mathcal{R}_i$ between VddH and VddL. Wire segments with different lengths might be driven by switches with different sizes.

3.2 LP Problem Formulation

Similar to [10], the net-based formulation is used, which partitions the constraints on path delay into constraints on delay across circuit elements or routing. Let $a(v)$ be the arrival time for vertex v in $\mathcal{G}$ and the timing constraints become

$$a(v) \leq T_{spec} \qquad \forall v \in \mathcal{PO} \tag{8}$$
$$a(v) = 0 \qquad \forall v \in \mathcal{PI} \tag{9}$$
$$a(u) + d(u,v) \leq a(v) \quad \forall u \in \mathcal{V} \wedge v \in \mathcal{FO}_u \tag{10}$$

where $\mathcal{V}$ is the set of vertices in $\mathcal{G}$, $d(u,v)$ is the delay from vertex u to v and $\mathcal{FO}_u$ is the set of fanout vertices of u.

The objective is to maximize interconnect power reduction given by the sum of (6) and (7). To incorporate them into mathematical programming, we introduce a variable $f_n(i,j)$ for j^{th} switch in $\mathcal{R}_i$ and some additional constraints. The new objective function after transformation plus the additional constraints can be expressed as

$$Maximize \sum_{i=0}^{N_r-1} 0.5 f_{clk} \Delta V dd^2 \sum_{j=0}^{N_s(i)-1} f_n(i,j) f_s(i,j) c_{ij}$$

$$+ \sum_{i=0}^{N_r-1} \sum_{j=0}^{N_s(i)-1} f_n(i,j) \Delta P_s(i,j) \quad (11)$$

s.t.

$$f_n(i,j) \leq \frac{s_{ik}}{C_{ik}} c_{ij} \quad 0 \leq i < N_r \wedge 0 \leq j < N_s(i) \wedge \forall k \in \mathcal{SL}_{ij} \quad (12)$$

The timing constraints (10) is then modified as follows. For the edges corresponding to routing in $\mathcal{G}$, the constraints considering slack can be expressed as

$$a(p_{i0}) + d(p_{i0}, p_{ik}) + S_{ik} \leq a(p_{ik})$$
$$0 \leq i < N_r \wedge \forall p_{ik} \in \mathcal{FO}_{p_{i0}} \quad (13)$$

where vertex p_{i0} is the source of $\mathcal{R}_i$ in $\mathcal{G}$, vertex p_{ik} is k^{th} sink of $\mathcal{R}_i$ in $\mathcal{G}$, S_{ik} is the slack allocated to k^{th} sink in $\mathcal{R}_i$ and $d(p_{i0}, p_{ik})$ is the delay from p_{i0} to p_{ik} in $\mathcal{R}_i$ using VddH. For the edges other than routing in $\mathcal{G}$, the constraints can be expressed as

$$a(u) + d(u,v) \leq a(v) \quad \forall u \in \mathcal{V} \wedge u \notin \mathcal{SRC} \wedge v \in \mathcal{FO}_u \quad (14)$$

where $\mathcal{SRC}$ contains vertices corresponding to routing tree sources.

The *time slack allocation problem* is formulated using objective function (11), additional constraints (12), slack upper bound constraints (3), and timing constraints (8), (9), (13) and (14). It is easy to verify that all the constraints are linear, and the objective function (11) is also linear.

THEOREM 1. *The time slack allocation problem is a linear programming (LP) problem.*

4. NETWORK FLOW BASED BUDGETING

The runtime of time slack allocation in EdTLC-LP can be very long for large circuits mainly due to the expensive computational time of linear programming. In this section, we formulate the time slack allocation problem as a min-cost network flow problem and present a new algorithm, *EdTLC-NW*, with significantly reduced runtime. Similar network flow formulation has been used for timing budgeting in high level synthesis [14].

4.1 Network Flow Formulation

We first deliberately distribute slack s_{ik} to the switches in the path from the source to to k^{th} sink in $\mathcal{R}_i$. As the min function in (5) cannot be efficiently handled by a network flow formulation, we define the sink with the minimum slack as the *critical sink* in the fanout cone of a switch with multiple sinks. We then use $s_{ik} c_{ij} / C_{ik}$ of this critical sink to replace the min operator over all sinks in the fanout cone of a switch. The dynamic power reduction (6) and leakage power reduction (7) for $\mathcal{R}_i$ can be rewritten as follows,

$$P_{dr}(i) = 0.5 f_{clk} \cdot \Delta V dd^2 \cdot \sum_{j=0}^{N_s(i)-1} S_{ik} \cdot [f_s(i,j) \cdot \frac{l_{ik} \cdot c_{ij}^2}{D_{ik} \cdot C_{ik}}] \quad (15)$$

$$P_{lr}(i) = \sum_{j=0}^{N_s(i)-1} S_{ik} \cdot [(\frac{l_{ik} \cdot c_{ij}}{D_{ik} \cdot C_{ik}}) \cdot \Delta P_s(i,j)] \quad (16)$$

The objective function can be rewritten as following by merging the coefficient of the slack S_{ik} of kth sink in $\mathcal{R}_i$ as follows,

$$Maximize \sum_{i=0}^{N_r-1} \sum_{k=0}^{N_k(i)-1} W_{ik} \cdot S_{ik} = \sum_{\forall Sink} W_{ik} \cdot S_{ik} \quad (17)$$

$$W_{ik} = \sum_{\forall j \in \mathcal{UBC}_{ik}} [0.5 f_{clk} \Delta V dd^2 c_{ij} f_s(i,j) + \Delta P_s(i,j)] \cdot \frac{c_{ij} \cdot D_{ik}}{(C_{ik} \cdot l_{ik})} \quad (18)$$

where set $\mathcal{UBC}_{ik}$ include all switches with k^{th} sink as the critical sink in $\mathcal{R}_i$.

Since $W_{ik} > 0$ for all sinks, we can restrict timing constraint (13) as the following equation to maximize the objective function,

$$S_{ik} = a(p_{ik}) - a(p_{i0}) - d(p_{i0}, p_{ik}), \quad 0 \leq i < N_r \wedge \forall p_{ik} \in \mathcal{FO}_{p_{i0}} \quad (19)$$

After substituting S_{ik} using (19) and rearrangement, objective function (17) can be expressed as,

$$Maximize \sum_{i=0}^{N_r-1} \sum_{k=0}^{N_k(i)-1} W_{ik} \cdot [a(p_{ik}) - a(p_{i0}) - d(p_{i0}, p_{ik})] \quad (20)$$

Similarly, slack bound constraint (3) can be rewritten as

$$a(p_{i0}) - a(p_{ik}) \leq -d(p_{i0}, p_{ik}) \quad (21)$$
$$a(p_{ik}) - a(p_{i0}) \leq d(p_{i0}, p_{ik}) + D_{ik} \quad (22)$$

We then merge the timing constraint (21) and (14) into the general expression (10).

Similar to [14], a virtual input node (SI) and a virtual output node (SO) are added into $\mathcal{G}$ to connect all nodes in $\mathcal{PI}$ and $\mathcal{PO}$, respectively. All edges connected to SI and SO have zero delay. We add a backward edge $e(p_{ik}, p_{i0})$ for each source sink pair in $\mathcal{R}_i$. A delay of $-(d(p_{i0}, p_{ik}) + D_{ik})$ is associated to $e(p_{ik}, p_{i0})$ to represent the slack upper bound. A virtual edge $e(SO, SI)$ with delay $-T_{spec}$ is then added. All constraints can now be represented by edges in $\mathcal{G}$. For example, edge $e(u,v)$ with delay $d(u,v)$ represents constraint $a(u) - a(v) < -d(u,v)$.

To represent the objective function (20) in $\mathcal{G}$, we associate a weight w_{uv} in each edge $e(u,v)$. For those edges $e(p_{i0}, p_{ik})$ corresponding to routing, let $w_{p_{i0} p_{ik}} = W_{ik}$. For other edges, let $w_{uv} = 0$. The objective function (20) can then be rewritten as,

$$Maximize \sum_{v \in V} a(v)(\sum_{u \in \mathcal{FI}_v} w_{uv} - \sum_{u \in \mathcal{FO}_v} w_{vu})$$
$$- \sum_{i=0}^{N_r-1} \sum_{k=0}^{N_k(i)-1} d(p_{i0}, p_{ik}) \quad (23)$$

where $\sum_{i=0}^{N_r-1} \sum_{k=0}^{N_k(i)-1} d(p_{i0}, p_{ik})$ is a constant and can be removed from the objective function (23), and $\mathcal{FI}_v / \mathcal{FO}_v$ is fanin/fanout set of vertex v.

For the optimization problem with constraints (10) and (22) and objective function (23), its dual problem is

$$\min \quad \sum_{e_{(i,j)} \in \mathcal{E}} (d(i,j) + D_{ij}) \cdot z_{ij} - d(i,j) \cdot y_{ij} \quad (24)$$

$$s.t. \quad \sum_{e(k,i) \in \mathcal{E}} (y_{ki} - z_{ki}) - \sum_{e(i,j) \in \mathcal{E}} (y_{ij} - z_{ij}) = \rho_i \quad (25)$$

$$\rho_i = \sum_{j \in \mathcal{FI}_i} w_{ji} - \sum_{k \in \mathcal{FO}_k} w_{ik} \quad (26)$$

$$y_{ij}, z_{ij} \in R_+ \quad (27)$$

170

To verify that the above dual problem is a min-cost network flow problem on $\mathcal{G}$, y_{ij} is the flow along $e(i,j)$ with cost $-d(i,j)$, z_{ij} is the flow along $e(j,i)$, which corresponds to routing and is associated with cost $d(i,j) + D_{ij}$. Obviously, no negative cycle is introduced by the backward edges. ρ_i is the demand in each vertex. Note that $\sum_{i \in \mathcal{V}} \rho_i = 0$ is satisfied as required in the min-cost network flow problem. Hence, we have the following theorem.

THEOREM 2. *The dual problem of the time slack allocation problem is a min-cost network flow problem.*

After solving the min-cost network flow problem, we can get the solutions for variables y_{ij} and z_{ij}. Similar to [14], we can calculate the solution of the primal problem. We first construct the residual graph $\mathcal{G}'(\mathcal{V}, \mathcal{E}')$ from the original $\mathcal{G}$. For any edge $e(i,j)$ in $\mathcal{G}'$ with non-zero flow, there are two edges $e(j,i)$ and $e(i,j)$ in $\mathcal{G}'$. The cost of each backward edge $e(j,i)$ is $d(i,j)$, and is equal to the complement of the forward edge cost. Let δ_i be the shortest distance from SI to vertex i in $\mathcal{G}'$. It has been proved in [14] that $a(i) = -\delta_i$ is an optimal solution to the primal problem. We use the push-relabel algorithm [15] for min-cost flow problem and Bellman-Ford algorithm [16] for shortest path problem.

4.2 Comparison Between EdTLC-LP and EdTLC-NW

The basic difference between EdTLC-LP and EdTLC-NW is the way to calculate the slack for each switch. EdTLC-LP chooses the minimum $s_{ik}c_{ij}/C_{ik}$ among all sinks while EdTLC-NW chooses the $s_{ik}c_{ij}/C_{ik}$ of the critical sink. Let k^{th} sink be the critical sink in the fanout cone of jth switch in $\mathcal{R}_i$. EdTLC-NW estimates number of VddL switches in $\mathcal{R}_i$ as

$$F_n(i) = \sum_{j=0}^{N_s(i)-1} \frac{s_{ik}}{C_{ik}} \cdot c_{ij} \qquad (28)$$

It has been proved in [10] that (5) can always give a lower bound of the number of VddL switches that can be achieved in $\mathcal{R}_i$ for uniform wire length. However, (5) cannot always give an infimum (the greatest lower bound) for the VddL switch number. Figure 1 (a) shows a simple example (uniform length of wire segments is assumed for simplicity).

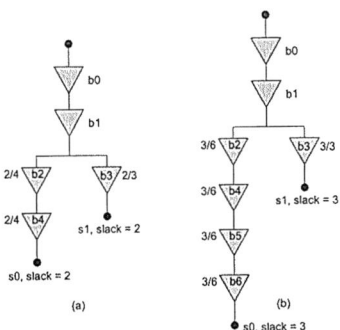

(a)   (b)

Figure 1: An example of estimated VddL switch #

Suppose $S1$ is the critical sink based on the current timing graph, and a slack of 2 is allocated to both $S0$ and $S1$. (5) estimates VddL switch number as $(2/4 + 2/4 +$

$2/3 + 2\min(2/4 + 2/3)) = 8/3$ while (28) gives an estimation as $(2/4 + 2/4 + 2/3 + 2 \cdot 2/3) = 3$. Obviously, we CAN achieve three VddL switches while satisfying the allocated slack $S0 = S1 = 2$. This indicates that EdTLC-LP cannot always give the infimum of VddL switch number.

On the other hand, EdTLC-NW always gives a greater estimation than EdTLC-LP while it might give an over-estimated result. As shown in Figure 1 (b), suppose $S1$ is still the critical sink. The slacks allocated to $S0$ and $S1$ are both 3. (5) gives an estimation as $(3/6 + 3/6 + 3/6 + 3/6 + 3/3 + 2\min(3/6 + 3/3)) = 4$ while (28) gives $(3/6 + 3/6 + 3/6 + 3/6 + 3/3 + 2 \cdot 3/3) = 5$. However, we cannot achieve five VddL switches while satisfying the allocated slacks $S0 = S1 = 3$. This indicates that EdTLC-NW may overestimate the VddL switch number in some cases.

In summary, EdTLC-LP may give a conservative estimation of power savings while EdTLC-NW may give an over-optimistic estimation. In the experiments to be presented in Section 5, the two formulations give a similar estimation for most cases due to the fact that the most critical sink is usually the one with the minimum $s_{ik}c_{ij}/C_{ik}$. In addition, EdTLC-NW can achieve results close to EdTLC-LP in terms of power reduction but with significantly shorter runtime.

5. EXPERIMENTAL RESULTS

5.1 Experimental Settings

We conduct the experiments on the largest MCNC benchmarks [17] including ten combinational circuits (group I in Table 2) and ten sequential circuits (group II in Table 2). We map them into FPGA with LUT size of 4 and cluster size of 10. We use the same Vdd-programmable logic blocks and interconnects in [10], but with a mix of different interconnect wire lengths. We use 60% length 4 wire and 40% length 8 wire for better performance and area tradeoff, as suggested in [18]. The unused interconnect switches are power-gated in all cases. Similar to [8], we customize the FPGA chip size for each benchmark circuit and use the smallest chip that fits each benchmark. We use 1.3v for VddH and 0.8v for VddL same as [10] in our experiments at $100nm$ technology node. We perform dual-Vdd assignment without delay increase compared to the circuit using only VddH in the rest of the paper.

We first use VPR [13] for single-Vdd placement and routing. Before applying budgeting algorithms to the Vdd programmable interconnects, a sensitivity based assignment [6] is first performed to assign Vdd-level for Vdd-programmable logic blocks without performance loss[1]. The cycle-accurate FPGA power simulator *fpgaEva-LP2* [11] is then used to calculate power.

5.2 Comparison of Interconnect Power

We first compare the the number of VddL switches achieved by EdTLC-LP and EdTLC-NW in Table 2, as it is a good indication of power reduction. The number of VddL switches is expressed in percent of used switch number. EdTLC-LP and EdTLC-NW achieve 84.43% and 84.05% VddL switches, respectively. Both achieve almost the same number of VddL switches.

---

[1]Both algorithms EdTLC-LP and EdTLC-NW do consider the fact that VddL logic blocks consume time slack.

171

We then present the interconnect power reduction achieved by EdTLC-LP and EdTLC-NW in Table 3. Single-Vdd FPGA with power-gating is used as the baseline for interconnect power. Columns 2-3 present the interconnect dynamic and leakage power for the baseline case. Columns 4-5 present the overall interconnect power reduction for EdTLC-LP and EdTLC-NW . Compared to the baseline case, EdTLC-LP and EdTLC-NW reduce total interconnect power by 52.71% and 52.03%, respectively. We also present the interconnect dynamic power and leakage power reduction in columns 6-9. Compared to the baseline, EdTLC-LP and EdTLC-NW reduce dynamic power by 52.03% and 51.69%, and leakage power by 62.71% and 62.51%, respectivly. Clearly, EdTLC-NW achieves as good results as EdTLC-LP .

For both algorithms, we also present the contribution of refinement step in Table 2 and Table 3. The refinement step achieves 3.49% and 4.17% VddL switches for EdTLC-LP and EdTLC-NW , respectively. Compared to baseline, the refinement step in EdTLC-LP /EdTLC-NW obtains 3.35%/ 1.03% more power reduction, 3.45%/0.96% more dynamic power reduction and 2.23%/2.57% more leakage power reduction, respectively. It is clear that the refinement step is effective to distribute surplus time slack and further reduce interconnect power. The first source of the surplus time slack is the difference between the continuous problem formulations (i.e., the allocated slack is continuous in budgeting) and the fact that Vdd-level assignment is discrete (i.e., the slack consumed by a VddL switches must be Δd). Secondly, the objective of our formulations is to maximize the estimated power reduction instead of the exact power reduction. The two may be different due to the fact that not all allocated slack is useful for power reduction.

	Circuit	Cluster#	EdTLC-LP	EdTLC-NW
	ex5p	123	68.32% (3.43%)	67.99% (4.87%)
	apex4	134	75.38% (6.65%)	73.27% (8.04%)
	misex3	153	73.77% (4.12%)	72.41% (9.11%)
	alu4	162	78.67% (4.89%)	78.08% (6.54%)
I	seq	198	70.49% (4.93%)	69.52% (6.63%)
	apex2	213	78.48% (3.98%)	78.09% (4.74%)
	des	218	82.13% (1.94%)	82.01% (1.86%)
	spla	399	78.46% (4.40%)	78.36% (6.01%)
	ex1010	493	78.77% (5.55%)	78.59% (6.84%)
	pdc	568	79.76% (4.19%)	79.29% (5.16%)
	tseng	131	97.02% (2.58%)	97.07% (0.71%)
	dsip	162	91.70% (1.10%)	91.67% (1.67%)
	diffeq	195	92.93% (2.58%)	92.90% (1.21%)
II	s298	256	89.62% (2.93%)	89.59% (1.41%)
	bigkey	294	81.43% (2.18%)	80.88% (7.27%)
	elliptic	421	98.72% (1.92%)	98.75% (0.28%)
	frisc	595	99.19% (4.74%)	99.19% (2.58%)
	s38584.1	704	97.77% (1.86%)	97.75% (0.65%)
	s38417	847	90.15% (2.53%)	89.86% (2.93%)
	clma	1358	85.93% (3.20%)	85.66% (3.38%)
	Ave	266	84.43% (3.49%)	84.05% (4.17%)

Table 2: Percentage of VddL switches achieved by EdTLC-LP and EdTLC-NW

5.3 Comparison of Runtime

Table 4 compares the runtime of EdTLC-LP and EdTLC-NW. Column "budget" refers to the runtime for distributing slack to each tree. The simplex method based LP solver [19] is used in EdTLC-LP . Column "total" presents the overall runtime. Column "speedup" shows the ratio of speedup achieved by EdTLC-NW compared to EdTLC-LP. EdTLC-NW achieves as good results as EdTLC-LP but runs $8X$ faster on average. It is clear that the speedup of the bud-

geting time increases for the larger circuits. For the largest circuit *clma*, which contains over 1k clusters, EdTLC-NW achieves up to $6000X$ speedup in time slack allocation and $20X$ speedup in overall runtime. The current commercial FPGA designs often contain 10k to 100k clusters [20], which indicates that the EdTLC-NW may have more speedup than that reported in this paper in practice. The min-cost network flow based algorithm EdTLC-NW takes negligible runtime in time slack allocation compared to EdTLC-LP . Clearly, the efficiency of EdTLC-NW makes our algorithm highly scalable, especially for the applications that need iterative budgeting for different time specifications.

	EdTLC-NW		EdTLC-LP		speedup(X)	
cir	budget	total	budget	total	budget	total
ex5p	1	33	1187	1254	118x	5x
apex4	1	33	184	228	184x	7x
misex3	1	38	155	227	155x	6x
alu4	1	36	101	131	101x	4x
seq	1	55	179	216	179x	4x
apex2	1	78	413	502	413x	6x
des	1	95	327	444	327x	5x
spla	1	300	837	1181	837x	4x
ex1010	2	346	2391	2804	1196x	8x
pdc	2	803	3633	4496	1817x	6x
tseng	1	32	83	91	83x	3x
dsip	1	44	181	223	181x	5x
diffeq	1	41	252	321	252x	8x
s298	1	96	371	494	371x	5x
bigkey	1	89	478	589	478x	7x
elliptic	1	229	877	1128	877x	5x
frisc	2	479	1823	2364	912x	5x
s38584	3	421	2305	2806	768x	7x
s38417	4	709	3719	4463	930x	6x
clma	9	2735	53712	56726	5968x	21x
Ave	1	335	3660	2775	3607x	8x

Table 4: Runtime (second) comparison between EdTLC-LP and EdTLC-NW

6. CONCLUSIONS

To reduce power in dual-Vdd FPGA, we have re-formulated the LP based time slack allocation problem to a min-cost network flow based problem and presented a new network flow based algorithm, EdTLC-NW , with significantly shorter run-time. Using single-Vdd FPGA with power-gating as the baseline, the linear programming (LP) based time budgeting algorithm EdTLC-LP [12] and EdTLC-NW reduce interconnect power by 52.71% and 52.52%, respectively. EdTLC-NW achieves as good results as EdTLC-LP but runs $8X$ faster on average. The speedup increases for larger circuits. For the largest circuit, EdTLC-NW achieves up to $6000X$ speedup in time slack allocation and $20X$ speedup in overall runtime. Clearly, the efficiency of EdTLC-NW makes our algorithm highly scalable, especially for the applications that may need iterative budgeting procedures. We expect that EdTLC-NW has more speedup in real designs than that reported in this paper since real designs are often bigger than the examples in this paper.

7. REFERENCES

[1] K. Poon, A. Yan, and S. Wilton, "A flexible power model for FPGAs," in *Proc. of 12th International conference on Field-Programmable Logic and Applications*, Sep 2002.

[2] F. Li, D. Chen, L. He, and J. Cong, "Architecture evaluation for power-efficient FPGAs," in *Proc. ACM Intl. Symp. Field-Programmable Gate Arrays*, Feb 2003.

1	2	3	4	5	6	7	8	9
	baseline power (W)		interconnect power reduction		dynamic power reduction		leakage power reduction	
	dynamic	leakage	EdTLC-LP	EdTLC-NW	EdTLC-LP	EdTLC-NW	EdTLC-LP	EdTLC-NW
ex5p	0.017687	0.000744	40.64% (3.25%)	40.42% (1.31%)	40.31% (3.28%)	40.08% (1.22%)	48.33% (2.55%)	48.47% (3.47%)
apex4	0.020157	0.000768	45.24% (4.92%)	44.19% (1.49%)	44.91% (4.93%)	43.87% (1.34%)	53.79% (4.81%)	52.38% (5.19%)
misex3	0.037557	0.000768	44.57% (3.23%)	44.02% (2.73%)	44.40% (3.23%)	43.86% (2.65%)	52.83% (2.89%)	52.14% (6.70%)
alu4	0.031968	0.000617	48.01% (6.09%)	47.92% (1.55%)	47.83% (6.14%)	47.73% (1.50%)	57.66% (3.34%)	57.45% (4.15%)
seq	0.053091	0.001042	45.43% (3.58%)	45.06% (1.65%)	45.31% (3.59%)	44.94% (1.60%)	51.71% (2.96%)	51.25% (3.95%)
apex2	0.053479	0.00121	49.71% (4.35%)	49.56% (0.95%)	49.51% (4.39%)	49.36% (0.90%)	58.65% (2.62%)	58.33% (2.97%)
des	0.069708	0.001174	50.18% (1.31%)	50.02% (0.74%)	50.02% (1.30%)	49.86% (0.73%)	59.73% (1.78%)	59.80% (1.78%)
spla	0.042551	0.002538	49.88% (3.74%)	49.75% (1.37%)	49.39% (3.80%)	49.27% (1.24%)	58.09% (2.65%)	57.83% (3.68%)
ex1010	0.028371	0.003003	46.08% (4.96%)	45.89% (1.38%)	44.74% (5.09%)	44.53% (1.03%)	58.77% (3.65%)	58.76% (4.76%)
pdc	0.055369	0.003998	47.87% (3.42%)	47.49% (1.08%)	46.99% (3.53%)	46.61% (0.97%)	60.06% (1.97%)	59.65% (2.71%)
tseng	0.006299	0.000545	62.39% (3.37%)	62.39% (0.06%)	61.44% (3.53%)	61.45% (0.03%)	73.29% (1.55%)	73.34% (0.39%)
dsip	0.049421	0.000842	55.71% (0.97%)	55.67% (0.72%)	55.50% (0.97%)	55.46% (0.71%)	68.00% (0.97%)	68.12% (1.38%)
diffeq	0.004697	0.000741	61.92% (2.93%)	61.93% (0.14%)	60.72% (3.13%)	60.73% (0.05%)	69.49% (1.64%)	69.57% (0.68%)
s298	0.010679	0.000884	57.06% (4.80%)	57.04% (0.29%)	56.26% (5.06%)	56.23% (0.25%)	66.74% (1.70%)	66.86% (0.78%)
bigkey	0.049461	0.000956	48.14% (0.98%)	47.84% (3.07%)	47.90% (0.97%)	47.60% (3.03%)	60.68% (1.57%)	60.54% (4.91%)
elliptic	0.015046	0.002065	62.71% (2.91%)	62.72% (0.01%)	61.04% (3.11%)	61.04% (0.01%)	74.94% (1.43%)	74.95% (0.04%)
frisc	0.012426	0.002612	64.32% (3.78%)	64.32% (0.01%)	62.05% (4.18%)	62.05% (0.00%)	75.11% (1.87%)	75.11% (0.05%)
s38584.1	0.058038	0.002767	62.00% (1.69%)	62.04% (0.09%)	61.44% (1.71%)	61.48% (0.09%)	73.89% (1.35%)	73.86% (0.25%)
s38417	0.056849	0.003581	54.86% (1.98%)	54.69% (1.47%)	54.07% (1.99%)	53.91% (1.43%)	67.36% (1.79%)	67.06% (2.15%)
clma	0.070958	0.008078	57.55% (4.70%)	57.39% (0.50%)	56.68% (5.06%)	56.53% (0.41%)	65.17% (1.51%)	64.93% (1.32%)
Ave.	0.037191	0.001947	52.71% (3.35%)	52.52% (1.03%)	52.03% (3.45%)	51.83% (0.96%)	62.71% (2.23%)	62.52% (2.57%)

Table 3: Total power achieved by EdTLC-LP and EdTLC-NW

[3] J. Lamoureux and S. J. Wilton, "On the interaction between power-aware FPGA CAD algorithms," in *Proc. Intl. Conf. Computer-Aided Design*, pp. 701–708, November 2003.

[4] J. H. Anderson, F. N. Najm, and T. Tuan, "Active leakage power optimization for FPGAs," in *Proc. ACM Intl. Symp. Field-Programmable Gate Arrays*, Februray 2004.

[5] F. Li, Y. Lin, L. He, and J. Cong, "Low-power FPGA using pre-defined dual-vdd/dual-vt fabrics," in *Proc. ACM Intl. Symp. Field-Programmable Gate Arrays*, Februray 2004.

[6] F. Li, Y. Lin, and L. He, "FPGA power reduction using configurable dual-vdd," in *Proc. Design Automation Conf.*, June 2004.

[7] A. Gayasen, K. Lee, N. Vijaykrishnan, M. Kandemir, M. J. Irwin, and T. Tuan, "A dual-vdd low power FPGA architecture," in *Proc. Intl. Conf. Field-Programmable Logic and its Application*, August 2004.

[8] Fei Li and Yan Lin and Lei He, "Vdd programmability to reduce FPGA interconnect power," in *Proc. Intl. Conf. Computer-Aided Design*, November 2004.

[9] Jason H. Anderson and Farid N. Najm, "Low-power programmable routing circuitry for FPGAs," in *Proc. Intl. Conf. Computer-Aided Design*, November 2004.

[10] Y. Lin and L. He, "Leakage efficient chip-level dual-vdd assignment with time slack allocation for FPGA power reduction," in *Proc. Design Automation Conf.*, June 2005.

[11] Y. Lin, F. Li, and L. He, "Power modeling and architecture evaluation for FPGA with novel circuits for vdd programmability," in *Proc. ACM Intl. Symp. Field-Programmable Gate Arrays*, Februray 2005.

[12] Y. Hu, Y. Lin, L. He, and T. Tuan, "Simultaneous time slack budgeting and retiming for dual-vdd FPGA power reduction," in *Proc. Design Automation Conf.*, July 2006.

[13] V. Betz, J. Rose, and A. Marquardt, *Architecture and CAD for Deep-Submicron FPGAs*. Kluwer Academic Publishers, Feb 1999.

[14] S. C. Soheil Ghiasi, Elaheh Bozorgzadeh and M. Sarrafzadeh, "A unified theory of timing budget management," in *Proc. Intl. Conf. Computer-Aided Design*, November 2004.

[15] A. V. Goldberg, "An efficient implentation of a scaling minimum-cost flow algorithm," *Journal of Algorithms*, vol. 22, pp. 1–29, 1997.

[16] C. L. R. Rivest, T. Cormen, *An Introduction to Algorithms*. MIT Press, 1990.

[17] S. Yang, "Logic synthesis and optimization benchmarks, version 3.0," tech. rep., Microelectronics Center of North Carolina (MCNC), 1991.

[18] D. Lewis and et al, "The stratix routing and logic architecture," in *Proc. ACM Intl. Symp. Field-Programmable Gate Arrays*, Feb 2003.

[19] M Berkelaar, *lp-solver: a public domain (MI)LP solver*. ftp://ftp.ics.ele.tue.nl/pub/lp_solve/.

[20] "Xilinx product datasheets," in *http://www.xilinx.com/literature*.

A Novel Approach for Variation Aware Power Minimization during Gate Sizing

V. Mahalingam
University of South Florida
Tampa, FL
mvenkata@cse.usf.edu

N. Ranganathan
University of South Florida
Tampa, FL
ranganat@cse.usf.edu

Justin E. Harlow III
University of South Florida
Tampa, FL
harlow@cse.usf.edu

ABSTRACT

Increasing dominance of process variations in the nanometer designs are posing significant challenges for circuit design and optimization. The variations in parameters such as channel length and the gate oxide thickness impacts circuit delay and power. In this paper, we propose a new gate sizing algorithm using fuzzy mathematical programming (FMP) in which the uncertainty due to process variations is modeled using fuzzy numbers. The variations in gate delay, which is a function of gate sizes and the fan-outs of the gate, are represented using triangular fuzzy numbers with linear membership functions. The variation aware gate sizing problem is formulated as a fuzzy mathematical program to perform a delay constrained power minimization in the presence of variations. Initially, a deterministic optimization is performed by fixing the fuzzy parameters to the worst and the average case values and the results are used to convert the fuzzy optimization problem into a crisp non-linear problem which is then solved using a non-linear optimization solver. The above model with delay and power as constraints, maximizes the robustness, i.e., the variation resistance of the circuit and thus the yield. The proposed approach was tested on ISCAS '85 benchmarks and the results were validated for timing yield using monte-carlo simulations. The fuzzy approach yields significantly better results compared to stochastic programming based gate sizing approach with a comparable runtime.

Categories and Subject Descriptors

B.6.3 [**Hardware**]: Logic Design—*Design Aids*

General Terms

Algorithms, Design

1. INTRODUCTION

Advances in VLSI technology have continued to shrink device geometries for improving performance of integrated

Permission to make digital or hard copies of all or part of this work for personal or classroom use is granted without fee provided that copies are not made or distributed for profit or commercial advantage and that copies bear this notice and the full citation on the first page. To copy otherwise, to republish, to post on servers or to redistribute to lists, requires prior specific permission and/or a fee.
ISLPED'06, October 4–6, 2006, Tegernsee, Germany.
Copyright 2006 ACM 1-59593-462-6/06/0010 ...$5.00.

circuits. This aggressive scaling however, has given rise to increased variations in device parameters. These variations can either be environmental or physical. Environmental variations are due to power supply voltage and temperature, while physical variations are due to discrepancy in transistor width, channel length, oxide thickness and interconnect dimensions. These variations can be attributed to difficulty in fabrication control and randomness in the number of dopants atoms in transistors. The end result is a loss in parametric yield due to a large difference in identically designed circuits. Hence, there is a strong need to consider these variations as a part of the design and optimization of VLSI circuits.

Circuit optimziation over the past three decades, has successfully improved performance and power of microprocessor designs. These optimizations achieve the improvement by making most paths of the circuits equally critical. However, with process variations, such optimization can worsen timing yield, defined as the percentage of chips meeting the timing specification, as any of these critical paths can fail [1]. A guarded approach to eliminate the effects of variability is to perform deterministic optimization at the worst case values of the varying parameters. Worst casing approach guarantess high yield, but leads to high overhead in terms of circuit area and power. On the other hand average case values for these parameters has less overheads, but results in unacceptable timing yield. It is clear that new methodologies are needed, which can guarantee high yield without a lot of area and power overhead. The problem of gate sizing can be formally defined as the process of finding the optimal drive strengths of individual gates of a circuit for a given objective function and constraints. For example, the objective function can be to minimze power or area for a specified timing target. Several solutions exist for the gate sizing problem, formulating it as unconstrained delay minimization or area and power minimization problem under delay constraints. In [7], authors use a iterative approach to find the most sensitive gate to size in the current iteration using posynomial delay models. Lagrangian relaxation [18] method and game theoretic approaches [13] have also been used to efficiently solve the gate sizing problem. However, none of these problems consider variability as a part of the optimization and hence are not suitable for optimizing designs in the nanometer regime, where optimization without variability is not effective.

Process variations can be categorized as inter-die and intra-die variations. Inter-die variations refers to variations across different dies and affects all the transistors in the chip in a similar fashion. Intra-die variations on the other hand, refers to variability within a single chip and for example can make gate length of some transistors larger and others smaller. Several studies in the literature have attempted to include process variations in timing analysis and optimization of digital circuits. Static timing analysis was replaced with statistical static timing analysis (SSTA) [5, 6], where continuous distributions are propogated instead of deterministic values to find closed form expressions for performance in presence of variations. More recently, statistical design optimization [1, 9, 12, 11] for improving power and area for an acceptable yield has also been attempted. In [1], the optimization uses a penalty function to improve the slacks of critical paths to improve yield. An SSTA engine is used in the iterative optimization framework [9] to find the most critical gate to size in terms of power/delay sensitivity. In [12, 11], a stochastic programming approach with chance constraints are used to incorporate yield in the gate sizing problem formulation. However the approaches using continuous distributions requires a number of complicated operations to be performed iteratively at each node and hence a prohibitive runtime [15, 10]. The stochastic programming based statistical optimization [11] technique on the other hand is reasonably fast, but is more conservative in terms of yield and hence lesser savings [4] in terms of area and/or power consumption.

In this paper, we propose a new gate sizing algorithm in presence of process variations using Fuzzy Mathematical Programming (FMP). The theory of fuzzy sets and systems have been proved to successfully model imprecision and an uncertainty in optimization problems. It has succesfully been applied in areas like pattern recognition and robotics [2]. In VLSI design automation, fuzzy logic has also been applied in high level synthesis operations like scheduling to model imprecise coefficients [10]. To the best of our knowledge, this is the first time concepts of fuzzy sets and systems and FMP are attempted to model uncertainty due to process variations in VLSI circuits. For simplicity, we use linear delay models [3] and linear membership functions [8], but our approach is applicable to solve problems including nonlinear or other posynomial delay models [14]. A genetic algorithm based approach for solving fuzzy nonlinear problems has been presented in [14].

The fuzzy optimization approach, initially performs a deterministic optimization assuming the worst and the average case values for the variation parameters. The results of these deterministic optimizations are used to convert the fuzzy optimization problem into a crisp non-linear problem using the symmetric relaxation method [2]. The crisp problem is then solved using a non-linear optimization solver. The crisp problem in general, has been proved to provide the most satisfying solution in presence of imprecision or variations in coeffecients of the constraints or objective function in the optimization problem [14]. In the context of variation aware circuit optimzation, the above crisp model with delay and power as constraints, maximizes the robustness, i.e., the variation resistance of the circuit and thus the yield. The proposed approach is tested on ISCAS '85 benchmarks and results indicate sizeable savings in power compared to the worst case deterministic gate sizing approach. The proposed fuzzy programming based gate sizing also provides better results than the stochastic programming approach both in power savings and execution time. The results are also validated using monte-carlo simulations, which indicate a high timing yield as well. The rest of the paper is organized as follows. In Section 2, we motivate why fuzzy programming is well suited for variation aware gate sizing. Some preliminaries of FMP is given section 3. The proposed fuzzy gate sizing approach is given in section 4 followed by experimental results and some conclusions in section 5 and 6.

2. MOTIVATION

The problem of gate sizing is a popular and well know topic in VLSI circuit optimization. A significant amount of research in this area have concentrated on gate sizing optimization with deterministic models. However, the increasing impact of process variations in the nanometer era, necessiates the need to develop design and optimization methods using non-deterministic models. In this section, we discuss why FMP is well suited for modeling uncertainty due to process variaitons in VLSI circuits. The impact of these variations are completely non-deterministic and the degree of uncertainty will be even higher in future generations [1].

The popular approach for handling these uncertainties due to process variations has been to use probabilistic models, which describes these uncertain parameters in terms of probability distributions. However, the probabilistic way of evaluating and optimizing these uncertainties is computationally expensive due to the need for complicated multiple integration techniques needed for continuous distributions [15, 10] or due to large number of scenarios for the discrete representation. Variation aware circuit optimization in VLSI gate sizing [9] using SSTA is also shown to have very high execution time. Furthermore, probabilistic modeling requires exhaustive description of uncertain parameters to build probabilistic distributions from historic (empirical) data and extensive details in some cases are not available. An alternative treatment of uncertainty is needed in the situations, where an expert can predict or obtain only the mean and worst case value of an uncertain parameter. FMP and interval arithmetic can be used to make decisions in the above conditions.

In addition to the above arguments, Buckley [4] has also shown that fuzzy programming based optimization guarantees solutions that is better or atleast as good as their stochastic counterparts. The author [4] compared stochastic and fuzzy programming methodologies using monte-carlo simulations. The fuzzy optimization in uncertain environments finds the best solution (supremum operation over all feasible solutions) as opposed to averaging (integrals over all feasible solutions) in stochastic programming based optimization. Hence, fuzzy programming selects a solution which is better or atleast as good as stochastic solution. The above arguments provided us the motivation to investigate fuzzy mathematical programming (FMP) approach to model uncertainty due to process variations in VLSI circuits. The performance of the proposed algorithm is compared with that of the stochastic programming based gate sizing in order to illustrate the efficiency of fuzzy programming for optimizing in presence of variations.

3. PRELIMINARIES: FMP

This section briefly discusses some of the relevant fuzzy mathematical programming and fuzzy numbers concepts. The reader is referred to [Sakawa 2002 [14] and Zadeh 1970[2]] for a detailed treatment of the concepts of fuzzy sets and systems. Zadeh (1965) introduced the concept of fuzzy sets and systems in which an element's belonging to a set need not be binary valued, but could be any value in between [0, 1]. The theory and methodology of fuzzy programming based optimization has been popular from the inception of decision making in fuzzy environments by Bellman and Zadeh [2]. Several models and approaches have been proposed for uncertainty management using fuzzy linear programming, fuzzy multi-objective programming, fuzzy dynamic programming, fuzzy integer programming, possibilistic programming and fuzzy nonlinear programming. An extensive list of references is provided in [14] and [17]. In order to solve a complex uncertain problem using fuzzy optimization, fuzzy modeling is very important. We model the uncertainty due to process variations using fuzzy numbers with linear membership functions. In recent works [9, 5], these variations are modeled as normally distributed random variables with zero mean and standard deviation σ. In this work, instead of using these distributions directly we extract the values of mean and 3σ to model these variations as interval valued fuzzy numbers. The 3σ value is assumed to be the deterministic worst case variation value, meaning all the random parameters are set to 3σ for maximum timing yield in worst case deterministic optimization.

Lofti A. Zadeh, in [2], defined fuzzy numbers using possibilistic distributions. In FMP the possibilistic distributions $\pi(.)$ are analogous to linear membership functions [10]. Triangular and trapezoidal are the commonly used possibilistic distributions in solving FMP problems. Other nonlinear membership functions could as well be used for improving modeling accuracy [14]. The triangular fuzzy number is denoted by a triple X = (x^m, x^l, x^u), where x^m is the most possible value or the mean value and x^l, x^u are the lower and upper bounds, denoting the pessimistic and optimistic value of the number. Depending on the context, the value x^l can be pessimistic or optimistic variation from the mean value x^m and the same holds for x^u. In the context of VLSI circuit optimization, the triple $L_{eff} = (L_{eff}^m, L_{eff}^l, L_{eff}^u)$ can be used to model variations in channel length. Since the general objective in circuit optimzation is to minimize delay or power, the pessimistic value in this context for effective channel length is L_{eff}^u, which is the sum of $L_{eff}^m + 3\sigma$. Similarly if we model the gate oxide thickness as a fuzzy triple, the pessimistic value is the upper bound value.

The fuzzy optimization problem is initially solved with fixing the varying parameters to their worst (Obj_1) and average case (Obj_2) values. Now with these bound values we can formulate a crisp problem, which will represent a satisficing solution in presence of variations. The objective function for the fuzzy programming problem takes values between this lower $Obj_l = \min(Obj_1, Obj_2)$ and upper $Obj_u = \max(Obj_1, Obj_2)$ bound values. Using these bound values and the symmetric defnition of fuzzy decision proposed by Bellman and Zadeh [2], the fuzzy problem can be de-fuzzified into a crisp nonlinear problem as shown in equation 1.

$$maximize \ \lambda \qquad (1)$$

$$\lambda(Obj_l - Obj_u) - \sum_{i=1}^{n} a_i x_i + Obj_u \leq 0,$$

$$\sum_{i=1}^{n} (b_{ji} + \lambda d_{ji}) x_i - c_j \leq 0, \qquad 1 \leq j \leq m$$

$$x_j \geq 0, \ 0 \leq \lambda \leq 1 \qquad 1 \leq i \leq n$$

Where, λ is the variation parameter introduced in the crisp problem which is to be maximized for an fuzzy optimal solution between the lower and upper bound values. The solution of this nonlinear programming problem can be interpreted as representing an overall degree of satisfaction in presence of varying parameters [14]. In the general formulation the variation parameter λ can take values between 0 and 1. However, for our gate sizing problem we can restrict it to a much smaller range. The smaller range does not affect the quality of the solution, but improves the runtime of the optimization process. In the next section, we explain the fuzzy linear programming formulation for the gate sizing problem in the presence of uncertainty due to process variations.

4. PROPOSED FUZZY GATE SIZING

In this section, we explain our formulation of the gate sizing problem in the presence of uncertainty due to process variations. Several formulations have appeared for the gate sizing problem over the years. The problem of gate sizing can be defined as finding the optimal drive strengths such that specified critical path timing is met and overheads are minimized. In this paper, we size the gates to minimize dynamic power with delay as constraints. We formulate the gate sizing problem as a linear programming problem, due to the simplicity of linear programming based modeling and wide availability of well developed fuzzy programming theory for linear programming based optimization. However, fuzzy programming based optimization can also solve gate sizing and other VLSI design optimization problems in a non-linear programming setup [14]. Next we discuss the power and delay models used in this work.

4.1 Power and Timing Models

The dynamic power consumption of a gate i is given as,

$$P_i = \frac{1}{2} f V_{dd}^2 E_i (C_i + C_{wire}) + P_{sc} \qquad (2)$$

where, P_i is the total dynamic power consumed by gate i, f is the clock frequency, V_{dd} is the supply voltage for the gate, E_i is the average swiching activity of the gate, C_i is the intrinsic gate capacitance internal to the gate and C_{wire} is the sum of all the interconnects that fanout from gate i. Thus, reducing the size (s_i) of the gate reduces the intrinsic gate capacitance of gate i, power consumption and fan-in load capacitances of the gate. A linear delay model [3] of a gate i is given by,

$$d_i = a_i - b_i s_i + c_i \sum_{j \epsilon fo(i)} s_j \qquad (3)$$

where, s_i refers to the size of gate i, $fo(i)$ is the set of gates that fanout from gate i, constant coefficients a_i, b_i, c_i are emprically determined by extensive SPICE simulations for each gate in the library for various sizes and fanout counts. The base (nominal) delay model used in this work, has also been used in a recent stochastic programming [11] based statistical gate sizing approach. The uncertain parameters in recent works has been modeled using the equation.

$$D = d_i + \sum_{j=1}^{n} d_j X_j + d_r X_r \qquad (4)$$

where, d_i is the nominal delay and X_j and X_r are the random parameters determining correlated and independent variations respectively. The magnitude of these variations is given by the variables d_j and d_r, which is determined from extensive simulations. The correlations of different areas of transistors are different and is usually affected by a small subset of all the gates in the chip. In this paper, we capture these variations using the concept of fuzzy numbers. We assume the delay of the gate as a interval with a lower and a upper bound value. In other words, every gate's delay is now a triangular value (average, low, high), instead of a discrete value. This interval delay value of the gate captures the uncertainty due to process variations. Each gate in the circuit can take any value in between low and high value. We focus on the intra-die variations, meaning, each transistor can have different amount of variation, however the method can be modified easily to incorporate the effect of inter-die variations. A extra fuzzy variable is needed and it will have negligible effect in the solution time. The uncertainty in the delay values are transfered to the coefficients b_i and c_i of the linear delay model shown in equation 3. These coefficients are the fuzzy numbers of the triangular form with a linearly varying membership function. The actual physical variability effects of these variables are not known, but the coefficient b_i approximates the effective channel length variations and c_i the gate oxide thickness variation. Next, we explain the proposed fuzzy gate sizing approach for optimization in presence of process variations.

4.2 Fuzzy Gate Sizing

In this paper, we use delay constrained dynamic power minimization for the gate sizing problem. If minimizing power is the only interest, then all the gates will be set to minimum size. However, the statement is to acheive minimum power for a specified timing target. Hence, the cost function of the deterministic optimization formulation include both delay and power as variables. Here, the cost function is simply defined as the problem interest of the optimization problem. The deterministic formulation of the sizing problem is given by,

$$min \sum_i P_i \qquad (5)$$

$$s.t. \quad D_p \leq T_{spec} \qquad \forall p \epsilon P$$

$$and \ D_p = \sum_{i \epsilon p}(a_i - b_i s_i + c_i \sum_{j \epsilon fo(i)} s_j)$$

where, T_{spec} is the specified timing target of the circuit, p denotes a particular path number in a circuit which belongs to the set of all paths P and D_p is the sum of the delays of all gates in path p. The summation of the dynamic power of all the gates is used as as the objective function. The dynamic power model in equation 2 is substituted here. The fuzzy version of the above deterministic optimization problem with uncertain parameters is given by,

$$min \sum_i P_i \qquad (6)$$

$$s.t. \quad D_p \leq T_{spec} \qquad \forall p \epsilon P$$

$$and \ D_p = \sum_{i \epsilon p}(a_i - \tilde{b}_i s_i + \tilde{c}_i \sum_{j \epsilon fo(i)} s_j)$$

where, s_i is bounded by minimum and maximum gate size, the coefficients $\tilde{b}_i$ and $\tilde{c}_i$ are the uncertain parameters. The uncertain parameters are modeled as fuzzy number triples of the form $(b_i, b_i - g_i, b_i + g_i)$ and $(c_i, c_i - h_i, c_i + h_i)$, where

g_i and h_i are the maximum variations for the coefficients b_i and c_i respectively. The coefficient b_i and c_i closely approximate the variation in effective channel length (L_{eff}) and oxide thickness (t_{ox}). The fuzzy gate sizing problem is then transformed into a crisp nonlinear problem using the following steps. A deterministic optimization is performed intially with the varying coefficients set to worst and average case values of the fuzzy number. In the worst case optimization, the fuzzy gate delay equations in the fuzzy problem is replaced with the following equation.

$$d_i = (a_i - (b_i - g_i)s_i + (c_i + h_i) \sum_{j \epsilon fo(i)} s_j) \qquad (7)$$

The gate delay in the above equation is most pessimistic estimate, resulting in the worst possible delay for the gate. It can also be seen that the worst case estimate corresponds to the lower bound in coeffecient b_i, since b_i is inversely proportional to the effective channel length and upper bound in coefficient c_i as it is directly proportional to gate oxide thickness. Similarly, the typical or nominal case of the gate delay is the case where the fuzzy numbers are fixed to their average case values. In the nominal case optimization, the fuzzy delay equations in the fuzzy problem is replaced with the following equation.

$$d_i = (a_i - (b_i)s_i + (c_i) \sum_{j \epsilon fo(i)} s_j) \qquad (8)$$

The deterministic optimization problem (equation 6) is solved with the delay equations set to these worst case and nominal case equations. MINOS optimization solver availabe through the NEOS optimization server is used to solve these linear programming problems. These results of these optimization correspond to worst case gate sizing (wc_{sizing}) and nominal case gate sizing (nc_{sizing}) values. Using these values and a new variation parameter λ the fuzzy optimization problem is transformed into a crisp nonlinear programming problem using the symmetric relaxation method [2]. The crisp nonlinear problem for gate sizing in the presence of process variations is given by,

$$maximize \ \lambda \qquad (9)$$

$$\lambda(nc_{sizing} - wc_{sizing}) - \sum_i P_i + wc_{sizing} \leq 0,$$

$$s.t. \quad D_p \leq T_{spec} \qquad \forall p \epsilon P$$

$$and \ D_p = \sum_{i \epsilon p}(a_i - (b_i - g_i * \lambda)s_i + (c_i + h_i * \lambda) \sum_{j \epsilon fo(i)} s_j)$$

where, the parameter λ is bounded by 0 and 1. The spatial correlations can be incorporated by making the coefficients g_i and h_i in the gate's delay equation as a function of the gate's location in the chip. The chip area can be paritioned into n areas as in [16] and gates in the same block will have same variation range. Eventhough, the parameter λ can take any values between 0 and 1, for the gate sizing problem, it can be easily bounded to a smaller value. In this paper, we bound the λ value to be between 0.5 and 0.75. We estimated that such a smaller bound is sufficient due to the dual requirement of high yield and low overhead for the gate sizing optimization in presence of variations. The smaller bound speeds up the fuzzy gate sizing procedure from 2-3 times, without affecting the final solution. The crisp optimzation problem has three variables in the cost function namely, power, delay and variation parameter λ.

The cost function is defined as the problem interest of the optimzation. If we consider the parameter λ as the variation resistance (robustnness) property of the circuit, meaning the ability to meet timing constraint even in the presence of variations. The problem tries to maximize variation resistance, constraints delay value even with variations to be less than specified timing and bounds the power value to be in between wc_{sizing} and nc_{sizing} values. Favoring the power value to be close to the nc_{sizing} value by maximizing variation resistance value. Hence, the crisp optimization problem tries to satisfy all the three requirements to the maximum degree. It has been proved for problems in other domain that the above formulation provides the most satisfying solution for optimization in presence of variations [14].

Finally, since the number of paths grow exponentially in the number of gates in the above optimization formulation, the path based formulation is converted to a node based optimization problem as in, [12, 11] . The transformation only introduces a sub-optimality close to 2% [11], for small circuits consisting of 20 levels of logic. The sub-optimality refers to the value of the dynamic power obtained when compared to the path based formulation. However, node based formulation significantly improve the run time and feasibility of optimizing large circuits. In the next section, we present the simulation steps and the experimental results of the fuzzy gate sizing approach tested on ISCAS benchmark circuits.

5. EXPERIMENTAL RESULTS

The proposed fuzzy linear programming optimization for gate sizing was tested on ISCAS '85 benchmark circuits. The complete simulation flow of the proposed approach is shown in Figure 1. First, the benchmark netlists are converted to Berkeley Logic Interchange Format (BLIF) using a C script. Sequential Interactive Synthesis (SIS) system, logic design tool from berkeley, takes this BLIF file as input and performs technology mapping to the user defined library (UDL). The UDL contains an inverter, a buffer, two-input nand and nor gates. SIS system's *read_blif* and *read_library* commands are used to read the benchmark input in BLIF format and the UDL in genlib format respectively. The benchmark circuit is then mapped using *map* command, which uses a tree covering algorithm for the covering phase. The mapped BLIF file is then converted to AMPL model using a C (*blif2ampl*) script. AMPL is a widely used modeling language for large scale mathematical programming problems. The equation coefficients for power and delay models are characterized for various gate sizes and fanouts using hspice simulations for a $0.18\mu m$ CMOS process. The fit is justified as the rms error is less than 7% for a restricted range of gate sizes (1x - 4x) [11]. The *blif2ampl* script uses these delay equations to generate the linear programming models for the benchmarks with delay coefficients set to mean and the maximum possible variation (worst case). The maximum variation in gate delay is assumed to be 25% from its mean value. This is translated into appropriate values for the coefficients b_i and c_i in the delay model, equation .

The linear optimization problems are solved using the Modular Incore Nonlinear Optimization Solver (MINOS), available through the NEOS server for optimization. The *blif2ampl* script uses the results of these optimizations and generates a fuzzy nonlinear AMPL model. The fuzzy nonlinear optimization problem is also solved using the MINOS

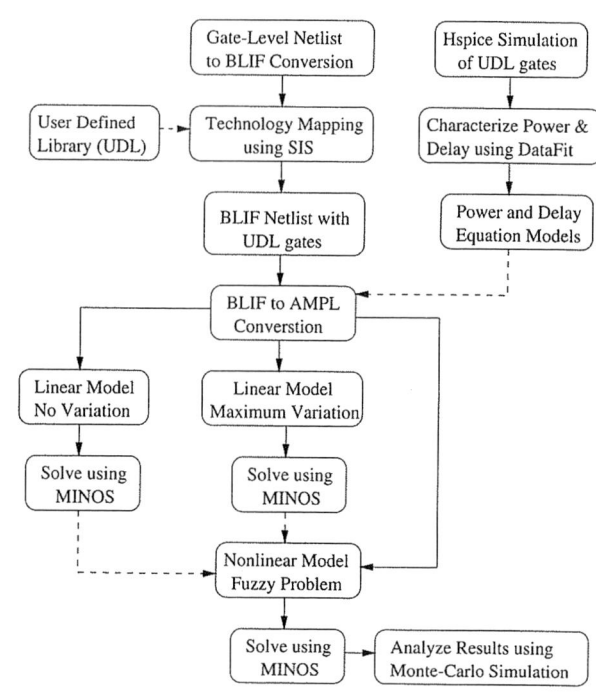

Figure 1: Fuzzy Gate Sizing: Simulation Flow

solver to find the optimal gate sizes in presence of variations in gate delay. The proposed fuzzy sizing approach is also compared with the stochastic programming based gate sizing under uncertainty [11]. The latter method is also implemented with the same setup, parameters and objective functions for fairness in comparison.

The power reduction achieved by the fuzzy sizing approach compared to worst case deterministic sizing and the stochastic programming approach is documented in Table 1. The worst case sizing results correspond to the delay coefficients set to their maximum variation case. The percentage improvement of fuzzy sizing compared to deterministic worst case sizing is calculated as,

$$PR_1 = \frac{Power_{DWC-GS} - Power_{F-GS}}{Power_{DWC-GS}} \star 100 \qquad (10)$$

Similarly, the percentage improvement of fuzzy sizing compared to stochasting sizing is calculated as,

$$PR_2 = \frac{Power_{S-GS} - Power_{F-GS}}{Power_{S-GS}} \star 100 \qquad (11)$$

It can be seen that there is a sizable savings in power by using the fuzzy sizing approach as compared to deterministic worst case gate sizing and stochastic programming approach. The execution time of the fuzzy optimization approach is also shown in Table 1. Once again fuzzy logic based optimization is seen to be better than stochastic programming approach in terms of execution time. Finally, to verify the result of the fuzzy sizing approach, we generate 10000 samples of the ISCAS benchmarks. The circuits are fixed with gate size outputs from the fuzzy sizing method and the other gate coefficients are assumed to have random variation value. The variation value is generated from a uniform distribution between minimum and maximum variation values used in the optimization. We then performed monte-Carlo simulation in these random samples to determine the fre-

Table 1: Experimental Results on Benchmark Circuits

ISCAS '85 Benchmark	Number of gates	Gate Sizing Power(μw)			% Reduction of F-GS over		RunTime (sec)	
		DWC-GS	S-GS	F-GS	DWC-GS	S-GS	S-GS	F-GS
c432	285	897	805	728	18.7	9.4	1.57	1.81
c499	567	1620	1487	1276	21.21	14.1	2.68	2.94
c880	441	1143	1090	1020	10.7	6.46	1.82	1.75
c1355	552	2090	1708	1540	26.3	9.8	2.25	2.42
c1908	773	2628	2252	1945	25.9	13.6	4.22	4.15
c2670	1190	3146	2595	2449	22.15	4.87	8.13	8.88
c3540	1553	3980	3426	3179	20.12	7.21	16.4	17.23
c5315	2534	5641	5387	5145	8.79	4.49	34.3	36.66
c6288	2400	5482	4864	4327	21.06	11.04	29.22	31.14

DWC-GS : Deterministic Worst Case Gate Sizing
S-GS : Stochastic Gate Sizing [11] and F-GS : Fuzzy Gate Sizing

quency of timing violations, i.e, number of times delay of the random circuit is greater than specified timing (T_{spec}). The fuzzy logic approach had an timing yield of around 99-100% for all the benchmark circuits. This confirms the fact that the FMP is a fast approach to design circuits with high yield and smaller area and power overheads.

6. CONCLUSION

In this paper, we proposed a new approach for gate sizing considering process variations using FMP. The variations in channel length and oxide thichkness are modeled as fuzzy numbers with linear membership functions. The proposed fuzzy gate sizing approach maximizes variation resistance (robustness) of the circuit, with delay and power as constraints in the formulation. Experimental results on ISCAS '85 benchmark circuits indicate an sizeable savings in power and a smaller execution time when compared with other statistical sizing approaches. The proposed results validated using Monte Carlo simulations confirms the high variation resistance of the circuits designed using the proposed approach.

7. REFERENCES

[1] Xiaoliang Bai, Chandu Visweswariah, N. Philip Strenski, and J. David Hathway. Uncertainty-Aware Circuit Optimization. *Design Automation journal*, pages 58–63, 2002.

[2] R. E Bellman and L. A Zadeh. Decision Making in Fuzzy Environment. *Management Science*, pages 141–164, 1970.

[3] M Berkelaar and J. Jess. Gate Sizing in MOS digital circuits with linear programming. *European Design Automation journal*, pages 217–221, 1990.

[4] J. J Buckley. Stochastic versus Possibilistic programming. *Fuzzy Sets and Systems*, pages 173–177, 1990.

[5] Srinivas Devadas, H. F Jyu, Kurt Keutzer, and Sharad Malik. Statistical Timing Analysis of Combinational Circuits. *International journal on Computer Design*, pages 38–43, 1992.

[6] Anirudh Devgan and Chandramouli Kashyap. Block-based Static Timing Analysis with Uncertainty. *International journal on Computer Aided Design*, pages 607–614, 2003.

[7] J. P Fishburn and A. E Dunlop. TILOS : A Posynomial Programming Approach to Transistor Sizing. *IEEE Transactions on CAD*, pages 326–336, 1985.

[8] N. Rafail Gasimov and Kursat Yenilmez. Fuzzy Linear Programming Problems with Fuzzy Membership Functions. *Mathematical Subject Classification*, pages 375–396, 2000.

[9] Masanori Hashimoto and Hidetoshi Onodera. A Performance Optimization Method by Gate Sizing using Statistical Static Timing Analysis. *International Symposium on Physical Design*, pages 111–116, 2000.

[10] Ireneusz Karkowski. Architectural Synthesis with possibilistic programming. *International journal on system sciences*, pages 14–22, 1995.

[11] M. Mani and M. Orshansky. A New Statistical Optimization Algorithm for Gate Sizing. *International journal on Computer Design*, pages 272–277, 2004.

[12] Murari Mani, Anirudh Devgan, and Michael Orshansky. An Efficient Algorithm for Statistical Minimization of Total Power under Timing Yield Constraints. *Design Automation journal*, pages 309–314, 2005.

[13] A. K Murugavel and N. Ranganathan. Gate Sizing and Buffer Insertion using Economic models for Power Optimization. *International journal on VLSI Design*, pages 195–200, 2004.

[14] Masatoshi Sakawa. Genetic Algorithms and Fuzzy Multiobjective optimization. *Kluwer Academic Publishers*, 2002.

[15] C. Schmidt and Grossmann I.E. The exact overall time distribution of uncertain task durations. *European journal of Operational Research*, 2000.

[16] Jaskirat Singh, Vidyasagar Nookala, Zhi-Quan Luo, and Sachin Sapatnekar. Robust Gate Sizing by Geometric Programming. *Design Automation journal*, pages 315–320, 2005.

[17] J.F Tang, D.W Wang, Y.K Fung, and Yung K.L. Understanding of Fuzzy Optimization: Theories and Methods. *journal of System Science and Complexity*, pages 117–136, 2004.

[18] Hiran Tennakoon and Carl Sechen. Gate Sizing using Lagrangian Relaxation combined with a fast gradient based preprocessing step. *ICCAD*, pages 395–402, 2002.

Adaptive Duty Cycling for Energy Harvesting Systems

Jason Hsu, Sadaf Zahedi, Aman Kansal, Mani Srivastava
Electrical Engineering Department
University of California Los Angeles
{jasonh,kansal,szahedi,mbs} @ ee.ucla.edu

Vijay Raghunathan
NEC Labs America
Princeton, NJ
vijay@nec-labs.com

ABSTRACT

Harvesting energy from the environment is feasible in many applications to ameliorate the energy limitations in sensor networks. In this paper, we present an adaptive duty cycling algorithm that allows energy harvesting sensor nodes to autonomously adjust their duty cycle according to the energy availability in the environment. The algorithm has three objectives, namely (a) achieving energy neutral operation, *i.e.,* energy consumption should not be more than the energy provided by the environment, (b) maximizing the system performance based on an application utility model subject to the above energy-neutrality constraint, and (c) adapting to the dynamics of the energy source at run-time. We present a model that enables harvesting sensor nodes to predict future energy opportunities based on historical data. We also derive an upper bound on the maximum achievable performance assuming perfect knowledge about the future behavior of the energy source. Our methods are evaluated using data gathered from a prototype solar energy harvesting platform and we show that our algorithm can utilize up to 58% more environmental energy compared to the case when harvesting-aware power management is not used.

Categories and Subject Descriptors

C.2.4 [**Computer Systems Organization**]: Computer Communication Networks—*Distributed Systems*

General Terms

Algorithms, Design

Keywords

Energy harvesting; low power design; energy neutral operation

1. INTRODUCTION

Energy supply has always been a crucial issue in designing battery-powered wireless sensor networks because the lifetime and utility of the systems are limited by how long the batteries are able to sustain the operation. The fidelity of the data produced by a sensor network begins to degrade once sensor nodes start to run out of battery power. Therefore, harvesting energy from the environment has been proposed to supplement or completely replace battery supplies to enhance system lifetime and reduce the maintenance cost of replacing batteries periodically.

However, metrics for evaluating energy harvesting systems are different from those used for battery powered systems. Environmental energy is distinct from battery energy in two ways. First it is an inexhaustible supply which, if appropriately used, can allow the system to last forever, unlike the battery which is a limited resource. Second, there is an uncertainty associated with its availability and measurement, compared to the energy stored in the

battery which can be known deterministically. Thus, power management methods based on battery status are not always applicable to energy harvesting systems. In addition, most power management schemes designed for battery-powered systems only account for the dynamics of the energy consumers (*e.g.,* CPU, radio) but not the dynamics of the energy supply. Consequently, battery powered systems usually operate at the lowest performance level that meets the minimum data fidelity requirement in order to maximize the system life. Energy harvesting systems, on the other hand, can provide enhanced performance depending on the available energy.

In this paper, we will study how to adapt the performance of the available energy profile. There exist many techniques to accomplish performance scaling at the node level, such as radio transmit power adjustment [1], dynamic voltage scaling [2], and the use of low power modes [3]. However, these techniques require hardware support and may not always be available on resource constrained sensor nodes. Alternatively, a common performance scaling technique is *duty cycling*. Low power devices typically provide at least one low power mode in which the node is shut down and the power consumption is negligible. In addition, the rate of duty cycling is directly related to system performance metrics such as network latency and sampling frequency. We will use duty cycle adjustment as the primitive performance scaling technique in our algorithms.

2. RELATED WORK

Energy harvesting has been explored for several different types of systems, such as wearable computers [4], [5], [6], sensor networks [7], *etc*. Several technologies to extract energy from the environment have been demonstrated including solar, motion-based, biochemical, vibration-based [8], [9], [10], [11], and others are being developed [12], [13]. While several energy harvesting sensor node platforms have been prototyped [14], [15], [16], there is a need for systematic power management techniques that provide performance guarantees during system operation. The first work to take environmental energy into account for data routing was [17], followed by [18]. While these works did demonstrate that environment aware decisions improve performance compared to battery aware decisions, their objective was not to achieve energy neutral operation. Our proposed techniques attempt to maximize system performance while maintaining energy-neutral operation.

3. SYSTEM MODEL

The energy usage considerations in a harvesting system vary significantly from those in a battery powered system, as mentioned earlier. We propose the model shown in Figure 1 for designing energy management methods in a harvesting system. The functions of the various blocks shown in the figure are discussed below. The precise methods used in our system to achieve these functions will be discussed in subsequent sections.

Harvested Energy Tracking: This block represents the mechanisms used to measure the energy received from the harvesting device, such as the solar panel. Such information is useful for determining the energy availability profile and adapting system performance based on it. Collecting this information requires that the node hardware be equipped with the facility to measure the power

Permission to make digital or hard copies of all or part of this work for personal or classroom use is granted without fee provided that copies are not made or distributed for profit or commercial advantage and that copies bear this notice and the full citation on the first page. To copy otherwise, or republish, to post on servers or to redistribute to lists, requires prior specific permission and/or a fee.
ISLPED '06, October 4–6, 2006, Tegernsee, Germany.
Copyright 2006 ACM 1-59593-462-6/06/0010...$5.00.

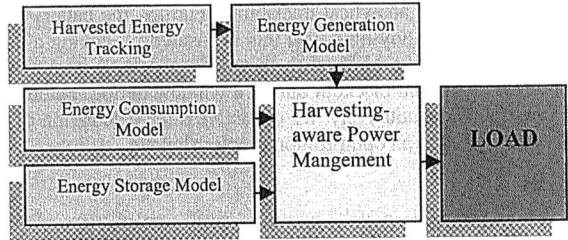

Figure 1. System model for an energy harvesting system.

generated from the environment, and the Heliomote platform [14] we used for evaluating the algorithms has this capability.

Energy Generation Model: For wireless sensor nodes with limited storage and processing capabilities to be able to use the harvested energy data, models that represent the essential components of this information without using extensive storage are required. The purpose of this block is to provide a model for the energy available to the system in a form that may be used for making power management decisions. The data measured by the energy tracking block is used here to predict future energy availability. A good prediction model should have a low prediction error and provide predicted energy values for durations long enough to make meaningful performance scaling decisions. Further, for energy sources that exhibit both long-term and short-term patterns (*e.g.*, diurnal and climate variations vs. weather patterns for solar energy), the model must be able to capture both characteristics. Such a model can also use information from external sources such as local weather forecast service to improve its accuracy.

Energy Consumption Model: It is also important to have detailed information about the energy usage characteristics of the system, at various performance levels. For general applicability of our design, we will assume that only one sleep mode is available. We assume that the power consumption in the sleep and active modes is known. It may be noted that for low power systems with more advanced capabilities such as dynamic voltage scaling (DVS), multiple low power modes, and the capability to shut down system components selectively, the power consumption in each of the states and the resultant effect on application performance should be known to make power management decisions.

Energy Storage Model: This block represents the model for the energy storage technology. Since all the generated energy may not be used instantaneously, the harvesting system will usually have some energy storage technology. Storage technologies (*e.g.*, batteries and ultra-capacitors) are non-ideal, in that there is some energy loss while storing and retrieving energy from them. These characteristics must be known to efficiently manage energy usage and storage. This block also includes the system capability to measure the residual stored energy. Most low power systems use batteries to store energy and provide residual battery status. This is commonly based on measuring the battery voltage which is then mapped to the residual battery energy using the known charge to voltage relationship for the battery technology in use. More sophisticated methods which track the flow of energy into and out of the battery are also available.

Harvesting-aware Power Management: The inputs provided by the previously mentioned blocks are used here to determine the suitable power management strategy for the system. Power management could be carried to meet different objectives in different applications. For instance, in some systems, the harvested energy may marginally supplement the battery supply and the objective may be to maximize the system lifetime. A more interesting case is when the harvested energy is used as the primary source of energy for the system with

the objective of achieving indefinitely long system lifetime. In such cases, the power management objective is to achieve energy neutral operation. In other words, the system should only use as much energy as harvested from the environment and attempt to maximize performance within this available energy budget.

4. THEORETICALLY OPTIMAL POWER MANAGEMENT

We develop the following theory to understand the energy neutral mode of operation. Let us define $P_s(t)$ as the energy harvested from the environment at time t, and the energy being consumed by the load at that time is $P_c(t)$. Further, we model the non-ideal storage buffer by its round-trip efficiency η (strictly less than 1) and a constant leakage power P_{leak}. Using this notation, applying the rule of energy conservation leads to the following inequality:

$$B_0 + \eta \int [P_s(t) - P_c(t)]^+ dt - \int [P_c(t) - P_s(t)]^+ dt - \int P_{leak} dt \geq 0 \quad (1)$$

where B_0 is the initial battery level and the function $[X]^+ = X$ if $X > 0$ and zero otherwise.

DEFINITION 1 $(\rho, \sigma_1, \sigma_2)$ function: A non-negative, continuous and bounded function $P(t)$ is said to be a $(\rho, \sigma_1, \sigma_2)$ function if and only if for any value of finite real number T , the following are satisfied:

$$\rho T - \sigma_2 \leq \int_T P(t) dt \leq \rho T + \sigma_1 \quad (2)$$

This function can be used to model both energy sources and loads. If the harvested energy profile $P_s(t)$ is a $(\rho_1, \sigma_1, \sigma_2)$ function, then the average rate of available energy over long durations becomes ρ_1, and the burstiness is bounded by σ_1 and σ_2. Similarly, $P_c(t)$ can be modeled as a (ρ_2, σ_3) function, when ρ_2 and σ_3 are used to place an upper bound on power consumption (the inequality on the right side) while there are no minimum power consumption constraints.

The condition for energy neutrality, equation (1), leads to the following theorem, based on the energy production, consumption, and energy buffer models discussed above.

THEOREM 1 (ENERGY NEUTRAL OPERATION): Consider a harvesting system in which the energy production profile is characterized by a $(\rho 1, \sigma 1, \sigma 2)$ function, the load is characterized by a $(\rho 2, \sigma 3)$ function and the energy buffer is characterized by parameters η for storage efficiency, and P_{leak} for leakage power. The following conditions are sufficient for the system to achieve energy neutrality:

$$\rho_2 \leq \eta \rho_1 - P_{leak} \quad (3)$$
$$B_0 \geq \eta \sigma_2 + \sigma_3 \quad (4)$$
$$B \geq B_0 \quad (5)$$

where B_0 is the initial energy stored in the buffer and provides a lower bound on the capacity of the energy buffer B. The proof is presented in our prior work [19].

To adjust the duty cycle D using our performance scaling algorithm, we assume the following relation between duty cycle and the perceived utility of the system to the user: Suppose the utility of the application to the user is represented by U(D) when the system operates at a duty cycle D. Then,

$$U(D) = 0, \qquad if\ D < D_{\min}$$
$$U(D) = k_1 D + \beta, \quad if\ D_{\min} \leq D \leq D_{\max}$$
$$U(D) = k_2\ , if\ D > D_{\max}$$

This is a fairly general and simple model and the specific values of $D_{\min}$ and $D_{\max}$ may be determined as per application requirements. As an example, consider a sensor node designed to detect intrusion across a periphery. In this case, a linear increase in duty cycle translates into a linear increase in the detection probability. The fastest and the slowest speeds of the intruders may be known, leading to a minimum and

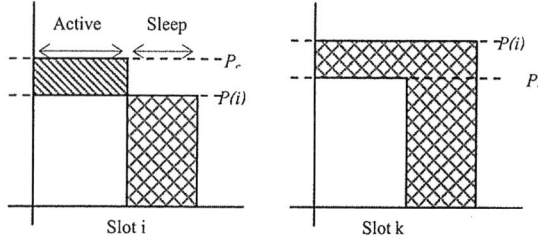

Figure 2. Two possible cases for energy calculations

maximum sensing delay tolerable, which results in the relevant D_{max} and D_{min} for the sensor node. While there may be cases where the relationship between utility and duty cycle may be non-linear, in this paper, we restrict our focus on applications that follow this linear model. In view of the above models for the system components and the required performance, the objective of our power management strategy is adjust the duty cycle $D(i)$ dynamically so as to maximize the total utility $U(D)$ over a period of time, while ensuring energy neutral operation for the sensor node.

Before discussing the performance scaling methods for harvesting aware duty cycle adaptation, let us first consider the optimal power management strategy that is possible for a given energy generation profile. For the calculation of the optimal strategy, we assume complete knowledge of the energy availability profile at the node, including the availability in the future. The calculation of the optimal is a useful tool for evaluating the performance of our proposed algorithm. This is particularly useful for our algorithm since no prior algorithms are available to serve as a baseline for comparison.

Suppose the time axis is partitioned into discrete slots of duration ΔT, and the duty cycle adaptation calculation is carried out over a window of N_w such time slots. We define the following energy profile variables, with the index i ranging over $\{1,..., N_w\}$: $P_s(i)$ is the power output from the harvested source in time slot i, averaged over the slot duration, P_c is the power consumption of the load in active mode, and $D(i)$ is the duty cycle used in slot i, whose value is to be determined. $B(i)$ is the residual battery energy at the beginning of slot i. Following this convention, the battery energy left after the last slot in the window is represented by $B(N_w+1)$. The values of these variables will depend on the choice of $D(i)$.

The energy used directly from the harvested source and the energy stored and used from the battery must be accounted for differently. Figure 2 shows two possible cases for $P_s(i)$ in a time slot. $P_s(i)$ may either be less than or higher than P_c, as shown on the left and right respectively. When $P_s(i)$ is lower than P_c, some of the energy used by the load comes from the battery, while when $P_s(i)$ is higher than P_c, all the energy used is supplied directly from the harvested source. The crosshatched area shows the energy that is available for storage into the battery while the hashed area shows the energy drawn from the battery. We can write the energy used from the battery in any slot i as:

$$B(i)-B(i+1)=\Delta TD(i)\left[P_c-P_s(i)\right]^+-\eta\Delta TP_s(i)\{1-D(i)\}-\eta TD(i)\left[P_s(i)-P_c\right]^+ \quad (6)$$

In equation (6), the first term on the right hand side measures the energy drawn from the battery when $P_s(i) < P_c$, the next term measures the energy stored into the battery when the node is in sleep mode, and the last term measures the energy stored into the battery in active mode if $P_s(i) > P_c$. For energy neutral operation, we require the battery at the end of the window of N_w slots to be greater than or equal to the starting battery. Clearly, battery level will go down when the harvested energy is not available and the system is operated from stored energy. However, the window N_w is judiciously chosen such that over that duration, we expect the environmental energy availability to complete a periodic cycle. For instance, in the case of solar energy harvesting, N_w could be chosen to be a twenty-four hour duration, corresponding to the diurnal cycle in the harvested energy. This is an approximation

since an ideal choice of the window size would be infinite, but a finite size must be used for analytical tractability. Further, the battery level cannot be negative at any time, and this is ensured by having a large enough initial battery level B_0 such that node operation is sustained even in the case of total blackout during a window period. Stating the above constraints quantitatively, we can express the calculation of the optimal duty cycles as an optimization problem below:

$$\max \sum_{i=1}^{N_w} D(i) \quad (7)$$

$$B(i)-B(i+1)=\Delta TD(i)\left[P_c-P_s(i)\right]^+-\eta\Delta TP_s(i)\{1-D(i)\}-\eta TD(i)\left[P_s(i)-P_c\right]^+ \quad (8)$$

$$B(1)=B_0 \quad (9)$$

$$B(N_w+1)\ge B_0 \quad (10)$$

$$D(i)\ge D_{min} \quad \forall i \in \{1,...,N_w\} \quad (11)$$

$$D(i)\le D_{max} \quad \forall i \in \{1,...,N_w\} \quad (12)$$

The solution to the optimization problem yields the duty cycles that must be used in every slot and the evolution of residual battery over the course of N_w slots. Note that while the constraints above contain the non-linear function $[x]^+$, the quantities occurring within that function are all known constants. The variable quantities occur only in linear terms and hence the above optimization problem can be solved using standard linear programming techniques, available in popular optimization toolboxes.

5. HARVESTING-AWARE POWER MANAGEMENT

We now present a practical algorithm for power management that may be used for adapting the performance based on harvested energy information. This algorithm attempts to achieve energy neutral operation without using knowledge of the future energy availability and maximizes the achievable performance within that constraint.

The harvesting-aware power management strategy consists of three parts. The first part is an instantiation of the energy generation model which tracks past energy input profiles and uses them to predict future energy availability. The second part computes the optimal duty cycles based on the predicted energy, and this step uses our computationally tractable method to solve the optimization problem. The third part consists of a method to dynamically adapt the duty cycle in response to the observed energy generation profile in real time. This step is required since the observed energy generation may deviate significantly from the predicted energy availability and energy neutral operation must be ensured with the actual energy received rather than the predicted values.

5.1. Energy Prediction Model

We use a prediction model based on Exponentially Weighted Moving-Average (EWMA). The method is designed to exploit the diurnal cycle in solar energy but at the same time adapt to the seasonal variations. A historical summary of the energy generation profile is maintained for this purpose. While the storage data size is limited to a vector length of N_w values in order to minimize the memory overheads of the power management algorithm, the window size is effectively infinite as each value in the history window depends on all the observed data up to that instant. The window size is chosen to be 24 hours and each time slot is taken to be 30 minutes as the variation in generated power by the solar panel using this setting is less than 10% between each adjacent slots. This yields $N_w = 48$. Smaller slot durations may be used at the expense of a higher N_w.

The historical summary maintained is derived as follows. On a typical day, we expect the energy generation to be similar to the energy generation at the same time on the previous days. The value of energy generated in a particular slot is maintained as a weighted average of the energy received in the same time-slot during all observed days. The weights are exponential, resulting in decaying contribution from older data. More specifically, the historical average maintained for each slot is given by:

$$\bar{x}_k = \alpha \bar{x}_{k-1} + (1-\alpha)x_k$$

where α is the value of the weighting factor, $\bar{x}_k$ is the observed value of energy generated in the slot, and $\bar{x}_{k-1}$ is the previously stored historical average. In this model, the importance of each day relative to the previous one remains constant because the same weighting factor was used for all days.

The average value derived for a slot is treated as an estimate of predicted energy value for the slot corresponding to the subsequent day. This method helps the historical average values adapt to the seasonal variations in energy received on different days. One of the parameters to be chosen in the above prediction method is the parameter α, which is a measure of rate of shift in energy pattern over time. Since this parameter is affected by the characteristics of the energy and sensor node location, the system should have a training period during which this parameter will be determined. To determine a good value of α, we collected energy data over 72 days and compared the average error of the prediction method for various values of α. The error based on the different values of α is shown in Figure 3. This curve suggests an optimum value of $\alpha = 0.15$ for minimum prediction error and this value will be used in the remainder of this paper.

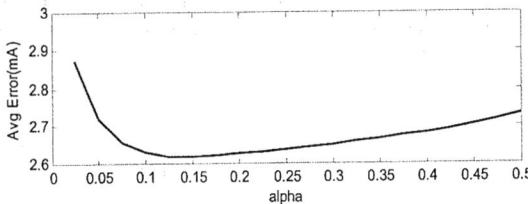

Figure 3. Choice of prediction parameter.

5.2. Low-complexity Solution

The energy values predicted for the next window of N_w slots are used to calculated the desired duty cycles for the next window, assuming the predicted values match the observed values in the future. Since our objective is to develop a practical algorithm for embedded computing systems, we present a simplified method to solve the linear programming problem presented in Section 4. To this end, we define the sets S and D as follows:

$$S = \{i \mid P_s(i) - P_c \geq 0\}$$
$$D = \{i \mid P_c - P(i)_s > 0\}$$

The two sets differ by the condition that whether the node operation can be sustained entirely from environmental energy. In the case that energy produced from the environment is not sufficient, battery will be discharged to supplement the remaining energy. Next we sum up both sides of (6) over the entire N_w window and rewrite it with the new notation.

$$\sum_{i=1}^{Nw} B_i - B_{i+1} = \sum_{i \in D} \Delta TD(i)[P_c - P_s(i)] - \sum_{i=1}^{Nw} \eta \Delta TP_s(i) + \sum_{i=1}^{Nw} \eta \Delta TP_s(i)D(i) - \sum_{i \in S} \eta \Delta TD(i)[P_s(i) - P_c]$$

The term on the left hand side is actually the battery energy used over the entire window of N_w slots, which can be set to 0 for energy neutral operation. After some algebraic manipulation, this yields:

$$\sum_{i=1}^{Nw} P_s(i) = \sum_{i \in D} D(i)\left(\frac{P_c}{\eta} + P_s(i)\left(1-\frac{1}{\eta}\right)\right) + \sum_{i \in S} P_c D(i) \quad (13)$$

The term on the left hand side is the total energy received in N_w slots. The first term on the right hand side can be interpreted as the total energy consumed during the D slots and the second term is the total energy consumed during the S slots. We can now replace three constraints (8), (9), and (10) in the original problem with (13), restating the optimization problem as follows:

$$\max \sum_{i=1}^{N_w} D(i)$$
$$\sum_{i=1}^{Nw} P_s(i) = \sum_{i \in D} D(i)\left(\frac{P_c}{\eta} + P_s(i)\left(1-\frac{1}{\eta}\right)\right) + \sum_{i \in S} P_c D(i)$$
$$D(i) \geq D_{min} \quad \forall i \in \{1,...,Nw\}$$
$$D(i) \leq D_{max} \quad \forall i \in \{1,...,Nw\}$$

This form facilitates a low complexity solution that doesn't require a general linear programming solver. Since our objective is to maximize the total system utility, it is preferable to set the duty cycle to D_{min} for time slots where the utility per unit energy is the least. On the other hand, we would also like the time slots with the highest P_s to operate at D_{max} because of better efficiency of using energy directly from the energy source. Combining these two characteristics, we define the utility co-efficient for each slot i as follows:

$$W(i) = \begin{cases} 1/P_c & for \quad i \in S \\ 1 \Big/ \left(\frac{P_c}{\eta} + P_s(i)\left(1-\frac{1}{\eta}\right)\right) & for \quad i \in D \end{cases}$$

where $W(i)$ is a representation of how efficient the energy usage in a particular time slot i is. A larger $W(i)$ indicates more system utility per unit energy in slot i and vice versa. The algorithm starts by assuming $D(i) = D_{min}$ for $i = \{1...N_w\}$ because of the minimum duty cycle requirement, and computes the remaining system energy R by:

$$R = \sum_{i=1}^{Nw} P_s(i) - \sum_{i \in D} D(i)\left(\frac{P_c}{\eta} + P_s(i)\left(1-\frac{1}{\eta}\right)\right) - \sum_{i \in S} P_c D(i) \quad (14)$$

A negative R concludes that the optimization problem is infeasible, meaning the system cannot achieve energy neutrality even at the minimum duty cycle. In this case, the system designer is responsible for increasing the environment energy availability (*e.g.*, by using larger solar panels). If R is positive, it means the system has excess energy that is not being used, and this may be allocated to increase the duty cycle beyond D_{min} for some slots. Since our objective is to maximize the total system utility, the most efficient way to allocate the excess energy is to assign duty cycle D_{max} to the slots with the highest $W(i)$. So, the coefficients $W(i)$ are arranged in decreasing order and duty cycle D_{max} is assigned to the slots beginning with the largest coefficients until the excess energy available, R (computed by (14) in every iteration), is insufficient to assign D_{max} to another slot. The remaining energy, R_{Last}, is used to increase the duty cycle to some value between D_{min} and D_{max} in the slot with the next lower coefficient. Denoting this slot with index j, the duty cycle is given by:

$$D(j) = \begin{cases} R_{Last}/P_c & if \quad j \in D \\ \dfrac{R_{Last}}{(P_s(j)-P_c)/\eta - P_s(j)} & if \quad j \in S \end{cases} + D_{min}$$

The above solution to the optimization problem requires only simple arithmetic calculations and one sorting step which can be easily implemented on an embedded platform, as opposed to implementing a general linear program solver.

5.3. Slot-by-slot continual duty cycle adaptation.

The observed energy values may vary greatly from the predicted ones, such as due to the effect of clouds or other sudden changes. It is thus important to adapt the duty cycles calculated using the predicted values, to the actual energy measurements in real time to ensure energy neutrality. Denote the initial duty cycle assignments for each time slot i computed using the predicted energy values as $D(i) = \{1, ...,N_w\}$. First we compute the difference between predicted power level $P_s(i)$ and actual power level observed, $P_s'(i)$ in every slot i. Then, the excess energy in slot i, denoted by X, can be obtained as follows:

$$X = \begin{cases} P_s(i) - P_s'(i) & if \quad P_s'(i) > P_c \\ P_s(i) - P_s'(i) - D(i)[P_s(i) - P_s'(i)](1-\frac{1}{\eta}) & if \quad P_s'(i) \leq P_c \end{cases}$$

Input: D: *Initial duty cycle*, X: *Excess energy due to error in the prediction*, P: *Predicted energy profile*, i: *index of current time slot*
Output: D: *Updated duty cycles in one or more subsequent slots*

```
AdaptDutyCycle()
 Iteration: At each time slot do:
  if X > 0
    W_sorted = W{1, ...,Nw} sorted in decending order.
    Q := indices of W_sorted
    for k = 1 to |Q|
        if Q(k) ≤ i or D(Q(k)) ≥ D_max   //slot is already passed
            continue
        if R(Q(k), D_max − D(Q(k))) < X
            D(Q(k)) = D_max
            X = X − R(j, D_max − D(Q(k)))
        else
        //X insufficient to increase duty cycle to D_max
            if P (Q(k)) > P_l
                D(Q(k)) = D(Q(k)) + X/P_l
            else
                D(Q(k)) = D(Q(k)) +      X
                                    ──────────────────
                                    P_c/η + P_s(Q(k))(1−1/η))

  if X < 0
    W_sorted = W{1, ...,Nw} sorted in ascending order.
    Q := indices of W_sorted
    for k = 1 to |Q|
        if Q(k) ≤ I or D(Q(k)) ≤ D_min
            continue
        if R(Q(k), D_max − D(Q(k))) > X
            D(Q(k)) = D_min
            X = X − R(j, D_min − D(Q(k)))
        else
            if P (Q(k)) > P_c
                D(Q(k)) = D(Q(k)) + X/P_c
            else
                D(Q(k)) = D(Q(k)) +      X
                                    ──────────────────
                                    P_c/η + P_s(Q(k))(1−1/η))
```

ALGORITHM 1 Pseudocode for the duty-cycle adaptation algorithm

The upper term accounts for the energy difference when actual received energy is more than the power drawn by the load. On the other hand, if the energy received is less than P_c, we will need to account for the extra energy used from the battery by the load, which is a function of duty cycle used in time slot i and battery efficiency factor η. When more energy is received than predicted, X is positive and that excess energy is available for use in the subsequent solutes, while if X is negative, that energy must be compensated from subsequent slots.

CASE I: X<0. In this case, we want to reduce the duty cycles used in the future slots in order to make up for this shortfall of energy. Since our objective function is to maximize the total system utility, we have to reduce the duty cycles for time slots with the smallest normalized utility coefficient, W(i). This is accomplished by first sorting the coefficient W(j) ,where j>i. in decreasing order, and then iteratively reducing Dj to Dmin until the total reduction in energy consumption is the same as X.

CASE II: X>0. Here, we want to increase the duty cycles used in the future to utilize this excess energy we received in recent time slot. In contrast to Case I, the duty cycles of future time slots with highest utility coefficient W(i) should be increased first in order to maximize the total system utility.

Suppose the duty cycle is changed by d in slot j. Define a quantity R(j,d) as follows:

$$R(j,d) = \begin{cases} P_l \cdot d & if \quad P_j > P_l \\ d\left(\dfrac{P_l}{\eta} + P_j\left(1 - \dfrac{1}{\eta}\right)\right) & if \quad P_j <= P_l \end{cases}$$

The precise procedure to adapt the duty cycle to account for the above factors is presented in Algorithm 1. This calculation is performed at the end of every slot to set the duty cycle for the next

slot. We claim that our duty cycling algorithm is energy neutral because an surplus of energy at the previous time slot will always translate to additional energy opportunity for future time slots, and vice versa. The claim may be violated in cases of severe energy shortages especially towards the end of window. For example, a large deficit in energy supply can't be restored if there is no future energy input until the end of the window. In such case, this offset will be carried over to the next window so that long term energy neutrality is still maintained.

6. EVALUATION

Our adaptive duty cycling algorithm was evaluated using an actual solar energy profile measured using a sensor node called Heliomote, capable of harvesting solar energy [14]. This platform not only tracks the generated energy but also the energy flow into and out of the battery to provide an accurate estimate of the stored energy.

The energy harvesting platform was deployed in a residential area in Los Angeles from the beginning of June through the middle of August for a total of 72 days. The sensor node used is a Mica2 mote running at a fixed 40% duty cycle with an initially full battery. Battery voltage and net current from the solar panels are sampled at a period of 10 seconds. The energy generation profile for that duration, measured by tracking the output current from the solar cell is shown in Figure 4, both on continuous and diurnal scales. We can observe that although the energy profile varies from day to day, it still exhibits a general pattern over several days.

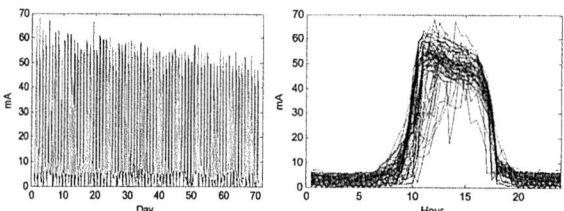

Figure 4 Solar Energy Profile (**Left:** Continuous, **Right:** Diurnal)

6.1. Prediction Model

We first evaluate the performance of the prediction model, which is judged by the amount of absolute error it made between the predicted and actual energy profile. Figure 5 shows the average error of each time slot in mA over the entire 72 days. Generally, the amount of error is larger during the day time because that's when the factor of weather can cause deviations in received energy, while the prediction made for night time is mostly correct.

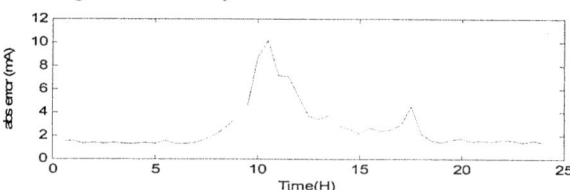

Figure 5. Average Predictor Error in mA

6.2. Adaptive Duty cycling algorithm

Prior methods to optimize performance while achieving energy neutral operation using harvested energy are scarce. Instead, we compare the performance of our algorithm against two extremes: the theoretical optimal calculated assuming complete knowledge about future energy availability and a simple approach which attempts to achieve energy neutrality using a fixed duty cycle without accounting for battery inefficiency.

The optimal duty cycles are calculated for each slot using the future knowledge of actual received energy for that slot. For the simple approach, the duty cycle is kept constant within each day and is computed by taking the ratio of the predicted energy availability and the maximum usage, and this guarantees that the senor node will never deplete its battery running at this duty cycle.

$$D = \eta \cdot \sum_{i \in \{1..Nw\}} P_s(i) \Big/ N_w \cdot P_c$$

We then compare the performance of our algorithm to the two extremes with varying battery efficiency. Figure 6 shows the results, using $D_{max} = 0.8$ and $D_{min} = 0.3$. The battery efficiency was varied from 0.5 to 1 on the x-axis and solar energy utilizations achieved by the three algorithms are shown on the y-axis. It shows the fraction of net received energy that is used to perform useful work rather than lost due to storage inefficiency.

As can be seen from the figure, battery efficiency factor has great impact on the performance of the three different approaches. The three approaches all converges to 100% utilization if we have a perfect battery ($\eta=1$), that is, energy is not lost by storing it into the batteries. When battery inefficiency is taken into account, both the adaptive and optimal approach have much better solar energy utilization rate than the simple one. Additionally, the result also shows that our adaptive duty cycle algorithm performs extremely close to the optimal.

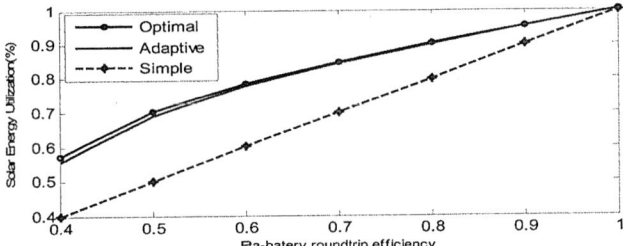

Figure 6. Duty Cycles achieved with respect to η

We also compare the performance of our algorithm with different values of D_{min} and D_{max} for $\eta=0.7$, which is typical of NiMH batteries. These results are shown in Table 1 as the percentage of energy saved by the optimal and adaptive approaches, and this is the energy which would normally be wasted in the simple approach. The figures and table indicate that our real time algorithm is able to achieve a performance very close to the optimal feasible. In addition, these results show that environmental energy harvesting with appropriate power management can achieve much better utilization of the environmental energy.

D_{max}	0.8	0.8	0.8	0.5	0.9	1.0
D_{min}	0.05	0.1	0.3	0.2	0.2	0.2

TABLE 1. Energy Saved by adaptive and optimal approach.

Adaptive	51.0%	48.2%	42.3%	29.4%	54.7%	58.7%
Optimal	52.3%	49.6%	43.7%	36.7%	56.6%	60.8%

7. CONCLUSIONS

We discussed various issues in power management for systems powered using environmentally harvested energy. Specifically, we designed a method for optimizing performance subject to the constraint of energy neutral operation. We also derived a theoretically optimal bound on the performance and showed that our proposed algorithm operated very close to the optimal. The proposals were evaluated using real data collected using an energy harvesting sensor node deployed in an outdoor environment.

Our method has significant advantages over currently used methods which are based on a conservative estimate of duty cycle and can only provide sub-optimal performance. However, this work is only the first step towards optimal solutions for energy neutral operation. It

is designed for a specific power scaling method based on adapting the duty cycle. Several other power scaling methods, such as DVS, sub-module power switching and the use of multiple low power modes are also available. It is thus of interest to extend our methods to exploit these advanced capabilities.

8. ACKNOWLEDGEMENTS

This research was funded in part through support provided by DARPA under the PAC/C program, the National Science Foundation (NSF) under award #0306408, and the UCLA Center for Embedded Networked Sensing (CENS). Any opinions, findings, conclusions or recommendations expressed in this paper are those of the authors and do not necessarily reflect the views of DARPA, NSF, or CENS.

REFERENCES

[1] R Ramanathan, and R Hain, "Toplogy Control of Multihop Wireless Networks Using Transmit Power Adjustment" in Proc. Infocom. Vol 2. 26-30 pp. 404-413. March 2000

[2] T.A. Pering, T.D. Burd, and R. W. Brodersen, " The simulation and evaluation of dynamic voltage scaling algorithms", in Proc. ACM ISLPED, pp. 76-81, 1998

[3] L. Benini and G. De Micheli, Dynamic Power Management: Design Techniques and CAD Tools. Kluwer Academic Publishers, Norwell, MA, 1997.

[4] John Kymisis, Clyde Kendall, Joseph Paradiso, and Neil Gershenfeld. Parasitic power harvesting in shoes. In *ISWC*, pages 132-139. IEEE Computer Society press, October 1998.

[5] Nathan S. Shenck and Joseph A. Paradiso. Energy scavenging with shoe-mounted piezoelectrics. *IEEE Micro*, 21(3):30ñ42, May-June 2001.

[6] T Starner. Human-powered wearable computing. *IBM Systems Journal*, 35(3-4), 1996.

[7] Mohammed Rahimi, Hardik Shah, Gaurav S. Sukhatme, John Heidemann, and D. Estrin. Studying the feasibility of energy harvesting in a mobile sensor network. In ICRA, 2003.

[8] ChrisMelhuish. The ecobot project. www.ias.uwe.ac.uk/energy autonomy/EcoBot web page.html.

[9] Jan M.Rabaey, M. Josie Ammer, Julio L. da Silva Jr., Danny Patel, and Shad Roundy. Picoradio supports ad hoc ultra-low power wireless networking. *IEEE Computer*, pages 42–48, July 2000.

[10] Joseph A. Paradiso and Mark Feldmeier. A compact, wireless, self-powered pushbutton controller. In *ACM Ubicomp*, pages 299–304, Atlanta, GA, USA, September 2001. Springer-Verlag Berlin Heidelberg.

[11] SE Wright, DS Scott, JB Haddow, andMA Rosen. The upper limit to solar energy conversion. volume 1, pages 384 – 392, July 2000.

[12] Darpa energy harvesting projects. http://www.darpa.mil/dso/trans/energy/projects.html.

[13] Werner Weber. Ambient intelligence: industrial research on a visionary concept. In *Proceedings of the 2003 international symposium on Low power electronics and design*, pages 247–251. ACM Press, 2003.

[14] V Raghunathan, A Kansal, J Hsu, J Friedman, and MB Srivastava, "Design Considerations for Solar Energy Harvesting Wireless Embedded Systems," (IPSN/SPOTS), April 2005.

[15] Xiaofan Jiang, Joseph Polastre, David Culler, Perpetual Environmentally Powered Sensor Networks, (IPSN/SPOTS), April 25-27, 2005.

[16] Chulsung Park, Pai H. Chou, and Masanobu Shinozuka, "DuraNode: Wireless Networked Sensor for Structural Health Monitoring," to appear in *Proceedings of the 4th IEEE International Conference on Sensors*, Irvine, CA, Oct. 31 - Nov. 1, 2005.

[17] Aman Kansal and Mani B. Srivastava. An environmental energy harvesting framework for sensor networks. In *International symposium on Low power electronicsand design*, pages 481–486. ACM Press, 2003.

[18] Thiemo Voigt, Hartmut Ritter, and Jochen Schiller. Utilizing solar power in wireless sensor networks. In *LCN*, 2003.

[19] A. Kansal, J. Hsu, S. Zahedi, and M. B. Srivastava. Power management in energy harvesting sensor networks. *Technical Report TR-UCLA-NESL-200603-02, Networked and Embedded Systems Laboratory*, UCLA, March 2006.

Power Reduction of Multiple Disks Using Dynamic Cache Resizing and Speed Control *

Le Cai
School of Electrical and Computer Engineering
Purdue University
lc@purdue.edu

Yung-Hsiang Lu
School of Electrical and Computer Engineering
Purdue University
yunglu@purdue.edu

ABSTRACT

This paper presents an energy-conservation method for multiple disks and their cache memory. Our method periodically resizes the cache memory and controls the rotation speeds under performance constraints. The cache memory stores the data from the disks for reuse. Enlarging the cache memory reduces disk accesses and disk utilization. This allows the disks to reduce their speeds and conserve energy because the disks' power consumption is quadratic to their speeds. However, the cache memory itself consumes power to retain data. Shrinking cache memory can save memory power while increasing disk accesses and degrading performance. Choosing proper cache sizes and rotation speeds can reduce the energy consumption of both memory and disks with satisfactory performance. We model cache resizing and speed setting as an optimization problem with minimizing the power consumption as objective and limiting disk utilization as constraints. We compare our method with the methods resizing cache based on request rates. The simulation results show that our method achieves better energy savings while limiting disk access latency.

Categories and Subject Descriptors

C.4 [**Performance of Systems**]: Performance attributes

General Terms

Design, performance

Keywords

Disk cache, disk rotation speed, power management

---

*This work is supported in part by National Science Foundation CNS-0347466, and Purdue Research Foundation. Any opinions, findings, and conclusions or recommendations expressed in this paper are those of the authors and do not necessarily reflect the views of the sponsors.

Permission to make digital or hard copies of all or part of this work for personal or classroom use is granted without fee provided that copies are not made or distributed for profit or commercial advantage and that copies bear this notice and the full citation on the first page. To copy otherwise, to republish, to post on servers or to redistribute to lists, requires prior specific permission and/or a fee.
ISLPED'06, October 4–6, 2006, Tegernsee, Germany.
Copyright 2006 ACM 1-59593-462-6/06/0010 ...$5.00.

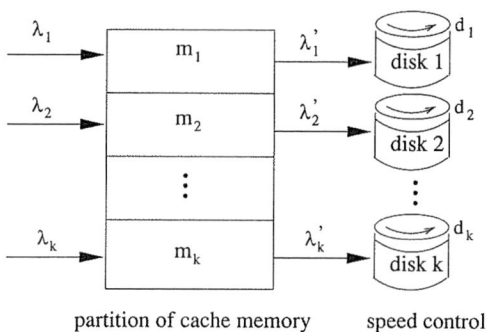

partition of cache memory   speed control

Figure 1: Caching based speed management (CBSM) for multiple disks.

1. INTRODUCTION

Energy conservation in storage system increasingly receives research attention [6, 8, 9, 14] because storage system, including cache, memory, and disks, consumes significant amounts of power. Most storage devices support low-power modes to reduce power consumption. For example, hard disks can switch to the standby or sleep mode when no disk accesses occur; memory banks can stay in the low-power modes after accesses. In addition to low-power modes, hard disks can reduce their power consumption by decreasing rotation speeds [4]. Adjusting disk rotation speeds achieves similar energy savings from the disks as frequency scaling from processors. When the rotation speeds drop, the power consumption of disks quadratically decreases while the disks' service time increases almost linearly [4, 13]. Therefore, energy savings can be achieved by lowering the rotation speeds.

Main memory is used as the cache of disks. More data cached in memory can eliminate more accesses to the disks so that they can be slowed down to save power. However, the memory itself consumes power to retain data. Using more memory consumes more power. Choosing proper cache memory sizes and disk rotation speeds can minimize the power consumption of both memory and disks. Allocating more memory to one disk reduces the cache memory used by other disks. Such cache memory resizing can save energy by allocating more cache memory to the disks which can achieve larger energy savings.

This paper presents a method to periodically adjust the cache sizes and the rotation speeds of multiple disks to achieve better energy savings while limiting disk utilization. We call the method caching based speed management (CBSM). The cache memory is distinguished from the rest of the main memory because the cache memory stores the

method	disk PM	memory PM	cache size/partition	feature	performance constraint
PB-LRU[16]	sleep	no	fixed/yes	caching	no
DRPM[4]	speed control	no	fixed/no	speed control	no
PDC[10]	speed control	no	fixed/no	file migration+speed control	no
Hibernator[15]	speed control	no	fixed/no	file migration+speed control	yes
CR[7]	sleep	no	fixed/no	code restructuring	no
CBSM	speed control	yes	variable/yes	caching+speed control	yes

Table 1: Various power management of disk arrays. PM: power management. No: not used. Yes: used. Sleep: transition disks to sleep or standby mode.

disk data for reuse. We study server systems because their data are often reused and the cache memory can considerably reduce disk accesses. Figure 1 shows the concept of CBSM. The cache size and the request rate of the ith disk are represented by m_i and λ_i, respectively. After caching, the request rate decreases to λ_i'. The ith disk runs at a rotation speed of d_i. The goal of CBSM is to determine m_i and d_i to reduce the total power consumption under performance constraints. The performance constraints are to limit the average disk utilization because high utilization causes longer latency [5]. The utilization is defined as the ratio between disk request rate and disk bandwidth. We model CBSM as an optimization problem. The power consumption and the disk utilization are expressed as the functions of m_i and d_i based on three relationships: (a) the cache size and the disk request rate, (b) the rotation speed and the disk's service time, and (c) the rotation speed and the disk's power consumption. We simulate the disk cache and the power manager to compare CBSM with other methods of resizing cache memory and controlling rotation speeds for multiple disks. The results show that CBSM consistently achieves energy savings with lower latency of disk accesses.

2. RELATED WORK

Researchers present many methods to reduce the power consumption of disk arrays. We compare these methods in Table 1. Zhu et al. [16] present a partition based LRU (PB-LRU) replacement to dynamically partition the cache memory for multiple disks. Their method enlarges the disks' idleness and saves power by switching them to low-power modes. However, in server systems, the disks experience short idleness even when the workload is light. Gurumurthi et al. [4] exploit multiple rotation speeds for power savings, called dynamic rotations per minute (DRPM). File migration can help DRPM to concentrate the workload on a small set of disks. When disks support only a few speeds, workload concentration makes a few disks running at high speeds while other disks can slow down and save power. Pinheiro et al. [10] propose popular data concentration (PDC) to move data across disks so that the first disk contains the most popular data, the second disk contains the second most popular data, and so on. Due to the limited bandwidth of a single disk, their method has to tradeoff power savings with performance degradation. Zhu et al. [15] improves PDC by dividing a disk array into groups and each group contains the same popular data. Their method, called Hibernator, models the relationship between the disk request latency and the rotation speed to limit performance degradation caused by speed control. Kandemir et al. [7] propose code restructuring (CR) to concentrate disk accesses on a small set of

symbol	description	unit
d	rotation speed	RPM
m	cache size of one disk	MB
p	memory static power	W/MB
e	memory dynamic energy	J/MB
r/R	min/max rotation speed	RPM
f	transfer rate at R	MB/s
h	rotation delay at R	second
λ	request rate	request/s
N	data set size	MB
α	Zipf parameter	
s	seek time	second
q	request size	MB
l	$h + \frac{q}{f}$	second
u	disk utilization	

Table 2: Symbols and their meaning. Symbols with subscript i: related to the ith disk.

disks so that the rest disks obtain longer idleness and more power-saving opportunities. Compared with these methods, our method has two advantages: (a) Using caching to manage workload avoids the complicated file migration or source codes revision. (b) Considering the energy consumed by the cache memory provides more energy-saving opportunities.

3. CACHING BASED SPEED MANAGEMENT

We model the power consumption of memory and disks to obtain proper cache sizes and disk speeds. We also estimate how they affect disk utilization. Table 2 lists the symbols used in our modeling.

3.1 Power and Performance Model

The disk's power consumption includes: static power when no accesses occur and dynamic power for accesses. We use the quadratic power model [4] to estimate the power consumption at different speeds. This is the latest model we can find for the power consumption of multiple-speed disks. The static power is $ad_i^2 + bd_i + c$, where d_i is within $[r, R]$. The values of r and R are 3600 and 12000 RPM. The coefficients a, b, and c are disk-specific constants. The dynamic power is proportional to the disk speed. The disk's idle power and peak power at R are 22.3W and 39W [4]. The dynamic power at different speeds is estimated using $u_i g d_i$, where u_i is the ith disk utilization. The value of g is $\frac{39 - 22.3}{12000} = 1.4 \times 10^{-3}$.

The memory's power consumption also includes static and dynamic parts. We use a power model of Rambus dynamic random-access memory (RDRAM) [11]. We consider that the RDRAM stays in the nap mode after memory accesses as proposed in [6, 8]. A 16MB memory chip consumes 10.5mW in the nap mode [11]. The static power per MB is computed: $p = \frac{10.5}{16} = 0.67$mW/MB. The dynamic power is estimated using the product of the data rate and the energy consumed to access one unit of data. When the RDRAM chip reaches its peak bandwidth 1.6 GB/second, its power consumption is 1325mW. The dynamic energy consumed to access one MB data (e) is estimated as $e = \frac{1325 - 10.5}{1.6 \times 1024} = 0.81$mJ/MB.

The disk utilization is defined as the ratio between request rate and disk bandwidth. The former is affected by the cache sizes while the latter varies with the rotation speeds. The disk's bandwidth is estimated using the reciprocal of the average disk's service time for one request. The service time includes three main parts: seek time (s_i), rotation delay, and transfer time. The rotation speed affects the latter two parts. At the peak speed R, the transfer rate is f_i and the rotation delay is h_i. We assume that the average transfer rate is proportional to the rotation speed and the rotation delay is inversely proportional to the rotation speed. Hence, at speed d_i, the service time is calculated as following.

$$s_i + \frac{R}{d_i}h_i + \frac{R}{d_i}\frac{q_i}{f_i} = s_i + \frac{R}{d_i}l_i, \quad l_i = h_i + \frac{q_i}{f_i} \quad (1)$$

3.2 Problem Formulation

In Figure 1, the request rates change from λ_i to λ_i' after the cache memory. We model the relationship between m_i and λ_i' assuming Zipf-like distribution of data popularity. Previous studies show that the distribution of data requests follows Zipf-like distributions for both web servers and multimedia servers [1, 3]. The probability of the jth most popular MB of data is proportional to $\frac{1}{j^{\alpha_i}}$. Here, α_i is the Zipf exponent. Its value is smaller than one ($0 < \alpha_i < 1$) [1, 3]. Larger α_i indicates that the accesses concentrate on a smaller portion of data. We use $v(j)$ to represent the probability of accessing the jth most popular MB of data. The value of $v(j)$ can be expressed as $v(j) = \frac{\Omega}{j^{\alpha_i}}$, where $\Omega = (\sum_{j=1}^{N_i} \frac{1}{j^{\alpha_i}})^{-1}$ and N_i is the data set size of the ith disk. We model the behaviors of the LRU replacement by assuming the cache always stores the most popular data. This simple model is used because we focus on the relationship between memory sizes and disk speeds. More accurate models can be found in [12]. The cache hit ratio is calculated as $\sum_{j=1}^{m_i} v(j)$. The request rate after caching is:

$$\lambda_i' = \lambda_i[1 - \sum_{j=1}^{m_i} v(j)] \simeq \lambda_i[1 - (\frac{m_i}{N_i})^{1-\alpha_i}] \quad (2)$$

The value of $\sum_{j=1}^{m_i} v(j)$ can be estimated using $(\frac{m_i}{N_i})^{1-\alpha_i}$ because an integral can approximate summation: $\sum_{j=1}^{N_i} \frac{1}{j^{\alpha_i}} \simeq \int_0^{N_i} \frac{1}{x^{\alpha_i}}dx = \frac{1}{1-\alpha_i}N_i^{(1-\alpha_i)}$. Hence, $\sum_{j=1}^{m_i} v(j) = \frac{\sum_{j=1}^{m_i} \frac{1}{j^{\alpha_i}}}{\sum_{j=1}^{N_i} \frac{1}{j^{\alpha_i}}} = (\frac{m_i}{N_i})^{(1-\alpha_i)}$. When m_i increases, $[1 - \sum_{j=1}^{m_i} v(j)]$ decreases so the disk request rate λ_i' decreases. The value of m_i is not larger than N_i because the disk cache cannot be larger than the disk's capacity. Hence, $\frac{m_i}{N_i} \leq 1$. As α_i grows, $[1 - (\frac{m_i}{N_i})^{1-\alpha_i}]$ and λ_i' decrease. Larger α_i means that the requests concentrate on a smaller amount of data. More requests are served from the cache memory and the number of disk requests decreases.

Based on the above models, we express the problem of

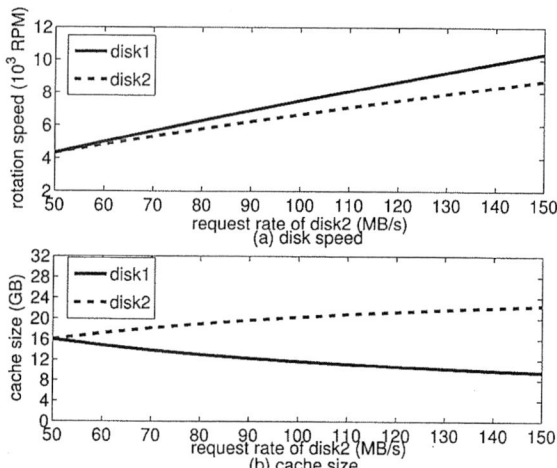

Figure 2: Request rates. The first disk's request rate is constant.

cache resizing and speed control for k disks as an optimization problem. The objective function is to minimize the power consumption of disks and their cache memory. Increasing request rates or reducing disk speeds enlarges the latency of each request and disk utilization. The constraints are (a) The sum of cache memory for each disk cannot exceed the installed memory size (M). (b) All disks' average utilization is smaller than the allowed utilization (U). The variables in the optimization problem are d_i and m_i. Other parameters are either constants or can be obtained at runtime.

$$\begin{array}{ll} \min & \sum_{i=1}^{k}(ad_i^2 + bd_i + c) + u_igd_i + pm_i + eq\lambda_i \\ & \sum_{i=1}^{k} m_i \leq M \\ & u_i = \lambda_i[1 - (\frac{m_i}{N_i})^{1-\alpha_i}](s_i + \frac{R}{d_i}l_i) \leq U \end{array} \quad (3)$$

In the objective function, $(ad_i^2 + bd_i + c)$ and u_igd_i are the static power and the dynamic power of the ith disk, respectively. The cache's static and dynamic power are represented by pm_i and $e\lambda_i$. The value of u_i is estimated using the product of disk request rate and disk service time. Their values are computed from (2) and (1). Enlarging m_i reduces u_i because the disk request rate decreases. Therefore, the disks' dynamic power decreases. When the decrease of the disks' dynamic power exceeds the increase of the cache memory's power, enlarging cache memory can save power.

3.3 Cache Sizes and Rotation Speeds

The formula (3) includes non-linear expressions. Although no explicit solutions are derived from (3), we conduct numeric analysis on proper cache sizes and disk speeds with different parameters. A two-disk array is used as an example to show the effects of request rate (λ) and disk utilization (U). Each disk stores a data set of 32GB with the Zipf exponent of 0.9. We use 32GB data sets to fit into a 33GB disk model [4]. Our method can be applied to larger data sets and the disks with larger capacity. Two disks have the same average seek time: $s_1 = s_2 = 3.4$ms. Their average rotation and transfer delays are same: $l_1 = l_2 = 3$ms. These numbers are obtained from practical workloads. We fix U as 50%. The first disk receives requests at 50MB/s while the second disk's request rate varies from 50MB/s to 150MB/s. Based on (3), we compute the proper disk speeds and cache

188

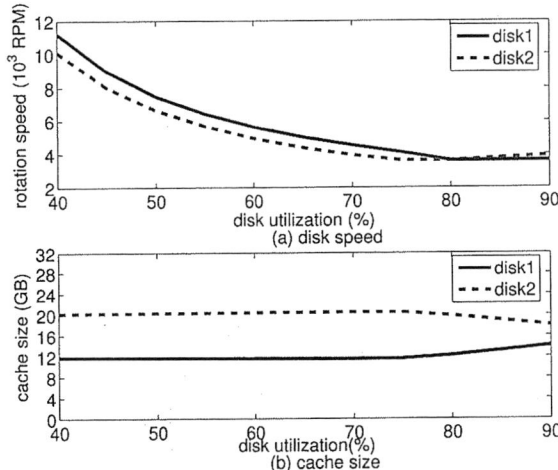

Figure 3: Cache size and rotation speed for different disk utilization. The first disk's request rate is constant.

sizes using *fmincon* in Matlab. The results are shown in Figure 2. The x-axes represent the request rate of the second disk. The y-axes are the disk rotation speeds and the cache sizes of two disks. When the request rate is 50MB/s, both disks use the same speed and cache size. As the second disk's request rate grows, the disk uses more cache memory to reduce disk accesses. As a result, the second disk runs at lower speeds than the first disk to save power. This result indicates that it is not most power-saving to make $\lambda'_1 = \lambda'_2$. Figure 2 (b) shows that the ratio between the cache sizes is not equal to the ratio of the request rates. For example, at 150MB/s, the rate ratio is $\frac{50}{150} = 0.33$ while the ratio of the cache sizes is $\frac{9.5}{22.5} = 0.42$. Allocating cache based on the request rates cannot achieve best power savings.

The allowed disk utilization (U) also affects proper disk speeds and cache sizes. In Figure 3, we fix the request rates as 50MB/s and 100MB/s. The value of U varies from 40% to 90%. Lower utilization cannot be achieved due to the request rates. In Figure 3, the disk speeds drop rapidly with the increase of U while the cache sizes remain almost constant. The ratio of cache sizes remains $\frac{12}{20} = 0.6$ because the request rates are fixed. Increasing utilization allows the disks to use lower speeds and smaller caches. However, reducing disk speeds save more energy. When U becomes 80%, the disk speeds reach minimum. In Figure 3 (b), further enlarging U changes the ratio of the cache sizes. More memory is allocated to the first disk because its dynamic power decrease more rapidly as the cache memory increases.

Our method adjusts the cache memory and the rotation speed of each disk every 10 minutes. We choose this period length to achieve fast adaption and small optimization overhead. When each period starts, the proper cache sizes and speeds are computed by solving the optimization problem (3). The solver is from Lindo system Inc. written in C. The values of the parameters in current period are predicted using the information obtained during previous period. By recording the accesses to the cache memory and disks, we obtain the request rates, the request sizes, and the sizes of working sets. We also count the access number of each file to estimate the Zipf exponent of data popularity. The disk seek time and the rotation delay are estimated using approxima-

tion because they are invisible to the power manager. We estimate the disk seek time using the product of the average single-track seek time and the seek track number. The track number is computed by multiplying the block distance between two consecutive disk requests and the average block number per disk track. We use the time to rotate half circle as the approximate for the average rotation delay.

4. SIMULATION AND RESULTS

We simulate the power manager and the disk cache to evaluate the energy savings and the performance impact of our method and the methods in comparison. SPECWeb99 is used to create web server workloads on a real machine. We use SPECWeb99 because web servers are one of most important server systems and SPECWeb99 is the state-of-art generator of web requests. We revise Linux kernel in the web server to record the access traces to the cache memory. The traces are adjusted for different data rates. The traces are processed by the simulator of the disk cache with LRU replacement. Its output is the traces of the disk accesses used by the disk simulator Disksim [2] and the simulator of the power manager. The former obtains the information of disk performance, such as disk access latency. The latter manages both memory and disks for power savings.

We compare our method with four power management methods. Each method includes power management schemes for memory and disks. For memory management, we use rate-based (RB) resizing: the cache sizes are proportional to the request rates. The memory size step is 16MB because different types of DRAM can support this granularity. For disk management, one of the four methods uses dynamic speed (DS) control based on the same performance model used in CBSM. The speed step is 1200 RPM as in [4]. The other three methods use fixed disk speeds (FS): 6000 RPM, 9000 RPM, and 12000 RPM. The four methods are named using the combination of cache resizing methods and disk speed control methods: RBDS, RBFS-6K, RBFS-9K, RBFS-12K. For CBSM and RBDS, the disk speeds are chosen to limit the disk utilization to 20%.

The workload varies in many characteristics, such as request rate and working set. CBSM can adapt cache sizes and disk speeds to the workload. In contrast, the methods using fixed speeds or cache sizes need many times of simulation to determine the proper cache sizes and disk speeds for different workloads. The methods resizing cache only based on one workload characteristics may not choose the proper cache sizes and disk speeds. In the simulation, we change the request rate of one disk to show how our method can save energy with low latency by adjusting cache sizes and disk speeds. We use a four-disk array with 64GB memory. Each disk stores a 32GB data set. The first three disks receive requests at constant rates: 5MB/s, 50MB/s, and 100MB/s, respectively. The fourth disk's request rate varies from 5MB/s to 150MB/s. Every ten minutes, the methods adjust cache sizes and disk rotation speeds. The disks' transfer rate (f_i) is 80MB/s at R.

Figure 4 shows the energy consumption and the disk request latency of the methods in comparison. The energy consumption includes that of cache memory and four disks. The x-axes represent the request rate of the fourth disk while the y-axes denote the energy consumption percentage and the percentage of disk request latency. The percentage is based on RBFS-12K because this method always

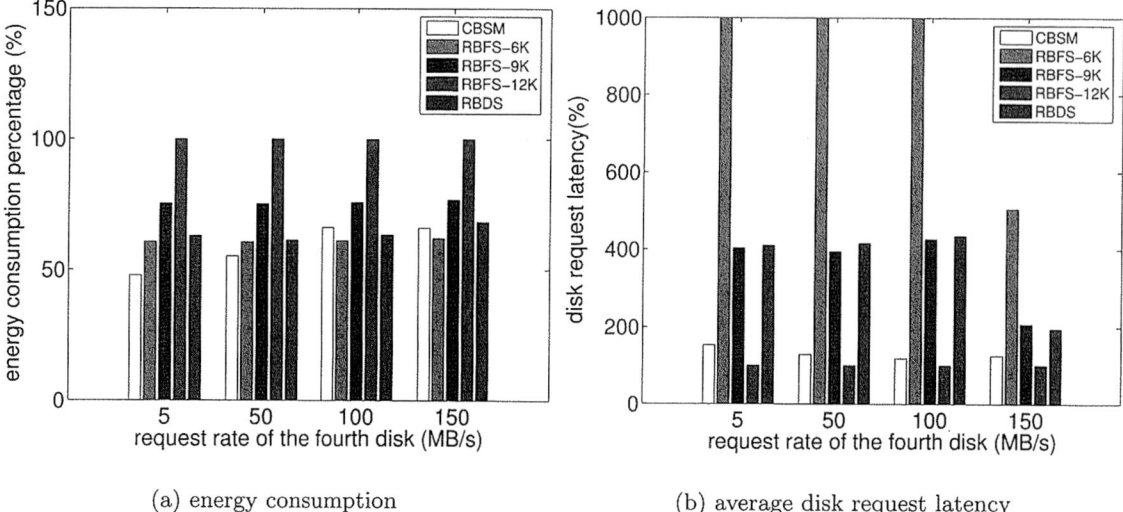

(a) energy consumption

(b) average disk request latency

Figure 4: Energy consumption and disk request latency of a four-disk array. The first three disks' request rates are constant: 5, 50, and 100MB/s. The latency percentage over 1000% is not shown.

keeps the highest rotation speeds to achieve better performance. Higher bars indicate more energy consumption or longer request latency. In Figure 4 (a), compared to RBFS-12K, CBSM saves 34% to 53% energy when the fourth disk receives different workloads. The energy savings increase as the request rate decreases because the fourth disk can use less cache memory and lower rotation speeds. CBSM consumes less energy than RBFS-9K since using fixed and high speeds cannot save power from disks when they receive different workloads. Although RBFS-6K save more energy than CBSM at 100MB/s and 150MB/s, RBFS-6K causes longer delay to disk requests. In Figure 4 (b), CBSM only increases the average latency by 17% to 53% while the average latency of RBFS-6K exceeds 1000% of RBFS-12K. CBSM's latency is also much shorter than RBFS-9K because CBSM adjusts disk speeds according to different rates of the disks. At most request rates, CBSM consumes less power and causes shorter latency than RBDS due to the joint management of cache memory and disk speeds. Overall, CBSM consistently achieves both energy savings and low request latency across different workloads.

5. CONCLUSION

This paper presents the joint power management of memory and multiple disks by adjusting the cache size and the rotation speed of each disk. The method is formulated as an optimization problem with constraints of limiting disk utilization. The method solves the optimization problem at runtime to obtain the proper cache sizes and rotation speeds. The simulation results show that our method can adapt to the workload variation and achieves both energy savings and low disk access latency for multiple disks. This paper can be extended to consider latency constraints and other cache replacement algorithms.

6. REFERENCES

[1] L. Breslau, P. Cao, L. Fan, G. Phillips, and S. Shenker. Web Caching and Zipf-like Distributions: Evidence and Implications. In *INFOCOM*, pages 126–134, 1999.

[2] J. S. Bucy and G. R. Ganger. The disksim simulation environment version 3.0 reference manual. http://www.pdl.cmu.edu/DiskSim/, 2003.

[3] C. Griwodz, M. Bar, and L. C. Wolf. Long-term movie popularity models in video-on-demand systems or the life of an on-demand movie. In *ACM International Conference on Multimedia*, pages 349–357, 1997.

[4] S. Gurumurthi, A. Sivasubramaniam, M. Kandemir, and H. Franke. DRPM: Dynamic Speed Control for Power Management in Server Class Disks. In *International Symposium on Computer Architecture*, pages 169–181, 2003.

[5] J. L. Hennessy and D. A. Patterson. *Computer Architecture: A Quantitative Approach*. Morgan Kaufmann, 3 edition, 2003.

[6] H. Huang, P. Pillai, and K. G. Shin. Design and Implementation of Power-Aware Virtual Memory. In *USENIX Annual Technical Conference*, pages 57–70, 2003.

[7] M. Kandemir, S. W. Son, and G. Chen. An evaluation of code and data optimizations in the context of disk power reduction. In *ISLPED*, pages 209–214, 2005.

[8] A. R. Lebeck, X. Fan, H. Zeng, and C. Ellis. Power Aware Page Allocation. In *International Conference on Architectural Support for Programming Languages and Operating Systems*, pages 105–116, 2000.

[9] Y.-H. Lu, E.-Y. Chung, T. Simunic, L. Benini, and G. D. Micheli. Quantitative Comparison of Power Management Algorithms. In *Design Automation and Test in Europe*, pages 20–26, 2000.

[10] E. Pinheiro and R. Bianchini. Energy Conservation Techniques for Disk Array-based Servers. In *International Conference on Supercomputing*, pages 68–78, 2004.

[11] Rambus Company. 128Mb RDRAM Split Bank Architecture Advanced Information, March 2003.

[12] G. S. Rao. Performance analysis of cache memories. *Journal of the ACM*, 25(3):278–395, 1978.

[13] E. Shriver. *Performance Modeling for Realistic Storage Devices*. PhD thesis, New York University, May 1997.

[14] T. Simunic, L. Benini, P. Glynn, and G. D. Micheli. Dynamic Power Management for Portable Systems. In *International Conference on Mobile Computing and Networking*, pages 11–19, 2000.

[15] Q. Zhu, Z. Chen, L. Tan, Y. Zhou, K. Keeton, and J. Wilkes. Hibernator: Helping disk arrays sleep through the winter. In *ACM Symposium on Operating Systems Principles*, October 2005.

[16] Q. Zhu, A. Shankar, and Y. Zhou. PB-LRU: A Self-tuning Power Aware Storage Cache Replacement Algorithm for Conserving Disk Energy. In *International Conference on Supercomputing*, pages 79–88, 2004.

Lifetime Aware Resource Management for Sensor Network Using Distributed Genetic Algorithm

Qinru Qiu   Qing Wu
Department of Electrical and Computer Engineering
Binghamton University
Binghamton, NY 13902
001-607-777-4918, 001-607-777-4536
{qqiu, qwu}@binghamton.edu

Daniel Burns   Douglas Holzhauer
Air Force Research Laboratory, Rome Site
26 Electronic Parkway
Rome, NY 13441
001-315-330-2335, 001-315-330-4920
{Daniel.Burns, Douglas.Holzhauer}@rl.af.mil

ABSTRACT

In this work we consider lifetime-aware resource management for sensor network using distributed genetic algorithm (GA). Our goal is to allocate different detection methods to different sensor nodes in the way such that the required detection probability can be achieved while the network lifetime is maximized. The contribution of this paper is twofold. Firstly, the resource management problem is formulated as a constraint optimization problem and is solved using a distributed GA. Secondly, empirical analysis results are provided that reveals the relationship between the configuration parameters and the quality of the search. A regression model is designed to estimate the runtime of the distributed GA given the configuration parameters. The model is utilized to find energy efficient configurations of the algorithm.

Categories and Subject Descriptors

J.7 [**Computer Applications**]: Sensors and Sensor Networks

General Terms

Experimentation

Keywords

Distributed Genetic Algorithm, Sensor Network, Energy Aware Design, Resource Management

1. INTRODUCTION

Due to the fast development of information technology, the networked distributed system is gradually replacing the conventional centralized system. It is a vision of the future that large numbers of low cost smart mobile devices will be integrated into the daily life of ordinary people. Accumulated, they provide the information processing capability that is equivalent to a high performance processing station. The emerging concept of Ambient Intelligence [1] and the recent developments of sensor networks [2], and wearable computers [3] reflect such vision. A distributed system consists of multiple heterogeneous networked processing elements, which are battery-powered and work on a set of tasks collaboratively. Each processing element has limited resources, such as battery energy, communication bandwidth, etc. It is a challenging task to efficiently utilize these resources to deliver required services during the runtime in a dynamic environment.

Copyright 2006 Association for Computing Machinery. ACM acknowledges that this contribution was authored or co-authored by an employee, contractor or affiliate of the U.S. Government. As such, the Government retains a nonexclusive, royalty-free right to publish or reproduce this article, or to allow others to do so, for Government purposes only.
ISLPED'06, October 4–6, 2006, Tegernsee, Germany.
Copyright 2006 ACM 1-59593-462-6/06/0010...$5.00.

Resource management is defined as the process that assigns tasks to different processing elements, schedules their start times and decides the level of service quality, which determines the resource usage, such as the energy dissipation and communication bandwidth, to run these tasks. The execution of each task represents a positive gain when measuring or quantifying the performance of the system. It also associates a cost, which represents the resource usage. The resource management problem can be formulated as a multi-objective optimization problem, i.e. maximizing the gain while minimizing the cost. It can also be formulated as a constraint optimization problem, i.e. maximizing the gain while satisfying the cost constraint or vise versa.

In this paper we focus on the management of the energy resource in an environment monitoring sensor network that is used to monitor, model and forecast physical processes, such as environment pollution, flooding, and fire etc. The basic configuration of each node in this network consists of a microprocessor, a wireless transceiver and an array of sensors such as light detector, barometer, humidity and thermopile sensors. A set of data acquisition and signal processing applications is available on each node. They provide the tradeoffs between detection quality and resource utilization. For example, increasing the sampling rate improves the probability of detecting an abnormal event however it increases the power consumption as well.

There is usually a significant cost associated with deploying an environment monitoring system. It is desirable that the system can work for a reasonably long time after it is deployed. A common approach is to incorporate certain level of redundancy in the system. More than one node usually will be deployed to cover the same region. These nodes may be turned on alternatively to extend the network lifetime or simultaneously to increase the detection probability. If the minimum detection accuracy is given as a user constraint, the resource management problem for the system is to determine which sensor nodes should be turned on to process which data acquisition and signal processing application such that the network lifetime can be maximized while meeting the required detection accuracy. This is a well known general assignment problem which has been proven to be NP-complete [4].

Most of the traditional resource optimization algorithms are solved in a centralized, off-line approach which is not suitable for a distributed system. In this paper we study the use of distributed genetic algorithm (GA) to solve the above mentioned optimization problem, potentially using processing capabilities residing on nodes of the distributed sensor network. One of the major characteristics of the GA is that it is "embarrassingly parallel", in the sense that, its workload can easily be evenly distributed among processors, making it an appropriate choice for solving optimization problems in distributed systems. The configurations of the distributed, multi-deme GA, such as the population size, the migration rate, and the parallelism, has a significant impact on the quality of the search [8]. Finding efficient configurations of the distributed GA is an important research topic. The contribution of this paper is twofold. Firstly, the resource management problem is formulated as a constraint optimization problem and is solved using a distributed GA. The simulation results show that the resulting task allocation scheme increases the system lifetime by 14.4% in average, comparing to heuristic approaches. Secondly, empirical analysis results are provided that reveals the relationship between the configuration parameters and the quality of the search. A regression model is presented that estimates the runtime of the distributed GA given the configuration parameters. This model is then used to find energy efficient configurations of the algorithm.

Many previous works on sensor network resource management and task allocation address network communication issues [5][6]. In these schemes the nodes are dynamically awakened to route a message. In reference [7] the resource allocation problem in a vehicle tracking system is modeled as a virtual market and solved using feedback control. This work focuses more on the tracking of a moving object rather than the collaborative detection of a static event. Therefore it cannot be applied in the environment monitoring system. Reference [9] focuses on task allocation on the gateways in a cluster-based sensor network. The problem is also formulated as a constraint optimization problem and is solved using simulated annealing, which is a centralized stochastic searching algorithm. Compared with reference [9], the resource management problem considered in this paper has a different set of constraints and objective functions and is solved using a distributed GA.

The rest of this paper is organized as follows. Section 2 introduces the sensor network architecture. Section 3 presents the distributed GA algorithm. Section 4 provides the empirical analysis of the relationship between the configuration parameters and the quality of search of GA and derives the regression model for runtime estimation. Section 5 discusses the utilization of the regression model to design energy efficient distributed GA. Sections 6 and 7 provide the experimental results and summaries, respectively.

2. SENSOR NETWORK ARCHITECTURE

We consider the sensor network that is deployed with a certain level of redundancy. The network can be partitioned into several clusters. Each cluster consists of p sensor nodes that are responsible for performing monitoring and hazard detection in the same region. Each sensor node is low cost and low quality; however combined together they provide very accurate detection. The nodes in the same cluster have direct communication with each other via wireless communication channels. The nodes in different clusters communicate with each other through gateways. In this work we assume that the clustering and routing scheme is provided. We also assume that each cluster has advanced data fusion capability so that the traffic of inter-cluster communication is low.

An array of w sensors is installed on each node. The reading from these sensors can be sampled by l different sampling frequencies. Obviously, higher sampling frequency leads to higher detection probability while consumes more energy. The sampled data from sensor i can be analyzed in x_i different ways. They provide different tradeoffs between accuracy and energy dissipation. A *detection method* (i.e. *task*) is considered as a combination of sensing function, sampling frequency and signal processing algorithm.

Each task-processor pair (i, k), $1 \leq i \leq n$ and $1 \leq k \leq p$, associates with two variables $pow_{i,k}$ and $prob_{i,k}$, which represent the power

consumption and the detection probability of task i when it is running on processor k. The $prob_{i,k}$ is a function of the location and the environment of the sensor node. We assume that this function is pre-calibrated and installed on each sensor node before its deployment. The sensor node will collect the environment information and calculate the detection probability using the provided function periodically. To improve the detection probability, the node is allowed to use more than one detection method at the same time. The combined detection probability of node k can be calculated as $1 - \prod_{i \in \Delta_k}(1 - prob_{i,k})$, where Δ_k is the set of tasks that are allocated to node k. The total node power consumption can be calculated as $\sum_{i \in \Delta_k} pow_{i,k}$. The detection probability $Prob$ of a cluster with p nodes is calculated as

$$Prob = 1 - \prod_{1 \leq k \leq p} \prod_{i \in \Delta_k}(1 - prob_{i,k}).$$

The goal of resource management is to find the Δ_k for each processor k so that the combined detection probability of the cluster is larger than the user defined constraint while the network lifetime is maximized. In this work, we define the network lifetime as the time from the deployment of the sensor network to the time when the first node runs out of battery energy. We assume that each sensor node is built with the smart battery Bus (SMBus) [10] which enables the system software to keep tracking of the remaining battery capacity and estimate the remaining lifetime.

3. RESOURCE MANAGEMENT USING DISTRIBUTED GA

A Genetic Algorithm (GA) is a stochastic search technique based on the mechanism of natural selection and recombination. It starts with an initial *population* of individuals, i.e. a set of randomly generated candidate solutions. The solutions are represented by *chromosomes*, which are collections of numbers or symbols that map onto parameters of the problem. Individuals are evolved from generation to generation, with *selection*, *mating*, and *mutation* operators that provide an effective combination of exploration of the global search space and pressure to converge to the global minimum. The solution quality is measured by a *fitness* function.

The Island multi-deme GA is one of the parallel GA models that are widely used [8]. In this model, the population is divided into several sub-populations and distributed on different processors. Each sub-population evolves independently for a few generations, before one or more of the best individuals of the sub-populations migrate across processors. The time between migrations is called *epoch*.

In this work the Island multi-deme GA is used to optimize the resource management for a cluster of sensor nodes. Each individual solution is a chromosome of n symbols, where n is the total number of tasks in the cluster. We assume that each task can only be selected by at most one sensor node in a cluster because multiple executions of the same task only generate redundant information. If the jth task is allocated to node x then the jth entry of the chromosome is equal to x. If the jth entry of the chromosome is -1 then this task is not allocated to any of the processors. Denote the user specified minimum detection probability as $prob_{th}$, the fitness function is:

$$fitness = \begin{cases} 0 & \text{if } Prob < Prob_{th} \\ \min_{1 \leq k \leq p}(B_k / \sum_{i \in \Delta_k} pow_{i,k}) & \text{otherwise} \end{cases} \quad (1)$$

where B_k is the remaining battery capacity of node k. The fitness of an individual is 0 if the corresponding resource management scheme cannot meet the user specified detection threshold; otherwise its fitness is equal to the minimum remaining lifetime of the nodes. Single point crossover mating function is used in our experiment. The mutation probability is set to 1%, and involves flipping bits in integer representations of the parameters stored in chromosomes.

The GA is running on np processors. Each sub-population is initialized randomly and its size is denoted as pop. The sub-population evolves independently for c generations, and then 5 of the best individuals are broadcast to all other processors. The three parameters, np, pop and c, will be referred as the *configuration parameters* in the rest of this paper. The value of the configuration parameters has significant impact on the convergence speed of the GA and the quality of the solution. An empirical analysis is next.

4. CONFIGURATION PARAMETERS

We are interested in understanding the effect of configuration parameters on the quality of the search of the distributed GA that is previously discussed. Some work has been carried out in this area [8]. However, most of these involve the analysis of simple optimization problems such as the fully deceptive function [12]. Whether their results can be applied to our problem is unknown. Due to the extremely large search space and very complicated stochastic behavior of the GA, we found that it is difficult to perform an analytical study. Therefore, extensive experiments have been simulated and the relation between the configuration parameters and the quality of search is derived empirically.

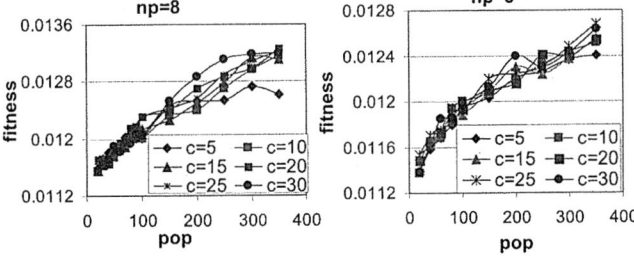

Figure 1 Normalized fitness vs. Sub-population size

Two sets of experiments have been carried out. In these experiments, we model a cluster of 10 sensor nodes. There are 100 tasks available. The GA is running on np sensor nodes with $np \leq 10$. The detection probability and the power consumption of each task are uniformly distributed random variables whose range is 1% ~ 25% and 0.1Watt ~ 10Watt respectively. The battery of each sensor has the capacity of 5000 Ampere·hour and the V_{dd} is 1V. Because GA is a stochastic algorithm, we run each simulation 50 times and report the mean value.

The first set of experiments is designed to find out the effect of the configuration parameters on the quality of the solution. We swept the np from 2 to 8, the pop from 25 to 350, and the c from 1 to 35. For each configuration, the distributed GA is simulated. The GA will stop when the fitness of the best individual does not improve for 2000 generations. The relation between the pop and the normalized fitness of the best individual is reported. Figure 1 shows two sets of data for np (i.e. the number of processors) equal to 8 and 3. The results show that increasing both the pop and the np improves the quality of the solution. However, varying c has very little impact on it. Therefore the quality of the solution is determined by the size of the total population which is the product of the pop and np.

The second set of experiments is designed to find out the effect of the configuration parameters on the runtime of the GA. The value of np, pop and c are swept in the same way as the first experiment. The GA stops when the fitness of the best individual exceeds the

threshold which is set to be 5 times of the expected fitness of a random individual. The number of generations that the GA has iterated is reported. Due to the iterative nature of GA, it is reasonable to assume that the runtime of each generation is approximately the same and it increases linearly as population size increases. Therefore we use the product of pop and the number of generations that the GA has iterated as a measure of the runtime.

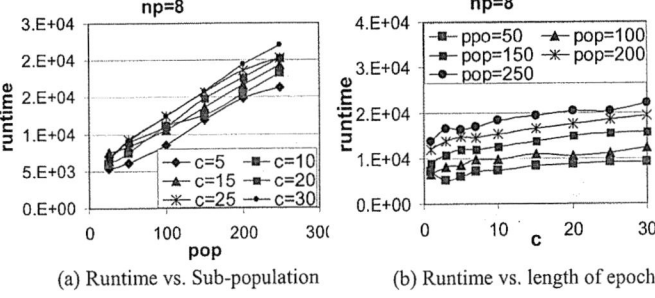

(a) Runtime vs. Sub-population   (b) Runtime vs. length of epoch

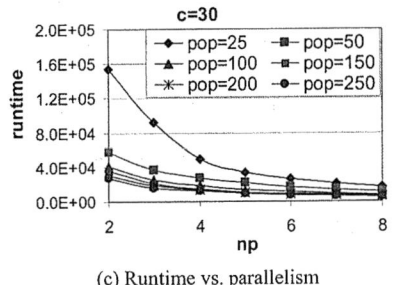

(c) Runtime vs. parallelism

Figure 2 Runtime vs. configuration parameters

The relation among pop, c, np and the runtime are extracted from the results of second experiment. Figure 2 (a)-(c) show some of the data that we have obtained. Several observations can be made from these data. First, when the size of the sub-population increases, the runtime increases linearly. Combined with the results from the first experiment we can see that if the goal of the GA is to find the best possible solution, then a large population should be used. However, if the goal of the GA is to find a good solution in a short time, then increasing the population size will not help. Instead a small population should be used. Second, reducing the migration rate will result an almost linear increase in solution time. The slope is the same for different sub population size. Third, increasing the number of processors will reduce runtime, and this effect is more dominant when the sub-population is small.

In order to consider the combined effect of all of the three configuration parameters, we introduce a new variable called *effective population* (*Epop*). The size of the effective population increases when the size of the sub-population, the parallelism or the migration rate increases. It can be calculated as the following:

$$Epop = pop + pop \cdot (np - 1) / c \qquad (2)$$

Given the effective population, the runtime of the distributed GA can be predicted. Let G denote the number of generations that the GA has iterated before it finds the solution with the required fitness. Figure 3 (a) gives the relation between $Epop$ and G. It shows that G is a continuous and differentiable function of $Epop$.

Based on the observation, we construct a prediction model to predict the number of generations that the GA has iterated.

$$G = a + a_0 / \sqrt{Epop} + \sum_{i=1}^{5} a_i / Epop^i \qquad (3)$$

The coefficients a, a_0 ..., a_5 are obtained using regression analysis. Note that the value of the coefficients will change if the experiment setup changes. Here the experiment setup includes the threshold of fitness and the distribution function of *prob* and *power*. For each new setup, regression analysis should be performed to obtain the values of the coefficients.

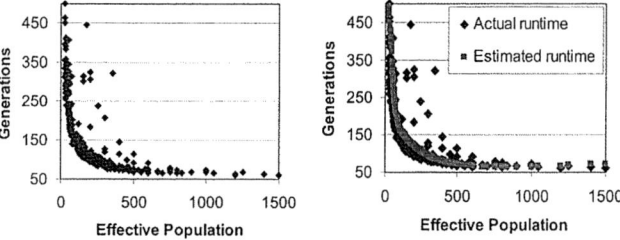

(a) *G* vs. effective population *G* (b) Comparing predicted and actual *G*

Figure 3 Runtime vs. effective population

The above model gives quite accurate prediction of the number of generations that GA has iterated given the configuration parameters. Figure 3 (b) compares the prediction model with the simulated results. The blue dots give the *G* value obtained from simulation and the magenta dots give the *G* value obtained using the prediction model. The runtime *T* is measured as the product of *pop* and the number of generations $T = G \cdot pop$.

5. ENERGY EFFICIENT CONFIGURATIONS OF DISTRIBUTED GA

Sensor nodes are energy constraint systems. Any application running on the sensor node should be designed carefully to achieve high speed and energy efficiency. In this section we will discuss how to select the configuration parameters to minimize the energy dissipation of the distributed GA.

In a computing system with fixed supply voltage (V_{dd}) and clock frequency, reducing the runtime of an algorithm leads to linear reduction of the energy dissipation if the processor can be turned off after the program finishes. More energy saving is possible by using *dynamic voltage and frequency scaling* (*DVFS*), which is one of the runtime power management approaches that is supported by many processors for the state-of-the-art mobile computing platforms. It is a property of CMOS digital circuit that reducing the V_{dd} can reduce the energy dissipation quadratically but increase the circuit delay linearly [11]. In a system with DVFS capability, the program is running at the minimum supply voltage and clock frequency so that it finishes just before the deadline. Due to the convex relation between the energy and the runtime, this gives more energy saving than running the program at the nominal speed and turn off the processor after the program finishes.

Population migration among the processors is an important feature in distributed GA. The communication energy to broadcast the best individuals must be considered. Under the assumption of a fixed transmission power and a constant transmission speed, the communication energy is proportional to the size of the transmitted data. The communication energy will not be affected by DVFS.

The computing energy is a product of the runtime and the power consumption of the processor. Therefore, the runtime model proposed in Section 4 is the key for the energy estimation of the distributed GA. While increasing the migration rate, decreasing the population size and increasing the parallelism reduce the runtime of

GA and consequently reduce the computing energy, frequent population migration leads to high communication energy. The configuration parameters must be selected carefully to minimize the overall system energy dissipation, which is the sum of computing energy and communication energy.

Let T_{nom} denote the process time for a single individual in each generation at nominal V_{dd} and let p_{nom} denote the power consumption of the processor at nominal V_{dd}. The energy dissipation of GA on a processor without DVFS can be calculated as:

$$E = T_{nom} \cdot p_{nom} \cdot T + G / c \cdot N \cdot E_{bit}, \qquad (4)$$

where *G* is the number of generations that GA has iterated, *T* is the runtime of the GA that is measured as the product of *pop* and *G*, *c* is the length of an epoch, *N* is the size of data that is broadcasted during each population migration, and E_{bit} is the energy to transmit one bit data. The first term in equation (4) is the computing energy and the second term is the communication energy. Furthermore, $T_{nom} \cdot p_{nom}$ represents the computing energy to processor one individual in each generation and $N \cdot E_{bit}$ represents the communication energy to broadcast the best individual during one migration. T_{nom}, p_{nom}, and E_{bit} are hardware related constant parameters. *N* is determined by the size of migrations which is also a constant value. Because we are not interested in calculating the absolute energy dissipation, we simplify equation (4) and consider a normalized energy dissipation which is calculated as the following,

$$E_{norm} = \frac{E}{T_{nom} \cdot p_{nom}} = T + \frac{G}{c} E_{mig}, \qquad (5)$$

where E_{mig} is the ratio of communication energy versus computing energy and it is calculated as $E_{mig} = \dfrac{N \cdot E_{bit}}{T_{nom} \cdot p_{nom}}$. As we can see the value of E_{mig} is determined by the system hardware configuration. For example, the power consumption of a Lucent ORiNOCO USB Wireless Adapter is 360mA in TX mode and 245mA in RX mode. The typical active power of an Intel XScale processor is 300mA. Assume that the data is transmitted at 1Mbit/s. If *N* equals to 1k bytes and T_{nom} equals to 5µs, which is the time to run 10k instructions at 200MHz clock, then E_{mig} is approximately 160.

E_{norm} is an increasing function of *pop* and a decreasing function of *np* because changing these two parameters only affects the computing energy. The only configuration parameter that affects both the computing and communication energy is *c*. Provided with the value of *pop*, *np*, E_{mig}, it is not difficult to find the optimal *c* that minimizes E_{norm} by solving the differential equation $\dfrac{\partial E_{norm}}{\partial c} = 0$.

Because GA is running on multiple sensor nodes, the total energy dissipation can be calculated as $E_{total} = np \cdot E_{norm}$ where *np* is the parallelism of the GA.

If the DVFS is available on the processor, then the computing energy can be scaled quadratically as the runtime decreases. The energy dissipation of GA on each processor can be calculated as the following,

$$EDVFS = T_{nom} \cdot p_{nom} \cdot T \cdot s^2 + G / c \cdot N \cdot E_{bit} \qquad (6)$$

Here *s* is the scaling factor and it is calculated as $s = \dfrac{T_{nom} \cdot T}{T_{req}}$, where

T_{req} is the deadline before which the GA must return a solution with the required fitness. Again, we simplify equation (6) and consider the normalized energy dissipation as the following,

$$EDVFS_{norm} = (T \cdot T_{nom} / T_{req})^3 + G / c \cdot E'_{mig}, \qquad (7)$$

E'_{mig} is calculated as $E'_{mig} = \dfrac{N \cdot E_{bit}}{p_{nom} \cdot T_{req}}$ which stands for the ratio of the communication energy for one migration versus the computing energy of the program if if takes exactly T_{req} time when running at the nominal V_{dd}. For the previous mentioned hardware system, which consists of Lucent ORiNOCO USB Wireless Adapter and Intel XScale processor, if the T_{req} is 1ms then E'_{mig} is approximately 0.8.

Again, the total energy dissipation can be calculated as $EDVFS_{total} = EDVFS_{norm} \cdot np$ and the optimal c that minimizes the energy dissipation can be found by solving the differential equation $\dfrac{\partial EDVFS_{norm}}{\partial c} = 0$.

Plug in the runtime estimation of G and T into equation (4)~(7), the energy dissipation of GA can be expressed as a function of the configuration parameters. Figure 4 (a) shows the relation between E_{norm} and c in a system without DVFS. The np and pop are set to 5 and 100 respectively. The E_{mig} varies from 90 to 180. As we can see from the figure, the energy is an increasing function of c for small E_{mig} and a decreasing function of c for large E_{mig}. Furthermore, when the E_{mig} falls into certain range, the energy is first a decreasing then an increasing function of c. In this case, we need to solve the previous mentioned differential equation to find the most energy efficient migration rate. When the parameter c gets larger, the E_{norm} under different E_{mig} approach to the same value. This is because the migration rate is so low that a small difference in the communication energy does not have a significant affect on the total energy.

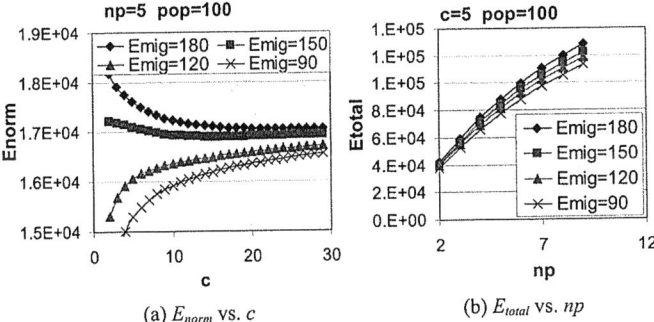

(a) E_{norm} vs. c (b) E_{total} vs. np

Figure 4 Energy vs. configuration parameters without DVFS

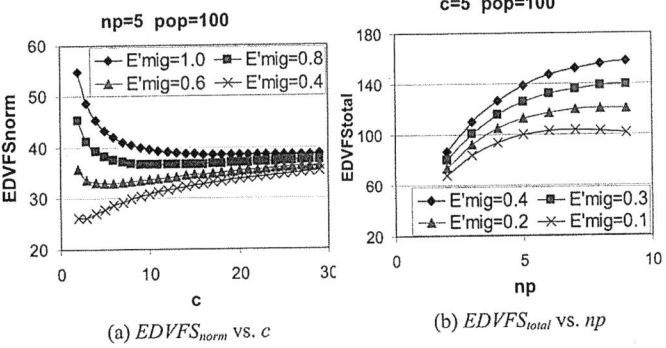

(a) $EDVFS_{norm}$ vs. c (b) $EDVFS_{total}$ vs. np

Figure 5 Energy vs. configuration parameters with DVFS

Figure 4 (b) shows the relation between E_{total} and np in a system without DVFS. The parameters c and pop are set to 5 and 100 respectively. E_{mig} varies from 90 to 180. It is interesting to note that

the total energy always increases no matter how we change the E_{mig}. This indicates that without DVFS the energy efficiency will decrease as the parallelism increases.

Figure 5 (a) shows the relation between $EDVFS_{norm}$ and c in a system with DVFS. The np is set to 5, the pop is set to 100 and the E'_{mig} varies from 1.0 to 0.4. In this figure, we see the similar trend as what has been shown for the system without DVFS. Figure 5 (b) shows the relation between the $EDVFS_{total}$ and np with E'_{mig} varies from 0.4 to 0.1. As we can see that for systems with $E'_{mig} \geq 0.3$, increasing the parallelism always increases the total energy dissipation. However, for systems with $E'_{mig} < 0.3$, increasing the parallelism will first increase then decrease the total energy. This is because increasing the parallelism reduces the overall computing energy quadratically and increases the overall communication energy linearly. Eventually the quadratic decreasing in computing energy will become dominant.

6. EXPERIMENTAL RESULTS

In order to evaluate the performance of the GA based resource management scheme, a C++ based software program is constructed to emulate the environment monitoring sensor network. The cluster consists of 10 low cost and low quality sensor nodes and 100 tasks. The battery of each sensor has the capacity of 5000 Ampere·hour. The detection probability and the power consumption of each task are randomly generated. Different distributions with different variances are tested in the experiment. Furthermore, to emulate the behavior of the real sensor network which is deployed in a dynamic environment, the detection probability of the sensors is constantly changing. Every 1000 hours, for a set of x sensor nodes, their detection probability $prob_i$, $1 \leq i \leq 100$ will be regenerated and reapplied to model the change of their environment. The x is set to be 1, 2, and 5.

The environment setup is named by a quintuplet (*distribution, prob variance, power variance, biased/unbiased, x*). The first field specifies the type of distribution that is used to generate the detection probability and power consumption of each task. It can either be uniform distribution or normal distribution. The second and third filed specifies the variance of the detection probability and the power consumption respectively. The fourth field is either biased or unbiased. When an environment setup is biased, half of the sensor nodes have lower power consumption than the others. This field is designed to model a heterogeneous network. The final field specifies the number of sensors whose detection probability changes due to changes in the environment. Table 1 column 1 gives the list of environment setups that were tested in our experiments. Note that the variance of power consumption is different for the biased and unbiased environment.

Our distributed GA algorithm which is presented in section 4 is denoted as *GA-lifetime*, since its objective is to maximize the lifetime of the sensor network. The program is distributed on 5 processors ($np = 5$). The subpopulation size is set to 100 ($pop = 100$) and the number of generations in each epoch is 5 ($c = 5$).

We designed two algorithms to compare with the GA-lifetime. The first one is also a distributed GA whose objective is to minimize the total power consumption of the cluster. Therefore it is denoted as *GA-power*. Instead of using equation (1), the GA-power uses a fitness function as the following.

$$fitness = \begin{cases} 0 & \text{if } Prob < \text{Prob}_{th} \\ 1 / \sum_{i \in \Delta_k} pow_{i,k} & \text{otherwise} \end{cases}$$

The second one is a heuristic algorithm which selects and allocates task based on the power versus detection probability ratio. For each task, it first selects the sensor node that has the highest power vs. detection probability ratio. Then it arranges the available tasks based on the descending order of this ratio. From the beginning of the list the algorithm selects the tasks one by one and assigns them to the sensor node, which is the most power efficient, until the overall detection probability of the cluster exceeds the user defined threshold $Prob_{th}$. In our experiment, the $Prob_{th}$ is set to 99.9%. The same threshold is applied to two other programs as well. We applied the above mentioned three resource management algorithms in the sensor network emulator. The lifetime of the network is recorded. The results are provided in Table 1. The first column specifies the environment setup and the last three columns specify the network lifetime (in hours) with different resource management algorithms.

ENVIRONMENT SETUP	GA-LIFETIME	GA-POWER	HEURIS TIC
uniform, 8.3, 7.1, biased, 1	44752.22	41930.85	40066.12
uniform, 8.3, 7.1, biased, 2	42158.9	35224.52	35028.29
uniform, 8.3, 7.1, biased, 5	42186.89	33207.64	31848.58
uniform, 8.3, 8.3, unbiased, 1	23874.2	26462.19	26776.26
uniform, 8.3, 8.3, unbiased, 2	26828.16	31983.17	32294.36
uniform, 8.3, 8.3, unbiased, 5	27656.94	29365.77	29794.95
normal, 2, 5.5, biased, 1	480000	430909.1	450000
normal, 2, 5.5, biased, 2	511200.4	436666.7	450000
normal, 2, 5.5, biased, 5	507500	404285.7	416923.1
normal, 2, 2, unbiased, 1	14531.39	14218.28	10358.73
normal, 2, 2, unbiased, 2	13892.64	10881.77	9143.943
normal, 2, 2, unbiased, 5	15708.92	16131.24	14628.62
normal, 1, 1, unbiased, 1	2788.086	2224.231	2255.613
normal, 1, 1, unbiased, 2	2759.893	2597.652	2248.155
normal, 1, 1, unbiased, 5	2857.993	2647.201	2647.037
normal, 1.5, 4.2, biased, 1	43315.71	40533.04	34495.43
normal, 1.5, 4.2, biased, 2	49774.59	52759.19	47284.2
normal, 1.5, 4.2, biased, 5	50744.86	50134.21	48893.41

Table 1 Network lifetime under different algorithms

Figure 6 shows the percent lifetime improvement of GA-lifetime relative to the heuristic algorithm. We can see that the GA-lifetime generally works better than the heuristic algorithm. The average lifetime improvement is 14.4%. The only case for which the heuristic algorithm works better than the GA-lifetime is when the detection probability and power consumption of the tasks are distributed uniformly and the network is unbiased. This is because, in this environment setup, the detection probability and power consumption have significant variety. Therefore, there exist some task-processor pairs that are much more power efficient than others. A similar reason can be used to explain why the GA-lifetime works relatively better in the environment setup with normal distribution.

Figure 7 shows the comparison between the GA-lifetime and the GA-power. The average lifetime improvement of GA-lifetime over GA-power is 6.5%. This indicates that merely reducing the power consumption is not a good way to improve the network lifetime. If a sensor node has more remaining battery, it should be allocated with more tasks even though it is not the most power efficient node that can be used to process these tasks. In another word, to extend the network lifetime, it is more important to evenly distribute the tasks.

We also observe that the GA-power outperforms the GA-lifetime when the environment setup is uniform and unbiased. This shows us that these two algorithms are complementary to each other, and they can be applied in different situations.

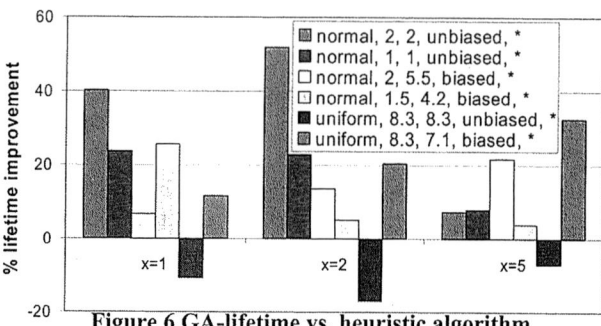

Figure 6 GA-lifetime vs. heuristic algorithm

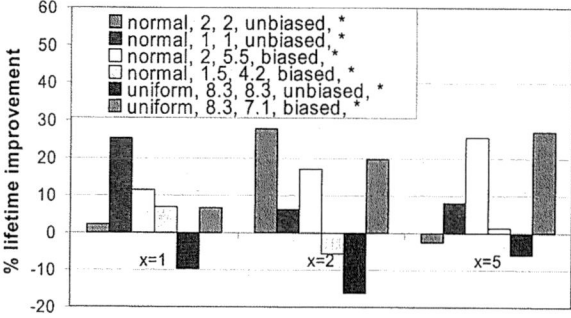

Figure 7 GA-lifetime vs. GA-power

7. CONCLUSIONS

In this paper we present a distributed GA algorithm that solves the resource management problem in a sensor network. A regression estimation model is presented that estimates the runtime of this algorithm. It is used to find the energy efficient configurations of the GA. The experimental results show that the proposed algorithm improves network lifetime by 14.4% in average.

8. REFERENCES

[1] ISTAG, "Ambient Intelligence: From Vision to Reality," Sept. 2003.
[2] I. F. Akyildiz, S. Weilian, Y. Sankarasubramaniam and E. Cayirci, "A Survey on Sensor Networks," *IEEE Communications Magazine*, Volume 40, Issue 8, pp. 102-114, Aug. 2002.
[3] E. R. Post and M. Orth, "Smart Fabric, or Wearable Computing," *Proc. First Int'l Symp. Wearable Computers*, pp. 167-168, Oct. 1997.
[4] H. Feltl and G. R. Raidl, "Evolutionary computation and optimization (ECO): An improved hybrid genetic algorithm for the generalized assignment problem," *Proceedings of the 2004 ACM symposium on Applied computing*, March 2004.
[5] W. Heinzelman, A. Chandrakasan, and H. Balakrishnan, "Energyefficient communication protocol for wireless microsensor networks," *Proceeding of International Conference on System Sciences (HICSS)*, Jan. 2000.
[6] C. Schurgers, V. Tsiatsis, S. Ganeriwal, and M. Srivastava, "Topology management for sensor networks: Exploiting latency and density," *Proceeding of International Symposium on Mobile Ad Hoc Networking and Computing*, 2002.
[7] G. Mainland, D. C. Parkes, and M. Welsh, "Decentralized, Adaptive Resource Allocation for Sensor Networks," *Symposium on Networked Systems Design and Implementation*, May 2005.
[8] E. Cantu-Paz, "A Survey of Parallel Genetic Algorithms," *Calculateurs Paralleles, Reseaux et Systems Repartis*, Vol. 10, No. 2.
[9] M. Younis, K. Akkaya, and A. Kunjithapatham, "Optimization of task allocation in a cluster-based sensor network," *Proceedings of IEEE International Symposium on Computers and Communication*, 2003.
[10] http://smbus.org/.
[11] M. Pedram, "Power Minimization in IC Design: Principles and Applications," *ACM Trans. on Design Auto. of Elec. Systems*, Vol. 1, No. 1, pp. 3-56, 1996.
[12] E. Cantu-Paz, "Markov chain models of parallel genetic algorithms," *IEEE Transactions on Evolutionary Computation*, Vol. 4, Issue 3, pp 216-226, Sep. 2000.

Everlast: Long-life, Supercapacitor-operated Wireless Sensor Node

Farhan Simjee and Pai H. Chou
Center for Embedded Computer Systems, University of California, Irvine, CA 92697-2625 USA
Email: {fsimjee,phchou}@uci.edu

ABSTRACT

This paper describes a supercapacitor-operated, solar-powered wireless sensor node called Everlast. Unlike traditional wireless sensors that store energy in batteries, Everlast's use of supercapacitors enables the system to operate for an estimated lifetime of 20 years without any maintenance. The novelty of this system lies in the feedforward, PFM (pulse frequency modulated) converter and open-circuit solar voltage method for maximum power point tracking, enabling the solar cell to efficiently charge the supercapacitor and power the node. Experimental results show that Everlast can achieve low power consumption, long operational lifetime, and high transmission rates, something that traditional sensor nodes cannot achieve simultaneously and must trade-off.

Categories and Subject Descriptors

C.3 [**Special-Purpose and Application-Based Systems**]: Real-time and embedded systems

General Terms

Management, Measurement, Performance, Design

Keywords

Maximum power point tracking, MPPT, supercapacitor, solar power, long life, wireless sensors, WSN

1. INTRODUCTION

Wireless sensor nodes are being deployed for new applications that are demanding higher data rates in large-scale networks with up to thousands of nodes. Today, many wireless sensor nodes achieve low power by low duty cycling. Going to higher data rates will dictate the use of high-capacity batteries and possibly solar cells to replenish the energy regularly. At the same time, large-scale deployment will demand lower cost per node.

Among the ultra low-power, low-cost radios with 1kbps data rate and a range of 3–10 meters, PicaRadio may run for several years without battery recharging or replacement [1, 2]. While useful for some applications, in order to achieve a 10–20 year operational lifetime, these nodes must operate at much lower duty cycles than one transmission per minute. The widely used Mica2 mote [3] uses a 38.4 kbps radio with a range of >80 meters in ground-level, line-of-sight communication. Powered by two AA batteries, it can run for a few hours in 100% duty cycle, or a few months when activated by a threshold detector with very infrequent events. With some external circuitry and a solar panel, the batteries can be recharged each day for significantly extended operations [4]. Due to the limited number of recharge cycles, the battery will require replacement after one to two years. Unfortunately, such recurring maintenance cost is likely to become very expensive or prohibitive if it must be done for thousands of deeply embedded nodes, which are likely to be difficult or expensive to access after deployment.

In both cases above, the battery is the primary limiting factor that prevents the node from operating maintenance-free for more than several years at non-trivial data rates. One solution that slows down battery aging is to place a supercapacitor in parallel with the battery so that transient power is delivered by the capacitor rather than the battery [5, 6, 7]. Fewer and shorter current pulses drawn from the battery allow more efficient use of battery capacity and increase the number of charge cycles possible. Also, a supercapacitor has been used to buffer solar energy for a few days to weeks to reduce the need for daily battery charging [8]. Unfortunately, even with less-frequent recharges the battery is still limited to holding charge for a few years before a reduction in energy capacity.

By removing the battery altogether and storing energy solely in the supercapacitor, there is now a viable option for achieving long-life operation. Supercapacitors have received wide attention recently due to their power density, low equivalent series resistance (ESR), and very low leakage current [9]. A typical supercapacitor offers more than half a million charge cycles and a 10-year operational lifetime until the capacity is reduced by 20%. At this point 80% of the useful energy is still available because the ESR is still very low unlike a battery whose useful energy drops to 50% at this point because the higher ESR causes premature end of life. By designing the node to operate on 50% energy capacity, the operational lifetime can be pushed out to 20 years.

The energy density of supercapacitors today is still less than batteries by an order of magnitude. However, when used in conjunction with a suitable solar cell, the supercapacitor should be able to sufficiently power the node when there is not enough sunlight in the environment. Aggressive power management techniques should also be able to close the energy density gap between supercapacitors and batteries. Given that a 350F capacitor with a capacity of 240 mAh from Maxwell Technologies costs $20 in single quantity, supercapacitors are also becoming competitive with rechargeable Li-ion batteries. Since a single supercapacitor will outlive half

---

Permission to make digital or hard copies of all or part of this work for personal or classroom use is granted without fee provided that copies are not made or distributed for profit or commercial advantage and that copies bear this notice and the full citation on the first page. To copy otherwise, to republish, to post on servers or to redistribute to lists, requires prior specific permission and/or a fee.
ISLPED'06, October 4–6, 2006, Tegernsee, Germany.
Copyright 2006 ACM 1-59593-462-6/06/0010 ...$5.00.

a dozen batteries, the drastically reduced maintenance costs and the anticipated price drop with volume are expected to significantly lower the overall cost of operating the sensor network.

The purpose of this work is to demonstrate the feasibility of such a sensor node operating on supercapacitors recharged by solar cells. One could not simply replace a battery with a supercapacitor because of the very different electrical characteristics and efficiency considerations in relation to the solar cells. We propose a feed-forward, pulse frequency modulated regulator that charges the supercapacitor at the optimal operating point for solar cells. We have constructed a complete wireless sensor node called Everlast with this solar/supercapacitor power circuitry and stress-tested its operation. Experimental results show that the Everlast node can achieve low power consumption, long operational lifetime, and high transmission rates simultaneously, without forcing the user to make trade-offs. This is expected to enable a whole new class of applications for low-cost, high performance, durable wireless sensor networks.

2. RELATED WORK

Charging a capacitor and drawing maximum solar power are both areas of active research. However, the authors are not aware of prior work on efficiently charging a capacitor using optimal solar power. The problem with capacitor charging is that a when large capacitor is attached to a solar cell, the component acts as a short and the solar voltage drops to the capacitor voltage. Although the solar cell will charge under this condition, it will not do so efficiently, because it will not be at the *maximum power point* (MPP), the voltage and current combination that maximizes power output under a given sunlight and temperature condition. A supercapacitor has been charged in this manner before but this reduces the solar efficiency by a factor of 2-5x depending on the solar cell topology [8].

A typical PWM regulator is not suitable because the ESR in the supercapacitor is so low that only a short pulse will prevent the supply voltage to fall sharply. For this reason, pulse power applications use resonant and Ward converters to efficiently and quickly charge capacitors [10, 11]. Unfortunately, the resonant converter requires an AC voltage that cannot be generated by the solar panel. The Ward converter inputs DC power, but the circuit is rather complex and requires a low impedance power source such as a battery or electrical outlet. Clearly, a new converter design is needed to meet the requirement of efficiently charging a capacitor from a high impedance power source.

The MPP also changes for a solar cell across temperature and sunlight. Numerous methods proposed to date for tracking the solar cell MPP include the hill climbing method, short-circuit method, and open-circuit voltage method [12, 13, 14, 15]. The hill climbing method of sweeping solar voltage while measuring the current requires a great deal of circuitry in the form of a microcontroller (MCU) to calulate the optimal power point. The power needed to run those chips are too high to keep the converter efficient. The short circuit method entails shorting the solar cell and measuring the short circuit current, which directly determines the MPP [16]. Finally, the open-circuit voltage method simply requires disconnecting the solar cell from any load and measuring the open-circuit solar voltage, which again is directly related to the MPP [17, 18]. The latter two methods are not as accurate as the hill climbing method, but are easy to implement with low power overhead.

3. PROBLEM STATEMENT

The problem is to build a power converter that can efficiently charge a large capacitor at the optimal voltage and current of the solar cell. Since the goals are cost reduction and circuit simplicity, the solution will require trade-offs among the efficiency of the power converter, cost, and the accuracy of the maximum power point tracking (MPPT) circuitry, and power overhead.

The regulator control circuitry must be able to both set the voltage or current to the optimal value and make corrections when the solar cell deviates from this value. Unlike a typical voltage regulator that uses feedback from the output, our converter requires the solar cell input to be fed forward into the control circuitry. To formulate the MPP problem, we define the following variables:

E : plane-of-array solar irradiance (W/m^2)

E_e : effective irradiance, or "suns"

AM_a : absolute air mass (dimensionless)

AOI : solar angle-of-incidence on module (degrees)

T_c, T_o : temperature of cells and reference temperature (Celsius)

I_{sco}, I_{mpo} : ref. short-circuit current and ref. max. power current

V_{oco}, V_{mpo} : ref. open-circuit voltage and ref. max. power voltage

$\alpha_{I_{sc}}, \alpha_{I_{mp}},$

$\beta_{V_{oc}}, \beta_{V_{mp}}$: temperature coefficients

$C_0 - C_4$: empirical coefficients

$K_{I_{mp}}, K_{V_{mp}}$: constants to calculate I_{mp} and V_{mp} from I_{sc} and V_{oc}

$$I_{sc}(E, T_c, AM_a, AOI) = (E/E_o) f_1(AM_a) f_2(AOI)$$
$$\cdot (I_{sco} + \alpha_{I_{sc}}(T_c - T_o)) \quad (1)$$

$$E_e = \frac{I_{sc}(E, T_c = T_o, AM_a, AOI)}{I_{sco}} \quad (2)$$

$$I_{mp}(E_e, T_c) = C_0 + E_e(C_1 + \alpha_{I_{mp}}(T_c - T_o)) \quad (3)$$

$$V_{oc}(E_e, T_c) = V_{oco} + C_2 \ln(E_e) + \beta_{V_{oc}}(T_c - T_o) \quad (4)$$

$$V_{mp}(E_e, T_c) = V_{mpo} + C_3 \ln(E_e) + C_4 (\ln(E_e))^2$$
$$+ \beta_{V_{mp}}(T_c - T_o) \quad (5)$$

$$\frac{\partial P_{solar}}{\partial V_{solar}} = \frac{\partial P_{solar}}{\partial I_{solar}} = 0 \quad (6)$$

$$V_{mp} = K_{V_{mp}} V_{oc} \quad (7)$$

$$I_{mp} = K_{I_{mp}} I_{sc} \quad (8)$$

The maximum power described in (3) and (5) is directly dependent upon the solar intensity and the temperature [19]. Under a given sunlight condition, (6) holds when the solar power output is maximized. Since it would be difficult to empirically find MPP at runtime without a PC and a correct solar model, finding the power peak will require the brute force hill-climbing technique and some computation. Fortunately, it also known that the voltage and current at MPP are directly proportional to V_{oc} and I_{sc}, as described by (7)(8), respectively [16, 17]. This method would simply require the regulator to shut down for a few moments to measure V_{oc} or I_{sc} to find the new MPP, and adjust the power converter. With these techniques, MPP can be reached more quickly since it requires only a single computation and a momentary shutdown of the converter.

4. EVERLAST SYSTEM DESIGN

Fig. 1 shows an overview of the Everlast design. The system is designed for three primary tasks: charging the capacitor using a pulse-frequency modulated (PFM) regulator, feeding the PFM controller the optimal operating point for the solar cell, and all the typ-

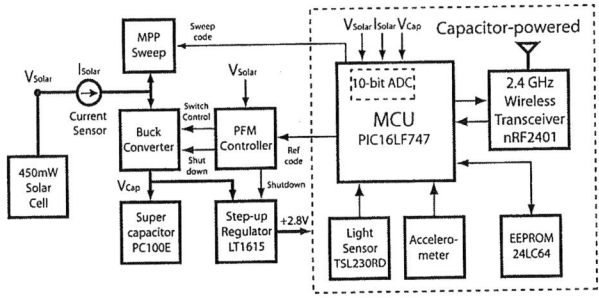

Figure 1: Everlast system block diagram

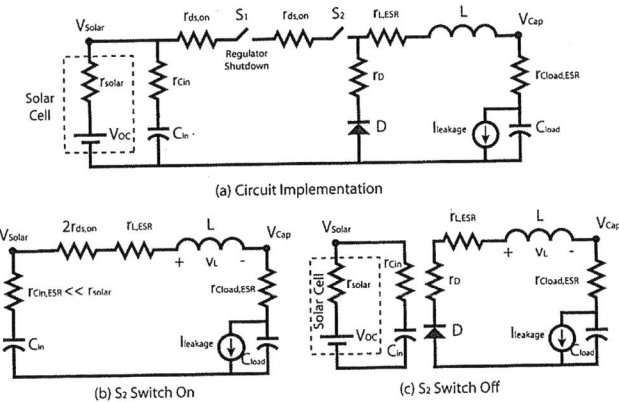

Figure 2: PFM Regulator

ical WSN functions including reading sensor data and communication with nodes and basestations through the wireless transceiver.

4.1 PFM Regulator

Since the capacitor is seen as a short to the solar cell when they are connected together, the traditional voltage regulators such as the PWM DC-DC converter and linear regulator would not function properly because V_{solar} would quickly fall to V_{cap} and greatly deviate from V_{mp}. With the feedback pin connected to the output load, the regulator would short the input and output until reaching the preset output voltage. As shown in Fig. 2, a PFM regulator is designed with the advantages of the switched-capacitor regulator and the buck converter to prevent shorting the input and output. At startup, the switch S_2 is open, and the solar cell charges the input capacitor. Once V_{solar} reaches the upper threshold of (9) which is set by the PFM Controller, S_2 is closed and C_{in} sends current to the inductor, L, and the supercapacitor, C_{load}. The current from the solar cell can be ignored in the on switch state, because the input

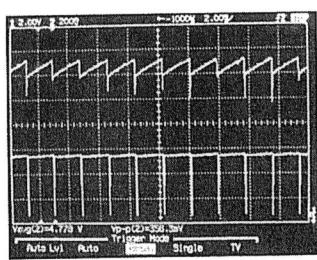

Figure 3: PFM Regulator Waveform: V_{solar} **(top),** S_2 **(bottom)**

capacitance ESR is much lower than the impedance of the solar cell. Since all these components have low resistive values, energy is transferred very quickly until the bottom threshold of (9) is met. Once switch S_2 is in the off state, C_{in} is once again recharged and the inductor transfers energy to C_{load} and thus charging the supercapacitor. Fig. 3 shows the solar voltage and pulses sent to the switch when the regulator is operating.

The circuit is similar to a switched-capacitor circuit in that it charges a capacitor and then transfers that energy to the output load capacitor. The only difference is that an inductor is added to buffer the voltage difference between V_{solar} and V_{cap}. The pulse frequency is primarily dependent upon the switching thresholds, input capacitance, and solar power (Eqn. 10).

The regulator efficiency is measured with a test setup that is similar to a deployed sensor node but at the same time is easily reproducible. A solar cell is placed under a high-power adjustable halogen lamp so that it can produce constant power. The solar cell is fed to the PFM regulator to charge a 10F supercapacitor. The efficiency is calculated by charging the capacitor from 0V to 2.4V and 1V to 2.4V and comparing the total solar energy generated with the change in the energy stored in the capacitor (Eqn. 11).

Fig. 4 shows the results across varying solar power, and generally over 20mW the efficiency plateaus. Charging from 0V is less efficient, because the higher voltage difference between V_{solar} and V_{cap} causes more parasitic losses. The efficiency curve with the capacitor already charged to 1V is a better indicator of real-world performance, as the sensor node is assumed to always be powered on.

C_{in} is also changed in the regulator to see how that affects operation. Since it is faster to charge a smaller capacitor, the frequency increases with smaller input capacitance. With $1\mu F$ C_{in}, the regulator can achieve 65% efficiency, because there is less resistive loss in the capacitor, MOSFET switch, and inductor at higher frequencies. As there is no sign of further efficiency loss due to dynamic power at higher frequencies, C_{in} can be further reduced. However, choosing an amplifier that modulates the switch with low supply current came at the cost of slower slew rate to drive the switch. If C_{in} is lowered, then the pulses will not fully turn on the FET. A future board revision will explore adding a FET gate driver and using power inductors, but as it stands, the regulator has been experimentally verified to operate in the 0.5Hz to 30kHz range across varying P_{solar} from 3mW to 300mW, respectively. No battery charger IC on the market can match this PFM regulator's ability to harvest low power. With a change of the passive components and switch, the regulator can also accommodate sub-mW piezoelectric generators and larger solar cells.

$$V_{mp} - V_{hyst} < V_{solar} < V_{mp} + V_{hyst} \tag{9}$$

$$f_{pulse} \propto \frac{1}{V_{hyst}} \frac{1}{C_{in}} \frac{1}{L} P_{solar}(t) \tag{10}$$

$$\eta = \frac{E_{solar}}{E_{cap,end} - E_{cap,start}} = \frac{P_{solar,avg} t_{chargetime}}{\frac{1}{2} C(V_{end}^2 - V_{start}^2)} \tag{11}$$

4.2 PFM Controller

The function of this PFM controller is to pulse the regulator every time V_{solar} exceeds the specified reference voltage, which is V_{mp} in this case. Fig. 5(a) shows V_{solar} being compared to a reference voltage to determine switching the regulator. Hysteresis is added to stabilize the amplifier and control the pulse frequency. The resistors are larger than the impedance coming from reference and solar cell so that V_{hyst} remains accurate (Eqn. 12,13). The volt-

199

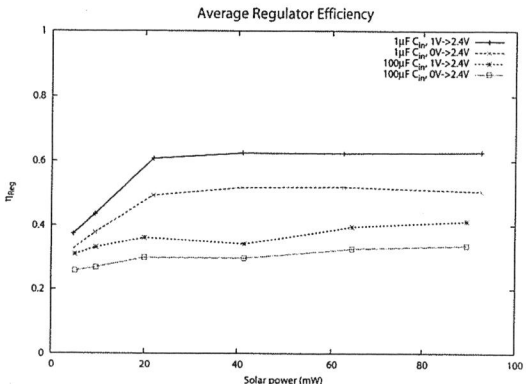

Figure 4: PFM Regulator Efficiency

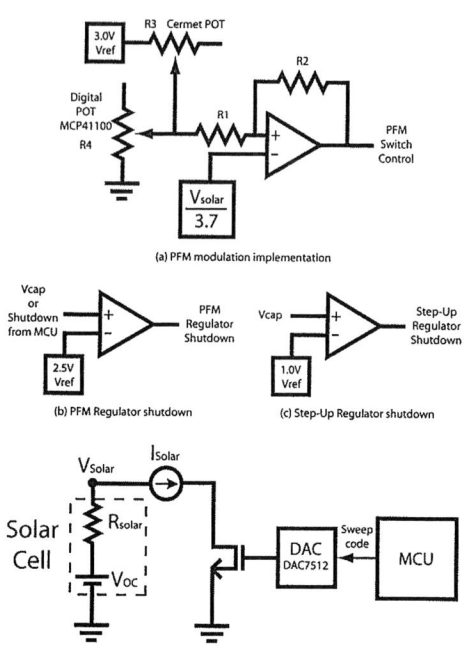

Figure 5: PFM Controller and DC sweep circuitry

age reference is divided by a cermet POT and an 8-bit digital POT, and the solar voltage is resistively divided so that the amplifier can compare it with the reference signal. The digital POT defaults to $100k\Omega$ and, along with the cermet POT, presets the solar voltage at which to regulate when the MCU is powered down.

The MCU fine-tunes the digital POT value until V_{solar} reaches V_{mp}. The MCU is used rather than some custom circuitry, because it is flexible enough to run any of the MPPT methods mentioned earlier. The hill climbing method can be achieved by sweeping the POT code while recording the voltage and current. The solar cell can be short-circuited momentarily by setting the POT value to 0 and thus closing the switch. Conversely, V_{oc} is measured by setting the POT value to 255 to open the switch. With the latter two, V_{mp} and I_{mp} can be easily calculated by the MCU using Eqns. (7)(8). With either of these values, the POT value can be adjusted until V_{solar} or I_{solar} reaches one of these respective values. Depending on whether an application requires accurate MPPT or low cost/complexity, the suitable method can be chosen and custom circuitry can be implemented.

As shown in Fig. 5(b), the functional block of the PFM controller also includes the ability to shut down the regulator when the capacitor is fully charged by comparing it with a 2.5V reference signal. The amplifier can be jumpered to also accept a signal from the MCU to shut down. In order to operate the sensor node from a dead start, the bootstrap circuitry in Fig. 5(c) keeps the step-up regulator shut down until V_{cap} reaches a high enough voltage that is set by voltage dividing from the 3.0V reference.

For testing purposes, the MPP sweep circuitry is included to sweep the circuit I_{solar} and measure V_{solar} to profile the solar cell as shown in Fig. 5(d). While the PFM regulator is shut down, the MCU sweeps the gate voltage of a MOSFET using a DAC. This sweeps the solar current, and both I_{solar} and V_{solar} are sampled after every DAC step. The MCU can then calculate P_{mp}, V_{mp}, and I_{mp} and take a light measurement to profile MPP at every light intensity level. The solar profile will be useful later to benchmark how accurately the MPPT methods can keep the solar cell at P_{mp}.

$$V_{hyst} = V_{ref}\frac{R_1}{R_1 + R_2} \qquad (12)$$

$$R_1, R_2 \gg R_3, R_4 \qquad (13)$$

4.3   WSN Functions

Although the primary goal of Everlast is to keep the solar cells at MPP and operate for 20 years, some of the latest technologies in

MCUs, sensors, and transceivers are adopted so that the board will operate with lower power requirements.

The PIC16LF747 is selected as the MCU because of its low supply current from $25\mu A$ at 31.25kHz to $930\mu A$ at 8MHz with V_{DD} at 3.0V using an internal RC oscillator. The MCU can operate from 2.0V to 5.5V and the internal RC oscillator can be controlled at runtime. There has been much work on dynamic voltage and frequency scaling (DVFS), and practically any of those techniques can be written into the firmware.

Most sensors today either produce an analog voltage, PWM signal, or PFM signal. The MCU can read the first using its internal 10-bit ADC or an external, possibly higher-precision, ADC. The latter two signals can be measured with an internal timer/counter or by polling. The TSL230RD light sensor improves upon traditional methods of measuring light intensity such as with a photodiode, because it has a pin-selectable light aperture for measuring a wider dynamic range of light. The light sensor draws 2mA when operating, but since it generates a square-wave output, the MCU can easily count the edges using the internal counter in a matter of 1–2 ms while consuming $12\mu J$ per sampling. The ADXL202 is a dual-axis $\pm$ 2g accelerometer consuming $600\mu A$. The supply current is low enough that the MCU I/O pin can power it during sampling and power it down otherwise. A few more ADC pins are drawn out to the board so that additional sensors can be added anytime. The board also includes an EEPROM that can be used for storing sensor data.

For wireless communication, we use the Nordic nRF2401 2.4GHz GFSK transceiver, a recent breakthrough in low power transceivers. Operating in the ISM band, it supports 125 channels and can support a dense network of sensor nodes. With the ability to send ShockBurst messages, the transceiver turns on to send a 256-bit packet in a 1Mbps datastream and goes back to standby. With 0dBm RF power, the transceiver has line of sight range of 300m according to the datasheet, and we have physically verified it for at least a 150m range. With integrated pseudo-MAC layer, the

Profile	MCU Core Freq. (MHz)	ADC Sampl./msg	Transceiver Msg/sec	I_{supply} (mA)
p1	Sleep	Off	Standby	.280
p2	8.0	Off	Standby	1.08
p3	8.0	Standby	Standby	1.14
p4	8.0	9	100	1.83
p5	4.0	9	100	1.49
p6	2.0	9	50	1.05
p7	1.0	9	25	.850
p8	8.0	Off	Rx mode	21.71

Table 1: Sample power modes, $V_{DD} = 2.8V$, Message = 256 bytes

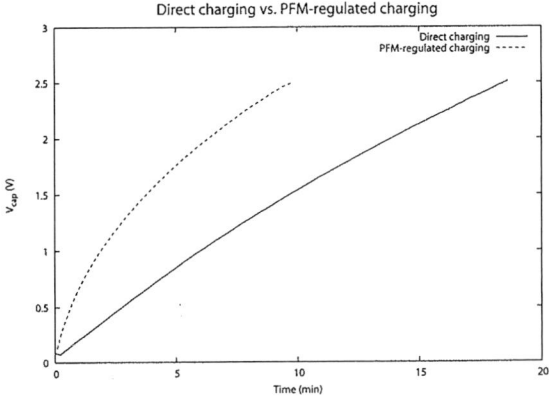

Figure 6: Comparison between direct capacitor charging and PFM-regulated charging

transceiver incorporates a 16-bit CRC and the destination address into the packet eliminating the need for error checking.

5. EXPERIMENTAL RESULTS

Three metrics determine the performance of the Everlast board: efficiently charging the supercapacitor, tracking the solar cell at MPP, and running continuously 24 hours a day.

Fig. 6 shows a comparison of charging a 10F supercapacitor from a fixed-intensity halogen light source with direct and PFM-regulated charging modes. With direct charging the solar cell is attached to the capacitor while PFM-regulated charging is done through the Everlast board. The latter charging scheme tracked the optimal power point to generate maximum solar power and charged the supercapacitor in about half the time. This makes more energy available for consumption by the sensor node, or enables the node to use a smaller, cheaper solar cell. With future improvements to the regulator efficiency, the gap between these two schemes should widen even further.

Using a solar profile generated from a DC sweep of the solar cell throughout the day, the voltage and current tracking constants, $K_{V_{mp}}$ and $K_{I_{mp}}$, respectively, are calculated and shown in Fig. 7. Voltage tracking is preferred, because under normal lighting conditions, $K_{V_{mp}}$ is between 0.68~0.72 and reaches up to 0.80 in low light conditions. This is much better than the wide swing range of the current tracking constant. The V_{oc} can also be easily measured by periodically shutting down the regulator with a timer and then sampling the voltage with an ADC or sample-and-hold circuit. Using this method, Fig. 8 shows that the voltage tracking method, with $K_{V_{mp}}$ fixed at 0.70, can track solar power very closely to the DC sweep (hill-climbing) method but without the need for a MCU/DSP and within a few milliseconds needed to sample the open-circuit

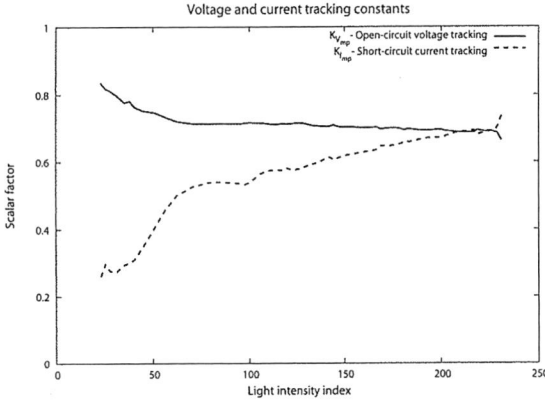

Figure 7: Voltage and current tracking constants calculated over one day

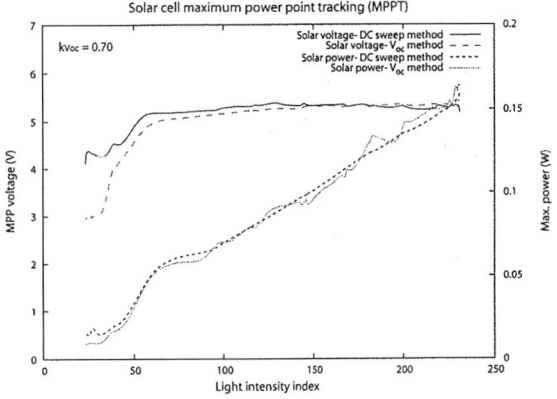

Figure 8: MPPT comparison between DC sweep and V_{oc} method

solar voltage. The voltage tracking method has a tracking error of less than 5% in normal lighting conditions and less than 11% in low lighting conditions in comparison to the DC sweep method. The tracking error increases in low lighting conditions, because the solar cell's $K_{V_{mp}}$ reaches 0.80 while the voltage tracking constant is still fixed at 0.70 in the tracking circuitry. This level of accuracy is acceptable for sub-Watt solar power generation, since the voltage tracking method can be implemented with a few inexpensive, discrete components, and the power requirements to operate a more accurate tracker would outstrip the increase in power generation.

The module must also be able to solely rely on stored energy in the absence of solar power. Creating a few sample modes in Table 1, it is worth noting that the transceiver can continuously transmit 25.6 kbps in profile p5, drawing only 1.49mA. Off a fully-charged 100F supercapacitor, p4 ran for 8.16 hours, p5 ran for 9.54 hours, and p6 ran for 11.1 hours. One point to mention is that the step-up regulator is only 40-50% efficient as it steps up V_{cap} of 1V–2.5V to 2.8V. One area to explore in the future will be to split the power supply rails to the components: 2.0V regulator supplying power to the MCU and transceiver, and 2.7V supplied to any ICs that require it. This will reduce the power consumption of the modules and increase efficiency, which should nearly double the autonomous operating time of each profile.

Fig. 9 is a stress test of Everlast running in p6 on a building rooftop draining the capacitor at night and recharging during the

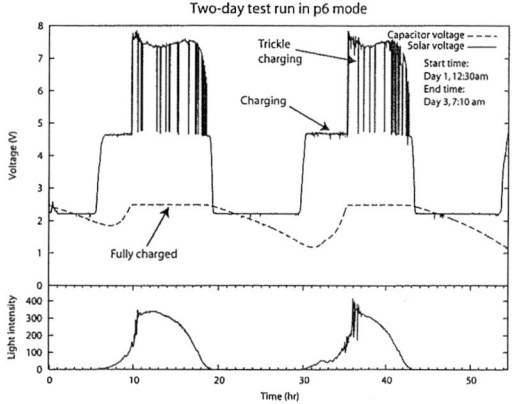

Figure 9: Two-day stress test

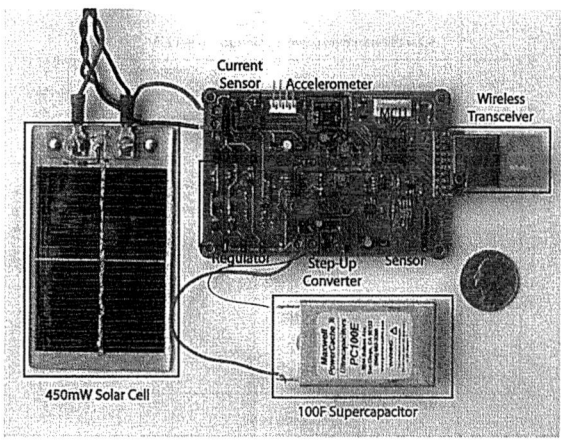

Figure 10: Everlast prototype board

day. In fact, all the data generated in the graph are sampled by the on-board ADC and transmitted from the sensor node itself to a base station. Taking a close look at the graph, throughout the night, the capacitor is drained, and during the morning, the regulator turns on and charges the capacitor. By high noon, the capacitor is fully charged at 2.5V, and the regulator is shut down, causing the solar voltage to jump to V_{oc}. During the day, the sensor node is still transmitting data and drains power from the capacitor. Once V_{cap} dips below 2.45V, the MCU restarts the regulator to recharge the capacitor back to full capacity. After the module is completely charged up, the regulator periodically "trickle charges" the capacitor back to full charge. As solar power decreases in the late afternoon, the power generation is lower than consumption and the capacitor starts to discharge.

6. CONCLUSION

The Everlast sensor node shatters the current preconceptions of designing a sensor node towards only a single goal of either long-life, high data rates, or low cost. With the the replacement of the battery with the supercapacitor, sensor nodes can operate for 20 years while maintaining high data rates. Using a feedforward PFM regulator with open-circuit voltage MPPT, the Everlast system charges the capacitor efficiently by enabling the solar cell to generate maximum power. More accurate MPPT and higher efficiency can be achieved with some fine-tuning of the present design.

Future work on the consumption side includes lowering supply rail voltages and exploring multiple capacitor configurations for an expected power reduction by a factor of 2.

Acknowledgments

This work was supported in part by the National Science Foundation grant CCR-0205712 and NSF CAREER Award CNS-0448668.

7. REFERENCES
[1] B. Warneke, M. Last, B. Liebowitz, and K.S.J. Pister. Smart Dust: communicating with a cubic-millimeter computer. *Computer*, 34:44–51, Jan. 2001.
[2] J.M. Rabaey, M.J. Ammer, J.L. da Silva, Jr., D. Patel, and S. Roundy. PicoRadio supports ad hoc ultra-low power wireless networking. *Computer*, 33:42–48, July 2000.
[3] Crossbow Technology. Mica2 datasheet. www.xbow.com.
[4] V. Raghunathan, A. Kansal, J. Hsu, J. Friedman, and M. Srivastava. Design considerations for solar energy harvesting wireless embedded systems. In *Proc. 4th Int. Conf. on Information Processing in Sensor Networks*, pages 457–462, Apr. 2005.
[5] P. Enjeti, J.W. Howze, and L. Palma. An approach to improve battery run-time in mobile applications with supercapacitors. In *IEEE 34th Annual Power Electronics Specialists Conference*, 2003.
[6] T.A. Smith, J.P. Mars, and G.A. Turner. Using supercapacitors to improve battery performance. In *Power Electronics Specialists Conference*, 2002.
[7] L. Gao, R.A. Dougal, and S. Liu. Active power sharing in hybrid battery/capacitor power sources. In *Eighteenth Annual IEEE Applied Power Electronics Conference and Exposition*, 2003.
[8] X. Jiang, J. Polastre, and D. Culler. Perpetual environmentally powered sensor networks. In *Proc. 4th Int. Conf. on Information Processing in Sensor Networks*, pages 463–468, Apr. 2005.
[9] Maxwell Technologies. BCAP0350 datasheet. http://www.maxwell.com.
[10] H. Pollock. High efficiency, high frequency power supplies for capacitor and battery charging. In *IEE Colloquium on Power Electronics for Demanding Applications*, pages 901 – 910, Apr. 1999.
[11] R.M. Nelms and J.E. Schatz. A capacitor charging power supply utilizing a ward converter. *IEEE Trans. Ind. Electronics*, 39:421–428, Oct. 1992.
[12] K.H. Hussein, I. Muta, T. Hoshino, and M. Osakada. Maximum photovoltaic power tracking: an algorithm for rapidly changing atmospheric conditions. In *IEE Proceeding Generation, Transmission and Distribution*, pages 59–64, Jan. 1995.
[13] C. Hua and C. Shen. Study of maximum power tracking techniques and control of DC/DC converters for photovoltaic power system. In *IEEE-PESC. Conf. Rec.*, pages 86–93, 1998.
[14] E. Koutroulis, K. Kalaitzakis, and N.C. Voulgaris. Development of a microcontroller-based, photovoltaic maximum power point tracking control system. *IEEE Trans. Power Electronics*, 16:46–54, Jan. 2001.
[15] Y. Kuo, T. Liang, and J. Chen. Novel maximum-power-point-tracking controller for photovoltaic energy conversion system. *IEEE Trans. Ind. Electronics*, 48:594–601, June 2001.
[16] T. Noguchi, S. Togashi, and R. Nakamoto. Short-current pulse-based maximum-power-point tracking method for multiple photovoltaic-and-converter module system. *IEEE Trans. Ind. Electronics*, 49:217–223, Feb. 2002.
[17] J.H.R. Enslin, M.S. Wolf, D.B. Snyman, and W. Swiegers. Integrated photovoltaic maximum power point tracking converter. *IEEE Trans. Ind. Electronics*, 44:769–773, Dec. 1997.
[18] Dong-Yun Lee, Hyeong-Ju Noh, Dong-Seok Hyun, and Ick Choy. An improved MPPT converter using current compensation method for small scale pv-applications. In *IEEE 18th Annual Applied Power Electronics Conf. and Expo.*, 2003.
[19] D.L. King, J.A. Kratochvil, and W.E. Boyson. Field experience with a new performance characterization procedure for photovoltaic arrays. In *In Proc. 2nd World Conf. and Exhib. on Photovoltaic Solar Energy Conversion*, 1998.

Model to Hardware Matching
For nano-meter Scale Technologies

Sani R. Nassif

IBM Research – Austin

11501 Burnet Road, Austin, TX 78758

nassif@us.ibm.com

ABSTRACT

With the semiconductor industry pushing past the 65nm node and forward to 45nm and beyond, a host of phenomena are becoming prominent. For some time now, manufacturing variability and its impact on power and performance has captured the attention of the CAD research community, and is now transitioning to the commercial EDA market. Simultaneously, however, our ability to reliably predict the outcome of a semiconductor manufacturing process has been steadily deteriorating. This is happening because the rapidly increasing process complexity which is introducing a host of systematic sources of variation, as well as a natural increase in core random variability due to scaling. These factors increase the error in our performance predictions, and thus lead to a gap in model to hardware matching.

In this tutorial, we will review the sources and impacts of model to hardware mismatch, and show examples of potential solutions to currently under development..

Categories and Subject Descriptors

B.7.0 [**Hardware**]: Integrated Circuits – *General.*

General Terms

Design.

Keywords

Model to hardware correlation; silicon manufacturing variations; integrated circuit modeling

1. What is a MOSFET

Until relatively recently (say in the 1990s) the definition of the layout features which determine the performance of a fabricated MOSFET were relatively simple. The intersection between the diffusion and polysilicon mask shapes determined the dimensions of the MOSFET, and those dimensions sufficed to characterize the behavior of the transistor. We do not mean to imply that the scaling of these dimensions did not make transistor modeling more difficult, but rather to point out that the behavior of the transistor was determined by local geometry and that its extraction from the layout was a straightforward task.

Permission to make digital or hard copies of all or part of this work for personal or classroom use is granted without fee provided that copies are not made or distributed for profit or commercial advantage and that copies bear this notice and the full citation on the first page. To copy otherwise, or republish, to post on servers or to redistribute to lists, requires prior specific permission and/or a fee.
ISLPED'06, October 4–6, 2006, Tegernsee, Germany.
Copyright 2006 ACM 1-59593-462-6/06/0010...$5.00.

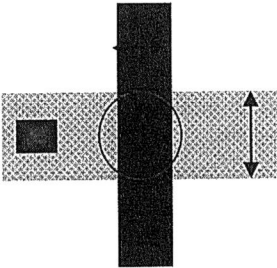

Figure 1: Geometry defining a MOSFET prior to the introduction of sub-wavelength RET.

As the industry migrated to sub-wavelength lithography, the difference between the drawn mask shapes and actual printed (wafer) regions began to increase. This resulted in the introduction of resolution enhancement techniques (RET) and optical pattern correction (OPC) to improve the fidelity of manufactured devices. In spite of advances bordering in many cases on the miraculous, it remains a fact that there is an ever growing gap between device layout as viewed by a designer, and final manufactured shapes as rendered in Silicon. This gap exhibits itself in two ways: (a) the precision with which the dimensions (length and width) of a device can be set are limited by the residual error that remains after RET is applied; and (b) the interaction between shapes on the mask due to interference, flare, and other lithography and illumination related issues means that the actual dimensions of the device are determined by all the shapes in the neighborhood of the device.

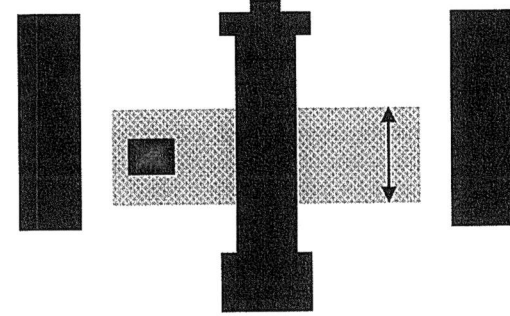

Figure 2: Geometry defining a MOSFET with RET assist features of various types.

At the 65nm node and below, assuming the current lithography roadmap remains as it is currently defined, the radius of influence that defines the neighborhood of shapes that play a part in determining the characteristics of a MOSFET is expected to increase from the current "nearest feature" to include one-removed features, auxiliary features, corners, vias, and other second order phenomena. This increase in the radius of influence will impact a number of areas:

• It will make the modeling of device behavior more difficult since it will be harder to define a canonical or typical device from which to perform model characterization. We will come back to this subject later in this paper.

• It will make the circuit extraction phase of design verification (where the layout is converted into a simulatable netlist) more complex since a larger number of geometries will need to be processed.

• It will severely reduce design composability, defined as the ability to compose a complex circuit function out of individual simple functions, e.g. building a multi-bit adder out of single-bit adders, which are in turn composed of individual NAND, NOR and similar logic gates.

The trends outlined thus far point to an increasing need to model and comprehend the interaction between design and manufacturing at the device level. A related and important trend is the increase in manufacturing variability, which we will discuss next.

2. Variability and Uncertainty

The performance of an integrated circuit is determined by the electrical characteristics of the individual linear and non-linear devices which constitute the circuit. Variations in these device characteristics cause the performance of the circuit to deviate from its intended range and can cause performance degradation and erroneous behavior. We will refer to this type of variability as "physical variability".

In addition, circuit performance is determined by the environment in which the circuit operates. This includes factors such as temperature, power supply voltage, and noise. Variations in operating environment can have a similar impact on circuit behavior as device variations. We will refer to this type of variability as "environmental variability".

It is tempting to think of physical variability as simply the result of systematic and random manufacturing fluctuations. But a variety of time-dependent wear-out mechanisms such as metal electro-migration and negative-bias threshold instability (NBTI) cause device characteristics to change over time –albeit with a time constant on the scale of months and years. In contrast, environmental variability is very much a function of time but at the same time-scale as that of the operation of the circuit, e.g. in the Nano-second range for a typical GHz design.

Whether physical or environmental, we can classify components of variability in various ways. For example, we can examine their temporal behavior (as we alluded to above). We can also examine their spatial distribution across chip, reticle, wafer, and lot. Most important, however, is the notion of whether we fully understand the interaction between the specific component of variability and the characteristics of the relevant design.

Consider the case where a physical or environmental component of variability is known to be a function of specific design characteristics. For example, it is well known that variation in the channel length of MOSFET devices is related to the orientation of the devices. With a suitable quantitative model relating the variation to design practice, a designer can make the appropriate engineering tradeoff and margin his design so as to

minimize the impact of the variations. Such phenomena are often systematic in nature, and we will refer to them simply as "variability".

Now consider the case where a physical or environmental phenomenon is not well understood, such that available information is limited to the magnitude of the variation, but without insight into its quantitative dependence on design. In such a case, the designer has no choice but to perform worst-case analysis, i.e. creating a large enough design margin to correct for the worst possible condition that may occur. Such large margins are usually wasteful in design resources and end up impacting the overall cost and performance. We will refer to these types of phenomena as "uncertainty".

Variability can be designed around and will typically cause small increases in design cost, while uncertainty needs to be margined against and will typically cause large increased in design cost. A key point to remember, however, is that the difference between the two is determined by our ability to understand and model the mechanisms as play, and that an investment in modeling and analysis can sometimes turn a source of uncertainty into a source of variability, thereby reducing design cost and/or improving design performance.

With increasing manufacturing process complexity, more and more phenomena are competing for limited modeling resources. This trend, unchecked, endangers our industry's ability to deliver future design improvements consistent with historical trends. Furthermore, many phenomena are increasing in magnitude as scaling continues [1], requiring more modeling and analysis resources.

The concrete understanding of variability, uncertainty, and their interaction with design depends on our ability to perform detailed design-space-oriented manufacturing process characterization. We will examine this topic next.

3. Characterization vs. Modeling

The prior two sections identified two trends: (1) an increase in the diversity of device implementation and interaction, and (2) an increase in random and systematic variability resulting from the manufacturing process as well as circuit operation. Putting aside the environmental sources of variability, which require a set of techniques for analysis, modeling, and optimization that are beyond the scope of this paper, we will focus on how our understanding of the manufacturing process is created. We will further focus on how this understanding of the process is used to generate simulatable models of circuits, acknowledging that Spice-level circuit simulation is usually the basis on which all modern design is built.

In earlier technologies, it was sufficient to create simple scribe-line test structures including a handful of MOSFETs with varying dimensions. The measured characteristics of the devices was used to extract device model parameters (e.g. for a BSIM [2] model). The resultant parameters were considered to be a complete encapsulation of the behavior of the technology. As technology scaled and device models became more complex (e.g. to handle phenomena such as short channel effects) the selection and number of devices that are included in the test structure increased somewhat.

In contrast, the impact of the local layout environment on device behavior in current technologies dramatically increases the number of factors that need to be considered, so the number of potential device layout variations that would need to be studied is much larger than the handful supported by current test structures and modeling strategies. Furthermore, since variability has become an important limiter of design performance, an assessment of within-die fluctuations is desirable. Such an assessment would require sufficient replication of identical structures to allow statistical characterization. Such a characterization may, for example, allow for the comparison of the spread in device behavior for different layout practices –an important degree of freedom for a designer attempting to reduce the impact of variations on a circuit.

Without models of the interaction between design implementation (layout) and manufacturing variability, we run the risk of excessive worst-casing as more and more of these phenomena emerge and challenge current design practices. Thus there is a need to change the manner in which we define the interaction between design and technology. When the characteristics of devices were substantially independent of the manner in which they were laid out, it was sufficient to merely characterize the manufacturing process. In current technologies, the focus needs to shift to modeling of the manufacturing process, i.e. creating quantitative relationships between design implementation and design performance. It is only through such models that we can convert uncertainty to variability, and reduce excessive margins and pessimism.

All of these facts point to the need for significantly more complex process modeling test structures. Some of desirable features of such structures are:

• Similarity to design. Structures should have layout features that are similar to realistic designs in order to ensure that the resulting data is meaningful.

• Spatial breadth. Since within-die and within-wafer variability often has strong spatial components, careful attention must be paid to the placement and size of structures in order to gather sufficient data.

• Statistical breadth. As pointed out above, we need a sufficient number of replicas to confidently measure distributions, as well as changes in distributions over time or for different design practices.

• Layout breadth. The structures should explore a sufficiently wide range of layouts to cover all or most of anticipated design practice.

• Ease of measurement. Crucial for both the timely delivery of initial models, as well as the continued monitoring of model to hardware matching as technology learning progresses.

No one structure can accommodate all the needs above, but a family of related structures can. Such a family might include large dense structures that are run infrequently for detailed modeling and characterization, and small scribe-line structures that are run often (perhaps on all wafers) and serve to monitor a smaller number of key variables.

At IBM we are making significant research investments in this area to produce easy-to-design test structures capable of

improving our understanding of manufacturing technology and its variations [6]. An example of such a structure is the device array shown in Figure 3, results from which are shown in Figure 4.

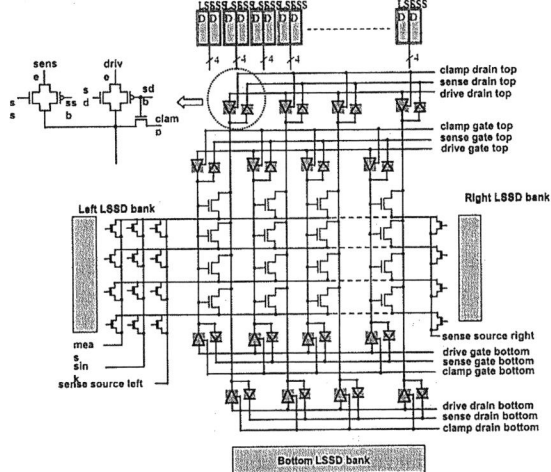

Figure 3: Schematic of an addressable FET characterization structure.

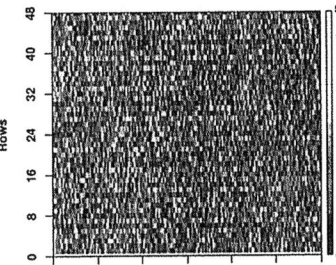

Figure 4: Spatial leakage distribution measured from device characterization test structure.

We believe this area will grow in importance and relevance with time, and will be a key enabler for future design/technology interaction.

4. Modeling And Virtual Fabrication

The previous section alluded to the need for improved models of technology performance and variability, since such models are crucial to the management of pessimism and design margin. But how are such models to be created, characterized and delivered? And what is the role of Computer-Aided Design (CAD) as well as Technology-CAD in the creation of such models?

The MOSFET circuit simulation device model (e.g. BSIM) has been the core definition of the design/technology interface. Such models are standardized, well understood, and are supported by both simulators as well as parameter extractors. The historical trend in this field has been for increasing model complexity (and corresponding number of model parameters) to capture an ever broader set of phenomena. This trend focuses on nominal accuracy of the model with respect to a "typical" or "golden" device, and can as easily be driven by measurements as by device characteristics generated from a traditional device simulator. Also, the complexity of these models makes for costly characterization cycles that require substantial experience to insure the delivery of robust, reliable and accurate models.

The trend to increased (a) layout dependence (radius of influence) and (b) variability is an opportunity to reexamine current modeling trends. In a regime where the within-die tolerance on key parameters is expected to reach a third or more of the mean value, we need to carefully understand the interplay between nominal accuracy (predicting the mean) vs. statistical accuracy (predicting the spread). This interaction is central to the understanding of the model to hardware matching issue, and needs to be driven by an understanding of the economic significance of misprediction on the potential profit of a design.

We strongly believe that there is a strong and immediate need for statistically accurate MOSFET models which would serve to allow the prediction and reduction of the impact of technology variations on design. Such models are not new, and have been suggested and implemented in the past. What is different now is the need for those models to be enhanced to include the interaction between device layout and device behavior. This is an area of research that can result in immediate benefit.

A statistical characterization methodology will also be required in order to insure that such models are easily created, validated, and updated as technology learning occurs. Such a methodology is not as well developed, but will be a corner stone of future design/technology interaction.

Generating early models of technology for exploratory design is crucial, especially due to the ever lengthening design cycles of large complex chips. Such models need to be created with little or no hardware available, and are often updated as more information becomes available. Such models currently focus on capturing the nominal behavior of technology, but need to be enhanced to capture variability. This presents a significant challenge to current methods, but also represents an opportunity for new algorithms and tools.

The concept of a "virtual factory" has existed for some time in the Technology-CAD community [3]. Perhaps it is time to revisit the concept of virtual fabrication and enhance it to allow the early modeling of variability as well as the interaction between layout and device performance. Such enhancement should be done such that the virtual factory meshes directly with advanced test structures such that it can readily adapt to new data, new physical mechanisms, and even new layout styles.

5. Model To Hardware Matching

The issues and efforts outlined thus far culminate in the problem of model to hardware matching. We use this term to denote our ability to predict, via modeling, the behavior of

hardware after fabrication. The confidence we have in such predictions is key to the economic success of the semiconductor business since performance and yield play a large part in determining the profitability of a design.

The increasing interaction between layout and device performance is one of many sources of additional potential disparity between models and reality. In fact, a large number of mechanisms, when not modeled with appropriate accuracy, can lead to deviation between the predicted and observed performance. A few examples include:

1.   Voltage and Temperature variability across the die due to differing power density [4]. Such variations, when not accounted for, can cause significant error in timing and static power (leakage) estimates.

2.   Metal thickness variability due to Chemical-Mechanical Polishing [5]. While physical models exist for such variability, it is still not widely applied in commercial CAD flows.

Considering that the increment in performance between technology generations has been reducing as CMOS nears maturity, and that design margin is directly related to our ability to accurately predict and bound the nominal and statistical performance of our designs, it is clear that the improvement of our modeling breadth and accuracy must have the highest priority.

6. ACKNOWLEDGMENTS
The author is indebted to the large design/technology interaction community within IBM, which is far too numerous to name, but especially to Anne Gattiker for her insights and collaboration.

7. REFERENCES
[1]   S. Nassif, *Delay Variability: Sources, Impacts and Trends*, Proceedings of 2000 IEEE ISSCC Conference.

[2]   B. Sheu, D. Scharfetter, P-K. Ko, and M-C. Jeng, *BSIM: Berkeley short-channel IGFET model for MOS transistors*, IEEE JSSC, Aug 1987.

[3]   P. Lloyd, *et. al. Technology CAD at AT&T*, Microelectronics Journal, March 1995.

[4]   H. Su, F. Liu, A. Devgan, E. Acar and S. Nassif, *Full Chip Leakage Estimation Considering Power Supply and Temperature Variations*", ISLPED 2003.

[5]   V. Mehrotra, S. Nassif and D. Boning, *Modeling the Effects of Manufacturing Variation on High-Speed Microprocessor Interconnect Performance*, 1998 IEEE IEDM Conference.

[6]   K. Agarwal, *et. al. A Test Structure for Characterizing Local Device Mismatches*. VLSI 2006.

Low Power Light-weight Embedded Systems*

Majid Sarrafzadeh[1], Foad Dabiri[1], Roozbeh Jafari[2], Tammara Massey[1], Ani Nahapetan[1]

Computer Science Department[1]
University of California, Los Angeles
{majid, dabiri, tmassey, ani}@cs.ucla.edu

Electrical Eng. Dept. / University of Texas at Dallas[2]
Electrical Eng. and Computer Sci. Dept. / UC Berkeley
rjafari@utdallas.edu

Abstract

Light-weight embedded systems are now gaining more popularity due to the recent technological advances in fabrication that have resulted in more powerful tiny processors with greater communication capabilities that pose various scientific challenges for researchers. Perhaps the most significant challenge is the **energy consumption concern and reliability**, mainly due to the small size of batteries. In this tutorial, we portray a brief description of low-power, light-weight embedded systems, depict several power profiling studies previously conducted, and present several research challenges that require low-power consumption in embedded systems. For each challenge, we highlight how low-power designs may enhance the overall performance of the system. Finally, we present a several techniques that minimize the power consumption in such systems.

Categories and Subject Descriptors

B.8.2 [**Performance and Reliability**]: Performance Analysis and Design Aids; C.3 [**Special-purpose and Application-based Systems**]: Real-time and embedded systems

General Terms

Design, Performance

Keywords

Light-weight embedded systems, sensor networks, power optimization

1. Introduction

Light-weight embedded systems, recently introduced due to the advancement of fabrication of powerful tiny processors, have the ability to revolutionize capture, processing and actuation in several collaborative and networked systems. The new class of tiny embedded systems has been widely utilized in several domains from medical monitoring applications [1][2] to collaborative object tracking systems [3]. Many systems similar to the aforementioned applications require low-profile, mobile and cost-effective devices. The physical size and the cost-effectiveness immediately deduce several constraints in processing power and communication bandwidth. In addition, it enforces restriction on the size of batteries. These unique limitations require rethinking and reinventing the design process

---

* Second to fifth author names appear in alphabetical order.

Permission to make digital or hard copies of all or part of this work for personal or classroom use is granted without fee provided that copies are not made or distributed for profit or commercial advantage and that copies bear this notice and the full citation on the first page. To copy otherwise, or republish, to post on servers or to redistribute to lists, requires prior specific permission and/or a fee.
ISLPED '06, October 4–6, 2006, Tegernsee, Germany.
Copyright 2006 ACM 1-59593-462-6/06/0010...$5.00.

in particularly light-weight embedded systems. Predominantly, power issues become a major concern in the design phase due to the unique properties of such systems. Optimization of power consumption in light-weight embedded systems is no longer just an objective function that is to be minimized. Power optimization is a tight constraint that must be accommodated to deliver a practical system.

2. System definition

An embedded system is a special-purpose system in which a computer is entirely encapsulated by the gadget it controls. Unlike a general-purpose computer, an embedded system performs pre-defined tasks, usually with very specific requirements and constraints [4]. Since the system is dedicated to a specific task, designers can optimize it, reducing the size and cost of the product. *Light-weight embedded systems* are often referred to low-profile, small size, unobtrusive and portable processing elements with **limited power resources**. Such systems typically incorporate sensing, processing and communications and are often manufactured to be simple and cost-effective. These **low profile** systems usually have limited computational capabilities, memory (storage), speed and I/O interfaces. Despite their low complexity, computationally intensive tasks impede light-weight embedded systems from being deployed in collaborative networks in large quantities. While their sensing capabilities allow for a seamless integration into the physical world, their processor architecture designs yield notable advantages such as reconfigurability and adaptability with various applications and environments. Consumers, on the other hand, constantly demand thinner, smaller and lighter systems with **smaller batteries** in which the battery life is enhanced to meet their lifestyle. Improving the performance of battery life, however, has been always a major scientific challenge for researchers. Due to its criticality, the battery life becomes an objective as opposed to being a constraint in traditional systems. In order to optimize the power consumption in such systems, researchers must understand the major sources of power consumption. Therefore, we present a power profiling study on light-weight embedded systems in the next section. In section 4, we portray a number of challenges with respect to power optimization in such systems. In sections 5 and 6, we discuss two problems that we selected in regards to low-power light-weight embedded system design in details. Finally, Section 7 concludes the paper.

3. Power profiling

In this section, we present a few widely used embedded systems deployed in several monitoring and mobile applications. Motes from CrossBow [5] are among popular candidates due to their tiny size and on-chip sensing. For applications where interaction with users is desired, Pocket PCs have also been broadly deployed [6]. Pocket PCs, in addition to their portability, incorporate relatively large displays, and communicate using commonly accepted wireless communication protocols such as Bluetooth and 802.11 Wi-Fi. These communication capabilities facilitate real-time and remote monitoring. Several power profiling studies have been conducted on motes [7] [8]. Pocket PCs behave differently mainly to due to their architecture dissimilarities with motes. Several studies are conducted on power profiling of handheld devices [9] [10] including Pocket PCs.

4. Challenges

In this chapter, we describe several challenges for low-power embedded system design. For each challenge, we begin with a general definition, followed by low-power solutions proposed for traditional systems and then we depict new techniques suitable for light-weight embedded systems suggested by researchers.

4.1 Scheduling for power management

Task scheduling on single or multiple processing elements is considered as one of the most common methods to achieve lower power consumption. In particular, in light-weight embedded systems, scheduling saves power by shutting down devices when they are not operating. Processing elements in embedded systems usually serve different requests at different times. Ordering task execution adjusts the lengths of idle periods and exploits the opportunities for power management [11][12][13]. Several approaches have been proposed for task scheduling on low-power embedded systems that consider highly constrained energy source and environmental sources [14][15][16][17].

4.2 Software power optimization

Software constitutes a major component of today's systems, and its role is projected to continue to grow. In traditional processors, instruction level analysis of a processor aids in developing power consumption models of software execution in processors. Software power evaluation also gives the designers the ability to optimize their programs in terms of power. Common techniques include code compression and coding [18][19][20]. Similar approaches have been applied to embedded systems with tiny processors. Light-weight embedded systems are highly constrained in terms of the memory size available to them. Most work on code compression thus far has focused mainly on the memory optimization. However, code compression has a significant effect on energy consumption. Since a compressed code is smaller in size, fewer accesses to the main memory is required resulting in less energy consumption. Meanwhile, reductions in memory accesses result in the reduction of power dissipation in the bus and interconnections [21][22]. This, in particular, gains more importance in systems with small batteries.

4.3 Low power communication

Communication on buses within a chip as well inter-device wireless communication has been always a major source for power consumption. On-chip bus communication becomes even more challenging in tiny-processors that does not feature advance architecture. A considerable amount of energy is consumed in on-chip interconnect and I/O buses. The main source of power dissipation occurs when the voltage swings in communication lines. Bus coding and encoding techniques can be used to reduce power consumption along with increasing the performance in terms of throughput and latency. These techniques reduce voltage swings along interconnect lines which can result in large power savings [23][24][25][26]. Wireless communication in light-weight embedded systems dominates the total power consumption [7].

Wireless communication distinguishes these systems from other traditional networks mainly due to the large of number of elements deployed, their power constraints, and their level of mobility. Several techniques have been proposed and various survey articles have been published in this area [27][28][29].

4.4 Low power security

Security protocols mostly involve complex computations and extensive communications. This challenge becomes even more critical in light-weight embedded systems mainly due to the constraints in processing power as well as communication bandwidth. In addition, the limitations in the battery size impede conducting complex tasks that require massive communications. Yet, due to the sensitivity of data communicated in many applications as well as the deployment of such applications in unattended and hostile environments, implementing quality, light-weight security protocols in embedded systems are imperative. Implementation of traditional security protocols in light-weight embedded systems, however, introduces several obstacles due to the resource constrained nature of such systems. More precisely, the constraints include the lack of large storage, the lack of powerful computational elements, a low bandwidth communication infrastructure, and limited power resources [30]. Therefore, various power-aware secure protocols have been proposed. Secure routing schemes have been investigated in [31][32][33] and [34]. Secure protocols also have been studied for data aggregation [35][36][37][38][39][40] and group formation [41][42][43].

4.5 Low power display

Classes of light-weight embedded systems in which interaction with users is essential require displays. The backlight in displays consumes a significant portion of available energy [44]. This becomes even more crucial in light-weight embedded systems where their available energy is highly constrained. Unfortunately, this issue has received little attention from researchers. [45] and [46] have proposed techniques for power reduction in displays. Techniques for low-power graphical user interface (GUI) and low-power human-computer interaction were suggested in [47] and [48].

4.6 Low power data management

Power optimization in data management has not been a major focus of researchers conventionally due to the availability of often unlimited energy source in most computation nodes. Yet, limited supply of energy, inadequate processing power, small memory size, power hungry communication and a low bandwidth communication infrastructure impose various challenges for data management in systems which are closely linked with physical environments. In addition, uncertainty in sensor readings due to environmental interference and faults in inexpensive embedded systems may result in the generation of inaccurate information. Several tree-based [49][50][51] and multi-path-based [52][53] query aggregation techniques have been proposed. In [54] researchers have addressed a combined method for in-network data processing. Overall, researchers have outlined several data management metrics in low-power distributed embedded systems which may enhance in-network data processing capabilities.

4.7 Fault tolerance and reliability

The most common approach to fault tolerance is the use of redundancy in systems. Adding a new redundant component creates a new source of power consumption in the system.

Therefore, in light-weight embedded systems, efficient fault tolerant techniques must be applied since the luxury of having redundant components can not be easily afforded. The other barrier is low bandwidth and expensive communication. Fault tolerance cannot be handled in a localized manner since it may not be feasible to collect information from all nodes. Finally, one major objective in manufacturing tiny embedded systems is the fabrication cost. The manufacturers tend to keep the systems cost-effective; therefore, the hardware may not be ultra reliable and reliability concerns mostly must be handled by the applications [55][56][57][58].

5. Minimum skew utilization

Previous research efforts have shown that communication dissipates significant amount of energy [7]. Consequently, design choices and protocols that affect the communication traffic have a great impact on system lifetime. For example, minimum-hop routing algorithms reduce the number of transmissions required to deliver a message at the destination and improve system energy dissipation [59]. In this section, we propose a technique called minimum skew utilization that aims at optimizing the power consumption and enhancing the system lifetime. This is done by evenly distributing node utilization and communication across the network. In our formulation, we attempt to minimize the skew in energy consumption (due to wireless communication) across highly congested nodes. We will show that our formulation will additionally yield a minimum skew distribution of energy consumption across all the nodes. The skew is defined as follows: There exists exponential number of paths connecting source to destination nodes. For every path, we form the following definition. There exists a node in every path that has the highest energy consumption rate (ideally if each path would be isolated from the rest of the network, the energy consumption rate due to wireless communication must be identical throughout the path, however, in reality, nodes may have incoming/outgoing edges from/to the rest of the network. Therefore, the energy consumption rate may not necessarily be uniformly distributed). In every path P_i connecting a source to a destination node, we identify the node with highest wireless communication traffic as p_{imax}. We define the skew of energy consumption as the difference between p_{imax} and p_{jmax} of paths i and j. In the next subsection, we will illustrate how our formulation minimizes the difference of highest traffic across every two paths. Even though, the number of paths is exponential, the upper bound on the number of links with highest traffic is still m (where m is total number of links). The definition of minimum skew is rephrased in Equation 1.

For every Pi and Pi connecting source(s) to destination(s)
$$\text{Minimize } |p_{imax} - p_{jmax}| \qquad (1)$$

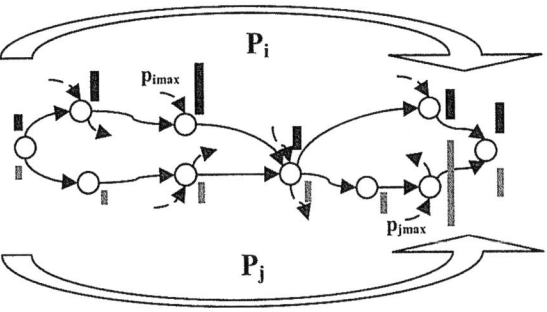

Figure 1. Min-skew definition

5.1 Methodology

Given a graph G_s for a network and a specific traffic pattern, the objective is to route the packets so that after the completion of packet routing, the maximum traffic across all nodes is minimized. This objective enhances the system lifetime. We assume that there is a specific node t in G_s serves as a gateway, base station, or destination node, and all packets have to be delivered to node t. Each packet transmitted from a source node to t can be viewed as a unit flow in the network G_s. More precisely, x_{ij} units of flow represents x_{ij} packets transmitted from v_i to v_j. The problem of packet routing is equivalent to finding a feasible network flow in G_s. Intuitively, the objective of minimizing the highest congested node is equivalent to minimizing the maximum flow passing through nodes. In the remainder of this section, we transform G_s to a new graph G_t in which, a formulation based on min-cost flow generates the optimal solution. We proceed to describe the transformation procedure, along with the mathematical properties of our technique.

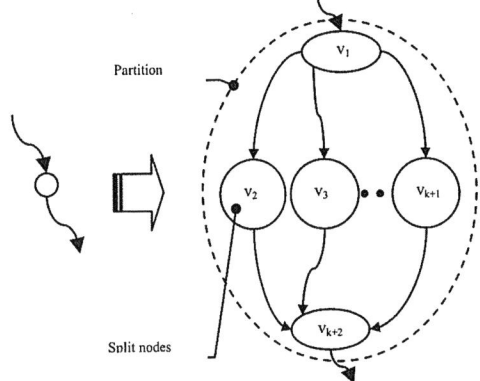

Figure 2. Node partitioning

We construct the network $G_t = (V_t, E_t)$ from the graph G_s according to the following rules. Each node in graph G_s is split into $k + 2$ nodes where k is a tunable parameter that controls the accuracy of the solution. We refer to the resulting set of nodes as a partition. In each partition, two nodes serve as receiver and transmitter (or input and output) and the rest of the nodes are called splits. Figure 2 illustrates an example partition in which, nodes v_1 and v_{k+2} are the receiver and the transmitter of the partition, respectively. Nodes v_2 thru v_{k+1} are the split nodes. The capacity of each node, v_i, is represented by u_i. In order to embed this constraints into the problem, we assign the flow upper bound of $u'_i = u_i/k$ to each of the split nodes. The upper bounds guarantee that a partition cannot pass more flow after relaying u_i units of flow, or equivalently, a node cannot exceed its capacity. The receiver, transmitter, and split nodes are assigned a special sequence of costs for passing the flow. The cost associated with receiver and transmitter nodes of a partition is zero. The costs associated with split nodes of a partition increase from left to right (Figure 2). This directs the min-cost flow technique to utilize the split nodes from left to right, when passing flow through a partition.

The cost on each split node is enforced such that it would be greater than the cumulative cost of the split nodes with smaller indices over all partitions. Intuitively, our cost assignment technique enforces the min-cost flow technique to utilize the split nodes with smaller indices, before trying to utilize a particular split node. This simple, yet effective idea exhibits the main property of our technique by which, we minimize the traffic across the highly congested links. For simplicity, we define cost rank where cost rank R_i has the following property:

$$R_i > \sum_{j<i} m.R_j \qquad (2)$$

The cost on all the other nodes from the original graph G_s is zero. The loss in edges may also be accommodated easily. For the sake of brevity, we remove the extensions for lossy edges.

We now prove that our cost assignment strategy implies that the min-cost flow technique in G_t corresponds to a routing scheme in G_s that minimizes the traffic across the most highly congested link. Let x_{ij} represent the amount of flow on edge (i, j). Let y_{il} represent the amount of flow going through split node l in partition i. Similarly, let c_{il} denote the associated cost of unit flow passing through split l of node i in G_t. The min-cost flow problem for graph G_t with the given supply and demand vector, can be written as:

$$\text{Minimize} \sum_{\forall i \forall l=1..k} c_{il} y_{il} \qquad (3)$$

The flow that passes through the receiver or transmitter nodes of a partition represents the total flow passing through that partition, or equivalently, it determines the traffic across the original link in G_s. Therefore, the objective is to minimize the maximum amount of flow passing through partitions.

The flow passing through any partition has to pass through its split nodes, and subsequently, min-cost flow solutions utilize the splits with lower costs before higher cost splits. We present three theorems; however, the proofs are omitted for brevity.

Theorem 1: The objective function in Equation 3 minimizes the maximum flow across the links of the graph $G_s = (V_s, E_s)$ with maximum error ε where $\varepsilon <$

$u'_i = u_i/k = 1/k$.

The intuition behind our proposed technique is that the cost assignment on the splits forces the network to route a flow from the lth split of link e_i, if it cannot be routed through any number of other nodes whose $(l-1)^{th}$ splits is empty.

Theorem 2: The solution L, generated by our technique, minimizes the difference of maximum flows across every two disjoint paths connecting a source to a destination node (with tolerance of $\varepsilon = 1/k$ - minimal-skew). In other words, it minimized the difference of traffic across the most highly congested links in the two paths.

Theorem 3: The lexicographically sorted solution of minimal-skew routing is unique.

5.2 Summary and open problems

We generated several benchmarks based on random geometric graphs with relatively a large number of nodes to illustrate the effectiveness of our technique. For simplicity, we assumed that communication capacity across all links is uniform. One hundred sensors are placed within areas of size 160×40, 160×60, 160×80, 200×40, 200×60, 200×80 and 200×100. In all networks, three source nodes are placed on far left side of the square area while the destination nodes/ gateways (three) are placed on the right side of the square. This particularly assists us to place the

source/destination nodes not within the close proximity of each other. The connectivity between nodes is determined by unit-radius disk model. The reception rate on communication links is chosen randomly between 80% and 100%. The locations of the nodes are generated conforming to a random uniform distribution over several size areas. For every particular area size, twenty benchmarks are generated with three source and three destination nodes. To compare our scheme against other routing algorithms, we consider a shortest path routing algorithm based on minimum cost flow (which is equivalent to the case where we have only one split ($k=1$)).

Overall, the average maximum communication traffic is reduced by a factor of 4 for $k=4$ compared to min-cost shortest path approach ($k=1$). The average delay of our scheme (for k =4) is about 20% greater than the min-cost shortest path. The average delay is not vastly increased due to existence of multiple edge disjoint paths between sources and destinations. The average delay also increases slightly as the k (number of splits) is increased. In general, highly connected networks provide a large number of parallel paths between nodes, which is of our interest and enhance the flexibility of data routing.

For the minimum skew utilization, a number of open problems must be solved to allow truly deployment of the technique in distributed and light-weight embedded systems. Firstly, the distributed version of this technique must be explored. Secondly, in highly dynamic networks where the quality of links may lively change, a fast optimal or sub-optimal solution is desired. Finally, the effect of several cost series on split nodes may be studied.

6. Static Voltage Scheduling

In this section we address the problem of static voltage scheduling in light-weight, low-power and high-performance systems. The scheduling problem refers to the assignment of a supply voltage level to each module in the architecture representing the system. The objective is to minimize the energy consumption for a given computation time or throughput constraints or both. Intuitively, the voltage scheduling problem can be stated as a timing management problem. In a given application with distinct constituting blocks, the maximum tolerable power reduction of individual blocks is desired, while the timing constraints of the system is not violated. These blocks are often modeled as nodes in a directed acyclic graph (DAG) where edges represent the dependency among modules.

6.1 Methodology

We present an optimal methodology for static voltage scheduling for low-power and high-performance systems. The main property of the methodology is the unified formulation with linear size number of constraints in the optimization problem as opposed to exponential number of constraints in previously proposed techniques. Our formulation can be applied to dynamic voltage scaling on single or multiple resources. Moreover, this problem formulation results a convex optimization problem.

An interesting observation is proven indicating that in the optimal voltage scheduling of a DAG, the delay between any node u and the output t is independent of the choice of the path taken between them and is unique. This property of the optimal solution leads us into generating only linear number of constraints. We prove this property in the following theorem.

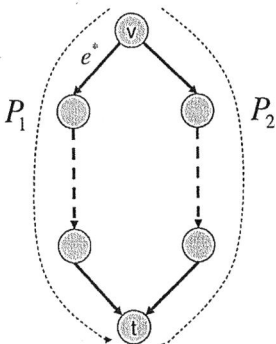

Figure 3. Figure for theorem 4

Theorem 4: In the optimal voltage scheduling of a DAG, the distance between any node v and the output t is independent of the choice of the path taken between them and is unique.

Proof: Suppose the claim is not true, i.e. there exists a node v where its distance to t through path P_1 is less then P_2 (see Figure 3). The intuition behind the proof is that, the larger delay assigned to an edge yields in more power saving which is the objective of the optimization problem. If P_1 is shorter than P_2, there exists an edge (e*) in P_1 that can be slowed down and still not violate the timing constraint because P_2 is on the critical path from v to t. One immediate candidate for e* is the first edge in P_1. Increasing the delay of e* by $d_{P2} - d_{P1}$ will not cause a timing violation and therefore, reduces the total power dissipation.

The following observation is immediately inferred from the above theorem: The delay of each path in the optimal solution from the primary input node s to primary output node t is equal to T. Now that the distance of every node to the destination is independent of the path taken, let t_i be a variable assigned to each node v_i that represents its distance to t. We call t_i the distance variable of node v_i. In other words, t_i is the delay of the system from node v_i to the output. Therefore, the delay of each node in the graph can be bounded by linear number of constraints.

In the next step, we show that the feasible solution space is in fact convex. With a convex objective and a convex feasible region, there can be only one optimal solution, which is globally optimal. All delay constraints are linear and can be viewed as planes, bounding the solution space. It is trivial that non-finite planes yield in a convex solution space. To show the convexity of the objective function which we prove that each term is convex. It can be easily shown that each term is proportional to the inverse of the delay squared which is trivially convex. Since the sum of convex functions is convex, the objective function is convex.

These voltages obtained from the optimal method might all have different values and therefore result in large number of power supplies with various voltage levels which may not be available. Current technologies, allow the designer to utilize only a few number of voltages. We propose to use the nearest neighbor mapping technique and map the optimal voltage of each node to its next available level. Experimental results show that this yields a very good power saving while the timing constraint is violated by a small fraction.

6.2   Summary and open problems

We evaluated the performance of our techniques on benchmarks from TGFF[60]. An average of 43.96% power reduction was gained for unbounded supply voltage assignment along with 40%

average power saving where discrete voltage levels are available. The results illustrate the efficiency of using the continuous optimal voltage scheduling for discrete case and observed that even limited number of voltage levels (less that 8) can provide us with near optimal power reduction.

In future, the optimum voltage scheduling method may be extended for dynamic voltage scheduling. Furthermore, developing design rules that assist developers with voltage scheduling at the design stage may be investigated. In addition, the effect of voltage level shifters on performance and their related optimization problems may be studied.

7.   Conclusion

Power consumption and battery life is among the most critical concerns in light-weight embedded systems. Improving the performance of battery life, however, has been always a major scientific challenge for researchers. In order to optimize the power consumption of such systems, researchers must first understand the major sources of power consumption. In this paper, we presented power profiling studies conducted on several embedded systems. We furthermore portrayed a number of challenges that require low-power solutions and depicted three techniques that enhances the power lifetime.

8.   References

[1] R. Jafari, A. Encarnacao, A. Zahoory, F. Dabiri, H. Noshadi, and M. Sarrafzadeh, "Wireless Sensor Networks for Health Monitoring," *MobiQuitous '05: Proc. Second Ann. Int'l Conf. Mobile and Ubiquitous Systems,* 2005.

[2] R. Jafari, F. Dabiri, P. Brisk, and M. Sarrafzadeh, "Adaptive and Fault Tolerant Medical Vest for Life-Critical Medical Monitoring," *SAC '05: Proc. 2005 ACM Symp. Applied Computing,* pp. 272-279, 2005.

[3] Aslam, J., Butler, Z., Constantin, F., Crespi, V., Cybenko, G., and Rus, D. 2003. Tracking a moving object with a binary sensor network. In *Proceedings of the 1st international Conference on Embedded Networked Sensor Systems* (Los Angeles, California, USA, November 05 - 07, 2003). SenSys '03. ACM Press, New York, NY, 150-161.

[4] Daniel Gajski, Frank Vahid, Sanjiv Narayan, and Jie Gong, *Specification and Design of Embedded Systems,* Prentice Hall, 1994.

[5] Crossbow technology inc. *http://www.xbow.com.*

[6] Hinckley, K., Pierce, J., Sinclair, M., and Horvitz, E. 2000. Sensing techniques for mobile interaction. In *Proceedings of the 13th Annual ACM Symposium on User interface Software and Technology* (San Diego, California, United States, November 06 - 08, 2000). UIST '00. ACM Press, New York, NY, 91-100.

[7] Victor Shnayder, Mark Hempstead, Bor rong Chen, Geoff Werner Allen, and Matt Welsh. Simulating the power consumption of large-scale sensor network applications. In *SenSys '04: Proceedings of the 2nd international conference on Embedded networked sensor systems,* pages 188-200, New York, NY, USA, 2004. ACM Press.

[8] Robert Szewczyk, Alan Mainwaring, Joseph Polastre, John Anderson, and David Culler. An analysis of a large scale habitat monitoring application. In *SenSys '04: Proceedings of the 2nd international conference on Embedded networked sensor systems,* pages 214-226, New York, NY, USA, 2004. ACM Press.

[9] Vijay Raghunathan, Trevor Pering, Roy Want, Alex Nguyen, and Peter Jensen. Experience with a low power wireless mobile computing platform. In *ISLPED '04: Proceedings of the 2004 international symposium on Low power electronics and design,* pages 363-368, New York, NY, USA, 2004. ACM Press.

[10] Marc A. Viredaz and Deborah A. Wallach. Power evaluation of a handheld computer. *IEEE Micro,* 23(1):66-74, 2003.

[11] Jing-Jang Hwang, Yuan-Chieh Chow, Frank D. Anger, and Chung-Yee Lee. Scheduling precedence graphs in systems with interprocessor communication times. *SIAM J. Comput.,* 18(2):244-257, 1989.

[12] Hesham El-Rewini and T. G. Lewis. Scheduling parallel program tasks onto arbitrary target machines. *J. Parallel Distrib. Comput.,* 9(2):138-153, 1990.

[13] Behrooz Shirazi, Mingfang Wang, and Girish Pathak. Analysis and evaluation of heuristic methods for static task scheduling. *J. Parallel Distrib. Comput.,* 10(3):222-2232, 1990.

[14] Yung-Hsiang Lu, Luca Benini, and Giovanni De Micheli. Low-power task scheduling for multiple devices. In *CODES '00: Proceedings of the eighth*

international workshop on Hardware/software codesign, pages 39-43, New York, NY, USA, 2000. ACM Press.

[15] Niraj K. Jha, Jiong Luo, "Battery-Aware Static Scheduling for Distributed Real-Time Embedded Systems," *Design Automation Conference* , pp. 444-449, 2001.

[16] Rakhmatov, D. and Vrudhula, S. 2003. Energy management for battery-powered embedded systems. *Trans. on Embedded Computing Sys.* 2, 3 (Aug. 2003), 277-324.

[17] Kansal, A., Potter, D., and Srivastava, M. B. 2004. Performance aware tasking for environmentally powered sensor networks. *In Proceedings of the Joint international Conference on Measurement and Modeling of Computer Systems* (New York, NY, USA, June 10 - 14, 2004). SIGMETRICS '04/Performance '04. ACM Press, New York, NY, 223-234.

[18] A. Peymandoust, T. Simunic, and G. de Micheli. Low power embedded software optimization using symbolic algebra. In *DATE '02: Proceedings of the conference on Design, automation and test in Europe*, page 1052, Washington, DC, USA, 2002. IEEE Computer Society.

[19] Eui-Young Chung, Luca Benini, and Giovanni De Micheli. Source code transformation based on software cost analysis. In *ISSS '01: Proceedings of the 14th international symposium on Systems synthesis*, pages 153-158, New York, NY, USA, 2001. ACM Press.

[20] Arpad Beszedes, Rudolf Ferenc, Tibor Gyimothy, Andre Dolenc, and Konsta Karsisto. Survey of code-size reduction methods. *ACM Comput. Surv.*, 35(3):223-267, 2003.

[21] Haris Lekatsas, Wayne Wolf, and Joerg Henkel. Arithmetic coding for low power embedded system design. In *DCC '00: Proceedings of the Conference on Data Compression*, page 430, Washington, DC, USA, 2000. IEEE Computer Society.

[22] A. Parikh, Soontae Kim, M. Kandemir, N. Vijaykrishnan, and M. J. Irwin. Instruction scheduling for low power. *J. VLSI Signal Process. Syst.*, 37(1):129-149, 2004.

[23] Mircea R. Stan and Wayne P. Burleson. Bus-invert coding for low-power i/o. *IEEE Trans. Very Large Scale Integr. Syst.*, 3(1):49-58, 1995.

[24] Benjamin Bishop and Anil Bahuman. A low-energy adaptive bus coding scheme. In *WVLSI '01: Proceedings of the IEEE Computer Society Workshop on VLSI 2001*, page 118, Washington, DC, USA, 2001. IEEE Computer Society.

[26] Peter Petrov and Alex Orailoglu. Low-power data memory communication for application-speci c embedded processors. In *ISSS '02: Proceedings of the 15th international symposium on System Synthesis*, pages 219-224, New York, NY, USA, 2002. ACM Press.

[25] Peter Petrov and Alex Orailoglu. Low-power instruction bus encoding for embedded processors. *IEEE Trans. Very Large Scale Integr. Syst.*, 12(8):812-826, 2004.

[27] I. Akyildiz, W. Su, Y. Sankarasubramaniam, and E. Cayirci. A survey on sensor networks, 2002.

[28] Suresh Singh, Mike Woo, and C. S. Raghavendra. Power-aware routing in mobile ad hoc networks. In *MobiCom '98: Proceedings of the 4th annual ACM/IEEE international conference on Mobile computing and networking*, pages 181-190, New York, NY, USA, 1998. ACM Press.

[29] Kemal Akkaya and Mohamed Younis. A survey on routing protocols for wireless sensor networks. *Ad Hoc Networks*, 2, 2004.

[30] Adrian Perrig, Robert Szewczyk, Victor Wen, David Culler, and J. D. Tygar. SPINS: Security protocols for sensor networks. In *Seventh Annual International Conference on Mobile Computing and Networks (MobiCOM 2001)*, Rome, Italy, July 2001.

[31] J. Deng, R. Han, and S. Mishra. Insens: Intrusion-tolerant routing in wireless sensor networks, 2002.

[32] Brad Karp and H. T. Kung. GPSR: greedy perimeter stateless routing for wireless networks. In *Mobile Computing and Networking*, pages 243-254, 2000.

[33] P. Papadimitratos and Z.J. Haas. Secure routing for mobile ad hoc networks.

[34] Sapon Tanachaiwiwat, Pinalkumar Dave, Rohan Bhindwale, and Ahmed Helmy. Poster abstract secure locations: routing on trust and isolating compromised sensors in location-aware sensor networks. In *SenSys '03: Proceedings of the 1st international conference on Embedded networked sensor systems*, pages 324-325, New York, NY, USA, 2003. ACM Press.

[35] Deborah Estrin, Ramesh Govindan, John S. Heidemann, and Satish Ku mar. Next century challenges: Scalable coordination in sensor networks. In *Mobile Computing and Networking*, pages 263-270, 1999.

[36] L. Hu and D. Evans. Secure aggregation for wireless networks, 2003.

[37] S. Madden, M. Franklin, J. Hellerstein, and W. Hong. Tag: a tiny aggregation service for ad-hoc sensor networks, 2002.

[38] Bartosz Przydatek, Dawn Song, and Adrian Perrig. Sia: secure information aggregation in sensor networks. In *SenSys '03: Proceedings of the 1st international conference on Embedded networked sensor systems*, pages 255-265, New York, NY, USA, 2003. ACM Press.

[39] Nisheeth Shrivastava, Chiranjeeb Buragohain, Divyakant Agrawal, and Subhash Suri. Medians and beyond: new aggregation techniques for sensor networks. In *SenSys '04: Proceedings of the 2nd international conference on Embedded networked sensor systems*, pages 239-249, New York, NY, USA, 2004. ACM Press.

[40] F. Ye, H. Luo, S. Lu, and L. Zhang. Statistical en-route detection and filtering of injected false data in sensor networks, 2004.

[41] Andreas Grlach, Andreas Heinemann, and Wesley W. Terpstra. Survey on location privacy in pervasive computing.

[42] Chris Karlof and David Wagner. Secure routing in wireless sensor networks: Attacks and countermeasures. *Elsevier's AdHoc Networks Journal, Special Issue on Sensor Network Applications and Protocols*, 1(2-3):293-315, September 2003.

[43] Sandro Rafaeli and David Hutchison. A survey of key management for secure group communication. *ACM Comput. Surv.*, 35(3):309-329, 2003.

[44] J. Flinn, K. Farkas, and J. Anderson. Power and energy characterization of the itsy pocket computer, 2000.

[45] Inseok Choi, Hojun Shim, and Naehyuck Chang. Low-power color tft lcd display for hand-held embedded systems. In *ISLPED '02: Proceedings of the 2002 international symposium on Low power electronics and design*, pages 112-117, New York, NY, USA, 2002. ACM Press.

[46] S. Pasricha and S. Mohapatra. Reducing backlight power consumption for streaming video applications on mobile handheld devices, 2003.

[47] Lin Zhong and Niraj K. Jha. Graphical user interface energy characterization for handheld computers. In *CASES '03: Proceedings of the 2003 international conference on Compilers, architecture and synthesis for embedded systems*, pages 232-242, New York, NY, USA, 2003. ACM Press.

[48] Lin Zhong and Niraj K. Jha. Energy efficiency of handheld computer interfaces: limits, characterization and practice. In *MobiSys '05: Proceedings of the 3rd international conference on Mobile systems, applications, and services*, pages 247-260, New York, NY, USA, 2005. ACM Press.

[49] Samuel Madden, Michael J. Franklin, Joseph M. Hellerstein, and Wei Hong. Tag: a tiny aggregation service for ad-hoc sensor networks. *SIGOPS Oper.Syst. Rev.*, 36(SI):131-146, 2002.

[50] Samuel R. Madden, Michael J. Franklin, Joseph M. Hellerstein, and Wei Hong. Tinydb: an acquisitional query processing system for sensor networks. *ACM Trans. Database Syst.*, 30(1):122-173, 2005.

[51] Y. Yao and J. Gehrke. Query processing in sensor networks, 2003.

[52] Jeffrey Considine, Feifei Li, George Kollios, and John Byers. Approximate aggregation techniques for sensor databases. In *ICDE '04: Proceedings of the 20th International Conference on Data Engineering*, page 449, Washington, DC, USA, 2004. IEEE Computer Society.

[53] Suman Nath, Phillip B. Gibbons, Srinivasan Seshan, and Zachary R. Anderson. Synopsis diffusion for robust aggregation in sensor networks. In *SenSys '04: Proceedings of the 2nd international conference on Embedded networked sensor systems*, pages 250-262, New York, NY, USA, 2004. ACM Press.

[54] Amit Manjhi, Suman Nath, and Phillip B. Gibbons. Tributaries and deltas: efficient and robust aggregation in sensor network streams. In *SIGMOD '05: Proceedings of the 2005 ACM SIGMOD international conference on Management of data*, pages 287-298, New York, NY, USA, 2005. ACM Press.

[55] Thomas Clouqueur, Kewal K. Saluja, and Parameswaran Ramanathan. Fault tolerance in collaborative sensor networks for target detection. *IEEE Trans. Comput.*, 53(3):320-333, 2004.

[56] F. Koushanfar, M. Potkonjak, and A. Sangiovanni-Vincentelli. Fault tolerance techniques in wireless ad-hoc sensor networks, 2002.

[57] Jonathan L. Bredin, Erik D. Demaine, MohammadTaghi Hajiaghayi, and Daniela Rus. Deploying sensor networks with guaranteed capacity and fault tolerance. In *MobiHoc '05: Proceedings of the 6th ACM international symposium on Mobile ad hoc networking and computing*, pages 309-319, New York, NY, USA, 2005. ACM Press.

[58] Murat Demirbas. *Scalable design of fault-tolerance for wireless sensor networks*. PhD thesis, The Ohio State University, 2004.

[59] Chalermek Intanagonwiwat, Ramesh Govindan, Deborah Estrin, John Heidemann, and Fabio Silva. Directed diffusion for wireless sensor networking. *IEEE/ACM Trans. Netw.*, 11(1):2–16, 2003.

[60] D. Rhodes R. Dick and W. Wolf. Tgff: Task graphs for free. In *CODES '98: Proceedings of the CODES*, pages 97–101, 1998.

Invited Talk

Low Power Design from Technology Challenge to Great Products

Barry Dennington
Philips Semiconductors
Senior Vice President
CTO/SoC Design Technology
barry.dennington@philips.com

ABSTRACT

Each generation of semiconductor process technology enables increased levels of integration and density on a single chip, Moores Law continues to prevail and the users of portable and hand held communications and entertainment products enjoy greater functionality and features. Lifestyles and user paradigms indicate that no matter how many features are added, service providers and manufacturers think of more and users cannot wait to acquire the latest products. Features, performance, fashion and, of course, fierce competition, drive the market and thereby set the challenge for the semiconductor designer.

The challenge for designers is to create systems on a single chip (SoC) to provide these features for the user and to enable service and content providers to realize new emerging market opportunities. This has been a challenge for many years but, now that nanometer process technologies form the enabling process technology, the design challenge is much greater. Nanometer design effects must be considered from the initial SoC architecture all the way through to manufacturing where design for manufacturing (DFM) effects must be overcome to enable reliable, high volume production. At the silicon level the features demanded by the users require extensive efforts to provide acceptable performance and reliability. Users of Cell Phones, PDAs and MP3 Players will be most familiar with the need for long battery life while, to achieve this, the SoC designer worries about how to design with lower supply voltages, higher leakage currents, on chip power density and reliability. Packaging techniques which assemble multiple chips to form Systems in Package (SiP) also create signal integrity and power dissipation issues. At the same time designers must be able to design with EDA design tools and methodology's that are still emerging and where no standards for low power design exist today to make the task easier.

This keynote will talk about the low power design techniques available to SoC designers, how they are implemented in SoCs and how they are implemented in existing and new designs. The talk will end with a view on the challenges coming up next and what needs to be done to prepare for them.

Categories and Subject Descriptors: B.7 INTEGRATED CIRCUITS, C.3 SPECIAL-PURPOSE AND APPLICATION-BASED SYSTEMS

General Terms: Management, Design

Keywords: SoC, Design for Low Power

Copyright is held by the author/owner(s).
ISLPED '06, October 4–6, 2006, Tegernsee, Germany.
ACM 1-59593-462-6/06/0010.

A Novel Dynamic Power Cutoff Technique (DPCT) for Active Leakage Reduction in Deep Submicron CMOS Circuits

Baozhen Yu, Michael L. Bushnell
Dept. of ECE and CAIP Research Center, Rutgers University
Piscataway, NJ USA 08854-808
baozhen@caip.rutgers.edu, bushnell@caip.rutgers.edu

ABSTRACT

Due to the exponential increase in subthreshold leakage and gate leakage with technology scaling, leakage power is becoming a major fraction of total VLSI chip power in active mode. We present a novel active leakage power reduction technique, called the *dynamic power cutoff technique* (DPCT). First, the switching window for each gate, during which a gate makes its transitions, is identified by static timing analysis. Then, the circuit is optimally partitioned into different groups based on the *minimal switching window* (MSW) of each gate. Finally, power cutoff transistors are inserted into each group to control the power connections of that group. Each group is turned on only long enough for a wavefront of changing signals to propagate through that group. Since each gate is only turned on during a small timing window within each clock cycle, this significantly reduces active leakage power. This technique can also save standby leakage and dynamic power. Results on ISCAS '85 benchmark circuits modeled using 70 *nm* Berkeley Predictive Models [1] show up to 90% active leakage, 99% standby leakage, 54% dynamic power, and 72% total power savings.

Categories and Subject Descriptors: B.7.1 [Integrated Circuits]: Types and Design Styles – Advanced technologies, VLSI

General Terms: Performance, Design

Keywords: Leakage current, standby current, dynamic power, power cutoff, stacking

1. INTRODUCTION

Leakage power is increasingly significant in CMOS circuits as the technology scales dow, due to the exponential increase of subthreshold and gate leakage currents with technology scaling. Many leakage reduction techniques have been proposed. Forced stacking [2] reduces leakage power by inserting an extra serially connected transistor in the gate pulldown or pullup path and turning it off in standby mode. Input vector control [3] uses the state dependence of leakage to apply a low-leakage input vector to the circuit in standby mode to save leakage power. The power cutoff technique [4], also called supply gating, reduces leakage by disconnecting the global supply voltage in standby

mode. One common problem of all techniques is that they can only reduce the circuit leakage power in standby mode.

Leakage is important in both standby and active operation modes. The leakage in active mode is significantly larger due to the higher die temperature in active mode. So, efficient leakage power reduction must target both standby and active leakage power. The dual V_{th} technique [5, 6] uses high-threshold voltage devices on noncritical paths to reduce leakage while using low-threshold devices on critical paths to maintain circuit speed. It reduces both active and standby leakage. However, this technique does not reduce the leakage on critical paths. Thus, it does not help much for practical circuits, whose paths are usually well balanced. Supply voltage scaling [7, 8], developed for switching power reduction, also reduces both active and standby leakage power. But level conversion is needed at the interface whenever an output from a low V_{DD} unit drives a high V_{DD} unit input. Bhunia *et al.* proposed dynamic leakage reduction using supply gating [9]. They use the Shannon expansion to identify the idle circuit parts and dynamically gate the supply to those parts to save active leakage power.

In this paper, we propose a novel active leakage power reduction technique using dynamic power cutoff, called the *dynamic power cutoff technique* (DPCT). We also target the idle part of the circuit when it is in active mode. However, instead of finding the idle circuit part by the Shannon expansion, we identify when a gate is idle from its *switching window*, the timing window during which the gate makes its transition within each clock cycle, using static timing analysis. Active leakage power is saved by turning on the power connections of each gate only within its switching window during each clock cycle. Standby leakage can also be reduced by turning off the power connections of all gates all of the time once the circuit is idle. This technique also reduces dynamic power by reducing the circuit glitches.

The paper is organized as follows. Section II briefly discusses the traditional power cutoff technique and its effect on performance. Section III gives the idea and implementation of our new *dynamic power cutoff technique* (DPCT). Section IV presents results and Section V concludes.

2. TRADITIONAL POWER CUTOFF TECHNIQUE

Power cutoff was originally proposed to reduce the standby leakage [4]. It inserts low V_t MOSFETs between the power connection of each logic gate and the global power line. Either pMOS or nMOS insertion is used to turn off either V_{DD} or GND of the circuit during idle mode to save leakage power. Figure 1 is a pMOS example.

When the pMOS cutoff transistor is turned off, the subthreshold leakage reduces dramatically due to the *stacking effect* [2]. Overall gate leakage also reduces because of a smaller voltage drop across gate oxides of transistors due to the dropped virtual V_{DD}. Overall leakage power is dominated by subthreshold and gate leakages, so power

---

Permission to make digital or hard copies of all or part of this work for personal or classroom use is granted without fee provided that copies are not made or distributed for profit or commercial advantage and that copies bear this notice and the full citation on the first page. To copy otherwise, to republish, to post on servers or to redistribute to lists, requires prior specific permission and/or a fee.
ISLPED'06, October 4–6, 2006, Tegernsee, Germany.
Copyright 2006 ACM 1-59593-462-6/06/0010 ...$5.00.

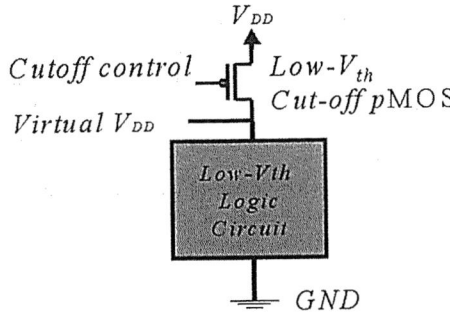

Figure 1: Concept of *pMOS* Insertion for Power Cutoff

cutoff is very effective to reduce the deep submicron leakage power. One problem is that data can be lost during the long sleep period due to the collapsed virtual V_{DD} signal. The power cutoff transistors have about ∼6% extra delay in 0.3 μm CMOS [7] and 3%-6% in 70 nm CMOS [9]. They also add chip area. More importantly, the traditional power cutoff technique only reduces standby leakage.

3. DYNAMIC POWER CUTOFF TECHNIQUE (DPCT)

Here, we propose a novel active leakage power reduction technique based on the power cutoff technique. We first identify when a gate is idle by finding its switching window using static timing analysis, and then turn off the power of each gate when it is idle within each clock cycle. We first introduce the basic idea of this *dynamic power cutoff technique* (DPCT), then discuss its implementation, and finally discuss its power savings.

3.1 Basic Idea of DPCT

Each gate only switches within a particular timing window, called the gate *switching window*, during each clock cycle even when the circuit is in active mode. If we turn on the power connection of each gate only during its switching window during each clock cycle, we save part of the active leakage power with very little effect on signals, usually a little extra delay in 70 nm CMOS technology. The active leakage power saved in a CMOS gate by doing this is proportional to the ratio of the gate power-off time to the clock period. The gate power-off time equals the clock period minus the gate switching window. This is shown in the basic DPCT architecture in Figure 2.

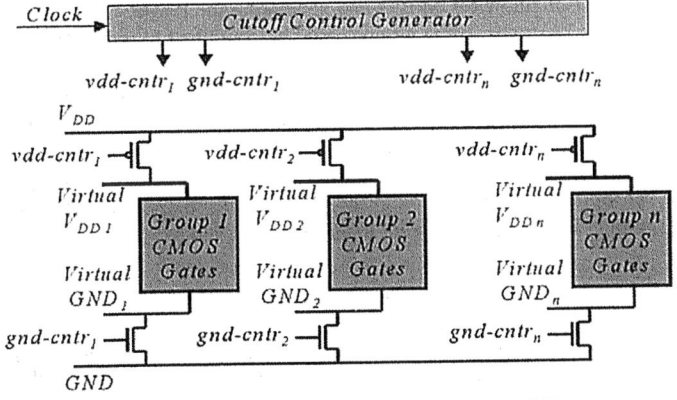

Figure 2: Architecture of a Circuit with DPCT

We use both *nMOS* and *pMOS* insertion to increase the leakage savings. If we left out the *GND* cutoff transistor, when a logic gate output is high, the *p*-tree is on and the *n*-tree is off. Therefore, a leakage path exists from the high output through the *n*-tree to *GND*. A similar argument holds for the V_{DD} cutoff transistor, so we need

both cutoff transistors. A circuit is partitioned into different groups based on the gate switching windows. Gates with the same switching window are treated as one group and the power connections of all gates within the same group are controlled by one pair of power cutoff MOSFETs, a *p*MOSFET and an *n*MOSFET. All such different groups make a partition of the circuit. There is one pair of cutoff control signals for each group, *vdd-cntr$_i$* and *gnd-cntr$_i$*, to control V_{DD} and *GND* of the gates in that group. The cutoff control signals are generated by the *cutoff control generator* using the global clock signal. They all have the same period as the global clock and are carefully tuned so that they turn on the power cutoff MOSFETs only during the switching window of that group within each clock cycle. For a global clock period of 1 GHz with a 50% duty cycle, the waveforms in Figure 3 show the relationship of the global clock and one pair of cutoff control signals, which control a group whose switching window is (60 ps, 180 ps).

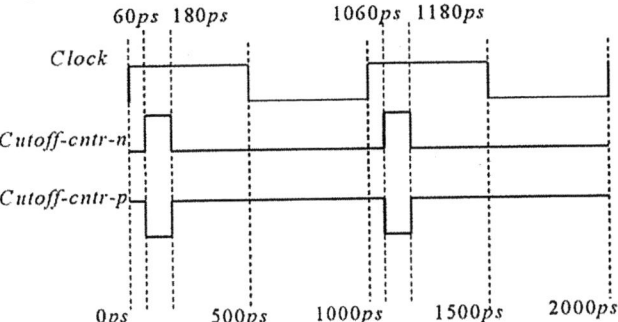

Figure 3: The Clock and One Pair of Cutoff Control Signals

3.2 Six Steps to Implement DPCT

There are several problems in implementing DPCT. First, the switching window widths of many gates are almost as big as the clock period. So, the possible power-off time of many gates is ∼ 0, which gives little leakage power savings by applying DPCT. Second, there may be thousands of switching windows within each circuit. It is very clumsy and expensive to add so many cutoff control signals and cutoff control MOSFETs in a circuit. So, we propose a six-step approach to implement DPCT, to optimize leakage power saving with minimal extra cost.

1st Step: Calculate the Minimal Switching Window of Each Gate by Static Timing Analysis

The *switching window* and *minimal switching window* of a gate are based on the timing window method of Raja *et al.* [10].

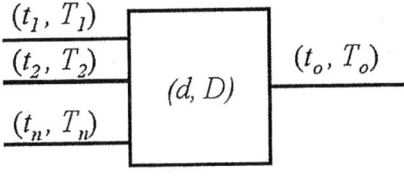

A CMOS Gate
Figure 4: Timing Window of a CMOS Gate

Switching Window Based on Traditional Timing Window. The timing window (t, T) for each circuit node is specified by two variables t and T. Here, t is the earliest time and T is the most delayed time of signal transition. Consider a CMOS gate with n inputs, best-case delay d and worst-case delay D in Figure 4. Each input has a timing window (t_i, T_i), and the output has a timing window (t_o, T_o). Then, the output node timing window is derived from the input timing windows and the gate delay:

$$T_o = max(T_i + D), \qquad t_o = min(t_i + d) \qquad (1)$$

Using Equation 1, we calculate the timing windows of all circuit nodes by a level-order traversal from *primary inputs* (PIs) to *primary outputs* (POs), if we know the delay of each gate and the timing window of each PI. The maximum T_o of all POs is the worst-case delay of the circuit. In a real circuit, the clock cycle is determined by the worst-case circuit delay. Usually, a 10% to 15% margin is added to make sure that the circuit can always finish its transitions even under the worst case.

Based on the timing window method discussed above, we define the *switching window* of a gate as $(min(t_i), T_o)$, where $min(t_i)$ is the earliest arrival time among all inputs, and T_o is the latest arrival of the signal at the output of the gate. The switching window of a gate defines a timing window from the earliest arrival time of its inputs to the end time of the latest possible transition the gate can make. A logic gate is in active mode only within its switching window during each clock cycle.

If we turn on the power of each gate only within its switching window during each clock cycle, we can save part of active leakage power without affecting its normal transition activity except for a little added delay. The percentage of active leakage power saving of a CMOS gate, PS_{gate}, is given by:

$$PS_{gate} = a \times t_{off}/T_{cycle} \qquad (2)$$

where T_{cycle} is the clock cycle period, t_{off} is the power-off time of that gate within each clock cycle, and $0 < a < 1$ is an efficiency factor proportional to t_{off}/T_{cycle}. From our experiments, we found noticeable leakage savings only when $t_{off}/T_{cycle} > 1/3$. This is because the virtual V_{DD} and GND take a little extra time to collapse after the cutoff transistors are turned off. Also, it takes some extra cost to operate the cutoff transistors.

However, the switching window of a gate will become much wider if the gate has very unbalanced minimal and maximal delays, or if its inputs come from different paths with big delay differences, or if some inputs already have wide switching windows. The wide switching windows of the gates will make the switching windows of their fanout gates even wider. The result is that the widths of many gates' switching windows are almost as big as the worst-case delay of the circuit. If we turn on the power of each gate within its switching window like this in each clock cycle, we cannot save much leakage power.

Minimal Switching Window. To solve the problem of the switching window, we propose another type of timing window, named the *minimal switching window* (MSW) of a CMOS gate, which is defined as the minimal timing window during which we can turn the gate on without affecting the logic function and worst-case circuit delay. It is represented by $((T_o - d), T_o)$, where T_o is the latest arrival of the signal at the output of the gate and d is the maximal delay of the gate. The idea is that we do not have to turn on the gate as early as the earliest input signal comes. As long as we turn on the gate d time units earlier than T_o, we can guarantee that the transition of its output happens no later than T_o. Because the worst-case delay of the circuit only depends on the latest transition time of each gate, not the earliest transition time, turning on the power of each gate only within its MSW during each clock cycle will not affect the function and the timing performance of the circuit. Of course, cutoff transistors will introduce some extra delay. But this extra delay always exists no matter which timing window we use.

The advantage of the MSW is that its width only depends on the maximal delay of the gate itself, which is usually less than $1/10$ of the worst-case circuit delay in big circuits. It does not blow up with the unbalanced delay of the gate and the delay differences of its inputs. By turning on each gate only within its MSW, we can save a large percentage of the active leakage of the circuit. Furthermore, as we only turn on the gate after all input signals are stabilized, the glitches caused by different input path delays are avoided. This leads to dynamic power savings.

To calculate each gate's MSW, we calculate each gate delay. Then static timing analysis calculates T_o of each gate. Finally, we apply $((T_o - d), T_o)$ to get the MSW of each gate. We use 70 nm CMOS Berkeley Predictive Models, BSIM3v3 models, for simulation. We model each CMOS gate as an RC network, as Wei *et al.* [6] do. The load capacitance C is calculated using the parameters and equations defined in the BSIM3v3 model manual. A look-up table based on SPECTRE^{TM} analog simulation is used to get the equivalent R of the n-tree or p-tree of a CMOS gate based on the gate type, the number of fanins, the number of fanouts, and the transistor sizes to compute the equivalent on-resistance. The delay calculation results from static timing analysis were verified on various benchmark circuits to be within 10% error compared with the results of SPECTRE^{TM} analog simulation. The delay calculation, static timing analysis, and MSW calculation are implemented as C programs.

To allow for this 10% delay estimation error and ensure that signals make full swings to logic 1 or 0, we doubled the MSW width to $((T_o - d) - 0.5 \times d, T_o + 0.5 \times d)$. We experimented with timing windows that were 1.0, 1.5, and 2.0 times the MSW width. The 1.0 figure gave output logic errors, but the 1.5 and 2.0 values worked correctly on all benchmarks, so we used 2.0 to provide a margin for process variations. This also gives some overlap of the power-on time between each gate with its fanin and fanout gates. This allows some early transitions to happen, which can reduce the potential delay cost of DPCT. The MSW width is usually less than $1/10$ of the clock period, so doubling it has little effect on the active leakage power savings. In the future, we will statistically derive the relationship between the timing window tolerance and the MSW width.

2nd Step: Greedy Partitioning by Dynamic Programming

Usually, different gates have different MSWs and there may be hundreds, even thousands, of different MSWs within a circuit. The number of MSWs within each ISCAS '85 benchmark circuit is shown in Table 1. Each MSW will need a pair of power cutoff MOSFETs and a pair of power cutoff control signals. Adding so many power cutoff control groups to a circuit will be quite expensive because of the extra cost of cutoff MOSFETs and generating the power cutoff control signals. Actually, some groups could be combined to reduce the cost with little effect on leakage power saving. The switching window of a combined group is the union of the MSWs of all gates within that group. For example, the switching window for a group combining n MSWs (ts_i, Te_i) will be $(min(ts_i), max(Te_i))$, $i = 1, ..., n$.

To optimize the leakage power saving and the extra cost, we propose an greedy partitioning algorithm based on dynamic programming. The function we want to optimize is given by:

$$OPT = pa \times PS - pb \times COST \qquad (3)$$

where PS is the estimated total active leakage power saving percentage under the current partitioning scheme, and $COST$ is the estimated indication of the total area and speed cost under the current partitioning scheme. Here, pa and pb are the weights of PS and $COST$, respectively. Adjusting the relative values of pa and pb allows us to choose whether we want to optimize for more power savings or for less cost.

The $COST$ is an indication of the area and delay cost of a partitioning scheme that we use for the greedy partitioning algorithm. It is proportional to the number of groups and the switching window width

Table 1: Greedy Partitioning Results on *ISCAS '85* Benchmarks

Circuit	Number of Gates	Number of Levels	Worst Case Delay (ps)	# of Groups		Average # Gates per Group		Estimated Active Leakage Saving (%)		Estimated Cost (%)	
				Before	After	Before	After	Before	After	Before	After
				Greedy Partitioning		Greedy Partitioning		Greedy Partitioning		Greedy Partitioning	
c432	160	18	982	41	13	3.9	12.3	95.4	83.2	20.4	11.1
c499	202	12	855	13	10	15.5	20.2	80.8	79.6	12.4	10.7
c880	383	25	819	210	15	1.8	25.5	88.1	74.7	86.4	14.2
c1355	546	25	830	28	23	19.5	23.7	89.4	88.6	13.2	11.5
c1908	880	41	1024	367	17	2.4	51.8	91.4	78.7	107.6	14.2
c2670	1193	33	1467	431	23	2.8	51.9	92.1	83.6	98.2	13.2
c3540	1669	48	1647	747	17	2.2	98.2	92.0	79.2	169.4	12.9
c5315	2307	50	1515	778	14	3.0	164.8	91.1	75.7	209.9	14.2
c6288	2416	125	4547	868	40	2.8	60.4	94.9	89.9	97.4	12.2
c7552	3512	44	1258	1366	12	2.6	292.3	92.2	80.1	442.6	15.5
Average	1105.7	35.1	1245.3	484.9	18.4	5.7	80.8	90.7	81.3	125.8	12.9

of each group. We define *COST* as:

$$COST = \sum_{k=1}^{N_{groups}} pcost \times Width_k / T_{cycle} \qquad (4)$$

where N_{groups} is the total number of groups under the current partitioning scheme, $Width_k$ is the width of the switching window of group k under the current partitioning scheme, and $pcost$ is the overall cost per group per unit time of switching window. We set $pcost = 0.1$ in our experiments to obtain the best match between the prediction from the optimizer and the results from the analog simulator. The simplified *COST* we defined here is not the real area cost or delay cost of DPCT, but it gives equally good results as the exact cost function, with much less computation.

To simplify the leakage power calculation, we assume that each gate consumes equal amounts of leakage. Although this is quite rough, it is good enough for our greedy partitioning algorithm. The total leakage saving of a partitioning scheme is:

$$PS = \sum_{k=1}^{N_{gates}} a \times (T_{cycle} - Width_k) / T_{cycle} \qquad (5)$$

where N_{gates} is the total number of gates within this circuit, $Width_k$ is the width of the switching window of the group where the gate k belongs under the current partitioning scheme, T_{cycle} is the clock period, and a is a parameter for estimation of leakage power saving. We calculate a by comparing the estimated active leakage power savings with the simulation results from NanoSim^{TM}. Based on our experiments, $a = 0.73$ is a good empirical value to match the analog simulation results.

The following is the greedy partitioning algorithm flow:

1. *Round all original MSWs into integer picosecond units.*

2. *Sort all MSWs into increasing order by start time.*

3. *Calculate the OPT of each individual group with the original MSW and record it in a table.*

4. *Calculate the OPT of each group that combines two consecutive MSWs. Compare it with the sum of the two individual OPTs. Record the larger value and the corresponding optimal grouping in the table as the optimal result.*

5. *Calculate the OPT of each group that combines three consecutive MSWs. Compare it with the OPT of all other possible combinations of the three MSWs. Record the largest value and the corresponding optimal grouping as the optimal result in the table.*

6. *Keep increasing the number of combined groups one by one until all MSWs are combined. Then we get the near optimal OPT and partitions for the entire circuit.*

Table 1 shows the number of groups, the number of gates per group, the estimated active leakage power saving, and the related cost before and after greedy partitioning. It also shows the number of gates and levels in each circuit. By our greedy partitioning, the average number of groups in a circuit reduces from 484.9 to 18.4 and the average number of gates per group increases from 5.7 to 80.8, while the corresponding average cost reduces from 125.8% to 12.9%. At the same time, the average active leakage power saving only changes from 90.7% to 81.3% after greedy partitioning. Thus, our greedy partitioning reduces the average cost greatly with little effect on power savings.

3rd Step: Insert Cutoff MOSFETs

After greedy circuit partitioning, a *p*MOS and an *n*MOS transistor are inserted into each group to control V_{DD} and GND signals of the gates of that group. To minimize the extra delay caused by the cutoff MOS-FETs, they are sized appropriately. The delay improvement becomes marginal beyond the size of 10× for the power cutoff transistor [9]. Also, not all gates switch at the same time within each group. In our experiments, all transistors in the original circuit are minimal size. The widths of the cutoff control MOSFETs are set to:

$$W = pw \times (10 \times L_{min}) \times n \qquad (6)$$

where L_{min} is the minimum feature size in a given process, which is 70 *nm* here; n is the number of gates within the group controlled by this cutoff MOSFET; and $0 < pw < 1$ is the maximal percentage of gates switching at the same time within this group, which is related to the PI signal activities and the circuit's architecture. We set the signal activities of all PIs to 0.5. We experimentally found the following empirical equations to set pw, which gives less than 6% delay penalty with less than 15% average chip area cost.

$$pw = 0.02 \quad if \quad n > 100, \qquad pw = 0.08 \quad if \quad 10 < n \le 50$$
$$pw = 0.06 \quad if \quad 50 < n \le 100, \quad pw = 1/n \quad if \quad n \le 10$$

4th Step: Generate Cutoff Control Signals

Cutoff control signals are used to control the power-on/off of a group based on the switching window of that group. One pair of cutoff control signals are required for each group, one to control the cutoff *n*MOSFET and the other to control the cutoff *p*MOSFET. All cutoff control signals have the same period as the global clock signal. Suppose that the clock period is 1 *GHz* with 50% duty cycle, and the *minimum switching window* (MSW) of a group (after greedy partitioning) is (60 *ps*, 180 *ps*). Figure 3 shows the waveforms of the clock and the two cutoff control signals for this group, *cutoff-cntr-n* to control the cutoff nMOSFET, and *cutoff-cntr-p* to control the cutoff pMOSFET.

We use clock stretchers [11] to generate the power cutoff control signals for each group. An example clock stretcher used to generate the cutoff control signals in Figure 3 is in Figure 5. It has three inverters and a *NAND* gate. The signal *cutoff-cntr-n* must rise at time *offset* from the rising *clk* edge, and remain high for time *width*, so that its partition is powered at the correct time, relative to *clk*, so that the wavefront of signals passes through it using minimal energy. Variable Δ_i indicates the logic gate's incremental output delay in the clock stretcher from the rising clock edge. We size the inverters and *NAND* gate in the clock stretcher so that their delays satisfy these conditions:

$$width = \text{MSW width} = (180-60)\,ps = 120\,ps \quad (7)$$
$$= \Delta_1 + \Delta_2 + \Delta_3 = t_{1f} + (t_{2r} - t_{2f}) + (t_{3f} - t_{3r})$$
$$offset = t_{2f} + t_{3r} = 60\,ps \quad (8)$$

where t_{ir} (t_{if}) is the rising (falling) delay of gate i. For gate 1, an *INVERTER*, $t_{1f} = 120\,ps$. For gate 2, a *NAND* gate, $t_{2r} = 30\,ps$ is the best case rising delay and $t_{2f} = 30\,ps$ is the worst case falling delay. For gate 3, $t_{3f} = t_{3r} = 30\,ps$. Figure 3 shows these control signals. As *cutoff-cntr-p* is just $\overline{\text{cutoff-cntr-n}}$, the maximal delay of gate 4, an inverter, is designed to be very small so that *cutoff-cntr-p* is delayed less than 10% of the width of the timing window, which is 12 *ps*.

We previously found 10% error in the static timing analysis compared with the analog simulator delay. We arbitrarily doubled the MSW for each gate, to make our method very insensitive to circuit delay variations due to various process corners. So, this allows up to 40% error in the rising and falling edge timings of cutoff control signals, so $delayerror = (10\% + 40\%) \times 2 = 100\%$. This greatly reduces the design complexity of the clock strechers. Analog simulation is used to verify the results to make sure that the cutoff control signals match our timing specifications. High V_{th} transistors should be used for all transistors in the clock stretchers to reduce their leakage power. Future work will improve on this by relating the error in the MSW width to actual clock stretcher design parameters and to process vari-

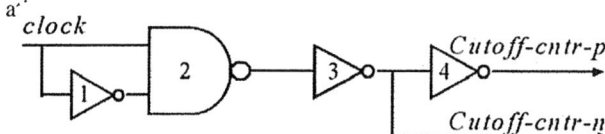

Figure 5: Clock Stretcher for Generating Cutoff Control Signals

5th Step: Add Latches to POs to Capture the Data

With traditional power cutoff, data can get lost during the long sleep period due to the collapsed virtual V_{DD} and virtual *GND* signals. With DPCT, however, the power of each gate is only turned off for a short time within each clock cycle. Also each gate shares some power-on time with its fanout gates. So, data at each intermediate gate can be passed to its fanout gates correctly before it collapses. To capture the data on POs, we add a latch to each PO. The signal on each PO is stored in the latch right before we turn off the power of the gate that drives that PO. We use the power cutoff control signals of that gate to control the corresponding latch. In a real circuit where each PO is usually followed by a flip-flop, these latches can be removed.

6th Step: Verify the DPCT Circuit Using Analog Simulation

Finally, the circuit with DPCT is simulated using Cadence SPECTRE^{TM} and compared with simulations of the original circuit without DPCT for the same test vectors. All POs are checked one by one to make sure that the circuit with DPCT functions correctly. Our DPCT method was verified and proved to be working correctly on all ISCAS '85 benchmarks.

3.3 Power Savings of DPCT

DPCT is mainly targeted for reducing active leakage power. However, it can also be used to reduce standby leakage power and dynamic power. By turning on each gate only within a small part of the entire clock cycle, DPCT significantly reduces *active leakage power*. When the circuit is in standby mode, we can save *standby leakage power* by turning off the power connections of all groups. By turning on the power of a gate only within its switching window, the gate can make transitions only when all of its inputs are ready. This automatically balances the delay differences between the inputs of each gate. Therefore, glitches, which are unnecessary transitions of the output due to different delays on inputs, are automatically eliminated. This results in *dynamic power savings*. In our DPCT method, we double the width of MSWs and combine the MSWs of some gates to reduce the extra cost of DPCT. A logic gate will have an output glitch if the path delays for an input transition from a PI to different inputs of the gate differ by an amount greater than the gate inertial delay. Combining MSWs of multiple gates, therefore, introduces glitches. But overall, circuits with DPCT have many fewer glitches compared with unmodified circuits, which may result in significant dynamic power savings.

4. EXPERIMENTAL RESULTS

We tested DPCT on the ISCAS '85 benchmarks in a 70 *nm* CMOS process modeled by Berkeley Predictive Models. For each benchmark circuit, the circuit without DPCT and the one with DPCT are running at the same frequency using the same test vectors. Random test vectors with 0.5 activities are used for all of the PIs. The clock period of the test vectors for each benchmark is chosen to be an integer about 10% larger than the worst-case circuit delay. V_{DD} is set to 1.0 *V*. The temperature is set to 90°C to reflect the real chip temperature when the circuit is active. Single low V_{th} MOSFETs are used, where the V_{th} voltages are 0.16 *V* and −0.19 *V* for nMOSFETs and pMOSFETs, respectively. All circuits are simulated using Synopsys Nanosim^{TM} to get their detailed power profile. The results are shown in Table 2.

Power Savings by DPCT. DPCT saves up to 90% active leakage power, up to 54% dynamic power, and up to 72% total power. The average active leakage saving is 84.4%, the average dynamic power saving is 9.7%, and the average overall power saving is 40.1%. The power savings of DPCT on bigger circuits are more significant than those on smaller circuits. As operating cutoff transistors introduce extra dynamic power, the dynamic power saving will be negative if the dynamic power saved by reducing glitches is smaller than the extra cost. That is why the dynamic power savings are small or negative on relatively small circuits, but quite significant on larger circuits such as c6288, where glitches are much more significant than in any other benchmark.

When the circuit is in standby mode, we can save standby leakage power by turning off the power to all groups. Our experimental results on ISCAS '85 benchmark circuits show more than 99% average standby leakage power savings.

Delay and Area Cost of DPCT. There are two costs of DPCT, delay and chip area. Just as with other power cutoff techniques, DPCT introduces about 6% delay. To minimize the delay, the power cutoff MOSFETs usually are more than 10 times larger than other transistors. Clock stretchers, used to generate cutoff control signals, also add extra chip area. These altogether introduce 15% area overhead, on average. Table 2 gives the area overhead of DPCT on ISCAS '85 benchmarks.

Table 2: Power Savings and Area Cost of DPCT on *ISCAS '85* Benchmarks

Circuit	Clock Frequency (Hz)	Total Power			Active Leakage Power			Dynamic Power			Area Cost (%)
		Without DPCT (μW)	With DPCT (μW)	Savings (%)	Without DPCT (μW)	With DPCT (μW)	Savings (%)	Without DPCT (μW)	With DPCT (μW)	Savings (%)	
c432	1G	75.06	50.44	32.8	35.76	6.93	80.6	39.30	43.51	−10.7	29.1
c499	1G	179.39	111.93	37.6	100.05	21.73	78.3	79.34	90.20	−13.7	12.1
c880	1G	140.72	114.13	18.9	65.09	10.81	83.4	75.63	103.31	−36.6	20.2
c1355	1G	209.83	151.51	27.3	101.39	15.93	84.3	108.44	135.51	−24.9	23.3
c1908	800M	345.59	242.75	29.8	141.27	22.98	83.7	204.32	219.76	−7.6	16.3
c2670	625M	495.85	275.57	44.4	240.80	29.27	87.8	255.05	246.30	3.4	13.7
c3540	500M	508.20	273.83	46.1	310.90	42.10	86.5	197.30	231.73	−17.5	9.2
c5315	625M	1064.60	625.57	41.2	509.00	88.64	82.6	555.60	536.93	3.4	6.0
c6288	200M	837.42	237.85	71.6	453.85	59.94	86.8	383.58	177.91	53.6	13.7
c7552	625M	1600.42	793.69	50.4	725.21	72.95	89.9	875.20	720.74	17.7	5.2
Average		545.71	287.73	40.1	268.33	37.13	84.4	277.38	250.59	9.7	14.9

5. CONCLUSIONS AND FUTURE WORK

We presented a novel low power design technique called DPCT that can reduce active leakage, standby leakage, and dynamic power by applying the dynamic power cutoff technique to a circuit. A six-step approach applies DPCT to a circuit automatically. The power savings and implementation costs of DPCT are discussed. Experimental results on ISCAS '85 benchmarks modeled by 70 *nm* Berkeley Predictive Models show great savings in active leakage, standby leakage, and dynamic power.

As DPCT is new, some of its effects are still under investigation, e.g., its effect on the noise margin, the power grid design, and the layout. Also, as the process variation is worsening, worst-case static timing analysis is too pessimistic. In the future, we will use statistical static timing analysis in DPCT to get more realistic design timings that account for process variations.

6. REFERENCES

[1] Y. Cao, T. Sato, M. Orshansky, D. Sylvester, and C. Hu, "New Paradigm of Predictive MOSFET and Interconnect Modeling for Early Circuit Design," in *Proc. of the Custom Integrated Circuits Conf.*, pp. 201–204, June 2000.

[2] M. Johnson, D. Somasekhar, and K. Roy, "Leakage Control with Efficient Use of Transistor Stacks in Single Threshold CMOS," in *Proc. of the Design Auto. Conf.*, pp. 442–445, June 1999.

[3] J. Halter and F. Najm, "A Gate-Level Leakage Power Reduction Method for Ultra Low Power CMOS Circuits," in *Proc. of the Custom Integrated Circuits Conf.*, pp. 475–478, 1997.

[4] H. Kawaguchi, K. Nose, and T. Sakurai, "A Super Cut-off CMOS (SCCMOS) Scheme for 0.5-V Supply Voltage with Picoampere Stand-By Current," *IEEE J. of Solid State Circuits*, vol. 35, pp. 1498–1501, Oct. 2000.

[5] Z. Chen, C. Diaz, J. Plummer, M. Cao, and W. Greene, "0.18 *μm* Dual V_t MOSFET Process and Energy-Delay Measurement," in *Proc. of the 1996 Int'l. Electron Devices Meeting*, pp. 851 – 854, Dec. 1996.

[6] L. Wei, Z. Chen, M. Johnson, K. Roy, and V. De, "Design and Optimization of Low Voltage High Performance Dual Threshold CMOS Circuits," in *Proc. of the Design Auto. Conf.*, pp. 489–494, June 1998.

[7] M. Takahashi *et al.*, "A 60-mw MPEG4 Video CODEC Using Clustered Voltage Scaling with Variable Supply-voltage Scheme," *IEEE J. of Solid-State Circuits*, vol. 33, pp. 1772–1780, Nov. 1998.

[8] T. D. Burd, T. A. Pering, A. J. Stratakos, and R. W. Brodersen, "A Dynamic Voltage Scaled Microprocessor System," *IEEE J. of Solid-State Circuits*, vol. 35, pp. 1571–1580, Nov. 2000.

[9] S. Bhunia, N. Banerjee, Q. Chen, H. Mahmoodi, and K. Roy, "A Novel Synthesis Approach for Active Leakage Power Reduction Using Dynamic Supply Gating," in *Proc. of the Design Auto. Conf.*, pp. 479–484, June 2005.

[10] T. Raja, V. Agrawal, and M. Bushnell, "CMOS Circuit Design for Minimum Dynamic Power and Highest Speed," in *Proc. of the 17th Int'l. Conf. on VLSI Design*, pp. 1035 – 1040, Jan. 2004.

[11] N. H. Weste and D. Harris, *CMOS VLSI Design: A Circuits and Systems Perspective*. Boston: Pearson Education/Addison-Wesley, 2005.

Analysis and Modeling of Subthreshold Leakage of RT-Components under PTV and State Variation

Domenik Helms[1], Günter Ehmen[1], and Wolfgang Nebel[2]
[1]OFFIS Research Institute, [2]University of Oldenburg
D - 26121 Oldenburg, Germany
helms@offis.de, ehmen@offis.de, nebel@offis.de

ABSTRACT

In this work we present a SPICE-based RTL subthreshold-leakage model analyzing components built in $70nm$ technology [1]. We present a separation approach regarding inter- and intra-die threshold variations, temperature, supply-voltage, and state dependence. The body-effect and differences between NMOS and PMOS introduce a leakage state dependence of one order of magnitude[2, 3]. We show that the leakage of RT-components still shows state dependencies between 20% and 80%. A leakage model not regarding the state can never be more accurate than this. The proposed state aware model has an average error of 6.7% for the RT-components analyzed.

Categories and Subject Descriptors:
B.8.2: Performance Analysis and Design Aids.

General Terms:
Design.

Keywords:
Leakage, Process Variation, State Dependence, Modeling.

1. INTRODUCTION

In recent years, a leakage paper motivation would have read like: *leakage will become the most important source of power consumption*. Today, leakage is the most important contributor to a system's power consumption and within the last 3 years there was an incredible amount of scientific approach in this area.

The physics of leakage are well understood and can be estimated with a sufficient accuracy at transistor level if the device geometry as the physical condition of the transistor is exactly known [4, 5].

Process variations randomly and unpredictably affecting the geometry are identified being the major hurdle of accurate leakage and performance estimation. Results of ongoing research have to be implemented to the EDA tools [3, 6, 7].

A huge amount of anti-leakage techniques exist which can be separated into 3 classes:

- Always applicable techniques, which if available will be used for each design i.E. well engineering [4] or high-k gate oxide [8].

- Leakage-performance tradeoff techniques [9], enabling the choice between fast and low leaking devices. The usefulness of this tradeoff depends on the design.

- Power management techniques offering a high performance and a low leakage mode in which the component is either slow or dysfunctional as power gating or adaptive body biasing - both reviewed in [10]. Both techniques are implemented on lowest levels of abstraction but have to be controlled system wide.

Both, the tradeoff and the power management techniques have to be evaluated at high level.

There are several approaches abstracting from accurate BSIM models [5] to much faster gate level models. But as the impact of tradeoffs and power management has to be evaluated at system level, gate level based models are still too complex for system level tools which usually do not even generate gate level details. Thus we developed two alternative RTL-leakage macromodels: The simulation based bottom-up model, abstracting from transistor level to RTL is described in [11]. The top-down characterization based model analytically describing the leakage of RT components is presented in this work.

Existing dynamic power, area and delay models for RT components have typical estimation errors in the order of 10%. To make leakage estimation accuracy comparable, our leakage models will have to regard all known parameters influencing leakage [8].

In the bottom-up model, the dynamic parameters temperature, supply voltage, and body voltage, and the variation parameters of channel length, oxide thickness, and channel doping are explicit input parameters. In the analytical model, only supply voltage, temperature and component state directly enter the model. Bulk voltage, deviating due to ABB indirectly enters our model by modifying the effective threshold voltage. Modeling a huge number of transistors in one model, random process variations (intra-die variations) only enter due to the non-linearity of the $I_{leak}(V_{th})$ relation which is captured inside the models as presented in [3, 7, 12]. Finally, the inter-die variations resulting in systematic (not statistic) deviations of the threshold voltage can be regarded by sequentially recomputing the models resulting in a probability density function of the leakage power distribution.

Permission to make digital or hard copies of all or part of this work for personal or classroom use is granted without fee provided that copies are not made or distributed for profit or commercial advantage and that copies bear this notice and the full citation on the first page. To copy otherwise, to republish, to post on servers or to redistribute to lists, requires prior specific permission and/or a fee.
ISLPED'06, October 4–6, 2006, Tegernsee, Germany.
Copyright 2006 ACM 1-59593-462-6/06/0010 ...$5.00.

In this paper we will thus address the emerging leakage problem by presenting an accurate leakage macro-model enabling leakage aware RT-synthesis. The problem of modeling PTV variations is discussed in several publications as shown in Section 2. After presenting the simulation environment in Section 3, the separation approach itself is only briefly described in Section 4. In Section 5, the major contribution of this work, the state dependent leakage model is developed and evaluation results are presented in Section 6.

2. RELATED WORK

Chen et al. [13] presented a subthreshold leakage estimation methodology for computing lower and upper bounds by regarding the stacking effect. [14] then presented a first useful separation approach, dividing leakage current into transistor count N, supply V_{DD}, leakage per device I_{dev} and a constant k as

$$I_{leak} = k \cdot N \cdot V_{DD} \cdot I_{dev}, \qquad (1)$$

where k gives the average leakage current per device at nominal voltage. In our approach, we first introduce data dependency to the parameter k, then we show that k accurately separates from the other parameters for the Berkeley Predictive Technology Model (BPTM) $70nm$ technology, and finally we introduce some corrections, as leakage does not perfectly separate into these 4 parts.

[2] analyses the leakage distribution for the 1 and 2 input gates reporting a substantial state dependency. There are many approaches at this level enabling accurate leakage current prediction if all relevant parameters are exactly known per device [5]. But some parameters are not accurately predictable on the lower levels. The effect of parameter variations on leakage was analytically investigated in [15], where [16] regards the distribution of these variations.

From a system level view, the dominant leakage parameters [8] described in Section 1 can be identified and predicted. In [17], a flow for modeling the thermal dependence of the leakage was presented, and in [18, 19], thermally dependent leakage estimation was combined with a chip-wide temperature prediction, thus also regarding the electro-thermal back-coupling introduced by the subthreshold current's thermal dependence.

[12] analyses the impact of threshold voltage variations in order to handle intra-die process variations. In [6] a methodology is presented, estimating the probability density function of the leakage current due to process parameter variation. In [3] parameter sensitivity analysis is introduced, enabling estimation of the effect of a variation on the average leakage. In [7] this sensitivity analysis regarding intra- and inter-die process variations is extended.

In order to combine the different parameters, an iterative approach is presented by [20], accurately modeling dynamic power and leakage power by regarding the interaction between temperature, supply voltage, and power consumption. They introduce a thermal system model handling the electro-thermal coupling as well as a supply-grid model handling the electro-electro coupling introduced by the finite capacitance of the supply system. The most complete high level leakage model was presented by [21], regarding all PTV variations, thus all parameters except for the state.

Best to our knowledge, the only other approach enabling PTV and state dependent RTL leakage analysis is our alternative approach, presented in [11]. As can be seen in the evaluation section, the strengths of the analytical model are easy model characterization and very fast model evaluation. The advantages of [11] are higher modeling accuracy and more direct model parameters.

3. DESCRIPTION OF THE SIMULATION ENVIRONMENT

Since there is no reliable higher level leakage estimation tool available, this work is based on the Berkeley SPICE simulator including the BSIM transistor model. The recent version BSIM4.40 is able to model various leakage effects like subthreshold current, gate tunneling and junction leakage including the Drain Induced Barrier Lowering (DIBL) effect. Unfortunately, in our experiments the recent transistor model failed to converge for several circuits as soon as they exceeded a size of 100-1000 transistors. The older BSIM3.52 version does not show these convergency problems, so we decided to use this version, even though it is not capable of estimating other leakage effects than subthreshold current. This limits the application of our model to technologies where subthreshold leakage is the dominating source of leakage. These are technologies down to $70nm$ structure size and smaller technologies having applied high-k gate dielectrics.

The MOSFET model was characterized using the BPTM card for the 70nm NMOS and PMOS [1]. The model-cards were created using the reference values for channel length, oxide thickness, threshold voltage and drain-source resistance.

For model characterization and evaluation, we need to have SPICE simulations of RT components, having each transistor properly configured. Hence, we synthesize each RT component using a commercial technology. Then we replace each gate of the component by a cloned SPICE version having the same behavior as the respective commercial gate. In Section 3.1 we present, how to clone a commercial technology to SPICE, and in Section 3.2 we describe the flow creating RTL SPICE.

3.1 Creation of a Generic SPICE Library

A commercial technology offers several hundred gates in different driving strengths. In order to limit the cloning effort, we prune the relevant gates of the commercial technology. As for instance, an AND gate has to be constructed using a NAND gate and an inverter, we pruned all logic gates which can be constructed from other cells resulting in the same transistor level description. The difference between a native AND gate and a constructed one is that the interconnect of the native one is optimized on layout level. But as long as the interconnect seems not to show relevant leakage effects, this pruning is valid for our work and reduces the cloning to 21 native gates in $4 - 10$ driving strengths.

As described in the next section, the gates of the SPICE library have to replace the gates of the commercial technology in a gate level description of an RT component. To make a replacement valid, our gates must have the same timing and power behavior. As accurate power figures for the commercial technology are not available, we can just ensure the same timing behavior. Thus, we create timing-equivalent RT-components for our model validation.

The only degree of freedom left in the BPTM regarding the timing is the transistor width and the sequence of serial

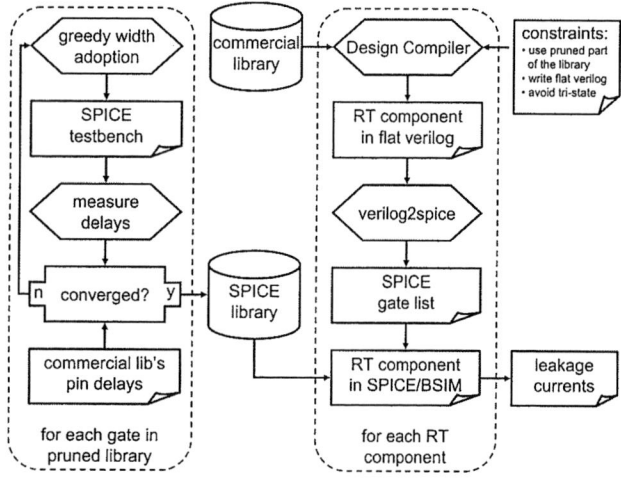

Figure 1: Adopting width of the 70nm BPTM technology transistors to generate gates having the same timing behavior. Right: RT components in Verilog can be simulated in SPICE using the SPICE library and the Verilog netlist.

transistors. Using a setup script interacting with SPICE and adopting transistor width of each transistor, we construct a SPICE sub-circuit for every pruned gate of the commercial library having the same delay behavior and the minimal total size (rf. Fig. 1 left side).

3.2 Synthesis of RT-Components

As presented on the right side of Fig. 1, we use Synopsys Design Compiler to synthesize the components. A script limits the synthesis to the pruned gates, forces non-hierarchical output as it eases up the conversion to SPICE, and avoids tri-state logic as this eases up SPICE conversion. We obtain a Verilog gate-netlist, which is automatically converted to SPICE, but now instantiating our SPICE library instead of the commercial technology's gates. We end up with professionally designed RT-components having the same timing behavior as the commercial technology and giving realistic leakage estimates.

4. PARAMETER SEPARATION

In this work we regard the impact of process, temperature and voltage variations (PTV-variations) together with state dependencies on leakage. This section will describe a separation approach splitting the influences of process variations, temperature, and supply voltage.

4.1 Variation of the Threshold Voltage

Since we will only regard sub-threshold currents, we reduce the process variations to the variation of the threshold voltage. We combine the separation approach [14] with a statistical process variation model [3]. For a transistor that is off, the subthreshold leakage results as

$$I_{sub} = kV_T^2 W/L \cdot \exp\left(V_{th}/nV_T\right) \qquad (2)$$

where $n = 1 + C_{dm}/C_{ox}$ is the subthreshold slope resulting from the ratio of the capacitances of the depletion layer and

the oxide. $V_T = k_B T/e$ is proportional to the absolute temperature T. Assuming that, due to process variations, the threshold voltage is Gaussian distributed

$$p\left(V_{th}\right) = \left(\sigma_V \sqrt{2\pi}\right)^{-1} \exp\left(-\frac{\left(V_{th} - \mu_V\right)^2}{2\sigma_V^2}\right), \qquad (3)$$

the average leakage due to the threshold variation then results as the expectation value $E(x) = \sum_i p(x_i)x_i$

$$\mu_I = \int_{-\infty}^{\infty} dV_{th}\, p\left(V_{th}\right) I\left(V_{th}\right)$$

$$= \frac{kV_T^2 W}{L\sigma_V \sqrt{2\pi}} \cdot \int_{-\infty}^{\infty} dV_{th}\, e^{\left(-\frac{\left(V_{th}-\mu_V\right)^2}{2\sigma_V^2} + \frac{V_{th}}{nV_T}\right)}$$

$$= \frac{kV_T^2 W}{L\sigma_V \sqrt{2\pi}} \cdot e^{\frac{\mu_V}{nV_T} + \frac{\sigma_V^2}{2n^2 V_T^2}} \cdot \int_{-\infty}^{\infty} dV_{th}\, e^{-\frac{\left(V_{th}-\mu_V+\sigma_V^2/(nV_T)\right)^2}{2\sigma_V^2}}$$

$$= kV_T^2 W/L \cdot e^{\frac{\mu_V}{nV_T} + \frac{\sigma_V^2}{2n^2 V_T^2}} \cdot \int_{-\infty}^{\infty} dV_{th}\, p\left(V_{th} + \frac{\sigma_V^2}{nV_T}\right)$$

$$= I_{sub}\left(\mu_V\right) \cdot f_P\left(\sigma_V, T\right) \cdot 1. \qquad (4)$$

Due to the nonlinear relation between temperature, threshold voltage and leakage, the threshold voltage can not be completely separated, but remains coupled with the temperature as

$$f_P\left(\sigma_V, T\right) = e^{\left(\sigma_V/nV_T\right)^2/2}. \qquad (5)$$

4.2 Separation of the Supply Voltage

In first order, the supply voltage influences the subthreshold current linearly, but due to the drain induced barrier lowering effect (DIBL), the threshold voltage depends on the supply voltage as

$$\Delta V_{th}\left(V_{DD}\right) = -\frac{V_{DD}}{2\cosh\left(L/l_c\right) - 2} \qquad (6)$$

with l_c being a technology constant. Thus, the supply voltage also influences the correction term (5), preventing an easy separation approach. We circumvent this by introducing a second order Taylor approximation at nominal values building an effective supply voltage function f_V. Using the fact, that $e^{\alpha(x+\delta)} \approx e^{\alpha x} \cdot (1 + \alpha\delta)$ for small $|\delta|$, the supply voltage dependence can be approximated as

$$f_V\left(V, T\right) = 1 + \alpha_V \cdot \frac{V - V^*}{T} + \beta_V \cdot \left(\frac{V - V^*}{T}\right)^2$$

$$\alpha_V = \frac{T^*}{V^*}\frac{\partial I}{\partial V}\bigg|_{T^*, V_{th}^*, V^*}, \quad \beta_V = \frac{T^{*2}}{2V^*}\frac{\partial^2 I}{\partial V^2}\bigg|_{T^*, V_{th}^*, V^*} \qquad (7)$$

The fitting parameters α_V and β_V can be easily characterized having SPICE simulation results. For the 70nm BPTM, they resulted as $\alpha_V \approx 188.5 KAV^{-2}$ and $\beta_V \approx 8460 K^2 A^2 V^{-3}$ (rf. Fig. 2).

4.3 Analysis of the Nonlinear Thermal Dependence

As can be seen in Table 1, the subthreshold leakage significantly depends on temperature, doubling with each $19K$

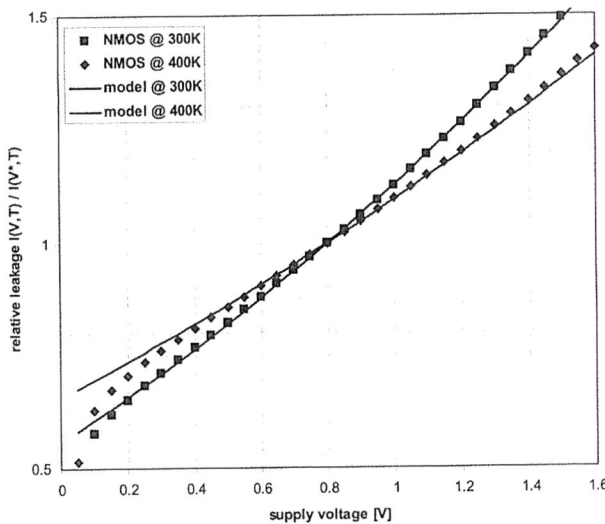

Figure 2: Evaluation of the voltage separation function $f_V(V, T)$. Within typical limits of supply voltage ($V_{DD} > 0.4V$) and temperature ($300K$ - $400K$), the separation error is below 1%

T [K]	I_{NMOS} [$nA/\mu m$]	I_{PMOS} [$nA/\mu m$]	error$_N$ [%]	error$_P$ [%]
300	2.23	0.12	0.0	-13.3
320	5.59	0.35	1.6	-6.1
340	12.5	0.89	2.1	0.0
360	25.7	2.05	1.9	5.0
380	48.8	4.34	0.8	9.2
400	86.6	8.56	-0.8	12.7

Table 1: Accuracy of the thermal leakage model for NMOS and PMOS devices. Negative errors mean an underestimation by the model.

(NMOS) and $16K$ (PMOS) of temperature increase. As we already handled the temperature-variation correlation in Section 4.1 and the temperature-voltage interaction in Section 4.2, the remainder of the thermal dependence of the subthreshold leakage strictly follows the analytical expression of Equation 2. The last two columns of Table 1 summarize the error, when trying to model subthreshold current of equation 2 using a characterized value of $n = 1.598$, thus approximating

$$f_T(T) = I_0 \cdot e^{V_{th}/1.598V_T}. \tag{8}$$

The subthreshold leakage's thermal dependence of an NMOS transistor can be approximated with less than 2% error between $300K$ and $400K$. The PMOS modeling is a little worse than this having a maximum error of 13% in this range. But as the subthreshold leakage for PMOS is approximately one order of magnitude smaller, the expected overall error for large circuits remains below 2%.

4.4 Combination of the Separated Models

The total leakage power of a component having N transistors can be modeled as

$$I_{sub} = I_0 \cdot k_{data} \cdot N \cdot f_P(\sigma_V, T) \cdot f_V(V_{DD}, T) \cdot f_T(T) \tag{9}$$

Component	N	I_{leak} [nA]	min. [%]	max. [%]	std. [%]
AddCla4	364	37.31	30.8	27.4	10.7
AddCla8	610	62.49	24.8	18.5	7.0
AddRpl4	330	34.50	35.2	29.6	14.2
AddRpl8	656	76.63	33.6	22.7	7.8
DecRpl4	54	8.18	31.9	56.5	22.1
DecRpl8	214	29.06	53.0	75.6	25.6
IncRpl4	56	8.48	62.6	36.8	28.9
IncRpl8	156	20.63	39.4	22.9	9.5
MultCsa4	758	73.09	17.5	21.1	6.9
MultWall4	1034	110.14	29.9	32.5	14.7
Mux2_4	50	13.06	65.1	86.7	51.1
Mux2_8	98	26.38	65.8	83.0	51.6
Mux3_4	104	10.26	71.0	68.9	40.2
Mux3_8	208	20.82	70.8	65.6	39.8
SubCla4	370	43.70	35.7	46.3	15.2
SubCla8	696	77.14	34.1	36.0	9.8
SubRpl4	274	32.92	36.3	39.2	13.9
SubRpl8	802	80.85	25.5	38.5	8.1

Table 2: Number of transistors N, average subthreshold leakage, percentage of deviation for minimum and maximum state and relative standard deviation for RT components up to 1000 transistors.

where the remainder k_{data} models the state dependency and will be discussed in the following section.

5. MODELING STATE DEPENDENCY

In this section, we will analyze the importance of state dependent leakage modeling and develop a k_{data} model.

5.1 Analysis of the State Dependence

Here, we perform a pure SPICE based minimum, maximum, and average leakage current analysis to evaluate the influence, the state has on the leakage of a huge RT structure. In 5.2, we then develop a model describing this state dependence.

Determining minimum and maximum leakage current is np-hard [22]. Hence, we limited our analysis to components with up to 17 inputs (2 times 8bit data plus 1 control) except for the multiplexers where minimum and maximum can easily be determined due to their symmetry. In addition we limited the analysis to components with less than 1000 transistors as simulation time in SPICE was reasonable then. Even with this limits, the total simulation time (for evaluation) was 6 weeks on a 8xPentium4 System running at $3GHz$, as a single BSIM evaluation needs $\approx 2.5ms$ on our system and we had to perform $1.5 \cdot 10^9$ evaluations.[1]

In Table 2, we summarize our results: On average, the minimum and maximum leakage are 39.5% and 41.9% away from the mean leakage. The average standard deviation was 19.2%, but relative standard deviation is sinking with the number of transistors. Monte Carlo simulations of a 3490 transistor 8 bit Wallace tree multiplier show a relative standard deviation of 4.3%.

---

[1]This high simulation effort was needed for detailed analysis of the state dependence. The final model is characterizable in a few seconds.

5.2 Development of a State Model

The state dependency of leakage is the least important one of all parameters introduced in Equation 9, but with 19.2% on average, it still needs to be considered.

We develop our model, starting with an RT-level soft-macro modeling the leakage state dependence by simplifying the transistor level equation for fixed temperature and supply, without variance and limiting the body effect to an effective width[2]:

$$I_{\text{sub}}(\text{data}) = \sum_{i=0}^{N-1} s_i \cdot (w_i^p I_p) + (1 - s_i) \cdot (w_i^n I_n), \quad (10)$$

where w_i^p and w_i^n are the effective widths (in μm) of the ith PMOS and NMOS device, s_i is the logic value and I_p and I_n are PMOS and NMOS leakage for an inverter with $1\mu m$ width. In order to reduce the complexity we replace all w_i^p and w_i^n by an average $\bar{w}_p$ and $\bar{w}_n$:

$$I_{\text{sub}} = N\bar{w}_p I_p \cdot \sum_{i=0}^{N-1} s_i + N\bar{w}_n I_n \cdot \sum_{i=0}^{N-1} (1 - s_i)$$

$$I_{\text{sub}} = N\bar{w}_n I_n + \sum_{i=0}^{N-1} s_i \cdot (\bar{w}_p I_p - \bar{w}_n I_n)$$

$$= N \left(\bar{w}_n I_n + p_{all}(\bar{w}_p I_p - \bar{w}_n I_n) \right) = N \left(\alpha^* + \beta^* p_{all} \right) \quad (11)$$

where p_{all} is the signal probability of all internal nodes. Assuming that the internal nodes are correlated to the inputs and can be linearly approximated from the input signal probability p_{input}, k_{data} can be modeled using

$$k_{\text{data}} = \alpha_D + \beta_D \cdot p_{input}, \quad (12)$$

where the data dependence parameters α_D and β_D can be fit by characterization. The number of transistors N resulted as a factor to the leakage, thus it can be separated as suggested in Equation 9. As there is no proof, that state dependence also separates from other parameters, we experimentally determined the separability. The term

$$I(\text{data}, T_A, V_A)/I(\text{data}, T_B, V_B)$$

varies less than 0.1% when randomly selecting T_A, T_B, V_A and V_B and evaluating the term for each possible input state.

6. EVALUATION

The set of characterization and the set of evaluation data both are obtained as follows: We create the RT Components shown in Table 2 and perform SPICE simulations measuring the leakage of these components while varying all parameters: Synthesizing the circuit, we randomly assign a threshold voltage variation between $0 \leq \sigma_{V_{th}} \leq 50mV$ with a mean value of $0.2V$ and $-0.22V$. We supply this circuit with a voltage randomly chosen from $0.6V \leq V_{DD} \leq 1.0V$ and set the ambient temperature to a random value from $300K \leq T \leq 400K$. Then we measure the subthreshold leakage of the component for different input states.

---

[2]In [11], we showed, that leakage reduction due to the body-effect can be accurately modeled with a state-dependent effective transistor width.

Compon.	w/o	w/D	Compon.	w/o	w/D
AddCla4	10.9	3.83	AddCla8	7.35	6.63
AddRpl4	14.4	7.40	AddRpl8	8.12	4.49
DecRpl4	22.2	8.27	DecRpl8	25.7	11.4
IncRpl4	29.0	18.1	IncRpl8	11.0	9.78
MultCsa4	7.26	2.98	MultWall4	14.9	6.02
SubCla4	15.4	4.60	SubCla8	10.1	3.00
SubRpl4	14.1	4.36	SubRpl8	8.43	2.86
			Average	14.2	6.69

Table 3: Evaluation of the relative standard deviation error of the parameter separation approach with no data awareness (w/o) and the data aware approach (w/D).

Component	#gates	std. [%]	max. [%]
AddCla4	83	4.11	12.9
AddRpl8	151	2.90	11.6
MultWall8	767	2.33	11.1

Table 4: Evaluation of the bottom-up model: Relative standard deviation and maximum estimation error of three $45nm$ RT components against statistical SPICE simulation.

6.1 Analytical Model Evaluation

We compute the relative standard deviation for three models. To evaluate how accurate the parameter-data separation is, we fix the input state and just vary all the other parameters. We characterize the separation function parameters using random sampling points. The resulting model error for all components is between 2% and 3%. For the next model, called 'w/o' in Table 3, we vary the data and all parameters. We do not characterize the data dependency but assume that $k_{\text{data}} = 1$. As the model error of the parameter separation alone is very low, the resulting error of the 'w/o' model is dominated by the data-variance of the component as presented in Table 2. The average error of the 'w/o' model is 14.1%. The final model, called 'w/D' in Table 3 uses all parts of Equation 9. Except for the smallest components (incrementer and decrementer), the standard deviation is always below 8%, the average error for all components is at 6.7%.

In comparison to the SPICE simulation time, the model evaluation time of each of our models is negligible as is is just evaluation of analytical functions. In order to characterize the models, we took 3000 sampling points needing one hour computation time for all components together.

In this evaluation, we neglect the effect of electro-thermal and electro-electro back-coupling, described in Section 2. But as back-coupling aware modeling means iteratively evaluating a model working at fixed parameters and recomputing these parameters afterwards, our model is well applicable to back-coupling approaches.

6.2 Comparison to the Bottom-Up Model

For comparison to our alternative model, Table 4 shows the evaluation results of the bottom-up model. As this model can estimate the effect of various parameter variations, the error of a single prediction is computed in comparison to a Monte Carlo SPICE simulation averaging over 1000 settings for intra-die variation of the parameters.

As both models show comparable accuracy, it strongly depends on the application, which is the better one. As the bottom-up model is characterized at transistor level, it avoids RTL spice simulation and directly models transistor geometry parameters (channel length and width, oxide thickness, channel doping) of which the variability is easier to predict. But for characterization, several thousand spice simulations are needed and have to be saved inside the model. The advantages of the top-down approach, presented here, are a) the few number of characterization data needed characterizing the few model parameters and b) the fast model evaluation.

7. CONCLUSION

On gate level, accurate leakage modeling is available, but as design for low leakage has to start at system level, models at higher abstraction are required. RTL models exist, but because state dependence has the smallest influence to leakage, it was not regarded so far on these higher levels. We showed that leakage can be modeled with 6.7% accuracy many orders of magnitude faster than SPICE could do. Without regarding state dependency, the accuracy would only be 14.1%. Even though, there are some further improvements possible:

We are currently working at an extension to the top-down model enabling estimation for all 3 important sources of leakage, subthreshold currents, gate leakage and junction leakage due to BTBT. Also modeling gate and junction leakage the variance of other parameters as oxide thickness and channel doping are regarded, too. Further work may also analyze the effect of temperature and supply voltage variations inside a component, as we assumed both being constant for all transistors inside one component.

Finally, the model proposed here has to be embedded into a high level power estimation tool determining the input parameters temperature and supply voltage by iteratively computing local energy consumption and the resulting thermal increase and voltage drop. Having RT level leakage estimation is the key enabler for leakage aware RT synthesis.

8. REFERENCES

[1] *Berkeley Predictive Technology Model:*
www-device.eecs.berkeley.edu/~ptm/

[2] S Mukhopadhyay, A Raychowdhury, K Roy: Accurate Estimation of Total Leakage Current in Scaled CMOS Logic Circuits Based on Compact Current Modeling. *DAC*, 2003.

[3] A Shrivastava, R Bai, D Blaauw, D Sylvester: Modeling and Analysis of Leakage Power Considering Within-Die Process Variations. *ISLPED*, 2002.

[4] K Roy, S Mukhopadhyay, H Mahmoodi-Meimand: Leakage Current Mechanisms and Leakage Reduction Techniques in Deep-Submicrometer CMOS Circuits. *Proc. of the IEEE Vol.91 No.2*, 2003.

[5] C Hu: BSIM Model for Circuit Design Using Advanced Technologies. *2001 Symposium on VLSI Circuit Digest of Technical Papers*, 2001.

[6] H Chang, S Sapatnekar: Full-Chip Analysis of Leakage Power under Process Variations, Including Spatial Correlations. *DAC*, 2005.

[7] R Rao, A Shrivastava, D Blaauw, D Sylvester: Statistical Estimation of Leakage Current Considering Inter- and Intra-Die Process Variation. *ISLPED*, 2003.

[8] D Helms, E Schmidt, W Nebel: Leakage in CMOS circuits - An Introduction. *PATMOS*, 2004.

[9] D Lee D Blaauw, D Sylvester: Static Leakage Reduction Through Simultaneous V_T/T_{ox} and State Assignment. *IEEE Tran on CAD of ICs and Systems Vol24 No7*, 2004

[10] S Narendra: Challenges and Design Choices in Nanoscale CMOS. *ACM Journal on Emerging Techn. in Computing Systems Vol1 No1*, 2005.

[11] D Helms, M Hoyer, W Nebel: Accurate PTV, State, and ABB Aware RTL Blackbox Modeling of Subthreshold, Gate, and PN-Junction Leakage. *PATMOS*, 2006.

[12] S Narendra, V De, S Borkar, D Antoniadis, A Chandrakasan: Full-Chip Sub-threshold Leakage Power Prediction Model for sum-$0.18\mu m$ CMOS. *ISLPED*, 2002.

[13] Z Chen, M Johnson, L Wei, K Roy: Estimation of Standby Leakage Power in CMOS Circuits Considering Accurate Modeling of Transistor Stacks. *ISLPED*, 1998.

[14] J A Butts, G S Sohi: A static power model for architects. *Proc. Int'l Symp. on Microarchitecture*, 2000.

[15] S Mukhopadhyay, K Roy: Modeling and Estimation of Total Leakage Current in Nano-scaled CMOS Devices Considering the Effect of Parameter Variation. *ISLPED*, 2003.

[16] S Bhardwaj, S Vrudhula: Leakage Minimization of Nano-Scale Circuits in the Presence of Systematic and Random Variables. *DAC*, 2005.

[17] Y Zhang, D Parikh, M Stan, K Sankaranarayanan, K Skadron: HotLeakage: A Temperature-Aware Model of Subthreshold and Gate Leakage for Architects. *Tech Report CS-2003-05, Univ. of Virginia Dept. of Computer Science*, 2003.

[18] W Liao, F Li, L He: Microarchitecture Level Power and Thermal Simulation Considering Temperature Dependent Leakage Model. *ISLPED*, 2003.

[19] K Banerjee, S-C Lin, A Keshavarzi, S Narendra, V De: A Self-Consistent Junction Temperature Estimation Methodology for Nanometer Scale ICs with Implications for Performance and Thermal Management. *IEEE*, 2003.

[20] H Su, F Liu, A Devgan, E Acar, S Nassif: Full chip leakage estimation considering power supply and temperature variations. *ISLPED*, 2003.

[21] S Borkar, T Karnik, S Narendra, J Tschanz, A Keshavari, V De: Parameter Variations and Impact on Microarchitecture. *DAC*, 2003.

[22] F A Aloul, S Hassoun, K A Sakallah, D Blaauw: Robust SAT-Based Search Algorithm for Leakage Power Reduction. *PATMOS*, 2002.

Power Optimization In A Repeater-Inserted Interconnect Via Geometric Programming

W. T. Cheung
Department of Electrical and
Electronic Engineering
The University of Hong Kong
Pokfulam Road, Hong Kong

wtcheung@eee.hku.hk

N. Wong
Department of Electrical and
Electronic Engineering
The University of Hong Kong
Pokfulam Road, Hong Kong

nwong@eee.hku.hk

ABSTRACT

We present an innovative geometric programming (GP) approach for minimizing the power dissipation of an interconnect with repeater insertion, subject to delay, bandwidth and area constraints. Repeater sizes and segment lengths are globally optimized in various technology nodes with respect to International Technology Roadmap for Semiconductors (ITRS). Relative power dissipation due to different power components is analyzed. We show that, on average, the power dissipation per unit length can be reduced by over 30% when the timing constraint is relaxed by 5%. The optimum number of repeaters is always given as an integer in our design flow. The relationships between power dissipation and respective design constraints are easily visualized in tradeoff curves. Additional design criteria, such as reliability of the interconnect delay against process variations, are easily incorporated into the optimization.

Categories and Subject Descriptors

B7.1 [Integrated Circuits]

General Terms

Design

Keywords

Interconnect, Optimization, Repeater, Power, Geometric Programming

1. INTRODUCTION

With the ever-shrinking feature size, delay caused by interconnects (on-chip wires) is ultimately becoming the bottleneck in the performance of a VLSI circuit. Low-power and minimal-delay interconnect design is critical in many system-on-a-chip (SoC) implementations. Repeater insertion is an effective means to significantly reduce the over-

Permission to make digital or hard copies of all or part of this work for personal or classroom use is granted without fee provided that copies are not made or distributed for profit or commercial advantage and that copies bear this notice and the full citation on the first page. To copy otherwise, to republish, to post on servers or to redistribute to lists, requires prior specific permission and/or a fee.
ISLPED'06, October 4–6, 2006, Tegernsee, Germany.
Copyright 2006 ACM 1-59593-462-6/06/0010 ...$5.00.

all delay in an interconnect. The leakage power dissipation in the repeaters, however, can account for more than 20% of total power consumption in nanometer technologies. Increased power dissipation raises the cost of packaging, and even causes reliability problems and failure of a chip [1]. Subsequently, the tradeoff between power dissipation and wire delay must be optimized when doing repeater insertion.

Previous work on optimizing repeater insertion, subject to delay and power constraints, can be found in [2,3]. However, in these approaches, power dissipation components due to leakage power and short-circuit power are not simultaneously considered. Analytical models for power consumption in interconnect with repeater insertion are discussed in [4], where the optimization of repeater sizes and (interconnect) segment lengths is formed into a set of nonlinear equations and solved numerically using Newton-Raphson method. Computational load is increased when delay variations are included into the optimization. Moreover, additional design constraints cannot be readily incorporated into the context. A new metric is defined in [5] to facilitate the tradeoff between power and delay, but analyses are based on theoretically optimal repeater sizes and segment lengths and do not capture the effect of power. Although geometric programming (GP) has been employed for optimizing the interconnect delay [6], the analysis does not consider repeater insertion and is only limited to the minimization of the Elmore delay.

The main contribution of this paper is the introduction of a flexible GP framework that allows accurate modeling of power dissipation in an interconnect with repeater insertion and its subsequent optimization. The approach exhibits high computational efficiency and permits easy addition of design constraints such as area, bandwidth and delay. The interaction between power dissipation and various design variables, such as repeater sizes and segment lengths, is easily visualized through tradeoff curves. Impacts on design variables with *delay fraction* (see Section 4) applied on timing constraints are analyzed. Robust analysis, incorporating the idea of process variations, is conveniently performed. The proposed design flow also automatically decides an integral number of repeater stages through error checks.

2. GEOMETRIC PROGRAMMING

Convex programming refers to the minimization of a convex objective function over a convex set. Geometric programming or GP [7,8] is a special type of convex program-

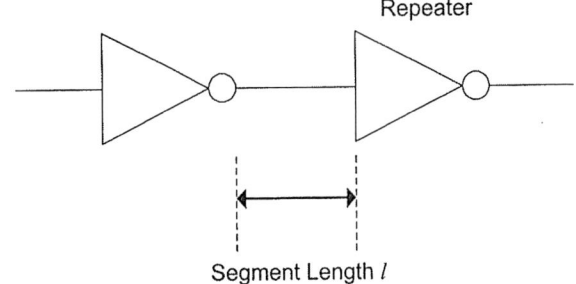

Figure 1: Interconnect with segment length l between two repeaters.

ming consisting of positive polynomials called *posynomials*. Let $x = (x_1, \cdots, x_n) \in \Re^n$ be a vector of real, positive variables. A function f is a posynomial of x if it has the form

$$f(x_1, x_2, \cdots, x_n) = \sum_{k=1}^{t} c_k x_1^{\alpha_{1k}} x_2^{\alpha_{2k}} \cdots x_n^{\alpha_{nk}}$$

where c_k's are positive real numbers and α_{ik}'s, called the exponents, are real. When $t = 1$, f is called a *monomial*. A GP optimization problem takes the form

$$\begin{aligned}
\text{minimize} \quad & f_o(x) \\
\text{subject to} \quad & f_i(x) \le 1 \quad, \quad i = 1, 2, \cdots, m, \\
& g_j(x) = 1 \quad, \quad j = 1, 2, \cdots, p, \\
& x_k > 0 \quad, \quad k = 1, 2, \cdots, n.
\end{aligned}$$

where f_o is the objective (posynomial) function, f_i are posynomials and g_j are monomials. The central property of a GP problem is that its global optimum, if exists, can be solved with great efficiency using standard interior-point algorithms [8–10] (e.g., in seconds on a PC for a problem with 10^3 variables and 10^4 constraints). When the problem cannot be solved due to mutually inconsistent constraints, an infeasibility certificate is given. Also, an initial starting point is irrelevant to the final solution and is generally unnecessary to a GP solver.

3. INTERCONNECT MODEL

3.1 Delay

First, we analyze the delay in an interconnect. The notations and expressions follow closely from those in [4, 11]. The interconnect model with repeaters inserted is shown in Fig. 1. We define the segment length l as the length of interconnect between two neighboring repeaters. When the repeater size is s, which is assumed uniform throughout the interconnect, and the segment length is l, the time constant τ of that segment is

$$\tau = r_s(c_0 + c_p) + \frac{r_s}{s}cl + rlsc_0 + \frac{1}{2}rcl^2 \qquad (1)$$

where r is the interconnect resistance per unit length, given by $r = \frac{\rho}{WT}$, with ρ being the metal resistivity, and W and T being the wire width and thickness which are assumed uniform throughout the interconnect. We use c to denote the interconnect capacitance per unit length, which can be

broken down into $c = c_a + c_b W$, with c_a being the fringing capacitance and c_b the parallel plate capacitance formed by the top and bottom metal layers. Also, c_0, c_p and r_s are defined as the input capacitance, output parasitic capacitance and output resistance, respectively, in a minimum-size repeater.

The delay in the line segment of length l is $\tau \ln 2$, which is the time difference between the input and output waveforms crossing half of their full-swing values. Thus the delay per unit length is given by

$$\frac{\tau}{l} \ln 2 = \left(\frac{r_s(c_0 + c_p)}{l} + \frac{r_s}{s}c + rsc_0 + \frac{1}{2}rcl \right) \ln 2 \qquad (2)$$

where (2) is minimized when

$$\left(\frac{\tau}{l} \right)_{opt} = 2\sqrt{r_s c_0 rc} \left(1 + \sqrt{\frac{1}{2}(1 + \frac{c_p}{c_0})} \right), \qquad (3)$$

attained at

$$l_{opt} = \sqrt{\frac{2r_s(c_0 + c_p)}{rc}} \text{ and } s_{opt} = \sqrt{\frac{r_s c}{rc_0}}.$$

3.2 Power Dissipation

CMOS inverters are used in repeater insertion. Assuming the inverters are symmetric, in which the effective output resistance is the same for both rising and falling edges. The power dissipation of a repeater consists of three components, namely, switching power (P_{sw}), short-circuit power (P_{sc}) and leakage power (P_{leak}), such that [4,12]

$$P_{repeater} = P_{sw} + P_{sc} + P_{leak}. \qquad (4)$$

The switching power, P_{sw}, is the power dissipated due to charging and discharging of the load capacitance, given by

$$P_{sw} = \alpha(s(c_p + c_0) + cl)V_{DD}^2 f \qquad (5)$$

where f is the clock frequency, V_{DD} the power supply voltage and α the switching factor which is the average fraction of repeaters switched during a clock cycle. We assume the typical value of $\alpha = 0.15$ [12].

The short-circuit power P_{sc} is the power consumed when direct current flows between power source and ground. Specifically, when input signal switches from V_{t_n} to $V_{DD} - V_{tp}$, where V_{t_n} and V_{tp} denote the threshold voltages of NMOS and PMOS, respectively, the NMOS and PMOS are momentarily both ON forming a path between the power rails. Suppose t_r is the time for the input voltage to rise from V_{t_n} to $V_{DD} - V_{t_p}$, the short-circuit power during this interval is

$$P_{sc} = \alpha t_r V_{DD} W_{n_{min}} s I_{sc} f \qquad (6)$$

where $t_r = \tau \ln 3$. $W_{n_{min}}$ is the width of an NMOS in a minimum sized inverter. Here we assume the case where the input switches from $V_{DD} - V_{tp}$ to V_{t_n} consumes the same amount of short-circuit power, and is already accounted for by α in (6). I_{sc} is approximately $65\mu A/\mu m$ across all technology nodes from SPICE simulation [4]. The value of α in (6) is assumed to be equal to that in (5).

The leakage power P_{leak} is due to subthreshold conduction of an inverter [1]. The simplified equation for it is

$$P_{leak} = V_{DD} I_{leak} = \frac{1}{2}V_{DD}(I_{off_n} + 2I_{off_p})W_{n_{min}}s \qquad (7)$$

where I_{off_n} and I_{off_p} are the subthreshold leakage current per unit transistor width in NMOS and PMOS, respectively.

Tech. node (nm)	130	90	65	45
width(nm)	335	230	145	103
thickness(nm)	670	482	319	236
ϵ_r	3.3	2.8	2.5	2.1
$c_a(fF/mm)$	70.95	58.24	65	52.08
$c_b(fF/\mu m^2)$	0.053	0.065	0.057	0.072
$c_c(fF)$	0.046	0.029	0.015	0.010
$r_s(k\Omega)$	6.23	9.04	9.6	13.2
$c_0(fF)$	1.33	1.1	1.03	0.9
$c_p(fF)$	3.32	2.04	1.22	0.6
$V_{DD}(V)$	1.1	1	0.7	0.6
$A_m(\mu m^2)$	0.248	0.125	0.061	0.034
$I_{off_n}(\mu A/\mu)$	2	3.56	20	35.5
$I_{off_p}(\mu A/\mu)$	1.34	2.38	13.4	23.83
$f(GHz)$	1.68	3.99	6.73	11.51

Table 1: Interconnect and other circuit parameters for top metal layer in different technology nodes based on ITRS2001 [14].

Also, we assume the width of a PMOS is twice that of an NMOS in an inverter. The factor $1/2$ in (7) arises from the approximation that along a long repeater-inserted interconnect, about half of the input signals are high.

Finally, consider a long interconnect of length L with n repeaters inserted, the total power dissipation is

$$P_{line} = nP_{repeater} = \frac{L}{l}P_{repeater}. \tag{8}$$

Therefore, the power dissipated in a repeater-inserted interconnect can be minimized by minimizing the product of the number of repeaters and the power consumed inside a repeater.

3.3 Area and Bandwidth

The total area of repeaters in a global interconnect with repeater insertion is proportional to the size of each repeater and inversely proportional to the segment length. Accordingly,

$$A = N\frac{L}{l}sA_m \tag{9}$$

where N is the number of global interconnects [13] and A_m is the value in Table 1. The bandwidth of an interconnect is inversely proportional to the delay of global interconnects, D, and directly proportional to the number of global interconnects, N, i.e.,

$$B \propto \frac{N}{D}. \tag{10}$$

In other words, the bandwidth is maximized when the delay is minimized.

4. OPTIMIZATION MODEL AND RESULTS

We present the GP model of a repeater-inserted interconnect based on equations in Section 3, namely,

$$\text{minimize } (P_{sw} + P_{sc} + P_{leak})/l \tag{11}$$

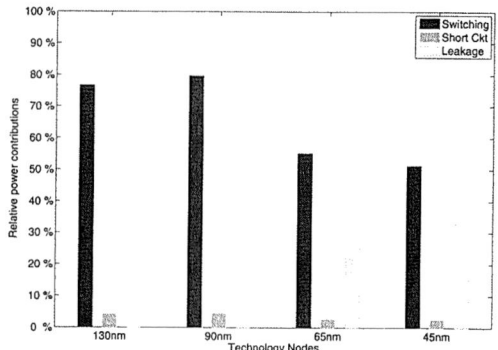

Figure 2: Relative contributions of power components with 5% delay fraction for different technology nodes.

fraction of delay	0%	5%	10%	15%	20%
s/s_{opt}	1	0.7249	0.6358	0.5757	0.5301
l/l_{opt}	1	1.3611	1.5442	1.6995	1.8410

Table 2: Ratio of repeater sizes and segment lengths to their optimal values in 65nm technology, where l_{opt} and s_{opt} are from (3).

subject to

$$
\begin{aligned}
s &\geq 1, \\
l &\leq L, \\
\alpha(s(c_p + c_o) + lc)V_{DD}^2 f &\leq P_{sw}, \\
\alpha t_r V_{DD} W_{n_{min}} s I_{sc} f &\leq P_{sc}, \\
\frac{1}{2}V_{DD}(I_{off_n} + 2I_{off_p})W_{n_{min}} s &\leq P_{leak}, \\
\frac{r_s(c_o + c_p)}{l} + \frac{r_s}{s}c + rsc_o + \frac{1}{2}rcl &\leq (1 + t_f)T_{req}.
\end{aligned}
$$

The objective function is the total power dissipated in a repeater per unit segment length, $P_{repeater}/l$. The delay requirement, T_{req}, is computed from $(\tau/l)_{opt}$ in (3). We introduce a *delay fraction* parameter, $t_f > 0$, for relaxing the upper bound in the delay constraint. The tuning of t_f then allows us to systematically investigate the tradeoff between delay and other design variables. We note in passing that additional constraints, such as area and bandwidth limits in GP-compatible forms, can be appended to the constraint list of (11) if necessary. In our experiments, (11) is solved with the freely available MOSEK [15] Matlab package under various technology nodes specified in Table 1. The computation consistently takes less than one second on a 3.2GHz PC with 512MB RAM.

Fig. 2 plots the relative power dissipation compositions with 5% delay fraction for various technology nodes. Although switching power is dominant in 130nm and 90nm technology nodes, the leakage power increases significantly when the feature size scales down. This increase of leakage power can mainly be attributed to the increasing subthreshold leakage. Nonetheless, the total power dissipated by a repeater-inserted interconnect decreases with reduced feature size due to smaller repeaters.

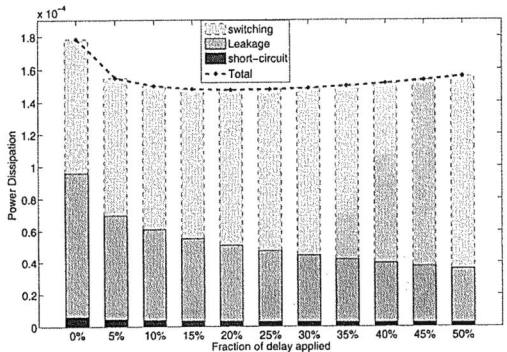

Figure 3: Optimized power dissipation in 65nm vs different delay fractions.

tech nodes (nm)	130	90	65	45
s/s_{opt}	0.6972	0.7094	0.7249	0.7398
l/l_{opt}	1.3076	1.3022	1.3636	1.3992
P/P_{opt}	0.9348	0.9771	0.8987	0.9111
$\frac{P}{l}/\frac{P_{opt}}{l_{opt}}$	0.7010	0.7319	0.6603	0.6556
$Area/Area_{opt}$	0.5332	0.5448	0.5316	0.5287

Table 3: Ratio of power per unit length to optimum values in various technology nodes with 5% delay fraction.

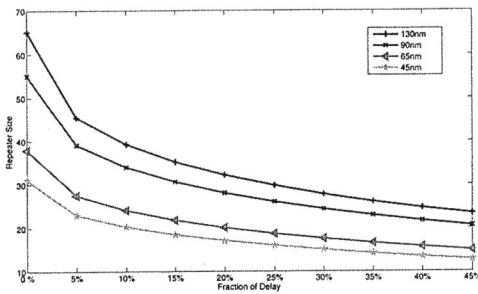

Figure 4: Tradeoff of repeater sizes against delay fractions under 130nm, 90nm, 65nm and 45nm technology nodes.

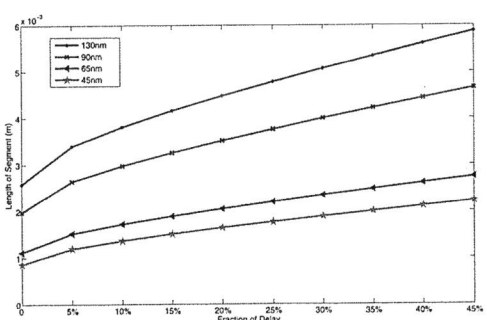

Figure 5: Tradeoff of segment lengths against delay fractions under 130nm, 90nm, 65nm and 45nm technology nodes.

Variables	Nominal	Robust
Repeater size, s	39.6092	39.9638
Segment length $(mm), l$	2.6082	2.5654
Width$(nm), W$	230	227.75
$T_{req}(ns)$	55.175	55.175

Table 4: Values obtained from nominal and robust designs in 90nm.

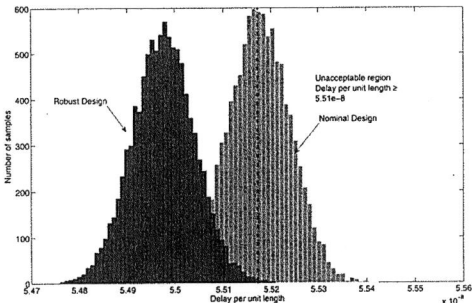

Figure 6: Histograms of delay per unit length for robust and nominal designs in 90nm.

Results of power consumption with various delay fractions in 65nm technology node are shown in Fig. 3. It is obvious that with an increasing delay fraction, the switching power gradually increases while the leakage power decreases. Generally, the total power dissipation reduces when the timing requirement is relaxed. In Tables 2 and 3, we compare the optimal repeater sizes and segment lengths from (11) with the values l_{opt} and s_{opt} in (3). As seen from Table 2, s/s_{opt} decreases while l/l_{opt} increases with an increasing delay fraction. Table 3 shows that power dissipation per unit length P/l is reduced in all nodes, with an average improvement of over 30%. By recovering the ratio s/l from s/s_{opt} and l/l_{opt} in Table 3 and substituting into (9), it can be verified that the area consumed by repeaters in various technology nodes is reduced by almost 50% with only a 5% delay fraction.

Relationships between design variables are easily observed in tradeoff curves. Figs. 3, 4 and 5 show the tradeoffs in total power dissipation, repeater sizes and segment lengths with various delay fractions. It can be observed that in all technology nodes, when the delay fraction increases, the optimal repeater size decreases while the optimal segment length increases. Also, the reduction in optimal repeater sizes is significant with only a slight increase of delay fraction from 0% to 5%.

5. ROBUST DESIGN

When fabrication technology scales below 100nm, process variations caused by manufacturing and lithography affect the circuit performance. To handle the impact of process variations, we apply the corner-based worst-case design optimization. In the analysis of process variation, we first assume that the mass of probability is concentrated in the interval $[\mu - 3\sigma, \mu + 3\sigma]$ when a variable is normally distributed with a mean μ and standard variation σ. For simplicity, we only consider W (the wire width) as a new design variable in the context of robust design, but extension to include other variables, such as the thickness T, is trivial. T_{req} is set to

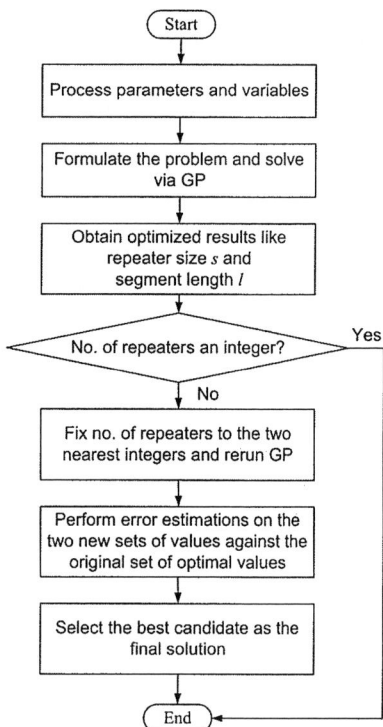

Figure 7: Overall design flow.

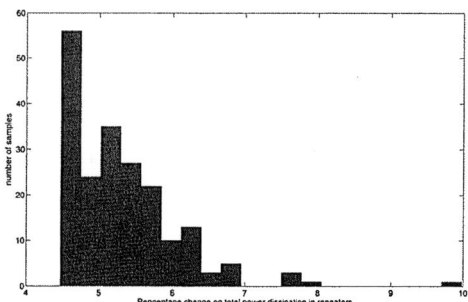

Figure 8: Histogram of percentage change in power dissipation.

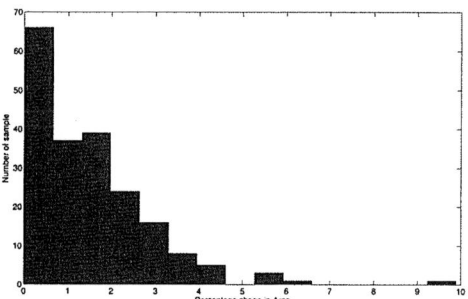

Figure 9: Histogram of percentage change in area.

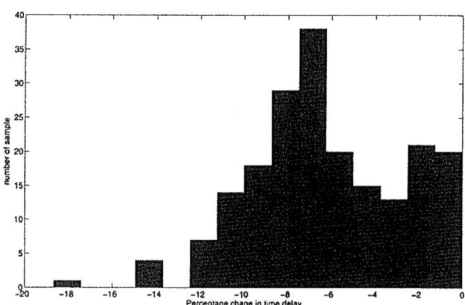

Figure 10: Histogram of percentage change in delay.

a worst-case figure 55.175ns, which is calculated from equation (3) with 5% delay fraction in 90nm technology node. Suppose W has a 95% confidence interval in normal distribution, where $W_{min} = \mu_w - \sigma_w$ and $W_{max} = \mu_w + \sigma_w$. The mean, μ_w, is the minimium wire width in a node and the standard deviation, σ_w, is 0.5% of minimium width. We formulate the time delay per unit length, t_i, as a function of s, l and W (cf. the last constraint in (11)),

$$t_i(s, l, W) \leq T_{req}, \quad \forall W \in [W_{min}, W_{max}], \qquad (12)$$

and optimize it at various design corners. Results obtained from such a robust GP are then compared with those from nominal GP, as listed in Table 4. A 10^4-point Monte-Carlo analysis is used to evaluate the variability. The histogram of delay time per unit length are shown in Fig. 6. Obviously, the overall delay from the robust design is smaller than that of the nominal one. Only a small portion of samples fall into the unacceptable region, i.e., a delay per unit length bigger than T_{req}. It can be concluded that the design using robust GP achieves higher reliability on repeater-inserted interconnect design, at the expense of a higher design cost (due to slightly bigger repeater sizes) compared with nominal design using GP.

6. DESIGN FLOW

Generally, the number of repeaters after GP optimization in Section 4 is given as a real number rather than an integer, and its conversion to an integer is necessary. The proposed optimization flow is shown in Fig. 7. The main idea in the repeater-stage quantization is to do an error check on power and a feasibility check on delay in the new designs evaluated at the two integers nearest to the (usually non-

Design Variables	s	l (mm)	P_{line} (mW)	Delay (ps)	Area (μm^2)
n_{opt}	39.058	2.640	4.495	101.3	92.48
n_{min}	41.531	2.778	4.722	106.6	93.445
Change(%)	6.333	5.240	5.045	5.240	1.038
n_{max}	38.942	2.632	4.696	101.0	92.487
Change(%)	-0.296	-0.299	4.469	-0.299	0.003

Table 5: Percentage change of design variables at nearest integral stages of repeaters when L= 50mm.

230

integral) optimal number of repeaters. Any infeasibility in the new solution is detected and the feasible design with less power dissipation is selected. Specifically, if the number of stages is a non-integer, it is rounded to two nearest integers, $n_{min} = \lceil x \rceil - 1$ and $n_{max} = \lceil x \rceil$, and the GP is rerun. The design with the smaller power consumption satisfying all other constraints is then taken as the final solution.

Figs. 8, 9 and 10 show the histograms of percentage changes in power, area and delay corresponding to the best candidate (the one with less power consumption among two) from our proposed method in 90nm technology node. 200 samples of interconnect length L with mean = 20mm and standard deviation = 5mm are used. It is observed that, compared to the optimal values in the first pass of GP, the mean increase in total power dissipation and mean deviation in total repeater area are 5.29% and 1.49%, respectively. Table 5 shows the results from another example obtained from optimization with interconnect length $L = 50$mm in 90nm technology node. It can be seen that when the number of stages is n_{min}, the solution is infeasible since the delay time is greater than the timing requirement. On the other hand, in the feasible design where n_{max} is used, power dissipation increases by about 4% with a slight increase in area.

7. CONCLUSION

We have presented a novel geometric programming (GP)-based optimization method for minimizing the power dissipation in a repeater-inserted interconnect. In the context of various technology nodes, the repeater sizes and segment lengths have been globally optimized. Their relationships to different delay fractions (timing relaxations) have been characterized by tradeoff curves. We have shown that the leakage power is insignificant in 130nm and 90nm processes. However, when technology scales down, leakage power becomes dominant. In the proposed design flow, the conversion of the number of repeaters from an optimized real number into an integer has been automated through an error check on power and a feasibility check on delay. Process variations of design variables have been examined and reliability of an interconnect, in terms delay, has been improved through a robust corner-case design procedure.

8. REFERENCES

[1] M. Mui, K. Banerjee, and A. Mehrotra, "Supply and power optimization in leakage-dominant technologies," *IEEE Trans. Computer-Aided Design*, vol. 24, no. 9, pp. 1362–1371, Sept. 2005.

[2] V. Adler and E. G. Friedman, "Repater design to reduce delay and power in resistive interconnect," *IEEE Trans. Circuits Syst. I*, vol. 45, pp. 607–616, May 1998.

[3] A. Nalamalpu and W. Burleson, "A practical approach to DSM repeater insertion: Satisfying delay constraints while minimizing area and power," in *Proc. 14th Annu. IEEE Int. ASIC/SOC Conf.*, 2001, pp. 152–156.

[4] K. Banerjee and A. Mehrotra, "A power-optimal repeater insertion methodology for global interconnects in nanometer designs," *IEEE Trans. Electron Devices*, vol. 49, no. 11, pp. 2001–2007, Nov. 2002.

[5] M. Mui, K. Banerjee, and A. Mehrotra, "A global interconnect optimization scheme for nanometer scale vlsi with implications for latency, bandwidth and power dissipation," *IEEE Trans. Electron Devices*, vol. 51, no. 2, pp. 195–203, Feb. 2004.

[6] L. Vandenberghe, S. P. Boyd, and A. E. Gamal, "Optimizing dominant time constant in rc circuits," *IEEE Trans. Computer-Aided Design*, vol. 17, pp. 110–125, Feb. 1998.

[7] R. J. Duffin, E. L. Peterson, and C. Zener, *Geometric Programming - Theory and Applications*. Wiley, 1967.

[8] S. Boyd, S. J. Kim, L. Vandenberghe, and A. Hassibi, "A tutorial on geometric programming," *to appear in Optimization and Engineering*, 2006.

[9] M. del Mar Hershenson, S. S. Mohan, S. P. Boyd, and T. H. Lee, "Optimization of inductor circuits via geometric programming," in *Design Automation Conference*, 1999, pp. 994 –998.

[10] M. Hershenson, S. P. Boyd, and T. H. Lee, "Optimal design of a CMOS op-amp via geometric programming," *IEEE Trans. Computer-Aided Design*, vol. 20, no. 1, pp. 1–21, 2001.

[11] H. B. Bakoglu, *Circuits, Interconnects and Packaging for VLSI*. Addision-Wesley, 1990.

[12] A. P. Chandrakasan and R. W. Brodersen, *Low Power Digital CMOS Design*. Norwell, MA: Kluwer, 1995.

[13] X. C. Li, J. F. Mao, H. F. Huang, and Y. Liu, "Global interconnect width and spacing optimization for latency, bandwidth and power dissipation," *IEEE Trans. Electron Devices*, vol. 52, no. 10, pp. 2272–2279, Oct. 2005.

[14] International technology roadmap for semiconductor(ITRS), semiconductor industry association, san jose, ca,2001. [Online]. Available: http://public.itrs.net/

[15] MOSEK optimization toolbox. [Online]. Available: http://www.mosek.com/

Input-specific Dynamic Power Optimization for VLSI Circuits

Fei Hu*
Intel Corporation
Folsom, CA 95630, USA
frank.hu@intel.com

Vishwani D. Agrawal
Auburn University
Auburn, AL 36849, USA
vagrawal@eng.auburn.edu

ABSTRACT

Literature proposes linear programming (LP) methods for glitch-less design of digital circuits. Considering the worst-case these methods ensure absence of glitches for any arbitrary state of primary input as well as internal signals. In this paper, we examine an unexplored aspect, i.e., glitch-free design with respect to a specific set of vectors (patterns). Introducing the logic-level concepts of glitch-generation patterns and glitch-generation probability, which are analyzable through logic simulation, we remove glitch filtering requirements from gates on which the given set of input vectors cannot produce glitches. We relax constraints of any existing LP either selectively or probabilistically. Such input-specific design from an LP model without process variation and another with process variation reduced the number of delay buffer overhead by up to 80% and 63%, respectively, while maintaining the power reduction and overall delay.

Categories and Subject Descriptors: J.6 [**Computer-Aided Engineering**]: Computer-aided design (CAD)

General Terms: Algorithms, Design

Keywords: Input specific, dynamic power optimization, glitch reduction

1. INTRODUCTION

Reduction of switching power dissipation of a circuit involves, among other things, glitch reduction. In conventional CMOS circuits, the spurious transitions at the output of a gate due to the differential delay of input paths are called *glitches* or *hazards*. Removal of such transitions can reduce the switching activity of a circuit and hence the switching power. The principal idea in glitch reduction is to find delay assignment for all gates in the circuit to reduce the differential path delays at gate inputs with respect to the inertial delays. Optimization techniques for glitch

---
*Formerly with Department of Electrical and Computer Engineering, Auburn University, Auburn, AL 36849, USA.

Permission to make digital or hard copies of all or part of this work for personal or classroom use is granted without fee provided that copies are not made or distributed for profit or commercial advantage and that copies bear this notice and the full citation on the first page. To copy otherwise, to republish, to post on servers or to redistribute to lists, requires prior specific permission and/or a fee.
ISLPED'06, October 4–6, 2006, Tegernsee, Germany.
Copyright 2006 ACM 1-59593-462-6/06/0010 ...$5.00.

reduction are the *balanced delay* [5] and *hazard filtering* [1] methods, implemented through a variety of algorithms such as *transistor sizing* [12, 14], *gate sizing* [3, 6], and *linear programming* [2, 7, 8].

A linear programming (LP) technique has the advantage that an LP solver derives a globally optimal solution in a relatively short time from the given model of the problem. Agrawal *et al.* [2] combined path balancing and hazard filtering in their LP model to determine the delay assignment for each gate. In subsequent work, their group proposed [8] an improvement reducing the complexity of the constraint set from exponential to linear in the circuit size. In another recent work, the use of a random delay model allows a robust glitch-free circuit design under given process-tolerance [7].

In all the previous LP models [2, 7, 8], the glitch optimization of the circuit is considered under arbitrary gate inputs. The LP solution ensures the absence of a glitch for any input vector sequence and for all input signal combinations at all gates. Such constraints result in an "overdesign" in the sense that glitches are virtually suppressed even for those signal states that are either impossible or only occur with very small probability. Restrictions on signal states can be due to two reasons, namely, circuit structure and functionally-relevant subsets of primary inputs. For a circuit where the total propagation delay is restricted, the conventional LP solution requires the insertion of many delay elements in non-critical paths. To reduce the additional power consumed by these elements, Raja *et al.* [9, 10, 11] have proposed new type of gates with different IO delays incorporating transmission-gates. Uppalapati *et al.* [13] used customized resistive feedthrough cells as delay elements, which consumed negligible amount of switching power. Howsoever small in number, delay elements [10, 13] add some capacitive loading that increases the per-transition dynamic power. Besides, they increase the total circuit area. Any reduction in the delay elements is therefore desirable.

In this paper, we explore a new aspect of the above problem, the application-specific circuit optimization. That is, we may only optimize the circuit for certain input sequences that will be applied to the circuit, e.g., functional vectors. Optimization of the circuit for these vector sequences ensures the low power dissipation when the circuit is in use and it can lead to a better solution because the optimization is customized to the application. In our experiments on ISCAS'85 benchmark circuits, when the input-specific optimization was considered in the previously published LP models of Raja *et al.* [8] and Hu [7], the number of required delay buffers (overhead) dropped by up to 80% and 63%, re-

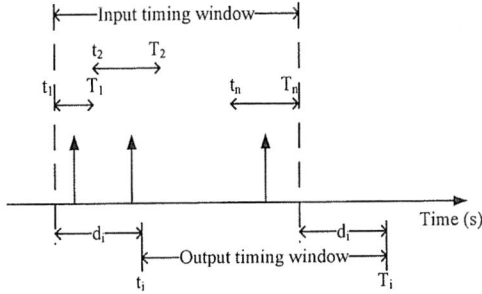

Figure 1: Illustration of timing window at gate i.

spectively, while maintaining the power reduction and overall circuit delay.

2. BACKGROUND AND MOTIVATION

Previous LP modeling [7, 8] considers the optimization of the circuit in the *worst-case*. As shown in Figure 1, a timing window of signal arrival time $[t_i, T_i]$ is propagated throughout the circuit [8], where t_i is the earliest arrival time and T_i is the latest arrival time for gate i. A constraint $d_i > T_i - t_i$ is imposed for each gate inertial delay d_i. Therefore, the LP solution ensures that the gate is free from glitches for *any* arbitrary signal transition at the inputs of the gate. However, we observe, this worst-case optimization may have introduced too much pessimism into the solution. For a circuit where the total propagation delay is restricted by design, the LP solution may require insertion of a large number of buffers on non-critical paths. As we know, the insertion of buffers is costly and their number should be kept as small as possible because it either increases the total power dissipation of the circuit (assuming conventional buffers) or the total area of the circuit (assuming resistance type of buffers [13]).

In general, the worst-case optimization could mean overdesign. We may not need the circuit to be optimized for *all* possible input sequences. On the contrary, we may only want the circuit be optimized for the set of input sequences that will actually be applied to the circuit while it is working, for example, the functional vectors. These input sequences can be a highly biased set depending on the system environment. Optimization of a circuit specific to such vector sequences ensures that the optimized circuit maintains the low power dissipation under the given system environment. At the same time, we are able to achieve a better solution with reduced overhead because the optimization is more customized.

3. GLITCH GENERATION

We discuss the generation of glitches and introduce the concepts of *glitch-generation pattern* and *glitch-generation probability*.

3.1 Glitch-generation pattern

Glitches and *hazards* refer to the spurious transitions at a gate output caused by differential delays of paths arriving at its inputs. Two factors are essential for glitch generation, i.e., transitions and path delays. Our treatment here is similar to that in path delay testing where only signal transitions and not the specific delays are considered [4]. We define a *glitch-generation pattern* for a gate as the input vector pair

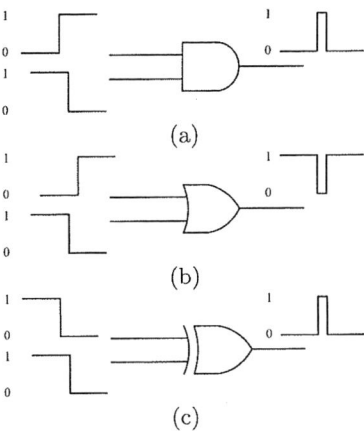

Figure 2: Glitch-generation in two-input gates.

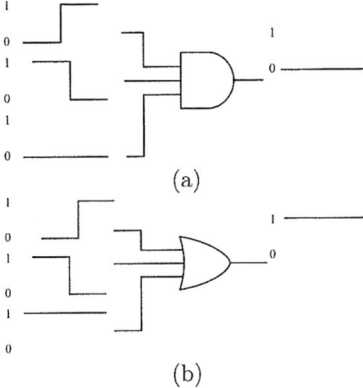

Figure 3: Glitch-suppression in multi-input gates by controlling value.

that can potentially generate a glitch at the output of the gate for some arbitrary input and inertial delays.

As shown in Figure 2, glitch-generation patterns for a two-input AND/OR gate are those vector pairs that produce two opposite transitions on different inputs. However, for a two-input XOR gate, a glitch can be potentially generated as long as both inputs have transitions.

For a gate with more than two inputs, a glitch cannot be generated if there is a steady controlling value (e.g., 0 for a AND gate) at any input of the gate. Therefore, the glitch-generation patterns for a multi-input AND gate will be those vector pairs that produce opposite transitions at any two inputs and no constant 0's at any other input. Similarly, the glitch-generation patterns for a multi-input OR gate will be those vector pairs that produce opposite transitions at any two inputs and no constant 1's at any other input. Since there is no controlling value for an XOR gate, the glitch-generation patterns for a two-input XOR gate are those vector pairs that produce transitions on both inputs. Figure 3 shows the effect of a controlling value on glitch generation.

3.2 Glitch-generation probability

We define *glitch-generation probability* P_g for a gate as the probability that a glitch-generation pattern-pair occurs at the inputs of that gate. The occurrence of a glitch means that the *steady-state* signal values during two consecutive clock periods at inputs of the gate match a glitch-generation

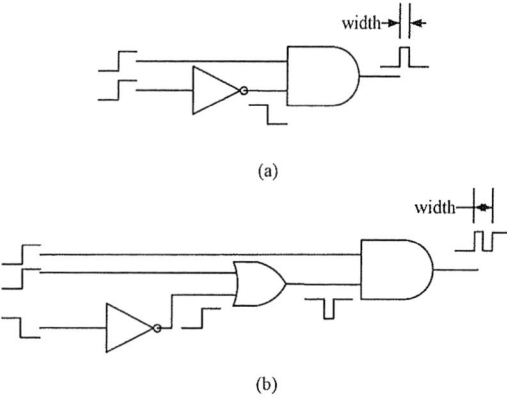

(a)

(b)

Figure 4: Hazard generation in logic circuits: (a) static hazard, (b) dynamic hazard.

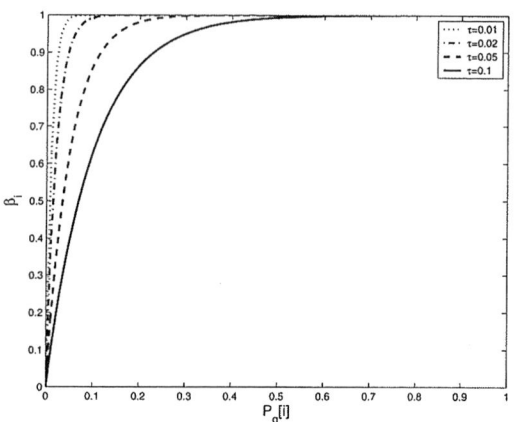

Figure 5: Function β_i for various selectivity factors τ.

pattern for that gate type. For a given set of N primary input vectors, glitch-generation probability for all gates can be obtained through zero-delay logic simulation of the circuit. Let us denote the number of times a glitch-generation pattern occurs at the input of gate i by $N_g[i]$, the glitch-generation probability for gate i, $P_g[i]$, is calculated as

$$P_g[i] = \frac{N_g[i]}{N} \qquad (1)$$

4. INPUT-SPECIFIC OPTIMIZATION

With the measure of glitch-generation probability, we can selectively relax the constraints for gates where glitches are unlikely to occur. This input-specific optimization technique is applied first to the basic LP model [8] and then to the process-variation-resistant LP model [7].

4.1 Application to the basic LP model

First, we apply the input-specific optimization to the previous basic LP model [8]. This will achieve a glitch-free circuit under the given set of input sequence. Our input-specific optimization is a "static" analysis, meaning that only probabilities (and not the signal values) of glitch generation are the basis for eliminating (relaxing) some LP constraints. As shown in Figure 4, glitches in a practical circuit can be either generated at a gate or propagated from the previous stages of the circuit. Our definition of glitch-generation probability only captures potential glitch generation and ignores possible glitch propagation from the previous stage.

Clearly, the accuracy of the glitch-generation probability to represent the chance that a glitch can be produced is strongly affected by the ratio of propagated glitches. Only when propagated glitch does not exist or has a negligible probability, can our glitch-generation probability represent the chance correctly. For the relaxed constraints, we assume that no (or a negligibly small number of) glitches are propagated from the previous stages of the circuit.

4.1.1 Selectively relaxed LP constraints

Assuming that no glitch is being propagated throughout the circuit, glitch-generation probability of a gate represents the chance that a glitch can be produced at the output of the gate if no proper path balancing or glitch filtering is done. For gates with zero glitch-generation probability, a

glitch-generation pattern will never be produced at the gate inputs by the given primary input vector sequence. It also means that a glitch will never occur no matter how path delays or gate delays change. Under this circumstance, we remove the glitch-filtering constraint for that gate from the LP.

The original glitch-filtering constraint for gate i has the form [8]:

$$d_i > T_i - t_i \qquad (2)$$

In the input-specific optimization, it is modified to

$$d_i > (T_i - t_i) \cdot \beta_i \qquad (3)$$

where $\beta_i \in \{0, 1\}$ is a constant determined by the glitch-generation probability of gate i:

$$\beta_i = \left\{ \begin{array}{ll} 0 & \text{if } P_g[i] = 0 \\ 1 & \text{if } P_g[i] > 0 \end{array} \right. \qquad (4)$$

This essentially retains the glitch-filtering constraints only for gates with non-zero glitch-generation probability. Note that such selective relaxation of constraints does not change the totally glitch-free property (i.e., no glitches are generated) in the resulting circuit because there is no need to suppress glitch propagation given that none is generated.

4.1.2 Probabilistically relaxed LP constraints

The selection of gates for glitch-elimination can be probabilistically generalized to allow even more relaxed constraints. The resulting LP solution will not guarantee that the circuit is totally glitch-free. However, it provides designers a trade-off between glitch power dissipation and cost (number of delay elements inserted) for a given critical delay requirement. We now replace the step function in Equation 4 with

$$\beta_i = 1 - e^{-P_g[i]/\tau} \qquad (5)$$

Here, β_i is an exponential function of the glitch-generation probability $P_g[i]$ with a selectivity factor τ. The function β_i with τ as parameter is illustrated in Figure 5. The adoption of an exponential function has two advantages. First, for gates where glitches are more likely to occur, the glitch-filtering constraint is strictly enforced ($\beta_i = 1$). Second, for gates where glitches are less likely to occur, the glitch-filtering constraint is relaxed accordingly. The fast rising slope of the exponential function for small $P_g[i]$ ensures that only a small number of glitches will be generated and propagated to the subsequent stages, which supports our assumption on neglecting the propagation of glitches.

By varying the selectivity factor τ, $0 \leq \tau \leq \infty$, a designer can adjust the slope of the function β_i. For a larger τ and milder slope of the function β_i, the circuit will consume relatively more power by allowing some glitches. At the same time, it will reduce the number of inserted delay elements for the same critical delay requirement. Designers can adjust the value of τ to obtain the desired solution according to their specific needs.

4.2 Application to process-variation LP model

Next, we apply the input-specific optimization to a process-variation-resistant LP model. The original LP formulation [7] considers intra-die variations of gate delays. Gate delays d_i are random variables and are assumed to have truncated normal probability distributions. A gate i has a nominal (also the mean) delay μ_{d_i} and standard deviation σ_{d_i}. All gates are assumed to have the same normalized standard deviation given by $r = \sigma_{d_i}/\mu_{d_i}$. The time window (Figure 1) at the output of a j-input gate i, $W_i = max\{T_j\} - min\{t_j\}$, is also a random variable with mean μ_{W_i} and standard deviation σ_{W_i}. The following inequality in the LP, which determines the nominal gate delays μ_{d_i}'s, ensures that gate i can produce a glitch with only a very small probability [7]:

$$\mu_{d_i} - \mu_{W_i} > 3 \cdot k(\sigma_{W_i} + r \cdot \mu_{d_i}) \cdot \alpha \qquad (6)$$

where k is a constant ($1/\sqrt{2} \leq k \leq 1.0$) whose value is taken as 0.85, and $\alpha \leq 1.0$ is an optimism factor. This glitch-filtering requirement for gate i ensures that the inertial delay to gate exceeds the timing window in spite of the process variation. We modify the Inequality 6 for all gates i as,

$$\mu_{d_i} > [\mu_{W_i} + 3 \cdot k(\sigma_{W_i} + r \cdot \mu_{d_i}) \cdot \alpha] \cdot \beta_i \qquad (7)$$

This glitch-filtering requirement on the delay of gate i is relaxed by a factor β_i. When $\beta_i = 0$, glitch-filtering constraint is altogether removed. As before, β_i is a function of $P_g[i]$ and can be chosen from Equation 4 or Equation 5.

It should be noted that constraints 6 and 7 require that the glitch-filter condition is satisfied even when the delays vary as much as three times standard deviation [7]. Such constraints are pessimistic and sometimes lead to "no solution" that meets the overall circuit delay budget. Values of $\alpha < 1.0$ reduce the pessimism to permit a solution while allowing a small number of glitches.

4.2.1 Optional tuning

Under process variation, the overall delay for an optimized circuit will not be a constant. The delays of critical paths are random variables and, therefore, the overall delay of the optimized circuit is a random variable with certain mean and variance. As illustrated below, under process variation, a solution to the input-specific optimization can lead to cases that must be avoided.

Consider the example shown in Figure 6. Under the input-specific optimization, glitch-filtering constraints for all AND and NAND gates are removed because the second PI to lower AND gate is always 0 in the specified input vectors. Delays for these AND/NAND gates are all set to the minimum value, $d_i = 1$, by the LP. The signal arrival time for the AND gate is between 20 and 40 due to the logic enclosed in the cloud. Given that the overall delay of the circuit

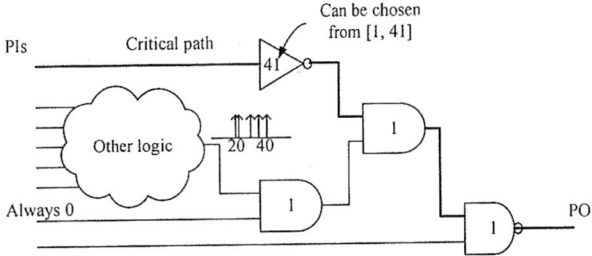

Figure 6: An undesirable solution under process variation when the input-specific optimization is applied directly. Bold lines indicate the critical path. The numbers on gates are their inertial delays.

should not exceed 43, the delay of the inverter can be chosen anywhere from a minimum value $d_i = 1$ to a maximum value of $d_i = 43 - 2 = 41$. However, in some cases, the LP solver will choose $d_i = 41$ if no constraint prevents it from doing so. This solution is undesired under process variation. The critical path PI, inverter and PO is unnecessary. This path will dominate the critical delay of the circuit under the process-variation and result in the degradation of critical delay distribution.

To avoid this undesirable solution, we include an additional term in the objective function. The original objective of the the LP model was to minimize the total buffer delays (which is a linear approximation for the number of delay buffers):

$$\text{Minimize } \Sigma_j \, d_j, \quad j \in \text{all delay buffers} \qquad (8)$$

For input-specific optimization under process-variation, this is replaced by

$$\text{Minimize } \Sigma_j \, d_j \; + \; \frac{TF}{N} \Sigma_i \, d_i, j \in \text{delay buffers}, \; i \in \text{gates} \qquad (9)$$

where the constant $TF \geq 0$ is a tuning factor, N is the total number of gates other than delay buffers.

When $TF > 0$, the tuning option is turned on. The value of TF is kept much smaller than 1.0 so that its impact on the overall optimization is minimized. However, as long as $TF > 0$, the LP solver is forced to minimize those gate delays that do not affect any constraints. With this tuning option, the gates on the dominating paths will be assigned minimum (rather than arbitrary) delays.

5. EXPERIMENTAL RESULTS

ISCAS'85 benchmark circuits were optimized by the input-specific optimization methods. Two input-specific optimization methods are illustrated. "IS-Opt1" is the input-specific optimization added to the previous basic LP model [8]. "IS-Opt2" is the input-specific optimization added to the previous process-variation-resistant LP model [7]. Results are compared to "un-optimized" circuits ("Un-opt") and "optimized" circuits from the basic LP model [8] ("Opt1") or the process-variation-resistant LP model [7] ("Opt2"). Same as in the published work [8], we use a unit-delay circuit as the un-optimized circuit, where each gate has a delay of one unit. Due to the space limitation, experimental results for the probabilistic relaxation method (Section 4.1.2) are not included.

235

Table 1: Input-specific optimization of ISCAS'85 benchmark circuits without process-variation.

Cir.	Max delay	Opt1 [8]			IS-Opt1		
		Avg. Pwr.	Cir. Delay	No. Buf.	Avg. Pwr.	Cir Delay	No. Buf.
c432	34	0.74	34	66	0.74	35	66
	68	0.74	68	58	0.74	69	41
c499	22	0.94	22	48	0.94	22	33
	33	0.94	33	0	0.95	33	0
c880	48	0.54	51	35	0.54	49	32
	120	0.54	121	30	0.54	122	24
c1355	48	0.93	48	192	0.93	48	113
	120	0.93	121	128	0.93	120	25
c1908	80	0.53	82	62	0.54	86	52
	200	0.54	203	34	0.53	204	3
c2670	64	0.74	65	34	0.74	66	30
	160	0.74	163	9	0.74	162	1
c3540	94	0.59	95	139	0.59	101	122
	235	0.59	239	78	0.59	239	73
c5315	98	0.56	100	167	0.56	104	170
	245	0.56	249	53	0.56	250	52
c6288	228	0.13	226	870	0.13	228	870
	620	0.13	620	857	0.13	620	853
c7552	86	0.52	89	91	0.52	88	84
	215	0.52	220	44	0.52	221	38

5.1 Input-specific optimization

The power dissipation and critical delay for "Opt1" and "IS-Opt1" are shown in Table 1. "IS-Opt1" adopted the selectively relaxed LP solution (Section 4.1.1). Similar to that in [8], circuits were simulated using a event-driven logic simulator with sequences of input vectors. Meaningful vectors would have been functional inputs. However, in the absence of such vectors or the functional information for these circuits, we either used test vectors or random vectors. For smaller circuits (i.e., c432 to c1355) complete gate level test vectors (with 100% stuck fault coverage) were used. For larger circuits, 50 random vectors with signal probability of 0.5 were used. Load capacitances for gates were assumed to be in proportion to the number of fanouts. Inserted delay buffers were assumed to be of resistance type and any additional power consumption by them was neglected. The average power for each circuit was normalized to the power dissipated by its un-optimized version, i.e., the un-optimized circuit has the power dissipation value of 1.

In Table 1, "Maxdelay" is the maximum specified critical delay parameter supplied to the LP. Clearly, the input-specific optimization is able to reduce the number of buffers inserted while maintaining the same performance in terms of power dissipation and critical delay. Depending on the vectors and circuits, a varying degree of improvement is achieved. In some cases the number of buffers inserted is reduced by up to 80%. Meanwhile, the power dissipation and critical delay values are the same or very close for "Opt1" and "IS-Opt1" in most cases.

5.2 Input-specific optimization under process-variation

5.2.1 Power analysis

Power dissipation and number of buffers inserted by "Opt2" and "IS-Opt2" are shown in Table 2. Under the process-variation, power dissipation of a circuit varies from sample to sample. Monte-Carlo simulation method is used where 1,000 sample cases of the optimized circuit were simulated. For each of these samples, as in [7], gate delays were in-

dependently sampled from normal distributions assuming 15% intra-die and 5% inter-die delay variation. In these experiments, "IS-Opt2" uses the selectively relaxed LP of Section 4. The tuning option of the objective Function 9 was turned on only for c1908, c3540, and c6288, where TF is chosen to be $\frac{1}{D_{max}}$. D_{max} is the maximum critical delay parameter.

In Table 2, "Nom. Pwr." represents the nominal power dissipation when no process-variation exists; "Mean Pwr." represents the mean value of the power distribution; and "Max Dev." represents the difference ratio between the maximum value of the power distribution and the power dissipation under no process-variation. "Max Dev." shows the degree of the deviation of average power from its design value due to the process-variation. All power values were normalized to the power dissipation of the un-optimized circuit.

We see that in all cases power dissipation of optimized circuits by "Opt2" and "IS-Opt2" is either the same or has only a slight difference. However, "IS-Opt2" achieves a solution with smaller number of delay buffers. Our technique, when compared to the reported input independent optimization [7] requires 40-60% fewer delay elements. The reduction of buffers is more obvious for larger D_{max} for each circuit. This is because for a smaller D_{max}, the optimization is more difficult. Removing of glitch-filtering constraint has a smaller effect on the reduction of buffers. Up to 63% reduction in the number of buffers is achieved for c2670 circuit.

Note that some of these examples use random inputs for demonstration. In an actual design, the input to combinational logic, when extracted from the system level simulation, will be further restricted. For example, the state-space of control logic may be much smaller than what is modeled by random inputs. Therefore, greater savings by this technique can be expected.

5.2.2 Delay analysis

The critical delays under process-variation are shown in Figures 7 and 8. "Nom. Delay" indicates the critical delay of the circuit under no process-variation, i.e., the nominal value of the critical path delay. We also show the maximum deviation ("Max. Dev.") of the critical delay from its intended value under the process-variation. We see that "Opt2" and "IS-Opt2" have equivalent performances in all cases. From the power dissipation results in Table 2 and these figures, we can conclude that the input-specific optimization method "IS-Opt2" achieves a better solution for a given input sequence. It maintains the same power and delay performance while reducing the overhead in terms of the number of delay buffers inserted.

6. CONCLUSION

In this paper, we have explored a new aspect of low-power optimization for VLSI circuits and proposed the input-specific optimization techniques. We consider optimizing the circuit for a given input sequence that may be specified for the circuit. We define the concept of glitch-generation probability. By observing the glitch generation probability for each gate, we can adaptively relax the glitch-filtering constraint. The experimental results show that we are able to obtain a better solution with fewer delay buffer insertions while maintaining similar power reduction and delay performance as before. Up to 80% and 63% reductions in delay buffer overheads have been achieved in our experiments.

Table 2: Power dissipations and number of delay buffers inserted for input-specific optimization of ISCAS'85 benchmark circuits under process variations.

Cir.	D_{max}	Un-opt Nom. Pwr.	Opt2 [7] Nom. Pwr.	Mean Pwr.	Max Dev. (%)	No. Buf.	IS-Opt2 Nom. Pwr.	Mean Pwr.	Max Dev. (%)	No. Buf.
c432	50	1.0	0.74	0.76	11.1	88	0.74	0.76	9.3	81
	99	1.0	0.74	0.74	3.7	106	0.74	0.74	3.3	76
c499	32	1.0	0.94	0.95	2.0	88	0.94	0.95	1.9	88
	48	1.0	0.94	0.95	1.0	129	0.94	0.95	1.8	58
c880	70	1.0	0.54	0.59	18.2	57	0.54	0.59	20.4	38
	174	1.0	0.54	0.55	8.6	62	0.54	0.56	9.0	38
c1355	70	1.0	0.93	0.98	10.2	305	0.93	1.01	13.1	253
	174	1.0	0.93	0.94	3.0	305	0.93	0.95	4.7	160
c1908	116	1.0	0.52	0.64	35.8	135	0.52	0.64	34.7	107
	290	1.0	0.52	0.58	21.4	190	0.52	0.57	18.4	104
c2670	93	1.0	0.74	0.80	13.6	249	0.73	0.79	11.3	186
	232	1.0	0.73	0.76	6.2	211	0.73	0.75	4.3	79
c3540	137	1.0	0.59	0.66	17.8	281	0.59	0.65	15.6	247
	341	1.0	0.59	0.62	10.1	311	0.59	0.61	7.4	188
c5315	143	1.0	0.55	0.63	20.8	399	0.55	0.63	21.0	389
	356	1.0	0.55	0.60	13.4	418	0.55	0.60	13.2	413
c6288	331	1.0	0.13	0.38	223.8	1121	0.13	0.38	225.2	1115
	899	1.0	0.13	0.26	125.3	1473	0.13	0.26	125.5	1243
c7552	125	1.0	0.52	0.59	18.7	481	0.52	0.58	18.1	389
	312	1.0	0.52	0.56	11.8	645	0.52	0.55	10.9	520

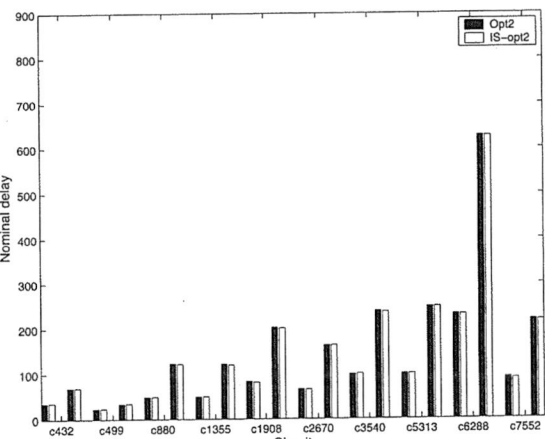

Figure 7: Nominal critical delay for optimized ISCAS'85 circuits.

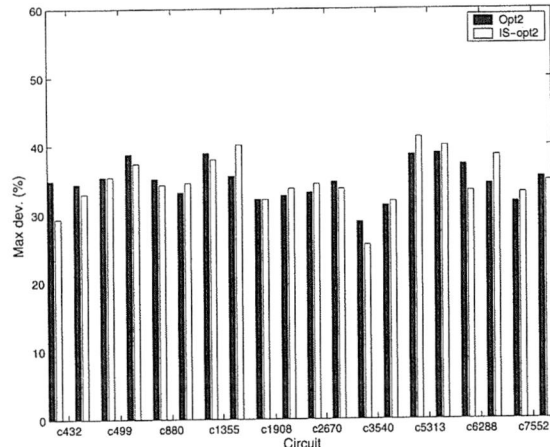

Figure 8: Maximum deviation of critical delay for optimized ISCAS'85 circuits.

7. REFERENCES

[1] V. D. Agrawal, Low Power Design by Hazard Filtering, In *Proc. Intl. Conf. on VLSI Design*, pp. 193–197, 1997.

[2] V. D. Agrawal, M. L. Bushnell, G. Parthasarathy and R. Ramadoss, Digital Circuit Design for Minimum Transient Energy and Linear Programming Method, In *Proc. Intl. Conf. on VLSI Design*, pp. 434–439, 1999.

[3] M. Berkelaar and E. Jacobs, Using Gate Sizing to Reduce Glitch Power, In *Proc. the ProRISC Workshop on Circuits, Systems and Signal Processing*, pp. 183–188, 1996.

[4] M. L. Bushnell and V. D. Agrawal, *Essentials of Electronic Testing for Digital, Memory & Mixed-Signal VLSI Circuits*, Springer, Boston, 2000.

[5] A. P. Chandrakasan and R. W. Brodersen, *Low Power Digital CMOS Design*, Kluwer Academic Publishers, Boston, 1995.

[6] S. Dutta, S. Nag and K. Roy, ASAP: A Transistor Sizing Tool for Area, Delay and Power Optimization of CMOS Circuits, In *Proc. IEEE ISCAS*, pp. 61–64, 1994.

[7] F. Hu, *Process-Variation-Resistant Dynamic Power Optimization for VLSI Circuits*, Ph.D. Dissertation, Auburn University, Dept. of ECE, Auburn, Alabama, May 2006.

[8] T. Raja, V. D. Agrawal and M. L. Bushnell, CMOS Circuit Design by a Reduced Constraint Set Linear Program, In *Proc. Intl. Conf. on VLSI Design*, pp. 527–532, 2003.

[9] T. Raja, V. D. Agrawal and M. L. Bushnell, CMOS Circuit Design for Minimum Dynamic Power and Highest Speed, In *Proc. Intl. Conf. on VLSI Design*, pp. 1035–1040, 2004.

[10] T. Raja, V. D. Agrawal and M. L. Bushnell, Variable Input Delay CMOS Logic for Low Power Design, In *Proc. Intl. Conf. on VLSI Design*, pp. 596–604, 2005.

[11] T. Raja, V. D. Agrawal and M. L. Bushnell, Transistor Sizing of Logic Gates to Maximize Input Delay Variability, *Journal of Low Power Electronics*, vol. 2, pp. 121–128, April 2006.

[12] C. V. Schimpfle, A. Wroblewski and J. A. Nassek, Transistor Sizing for Switching Activity Reduction in Digital Circuits, In *Proc. the Euro. Conf. on Theory and Design*, 1999.

[13] S. Uppalapati, M. L. Bushnell and V. D. Agrawal, Glitch-Free Design of Low Power ASICs using Customized Resistive Feedthrough Cells, In *Proc. VLSI Design and Test Symp.*, pp. 41–48, 2005.

[14] A. Wróblewski, C.V. Schimpfle, O. Schumacher and J. A. Nossek, Minimizing Spurious Switching Activities with Transistor Sizing, *J. of VLSI Design*, vol. 15, pp. 537–546, no. 2, 2002.

Two-phase Fine-grain Sleep Transistor Insertion Technique in Leakage Critical Circuits

Yu Wang, Yongpan Liu, Rong Luo, Huazhong Yang, Hui Wang
E.E. Dept., Tsinghua University, Beijing, P.R.China
(8610)62772966
{wangyuu99, ypliu99}@mails.tsinghua.edu.cn

ABSTRACT

Multi-threshold CMOS is a valuable leakage reduction method in circuit standby mode. Reducing leakage current through fine-grain sleep transistor insertion (*FGSTI*) makes it easier to guarantee circuit functionality and improves circuit noise margins. In this paper, we first indicate the negligible dependence of ST size on the amount of leakage saving which makes the two-phase *FGSTI* reasonable based on our leakage current and delay models. Then we introduce a novel two-phase *FGSTI* technique: a) ST placement and b) ST sizing, which are formally modeled as two linear programming (LP) models respectively. Our experimental results show that the two-phase *FGSTI* technique can achieve 78.91%, 92.55%, 97.97% leakage saving when the circuit slowdown is 0%, 3%, 5% respectively. Comparing to the simultaneous ST placement and sizing method using mix integer linear programming (MLP) [1], our technique leads to on average 2% more leakage current reduction while at least 10X runtime saving since fewer variables and constraints with less approximation are used in the LP models. When the circuit slowdown is large enough to perform conventional fixed slowdown method, our technique can still achieve 75.48% ST area saving. Moreover, we show that when the circuit slowdown is 0%, it should be carefully considered to use *FGSTI* technique due to a large amount of leakage feedback gates.

Categories and Subject Descriptors:

J.6 [**Computer Aided Engineering**]: Computer aided design (CAD), B.6.3 [**Design Aids**]: Optimization

General Terms

Algorithms, Design

Keywords

Leakage current reduction, two-phase fine-grain sleep transistor insertion, mixed integer linear programming.

1. INTRODUCTION

With the development of the fabrication technology, leakage power dissipation has become comparable to switching power dissipation [2]. As we all know, the total power dissipation consists of dynamic power, short circuit power and leakage power. The behavior of the short circuit power dissipation remains at

Permission to make digital or hard copies of all or part of this work for personal or classroom use is granted without fee provided that copies are not made or distributed for profit or commercial advantage and that copies bear this notice and the full citation on the first page. To copy otherwise, or republish, to post on servers or to redistribute to lists, requires prior specific permission and/or a fee.
ISLPED'06, October 4–6, 2006, Tegernsee, Germany.
Copyright 2006 ACM 1-59593-462-6/06/0010...$5.00.

around 10% of the total power dissipation [3]. At the 90nm technology node, leakage power may make up 42% of total power [4]. Leakage power reduction techniques can be broadly categorized into two main categories [5]: process level and circuit level techniques. The circuit level techniques consist of adapt body bias [6], DVTS [7], input vector control [8], dual-V_t assignment [9-11] and Multi-Threshold CMOS (ST insertion) [1] [12-18]. Among these, Multi-Threshold CMOS (MTCMOS) technique is essentially placing a ST between the gates and the power/ground (P/G) net in a circuit in order to put it into sleep mode when the circuit is standby.[*]

The most popular MTCMOS technique is gating the power of sizable blocks using large sleep transistors which is concluded as *block based ST insertion* (*BBSTI*) technique. In *BBSTI* techniques, all the gates in the block are assumed to have a fixed slowdown, so it is also called fixed slowdown method. The existing literatures on *BBSTI* techniques [12-16] present some details in clustering gates into blocks in order to optimize the leakage current and ST size. All these literatures focus on how to reduce the ST area penalty along with a remarkable leakage saving: [12] first gives out a mutual exclusion method; [13] [14] present several fast heuristic techniques for efficient gate clustering; [15] [16] propose a distributed sleep transistor network (DSTN) approach which assumes that all the sleep devices are connected to further reduce the area penalty.

Although *BBSTI* techniques greatly reduce the area penalty, they induce large ground bounce in the P/G network which has adverse effects on circuit speed and noise immunity [18]. What is more, ST size is determined by the worst case current of the clustering block which is quite difficult to determine without comprehensive simulation [12]. Thus it is harder to guarantee circuit functionality for large blocks with only one ST [17].

In recent years gate level ST insertion, which can be also called *fine-grain ST insertion* (*FGSTI*) technique [1] [17] [18] (as shown in figure 1 (a)) shows some advantages over the *BBSTI* technique (as shown in figure 1 (b)). It is easier to guarantee circuit functionality in an *FGSTI* technique as ST sizes are not determined by the worst case current of large circuit blocks. And *FGSTI* technique leads to a smaller simultaneous switching current when the circuit changes between standby mode and active mode, thus improves circuit noise margins. Furthermore, better circuit slack utilization can be achieved as the slowdown of each gate is not fixed, and then leads to a further reduction of leakage and area. As shown in [18], *FGSTI* technique corresponds to an area penalty of roughly 5% using standard cell placement.

---

[*] This work is sponsored by the grants from National 863 project of China (No. 2005AA1Z1230) and NSFC (No. 60506010).

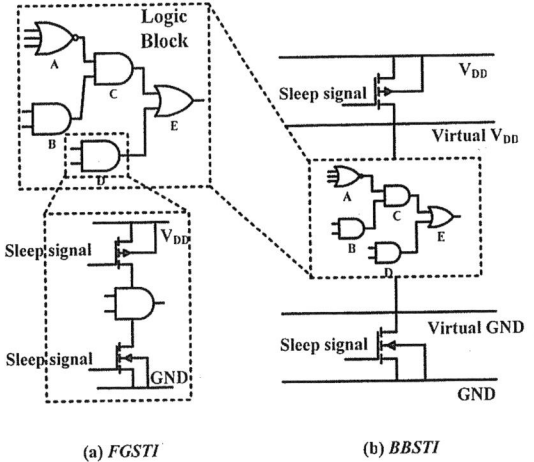

Figure. 1 *Fine-grain ST insertion (FGSTI) vs Block based ST insertion (BBSTI)*

In [17], a fine-grain MTCMOS design methodology and several design rules are proposed. The authors also make a comparison between local and global sleep devices. In [18], a selectively ST insertion methodology with better utilization of circuit slack is proposed in detail. They use a heuristic method to determine ST placement and sizing when circuit slowdown is 3%. When the circuit slowdown exceeds 5%, they solve an LP model to get optimal ST size. Although their method can give out an optimal sizing result, the heuristic step may lead to a local optimal point because ST placement is affected by ST sizing under their leakage current model assumption. In [1], a simultaneous fine-grain ST placement and sizing method using MLP is presented. The MLP model leads to an accurate result, but its computation time is considerably long.

This paper presents a novel two-phase *FGSTI* technique which has three contributions to leakage reduction:

(a)   Simple leakage current and delay models of a single gate are proposed. Our model analysis is the first to provide the designer the negligible dependence of ST size on the amount of leakage saving which makes the two-phase *FGSTI* reasonable.

(b)   The novel two-phase *FGSTI* technique: a) ST placement and b) ST sizing are modeled as two simple LP models respectively. Fewer variables and constraints with less approximation are used in our models, thus our two-phase *FGSTI* technique is more accurate and faster compared to simultaneous ST placement and sizing method [1]. Our ST placement can achieve an impressive leakage saving when the conventional fixed slowdown method can not be performed. Furthermore, if the circuit slowdown is large enough to use conventional fixed slowdown method, our ST sizing still leads to a much smaller total ST size.

(c)   We show that when the circuit slowdown is 0%, it may be inappropriate to use *FGSTI* technique due to the usage of different type ST to avoid floating states. A large amount of buffers bring not only additional area, but also considerable dynamic power penalty.

The paper is organized as follows. In Section 2, our leakage current and delay models are given out and analyzed to prove the rationality of our two-phase *FGSTI* technique. The two-phase *FGSTI* technique is proposed in Section 3. The implementation and experimental results are presented and analyzed in Section 4. In Section 5, we conclude this paper.

2. PRELIMINARIES

In this section, the leakage current and delay models used in our two-phase *FGSTI* technique are given out and analyzed to prove that a *FGSTI* design can be performed in two phases. ST is used with variable size which is decided by the process technology in our two-phase *FGSTI* design. A combinational circuit is represented by a directed acyclic graph (DAG) $G = (V, E)$. A vertex $v \in V$ represents a CMOS gate from the given library, while an edge $(i, j) \in E$, $i, j \in V$ represents a connection from vertex i to vertex j.

2.1 Leakage current model

For the gates without ST, a leakage lookup table is created by simulating all the gates in the standard cell library under all possible input patterns. Thus the leakage current $I_l^{w/o}(v)$ can be expressed as:

$$I_l^{w/o}(v) = \sum_{IN} I_l(v, IN) \times PB(v, IN) \tag{1}$$

Where $I_l(v, IN)$ and $PB(v, IN)$ are the leakage current and the probability of gate v under input pattern IN.

We simply use a linear model to represent leakage current $I_l^{ST}(v)$ based on HSPICE simulation results:

$$I_l^{ST}(v) = A(v) \times (W/L)_v \tag{2}$$

where $A(v)$ is constant and decided by the gate type. Here we assume all the input patterns have same probability and estimate every $A(v)$ for all the standard cells in the library. Consider two standard cells: NOR2XL and NAND4XL in the TSMC $0.18\mu m$ standard cell library, the largest error is about 52% as shown in table 1. The error of linear approximation may be neglected in *FGSTI* due to Law of large numbers [19] with the growing circuit size. As we will mention in Section 2.3, the influence of the linear model error on the *FGSTI* technique will be diminished by the large difference between leakage current of a gate with or without ST.

Table 1. Leakage current in NOR2XL and NAND4XL

	Leakage current in NOR2XL (fA) $A(v)$=1.60495			Leakage current in NAND4XL (fA) $A(v)$= 2.97335		
	Hspice	Our model	Error	Hspice	Our model	Error
w/o ST	14606.8	N/A	N/A	12261.3	N/A	N/A
(W/L) =2	5.3899	3.2099	-40.4%	10.19664	5.9467	-41.7%
(W/L) =4	5.9464	6.4198	7.96%	10.9738	11.8934	8.38%
(W/L) =8	8.5931	12.8396	49.4%	15.58464	23.7868	52.6%
(W/L) =16	27.6482	25.6792	-7.12%	51.37328	51.3733	-7.40%

2.2 Delay model

As shown in [20], gate delay is influenced by the ST insertion. The load dependent delay $d^{w/o}(v)$ of gate v without ST is given by:

$$d^{w/o}(v) = \frac{KC_L V_{DD}}{(V_{DD} - V_{THlow})^\alpha} \tag{3}$$

where C_L, V_{THlow}, α, K are the load capacitance at the gate output, the low threshold voltage, the velocity saturation index and the proportionality constant respectively.

The propagation delay $d^{ST}(v)$ of gate v with ST can be expressed as:

$$d^{ST}(v) = \frac{KC_L V_{DD}}{(V_{DD} - 2V_x - V_{THlow})^\alpha} \tag{4}$$

where V_x is the V_{ds} of the ST, that is to say the voltage drop from V_{DD} to the virtual V_{DD}. $\Delta D(v)$ is derived from the above equations:

$$\Delta d(v) = d^{ST}(v) - d^{w/o}(v) = \left(\left(1 - \frac{2V_x}{V_{DD} - V_{THlow}}\right)^{-\alpha} - 1 \right) d^{w/o}(v) \tag{5}$$

239

$I_{ON}(v)$ is the current flowing through ST in gate v during the active mode, which can be expressed as given by [18]:

$$I_{ON}(v) = \mu_n C_{ox}(W/L)_v((V_{DD} - V_{THhigh})V_x - \frac{V_x^2}{2}) \tag{6}$$

$$= \mu_n C_{ox}(W/L)_v(V_{DD} - V_{THhigh})V_x$$

Thus the voltage drop V_x in gate v due to ST insertion can be expressed as:

$$V_x = \frac{I_{ON}(v)}{\mu_n C_{ox}(V_{DD} - V_{THhigh})} \times \frac{1}{(W/L)_v} \tag{7}$$

Refer to equation (3) and (4), V_x in gate v due to ST insertion can also be given out as:

$$V_x = \frac{1}{2}\left(1 - \left(\frac{d^{w/o}(v)}{d^{ST}(v)}\right)^{1/\alpha}\right)(V_{DD} - V_{THlow}) \tag{8}$$

2.3 Relationship between ST placement and sizing

From previous part, we find that a linear leakage current model may have an error as large as 50% comparing with the HSPICE simulation results. Refer to [18], the leakage current for a gate with ST is also modeled as a linear function from [21]:

$$I_l^{ST}(v) = \mu_n C_{ox}(W/L)_v e^{1.8} V_T^2 e^{\frac{V_{gs} - V_{THhigh}}{nV_T}}(1 - e^{\frac{-V_{dx}}{V_T}}) \tag{9}$$

where μ_n is the N-mobility, C_{ox} is the oxide capacitance, V_{THhigh} is the high threshold voltage, V_T is the thermal voltage, n is the sub-threshold swing parameter, $(W/L)_v$ represents the ST size of gate v. Notice that their model is also linear by assuming parameters except $(W/L)_v$ are constant decided by process information and gate structure. Such a linear model will also consume comparative error as our leakage current model.

However, as we explore the leakage current model further, the leakage current of a gate without ST is much larger than that of a gate with ST as shown in Table 2, so that the error of the linear model can be neglected in the *FGSTI* procedure. In table 2, the leakage current of cells in the TSMC $0.18\mu m$ standard cell library under two different ST conditions: with ST or without ST are compared. As shown in table 1, the leakage current of a gate with ST become larger with a larger (W/L). Therefore, the largest leakage current of a gate with ST is derived by setting the (W/L) of a ST to 16 which is the maximum ratio of ST in our *FGSTI* technique.

Table 2. Leakage current comparison of standard cells (fA)

Cell Name	$I^{w/o}$	I^{ST}	$I^{w/o}/I^{ST}$	Cell Name	$I^{w/o}$	I^{ST}	$I^{w/o}/I^{ST}$
NAND2XL	14076.3	45.03	313	AND3X4	54900.3	53.4	1028
NAND2X4	84392.0	45.5	1854	BUFX4	80876.7	53.4	1513
INVXL	14213.2	36.9	388	NOR2X1	16261.1	27.6	516
NOR2XL	14606.8	27.6	528	CLKINVX4	38763.4	37.2	1043
XOR2XL	95853.9	53.4	1794	NAND4X4	72554.2	51.4	1411
NAND4XL	12261.2	51.4	239	AND2XL	26956.7	53.4	505
NAND3XL	14186.1	49.3	288	AND4XL	13768.6	53.4	258
AND2X4	60305.1	53.4	1129	OR4XL	33827.5	53.4	633
AND4X4	48899.1	53.4	915	CLKINV8	69175.0	37.7	1833

As shown in table 2, the leakage current difference is at least 238X. Referring to equation (5) and (7), the delay difference is less than 20% of the original gate delay under the same condition. However, the delay difference of a gate with different ST size is much larger, for example, setting the (W/L) of a ST to 1 will lead to about 140% additional delay comparing with the original gate without ST. Also from table 1, the leakage current difference of a gate with different ST size is less than 1% of the original gate leakage. Hence, the leakage current variation range due to the change of ST size can be neglected since it is much smaller

compare to the leakage saving of changing a gate's ST condition. In a word, although leakage current *is* reduced by sizing the ST, ST placement is not affected by ST sizing due to the large gap between their effects on leakage saving.

With technology scaling down, the leakage current difference may be smaller under different ST condition, but it will still be very large due to high V_{th} ST and stacking effect. Hence we can draw a conclusion that *the leakage reduction depends on where to insert ST and the leakage difference of each gate under different ST condition; while the area penalty is decided by the ST sizing procedure.*

We further assume that ST placement and sizing are *independent* in a *FGSTI* design. Therefore, we develop a two-phase *FGSTI* technique: first, ST placement can be performed to decide where to put the ST and achieve most of the leakage saving; and then ST sizing can be used to reduce the area overhead along with further leakage current reduction.

3. TWO-PHASE *FGSTI* TECHNIQUE

In this section two-phase *FGSTI* technique is modeled using linear programming. First ST placement phase shows how to place the ST as many as possible in order to reduce the total leakage, and then an optimal sizing method is given out for ST sizing phase to reduce the area overhead based on the ST placement information. At the end of this section, the simultaneous placement and sizing method [1] is briefly reviewed for comparison.

3.1 ST placement

A novel ST placement method is proposed that tries to maximize the leakage saving in the circuits through mixed integer linear programming (MLP). First, the object function for the total leakage current is constructed as below:

$$I(G) = \sum_{v \in V}\left(I_l^{w/o}(v) \times (1 - ST(v)) + I_l^{ST}(v) \times ST(v)\right) \tag{9}$$

where $ST(v)$ is a binary variable to represent gate v's ST condition, $ST(v) = 1$ means gate v has ST inserted and $ST(v) = 0$ means gate v is without ST. As ST size is not considered, we choose the largest ST size $(W/L)_{max}$ in equation (2) to obtain the minimum delay overhead. The leakage current of gate v with ST is given by:

$$I_l^{ST}(v) = A(v) \times (W/L)_{max} \tag{10}$$

It can be derived that $I_l^{ST}(v)$ is a constant for each gate to simplify the MLP model further.

The timing constraints of $G(V, E)$ can be expressed as:

$$t_a(m) = 0 \qquad m \in PI \tag{11}$$

$$t_a(n) + d(n) \le T_{req} \qquad n \in PO \tag{12}$$

$$t_a(i) + d(i) \le t_a(j) \qquad \forall (i,j) \in E, \ i,j \in V \tag{13}$$

where PI and PO refer to the primary input and primary output gates of the circuit; $t_a(v)$ represents the arrival time of gate v, T_{req} is the overall circuit delay; $d(v)$ represents the gate delay which can be expressed as using equation (5) and (7):

$$d(v) = d^{w/o}(v) + \Delta d(v) \times ST(v)$$

$$= d^{w/o}(v) + \left(\left(1 - \frac{2V_x}{V_{DD} - V_{THlow}}\right)^{-\alpha} - 1\right)d^{w/o}(v) \times ST(v)$$

$$= d^{w/o}(v) + \left(\left(1 - \frac{2\frac{I_{ON}(v)}{\mu_n C_{ox}(V_{DD} - V_{THhigh})} \times \frac{1}{(W/L)_{max}}}{V_{DD} - V_{THlow}}\right)^{-\alpha} - 1\right)d^{w/o}(v) \times ST(v) \tag{14}$$

$$= d^{w/o}(v) + \Gamma d^{w/o}(v) \times ST(v)$$

where $d^{w/o}(v)$ and Γ are constant for each gate v. Similarly we choose the largest ST size $(W/L)_{\max}$ to get the minimum delay overhead.

Minimize:

$$I(G) = \sum_{v \in V} \left(I_l^{w/o}(v) \times (1 - ST(v)) + A(v) \times (W/L)_{\max} \times ST(v) \right)$$

Subject to:

{Timing constraints}

$$t_a(m) = 0 \qquad\qquad m \in PI$$

$$t_a(n) + d(n) \leq T_{req} \qquad\qquad n \in PO$$

$$t_a(i) + d(i) \leq t_a(j) \qquad \forall (i,j) \in E,\ i,j \in V$$

$$d(v) = d^{w/o}(v) + \Gamma d^{w/o}(v) \times ST(v) \qquad v \in V$$

{Variable bounds}

$$ST(v)\ \text{are binary variables}$$

Figure 2. MLP model for leakage minimization through ST placement

---

The general form of our MLP model for ST placement is shown in figure 2. ST placement is similar as dual V_{th} assignment with fixed high and low V_{th} values, thereby it can also be solved by sensitive based heuristic algorithms which are previously dealing with dual Vth assignment [9-11].

3.2 Optimal ST sizing

After the ST condition for each gate v is decided, we use linear programming to get the optimal ST size. First the object function for optimal ST sizing is given out as below:

$$Area(ST) = \sum_{v \in V} \left((W/L)_v \times ST(v) \right) \qquad (15)$$

where $ST(v)$ is a binary value given out in the ST placement phase; $(W/L)_v$ is a continuous variable. Because the transistor length for ST is assumed to be a constant: the minimum length, $(W/L)_v$ is used instead of $W \times L$ to represent the ST area. Moreover, the expression for $(W/L)_v$ can be derived from equation (7) and (8) as below:

$$(W/L)_v = \frac{I_{ON}(v)}{\mu_n C_{ox}(V_{DD} - V_{THhigh})} \times \frac{1}{V_x} \qquad (16)$$

$$= \frac{I_{ON}(v)}{\mu_n C_{ox}(V_{DD} - V_{THhigh})} \times \left(\frac{1}{2}\left(1 - \left(\frac{d^{w/o}(v)}{d^{ST}(v)}\right)^{1/\alpha}\right)(V_{DD} - V_{THlow}) \right)^{-1}$$

The timing constraints can also be expressed as equation (11), (12) and (13). The propagation delay $d^{ST}(v)$ of gate v with ST can be rewrite using equation (5) and (7) as:

$$d^{ST}(v) = d^{w/o}(v) + \Delta d(v) \qquad (17)$$

$$= d^{w/o}(v) + \left(\left(1 - \frac{2V_x}{V_{DD} - V_{THlow}}\right)^{-\alpha} - 1 \right) d^{w/o}(v)$$

$$= d^{w/o}(v) + \left(\left(1 - \frac{2\frac{I_{ON}(v)}{\mu_n C_{ox}(V_{DD} - V_{THhigh})} \times \frac{1}{(W/L)_v}}{V_{DD} - V_{THlow}}\right)^{-\alpha} - 1 \right) d^{w/o}(v)$$

With a given boundary of $(W/L)_v$: $[(W/L)_{\min}, (W/L)_{\max}]$, the boundary of $d^{ST}(v)$ can be easily gained: $[d_{\min}^{ST}(v), d_{\max}^{ST}(v)]$ using equation (17). Consequently, the general form of our LP model for ST sizing is show in figure 3.

3.3 Simultaneous ST placement and sizing

The object function of simultaneous ST placement and sizing is very similar to ST placement as shown in equation (9):

$$I(G) = \sum_{v \in V} \left(I_l^{w/o}(v) \times (1 - ST(v)) + A(v) \times (W/L)_v \times ST(v) \right) \qquad (18)$$

Minimize:

$$Area(ST) = \sum_{v \in V} \left(\frac{I_{ON}(v)}{\mu_n C_{ox}(V_{DD} - V_{THhigh})} \times \left(\frac{1}{2}\left(1 - \left(\frac{d^{w/o}(v)}{d^{ST}(v)}\right)^{1/\alpha}\right)(V_{DD} - V_{THlow}) \right)^{-1} \times ST(v) \right)$$

Subject to:

{Timing constraints}

$$t_a(m) = 0 \qquad\qquad m \in PI$$

$$t_a(n) + d(n) \leq T_{req} \qquad\qquad n \in PO$$

$$t_a(i) + d(i) \leq t_a(j) \qquad \forall (i,j) \in E,\ i,j \in V$$

$$d(v) = d^{w/o}(v) + (d^{ST}(v) - d^{w/o}(v)) \times ST(v) \qquad v \in V$$

{Variable bounds}

$$d_{\min}^{ST}(v) \leq d^{ST}(v) \leq d_{\max}^{ST}(v)$$

Figure 3. LP model for optimal ST sizing

---

where $ST(v)$ and $(W/L)_v$ are variables which decide where to put ST and how to size ST respectively.

The timing constraints also follow equation (11), (12) and (13). Refer to equation (14), gate delay $d(v)$ for gate v can be derived as:

$$d(v) = d^{w/o}(v) + \left(\left(1 - \frac{2\frac{I_{ON}(v)}{\mu_n C_{ox}(V_{DD} - V_{THhigh})} \times \frac{1}{(W/L)_v}}{V_{DD} - V_{THlow}}\right)^{-\alpha} - 1 \right) d^{w/o}(v) \times ST(v) \qquad (19)$$

$$= d^{w/o}(v) + d^{w/o}(v) \times \Phi((W/L)_v) \times ST(v)$$

As we can see from equation (18) and (19), this problem is actually a non-linear programming model. In [1], Taylor series expansion and piecewise linear approximation technique are used to get a mixed integer linear programming model. Some dummy variables are needed for linear approximation and more linearization constraints are added in the MLP model for each dummy variable. Unfortunately, the model size becomes extremely large with the increasing gate number in the circuit.

4. IMPLEMENTATION AND EXPERIMENTAL RESULTS

4.1 Implementation

All ISCAS85 benchmark circuit netlists are synthesized using Synopsys Design Compiler and a TSMC $0.18\mu m$ standard cell library. A leakage current look up table of all the standard cells without ST is generated using HSPICE. In addition, every $A(v)$ in equation (2) for all the standard cells is estimated using HSPICE simulation results under different $(W/L)_v$. The values of various transistor parameters are taken from the TSMC $0.18\mu m$ process library. $V_{DD}{=}1.8V$, $V_{THhigh}{=}500mV$, $V_{THlow}{=} 300mV$, and $I_{ON}{=} 200\mu A$ for all the gates in the circuit. The timing constraints are set up with a specialized static timing analysis (STA) tool [10], and the MLP and LP models for ST placement and sizing are automatically generated. We use an LP solver named lp_solve [22] to solve the models.

We assume $1 \leq (W/L)_v \leq 16$, corresponding to a least delay variance of 6% if ST is assigned to every gate in the circuit. We perform our two-phase $FGSTI$ technique by first using the MLP model to get $ST(v)$ for all the gates in the circuit and then solve the LP model to get the optimal $(W/L)_v$ based on the results of $ST(v)$. The MLP model to simultaneously determine ST placement and sizing are also solved using the same LP solver under a same set of parameters for comparison with the two-phase $FGSTI$ technique.

4.2 Results for two-phase $FGSTI$ technique

For 0%, 3%, 5% circuit slowdown, we can not get a valid solution from conventional fixed slowdown method [12]. Thus the

241

leakage current saving for 0%, 3%, 5% circuit slowdown are compared between our two-phase *FGSTI* technique and MLP method [1]. As shown in table 3, our two-phase *FGSTI* technique can achieve 78.91% leakage saving even when the circuit slowdown is 0%. When the circuit slowdown is 3%, 5%, the leakage saving of our two-phase *FGSTI* technique is 92.55%, 97.97% respectively. Because of less approximation in ST placement phase, more ST's can be assigned to different gates and additional leakage saving is achieved. The leakage saving is about on average 2% more than the MLP method [1].

In table 4, we show that our two-phase *FGSTI* technique can achieve a very impressive runtime saving. As the LP model for ST sizing only need seconds to solve, the runtime saving is largely due to two reasons: one is the two-phase procedure of *FGSTI* technique and the other is less variables and constraints used in MLP model for ST placement. For circuit C432, there are only 271 constraints and 338 variables in our MLP model for ST placement; however, in [1] there are 2975 constraints and 1183 variables. Although the MLP problem still need a long time to solve, as we can see from some of the benchmark, especially the small ones, our two-phase *FGSTI* technique can achieve at least 10X runtime saving. We only list the results of 4 benchmarks, because other benchmarks take hours to get the optimal results. The stopping time criteria is set to 4 hours for larger circuits. Heuristic algorithms can get near optimal results with a very fast speed, but as we all know the heuristic may lead to local optimal and can not guarantee the optimality of the result; thus we can use the results of LP models as a reference.

When the circuit slowdown is larger than 6%, ST can be assigned to all the gates in the circuits, the two-phase procedure of *FGSTI* technique is changed to one: ST sizing, while ignoring the ST placement. Our LP model for ST sizing leads to a same result as optimal sizing method in [16]. We compare the area penalty with the fixed slowdown method and the MLP method in table 5. With 7% circuit slowdown, our ST sizing LP model causes 75.48% ST area saving compared to fixed slowdown method and the result is almost the same with MLP method. In table 5, ST

area is calculated using equation (15), just summing up all the $(W/L)_v$, since the length of ST is a constant.

Table 5. ST sizing results comparison with MLP and fixed slowdown method

ISCAS85 benchmark circuits	7% circuit slowdown			9% circuit slowdown		
	ST sizing	MLP	Fixed slowdown	ST sizing	MLP	Fixed slowdown
C432	714	714	2317.72	596	597	1802.67
C499	1146	1146	2797.72	959	959	2176.01
C880	876	876	5252.58	780	780	4085.35
C1355	3365	3364.	7515.44	2719	2720	5845.35
C1908	2355	2355	12493.73	2081	2081	9717.36
C2670	2087	2088	17540.59	1937	1937	13642.71
C3540	3371	3371	23300.60	3160	3160	18122.72
C5315	4292	4293	31940.60	3917	3918	24842.74
C6288	11733	11733	33558.89	9610	9611	26101.41
C7552	8980	8981	48905.19	8197	8197	38037.45
Area saving	75.48%	N/A	N/A	73.0%	N/A	N/A

4.3 ST type Consideration

From table 3, when the circuit slowdown is below 6%, not all the gates in the circuit can be assigned with ST. FGSTI technique can cause a gate with ST to drive a gate without ST which leads to a floating state at the output of the gate with ST and large power dissipation in the gate without ST. As mentioned in [18], leakage feedback gate structure [23] shown in figure 4 is used in order to avoid the floating states. We assume that the leakage feedback structure can achieve the same delay as the normal ST insertion with a larger area and dynamic power consumption penalty. We examine all the gates with ST in the circuits and find out how many gates with ST should be changed into leakage feedback structure.

In table 6, when the circuit slowdown is 0%, about 82.75% of the total gates can change into gate with ST, and about 37.10% of the gates with ST should change into leakage feed back structure. When circuit slowdown is 3% and 5%, about 93.47% and 98.03% of the total gates can be changed into gates with ST, and only 19.78% and 9.90% of them should be changed into leakage feedback structure respectively. When the circuit slowdown is 0%, some of the benchmarks, such as C499, C1355, need to change 80.41% and 66.41% of original ST into leakage feedback struct-

Table 3. Leakage current comparison between two-phase *FGSTI* and MLP method

ISCAS85 benchmark circuits	Original I_{leak} (pA)	Total gate Num.	0% circuit slowdown				3% circuit slowdown				5% circuit slowdown			
			Two-phase FGSTI		MLP		Two-phase FGSTI		MLP		Two-phase FGSTI		MLP	
			I_{leak} (pA)	ST gate Num.	I_{leak} (pA)	ST gate Num.	I_{leak} (pA)	ST gate Num.	I_{leak} (pA)	ST gate Num.	I_{leak} (pA)	ST gate Num.	I_{leak} (pA)	ST gate Num.
C432	4609.417	169	1759.279	130	1964.911	127	463.719	151	463.719	151	205.459	157	205.459	157
C499	21374.953	204	14479.83	97	14587.294	101	805.901	189	1451.785	164	105.253	200	757.367	189
C880	9261.315	383	619.241	352	619.241	352	232.245	370	364.903	365	126.149	375	126.149	375
C1355	11874.533	548	6417.456	308	6712.258	287	5099.370	402	5220.438	386	945.191	535	4382.772	417
C1908	23418.219	911	2498.797	830	3177.773	831	590.002	878	1296.372	882	224.510	900	258.150	900
C2670	35191.285	1279	1356.567	1235	1382.020	1235	364.575	1264	667.634	1260	161.849	1274	269.792	1270
C3540	40369.652	1699	2060.401	1617	2251.211	1612	1020.377	1658	1558.656	1637	270.415	1690	611.115	1675
C5315	56292.203	2329	1660.947	2253	1841.916	2254	788.566	2293	1025.466	2283	433.763	2312	593.592	2305
C6288	40968.834	2447	7427.753	1948	8083.815	1903	2545.587	2282	3042.084	2248	977.670	2385	1088.476	2382
C7552	85523.934	3566	3012.412	3415	4190.660	3385	1320.156	3504	2004.873	3471	682.362	3539	975.917	3519
Leakage saving	N/A	N/A	**78.91%**	N/A	77.49%	N/A	**92.55%**	N/A	91.24%	N/A	**97.97%**	N/A	94.55%	N/A
Additional Leakage saving (MLP-two-phase)/MLP			6.31%				14.95%				62.75%			

Table 4. Runtime comparison between two-phase *FGSTI* and MLP method (Time in s)

ISCAS85 benchmark circuits	0% circuit slowdown				3% circuit slowdown				5% circuit slowdown			
	Two-phase FGSTI			MLP	Two-phase FGSTI			MLP	Two-phase FGSTI			MLP
	ST placement	ST sizing	Total		ST placement	ST sizing	Total		ST placement	ST sizing	Total	
C432	0.491	1.234	1.725	32.047	1.502	1.593	3.095	1905.094	0.551	1.625	2.176	635.016
C499	3.856	1.454	5.31	75.0	2400.802	2.390	2403.19	154825.17	23.303	2.938	26.241	99610.59
C880	0.351	7.619	7.97	134.109	0.501	6.578	7.079	2973.734	0.321	6.344	6.665	958.766
C2670	43.61	107.797	151.407	22121.922	42.532	221.734	264.266	27438.11	4.25	176.453	180.703	57462.64

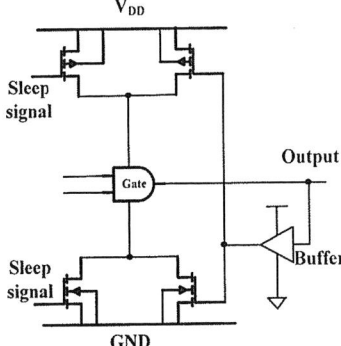

Figure 4. Leakage feedback structure

ure. This will lead to a large area increasing due to large number of high V_{th} feedback buffers. Therefore, if *FGSTI* technique will be used when there is no circuit slowdown, the penalty of using leakage feedback structure should be carefully examined because of the extra area and dynamic power consumption.

Table 6. Consideration of different ST type

ISCAS85 benchmark circuits	Total gate number	0% circuit slowdown		3% circuit slowdown		5% circuit slowdown	
		Total ST number	Leakage feed-back	Total ST number	Leakage feed-back	Total ST number	Leakage feed-back
C432	169	130	37	151	31	157	19
C499	204	97	78	189	48	200	32
C880	383	352	105	370	68	375	46
C1355	548	308	206	402	134	535	64
C1908	911	830	261	878	171	900	83
C2670	1279	1235	293	1264	193	1274	98
C3540	1699	1617	368	1658	246	1690	122
C5315	2329	2253	428	2293	294	2312	148
C6288	2447	1948	800	2282	485	2385	213
C7552	3566	3415	937	3504	577	3539	256
Average	N/A	82.75%	37.10%	93.47%	19.78%	98.03%	9.90%

5. Conclusions

In this paper, we present a novel two-phase *FGSTI* technique to reduce the leakage current using MTCMOS scheme. Simple leakage current and delay models for our two-phase *FGSTI* technique are proposed and analyzed to prove the rationality of our method. ST placement and sizing are modeled using MLP and LP models respectively. Our experimental results show that the two-phase *FGSTI* technique can achieve 78.91%, 92.55%, 97.97% leakage saving when the circuit slow down is 0%, 3%, 5% respectively. Moreover, two-phase *FGSTI* technique leads to 2% more leakage saving and at least 10X runtime saving comparing with simultaneous ST placement and sizing method using MLP. When the circuit slowdown is larger than 6%, the two-phase *FGSTI* can achieve 75.48% ST area saving comparing with fixed slowdown method. In conclusion, two-phase *FGSTI* technique is reasonable from our results. However, we show that the penalty of using leakage feedback structure during *FGSTI* technique should be carefully examined when the circuit slowdown is below 6%.

There are still some unsolved problems in *FGSTI* technique as our future work. Fast heuristic algorithms are needed for ST placement phase because the MLP model is very time consuming and can not handle large circuits. Furthermore, the detailed comparison between *FGSTI* and *BBSTI* techniques should be carefully examined in the physical level, such as place and routing penalty.

6. Reference

[1] Y. Wang, H. Lin, H.Z. Yang, R. Luo, H. Wang, "Simultaneous Fine-grain Sleep Transistor Placement and Sizing for Leakage Optimization," in *Proc. of ISQED '06*, 2006, pp. 723-728.

[2] G. Moore, "No exponential is forever: But forever can be delayed," in *IEEE ISSCC* Dig. Tech. Papers, 2003, pp. 20 - 23.

[3] D. Duarte, N. Vijaykrishnan, M. J. Irwin, and M. Kandemir, "Formulation and validation of an energy dissipation model for the clock generation circuitry and distribution networks," in *Proc. of VLSI Design*, 2001, pp. 248 - 253.

[4] J. Kao, S. Narendra, A. Chandrakasan, "Subthreshold Leakage modeling and reduction techniques", in *Proc. of ICCAD*, 2002, pp 141 – 149.

[5] K. Roy, S. Mukhopadhay, H. Mahmoodi-Meimand, "Leakage Current Mechanisms and Leakage Reduction Techniques in Deep-Submicrometer CMOS Circuits", in *Proc. of the IEEE*, Vol. 91, No. 2, Februray 2003 pp 305 – 327.

[6] S. Narendra et.al, "Forward body bias for microprocessors in 130-nm technology generation and beyond", in *IEEE JSSC*, Vol. 38 , No. 5 ,May 2003 pp. 696 - 701.

[7] C.H. Kim, K. Roy, "Dynamic VTH scaling scheme for active leakage power reduction", in *Proc. of DATE* 2002 pp.163 - 167.

[8] S. Mukhopadhyay et. al., "Gate Leakage Reduction for Scaled Devices Using Transistor Stacking", in *IEEE TVLSI*, Vol. 11, No. 4, Aug. 2003, pp. 716 - 729.

[9] L. Wei, Z. Chen, and K. Roy, "Design and Optimization of Dual Threshold Circuits for Low Voltage, Low Power Applications", in *IEEE TVLSI*, Vol.2. 17, NO. 1, 1999, pp. 16-24.

[10] Y. Wang, H.Z. Yang, H. Wang, "Signal-path Level Dual-Vt Assignment for Leakage Power Reduction," in Journal of Circuits, System and Computers, 2006, Vol. 15, No. 2, pp:197-216.

[11] Qi Wang; Vrudhula, S.B.K.; "Algorithms for minimizing standby power in deep submicrometer, dual-Vt CMOS circuits," in *IEEE TCAD*, Vol: 21 ,Issue: 3 ,March 2002 pp. 306 - 318 .

[12] J. Kao, S. Narendra, and A. Chandrakasan, "MTCMOS hierarchical sizing based on mutual exclusive discharge patterns," in *Proc. of DAC*, 1998, pp. 495–500.

[13] M. Anis, S. Areibi, and M. Elmasry, "Dynamic and leakage power reduction in MTCMOS circuits using an automated efficient gate clustering technique," in *Proc. of DAC*, 2002, pp. 480–485.

[14] W. Wang, M. Anis, S. Areibi, "Fast techniques for standby leakage reduction in MTCMOS circuits" in *Proc. of IEEE SOC*, 12-15 Sept. 2004 pp 21 – 24.

[15] C. Long; L. He; "Distributed sleep transistors network for power reduction" in *Proc. of DAC*, 2-6 June 2003 pp. 181 – 186.

[16] C. Long; L. He; "Distributed sleep transistor network for power reduction" in *IEEE TVLSI*, Volume: 12, Issue: 9, Sept. 2004 pp. 937 – 946.

[17] B. H. Calhoun, F. A. Honoré, and A. P. Chandrakasan, "A Leakage Reduction Methodology for Distributed MTCMOS," in *IEEE JSSC* Vol. 39, No. 5, May 2004, pp. 818 - 826.

[18] V. Khandelwal, A. Srivastava; "Leakage Control Through Fine-Grained Placement and Sizing of Sleep Transistors," in *Proc. of ICCAD* 2004, pp 533 - 536.

[19] W. Feller, "The Strong Law of Large Numbers," in An Introduction to Probability Theory and Its Applications, Vol.1, 3rd ed. New York: Wiley, pp.243-245,1968.

[20] S. Mutoh et al. "1-V Power Supply High Speed Digital Circuit Technology with Multithreshold Voltage CMOS," in *IEEE JSSC*, Vol. 30, No. 8 August 1995.

[21] S. Mukhopadhyay, K. Roy, " Modeling and Estimation of Total Leakage Current in Nano-scaled CMOS Devices Considering the Effect of Parameter Variation," in *Proc of ISLPED* Aug. 2003.

[22] http://groups.yahoo.com/group/lp_solve/

[23] J. Kao, A. Chandrakasan, " MTCMOS Sequential Circuits," in *Proc. of ESSDERC*, Sept 2003.

Register File Caching for Energy Efficiency

Hui Zeng and Kanad Ghose
Department of Computer Science
State University of New York at Binghamton
{hzeng, ghose}@cs.binghamton.edu

ABSTRACT

With the use of faster clocks and larger instruction windows in high-end superscalar processors, the physical register files (RFs) can no longer be accessed in a single cycle. To combat the consequential performance penalty, the RFs employ multiple levels of bypassing. Register file caching, which caches a small subset of the registers in a faster, smaller structure called the register file cache (RFC) has also been proposed as a remedy for this problem. We introduce a relatively simple RFC design that partitions the RFC into two separate components: a FIFO queue for holding register values that are used over a short duration following their writeback and another small set-associative cache holding values that are likely to be used over a longer duration. Results written to the RFC are easily classified into these categories and the classification bit is also used to predict the nature of the result for the next execution of the same instruction. We show that significant energy savings – about 38% on the average – occurs in accessing register operands when a 28-entry RFC is used, together with a 96-entry RF with no additional bypassing when compared with a base case design that has 128 registers with a 2 cycle access time and having one additional level of bypassing. The performance drop compared against the base case is also negligible (0.3% drop).

Categories and Subject Descriptors

C.1 [**Processor Architectures**]: Other Architecture Styles – *Pipeline processors.*

General Terms: Performance, Design

Keywords: Register Files, Register Caching, Energy-Efficiency

1. INTRODUCTION

The almost ubiquitous improvements in device technology have enabled the use of higher speed clocks. At the same time, high-end superscalar microprocessors that use out-of-order instruction processing have relied on the use of bigger and bigger register files to expose and exploit instruction level parallelism.

Consequently, the register file access time has gone up beyond a single cycle. Processor designers have addressed the performance

Permission to make digital or hard copies of all or part of this work for personal or classroom use is granted without fee provided that copies are not made or distributed for profit or commercial advantage and that copies bear this notice and the full citation on the first page. To copy otherwise, or republish, to post on servers or to redistribute to lists, requires prior specific permission and/or a fee.
ISLPED'06, October 4–6, 2006, Tegernsee, Germany.
Copyright 2006 ACM 1-59593-462-6/06/0010...$5.00.

problems inherent with the use of a multi-cycle access register files by using multiple levels of bypassing or by using alternative register file designs or by using a register file cache (RFC or simply a register cache) that permits a single cycle access to store register values that are expected to be accessed by the issued instructions [BDA 01, CGTV 00, HM 00, TA 03, BuSo 04]. The RFC is backed up by the larger register file (RF). A number of register caching schemes have been proposed to avoid the performance penalty associated with multi-cycle register accesses [BDA 01, CGTV 00, LVA 95, HM 00, BuSo 04]. Some of these schemes can get quite sophisticated and complex [BuSo 04].

This paper introduces a relatively simple register caching scheme whose primary goal is to reduce the energy spent in reading and writing a large, multi-cycle register file. As seen from our results given later, our register caching mechanism avoids more than 86% of the operands reads and about 35% of the writes that would otherwise take place in a superscalar processor that does not have a register cache. Furthermore, the proposed scheme a 28-entry RFC and a 96 entry register files provide performance level that comes within 0.3% of that provided by a base design that uses a 128-entry register file *with one additional level of bypassing*. (If this additional level of bypassing is not used in the base case, our scheme provides a significant performance gain; as this is not a fair comparison, we assume that the base case uses one additional level of bypassing.) Compared to the base case, the power spent in register file accessing is reduced by 38% on the average compared to the base case.

A detailed microarchitecture-level, cycle-by-cycle simulator [Sh 05] is used for the studies reported in this paper. This simulator extends the well-known Simplescalar simulator [Sim 06] in a number of ways that permit accurate power estimation of the various datapath components. This simulator models an out-of-order processor that uses register renaming for handling the true data dependencies and for eliminating anti and output dependency in the renamed instruction stream. As an instruction targeting a result into a register is dispatched, the source architectural registers are remapped to the physical registers representing their most recent instances using a register alias table (RAT) and a new physical register is assigned to the destination architectural register. A reorder buffer (ROB) is used to implement precise interrupts in the face out out-of-order instructions completions. The committed register values within the physical register file are pointed to by a retirement register alias table (R-RAT). The basic datapath thus has a strong resemblance to the Intel P4 style datapath. We modify this simulator to incorporate the proposed RFC design and to study the energy-performance tradeoffs.

2. Related Work

There is a large body of work that targets the reduction of register file ports. Alternative register file organizations have been explored primarily for reducing the access time (which goes up with the number of ports and registers), particularly in wire-delay dominated circuits [BTME 02, CGTV 00, TA 03, LVA 95]. Replicated register files in a clustered organization have been used in the Alpha 21264 processor [Ke 99] to reduce the number of ports in each replica and also to reduce delays in the connections in-between a function unit group and its associated register file.

Multi-banked register files with dissimilar banks, organized as a two-level structure (cached RF) or as a single-level structure, with dissimilar components were used to reduce the register file complexity and to reduce the effective RF access time in [CGTV 00]. In [BDA 01], a two-level RF implementation is proposed, along with the use of multiple register banks. The complexity/access delay reduction in this technique comes from the use of banks with a single read port and a single write port in each bank. The two-level structure allows the first level (banked) register file to be kept small (and fast), with the higher speed compensating for IPC drops arising from limiting the number of ports. In any design that uses a multi-bank structure with fewer ports on each bank, additional complexities, delays and energy dissipations are introduced by the connections and arbitration logic needed by the banks to arbitrate for accessing shared connection paths. The idea of caching recently produced values was used in [HM 00] (hereafter called "the VAB scheme"). At the time of instruction writeback, function units write results into a cache called Value Aging Buffer (VAB). The register file, holding both speculative and committed register values, is updated only when entries were evicted from the VAB. Furthermore, when the required value is not in the VAB, a read of the register file is needed, requiring at least some read ports for reading the source operands. Unless a sufficient number of register file ports is available or the number of entries in the VAB is sufficiently large, the performance degradation can be considerable. In addition, in the VAB scheme, the multi-cycle register file access is still an intimate part of the issue process. In [BuSo 04], a variety of register cache organizations, allocations and replacement policies are studied. The allocation and replacement policies studied in [BuSo 04] require keeping track of the number of active consumers of an instruction. This study, in contrast, requires no such bookkeeping and looks at alternative techniques for boosting the performance and/or energy efficiency of a RFC.

3. The Proposed Register File Cache

It is well known that most register values (results targeting a destination register) are consumed within a few cycles of their generation, either in the course of bypassing or very shortly thereafter [FS 92, LoGa 95, Po 03]. We will hereafter refer to such values as *short-term* values. However, there are some register values that are accessed over a more prolonged period. Examples of such values – called *long-term values* hereafter – are updates generated for the stack or environment pointer, values whose consumers are delayed due to cache misses or dependencies on long latency operation. A register value is thus either short-term or long-term.

We introduce a relatively simple register caching mechanism in this paper that targets both types of register values – short-term as well as long-term. The proposed register design actually uses two separate forms of register caches, one geared towards short-term values and the other targeting long-term values. Figure 1 depicts the overall structure of our register file cache (RFC) and the associated components and connections.

Short-term values are provided from a FIFO queue. The depth of this queue is chosen to cover the majority of consumed values. The FIFO queue is implemented as a circular queue, within a small register file. All results generated by the execution units – whether they are short-term or long-term are written into this FIFO part of the register cache. We will refer to the FIFO part of the cache as the SRFC (short-term RFC). For the configuration studied, the FIFO queue has a total of 12 entries. Note that if the main register file requires 3 cycles for accessing, the first 2 levels of this FIFO queue effectively form the multi-level bypass logic that will be used to avoid the full 3-cycle access penalty that would otherwise be incurred in accessing a result whose writing was initiated one cycle earlier.

The second part of the RFC is called the long-term RFC, LRFC. The LRFC is a small set-associative structure (4 ways, 4 sets for a total of 16 entries). The LRFC uses a LRU replacement to cache long-term values as they are evicted from the FIFO SRFC. (Recall that all results, whether they are short-term or long-term, always go through the FIFO SRFC).

As seen from the experimental results presented later, the segregated RFC design introduced here provides a hit rate of 86% on the average across all SPEC 2000 benchmarks. On a RFC hit, the register values are retrieved in a single cycle, as the RFC is small structure. On a RFC miss additional cycles are spent in accessing the register operand from the slower and larger RF. The proposed RFC design not only saves a significant part of reads to the main register file but also saves some writes. Approximately 35% of the values evicted from the RFC (SRFC or LRFC) do not need to be written to the RF because they are committed and deallocated by the time such writes to the RF are needed.

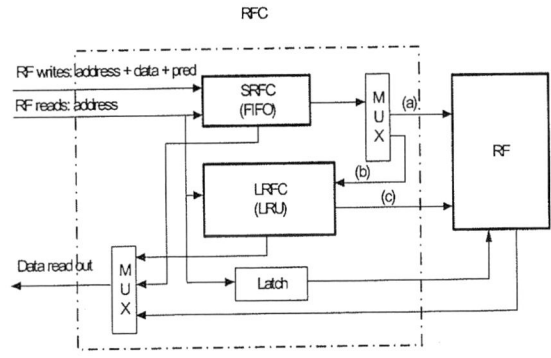

Figure 1. The Proposed Register File Cache Organization

To handle branch mispredictions and the associated squashings, all values – whether they are in the SRFC, LRFC are tagged with the branch tag – as in a traditional design. *The energy overhead for the branch tag within our RFC is accounted for in our studies along with all other components of the overhead.* The other

245

components of the RFC shown in Figure 1 will be explained later. We now discuss how we very simply identify and predict a result to be either short-term or long-term.

It is worth noting that the 28-entry RFC design shown in Figure 1 is far simpler in complexity compared to a fully associative 28-entry RFC design. Furthermore, the segregation of the RFC into the SRFC and the LRFC permits short term values to exit the cache quickly instead of occupying valuable cache space, while the set-associative LRFC handles long term values without resorting to the use of full associative addressing.

3.1 Implementation of the SRFC and LRFC

As indicated earlier, the SRFC is implemented as a circular queue. To enable instructions to access an operand stored in this queue, each entry in the queue is associatively addressed using the address of the physical register to which the value is destined for. Note that the first few levels of this FIFO queue are identical to the facilities provides by a multi-level bypass logic that allows register values to be accessed in between the time at which they are written to the register file and the time at which the multi-cycle write operation ends. Such bypass logic also require full comparators, so in essence the additional comparators needed for the SRFC in our scheme are only the ones used in the last few levels of the FIFO. Specifically, for the 4-way superscalar configuration studied in this paper, if a 3 cycle access register file is used, 2 additional levels of bypassing are needed. With the use of these additional levels of bypassing, up to 8 results, written in two consecutive cycles can be bypassed to one of the two sources of up to 4 instructions that can issue in a cycle. So the number of comparators needed in the bypass network of the base case is 8 X 4 X 2 = 64. The LRFC has 6 read ports and 4 write ports. The 12-entry SRFC to support reads on all of its 6 read ports requires a total of 12 X 6 = 72 comparators. So the additional number of comparators used in our scheme for the FIFO SRFC is only 8.

The LRFC is our design is a 4-way set associative with 4 sets for a total of 16 entries. The LRFC has 6 read ports and 4 write ports. The LRFC is accessed (set) associatively using the physical register address, as in the case of the SRFC. The LRFC is implemented as a 4-way set associative design, requiring 6 X 4 = 24 comparators itself. Thus compared to the 64 comparators used in the 2 additional levels of full bypassing used in a 3 cycle access register file design, the total number of comparators needed to support associative addressing of the SRFC and LRFC is 96, i.e., 32 more compared to the 3 cycle register file design.

3.2 Detecting and Predicting a Long Term Result

In our scheme, each result carries a single bit tag that indicates if it is short term or long term in nature. By default, a register value is tagged as short term. Short term values are written to the FIFO SRFC and migrate from there to the main register file (RF), as they are shifted out of the SRFC (path (a) in Figure 1). If subsequently a register value marked as short term is accessed from the main register file, it is retagged as long term. Similarly, a value tagged as long-term gets retagged as a short-term value if it is never accessed from the LRFC.

A result that is predicted to be long term is first written to the SRFC (just like the short term values) but migrates to the LRFC when it is shifted out of the SRFC (Path (b) in Figure 1). When this value is evicted from the LRFC as a victim for replacement, it is written to the main register file (path (c) in Figure 1).

When the register value is retired into an architectural register file, the associated 1-bit tag and information to locate the instruction generating this result within the I-cache (available in the reorder buffer entry of the instruction) is stored away. (Appropriate details are given later in Section 2.4).

When the register is eventually deallocated, this stored information is used to update the I-cache entry of the instruction with the updated tag bit. This tag bit serves as the prediction made for the type of the result (short-term or long-term) generated by the instruction when it is *subsequently* executed.

3.3 Optimizing Accesses to Values Recently Reclassified as Long-Term

As described above, a short-term value accessed from the main register file has to be retagged as a long-term value. As this value is read out from the RF, a copy is also installed into the LRFC. Within a single cycle, we allow only one such data movement from the RF to the LRFC to conserve the number of write ports to the LRFC. A single bit flag, set on a transfer from the RF to the LRFC and cleared at the end of a clock cycle is used to implement this limiting mechanism. We have imposed this restriction to reduce then number of additional write ports into LRFC. By doing this simple optimization, we could see an improvement of 21% in the hit rate of the RFC (i.e., the SRFC and LRFC combined).

3.4 Storing, Updating and Using the Prediction

Each instruction is tagged with a single bit to indicate if the result it generates into a register is short-term or long-term. This bit, in essence, is the prediction made for the result generated by a subsequent execution of the instruction. This "prediction" bit is stored in the I-cache and effectively increases the width of the cache lines by a factor if 1/32, as each instruction is 32 bits long. (This bit has no significance for instructions like stores and branches that do not generate a result into an addressable register.)

Updating a prediction from short-term to long-term or vice-versa requires the I-cache entry of the associated instruction to be updated – this is done as follows. At the time of instruction fetching, bits indicating the cache way that provided the instruction is propagated with the instruction as it is issued. Since the I-cache used in this study is a 2-way set associative cache, only 1 bit is needed for our design. As a result is generated by this instruction, this bit moves along with the result in the rest of the pipeline. When the instruction is retired, the retirement register alias table (R-RAT) entry pointing to the committed value is augmented to hold this "way indicator" bit and parts of the instruction address (available from the reorder buffer entry of the instruction at the time of retirement) that locates the cache set and the instruction within the cache line. The prediction made for the result is also stored within the augmented part of the R-RAT entry. For the design studied, the R-RAT entries are extended by 13 bits (1 way bit, 1 prediction bit, 3 bits for instruction offset within line and 9 bits for the I-cache set address). At the time of deallocating this register, the information in the extended R-RAT entry is used to write the prediction information into the I-cache, only if the prediction needs to be revised. Our simulation results indicate that such updates are needed only for about 15% of the committed instructions. Note that a I-cache lookup is not performed during this update, so if the contents of the cache line got replaced, some other instruction predictions are inadvertently updated. Our simulations show that the overall impact of such inadvertent updates to the prediction is negligible because of the very high I-cache hit rates, therefore justifying a direct update approach to a full-fledged cache lookup based update.

As an instruction targeting a physical register is dispatched, its associated prediction is also stored in its issue queue (IQ) entry. When this instruction issues, the associated prediction bit moves with the instruction. When the result is written into the SRFC, this bit then moves with the result and decides where the result moves to – either the RF or the LRFC – as it shifts out of the SRFC.

3.5 Handling Register Cache Misses

On a RFC miss, which is discovered at the end of the cycle in which the instruction is issued, the instruction needs to be replayed (that is, re-issued) to permit the operand to be fetched from the slower RF. To simplify the replay mechanism, we do not release the issue queue entry of an instruction till it is successful in reading its operand. We use the usual bit that marks an issue queue entry as valid to delay the deallocation of the entry. Note that dependents of an instruction that experiences a RFC miss may actually get selected for issue before the miss is discovered, as the broadcast of a wakeup signal generally precedes the availability of the corresponding result by at least one cycle in a high-end datapath. These dependants of the instruction that had a RFC miss (as well as dependants of the dependants) are handled in the same manner - using replays. In the datapath used for this study, where a RF access requires 2 additional cycles for accessing the RF following the detection of a RFC miss, the replays discussed above can delay the release of 8 issue queue entries in the absolute worst case, given a 4-way dispatching. The actual number of dependants are significantly lower than this and in reality no more than 1 to 2 issue queue entries are locked up. In any case, this overhead is necessary for any modern machine that speculates on data or load/store dependencies.

4. Energy/Power Estimation

The overall energy spent in accessing register operands with our register file caching scheme can be broken down as follows:

- Ewrite,rf: energy spent in writing results from the RFC to the RF

- Eread,rf: energy spent in accessing the main register file on a RFC miss

- Eread, rfc: energy spent in probing and reading the SRFC and LRFC in parallel in the course of a source register access.

- Ewrite, srfc: energy spent in writing the results into the SRFC.

- Ewrite, lrfc: energy spent in writing results from the SRFC or the RF to the LRFC

- Eiq: additional issue queue energy spent in propagating the prediction bit and way bit in the IQ

- Er-rat: energy spent in writing and reading the extended R-RAT entries for saving and updating the prediction information (Section 3.2).

- Eicache: energy spent in storing, updating and reading out the prediction bit in the I-cache (Section 3.2).

- Erfc, inv: energy spent in invalidating entries from the RFC on branch mispredictions or retiring physical registers.

- Erp: energy spent in replaying instructions on a RFC miss.

With the RFC present, (Ewrite, rf + Eread, rf) represents the energy expended in accessing the main register file, while (Eread, rfc + Ewrite, lrfc + Ewite, srfc) represents the energy dissipated in the RFC (SRFC, LRFC and the other components as shown in Figure 1). The last five energy components together represent the energy overhead for our scheme.

These energy components given above are weighted with access data derived from the microarchitectural simulations and added up to obtain the overall energy spent in accessing register operands in our design. For the base case, the energy expended from accessing register operands comes from the register file reads and writes and from the energy expended in the additional level of full bypassing needed.

We use 0.18 micron CMOS layouts for all major datapath components relevant to this study. We assume that all comparators used in the bypass networks (including the base case) and for associative addressing within the RFC to dissipate energy only on an exact match and use the design introduced in [EGKP 02]. We use SPICE measurements to derive the energy coefficients for different types of activities/transitions. These are combined with the data collected from the microarchitectural simulations to obtain the average register operand access energy per cycle in the base case (RF accesses) and the average energy spent in accessing a register operand per cycle with the RFC (RFC accesses on a hit and RF accesses on a miss PLUS the energy overhead components, as described earlier).

Table 1. Configuration of the Simulated Processor

Parameter	Configuration
Machine width	4-wide fetch, 4-wide issue, 4 wide commit
Window size	64-entry issue queue, 64 entry load/store queue,
Registers	Various sizes studied, as specifically indicated in
Function Units and Latency (total/issue)	4 Int Add (1/1), 1 Int Mult (3/1) / Div (20/19), 2 Load/Store (2/1), 4 FP Add (2), 1FP Mult (4/1) / Div (12/12) / Sqrt (24/24)
L1 I–cache	32 KB, 2–way set–associative, 32 byte line, 1 cycles
L1 D–cache	32 KB, 4–way set–associative, 32 byte line, 2 cycles
L2 Cache unified	512 KB, 4–way set–associative, 128 byte line, 8
BTB	1024 entry, 4–way set–associative. Mimimum branch misprediction penalty – 10 cycles
Branch Predictor	Combined with 1K entry Gshare, 10 bit global history, 4K entry bimodal, 1K entry selector
Memory	128 bit wide, 120 cycles first chunk, 2 cycles
TLB	64 entry (I), 128 entry (D), fully associative

5. Experimental Results

Table 1 shows the configuration of the baseline machine ("Base Case"), which uses two separate integer and floating point physical register files, each with 128 registers and with an access time of 2 cycles. The base case is described as the **128/2** configuration in the results (128 registers in the integer or floating point register file, with a 2 cycle register access time.) *We*

247

assume that the base case also uses an additional level of bypassing to keep comparisons fair. For performance comparison, we also have an "Ideal Case" that also uses a configuration identical to the base case but with a 1-cycle register access time, a configuration described as **128/1** in the results. In both of these cases, the register files are assumed to be fully ported, with 8 read ports and 4 write ports. The machine configurations used with the RFC are as follows:

RFC+96/2: This configuration uses a RFC with a 12-entry SRFC, 16 entry (4sets, 4 ways) LRFC and 96-entry register files with 6 read ports and 4 write ports each and a cycle time of 2 cycles. Note that in this configuration, the register files have fewer ports compared to the base and ideal cases.

RFC+128/2: This configuration is identical to the RFC+96/2 configuration except that each of the integer and floating point registers has 128 registers.

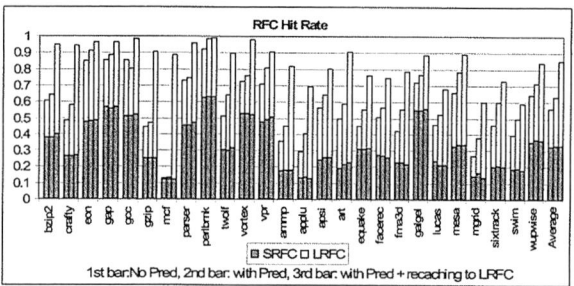

Figure 2. RFC hit rates.

We first show how the proposed RFC design eliminates a large majority of accesses to the main register file. Figure 2 shows the hit rates to the RFC for the RFC+96/2 configuration across the execution of 100 million instructions following the initialization and warmup phase for the SPEC 2000 benchmark programs as in SimPoint [SPHC 02]. There are 3 sets of bars for each benchmark, and each bar breaks down the hits to the SRFC and the LRFC individually. The first bar shows the hit rates achieved without the use of any predictions (i.e., with a default prediction of long-term for all results). The second bar shows the hit rates obtained with the use of a single bit prediction as described in Section 2.2, while the 3rd bar shows the hit rates realized using the optimization of Section 3.3 where results reclassified as long-term due to a RF access are also moved into the LRFC. For some benchmarks, like mcf, that have a substantial delay between initial set of consumers and later consumers, the optimization of Section 3.3 pays off handsomely. The prediction and the additional optimization essentially classify and exploits long-term results, respectively, so for each benchmark, the LRFC hit rates go up progressively with the prediction and the optimization. The average hit rate achieved across all benchmarks is 86% with prediction and the optimization. Even benchmarks that have a low reference locality (such as mcf and applu) achieve a high RFC hit rate. The simple RFC mechanism proposed in this paper is thus quite effective at eliminating RF accesses. In all of the subsequent results, we use a RFC with prediction and the optimization of Section 3.2.

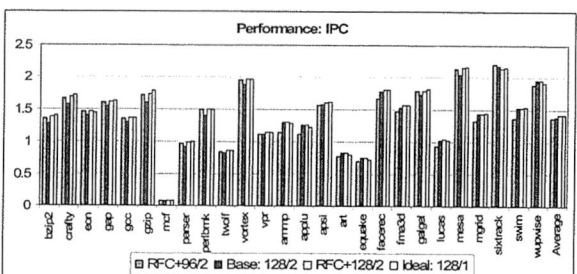

Figure 3. Performance (IPC) for the 4 CPU configurations

Figure 3 shows the IPCs realized with the four configurations, including the ideal case (128/1). There are four bars for each SPEC 2000 benchmark, with one bar per CPU configuration. On average, the performance realized by the RFC+96/2 configuration comes within 0.3% of the base case even though smaller register files (96 vs. 128) and fewer ports (8 read ports + 4 write ports vs. 6 read ports + 4 write ports) are used. The RFC+128/2 configuration actually shows a small performance gain around 2% over base case (128/2) on the average across the benchmarks. This configuration also provides a performance level very close to that of the ideal case (128/1). Note again that for the 128/2 configurations, we assumed that the RF has an additional level of bypassing; the configurations with the RFC do not need such additional bypassing as the front end of the SRFC effectively serves as a multi-level bypass network.

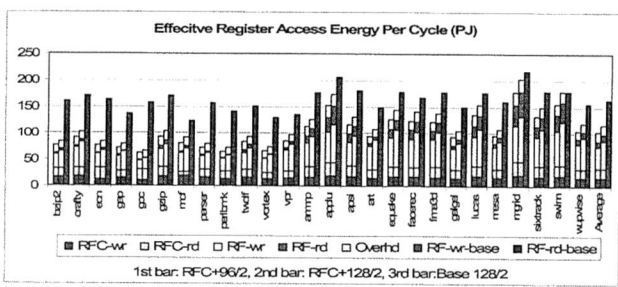

Figure 4. Average energy spent per cycle in accessing registers operands for different CPU configurations.

The average energy spent per cycle for accessing the register operands for the various configurations for the various benchmarks are shown in Figure 4. Again there are three bars for each benchmark, one per configuration. For the configurations without the RFC, Base 128/2, the bars are split into two parts – one part corresponding to the average read energy per cycle and the other corresponding to the average write energy per cycle. For the configurations that use the RFC, the bars are segmented into 5 parts: one for the average RFC read energy, a part for the average RFC write energy, a part for the RF read energy (on RFC miss), a part for the RF write energy (writebacks from the RFC to the RF) and a final part accounting for the overhead of the register file caching scheme, as described in Section 3. The results shown in Figure 4 confirm what one expects from the RFC hit rates shown in Figure 2. On the average, the RFC+96/2 configuration provides close to 38% energy savings on the average compared to the base case (128/2). The energy savings with the RFC+128/2 configuration is somewhat lower (about 34%) on the average because we have a bigger register file; this configuration provides a performance comparable to the ideal 128/1 case, within 0.1%. Note also that for benchmarks like mcf and applu that have bad reference locality, the energy savings are somewhat less pronounced even though they have a reasonable RFC hit rate. For swim and mgrid, the power savings are rather small as they have a low cache hit rate, as seen from Figure 2. As expected, the RFC read energy dominates the RFC write energy, as the RFC reads use associative lookup. The use of dissipate on a full match comparator design brings down the RFC read energy component; this component would have been far higher if traditional pulldown-on-mismatch comparators are used. The energy components shown in Figure 4 also clearly demonstrate how RF read and write access energies are reduced with the use of the RFC.

6. Conclusions

We introduced a register file cache (RFC) design that provides significant dynamic power savings in accessing register operands with negligible performance loss (about 0.3% on the average) for a superscalar datapath that has a register file that requires multi-cycle access. The basic concept behind our RFC is the segregation of the cache into two parts – a FIFO cache called the SRFC for caching short term values that are consumed within a few cycles of their production and a very small set associative cache called the LRFC for caching values that are likely to be consumed over a longer period. The overall implementation has a lower complexity compared to a fully associative design and turns out to be very effective in caching long term values. The proposed scheme also detects and predicts register values as short term or long term. The prediction information is stored in the I-cache entry of the instruction that produces the value and is used to determine if a value should be kept in the LRFC as it is shifted out of the SRFC. The technique proposed achieves a RFC hit rate of 86% on the average across all SPEC benchmarks and reduces the energy spent in accessing register operands by 40% compared to a base case that has no RFC but has one additional level of bypassing. Furthermore, the performance achieved with the RFC is almost identical to that of the base case on the average.

Although we have not quantified the leakage power tradeoffs in our design, the leakage introduced by the additional structures is somewhat offset by the savings in the register file area – with the use of the RFC, the main RF can be smaller compared to a base case design that has comparable performance and requires a larger RF with more ports. The leakage in the R-RAT extensions can be reduced by making the bitcell designs stacked in the extension part of the R-RAT entries. Any increase in the switching delays can be tolerated as these structures are off the critical path (specifically, in the retirement part of the pipeline). The leakage introduced by the extra bit needed for each instruction in the I-cache is similarly addressable using stacked or sleep devices. In fact if leakage power becomes a dominant factor, the I-cache itself has to be made resilient against leakage and the techniques used will extend naturally to the extra bit needed for each instruction.

The proposed RFC design is thus quite effective in making register accesses energy efficient.

7. ACKNOWLEDGMENTS

We would like to thank Joe Sharkey for his contributions on simulation framework used in this work. This research was supported in part by the National Science Foundation, award numbers CNS 0454298 and EIA 9911099, and by the Integrated Electronics Engineering Center at SUNY Binghamton.

8. REFERENCES

[BDA 01] Balasubramonian, R., Dwarkadas, S., Albonesi, D., "Reducing the Complexity of the Register File in Dynamic Superscalar Processor", in Proc. MICRO-34, 2001, pp 237 - 248.

[BaSo 03] Balakrishnan, S., Sohi, G. S., "Exploiting Value Locality in Physical Register Files", in Proc. MICRO-36, Dec. 2003, pp 265 – 276.

[BuSo 04] Butts, J.A. and Sohi, G. S., "Use-Based Register Caching with Decoupled Indexing", Proc. ISCA-31, 2004, pp 302.

[BTME 02] Borch, E., Tune, E., Manne, S., Emer, J., "Loose Loops Sink Chips", in Proceedings of Int'l. Conference on High Performance Computer Architecture (HPCA-02), 2002, pp 161 - 170.

[CGTV 00] Cruz, J-L., Gonzalez, A., Topham, N. and Valero, M., "Multiple-Banked Register File Architecture", in Proceedings ISCA-27, 2000, pp. 316-325.

[EGKP 02] Ergin, O. Ghose, K., Kucuk, G. and Ponomarev, D.P. "A Circuit-Level Implementation of Fast, Energy-Efficient CMOS Comparators for High-Performance Microprocessors, in Proc. 20th ICCD, 2002, pp.118-121.

[FS 92] Franklin, M.,and Sohi, G.S., "Register Traffic Analysis for Streamlining Inter-Operation Communication in Fine-Grain Parallel Processors", in Proc. MICRO-25, 1992, pp 236 - 245.

[HM 00] Hu, Z. and Martonosi, M., "Reducing Register File Power Consumption by Exploiting Value Lifetime Characteristics", in Workshop on Complexity-Effective Design, 2000.

[Ke 99] Kessler, R.E., "The Alpha 21264 Microprocessor", IEEE Micro, 19(2) (March 1999), pp. 24-36.

[LVA 95] Llosa, J., Valero, M. and Ayguade, E., "Non-consistent Dual Register Files to Reduce Register Pressure", in Proceedings of HPCA, 1995, pp. 22-31.

[LoGa 95] Lozano, G. and Gao, G., "Exploiting Short-Lived Variables in Superscalar Processors", in Proc. MICRO-28, 1995, pp 292 - 302.

[Po 03] Ponomarev, D., et. al."Reducing Datapath Energy Through the Isolation of Short-Lived Operands", in Proc. of PACT-12, 2003, pp 258- 268.

[Sim 06] Simplescalar simulator code and documentation available from Simplescalar LLC at: www.simplescalar.com.

[Sh 05] Sharkey, J., "M-Sim: A Flexible, Multi-threaded Simulation Environment." Tech. Report CS-TR-05-DP1, Department of Computer Science, SUNY Binghamton, 2005.

[SPHC 02] Sherwood, T., Perelman, E., Hamerly, G. and Calder, B., "Automatically characterizing large scale program behavior", In 10th International Conference on Architectural Support for Programming, October 2002, pp 45 - 57.

[TA 03] Tseng, J.H. and Asanovic, K., "Banked Multiported Register Files for High-Frequency Superscalar Processors", in Proc. ISCA-30, 2003, pp 62 - 71.

L-CBF: A Low-Power, Fast
Counting Bloom Filter Architecture

Elham Safi, Andreas Moshovos, and Andreas Veneris
Electrical and Computer Engineering Department
University of Toronto
{elham, moshovos, veneris@eecg.utoronto.ca}

ABSTRACT

We study the energy, latency and area characteristics of two Counting Bloom Filter implementations using full custom layouts in a commercial 0.13μm technology. The first implementation, S-CBF, uses an SRAM array of counts and a shared counter. The second, L-CBF, utilizes an array of up/down linear feedback shift registers. Circuit level simulations demonstrate that for a 1K-entry CBF with a 15-bit count per entry, L-CBF is 3.7 or 1.6 times faster than the S-CBF depending on the operation. The L-CBF requires 2.3 or 1.4 times less energy per operation compared to the S-CBF. However, the L-CBF requires 3.2 times more area. We demonstrate that for one application of CBFs (early hit/miss detection for L1 caches [12] for an aggressive dynamically-scheduled superscalar processor) the energy consumed by the L-CBF is 60% of the energy consumed by the S-CBF for most of the SPEC CPU 2000 benchmarks.

Categories and Subject Descriptors
B.7.1 Integrated Circuits , C.1 Processor Architecture.

General Terms
Design, Measurement, Performance, Experimentation.

Keywords
Counting Bloom Filters, Processors, Delay, Energy per Operation.

1. INTRODUCTION

An increasing number of architectural techniques rely on hardware counting bloom filters (CBFs) to improve upon the power, latency and complexity of various key processor structures. For example, CBFs have been proposed to improve performance and power in snoop-coherent multiprocessor or multicore systems [8,9], to improve the scalability of load/store scheduling queues [10] and to reduce replays by assisting in early hit/miss determination at the L1 data cache [12]. In these proposals CBFs eliminate broadcasts over the interconnection network in multiprocessor systems [8], or accesses to much larger

Permission to make digital or hard copies of all or part of this work for personal or classroom use is granted without fee provided that copies are not made or distributed for profit or commercial advantage and that copies bear this notice and the full citation on the first page. To copy otherwise, or republish, to post on servers or to redistribute to lists, requires prior specific permission and/or a fee.
ISLPED'06, October 4–6, 2006, Tegernsee, Germany.
Copyright 2006 ACM 1-59593-462-6/06/0010...$5.00.

and thus much slower and power-hungry content addressable memories [10], or cache tag arrays [8,9,12].

In all aforementioned hardware applications, the CBF improves the energy and latency of membership tests (e.g., whether a memory block is currently cached). It does so by providing a definite answer for most but not all tests. Thus, the CBF does not replace the underlying conventional mechanism (e.g., cache tags). Instead, the CBF dynamically bypasses the conventional mechanism as frequently as possible. Accordingly, the benefits obtained through the use of a CBF depend on how frequently it can be utilized and on the CBF's energy and latency characteristics. The more tests are serviced by the CBF alone and the lower the power and latency of the CBF, the higher the benefits. Architectural techniques and application behavior determine how many tests can be serviced by the CBF. In this work we are only concerned with implementations of CBFs that improve on energy and latency.

Conceptually, a CBF is an array of counts for which three operations are defined: increment by one, decrement by one, and test if zero. We will refer to the first two operations as updates and to the third as a probe. Previous work assumed a straightforward SRAM-based implementation which we will refer to as S-CBF (see Section 2.1). In this work we investigate the energy, latency and area of this implementation in a commercial 0.13 μm CMOS technology. However, the key contribution of this work is L-CBF, a novel implementation of CBFs that relies on up/down linear feedback shift registers (LFSRs). We demonstrate that this implementation is significantly faster, and that it requires significantly less energy than the S-CBF implementation. Using architecture level simulation of most of the SPEC CPU 2000 programs we demonstrate that L-CBF can significantly reduce power for the early detection of L1 data cache misses [12].

In more detail, the contributions of this work are as follows:

- L-CBF, a LFSR-based counting bloom filter architecture, is proposed.

- The energy, latency and area of L-CBF and S-CBF are compared using their circuit level implementation and full-custom layouts in 0.13μm fabrication technology.

- The relative energy dissipation of L-CBF and S-CBF is compared for the early detection of L1 data cache misses for most SPEC CPU 2000 programs [12].

To the best of our knowledge this is the first work that investigates the energy, latency and area of full-custom implementations of CBFs using a commercial fabrication technology. The idea of using LFSRs for the design of CBF has been proposed before but no design or evaluation of its characteristics was reported [8].

The rest of this paper is organized as follows. Section 2 reviews CBFs and the previously assumed S-CBF implementation. Section 2.2 presents the L-CBF design. Section 3 demonstrates the experimental results. Section 4 summarizes our findings.

2. COUNTING BLOOM FILTERS

Without the loss of generality, we restrict our attention to utilizing CBFs for the early detection of L1 data cache misses [12]. The concepts and implementations presented are directly applicable to other CBF applications.

In this application, the CBF determines whether a particular block of memory is currently cached in the L1 data cache. Given a block address A, the CBF reports whether A appears in any of the tags of the data cache. The CBF provides two possible answers: (1) "no, the block is definitely not cached", and (2) "unknown, may be the block is cached". In the first case, we determine that A is not cached and hence this access will result in a miss. Provided that the CBF is much faster and dissipates much less power than the L1 tag arrays, we manage to obtain the desired answer much faster and to save power. In the second case, the CBF cannot provide a definite answer and thus we do have to access the L1 tags. In this case, we incur a power penalty since we had to also access the CBF. We may also incur a latency penalty if the CBF and the L1 tag accesses are serialized (we may avoid this latency penalty if we probe the CBF and the L1 tags in parallel, in which case power benefits will be possible only if we can terminate the L1 tag access in progress when the CBF provides a definite answer).

As shown in Figure 1, the CBF can be thought of as an array of counts that is indexed via a hash function of the address A, and where three operations are defined: (1) increment count, (2) decrement count, and (3) test if count is zero, or probe. The first two operations increment or decrement the corresponding count by one, while the probe operation tests if the count is zero and returns true or false (single bit output). Simply using a portion of the address and not a more elaborate hash function has been shown to work well [8,12].

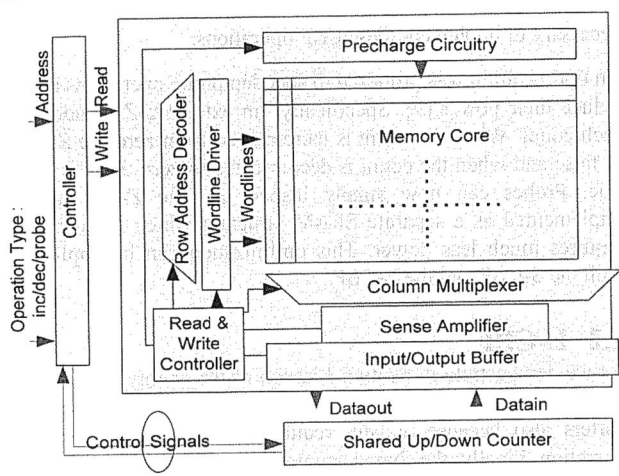

Figure 2: S-CBF architecture: An SRAM holds the CBF counts and updates are implemented as read-modify-write sequences.

Initially, all CBF counts are zero and the L1 is empty. When a block is allocated into the L1, the corresponding CBF entry is incremented by one. When a block is evicted from the L1, the corresponding CBF entry is decremented by one. To test whether A currently exists in the L1, we inspect the corresponding CBF count. If the count is zero then A is definitely not in the L1 since we would have incremented the count the moment it was cached. If the count is non-zero then it is unknown whether A is cached. Since many blocks can map onto the same CBF count, it is possible that some other cache block incremented the count[1]. Therefore, in this case we need to check the L1 tags to determine whether A is cached. It is for the latter reason that a CBF is an imprecise representation of the cached blocks. Specifically, the CBF represents a superset of the cached blocks.

A CBF is characterized by the number of entries it contains and the width of the count of each entry. Multiple CBFs with different hash functions can be used to improve accuracy [9,12]. In addition, count values are bounded. Since the same count entry is incremented and decremented on a block's allocation and eviction respectively, a count can never become negative and can never exceed the number of the total cache blocks.

2.1 S-CBF: SRAM-Based CBF

Previous work assumed a straightforward CBF implementation comprising an SRAM array to hold the counts, a shared up/down counter, a zero-comparator and a small controller [9]. This implementation is shown in Figure 2. Updates are implemented as read-modify-write sequences as follows: (1) the count is read from the SRAM, (2) it is adjusted using the counter, and (3) it is written back to the SRAM. The probe operation is implemented as a read from the SRAM and then a comparison with zero using

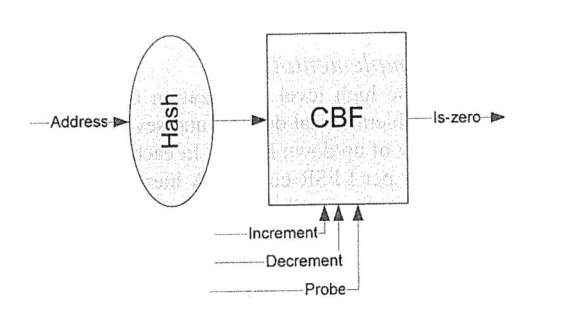

Figure 1: A CBF for cache block address membership test.

---

[1] Ideally, a separate entry would exist for every possible block address A. However, this would result in a prohibitively large table (e.g., a table with 32 million entries for a processor with a 4Gbyte address space and 32-byte cache blocks) and would negate any benefits. Accordingly, a small table is used and addresses are hashed onto the table. Hence multiple addresses may map onto the same table entry.

251

the zero-comparator. A small controller coordinates all actions necessary to implement these CBF operations.

An optimization was proposed to speedup probe operations and to reduce their power [9]. Specifically, an extra bit, Z, is added to each count. When the count is incremented from zero the Z is set to false and when the count is decremented to zero the Z is set to true. Probes can now simply inspect Z. The Z-bits can be implemented as a separate SRAM structure which is faster and requires much less power. This optimization can be applied to both the S-CBF and the L-CBF.

2.2 L-CBF

As we demonstrate in Section 3, much of the energy in S-CBF is consumed on the bitlines and wordlines. Latency and energy suffers also because updates require two SRAM accesses per operation. Finally, the shared counter further increases energy and latency.

We could avoid accesses over long bitlines by building an array of up/down counters. Then, that would eliminate the need to read the value of each counter and updates would be localized. Unfortunately, up/down arithmetic counters require many gates and are slow [11]. We make the following two observations: (1) the actual count sequence used in a CBF is not important, and (2) externally, we only care whether a count is "zero" or "non-zero". The L-CBF exploits these two properties to offer the benefits that are possible with an array of up/down counters while avoiding the overheads associated with using arithmetic counters. L-CBF uses up/down LFSRs which offer a better latency, power and complexity trade-off than other non-arithmetic counters. As we demonstrate in Section 3, L-CBF significantly reduces energy and latency compared to S-CBF albeit at the expense of increased area. However, this is a minor concern in modern processor designs for two reasons: (1) there is an abundance of resources, and (2) the CBF is tiny compared to most other processor structures (e.g., caches and branch predictors). It is unlikely that the same resources could improve performance if applied to other processor structures that are already much larger and optimized.

In the rest of this section, we review LFSRs, the construction of up/down or reversible LFSR counters, and present the organization of L-CBF.

2.2.1 Linear Feedback Shift Registers

In this section, we review LFSRs and the construction of up/down LFSR counters. A maximum-length n-bit LFSR counter sequences through 2^n-1 states. Without the loss of generality we restrict our attention to the Galois configuration of LFSRs [1]. Figure 3 shows a maximum-length 8-bit LFSR. The LFSR comprises a shift register and a few XNOR gates. Each bit of the shift register is either shifted as-is to the next bit (no tap) or is XNORed with the output of bit 7 (tap). By appropriately selecting

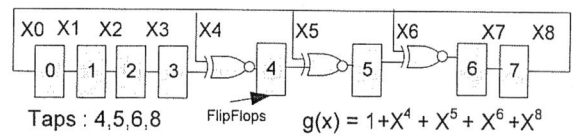

Figure 3: An 8-bit maximum-length LFSR.

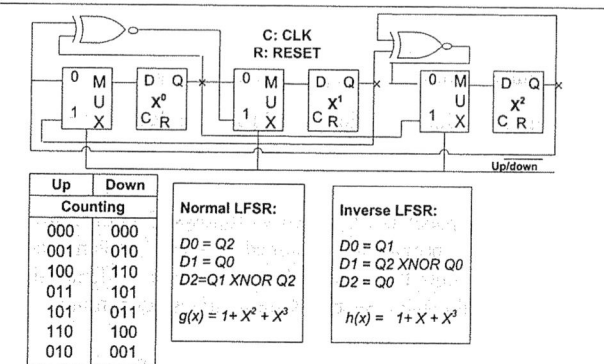

Figure 4: A 3-bit maximum-length up/down LFSR.

the tap locations it is always possible to build a maximum-length LFSR of any width that has either two or four taps [1,5]. Furthermore, ignoring wire length delays and the fan-out of the feedback line, the delay of the maximum-length LFSR is independent of its size [11]. As we show in Section 3.2 , latency increases only slightly as the number of bits increases, primarily as a result of increased capacitance on the control lines.

The tap locations for a maximum-length n-bit LFSR can be represented as a primitive polynomial $g(x)$. Figure 3 shows an example of such a polynomial. In general, an LFSR can be expressed as:

$$g(x) = \sum_{i=0}^{n} C_i \times X^i \quad (C_0 = C_n = 1)$$

where X^i corresponds to the output of the i-th bit of the shift register and the constants C_i are either 0 (no tap) or 1 (tap). This formula represents a uni-directional ("up") LFSR. If the primitive polynomial for a maximum-length n-bit LFSR is $g(x)$ (as defined by the preceding formula), then the primitive polynomial $h(x)$ of an LFSR that generates the reverse sequence is [5]:

$$h(x) = \sum_{i=0}^{n} C_i \times X^{n-i} \quad (C_0 = C_n = 1)$$

The superposition of the two LFSRs (the original and its reverse) forms a reversible ("up/down") LFSR. This reversible LFSR can be implemented using the same shift register, a 2-to-1 multiplexer per bit to control the direction of the shift and several XNOR gates. Figure 4 shows the construction of a 3-bit up/down LFSR. In general, it is possible to construct a maximum-length up/down LFSR of any width with either two or six XNOR gates (i.e., four or eight taps).

2.2.2 L-CBF Implementation

Figure 5 shows the high level organization of the L-CBF. The L-CBF includes a hierarchical decoder and several partitions each containing an array of up/down LFSRs. In each partition there is a local zero detector per LFSR counter. A hierarchical multiplexer collects these local "is-zero" signals and provides the single "is-zero" output. The L-CBF accepts three inputs and produces a single output "is-zero". The 2-bit input operation encodes any of the three possible operations or none. The address lines are used to specify the address in question and the reset signal is used to initialize all LFSRs to the "zero" state. An external clock source

252

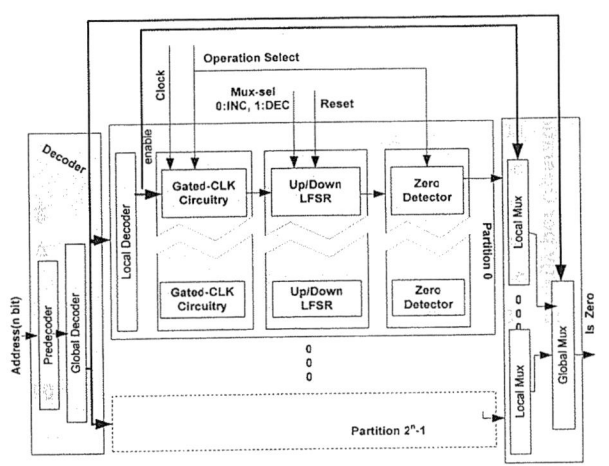

Figure 5 : L-CBF architecture.

is needed. The LFSRs use two non-overlapping phase clocks that are generated internally from the external clock signal.

We use a hierarchical decoder for decoding the address to minimize the energy and delay [2]. The decoder consists of a pre-decoding stage, a global decoder that selects the appropriate partition, and a set of local decoders, one per partition. Each partition contains an array of up/down LFSRs. Each row in each partition contains an up/down LFSR and its zero detector. Finally, we use a hierarchical multiplexer for selecting the appropriate zero-detector output for the "is-zero" operation. Figure 6 shows the basic cells we used to implement each LFSR and the zero-detector. Shown are the flip-flop for the shift register cells, the multiplexers that control the direction of change ("up"/"down"), the XNOR gate, and a bit-slice of the zero-detector. Due to space limitations we do not provide additional details on the L-CBF implementation.

3. EXPERIMENTAL RESULTS

In this section, we compare the energy consumption, delay, and area of the S-CBF and L-CBF implementations. We first compare the designs on a per operation basis and then report energy savings with L-CBF over S-CBF for L1 hit/miss detection using architectural simulation of several SPEC 2000 benchmarks. We

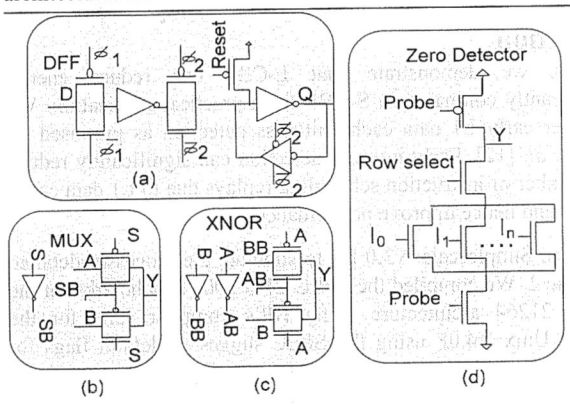

Figure 6: The cells used to implement each up/down LFSR: (a) the two-phase flip-flop (b) the 2-to-1 mux (c) XNOR gate, (d) a bit-slice of the embedded zero detector .

implemented all designs using Cadence(R) tools in a commercial 0.13μm fabrication technology. We did not use an automated process to generate the designs. Instead, we used full-custom design and attempted to optimize the energy and latency of both designs as much as possible. We used the Spectre simulator for circuit simulations. This is the vendor recommended simulator for design validation prior to manufacturing.

The rest of this section is organized as follows. We initially consider a 1K-entry CBF with 15-bit entries as it is representative of the CBFs used in previous proposals [9,12]. In Section 3.1 we compare the energy, delay and area of the two designs for each of the three operations (increment, decrement and probe). In Sections 3.2 we study how energy and delay change as we vary the number of entries, the width of the counters and the number of taps. In Section 3.3 we demonstrate that L-CBF can reduce energy to 60% compared to S-CBF when used for early L1 cache hit/miss determination.

3.1 Delay and Energy per Operation

We compare implementations of a 1K-entry, 15-bit count per entry CBF. For S-CBF, we use an SRAM with a total capacity of 15Kbits. We partitioned the SRAM in order to minimize its power/delay product. For the S-CBF we do not consider the delay and energy overhead of the shared counter since our goal is to demonstrate that the L-CBF consumes less energy and it is also faster. To further reduce energy for probes in the S-CBF design, we introduce an extra bit per entry which is updated only when the count changes from or to zero as described in Section 2.1 (Z-bits). On a probe, we only read this bit. Furthermore we applied a number of latency and power optimizations on the S-CBF [2,3,7,4]. The Divided Word Line (DWL) technique which adopts a two-stage hierarchical row decoder structure was used to improve speed and power [3,7]. Power was further reduced via pulse operation techniques for the word-lines, the periphery circuits and the sense amplifiers [7]. We also used multi-stage static CMOS decoding [4] and current-mode read and write operations for further power reduction [7]. For the L-CBF implementation we use 16-bit LFSRs so that the LFSR can count at least 2^{15} values.

Table 1 shows the delay in picoseconds, the energy (static and dynamic) per operation in picojoules and the area in square micrometers for both the L-CBF and the S-CBF. The last column reports the ratio of S-CBR over L-CBF per metric. We report two rows per category, one for the update and one for the probe operation. For delay and energy we report the worst case which we measured selecting appropriate input vectors. Given that we do not consider the overhead (latency and energy) of the shared counter, the measurements for the S-CBF are optimistic.

Table 1. Energy, delay and area of the S-CBF and L-CBF implementations of a 1K-entry, 15-bit CBF.

	Operation	L-CBF	S-CBF	S-CBF/ L-CBF
Delay (ps)	inc/dec	447.26	1670	3.7
	probe	580.32	910.12	1.6
Energy (pj)	inc/dec	38.73	88.98	2.3
	probe	30.36	41.02	1.4
Area (um²)		945825	295570	0.31

253

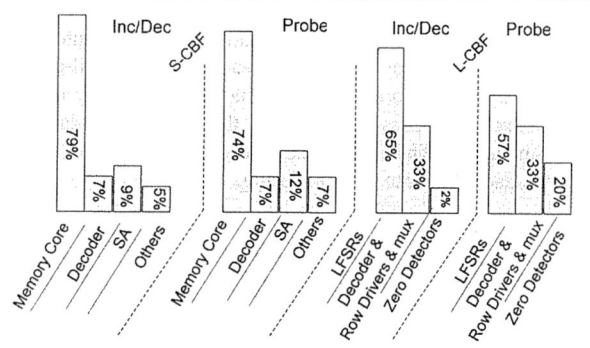

Figure 7: Per component energy consumption for the S-CBF and the L-CBF designs. Two sets of results are shown per design, one for the update operations (Inc/Dec) and one for the probe operation.

The L-CBF is 3.7 and 1.6 times faster than the S-CBF during updates and probes respectively. In addition, the L-CBF consumes 2.3 and 1.4 times less energy compared to the S-CBF for updates and probes respectively. These significant gains in speed and energy consumption come at the expense of increased area. The L-CBF is about 3.2 times larger than the S-CBF. As mentioned in Section 2.2 this is less of a concern in modern processor designs.

Figure 7 shows a per component breakdown of energy consumption for the two designs and for the two operation categories. For the S-CBF, we can observe that most of the energy (79% and 74% respectively for updates and probes) is consumed by the memory core (worldlines, bitlines and SRAM cells). The decoder and the sense-amplifiers consume considerably less energy. This is expected as we applied aggressive energy and latency optimizations to these components. Finally, a small percentage of the overall energy is consumed by peripheral circuitry such as the precharge and write logic.

For the update operations of the L-CBF, 2% of the total power is dissipated due to the leakage in the zero-detectors that are inactive during updates. Same reasoning applies to 33% of the total power that is dissipated in the inactive parts during the probe.

3.2 Sensitivity Analysis

Thus far we have focused on a specific CBF. In this section we vary the number of entries and the width of the counts. Figure 8 reports the energy per operation for CBFs of 64 through 1K entries in power of two steps. We observe that the L-CBF always consumes less energy than the S-CBF and that the relative difference increases slightly for larger entry counts.

Figure 9 reports the energy per operation as a function of count width in the range of 4 to 16 bits. In this experiment we limit our attention to a 64-entry CBF. Along the L-CBF measurements we also report the number of taps needed by each count width (either four or eight). We observe that L-CBF's energy scales better than S-CBF's. L-CBF energy increases slightly for wider counts. Communication in the L-CBF is primarily between adjacent cells. For this reason, increasing the number of cells does not impact overall energy significantly. S-CBF's energy increases at a greater rate because additional bitlines and sense amplifiers are introduced and to a lesser extent because the wordlines become

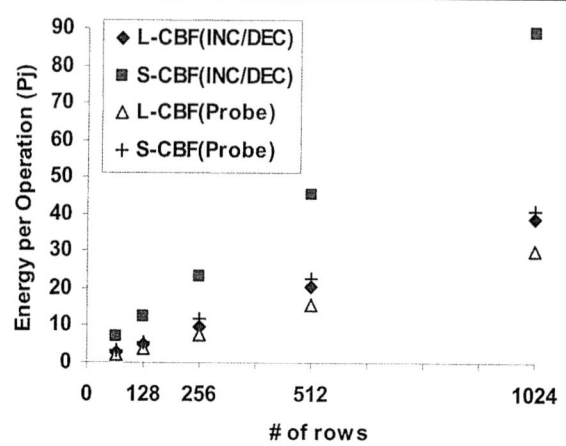

Figure 8: Energy per operation as a function of entry count for L-CBF and S-CBF for 15-bit counts.

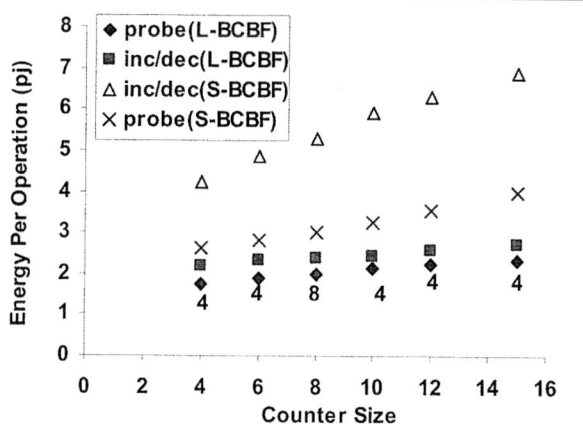

Figure 9: Energy per operation as a function of count width for a 64-entry CBF.

longer. As it can be seen in Figure 9 changing the number of taps in the L-CBF does not significantly impact energy.

3.3 Energy Savings for Early Hit/Miss Detection

Finally, we demonstrate that L-CBF can reduce energy significantly compared to S-CBF for a practical application. We consider early L1 data cache hit/miss detection as proposed by Peir *et. al.* [12]. Early hit/miss detection can significantly reduce the number of instruction scheduling replays due to L1 data cache misses and hence improve performance.

We used Simplescalar v3.0 [6] to simulate the processor detailed in Table 2. We compiled the SPEC CPU 2000 benchmarks for the Alpha 21264 architecture using HP's compilers and for the Digital Unix V4.0F using the SPEC suggested default flags for peak optimization. We used a reference input data set all the benchmarks. To obtain reasonable simulation times, samples were taken for five billion committed instructions per benchmark. We skipped 100 billion committed instructions prior to collecting

Table 2. Base processor configuration

Branch Predictor	Fetch Unit
8K-entry GShare and 8K-entry bi-modal 16K selector 2 branches per cycle	Up to 8 instr. per cycle 64-entry Fetch Buffer Non-blocking I-Cache
Issue/Decode/Commit	**Scheduler**
any /8 instr. per cycle	128-entry 64-entry LSQ
FU Latencies	**Main Memory**
Default simplescalar values	Infinite, 200 cycles
L1D/L1I Geometry	**UL2 Geometry**
64KBytes, 4-way set-associative with 64-byte blocks	2Mbytes, 8-way set-associative with 64-byte blocks
L1D/L1I/L2 Latencies	**Cache Replacement**
3/3/16 cycles	LRU
Fetch/Decode/Commit Latencies	
4 cycles + cache latency for fetch	

measurements for all benchmarks except for art and parser for which we only skipped 20 billion instructions.

We simulate a 512-entry CBF with 11-bit counts. The CBF is indexed using nine continuous address bits starting immediately after the last bit that is used as an offset within a cache block. Figure 10 shows the ratio of the energy consumed by the L-CBF over the energy consumed by the S-CBF for this application. A breakdown also in terms of updates and probes is shown. Overall, L-CBF reduces energy by about 40%. Should a larger CBF was used; the energy savings would be higher. Moreover, in this experiment we do not consider any energy savings that would be possible by voltage scaling in the L-CBF. Because the L-CBF is faster than the S-CBF it may be possible to reduce power further by scaling its voltage supply.

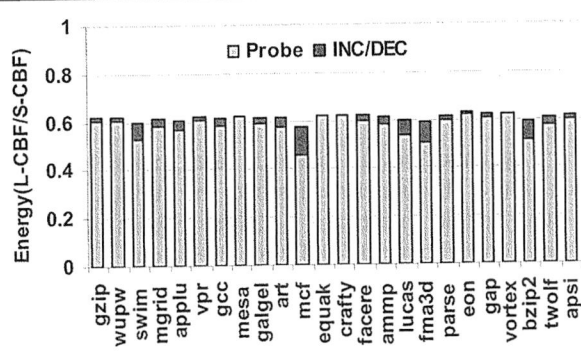

Figure 10: Energy ratio of L-CBF over S-CBF for a 512-entry, 11-bit count CBF for early L1 data cache hit/miss detection.

4. SUMMARY

We presented two designs of CBFs, one based on a SRAM array of counts and one based on an array of linear feedback shift register counters. We evaluated the energy, latency and area of the two implementations of CBFs using a commercial semiconductor technology. Finally, we studied energy consumption for a practical application of CBFs using architectural simulation. The LFSR-based CBF design is superior to the SRAM-based CBF design in both latency and energy at the expense of more area.

ACKNOWLEDGMENTS

We would like to thank Farid Najm and the anonymous reviewers for their comments. We are grateful to Mohammad Haji rostam for his assistance in circuit design and simulations and Navid Azizi for his help with SRAM design. This work was supported by an NSERC Discovery Grant, a Canada Foundation for Innovation Equipment Grant, an Intel Research Council Grant and by Semiconductor Research Corporation under contract #901.001.

REFERENCES

[1] P. Alfke, "Efficient Shift Registers, LFSR Counters, and Long Pseudo-Random Sequence Generators", Xilinx, Application Note 052, Jul. 1996.

[2] B. S. Amrutur and M. A. Horowitz, "Fast Low-Power Decoders for RAMs", IEEE Journal of Solid-State Circuits, 36(10):1506- 1515, Oct. 2001.

[3] B. S. Amrutur, "Design and Analysis of Fast Low Power SRAMs", Ph.D. dissertation, Electrical Engineering Department, Stanford University, 1999.

[4] B. S. Amrutur and M. A. Horowitz, "Speed and Power Scaling of SRAM's", IEEE Journal of Solid-State Circuits, 35(2):175-185, Feb. 2000.

[5] P. H. Bardell, W. H. McAnney, and J. Savir, "Built-In Test for VLSI: Pseudorandom Techniques", John Wiley & Sons, Inc., 1987.

[6] D. Burger and T. Austin. "The Simplescalar Tool Set v2.0", Technical Report UW-CS-97-1342, Computer Sciences Department, University of Wisconsin-Madison, Jun. 1997.

[7] M. Margala, "Low-power SRAM Circuit Design", In Proc. of IEEE Workshop on Memory Technology, Design and Testing, 115 - 122, Aug. 1999.

[8] A. Moshovos, "RegionScout: Exploiting Coarse-Grain Sharing in Snoop-Coherence", In Proc. Annual International Symposium on Computer Architecture, Jun. 2005.

[9] A. Moshovos, G. Memik, B. Falsafi, and A. Choudhary, "Jetty: Filtering Snoops for Reduced Energy Consumption in SMP Servers", In Proc. of the Annual International Conference on High-Performance Computer Architecture, 85–96, Feb. 2001.

[10] S. Sethumadhavan, R. Desikan, D. Burger, C.R. Moore, S.W. Keckler, "Scalable Hardware Memory Disambiguation for High-ILP Processors", IEEE Micro, 24(6):118 - 127, Nov. 2004.

[11] M. R Stan, "Synchronous Up/Down Counter with Clock Period Independent of Counter Size", In Proc. IEEE Symposium on Computer Arithmetic, 274-281, Jul. 1997.

[12] J. K. Peir, S.C. Lai, S.L. Lu, J. Stark, and K. Lai, "Bloom Filtering Cache Misses for Accurate Data Speculation and Prefetching", In Proc. Annual International Conference on Supercomputing, Jun. 2002.

A Low Power SRAM Architecture Based on Segmented Virtual Grounding

Mohammad Sharifkhani, Manoj Sachdev

University of Waterloo, 200 University Ave. W., Waterloo, On, N2L3G1, Canada

+519-888-4567 (ext.2033), msharifk@vlsi.uwaterloo.ca

ABSTRACT

A novel architecture for the reduction of both dynamic and static power consumption of static random access memories (SRAM) is presented. The scheme is based on the segmented virtual grounding (SVGND) of the SRAM cells. Substantial leakage reduction is achieved by increasing the threshold voltage of the cell transistors through body effect. The write and read energy consumptions are reduced significantly by decreasing the bitline voltage swing and the number of bitlines affected in each transaction. Unlike recently reported low-power schemes, SVGND allows multiple words to be placed in each row while keeping the dynamic power low. This feature is achieved by introducing an additional operation mode to the SRAM cells. The architecture is implemented in a 130nm CMOS technology. Using this scheme, the read and write array energy consumption can be saved by 44% and 84% respectively. Measurement results portraits 15 times leakage reduction compared to the conventional scheme.

Categories and Subject Descriptors

B.3.2 [**Hardware**]: Memory units – design style.

General Terms: Design.

Keywords

Low-power, static-random access memory, SRAM, Leakage reduction, write power reduction.

1. INTRODUCTION

The explosive growth of battery operated applications has compelled circuit designers to move toward low-power circuits and architectures. Scaled technologies with increased leakage current demand novel low-power techniques to combat the power drain in the standby mode. In addition, with increased capacitance of the interconnect wires and bitlines, restraining the dynamic power consumption is becoming a challenging task. SRAM units with long and heavily loaded interconnects and numerous leakage paths impose a major obstacle in power reduction.

Permission to make digital or hard copies of all or part of this work for personal or classroom use is granted without fee provided that copies are not made or distributed for profit or commercial advantage and that copies bear this notice and the full citation on the first page. To copy otherwise, or republish, to post on servers or to redistribute to lists, requires prior specific permission and/or a fee.

ISLPED '06, October 4–6, 2006, Tegernsee, Germany.

Copyright 2006 ACM 1-59593-462-6/06/0010...$5.00.

Due to their low activity and high density nature, SRAM cells, play the most important role in the DC current of the unit. On the other hand, the voltage variation of the heavily loaded bitlines takes a significant share of the dynamic power consumption. In particular, dynamic power consumption is important in write operation since it requires high voltage swings over the bitlines. Recently, with the expansion of the SRAM arrays on CMOS chips, power reduction of these devices has received significant attention by many researchers.

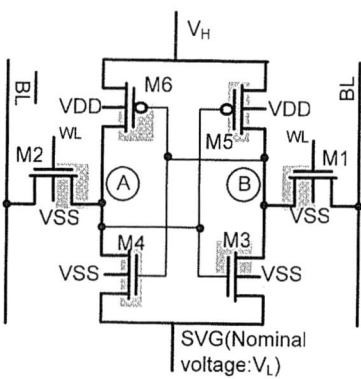

Figure 1 Cell with a virtual ground, nominally connected to V_L; The gray areas shows the internal capacitances of the cell affected by variation of the virtual ground voltage when $V_A = V_L$, $V_B = V_H$.

Itoh provides an excellent review on low power SRAM design strategies including leakage reduction methods [1]. Reduction of the dynamic power consumption especially power consumption associated with the write operation is the main focus of [2, 3, 4].

Low dynamic power methods are mostly based on reduction of the signal swings over the bitlines. Reduction of the bitline precharge voltage reduces the power dissipation in both read and write operations. This method has been utilized in [3] at the expense of noise margin degradation in read and write operations and some design overheads. In this scheme, the bitline voltage swing is reduced to $V_{dd}/2$. Recently, another write power reduction scheme has been reported in [4] which is based on hierarchical bitline and local sense amplification. In this scheme the voltage swing of the bitlines for the write operation is reduced to Vdd/10 at the expense of two additional full swing control signals in the array that run in parallel to the wordline. In this scheme both read and write operations require switching of these signals. Aside from complexity and area overhead, having additional full-swing

256

control signals imposes substantial limitation on the power reduction efficiency, especially if more than one word is stored in a row. In addition, this configuration is destructive to the neighboring words in the same row in write operation. Therefore, this scheme can be categorized as a single word per row architecture similar to [2].

Figure 1 shows an SRAM cell featured with a virtual ground. Reduction of the write power consumption has been successfully implemented using virtual grounding methods [2, 5, 6]. In these schemes, the voltage swing of the bitlines is reduced significantly by reducing the write noise margin of the cell. Reduction of the write noise margin is achieved by reducing the cell supply voltage during the write operation which results in the destruction of the data stored in a cell. During the read operation, however, the supply voltage increases to provide sufficient overdrive for discharging the bitline. [2] and its predecessor [4] have reduced the write power consumption significantly by floating the source line of the cells on the same row and reducing the required bitline variation to Vdd/6 with a small area overhead. This scheme, however, drives the non-accessed neighboring cells on the shared source line to an unstable floating region during the write operation, destroying the data of the words located on the same row. This effect limits the application of this technique to high bitwidth applications in which one row represents only one word. A similar write power reduction method has been reported in [6] with the source line of the cells on the same column are shared instead of the source line of the rows, saving the power consumption of the neighboring bitlines. However, in that scheme, the voltage swing of the cell internal nodes of the accessed column increases the read power consumption significantly.

Due to higher leakage currents in the scaled technologies, DC power reduction of the SRAM cells has received significant attention. SRAM cells with multiple threshold voltages have been suggested in [7]. In [8], variation of the supply voltage is adopted to cut the leakage current of a whole array in a cache memory. In general, variation of the supply voltage or the virtual ground voltage of the SRAM cells requires careful power trade-off considerations. That is because the power supply and virtual ground nodes are highly capacitive and excessive voltage variations can impose significant dynamic power consumption.

This paper proposes a segmented virtual grounding scheme to reduce the power consumption of the SRAM. It investigates the architectural issues for the scheme. The rest of the paper is organized as follows. Next section discusses the operation modes of an SRAM cell and the corresponding voltage settings in the proposed scheme. Section three elaborates the architecture that enables the said voltage settings under different operational modes. Section four presents the implementation and the results of

an SRAM array that is based on the SVGND scheme. Section five elaborates the dynamic power efficiency using a numerical example. Finally, section six concludes the paper. This architecture is implemented in a CMOS 130 nm technology with the nominal supply of 1.2V.

2. Cell Voltage Settings and Operation

Reduction of the leakage current of an SRAM cell, during the data retention mode, is the prominent motivation in a low DC power configuration. There are three leakage currents in a cell. In figure 1 lets assume node A carries a logic '0' at V_L while node B carries a logic '1' at V_H. In the conventional scheme, V_L and V_H are equal to VSS and VDD, respectively. Two leakage currents pass through the two internal inverters. The currents are determined by the leakage current of the transistors that are off (i.e., $V_{gs}=0$), namely M3 and M6. Assuming that the bitline voltages are at V_H the third leakage current takes path from M2 toward M4. The amount of this current is determined by M2. The leakage current through M1 is negligible since its V_{ds} is zero. According to [9] the leakage current has an exponential relationship with the threshold voltage of the leaking transistors:

$$I_s = I_0 e^{(V_{gs}-V_{th})/nV_T}(1-e^{-V_{ds}/V_T}) \qquad (1)$$

where $V_T = q/kT$ and $I_0 = \mu_0 C_{ox}(W_{eff}/L_{eff})V_T^2 e^{1.8}$. It is evident that, for supply voltage larger than V_T, plain reduction of the supply voltage induces a minor effect on the leakage current as V_{gs} remains unchanged and equal to zero for M2, M3 and M6. However, the leakage current drops exponentially once the threshold voltages of the leaking transistors are increased. This is achieved by exploiting the body effect of the transistors in SVGND scheme.

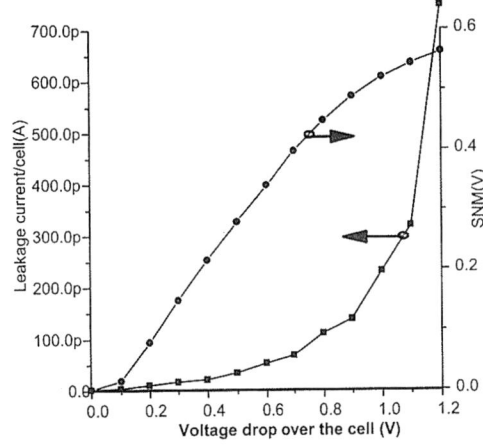

Figure 2 Measured leakage and simulated SNM as functions of voltage across the cell when V_H and V_L digress from the mid-rail(Vdd/2)

In the proposed scheme, the bodies of the load transistors are connected to VDD the highest voltage available on the chip while the body of the NMOS transistors and the access transistors are connected to the ground, VSS. When a cell is not accessed, the wordline is connected to the ground, VSS. The source of the PMOS transistors are connected to V_H, the high voltage of the cells; and the source of the drive transistors are at V_L, the virtual ground voltage of the cell. Evidently, all transistors are reverse body biased and offer a higher threshold voltage. However, reduction of the voltage across the cell (V_H-V_L) affects the static noise margin (SNM).

Figure 2 shows the measured leakage current of the SRAM cell when both V_H and V_L digress from VDD/2 towards VDD and VSS, respectively. The measurement results are based on a 130 nm, 1.2V, CMOS technology with nominal threshold voltages of 0.3V(Vbs = 0) for typical cell transistor sizes at 25°C. It can be seen that the leakage current increases exponentially when the effective voltage over the cell exceeds one V_{th} of the transistors (i.e., 0.4V when it is affected by body effect) as predicted in [7]. Therefore, the nominal voltage drop over the cell is chosen to be close to V_{th}, keeping the drive and load transistors in weak inversion region. The second curve in the figure depicts the SNM with respect to the effective voltage across the cell. It is apparent that the SNM degrades gracefully, and at 400mV it is approximately 220mV for this cell which is adequate to retain the data in data retention mode.

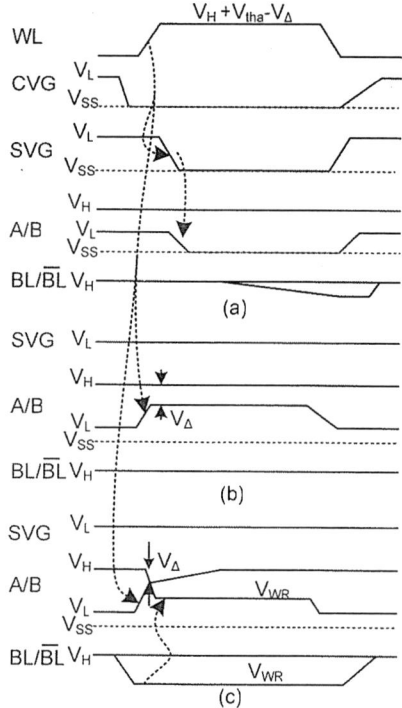

Figure.3 conceptual waveforms of cell internal nodes under read (a), AR-more(b), and write(c) operations

Figure 3.a shows the waveform of the cell internal nodes during the read operation. A successful read operation asks for sufficient voltage drop over the bitlines which are highly capacitive loads. Drive transistors, M3 and M4, need to gain a high drive capability during the read operation to discharge the bitlines. Hence, the virtual ground voltage of a cell drops to VSS in read operation. This action reduces the threshold voltage and increases the overdrive voltage V_{gs} of M3 and M4 simultaneously which results in sufficient voltage drop over the bitlines.

Depending on the number of words located in a row, only a few cells in a row need to discharge the bitlines in order to be read. Therefore, the virtual ground voltage variation is restricted only to the cells that are read in a row. That leaves the non selected cells on the same row at the nominal virtual ground voltage level of V_L during the operation. This operational mode is called *accessed retention mode* (AR-mode). Introduction of the AR-mode drops the dynamic power significantly because of two reasons: One; the variation of the virtual ground voltage of a cell associates with the (dis)charge of the internal capacitances of the cell. Shown in gray in figure 1, the said capacitance is 3-4 times larger than the capacitive load imposed to the bitline by a cell. The selective discharge of the SVG node to VSS of only to-be-read cells in a row prevents the unnecessary discharge of the internal capacitances of the non selected cells on the same row which are at the AR-mode. Two; since the virtual grounds of the non selected cells remain at the nominal voltage of V_L in AR-mode, those cells have limited drive capability. Therefore, the non selected cells on the same row do not discharge the corresponding bitlines because of AR-mode voltage setting. This results in a significant power reduction.

Conceptual waveforms of the cell internal nodes during AR-mode are shown in figure 3.b. Proper wordline voltage setting guarantees the data stability of the cell during the AR-mode. The wordline voltage is limited to $V_{wl}=V_H+V_{tha}-V_\Delta$ for the whole row; V_{tha} is the threshold voltage of the access transistors and V_Δ is close to the SNM of the cell in the retention mode in the said equation. In light of figure 3.b, the cell's behavior can be explained using figure 1 assuming node A carries a zero($V_A=V_L$) and node B carries a one ($V_B=V_H$) before the wordline becomes active. Since node B is at V_H , V_{gs} of M1 is less than V_{tha} when the cell is accessed. Therefore, M1 remains off and has no effect on node B. Node A however, raises to V_H - V_Δ where V_{gs} of M2 is equal to V_{tha} and M2 turns off. Therefore, all six transistors remain off when the cell goes to the AR-mode, leaving V_Δ voltage difference between nodes A and B. This is due to the fact that under the nominal voltage setting the overdrive voltage of M4 is too small to overcome M2 when M2 is on ($V_{gs} > V_{tha}$). So M2 lifts up the voltage of node A until this voltage reaches to V_H-V_Δ where M2 turns off(V_{gs}

$= V_{tha}$). The cell is able to recover its original state once the access transistors are deactivated. It is known that as long as the initial differential voltage between node A and node B (i.e., V_Δ) is larger than the offset voltage between the inputs of the two internal inverters, the cell can recover its original state [10]. That is, when the access transistors are deactivated, the V_{gs} of M4 is V_H-V_L which pulls node A towards V_L. In this design, V_Δ is chosen to be 200mV.

Figure 3.c shows the conceptual waveforms during the write operation. The write operation requires less bitline voltage swing in this scheme compared to the conventional write operation which needs a full VDD swing over the bitlines. The cell keeps its nominal supply and virtual ground voltages during the write operation which results in weak drive transistors. Reduction of the corresponding bitline voltage from precharge voltage of V_H down to V_{WR} is enough to set the logic state of the cell if V_{WR} is chosen to be sufficiently below V_H-V_Δ to impose proper differential voltage between nodes A and B. Simulation results suggests that 300mV (i.e., $V_H -V_{WR} > 300mV$) variation over the bitlines is sufficient to flip the cell. Such a small bit-line voltage variation over the bitlines reduces the write power consumption significantly.

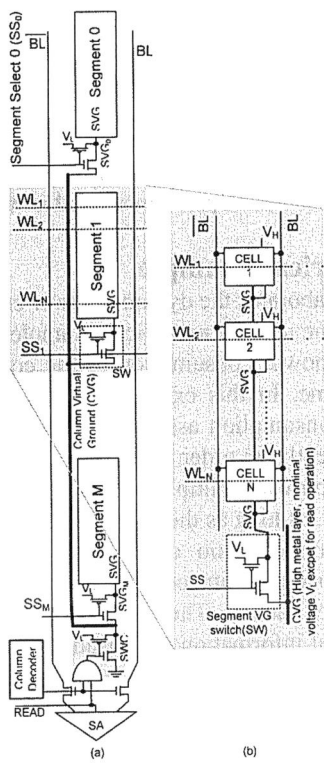

Figure 4 Architecture of a column (a), and a segment (b)

3. SVGND Architecture

SVGND scheme constitutes the operational modes and voltage settings that were discussed in the previous section. Figure 4 shows the concept of segmentation of the cells in

a column. A segment is defined as a set of N cells on the same column with a shared segment virtual ground (SVG). A virtual ground switch (SW) connects the virtual ground of the segment to the virtual ground of the column (CVG). The CVG is a high metal layer node shared between all virtual ground switches on the same column. If the virtual ground switch(SW) of a segment is activated by turning signal SS to a high voltage, the virtual ground voltage of the segment is equal to the virtual ground voltage of the column; otherwise, the virtual ground of the segment keeps its nominal voltage, V_L. The logic of the virtual ground switch (SW) is an inverter which drives the SVG node either to V_L or to the voltage of CVG node depending on the control signal, SS.

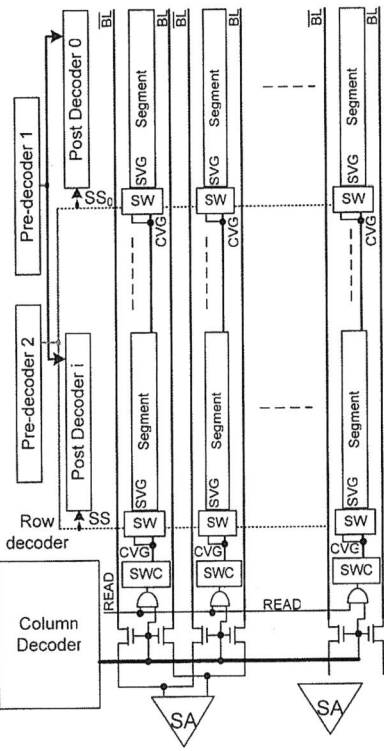

Figure 5 A memory array based on SVGND scheme

Figure 5 depicts an SRAM architecture with the SVGND scheme. In this figure, wordlines are removed for clarity. The activating signal for the virtual ground switch (SW) which is segment select signal (SS_i) is generated by the predecoder of the row decoder at no additional hardware cost. If a cell is accessed for either read or write operation, the (SS_i) signal of that segment and the neighboring segments on the same row is also active. On the other hand, the column virtual ground(CVG) is lowered to VSS only if one of the cells on the column is to be *read*. CVG of other columns remains at V_L under read operation. This operation is achieved by introducing column virtual ground switch(SWC) as shown in figure 4. Lowering the CVG of a column allows the virtual ground of the accessed cell on that column to be lowered to VSS. When the virtual ground of a cell is pulled down to the ground, the drive transistors become stronger . That is due to the elimination the body effect and increased V_{gs}. However, other cells on the same row and in the non selected columns will go to the AR-mode and do not discharge the corresponding bitlines during read or write operation.

4. Realization and Results

Figure 6 shows the layout of a 2048x22bit SRAM unit based on SVGND. The architecture is implemented in a 130nm CMOS technology on a separate chip. The unit is divided into four arrays where each array takes 150um x 410um. The area overhead associated with the segment virtual ground switch is less than 8% since N is chosen to be equal to 8. Clearly, unlike source line switch reported in [2], since only one cell in a segment discharges the bitline, the segment virtual ground switch can be as small as half of a cell. Unlike recently published low-power SRAMs of [2,4], SVGND is able to accommodate more than one word on a row which reduces the area and power of the address decoders. In this design, each row accommodates four 22-bit words in each array.

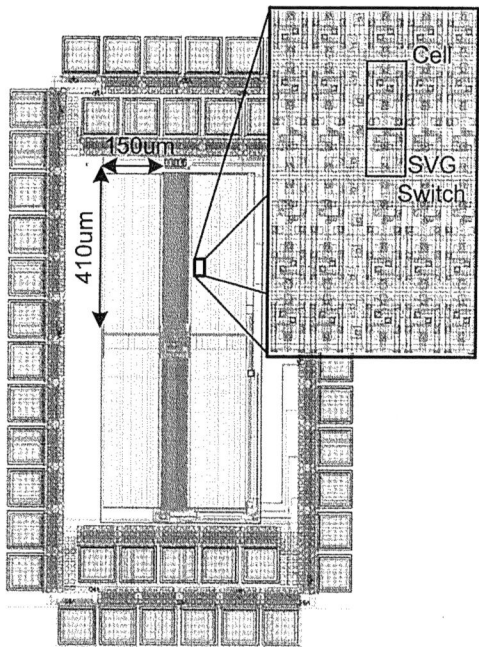

Figure 6 Chip layout

Figure 7 shows the spice simulated waveforms of the cell internal nodes of the post-layout extracted circuit. In this figure, V_H and V_L are chosen to be 0.9V and 0.5V respectively. Since the wordline voltage of $V_{wl} = 1.2V$ is chosen, the resulting V_Δ of 0.2V appears on the internal nodes when the cell goes to the accessed retention mode. In this figure, V_{wr} is 0.6V which is sufficient to flip the cell's state. This scheme drops the leakage current to 27pA/Cell.

The array write power consumption is estimated to be 84% less than the conventional write power consumption. The read power consumption is also reduced by 56% because the non-selected bitlines are not discharged in the read operation and precharge voltage is reduced to V_H.

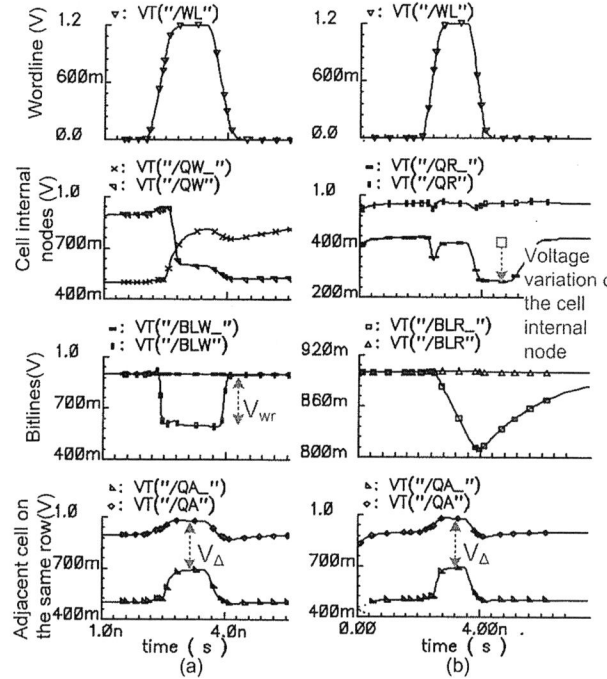

Figure 7 Spice simulated waveforms of the post-layout extracted circuit in write(a) and read(b) operation, the bottom waveform in both plots show a cell in AR-mode

5. A Numerical Example

This section elaborates the dynamic power efficiency of the SVGND scheme using a numerical example. This example compares the power consumption of the conventional and SVGND scheme. In this example we derive the column read energy consumption associated with one bit for both schemes. Evidently, in order to find the overall array power consumption, on can multiply the said value to the word size. Therefore, as long as the comparison is concerned, the energy consumption ratio of one bit is sufficient. The values of the load capacitances of different nodes are based on the layout extraction. The voltage variation values and the architectural information are based on the implemented chip. The architectural information and voltage levels are assumed to be the same for both schemes to make a fair comparison.

In this example, the number of words in a row, P, is chosen to be equal to 4 while the number of cells on a bitline, M, is 128. Therefore in order to access a single bit(cell) of a word, P bits(cells) are accessed. The bitline precharge voltage is $V_{pre}=0.9V$ and there is a voltage swing of $V_{read} =0.2V$ over the bitline(s) in the read access. Therefore, the column energy consumption associated with the conventional scheme is:

$$E_{CONV} = P \times V_{read} \times V_{pre} \times C_{BL} \times M$$

where C_{BL} is the bitline capacitance associated with one cell. Note that in conventional scheme all P bitlines are discharged during the read operation. In the SVGND scheme, the number of cells in a segment, N, is 8. The V_L swing of 0.5Vis applied to the cell internal node voltages as well as CVG of the to-be-read column. Therefore, the column energy consumption of SVGND scheme is:

$$E_{SVGND} = V_{read} \times V_{pre} \times C_{BL} \times M + V_L^2 \times C_{CVG} \times M + V_L^2 \times C_{INT} \times N$$

where C_{CVG} is the capacitive load imposed to the CVG by one cell and C_{INT} is the capacitive load of the internal nodes of a cell seen from SVG. In SVGND, in order to access one bit, only one bitline is discharged in addition to a CVG and the internal cell capacitance of one segment. C_{CVG} is 4 times smaller than C_{BL} since the former wire does not carry the capacitance of access transistors and also it is in a high metal layer with far neighboring parallel wires in contrast to the bitlines. On the other hand, as it was mentioned before, C_{INT} is 3 times bigger than C_{BL} because of the internal gate oxide capacitance (See figure 1). The effect of 8% bitline wires extension associated with SW is negligible as SW does not impose a junction capacitance to the bitlines and the additional interconnect capacitance is small(<4%). Dividing E_{SVGND} to E_{CONV} portraits the read power saving efficiency: E_{SVGND}/E_{CONV}=0.44.

Evidently, the write power saving is more since conventional scheme asks for full swing variation over the bitlines. Also, if the number of words in a row increases or bitlines length is doubled, the power saving will be even more.

6. Conclusion

A low power SRAM architecture based on segmented virtual grounding scheme is introduced. The scheme offers low power, low energy consumption in both dynamic and static perspectives. Leakage power is reduced by exploiting the body effect and increasing the threshold voltage of the drive and access transistors. 15 times leakage reduction is achieved in this scheme which is comparable to the 16 times leakage reduction technique recently reported in [11].

The scheme portraits low dynamic power in read and write operation. Read power reduction is reduced by preventing the non selected cells on the accessed row from discharging their bitlines. Write power is saved by the significant reduction of the bitline voltage swing. Unlike the recently published low power SRAMs, SVGND allows multiple words to be fit in a row while keeping the power consumption of non selected bitlines low when a word is

accessed. This feature reduces the area and power overhead of the row decoders.

The scheme requires multiple mid-rail DC voltages. The power consumption of these reference voltage generators can be substantial especially when the array is not active; the generator current becomes larger than the total leakage current of the array. This calls for agenerator-current reduction through circuit techniques that are familiar to DRAM designers [12].

7. REFERENCES

[1] K. Itoh, K. Sasaki, and Y. Nakagome, Trends in low-power SRAM circuit technologies, *Proc. IEEE, vol. 83*, pp. 524-543, 1995.

[2] K. Kanda, S. Hattori, and T. Sakurai, 90 % write power-saving SRAM using sense amplifying memory cell, *IEEE J. Solid-State Circuits, vol. 93*, pp. 929-933, 2004.

[3] K. W. Mai et al., Low-power SRAM design using half-swing pulse-mode techniques, *IEEE J. Solid-State Circuits, vol. 33*, pp. 1659-1671, 1998.

[4] B. Yang and L. Kim, A low power SRAM using hierarchical bit-line and local sense amplifiers, *IEEE J. Solid-State Circuits, vol. 40*, pp. 1366-1376, June 2005.

[5] H. Mizuno and T. Nagano, Driving source-line cell architecture for sub-1-v high speed low-power applications, *IEEE J. Solid-State Circuits, vol. 31*, pp. 552-557, 1996.

[6] N. Shibata, A switched virtual-gnd level technique for fast and low power SRAMs, *IEICE Trans. Electron., vol. E80-C*, p. 1598-1607, 1997.

[7] Y. Nakagome, M. Horiguchi, T. Kawahara, and K. Itoh, Review and prospect of low-voltage RAM circuits, *IBM J. Res. and Dec., vol. 47, no. 5/6*, pp. 525-552, Sept. 2003.

[8] K. Zhang, et al, a 3-GHz 70Mb SRAM in 65nm CMOS technology with integrated column-based dynamic power supply, *Proc. of IEEE International Solid-State Circuit Conference (ISSCC'05)*, pp.474-475, 2005.

[9] R. Gu and M. Elmasry, Power dissipation analysis and optimization of deep submicron CMOS digital circuits, *IEEE J. Solid-State Circuits*, vol. 31, pp. 707-713, May 1996.

[10] H. Veendrick, The behavior of flip-flops used as synchronizers and prediction of their failure rate, *IEEE J. Solid-State Circuits, vol. SC-15, no. 2*, pp. 169-176, Apr. 1980.

[11] R. Islam, A. Brand, D. Lippincott, Low power SRAM techniques for handheld products, Proc. of International Symposium on Low Power Devices (ISLPED'05), pp.198-202.

[12] K. Itoh, VLSI Memory Chip Design. Springer-Verlag, 2001.

Process Variation Aware Cache Leakage Management

Ke Meng
k-meng@northwestern.edu

Russ Joseph
rjoseph@eecs.northwestern.edu

Department of Electrical Engineering and Computer Science
Northwestern University
Evanston, IL 60208

ABSTRACT

In a few technology generations, limitations of fabrication processes will make accurate design time power estimates a daunting challenge. Static leakage current which comprises a significant fraction of total power due to large on-chip caches, is exponentially dependent on widely varying physical parameters such as gate length, gate oxide thickness, and dopant ion concentration. In large structures like on-chip caches, this may mean that one portion of a cache may consume an order of magnitude larger static power than equivalently sized regions.

Under this climate, egalitarian management of physical resources is clearly untenable. In this paper, we analyze the effects of within-die and die-to-die leakage variation for on-chip caches. We then propose *way prioritization*, a manufacturing variation aware scheme that minimizes cache leakage energy. Our results show that significant average power reductions are possible without undue hardware complexity or performance compromise.

Categories and Subject Descriptors

B.3.2 [**Memory Structures**]: Design Styles - *cache memories*

General Terms

Design, Performance, Experimentation

Keywords

process variation, low power, leakage, cache management, Gated-VDD, selective cache ways

1. INTRODUCTION

In future technology generations, the promise of burgeoning transistor budgets will be tempered by a sobering fact: foundries will have increasingly limited control of transistor quality. ITRS lists several fundamental challenges for chip manufacture which limit fabrication accuracy but have no known solutions [15]. These quality control issues manifest themselves as randomly distributed variations in characteristics such as transistor length and gate oxide thickness [4].

Permission to make digital or hard copies of all or part of this work for personal or classroom use is granted without fee provided that copies are not made or distributed for profit or commercial advantage and that copies bear this notice and the full citation on the first page. To copy otherwise, to republish, to post on servers or to redistribute to lists, requires prior specific permission and/or a fee.
ISLPED'06, October 4–6, 2006, Tegernsee, Germany.
Copyright 2006 ACM 1-59593-462-6/06/0010 ...$5.00.

Ultimately, these variations affect power and performance by altering circuit behavior, leading to physical implementations which may no longer meet design specifications

Leakage current, well known as a prominent threat for deep submicron design, is highly susceptible to manufacturing variations and will consequently be more difficult to optimize [7]. Recent industry data show as much as a 20× variation in total chip leakage power for dies from the same wafer [4]. The majority of this variation comes from deviations in leakage power. As a result of the exponential dependence on gate length (L) and threshold voltage (v_t), which vary on a die-to-die and within-die basis, even schematically equivalent cells on the same chip can have dramatically different leakage power.

Due to the huge potential variances in device characteristics, microarchitectures can no longer be designed with an egalitarian view of hardware resources. In particular, fixed parameter leakage models and optimizations will become increasingly less effective because they can only target the fraction of chips which have no imperfections, ignoring the majority of chips which have significant on-chip variations. The consequences will be significant for cache memories due to their large on-chip area. First, they will account for a large portion of the total chip leakage. Second, they will be exposed to significant amounts of spatially dependent parameter variation. Although techniques like adaptive body biasing ([16]) have potential for reducing leakage, they necessitate use of complicated circuit techniques which may not be feasible for fine grain spatial leakage control.

In this work, we examine the effects that leakage variation can have on cache memories. This paper makes two primary contributions:

- We develop architecture-level statistical models and methodology for studying cache leakage under manufacturing variations. We are among the first to survey the local and global effects of manufacturing variation on cache leakage.

- Based on our analysis, we introduce *way prioritization*, a leakage reduction strategy that appropriately sizes caches to reduce the average and worst case leakage power without compromising performance. Overall, we show a 28% reduction in L3 leakage power consumption versus a variation oblivious resizing strategy.

The remainder of this paper is organized as follows: Section 2 introduces the statistical models used in this paper. In Section 3, we propose a cache organization which accounts for process variation to improve leakage management. Experimental methodology and evaluation of way prioritization follow in Sections 4 and 5. Finally, we conclude with a summary in Section 6.

2. MODELING PROCESS VARIATION

In this paper, we leverage existing stochastic techniques common in the circuit design community to develop architectural models for on-chip caches. Our approach, depicted in Figure 1, uses two phases. In the first, we construct a statistical model for a cache based on organizational parameters such as capacity and block size, as well as physical parameters such as geometric position on chip and overall die size. For a fixed set of manufacturing quality parameters, this model captures the local leakage variation within different regions of the cache and the total cache leakage variation across dies. In the second phase, we conduct Monte Carlo analysis by generating multiple random samples which have the statistical properties described by our model. While Monte Carlo simulation is not as elegant as some recently proposed analytic strategies for computing total leakage [6, 14], our methodology experiments allow us to examine several spatial aspects of the problem that the analytic techniques do not support.

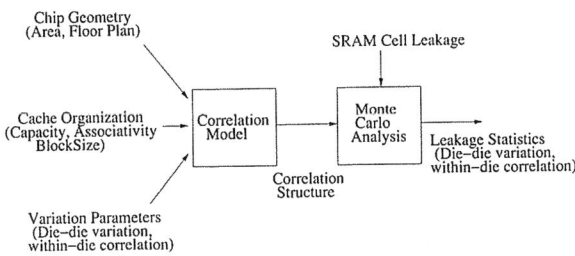

Figure 1: Leakage variation modeling approach used in this work.

2.1 Leakage Model for An Individual SRAM Cell

A six-transistor (6-T) SRAM cell serves as the basic construction block of a standard cache design. We develop a 6-T leakage model as the basis of our static power analysis. In current CMOS technologies, leakage current is composed of three major sources: subthreshold leakage, gate leakage and substrate leakage. SPICE simulation with PTM [5] 32nm technology predictive model card identifies subthreshold leakage as the dominant source for near future technologies. In the rest of this paper, we focus on subthreshold leakage, but the methodology is readily extendable to other types of leakage.

The basis of a 6-T cell is a pair of cross-coupled inverters. Subthreshold leakage occurs on a PMOS transistor and an NMOS transistor from each inverter regardless of the value stored in the cell. As shown by Rao et al., gate length variation plays a dominant role in subthreshold leakage [14]. In this work, we focus on the leakage current deviations caused by gate length variation.

Rao et al. took the following empirical equation to model the subthreshold leakage current of a single transistor with respect to the transistor gate length [14]:

$$I = p_1 e^{(p_2 L + p_3 L^2)} \qquad (1)$$

We extend the usage of the above equation to a SRAM cell. First we create a 6-T SPICE model, then we collect its leakage current during a series of simulations where we sweep transistor gate lengths over the $+/-10\%$ range of their

nominal values. We then apply curve fitting to identify the constants. In our simulations, we assume a perfect correlation between gate lengths in an individual SRAM cell due to their proximity. These are common assumptions found in the literature [1, 14].

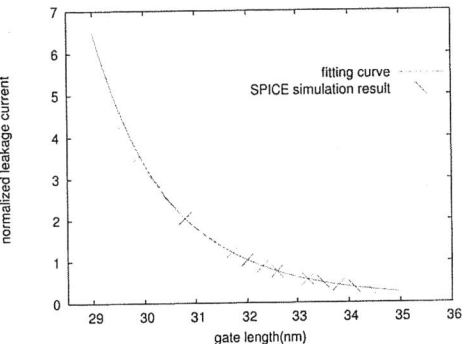

Figure 2: SPICE simulation and Curve fitting results.

Figure 2 shows the comparison between the leakage value obtained from SPICE simulation and the fitting curve. The fit matches the simulation results accurately over a wide range.

The fabricated gate length is the sum of nominal value and the variation:

$$L_{total} = L_{nominal} + L_{intradie\_variation} + L_{interdie\_variation} \qquad (2)$$

Intra-die variation expresses the difference in gate length from a transistor to another transistor on a single die, while inter-die variation refers to the die-to-die variation. We model both variations as normally distributed random variables $L_{interdie\_variation}$ and $L_{intradie\_variation}$. In the following section, we discuss the gate length correlation patterns which form the basis for the variation models.

2.2 Spatial Correlation Between SRAM Cells

Experimental results from physical measurements in [10] show that strong spatial correlation exists on intra-die gate length variations. For two identical transistors, the correlation factor of their gate length drops almost linearly as the distance between the two devices increases. The correlation nears zero as the separation approaches half the width of the die. This spatial correlation is also independent of chip size if one chip is printed per field. The consequence is that there is physical-spatial locality for leakage in regular array structures such as caches.

To model the spatial correlation in a die, we use the hierarchical correlation modeling method introduced in [1] to capture within-die variation. Figure 3(a) illustrates how the hierarchy corresponds to physical spacing for a small example. At the top level, we generate a random variable which is an additive term present in the gate length of all devices. We recursively divide the area into quadrants and generate random additive terms for each quadrant. The effective gate length for specific device can be found by adding the random variables in its hierarchy to the nominal gate length. In our experiments, we choose a grid with an area of 32 6-T cells as

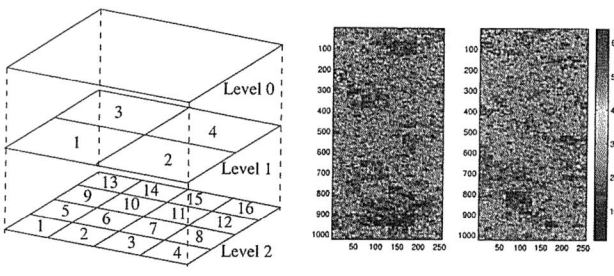

(a) Hierarchal correlation matrix   (b) Correlated leakage variation

Figure 3: Construction of correlation models used in this work and resulting leakage variation for two samples.

the terminal size and end the recursion there. Consequently, for a 25% chip area 16MB data array, we have a total of 12 levels and 4,194,304 atomic macro cells over the cache data array. By scaling the additive terms as the grid size decreases, we can match the spatial correlation presented in [10].

Figure 3(b) shows an example of how severe the leakage variation can be in cache structures. The same batch of simulation runs also featured a maximum 20× cache leakage variation from die-to-die. The implication is that it may no longer be wise to treat two chips, two processor cores on the same chip, or even two sub-arrays in the same cache identically. It is important to recognize the physical differences rooted in process variation. Additional power savings can be gained if we can find ways to manage a processor on a granularity fine enough to isolate spatial variations.

3. PROCESS VARIATION AWARE CACHE SIZING

Because caches normally occupy a large portion of on-chip area, they are a major target for static power control. The regularity of caches provide a large design space and flexibility for cache structure design. In this section, we propose a cache design with leakage variation awareness.

3.1 Enabling Techniques

Previous work on power aware processor design has examined several methods for reducing both dynamic and static power for caches [2, 11, 13, 9]. Our approach extends two widely known techniques, which we briefly summarize:

- **Selective Cache Ways [2]** - In an n-way set associative cache, one or more ways (regular sub-arrays) can be disabled, effectively reducing both the associativity and the overall size of the cache. Power saving modes can then be applied to decoders, precharge structures, sense amplifiers, and SRAM cells in the disabled ways.

- **Gated VDD [13]** - Portions of logic or memory can be selectively enabled or disabled by placing a controlling transistor on conductive paths to VDD/VSS. Leakage current can be effectively eliminated in SRAM cells via this technique.

We use cache ways as the granularity at which to apply VDD gating. With these two technologies, caches can be effectively sized to meet workload demands.

3.2 Way Prioritization

Conventional cache sizing strategies do not differentiate cache ways, despite the fact that some portions of the cache will be leakier than others. This egalitarian view of the cache may not achieve good energy savings if one of the enabled ways happens to have a high overall leakage current. Ideally, we wish to only use cache ways that have energy costs which are commensurate with the performance benefit from having increased cache size.

We propose *way prioritization* as a technique to enable the appropriate number of cache ways and select which subset of the cache array to make active. Way prioritization achieves good overall energy savings for a given performance level by accounting for the effects of within-die and die-to-die variation. Just as in the standard selective ways approach, the cache needs to be appropriately sized for a given workload. Furthermore, way prioritization can be applied in concert with circuit-level leakage and variation aware design techniques such as adaptive body-biasing [16] and multiple v_t assignment to maximize leakage savings. In this section, we focus on the mechanisms that allow us to isolate and prioritize ways. We discuss tradeoffs in sizing policies in the subsequent section. Under way prioritization the SRAM cells belonging to the same way are organized into a sub-array on the chip. In our case, this has the additional benefit of taking advantage of the spatial leakage correlation.

The key hardware difference between a standard selective cache ways implementation and a way prioritized one is a set of hardware registers which identify the leaky cache ways and make cache sizing effective. Figure 4 depicts the hardware differentiation and highlights the PRIORITY and DEGREE registers. The PRIORITY register consists of n entries of $log_2 n$ bit size for an n-way set associative cache. The entries in the PRIORITY register correspond to physical ways sorted by descending leakage power. The DEGREE register supplements this information by tracking the absolute leakage of the corresponding physical way. When the cache is being resized for a particular workload, these registers can be queried to determine *how many* ways should be enabled and *which specific* ways should be enabled.

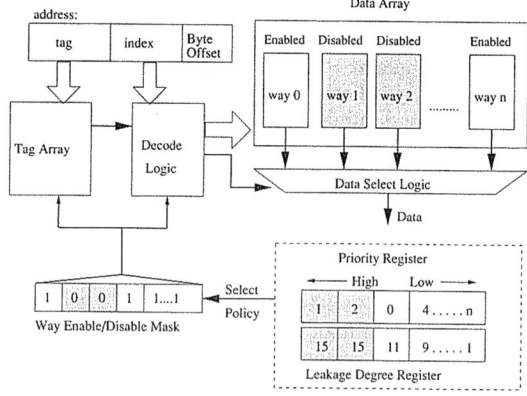

Figure 4: Hardware organization for prioritized cache ways

The measurements needed to populate the leakage registers can be collected off-line during the manufacturing test phase. Individual cache ways can be independently enabled as part of a built in self-test (BIST) sequence while the rest

264

of the processor is left idle. The leakage current for each way can be calculated from ammeter readings of total chip current draw. Collected data can be quantized, physical ways can be sorted by their leakage power, and the resultant information can be kept in non-volatile near-chip storage. At boot time, the PRIORITY and DEGREE registers can be configured based on the previously determined values.

3.3 Cache Sizing Policies for Individual Chips and Workloads

Way prioritization allows the cache to be sized and configured on a per workload and per chip basis. Given knowledge of how application performance varies with increasing total cache size, we can either chose a sizing which minimizes power for a fixed performance level or we can target a more flexible power/performance optimization metric (e.g. energy-delay product). We assume that static profiling [2] or dynamic working set analysis [8] can identify how performance scales with cache size. The remainder of this discussion focuses on policies.

For a fixed performance level, for example 2% slowdown, a working set profile provides an appropriate cache size, or equivalently, k, the optimal number of active ways. Under way prioritization, the cache can be resized by enabling the last k physical ways held in the PRIORITY register and disabling all other ways. Any dirty data held in a newly disabled way is written back to the next level in the memory hierarchy before VDD is finally gated. Because the physical ways are pre-sorted by leakage, this cache configuration represents the organization which would produce the minimal leakage power for that given performance level.

For optimizations which allow a variable amount of performance degradation, we need to know the incremental energy cost of enabling each additional way. Different physical chips may have different total leakage or different ratios of leakage between ways. The DEGREE register tracks how much leakage energy each additional way contributes. When reconfiguring the cache to minimize energy-delay, for example, the optimal value can be found by iterating through the PRIORITY and DEGREE registers. Initially, the core power would be added to the leakage power for the least leaky way and that sum would be multiplied by $slowdown^2$ for the minimal cache size. For each additional way, the total chip power increases, but the delay (obtained via profiling) decreases. The minimal energy-delay product can be found by repeating this process. If core power increases and delay decreases monotonically with cache size, the search process need not be exhaustive: an increase in energy-delay signals that the optimal sizing has already been reached. In the worst case, none of the ways are disabled, and a total of n iterations are performed. Note that the pre-sorting of the PRIORITY register means that we do not need to identify all possible combinations of physical ways. Because this sizing process is slightly more complicated than trying to meet a fixed performance goal, it may be preferable to implement it in microcode or the operating system.

In addition to manufacturing variations in gate length, runtime parameters like temperature can also have an influence on leakage power. Our work focuses primarily on large L3 caches which typically have lower peak temperatures, so we believe that the thermal contribution to variation will be relatively small. However, on-chip thermal sensors could be combined with structures like the DEGREE register to adjust power estimates for runtime conditions. We plan to investigate this approach in future work.

4. EXPERIMENTAL METHODOLOGY

4.1 Processor Model

Our experiments model a high-performance server class processor comparable to the Intel Montecito [12]. The processor features two symmetric cores, each of which has private L1, L2, and L3 caches. Table 1 summarizes the pipeline configuration and cache organization for our processor. The processor executes the Alpha ISA.

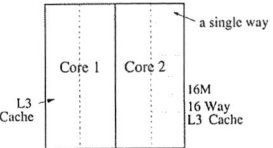

Figure 5: The base floorplan for performance simulation

We target this design for a future 32nm technology process where leakage and process variation will play prominent roles. We devise the floorplan for this design by assuming that the footprint for a single processor core is equivalent to that of a single core for Montecito in a 90nm process. We then scale the core to 32nm and assume a base floorplan for our experiments as presented in Figure 5. The cache structure studied is a re-sizable 16M L3 Cache which occupies 25% of the total chip area.

The processor is modeled via the M5 Full System simulator [3] which includes detailed models of pipelines, caches, buses, and off-chip memory. The simulator runs a slightly modified version of the linux-2.6.8.1 kernel and captures all of the performance effects of the multiprogrammed workloads used in this study.

Single Core		Workload	
		Name	Apps
Clock	2.5 GHz	int.2.1	bzip2, crafty
Fetch/Decode	4 inst	int.2.2	eon, twolf
Issue	6 inst, out-of-order	fp.2.1	ammp, art
IQ/LSQ/ROB	32/40/80 entries	fp.2.2	equake, mesa
Func Units	4 IntALU,1 IntMul	mix2.1	ammp, bzip2
	1 FPALU,1 FPMul	mix2.2	art, crafty
	2 MemPorts	mix2.3	equake, eon
L1 I-Cache	16KB 4-w 1 cyc	mix2.4	mesa, twolf
L1 D-Cache	16KB 4-w 2 cyc		
L2 I-Cache	1MB 8-w 7 cyc		
L2 D-Cache	256KB 4-w 6 cyc		
L3 Cache	16MB 16-w 20 cyc		

Table 1: Processor parameters for a single core and workloads used in this study.

4.2 Workloads

To evaluate the efficacy of our leakage management approach, we use several multiprogrammed workloads which showcase a variety of memory usage patterns. Individual applications are taken from the SPEC CPU2000 benchmark suite. To reduce the total number of simulations, we identify a subset of SPEC applications which exhibit a range of performance characteristics and then group them into eight mix sets. Table 1 also shows the workload groupings used in the experiments. The benchmarks were compiled with `gcc-4.0.1` for `alpha-linux` using `-O4 -ffast-math -funroll-loops` flags. All of our workloads run on a single core of our dual core design. We report performance and power savings with respect to a single core.

5. RESULTS

5.1 Leakage Variation in Cache Structures

We used our Monte Carlo method to evaluate spatial leakage variation under several different cache configurations. Each simulation point consists of 10,000 samples. In all cases we assumed normal distribution on gate length intra-die variation and the 3σ value of the distribution was set to 9.4% of the nominal value.

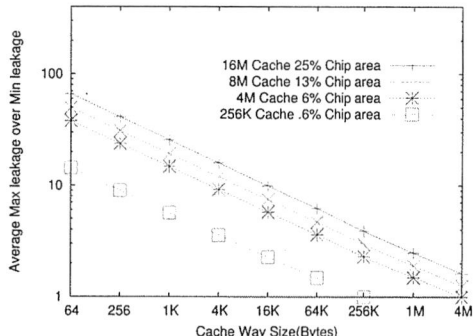

Figure 6: Average max leakage over min leakage with respect to way sizes.

Figure 6 shows the dramatic difference in cache leakage for regions chosen from different locations in the data array. This log-log plot tracks the ratio of maximum to minimum leakage power for cache regions of varying size. For each region size, we first select square sub-arrays of the cache which have maximum and minimum leakage, and then we compute their ratio. We draw these curves for a wide range of cache sizes. From Figure 6, we first see that the leakage ratio decreases rapidly as the region size increases. This is due to the fact that when the regions are small, there are many distant sections to choose from, increasing the chance that the regions do not have similar parameter sizes. As the regions grow, both the maximum and minimum leakage regions tend towards mean values, and the distance between the regions decreases. The second trend is that increasing cache sizes boost the max/min ratio. This is due to the fact that larger caches have a larger population of 6-T cells and hence longer "tails" on the distribution. We can extrapolate the benefits of way prioritization if we consider cache ways to be our regions. First, the savings are likely maximized on very large caches. Second, greater associativity and hence a larger number of small cache ways increase the potential for power savings. In the remainder of this paper, we focus our analysis on a large, highly associative cache (16-way 16M L3).

Again employing our Monte Carlo approach, we create 10,000 random samples of the L3 cache described in Section 4, assuming zero global inter-die variation. For each run, we

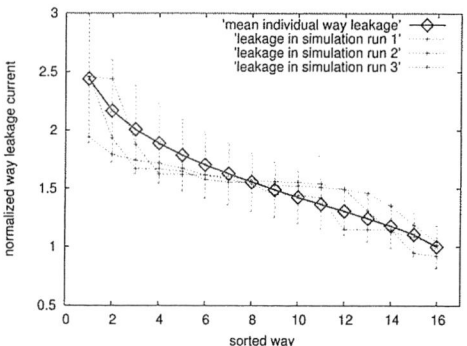

Figure 7: Normalized mean leakage current from most leaky to least leaky ways.

Inter-die Variation 3σ Perc. of Nominal Value		Normalized Way Leakage Rank(Decreasing Leakage)			
		1	6	10	16
0%	mean	2.44	1.70	1.43	1.00
	stdev	0.55	0.29	0.23	0.18
2.8%	mean	2.59	1.80	1.50	1.05
	stdev	1.10	0.67	0.54	0.37
5.6%	mean	2.75	1.90	1.58	1.10
	stdev	1.57	0.98	0.79	0.52
9.4%	mean	3.05	2.09	1.73	1.19
	stdev	2.40	1.50	1.20	0.78

Table 2: The influence of inter-die variation on way leakage current distribution.

sorted the cache ways by decreasing leakage current. Figure 7 shows the normalized mean leakage current for the ranked cache ways and the standard deviation spread. We can see that on average the leakiest way consumes 2.44 times more static power than the least leaky way. Figure 7 also shows way leakage current for several samples. Clearly, different chips can have divergent leakage profiles for their ways.

Table 2 lists the normalized mean leakage currents and standard deviation of the leakage currents calculated from the simulation results of leakage-severity-sorted ways. This time the 3σ value of inter-die variation changes from 0% to 9.4% and both the mean leakage and standard deviation show an increase as the inter-die variation increases. The increase for mean leakage is quite limited while the standard deviation sees a significant jump. It indicates that we can expect much more variance in way leakage profiles and a better savings from way prioritization.

5.2 Energy Savings of Variation Aware Sizing

Figure 8 shows the benefits of way prioritization as applied to the L3 cache for multiprogrammed workloads. In these experiments, we again assume zero inter-die variation and a 3σ 9.4% intra-die variation. All the benchmarks place small to moderate pressure on the L3 Cache. We assume a static cache sizing policy which chooses the maximum number of ways to disable while not reducing performance by more than 2%.

In Figure 8, the y-axis represents the static power consumed by working ways as a percentage of the total static power of the L3 Cache. Clearly, our method provides satisfying power saving over standard selective ways (for both the average and worst case). Prioritized ways consume an average 28% and 48% less cache static power over the non-variation aware mean and worst-case scenarios.

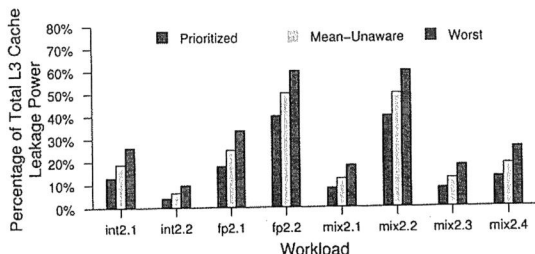

Figure 8: Cache leakage energy of way prioritization (prioritized) versus variation unaware selective cache ways in average (mean-unaware) and pathologically bad (worst) cases.

Results based on the energy-delay metric are presented in Figure 9. We assume that the cache static leakage power comprises 20% of the total core power consumption and that the dynamic power of the whole core is unchanged when cache ways are closed. We normalize the results to the full cache leakage power. In theory, the best energy-delay product achievable is 0.8, when all ways are closed and the performance does not suffer. The results in Figure 9 indicate we can achieve better energy-delay product than mean-unaware case in which all ways are assumed to consume the same amount of power and the worst-case scenario in which the most leaky ways are chosen to stay on.

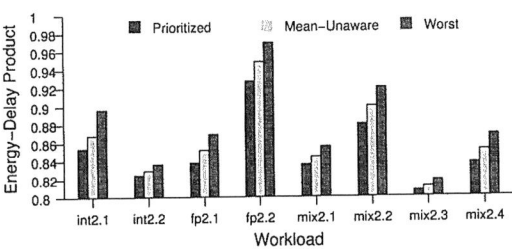

Figure 9: Energy-Delay product of way prioritization (prioritized) versus variation unaware average (mean-unaware) and pathologically bad (worst) cases.

A relevant feature not captured in the above figures is that a single mean leakage current profile cannot represent actual leakage profiles. In certain situations a variation oblivious cache-sizing method could make a faulty decision in closing ways. In addition to picking the wrong physical ways to close, unaware sizing risks choosing the wrong number of ways to close. By taking into account the individual leakage profile of a chip, way prioritization always chooses the best energy-delay sizing.

6. CONCLUSION

In future technologies, leakage power will comprise a significant portion of the total chip power and the effects of manufacturing variations will be pronounced. This paper is among the first to examine the effects that parameter variation have on cache leakage. Our findings show that spatial variation can have significant influence on the cache leakage profile. In particular, equal areas of the cache may have leakage factors that differ by more than an order of magnitude. We propose way prioritization, a cache organization that considers the leakage profile of a chip and resizes the cache by closing sub-arrays that have higher leakage factors. The result is a substantial energy savings with little impact on complexity or performance. Furthermore, because way prioritization works at the microarchitectural level, it is complementary to existing circuit-level leakage reduction and variation tolerance schemes.

7. REFERENCES

[1] A. Agarwal, D. Blaauw, S. Sundareswaran, V. Zolotov, M. Zhou, K. Gala, and R. Panda. Path-based statistical timing analysis considering inter and intra-die correlations. In *Proc. of TAU*, 2002.

[2] D. H. Albonesi. Selective cache ways: On-demand cache resource allocation. In *Proc. of the 32nd Annual IEEE/ACM Int. Symp. on Microarchitecture*, pages 248–259, Nov. 1999.

[3] N. L. Binkert, E. G. Hallnor, and S. K. Reinhardt. Network-oriented full-system simulation using m5. In *6th Workshop on Computer Architecture Evaluation using Commercial Workloads (CAECW)*, Feb. 2003.

[4] S. Borkar et al. Parameter variations and impact on circuits and microarchitecture. In *Proc. of the 40th DAC*, 2003.

[5] Y. Cao, D. S. T. Sato, M. Orshansky, and C. Hu. New paradigm of predictive mosfet and interconnect modeling for early circuit design. In *Proc. of CICC*, pages 201–204, 2000. http://www.eas.asu.edu/ptm.

[6] H. Chang and S. Sapatnekar. Full-chip analysis of leakage power under process variations - including spatial correlations. In *Proceedings of the ACM/IEEE Design Automation Conference*, 2005.

[7] A. Devgan and S. Nassif. Power variability and its impact on design. In *Proceedings of the 18th International Conference on VLSI Design (VLSID-05)*, 2005.

[8] A. Dhodapkar and J. E. Smith. Managing multi-configuration hardware via dynamic working set analysis. In *Proc. of 29th Int. Symp. on Computer Architecture (ISCA-29)*, May 2002.

[9] K. Flautner, N. Kim, S. Martin, D. Blaauw, and T. Mudge. Drowsy caches: Simple techniques for reducing leakage power. In *Proceedings of 29th International Symposium on Computer Architecture (ISCA-29)*, May 2002.

[10] P. Friedberg, Y. Cao, J. Cain, R. Wang, J. Rabaey, and C. Spanos. Modeling within-die spatial correlation effects for process-design co-optimization. In *Proc. of the 6th Int. Symp. on Quality Electronic Design*, 2005.

[11] S. Kaxiras, Z. Hu, and M. Martonosi. Cache decay: Exploiting generational behavior to reduce cache leakage power. In *Proceedings of the 28th International Symposium on Computer Architecture (ISCA-28)*, June 2001.

[12] C. McNairy and R. Bhatia. Montecito: A dual-core dual-thread itanium processor. *IEEE Micro*, 25:10–20, Apr. 2005.

[13] M. Powell, S.-H. Yang, B. Falsafi, K. Roy, and T. Vijaykumar. Gated-Vdd: A circuit technique to reduce leakage in deep-submicron cache memories. In *ACM/IEEE ISLPED*, 2000.

[14] D. B. Rajeev Rao, Ashish Srivastava and D. Sylvester. Statistical analysis of subthreshold leakage current for vlsi circuits. *IEEE Trans. on VLSI Systems*, 12:131–139, Feb. 2004.

[15] Semiconductor Industry Association. International Technology Roadmap for Semiconductors, 2003. http://public.itrs.net/Files/2003ITRS/Home.htm.

[16] J. W. Tschanz, J. T. Kao, S. G. Narendra, R. Nair, D. A. Antoniadis, Fellow, A. P. Chandrakasan, and V. De. Adaptive body bias for reducing impacts of die-to-die and within-die parameter variations on microprocessor frequency and leakage. *IEEE JOURNAL OF SOLID-STATE CIRCUITS*, 37:1396–1402, Nov. 2002.

Substituting Associative Load Queue with Simple Hash Tables in Out-of-Order Microprocessors*

Alok Garg[†], Fernando Castro, Michael Huang[†], Dani Chaver,
Luis Piñuel, and Manuel Prieto
[†]University of Rochester and Universidad Complutense Madrid
[†]{garg, huang}@ece.rochester.edu, fcastror@fis.ucm.es, {dani02, lpinuel, mpmatias}@dacya.ucm.es

ABSTRACT

Buffering more in-flight instructions in an out-of-order microprocessor is a straightforward and effective method to help tolerate the long latencies generally associated with off-chip memory accesses. One of the main challenges of buffering a large number of instructions, however, is the implementation of a scalable and efficient mechanism to detect memory access order violations as a result of out-of-order scheduling of load and store instructions. Traditional CAM-based associative queues can be very slow and energy consuming. In this paper, instead of using the traditional age-based load queue to record load addresses, we explicitly record age information in address-indexed hash tables to achieve the same functionality of detecting premature loads. This alternative design eliminates associative searches and significantly reduces the energy consumption of the load queue. With simple techniques to reduce the number of false positives, performance degradation is kept at a minimum.

Categories and Subject Descriptors

C.1.0 [**Processor Architectures**]: General

General Terms

Design, Experimentation, Measurement

Keywords

LSQ, Memory disambiguation, Hash table, Scalability

1. INTRODUCTION

With high operation frequency, modern out-of-order processors often need to buffer a very large amount of instructions to be able to overlap useful processing with relatively long latencies associated with accesses to lower levels of the memory hierarchy. Processor features such as multi-threading further increase the demand on the instruction buffering capability. Increasing the number of in-flight instructions, however, requires scaling up different microarchitectural structures. This can cause significant increase in

energy consumption, especially if the structure is accessed associatively. One such example is the logic that enforces correct memory-based dependences, generally referred to as the load-store queue (LSQ) and implemented as two separate queues: the load queue (LQ) and the store queue (SQ). Conventional implementations of these queues are age-based, containing complete addresses. Memory instructions need to update one queue and associatively check the other. The associative search operation is a major concern for the scalability of these queues as not only energy consumption increases, the latency of accesses also worsen with the increase of queue size and may present complications in the logic design. As such, unconventional implementations that avoids associative searches should be explored.

In this paper, we focus on the LQ and propose an alternative implementation: instead of explicitly expressing the full address in the queue and implicitly encoding the age information in the allocation order of the entries, we explicitly maintain the age information and implicitly track the address via a hash table. The resulting structure requires only index-based access and thus can be much more energy-efficient and easy-to-scale. We do, however, incur false positive memory order violation detections and thus extra replays. We show that with a few very simple mitigation techniques, the number of replays can be drastically reduced. Furthermore, because our implementation does not place a limit on the number of in-flight loads, we can reduce the chance of stalling processor because the LQ is full. This can offset the performance degradation caused by extra replays. We show that our design can drastically reduce the energy consumption of the LQ at a slight performance cost: on average, about 85% of LQ energy is saved. Performance degradation is around 1% compared to conventional design with optimally sized LQ ignoring any latency issues of a large LQ. Overall, processor-wide energy savings range from 1-4%.

The rest of the paper is organized as follows: Section 2 discusses the issues of the conventional LQ design and our alternative implementation; Section 3 describes our experimental methodology; Section 4 presents quantitative evaluation of our design; Section 5 summaries related work; and Section 6 concludes.

2. TRACKING LOADS WITH HASH TABLES

2.1 Conventional design

In a typical out-of-order core, the result of a store instruction is not released to the memory hierarchy until the store commits so as to easily support precise exception. A

---

*This work is supported in part by National Science Foundation under the grant 0509270, the Spanish goverment through research contract TIN 2005-05619, and by the Hipeac European Network of Excellence.

Permission to make digital or hard copies of all or part of this work for personal or classroom use is granted without fee provided that copies are not made or distributed for profit or commercial advantage and that copies bear this notice and the full citation on the first page. To copy otherwise, to republish, to post on servers or to redistribute to lists, requires prior specific permission and/or a fee.
ISLPED'06, October 4–6, 2006, Tegernsee, Germany.
Copyright 2006 ACM 1-59593-462-6/06/0010 ...$5.00.

load instruction, on the other hand, is allowed to access the memory hierarchy as soon as the address is available. This speculative access is premature if a store older in program order executes later and modifies (part of) the data being loaded. In conventional implementations, when a store executes, its age and address information is compared to that of the loads to find out whether there is any premature load. The LQ implicitly stores the age information by using an age-ordered queue. If the portion of the queue belonging to loads younger than the store registers a match, a store-load replay is triggered [7]. Although the high-level concept of the LQ is not overly complex, practical designs have to deal with various issues such as handling of partial overlap between loads and stores. The (physical) address is explicitly stored in the LQ and compared. Unfortunately, address is a very wide operand (more than 32 bits in current generation products) and the width continues to grow. Searching the queue associatively is not only slow but also power-consuming.

2.2 Hash table-based tracking

We propose an alternative design that eliminates associative search and the capacity issue of the LQ. The first key difference between our design and the conventional design is that we explicitly assign and track the age of loads. Recall that to determine whether a store-load replay is needed requires two pieces of information: address and age. Conventional design allocates an LQ entry for each load at dispatch in program order thereby implicitly encoding the age information within the position of the entry. By explicitly encoding and tracking the age, we are no longer bound to perform entry allocation at the early stage of dispatch and therefore can choose from a much wider variety of implementation options. For example, the LQ can be split into multiple smaller ones that are word interleaved.

The second difference is that we use the simple, oft-used indexing table or a hash table to avoid associative comparison of address: each load, upon execution, will use the address to hash into the table and record its age and optionally some address information. (We refer to this table as the load table.) Because the hashing already carries some address information, we only need to record partial information about address – just like the way cache tag keeps partial information. However, one important difference between maintaining a cache and keeping track of loads for order violation detection is that we do not necessarily need to keep the full address. If there is a partial address match between a store and a younger load, we can conservatively assume an address match has occurred. This only affects the efficiency of the system, not correctness. This observation leads to the third difference: in our design, we choose not to record any additional address information (beyond hashing). When two addresses map to the same entry in the load table, we simply assume that they are accessing the same address.

Clearly, multiple loads can hash into the same entry of the load table. We simply keep the age of the *youngest* load. This is because the *existence* of an order violation is all-important, whereas the identity of the load(s) involved is dispensable. Clearly, keeping the age of the youngest load is sufficient to detect the existence of order violation. When a replay is needed, we can simply replay from the instruction following the store in program order – rather than starting from the oldest load among all triggering loads. In fact, trying to identify the oldest load that needs to replay requires additional circuit complexity existing processors such as the

IBM POWER 4 chose to avoid [16]. It, too, replays from the store onward.

To summarize, when a load executes, it accesses the load table based on the address and if the age currently stored in the entry is older, it updates the entry with its own age, otherwise, the entry remains unchanged. When a store executes, it similarly accesses the table to read the current age. If the age is younger its own age, a replay (from the instruction following the store) is triggered.

Representing age: To represent the age of memory instructions, we simply take their ROB ID and augment it with a 2-bit prefix to handle wrap-arounds. This prefix increments every time the write (dispatch) pointer of the ROB wraps around. Because at any moment two in-flight instructions can only differ by one in the prefix, we cycle the prefix from 1 to 3 and when comparing two age IDs, the prefix part follows the fixed rule $1 < 2$, $2 < 3$, and $3 < 1$.

When the read (commit) pointer of ROB wraps around, all entries in the load table with the old prefix represent instructions that are committed. We clean up those entries with the old prefix by flash-resetting their prefix to 0 to indicate invalid age. If we do not perform the clean-up, these age IDs will be misinterpreted as future ages. Note that depending on the actual implementation, the flash reset can take multiple cycles. During this clean-up action, we can not start assigning new prefixes, *i.e.*, the ROB write pointer can not wrap around. This guarantees that at any moment, there are only two neighboring prefixes in the load table.

Handling coherence and consistency: The LQ also serves the purpose of maintaining load-load ordering for coherence and consistency. First off, a cache-coherent design requires write serialization: all writes to the same location have to appear in the same order to all processors. As a result, if a younger load obtained some data, subsequently, the data is updated by a store from another processor, then an older load can not obtain the new data. Note that write-serialization is quite a standard requirement even in a uniprocessor environment. This is because even in uniprocessor systems, DMA can also assume the role of a bus master and write to the memory. To simplify software, (almost) all current commercial processors maintain cache coherence with DMA, which implies write serialization.

In a conventional design, an invalidation searches through the entire LQ to find matching addresses. When matches are found, they are marked in the LQ. Every load, upon execution, also searches in the LQ associatively (just like a store). If there is a match of the address with a younger load whose entry is marked (by an invalidation), this suggests that the younger load has consumed the old data and therefore a load-load replay is triggered [7].

The guarantee of write serialization can be supported by the hash table implementation as well. An invalidation will also hash into the load table and mark the entries corresponding to the invalidated addresses. When a load executes, we inspect the invalidation mark in the entry. If the mark is set and the age recorded in the table is younger than that of the current load, we trigger a load-load replay. Note that for load instructions, we already need to perform a read of the hash table to determine whether the age needs to update. Thus, checking of load-load replay is essentially for free in our design. Similar to the store-load replay, we can not pin-point the identity of the younger loads and have to replay from the instruction following the older load. Again, such a more conservative range of replay is already adopted in existing processors for circuit simplicity [16].

LQ can also be used to ensure that load speculation does not violate the memory consistency model. For example, in MIPS R10000, which implements sequential consistency, loads actually execute, speculatively, before they are allowed to by the consistency model. To guard against mis-speculation, an external invalidation searches the LQ and marks matching entries. When the marked loads reach the commit stage, a "soft" exception – essentially a replay – occurs. This type of replay can also be implemented by the load table. However, because the load table does not provide the identity of loads, only the existence of a match, we can only conservatively replay from the oldest load instruction still in-flight if there is a match. While this would be functionally correct, it is perhaps an inefficient implementation. In this paper, we focus on uni-processors. We leave the study of our design in multiprocessor domain and potential optimizations in that environment to future work.

2.3 Handling multiple data sizes

Unfortunately for memory order tracking, accesses come at different sizes: from byte to 8 bytes or even larger. This creates a challenge to accurately and efficiently track accesses: If tracking happens at a fine granularity (*e.g.*, byte), a wider access needs to simultaneously mark multiple entries, clearly energy-inefficient and complicated circuit-wise. If, on the other hand, tracking is done at a coarse granularity (*e.g.*, 8-byte word), we lose the ability to distinguish two finer-grain accesses to two different portions within the same data chunk. This can potentially result in pathological scenario and create numerous spurious replays: If a loop performs a read-modify-write cycle on an array of bytes, and the issue logic favors loads over stores, then there can be many re-orderings of stores and loads from consecutive iterations to neighboring bytes. These would be incorrectly construed by the coarse-grain tracking logic as ordering violation to the same memory location.

To handle multiple access widths efficiently and without complicated circuitry, we use a main table to track the predominant access width (64 bits, or quad-word in our experimental system) and use a "side" table to handle other widths. Each entry in the main table contains two bits to encode the current tracking width. When a load-quad-word instruction executes, it only accesses the main table. When a load instruction with narrower width executes, it first accesses the main table, if the entry is invalid (the prefix part of the age is 0), it will set the entry's width field to its own width and proceed to the side table to update it. If the main table's entry already has a valid age, then that age will be updated if the load's age is younger. (This way, the main table *always* contain the age of the youngest load that accessed *any* part of the quad-word.) Comparing the entry's existing width and that of the load, there are two possible cases: (a) the entry's width is the same or wider than that of the load, or (b) the entry's width is narrower.

In case (a), because we are already tracking at a coarser granularity, finer-grain information is useless. So we simply "upgrade" the width of the load to the same as the entry's and proceed to the side table if necessary. For example, if the entry's width is double word (4 bytes), then we take the load's address, ignore the last two bit (to get double-word

address) and use that to hash into the side table and update the age if necessary.

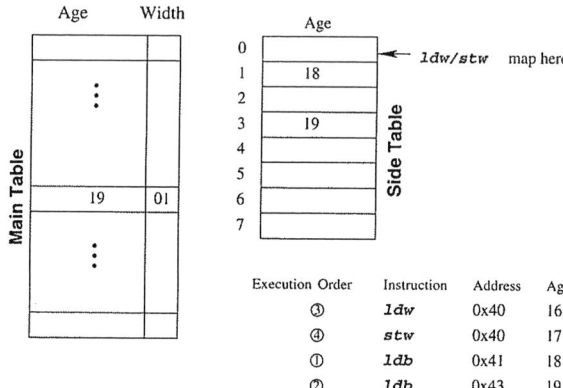

Figure 1. Example of table update with width and age upgrades. *ldw*, *stw*, and *ldb* stand for load-word, store-word, and store-byte, respectively. Example shows an 8-entry side table with a hashing function that takes the 3 least significant bits of byte, word, or double-word addresses, depending on the access width.

In case (b), because we do not want the complexity of having to access and update multiple entries, we forgo the fine-grain information we have accumulated so far and upgrade the entry's width to that of the load and proceed to update the side table. If the upgraded width is quad-word, the side table is not accessed. Before we access the side table, we upgrade the load's age to the elder of its own age or that of the entry in the main table. The reason to upgrade the age is best explained by an example. In Figure 1, we show a sequence of 4 instructions in which the two younger *ldb* instructions execute first. The figure shows the state of the two tables after their execution. When *ldw* executes, the width of the main table entry will upgrade, indicating that the side table will start to track this quad-word at word granularity instead. *ldw* will map into entry 0 of the side table and the new age needs to reflect all accesses to the entire word starting at address 0x40. Rather than traversing multiple entries of the side table to figure out the new age, we adopt the simple, if conservative, solution: to take the age in the main table which reflects the age of the youngest load accessing anywhere in the quad-word.

When a store executes, it compares its age to that in the corresponding entry in the main table. If the store's age is younger, then no replay is needed. Otherwise, we may need to replay. However, if the store has a narrower width, we may have more fine-grained tracking information from the side table that can rule out violation. Depending on the width of the store and that of the entry, there are also two cases: (a) the store's width is the same or narrower than the corresponding entry's in the main table, or (b) the store's width is wider. In case (a), we simply treat the store as a wider access (with the same width as the entry) and consult the side table to determine whether we replay. In case (b), though the side table contains the age information, it does so in a "fragmented" way, and we need to access the side table multiple times to find out the age of the youngest load that overlaps with the store. To avoid complexity, we do not do so. We simply ignore the side table in this case.

270

In summary, the side table essentially provides some extra space to allow us to "zoom in" and track select quad words at a finer granularity. At any time, a single quad-word is tracked with only one data width. However, that width differs from quad-word to quad-word. The entries in the side table, therefore, are tracking different widths. Hashing conflicts may incur spurious replays, but does not affect correctness.

2.4 Mitigating the effect of pollution

Because of hashing conflicts, we may have spurious replays that an age-ordered LQ recording full address does not have. We also have another and more serious source of spurious replays – table pollution. Recall that updates to the load tables (main table and side table) occur at execution time when the instructions are speculative and may be on the wrong path. Later on, when a recovery of misprediction or replay takes place, the processor rolls back and starts to re-assign older ages to memory instructions on the right path. However, the tables already have many (incorrect) future ages making apparent order violation very likely. Unfortunately, we can not easily undo the updates from the load tables. It is certainly possible to maintain a log of updates and walk through the log to undo updates after a recovery or a replay. However, it is clearly inefficient and undesirable.

We propose a technique to mitigate the effect using a simple, global register. This register can be thought of as a degenerate main table that has only one entry. When a load executes, if its age is younger than that recorded by the register, the register is updated. Because this degenerate table has only one entry, it is very easy to roll back the age to that of the branch upon branch misprediction recovery, or to that of the first instruction to be replayed. If this single register can rule out the possibility of order violation, we do not check the larger tables. This actually has a filtering effect that reduces the energy consumption of accessing the bigger tables.

In practical implementations, we are not limited to a single register. We can use a small number of registers (essentially making a table of a few entries). Intuitively, we get diminishing returns as we increase the number registers. In our limited exploration, adding a second registers can further reduce the number of spurious replays and seems to be a good choice. In this paper, we use a pair of such registers.

Alternatively, we can use a monotonically increasing counter (with wrap-arounds) to replace the ROB ID as the age. With such an age mechanism, when a misprediction recovery or replay happens, we continue to increment age counter rather than reuse already-assigned age IDs and this can significantly cut down spurious replays due to table pollution. The disadvantage is the extra bits needed to store the age. In contrast, ROB ID is already used by the issue logic to update instruction execution status and thus an ROB ID-based age representation is nearly for free. However, as we will show later, the improvement in replay reduction is significant and could easily justify the extra cost in the design.

3. EXPERIMENTAL SETUP

We evaluate our scheme using a heavily-modified SimpleScalar [3] 3.0d tool set with Wattch extension [2]. The Wattch model is extended to include the energy consumption of the hash tables. Besides implementing separate ROB, issue queue, and register files, we have added support to more faithfully model modern microprocessors. Specifically,

we allow speculative issue of load instructions when there are prior unresolved stores; we faithfully model store-load replays [7] (instructions are re-fetched and re-issued); we model load rejections [16]; we issue dependent instructions of the load speculatively (assuming a cache hit) and perform scheduler replays [7] if the load misses in the cache.

Our simulator models only uniprocessor and we do not have a source to faithfully model invalidation messages in a uniprocessor. Thus we do not model load-load ordering search in the LQ. This favors the conventional design by reducing the energy expenditure – for our hash table-based design, this checking is for free.

To evaluate our design we use three processor configurations (config1, config2, and config3) as shown in Table 1. *Config1* closely mimics POWER 4 processor whereas *Config2* and *Config3* are scaled-up versions of config1. To focus on energy consumption of the baseline load queue we size it optimally relative to ROB size for all three configurations.

We use highly optimized alpha binaries of all 26 SPEC CPU2000 benchmarks. We simulate 500 million instructions after fast-forwarding 1 billion instructions.

Processor core			
Issue/Decode/Commit width	8 / 8 / 8		
Functional units	INT 8+2 mul/div, FP 8+2 mul/div		
Branch predictor	Bimodal and Gshare combined		
- Gshare	8K entries, 13 bit history		
- Bimodal/Meta table/BTB	4K/8K/4K (4 way) entries		
Branch misprediction penalty	at least 7 cycles		
Memory hierarchy			
L1 instruction cache	64KB, 1-way, 128B line, 2 cycles		
L1 data cache	32KB, 2-way, 128B line, 2 cycles, 2 ports		
L2 unified cache	1MB, 8-way, 128B line 15 cycles		
Memory access latency	120 cycles		
	Config1	Config2	Config3
Issue queue (INT, FP)	(32, 32)	(48, 48)	(64, 64)
ROB	128	256	512
Register (INT, FP)	(100, 100)	(200, 200)	(400, 400)
LSQ(LQ,SQ) 2 search ports	80 (48,32)	144 (96,48)	256 (192,64)
Table size (main, side)	(512, 128)	(1024, 256)	(2048, 512)

Table 1. System configuration.

4. EVALUATION

In this section, we perform some quantitative analysis on the proposed design and the optimizations. For brevity, we report data in a compact format. We group applications into integer (INT) and floating-point (FP) groups. All metrics are first normalized to the result of the conventional configuration (baseline) and then averaged within the groups. We show the average in solid bars and the minimum and maximum for the group in superimposed I-beams.

4.1 Energy impact of hash table-based tracking

We first inspect energy savings on the LQ. Intuitively, energy reduction will be significant because we no longer have a wide operand CAM structure and under most circumstances, each access consists of just one main table access. We see in Figure 2-(a) that this is indeed the case. Across configurations and applications, the energy savings consistently fall in the range of 77-94%. The averages are about 83-88% depending on the configuration. Recall that if load-load order checking is performed, the baseline could spend even more energy.

Next, we look at performance degradation. The performance is less straightforward to compare because the differ-

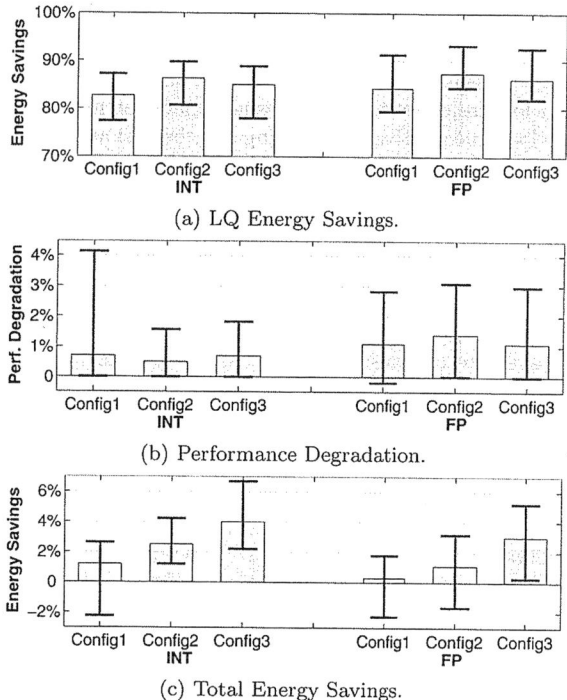

(a) LQ Energy Savings.

(b) Performance Degradation.

(c) Total Energy Savings.

Figure 2. Average energy savings from the LQ (a), performance degradation (b), and processor-wide energy savings (c) over the conventional systems.

ence is sensitive to the LQ size of the baseline. Here we size the LQ in the baseline system to keep stalling due to LQ-fillup very low. Reducing the LQ size even moderately would result in noticeable slowdown for the baseline configuration, especially for floating-point applications. In the scaled-up configurations, we use ever larger LQs, ignoring any practicality issues. With these in mind, we can see that slowdown induced by our design is very small, about 1%. Even the worst-case slowdown is insignificant. The minimum degradation is almost always 0%. The overall processor-wide energy savings are shown in Figure 2-(c). We see that for the smallest configuration (Config1), we already make a net energy gain. As we scale to larger configurations, our gain increases. This is because larger LQs are increasingly inefficient – if they can be built to start with. Note that energy consumption is spread out over different components. Thus it is expected that the drastic energy reduction in the LQ would translate into much smaller global energy savings. Furthermore, the benefit of a table-based tracking mechanism goes beyond energy savings. Large CAM structures are best avoided. Our results have clearly showed that the more sensible table-based implementation can competently and energy-efficiently track memory access order in large scale.

In these statistics, we use monotonically increasing age counter to mitigate pollution effect. If we choose ROB ID-based age and use the alternative of global register for pollution mitigation, the performance will degrade 1.1% and 0.5% on average for the integer and floating-point applications, respectively.

4.2 Understanding the optimization techniques

To understand the effect of the optimization techniques, we show the breakdown of the number of replays and the ef-

fect of the optimization techniques in Figure 3. For this analysis, we only show the averages for one configuration (Config2) while the results from other configurations are very similar. We break down the number of replays according to their type, from bottom up: true replays, false replays due to the mismatch between the access width and the tracking width, those due to hashing conflict, and finally, those due to table pollution caused by squashed instructions. We show the results of using four different configurations: using just the main table (M), using the main and the side table (MS), using both tables plus the pair of global registers (MSG), and using the age counter with main and side table (MSC).

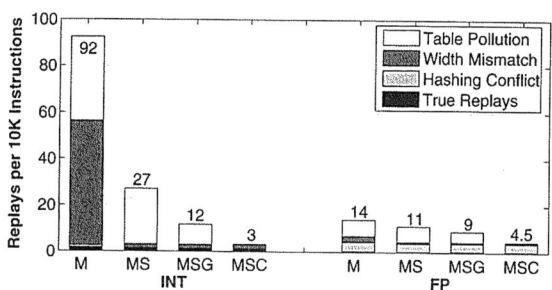

Figure 3. Breakdown of replay rates.

We see that the results are rather intuitive. (1) Integer applications tend to have many narrow-width data accesses, which cause significant number of false replays when we track access order only using the main table at quad-word granularity. Floating-point applications, on the other hand, have far fewer replays due to width mismatch when we only use the main table. With the use of the side table, these false replays are almost completely eliminated. (2) Integer applications tend to suffer more from table pollution due to the higher branch misprediction rate. Floating-point applications suffer far less from table pollution, but the effect is still visible. With the global registers, we are able to remove a significant portion of false replays: about 63% and 32% for integer and floating-point applications, respectively. With age counter, this source of replay is all but completely eliminated. (3) Hashing conflict-induced replays are almost negligible in integer applications, whereas they are more noticeable in floating-point applications because the working set is generally larger. The techniques aimed at reducing other false replays have little impact on these conflict-induced replays. A better hashing function may further improve our design for floating-point applications.

5. RELATED WORK

Recognizing the scalability issue of the LSQ, many different proposals have emerged recently. A large body of work adopts a two-level approach to disambiguation and forwarding. The guiding principle is largely the same. That is to make the first-level (L1) structure small (thus fast and energy efficient) and still able to perform a large majority of the work. This L1 structure is backed up by a much larger second-level (L2) structure to correct/complement the work of the L1 structure [1, 8, 17]. A variation of a two-level approach is the dual-queue approach where memory instructions predicted to have in-flight dependence are kept in an associative queue with conventional searching capabilities and others are kept in less power-hungry FIFOs [5, 6].

Another body of work only uses a one-level structure (for stores) but reduces check frequency through clever filtering or prediction mechanisms such as bloom filter [11, 13].

Exact memory dependence prediction is an alternative to address-based dependence-checking mechanism. Ambitious and complex dependence prediction is used in [14, 15]. Unfortunately, achieving highly accurate prediction of the actual communicating pairs of memory instructions requires a large number of tables, some of which highly-ported. Furthermore, enforcing the predicted dependence requires non-trivial support from the issue logic. Value-based re-execution presents a new paradigm for memory disambiguation. In [4], the LQ is eliminated altogether and loads re-execute to validate the prior execution. To address the energy increases due to re-execution, and the performance impact due to increased memory pressure, various filters are developed to reduce the re-execution frequency [4, 12].

Cooperative disambiguation uses compiler-based analysis to rule out the possibility for certain loads to cause dependence violation and therefore allow them to completely bypass the LQ [10].

Finally, slackened memory dependence enforcement adopts a different philosophy of enforcing memory dependence. Two decoupled executions of all memory instructions are used, the leading front-end execution focuses on speed and efficiency of common-case communication and the trailing execution follows program order to ensure correctness and resets the front-end execution when it failed to correctly enforce dependence [9].

In contrast to this body of prior work, our approach still maintains the conventional address-based memory disambiguation strategy, *i.e.*, we do not rely on dependence prediction or re-execution for validation. Compared to the solutions that still use the LQ, our approach is different in that we explicitly represent age and use hashing functions to implicit represent the address. This allows us to eliminate expensive associative comparisons of wide address operands (32 bits or more) and replace them with a single comparison of the much narrower age ID (about 10 bits) for most memory accesses.

6. CONCLUSIONS

In this paper, we have proposed a novel approach of tracking memory access order violation. The design uses a hash table-based tracking methodology maintaining explicit age information in the table. This data structure provides a number of benefits: First, detecting order violation involves only table indexing and a single age comparison per table. Compared to the fully-associative matching of the much wider address operand, this drastically reduces energy consumption; Second, thanks to the absence of large CAM structures, the access latency is low and does not increase much when scaling up the tables, making them a much better choice for scalable microarchitectures; Third, the design does not place a limit on the number of in-flight load instructions that can be tracked, potentially removing a bottleneck on the instruction-buffering capability. A straightforward implementation of the design, however, does introduce spurious replays due to different data access widths and table pollution as a result of branch misprediction and replays. We have presented two effective mitigation techniques. First, we use a side table to provide finer-grain tracking of non-predominant access widths. Second, we provide

two alternatives – a pair of global registers or a monotonically increasing age assignment logic – to filter out a large fraction of pollution-induced false replays. We have shown that the overall implementation is very effective: with very little slowdown, depending on the configuration, we cut the energy consumption of the LQ by an average of about 83-88%. Taking into account the energy cost of extra replays, the processor still makes a net gain in energy on average.

7. REFERENCES

[1] H. Akkary, R. Rajwar, and S. Srinivasan. Checkpoint Processing and Recovery: Towards Scalable Large Instruction Window Processors. In *International Symposium on Microarchitecture*. December 2003.

[2] D. Brooks, V. Tiwari, and M. Martonosi. Wattch: A Framework for Architectural-Level Power Analysis and Optimizations. In *International Symposium on Computer Architecture*. June 2000.

[3] D. Burger and T. Austin. The SimpleScalar Tool Set, Version 2.0. Technical report 1342, Computer Sciences Department, University of Wisconsin-Madison, June 1997.

[4] H. Cain and M. Lipasti. Memory Ordering: A Value-based Approach. In *International Symposium on Computer Architecture*. June 2004.

[5] F. Castro, D. Chaver, L. Pinuel, M. Prieto, M. Huang, and F. Tirado. A Power-Efficient and Scalable Load-Store Queue Design. In *International Workshop on Power And Timing Modeling, Optimization and Simulation*. September 2005. Lecture Notes in Computer Science Vol. 2236(8):1-9.

[6] F. Castro, D. Chaver, L. Pinuel, M. Prieto, M. Huang, and F. Tirado. Load-Store Queue Management: an Energy Efficient Design based on a State Filtering Mechanism. In *International Conference on Computer Design*. October 2005.

[7] Compaq Computer Corporation. *Alpha 21264/EV6 Microprocessor Hardware Reference Manual*, September 2000. Order number: DS-0027B-TE.

[8] A. Gandhi, H. Akkary, R. Rajwar, S. Srinivasan, and K. Lai. Scalable Load and Store Processing in Latency Tolerant Processors. In *International Symposium on Computer Architecture*. June 2005.

[9] A. Garg, M. Rashid, and M. Huang. Slackened Memory Dependence Enforcement: Combining Opportunistic Forwarding with Decoupled Verification. In *International Symposium on Computer Architecture*. June 2006.

[10] R. Huang, A. Garg, and M. Huang. Software-Hardware Cooperative Memory Disambiguation. In *International Symposium on High-Performance Computer Architecture*. February 2006.

[11] I. Park, C. Ooi, and T. Vijaykumar. Reducing Design Complexity of the Load/Store Queue. In *International Symposium on Microarchitecture*. December 2003.

[12] A. Roth. Store Vulnerability Window (SVW): Re-Execution Filtering for Enhanced Load Optimization. In *International Symposium on Computer Architecture*. June 2005.

[13] S. Sethumadhavan, R. Desikan, D. Burger, C. Moore, and S. Keckler. Scalable Hardware Memory Disambiguation for High ILP Processors. In *International Symposium on Microarchitecture*. December 2003.

[14] T. Sha, M. Martin, and A. Roth. Scalable Store-Load Forwarding via Store Queue Index Prediction. In *International Symposium on Microarchitecture*. December 2005.

[15] S. Stone, K. Woley, and M. Frank. Address-Indexed Memory Disambiguation and Store-to-Load Forwarding. In *International Symposium on Microarchitecture*. December 2005.

[16] J. Tendler, J. Dodson, J. Fields, H. Le, and B. Sinharoy. POWER4 System Microarchitecture. *IBM Journal of Research and Development*, Vol. 46(1):5-25, January 2002.

[17] E. Torres, P. Ibanez, V. Vinals, and J. Llaberia. Store Buffer Design in First-Level Multibanked Data Caches. In *International Symposium on Computer Architecture*. June 2005.

A Novel Power Optimization Technique for Ultra-Low Power RFICs

Amin Shameli, Payam Heydari
Department of EECS, University of California, Irvine, CA 92697, USA
{ashameli,payam}@uci.edu

ABSTRACT

This paper presents a novel power optimization technique for ultra-low power (ULP) RFICs. A new figure of merit, namely the $g_m f_T$-to-current ratio, $(g_m f_T/I_D)$, is defined for a MOS transistor, which accounts for both the unity-gain frequency and current consumption. It is demonstrated both analytically and experimentally that the $g_m f_T/I_D$ reaches its maximum value in moderate inversion region. Next, using the proposed method, a power optimized common-gate low-noise amplifier (LNA) with active load has been designed and fabricated in a CMOS 0.18μm process operating at 950MHz. Measurement results show a noise-figure (NF) of 4.9dB and a small signal gain of 15.6dB with a record-breaking power dissipation of only 100μW.

Categories and Subject Descriptors

B.7.1 [Integrated Circuits]: Types and Design Styles; B.7.m [Integrated Circuits]: Miscellaneous

General Terms

Design, Measurement

Keywords

Ultra-low power, radio-frequency integrated circuit, low-noise amplifier, CMOS

1. INTRODUCTION

Power minimization of the RF front-end continues to be one of the most important challenges in the design of wireless transceivers being employed in emerging technologies such as body implanted sensors, radio frequency identification systems (RFID), and sensor networks. All these examples share a common concern in that they entail extremely low power consumption in sub-milliwatt range so as to enable the constituent transceiver to operate either without integrated power supply or at long battery-life.

Due to its high integration capability and its continuously scaled feature size, the scaled CMOS technology remains a viable candidate for implementation of ultra-low power (ULP) integrated circuits. In the meantime, a rich prior research on CMOS ULP digital IC design has proved the use of CMOS technologies to design micropower baseband systems. Heavily supported by the

Permission to make digital or hard copies of all or part of this work for personal or classroom use is granted without fee provided that copies are not made or distributed for profit or commercial advantage and that copies bear this notice and the full citation on the first page. To copy otherwise, or republish, to post on servers or to redistribute to lists, requires prior specific permission and/or a fee.
ISLPED'06, October 4–6, 2006, Tegernsee, Germany.
Copyright 2006 ACM 1-59593-462-6/06/0010...$5.00.

concept of system-on-chip (SOC), the design of ULP RFICs in CMOS technologies, however, entails several design challenges.

One of the main limitations in the ULP CMOS RFIC design is the low value of transistor's transconductance, g_m, due to the low bias current. Having already examined in low-power analog/digital ICs, an effective way of minimizing power consumption is to bias the transistor(s) in weak inversion region where the transistors achieve maximum value of g_m/I_D [1], [2]. Nonetheless, a weakly inverted transistor exhibits poor frequency response, and therefore, may not be used extensively in RFIC design. In this paper, we will investigate the transistor's optimum region of operation to achieve ultra-low power consumption in RF frequencies.

The remainder of the paper is organized as follows. Section 2 gives an overview of the MOS transistor regions of operation. Section 3 introduces a new figure of merit, the $g_m f_T$-to-current ratio $(g_m f_T/I_D)$. Section 4 explains the design of a common-gate low noise amplifier (LNA). Section 5 illustrates the measurement results of the designed LNA. Finally, Section 6 provides the summary and conclusions.

II. AN OVERVIEW OF MOSFET REGIONS OF OPERATION

MOS transistors in the linear RF blocks are usually biased in strong inversion. However, at least from a conceptual perspective, these blocks can be realized by biasing the transistors at weak or moderate inversion. Assuming that the transistor is operating in the edge of the saturation region, the channel length modulation will be negligible [3], and the drain-source current will thus be expressed by Eq. (1):

$$I_D = 2nU_T^2 \frac{W}{L} \mu C_{ox} \ln^2(1 + e^{(V_{GS}-V_T)/2nU_T}) \qquad (1)$$

where C_{ox} is the gate-oxide capacitance per unit area, μ is the carrier mobility, U_T defined as $U_T = kT/q$ is the thermal voltage, and n defined as $n = 1+0.5\gamma(2\Phi_F+V_{SB})^{1/2}$ is the subthreshold slope factor whose value depends on the process, and is varying from 1.1 to 1.9. Eq. (1) is valid in all regions of inversion [3].

To take into account the carrier's velocity saturation effect in MOS transistor, the effective carrier mobility, μ_{eff}, defined in [3]–[5] and reiterated in Eq. (2), is utilized in Eq. (1):

$$\mu_{eff} = \frac{\mu_0}{[1+\theta_1(V_{GS}-V_T)+\theta_2(V_{GS}-V_T)^2]\cdot(1+V_{DS}/E_cL)} \qquad (2)$$

where μ_0 is the low-field mobility, E_c is the electric field at which the mobility saturates, L is the channel length of the transistor, and θ_1 and θ_2 are process dependent parameters. The value of θ_2, nevertheless, is very small and can be neglected in many applications [3], [6]. To determine the transistor's region of operation, irrespective of its aspect ratio, the normalized current $I_N = I_D/I_Z$ is employed in forthcoming analytical developments. I_Z captures the transistor's aspect ratio and is defined as [3], [7]:

$$I_Z = 2nU_T^2 \frac{W}{L}\mu_0 C_{ox} \quad (3)$$

To determine the transistor's region of operation, one should also look at the carrier concentration at the interface between the silicon substrate and the gate oxide. The MOS transistor makes a transition from weak to moderate inversion region if both the minority and the majority carrier concentrations become equal [3], [6]. The gate-source potential at which the carrier concentration in the inverted channel becomes equal to the carrier concentration of the silicon substrate is commonly referred to as threshold voltage. The upper limit of moderate inversion, on the other hand, is not clearly defined in [3]. It is loosely defined as the voltage below which the strong inversion equations are not valid. Looking at Eq. (1), for $V_{GS} - V_T \gg 2nU_T$, the exponential part is approximated as $(V_{GS} - V_T)^2$. Eq. (1) is thus simplified to the I-V relationship of the MOS transistor in strong inversion. Therefore, the upper limit of moderate inversion region is roughly defined as the gate-source voltage at which $\exp[(V_{GS} - V_T)/2nU_T] = 10$. Solving this criteria for overdrive voltage yields $V_{GS} - V_T = 4.6nU_T$. Table 1 summarizes the voltage and current criteria for transition from one region of operation to another.

Table 1: Bounds for transition from weak-to-moderate and from moderate-to-strong inversion

	Transition from Weak to Moderate inversion	Transition from Moderate to Strong Inversion
$V_{GS} - V_T$	0	$4.6nU_T$
$I_N = I_D/I_Z$	0.48	5.75

3. $g_m f_T$-TO-CURRENT RATIO

Power consumption is becoming as equally important as performance (e.g., bandwidth and speed of operation) in next generation wireless communication systems. To capture both performance and DC power consumption of a MOS transistor in any region of operation, we define a new figure of merit, the $g_m f_T$-to-current ratio ($g_m f_T/I_D$). By taking into account both g_m and f_T, maximizing the $g_m f_T/I_D$ for a fixed bias current leads to the maximum achievable gain-bandwidth-product (GBW). This unique attribute makes the $g_m f_T/I_D$ a proper objective function for the optimization of the ULP RF/analog circuits. As illustrated later in this section, the $g_m f_T/I_D$ will be expressed as a function of the normalized current I_N. This will allow us to readily show that to maximize the $g_m f_T/I_D$, the MOS transistor must operate in moderate inversion region.

In the following, we will express the $g_m f_T/I_D$ as a function of the normalized current by deriving both transconductance-to-current ratio and cut-off frequency in terms of I_N. To derive a general expression for transconductance of the MOS transistor, which remains valid for all regions of operation, one approach is to calculate the $\partial i_D/\partial v_{GS}$ using the general I-V relationship. Empirical

models for the MOS transistors, however, overlook the moderate inversion, thereby leading to inaccurate results [3]. The interpolation models such as Eq. (1) can also lead to an error in moderate inversion due to nonphysical (i.e., interpolative) nature of these models. Using the simplified charge sheet model presented in [3], on the other hand, leads to a closed-form expression for g_m/I_D in terms of the normalized current I_N that has acceptable level of accuracy in weak, moderate, and strong inversion regions:

$$\frac{g_m}{I_D} = \frac{1}{nU_T} \cdot \frac{2}{1 + \sqrt{4I_N + 1}} \quad (4)$$

S-parameter measurements were carried out on an NMOS transistor with W=10μm and L=0.18μm. The transistor was fabricated in Jazz Semiconductor's 0.18μm CMOS process to experimentally evaluate the accuracy of Eq. (4). The transistor's transconductance was derived from the measured S-parameters of the transistor [8]. Fig. 1 compares Eq. (4) with the actual measurement. The measurement result approximately follows Eq. (4) for low values of I_N, as shown in Fig. 1. However, this equation overestimates the g_m/I_D for I_N values greater than 10 (i.e., in strong inversion) due to non-idealities such as series source resistance. As will be shown later in this section, even by overestimating the g_m/I_D value in strong inversion region, the $g_m f_T/I_D$ will still reach its maximum value in the moderate inversion.

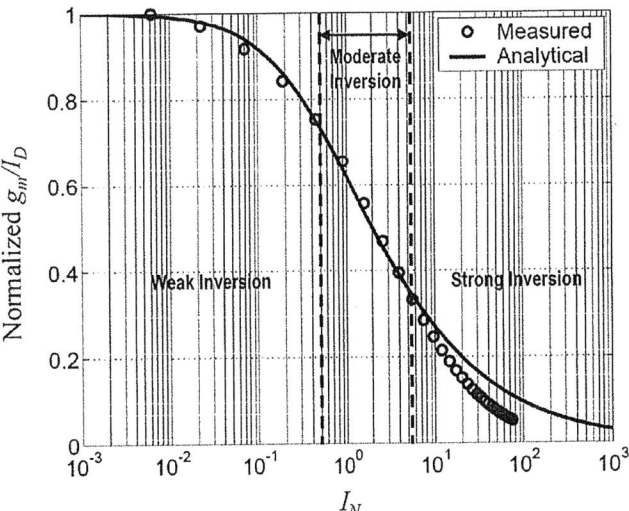

Fig. 1: Normalized g_m/I_D versus I_N for both analytical expression and the measurement result of a fabricated NMOS with W=10μm and L=0.18μm.

Now, we need to derive the transistor's unity-gain frequency f_T as a function of the normalized current by deriving the gate-source and gate-bulk intrinsic capacitances, C_{GS} and C_{GB}, in all inversion regions of operation. These parameters are defined for a transistor biased in saturation in Eqs. (5)-(7) in terms of I_N [3].

$$C_{GS} = WLC_{ox}P(I_N) \quad (5)$$

$$C_{GB} = WLC_{ox}\frac{n-1}{n}[1 - P(I_N)] \quad (6)$$

where:

$$P(I_N) = \left[\frac{3}{2} + \frac{1 + \sqrt{4I_N + 1}}{2I_N} \right]^{-1} \quad (7)$$

By substituting (4), (5), and (6) into $f_T = g_m/[2\pi(C_{GS} + C_{GB})]$, the unity-gain frequency is expressed in terms of the normalized current, i.e.,

$$f_T = \frac{I_D}{\pi W L C_{ox} U_T} \cdot \frac{[n - 1 + P(I_N)]^{-1}}{1 + \sqrt{4I_N + 1}} \quad (8)$$

In Eq. (8), I_D can be replaced by $I_N I_Z$. The result is as follows:

$$f_T = \frac{2n\mu U_T}{\pi L^2} \cdot \frac{I_N [n - 1 + P(I_N)]^{-1}}{1 + \sqrt{4I_N + 1}} \quad (9)$$

To take into account the velocity saturation in short-channel MOS devices, the mobility, μ, needs to be replaced by the effective mobility, μ_{eff}, defined in Eq. (2). By expressing the overdrive voltage in Eq. (2) with respect to the normalized current using Eqs. (1) and (3), and by neglecting the second- and third-order terms in the dominator of Eq. (2), the effective carrier mobility is approximately expressed as:

$$\mu_{eff} = \frac{\mu_0}{1 + \theta\sqrt{I_N}} \quad (10)$$

where the parameter θ is process dependent and is varying from 0.5 to 2. Substituting Eq. (10) into Eq. (9) will result in the following expression for the transistor's unity-gain frequency:

$$f_T = \frac{2n\mu_0 U_T}{\pi L^2} \cdot \frac{I_N [n - 1 + P(I_N)]^{-1}}{(1 + \theta\sqrt{I_N})(1 + \sqrt{4I_N + 1})} \quad (11)$$

Eq. (11) is compared with the actual measurement on f_T of the same NMOS transistor fabricated in a 0.18μm CMOS process. As indicated in Fig. 2, for low values of I_N where the transistor is in either weak or moderate inversion, the proposed analytical model closely follows the measurement data. However, as I_N increases, the measured f_T saturates and then starts to decrease. The reason for this frequency drop is that for large bias currents, the series source resistance becomes comparable to the input resistance of the transistor in common-gate configuration.

The voltage drop across the series source resistance causes the transistor to enter the triode region; hence, g_m will drop. On the other hand, the gate capacitance increases as the transistor enters the triode region [3]. A decrease in g_m and an increase in C_{GS} result in a drop in f_T. Using Eqs. (4) and (11), the $g_m f_T$-to-current ratio is:

$$\frac{g_m f_T}{I_D} = \frac{4\mu_0}{\pi L^2} \cdot \frac{I_N [n - 1 + P(I_N)]^{-1}}{(1 + \theta\sqrt{I_N})(1 + \sqrt{4I_N + 1})^2} \quad (12)$$

Eq. (12) is employed to find the optimum value of the normalized current at which the $g_m f_T/I_D$ reaches its maximum value. The optimum I_N is plotted in Fig. 3 for different values of n and θ.

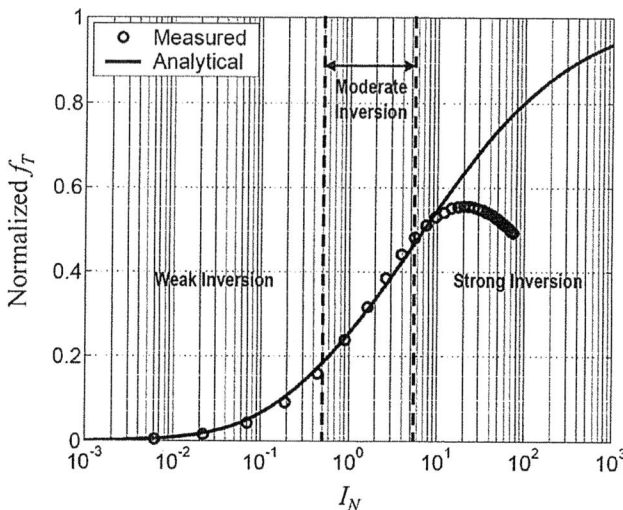

Fig. 2: Normalized f_T versus I_N for both theoretical expression and the measurement result of a fabricated NMOS with W=10μm and L=0.18μm.

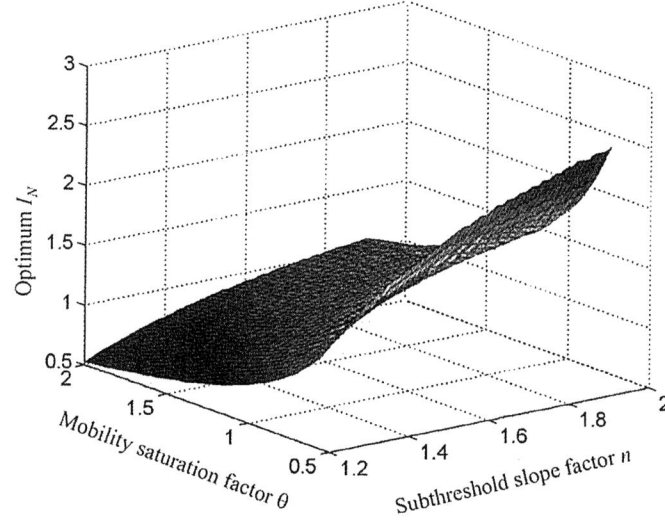

Fig. 3: Optimum I_N leading to the maximum $g_m f_T/I_D$, with respect to different values of n and θ

The optimum value of I_N in Fig. 3 varies from 0.5 to 2.5 for different values of n and θ. According to Table 1, the transistor stays in moderate inversion region for $0.48 < I_N < 5.75$, and therefore, the maximum $g_m f_T/I_D$ occurs in moderate inversion region. Fig. 4 demonstrates the variation of the $g_m f_T/I_D$ with respect to I_N, which was obtained using both the analytical model of Eq. (12) and the measurement. Fig. 4 verifies that the maximum value of the $g_m f_T/I_D$ occurs in moderate inversion region.

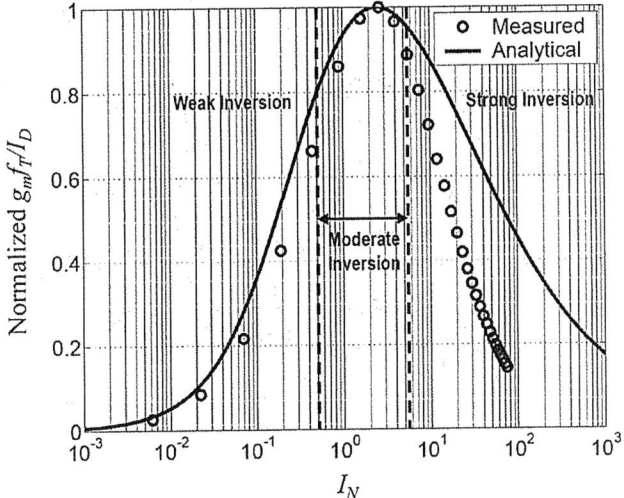

Fig. 4: Normalized $g_m f_T/I_D$ versus I_N for both theoretical expression and measurement result of a fabricated NMOS with W=10μm and L=0.18μm.

4. LNA DESIGN

Shown in Fig. 5 is a power optimized common-gate low noise amplifier (LNA) operating at 950MHz RF frequency. Using the power optimization technique illustrated in Section 3, the LNA is optimized to consume only 100μW power from a 1V supply voltage. The input transistor M_1 of the power-optimized LNA is biased in the moderate inversion region to maximize the gain-bandwidth product (GBW) for the circuit at the 100μA dc bias current. Due to the extremely low bias current, the transistor's g_m is low, and therefore, the active load structure has been employed to improve gain. The bias point of the transistor M_1 is provided by the self-biasing resistor R_F. It is readily shown that the thermal noise of the resistor R_F to the output can be modeled as a current noise source with power spectral density of $4kT/R_F$. Consequently, the noise contributed by the self-biasing resistor decreases as R_F increases.

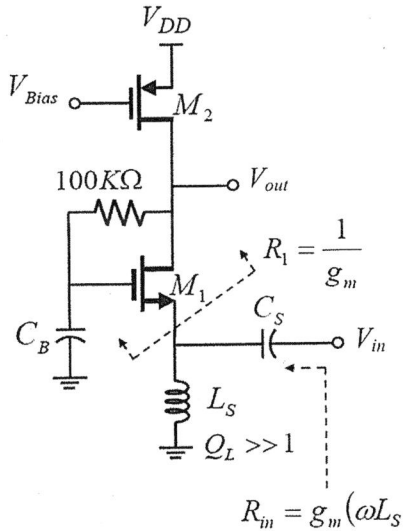

Fig. 5: Common-Gate low noise amplifier with active load and self bias resistor

The input matching is achieved using the LC tuned circuit, as shown in Fig. 5. The matching circuit, on the other hand, introduces additional voltage gain, which will improve the overall gain of the circuit.

Being dominant at lower frequencies, the input-referred noise corresponding to the channel thermal noise and the flicker noise are expressed as follows [3], [10]:

$$\overline{V_{nd}^2} = \frac{4kT\gamma}{g_m} \tag{13}$$

$$\overline{V_{fg}^2} = \frac{K_{1/f}}{WLC_{ox}^2 f} \tag{14}$$

where f is the operating frequency, γ is a bias dependent parameter, and $K_{1/f}$ is a process dependent coefficient. γ, however, decreases as the transistor's bias moves from strong to weak inversion region [9].

In addition, the thermal noise due to the polysilicon gate resistance is important in RFICs as devices with large aspect ratios are used. It can, however, be reduced by layout optimization techniques such as using multi-finger transistor layout.

Furthermore, the gate-induced noise expressed as [10]

$$\overline{I_{ng}^2} = \frac{4kT\delta(2\pi f C_{GS})^2}{5g_m} \tag{15}$$

becomes an important noise component at multi-GHz frequencies [3], [10]. In Eq. (15), δ is a bias dependent parameter and is roughly equal to 2γ. The gate induced noise, however, is correlated with the channel noise with correlation coefficient of approximately $j0.395$ [10]. Nonetheless, the gate-induced noise at the 950MHz RF frequency of interest is negligible. For the LNA circuit of Fig. 5, the channel thermal noise is thus the dominate noise source.

In order to calculate the noise figure (NF) of the circuit, we must calculate the signal-to-noise ratio (SNR) at the input and the output of the power-optimized common-gate LNA. As mentioned above, both passive input matching circuit and the core active components contribute to the LNA's gain. Assuming the matching circuit to be lossless and transforming the $1/g_{m1}$ seen from the source terminal of transistor M_1 to the source resistance of $R_s = 50\Omega$, the input power at the input port is equal to the power at the source terminal of the NMOS transistor, i.e.,

$$\frac{V_s^2}{1/g_{m1}} = \frac{V_{in}^2}{R_s} \tag{16}$$

where V_s is the voltage at the source node of the transistor M_1, V_{in} is the input voltage from the 50Ω source. Using Eq. (16), the voltage gain of the matching circuit can be expressed as Eq. (17)

$$G_p = \sqrt{\frac{1}{R_s g_m}} \tag{17}$$

The noise current at the output node can be expressed as Eq. (18).

$$\overline{I_{n,out}^2} = 4kT\gamma_n g_{m1} + 4kT\gamma_p g_{m2} + 4kTR_s G_P^2 g_{m1}^2 \tag{18}$$

where first and second terms are thermal noise of transistors M_1 and M_2, and the third term is the noise contribution of the input source. The current due to the input signal at the output node can be expressed as

$$\overline{I^2_{signal,out}} = G^2_P g^2_{m1} \overline{V^2_{in}} \qquad (19)$$

Using Eqs. (17) to (19), the output SNR of the LNA circuit is expressed as

$$SNR_{out} = \frac{\overline{I^2_{signal,out}}}{\overline{I^2_{noise,out}}} = \frac{1}{4kTR_s} \cdot \frac{g_{m1} \cdot \overline{V^2_{in}}}{(1+\gamma_n)g_{m1} + \gamma_p g_{m2}} \qquad (20)$$

The input SNR is expressed as

$$SNR_{in} = \frac{\overline{V^2_{in}}}{4kTR_s} \qquad (21)$$

Using Eqs. (20) and (21), the NF of the LNA circuit is calculated as follows

$$NF = \frac{SNR_{in}}{SNR_{out}} = 1 + \gamma_n + \gamma_p \frac{g_{m2}}{g_{m1}} \qquad (22)$$

Eq. (22) predicts that the circuit's NF will increase as g_{m2} increases. However, as the transistor's bias point moves toward the weak inversion from the strong inversion region, the γ value will decrease and the value of g_{m1} will increase for constant bias current, which will reduce the circuit's NF.

5. MEASUREMENT RESULTS

The LNA circuit has been fabricated using Jazz Semiconductor CMOS 0.18μm process. Fig. 6 shows the die photo of the circuit occupying an area of 1057μm×865μm. A source follower buffer was placed at the output of the amplifier to avoid the loading effect of the measurement devices at the high impedance output of the LNA circuit. The buffer was also fabricated separately and the final measurement results were derived by de-embedding the effect of the source follower. The de-embedding was achieved by measuring the NF and gain of the buffer separately.

The input return loss of the circuit was measured using Agilent E8358A Network Analyzer. The circuit exhibits a minimum -25dB return loss at the 950MHz RF frequency.

The NF and gain of the LNA are shown in Fig. 8. The results were calculated from the measured data for the combination of the LNA and buffer, and by de-embedding the buffer effect from the measurement result.

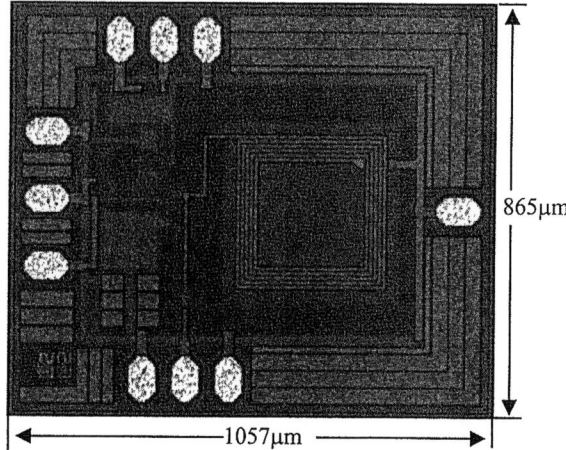

Fig. 6: Die photo of the common-gate Low noise amplifier fabricated in CMOS 0.18μm process

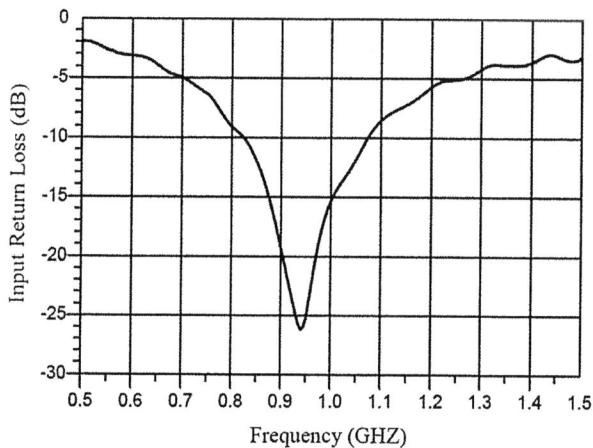

Fig. 7: Measured input return loss of the common-gate LNA

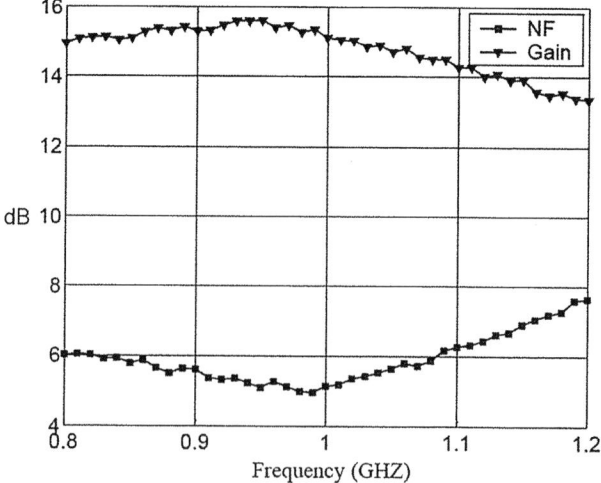

Fig 8: Measured noise figure and gain of the common-gate LNA

The circuit exhibits a NF of 4.9dB and a small-signal voltage gain of 15.6dB at the RF frequency of 950MHz. Results of the two-tone test and gain compression for the LNA are shown in Figs. 9 and 10, respectively. The LNA exhibits an IIP3 of -13.7dBm and an input-referred 1-dB compression point of -21.8dBm. The summary of the circuit performance is provided in Table 2. Compared to the prior work presented in [11]–[13], the circuit exhibits lower power consumption, higher gain, and slightly better NF.

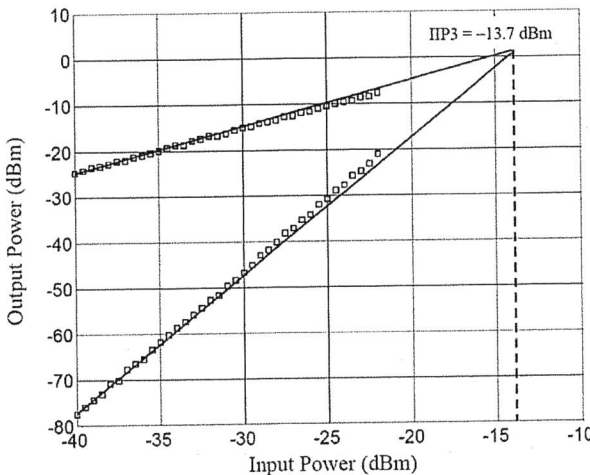

Fig. 9: Measured 3rd order input intercept point (IIP3)

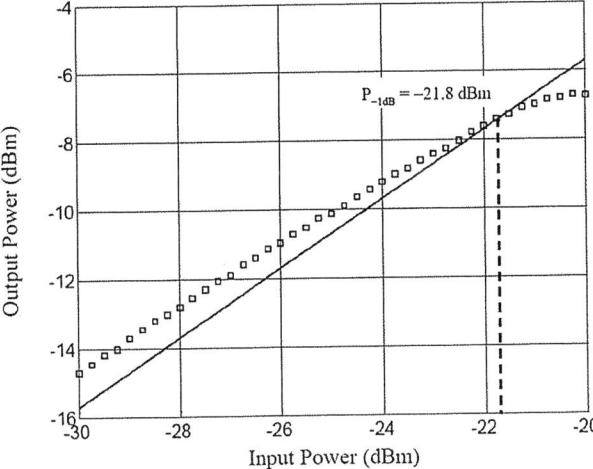

Fig. 10: Measured input referred 1-dB compression point

Table 2: Summary of the common-gate LNA performance

Parameter	Value
Supply Voltage	1 V
DC Current	100 μA
Operational Frequency	950MHz
S11 (dB)	-25 dB
IIP3	-13.7 dBm
P_{-1dB}	-21.8 dBm
Gain	15.6 dB
NF	4.9 dB

6. CONCLUSIONS

A closed-form expression for the $g_m f_T$-to-current ratio ($g_m f_T/I_D$), as a new figure of merit for performance optimization of ULP RF circuit was derived. It was shown both analytically and experimentally that $g_m f_T/I_D$ reaches its maximum value in moderate inversion region. A power optimized common-gate low-noise amplifier (LNA) with active load was fabricated in a CMOS 0.18μm process operating at 950MHz. The actual measurement results have shown a 4.9dB NF and a small signal gain of 15.6dB with a record-breaking power dissipation of only 100μW.

7. ACKNOWLEDGMENT

This project was supported by NSF CAREER Award under contract ECS-0449433 and by Fujitsu Labs of America through a UC-Micro project. The authors would like to thank Jazz Semiconductor, particularly Juan Cordovez, for providing the device measurement data.

8. REFERENCES

[1] F. Silveira, D. Flandre, P. G. A. Jespers, "A g_m/I_D Based Methodology for the Design of CMOS Analog Circuits and Its Application to the Synthesis of a Silicon-on-Insulator Micropower OTA," *IEEE J. of Solid-State Circuits*, vol. 31, no. 9, pp. 1314-1319, Sep. 1996.

[2] D. J. Comer, D. T. Comer, "Using the Weak Inversion Region to Optimize Input Stage Design of CMOS Op Amps," *IEEE Trans. on Circuit and Systems-II*, vol. 51, no. 1, pp. 8-14, Jan. 2004.

[3] Y. Tsividis, *Operation and Modeling of the MOS Transistor*, 2nd edition, McGraw Hill, 1999.

[4] W. Liu, *MOSFET Models for SPICE Simulation including BSIM3v3 and BSIM4*, John Wiley & Sons, 2001.

[5] J. R. Hauser, "A New and Improved Physics-Based Model for MOS Transistors," *IEEE Trans. on Electron Devices*, vol. 52, no. 12, pp. 2640-2647, Dec. 2005.

[6] B. G. Streetman, S. Banerjee, *Solid State Electronic Device*, 5th edition, Prentice Hall, 2000.

[7] C. C. Enz, Y. Cheng, "MOS Transistor Modeling for RF IC Design," *IEEE J. Solid-State Circuits*, pp. 186-201, vol. 35, no. 2, pp. 186-201, Feb. 2000.

[8] A.F. Tong, K.S. Yeo, L. Jia, C.Q. Geng, J.-G. Ma and M.A. Do, "Simple and Accurate Extraction Methodology for RF MOSFET Valid up to 20GHz," *IEE Proc. Circuits, Devices and Systems*, vol. 151, no. 6, pp. 587-592, Dec. 2004.

[9] C. C. Enz, "An MOS Transistor Model for RF IC Design Valid in All Regions of Operation," *IEEE Trans. Microwave Theory and Techniques*, vol. 50, no. 1, pp. 342-359, Jan. 2002.

[10] T. H. Lee, *The Design of CMOS Radio-Frequency Integrated Circuits*, 2nd edition, Cambridge University Press, 2004.

[11] B. G. Perumana, S. Chakraborty, C. H. Lee, J. Laskar, "A Fully Monolithic 260-μW, 1GHz Subthreshold Low Noise Amplifier," *IEEE Microwave and Wireless Components Letters*, vol. 15, no. 6, pp. 428-430, June 2005.

[12] H. H. Hsieh, L. H. Lu, "A CMOS 5-GHz Micro-Power LNA," *IEEE Radio Frequency Integrated Circuit Symposium*, June 2005, pp. 31-34.

[13] A. Molnar, B. Lu, S. Lanzisera, B. W. Cook, K. S. J. Pister, "An Ultra-low Power 900MHz RF Transceiver for Wireless Sensor Networks," *IEEE Custom Integrated Circuits Conf.*, Oct. 2004, pp. 401-404.

A CMOS Analog Frontend for a Passive UHF RFID Tag

Alessio Facen
University of Parma
Dipartimento di Ingegneria dell'Informazione
Parco Area delle Scienze 181/A
Parma, Italy
alessio.facen@nemo.unipr.it

Andrea Boni
University of Parma
Dipartimento di Ingegneria dell'Informazione
Parco Area delle Scienze 181/A
Parma, Italy
andrea.boni@unipr.it

ABSTRACT

The paper discusses the design of the analog frontend of a passive UHF RFID tag, compatible with ISO/IEC 18000-6b standard. An efficient ESD-protected power retrieving circuit, based on the antenna features, a rectifier bridge and a charge pump, is introduced, as well as an auto-calibrated clock generator. The chip, implemented in a 0.18μm digital CMOS technology, does not need any post-fabrication trimming or external component besides the antenna; according to simulations, a correct communication is achieved at a distance of several meters between reader and tag.

Categories and Subject Descriptors

B.7 [Integrated Circuits]: Miscellaneous

General Terms

Design

Keywords

UHF integrated circuits, electromagnetic coupling

1. INTRODUCTION

Industries looking for a reliable, hard-to-falsify, tiny and especially cheap way to stock and manage goods are nowadays abandoning optical bar reading architectures in favor of UHF passive RFID tags. Indeed, these devices allow a reading range of few meters and don't require any external battery or power generator, but obtain the power supply directly from the RF signal, providing, therefore, a great advantage in terms of both size and cost.

An ISO/IEC 18000-6b compliant UHF tag has been designed, using a digital 0.18μm CMOS technology. The standards [2][1] dictate a communication with a 40kHz ASK signal modulated with a 869.5MHz carrier. Modulation depth in reader-to-tag transmission is set to 18% or 100% with a Manchester encoding, while tag-to-reader backscatter answer is FM0 encoded. Simulations show that the designed

Permission to make digital or hard copies of all or part of this work for personal or classroom use is granted without fee provided that copies are not made or distributed for profit or commercial advantage and that copies bear this notice and the full citation on the first page. To copy otherwise, to republish, to post on servers or to redistribute to lists, requires prior specific permission and/or a fee.
ISLPED'06, October 4–6, 2006, Tegernsee, Germany.
Copyright 2006 ACM 1-59593-462-6/06/0010 ...$5.00.

chip, connected to an unity-gain adapted antenna, can establish a correct communication with a 500mW ERP interrogator within a mutual distance higher than 5m.

The architecture of the system and the requirements of the antenna are presented in sect. 2. Sect. 3 focuses on the rectifier bridge, showing the advantages of the proposed structure with respect to traditional ones. Sect. 4 handles the protections of the chip against ESD and overvoltage events. The charge pump is discussed in sect. 5, while the voltage regulation system is explained in sect. 6. The clock generation and the demodulation circuits are presented, respectively, in sect. 7 and 8. Finally, conclusions are drawn in sect. 9.

2. SYSTEM ARCHITECTURE

The designed RFID tag is made up of a differential antenna and a chip whose only two I/O pins are connected to the antenna terminals. The input section of the chip can be modelled with the R-C series equivalent, which together with the external inductance, L, and the antenna radiation resistance, R_{ANT} forms an RLC resonator, Fig. 1. It is worth to point out that a convenient equivalent inductance can be obtained by acting on the antenna, without resorting to discrete components, by shifting its impedance resonance frequency slightly above the UHF carrier one. Therefore, the peak voltage available at the chip input, under power matching conditions, can be estimated as:

$$V_{INp} \approx \sqrt{2 \frac{P_{AV}}{R_{ANT}}} \frac{1}{\omega C} \quad (1)$$

where P_{AV} is the available power at the chip input, estimated according to the Friis transmission equation. Eqn. (1) suggests that, in order to enhance the amplitude of the input RF signal voltage, both the antenna equivalent resistance and the chip input capacitance must be minimized. However, the antenna radiation resistance is limited by by the maximum achievable antenna efficiency, while a high inductance negatively impacts the antenna size.

The architecture of the analog front-end of the chip is shown in Fig. 2. The RF terminals are connected to a rectifier bridge and a voltage multiplier, providing the power supply to the analog blocks. In order to mantain a sufficient power supply level during the interrogation and backscatter phases, an on-chip 1nF capacitor is employed as energy-storage element: its charging occurs during the available time interval (400μs) that the standards provide for the RFID powering up. The charge pump output voltage is then regulated by means of a suitable circuit, providing the

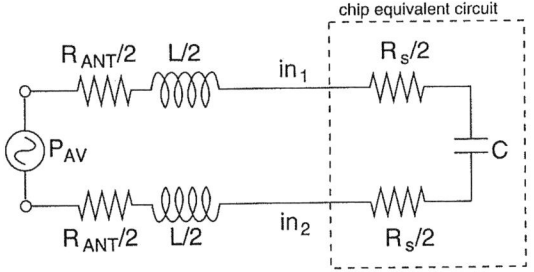

Figure 1: Passive tag equivalent schematic.

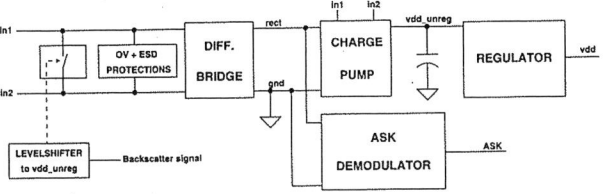

Figure 2: Frontend block diagram.

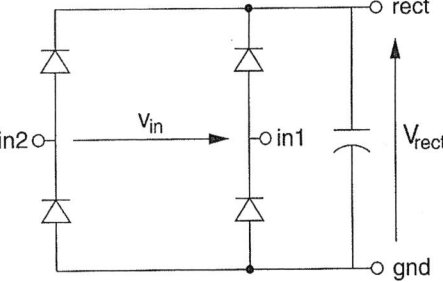

Figure 3: Full-wave bridge rectifier.

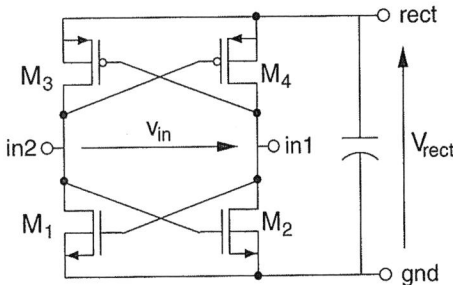

Figure 4: Proposed rectifier.

power supply level for the digital core and the self-regulated clock generator.

A demodulator translates the envelope of the RF signal at the end of the rectifier into CMOS levels and provides the ASK signal to the digital core. The latter, after processing the signal, generates a CMOS output that is used to drive the backscatter switch placed between the chip inputs, thus modulating the radar cross section of the tag. Since the switch is implemented with a MOS device, better performance in terms of difference of effective radiating area is achieved by driving the transistor with the highest available voltage, corresponding to the unregulated charge-pump output.

Between the chip inputs a protection circuit has been introduced against ESD and overvoltage events, gaining robustness compatibility with production requirements. It is worth noticing that a recently reported implementation avoids the use of such a protection in favor of lower capacitive load at the input [4].

The required 40kHz frequency reference for the digital circuitry is obtained from an integrated ring-oscillator with digital calibration, which allows to achieve the 15% precision on the clock period as dictated by the standard, without resorting to expensive trimming procedures at the wafer sort.

3. DIFFERENTIAL RECTIFIER

The rectifier used in Fig. 2 is used both to provide the envelope of the RF signal to the ASK demodulator and to feed the first stage of the charge pump with a DC voltage. Input unmodulated voltage at the rectifier terminals can be written as

$$v_{in} = V_{IN} sin(\omega t) = V_{IN} sin(\frac{2\pi}{T_c} t) \qquad (2)$$

T_c being the period of the RF carrier.

The well-known full-wave bridge rectifier (Fig. 3) can be a useful solution when the available power is sufficiently high to neglect the voltage drop across each diode. Reported

solutions [6][8], in order to reduce the threshold, make use of Shottky diodes; another implementation [4], based on the Greinacher rectifier, uses SOI process. In the present project the emphasis was on a low-cost implementation, therefore a digital CMOS $0.18\mu m$ technology has been employed, and low-threshold MOS transistors have been used as a convenient alternative to Shottky diodes.

At the operating frequencies, assuming a current consumption of $10\mu A$ and 1pF equivalent capacitance, the voltage drop at the output due to the capacitor discharge is estimated in few mV, so it can be neglected in the calculations. Therefore, the rectified voltage after the bridge can be approximated as

$$V_{rect} = V_{IN} - 2V_T \qquad (3)$$

V_T being the diode threshold.

This solution loses its convenience when the available input power becomes low enough to consider V_T no more negligible with respect to V_{IN}. In fact, no DC voltage can be obtained if $V_{IN} < 2V_T$; furthermore, complex circuitry (i.e. high power consumption) is needed to demodulate the signal when V_{rect} does not reach a sufficient level, that is when V_{IN} is not some hundred millivolts above $2V_T$.

This problem is overcome with the solution in Fig. 4.

The MOS's are employed as switches: assuming their on-resistance negligible and the same threshold V_T both for PMOS and NMOS devices, in the intervals where $|v_{in}| > V_T$, rect (the rectifier output pin) is connected through a PMOS to the input pin featuring the highest voltage, while an NMOS sets gnd to the lowest potential. On the other hand, in the intervals where $|v_{in}| < V_T$ the switches are turned off, and (still considering negligible the voltage drop due to the capacitor discharge) $V_{rect} = V_T$. Waveforms showing this behavior are presented in Fig. 5.

281

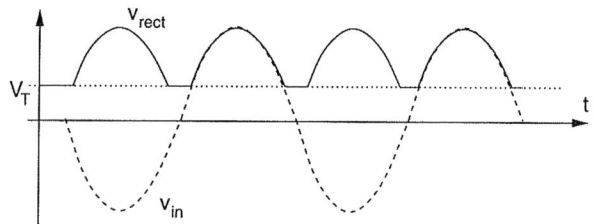

Figure 5: Proposed rectifier: waveforms when $r_{on}=0$.

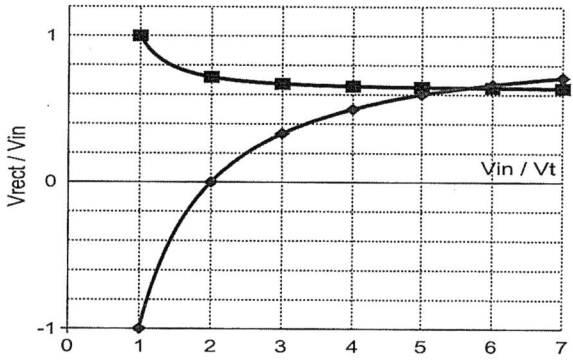

Figure 6: Comparison between proposed (blue ■) and full-wave (red ♦) outputs.

With simple calculations, it can be shown that the average value of the output voltage is

$$\overline{V_{rect}} = 4V_T \frac{T_1}{T_c} + \frac{2V_{IN}}{\pi} cos(2\pi \frac{T_1}{T_c}) \qquad (4)$$

where $T_1 = \frac{T_c}{2\pi} arcsin(\frac{V_T}{V_{IN}})$.

The minimum value of V_{IN} that produces a positive output voltage is V_T. A comparison between the two solutions, assuming the same V_T for both circuits, shows that for very low input power levels the adopted one exhibits superior efficiency, as reported in Fig. 6: the chosen structure is more convenient for $V_{IN} < 5.65V_T$. Yet it is worth noticing that in presence of elevate input power levels, where the proposed rectifier exhibits lower performance, the charge pump output voltage is in any case far higher than the minimum level needed to power the chip core.

In order to achieve a stable rectified voltage, the average on-resistance of the switches is increased by changing the transistors size ratio, thus obtaining an RC low-pass filter which, however, must be still able to track the envelope variations during the interrogation phase. Even if increasing the resistor value should negatively impact the quality factor of the RLC resonator, this solution does not worsen the circuit performances: analysis and simulations show that the equivalent series-resistor of the tank is mostly determined by the charge pump [9].

4. ESD & OVERVOLTAGE PROTECTIONS

Since only two I/O pins are present in the whole chip, the ESD protection circuit should provide a safety low-resistance path only between in_1 and in_2. The use of simple clamping diodes has the major drawback of introducing a high capacitive load at the inputs, thus reducing the quality factor of the RLC tank. In fact, in order not to alterate the behavior of the circuit when the available power is very low, a series of more diodes should be positioned between the inputs, thus these devices should be large-sized enough to absorbe the ESD event. Moreover, such a way of clamping the RF signal would cause the loss of the amplitude modulation when the input power is sufficiently high to turn on the diodes.

The proposed ESD protection circuit is shown in Fig. 7. In its simpler version, the circuit has the gates of transistors M_1 and M_2 tied to gnd: the structure formed by M_1, M_2, M_3 and M_4 is equivalent to the full-wave rectifier of Fig. 3, and allows the gnd and $rect$ potentials to track fast impulsive voltage peaks, which would be ignored by the RC filter of the main rectifier. The voltage step on v_{rect} caused by an ESD event is then reported on the gate of M_5 by C_C, turning M_5 on and allowing the massive current peak to flow through $M_1 - M_5 - M_4$ or $M_2 - M_5 - M_3$: each MOS transistor has to be consequently sized. Once exhausted the ESD event, the resistor R_C allows to turn off M_5.

A second protection circuit is introduced in order to avoid the breakdown of the devices that would happen if the tag were too close to the interrogator. In fact, the employed technology allows a maximum voltage of 1.8V between two terminals of the same transistor: simulations show that this condition is reached in the bridge MOS's for a tag reader distance of 1.2m under typical conditions. With the circuit shown in the left part of Fig. 7, this distance is reduced to 30cm (although it should be noticed that the far-field approximation used in the calculations of the available power at the tag antenna is no more valid in such region). The circuit works as follows: if the rectified level is lower than the threshold of the two diodes, the node A is approximately grounded and the circuit will not affect the rectifier normal behavior. If the rectified voltage is high enough to turn on the diodes, $V_A = V_{rect} - 2V_{TD}$, V_{TD} being the diode threshold, and the pair $M_{1,2}$ is turned on. The higher the rectified voltage, the higher the current drawn by $M_{1,2}$, causing a current flow from in_1 to in_2 and a consequent input voltage reduction due to the change of the equivalent impedance. The $R_F - C_F$ filter is needed to prevent the overvoltage protection to suppress the AM modulation.

It should be noticed that the modified ESD protection circuit exhibits better performance with respect to the A-grounded version, since an ESD event affecting $rect$ also causes the overvoltage section to turn on M_1 and M_2, thus creating an additional path for the current to flow through M_1 and M_2.

5. VOLTAGE MULTIPLIER

The differential bridge can not provide a sufficient supply level to the digital core when the tag is located far from the interrogator, even resorting to the higher quality factor that can be obtained from the RLC equivalent circuit. A voltage multiplier is thus introduced; the schematic of the circuit is presented in Fig. 8.

The multiplier is based on a three-stages differential Dickson charge pump [5], having the RF inputs in_1 and in_2 as complementary clocks. As a matter of fact, these are sinusoidal signals, so the efficiency of the circuit is far lower than the one predicted by using square-waves; on the other hand,

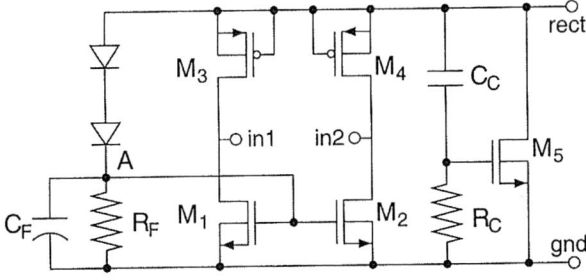

Figure 7: ESD and overvoltage protection circuit.

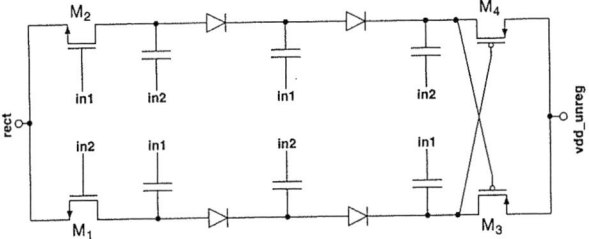

Figure 8: Charge pump.

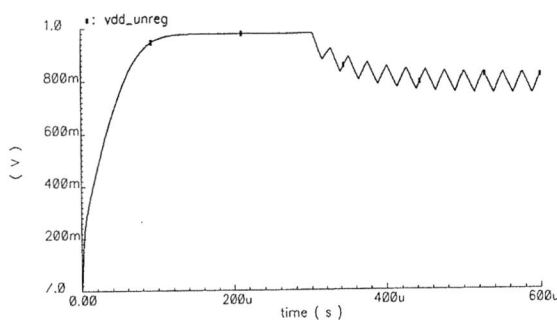

Figure 9: Charge pump output level.

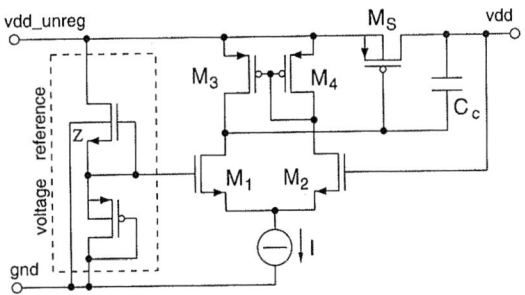

Figure 10: Power supply regulator.

it is not possible to obtain a squared RF voltage from in_1 and in_2 using the rectified voltage, since the required buffers would introduce an extra capacitive load to the inputs (lowering the quality factor) and V_{rect} may not be sufficiently high to bias the buffer itself.

The diodes employed in second and third stages are implemented with low-threshold diode-connected NMOS transistors, while the last stage is a peak detector like the one used in the bridge. Before the last structure, where the voltage peaks are higher than in the rest of the circuit, a further overvoltage protection is introduced, forcing a current to flow towards gnd if safety-level potentials are exceeded.

As explained in sect. 3, V_{rect} is lower than the input peak level, V_{IN}. Since the adopted technology provides native zero-voltage-threshold NMOS transistors, the first stage has been realized using two of these devices, M_1 and M_2, with the gates driven by the input signals complementary to the ones provided to the respective first capacitor. This solution allows to overcome the threshold loss that would have occurred using diodes as in the successive stages.

A simulation (inclusive of layout parasitics) showing the charge pump output level in the worst power-consumption corner is reported in Fig. 9. The estimated current request of the digital core is $3\mu A$ at high temperature, assuming a distance of 5 meters between a 500mW ERP interrogator and the tag, a unity-gain adapted antenna and a 100% modulation depth starting after the capacitor charge. It can be noticed that the output level always remains above 0.75V. The worst-case simulated efficiency of the on-chip power generation circuit, made up by rectifier and charge pump, for a current request of $3\mu A$ and the same reader-tag configuration as above is 15%.

6. REGULATOR

The supply voltage provided at the output of the charge pump is strongly dependent on process variations and tem-

perature. Furthermore, this potential is determined also by the current consumption of the circuit and the distance from the reader: moving the tag with respect to the interrogator causes a variation of vdd_unreg that could be noticeable along the whole reader-tag transmission stream.

While using the unregulated supply voltage proves useful to drive the backscatter transistor in order to maximize the radar cross section variation, it prevents from designing a stable clock generation circuit: since in the proposed chip a ring oscillator has been employed, whose output frequency strongly depends on the supply level, the variability of the clock frequency with vdd_unreg would be unacceptable.

It is therefore mandatory to introduce a voltage-regulation system, allowing also to achieve a better control of the digital gates behavior. The regulator, whose schematic is presented in Fig. 10, has the well-known structure of a series transistor on the power path driven by an op-amp.

The designed voltage reference, left part of Fig. 10, is made up of a diode-connected PMOS biased with the current given by a depletion NMOS transistor (zero threshold voltage) whose gate-source voltage is set to zero. Setting the aspect ratios of the MOS devices allows to achieve relatively small variations with temperature, while employing channel lengths far higher than the minimum one grants both good supply independency and very low power consumption. The latter feature and the tiny area required are the main advantages of the proposed reference generator with respect to the traditional bandgap based ones [3], but suffer far more of the process variability, although two intrinsic feedback mechanisms are present to reduce the effect of this dependence. In fact, the employed PMOS has the same electrical model of the devices used in the digital section of the circuit, so

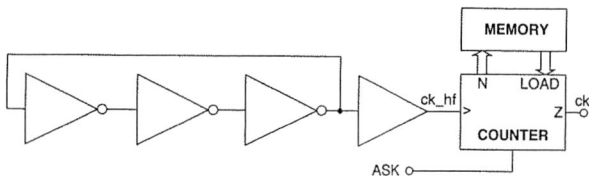

Figure 11: Clock generation circuit.

the larger current request of the latter due to a PMOS fast process is reduced by the lower regulated supply voltage; the second feedback is due to the body effect that raises the threshold of the ZVT NMOS as the reference voltage increases, thus limiting the reference current.

According to simulations, the designed voltage reference ranges from 0.5 to 0.7V with variations of less than 2% over a 0÷85°C temperature span, and the regulator exhibits a PSRR higher than 30dB for an unregulated power supply ranging from 0.75 to 1.8V.

7. CLOCK GENERATION

Most commercial RFID tags make use of RC oscillators, tuned to the correct modulation frequency (40kHz) with a post-fabrication trimming procedure. This kind of solution, besides its expensiveness, exhibits both a considerable power consumption and area occupation given the low desired output frequencies. For the same reason it is not possible to employ integrated inductors; moreover, the low power supply raises several difficulties in the design of LC-tank based circuits. Ring oscillators are by far the smallest and least power-hungry circuits, but they suffer of high frequency spread over process and temperature variations, even in presence of a regulated voltage supply.

The ISO/IEC 18000-6b standard-regulated reader-to-tag communication provides a preamble field, made up of 9 Manchester-encoded zeros. The basic idea used in the proposed design is generating a high-frequency clock and, by synchronizing it with the preamble ASK demodulated signal, obtaining the 40kHz clock by means of a digital counter. The measure of the period is done by averaging the number of clock periods counted during the single bits of the preamble field, then the resulting value is stored and used to define the final value of the counter in the successive phases. A block diagram of the clock generation circuit is shown in Fig. 11.

The tolerance on the generated signal is set basically by the one dictated from the standard specification on the return link transmission data-rate, that is ±15% of the period $T_L = 25\mu s$: in order to fulfill the requirement, the minimum frequency f_{ckH} of the HF-clock can be obtained from eqn. (5):

$$\frac{\Delta T_{L,max}}{T_L} = 1 - \frac{int(T_L f_{ckH})}{T_L f_{ckH}} \qquad (5)$$

The minimum acceptable value of the high-frequency clock is $f_{ckH,min}$ =240kHz, suggesting at least a three-bits counter to generate the 40kHz low-frequency clock.

The main feature requested to the desired oscillator is the low power consumption. Therefore, the ring oscillator architecture has been chosen, with the minimum number of

stages (three). In order to minimize the current request, no low-threshold MOS transistors have been used: given the regulated voltage supply, it happens that $|V_{THp}|, V_{THn} > V_{DD}/2$. During the transitions there is no time interval where both the MOS are in strong inversion, therefore saving considerable amounts of dynamic power.

The propagation time of a single inverter, loaded by an equal port, can be calculated on the basis of the analysis performed in [7], assuming that the current of the MOS working in the subthreshold region affects negligibly the charge of the loading capacitor. For example, t_{pHL} can be expressed as

$$t_{pHL} = \left(\frac{\frac{V_{THn}}{V_{DD}} + \alpha}{1 + \alpha} - \frac{1}{2} \right) t_T + \frac{C_L V_{DD}}{I_{D0}} \qquad (6)$$

$$t_T = \frac{C_L V_{DD}}{I_{D0}} \left(\frac{0.9}{0.8} + \frac{V_{D0}}{0.8 V_{DD}} ln \frac{10 V_{D0}}{e V_{DD}} \right) \qquad (7)$$

where α is the velocity saturation parameter, V_{D0} and I_{D0} are, respectively, the drain-source saturation voltage and the maximum drain current of the MOS when the inverter input voltage is V_{DD}, while C_L is the loading capacitor, including the layout parasitics. It should be pointed out that the capacitor value is strongly dependent from the output voltage, since the transition is roughly centered around the MOS threshold voltage: an average value has been used in calculations. The same analysis can be applied for the calculation of t_{pLH}.

Since C_L and I_{D0} are simple functions of the transistors length and width, the MOS's dimensions can be easily obtained from (6) and (7), given the desired output frequency.

It is worth noticing that the inverter delay suffers far more of temperature and process variations with respect to the typical situation, as the MOS threshold is no more far lower than the supply voltage: a wide spread of the output frequency over PVT space can be expected. In fact, setting the transistors size to obtain a minimum frequency of 240kHz over the corners reflects into the maximum frequency being around 1 MHz: this causes the extension of the counter to 5 bits in order not to experience an overflow. Furthermore, since a 5-bits counter can manage an input frequency of 1.24MHz for an output frequency of 40kHz, the span of the desired oscillator has been traslated to 340÷1200kHz, thus reducing the maximum error on the low-frequency clock period to 10.9% (for a frequency of 359kHz divided by 8). The maximum average current requested by the oscillator, including the buffer, is 330nA.

8. ASK DEMODULATOR

The envelope of the AM signal obtained from the bridge (*rect*) is translated into CMOS levels with the circuit shown in Fig. 12. In order to provide two suitable inputs to a comparator, the average level of *rect* is retrieved by means of an RC low pass filter able to track long-term variations of the rectified level. Not needing particular precision on the pole frequency, the large capacitor and resistor needed to reject the 40kHz Manchester modulation can be implemented by MOS devices: nominally, C=5pF and R=2MΩ. Such a large resistance is obtained with a series of 2 diode-connected native zero-threshold-voltage NMOS's with very low W/L.

It should be noticed that the designed filter pole frequency is about 16kHz, not far lower than the 40kHz ASK one. In fact, the filter should reach the average level in few clock pe-

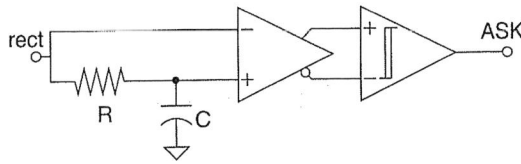

Figure 12: ASK demodulator.

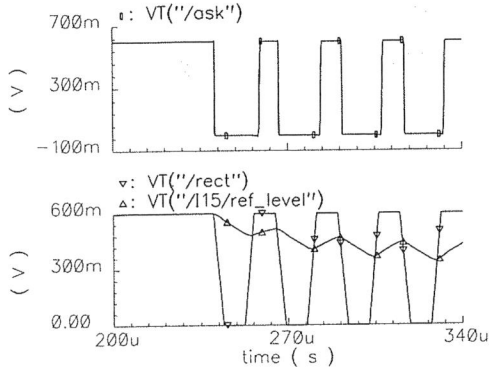

Figure 13: Preamble field: first demodulated bits

	WC	TC
Maximum operating distance	5.3m	>8m
Bridge+multiplier efficiency @ 5m	15%	17%
Regulated output voltage	0.5V	0.6V
Clock period error on 40kHz	10.9%	8%
Oscillator current consumption	330nA	280nA

Table 1: Chip features: worst & typical corners.

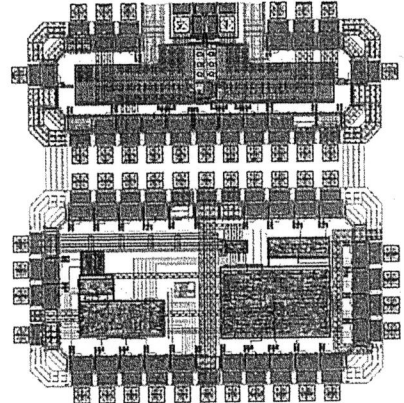

Figure 14: Test chip layout

riods after the powering up, in order to provide a precise reference period to the clock synchronization system explained in sect. 7: the waveforms in Fig. 13 show that, after the powering-up time, the settling of the average level after the filter causes an error in the estimation of the period of the preamble first bits. Since there are 9 zeros in the preamble field, in the estimation of the period the first bits have to be skipped while the filter output is settling. Nonetheless the issue is not so critical while demodulating the main communication, since the Manchester encoding prevents the average level to exhibit great variations.

Finally, the difference between the *rect* signal and its average level has been amplified before being analyzed by the comparator, since the large offset of the latter could cause a wrong demodulation when the modulation depth is 18%.

9. CONCLUSIONS

The analog frontend of a passive UHF RFID tag has been designed, suitable to achieve a correct communication based on ISO/IEC 18000-6b standard within a distance of several meters between reader and tag. The circuit, comprehensive of ESD and overvoltage protections, is made up of a low-activation power retrieving section, a demodulator and a self-calibrated on-chip clock generator. The main features of the designed chip are summarized in Table 1.

A first test-chip will soon be available to perform experimental measurements of the designed tag: the layout, comprehensive of probe-points pads, analog frontend and digital core, is shown in Fig. 14, and its size is $1500 \times 1500 \mu m^2$, while the single tag occupies only $700 \times 600 \mu m^2$.

10. ACKNOWLEDGEMENTS

The work is supported by IMEM CNR, Commessa "Nuovi sistemi elettronici a iperfrequenze".

11. REFERENCES

[1] Electromagnetic compatibility and radio spectrum matters (erm); short-range devices (srd) intended for operation in the 862 mhz to 870 mhz band; system reference document for radio frequency identification (rfid) equipment. *ETSI TR 101 445 V1.1.1*, April 2002.

[2] Iso/iec 18000-6:2004(e) international standard. August 2004.

[3] A. Boni. Op-amps and startup circuits for cmos bandgap references with near 1-v supply. *IEEE Journal of Solid-State Circuits*, 37(10).

[4] J. Curty, N. Joehl, C. Dehollain, and M. J. Declercq. Remotely power addressable uhf rfid integrated system. *IEEE Journal of Solid-State Circuits*, 40(11).

[5] J. F. Dickson. On-chip high-voltage generation in mnos integrated circuits using an improved voltage multiplier technique. *IEEE Journal of Solid-State Circuits*, SC-11(3).

[6] U. Karthaus and M. Fischer. Fully integrated passive uhf rfid transponder ic with 16.7-uw minimum rf input power. *IEEE Journal of Solid-State Circuits*, 38(10).

[7] T. Sakurai and A. R. Newton. Alpha-power law mosfet model and its applications to cmos inverter delay and other formulas. *IEEE Journal of Solid-State Circuits*, 25(2).

[8] K. Seemann, F. Cilek, G. Hofer, and R. Weigel. Single-ended ultra-low-power multistage rectifiers for passive rfid tags at uhf and microwave frequencies. In *IEEE Radio and Wireless Symposium 2006 Proceedings*.

[9] G. D. Vita and G. Iannaccone. Design criteria for the rf section of uhf and microwave passive rfid transponders. *IEEE Transactions on Microwave Theory and Techniques*, 53(9).

High-Speed Low-Power Frequency Divider with Intrinsic Phase Rotator

Stephan Henzler
Advanced Systems and Circuits
Infineon Technologies AG
81726 Munich, Germany
henzler@ieee.org

Siegmar Koeppe
Advanced Systems and Circuits
Infineon Technologies AG
81726 Munich, Germany
siegmar.koeppe@infineon.com

ABSTRACT

A CMOS divider concept without static power consumption, except leakage power, is proposed. The circuit divides an input signal by two and generates four phases with highly accurate phase skew of 90 degrees. In a 90nm low-power CMOS technology, the maximum operation frequency is 11.6 GHz for a supply voltage of 1.5V slow process and worst case operation parameters. Higher frequencies can be achieved by a hybrid approach where the signal is first divided by a factor of two in a single CML stage and then by the proposed circuit by another factor of two for the generation of the four phases. The divider is applied to dual modulus pre-scalers and IQ receivers. A variant of the circuit contains an intrinsic phase-rotator, so the power consumption of the pre-scaler is not only reduced due to the logic style but also by a simplified architecture of the overall pre-scaler.

Categories and Subject Descriptors

B.4.1 [**Input/Output and Data Communications**]: Data Communications Devices

General Terms

Design

Keywords

Low-Power, Divider, Phase-Rotator, Pre-Scaler

1. INTRODUCTION

High-speed frequency dividers are important building blocks in radio frequency signal processing circuits, frequency synthesizers and fast sample-and-hold circuits. Multiple output signals with fixed phase-skew are beneficial for phase rotators, IQ-signal processing, dynamic logic, and high speed mixed-signal circuits. In fractional-N PLLs (phase-locked-loop) and dual modulus pre-scalers, respectively, it is desirable to skip a phase of 90 degrees without the occurrence

Permission to make digital or hard copies of all or part of this work for personal or classroom use is granted without fee provided that copies are not made or distributed for profit or commercial advantage and that copies bear this notice and the full citation on the first page. To copy otherwise, to republish, to post on servers or to redistribute to lists, requires prior specific permission and/or a fee.
ISLPED'06, October 4–6, 2006, Tegernsee, Germany
Copyright 2006 ACM 1-59593-462-6/06/0010 ...$5.00.

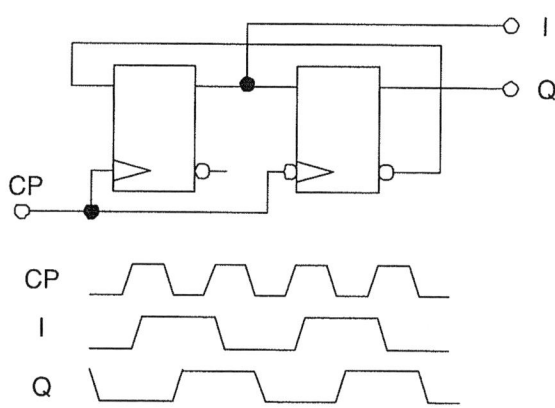

Figure 1: Basic structure of an IQ-divider. The maximum frequency is limited by the data-to-q delay of the flipflops.

of glitches. Small phase noise is an additional requirement for good divider circuits. Current mode logic (CML) is the state-of-art approach for very high input frequencies, but the static current sources within these circuits cause high power consumption independent of the input frequency. The small signal swing makes this analog approach sensitive to variations of process and environmental conditions. For a low power consumption of the system it is desirable to replace CML circuits completely by conventional CMOS logic, or to shift the interface between CML and CMOS to the highest possible frequency.

A conventional factor of two divider generating output signals with a phase skew of 90 degrees is shown in Fig. 1. For very high clock frequencies, edge triggered flipflops are not applicable as the data-to-q delay must be less than half the input period. Conventional master-slave latch pairs are at risk of race conditions especially if the skew between the two clock phases is subject to variations. The single ended structure is disadvantageous for applications which require precisely aligned output signals. In this paper we propose a fully differential high-speed and low-power frequency divider based on CMOS logic styles, i.e. without any power demanding current sources like in CML. The circuit architecture is discussed in section 2. The maximum frequency and the power consumption is investigated in section 3 and compared to prior-art. An intrinsic phase rotator which simplifies the implementation of dual-modulus pre-scalers is

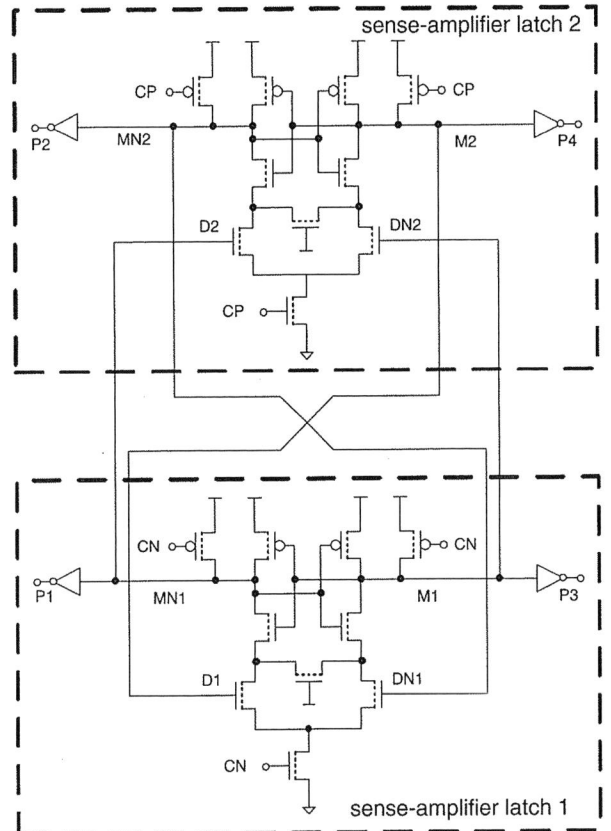

Figure 2: First stage of the divider circuit. The precharged internal nodes are used to couple the two complementarily clocked sense-amplifiers. Pulse signals occur at the outputs of the inverters, and are passed to the second divider stage for conditioning and post-processing.

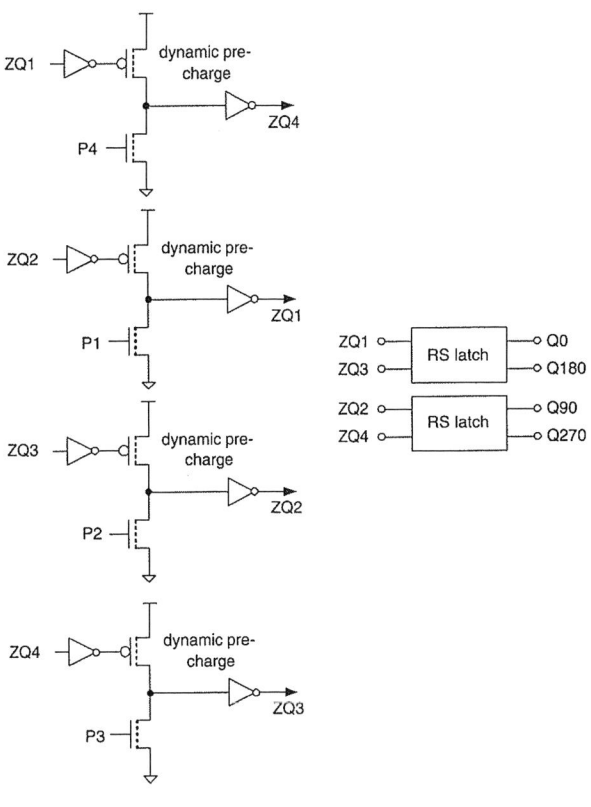

Figure 3: Post-processing stage consisting of circularly precharged dynamic level-converters. Even short pulses without full signal swing are reliably conditioned and passed to subsequent circuit blocks. A symmetric duty cycle is achieved by subsequent RS-latches.

presented in section 4. Finally, the accuracy of the output phases under process variations is discussed in section 5.

2. DIVIDER ARCHITECTURE

The intension of the circuit is to divide an input frequency by a factor of two and to provide complementary output pairs with an highly accurate phase skew of 90 degrees. The first part of the circuit, which can be partitioned in two stages, is the pulse generator as shown in Fig. 2. Two differential sense-amplifiers [5] with precharged internal nodes are connected corresponding to Fig. 1. However, the sense-amplifiers sample input data on the rising edge of the respective clock signal, so races through the latches are impossible. The internal precharged nodes M and MN are used directly to couple the two sense-amplifiers. This minimizes the interconnect latency and allows for extremely high input frequencies. Due to the differential feedback structure inside the latches, data is sampled correctly even without full swing input signals. Although the node capacitancies of the internal storage nodes M and MN are critical for the speed of the sense-amps, it is justified to use these signals for the coupling since the whole pulse generator is considered as a full custom macro cell

Inverters are used to decouple the second divider stage from the pulse generator. For high input frequencies and/or parameter variations, the pulse signals $P_1 - P_4$ at the outputs of these inverters may not reach the full V_{DD} level and may varay in their pulse width. To avoid cross currents and enable reliable signal conditioning, dynamic level converters are used to interface the pulse generator and subsequent circuit blocks. The principle is depicted in Fig. 3 and a signal-time diagramm is given in Fig. 4. The pulse signals $P_1 - P_4$ are used to discharge dynamic precharged nodes. A signal derived from the dynamic node which is discharged by the following pulse, i.e. 90 degrees later, is used to precharge the dynamic node again. Contentions are avoided as the circular control assures that each node is disconnected from V_{DD} when the corresponding pulse occurs. The full swing signals $ZQ_1 - ZQ_4$ at the outputs of the dynamic level converters have a pulse width of half of the input period plus two inverter delays, i.e. approximately one quarter of the output period. If output signals with a 50% duty cycle are required, conventional set/reset latches can be used to generate symmetric output signals.

Fig. 5 shows SPICE simulations of the pulse signal P_1 and the output signal ZQ_1 of the respective level converter for the nominal case and the slow and the fast process corner,

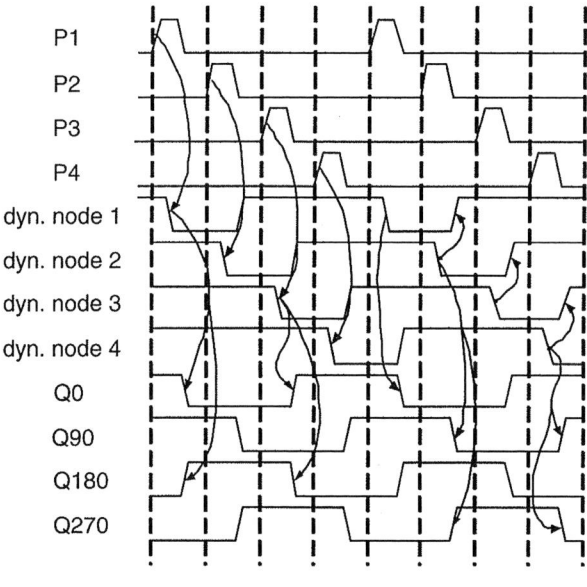

Figure 4: Signal diagram describing the circular precharge in the post-processing stage.

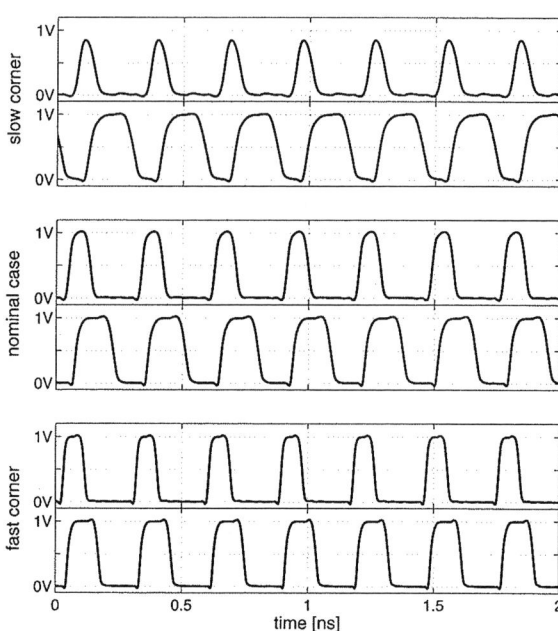

Figure 5: Pulse signal (top) and output signal of the dynamic levelshifter stage (bottom) for the best and worst process corner compared to the nominal case at $1.0V$ supply voltage and $7.0GHz$ input frequency.

respectively. The supply voltage is $1.0V$ and the input frequency is set to a value of $7.0GHz$ (maximum value of target application). In the worst case the pulse width and the peak value are significantly reduced, but the output signal of the level shifter has a full swing again.

In contrast to divider approaches in current mode logic, the complete circuit is intended to generate full-swing signals at each node and does not contain any current sources which cause static power consumption. However, to operate properly, the circuit does not require full swing signals at the internal nodes. For the sense-amplifiers in the pulse generator the voltage difference of the differential input signals is more essential than the absolute voltage value. The dynamic approach makes the level converters in the second stage relatively insensitive to the input level. The accuracy of the 90 degrees phase skew is not affected by these level issues, as the circuit is absolutely symmetric. Weak signals may increase the latency from the input signal to the output signals. However, the phase relation of the four output signals will not be affected as the signal processing is identical in each path.

3. PERFORMANCE EVALUATION

The high accuracy of the phase skew discussed in the previous section can be achieved only if the intrinsic symmetry of the circuit is maintained in the layout of the divider block. Fig. 6 shows the implementation of the divider in a 90nm low-power CMOS technology with medium threshold voltage. The symmetry was the number one design criterion for the internal structure of the sense-amplifiers but also for the overall structure and the wiring. The two sense-amplifiers are located at the borders of the module. The pulse signals of the sense-amp divider are transmitted across identical

wiring to the shifter core which is located in the center of the circuit. The set/reset latches lie in-between the sense-amps and the shifter core. The circuit is designed for a maximum operating frequency of $7GHz$ for $V_{DD} = 1.0V$, $T = 130°C$ and worst case process parameters. The maximum frequency and the power consumption in dependence of the supply voltage is shown in Fig. 7. All simulations are based on RC extracted layouts and silicon proven device models. The divider consumes $0.36\frac{mW}{GHz}$ at $V_{DD} = 1.0V$ and $1.02\frac{mW}{GHz}$ at $V_{DD} = 1.6V$ at a maximum operation frequency of $12.4GHz$. For other target frequencies the power can be optimized by dedicated device dimensions within the circuit.

4. INTRINSIC PHASE ROTATOR AND APPLICATION IN DUAL-MODULUS PRE-SCALERS

Multi modulus pre-scalers are important building blocks of high-speed frequency synthesizers. The phase switching pre-scaler architecture proposed in [1] is a smart technique to realize a division by N and $N+1$. The pre-scaler consists of a series connection of asynchronous sub-dividers. One of these dividers generates multiple clock phases. A phase-rotator selects one phase and has the ability to circularly switch to the proximate phase. Skipping one phase means that the next clock edge at the output of the phase-rotator is delayed by a phase of $\frac{360°}{N}$ where N is the number of phases. Refering to the input signal this corresponds to a phase of $\frac{360°}{N}k$ where k is the divider factor between the clock input and the phase rotator. In conventional approaches [9] the phase generator and the rotator are distinct circuit blocks.

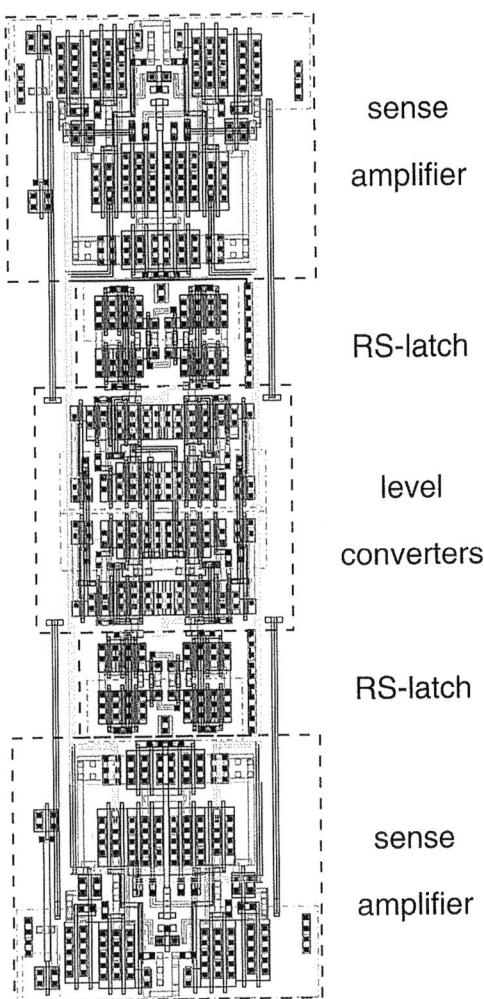

sense

amplifier

RS-latch

level

converters

RS-latch

sense

amplifier

Figure 6: Symmetric layout of the proposed IQ-divider. The symmetry slightly increases the area consumption but translates in excellent phase relations.

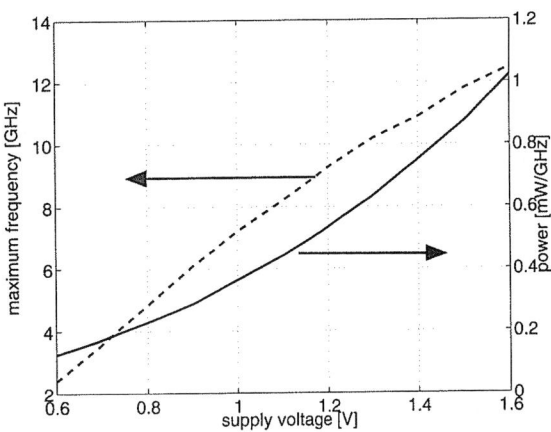

Figure 7: Power consumption and maximum frequency of the IQ divider in dependence on the supply voltage under worst case process conditions with $10fF$ load capacitances.

ref.	tech $[\mu m]$	V_{DD} $[V]$	f_{max} $[GHz]$	power $[mW/GHz]$
[2]	0.35	3.3	2.7	5.3
[3]	bip.	2.3	2.5	1.5
[4]	0.8	5	1.9	9.5
[6]	0.35	3	2.7	1.8
[7]	0.8	5	1.59	7.9
[8]	0.25	1.8	2	0.24
[10]	0.12	1.5	15	7.7
[11]	0.8	5	1.8	29.4
[12]	0.18	1.8	10	1.5
this work:				
slow mode	0.09	0.6	2.4	0.19
fast mode	0.09	1.5	11.6	1.3

Table 1: Power and performance of various pre-scaler architectures reported in literature.

A synchronization circuit is required to avoid glitches during the transition from one phase to the subsequent phase. This increases the overall power and area and causes systematic phase errors due to inevitable wiring asymmetries.

The proposed divider with a modified pull-down network in the dynamic coupling stage, combines the phase generation and the rotator in one circuit block. This reduces the power consumption and possible asymmetries to a minimum. An exemplary implementation is shown in Fig. 8. In principle, each dynamic node can be discharged by each pulse generated by the pulse generator, but only one of the four available pulses is activated, determined by one-hot-coded control signals $E1 - E4$. The circular structure is maintained, i.e. each dynamic node is discharged 90 degrees after its preceding node and recharges the latter again. When a phase skip is requested, a simple control circuit reconfigures the post-processing stage by disabling the current discharge path and enabling the path corresponding to the subsequent pulse. An exemplary circuit based on edge triggered flipflops is depicted on the top of Fig. 8. In contrast

to conventional approaches, this can be done without any synchronization, as the dynamic push-pull technique is sort of self-synchronized and avoids glitches completely. Even if the skip request occurs before a certain node is completely discharged no glitches occur because the respective node is definitely discharged 90 degrees later. Multiple buffers in the output path avoid meta-stable states at the output.

In the 128/129-pre-scaler of Fig. 10, the integrated divider/ phase-rotator is used as the first divider stage. The signal $ZQ0$ and the complementary signal $ZQ180$ are used to drive the clock and the inverse clock input of the subseqent toggle flipflop directly. This flipflop and all further flipflops are implemented as master-slave toggle flipflops with a differential slave stage. The differential slave provides well aligned complementary output signals which are directly used as clock signals for the next stage. If the divider factor 129 is selected by the mode signal, the input signal of the last divider stage is used to trigger a 90 degrees phase shift in the rotator. Each phase shift corresponds to half of an input clock cycle. Pre-scalers based on this integrated divider/phase-rotator approach allow for high-speed CMOS pre-scalers with low power consumption. A comparison with published pre-scalers

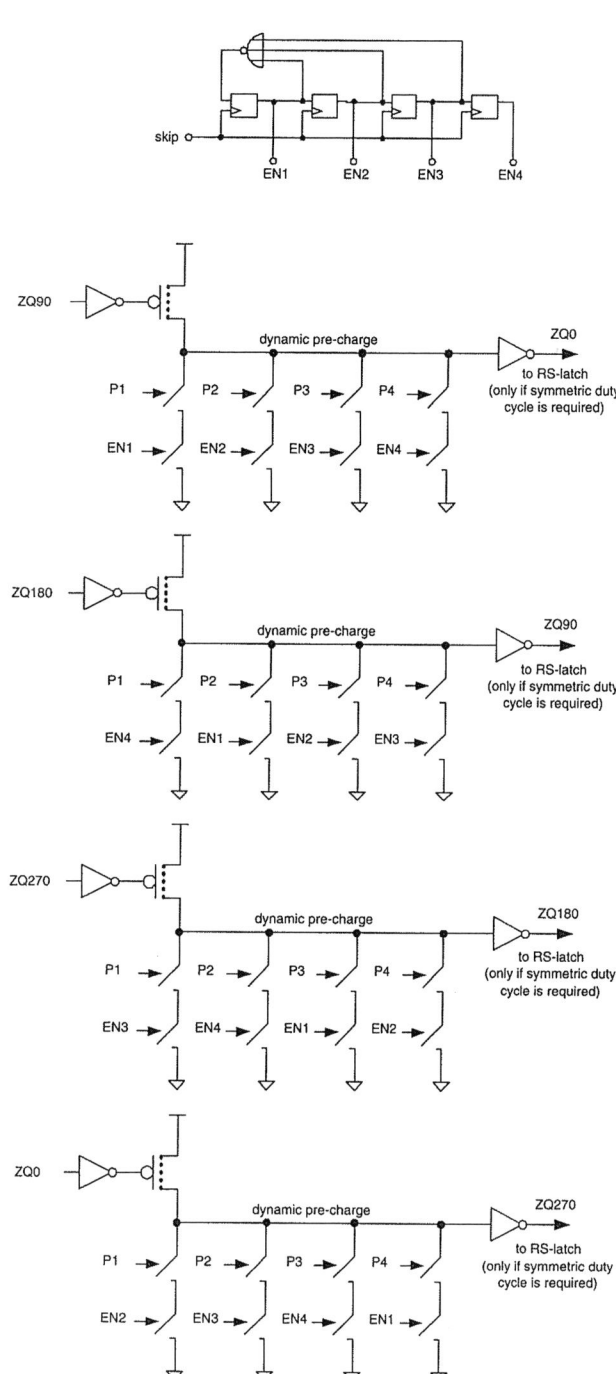

Figure 8: Modified post-processing stage: An intrinsic phase-rotator is obtained by additional discharge branches which are controlled asynchronously by the one-hot coded signals $EN1 - EN4$ of the cyclic pointer circuit (top).

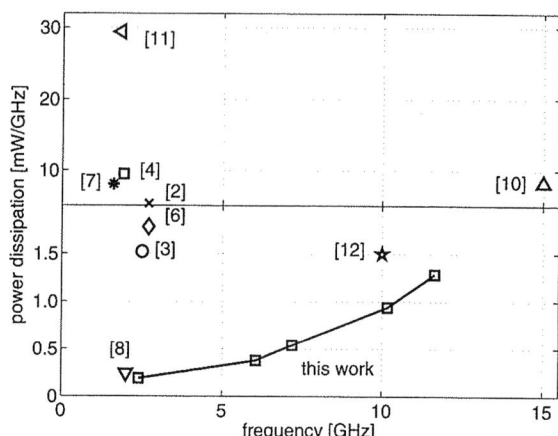

Figure 9: Comparison of prior-art with various operation modes of the pre-scaler using the proposed IQ-divider.

is summarized in Tab. 1 and illustrated in Fig. 9. The pre-scaler proposed here combines the lowest power consumption with input frequencies up to $12.4GHz$. If higher frequencies are required, a first divide-by-2 stage in current mode logic (CML) can be used. In contrast to conventional approaches [1, 9] only one full-speed CML register is required and only two instead of four signals have to be converted from the CML to the CMOS domain.

5. SIGNAL GENERATION FOR IQ-MIXERS

IQ-signal processing circuits require fast clock dividers which generate output signals with a precise phase skew of 90 degrees. Even small deviations translate directly into an increased bit error rate. Especially for mobile applications a low power consumption is an additional requirement. Fast low-power CMOS divider circuits based on dynamic true-single-phase flipflops [13] have been proposed [2, 8, 6, 4]. However, most IQ circuit blocks like Gilbert-multipliers require complementary clock signals. The single ended divider structure of prior-art approaches is a considerable drawback if skewless complementary output clock signals are required. The proposed divider circuit, however, is completely symmetric, and complementary output signals can be provided simultaneously. To characterize the phase-mismatch of the divider under 3σ process variations, Monte-Carlo simulations have been carried out for a supply voltage of $V_{DD} = 1.1V$ and an input frequency of $f_{in} = 6.0GHz$. Global and local process variations result in a variation of the phase relation with a standard deviation of $\sigma_{phase} = 0.87°$. For $f_{in} = 8.2GHz$ the phase variation is $\sigma_{phase} = 1.32°$. Hence the divider is well suited in all existing wireless applications especially if the circuit is designed with some margins, i.e. if the circuit is not operated at its absolute frequency limit.

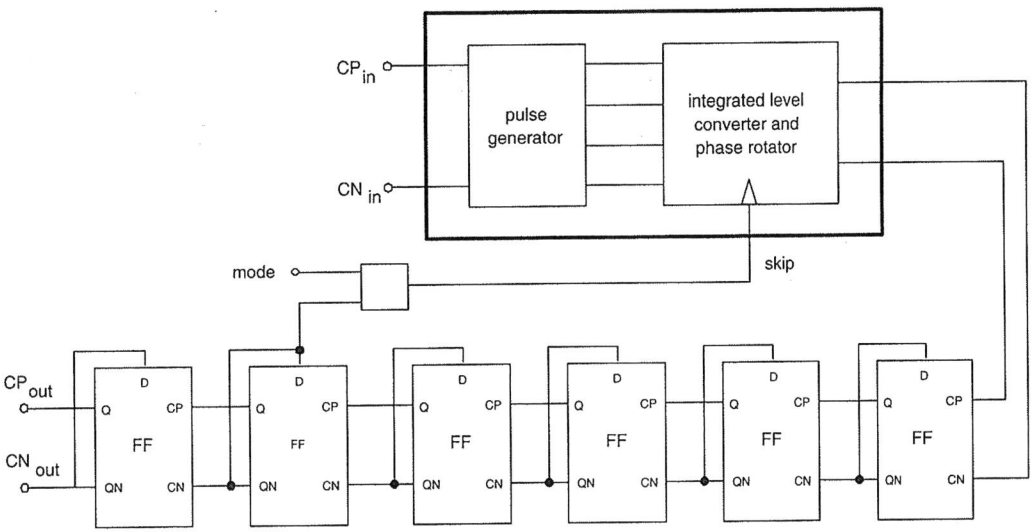

Figure 10: Block diagram of a 128/129 pre-scaler using the proposed IQ-divider as first divider stage.

6. CONCLUSIONS

A high-speed low-power divider topology without static current sources has been proposed for a $90nm$ low-power CMOS technology. A maximum input frequency of $12.4GHz$ is achieved with a maximum power consumption of $1.02\frac{mW}{GHz}$. The fully symmetric circuit allows for the generation of output signals with a highly precise phase skew of $90°$. This makes the divider predestinated for low-power IQ-transceivers. The intrinsic phase rotator enables multi-modulus pre-scalers with small power and area overhead. A glitch-free operation is achieved without any synchronization of the phase skip signal.

7. REFERENCES

[1] Jan Craninckx and Michiel S. J. Steyaert. A 1.75-GHz/3-V Dual Modulus Divide-by-128/129 Pre-Scaler in 0.7-um CMOS. Journal of Solid State Circuits, Vol. 31, No. 7, 890–897, Jul. 1996.

[2] Chun-Lung Hsu and Wu-Hung Lu. Glitch-Free Single-Phase D-FF for Dual-Modulus Prescaler. Conference on ASIC, 711–714, Oct. 2003.

[3] Herbert Knapp, Wilhelm Wilhelm, Mira Rest and Hans-Peter Trost. A 3.8-mW 2.5GHz Dual Modulus Pre-Scaler in 0.8um Silicon Bipolar Technology. International Symposium on Lower Power Electronics and Design, 23–20, Aug. 1998.

[4] Patrik Larsson. High-Speed Architecture for a Programmable Frequency Divider and a Dual-Modulus Prescaler. Journal of Solid State Circuits, Vol. 31, No. 5, 744–748, May 1996.

[5] Masataka Matsui, Hiroyuki Hara, Yoshiharu Uetani, Lee-Sup Kim, Tetsu Nagamatsu, Yoshinori Watanabe, Akihiko Chiba, Kouji Matsuda and Takayasu Sakurai. A 200MHz 13mm2 2-D DCT Macrocell Using Sense Amplifying Pipeline Flip-Flop Scheme. Journal of Solid State Circuits, Vol. 29, No. 12, 1482–1490, Dec. 1994.

[6] Ram Singh Rana. Dual-Modulus 127/128 FOM Enhanced Prescaler Design in 0.35-um CMOS Technology. Journal of Solid State Circuits, Vol. 40, No. 8, 1662–1670, Aug. 2005.

[7] Soares J. Navarro and W. A. M. Van Noije. A 1.6-GHz Dual Modulus Pre-scaler Using Extended True-Single-Phase-Clock CMOS Circuit Technique (E-TSPC). Journal of Solid State Circuits, Vol. 34, No. 1, 97–102, Jan 1999.

[8] Marc Tiebout. A 480uW 2GHz Ultra Low Power Dual-Modulus Pre-Scaler in 0.25um Standard CMOS. International Symposium on Circuits and Systems, 741–744, May 2003.

[9] Marc Tiebout, Christoph Sandner, Hans-Dieter Wohlmuth, Nicola Da Dalt and Edwin Thaller. A Fully Integrated 13GHz Delta Sigma Fractional-N PLL in 0.13um CMOS. International Solid State Circuits Conference, 386–387, Feb. 2004.

[10] Hans-Dieter Wohlmuth and Daniel Kehrer. A 15GHz 256/257 Dual-Modulus Pre-Scaler in 120nm CMOS. European Solid State Circuits Conference, 77–80, Sept. 2003.

[11] Ching-Yuan Yang, Guang-Kaai Dehng, June-Ming Hsu and Shen-Iuan Liu. New Dynamic Flip-Flops for High-Speed Dual-Modulus Pre-scaler. Journal of Solid State Circuits, Vol. 33, No. 10, 1568–1571, Oct. 1998.

[12] X. P. Yu, M. A. Do, J. G. Ma, K. S. Yeo, R. Wu and G. Q. Yan. Low Power High-Speed CMOS Dual-Modulus Pre-Scaler Design with Imbalanced Phase-Switching Technique. Proceedings on Circuits, Devices and Systems, Vol. 152, No. 2, 127–132, Apr. 2005.

[13] Jiren Yuan and Christer Svensson. High-Speed CMOS Circuit Technique. Journal of Solid State Circuits, Vol. 24, No. 1, 62–70, Feb 1989.

An Optimal Analytical Solution for Processor Speed Control with Thermal Constraints*

Ravishankar Rao, Sarma Vrudhula,
Chaitali Chakrabarti
Consortium for Embedded Systems
Arizona State University
Tempe, AZ 85281, USA

{ravirao, vrudhula, chaitali}@asu.edu

Naehyuck Chang
Computer Systems Lab
School of Computer Science and Engineering
Seoul National University
Seoul, 151-010, South Korea

naehyuck@cslab.snu.ac.kr

ABSTRACT

As semiconductor manufacturing technology scales to smaller device sizes, the power consumption of clocked digital ICs begins to increase. Dynamic voltage and frequency scaling (DVFS) is a well-known technique for conserving energy. Recently, it has also been used to control the CPU temperature as part of Dynamic Thermal Management (DTM) techniques. Most works in these areas assume that the optimum speed profile (for either minimizing energy or maximizing performance) is a constant profile. However, in the presence of thermal constraints, we show that the optimal profile is in general, a time-varying function. We formulate the problem of maximizing the average throughput of a processor over a given time period, subject to thermal and speed constraints, as a problem in the calculus of variations. The variational approach provides a powerful framework for precisely specifying and solving the speed control problem, and allows us to obtain an exact analytical solution. The solution methodology is very general, and works for any convex power model, and simple lumped RC thermal models. The resulting speed profiles were found to consist of up to three segments, of which one of them is a decreasing function of time, and the others are constant. We analyze the effect of different parameters like the initial temperature, thermal capacitance and the maximum rated speed on the nature and the cost of the optimum solution. We also propose a two-speed solution that approximates the optimal speed curve. This solution was found to achieve a performance close to that of the optimum, and is also easier to implement in real processors.

Categories and Subject Descriptors

C.4 [**Performance of systems**]: modeling techniques, performance attributes; G.1.6 [**Optimization**]: constrained optimization

---

*We gratefully acknowledge the support for this work by the Consortium for Embedded Systems at the Arizona State University and by a grant from the National Science Foundation, grant number CNS-0509540.

Permission to make digital or hard copies of all or part of this work for personal or classroom use is granted without fee provided that copies are not made or distributed for profit or commercial advantage and that copies bear this notice and the full citation on the first page. To copy otherwise, to republish, to post on servers or to redistribute to lists, requires prior specific permission and/or a fee.
ISLPED'06, October 4–6, 2006, Tegernsee, Germany.
Copyright 2006 ACM 1-59593-462-6/06/0010 ...$5.00.

General Terms

algorithms, performance, theory

Keywords

optimal control, thermal management, temperature, DTM, DVFS

1. INTRODUCTION

Over the past decade, microprocessor power consumption has increased exponentially [5]. Higher power consumption means greater energy costs, both for powering the chip, and for cooling it. Further, as chip form factors shrink, the power density increases too. This makes it difficult to design thermal solutions that ensure that the chip temperature is always within safe limits. Even if achieved, designing for the worst case could result in prohibitively expensive or bulky systems. Hence, thermal designers target a power consumption value smaller than the worst case power consumption – they ensure that the case temperature θ remains below a specified maximum value θ_{max} whenever the processor is operating at or below the Thermal Design Power (TDP) P_m. Such a design allows the processor to operate temporarily at higher speeds that dissipate more power than the TDP. If the TDP is exceeded for too long, however, the chip could heat up dangerously beyond θ_{max}, and the thermal solution, by design, cannot prevent it.

To handle such scenarios, processor manufacturers usually provide a hardware mechanism to quickly reduce the power consumption whenever an on-chip temperature sensor detects a thermal emergency [1, 5, 7]. These mechanisms include varying the duty cycle of the clock, throttling the instruction fetch unit of the processor (fetch throttling), and reducing the clock frequency and/or voltage scaling (DVFS). They constitute the simplest of a class of techniques known as Dynamic Thermal Management (DTM). Brooks and Martonosi [2] provide a comprehensive overview and comparison of different DTM techniques. All these techniques involve some kind of processor slowdown, and as pointed out by Skadron *et al* [14], the challenge of DTM is to adaptively control the processor speed to minimize the slowdown but still meet thermal constraints.

The DTM techniques implemented in current processors [1, 7] are fixed-response reactive heuristics i.e. (i) the response does not vary in proportion to the thermal emergency, and (ii) they are invoked only when the temperature crosses a certain threshold. Skadron *et al* proposed a feedback control system that adaptively varies the fetch toggling rate of the processor to maintain the chip temperature just under the maximum allowed temperature T_{max}. This allowed the processor to achieve safe thermal operation while suffering significantly lower performance penalty. Srinivasan and Adve [15] proposed predictive DTM algorithms for multimedia applications.

These algorithms are able to extract higher performance than reactive DTM techniques by using profiled temperature information to achieve maximum performance under thermal constraints.

A problem central to dynamic thermal management that has not been addressed so far, is the derivation of the shape of the performance optimal speed profile (a function of time), while satisfying specified thermal constraints. Consider the following simple problem in processor speed control: (X-MAX): maximize a processor's performance (measured in number of clock cycles executed) over a given time duration.[1] The optimal solution for this problem is not a function of time – one would simply operate at the largest supported speed $u_{\max}$. However, this solution does not account for the effect of thermal constraints on the optimal speed profile.

An ideal speed control technique should to be able to achieve a desired combination of performance, energy consumption, temperature, acoustic noise (from cooling fans), and reliability. This paper takes an important first step towards that goal by revisiting the simplest problem in DTM mentioned above (X-MAX), but this time, with the thermal constraint $T \leq T_{\max}$. Using the theory of the calculus of variations, we obtained an exact analytical solutions to this problem. The thermal constraints were observed to cause the optimal speed profile to be, in general, a piecewise non-linear function of time. The solution to this problem provides a theoretical upper bound to the maximum performance improvement that can be achieved by using speeds that dissipate power larger than the TDP. We also devised a two-speed (piecewise constant) profile that achieves nearly the same performance as the continuously time-varying performance optimum curve. This result makes it easier to implement the performance-optimum speed profile and to extend it for more complex scenarios.

2. NOTATION, MODELS, AND PROBLEM FORMULATION

Throughout this paper, the notation $x' = dx/dt$ will be used to denote the (total) derivative, always w.r.t. time t. Different glyphs will be used to differentiate between a variable (e.g. P) and a function assigned to a variable (e.g. $P = P(u)$).

2.1 Processor model

In this work, we restrict ourselves to a single DTM mechanism, namely DVFS, so that there are potentially two control variables, the supply voltage and clock speed u. Our processor model is similar to that of many works on DVFS like [11]. The processor can attain any speed u between 0 and $u_{\max}$ cycles/s. At each speed, the minimum feasible voltage [8, 12, 13] is assumed to be used. Hence, the processor's power consumption P is a function of only the speed. This function $P(u)$, is assumed to be increasing, strictly convex, and once differentiable. The power slope is defined as $S(u) = dP(u)/du$. *Although it is usually not emphasized in the notation, u, P, and S are understood to be functions of time.*

In recent times, there has been considerable research into developing accurate processor power models, particularly for modeling leakage currents [8, 12] and leakage dependence on temperature [10]. Our work is independent of the specific form of the power model, and is applicable to most leakage models including [8, 12]. Currently, however, our work does not include the effect of temperature on the leakage currents. The performance of a processor x over a given time interval $[t_i, t_f]$ is defined to be the number of clock cycles (amount of work done) executed by the processor over that interval and is given by $x = \int_{t_i}^{t_f} u\, dt$. Note that $x' = u$.

---

[1] The number of clock cycles executed over a given duration is proportional to the average throughput, which, for a given Instructions Per Clock (IPC) is proportional to the clock frequency [10].

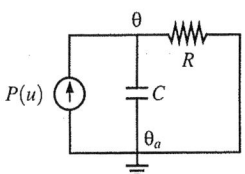

Figure 1: Equivalent thermal RC circuit.

2.2 Thermal model

Using a well-known duality between heat transfer and electrical phenomena [3], the relationship between the chip temperature θ and the processor power P can be modeled by an equivalent thermal R-C circuit. A second-order RC model is usually adequate to accurately model the transient thermal behavior of the chip package and heat sink [4]. In the interest of clarity however, we have used a lumped first order lumped RC model shown in Figure 1. This thermal model provides the following relationship between the power consumption P and the chip junction temperature θ (measured relative to the fixed ambient temperature):

$$RC\theta' + \theta - RP(u) = 0 \qquad (1)$$

where R (unit: $^\circ C/W$) is the thermal resistance from junction to ambient, and C (unit: $J/^\circ C$) is the thermal capacitance.

The thermal resistance is a measure of the temperature θ_{ss} to which the chip junction will heat up to in steady state, for a given constant power dissipation P, i.e. $R = \theta_{ss}/P$. In most electronic systems, the main modes of heat transfer are through conduction (from chip junction through the packaging and heat sink to the heat sink's outer surface) and convection (from heat sink to ambient). Each mode has its own equivalent thermal resistance, or $R = R_{\text{cond}} + R_{\text{conv}}$. Typically, R_{cond} is determined by the thermal conductivity and geometry of the packaging and heat sink (fixed) while R_{conv} by the cooling mechanism [3, 6]. If an adaptive cooling mechanism like a fan is used, R_{conv} is not fixed [3]. As we are interested in controlling only the processor speed in this work, we assume that the cooling solution operates at a fixed "speed", so that R_{conv} and hence R does not vary with time.

The thermal capacitance C is a measure of the heat storage capacity of the system. It is chiefly determined by the mass and specific heat capacity of the packaging and heat sink (fixed). Larger values of C allow the temperature to rise more slowly for the same power dissipation. This is an important factor in optimal speed control under thermal constraints as it determines the amount of time a chip can operate at powers larger than the Thermal Design Power (TDP). For a given initial temperature θ_0, and a constant speed u, the time taken t_m to reach the maximum temperature can be obtained using (1) as

$$t_m(u) = -RC \log \left[\frac{RP(u) - \theta_{\max}}{RP(u) - \theta_0} \right] \qquad (2)$$

The temperature profile generated by a constant speed u can be found using (1) and is given by

$$\theta = RP(u) + \left[\theta_0 - RP(u)\right] e^{-t/(RC)}. \qquad (3)$$

If $\theta_0 < P(u)$, the temperature profile is increasing, else, non-increasing. In either case, the steady state temperature $\theta_{ss} = P(u)R$. To ensure that the chip reliability is not compromised, it is essential to always maintain the chip temperature below a maximum value $\theta_{\max}$ that is specified by the processor manufacturer [1, 7, 14]. As $P(u)$ is an increasing function, there is a unique solution to the equation $P(u)R = \theta_{\max}$, which is denoted as u_m. The corresponding power consumption $P(u_m)$ is denoted as P_m. The speed u_m

is the maximum speed that can, *in steady state*, maintain the chip temperature below the maximum temperature θ_{max}, i.e. if $u > u_m$, $\theta_{ss}(u) > \theta_{max}$.

Note that if at any instant, the temperature reaches θ_{max}, reducing the speed immediately to u_m will ensure that the temperature remains at θ_{max}. Hence, we assume that an ideal DTM technique will scale down the processor speed to u_m at the onset of a thermal emergency $\theta = \theta_{max}$. The thermal RC system equation (1), together with the constraint $\theta(t) \leq \theta_{max}$ then constitute the thermal constraints that need to be added to the original speed control problem (X-MAX).

2.3 Problem formulation

The performance optimization problem under thermal constraints (henceforth called X-MAX-T) involves controlling the speed profile u to maximize the number of clock cycles executed over a time duration $[0, T]$ subject to (1), the temperature upper bound, and speed constraints. It can be expressed as

$$\max_u \quad X = \int_0^T u \, dt \quad (4)$$

$$\text{subject to} \quad RC\theta' + \theta - RP(u) = 0, \quad \theta(0) = \theta_0 \quad (5)$$

$$\theta \leq \theta_{max}, \quad (6)$$

$$\text{and} \quad 0 \leq u \leq u_{max}. \quad (7)$$

The above formulation involves computing an unknown function $u(t)$ that optimizes a function of this unknown function ($\int_0^I u \, dt$), subject to constraints on other functions of u. This kind of optimization problem does not fit the traditional realm of optimization problems because the unknowns are functions, and not simple variables. Instead, they are best approached using the theory of the calculus of variations [16] that was developed particularly for solving such problems. Using this theory, we now show how the optimal speed profile can be obtained for X-MAX-T.

3. THE OPTIMAL SPEED PROFILE

Clearly, if the thermal constraints (5) and (6) were omitted, the optimum solution to the problem is $u = u_{max}, 0 \leq t \leq T$. We call this the *unconstrained solution*. As the chip is not designed for worst case power dissipation by assumption, $u_{max} > u_m$, or $RP(u_{max}) > \theta_{max} \geq \theta_0$. This means that the temperature will eventually reach θ_{max} in a time $\tilde{t}_m \equiv t_m(u_{max})$ given by (2). If $\tilde{t}_m \geq T$, the unconstrained solution is feasible. Otherwise, the unconstrained solution violates the thermal constraint $\theta \leq \theta_{max}$ over the interval $[\tilde{t}_m, T]$. Then, we use a result from [16] that states that in the true (constrained) solution, the thermal constraint is binding over a subinterval $[t_m, T]$ of $[\tilde{t}_m, T]$, or $u = u_m$ for $t_m \leq t \leq T$ and the optimum speed profile has two segments. We now need to find the optimum speed profile u over the segment $[0, t_m]$ and also find the optimum "switch point" t_m between the segments.

3.1 Solution without binding speed constraints

We first assume that the speed constraint $u \leq u_{max}$ is not binding over any portion of the speed profile. The original problem can now be recast as a variational calculus problem over the interval $[0, t_m]$ with an unknown endpoint t_m and a terminal cost $u_m(T - t_m)$, as follows:

$$\min_u \quad J = -\left[\int_0^{t_m} u \, dt + u_m(T - t_m) \right] \quad (8)$$

$$\text{subject to} \quad \psi = RC\theta' + \theta - RP(u) = 0, \quad (9)$$

$$\text{and} \quad \theta(0) = \theta_0, \quad \theta(t_m) = \theta_{max}. \quad (10)$$

The solution methodology for this problem is detailed in the Appendix. We summarize the final optimal solution here:

$$u^*(t) = \begin{cases} S^{-1}\left((1/(\alpha R)) e^{-t/(RC)} \right), & 0 \leq t < t_m \\ u_m, & t_m \leq t \leq T, \end{cases} \quad (11)$$

where the parameters α and t_m are determined using the continuity conditions

$$u^*(t_m^-) = u_m, \quad \theta^*(t_m^-) = \theta_{max}. \quad (12)$$

The optimum solution consists of a time-varying, and a constant segment. The time-varying portion is expressed in terms of the inverse power slope function S.[2] At t_m, the optimum solution switches to the constant segment $u = u_m$. This has the effect of maintaining the temperature at θ_{max} during $[t_m, T]$. If the solution process produces a value of t_m greater than T, we must re-solve the problem by setting $t_m = T$. The optimum solution will then have a single time-varying segment. The optimum value of the parameter α is now determined by the temperature continuity condition $\theta(t_m) = \theta(T) = \theta_{max}$.

3.2 Solution with binding speed constraints

If the solution obtained above violates the speed constraint $u \leq u_{max}$, it must do so over an interval $[0, \tilde{t}_u]$ where $\tilde{t}_u < t_m$. This is because the optimum speed profile is a non-increasing function of time. Using the same reasoning as before, we must have the true optimum solution such that $u = u_{max}$ over some interval $[0, t_u]$ such that $t_u \leq \tilde{t}_u$. To find the optimum switch point t_u and compute the true optimum solution, we must now re-solve the problem described by (8)–(10) with the new objective

$$\min_u \quad -\left[u_{max} t_u + \int_{t_u}^{t_m} u \, dt + u_m(T - t_m) \right]. \quad (13)$$

It can be shown that[3] the optimum solution to this problem is

$$u^*(t) = \begin{cases} u_{max}, & 0 \leq t \leq t_u, \\ S^{-1}\left((1/(\alpha R)) e^{-(t-t_u)/(RC)} \right), & t_u < t \leq t_m, \\ u_m, & t_m < t \leq T, \end{cases} \quad (14)$$

where the unknown parameters α, t_u and t_m are given by the continuity conditions

$$u^*(t_u^+) = u_{max}, \quad u^*(t_m^-) = u_m, \quad \theta^*(t_m^-) = \theta_{max}. \quad (15)$$

Now, if the above solution process results in a value of t_u that exceeds $t_m(u_{max})$, the optimum solution is a two-speed profile $u^*(t) = u_{max} \ \forall \ 0 \leq t \leq t_m(u_{max})$, and u_m otherwise.

3.3 A two-speed approximate solution

As real processors cannot change speed smoothly, we propose a piecewise constant speed policy, CONST that approximates the optimal solution. This speed policy operates at a constant speed u_1 until $\theta = \theta_{max}$ and then changes to u_m. The optimum value of u_1 must maximize $x(u_1) \equiv u_1 t_m(u_1) + u_m(T - t_m(u_1))$. It is found by solving the following optimization problem:

$$\max_{u_1} \quad -(u_1 - u_m)RC \log\left[\frac{RP(u_1) - \theta_{max}}{RP(u_1) - \theta_0} \right], \quad (16)$$

$$\text{subject to} \quad 0 \leq u_1 \leq u_{max}. \quad (17)$$

---

[2] As the power is a strictly convex function of speed, the power-slope is an increasing function of speed. Hence, the optimum speed profile in this segment is a decreasing function of time.

[3] The solution proceeds on similar lines as the derivation in the Appendix. We have omitted the details for brevity.

This is a non-linear optimization problem in one variable with two simple linear constraints. The logarithm and the non-linear nature of the P(u) function make it difficult to obtain an analytical solution. However, the problem is easily solved numerically.

4. EXPERIMENTAL SETUP

We obtained an analytical power model for the full chip power consumption of a 70 nm CMOS processor based on data from [8, Table 1]. We made three changes to the parameters used in [8] to model a high performance high power processor. (i) we chose L_g, the number of devices as 200 million based on the transistor count of typical 90 nm and 65 nm processors, (ii) we set P_{on}, the power required to keep the processor on as 10 W, and (iii) we computed the coefficient C_{eff}, the switched capacitance per device as 1×10^{-16} J V^{-2} cycle^{-1}. This was obtained by dividing the value for C_{eff} of a 10 inverter chain (20 devices) from [12, Table 1] by 20.

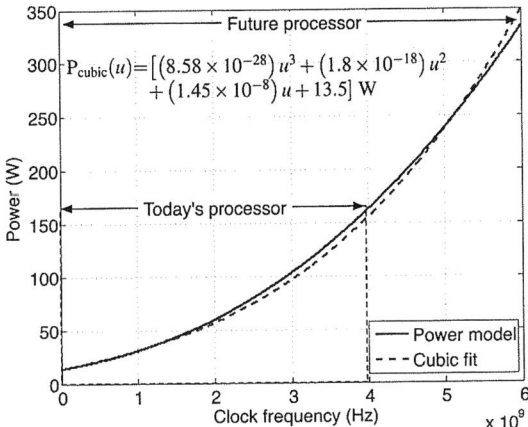

Figure 2: Processor power model, and a cubic approximation.

Our aim was to obtain a representative model of power consumption in existing high-end processors to illustrate our solution methodology, which by itself, is independent of the actual power model. We also wish to extrapolate this power model to model a future high-performance processor. To achieve this, and to be able to make a fair comparison to the power model of the current processor, we use the same power speed relationship for both, but limit the maximum speeds to 4 GHz for the current processor and 6 GHz for a future processor.

Figure 2 shows a plot of this power-speed relationship (solid curve). The power model P(u) presented in [8] has a power-slope function $S(u) = dP/du$ that is not invertible. To illustrate our analytical solution, it is convenient to approximate the power-speed relationship to a simpler form (we chose a cubic) that has an invertible power slope. Figure 2 shows the cubic fit to the earlier power model. The RMS error over the range of frequencies 0 to 6 GHz was found to be 5.39 W.

5. NUMERICAL RESULTS

In this section, we study the effect of the initial temperature θ_0, and the maximum speed u_{max} on the nature and cost of the optimum solution. We also compare the performance of the optimum profile with the two-speed approximation and a "safe" speed policy that always operates within the TDP.

5.1 Effect of initial temperature

The initial temperature has a strong effect on the performance of the optimal solution. Figure 3 plots the optimum speed profiles for

different values of θ_0, for a fixed maximum speed $u_{max} = 4$ GHz. For the lowest possible value $\theta_0 = \theta_{ambient}$, the thermal limit is not reached over the duration $T = 140$ s, and the unconstrained optimum solution $u(t) = u_{max}$ is the optimum solution. This constitutes the upper bound on the achievable performance. As the initial temperature rises, the optimum speed profile is forced to scale down towards u_m, which happens faster as θ_0 increases.

The resulting performance degradation is almost linear in θ_0. This translates roughly to a reduction in average throughput of 33.3 million cycles/s per °C increase in θ_0. This dependence of performance on θ_0 presents opportunities for balancing the temperature over successive workloads executed over a processor. As one workload's final temperature is the initial temperature of the next, optimizing the performance or energy consumption over a set of workloads must consider the effect of each tasks final temperature on the cost of future tasks.

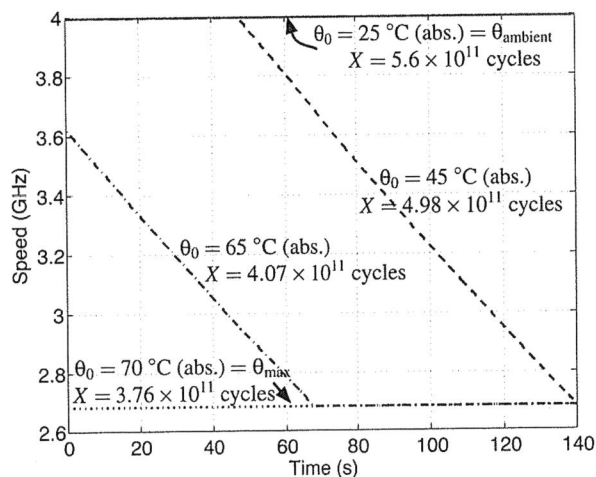

Figure 3: Effect of initial temperature on optimum solution.

5.2 Effect of the maximum speed

If the maximum speed of a processor were arbitrarily large, the resulting performance optimum solution would constitute an upper bound on the performance that can be achieved by using speeds larger than u_m. The maximum speed used by such a speed profile can be valuable for a designer as speeds larger than that speed will not help achieve higher performance. Figure 4 shows the optimal speed profiles for different values of u_{max}, for a fixed initial temperature of 45°C (abs). It can be seen that for $u_{max} > 4.63$ GHz, the maximum speed no longer constrains the optimum performance profile. The corresponding performance value is the maximum performance that can be achieved using the given power-speed relationship if all parameters other than u_{max} are kept constant.

Currently, the maximum power consumption allowed in a processor is about 10% to 40% above the TDP. This results in an even smaller difference between u_{max} and u_m. For the processor model we study here, we can see from Figure 4 that as u_{max} gets closer to u_m, the optimum performance sharply falls by up to 15%. While the power consumption of processors is rising quickly with each technology generation, the thermal solutions do not scale correspondingly. Hence, a practical solution for future processors may be to design speeds $> u_m$ that can only be used temporarily. This then creates a need for developing intelligent ways to use those larger speeds to extract maximum performance.

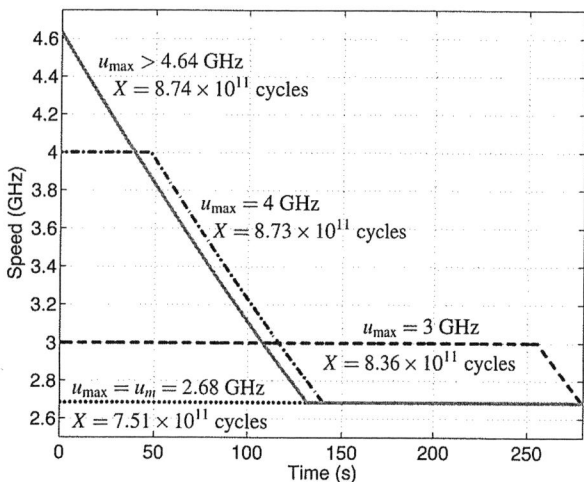

Figure 4: Effect of maximum speed on the optimum solution.

5.3 Effect of thermal capacitance

The thermal capacitance of the package-heat sink combination is an important factor to determine the time taken to reach the maximum temperature after which a forced slowdown to u_m occurs. As seen from (2), this time is directly proportional to the thermal capacitance. To study the effect of thermal capacitance on the optimum solution, we computed the optimum profiles for different values of the thermal capacitance, by keeping $u_{max} = 5$ GHz, and $\theta_0 = 45°C$ (abs.). The resulting solutions are plotted in Figure 5. The value of 340 J/K that is typical of most processor-heat sink packages is able to stay above u_m for only two minutes.

There exist novel packaging solutions using removable aluminum plates [9] and phase-change material based heat sinks [17] that can provide much larger thermal capacitances. The use of such heat sinks for smoothing out spikes in electronic workloads has been studied by the electronic packaging community [9, 17]. From Figure 5, it can be seen that using such solutions for increasing the thermal capacitance can increase both, the time taken to reach θ_{max} and the performance. The increase in both cases is linear. The performance increase translates to an increase in average throughput of about 0.5 million cycles/s per J/K increase in C.

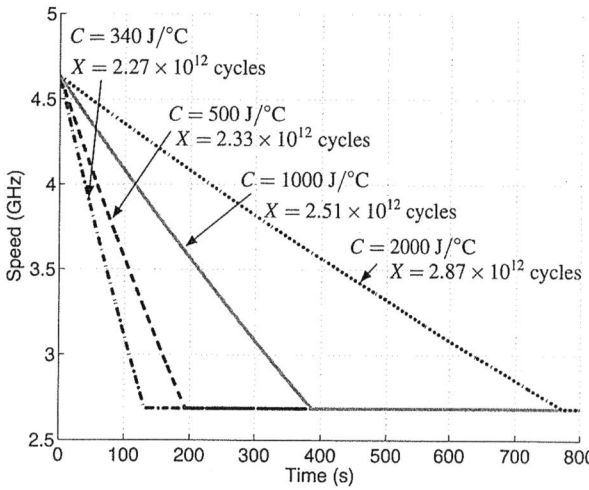

Figure 5: Effect of thermal capacitance on optimum solution.

5.4 Comparison of OPT and CONST policies

We wish to compare the performance of the optimum speed profile (OPT) with the two-speed approximation (CONST) described in Section 3.3. We also examine the policy SAFE, which avoids thermal emergencies altogether by always operating at the speed u_m. Clearly, SAFE will have a lower performance than MAX. But we wish to see how much performance improvement can be gained over it by operating at "unsafe speeds", for a modern processor with $u_{max} = 4$ GHz.

Figure 6 shows the optimum speed and temperature profiles for OPT, CONST and SAFE for a time duration $T = 120$ s. The analytical solution for the optimum speed profile OPT is given by,

$$
u^*(t) = \begin{cases} 4.00 \text{ GHz}, & 0 \le t < 16.44 \text{ s}, \\ \left(-0.7059 + 2.2768\sqrt{-1 + 5.7107e^{-0.0049*t}}\right) \text{ GHz}, & \\ & 16.44 \text{ s} < t \le 108.80 \text{ s}, \\ 2.68 \text{ GHz}, & 108.8 \text{ s} < t \le 120 \text{ s} \end{cases}
$$

(18)

while the optimum solution for the two-speed profile CONST is given by

$$
u^*_{CONST} = \begin{cases} u_1^* = 3.75 \text{ GHz}, 0 \le t \le 73.98 \text{ s} \\ u_m = 2.68 \text{ GHz}, 73.98 \text{ s} < t \le 120 \text{ s}. \end{cases}
$$

(19)

The resulting performance for the three speed policies are $X_{SAFE} = 3.221 \times 10^{11}$ cycles, $X_{CONST} = 4.003 \times 10^{11}$ cycles, and $X_{OPT} = 4.010 \times 10^{11}$ cycles. *An intelligent speed control policy like OPT could extract up to 25% more performance than SAFE.* However, the continuous speed variation required by OPT is not practical. The two-speed solution CONST was able to achieve up to 0.62% of the performance of the optimum policy. Additionally, this solution has the advantage that it is easier to implement in a real processor. The optimum speed value u_1^* can be pre-computed for different values of the time duration T by numerically solving the single-variable optimization problem described in 3.3. The resulting solutions can then be stored in a look-up table for easy access at run-time.

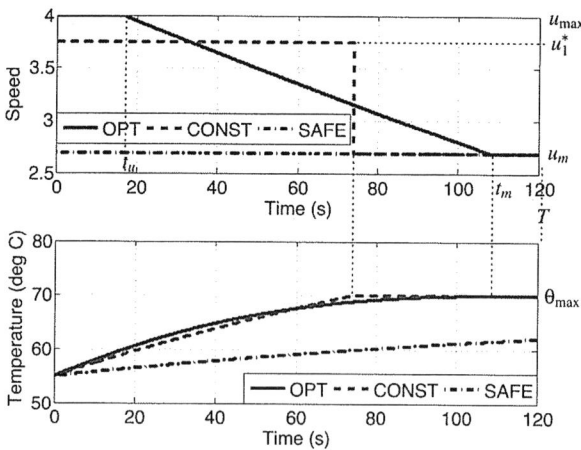

Figure 6: Optimum speed profile of different policies.

6. CONCLUSION

The disparity between thermal solutions and chip heat dissipation is growing just like the one between battery capacity and chip power consumption. The technological solutions (battery capacity increase, improved thermal design) alone cannot support the

296

increased capabilities of future processors. In the case of batteries, researchers responded by devising high-level battery models to account for non-linear effects and developed system-level load shaping techniques that varied the current demand from the battery to extract maximum charge. We need an analogous approach to the problem of system-level thermal management. While accurate thermal and power models already exist, the optimum manner to control the processor speed to address thermal concerns has only recently begun to be studied [2, 14, 15].

In this paper, we address a fundamental problem in system-level thermal management: What is the best way to control the speed of a processor to get maximum performance, while satisfying thermal and speed constraints? We used a novel formulation and solution methodology based on the calculus of variations, and were able to obtain an analytical solution. This solution serves as an upper bound for the maximum performance that can be extracted from a processor by speed control. We obtained a cubic power-speed relationship based on the literature, and studied the effect of different parameters like the initial temperature, thermal capacitance and maximum speed on the optimum solution. The performance was found to drop almost linearly with rising initial temperatures, and increase linearly with increasing thermal capacitance. Allowing larger maximum clock speeds helps improve the performance, but only up to a certain limit. The cubic power model also enabled us to solve for the optimum speed profile in closed form. For future work, we are working on extending the proposed solution for higher order thermal models, and including terminal temperature constraints ($\theta(T) \le \theta_f$) in our formulation.

7. REFERENCES

[1] Advanced Micro Devices. *CPU Thermal Management: Application Note.*

[2] D. Brooks and M. Martonosi. Dynamic thermal management for high-performance microprocessors. In *Intl' Symp. High Speed Computer Arch. (HPCA)*, pages 171–182, 2001.

[3] Y. A. Cengel. *Heat Transfer*. McGraw Hill, 2 edition, 2002.

[4] B. M. Guenin. Calculation corner: Simplified transient model for IC packages. *Electronics Cooling*, 8(3), August 2002.

[5] S. H. Gunther, F. Binns, D. M. Carmean, and J. C. Hall. Managing the impact of increasing microprocessor power consumption. *Intel Technology Journal*, Quarter 1 2001.

[6] W. Huang, M. R. Stan, K. Skadron, K. Sankaranarayanan, S. Ghosh, and S. Velusamy. Compact thermal modeling for temperature-aware design. In *Proc. Design Automation Conf. (DAC)*, pages 878–883, 2004.

[7] Intel Corp. *Intel Pentium 4 Processor 6x1 Sequence: Datasheet*, January 2006.

[8] R. Jejurikar, C. Pereira, and R. Gupta. Leakage aware dynamic voltage scaling for real-time embedded systems. In *Proc. Design Automation Conf. (DAC)*, pages 275–280, 2004.

[9] S. Krishnan and S. V. Garimella. Thermal management of transient power spikes in electronics—phase change energy storage or copper heat sinks? *Journal of Electronic Packaging*, 126(3):308–316, 2004.

[10] W. Liao, L. He, and K. M. Lepak. Temperature and supply voltage aware performance and power modeling at microarchitecture level. *IEEE Trans. Computer-Aided Design*, 24(7):1042–1053, July 2005.

[11] J. R. Lorch and A. J. Smith. PACE: A new approach to dynamic voltage scaling. *IEEE Trans. Computers*, 53(7):856–869, July 2004.

[12] S. M. Martin, K. Flautner, T. Mudge, and D. Blaauw. Combined dynamic voltage scaling and adaptive body biasing for lower power microprocessors under dynamic workloads. In *Proc. Intl' Conf. Computer Aided Design (ICCAD)*, pages 721–725, 2002.

[13] J. M. Rabaey, A. Chandrakasan, and B. Nikolic. *Digital Integrated Circuits*. Prentice Hall, 2 edition, 2002.

[14] K. Skadron, T. Abdelzaher, and M. R. Stan. Control-theoretic techniques and thermal-RC modeling for accurate and localized

dynamic thermal management. In *Proc. Intl' Symp. High Perf. Comp. Arch. (HPCA)*, pages 17–28, 2002.

[15] J. Srinivasan and S. V. Adve. Predictive dynamic thermal management for multimedia applications. In *Proc. Intl' Conf. Supercomputing (ICS)*, pages 109–120, 2003.

[16] F. Y. M. Wan. *Introduction to the Calculus of Variations and its Applications*. Chapman and Hall, 1995.

[17] D. won Yoo and Y. K. Joshi. Energy efficient thermal management of electronic components using solid-liquid phase change materials. *Device and Materials Reliability, IEEE Transactions on*, 4(4):641–649, 2004.

APPENDIX

Derivation of the performance optimal speed profile: A necessary condition to extremize the objective (8) is that its first variation δJ must vanish.[4] As the system equation (9) is satisfied at every instant over $[0, t_m]$, its variation must also vanish, or $\delta\psi = 0$. Now, we can use a Lagrangian multiplier function $\lambda(t)$ to form the augmented performance index

$$I = J - \int_0^{t_m} \lambda\psi\,dt = \int_0^{t_m} (-u - \lambda\psi)\,dt - u_m(T - t_m) \quad (20)$$

As $\delta J = \delta\psi = 0$, we must have $\delta I = 0$, or

$$\delta I = \int_0^{t_m} \left[\{-\delta u\} - \lambda\{ -RS\,\delta u + RC\,\delta\theta' + \delta\theta \} \right] dt + u_m\,dt_m = 0$$

It can be shown that on integrating by parts, and noting that $\delta\theta' = (\delta\theta)'$, the above equation reduces to

$$\delta I = \int_0^{t_m} \left[\{ -1 + \lambda RS \}\,\delta u + \{ -\lambda - (-RC\lambda') \}\,\delta\theta \right] dt - u_m\,dt_m$$
$$+ \left[-u(t_m) - \lambda(t_m)\{ \theta_{\max} - RP(u(t_m)) \} \right] dt_m = 0$$

where the last term with dt_m arises because t_m is a variable endpoint [16]. As δI must vanish for all variations δu, $\delta\theta$ and dt_m, we require their respective coefficients to vanish. This results in the following the two Euler-Lagrange equations

$$S(u) = 1/(R\lambda), \qquad \lambda = \alpha e^{t/(RC)},$$

and a transversality condition

$$u(t_m) - u_m = \alpha \left[RP(u(t_m)) - \theta_{\max} \right] e^{t_m/(RC)} \quad (21)$$

The two Euler-Lagrange equations reduce to

$$S(u(t)) = [1/(\alpha R)]\,e^{-t/(RC)}, \quad (22)$$

which gives us the form of the optimum solution. To determine the actual optimum speed profile, we must determine two unknowns α and t_m, and hence, we need two equations. One equation is provided by the transversality condition (21), and another is obtained by the condition [16] that the variable involved in the inequality constraint θ must be continuous at the switch point, or $\theta(t_m) = \theta_{\max}$.[5] Note that $u(t_m) = u_m$ always satisfies the transversality condition, as $P(u_m) = P_m = \theta_{\max}/R$.

---

[4]The variation can be considered the variational analogue of the differential operator in differential calculus. The theoretical underpinnings of the variational operator have been explained in [16].

[5]This makes intuitive sense as the temperature (unlike the speed) cannot vary suddenly due to the non-zero thermal capacitance.

Temperature-Aware Floorplanning of Microarchitecture Blocks with IPC-Power Dependence Modeling and Transient Analysis *

Vidyasagar Nookala   David J. Lilja   Sachin S. Sapatnekar

ECE Dept, University of Minnesota, Minneapolis, MN

{vidya,lilja,sachin}@ece.umn.edu

ABSTRACT

Operating temperatures have become an important concern in high performance microprocessors. Floorplanning or block-level placement offers excellent potential for thermal optimization through better heat spreading between the blocks, but these optimizations can also impact the throughput of a microarchitecture, measured in terms of the number of instructions per cycle (IPC). In nanometer technologies, global buses can have multicycle delays that depend on the positions of the blocks, and it is important for a floorplanner to be microarchitecturally-aware to be sure that thermal and IPC considerations are appropriately balanced. This paper proposes a methodology for thermally-aware microarchitecture floorplanning. The approach models the interactions between the IPC and the temperature distribution, and incorporates both factors in the floorplanning cost function. Our approach uses transient modeling and optimizes both the peak and the average temperatures, and employs a design of experiments (DOE) based strategy, which effectively captures the huge exponential search space with a small number of cycle-accurate simulations. A comparison with a technique based on previous work indicates that the proposed approach results in good reductions both in the average and the peak temperatures for a range of SPEC benchmarks.

Categories and Subject Descriptors: B.7.2 [Hardware]: Integrated Circuits - Design Aids

General Terms: Performance, Experimentation

Keywords: Microarchitecture, Floorplanning, Transient Analysis

1. INTRODUCTION

Due to rapid increases in on-chip power and integration densities, operating temperatures have become an important concern in high performance integrated circuits in nanometer technologies. A high temperature can affect the reliability of a circuit, thus reducing its lifetime [1], through phenomena such as electromigration and Negative Temperature Bias Instability (NBTI). With every process generation, circuit performance becomes more sensitive to thermal effects due to the decreasing limits on the maximum junction temperature [2]. In addition, the temperature dependence of the leakage power results in an undesirable positive feedback, commonly

---

*This work was supported in part by a gift from Intel Corporation, by the NSF under award CCCR-0205227, by the Minnesota Supercomputing Institute, and by the University of Minnesota Digital Technology Center.

Permission to make digital or hard copies of all or part of this work for personal or classroom use is granted without fee provided that copies are not made or distributed for profit or commercial advantage and that copies bear this notice and the full citation on the first page. To copy otherwise, to republish, to post on servers or to redistribute to lists, requires prior specific permission and/or a fee.
ISLPED'06, October 4–6, 2006, Tegernsee, Germany.
Copyright 2006 ACM 1-59593-462-6/06/0010 ...$5.00.

---

referred to as *thermal runaway*, which could even lead to catastrophic chip failures. While advanced [3] packaging solutions can result in enhanced heat removal capabilities, the costs associated with these solutions are typically prohibitive. Therefore, it is important to develop temperature-conscious design techniques that alleviate on-chip thermal problems.

On-chip temperature distributions depend not only on the total power dissipation, but also on the spatial distribution of the power sources and the material properties of the medium that permit vertical and horizontal heat transfer in a chip. Physical design methods, such as floorplanning and placement, can impact the thermal profile of a chip by altering the spatial distribution of power sources, indicating a scope for improvement through better heat spreading that evens the temperature distribution on the chip. In addition, physical design optimizations can complement other thermal- and power-aware design [4] techniques implemented at a higher, architecture level such as Dynamic Thermal Management (DTM) [5].

The topic of thermally-aware floorplanning/placement has attracted some attention in the last few years, both at the circuit and microarchitecture levels. The primary difference between circuit and architecture level treatments is the level of knowledge about the spatial distribution of power. At the architectural level, the circuit is defined only in terms of large functional blocks and coarse estimates of power are available, while at the circuit level [6, 7, 8], the power consumptions of individual macro cells or blocks are all well known, and more accurate estimations are possible. However, there are many more flexibilities at the architectural level that permit significant design changes that reduce the overall power and temperature distribution.

This work focuses on the interactions between microarchitecture design and physical design, in particular, floorplanning, to explore performance-temperature tradeoffs. In the nanometer regime, the choice of a floorplan can significantly affect the performance of a processor, measured in terms of the number of instructions per cycle (IPC) [9, 10, 11, 12]. The chief culprit is the delay associated with global wires, such as buses, which can have multicycle delays [13] , thus requiring *wire-pipelining* [14] in order to support high operating frequencies. Moreover, the fluctuations in the IPC can change the activity patterns of the blocks, resulting in variations in the power densities. In other words, floorplanning can affect the temperature profile not only through heat spreading but also because the spatial and temporal distributions of power densities vary due to wire-pipelining. A good floorplanning strategy must therefore consider such interaction between IPC and power (and hence temperature) and jointly optimize both the performance and temperature objectives.

A few recent works [15, 16, 17, 18] propose techniques for thermal-aware microarchitecture floorplanning. While these indicate a welcome progress, they suffer from two drawbacks:

- They do not model the IPC-power interaction in the floorplanning step and assume that the block power consumptions are layout independent. Specifically, the power densities that are obtained for a zero-bus-latency scenario, which typically represents the worst case for dynamic power (and the best case for IPC), are assumed to be valid for all floorplans irrespective of the amount of pipelining required by the buses, and this can result in overestimation of the temperature.

- They attempt to minimize the steady-state temperature of a chip. However, steady-state can only occur when the power dissipation is constant, which may not be true in general since programs tend to exhibit phases of varying activities [19]. In such a case, a transient modeling [20] provides a better picture of the thermal behavior of the chip: the execution times of the standard benchmarks that are used in simulations, such as SPEC [21], are typically in the range of seconds, which are significantly larger than typical thermal time constants, making it imperative to model transients. In addition, transient modeling also captures an accurate depiction of the dependence of leakage current on temperature.

A better strategy may be to focus on minimizing the peak transient temperature over the entire execution time of a program. Furthermore, besides the peak temperature, it is useful to capture the temporal average of the temperature distribution, since many reliability mechanisms depend on this.

Although some of the previous approaches do consider the temperature transients, the emphasis is on modeling the impact of temperature on leakage power, only a small portion of the execution time is considered for analysis, and the goal of floorplanning is to minimize the steady-state temperature.

In this paper, we propose a methodology for multiobjective microarchitecture floorplanning, where the objectives are minimizing the temperature (both average and peak), based on transient analysis, and maximizing the performance (IPC). Our approach models the impact of wire-pipelining (i.e., changes in the IPC, on power densities in the floorplanning step) and temperature-leakage power dependencies. For the purposes of a complete transient analysis that considers the entire execution times of the programs, we use a larger timestep than those employed in the limited-time analyses of [15, 16, 17, 18]. Since the floorplanning that we address involves big microarchitecture blocks, which have larger time constants than ordinary cells, the temperatures change at a slow rate, in which case, a large timestep, which reduces the analysis time by a tremendous amount, can be chosen without much loss in accuracy.

2. THERMAL ESTIMATION

A key component of a thermally-aware design methodology is a framework to estimate the temperature distribution of a chip. In the thermal analysis context, a chip can be viewed as a multilayered grid network, essentially a discretization of the chip geometry, where the nodes of the network correspond to the centers of the grids, and the connections between the nodes represent the heat flow paths in the chip. In such a set-up, the power sources $\vec{P}$ are located at the nodes of the network and based on the duality of electricity and heat transfer, the temperature distribution of the network is governed by the following differential equation:

$$\vec{C} \cdot \frac{d\vec{T}}{dt} + G \cdot \vec{T} = \vec{P} \qquad (1)$$

where G is the thermal conductance matrix of the network, $\vec{T}$ is the temperature distribution of the nodes of the network. The first term on the LHS of (1) represents the transient behavior of the temperature, with $\vec{C}$ modeling the thermal capacitances. Several techniques for thermal analysis have been proposed in the past, some of which can be found in [22].

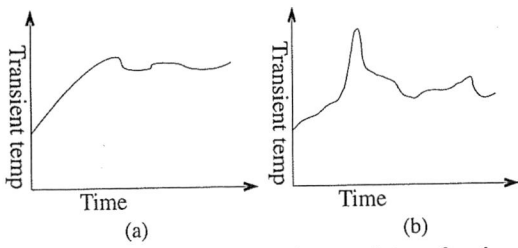

Figure 1: The results of two transient analyses of a circuit under two different implementations.

3. AVERAGE TEMPERATURE

Figure 1 shows two possible transient scenarios for a circuit, where the maximum transient temperature of the circuit is plotted against time elapsed. Although the curve of Figure 1(a) has a lower peak than that of Figure 1(b), Figure 1(b) offers a better average, where the curve is below that of Figure 1(a) for a majority of the time. As noted in [1], the reliability or mean time to failure (MTTF) decreases exponentially with temperature. Therefore, Figure 1(b) may represent a higher reliable case than Figure 1(a). In such a scenario, attempting to minimize the peak temperature can result in suboptimal thermal profiles. Nevertheless, a higher peak, seen in Figure 1(b), is not desirable due to the constraints it places on the package hardware. Therefore, a better approach may be to consider both the peak and the average temperatures in the optimization objectives, and we do this in our floorplanning methodology.

4. FLOORPLANNING FLOW

Figure 2 shows the flow of the proposed temperature-aware microarchitecture floorplanning methodology. The approach accepts a microarchitecture block configuration, a set of buses, benchmarks and a target frequency as inputs and generates a floorplan of the blocks that is both optimal in both IPC and temperature.

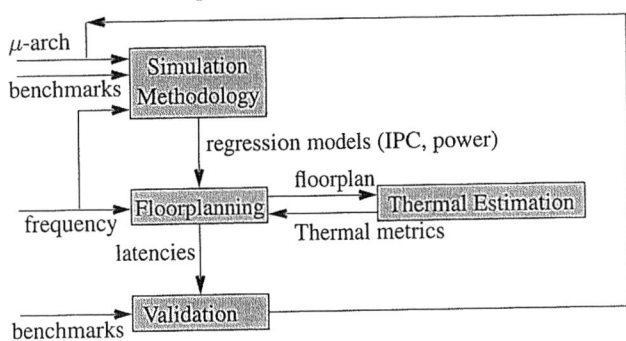

Figure 2: Thermal-aware floorplanning: design flow.

An important issue of the design flow is estimating the IPC and the block power dissipations required to generate the temperature distribution of the microarchitecture layout. In particular, the number of pipelined latencies required by each bus of the microarchitecture is proportional to its length, and therefore for every floorplan, there is a corresponding bus-latency configuration, and consequently an IPC and a power (and temperature) distribution. However, the large search space explored during floorplanning makes it virtually impossible to use simulations for each floorplan that is to be evaluated. Specifically, if each of n wires on a layout can have k possible latencies, then the cycle-accurate simulator may have to perform up to n^k simulations to fully explore the search space. We use a simulation strategy, first proposed in [12] for IPC-aware floorplanning, that is based on design of experiments (DOE) to limit the number of cycle-accurate simulations to a practical level. This approach, which reduces the number of simulations to a linear function of n, forms the preprocessing step of the flow.

Unlike [17, 18] and also our previous work on IPC-aware floorplanning [12], where the purpose of the simulations is to characterize the variations in the IPC in terms of changes in the bus latencies, the objective of the simulation strategy of Figure 2 is to model the variations in both IPC and power densities, and thus capture the IPC-power dependence. The variations are encapsulated in the form of regression functions, with the bus latencies as variables, both for IPC and power.

The floorplanner is based on a simulated annealing (SA) framework and uses the regression models to optimize a cost function that is a weighted sum of, besides traditional objectives such as area and aspect ratio, the IPC[1] and the thermal terms, both the peak and average temperatures, as described in section 3.

After every SA move, the floorplanner estimates the block power densities from the regression models and passes them along with the corresponding floorplan to the thermal simulator, which in turn returns the thermal metrics that are part of the cost function. The performance and thermal profile of the resultant layout can then be determined from cycle-accurate simulations. In addition, the entire design flow of Figure 2 may be repeated for several microarchitectural block configurations to identify the optimal configuration.

4.1 Microarchitecture and simulator

The microarchitecture that we employ in this work is based on the DLX architecture [23] and resembles a real processor, Alpha 21362 [24]. The configuration and the corresponding functional blocks are shown in Table 1 and Figure 3, respectively. The instruction fetch and decode blocks are labeled as fet and dec, respectively, while $il1$ and $dl1$ are the level-1 instruction and data caches, respectively. The instruction and data translation look-aside buffers (TLB) are indicated as $itlb$ and $dtlb$, respectively, while $l2$ is the unified level-2 cache. The block ruu is the register update unit, which contains the reservation stations and issue logic, while lsq represents the load store queue. The register file is shown as reg, whereas $bpred$ is the branch predictor. The blocks $iadd1$, $iadd2$, $iadd3$, $imult$, $fadd$ and $fmult$ are the instruction execution units. The figure also shows the 22 buses that can impact the performance (IPC) and block power densities of the processor, when pipelined.

Parameter	Value
Fetch width	8 instrs/cycle
Issue width	8 instrs/cycle
Commit width	8 instrs/cycle
RUU entries	128
LSQ entries	64
IFQ entries	16
Branch pred	comb, 4K table 2-lev 2K table, 11-bit 2K BHT
BTB	512 sets, 4-way
IL1	64K, 64B, 2-way LRU, latency: 1
DL1	32K, 32B, 2-way LRU, latency: 1
L2	2M, 128B, 4-way latency: 12
ITLB, DTLB	128 entries Miss latency: 200

Table 1: Block configuration of the processor.

For estimating the IPC and power data, we use Wattch [25], which is based on `sim-outorder` [23] simulator. The impact of the bus latencies is modeled as dummy pipeline stages in the simulator and the latencies are made configurable. The remainder of this section explains each step of the flow of Figure 2 in detail, and we tie the description to the microarchitecture of Figure 3.

4.2 Simulation strategy

Statistical *design of experiments* is an approach that characterizes the response of a system in terms of changes in the factors which influence the response of the system. The basic idea is to conduct

a set of experiments, in which all factors are varied systematically over a specified range of acceptable values, such that the experiments provide an appropriate sampling of the entire search space. The subsequent analysis of the resulting data will identify the critical factors, the presence of interactions between the factors, etc. In this work, the system is a microarchitecture, such as that shown in Figure 3, the response is the IPC/power, and the factors are the latencies of the buses of the microarchitecture.

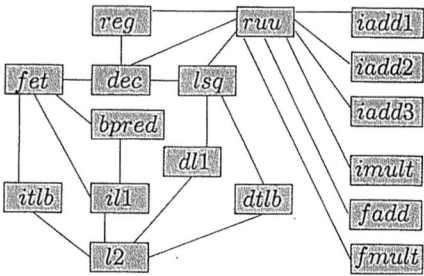

Figure 3: The functional blocks of the microarchitecture. The lines between the blocks represent the buses of the processor.

Since it is impractical to fully explore the exponential search space, even when the number of factors (buses) is small ($n = 22$), we employ a *fractional factorial* design [26] to reduce the number of simulations. In addition, as both power and IPC depend on the same set of variables, i.e., bus latencies, a single design can be used to characterize both responses.

An important advantage of factorial designs is the ability to model and estimate interactions between the factors. We have identified a few potential significant interactions, which resulted from the nature of wire-pipelining models integrated into the simulator:

- We have incorporated functional unit scheduling in the simulator. Specifically, the number of latencies inserted on the three buses between the register update unit and the three integer adders can be different, and while issuing an integer add instruction, of all the available units, the one with the least latency is chosen. This indicates possible significant (two and three factor) interactions, which need to be estimated.
- In the decode stage, the number of extra pipeline stages to be inserted is modeled as a maximum function of the latencies of the buses $dec - reg$, $dec - ruu$ and $ruu - reg$ (refer to Figure 3). Such a nonlinear function implies significant (two and three) factor interactions among these three factors.

In this work, we use a two-level resolution III fractional factorial design [26], where the two levels correspond to the lowest and highest values (extremes) for the bus latencies, which can be obtained by assuming worst-case and best-case scenarios for the corresponding wire lengths. For n factors, the number of experiments required is equal to the nearest highest power of two, which, since $n = 22$, turns out to be 32 for our work[2].

The floorplanning approaches of [17, 18], although do not model the dependence of power on bus latencies, propose simulation strategies to capture the IPC impact of bus latencies. The method of [17] constructs linear regression models using simulations by varying each latency independently, whereas [18] uses latency-independent models to capture the IPC variations. While these have been demonstrated to work well for IPC since a reasonably accurate relative ordering of variables is sufficient [27], such one-at-a-time approaches may not effectively track absolute variations, required in the case of power, as compared to the DOE approach [28] used in this work. The reason for the requirement of "absoluteness" is that the power

---

[1] The cost function actually includes the CPI, or Clocks Per Instruction, the reciprocal of IPC, since the objective is to maximize the IPC.

[2] Although resolution III designs work under the notion of negligible interactions, since n is less than the number of simulations, we have an *unsaturated* design, and this allows the estimation of a few interactions by projecting them as additional factors.

and temperature may not have a perfect correlation [29], and power-criticality does not necessarily imply temperature-criticality. This lack of fidelity[3], coupled with the dependence of leakage current on temperature, indicates that any error in power estimation can result in significant inaccuracies in the temperature computations.

4.2.1 Reducing simulation times

Cycle-accurate simulations are inherently slow and most SPEC benchmarks with `reference` input sets, when simulated can take days to complete. Therefore, although the resolution III design strategy of section 4.2 requires a small number of simulations, the run time of each simulation is still an issue. To speed up the simulations, we utilize SMARTS [30], a periodic sampling technique, which works well both for throughput (IPC) and power/energy, particularly for the SPEC benchmarks.

4.2.2 Power/IPC regression models

The SMARTS technique involves fastforwarding program segments between successive samples chosen for detailed simulation. However, the transient modeling requires that the block power densities be collected periodically for every timestep. For this, we extrapolate the power data collected for each sample for the succeeding fastforwarded portion. While we do not offer a proof, the concept of periodic sampling is inherently based on this assumption, and there is empirical evidence that it works well at least for average power/energy estimation [30].

The total execution time obtained from a simulation is then segmented into slots of size equal to the transient analysis timestep. Therefore, the data collected from the simulation can be arranged as an array P indexed by the timestep and the block number, i.e., the entry $P(a, b)$ of the array corresponds to the power consumption of block b (one of the 17 blocks of Figure 3) during timestep a. Since 32 simulations performed (per benchmark), there are 32 such tables. For each entry $P(a, b)$ (per benchmark), a regression model is constructed from the 32 values [26], based on least-squares approximation, where the variables are the bus latencies. Equation (2) shows one such a model, constructed to estimate the power dissipation at entry $P(a, b)$, where β_is represent the regression coefficients computed from the 32 values obtained for the correspond entry (a, b). Each x variable in (2), say x_i, represents an encoding of the latency of bus i, l_i, where the minimum and the maximum latencies are coded as -1 and +1, respectively, and $\mathcal{I}$ is the set of interactions described in section 4.2.

$$x_i = -1 + \left(\frac{2 \cdot l_i}{\min(i) + \max(i)} \right), \quad 1 \le i \le 22$$

$$P(a, b) = \beta_0 + \sum_{i=1}^{22} \beta_i \cdot x_i + \sum_{(ij) \in \mathcal{I}} \beta_{ij} \cdot x_i \cdot x_j + \sum_{(ijk) \in \mathcal{I}} \beta_{ijk} \cdot x_i \cdot x_j \cdot x_k \quad (2)$$

An IPC regression model is similarly constructed for each benchmark from the statistics gathered from the 32 simulations. In addition, although we construct separate regression functions for IPC and power, since the associated variables are the same, a direct relation between the power and the IPC estimates can be obtained by composition of the regression functions.

4.3 Temperature estimation

We use HotSpot [29] in this work for thermal analysis. In this approach, the nodes of the multi-layered thermal network described in section 2 are the centers of the blocks of the microarchitecture. The tool also provides a framework for transient modeling, and accepts a floorplan, the length of the timestep, and the block power

---

[3] A well known case where the property of fidelity holds is Elmore delay modeling: although the estimated delays may be inaccurate, the metric accurately tracks the variations in the delays.

---

dissipations averaged over each timestep as inputs. The differential equation (1) is solved at each timestep to estimate the new set of temperatures (with the initial conditions being those of the previous timestep). The leakage power component of the succeeding timestep can then be updated using the new temperatures.

Choice of timestep

In general, the smaller the timestep, the higher is the accuracy of the transient analysis. It is clearly impractical to perform the analysis for every clock cycle of execution, and the authors of HotSpot suggest a size of about 10000 clock cycles at a frequency of 3GHz, i.e., a timestep of about $3.3\mu s$. Although this reduces the analysis time by a significant factor, it still makes it prohibitive to incorporate transient analysis into the iterative scheme of the floorplanning step, where thousands of floorplans are evaluated.

To solve this issue, we choose an interval of one million clock cycles, which amounts to about a few hundreds of microseconds for gigahertz frequencies, and this can possibly affect the accuracy of the computations. However, since the focus of the optimizations involves relatively larger microarchitecture blocks (than the macro cells considered in circuit level optimizations), the thermal RC constants tend to be higher, typically in the range of tens of milliseconds, and this indicates a minimal loss of accuracy since each time constant still involves a high number of timesteps. For instance, ruu, a medium sized block of the microarchitecture of Figure 3, has a time constant of about 120ms. As noted in [29], the temperatures rise slowly, and it takes more than 100,000 clock cycles to observe an increase of as small as 0.1°C in the temperature. In addition, we use a single iteration to solve the differential equation of (1) during each timestep.

4.4 Floorplanning cost function

The floorplanner is based on simulated annealing (SA), which uses the power and IPC regression models built out of the simulation methodology described in section 4.2 in the cost function. We use PARQUET [31], a floorplanner available in the public domain.

The cost function C is a weighted sum of, besides the chip area ($Area$) and the aspect ratio (AR), the average (T_{avg}), and the peak (T_{peak}) transient temperatures, as shown below:

$$C = W_1 \cdot Area + W_2 \cdot AR + W_3 \cdot CPI + W_4 \cdot (T_{avg} + T_{peak}) \quad (3)$$

where the Ws represent the relative weights of the optimization terms. It can be seen that the cost function actually contains CPI, the reciprocal of IPC, since the objective is maximizing IPC. If N_t is the number of timesteps in the transient analysis and T_i is the maximum of the block temperatures at timestep i, the average and the peak temperatures are determined as follows:

$$T_{avg} = \frac{1}{N_t} \sum_i T_i \text{ and } T_{peak} = \max_i T_i \quad (i = 1, 2, \cdots, N_t)$$

5. VALIDATION

5.1 Benchmarks

We choose a set of eight SPEC 2000 benchmarks, which, along with the corresponding instruction counts of the `reference` input sets, are shown in Table 2. The benchmarks are chosen because of their distinct instruction mixes. For instance, `mesa` has a high percentage of conditional branches, while `gcc` has a very large number of memory operations. All benchmarks are complied at optimization level O3 using the SimpleScalar version of gcc.

5.2 Experimental set up

The areas of the blocks of Figure 3 are estimated using [32]. The total area of the chip is about $2cm^2$ at 90nm technology, with the L2 cache consuming about 70% of the area. Only the chip core that also includes the L1 caches is considered during floorplanning, and the L2 cache is wrapped around the core floorplan, just as is done in [17] and Alpha 21362 [24]. We choose a frequency of 4GHz for our experiments, and therefore, a timestep of $250\mu s$. For the bus

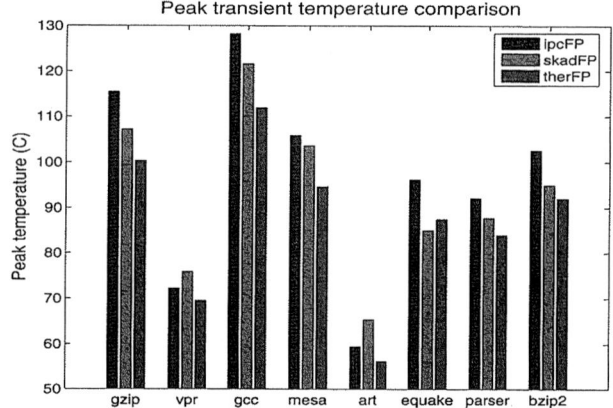

Figure 4: Comparison of the peak transient temperatures obtained from the three floorplanning approaches.

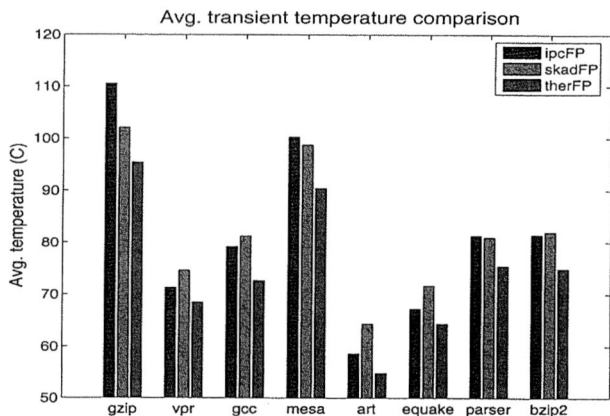

Figure 5: Comparison of the average temperature metric for the three floorplanning scenarios.

Benchmark	Type	Instr. (B)
gzip	Integer	63
vpr	Integer	11
gcc	Integer	35
mesa	Floating-point	305
art	Floating-point	54
equake	Floating-point	175
parser	Integer	301
bzip2	Integer	94

Table 2: Benchmarks from the SPEC 2000 suite, along with the reference instruction counts.

latency ranges that are to be used in the resolution III design, the low value is chosen to be 0, depicting the best case placement of the connecting blocks. The high value is chosen to equal the corner-to-corner latency of the chip core, which is found to be 6 clock cycles at 4GHz, based on the computations of [17].

For each of the eight SPEC benchmarks of Table 2, 32 cycle-accurate simulations are performed, as prescribed by the resolution III design. Although the floorplan can be optimized for each of the benchmarks, in practice, a processor must be optimized so that it performs well over a range of benchmarks. In other words, one must generate a single floorplan for the processor that is, on average, optimal over all benchmarks. For this purpose, the IPC and power regression coefficients are averaged over the eight benchmarks to generate a new set of regression models that are used in the optimization process to generate a single floorplan. In addition, for the purposes of transient analysis, we use an initial temperature of 40°C for the blocks of the architecture.

We integrate HotSpot with Wattch to enable thermal analysis during simulations. *Although we use SMARTS to speed up the simulation strategy of section 4.2, detailed cycle-accurate simulations, without fastforwarding any program portions, for the entire execution times of the benchmarks are performed for validating the floorplanning solutions.* In addition, we use a relatively smaller timestep of 10000 clock cycles, as compared to that of 1000000 cycles used during optimization, for transient analysis, i.e., the power data are averaged over every 10000 clock cycles and are provided to the HotSpot solver to determine the set of temperatures.

5.3 Results

We compare our proposed thermal floorplanning technique with two other approaches. The long run times of the simulations is the main obstacle that limits the number of comparisons that can be made. The floorplanners that are compared listed below:

- **ipcFP:** IPC only floorplanning, the cost function of the floorplanning does not consider any thermal issues.
- **therFP:** Our proposed temperature-aware floorplanning, where the cost includes IPC and both the average and peak transient temperatures, along with the core area and aspect ratio.

- **skadFP:** A temperature-aware floorplanning approach based on [17]: the block power densities are assumed to be independent of the bus latencies. In addition, the cost includes only the peak transient temperature, along with the IPC, area and aspect ratio[4].

For **therFP** and **skadFP**, we choose a weight of 0.4 for both IPC and temperature, and 0.1 for area and aspect ratio, i.e., $w_1 = w_2 = 0.1, w_3 = w_4 = 0.4$ in (3). For the IPC-only floorplanner **ipcFP**, we have $w_1 = w_2 = 0.1, w_3 = 0.8, w_4 = 0$. The idea is to provide a greater emphasis on the primary issues, the IPC and the temperature, while still attempting to limit the total area.

The white spaces (WS) and the aspect ratios (AR) of the floorplans obtained using the three approaches, shown in Table 3, imply that all of the three result only in a small increase in the area. For instance, a core WS of about 6% in **therFP** indicates an overall increase of 1.5% in the chip area (equivalent to 2.03cm^2). Besides, both **skadFP** and **therFP** produce floorplans of almost perfect AR.

Case	Core WS (%)	Core AR
ipcFP	5.33	1.15
skadFP	7.60	1.02
therFP	6.21	1.03

Table 3: Comparison of white space (WS) and aspect ratio (AR) for the three floorplanners.

Figure 4 plots the peak transient temperatures obtained using the three floorplanners for various benchmarks. The graphs show that, for a majority of the benchmarks, both our **therFP** and **skadFP** obtain good reductions in the peak temperatures when compared to **ipcFP**, and this is particularly true for those which exhibit high temperatures. In addition, **therFP** outperforms **skadFP** for almost all benchmarks despite not explicitly attempting to minimize the peak temperature as is done in **skadFP**. For instance, for the benchmark gcc, the floorplan generated by **therFP** reduces the peak by about 16°C as compared to **ipcFP**, while it is about 7°C for **skadFP**.

Figure 5 compares the average transient temperatures obtained using the three approaches. The plots indicate that **therFP** outperforms both **ipcFP** and **skadFP** by significant amounts for all benchmarks. Reductions of about 9°C and 6°C are obtained over **ipcFP** and **skadFP**, respectively, for gcc. In addition, since the floorplans are optimized for the average cases and not specifically for each benchmark, the optimization potential for each benchmark may not be fully exploited. Furthermore, benchmarks that have low power profiles such as art and vpr do not offer much scope for optimization, the resultant improvements tend to be small, and in

---

[4]We choose to include the peak transient temperature in our implementation of [17] for convenience. Moreover, although the original implementation attempts to minimize the steady-state temperature, the authors use peak transient temperature as a metric of their validation process.

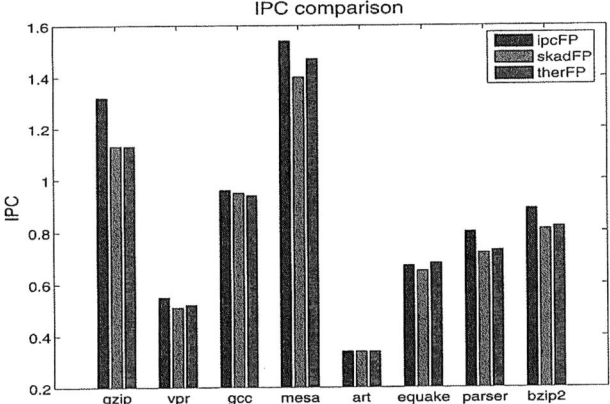

Figure 6: Comparison of the throughput (IPC) metric for the three floorplanning scenarios.

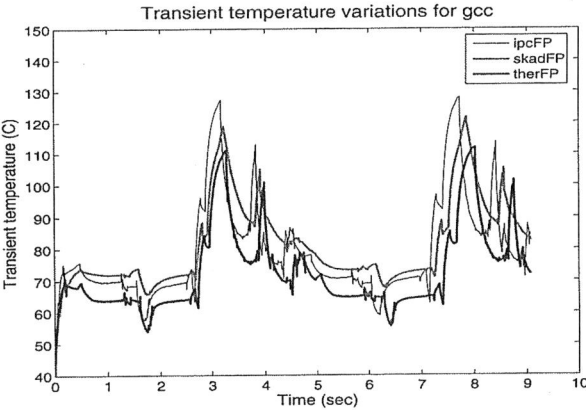

Figure 7: Transient curves for the benchmark `gcc` obtained using the three floorplans.

fact, **skadFP** worsens the thermal profiles obtained for `art` and `vpr`, where both the average and the peak temperatures are higher than those of **ipcFP**, as shown in Figures 4 and 5.

Finally, Figure 6 depicts the performance (IPC) degradation obtained in **therFP** and **skadFP** due to the inclusion of thermal issues in the cost function, besides performance. On an average, both **therFP** and **skadFP** result in almost identical IPCs, about 6% less than **ipcFP**, where no thermal metrics are considered in the cost.

Figure 7 depicts the temporal distributions of the temperature for the entire execution time of a benchmark, `gcc`, for the three floorplanners. The figure shows that **therFP**, where the transient curve is below those of the other cases, produces the best profile among the three cases, as is also shown in Figures 4 and 5. In addition, two observations can be made from this figure:

- A steady-state never occurs in all cases, even after a time as long as 10 seconds, and this is true for most of the benchmarks, except `art` and `vpr`, which have low averages, as seen in Figure 5, and do not exhibit significant variations.

- Although **skadFP** exhibits a lower peak than **ipcFP**, the corresponding curve is consistently higher than that of **ipcFP**, resulting in a higher average. This discrepancy, which is also observed for the benchmark `equake` as illustrated in Figures 4 and 5, indicates that the peak and average temperatures may not have a perfect correlation, which underlines the importance of including temporal average in the objectives.

6. CONCLUSION

Thermal issues have become an important concern in microprocessors designed in nanometer technology nodes. This paper presented a strategy for thermally-aware floorplanning for microprocessors, where the optimization objectives also include the throughput (IPC) issues. The approach also models the IPC-power interaction, and uses a complete transient analysis that captures a thermal profile of a chip in a better way than the steady-state approach, during the floorplanning optimization. The results indicate good improvements both in the average and peak temperatures when compared to an approach derived from a previous work.

7. REFERENCES

[1] J. Srinivasan *et al.*, "The impact of technology scaling on lifetime reliability," in *Proc. Dependable Systems and Networks*, pp. 70–80, Jun. 2004.

[2] Semiconductor Industry Association, "International Technology Roadmap for Semiconductors," 2001. Available at http://public.itrs.net.

[3] S. H. Gunther *et al.*, "Managing the impact of increasing microprocessor power consumption," *Intel Tech. Journ.*, vol. 5, pp. 1–9, May 2001.

[4] D. M. Brooks *et al.*, "Power-aware microarchitecture: Design and modeling challenges for next-generation microprocessors," *IEEE Micro*, vol. 20, pp. 26–44, Aug. 2000.

[5] D. Brooks and M. Martonosi, "Dynamic thermal management for high-performance microprocessors," in *Proc. ACM HPCA*, pp. 171–182, Feb. 2001.

[6] W.-L. Hung *et al.*, "Thermal-aware floorplanning using genetic algorithms," in *Proc. IEEE ISQED*, pp. 634–639, Mar. 2005.

[7] J. Cong *et al.*, "Thermal-driven floorplanning algorithm for 3D ICs," in *Proc. IEEE ICCAD*, pp. 306–313, Nov. 2005.

[8] B. Goplen and S. S. Sapatnekar, "Efficient thermal placement of standard cells in 3D ICs using a force directed approach," in *Proc. IEEE ICCAD*, pp. 86–89, Nov. 2003.

[9] M. Ekpanyapong *et al.*, "Profile-guided microarchitectural floorplanning for deep submicron processor design," in *Proc. ACM DAC*, pp. 634–639, Jun. 2004.

[10] C. Long *et al.*, "Floorplanning optimization with trajectory piecewise-linear model for pipelined interconnects," in *Proc. ACM DAC*, pp. 640–645, Jun. 2004.

[11] A. Jagannathan *et al.*, "Microarchitecture evaluation with floorplanning and interconnect pipelining," in *Proc. ACM ASPDAC*, pp. 1–8, Jan. 2005.

[12] V. Nookala *et al.*, "Microarchitecture-aware floorplanning using a statistical design of experiments approach," in *Proc. ACM DAC*, pp. 579–584, Jun. 2005.

[13] P. Saxena *et al.*, "Repeater scaling and its impact on CAD," *IEEE Trans. Computer-Aided Design Integr. Circuits Syst.*, vol. 23, pp. 451–463, Apr. 2004.

[14] P. Cocchini, "Concurrent flip-flop and repeater insertion for high performance integrated circuits," in *Proc. IEEE ICCAD*, pp. 268–273, Nov. 2002.

[15] Y. Han *et al.*, "Temperature aware floorplanning," in *Proc. Temperature-Aware Computer Systems*, Jun. 2005.

[16] Y. W. Wu *et al.*, "Joint exploration of architectural and physical design spaces with thermal consideration," in *Proc. ACM ISLPED*, pp. 123–126, Aug. 2005.

[17] K. Sankaranarayanan *et al.*, "A case for thermal-aware floorplanning at the microarchitectural level," *Journ. ILP*, vol. 8, pp. 8–16, Oct. 2005.

[18] M. Ekpanyapong *et al.*, "Thermal-aware 3D microarchitectural floorplanning," Technical Report GIT-cercs-04-37, Georgia Institute of Technology, Atlanta, Apr. 2004.

[19] C. Iski and M. Martonosi, "Phase characterization for power: Evaluating control-flow-based and event-counter-based techniques," in *Proc. ACM HPCA*, pp. 281–291, Feb. 2003.

[20] T. Wang and C. C. Chen, "3-D thermal-ADI: A linear-time chip level transient thermal simulator," *IEEE Trans. Computer-Aided Design Integr. Circuits Syst.*, vol. 21, pp. 1434–1445, Dec. 2002.

[21] J. L. Henning, "SPEC CPU 2000: Measuring CPU performance in the new millennium," *IEEE Trans. Comput.*, vol. 33, pp. 28–35, Jul. 2000.

[22] Y. Zhan and others, "Electrothermal analysis and optimization techniques for nanoscale integrated circuits," 2006. Available at http://www.aspdac.com/aspdac2006/archives/2d/index.html.

[23] D. C. Burger and T. M. Austin, "The SimpleScalar tool set, version 2.0," Technical Report CS-TR-97-1342, The University of Wisconsin, Madison, Jun. 1997.

[24] P. Bannon, "Alpha 21364: A scalable single-chip SMP," 1998. Available at http://www.digital.com/alphaoem/microprocessorforum.htm.

[25] D. Brooks *et al.*, "Wattch: A framework for architectural-level power analysis and optimizations," in *Proc. ACM ISCA*, pp. 83–94, Jun. 2000.

[26] D. C. Montgomery, *Design and analysis of experiments*. New York, NY: John Wiley, 1991.

[27] V. Nookala *et al.*, "Comparing simulation techniques for microarchitecture-aware floorplanning," in *Proc. IEEE ISPASS*, pp. 80–88, March 2006.

[28] V. Czitrom, "One-factor-at-a-time versus designed experiments," *The American Statistician*, vol. 53, pp. 126–131, May 1999.

[29] K. Skadron *et al.*, "Temperature-aware microarchitecture," in *Proc. ACM ISCA*, pp. 2–13, Jun. 2003.

[30] R. E. Wunderlich *et al.*, "SMARTS: Accelerating microarchitecture simulation via rigorous statistical sampling," in *Proc. ACM ISCA*, pp. 84–97, Jun. 2003.

[31] S. N. Adya and I. L. Markov, "Fixed-outline floorplanning through better local search," in *Proceedings of IEEE ICCD*, pp. 228–334, Oct. 2001.

[32] M. Steinhaus *et al.*, "Transistor count and chip-space estimation of simplescalar based microprocessor models," in *Proc. Complexity-Effective Design*, Jun. 2001.

Power Efficiency for Variation-Tolerant Multicore Processors

James Donald and Margaret Martonosi
Department of Electrical Engineering
Princeton University
Princeton, NJ
{jdonald, mrm}@princeton.edu

ABSTRACT

Challenges in multicore processor design include meeting demands for performance, power, and reliability. The progression towards deep submicron process technologies entails increasing challenges of process variability resulting in timing instabilities and leakage power variation. This work introduces an analytical approach for ensuring timing reliability while meeting the appropriate performance and power demands in spite of process variation. We validate our analytical model using Turandot to simulate an 8-core PowerPC™ processor. We first examine a simplified case of our model on a platform running independent multiprogrammed workloads consisting of all 26 of the SPEC 2000 benchmarks. Our simple model accurately predicts the cutoff point with a mean error less than 0.5 W. Next, we extend our analysis to parallel programming by incorporating Amdahl's Law in our equations. We use this relation to establish limit properties of power-performance for scaling parallel applications, and validate our findings using 8 applications from the SPLASH-2 benchmark suite.

Categories and Subject Descriptors

C.1 [**Processor Architectures**]: Parallel Architectures; C.4 [**Performance of Systems**]: Performance Attributes

General Terms

Performance

Keywords

multicore, variation, power, parallel applications

1. INTRODUCTION

Process variation is an ever-increasing challenge in microprocessor design. Deep submicron technologies pose significant risks for wider spread in timing paths, as well as variations in leakage power. Within a few technology generations, it is expected that within-die variations will become more significant than die-to-die variations [4], and manifest in multicore chips as core-to-core variations [14]. Architects must design these chips with appropriate options to ensure reliability while still meeting appropriate performance and power requirements. This involves fallback modes at the circuit, architectural, and system level.

Some post-silicon circuit techniques can be applied to ensure valid timing, but these entail non-trivial costs in terms of dynamic and leakage power. Our starting parameter is P_{excess}, the amount of excess power on a core resulting from inherent leakage variation or as a side effect of circuit techniques to ensure proper timing. For maximum reliability, two techniques that ensure accurate timing on process-variant cores are adaptive body bias (ABB) and V_{DD} adjustment [34]. ABB refers to toggling an additional voltage between the base and source.

This allows various timing paths to be sped up at the cost of possibly significant leakage power increase. V_{DD} adjustment generally does not increase the leakage as much, but has repercussions on both leakage and dynamic power. Furthermore, these effects on core power occur on top of inherent variations in leakage power, which can vary significantly across cores [14].

Our work takes into account these power discrepancies to formulate policies for post-silicon adaptivity to meet desired performance and power budgets. Modern platforms often have overall goals or configurable power modes that focus on maximizing the ratio of *performance/watt* rather than performance alone. In studying methods toward this goal, our approach is to turn off cores that are particularly expensive in terms of power consumption. For this it is necessary to establish the appropriate optimal tradeoff points. We seek to quantify these cutoffs depending on the level of variation and execution characteristics of the applications. By giving the system knowledge of power variation traits through diagnostics, these bounds can be known or calculated at system runtime and then used to properly tune performance and power to sustain desired user demands.

Since chip multiprocessors are becoming a widespread basis for platforms in the server, desktop, and mobile sectors, we tailor our analysis toward multicore designs. Modern devices now run a wide variety of applications, often concurrently, and must efficiently operate depending on their tasks at hand. We first derive an appropriate performance/power tradeoff point for the case of running 8 programs simultaneously through multiprogramming. We then extend our analysis to parallel programs, since multicore designs have become a major motivation factor toward seeing widespread use of parallel applications in all sectors.

We propose that multicore-based systems can adapt readily to meet power-performance requirements. Specifically, the core power ratings may be identified at the time of system integration, and can even be reconfigured through system diagnostics after power ratings change due to long-term depreciation effects. With knowledge of these variations, the system can choose how to efficiently allocate cores to particular tasks and put cores to sleep if potential additional performance is not power-efficient. Our specific contributions are as follows:

- We derive an analytical bound for estimating the amount of tolerable process variation for multicore policies seeking to maximize *performance/watt*. These excess power cutoff values, ranging from 1 to 6 W, are used to decide when to turn off extra power-consuming cores.

- We introduce PTCMP, a fast multicore simulation environment, and use this to validate our analysis using workloads formed from the SPEC 2000 suite.

- We extend our equations to the problem of parallel programming by incorporating Amdahl's Law, and use the new derived relation to establish limit properties for power-efficient parallel scaling.

- Using extended features of PTCMP we demonstrate these properties using 8 applications from the SPLASH-2 benchmark suite. The high parallel efficiency of raytrace, for example, allows it to increase in performance/power ra-

Permission to make digital or hard copies of all or part of this work for personal or classroom use is granted without fee provided that copies are not made or distributed for profit or commercial advantage and that copies bear this notice and the full citation on the first page. To copy otherwise, to republish, to post on servers or to redistribute to lists, requires prior specific permission and/or a fee.
ISLPED'06, October 4–6, 2006, Tegernsee, Germany.
Copyright 2006 ACM 1-59593-462-6/06/0010 ...$5.00.

Global Design Parameters	
Process Technology	$35nm$
Target Supply Voltage	0.9 V (selectively adjusted for variation)
Clock Rate	2.4 GHz
Organization	8-core, shared L2 cache
Core Configuration	
Reservation Stations	Int queue (2x20), FP queue (2x5), Mem queue (2x20)
Functional Units	2 FXU, 2 FPU, 2 LSU, 1 BXU
Physical Registers	80 GPR, 72 FPR, 60 SPR, 32 CCR
Branch Predictor	16K-entry bimodal, gshare, selector
Memory Hierarchy	
L1 Dcache	32 KB, 2-way, 128 byte blocks, 1-cycle latency, 15-cycle snoop latency
L1 Icache	64 KB, 2-way, 128 byte blocks, 1-cycle latency
L2 cache	8 MB, 4-way LRU, 128 byte blocks, 9-cycle latency
Main Memory	80-cycle latency

Table 1: Design parameters for modeled 8-core CPU.

tio by running on as many as 7 cores although this maximum is easily offset by power variation beyond 0.28 W.

The next section describes our experiment and simulation methodology. Section 3 provides our analytical model and simulations for multiprogrammed workloads, while Section 4 extends our analysis and validation to parallel programs. Section 5 covers related work and Section 6 offers our conclusions.

2. EXPERIMENT METHODOLOGY

2.1 Architectural Model

We use an enhanced version of Turandot [26] and PowerTimer [5] to model performance and power of an 8-core PowerPC^{TM} processor. Our cycle-level simulator also incorporates HotSpot version 2.0 [13, 32] in order to model the temperature-dependence of leakage power from various components. Our process and architectural parameters are given in Table 1.

We assume ABB and adaptive V_{DD} can be applied at the granularity of individual cores, ensuring timing correctness while bringing some cores beyond their normally allowed power specification.

2.2 Simulation Setup

This work introduces *Parallel Turandot CMP* (PTCMP), our cycle-level simulator. Unlike its predecessor Turandot CMP [21], PTCMP is programmed with POSIX threads rather than process forking to achieve lightweight synchronization and parallel speedup. This infrastructure is an alternative to Zauber [22], which also avoids slowdown in the fork-based Turandot CMP, but by using a non-cycle-accurate approximation. Our method maintains all necessary cycle-level communication. PTCMP is able to test various combinations of CMP and SMT configurations without limits on the number of cores that have been vexing for some prior simulators. The maintained cycle-level communication not only aids with accurate modeling of shared cache contention, but is also necessary for other enhancements described below.

Like its predecessors, PTCMP also incorporates online calculations for power and temperature through integration with PowerTimer and HotSpot.

2.3 Benchmarks

We use all 26 benchmarks from the SPEC 2000 benchmark suite [12] to formulate several 8-program workloads. These programs are traced with Aria [26] in their appropriate SimPoint [30] intervals. The SPLASH-2 applications, on the other hand, are traced using the Amber tool on Mac OS X [2] from the beginning to end of their complete algorithm executions.

2.4 Modeling Thread Synchronization and Coherence

There are two main issues in our extensions to Turandot for parallel program simulation: synchronization and coherency.

Fortunately, from an implementation perspective these can be dealt with independently. We implement modeled lock synchronization (not to be confused with the implementation's internal synchronization) and a MESI cache coherence protocol [28] to maintain memory consistency across each core's local cache with respect to the shared L2 cache.

We use Amber's thread synchronization tracing system in order to accurately track the status of pthread-based mutexes and condition variables. Our trace-driven simulator then models stalls for individual threads when such thread-communication dependencies are detected.

For shared memory coherence we implement a MESI [28] cache-coherence protocol to allow private copies of data in each core's local data cache. The data coherence is with respect to the shared L2 cache. We have correspondingly extended PowerTimer to account for the energy cost of cache snoop traffic.

2.5 Metrics

We are most interested in the metric of performance per watt. Since we use SimPoint-generated [30] subsections of the SPEC 2000 suite and some benchmarks may complete in different proportions depending on system properties, we use the weighted speedup metric [33] in order to measure performance. This effectively takes the sum of the programs' executions relative to their baseline single-threaded performance on our processor.

For our parallel program experiments, we use complete executions of 8 benchmarks from the SPLASH-2 suite. Because these are run to completion, in Section 4 we simply use the speedup ratio relative to the performance of a single node.

3. POWER-PERFORMANCE OF MULTIPROGRAMMED WORKLOADS

3.1 Analysis

We seek to maximize the throughput/energy ratio in spite of excess power on cores due to process variation, defined as P_{excess}. This excess power can arise due to inherent leakage variation but also as an after-effect of circuit techniques applied to ensure timing reliability. Specifically, cores that do not meet timing requirements at the time of manufacture can be receive ABB or V_{DD} adjustments, but this may cause these cores to go beyond its specified power limit [34].

ABB, when applied in forward mode, involves placing a positive bias between the body and source. Thus, these two techniques may be best used in combination to ensure timing requirements. This does not increase the dynamic switching power, but increases leakage power significantly more than V_{DD} adjustment [34]. Thus, these techniques may be best used in combination to ensure timing requirements.

For a given timing adjustment the supply voltage must be scaled up roughly linearly, resulting in an approximately linear increase in leakage power and quadratic increase in dynamic power. For cores which already meet their timing requirements with sufficient slack, we may also apply ABB in reverse (known as reverse body bias, RBB) or lower V_{DD} in order to save power. Even in a fortunate scenario where all cores have some timing slack, the degree to which ABB or V_{DD} adjustment can be applied will differ across cores. This combined with inherent leakage variation results in a set of cores on one die with possibly very different power characteristics.

For the purpose of managing these resultant power variations, the metric we aim to maximize is the ratio of *performance/watt*, a current focus for modern server applications [18] and one of the primary concerns for mobile platforms. We have also considered some more complex scenarios such as minimizing power for a fixed performance deadline or maximizing performance for a fixed power budget. These other analysis routes are interesting areas for future study, but for simplicity we have chosen to focus on maximizing the performance/power ratio.

305

Our approach is to find the appropriate cutoff point such that a system may decide to turn a power-hungry core off. There may often be a benefit to retaining an extra power-hungry core, since more running cores can help amortize power cost of dynamic and leakage power from shared resources such as the L2 cache.

We wish to see the appropriate cutoff point for when this core offers enough performance to make its wattage worthwhile, versus when we should put this core into sleep mode and make do with the remaining resources. Our criteria for when a core should be disabled can be stated in terms of an inequality relating the performance/power ratio of N cores to $N-1$ cores, as follows:

$$\frac{perf_N}{power_{N\&excess}} \leq \frac{perf_{N-1}}{power_{N-1}} \quad (1)$$

Here, $perf_N$ and $power_{N\&excess}$ represent the performance of power of the full processor with all N cores used, including the core with excess power. The corresponding $perf_{N-1}$ and $power_{N-1}$ represent those values when the Nth core (sorted from lowest to highest excess power) is turned off. We then expand the condition of Equation 1 as such:

$$\frac{N perf_1 \alpha_N}{N P_{core} + P_{excess} + P_{shar,N}} \leq \frac{(N-1)perf_1 \alpha_{N-1}}{(N-1)P_{core} + P_{shar,N-1}} \quad (2)$$

where N represents the number of active cores, P_{core} denotes average core power, P_{shar} denotes power shared among cores, such as power consumed by the L2 cache, and α represents a speed factor to take resource contention into account. Various elements such as P_{shar} are subscripted to indicate they have a specific value for different core configurations, while others such as P_{core} are taken as constant across different values of N. In fact we assume only a single value of P_{core}, since core power tends to vary significantly less than cache and interconnect power.

α is a factor typically less than 1. It represents the slowdown caused by contention on shared resources, a critical design element of CMPs. If there is little shared cache or memory contention, it becomes likely that $\alpha \approx 1$ [21]. This property does not hold for memory-intensive benchmarks, so we use the much weaker assumption that $\frac{\alpha_N}{\alpha_{N-1}} \approx 1$, which says that the incremental contention from an additional core is reasonably small for moderately sized N.

Solving for P_{excess} under these conditions, this results in:

$$P_{excess} \geq (\frac{N}{N-1})P_{shar,N-1} - P_{shar,N} \quad (3)$$

This equation states our criterion in a relatively simple manner, by depending only on the power cost of shared resources but not the baseline core power nor contention factors. In essence, we plan to turn off any cores for which Equation 3 is true. This is one of our key insights that we utilize and validate.

If the condition of Equation 3 were to be checked and acted upon dynamically, this would require knowing the value of P_{excess}, which is calculated per core relative to the average across all cores, and P_{shar}. Direct power measurement, such as used in Intel's Foxton technology [27], could be done individually on all cores and the shared cache in order to provide the numerical input for these calculations.

While not readily apparent from Equation 3, a general characteristic of the cutoff point is that it increases roughly linearly with respect to the power cost of shared resources. This can be seen more clearly in the simplest case. When P_{shar} does not vary significantly with respect to N, the expression in Equation 3 simplifies down to:

$$P_{excess} \geq \frac{P_{shar}}{N-1} \quad (4)$$

integer-only	bzip2, crafty, eon, gcc, parser, perlbmk, vortex, vpr
FP-only	applu, equake, galgel, mesa, mgrid, sixtrack, swim, wupwise
memory-bound	ammp, applu, art, gap, lucas, mgrid, swim, twolf
CPU-bound	apsi, eon, equake, fma3d, gcc, gzip, mesa, sixtrack
mix1	ammp, crafty, facerec, galgel, mcf, parser, vpr, wupwise
mix2	applu, apsi, eon, gap, gcc, gzip, lucas, twolf

Table 2: Various 8-program workloads consisting of SPEC benchmarks for our experiments.

In this case, the amount of room for power variation increases linearly with P_{shar}. Similarly, with N in the denominator, increasing the number of active cores takes care of amortizing these costs and hence reduces room for large values of P_{excess}. In our validation experiments, we use Equation 3 to account for variation in shared cache power, but our results in Section 3.3 do confirm this insight of amortized shared power costs with varying N.

3.2 Test Workloads

We examine six characteristic workloads for the purpose of confirming our power-performance tradeoff analysis in the multiprogramming case. Our specified workloads are grouped as shown in Table 2. The first four workloads have benchmarks selected for the purposes as listed in the table. The remaining two workloads are formulated to include all remaining SPEC benchmarks.

3.3 Results

Our validation uses the workloads listed in Table 2. However, when a lower power mode is entered by putting one core to sleep, one corresponding benchmark must be removed from the workload. In each case, we remove the median benchmark in terms of core power consumption, in order to ensure its value of P_{core} reasonably matches the average core power. For example in the mix1 workload, ammp is moderate in terms of power consumption and so is absent in the $N = 7$ case.

Modern process technologies create a roughly Gaussian distribution of product bins with nontrivial quantities of processors in bins for poor timing performance when using default voltage settings. Thus, our analysis needs to explore power excesses ranging from zero to as much as 50% in excess of the target core power. Figure 1 shows the performance/power ratio across various values of P_{excess} for $N = 6$ and $N = 8$, respectively. Each arrow represents the change in overall performance/power ratio due to turning off one core. Specifically, the lengths of the arrows represent the magnitude of gains or losses in the ratio due to dropping to a configuration of $N - 1$ cores. The boxes and cross-marks distinguish whether the analytical criterion in Equation 3 is satisfied or not. In these two figures, all non-satisfying cases result in a decrease in performance/power while almost all criterion-satisfied cases result in an improvement. The Y axes shows a relevant subrange for which the ratio rises or falls in this region.

Notably, the observed and predicted safe ranges for P_{excess} are somewhat larger in the 6-program case than the 8-program one. This is an example of the difference in amortized shared power costs as we have described in Section 3.1. The smaller N configuration gives a larger safe range for P_{excess}, and we observe this in most cases of reduced N for all workloads. Furthermore, for a given 8-core processor the probability of having a dangerously large excess power remaining after shutting down 3 cores is unlikely as those cores were already likely to be the most power-hungry ones. The obvious drawback in the mode with less active cores, as shown in Figure 1(a), is that this configuration has a lower overall performance/power ratio. Thus, our most relevant cases lie with using all 8 cores.

For clarity we have performed our tests assuming only a single process-variant core, but our equations can also apply to variance across multiple cores. In the case of multiple variations, P_{core} would refer to the average power of the $N - 1$ lower power cores, while P_{excess} still represents the difference

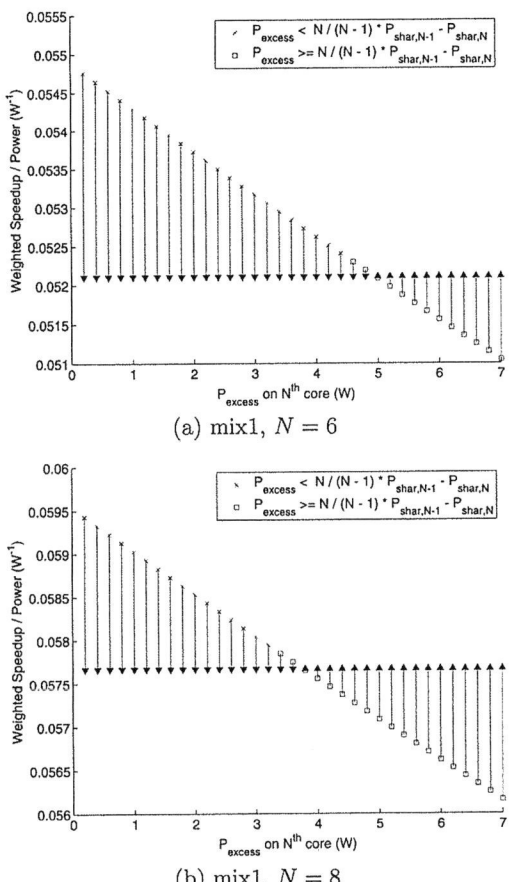

(a) mix1, $N = 6$

(b) mix1, $N = 8$

Figure 1: Performance/power ratio impacts due to sleeping an additional core when running the mix1 workload.

workload ($N = 8$)	P_{excess} cutoff	model agreement
integer-only	1.25 W	−0.04 W
FP-only	3.88 W	+0.95 W
memory-bound	5.61 W	−0.09 W
CPU-bound	3.48 W	−0.01 W
mix1	3.77 W	+0.39 W
mix2	2.06 W	−0.29 W

Table 3: Summary of P_{excess} cutoffs and respective agreement with analytical model for multiprogrammed workloads (all $N = 8$).

with multiple power modes, and feasibly a highest power mode may seek to maximize performance and still have use for the core in some situations, while a lower power mode may aim to maximize performance/watt as has been the goal in our work.

4. POWER-PERFORMANCE OF PARALLEL APPLICATIONS

4.1 Analysis

Our analysis here uses much of the same methodology as in Section 3.1. A key difference is that our speed factor is no longer $\alpha_N N$, which is effectively linear, but rather dictated strongly by Amdahl's Law for parallel computation. The vast topic of parallel computation can certainly entail many complex versions of Amdahl's Law [11, 24, 31], but we choose the most basic form as sufficient for our analysis:

$$speedup_N = \frac{1}{s + \frac{1-s}{N}} \qquad (5)$$

where s represents the fraction of sequential/serial computation that cannot be parallelized, and likewise $(1-s)$ represents that fraction that can be parallelized. The value of s is an important important application characteristic in deciding optimal performance tradeoffs. Using this, we perform an analysis similar to that used in Section 3.1. We evaluate Equation 1 and solve for P_{excess} as follows:

$$\frac{\frac{1}{s+\frac{1-s}{N}} perf_1}{NP_{core} + P_{excess} + P_{shar,N}} \leq \frac{\frac{1}{s+\frac{1-s}{N-1}} perf_1}{(N-1)P_{core} + P_{shar,N-1}} \qquad (6)$$

$$P_{excess} \geq \frac{s + \frac{1-s}{N-1}}{s + \frac{1-s}{N}}[(N-1)P_{core} + P_{shar,N-1}] - NP_{core} - P_{shar,N} \qquad (7)$$

The above relation is more complex than the properties found in Section 3, but we can use it to study several properties of parallel programs. Because this criterion relies heavily on s, which is an empirical constant that varies not only from benchmark to benchmark but even within different configurations for a single benchmark, it cannot used to accurately predict cutoff points as we had done in Section 3. It does, however, have distinct limit properties that give us much insight to the general power behavior of parallel applications.

First, Equation 7 is actually a special case of its analog for the multiprogram analysis. When $s = 0$, this represents that the application is completely parallelizable with no penalty of dependencies or contention. Substituting $s = 0$ into the expression and simplifying yields exactly Equation 3.

On the other hand, the case of $s = 1$ represents the worst case of limited parallel speedup, where a program will not run any faster on multiple cores as compared to just one. If we substitute in $s = 1$, the expression evaluates to $P_{excess} \geq -P_{core}$, which is always true and confirms that in such a specific situation reducing the execution down to a single core will always increase the performance/power ratio. Thus, in order to seek a P_{excess} cutoff that is greater than zero, it helps if $s \ll 1$.

Next, always of interest in parallel programming problems are the limits of scaling up to large values of N. Taking the limit for large N with a nonzero value of s results in the fol-

between the default power consumption of the most consuming core and that average.

We have summarized the cutoff points and our analytical criterion's margin of error for all workloads in Table 3. Error in this sense refers to the small range where our analytical criterion would make an incorrect system decision, such as turning off a core but resulting in reduced performance/power or not disabling a core when it should have. Equation 3 predicts the correct P_{excess} cutoff points with a mean squared error of less than 0.5 W. For perspective, the target design power for the entire processor is around 90 watts, where each core consumes about 8 watts. Total L2 cache and bus power ranges from 10 watts to 40 watts. At low access rates, the shared cache primarily consumes leakage power, while at high cache activity the total L2 power consumption is mainly reflected by the rate of cache accesses.

3.4 Underlying Themes

Our validation experiments have focused on only CPU power, but system designers may wish to include the full power cost of other resources including RAM, chipsets, memory-buffering add-ons, and other essentials. Equation 4 describes the general trend of how including all these elements generally serves to raise the cutoff point, due to better amortizing shared costs. This is one way to confirm general intuition regarding power efficiency of multicore designs.

On the flip side, Equation 4 also denotes an inverse relation between the P_{excess} cutoff and the number of cores. In the future, if multicore designs can feasibly can scale up to many more cores, this increases the chance that one core may become no longer worthwhile to use in the goal of maximizing performance/power. Even so, modern mobile platforms are designed

307

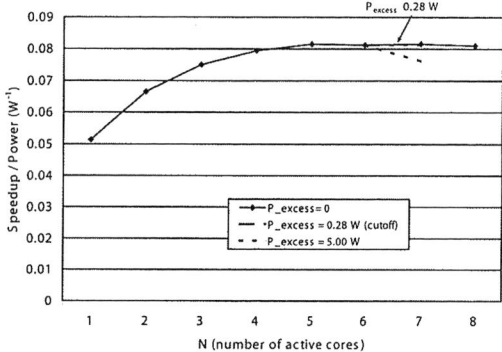

(a) `raytrace`, best performance/power ratio at $N = 7$.

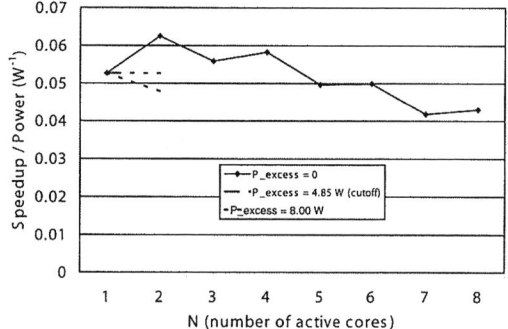

(b) `cholesky`, best performance/power ratio at $N = 2$.

Figure 2: Performance/power for two benchmarks across varying N and varying P_{excess} at the value of N providing the highest performance/power ratio for each benchmark.

lowing suboptimality condition:

$$P_{excess} \geq P_{shar,N-1} - P_{shar,N} - P_{core} \qquad (8)$$

Since typically $P_{shar,N} \geq P_{shar,N-1}$ this relation almost always also evaluates to a negative cutoff value for P_{excess}, meaning that the performance/power ratio only decreases after going beyond some finite N. Intuitively, we would thus expect a plot of this ratio vs increasing N to be a concave-down function that begins to decrease after leveling off. Furthermore, if the performance/power ratio grows only slowly before reaching this peak, we would expect a smaller allowable range for P_{excess}, as compared to the values found in Section 3 where the ratio would continue growing even up through $N = 8$.

4.2 Results

We use 8 of the 12 benchmarks from the SPLASH-2 benchmark suite [36]. Although we have actually conducted experiments with all 12 programs, three of the benchmarks—`ocean`, `fft` and `radix`—are algorithmically restricted from running a number of threads that is not a power of 2. To show clear tradeoffs across the number of cores we have focused only on other benchmarks which do provide this flexibility. Among the remaining 9 benchmarks, `volrend`'s runtimes are an order of magnitude longer than that of other programs, so we have focused on the remaining 8.

We use only the true execution phase of each SPLASH-2 benchmark run for our timing and power measurements. This phase begins after creation of all child threads and ends upon their completion, but does not include any long initialization phases beforehand nor the section of code at the end of each benchmark that generates a summary report.

Figure 2 gives examples of varying performance/power ratios with respect to N and P_{excess}. In each case, the main curve spanning all core counts assumes zero power excess on all cores. The two additional lines represent the change in

application	s	max N	P_{excess} cutoff
barnes	0.039	7	2.02 W
cholesky	0.254	2	4.85 W
fmm	0.066	5	1.41 W
lu	-0.009	8	1.34 W
radiosity	0.083	6	6.04 W
raytrace	0.044	7	0.28 W
water-nsquared	0.025	8	2.03 W
water-spatial	0.019	4	9.15 W

Table 4: Experimental results for SPLASH-2 benchmarks, showing most power-efficient N, cutoff for P_{excess} at that configuration, and each benchmark's corresponding s value.

power efficiency for possible values of P_{excess} at the otherwise optimal performance/watt point. In the first example (`raytrace`), we see a best configuration at $N = 7$, with only a small allowable range of power variation. In the second example (`cholesky`), we see good power efficiency occurring not beyond 2 cores. This is largely in part due to this algorithm's large serial portion [36].

The non-smooth patterns in the `cholesky` graph reveal other notable effects. In particular, core configurations that are not set as a power of 2 each take an additional performance reduction, consequently resulting in poorer performance/power. In fact, most of our benchmarks were found to show some degree of performance preference towards power-of-2 thread counts. This can be explained by non-ideal realities such as cache alignment. Such additional factors affecting performance have a rather direct effect on the performance/power ratio.

One difference between our results here as compared to Section 3 is that limitations are reached at a finite number of cores, as formulated by our analysis in Section 4.1. Our results for all parallel programs are summarized in Table 4. Individual s values used in our calculations for various benchmarks are calculated from the best fit according to speedups obtained through our simulations. However, we have also tested these programs on a real 8-way SMP system to confirm similar respective parallel speedup characteristics.

There are a few interesting cases shown. For one, `lu`'s characteristics best fit a negative value of s, meaning it received super-linear speedup with respect to N in some cases. This is unusual, although not impossible, as many complex effects such as improved cache hit ratios can combine for such a result.

Second, `radiosity` and `water-spatial` have unusually high allowable P_{excess} ranges for their optimal core configurations. The reason for this seems to be due to an interplay of effects that just happen to cause peak performance—possibly more due to negative effects on the adjacent configurations—at these choices of N for these two benchmarks.

Overall, these results confirm much of the intuitive limit behavior specified by our analytical formulation. However, unlike in Section 3, this formulation cannot accurately predict the numerical value of P_{excess} at any given finite point. These deviations from the analytical prediction are due to many application-specific non-ideal factors. A power-efficient multicore system design can take advantage of estimates based on the limit behaviors we have formulated, but an interesting topic for future study would be how a dynamic policy would adjust estimates to take into account application-specific special cases. Such a policy would not only utilize direct measurement of core power, as needed in the multiprogrammed case of Section 3, but also involve performance monitoring to track parallel efficiencies.

5. RELATED WORK

Much prior work has examined power and performance characteristics of multicore architectures when running multiprogrammed workloads [7, 10, 16, 17, 21, 22, 29] as well as parallel applications [3, 8, 15, 19, 20]. To the best of our knowledge, however, ours is the first to examine these problems in the context of process variation.

Although the majority of past research on variation tolerance has been at the circuit and device levels, recently a number of architectural approaches have been proposed. These include variation-tolerant register files [23], caches [1, 25], and pipeline organizations [9, 35]. Furthermore, Humenay et al. propose a model for variations in multicore architectures [14] while Chandra et al. provide a methodology for modeling variations during system-level power analysis [6].

6. CONCLUSIONS

Our work presents a foundation for power-performance optimization in the face of process variation challenges. We have formulated a simple analytical condition relating the shared power costs to predict an optimal cutoff point for turning off extra cores. Using PTCMP to model an 8-core processor, we have shown our model agrees on average within 0.5 W for the basic case of multiprogrammed workloads. P_{excess} cutoff values on our simulated processor range from 1 to 6 W depending on the workloads at hand.

Our analysis has been extended to parallel programming, a more complex problem that is relevant as software in the server, desktop, and mobile sectors all toward more common use of multithreaded applications. We have shown that our equations can be augmented to incorporate Amdahl's Law. Using the parameter s to represent each application's fraction of sequential execution, we have formulated a model to predict limit property trends across a range of parameters and demonstrated these properties on the SPLASH-2 benchmarks.

The purpose for finding these appropriate tradeoff points comes from a system design perspective. If a system is aware of its inter-component dynamic and leakage power excesses, it can make variation-aware decisions for allocating cores in a power-efficient manner. For future work, we intend to combine our on-off mechanism with dynamic voltage and frequency scaling and dynamic ABB to cover a more exhaustive power management design space. In an age when portable devices may execute different types of applications, such techniques are necessary to provide appropriate tradeoffs in performance and power in spite of different application characteristics and process variations.

7. ACKNOWLEDGEMENTS

We thank the Architecture/Performance Group at Apple and Jonathan Chang for their assistance with various tools. We also thank Chris Sadler and the anonymous reviewers for their helpful comments. This work is supported in part by grants from NSF, Intel, SRC, and the C2S2/GSRC joint microarchitecture thrust.

8. REFERENCES

[1] A. Agarwal et al. A Process-Tolerant Cache Architecture for Improved Yield in Nanoscale Technologies. *IEEE Transactions on VLSI Systems*, 13(1), Jan. 2005.

[2] amber(1) manual page. BSD General Commands Manual, Dec. 2005.

[3] M. Annavaram, E. Grochowski, and J. Shen. Mitigating Amdahl's Law Through EPI Throttling. In *ISCA '05: Proc. of the 32nd Intl. Symp. on Computer Architecture*, June 2005.

[4] S. Borkar et al. Parameter Variations and Impact on Circuits and Microarchitecture. In *DAC '03: Proc. of the 40th Design Automation Conf.*, June 2003.

[5] D. Brooks et al. Power-Aware Microarchitecture: Design and Modeling Challenges for Next-Generation Microprocessors. *IEEE Micro*, 20(6):26–44, Nov/Dec. 2000.

[6] S. Chandra et al. Considering Process Variations During System-Level Power Analysis. In *ISLPED: Proc. of the Intl. Symp. on Low Power Electronics and Design*, Oct. 2006.

[7] J. Donald and M. Martonosi. Temperature-Aware Design Issues for SMT and CMP Architectures. In *WCED-5: Proc. of the 5th Wkshp. on Complexity-Effective Design*, June 2004.

[8] M. Ekman and P. Stenstrom. Performance and Power Impact of Issue-width in Chip Multiprocessor Cores. In *ICPP '03: Proc. of the 32nd Intl. Conf. on Parallel Processing*, Oct. 2003.

[9] D. Ernst et al. Razor: A Low-Power Pipeline Based on Circuit-Level Timing Speculation. In *MICRO-36: Proc. of the Intl. Symp. on Microarchitecture*, Dec. 2003.

[10] S. Ghiasi and D. Grunwald. Design Choices for Thermal Control in Dual-Core Processors. In *WCED-5: Proc. of the 5th Wkshp. on Complexity-Effective Design*, June 2004.

[11] J. L. Gustafson. Reevaluating Amdahl's Law. *Communications of the ACM*, 31(5):532–533, May 1988.

[12] J. L. Henning. SPEC CPU2000: Measuring CPU Performance in the New Millennium. *IEEE Computer*, 33(7):28–35, July 2000.

[13] W. Huang et al. Compact Thermal Modeling for Temperature-Aware Design. In *DAC-41: Proc. of 41st Design Automation Conf.*, June 2004.

[14] E. Humenay et al. Impact of Parameter Variations on Multicore Architectures. In *ASGI: Proc. of the First Wkshp. on Architectural Support for Gigascale Integration*, June 2006.

[15] I. Kadayif, M. Kandemir, and U. Sezer. An Integer Linear Programming Based Approach for Parallelizing Applications in On-Chip Multiprocessors. In *DAC '02: Proc. of the 39th Design Automation Conf.*, June 2002.

[16] S. Kaxiras et al. Comparing Power Consumption of an SMT and a CMP DSP for Mobile Phone Workloads. In *CASES '01: Proc. of the 2001 Intl. Conf. on Compilers, Architecture, and Synthesis for Embedded Systems*, Nov. 2001.

[17] R. Kumar et al. Processor Power Reduction via Single-ISA Heterogeneous Multicore Architectures. *Computer Architecture Letters*, Apr. 2003.

[18] J. Laudon. Performance/Watt: The New Server Focus. In *dasCMP '05: Proc. of the Wkshp. on Design, Analysis, and Simulation of Chip Multiprocessors*, Nov. 2005.

[19] J. Li and J. F. Martínez. Power-performance implications of thread-level parallelism on chip multiprocessors. In *ISPASS: Proc. of the 2005 Intl. Symp. on Performance Analysis of Systems and Software*, Mar. 2005.

[20] J. Li and J. F. Martínez. Dynamic Power-Performance Adaptation of Parallel Computation on Chip Multiprocessors. In *HPCA '06: Proc. of the 12th Intl. Symp. on High-Performance Computer Architecture*, Feb. 2006.

[21] Y. Li et al. Performance, Energy, and Thermal Considerations for SMT and CMP Architectures. In *HPCA '05: Proc. of the 11th Intl. Symp. on High-Performance Computer Architecture*, Feb. 2005.

[22] Y. Li et al. CMP Design Space Exploration Subject to Physical Constraints. In *HPCA '06: Proc. of the 12th Intl. Symp. on High-Performance Computer Architecture*, Feb. 2006.

[23] X. Liang and D. Brooks. Latency Adaptation of Multiported Register Files to Mitigate Variations. In *ASGI: Proc. of the First Wkshp. on Architectural Support for Gigascale Integration*, June 2006.

[24] Massively Parallel Technologies. http://www.massivelyparallel.com/, 2006.

[25] K. Meng and R. Joseph. Process Variation Aware Cache Leakage Management. In *ISLPED: Proc. of the Intl. Symp. on Low Power Electronics and Design*, Oct. 2006.

[26] M. Moudgill, J.-D. Wellman, and J. H. Moreno. Environment for PowerPC Microarchitecture Exploration. *IEEE Micro*, 19(3):15–25, May/June 1999.

[27] S. Naffziger. Dynamically Optimized Power Efficiency with Foxton Technology. In *Proc. of Hot Chips 17*, Aug. 2005.

[28] M. Papamarcos and J. H. Patel. A Low-Overhead Coherence Solution for Multiprocessors with Private Cache Memories. In *ISCA '84: Proc. of the 11th Intl. Symp. on Computer Architecture*, June 1984.

[29] M. D. Powell, M. Gomaa, and T. N. Vijaykumar. Heat-and-Run: Leveraging SMT and CMP to Manage Power Density Through the Operating System. In *ASPLOS-XI: Proc. of the 11th Intl. Conf. on Architectural Support for Programming Languages and Operating Systems*, 2004.

[30] T. Sherwood et al. Automatically Characterizing Large Scale Program Behavior. In *ASPLOS-X: Proc. of the 10th Intl. Conf. on Architectural Support for Programming Languages and Operating Systems*, Oct. 2002.

[31] Y. Shi. Reevaluating Amdahl's Law and Gustafson's Law. Computer Sciences Department, Temple University (MS:38-24), Oct. 1996.

[32] K. Skadron et al. Temperature-Aware Microarchitecture. In *ISCA '03: Proc. of the 30th Intl. Symp. on Computer Architecture*, Apr. 2003.

[33] A. Snavely and D. Tullsen. Symbiotic Jobscheduling for a Simultaneous Multithreading Architecture. In *ASPLOS IX: Proc. of the 8th Intl. Conf. on Architectural Support for Programming Languages and Operating Systems*, Nov. 2000.

[34] J. Tschanz et al. Effectiveness of Adaptive Supply Voltage and Body Bias for Reducing Impact of Parameter Variations in Low Power and High Performance Microprocessors. *IEEE Journal of Solid-State Circuits*, 38(5), May 2003.

[35] X. Vera, O. Unsal, and A. González. X-Pipe: An Adaptive Reslient Microarchitecture for Parameter Variations. In *ASGI: Proc. of the First Wkshp. on Architectural Support for Gigascale Integration*, June 2006.

[36] S. C. Woo et al. The SPLASH-2 Programs: Characterization and Methodological Considerations. In *ISCA '95: Proc. of the 22nd Intl. Symp. on Computer Architecture*, June 1995.

Power-Conscious Configuration Cache Structure and Code Mapping for Coarse-Grained Reconfigurable Architecture

Yoonjin Kim[†], Ilhyun Park[†], Kiyoung Choi[†], Yunheung Paek[‡]

[†] Design Automation Laboratory, [‡] Software Optimization & Restructuring Laboratory
School of EECS, Seoul National University, Seoul, South Korea

{awgsize1, pih102, kchoi, ypaek}@snu.ac.kr

ABSTRACT

Coarse-grained reconfigurable architecture aims to achieve both performance and flexibility. However, power consumption is no less important for the reconfigurable architecture to be used as a competitive processing core in embedded systems. In this paper, we show how power is consumed in a typical coarse-grained reconfigurable architecture. Based on the power breakdown data, we suggest a power-conscious configuration cache structure and code mapping technique, which reduce power consumption without performance degradation. Experimental results show that the proposed approach saves much power even with reduced configuration cache size.

Categories and Subject Descriptors

C.1.3[**Processor Architecture**]: Other Architecture Styles – *Adaptive architectures, heterogeneous systems, pipeline processors.*

General Terms

Design, Performance, Experimentation, Verification.

Keywords

System-on-Chip (SoC), Low Power, Coarse-Grained Reconfigurable Architecture (CGRA), Configuration Cache, Loop Pipelining, Context Pipelining, Temporal Mapping, Spatial Mapping.

1. INTRODUCTION

With the growing demand for high quality multimedia, especially over portable media, efficient algorithms for audio and/or video data transfer and processing have been developed. These algorithms have the characteristics of data-intensive computation of high complexity. For such applications, we can consider two extreme approaches to implementation: software running on a general purpose processor and hardware in the form of ASIC. In the case of general purpose processor, it is flexible enough to support various applications but may not provide sufficient performance to cope with the complexity of the applications. In the case of ASIC, we can optimize best in terms of power and performance

but only for a specific application. With a coarse-grained reconfigurable architecture (CGRA), we can take advantage of the two approaches. This architecture has higher performance level than general purpose processor and wider applicability than ASIC.

A typical CGRA consists of an array of processing elements (PEs) and configuration cache memory. Each word (also called context word) of the configuration cache determines the functionality of a processing element and the interconnections to other PEs. By loading the words from the configuration cache into the array, we can dynamically change the configuration of the entire array within just one cycle. If the cache is not big enough to hold all contexts required for an application, then we may have to fetch them from the main memory. Therefore configuration cache size and structure can affect performance and flexibility of CGRA. They also affect power consumption since frequent reconfiguration of CGRA causes many cache-read operations and in turn causes much power consumption. Therefore configuration cache design is important in the trade-off among performance, flexibility, and power.

In this paper, we suggest a novel power-conscious configuration cache structure called hybrid configuration cache, which consists of spatial cache and temporal cache. Spatial cache has a few layers of memory to store contexts for spatial mapping. Each PE has its own spatial cache. Temporal cache has many layers for temporal mapping but only on one column. Other columns are configured through context pipelining. By combining spatial mapping and temporal mapping properly on this structure, we can save much power while keeping the performance.

This paper is organized as follows. After mentioning the related work in Section 2, we describe our reconfigurable architecture and power-breakdown in Section 3. In Section 4, we propose a power-conscious configuration cache structure and code mapping. We show the experimental results in Section 5 and conclude in the last section.

2. RELATED WORK

There have been many researches on CGRA as summarized in [1]. Most of the researches have been carried out in three different aspects: architecture exploration, code compilation and mapping, physical implementation. Power/energy consumption is yet another aspect of CGRA research.

Various architecture exploration flows have been suggested [2][3][4] and all these flows generate a good instance of CGRA considering area cost and performance. In the aspect of energy, interconnect architecture explorations have been suggested for low energy [5][6]. Because CGRA has complex and heavy inter-

---

Permission to make digital or hard copies of all or part of this work for personal or classroom use is granted without fee provided that copies are not made or distributed for profit or commercial advantage and that copies bear this notice and the full citation on the first page. To copy otherwise, or republish, to post on servers or to redistribute to lists, requires prior specific permission and/or a fee.
ISLPED'06, October 4–6, 2006, Tegernsee, Germany.
Copyright 2006 ACM 1-59593-462-6/06/0010...$5.00.

connection for performance and flexibility, power consumption of interconnection is crucial. In [5] the authors have proposed energy-aware interconnection exploration that aims to minimize energy by changing the topology between global register file and function units. However, this exploration only provides the trade-off between performance and energy. In [6] the authors have suggested hierarchical generalized mesh structure exploration that continues to exploit locality while reducing the cost of long connections but it has been only evaluated for specific reconfigurable DSPs. In the case of code compilation and mapping, loops have been focused on mainly for performance [7][8][9]. To the best of our knowledge, however, energy/power aware mapping has not been studied yet.

Many reconfigurable architectures have been implemented physically with various technologies [10][11][12][13][14][15]. Most of the researches in this aspect have focused on efficient design with respect to small area and high performance. In [11][12][15], even though authors have presented power estimation data of the implemented architectures, these are only accessorial results and don't mean power/energy-aware implementation. In [16][17], authors have emphasized that the implemented architectures are power-efficient compared with fine-grained architecture such as FPGA in specific applications but the architectures are not general CGRA but specific types for running some applications with low power.

3. PRELIMINARIES
3.1 Coarse-Grained Reconfigurable Architecture

A typical coarse-grained reconfigurable architecture consists of a microprocessor, a reconfigurable array, and their interface. There are three ways of connecting the reconfigurable array to the processor [18]. First, the array can be connected to a bus as an 'Attached IP' as shown in Figure 1 (a). Secondly, the array can be placed next to the processor as a 'Coprocessor' as shown in Figure 1 (b). In this case, the communication is done using a protocol similar to those used for floating point coprocessors. Finally, the array can be placed inside the processor like a 'FU (Functional Unit)' as shown in Figure 1 (c). In this case, the instruction decoder issues special instructions to perform specific functions on the reconfigurable array as if it were one of the standard functional units of the processor.

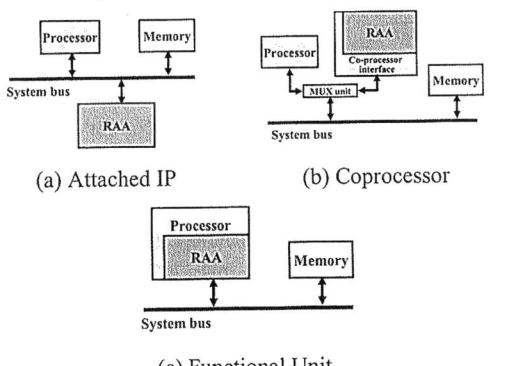

(a) Attached IP   (b) Coprocessor

(c) Functional Unit

Figure 1. Basic types of reconfigurable array coupling [17].

We have implemented the first type of reconfigurable architecture connecting the reconfigurable array as an Attached IP. It consists of a RISC processor, a main memory block, a DMA controller, and a coarse-grained Reconfigurable Core Module (RCM), which is similar to the Morphosys architecture [10]. The communication bus is AMBA AHB [19], which couples the RISC processor and the DMA controller as master devices and the RCM as a slave device. The RISC processor executes control intensive, irregular code segments and the RCM executes data-intensive kernel code segments. The block diagram of the entire reconfigurable architecture is shown in Figure 2.

3.2 Power Breakdown of Coarse-Grained Reconfigurable Architecture

We have implemented the architecture shown in Figure 2 at the RT-level with VHDL [13][14]. It contains 8x5 reconfigurable array of PEs, which we think is big enough for most of the applications considered in our experiments. We have analyzed the power breakdown of the reconfigurable architecture. The architecture has been synthesized using Design Compiler [20] with technology of DongbuAnam [21] 0.18 μm. We have used SRAM Macro Cell library for the frame buffer and configuration cache. ModelSim [22] and PrimePower [20] have been used for gate-level simulation and power estimation. To obtain the power breakdown data, we have used 2D-FDCT as the kernel for simulation with operation frequency of 100MHz and typical case of 1.8V Vdd and 27℃.

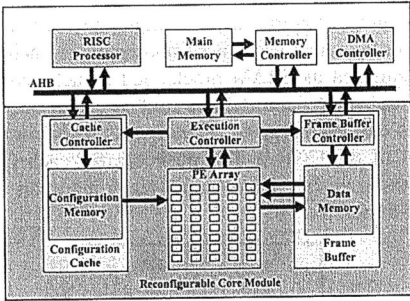

Figure 2. Block diagram of our coarse-grained reconfigurable architecture.

As can be observed from Figure 3, the reconfigurable architecture spends about 48% of total power in the PE array that is composed of many computational logics such as ALU, divider, multiplier, shifter, and register files. The PE array occupies most of the power consumption, which is natural because coarse-grained architecture aims to achieve high performance and flexibility with plenty of resources.

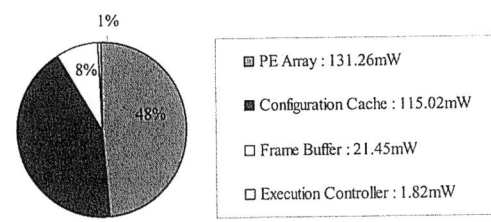

Figure 3. Power breakdown for CGRA running 2D-FDCT.

311

In the case of configuration cache, it spends about 43% of overall power taking the second largest portion. Even though the frame buffer uses the same kind of SRAM as the configuration cache and the size is about the same, it consumes much less power (8%). This is because the configuration cache performs read operations frequently to load the context words, one for each PE, whereas the frame buffer performs load/store operations less frequently to access data on row basis rather than for every PE.

3.3 Configuration Cache and Code Mapping

We can consider two different models for mapping loops onto coarse-grained reconfigurable architecture – SIMD and loop pipelining. SIMD computation model is efficient for computation intensive, data-parallel applications requiring less context words to configure the PE array [10]. Since data load and computation are temporarily separated in this model, array elements are not efficiently utilized. In the case of loop pipelining, different operations in a loop can be executed simultaneously in a pipeline [8]. With this flexibility, data load and computation can be simultaneously executed and all reconfigurable array elements can be efficiently used. In some loops, the performance of pipelining is roughly the same as the performance of SIMD. However, if a loop has frequent memory operations, the pipelining will render much higher performance. Therefore, we apply the loop pipelining technique to the mapping we propose in this paper.

The configuration cache of SIMD model is designed to broadcast the configuration. So PEs in the same row or column share the same context word for SIMD operation [10]. However, in the case of loop pipelining, each PE can be configured by different context word. Our configuration cache is composed of 40 Cache Elements (CEs) implemented as SRAM blocks and a cache controller for controlling each CE (Figure 4). The context register between a PE and a CE is used to keep the cache access path from being the critical path of the CGRA. It is desirable for each CE to have enough layers to enable fast dynamic reconfiguration for various context switching.

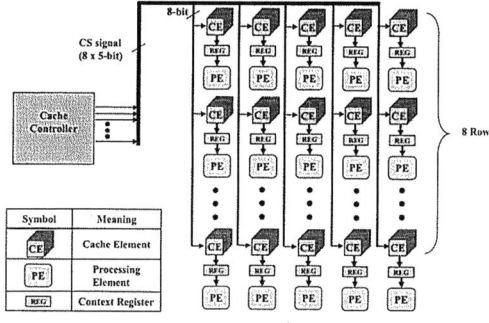

Figure 4. Distributed configuration cache structure.

4. AN APPROACH TO REDUCING POWER IN CONFIGURATION CACHE

As mentioned in the previous section, power consumption in configuration cache is critical. In this section we propose an approach to reducing the power in configuration cache. It achieves its goal by combining power conscious code mapping and a novel cache structure.

4.1 Spatial Mapping and Temporal Mapping

To illustrate the power-conscious code mapping, we assume a general mesh-based coarse-grained reconfigurable array of PEs, where a PE is a basic reconfigurable element composed of an ALU, an array multiplier, shifter, etc. and the configuration of each PE is controlled by a CE. Each row of the array shares read/write-buses. Figure 5 shows the case of 4x4 array with two read buses and one write-bus connecting the array to the frame buffer.

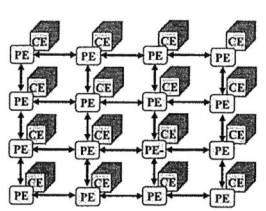

(a) Array organization

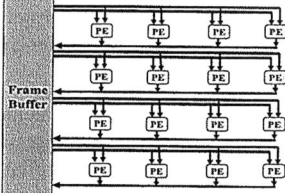

(b) Data bus structure

Figure 5. 4x4 reconfigurable array.

When mapping kernels onto the reconfigurable architecture with loop pipelining, we can consider two mapping techniques: spatial mapping and temporal mapping. Figure 6 shows the difference between spatial mapping and temporal mapping. In the case of spatial mapping, each PE executes a fixed operation with static configuration. This means that loop body is spatially mapped onto the reconfigurable array. The advantage of spatial mapping is that it may not need reconfiguration during execution of a loop because of fixed functionality of each PE. However one disadvantage is that spreading all the operations of the loop body over the limited reconfigurable array may require too many resources. Moreover, data dependencies between the operations should be taken care of by allocating interconnect resources to them and we may have to use PEs as delay registers to make every stage to have the same latency in number of cycles.

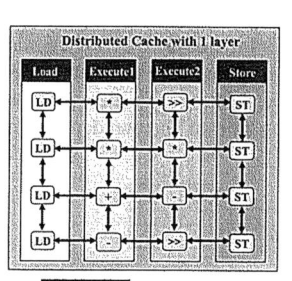
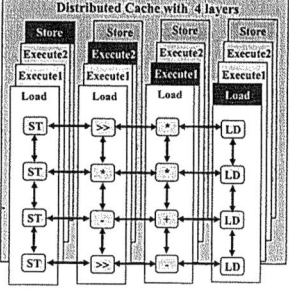

Operation : Operation executed in the current cycle

(a)Spatial mapping   (b) Temporal mapping

Figure 6. Mapping techniques.

Therefore, if the loop is simple enough to map the loop body to the limited reconfigurable array and there is not much data dependency between the operations, the loop can be mapped onto the reconfigurable array with small configuration data. In the case of temporal mapping (Figure 6(b)), a PE executes multiple operations by changing the configuration dynamically within a loop. Therefore, complex loops having many operations with heavy data dependencies can be mapped better in temporal fashion, pro-

312

vided that the configuration cache has sufficient layers to execute the whole loop body.

4.2 Spatial Mapping with Context Reuse

Because most power consumption in the configuration cache is due to memory read-operations, one of the most effective ways to achieve power reduction in the configuration cache is to reduce the frequency of read operations. Even though temporal mapping is more efficient in mapping complex loops onto the reconfigurable array, it requires many configuration cache layers for each PE and performs power consuming read-operation every cycle. On the other hand, spatial mapping can consume less power than temporal mapping because each PE executes a fixed operation within a loop and therefore it is not necessary to read a new context word from the cache every cycle.

As shown in Figure 7, since we have a context word register between a PE and the cache, one read-operation is enough in spatial mapping to configure PEs for static operations throughout the execution of a loop. In summary, spatial mapping with context reuse is more efficient than temporal mapping from the viewpoint of power consumption of configuration cache. However, all kinds of loops cannot be spatially mapped because of the limitation of the spatial mapping. Moreover, if we consider performance, temporal mapping is a better choice for loops having long and complex loop body. In the next subsection, we propose a new cache structure and mapping technique that reduce power consumption while retaining the merits of temporal mapping.

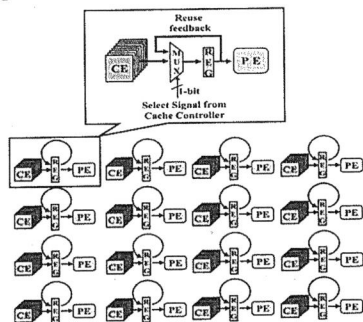

Figure 7. Configuration cache structure for context reuse.

4.3 Temporal Mapping with Context Pipelining

As shown in Figure 6 (b), in temporal mapping with loop pipelining, operations flow column by column from left to right. In Figure 6 (b) for example, the first column executes 'Load' in the first cycle and then in the second cycle, the second column executes 'Load' while the first column executes 'Execute1'. In temporal mapping, there is no need for a PE to have a CE.

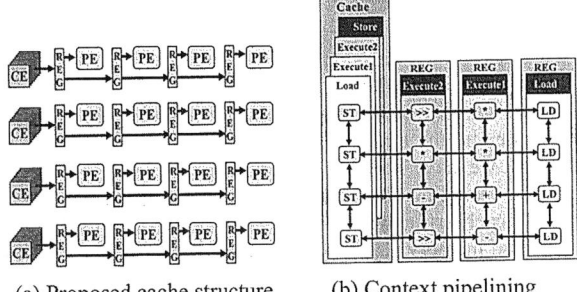

(a) Proposed cache structure (b) Context pipelining
Figure 8. Cache structure for context pipelining.

Instead, the context word can be fetched from the left neighboring column. By organizing a pipelined cache structure as shown in Figure 8, we can propagate the context words column by column through the pipeline. In this way, we can remove most of the CEs from the array, thereby saving power consumption without performance degradation. Compared with the distributed cache structure in Figure 4, the number of cache read operations is reduced to 1/5.

4.4 Hybrid Approach

Because spatial mapping with context reuse and temporal mapping with context pipelining are both power-efficient, we propose power-conscious configuration cache structure that supports both mapping techniques – we call it hybrid configuration cache structure. Figure 9 shows the connection structure for only context reuse and the one for both context reuse and context pipelining. Figure 10 shows the hybrid configuration cache structure. It is composed of cache controller, spatial cache, temporal cache, multiplexer, and de-multiplexer. Cache controller supports the same functions as the previous controller and in addition controls the selection between spatial cache and temporal cache. Compared with the temporal cache, the spatial cache has much less number of layers since it is not necessary to change the context at every cycle. Although the temporal cache has many layers, we keep only one column of cache memory thanks to context pipelining.

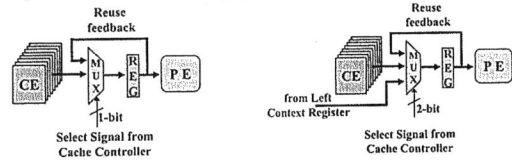

(a) Structure for only context reuse (b) Structure for context reuse and context pipelining.

Figure 9. Connection between CE and PE.

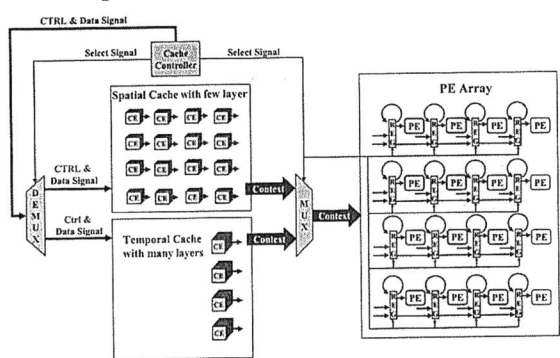

Figure 10. Hybrid configuration cache structure.

Our approach of power-conscious code mapping based on hybrid configuration cache structure exploits both spatial and temporal mapping to save power consumption. When we select spatial mapping, we save power because the number of layers has been reduced a lot. When we select temporal mapping, we save power since many memory read-

313

operations are replaced with register shift operations. The proposed approach also saves cache area since we keep only one column of temporal CEs and much less layers of spatial CEs. As mentioned previously, the approach does not incur any performance degradation.

5. EXPERIMENTS

5.1 Experimental Setup

5.1.1 Hardware design and power estimation

We have implemented reconfigurable architecture with power-conscious configuration cache based on the 8x5 reconfigurable array model mentioned in section 3. The proposed reconfigurable architecture with new configuration cache has been designed at the RT-level with VHDL. The spatial cache of the proposed architecture has 16 layers, which is half the size of the previous one and temporal cache has 32 layers, which is the same as the previous one. The proposed architecture has been synthesized with Design Compiler [20] using technology of DongbuAnam [21] 0.18 μm. ModelSim [22] and PrimePower [20] are used for gate-level simulation and power estimation. Simulation conditions are the same as the previous one mentioned in subsection 3.2 (operation frequency of 100MHz and typical case of 1.8V Vdd and 27℃).

5.1.2 Application kernels

We have applied several kernels of Livermore loops benchmark [23], DSPstone [24], and representative loops in MPEG-4 AAC decoder, H.263 encoder, and H.264 decoder to the previous and proposed architectures. Table 1 shows the mapping feasibility of the kernels. Each circle marking in the table indicates the best mapping technique for the corresponding kernel. The criterion is performance. If the performance of spatial mapping is the same as temporal mapping, the choice is spatial mapping. Hybrid mapping means that both mapping techniques are used. In some case, temporal mapping is necessary for performance and spatial mapping is also indispensable to complete mapping of the kernel, which is classified as hybrid mapping.

Table 1. Mapping feasibility of evaluated kernels

Kernels	Mapping feasibility		
	Spatial	Temporal	Hybrid
*First_Diff(120[†])	O		
*Tri-Diagonal(60[†])	O		
*Hydro(20[†])	O		
*ICCG(30[†])		O	
**Dot_Product(40[†])	O		
**24-Taps FIR(5[†])		O	
Complex_Mult in MPEG-4 AAC dec.(30[†])	O		
ITRANS in H.264 dec.			O
2D-FDCT in H.263 enc.		O	
SAD in H.263 enc			O
Matrix(10x8)-Vector(8x1) Multiplication(MVM)		O	

* Livermore loop benchmark suite, ** DSPstone benchmark suite,
[†] : Iteration number

5.2 Results

To demonstrate the effectiveness of our approach, we have compared the previous and proposed architectures for spatial and temporal mapping. Table 2 shows comparison of power consumption between the two mapping models. The kernels in Table 2 have been selected based on feasibility of spatial techniques. Compared to the previous architecture, we have saved up to 71.3% of the total power consumed in the configuration cache and 28% of that in the entire architecture using only temporal mapping. We have saved more power using spatial mapping. For the configuration cache and the entire architecture we have saved up to 83.7% and 32.9% respectively. We see from these experiments that we should use spatial mapping whenever it is feasible for the kernel.

Table 2. Power comparison between mapping techniques

Kernels	Prev Spatial Power (mW)	Prop					
		Temporal			Spatial		
		Power (mW)	Reduced(%) Cache	Entire	Power (mW)	Reduced(%) Cache	Entire
First_Diff	360.63	259.51	71.3	28.0	241.85	83.7	32.9
Tri-Diagonal	340.05	257.95	67.7	24.1	240.20	81.9	29.4
Dot_Product	321.05	253.78	65.9	21.0	244.90	74.7	23.7
Complex_Mult	371.13	285.75	67.8	23.0	269.13	81.9	27.5
Hydro	295.20	225.04	65.9	23.8	211.10	78.2	28.5

Table 3. Power measurement of temporal and hybrid mapping

Kernels	Power(mW)		Reduced(%)	
	Prev	Prop	Cache	Entire
ICCG	326.73	241.05	67.4	26.2
24-Taps FIR	330.15	231.56	69.2	29.9
MVM	309.54	226.95	68.0	26.7
ITRANS	246.88	168.17	61.4	31.9
2D-FDCT	269.55	196.12	64.5	27.2
SAD	237.86	181.99	47.6	23.5

Power: Overall power of the architecture
Prev: Previous architecture, **Prop:** Proposed architecture
Reduced Cache: Power reduction ratio compared with previous architecture in configuration cache
Reduced Entire: power reduction ratio compared with previous architecture in entire architecture

Table 4. Memory size evaluation

Architecture	Memory Element(Byte) SRAM in config' cache	Reduced(%)
Previous	5120	0
Proposed	3584	30.0

Table 3 shows comparison of power consumption between two architectures. The kernels in Table 3 have been selected based on the feasibility of temporal or hybrid mapping. Compared to the previous architecture, we have reduced the power by up to 69.2% for the configuration cache and 31.9% for the entire architecture using temporal and hybrid mapping.

Table 4 shows memory size evaluation between the previous architecture and the proposed one. Compared to the previous architecture, we have reduced the memory by up to 30%. This means that reconfigurable architecture with power-conscious configuration cache structure is more efficient than previous one in the aspect of memory size as well as power saving.

314

6. CONCLUSION

Coarse-grained reconfigurable architectures are considered to be appropriate for embedded systems because it can satisfy both flexibility and high performance. However, power consumption is also crucial for the reconfigurable architecture to be used as a competitive processing core in embedded systems. Most reconfigurable architectures have a configuration cache for dynamic reconfiguration, which consumes much power. In this paper we suggest a novel hybrid configuration cache structure that combines spatial cache and temporal cache. For spatial mapping, each PE in the reconfigurable architecture has a spatial cache with a few layers. For temporal mapping, only the PEs on the first column have temporal caches with many layers. Other columns are configured through context pipelining. Our architecture can be used to achieve much power-saving in a reconfigurable architecture while maintaining the same performance as the previous approach.

In the experiments, we show that our approach saves much power even with reduced configuration cache size. In the case of spatial mapping, power reduction ratios in the configuration cache and the entire architecture are up to 83.7% and 32.9% respectively compared with previous architecture. In the case of temporal and hybrid mapping, we have reduced the power by up to 69.2% and 31.9% respectively.

7. ACKNOWLEDGEMENTS

This work is supported by IT R&D Project funded by Korean Ministry of Information and Communications and by the IC Design Education Center (IDEC).

8. REFERENCES

[1] Reiner Hartenstein, "A decade of reconfigurable computing: a visionary retrospective," in *Proc. of Design Automation and Test in Europe Conf.*, pp. 642-649, Mar. 2001.

[2] Reiner Hartenstein, M. Herz, T. Hoffmann, and U. Nageldinger, "KressArray Xplorer: a new CAD environment to optimize reconfigurable datapath array architectures," in *Proc. of Asia and South Pacific Design Automation Conf.*, pp. 163-168, Jan. 2000.

[3] Bingfeng Mei, Serge Vernalde, Diederik Verkest, and Rudy Lauwereins, "Design methodology for a tightly coupled VLIW/reconfigurable matrix architecture: a case study," in *Proc. of Design Automation and Test in Europe Conf.*, pp. 1224-1229, Mar. 2004.

[4] Nikhil Bansal, Sumit Gupta, Nikil D. Dutt, and Alex Nicolau, "Analysis of the performance of coarse-grain reconfigurable architectures with different processing element configurations," in *Proc. of Workshop on Application Specific Processors*, Dec. 2003.

[5] Andy Lambrechts, Praveen Raghavan, Murali Jayapala, "Energy-Aware Interconnect-Exploration of coarse-grained reconfigurable processors," in *Proc. of Workshop on Application Specific Processors*, Sept. 2005.

[6] Hui Zhang, Marlene Wan, Varghese George, and Jan Rabaey, "Interconnect architecture exploration for low-energy reconfigurable single-chip DSPs," in *Proc. of VLSI' 99*, April 1999.

[7] Frank Hannig, Hritam Dutta, and Jurgen Teich, "Mapping of regular nested loop programs to coarse-grained reconfigurable arrays – constraints and methodology," in *Proc. of EEE*

Int. Parallel & Distributed Processing Symp., p. 148a, April 2004.

[8] Jong-eun Lee, Kiyoung Choi, and Nikil D. Dutt, "Mapping loops on coarse-grained reconfigurable architectures using memory operation sharing," in Technical Report 02-34, Center for Embedded Computer Systems(CECS), Univ. of California Irvine, Calif., 2002.

[9] Minwook Ahn, Jonghee W. Yoon, Yunheung Paek, Yoonjin Kim, Mary Kiemb, and Kiyoung Choi, "A spatial mapping algorithm for heterogeneous coarse-grained reconfigurable architectures," *Proc. of Design Automation and Test in Europe Conf.*, Mar. 2006.

[10] Hartej Singh, Ming-Hau Lee, Guangming Lu, Fadi J. Kurdahi, Nader Bagherzadeh, and Eliseu M. Chaves Filho, "MorphoSys: an integrated reconfigurable system for data-parallel and computation-intensive ap-plications," *IEEE Trans. on Computers*, vol. 49, no. 5, pp. 465-481, May 2000.

[11] Jurgen Becker and Martin Vorbach, "Architecture, memory and interface technology integration of an indutrial/academic configurable sys-tem-on-chip (CSoC)," in *Proc. of IEEE Computer Society Annual Symp. on VLSI*, 2003.

[12] Francisco-Javier Vererdas, Michael Scheppler, Will Moffat, and Bingfeng Mei, "Custom implementation of the coarse-grained reconfigurable ADRES architecture for multimedia purposes," in *Proc. of Int. Conf. on Field Programmable Logic and Applications*, pp. 106-111, Aug. 2005.

[13] Yoonjin Kim, Chulsoo Park, Shinwon Kang, Hyunjik Song, Jinyong Jung, and Kiyoung Choi, "Design and evaluation of coarse-grained reconfigurable architecture," in *Proc. of Int. SoC Design Conf.*, pp. 227-230, Oct. 2004.

[14] Yoonjin Kim, Mary Kiemb, Chulsoo Park, Jinyong Jung, and Kiyoung Choi, " Resource sharing and pipelining in coarse-grained reconfigurable architecture for domain-specific optimization," in *Proc. of Design Automation and Test in Europe Conf.*, pp 12-17, Mar. 2005.

[15] Marco Lanuzza, Martin Margala, and Pasquale Corsonello, "Cost-effective low-power processor-in-memory-based reconfigurable datapath for multimedia applications," in *Proc. of Int. Symp. on Low Power Electronics and Design*, pp. 161-166, Aug. 2005.

[16] Sami Khawam, Tughrul Arslan, and Fred Westall, "Synthesizable reconfigurable array targeting distributed arithmetic for system-on-chip applications," in *Proc. of IEEE Int. Parallel & Distributed Processing Symp.*, p. 150, April 2004.

[17] Francisco Barat, Murali Jayapala, Tom Vander Aa Henk Corporaal, Geert Deconinck, and Rudy Lauwereins, "Low power coarse-grained reconfigurable instruction set processor," in *Proc. of Int. Conf. on Field Programmable Logic and Applications*, pp. 230-239, Sept. 2003.

[18] Francisco Barat and Rudy Lauwereins, "Reconfigurable instruction set processors: a survey," in *Proc. of Int. Workshop on Rapid System Prototyping*, pp. 168-173, April 2000.

[19] ARM Corp. : http://www.arm.com/arm/AMBA

[20] Synopsys Corp. : http://www.synopsys.com

[21] DongbuAnam Semiconductor : http://www.dsemi.com

[22] Model Technology Corp. : http://www.model.com

[23] http://www.netlib.org/benchmark/livermorec

[24] http://www.ert.rwth-aachen.de/Projekte/Tools/DSPSTONE

Dynamic Thermal Management for MPEG-2 Decoding

Wonbok Lee
Department of Electrical Engineering
University of Southern California
Los Angeles CA 90089
(213) 821-4206
wonbokle@usc.edu

Kimish Patel
Department of Electrical Engineering
University of Southern California
Los Angeles CA 90089
(213) 821-4206
kimishpa@usc.edu

Massoud Pedram
Department of Electrical Engineering
University of Southern California
Los Angeles CA 90089
(213)-740-4458
pedram@usc.edu

ABSTRACT

In this paper, we propose an effective dynamic thermal management (DTM) scheme for MPEG-2 decoding by allowing some degree of spatiotemporal quality degradation. Given a target MPEG-2 decoding time, we dynamically select either an intra-frame spatial degradation or an inter-frame temporal degradation strategy in order to make sure that the microprocessor chip will continue to stay in a thermally safe state of operation, albeit with certain amount of image/video quality loss. For our experiments, we use the MPEG-2 decoder program of MediaBench and modify/combine Wattch and HotSpot for the power and thermal simulations and measurements, respectively. Our experimental results show that we achieve thermally safe state with spatial quality degradation of 0.12 Root Mean Square Error (RMSE) and with frame drop rate of 12.5% on average.

Categories and Subject Descriptors

B.7.2 [Hardware]: Design Aids

General Terms: Design, Reliability

Keywords

Thermal model, temperature-aware design, MPEG-2 decoding.

1. INTRODUCTION

Peak power dissipation and resulting temperature rise have become the dominant limiting factor to processor performance and a significant component of its cost. Expensive packaging and heat removal solutions are needed to achieve acceptable substrate and interconnect temperatures in high-performance microprocessors. The heat flux in state-of-the-art microprocessors chips is currently in the range of 10-20 W/cm^2, which is al-ready exceeding the confines of air cooling. Current thermal solutions are designed to limit the peak processor power dissipation to ensure its reliable operation under worst-case scenarios. However, the peak processor power and ensuing peak temperature are hardly ever observed. Dynamic thermal management (DTM) has been proposed as a class of micro-architectural solutions and software strategies to achieve the highest processor performance under a peak temperature limit. Furthermore, it is known that power density across the chip is non-uniform, resulting in localized hot spots. DTM solutions must address this phenomenon as much as they tackle system-wide temperature violations. When

Permission to make digital or hard copies of all or part of this work for personal or classroom use is granted without fee provided that copies are not made or distributed for profit or commercial advantage and that copies bear this notice and the full citation on the first page. To copy otherwise, or republish, to post on servers or to redistribute to lists, requires prior specific permission and/or a fee.
ISLPED '06, October 4-6, 2006, Tegernsee, Germany.
Copyright 2006 ACM 1-59593-462-6/06/0010...$5

the chip approaches the thermal limit, a DTM controller initiates hardware reconfiguration, slow-down, or shutdown to lower the chip temperature.

Traditionally, thermal issues within a chip have been handled at the package level. Chip manufacturers have devised sophisticated, albeit expensive, packaging and cooling assemblies, i.e., heat sinks and micro-fluidic conduits, to the processor chips so as to efficiently transfer heat generated within a chip to the ambient environment. However, packaging and cooling systems without knowledge about the resource utilization and power dissipation demands of a software program running on a micro-processor chip have some major limitations. As such, micro-architecture level solutions can result in changes to the dynamic temperature profile of a chip so as to avoid the worst-case power density and temperature conditions.

In order to reduce chances of creating a major thermal problem at the architectural level, a number of DTM strategies have been proposed [1]-[9]. These techniques rely on two pre-defined levels of thermal limits: a trigger temperature and an emergency temperature. The trigger temperature is a thermal limit over which a dynamic predictive/reactive thermal management schemes will be initiated whereas the emergency temperature is a thermal limit over which the chip may be damaged and hence must be avoided at all cost. Those DTM schemes include architectural adaptations such as fetch-toggling [1] (instruction fetching is stalled for next N cycles), instruction cache throttling [2] (throttle the instruction forwarding from the instruction cache to the instruction buffer), activity migration [3] (dispatching computations to different locations on the die) and dynamic voltage and frequency scaling [4] (DVFS). Those DTM schemes are application-independent schemes. On the contrary, in this paper we propose an application-specific DTM technique, specifically designed for an MPEG-2 decoding program running on a general purpose microprocessor chip.

As computers become faster, absolute decoding time of a frame in MPEG-2 video stream becomes smaller. However, MPEG-2 standard prescribes a fixed frame rate of 29.97framess/sec (NTSC) and 25frames/sec (PAL) [10]. The frame rate is determined in consideration of slow trace/pursuit nature of human visual system (HVS). A frame rate higher than this 25-30 does not effectively improve the perceived quality of image/video streams to human eyes. Hence this, frame rate and processor speed dependent, available residual time from the given frame decoding time can be used to achieve thermally safe state of the processor.

This paper is organized as follows. In section 2, current state-of-the-art in DTM is reviewed. Section 3 introduces the motivational example of our DTM for MPEG-2 decoding while section 4 covers the theoretical parts of our DTM scheme. Our simulation environment, benchmark programs and implementation details will be followed in section 5. In section 6, we show the experimental results of our work and section 7 is the conclusions.

2. PRIOR WORK

Many of the requisite performance features of microprocessor such as real-time processing and mean-time to failure are significantly affected by the power dissipation and resulting temperature. Hence, dynamic thermal management (DTM) has been proposed as a class of micro-architectural solutions and software strategies to achieve the highest processor performance under a peak temperature limit. To this end, there have been a number of studies that address this problem as explained in the following discussion.

Recently, a number of architectural adaptations based DTM solutions have been proposed. In [1]-[3], the authors propose fetch toggling, instruction cache throttling, and activity migration, respectively. In [4], the authors consider instruction window resizing and switching among active functional units as DTM techniques for multimedia applications. In [6]-[8], Skadron et al. introduce several DTM methods. In [6], they introduce temperature-tracking based frequency scaling, localized toggling and computation migration to spare hardware units. In [7], they propose a hybrid DTM technique that combines fetch gating and DVS. In [8], they propose a formal feedback control theory and use DTM as a test vehicle. DTM is invoked in response to the localized hotspot rather than chip-wide temperature.

In [5], Mircea et al. propose HotSpot which is an accurate yet fast thermal model based on thermal resistance and thermal capacitance at micro-architectural functional block level as well as two dimensional grid levels. Their model is implemented and widely used as a popular thermal simulator [11] . In [9], Lee et al. propose a software solution for temperature sensing that utilizes the built-in performance monitoring unit (PMU) to generate performance related information such as I/D cache access, number of instructions, number of stalls, etc. They derive temperature behavior by associating these performance figures with power.

Many power management (PM) schemes for MPEG have been proposed. In [12], Son et al. propose a dynamic voltage scaling (DVS) on MPEG decoding. Basically, they apply two dynamic voltage scaling schemes on MPEG decoding. One is based on delay and drop rate minimization algorithm and the other is based on predictive (per Group of Picture, GOP) decoding time algorithm. Their delay and drop rate minimization algorithm regulates the system voltage depending on the system clock speed and the current decoding status. The proposed algorithm assumes MPEG decoding in a low-performance machine in which the frame rate is less than 30frame/second.

In [13][14], Choi et al. propose an off-chip latency driven dynamic voltage and frequency scaling (DVFS) for MPEG decoding. In designing their DVFS strategy, the authors utilize the frame dependent versus frame independent parts within MPEG decoding process in [13] and the on-chip versus off-chip (CPU versus memory) dependent workloads within the frame decoding process in [14]. Their schemes are effective in slow machine which has a frame rate of 10 ~ 15.

All of the aforementioned schemes make use of the available, frame rate dependent, slack time to employ various low power strategies, but none of them make use of it for DTM. Since CPU speed and computational power are increasing rapidly, we have more slack time available during the decoding of an MPEG frame (with a fixed deadline of say 33ms). In this paper we propose to gather and distribute this available slack time to achieve thermally safe state of the microprocessor chip during MPEG-2 decoding. The safe thermal state comes at the expense of image/video quality.

Table 1. Baseline Configuration of Simulated Processor

Main Memory Latency	100 cycles/10 cycles
L1 I/D Cache	64KB 2-way 32Byte block
	1 cycle hit latency
I/D-TLB	Fully associate, 128 entries
	30 cycles miss latency
Branch Predictor	4K Bimodal
Functional Units	4 IntegerALU,1 IntegerMULT/DIV
	2 FP ALU. 1 FP MULT/DIV
RUU/LSQ size	64/32
Instruction Fetch Queue	8
In order Issue	False
Wrong Path Execution	True
Issue Width	6 instruction per cycles

Table 1 summarizes the architectural parameters that we use in our simulation.

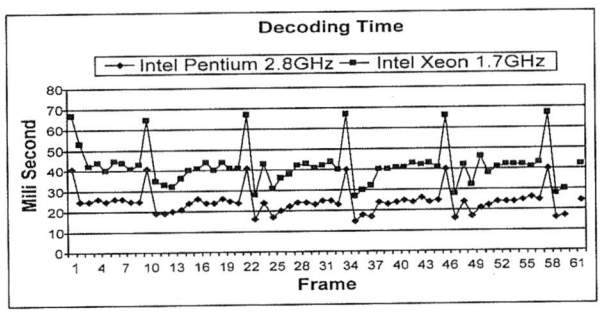

Figure 1 Actual MPEG-2 Decoding Time

3. MOTIVATIONAL OBSERVATION

Figure 1 reports per-frame decoding time variation of a MPEG-2 video stream decoded with MediaBench MPEG-2 decoder program [15]. The video stream has 60 frames and 704x480 resolutions. We run the decoder with two different machines: An Intel Xeon 1.7GHz and an Intel Pentium IV 2.8GHz. The OS is Linux-2.6.15. Note that the actual decoding time varies depending on the types of frame (I, P and B). Since I-frame is computation intensive compared to the other two frame types, its decoding time is longer than the others. In each machine, the average decoding times are 42.01msec and 24.01msec, respectively. Since the MPEG-2 standard specifies its frame rate as 29.97fps (which corresponds to approximately 33msec/frame), we clearly cannot finish the decoding of a frame with this frame rate in Intel Xeon chip. For this reason, MPEG standard has a frame discarding scheme [16] whereby it can drop B frame, P frame and I frame in stepwise manner, depending on the machine's clock speed.

In Pentium IV, on the contrary, the actual decoding time takes less than 33msec/frame. For such a case, MPEG standard has its frame rate control scheme that waits for some time to display the frame in a regular interval. For example, Berkeley MPEG-1 [17] uses 'select' function call to slow down displaying frames in a fixed rate of 29.97fps. Since the current state-of-the-art processors are much faster, it is expected that more residual time will be available within the allowed frame decoding time.

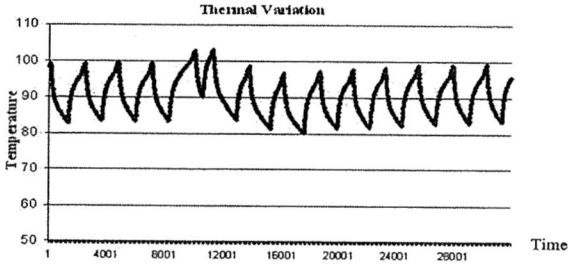

Figure 2 Thermal Variations & Violations in the Simulator

In Figure 2, we simulate this ever-decreasing actual decoding time in our simulator. (Section 5 will explain the simulator in detail) Simply speaking, we assume some fixed number of cycles that correspond to the given decoding time deadline (33msec). If the actual decoding finishes earlier than these many cycles (deadline), we stall the processor inside the simulator for the rest of the cycles until the deadline is reached and only then, we start the decoding of the next frame. The corresponding results are shown in Figure 2 where the X-axis plots the simulation cycles in 10K granularity and the Y-axis plots the temperature in Celsius. As shown in the figure, the temperature starts to decrease when the actual decoding finishes any time earlier than the given frame decoding time. Note that the peak temperature goes up to 103°C, which can invoke logical or timing error in the chip.

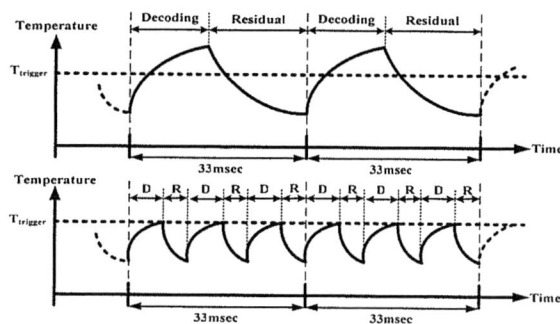

Figure 3 Overview of our DTM Scheme

In order to avoid thermal violation (in Figure 2), we propose a new DTM scheme, especially for the MPEG-2 decoding. Figure 3 shows the basic idea behind: Given a deadline for frame decoding, the conventional MPEG-2 decoder uses the first part of the decoding time to finish the decoding task while it rests in the second part. Unfortunately, the trigger temperature/emergency temperature of the chip may be exceeded in the first part. In our strategy, short periods of decoding are interleaved with short

periods of processor stalls so that the chip temperature never exceeds the trigger temperature, yet the decoding task is completed before the deadline.

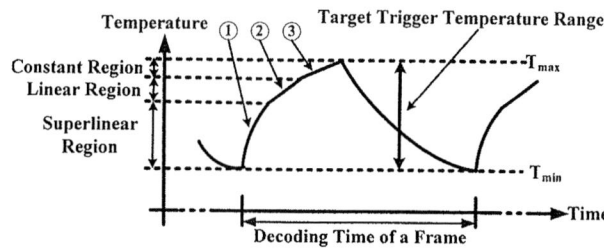

Figure 4 Detailed Analysis of Temperature Variation

4. DTM METHODOLOGIES

4.1. Thermal Models

We use a thermal model developed by Skadron et al. in [8]. In this model, the temperature increase in the chip is represented by:

$$\Delta T = (\frac{P}{C_{th}} - \frac{T_{old}}{R_{th} \cdot C_{th}}) \cdot \Delta t \qquad (1)$$

where Δt is a time interval, P is the average power dissipated in the interval, R_{th} is a thermal resistance, C_{th} is a thermal capacitance and T_{old} is the initial temperature, respectively. After a time interval, the new temperature is:

$$T_{new} = T_{old} + \Delta T \qquad (2)$$

Let $t_{initial}$ and t_{final} denote two instances of time (and their difference denoted by Δt), respectively. Moreover, assume that the power dissipation is non-zero when the processor is running. Now, the thermal rising gradient with respect to time is calculated as:

$$\text{Rising: } \frac{\Delta T_r}{\Delta t} = (\frac{P}{C_{th}} - \frac{T_{old}}{R_{th} C_{th}}) \qquad (3)$$

In contrast, when the processor is stalled (it is put in the standby mode) during the residual time, assume that the chip power dissipation is negligible compared to active power i.e., $P=0$. Then, the thermal falling gradient with respect to time is calculated as:

$$\text{Falling: } \frac{\Delta T_f}{\Delta t} = (-\frac{T_{old}}{R_{th} C_{th}}) \qquad (4)$$

In Figure 4, we model this thermal gradient over time as three piecewise linear functions. Since the amount of decoding workloads/steps within a MPEG-2 frame decoding is frame dependent yet more or less the same, we specify the temperature variation during MPEG-2 frame decoding on a DTM-ignorant machine as T_{max} and T_{min} peaks and valleys, respectively. Notice that T_{max} and T_{min} are mostly invariant when a program is in a steady-state but may slightly vary when a program behavior changes. Moreover, obtaining T_{max} and T_{min} is not always possible in the actual system but is always feasible in the thermal simulator, which is aimed at anticipating application's real on-chip thermal behavior. Then, the thermal behaviors of MPEG-2 decoding program are divided into three regions:

1. **Super-linear Region:** $\frac{P}{C_{th}} \gg \frac{T_{old}}{R_{th} C_{th}}$ In this region,

    $\frac{\Delta T_r}{\Delta t}$ changes dramatically and the power term

    dominates the other temperature term in equation 1. Since the thermal gradient during the rising of the curve

318

is higher than falling counterpart, a longer processor stall time is needed compared to the time it took for the temperature to rise to the same level.

2. **Linear Region:** $\dfrac{P}{C_{th}} > \dfrac{T_{old}}{R_{th}C_{th}}$ In this region, $\dfrac{\Delta T_r}{\Delta t}$ changes almost linearly and the power term is relatively larger than the temperature term in equation 1. The thermal gradients during the rising and the falling of the curve are comparable and both take almost same amount of time.

3. **Constant Region:** $\dfrac{P}{C_{th}} \approx \dfrac{T_{old}}{R_{th}C_{th}}$ In this region, $\dfrac{\Delta T_r}{\Delta t}$ becomes almost zero and the power term is comparable to the other temperature term in equation 1. Since the thermal gradient during the falling of the curve is higher than the rising counterpart, a shorter processor stall time is needed compared to the time it took for the temperature to rise to the same level.

Unfortunately, those three regions are not sharply bounded in the thermal gradient curve. Moreover, the trigger temperature is a material/architectural parameters dependent value whereas T_{max}/T_{min} is MPEG-2 input file dependent one, the trigger temperature can be located at any level within the thermal gradient curve. Hence, we carry out the following steps in order to build our DTM framework for MPEG-2 decoding.

1) Run a MPEG-2 video stream in MPEG-2 decoder program and get both T_{max} and T_{min} on the machine without any DTM policy applied.

2) Check $T_{trigger}$ of the processor. If $T_{trigger} > T_{max}$, machine is thermally safe and no DTM policy is needed to be applied.

3) If $T_{trigger} < T_{min}$, this means decoding workload is large and DTM policy must do significant quality degradation to achieve thermally safe state.

4) If $T_{min} < T_{trigger} < T_{max}$, which we show as a target trigger temperature range in Figure 4, then, if $T_{trigger}$ lies in the constant region, thermally safe state can be achieved with little or no quality degradation, whereas if it lies in the linear region some quality degradation must be accepted to achieve thermally safe state. Finally, if $T_{trigger}$ lies in the super-linear region, which is the worst case, then thermally safe state can be achieved only at the cost of significant image/video quality degradation.

4.2. DTM Policy

During the residual time, as shown in Figure 3 (top figure), the thermal gradient is high during its initial phase and slowly decreases afterwards. Even though significant amount of time is spent in 'stalling the processor', i.e. doing nothing but waiting for the arrival of frame display time, the drop in the temperature is relatively small. This leads us to the idea depicted in Figure 3 (bottom figure): Stall the processor only for the duration of time till the thermal falling gradient remains steep.

From the analysis presented in the previous section, high thermal falling gradient occurs only when the chip operates in either the linear or the constant regions. Our experiments with a set of input files with a trigger temperature of 81.8°C show that the trigger temperature is positioned in the 'linear region' where thermal rising/falling gradients are comparable. Based on this simulation results and extensive experiments, we choose an empirical value of 1 million cycles to stall the processor every time we reach the triggering temperature. This gives us comparable rising and thermal falling gradients

Note we may end up missing deadlines for frames unless we do have enough residual time. In order to avoid these deadline misses, we collect this slack time (actual decoding time subtracted from given decoding time) for the future use. This slack time saving is accomplished by having a buffer in the main memory which has the size of 3 frames. Every time we finish actual decoding before the given decoding time and the buffer has space for frames, we write the decoded frame to the buffer and claim the remaining slack time for the future use. If the buffer does not have space, we wait for the buffer to have a space. If we either miss or predict to miss the deadline for the frame being decoded, we resort to either spatial or temporal quality degradations to meet the deadlines. This deadline satisfaction and slack time collection continue over the whole execution time.

Spatial quality degradation: After enough amount of frame decoding time, e.g. around 100msec, thermal behavior of MPEG-2 decoding program becomes monotonous. Hence in decoding of all subsequent frames, we can predict how many times the thermal curve will reach to the trigger temperature for the following frame decoding. Since we allow stalling 1 million cycles every time we reach the trigger temperature, the total number of stall cycles can be easily predicted. If the predicted stall cycles are larger than the available slack time that we collected, deadline miss is expected hence activation of spatial quality degradation (as explained latter) is triggered.

Temporal quality degradation: Note that spatial quality degradation does not guarantee the deadline satisfaction, i.e., we may run out of slack times to meet the deadline. If deadline miss occurs, we will drop the next frame. The rationale is that this frame has already missed its deadline and it is going to be displayed at the time when the next frame is supposed to be displayed. In other words, instead of delaying the display of the whole sequence of frames, we decide to drop the next frame so that the second to the next frame can be decoded before its deadline.

4.3. Spatiotemporal Quality Degradation

In our definition, spatial quality degradation is the ratio of how the modified frame differs from the original frame. We use root mean square error (RMSE) as a measurement metric of the spatial image quality degradation. The temporal quality degradation, in our definition, is the ratio of how the modified video stream differs from the original video stream. We use the number of skipped/dropped frames as a measurement metric of the temporal video quality degradation. Clearly, the spatial quality degradation is an intra-frame level image distortion whereas the temporal quality degradation is an inter-frame level video distortion. Note that the time saved due to spatial quality degradation is smaller than the temporal quality degradation.

4.3.1. Spatial Quality Degradation

In order to find the best decoding steps to minimally distort the frame image quality, we analyze the typical MPEG-2 decoding sequence shown in Figure 5. Frame decoding in MPEG-2 has several major steps: Variable Length Decoding (VLD), Inverse Quantization (IQ), motion compensation, Inverse Discrete Cosine

Transformation (IDCT), dither, display, etc. Among those steps, we observe the SNR scalability and the saturation control, since they are employed to enhance the image quality. SNR scalability provides the enhancement of video quality by means of enhancement layer. Basically, it has two levels of layers: a base layer and an enhancement layer. Base layer includes the coarse level of DCT coefficients and the enhancement layer includes the finer level of DCT coefficients. On the contrary, the saturation control is clipping the results of IQ. Those two fine granularity scalability (FGS) techniques in MPEG-2 are initially introduced to cope with the time-varying bandwidth for the smooth image quality degradation. From the experiments, we obtain that these two steps consume approximately 10% of the total frame decoding time. Since they are spatial quality related metrics and relatively easy to be divided from MPEG-2 decoding steps, we choose them to do the spatial quality degradation in our DTM framework.

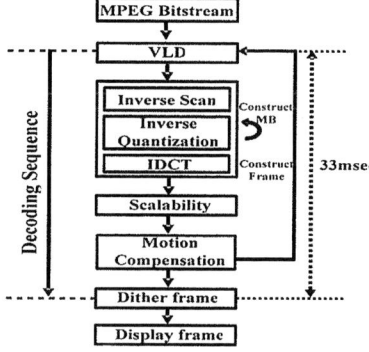

Figure 5 Typical MPEG-2 Decoding Steps

4.3.2. Temporal Quality Degradation

As mentioned earlier, temporal quality degradation is done by dropping the frames when deadline miss occurs. Note that not all the frames can be dropped arbitrarily. If a P frame is dropped, then all the subsequent P frames must be dropped till the next I frame. Whereas a B frame can be dropped arbitrarily since the next B frame does not depend on the B frame currently being dropped. In this paper hence we decide to drop only B frames so if the next frame to be dropped is not B then we keep decoding other I and P frames till we get the B frame.

Table 2 MPEG-2 Input Files Used in the Experiments

Input files	No. of frame	Frame resolution	I: P: B frame distribution
gitape.m2v	14	720 x 480	1: 4: 9
mei60f.m2v	50	704 x 480	5: 13: 32
hhilong.m2v	45	720 x 576	3: 8: 34
time_015.m2v	50	704 x 480	5: 12: 33
soccer_015.m2v	51	640 x 480	4: 14: 33
tens_015.m2v	47	352 x 192	5: 12: 30
cact_015.m2v	50	352 x 192	5: 12: 33

5. SIMULATION ENVIRONMENT

For our experiments, we modify and combine Simplescalar [18] Wattch [19] and HotSpot [11]. The simulated microprocessor model is based on ALPHA 21364, which has the feature size of 0.18μ, V_{dd} of 1.6V and a clock speed of 1GHz. The

power model used in the simulation does not model leakage power. In order to avoid modifying the default floor-plan in HotSpot, we use the same feature size and linearly scale both V_{dd} to 1.8V and a clock speed to 1.2GHz. The trigger and emergency temperature are set to 81.8 and 85.0°C, respectively [6], whereas the ambient and initial temperatures are set to 40.0 and 60.0°C respectively. Our combined simulator generates thermal results of each functional unit every 10K cycles.

For the application programs, we use MPEG-2 decoder program of MediaBench benchmark suite [15]. Table 2 summarizes the MPEG-2 video input files used in our experiments, which we mostly obtained from [20]. We add a few custom made files for the better comparison and limit the total number of frame to 51 in all experiments. Since 0.1°C rise/fall of temperature may take 100K cycles [6], our profile of thermal behavior of a program has long enough time. Our DTM policy is implemented in the MPEG-2 decoder program such that it interacts with our combined simulator.

Table 3 Thermal Behaviors in the Hottest Functional Unit

Input files	real decoding time (msec)	Max/Min temperature (°C)	
		DTM-ignorant	DTM
gitape	21.5	101.5 / 85.5	81.8 / 80.5
mei60f	19.6	99.6 / 83.8	81.8 / 80.5
hhilong	17.2	97.2 / 81.9	81.8 / 80.5
time_015	11.8	91.5 / 76.2	81.8 / 80.5
soccer_01	8.5	82.5 / 70.5	81.8 / 72.4
tens_015	4.0	73.4 / 63.2	N/A
cact_015	4.0	73.4 / 64.1	N/A

6. EXPERIMENTAL RESULTS

In Table 3, we summarize the experimental results. The left column shows the actually measured decoding time in Intel Pentium IV 2.8GHz and right two columns compare the maximum/minimum temperatures without and with DTM scheme, respectively. As you see, maximum temperature in DTM-ignorant system shows that thermal crisis can occur in some cases. Note that the maximum-minimum temperatures for the input files with same resolution are similar since they have approximately the same decoding workload. When the resolution becomes smaller, the maximum-minimum temperatures are both decreased. The N/A parts mean that no DTM scheme is necessary for those input files.

Table 4 Spatial/Temporal Quality Degradation

Input files	Image/Video Quality Degradation			
	Spatial		Temporal	
	Scaled frames	RMSE	Dropped frames	Frame drop ratio (%)
gitape	5	0.119	5	35.7
mei60f	8	0.125	15	30.0
hhilong	0	N/A	8	8.8
time_015	0	N/A	0	0
soccer_015	0	N/A	0	0
tens_015	0	N/A	0	0
cact_015	0	N/A	0	0

In Table 4, we summarize the experimental results for the image/video quality degradation. For the measurement of spatial quality degradation, we use root mean square error (RMSE) of the luminance (Y) values of frames. Note that this RMSE values are not calculated among all frames but for the spatially scaled frames only. For the temporal quality degradation, we show the number of frames dropped. As shown, when the resolution of a frame becomes smaller, number of dropped frames reduces since we have enough amount of residual time and become non-aggressive, i.e., scale frame instead. As expected, frames with a large resolution have high frame drop ratio.

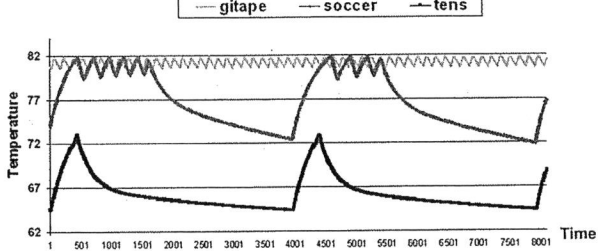

Figure 6 Comparison of the Thermal Variations

In Figure 6, we show the run time thermal behavior during the simulation. For simplicity, we use three input files which have different resolutions: gitape, soccer and tens. Each point in the X-axis is the measurement step in 10K cycles, Y-axis is the temperature in Celsius, and all measurements are made when programs reach to the thermally steady state. We categorize input files into three groups based on their workload, i.e. actual decoding time: Files with heavy workload execute DTM aggressively (top graph), files with medium workload execute DTM non-aggressively (middle graph), and files with light workload never execute DTM (bottom graph). Clearly, actual decoding time (in turn, the residual time) has strong relationship with the necessity of DTM scheme and our scheme shows that we achieve thermally safe state with different image/video quality degradation, i.e., frame drop ratio and the number of scaled frames are smaller in soccer than frame dropped/scaled in gitape.

7. CONCLUSIONS

In this paper, we propose an effective DTM scheme for MPEG-2 decoding with the spatiotemporal quality degradation. Our DTM algorithm makes use of the ever-decreasing actual frame decoding time and utilizes the residual time within a frame decoding by distributing it to achieve thermally safe state. As a consequence, quality degradation is observed. Our experimental results show that we achieve thermally safe state with spatial quality degradation of 0.12 in terms of RMSE value and with a frame drop rate of 12% on average. Our future research will carry out the analysis of the other steps of MPEG decoding in order to do more stepwise quality degradation with respect to decoding time.

8. REFERENCE

[1] D. Brooks, M. Martonosi, "Dynamic Thermal Management for High-Performance Microprocessors," *Proceedings of International Symposium on High Performance Computer Architecture (HPCA)*, January, 2001.

[2] H. Sanchez, B. Kuttanna, T. Olson, M. Alexander, G. Gerosa, R. Philip, J. Alvarez, "Thermal Management System for High Performance PowerPC Microprocessors," *Proceedings of IEEE Computer Society International Conference (COMPCON)*, 1997.

[3] S. Heo, K. Barr, K. Asonovic, "Reducing Power Density through Activity Migration," *Proceedings of International Symposium on Low Power Electronics and Design (ISLPED)*, August, 2003.

[4] Jay Srinivasan, Sarita V. Adve, "Predictive Dynamic Thermal Management for Multimedia Applications," *Proceedings of International Conference on Supercomputing (ICS)*, June, 2003.

[5] Mircea R. Stan, K. Skadron, M. Barcella, W. Huang, K. Sankaranarayanan, S. Velusamy, "HotSpot: a Dynamic Compact Thermal Model at the Processor-Architecture Level," *Microelectronics Journal: Circuit and Systems*, 2003.

[6] K. Skadron, Mircea R. Stan, W. Huang, S. Velusamy, Kathik Sankaranarayanan, David Tarjan, "Temperature-Aware Microarchitecture," *Proceedings of International Symposium on Computer Architecture (ISCA)*, June, 2003.

[7] K. Skadron, "Hybrid Architectural Dynamic Thermal Management," *Proceedings of the Design Automation and Test in Europe (DATE)*, 2004.

[8] K. Skadron, T. Abdelzaher, Mircea R. Stan, "Control-Theoretic Techniques and Thermal-RC Modeling for Accurate and Localized Dynamic Thermal Management," *Proceedings of International Symposium on High-Performance Computer Architecture (HPCA)*, 2002.

[9] K. J. Lee, K. Skadron, "Using Performance Counters for Runtime Temperature Sensing in High-Performance Processors," *Proceedings of High Performance, Power-Aware Computing (HP-PAC)*, April, 2005.

[10] MPEG-2 Standard: International Organization for Standardization/International Electro-technical Commission (ISO/IEC) 13818-2.

[11] HotSpot at http://lava.cs.virginia.edu/HotSpot

[12] D. Son, C. Yu, H. Kim, "Dynamic Voltage Scaling on MPEG Decoding," *Proceedings of International Conference on Parallel and Distributed Systems (ICPADS)*, 2001.

[13] K. Choi, R. Soma, Massoud Pedram, "Off-chip Latency Driven Dynamic Voltage and Frequency Scaling for an MPEG Decoding," *Proceedings of Design Automation Conference (DAC)*, June, 2004.

[14] K. Choi, K. Dantu, W. C. Cheng, Massoud Pedram, "Frame-Based Dynamic Voltage and Frequency Scaling for a MPEG Decoder," *Proceedings of International Conference on Computer Aided Design (ICCAD)*, 2002.

[15] MediaBench at: http://euler.slu.edu/~fritts/mediabench

[16] M. Verderber, A. Zemva, A. Trost, "HW/SW Codesign of the MPEG-2 Video Decoder," *Proceeding of International Parallel and Distributed Processing Symposium*, 2003.

[17] Berkeleympeg http://bmrc.berkeley.edu/frame/research/mpeg

[18] Simplescalar turorial at http://www.simplescalar.com

[19] D. Brooks, V. Tiwari, M. Martonosi, "Wattch: A Framework for Architectural-Level Power Analysis and Optimizations," *Proceedings of International Symposium on Computer Architecture (ISCA)*, 2000.

[20] MPEG2 streams at http://www.mpeg2.de/video/streams

A Low-Power Active Substrate-Noise Decoupling Circuit with Feedforward Compensation for Mixed-Signal SoCs

Song Guo and Hoi Lee
Mixed-Signal & Power IC Laboratory
Department of Electrical Engineering
The University of Texas at Dallas
Richardson, TX 75083-0688
{sxg041200, hoilee}@utdallas.edu

ABSTRACT

This paper presents a low-power active-decoupling circuit using feedforward-compensation technique for SoC substrate-noise reduction. The proposed feedforward technique not only generates a left-half-plane zero for the decoupling circuit achieving stability and wider bandwidth in low-power condition, but also increases the dynamic current during transient. As a result, substrate-noise suppression has been significantly improved with larger-amplitude and higher-frequency noise sources. In a standard 0.13μm CMOS process, simulation results show that the proposed feedforward technique enhances both the bandwidth and dynamic current of the decoupling circuit by 2 and 79 times, respectively, without additional static power consumption. The decoupling circuit thus improves the crosstalk noise suppression from 3.6 to 6.6 times with a 1GHz noise-source amplitude increasing from 100mV to 500mV.

Categories and Subject Descriptors

B.7.1 [Integrated Circuits]: Types and Design Styles– *VLSI (very large scale integration)*

General Terms

Design, Performance

Keywords

Active substrate noise decoupling, feedforward frequency compensation, substrate noise suppression, system-on-a-chip.

1. INTRODUCTION

The monolithic integration of digital, analog and RF systems to realize mixed-signal systems-on-a-chip (SoC) design is continually driven by scaling of CMOS technologies for reducing chip size, power consumption and cost. One of the major bottlenecks of SoC is the substrate coupling between digital circuits and sensitive analog and RF circuits through their shared substrate [1]-[3]. The switching noise transient from digital circuits can flow through the substrate and severely degrades the performances of the RF and analog circuits. In fact, this issue has a growing impact on today's submicron CMOS mixed-signal integrated circuits with gigahertz speed. Different techniques for reducing the substrate noise have been reported [4]-[8]. For

Permission to make digital or hard copies of all or part of this work for personal or classroom use is granted without fee provided that copies are not made or distributed for profit or commercial advantage and that copies bear this notice and the full citation on the first page. To copy otherwise, or republish, to post on servers or to redistribute to lists, requires prior specific permission and/or a fee.
ISLPED '06, October 4–6, 2006, Tegernsee, Germany.
Copyright 2006 ACM 1-59593-462-6/06/0010...$5.00.

example, decoupling capacitor or guard band has been adopted to reduce the substrate noise [4]-[7]. However, when the frequency of substrate noise is over several hundred MHz, these techniques become less effective due to the parasitic line inductance. Active-decoupling circuit is thus developed to suppress high-frequency substrate noise by realizing a Miller capacitance [8], which in turns depends on the gain of its amplifier $A_v(\omega)$. However, $A_v(\omega)$ is frequency dependent and the decoupling effect will then drop quickly when the frequency of the substrate noise exceeds -3dB bandwidth of the amplifier. In general, the -3dB bandwidth of the amplifier increases with the power consumption and there are usually many active-decoupling circuits in SoCs for minimizing the substrate noise. The overall chip power consumption can be greatly increased if the power consumption of each active-decoupling circuit is not optimized. Unfortunately, there is still lack of discussion on simultaneously optimizing both -3dB bandwidth and power consumption of the active-decoupling circuit. It is also lack of study on the capability of the active-decoupling circuit to suppress large-amplitude substrate noise. This capability is critical as wide -3dB bandwidth of the amplifier is not sufficient to reduce the large-amplitude substrate noise.

Motivated by above concerns, the study on amplifiers for the conventional active-decoupling circuit is discussed in Section II. Section III addresses the design of the amplifier using the proposed feedforward compensation for the active-decoupling circuit. The proposed feedforward compensation not only significantly improves -3dB bandwidth of the active-decoupling circuit without extra power consumption, but also increases dynamic current during transient for suppressing large-amplitude substrate noise. Simulations and discussions of substrate-noise decoupling are presented in Section IV. Finally, conclusions are given in Section V.

2. STUDY ON AMPLIFIERS FOR ACTIVE-DECOUPLING CIRCUIT

If the amplifier of the active-decoupling circuit is single-stage design, its structure can be modeled as Fig. 1(a). The amplifier load capacitance C_L of around 10pF is used to model the decoupling capacitor in the active-decoupling circuit. The -3dB bandwidth ω_{-3dB} of the single-stage amplifier $\omega_{-3dB}=g_{m1}/(A_v \cdot C_L)$ is inversely proportional to its gain A_v. Therefore, both high gain and wide bandwidth cannot be achieved simultaneously. In order to improve both gain and -3dB bandwidth of the single-stage amplifier, a conventional voltage buffer using the source follower is adopted to connect after the gain stage to drive C_L [8] and its structure is shown in Fig. 1(b). Since the gain of the amplifier is mainly contributed by the first stage, we can assume

$A_v = A_1 = g_{m1} r_{o1}$ and the gain of the source follower equals to 1. There are two poles in the amplifier as follows:

$$\omega_{-3dB} = \frac{g_{mL}}{C_L} \qquad (1)$$

$$\omega_2 = \frac{g_{m1}}{A_v \cdot C_{o1}} \qquad \qquad (2)$$

Based on (1), the -3dB bandwidth of the amplifier is improved by A_v times compared to the single-stage counterpart by assuming both amplifiers have the same transconductances. However, for achieving the phase margin of 45°, ω_2 should be equal to $A_v \cdot \omega_{-3dB}$. This implies $g_{m1} = A_v^2 g_{mL}(C_{o1}/C_L)$. Large g_{m1} is thus needed to ensure the amplifier to be stable, which corresponds to large power consumption. As a result, this structure is not favorable to low-power design for stability concern. In addition, the dynamic current during transient is limited by the static current I to charge and discharge C_L with the use of the source follower output stage.

3. PROPOSED AMPLIFIER WITH FEED-FORWARD COMPENSATION

Fig. 1(c) shows the proposed amplifier with feedforward compensation for the active-decoupling circuit. There are two poles and one zero in the amplifier given as

$$\omega_{-3dB} = \frac{g_{m1}}{A_1 C_{o1}} \qquad (3)$$

$$\omega_2 = \frac{g_{mL}}{C_L} \qquad (4)$$

$$\omega_z = \omega_{-3dB}(A_1 \frac{g_{mL}}{g_{mf}} + 1) \approx \omega_{-3dB} \cdot A_1 \frac{g_{mL}}{g_{mf}} \qquad (5)$$

where A_1 is the amplifier first-stage gain and g_{mf} is the transconductance of the feedforward stage. Due to small C_{o1}, ω_{-3dB} can be located at high frequency. In addition, the feedforward stage generates a left-half-plane zero ω_z, which provides positive phase shift to partly cancel the negative phase

shift of ω_2 for better stability. As a result, ω_2 is located at the frequency smaller than the unity-gain frequency of the amplifier to reduce the value of g_{mL}. As a result, the proposed structure is favorable to low-power design. In addition, the dc gain of the proposed amplifier is given as

$$A_v \approx -\left(A_1 + \frac{g_{mf}}{g_{mL}}\right) \qquad \qquad (6)$$

As $g_{mf}/g_{mL} > 1$, the magnitude of A_v of the proposed amplifier is increased compared to the conventional amplifier shown in Fig. 1(b), which improves the noise suppression capability. Moreover, the proposed feedforward stage realizes a push-pull output stage to increase the dynamic current such that the ability of the active-decoupling circuit to suppress large-amplitude noise can be enhanced.

Fig. 2 shows the core circuit implementation of the proposed amplifier for the active-decoupling circuit. Two input capacitors of 10pF each are used for decoupling the dc voltage at V_{in+} and V_{in-}, while high-frequency substrate noise at V_{in-} can be coupled to the gate of M0 and Mf through the input capacitor. It should be noted that transconductances g_{m1}, g_{mL} and g_{mf} are realized by nMOS transistors (M1, M0), ML and Mf, respectively. Generally, nMOS transistor dissipates less current to realize the same transconductance due to larger mobility compared to the pMOS counterpart. As a result, the implementation of the proposed amplifier is suitable for low-power design.

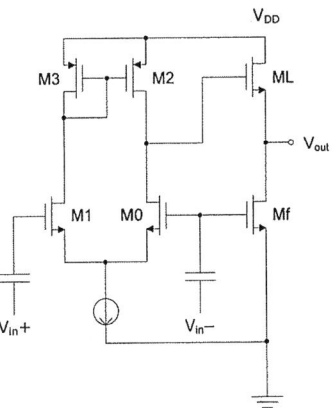

Fig. 2: Core schematic of the proposed amplifier for active decoupling.

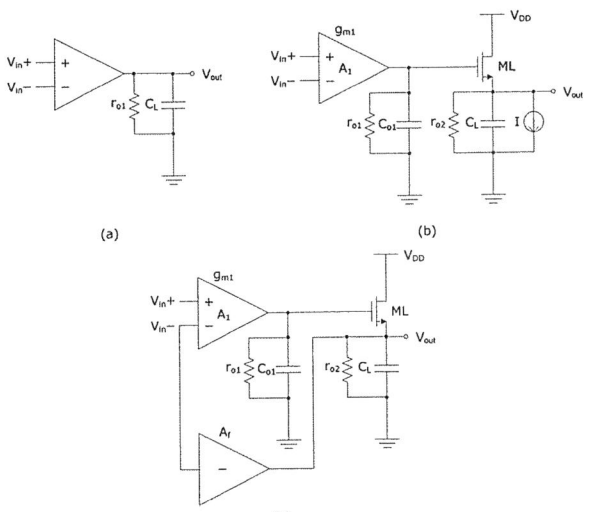

Fig. 1: Structures of the (a) single-stage, (b) conventional two-stage [8] and (c) proposed amplifiers for active-decoupling circuit.

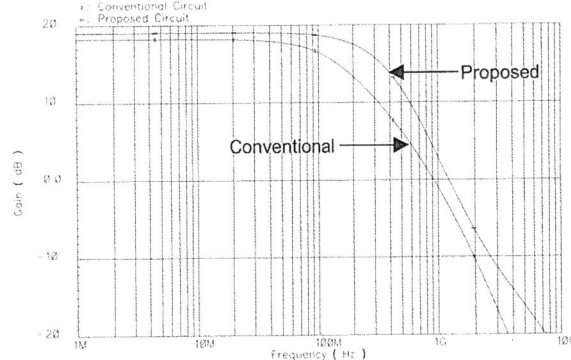

Fig. 3: Simulated frequency responses of conventional and proposed amplifiers for the active-decoupling circuit.

323

Table 1: Performance summary of different amplifiers

	Conventional	Proposed
Power Supply	1.2V	1.2V
Load Capacitor (C_L)	10pF	10pF
-3dB Bandwidth	138MHz	288MHz
Phase Margin	60°	60°
Power Consumption	4.8mW	4.8mW
DC Gain	18.2dB	19dB
Maximum Dynamic Current	398µA	31.5mA

Both the proposed and conventional (structure shown in Fig. 1(b)) amplifiers have been implemented in a 0.13-µm CMOS n-well process. Fig. 3 shows simulated frequency responses of both amplifiers and Table I provides the detailed simulation results. In order to give clearer performance comparisons on -3dB bandwidth and dc gain, both amplifiers are intentionally designed to dissipate the same power consumption of 4.8mW and provide the same phase margin of 60°. The proposed amplifier with feedforward compensation achieves 2-fold improvement in -3dB bandwidth compared to the conventional counterpart. The ability of providing the dynamic current is tested by applying an ideal pulse of 0.5V with 1ps rise and fall time at V_{in}. in order to model large-amplitude noise change. The maximum dynamic current of the proposed amplifier is improved by 79 times. These results indicate that the proposed amplifier with feedforward compensation can be better adopted into the active-decoupling circuit to suppress both high-frequency and large-amplitude substrate noise.

4. SIMULATIONS & DISCUSSIONS OF SUBSTRATE-NOISE DECOUPLING
4.1 Simulation Setup
Fig. 4 shows the schematic of the simulation circuit for both small-signal and transient analysis of the substrate-noise suppression. All p-substrate and n-well are modeled by using several resistors R1 (5Ω) and a capacitor (Cp = 5pF). L1 and L2 are line parasitic inductances with typical values of 3nH and 5nH, respectively. Similarly, R2 and R3 are line parasitic resistors of the power supply and the reference voltage of 5Ω and 15Ω, respectively. The noise source is modeled as an ac signal voltage for the ac analysis to evaluate the decoupling effect and a sinusoidal voltage for the transient analysis to evaluate large-amplitude noise suppression. For performance comparison, three types decoupling circuits are simulated including a pure capacitive decoupling circuit, the conventional active-decoupling circuit, and the proposed active-decoupling circuit with the amplifier using feedforward compensation. It should be also noted that the decoupling capacitor C in all three decoupling circuits is 30pF.

4.2 Decoupling Effect Simulations
The decoupling effect simulations are performed by taking the difference between voltages at V_c using different decoupling circuits and the voltage without the decoupling circuit at different frequencies. As shown in Fig. 5, both active-decoupling circuits achieve better decoupling effect especially at higher frequencies, compared to the capacitive decoupling circuit. In particular, the proposed active-decoupling circuit achieves the best decoupling effect to suppress high-frequency substrate noise due to the result of wider -3dB bandwidth achieved by the amplifier with feedforward compensation.

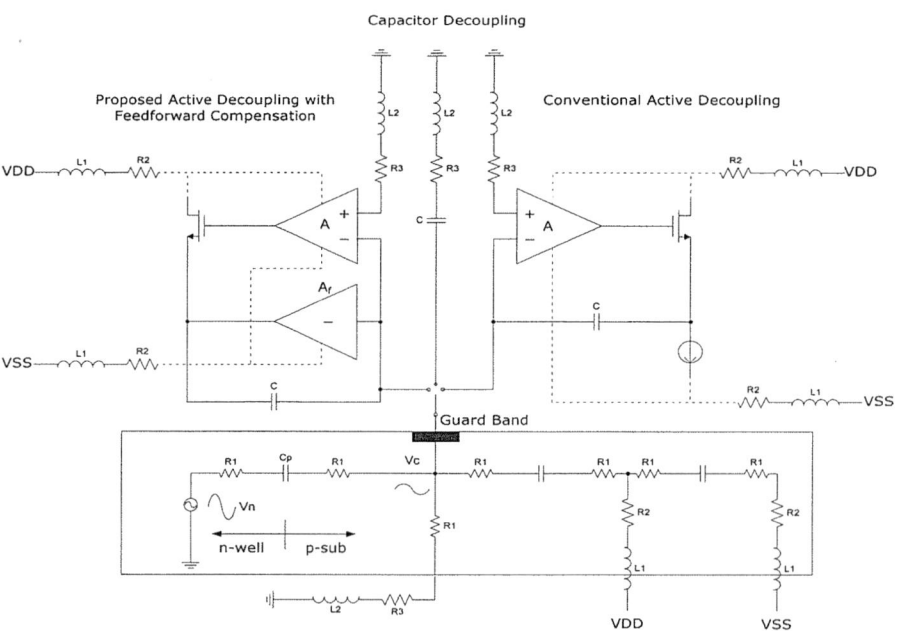

Fig. 4: Simulation circuit for the substrate noise decoupling effect.

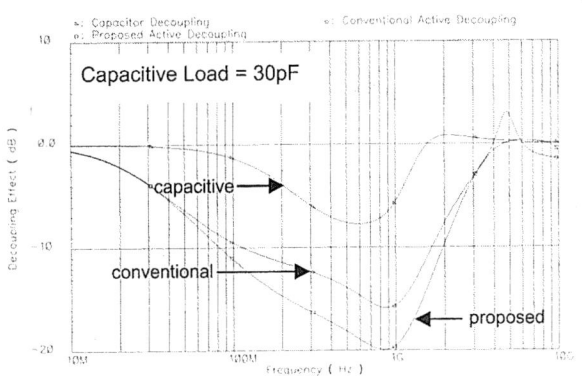

Fig. 5: Simulated decoupling effect.

4.3 Large-Amplitude Noise Suppression

Figs 6(a) and 6(b) show the simulated V_c at noise-source amplitudes of 100mV and 500mV under different frequencies changing from 10MHz to 2GHz. Under both noise amplitudes of 100mV and 500mV, the value of V_c is always smaller from 100MHz to 2GHz by using the proposed active-decoupling circuit. The rate of V_c increase using the proposed active-decoupling circuit is also always slower than that using the conventional counterpart. In particular, the proposed active-decoupling circuit improves the substrate-noise suppression from 3.6 to 6.6 times as compared to the conventional active-decoupling circuit when the 1 GHz noise-source amplitude increases from 100mV to 500mV. This is due to the fact that the capability of suppressing large-amplitude high-frequency substrate noise is not only dependent on the -3dB bandwidth of the amplifier, but also the amount of dynamic current during transient to charge or discharge the decoupling capacitor. Since the feedforward stage in the proposed amplifier realizes a push-pull output stage, the dynamic current of the active-decoupling circuit is significantly increased during transient. Furthermore, under 500mV 1GHz noise source, the proposed active-decoupling circuit can reduce the substrate noise to 71.6mV, while almost no noise suppression is provided by using the conventional active-decoupling circuit.

5. CONCLUSION

A low-power active-decoupling circuit with feedforward compensation for mixed-signal SoCs has been presented. The problem of simultaneously achieving wide -3dB bandwidth, stability, low-power and large dynamic current in the amplifier for

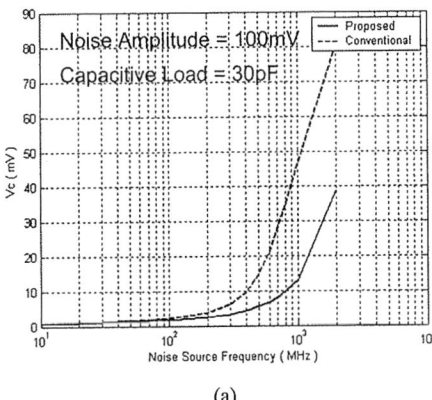

(a)

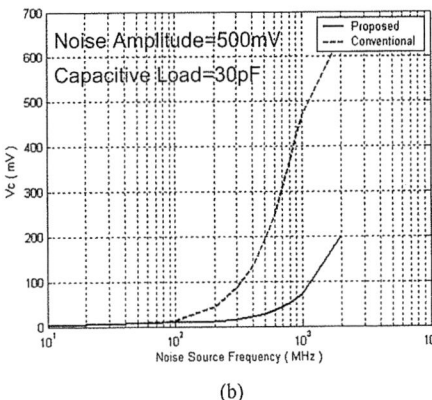

(b)

Fig. 6. Simulated V_c with noise amplitudes of (a) 100mV and (b) 500mV.

active decoupling has been addressed. A new amplifier with feedforward compensation is proposed to be adopted in the active-decoupling circuit for improving both -3dB bandwidth and dynamic current without extra power consumption. With the proposed amplifier, simulations verify that the capability of the active decoupling circuit to suppress higher-frequency and larger-amplitude substrate noise has been significantly enhanced.

6. REFERENCES

[1] D. K. Su, M. J. Loinaz, S. Masui, and B. A. Wooley, "Experimental results and modeling techniques for substrate noise in mixed-signal integrated circuits," *IEEE J. Solid-State Circuits*, vol. 28, pp. 420–430, Apr. 1993.

[2] M. Van Heijningen, J. Compiet, P. Wambacq, S. Donnay, M. G. E. Engels, and I. Bolsens, "Analysis and experimental verification of digital substrate noise generation for epi-type substrates," *IEEE J. Solid-State Circuits*, vol. 35, pp. 1002–1008, Jul. 2000.

[3] B. E. Owens, S. Adluri, P. Birrer, R. Shreeve, S. K. Arunachalam, K. Mayaram, and T. S. Fiez, "Simulation and measurement of supply and substrate noise in mixed-signal ICs," *IEEE J. Solid-State Circuits*, vol. 40, pp. 382-391, Feb. 2005.

[4] M. Nagata, J. Nagai, K. Hijikata, T. Morie, and A. Iwata, "Physical design guides for substrate noise reduction in CMOS digital circuits," *IEEE J. Solid-State Circuits*, vol. 36, pp. 539–549, Mar. 2001.

[5] M. Badaroglu, M. van Heijningen, V. Gravot, J. Compiet, S. Donnay, G. Gielen, and H. De Man, "Methodology and experimental verification for substrate noise reduction in CMOS mixed-signal IC's with synchronous digital circuits," *IEEE J. Solid-State Circuits*, vol. 37, pp. 1383–1395, Nov. 2002.

[6] M. S. Peng and H. S. Lee, "Study of substrate noise and techniques for minimization," *IEEE J. Solid-State Circuits*, vol. 39, pp. 2080–2086, Nov. 2004.

[7] H. Kaul, D. Sylvester, and D. Blaauw, "Clock net optimization using active shielding," in *Proc. ESSCIRC*, vol. C13.3, Sep. 2003, pp. 265-268.

[8] T. Tsukada, Y. Hashimoto, K. Sakata, H. Okada, and K. Ishibashi, "An on-chip active decoupling circuit to suppress crosstalk in deep-submicron CMOS mixed-signal SoCs," *IEEE J. Solid-State Circuits*, vol. 40, pp. 67–79, Jan. 2005.

Power-Efficient Pulse Width Modulation DC/DC Converters with Zero Voltage Switching Control

Changbo Long, Sasank Reddy, Sudhakar Pamarti, Lei He
EE Dept., UCLA
longchb@synopsys.com,{sasank,spamarti,lhe}@ee.ucla.edu

Tanay Karnik
Intel
tanay.karnik@intel.com

ABSTRACT

This paper proposes a power-efficient PWM DC/DC converter design with a novel zero voltage switching (ZVS) control technique. The ZVS control is realized by an inner feedback loop which is implemented by simple digital circuitry between the input and output of the power transistors and achieves real-time zero voltage switching (ZVS) for various loading and device parameters with power efficiencies over 90.0%. In addition, an outer feedback loop is used to ensure that the output precisely tracks a reference voltage level. We have also built the relationship between the output voltage ripple and the speed of the voltage comparators which has shown to introduce new low-frequency signals to the loops and cause significant output voltage ripples. Experiment results show that the output ripple could be reduced by 4x by carefully handling the generation and propagation of these low frequency signals.

Categories and Subject Descriptors

B.7.1 [**Integrated Circuits**]: Types and Design Styles

General Terms

Design.

Keywords

DC/DC conversion, zero voltage switching.

1. INTRODUCTION

Power consumption has become one of the most important issues in modern electronics due to increased complexity and speed of the system. In order to curb the effect of power on a system as a whole, multiple power domains have been proposed as an architecture scheme for low power design. To support multi-Vdd, an array of supply voltages need to be generated. DC/DC converters can be integrated on chip and convert the input voltage to different voltage levels internally. Recently, a great deal of research [1–5] has been devoted to improving the power efficiency and reducing the area cost of on-chip DC/DC converters. However, there are

Permission to make digital or hard copies of all or part of this work for personal or classroom use is granted without fee provided that copies are not made or distributed for profit or commercial advantage and that copies bear this notice and the full citation on the first page. To copy otherwise, to republish, to post on servers or to redistribute to lists, requires prior specific permission and/or a fee.
ISLPED'06, October 4–6, 2006, Tegernsee, Germany.
Copyright 2006 ACM 1-59593-462-6/06/0010 ...$5.00.

still many unsolved problems. For instance, the basic linear regulator and the charge-recycling voltage regulator are designs that have been looked at as candidates for on-chip integration because there are no filter elements. However, the relative low power efficiency, typically less than 80% [1], of these designs has limited their application.

In this paper, we propose design techniques and analysis to address the above problems for high frequency PWM buck converters. We first introduce a real-time ZVS technique which relies on a feedback loop as opposed to tuning device parameters to achieve ZVS during design time as in traditional methods [4]. Our experiment results show that the feedback mechanism guarantees ZVS under different loading and device parameters. Furthermore, using the real-time ZVS technique we are able to achieve power efficiencies over 90.0%. We then study close loop design and analysis of PWM buck converters. Our experiment results show that output voltage ripple in a closed loop PWM buck converter can be reduced up to 4x by correctly analyzing and optimizing the sources that generate and propagate low frequency signals.

2. BACKGROUND AND DESIGN OVERVIEW

2.1 Principles of PWM Buck Converters

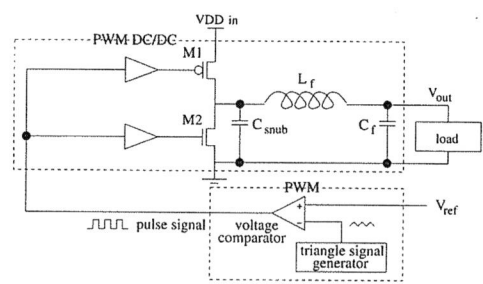

Figure 1: Schematic of a PWM buck converter.

Compared to other designs, such as linear regulators, PWM buck converters consume more area but have higher power efficiency. The schematic of a PWM buck converter is shown in Fig. 1. It consists of two power transistors, $M1$ and $M2$, with their drivers, a low-pass LC filter consisting of L_f and C_f, a snubber capacitor C_{snub}, and a pulse width modulator.

The output voltage level V_{out} is the DC component of the pulse signal generated by the PWM, and it is

$$V_{out} = Vdd_{in} \cdot D, \qquad (1)$$

where D is the duty cycle of the pulse signal, which is controlled by V_{ref} as an input of the PWM. In fact,

$$D = \frac{V_{ref}}{Vdd_{in}}. \quad (2)$$

Therefore, we have

$$V_{out} = V_{ref}. \quad (3)$$

As shown in [6], the output voltage ripple of a PWM buck converter can be expressed as

$$\Delta V_{out} = \frac{Vdd_{in}(1-D)D}{8L_f C_f f^2}, \quad (4)$$

where L_f and C_f are the inductance and capacitance of the LC filter and f is the frequency of the pulse signal. f is also called the operation frequency of the buck converter.

Equation (4) shows that to keep ΔV_{out} at a low level, L_f and/or C_f has to be large if the operation frequency f is low. In other words, an effective way to reduce the area of the LC filter in the buck converter is to use a high operation frequency [2]. However, a high operation frequency leads to a high switching power loss. To reduce the switching power loss, a technique called zero voltage switch (ZVS) has been widely adopted. As shown in [4], ZVS ensures that both power transistors switch under a zero voltage drop between source and drain.

2.2 Overview of the Proposed Topology

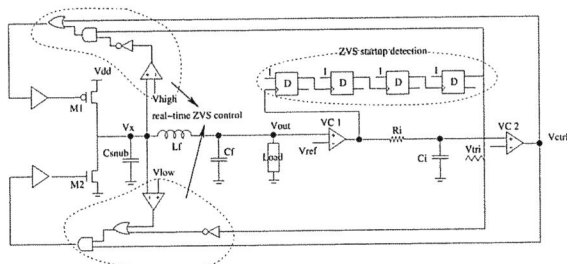

Figure 2: Overview of the proposed circuit topology.

Our study in this paper is based on the circuit topology shown in Figure 2. This circuit contains two feedback loops. The outer loop consists of two power transistors M_1 and M_2, LC filter L_f and C_f, voltage comparator $VC1$, RC integrator R_i and C_i, pulse modulation element $VC2$, and real-time ZVS circuitry. The inner loop starts from the output of the two power transistors, passes through the ZVS circuitry and ends at the input of the two power transistors.

The outer loop ensures that the output voltage level V_{out} tracks the reference voltage level V_{ref}. It is considered a negative feedback loop since there are only one set of negative components, the power transistors, in the entire looop. For example, if V_{out} is higher than V_{ref}, the output of voltage comparator $VC1$ is high, C_i is charged, and the voltage level at the positive input of $VC2$ is increased, which increases the duty cycle of the switching signal and therefore decreases V_{out}.

The design of voltage comparator is adopted from [7], which has shown high resolution and low power consumption. The topology of the voltage comparator is shown in Figure 3, which is composed by an input amplification stage, two flip flops and a RS latch. Two clock signals are used to clear up previous results and evaluations. More details are described in [7].

The real-time ZVS technique is achieved by the inner feedback loop. To avoid the influence of startup strike, a detection sub-circuit is included as shown in Figure 2. The idea is not to startup the ZVS circuitry until V_{out} has been stabilized. More details will be described in Section 4.

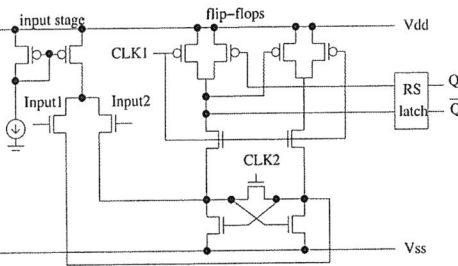

Figure 3: Topology of the voltage comparator.

3. CLOSE LOOP DESIGN AND ANALYSIS

3.1 Analysis on output voltage ripple

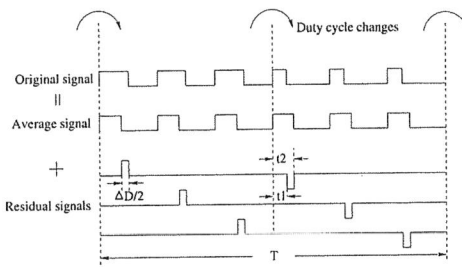

Figure 4: Illustration of changes in duty cycle in the switching signal.

Our study is based on the design shown in Figure 2. The voltage comparator $VC2$ in the figure modulates the duty cycle of the switching signal feeding the two power transistors and this modulation introduces a signal that contains a wide range of frequencies. Low frequency components of this signal may pass the LC filter and cause a voltage ripple with a frequency approximately equal to the frequency of the LC filter, $1/(2\pi\sqrt{L_f C_f})$, at the output terminal. Note that the voltage comparator $VC1$ may also bring in low frequency components but it is filtered by the $R_i C_i$ integrator and has a smaller impact to the overall output ripple.

Figure 4 illustrates the behavior of the operating switching signal during steady state. In the figure, this signal is decomposed into an average signal, which maintains a constant duty cycle and a set of residual signals. The average signal contains only high frequency components, and the residual signals contain a wide range of frequency components. The duty cycle switches between two states D_1 and D_2 has a frequency the same as the output voltage ripple, which is roughly equal to frequency of the LC filter, i.e.,

$$T = 2\pi\sqrt{L_f C_f}. \quad (5)$$

By Fourier transformation, the residual signals can be expressed as follows,

$$
\begin{aligned}
f(t) =& \frac{2V_{dd}}{\pi}\Big[\sum_{n=0}^{\infty}\Big(\frac{sin(2n+1)\omega t_2 - sin(2n+1)\omega t_1}{2n+1}\Big)cos(2n+1)\omega t \\
&+ \sum_{n=0}^{\infty}\Big(\frac{cos(2n+1)\omega t_1 - cos(2n+1)\omega t_2}{2n+1}\Big)sin(2n+1)\omega t\Big] \\
=& \frac{2V_{dd}}{\pi}\sum_{n=0}^{\infty}\frac{\sqrt{2-2cos(2n+1)\omega(t_2-t_1)}}{2n+1}cos((2n+1)\omega t + \phi) \\
\approx& \frac{2V_{dd}}{\pi}\sum_{n=0}^{\infty}\omega(t_2-t_1)cos((2n+1)\omega t + \phi),
\end{aligned}
$$

where

$$\omega = \frac{1}{\sqrt{L_f C_f}}. \quad (6)$$

Notice that $t_2 - t_1 = T \cdot \frac{\Delta D}{2}$, we have

$$f(t) = \frac{V_{dd} T}{\pi \sqrt{L_f C_f}} \Delta D \sum_{n=0}^{\infty} cos((2n+1)\omega t + \phi). \quad (7)$$

From (6) we can see that ω is the 3db frequency of the LC filter which implies that the magnitude of a signal at this frequency becomes half when it passes through the LC filter. Also, equation (7) implies that the residual signals contain frequency components of ω, 3ω, $\cdots$, etc and the magnitudes of these components are proportional to ΔD. Given that ω is the 3db frequency of the LC filter, when the residual signals enter into the LC filter, the low frequency components, such as ω, 3ω and 5ω, can pass through the LC filter with significant magnitudes and appear as a ripple at the output terminal. Because the magnitudes of the low frequency components of ω, 3ω and 5ω etc. are proportional to ΔD, the magnitude of the ripple is proportional to ΔD too. Practically every voltage comparator has a smallest value of ΔD. In slower voltage comparators this ΔD are generally larger than in faster comparators. Therefore, we suggest to use fast voltage comparators to avoid large ripple at the output.

4. REAL-TIME ZVS TECHNIQUE

4.1 Traditional ZVS techniques

A representative traditional design-time ZVS technique without feedback is presented in [4]. To illustrate the ZVS conditions, each cycle of the internal switching signal is divided into four time fragments, T_1, T_2, T_3 and T_4 in the figure. During T_1 and T_3, the NMOS power transistor M_2 and PMOS power transistor M_1 close, respectively. During T_2 and T_4, both transistors are open. These two time fragments are called deadtimes. V_x drops to zero at the beginning of T_1 when M_2 starts to close. Also, V_x reaches Vdd when M_1 starts to close. Thus, M_1 and M_2 switch at a zero voltage drop between the source and drain i.e, zero voltage switching. During T_2, the inductance current I_{L_f} charges C_{snub} and V_x increases to Vdd. Similarly during T_1, the inductance current I_{L_f} discharges C_{snub} and V_x decreases to zero.

One can see that the conditions to achieve ZVS are very restrictive. Parameters have to be re-tuned when loading current changes. Design-time ZVS techniques without feedback are vulnerable to variations of devices and loading.

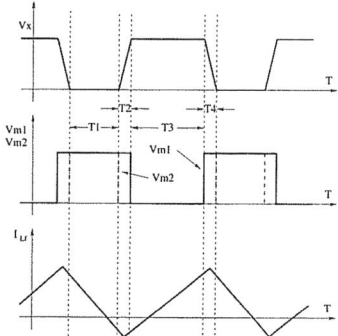

Figure 6: Conditions to achieve zero voltage switching.

4.2 Real-time ZVS technique

Assuming V_{ctrl} as the control signal of the power transistors, without considering ZVS, the control signals for the PMOS power transistor, M_1, and NMOS power transistor, M_2, are the same as V_{ctrl}, i.e.,

$$V_{m1} = V_{m2} = V_{ctrl}. \quad (8)$$

To achieve ZVS, V_{m1} and V_{m2} can be expressed as

$$V_{m1} = V_{ctrl} + \overline{V_x}, \quad (9)$$

and

$$V_{m2} = V_{ctrl} \cdot \overline{V_x}, \quad (10)$$

where V_x is the output voltage level of the power transistors. Equation (9) ensures that V_{m1} is 1 whenever V_x is 0, which implies that M_1 turns on (V_{m1} changes from 1 to 0) only when V_x is 1, i.e., the voltage drop between the source and drain of M_1 is zero. Similarly, (10) ensures that V_{m2} is 0 whenever V_x is 1, which implies that M_2 turns on (V_{m2} changes from 0 to 1) only when V_x is 0, i.e., the voltage drop between the source and drain of M_2 is zero.

Equation (9) and (10) explain the principle to achieve ZVS. The implementation of this principle, however, needs more careful analysis. To ensure that V_x stays at the perfect Vdd/zero level when M_1 and M_2 start to open, we use voltage comparators instead of inverters to implement $\overline{V_x}$ as shown in Figure 2. To drive M_1, V_x is compared with $V_{high} = VDD - \Delta v$ and only when V_x is is higher than V_{high} M_1 can open. Similarly, to drive M_2, V_x is compared with a low voltage level $V_{low} = \Delta v$ and only when V_x is lower than V_{low}, M_2 can open. In order to reduce the overshoot on V_x, we use a Δv slightly larger than zero (0.3v). However, as shown in our experiment results, it is impossible to fully eliminate overshoot.

Overall, the real-time ZVS scheme presented as part of this design is very unique in the fact that it does not rely on manually calculating the duty cycle delays for a particular design. Although there are designs that do automatic ZVS control, they do not meet the power and area requirements for on-chip high frequency applications [8,9].

4.3 ZVS startup detection

Initially, when the circuit is cold started, there is a period of time where the output signals have abnormally large variation. During this startup period there is no need to turn on the automatic ZVS calibration control for the system. In order to avoid tuning ZVS during the startup period, a series of D flip-flops are used as delay elements to prevent the startup of the ZVS calibration scheme until after the initial transient spikes, as shown in Fig. 2.

5. EXPERIMENT RESULTS

5.1 Close loop design and analysis

To verify the idea that the output voltage ripple is proportional to the smallest duty cycle change ΔD, we have implemented the voltage comparator, as in Figure 3, at two different clock rates. One voltage comparator operates at 400MHz and the other at 1.6 GHz frequencies. We compare the output voltage ripple of the PWM buck converters implemented with these two comparators in Figure 5. Note that 130nm technology is used, the Vdd voltage is 1.3 Volt, and the reference voltage level is 0.85 Volt. As shown in the figure, the voltage ripple of the 400 MHz comparator is about 11.7% as compared to the reference voltage, while the voltage ripple of the 1.6 GHz comparator is 2.9%. By increasing the frequency of the comparator by 4X, we notice a reduction in ripple by around 4X. These experiment results show that when designing the closed loop PWM buck converter, the elements of the closed loop need to be carefully considered so that the output voltage ripple is minimized.

328

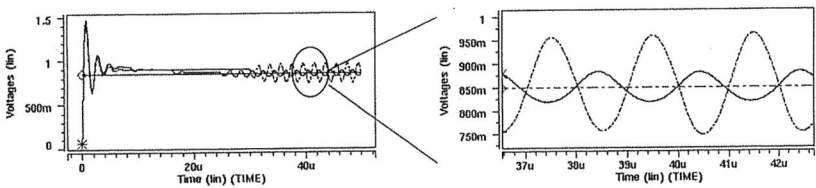

Figure 5: Comparison of voltage ripple between buck converters implemented by a 0.4GHZ voltage comparator and 1.6GHz voltage comparator.

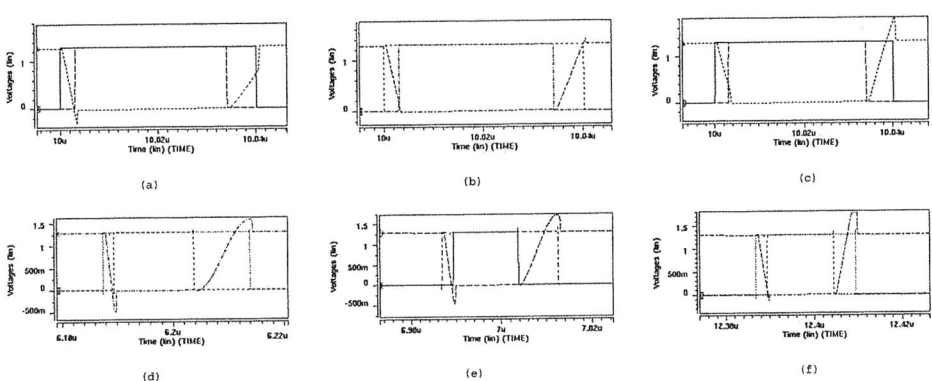

Figure 7: Design-time ZVS under loading of 5 ohms (a), 10 ohms (b), and 50 ohms (c) and real-time ZVS with feedback loop under loading of 5 ohms (d), 10 ohms (e), and 50 ohms (f). (A) and (c) fails ZVS.

5.2 ZVS control

One of the strong points of the design presented in this paper is the fact that the power of the transistors is consistent over a multitude of loads. This is mainly due to the fact that there is a real-time ZVS control system established in the circuit.

Figure 7 contains a comparison between design-time ZVS without feedback and real-time ZVS with feedback. In the case of the load of 5 and 50 ohms with design-time ZVS, one can see that the circuit is out of the ZVS mode of operation. With a load of 5 ohms, the C_{snub} capacitor is charged too slow and the voltage of V_x does not transition from the lower state to the higher state before the M_1 transistor switches on. On the other hand with a load of 50 ohms, the design-time ZVS circuit discharges the C_{snub} too slowly and the circuit is in a non optimal energy efficiency state once again. The real-time circuit always stays in the ZVS mode of operation with the various loads. Although, there is an overshoot (refer to Section 4.2) for the real-time ZVS circuit, this does not affect the energy usage of the power transistors.

Overall, not having a proper ZVS control system will cause problems in terms of power usage of the switch transistors. In terms of overall power efficiency of the circuit, both ZVS operation and the actual magnitude of the current through the load play significant roles. But our experiments show that if the proper load current is used with real-time ZVS, power efficiencies are greater than 90% consistently for various loads.

6. CONCLUSION

In this paper, we have designed a novel PWM circuit that can be used to provide a range of Vdd levels for a variety of loads by two feedback loops. With the use of an inner feedback loop between the output and input of the power transistors, we are able to ensure real-time zero voltage switching. This enables the reduction of power consumed by these transistors and achieves power efficiencies over 90% for a va-

riety of loads. Also, an outer feedback loop is employed in the PWM circuit to track the reference voltage level. We show that this closed loop should be propriately modeled and designed to ensure a low output voltage ripple.

7. REFERENCES

[1] S. Rajapandian, Z. Xu, and K. Shepard, "Charge-recycling voltage domains for energy-efficient low-voltage operation of digital cmos circuits," in *Computer Design, 2003. Proceedings. 21st International Conference on*, pp. 98–102, 2003.

[2] V. Kursan, S. G. Narendra, V. K. De, and E. G. Friedman, "Analysis of buck converters for on-chip integration with a dual supply voltage microprocessor," *IEEE Trans. VLSI Syst.*, vol. 11, pp. 514–522, June 2003.

[3] G. Schrom, P. Hazucha, J.-H. Hahn, V. Kursun, D. Gardner, S. Narendra, T. Karnik, and V. De, "Feasibility of monolithic and 3d-stacked dc-dc converters for microprocessors in 90nm technology generation," in *Proc. Intl. Symp. Low Power Electronics and Design*, pp. 263–268, 2004.

[4] A. J. Stratakos, S. R. Sanders, and R. W. Brodersen, "A low-voltage cmos dc-dc converter for a portable battery-operated system," in *Power Electronics Specialists Conference, PESC '94 Record., 25th Annual IEEE*, pp. 619–626, June 1994.

[5] A. J. Stratakos, "High-efficiency low-voltage dc-dc conversion for portable applications," *PhD dissertation, University of California, Berkeley*, 1998.

[6] D. W. Hart, ed., *Introduction to Power Electronics*. Prentice Hall, Upper Saddle River, N.J., 1997.

[7] G. Yin, F. Eynde, and W. Sansen, "A high-speed CMOS comparator with 8-b resolution," *IEEE Journal of Solid-state Circuits*, vol. 27, pp. 208–211, Feb. 1992.

[8] Y. Jang and M. Jovanovic, "A new zvs-pwm full-bridge converter," in *IEEE International Telecommunications Energy Conference*, pp. 232–239, Septemeber 2002.

[9] J. Cho, J. Sabate, G. Hua, and F. Lee, "Zero-voltage and zero-current-switching full bridge pwm converter for high power application," in *IEEE Transactions on Power Electronics*, pp. 102–108, May 1994.

Behavioral Modeling of Opamp Gain and Dynamic Effects for Power Optimization of Delta-Sigma Modulators and Pipelined ADCs

Anas A. Hamoui, T. Alhajj, and M. Taherzadeh-Sani

Department of Electrical & Computer Engineering, McGill University, Montreal, Canada
anas.hamoui@mcgill.ca

ABSTRACT

This paper proposes a simple, yet accurate, analytical model for the effect of opamp gain and dynamics (slew rate and bandwidth) on the transfer function of switched-capacitor (SC) amplifiers and integrators. Furthermore, it demonstrates the detrimental effects of: a) the nonlinear variation in the opamp dc gain; and b) the feedforward transmission of the feedback capacitor, on the harmonic distortion and settling behavior of these SC stages. These effects, typically ignored in the behavioral simulations of SC stages, are analyzed and modeled. Thus, accurate behavioral simulations of $\Delta\Sigma$ modulators or pipelined analog-to-digital converters (ADCs) can be performed in SIMULINK, using the proposed models for their SC building blocks (integrators or amplifiers). The proposed behavioral models are validated in HSPICE. Behavioral simulation examples are presented to illustrate the importance of such accurate modeling for low-power design.

Categories and Subject Descriptors: B.7.1 [Integrated Circuits]

General Terms: Design, Performance, Theory

Keywords: Behavioral modeling, synthesis, discrete-time systems, analog-to-digital conversion, sigma-delta ($\Sigma\Delta$) modulation.

1. INTRODUCTION

Switched-capacitor (SC) integrators and amplifiers (Fig. 1) are the basic building blocks of $\Delta\Sigma$ modulators and pipelined ADCs, respectively. One of the most challenging aspects in the behavioral simulation of an SC stage (amplifier or integrator) is the modelling of the circuit nonidealities in its opamp. Accordingly, this paper proposes an analytical model for the behavioral simulation of the effect of opamp gain and dynamics on the gain of an SC amplifier and the transfer function of an SC integrator.

Several tools for the high-level synthesis and behavioral simulation of pipelined ADCs [1] and $\Delta\Sigma$ modulators [2,3] have been developed. However, this paper focuses on modeling specific building blocks (amplifiers and integrators) for behavioral simulations in the widely-used and versatile SIMULINK tool. The goal is to develop analytical models which are reasonably accurate, while being simple, tractable, and meaningful to circuit designers (*only circuit-design parameters, no empirical values*).

Several SIMULINK-based models for the behavioral simulation of the gain of an SC amplifier [4] or the transfer function of an SC integrator [5,6] have been previously reported. Opamp dynamics in SC integrators have also been previously modeled [7]. However, in modeling the opamp circuit nonidealities and feedback-system behavior, many of these previously-reported models did not account for: a) the nonlinear variations in the opamp dc gain with its output voltage; and/or b) the feedforward transmission of the feedback capacitor C_F in an SC stage (Fig. 1).

Permission to make digital or hard copies of all or part of this work for personal or classroom use is granted without fee provided that copies are not made or distributed for profit or commercial advantage and that copies bear this notice and the full citation on the first page. To copy otherwise, or republish, to post on servers or to redistribute to lists, requires prior specific permission and/or a fee.
ISLPED'06, October 4-6, 2006, Tegernsee, Germany.
Copyright 2006 ACM 1-59593-462-6/06/0010...$5.00.

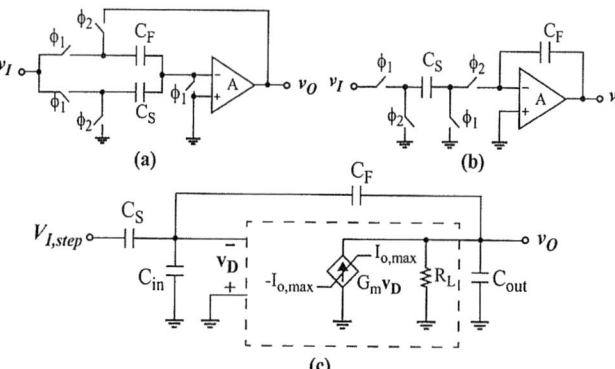

Fig. 1 (a) SC amplifier; (b) SC integrator; c) Configuration of (a) and (b) during the charge-transfer clock phase ϕ_2, with the opamp equivalent-circuit model shown. Here, C_S and C_F are the sampling and feedback capacitors; C_{in} and C_{out} are the total capacitances, including any parasitic, at the opamp's input and output nodes.

Both of these effects have a detrimental impact on the opamp settling behavior and harmonic distortion in an SC stage, especially when designing for low-power in nano-scale CMOS technologies (Section 6.2). In this paper, these effects are analyzed and modeled.

Accordingly, in Section 2, this paper develops a feedback model for an SC stage during its charge transfer phase. In Sections 3 and 4, it models the effect of the following opamp nonidealities on the transfer function of SC stages (Fig. 1):

1) finite dc gain A_0
2) nonlinear variations in dc gain A_0 with output voltage v_O
3) limited output-signal swing (output saturation voltage V_{Osat})
4) dynamic effects (finite bandwidth ω_{3dB} and slew rate SR)
5) parasitic capacitances (C_{in} and C_{out})
6) feedforward transmission of feedback capacitor C_F

In Section 5, SIMULINK models for SC amplifiers and integrators are proposed. In Section 6, simulation results are presented to validate the proposed behavioral models and demonstrate the importance of using these models for the accurate behavioral simulation and, hence, power optimization of ADCs.

2. EQUIVALENT FEEDBACK MODEL

2.1 Open-Loop Opamp Characteristics

The following analysis assumes a dominant-pole load-compensated opamp with a first-order transfer function

$$A(s) \equiv A_0/(1 + s/\omega_p) \qquad (1)$$

where $A_0 \equiv G_m R_L$ and $\omega_p \equiv 1/(R_L C_L)$ are the opamp's open-loop dc gain and dominant pole, respectively. Here, G_m is the short-circuit transconductance, R_L is the equivalent load resistance, and C_L is the equivalent load capacitance of the opamp (Fig. 1c).

2.2 Closed-Loop SC-Stage Characteristics

Consider an SC stage, as configured during its charge-transfer phase (Fig. 1c). During this clock phase, the feedback capacitor C_F provides not only signal feedback, but also signal feedforward.

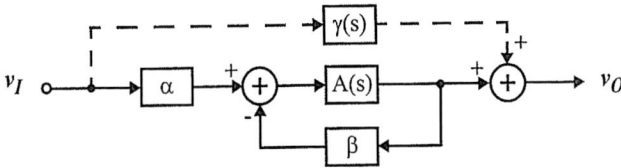

Fig. 2 Equivalent feedback model of an SC stage during its charge-transfer phase (Fig. 1c): a) series-shunt model ($\gamma=0$); b) return-ratio model (forward transmission γ modeled).

Typically, a *series-shunt feedback model* is used to represent the SC stage during its charge-transfer phase. This corresponds to the feedback model in Fig. 2, *without* the feedforward path (dashed line). Hence, in a series-shunt feedback model, the forward transmission due to feedback capacitor C_F is neglected compared to the larger forward transmission of the opamp [8]. However, this leads to detrimental modeling inaccuracies (Section 6.2.2).

In this paper, a *return-ratio feedback model* is utilized to more accurately model an SC stage during its charge-transfer phase. This corresponds to the feedback model in Fig. 2, with the feedforward path (dashed line) utilized to account for the forward transmission $\gamma(s)$ due to feedback capacitor C_F [9]. Accordingly, by considering the definitions and equivalent circuit of each term in the return-ratio feedback model, the closed-loop transfer function of the SC stage (Fig. 1c) can be expressed as:

$$A_{CL}(s) \equiv \frac{\alpha A(s)}{1 + \beta A(s)} + \gamma(s) = \alpha_0 K \frac{(1 - s/\omega_z)}{(1 + s/\omega_{3\text{dB}})} \quad (2)$$

α	$= C_S/(C_S + C_{in} + C_F)$	*Forward Factor*
β	$= C_F/(C_S + C_{in} + C_F)$	*Feedback Factor*
α_0	$= (A_0\beta)/(1 + A_0\beta)$	*Gain Factor due to Finite A_0*
K	$= C_S/C_F = \alpha/\beta$	*Nominal Capacitor Ratio*
$\omega_{3\text{dB}}$	$= \beta\,(G_m/C_L)/\alpha_0$	*Closed-Loop Dominant Pole*
ω_Z	$= G_m/C_F$	*Closed-Loop Transmission Zero*
C_L	$= C_{out} + \beta(C_S + C_{in})$	*Equivalent Open-Loop Load Cap.*

Hence, the forward transmission $\gamma(s)$ due to feedback capacitor C_F introduces a transmission zero ω_Z in the transfer function.

3. OPAMP DYNAMICS

Consider an SC stage, as configured during its charge-transfer phase (Fig. 1c). Its step response depends on the step response of its opamp. As depicted in Fig. 3, the opamp step response typically includes a slewing (nonlinear settling) region due to finite slew rate SR, followed by a linear settling region due to finite bandwidth $\omega_{3\text{dB}}$. In the following, the closed-loop step response $v_O(t)$ of the SC stage is derived, assuming a step voltage of height $V_{I,step}$ is applied at its input at time $t = 0$.

3.1 Linear Settling

During linear settling, the opamp output approaches its final value exponentially. Here, the step response of the SC stage is

$$V_O(s)\big|_{\text{LIN}} = \frac{V_{I,\text{step}}}{s} A_{CL}(s) \quad (3)$$

where $A_{CL}(s)$ is the closed-loop transfer function in (2). By taking the inverse-Laplace transform of (3), the step response of the SC stage can expressed in the time domain as

$$v_O(t)\big|_{\text{LIN}} = v_O(0^-) + V_{O,step}\,[1 - K_Z e^{-t/\tau}] \quad (4)$$

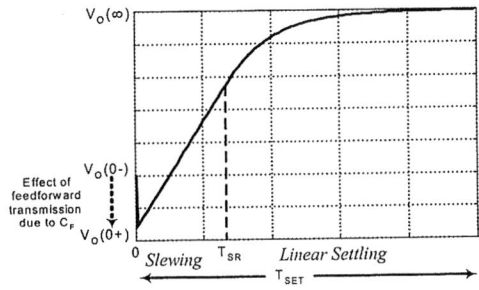

Fig. 3 Typical settling behavior of an SC stage in its charge-transfer phase (Fig. 1c), when a step voltage is applied at its input.

Here,

$V_{O,\text{step}}$	$\equiv v_O(\infty) - v_O(0^-) = \alpha_0 K V_{I,\text{step}}$	*Nominal Output Step*
K_Z	$\equiv 1 + \omega_{3\text{dB}}/\omega_Z$	*Transmission-Zero Factor*
τ	$\equiv 1/\omega_{3\text{dB}}$	*Closed-Loop Time Constant*

where $v_O(0^-)$ is the value immediately before $t = 0$ and $v_O(\infty)$ is the asymptotic final value of the opamp output.

Based on equation (4), the value of the opamp output immediately after $t = 0$ is

$$v_O(0^+) = v_O(0^-) + (1 - K_Z)V_{O,step} \quad (5)$$

Therefore, at the start of the charge-transfer phase, since $K_Z > 1$, the feedforward transmission due to feedback capacitor C_F causes a step change in the output in a direction opposite to its final value (Fig. 3). Hence, the *actual* output change required for complete settling is increased from $|V_{O,step}|$ to $|K_Z V_{O,step}|$. The resulting adverse effect on the settling and distortion errors is shown in Section 6.2.2.

3.2 Slewing

During slewing (nonlinear settling), the maximum rate at which the opamp output changes is limited by its slew rate SR. Here, the step response of the SC stage is

$$v_O(t)\big|_{\text{SR}} = v_O(0^+) + SR_0\, t \quad (6)$$

where $SR_0 \equiv SR \cdot sign(V_{O,step})$ to account for both rising and falling outputs.

3.3 Slewing followed by Linear Settling

Assume that the opamp step response has a slewing behavior followed by linear-settling (Fig. 3), as is typically the case in an SC stage. Let T_{SR} denote the slewing period. Then,

- For $0 < t \le T_{\text{SR}}$: slewing occurs. Hence, the step response is

$$v_O(t)\big|_{\text{SR/LIN}} = v_O(0^+) + SR_0\, t \quad (7)$$

- For $t \ge T_{\text{SR}}$: linear settling occurs and the output approaches $v_O(\infty)$ exponentially. Hence, the step response is

$$v_O(t)\big|_{\text{SR/LIN}} = v_O(T_{\text{SR}})\big|_{\text{SR}}\, e^{(T_{\text{SR}} - t)/\tau} + v_O(\infty)[1 - e^{(T_{\text{SR}} - t)/\tau}] \quad (8)$$

Based on the condition of continuity of the derivatives of $v_O(t)\big|_{\text{SR/LIN}}$ in (7) and (8) at $t = T_{\text{SR}}$, the slewing period is

$$T_{\text{SR}} = \frac{K_Z|V_{O,step}|}{SR} - \tau \quad (9)$$

Observe that, during linear settling, the maximum slope of the step response in (4) occurs at $t = 0$ and has a value of $K_Z|V_{O,step}|/\tau$. Therefore, slewing occurs if $K_Z|V_{O,step}|/\tau > SR$.

3.4 Modeling of Opamp Dynamics in SC Stages

In an SC stage, opamp dynamics cause errors in charge transfers between capacitors, during the charge-transfer phase. Let T_{SET} denote the settling period (the time available for charge transfer). Let $v_O(n)$ denote the output of the SC stage, *assuming no opamp dynamic limitations*. To account for opamp dynamics, the output of the SC stage must be expressed as $\hat{v}_O(n)$ with

- if $\dfrac{K_Z |V_{O,step}|}{\tau} < SR$ (no slewing): $\hat{v}_O(n) = v_O(T_{SET})\big|_{LIN}$ (10)

- else if $T_{SR} > T_{SET}$ (only slewing): $\hat{v}_O(n) = v_O(T_{SET})\big|_{SR}$ (11)

- else (slewing+linear): $\hat{v}_O(n) = v_O(T_{SET})\big|_{SR/LIN}$ (12)

Here, index n denotes time $t = n\,T_S$, where T_S is the clock period.

4. NON-LINEAR OPAMP DC GAIN

In a CMOS opamp, the dc gain varies with the output voltage due to the dependency of the output resistance r_{ds} of a MOS transistor on its drain-to-source voltage v_{DS}. The nonlinear variations in the opamp dc gain A_0 can be modeled as a function of the opamp output voltage v_O using [10]:

$$A_0(v_O) = \begin{cases} A_{0max}\left[1 - \left(\dfrac{v_O}{V_{Osat}}\right)^2\right] & \text{for } v_O \le V_{O,sat} \\ 0 & \text{for } v_O > V_{O,sat} \end{cases} \quad (13)$$

where A_{0max} is the maximum dc gain and V_{Osat} is the output saturation voltage of the opamp. In the behavioral modeling of an SC stage (in Section 5), equation (13) will be utilized to express the dc gain A_0 as a function of the SC-stage output voltage $\hat{v}_O(n)$.

5. BEHAVIORAL MODELING OF SC STAGES

The output of an SC amplifier (Fig. 1a) can be expressed as

$$v_O(n) = \alpha_0\,(1 + K)\,v_I\!\left(n - \frac{1}{2}\right) \quad (14)$$

Here, $1 + K = 1 + C_S/C_F$ is the amplifier gain and $v_I(n)$ is the amplifier input. The parameter α_o, as defined in (2), is the gain error due to the opamp's finite nonlinear dc gain A_0.

The output of an SC integrator (Fig. 1b) is expressed as [10]:

$$v_O(n) = \alpha_0\,K\,v_I(n-1) + \beta_0\,v_O(n-1) \quad (15)$$

where

$$\beta_0 = \alpha_o\left[1 + \frac{(1 + C_{in}/C_F)}{A_0}\right] \quad (16)$$

Here, $K = C_S/C_F$ is the integrator gain and $v_I(n)$ is the integrator input. Parameter α_o, as defined in (2), is the gain factor and parameter β_0 is the phase factor (shifted pole frequency) in the SC integrator, due to its opamp's finite nonlinear dc gain A_0.

Accordingly, the SIMULINK block models in Fig.4 and Fig. 5 can be utilized for the behavioral simulation of SC amplifiers (Fig. 1a) and integrators (Fig. 1b), respectively. They account for the effects of all opamp nonidealities listed in Section 1 on the amplifier gain in (14) and on the gain and phase of the integrator transfer function in (15), as described above.

In Fig. 4 and Fig. 5:

- The function block *opamp_gain* computes the opamp dc gain A_0 as a function of the SC stage output $\hat{v}_O(n)$, as per equation (13). The function blocks *alpha_0* and *beta_0* compute, respectively, α_o and β_o based on A_0 and the SC stage capacitances, as per equations (2) and (16).

- A saturation (limiter) block is placed at the SC stage output to model the output saturation level V_{Osat} of its opamp.

- The function block labelled *SR/BW* models the opamp dynamics, as per expressions (10)-(12). Its input variables are $v_O(n)$, $v_O(0^-)$, and α_o. Observe that:
  - In an SC amplifier, the output is reset during the sampling phase. Therefore, $v_O(0^-) = 0$ is used in Fig. 4.
  - In an SC integrator, capacitor C_F is not reset during the sampling phase, but holds the previous integration-phase charge. Therefore, $v_O(0^-) = \hat{v}_O(n-1)$ is used in Fig. 5.

The SIMULINK model in Fig. 5 models a non-inverting, delaying SC integrator. For a non-delaying integrator, simply delete the delaying block at the integrator input. For an inverting integrator, replace gain K with $-K$.

6. BEHAVIORAL SIMULATION RESULTS

6.1 Gain Error

Assume a step voltage of height $V_{I,step}$ is applied at the input of an SC stage during its charge-transfer phase (Fig. 1c). With an ideal opamp, the resulting change at the output is $K V_{I,step}$. Therefore, the gain error of the SC stage can be defined as

$$\delta g \equiv 1 - \frac{v_O(T_{SET}) - v_O(0^-)}{K V_{I,step}} \quad (17)$$

Typically, the step response of an SC stage is characterized by slewing followed by linear settling. Therefore, the gain error can be expressed, using (8) and (17), as

$$\delta g = \frac{1}{1 + \beta A_0} + \frac{SR\tau}{K V_{I,step}} e^{(T_{SR} - T_{SET})/\tau} \quad (18)$$

Observe that the left-hand term accounts for the gain error due to the opamp finite gain, while the right-hand term represents the gain error due to the opamp dynamics (finite bandwidth and slew rate).

To validate the accuracy of the proposed behavioral models, the equivalent circuit model in Fig. 1c was simulated in HSPICE, with the opamp's slew rate modeled by limiting the maximum current of its transconductance G_m. The initial output voltage was set to zero.

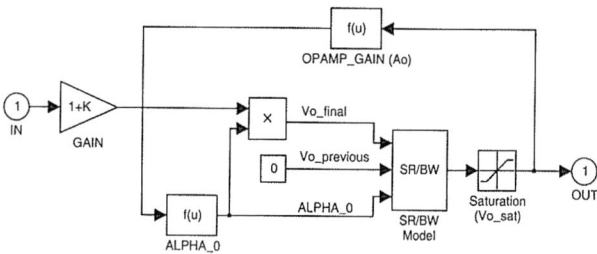

Fig. 4 SIMULINK model of an SC amplifier (Fig. 1a).

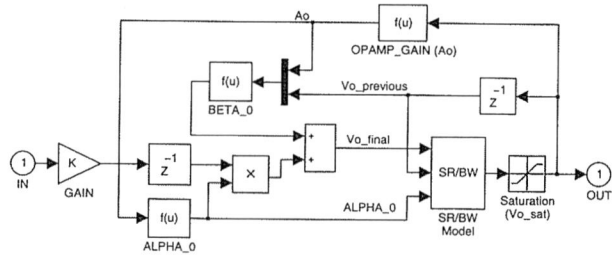

Fig. 5 SIMULINK model of a non-inverting delaying SC integrator (Fig. 1b)

Fig. 6 shows the results for δg versus $V_{I,step}$, computed based on HSPICE simulations and on the proposed behavioral models in equation (18). The excellent agreement between these results confirms the accuracy of the proposed behavioral models.

6.2 ADC Power Optimization

6.2.1 Opamp Nonlinear DC Gain vs. Harmonic Distortion

Consider a 3rd-order 5-bit $\Delta\Sigma$ modulator, with its noise transfer function having 1 zero at dc and 2 complex-conjugate zeros optimally placed within the signal band [11]. The modulator was simulated in SIMULINK with its 1st integrator modeled using the proposed behavioral model (Fig. 5), while the 2nd and 3rd integrators were assumed to be ideal.

Fig. 7 shows the signal-to-noise-plus-distortion ratio (SNDR) of the $\Delta\Sigma$ modulator versus the maximum dc gain A_{0max} of its opamp, when: a) the dc gain is constant ($A_0 = A_{0max}$); and b) the dc gain varies nonlinearly with the opamp output, as per (13). Accordingly, when designing using moderate-gain opamps (~50dB), the SNDR is significantly degraded (>8dB) due to nonlinear dc-gain variations. Thus, the effect of nonlinear dc gain cannot be neglected in behavioral simulations, especially when designing using moderate gain opamps: For example, in nano-scale CMOS technologies, high dc gains for the opamps are difficult to achieve at low power, due to the low supply voltages and the poor intrinsic gains of the MOS transistors.

6.2.2 Feedforward Transmission of C_F vs. Power Dissipation

A 2nd-order 1-bit $\Delta\Sigma$ modulator was simulated in SIMULINK, with its 1st integrator modeled using the proposed behavioral model (Fig. 5) to account for the opamp nonidealities. Assume:

- opamp's short-circuit transconductance: $G_m = I_{BIAS}/V_{OV}$
- opamp's slew rate: $SR = I_{BIAS}/C_L$;
- integrator closed-loop time constant: $\tau = (C_L/\beta)(V_{OV}/I_{BIAS})$

where I_{BIAS} is the opamp bias current and V_{OV} is the overdrive voltage of the opamp transistors. Accordingly, by expressing τ and SR in terms of I_{BIAS} in the proposed models, behavioral simulations can be performed to find the optimum I_{BIAS} required to achieve a given linearity and resolution.

Fig. 8 shows the modulator SNDR versus its opamp bias current I_{BIAS}, with: a) $K_Z = 1$ to neglect the effect of feedforward transmission due to the feedback capacitor C_F; and b) $K_Z = 1 + \omega_{3dB}/\omega_Z$. Accordingly, the estimated I_{BIAS} for achieving a given SNDR can be significantly underestimated and, more drastically, the modulator can be unstable when biased at this estimated current level. For example, with the effect of feedforward

transmission due to C_F modeled, behavioral simulations show that a normalized $I_{BIAS} = 1$ is required to achieve an SNDR = 71 dB. Furthermore, a normalized $I_{BIAS} \geq 0.75$ is required to ensure that the $\Delta\Sigma$ modulator is stable. However, behavioral simulations with $K_Z = 1$ would have erroneously predicted that a normalized $I_{BIAS} = 0.45$ is sufficient to achieve an SNDR = 71dB, while in practice the $\Delta\Sigma$ modulator is unstable at this bias-current level.

7. CONCLUSION

The effects of opamp gain and dynamics on SC amplifier gain and SC integrator transfer function were analyzed and modeled. SIMULINK models of SC amplifiers and integrators were also developed and validated using HSPICE simulations. It was shown that the nonlinear dc gain variations in the opamp and the feedforward transmission of the feedback capacitor in an SC stage, typically ignored in the behavioral simulations of SC stages, have detrimental effects on the harmonic distortion and settling behavior. SIMULINK simulation results were presented to show the need for the proposed accurate models in low power design.

8. REFERENCES

[1] J. Ruiz-Amaya et. al. , "Behavioral modeling, simulation and high-level synthesis of pipeline A/D converters," in *Proc. IEEE Int. Symp. Circuits Syst.*, May 2005, pp. 5609 - 5612.

[2] K. Francken and G. Gielen, "A high-level simulation and synthesis environment for $\Delta\Sigma$ modulators," *IEEE Trans.Comput.-Aided Des. Integr. Circuits Syst.*, vol. 22, no. 8, pp. 1049-1061, Aug. 2003.

[3] J. Ruiz-Amaya et al. , "High-level synthesis of switched-capacitor, switched-current and continuous-time $\Sigma\Delta$ modulators using SIMULINK-based time-domain behavioral models," *IEEE Trans. Circuits Syst. I*, vol. 52, no. 9, pp. 1795-1810, Sept. 2005.

[4] E. Bilhan et. al , "Behavioral model of pipeline ADC by using SIMULINK," in *Proc. Southwest Symp. Mixed-Signal Design*, Feb. 2001, pp. 147-151.

[5] P. Malcovati et al., "Behavioral modeling of switched-capacitor sigma-delta modulators," *IEEE Trans. Circuits Syst. I*, vol. 50, pp. 352-364, Mar. 2003.

[6] H. Zare-Hoseini, I. Kale, and O. Shoaei, "Modeling of switched-capacitor delta-sigma modulators in SIMULINK," *IEEE Trans. Instrumentation and Measurement*, vol. 54, no. 4, Aug. 2005, pp. 1646 - 1654.

[7] R. del Rio et al., "Reliable analysis of settling errors in SC integrators - application to the design of high-speed modulators," in *Proc. IEEE Int. Symp. Circuits Syst.*, May 2000, vol. 4, pp. 417-420.

[8] A. S. Sedra & K. C. Smith, *Microelectronic Circuits*, 5th ed. Oxford: 2003.

[9] P. R. Gray, P. J. Hurst, S. H. Lewis, and R. G. Meyer, *Analysis and Design of Analog Integrated Circuits*, 4th ed. Wiley: 2001.

[10] A. A. Hamoui and K. Martin, *Delta-Sigma Data Converters in Low-Voltage CMOS for Broadband Digital Communication*. Dordrecht, Netherlands: Springer, to be published in 2007.

[11] A. A. Hamoui and K. Martin, "High-order Multibit Modulators and Pseudo Data-Weighted-Averaging in Low-Oversampling $\Delta\Sigma$ ADCs for Broad-band Applications," *IEEE Trans. Circuits Syst. I*, vol. 51, no. 1, pp. 72-85, Jan. 2004.

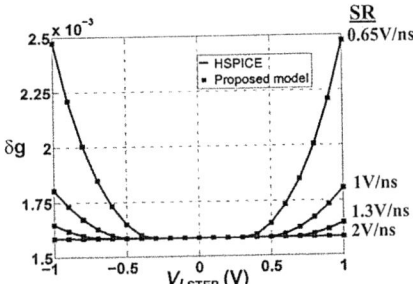

Fig. 6 Gain error vs. input step voltage for various SR values ($C_S = C_F = C_{out} = 2$pF; $C_{in} = 0.2$pF; $R_L = 200$kΩ; $G_m = 10$mA/V).

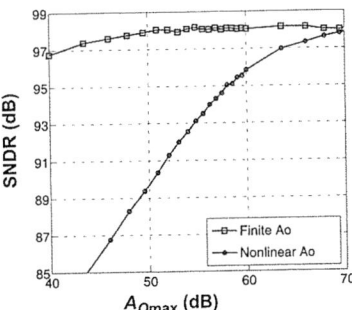

Fig. 7 SNDR vs. maximum dc gain A_{0max}

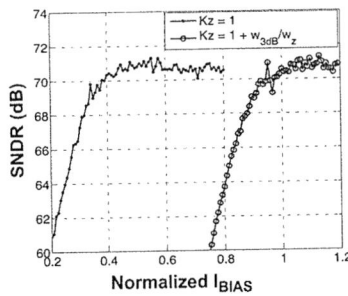

Fig. 8 SNDR vs. normalized bias current I_{BIAS}

Low-power Fanout Optimization Using MTCMOS and Multi-Vt Techniques

Behnam Amelifard
Department of EE-Systems
University of Southern California
Los Angeles, CA
(213) 740-9481
amelifar@usc.edu

Farzan Fallah
Fujitsu Laboratories of America
Sunnyvale, CA
(408) 530-4544
farzan@fla.fujitsu.com

Massoud Pedarm
Department of EE-Systems
University of Southern California
Los Angeles, CA
(213) 740-4458
pedram@ceng.usc.edu

ABSTRACT
This paper addresses the problem of low-power fanout optimization. We show that due to neglecting short-circuit current, previous analytical techniques proposed to optimize the area of a fanout tree may result in excessive power consumption. This shows to achieve a low-power fanout tree, an accurate power consumption model should be used as the objective function. Moreover, we propose an efficient method to minimize the total power consumption of a fanout tree by using MTCMOS and Multi-Vt techniques. Experimental results show that depending on the activity factor of the circuit, the proposed technique can reduce the power consumption of the fanout tree 18% to 45%.

Categories and Subject Descriptors
B.6.3 [**Design Aids**]: Automatic synthesis, Optimization

General Terms
Algorithms, Design, Performance

Keywords
Low-power design, fanout optimization, fanout tree, buffer chain

1. INTRODUCTION
Fanout optimization, an operation performed in logic synthesis, is the problem of building an inverter tree topology between a source and some sinks and sizing the inverters in this topology so that the driving capacitance at the source is less than an upper bound and the timing constraints at sinks are met [1][2]. Different objective functions have been considered for the fanout optimization problem such as minimizing area [2][3][4], minimizing power consumption [3][5], and minimizing load on the source [6]. In this paper we minimize the total power consumption. Since both dynamic and leakage power dissipation of an inverter chain are proportional to its area, it has been widely accepted that power minimization of the fanout tree is equivalent to its area optimization [3][5]. In this paper, however, we show that due to short-circuit power dissipation, minimizing area does not necessarily result in a minimized power dissipation solution and the solution obtained from an area optimization technique may dissipate excessive short-circuit power.

To reduce both the active power and the standby leakage power, we utilize multi-Vt and MTCMOS techniques in a fanout tree. For doing this, at the first step, we use high-threshold voltage inverters in the fanout tree to reduce the leakage power

consumption in both active and standby modes. If due to a high delay penalty high-Vt inverters cannot be used in the chain, then by using the MTCMOS technique, we try to reduce the leakage power consumption in the standby mode.

The remainder of the paper is organized as follows. After presenting the preliminaries in Section 2, in Section 3 the problem of low-power fanout optimization with one sink, i.e., an inverter chain, is formulated. Section 4 shows how a low-power fanout tree can be designed from the power-optimized inverter chains. Simulation results are given in Section 5, while Section 6 concludes the paper.

2. PRELIMINARIES
2.1 Delay Model
In logical effort, the delay of a gate with input capacitance C_{in}, which drives the load capacitance C_L, is modeled as,

$$D = \tau_0(p + gh) \qquad 1$$

where τ_0 is a technology-dependent parameter, g is the logical effort of the gate, $h = C_L/C_{in}$ is the electrical effort and p is the parasitic delay of the gate. τ_0 is a constant and without losing generality it can be assumed to be one. For an inverter, the value of logical effort g equals one and p is the ratio of diffusion to input capacitance of the template inverter, denoted by p_0, i.e., $p_0 = C_{diff,T}/C_{in,T}$. Since both input and diffusion capacitances of an inverter are scaled linearly when changing the size of inverter, for a scaled inverter the ratio of diffusion-to-gate capacitance remains constant. i.e.,

$$p_0 = C_{diff} / C_{in} \qquad 2$$

where C_{diff} is the diffusion capacitance at the output.

In a multi-Vt technology, the values of the logical effort and parasitic delay change as follows [5][8],

$$g_h = \frac{(V_{dd} - V_{t,0})^\alpha}{(V_{dd} - V_{t,h})^\alpha}, \quad p_h = p_0 \frac{(V_{dd} - V_{t,0})^\alpha}{(V_{dd} - V_{t,h})^\alpha} \qquad 3$$

where g_h and p_h are the logical effort and parasitic delay for an arbitrary $V_{t,h}$ threshold voltage, $V_{t,0}$ is threshold voltage of the template inverter and V_{dd} is the supply voltage and α is a technology parameter which is around 1.3 for short-channel devices.

In an MTCMOS circuit when the sleep transistor is ON, it can be modeled as a resistor whose resistance is inversely proportional to its width; hence, for an inverter [7],

$$g^r = 1, \quad p^r = p_0 \qquad 4$$

and,

$$g^f = (1 + \kappa \frac{C_{in}}{w}), \quad p^f = (1 + \kappa \frac{C_{in}}{w})p_0 \qquad 5$$

where g^r and p^r (g^f and p^f) are the logical effort and parasitic delay for the rise (fall) delay, w is the width of the sleep transistor and κ is a constant which depends on the technology and the threshold voltage of the sleep transistor.

Permission to make digital or hard copies of all or part of this work for personal or classroom use is granted without fee provided that copies are not made or distributed for profit or commercial advantage and that copies bear this notice and the full citation on the first page. To copy otherwise, or republish, to post on servers or to redistribute to lists, requires prior specific permission and/or a fee.
ISLPED'06, October 4–6, 2006, Tegernsee, Germany.
Copyright 2006 ACM 1-59593-462-6/06/0010...$5.00.

2.2 Power Dissipation Model

The power dissipation of an inverter has three components: dynamic power, short circuit power, and leakage power. The dynamic power is equal to,

$$P_{dyn} = \chi f V_{dd}^2 C \qquad 6$$

where χ is the switching activity of the inverter, f is the frequency, and C is the sum of the input gate capacitance and output diffusion capacitance of the inverter, i.e., $C = C_{diff} + C_{in}$. By using (2), (6) can be re-written as,

$$P_{dyn} = \chi f V_{dd}^2 (1 + p_0) C_{in} \qquad 7$$

The second source of power dissipation in digital circuit is due to the short-circuit current. Several techniques have been proposed to address the problem of short circuit power estimation [10], but due to their complexity, they may not be very useful during a gate-level optimization process. In this paper, by observing the fact that short-circuit power dissipation of an inverter is a linear function of its size and input transition time [10], and also the fact that input transition time itself can be approximated as a linear function of the electrical effort of the previous stage in the chain, the short-circuit power dissipation of the ith inverter in a chain is modeled as,

$$P_{sc} = K_{sc} h_{i-1} C_{in} \qquad 8$$

where K_{sc} is a technology-dependent parameter, h_{i-1} is the electrical effort of the $i\text{-}1^{th}$ inverter and C_{in} is the input capacitance of the ith inverter. Transistor level SPICE simulations show this technique, despite its simplicity, is accurate enough to be used in a gate-level optimization technique.

The third source of power dissipation is leakage. In current technologies, the major components of leakage current are subthreshold and gate-tunneling currents [9]. The total leakage power dissipation of an inverter can be modeled as

$$P_{leak} = (K_{sub} + K_{ox}) C_{in} \qquad 9$$

where K_{sub} and K_{ox} are technology parameters which depend on the effective channel length, oxide thickness, temperature, and supply voltage. Moreover, K_{sub} is also a function of the threshold voltage.

Having had different components of the power consumption, the total power dissipation of inverter i in a chain can be expressed as,

$$P = P_{dyn} + P_{leak} + P_{sc} = C_i (K_{dyn} + K_{sub} + K_{ox} + K_{sc} h_{i-1}) \qquad 10$$

where C_i is the input capacitance of inverter i and $K_{dyn} = \chi f V_{dd}^2 (1 + p_0)$.

3. LOW-POWER INVERTER CHAINS

In our approach, to construct the low-power fanout tree topology and size the inverters in the tree, the problem is decomposed into sub-problems in the forms of inverter chains, and each sub-problem is separately solved for each sink. The solutions to the sub-problems are then merged to find the solution to the main problem. So, in this section we formulate the problem of minimizing power dissipation of an inverter chain under timing and input capacitance constraints, i.e.,

$$\begin{cases} \min & Power \\ s.t & (1) \quad Delay \leq T \\ & (2) \quad C_1 \leq C_{\max} \end{cases} \qquad 11$$

where T is the timing constraint on the sink, C_1 is the input capacitance of the first inverter, and $C_{\max}$ is the maximum tolerable load on the source.

Since both dynamic and leakage power dissipation of an inverter are proportional to its size, if short-circuit power consumption is ignored, the problem of finding the minimum power consumption inverter chain is the same as finding the minimum area inverter chain. In [2] the problem of minimizing the area of an inverter chain given a constraint on the delay of the chain and a constraint on the load of the source has been formulated using logical effort. By using Lagrangian relaxation technique [11], it can be shown that when the input capacitance constraint of the fanout chain is "loose", i.e., in the optimal solution $C_1 < C_{\max}$, such a formulation results in a solution in which the following relation holds among the sequence of electrical efforts,

$$h_{i+1} = h_i (h_i - h_{i-1} + 1) \qquad 12$$

where h_0 is defined as 0 and h_1 can be found from solving this polynomial equation,

$$\sum_{i=1}^n h_i = T - n p_0 \qquad 13$$

It can be verified that in (12),

$$h_{i+1} > h_1^{2^i} \qquad 14$$

From (14), it is easy to see the value of h_i's grow exponentially and based on (8), the short circuit power dissipation of the inverters grows very fast.

Based on the above fact, we give a precise objective function for minimizing the total power dissipation of an inverter chain. In order to simplify the equation, without losing generality, we assume the driver and load of the chain are fixed-sized inverters. The driver is called 0th inverter, while the load is called $n+1^{th}$ inverter. Hence, the power dissipation of the inverter chain (i.e., the objective function of (13)) can be modeled as,

$$Power = \tilde{C}_L \sum_{i=1}^n \frac{1 + k_\phi h_{i-1}}{\prod_{j=i}^n h_j} + \tilde{C}_L k_\phi h_n \qquad 15$$

where $\tilde{C}_L = C_L (K_{dyn} + K_{sub} + K_{ox})$ and $k_\varphi = K_{sc} / (K_{dyn} + K_{sub} + K_{ox})$.

Since the size of the load is fixed, the dynamic and leakage power dissipation of the load inverter are constant; however, the short-circuit power consumption of this inverter is a function of the electrical effort of the last stage in the chain. Therefore, we have included the short-circuit power dissipation of the load into the objective function as the last term.

The constraints of (11) in logical effort notion are similar to those in [2], i.e., the delay constraint can be expressed as,

$$Delay = \sum_{i=1}^n (p_0 + h_i) \leq T \qquad 16$$

while the input capacitance constraint can be written as,

$$C_1 = C_L / \prod_{i=1}^n h_i \leq C_{\max} \qquad 17$$

Therefore, problem (11) is a minimization of a posynomial function with posynomial inequality constraints that can be easily solved in polynomial time [11]. Notice that to find the minimum inverter chain, the abovementioned mathematical program should be solved for different values of n. The upper and lower bounds of n are similar to those in [2] and [5]; however, based on the polarity of the sink node, only even or odd numbers of inverters between these bounds are considered when searching for the optimum solution [5].

Although by solving the above mathematical problem the total power consumption in the active mode is reduced, the standby leakage power consumption is weakly decreased. Many techniques have been proposed to reduce the standby leakage power, while maintaining high performance in the active mode. A combination of MTCMOS [12] and multi-Vt techniques has been shown to be very effective in reducing the standby leakage power dissipation [13]. In this scheme, by using high-Vt transistors in the non-critical paths their active and standby leakage power consumption is reduced. For the gates on the critical path low-Vt transistors are used to achieve the high performance but MTCMOS technique is applied to these gates to reduce the standby subthreshold current. We use a similar technique to suppress standby-mode leakage power consumption of our fanout trees.

Notice if the threshold voltage of all inverters in the inverter chain increases to $V_{t,h}$, (16) should be modified to,

$$Delay = g_h\left(\sum_{i=1}^{n} p_0 + h_i\right) \leq T \qquad 18$$

where g_h is obtained from (3). Moreover, due to exponential reduction of the subthreshold current, $\tilde{C}_L$ in (15) should be changed to,

$$\tilde{C}_L = C_L(K_{dyn} + K_{sub}\exp(V_{t,0} - V_{t,h}) + K_{ox}) \qquad 19$$

In the following sub-section we show how to modify this mathematical program for the case that MTCMOS technique is used.

3.1 Low-power MTCMOS inverter chain

Figure 1 shows three different ways to build an MTCMOS inverter chain. Although the techniques shown in Figure 1.b and 1.c seem to be more area-efficient, they are not compatible with the merge transformations that we are going to use to build the fanout tree from the inverter chains. Hence, in the remainder of this paper we assume the structure of Figure 1.a for the MTCMOS inverter chain.

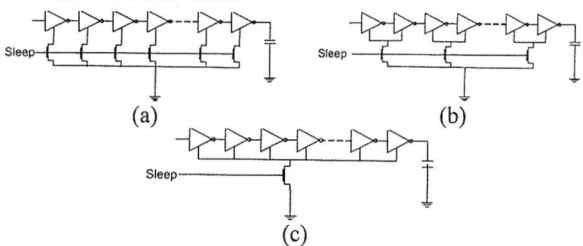

(a)   (b)

(c)

Figure 1. Different scenarios for implementing MTCMOS inverter chains

By using (3) and (4), the rise and fall delays of the inverter chain from sink to source, d^r and d^f, can be expressed as functions of the electrical efforts of the inverters in the chain. Since d^f and d^r are not equal, we define the delay of the inverter to be the maximum of the fall and rise delays. To minimize the total power consumption of the MTCMOS inverter chain, the delay constraint in (11) must be modified as,

$$Delay = \max\{d^f, d^r\} \leq T \qquad 20$$

The objective function of (11) also needs to be modified to model the gate-tunneling current of the sleep transistors in the active mode. Thus, (15) should be modified as,

$$Power = \sum_{i=1}^{n}\left(\tilde{C}_L \frac{1 + k_\phi h_{i-1}}{\prod_{j=i}^{n} h_j} + \tilde{K}_{ox} w_i\right) + \tilde{C}_L k_\phi h_n \qquad 21$$

On the other hand, in practice there is a budget for the total size of sleep transistors in the chain. So, in the mathematical program (11) a third constraint should be added to limit the total size of the sleep transistors. With these modifications, the low-power MTCMOS inverter chain optimization can be expressed as,

$$\begin{cases} \min \quad Power \\ s.t \quad (1) \quad Delay \leq T \\ \qquad (2) \quad C_1 \leq C_{\max} \\ \qquad (3) \quad Sleep\ Transistor\ Size \leq W_0 \end{cases} \qquad 22$$

where W_0 is the budget on the size of the sleep transistor.

4. BUILDING A FANOUT TREE

In this section we show how to build a fanout tree with more than one sink. The typical fanout tree we want to build is shown in Figure 2, where the first m sinks are not on the critical path

and hence the corresponding tree can be designed using high-Vt devices, while the next k sinks are on the critical path and hence

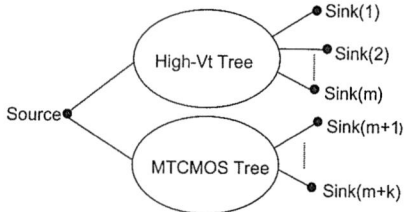

Figure 2. MTCMOS and High-Vt fanout trees

MTCMOS technique should be used for them. [6] introduced two transformations that could be performed on a fanout tree, namely merging and splitting and it showed that these transformations maintain the same area, delay, and input capacitance. We have extended the merging and splitting techniques, as shown in Figure 3, to handle MTCMOS fanout trees.

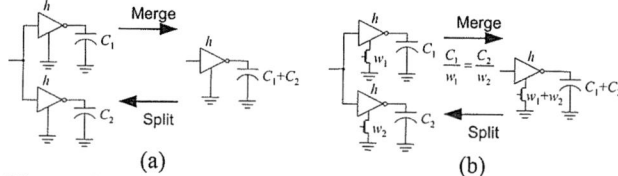

(a)   (b)

Figure 3: Split and merge transformations (a) original transformations (b) extended transformation for MTCMOS inverters

It can be shown that split/merge transformations and their extended versions applied to a fanout tree preserve the delay, input capacitance, and power dissipation values of the tree. By using these transformations, any fanout optimization problem with N sink nodes can be converted to N inverter chain optimization problems, whose respective power dissipation will be the same. To apply such transformations two issues should be addressed. The first issue is the input capacitance allocation to different chains in a decomposed fanout tree. It was shown in [2] that this problem is NP-complete and a heuristic, which we use in this paper, has been developed to allocate the input capacitance. The second issue is finding the inverter chains for which the high-Vt can be used and allocating the sleep transistor width among other chains. To address these problems, after the input capacitance allocation, we determine which inverter chains can work with the high-Vt. For an inverter chain whose allocated input capacitance is $C_{\max,i}$, the load capacitance is $C_{L,i}$, and the required time at the sink is T_i, this can be done by checking if the following relation holds for at least one even or odd n depending on the polarity of the sink,

$$S_{i,n} = T_i - \left(ng_h(C_{L,i}/C_{\max,i})^{1/n} + ng_h p_0\right) \geq 0 \qquad 23$$

Note the value inside the parentheses is the minimum delay of a high-Vt inverter chain with n inverters. In this case, we say the sink is *non-critical* and the corresponding chain can work with high-Vt devices without violating the timing constraint of the sink. On the other hand, if (23) does not hold for any n, the sink is called *critical*. In this case an MTCMOS inverter chain should be used in the corresponding chain. However, it should be noticed the circuit at the end of the MTCMOS chain needs to be in standby mode whenever the chain is in standby mode; otherwise, very high short circuit current flows through the circuit. If the critical sink does not drive an MTCMOS gate, only low-Vt inverters (without sleep transistors) are used in the corresponding chain.

To allocate the sleep transistor width to different MTCMOS inverter chains, we use the following heuristic. From the set of constraints of (22) it can be seen that in an MTCMOS inverter chain with n inverters, the power cost is a decreasing function of

Table 1: Comparison between *MinPowerFO* and *MinAreaFO* for a few inverter chains

Circuit	Circuit Specification				Fanout Type	ΔA (%)	ΔP (%)				
	C_{in}	C_{out}	T	P			$\acute{a}$=10%	$\acute{a}$=30%	$\acute{a}$=50%	$\acute{a}$=70%	$\acute{a}$=90%
C1	1	100	23	+	high-Vt	54.11	53.78	29.55	19.75	14.44	11.12
C2	2	135	20	+	high-Vt	154.81	44.57	30.20	25.73	23.56	22.27
C3	2	100	21	-	high-Vt	152.11	43.34	33.97	31.47	30.32	29.65
C4	2	100	17	-	MTCMOS	196.81	57.59	34.05	24.82	19.88	16.81
C5	2	70	15	+	MTCMOS	166.49	48.14	26.07	18.67	14.95	12.72
C6	4	550	20	-	MTCMOS	204.94	49.18	27.37	20.01	16.32	14.09
Average						154.87	49.43	30.20	23.41	19.91	17.78

Table 2: Comparison between *MinPowerFO* and *MinAreaFO* for a few fanout optimization problems

Circuit	Circuit Specification						ΔA (%)	ΔP (%)				
	C_{in}	$C_{out,max}$	T_{min}	T_{max}	$P+$	$P-$		$\acute{a}$=10%	$\acute{a}$=30%	$\acute{a}$=50%	$\acute{a}$=70%	$\acute{a}$=90%
T1	25	550	15	23	2	3	87.33	48.35	38.79	29.23	19.67	10.12
T2	20	1100	14	50	3	3	189.34	39.95	34.52	29.09	23.66	18.23
T3	17	135	40	90	4	1	163.44	52.34	46.38	40.43	34.47	28.52
T4	14	550	9	32	1	5	209.43	34.55	30.74	26.94	23.14	19.34
T5	10	70	12	52	2	6	121.22	57.77	47.18	36.60	26.01	15.43
T6	14	100	12	21	7	3	178.91	39.44	34.88	30.32	25.76	21.21
Average							158.27	45.40	38.74	32.10	25.45	18.80

the available slack defined as (23). Since using sleep transistors in the chain incurs delay overhead and reduces the available slack, we allocate the sleep transistor budget in a way that a larger transistor width is assigned to a chain with less slack, i.e.,

$$W_i = \frac{1/S_i}{\sum_{j=1}^{k_0} 1/S_j} W_{tot} \qquad 24$$

where $k_0 \leq k$ is the number of MTCMOS chains, W_i ($1 \leq i \leq k_0$) is the width of the sleep transistor allocated to the ith chain, $S_i = \max_n \{S_{i,n}\}$ is the slack of the ith chain, and W_{tot} is the total budget for sleep transistor width.

5. SIMULATION RESULTS

The proposed technique in Sections 3 and 4, which we call *MinPowerFO*, has been developed in MATLAB optimization toolbox. To study the efficiency of our technique in reducing the power consumption of the fanout trees, we performed a set of experiments and compared the results of *MinPowerFO* with the results of *MinAreaFO*, which minimizes the area of the fanout tree [2]. The technology parameters we used in these sets of experiments are based on a 65nm technology node [14] and have been obtained by transistor level simulation of devices. In this technology, the supply voltage is 1.0V and the values of low and high threshold voltages are 0.2V and 0.3V, respectively. Moreover, the oxide thickness of both NMOS and PMOS transistors is 17A^0. Simulation results for a few random problems, in the form of inverter chains, are shown in Table 1. In this table, C_{in} is the maximum allowed capacitance at the input of the inverter chain, C_{out} is the sink load, T is the required time at the sink, and P is the polarity of the sink. In each case, the constraint on the size of the sleep transistor has been assumed to be half of the total size of inverters in the minimum area solution. The power dissipation of circuits using these techniques has been compared for different activity factors $\acute{a}$ (i.e., the percentage of the time the circuit is in the active mode). In this table, ΔA is the area increase of the *MinPowerFO* compared to that of *MinAreaFO*, while ΔP is the power reduction of the *MinPowerFO* technique compared to that of *MinAreaFO*.

In the second set of experiments, the fanout optimization problem is solved for a group of arbitrary problems. Each problem states one source and multiple sinks with capacitive load, required time, and polarity constraints specified for each sink. The specification of each circuit, including the maximum input capacitance (C_{in}), the number of sinks with positive and negative polarities ($p+$ and $p-$), the maximum and minimum required times of all sinks (T_{max}

and T_{min}), and the maximum sink capacitances ($C_{out,max}$), are shown in Table 2. From the table, one can see depending on the activity factor of the fanout circuit, the average power reduction ranges from 18% to 45%.

6. CONCLUSION

In this paper we showed that the fanout optimization with area and power objective functions are not the same and a fanout tree optimized for area may dissipate excessive short-circuit power. By modeling all components of power dissipation, we formulated the fanout optimization problem as a geometric program for a circuit with one sink. To reduce standby power consumption, we proposed using multi-Vt and MTCMOS fanout trees, where high-Vt fanout tree is constructed for the sinks on the non-critical paths, while the MTCMOS fanout tree is constructed for the sinks on the critical paths. Experimental results show the proposed technique is very effective in reducing the total power consumption of fanout trees for various activity factors.

7. REFERENCES

[1] Salek, A., *et al.* Hierarchical buffered routing tree generation. *IEEE Trans. on CAD*, 21, (May 2002), 554-567.
[2] Rezvani, P., *et al.* A fanout optimization algorithm based on the effort delay model. *IEEE Trans. on CAD*, 22, (Dec. 2003), 1671-1677.
[3] Zhou, D., Liu, X. Minimization of chip size and power consumption of high-speed VLSI buffers. In *Proc. ISPD*, 1997, 186-191.
[4] Singh, K. J., *et al.* A heuristic algorithm for the fanout problem. In *Proc. DAC*, 1990, 357–360.
[5] Amelifard, B., *et al.*, Low-power fanout optimization using multiple threshold voltage inverters. In *Proc. ISLPED*, 2005, 95-98.
[6] Kung, D. S. A fast fanout optimization algorithm for near-continuous buffer libraries. In *Proc. DAC.*, 1998, 352-355.
[7] Sutherland, I., *et al. Logical Effort: Designing Fast CMOS Circuits.* Morgan Kaufmann, San Fransisco, CA, 1999.
[8] Sakurai, T., *et al.* A simple MOSFET model for circuit analysis. *IEEE Trans. Electron Device*, 38 (Apr. 1991), 887-894.
[9] De, V., *et al.* Techniques for leakage power reduction. in *Design of High-Performance Microprocessor Circuit, Circuits*, Chandrakasan, A., *et al.*, IEEE , Piscataway , NJ, 2001.
[10] Pedram, M. Power minimization in IC design: Principles and applications. *ACM Trans. on Design Automation of Electronic Systems*, 1,1 (Jan. 1996), 3-56.
[11] Gill, P. E., *et al. Practical Optimization*, Academic Press, New York, 1981.
[12] Anis, M., *et al.* Dynamic and leakage power reduction in MTCMOS circuits using an automated efficient gate clustering Technique. In *Proc. DAC*, 2002, 480-485.
[13] Usami, K., *et al.* Automated selective multi-threshold design for ultra-low standby applications. In *Proc. ISLPED*, 2002, 202-206.
[14] http://www.eas.asu.edu/~ptm/

A New Technique for Jointly Optimizing Gate Sizing and Supply Voltage in Ultra-Low Energy Circuits

Scott Hanson   Dennis Sylvester   David Blaauw

University of Michigan

{hansons,dmcs,blaauw}@umich.edu

ABSTRACT

Mobile applications with battery lifetimes on the order of thousands of days have placed stringent energy requirements on circuits. In this paper, we propose a new energy optimization technique for ultra-low energy circuits operating in the subthreshold regime. Our technique uses simultaneous gate sizing and supply voltage scaling to reduce energy. We demonstrate the effectiveness of our technique on benchmark circuits and offer insight on the roles of the timing distribution and wire capacitance in determining the achievable energy reductions.

Categories and Subject Descriptors: B.6.3 [**Design Aids**]: Optimization

General Terms: Algorithms, Performance, Design

Keywords: Subthreshold circuits, gate sizing, voltage scaling

1. INTRODUCTION

The rise of mobile computing has moved energy and power optimization to the forefront of the semiconductor industry. For a growing class of applications, energy minimization is the overriding priority. The ZigBee Alliance, for example, has specified a low power wireless standard for applications ranging from medical sensing to home security to home environment controllers [1]. While the performance demands of these applications are low, the target battery life is on the order of hundreds or thousands of days. It is therefore very important to investigate techniques that minimize energy, potentially at the expense of performance. In this paper, we describe a new technique that uses simultaneous voltage scaling and gate sizing to achieve optimal energy. We emphasize that the focus of this paper is energy minimization rather than power minimization since energy is a more relevant metric when battery life is the primary concern.

Aggressive voltage scaling has emerged during the past few years as an extremely effective solution to the power and energy minimization problems. Dramatic energy reductions are possible, particularly when voltage is allowed to scale into the subthreshold ($V_{dd} < V_{th}$) regime. Much of the previous work focused on the functional limits of voltage scaling and techniques for extending those limits [2][3][4]. However, recent work has shown that the operating voltage that results in minimum energy consumption is well above the minimum functional voltage [5][6]. As voltage is scaled to subthreshold voltages, leakage

---

Permission to make digital or hard copies of all or part of this work for personal or classroom use is granted without fee provided that copies are not made or distributed for profit or commercial advantage and that copies bear this notice and the full citation on the first page. To copy otherwise, or republish, to post on servers or to redistribute to lists, requires prior specific permission and/or a fee.

ISLPED'06, October 4–6, 2006, Tegernsee, Germany.

Copyright 2006 ACM 1-59593-462-6/06/0010...$5.00.

---

energy increases significantly (due to a rapid increase in circuit delay) and places a limit on the energy efficiency of further voltage scaling. Design techniques for maximizing energy efficiency in the subthreshold regime remain largely unexplored and deserve attention.

In the superthreshold regime ($V_{dd} > V_{th}$), it is widely known that minimum energy operation is obtained by setting gates to their minimum sizes, thereby reducing the dynamic energy as much as possible. However, in this paper we find that this is typically not true in the subthreshold regime. We show that increasing the sizes of certain gates in a circuit will reduce the overall leakage energy of the circuit. As a result, the minimum energy operating voltage can be reduced, thereby improving overall energy efficiency. We therefore propose a new optimization technique that alternately sizes gates and scales voltage to achieve minimum energy operation. It is important to note that we are not strictly decreasing leakage. We use leakage reductions to enable supply voltage reductions. This new sizing technique is unique in two respects: it reduces total energy by *increasing* gate sizes and it simultaneously sizes gates and scales supply voltage. Our sizing tool reduces energy by up to 15% in benchmark circuits compared to the case when only voltage scaling is used.

The remainder of this paper is organized as follows. In Section 2, we describe the key implications of low voltage design. We then describe our energy optimal gate sizing/voltage scaling tool in Section 3 and present a detailed analysis of the performance of the tool on a set of benchmark circuits in Section 4. Finally, we summarize the key conclusions of this paper in Section 5.

2. OPTIMIZATION OPPORTUNITIES

We begin our exploration of low voltage optimization opportunities by considering a chain of 50 inverters in a 130nm technology with an activity factor (α) of 0.2. Complex circuits behave similarly to the simple inverter chain, so our discussion is relevant for circuits of varying complexity. Figure 1 shows the energy consumed by the inverter chain per cycle as a function of V_{dd}. As predicted in [5][6], energy reaches a minimum at a voltage (called V_{min}) due to the rise in leakage energy but continues to function below 100mV. The rise in leakage energy is a result of the rapid increase in delay when V_{dd} drops below V_{th}. For this circuit, leakage accounts for 33% of the total energy at V_{min} and offers a unique optimization opportunity.

Reducing leakage results in two benefits, both illustrated in the inset of Figure 1. On one hand, leakage at V_{dd}=266mV is reduced. More importantly, V_{min} is reduced, enabling further energy savings. It is evident that dynamic energy and leakage energy must be optimized simultaneously. Optimization of only dynamic energy (via reduction of V_{dd}) yields a circuit that is very sensitive to leakage. Addressing leakage is consequently a high priority for energy optimality at low V_{dd}.

The energy consumed by a circuit per clock cycle may be represented as the sum of dynamic and leakage energies as shown in Equation 1. C_s is the switched capacitance, I_{leak} is the leakage current for the circuit, and T_{CLK} is the clock period.

$$E = E_{DYN} + E_{LEAK} = \frac{1}{2} \cdot C_s \cdot V_{dd}^2 \cdot \alpha + I_{leak} \cdot V_{dd} \cdot T_{CLK} \quad \text{(Eq. 1)}$$

By reducing T_{CLK} through gate sizing, the designer can decrease the amount of time that a circuit leaks per instruction. As long as the energy overhead of the sizing technique is low and a relatively small fraction of the total paths are critical, a reduction in T_{CLK} can yield significant leakage energy reductions. This leakage reduction, in turn, will drive further reduction in V_{min}. The notion of increasing gate sizes to reduce energy consumption is counterintuitive. Low power designers have typically chosen minimum-sized gates in order to reduce the dynamic energy consumed by a circuit. The authors of [4] point out that gate sizes may be increased to achieve energy reduction if there are few critical paths but suggest that this is a special case. As we discuss further in Section 4.1, we find that a skewed, unbalanced timing distribution with relatively few critical paths is common for a minimum sized design.

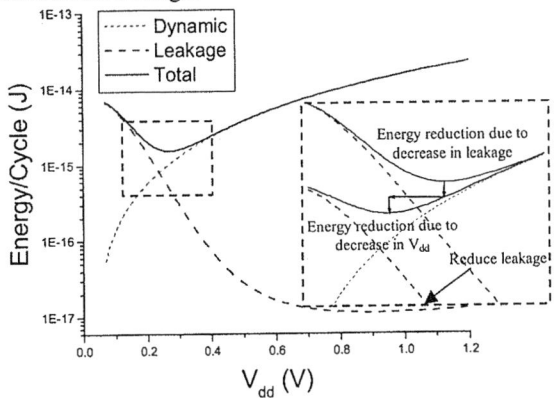

Figure 1. Energy in an inverter chain (n=50, α=0.2) is minimized at V_{dd}=266mV due to the rise in E_{leak}

3. A LOW VOLTAGE SIZING TOOL

The simultaneous scaling of V_{dd} and sizing of gates is a difficult problem since the two degrees of freedom (gate sizes and V_{dd}) are interdependent. A change in the sizing of the circuit causes the energy optimal supply voltage (V_{min}) to change. Conversely, changing V_{dd} alters the timing and leakage characteristics of a circuit and affects the energy optimal sizing point. The problem may be formulated as a multi-dimensional constrained optimization as shown in Figure 2(a), where W is the set of all gate sizes and W_i is the size of gate i. $W_{L,i}$ and $W_{U,i}$ are the lower and upper bounds for the size of gate i. Within those bounds, gate sizes may only assume discrete values set by the standard cell library. We do not initially place a constraint on total circuit area. We will investigate the implications of this decision in Section 4.3. V_L and V_U are the lower and upper bounds on V_{dd}.

A general nonlinear optimizer [7] could be used to solve the constrained optimization problem in Figure 2(a). However, such general optimization methods often incur high runtimes, making optimization of large circuits impractical. In this paper, we re-

formulate the problem as two simpler sub-optimization steps that are performed iteratively as shown in Figure 2(b). The two sub-optimizations are: (1) supply voltage optimization, and (2) sizing optimization. Both sub-optimization problems are well known and a number of different methods are available for solving them efficiently. For the supply voltage optimization, we use a binary search and for the sizing optimization we use a simple sensitivity based method similar to the approach in [8].

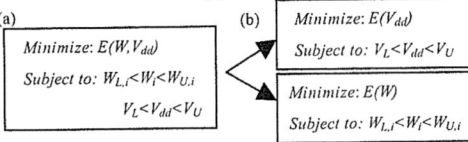

Figure 2. (a)The simultaneous optimization of gate size and supply voltage may be formulated as a two dimensional constrained optimization. (b) We simplify the problem by breaking it into two simpler optimization steps.

Figure 3 shows pseudo-code describing the operation of the sizing algorithm. The top-level optimization iterates between the two sub-optimizations to converge on the optimal solution. The algorithm begins with a circuit composed entirely of minimum-sized gates operating at a sufficiently high voltage (i.e. well above V_{min}). Using this initial approximate solution, the first sub-optimization is solved to find the energy optimal supply voltage (V_{min}) for the unsized circuit using SPICE-generated characterization data. At this voltage, leakage represents a significant portion of total energy, and the sensitivity of leakage to gate sizing is very high. A sizing optimization is performed at this $V_{dd}=V_{min}$. Each gate in the circuit is evaluated using a sensitivity metric to determine the change in energy that would result from a unit increase in gate size. The gate with the highest sensitivity is then sized up. Sizing continues at the same voltage until gate sizing no longer results in an energy improvement. At this point, a new supply voltage optimization is performed and V_{min} is again determined for the circuit. Iteration between voltage optimization and sizing optimization continues until convergence. We now show that this iterative formulation will converge to an optimal solution.

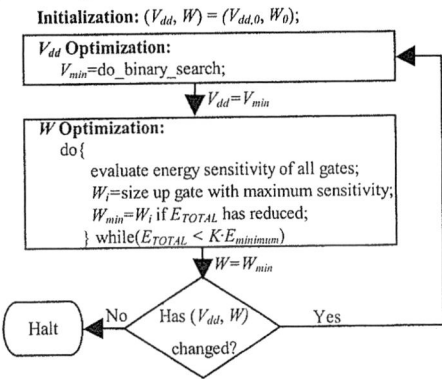

Figure 3. Pseudo-code for a low voltage sizing tool.

To guarantee convergence, we assume that the two sub-optimizations are ideal. In other words, we require: 1) that the V_{dd} optimizer returns the energy optimal supply voltage, V_{min}, for a fixed set of gate sizes, W and 2) that the gate size optimizer returns the energy optimal set of gate sizes, W_{min}, for a fixed supply voltage, V_{dd}. It is also assumed that any change in V_{dd} or

W suggested by a sub-optimizer reduces the total energy. Furthermore the outputs of the V_{dd} optimizer and gate size optimizer may be described by functions $f(W)$ and $g(V_{dd})$, respectively, where $f(W)$ gives V_{min} for a particular set of gate widths and $g(V_{dd})$ gives the set of W_{min} for a particular V_{dd}.

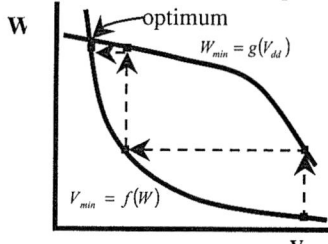

Figure 4. The hypothetical output of the two sub-optimizers is plotted (solid lines). Dashed lines show the path traversed by the top level optimization as the optimum is approached.

The characteristic curves for $f(W)$ and $g(V_{dd})$ are plotted in Figure 4. For clarity of illustration, the total width is shown on the Y-axis, although the figure can be extended to a multi-dimensional space where each individual gate size is an individual gate width. The solution to the top level optimization lies at the point where $f(W)$ and $g(V_{dd})$ intersect. At this point, both gate sizing and supply voltage are optimal. Our optimization algorithm will iterate back and forth between $f(W)$ and $g(V_{dd})$ until convergence as shown with arrows on the diagram. The optimization will converge as long as the following conditions are satisfied:

$$\frac{\partial f(W)}{\partial W} = \frac{\partial V_{min}}{\partial W} \leq 0, \quad \frac{\partial g(V_{dd})}{\partial V_{dd}} = \frac{\partial W_{min}}{\partial V_{dd}} \leq 0, \quad \text{(Eq. 2)}$$

$$\frac{\partial^2 f(W)}{\partial W^2} = \frac{\partial^2 V_{min}}{\partial W^2} \geq 0, \quad \frac{\partial^2 g(V_{dd})}{\partial V_{dd}^2} = \frac{\partial^2 W_{min}}{\partial V_{dd}^2} \leq 0$$

Note that the first two conditions in Equation 2 tend to push optimization toward larger gate sizes and lower V_{dd}. In other words, a reduction in V_{dd} causes an increase in W, which causes further reduction in V_{dd}. The first condition requires that an increase in W always leads to a reduction in V_{min}. It is clear an increase in W results in larger dynamic energy consumption. However, we know that the total energy consumption is reduced by such a gate size increase since this is the objective of the optimizer. Consequently, an increase in W results in a higher dynamic energy/leakage energy ratio. In [5] and [6], it was shown that a circuit with a larger dynamic/leakage energy ratio tends to have a lower V_{min}. This shows that an increase in transistor width by the sizing optimizer will result in a V_{min} that is lower, satisfying the first condition in Equation 2. The second condition means that a reduction in V_{dd} always leads to an increase in W_{min}. We saw in Section 2 that a reduction in V_{dd} causes leakage energy to increase. To mitigate this leakage, the gate size optimizer will reduce T_{CLK} further by increasing the sizes of the gates along the critical path. Hence, a V_{dd} reduction will lead to a gate size increase, which satisfies the second equation in Equation 2. The third and fourth conditions, which place requirements on the second derivatives of $f(W)$ and $g(V_{dd})$ are satisfied because the energy overheads of gate sizing and V_{dd} scaling force $f(W)$ and $g(V_{dd})$ to saturate near the optimum.

To model delay and energy in our sizing algorithm, we use SPICE characterization data for a 130nm standard cell library at twelve voltage points ranging from 130-350mV. This range of

voltages is sufficient to contain all V_{min} values for a variety of benchmarks under widely varying switching activity conditions. We verify our sizing algorithm using a set of benchmarks that includes the ISCAS85 benchmarks as well as two more complex circuits. The standard cell library contains 38 cells including inverters, 2 and 3-input NAND gates, 2 and 3-input NOR gates, and buffers of various sizes. Each benchmark is originally synthesized using only minimum-sized gates. The average switching activity values for each node are extracted from a Verilog simulation assuming an input switching activity of 0.2. In addition, we use a simple wireload model of the form $k(1+0.4(FO-1))$, where k represents the wire capacitance for a gate with one fanout, and FO is the number of fanout gates at the node of interest [9]. For our technology we choose $k=5fF$ which corresponds to a wire length of approximately 20μm.

Table 1. Results of sizing various benchmarks

Benchmark	Number of Gates	ΔE_{TOTAL} (%)	$\Delta E_{DYN} / \Delta E_{TOTAL}$ (%)	ΔArea (%)
c432	161	8.6	85	39.9
c499	544	5.9	99	20.1
c880	366	6.1	71	32.1
c1908	507	5.6	91	26.7
c1355	582	8.4	96	30.4
c2670	860	11.4	80	27.9
c3540	984	7.6	80	15.5
c5315	1668	11.7	84	19.8
c6288	2480	5.7	80	21.8
c7552	2087	15.0	81	39.8
SOVA	17559	14.7	85	47.2
R4	35039	13.9	96	60.9

4. RESULTS

In this section, we examine the energy reductions achieved using our new optimization technique. Table 1 summarizes the performance of our sizing tool on a number of benchmarks. The column labeled "ΔE_{TOTAL}" lists the energy reductions achieved using our technique. It compares the energy of the unsized design at $V_{dd}=V_{min,unsized}$ and the energy of the sized-up design at $V_{dd}=V_{min,sized-up}$. Table 1 also includes a column labeled "$\Delta E_{DYN} /\Delta E_{TOTAL}$." This quantity represents the fraction of energy savings attributed to dynamic energy improvement. Though we initially target leakage reduction, most of the energy benefits are a result in V_{min} reduction. We also list the area penalty (ΔArea), which is the increase in total transistor width as compared to the unsized design. Across all of the benchmarks, we observe that energy improves by 5.6% to 15% with the area penalty ranging from 15.6% to 60.9%. In the remainder of this section we look closely at the key factors affecting the efficiency of our algorithm.

4.1 The Effect of the Timing Distribution

The proposed sizing technique ultimately requires some timing slack to be effective. Timing slack is not easily exploited in a well-balanced circuit with many near-critical paths since many gates must be sized up to achieve small changes in T_{CLK}. As a result, as with traditional power optimization techniques our approach is more effective at reducing energy when a circuit has few critical paths. Figure 5 shows the timing distributions for c499 and c7552 before and after the completion of all gate sizing. The "before" and "after" distributions are measured at different V_{dd} but are normalized to their respective T_{CLK} values to facilitate

340

comparison. It is obvious from the initial timing distributions that c499 has many more critical timing paths than c7552 since the average relative path delay in the c499 initial distribution is much higher than that of the c7552 distribution. The shapes of the distributions suggest that c499 is well balanced compared to c7552. After gate sizing, the timing distribution of c499 moves slightly, but the shift in the c7552 distribution is far more significant. It is not surprising that the energy reduction achieved in c499 (5.9%) is much lower than in c7552 (15.0%). These observations show that the shape of the timing distribution is a strong indicator of whether or not our technique is effective at reducing energy. We can confirm this observation by finding the correlation between the "criticality" of a timing distribution and the observed energy savings. We quantify the "criticality" of an initial timing distribution using the average path delay and find that energy savings and average path delay are related with a correlation coefficient of 0.62. Only nine benchmarks are included in these calculations because the runtime to generate path distributions is prohibitive for large circuits.

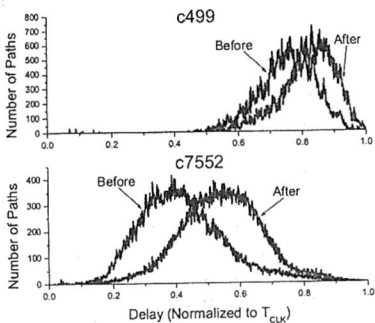

Figure 5. Timing distributions for c499 and c7552 behave differently. Delay is normalized to the clock period (T_{CLK}).

4.2 The Role of Wire Capacitance

The dynamic, leakage and total energy savings achieved for the c7552 benchmark are plotted as a function of wire capacitance in Figure 6(a). The change in transistor width after gate sizing is also plotted. At small wire capacitances, the change in total width is small (<20%). In this region, gate capacitance tends to dominate total load capacitance, minimizing the benefit of gate sizing. As wire capacitance grows, gate sizing becomes beneficial, as evidenced by the increase of both transistor width and energy savings in Figure 6(a). Designs with long routes will therefore benefit from combined V_{dd} scaling and gate sizing. Most of the energy reduction in Figure 6(a) is due to dynamic energy reduction. This is consistent with the results of Table 1, which shows that reductions in dynamic energy are responsible for 71-99% of total energy reductions. Put another way, the primary target of simultaneous V_{dd} scaling and gate sizing is reduced dynamic energy rather than reduced leakage energy.

4.3 Mitigating the Area Penalty

Though we assume energy to be the most important metric, we cannot ignore the area of the circuit. An increase in area is accompanied by an increase in cost. For many low energy applications (for example, a widespread sensor network) cost and area are overriding priorities [10]. The area penalties listed in Table 1, ranging from 15.6-60.9%, may result in an intolerable cost for such applications. By adding an area constraint to the minimization problem expressed in Section 3, the cost of sizing

can be reduced. Figure 6(b) highlights the effectiveness of our sizing technique when an area constraint is asserted for several benchmarks. In the case of the R4 benchmark, the area penalty can be reduced from 60.9% to 26.4% with only 10% reduction in the energy savings. The assertion of an area constraint is simple and should be considered by designers with area limitations.

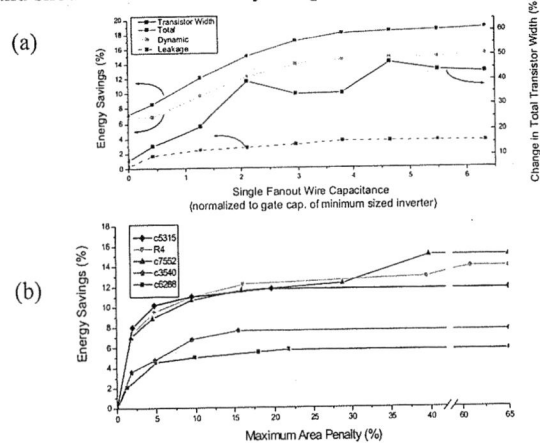

Figure 6. (a) Energy savings (% of total energy) and change in total transistor width as a function of wire capacitance. (b) Energy savings (% of total energy) as a function of the maximum allowable area penalty.

5. CONCLUSIONS

In this paper, we describe a technique that uses a combination of gate sizing and V_{dd} scaling to reduce energy. Our technique uses leakage energy reductions to drive extended voltage scaling and achieves total energy reductions of up to 15% across a set of benchmark circuits. Our results show that different design types respond differently to our technique. We find that designs with a wide timing distribution and few critical paths experience significant energy reductions using our technique. Since V_{dd} reduction ultimately drives energy reduction, we also find that the energy consumption in designs with wire-dominated load capacitances may be improved substantially using our technique. Finally, we show that the area penalty of our technique can be mitigated by applying an area constraint during optimization.

References

[1] ZigBee Alliance. http://www.zigbee.org/.

[2] J. Meindl, J. Davis, "The Fundamental Limit on Binary Switching Energy for Terascale Integration (TSI)," *IEEE Journal of Solid-State Circuits* 35, No. 10, 1515-1516 (October 2000).

[3] A. Wang, A. Chandrakasan, "A 180mV FFT Processor Using Subthreshold Circuit Techniques," *Int. Solid-State Circuits Conf. (ISSCC)*, pp. 292-529, 2004.

[4] B. Calhoun, A. Wang, A. Chandrakasan, "Device Sizing for Minimum Energy Operation in Subthreshold Circuits," *Custom Integrated Circuits Conf. (CICC)*, pp. 95-98, 2004.

[5] B. Zhai, D. Blaauw, D. Sylvester, K. Flautner, "Theoretical and Practical Limits of Dynamic Voltage Scaling," *Design Automation Conf. (DAC)*, pp. 868-873, 2004.

[6] B. Calhoun, A. Chandrakasan, "Characterizing and Modeling Minimum Energy Operation for Subthreshold Circuits," *Int. Symp. on Low Power Electronics and Design (ISLPED)*, pp. 90-95, 2004.

[7] B.A. Murtagh, M.A. Saunders, "MINOS 5.4 User's Guide, Report SOL 80-20R," *Systems Optimization Lab., Stanford University*, Dec. 1983 (revised Feb. 1995).

[8] J.P. Fishburn, A.E. Dunlop, "TILOS: A Posynomial Programming Approach to Transistor Sizing," *Int. Conf. on Comp. Aided Design (ICCAD)*, pp. 326-328, 1985.

[9] G.A. Sai-Halasz, "Performance Trends in High-End Processors," *Proc. of the IEEE 83*, 20-36 (Jan. 1995).

[10] J.M. Kahn, R.H. Katz, and K.S.J. Pister, "Emerging Challenges: Mobile Networking for Smart Dust," *Journal of Comm. And Networks* 2, No. 3, 188-196 (Sep. 2000).

Considering Process Variations During System-Level Power Analysis

Saumya Chandra[†‡]   Kanishka Lahiri[‡]   Anand Raghunathan[‡]   Sujit Dey[†]

[†]Dept. of ECE, University of California, San Diego (saumya,dey@ece.ucsd.edu)
[‡]NEC Laboratories America, Princeton, NJ (klahiri,anand@nec-labs.com)

ABSTRACT

Process variations will increasingly impact the operational characteristics of integrated circuits in nanoscale semiconductor technologies. Researchers have proposed various design techniques to address process variations at the mask, circuit, and logic levels. However, as the magnitude of process variations increases, their effects will need to be addressed earlier in the design cycle.

In this paper, we propose techniques for accurately and efficiently incorporating the effects of process variations into system-level power estimation tools. To motivate our work, we first study the impact of process variations on the power consumption of an example System-on-Chip (SoC). We consider simple extensions of current approaches to system-level power estimation (spreadsheet-based and simulation-based power estimation), and demonstrate their limitations in performing variation-aware power estimation. We propose a system-level power estimation methodology that can accurately and efficiently analyze the impact of process variations on SoC power. The proposed methodology combines efficient trace-based analysis, power-state based leakage modeling, and Monte Carlo sampling. The key benefit of the proposed methodology is that it captures the necessary inter-dependencies while avoiding iterative system-level simulation. Our implementation of the proposed techniques within an in-house system-level power estimation framework indicates 2-5 orders of magnitude efficiency gains, with negligible loss in accuracy, compared to direct Monte Carlo techniques that require iterative system simulation.

Categories and Subject Descriptors: C.5.4 [VLSI Systems]

General Terms: Algorithms, Design, Experimentation

Keywords: Process variations, Power analysis, Power Estimation, Low Power Design, System-on-Chip

1. INTRODUCTION

Integrated circuits (ICs) fabricated using nanoscale technologies are expected to be increasingly prone to manufacturing induced variations, which cause the characteristics of devices to vary both within a die and across dies [1]. These variations cause the performance and power consumption of circuits, and hence the systems that contain them, to display statistical, rather than deterministic behavior. Traditional design methodologies, based on typical and worst-case circuit models, break down in the face of increas-

---

ing variations, resulting in over-design, increased design effort, decreased yield, or an inability to meet design goals.

Recognizing the above challenges, various techniques at different stages of the design flow are being researched to aid in IC design in the presence of variations. The efforts in this area have thus far focused on addressing the problem at the mask [2], circuit [1, 3], and logic levels [4, 5, 6]. While these techniques have shown promise (and are already being incorporated into commercial IC design flows), they cannot completely address the problem, especially in the face of continually increasing process variations. Variation-aware design has only recently started to receive attention at higher levels of the design flow, namely, at the architecture and system levels. Statistical models to incorporate the impact of variations on microprocessor performance and power consumption were proposed in [7]. Tradeoffs between throughput, power, and area in parallel architectures under process variations were studied in [8]. The effects of process variations on embedded SRAM memory architectures were studied in [9] and reconfigurable buffers were exploited to optimize power and performance under variations.

Key to designing variation-tolerant systems is the availability of accurate and efficient analysis tools that predict the impact of variations on design metrics, such as the system-level power consumption. However, this area has not received much attention. In this work, we propose an efficient methodology based on efficient trace analysis, power-state based leakage modeling, and Monte Carlo sampling to provide SoC designers with feedback about power consumption under process variations.

2. SOC POWER UNDER VARIATIONS

In this section, we consider the example SoC architecture illustrated in Figure 1(a). The SoC implements an image processing application that consists of software running on an ARM946 processor [10], and dedicated hardware (Filter_HW) that accelerates pixel-level filtering operations. In addition, the SoC contains the AHB on-chip bus [11], an integrated memory controller, and an interrupt controller. The SoC is implemented using a commercial 90 nm standard cell library [12] and operates at a frequency of 206 Mhz and a voltage of 1 V. An in-house, cycle-accurate, simulation-based system-level power analysis tool [13] was used to generate dynamic and leakage power traces for each component while executing a specific test-bench. We used Monte Carlo techniques (described in Section 3) to estimate the impact of variations in leakage power, due to chip-to-chip variations in effective channel length (L_{eff}). For this study, we assumed that L_{eff} follows a normal distribution with $\mu = 90$ nm and $3\sigma/\mu = 30\%$. The power variations thus estimated for each component are captured using a box-whisker representation as shown in Figure 1(b). The lower and upper extremities of each box represent the 25th and 75th percentiles of a component's average power consumption (including both dynamic and leakage power). The whiskers denote the minimum and maximum values. For example, for the ARM processor, the inter-quartile range (the box height) is 25% of its average power, suggesting that variations significantly impact its power charac-

---

Permission to make digital or hard copies of all or part of this work for personal or classroom use is granted without fee provided that copies are not made or distributed for profit or commercial advantage and that copies bear this notice and the full citation on the first page. To copy otherwise, to republish, to post on servers or to redistribute to lists, requires prior specific permission and/or a fee.
ISLPED'06, October 4–6, 2006, Tegernesee, Germany.
Copyright 2006 ACM 1-59593-462-6/06/0010 ...$5.00.

teristics. However, for the AHB it is only 8%. This difference is explained as follows. While leakage power is highly sensitive to channel length variations, dynamic power is relatively immune. Hence, components for which leakage accounts for a greater portion of their total power consumption display higher total power variations. The breakdown of a component's total power into leakage and dynamic power is determined by how much time it spends in its different power-states (active, idle, sleep, deep-sleep, *etc.*). In the example, the ARM processor spends a significant amount of time in the idle state waiting for the `Filter_HW` and therefore, is more affected by variations. The AHB is almost always active, serving requests from either the ARM processor or other HW. Its power consumption consists largely of dynamic power, and is hence less susceptible to variations. This example suggests that the extent to which variations affect individual component power characteristics depends critically on *component workload profiles* and *power-states*. Accounting for such inter-dependencies accurately and efficiently is a key objective of the proposed methodology.

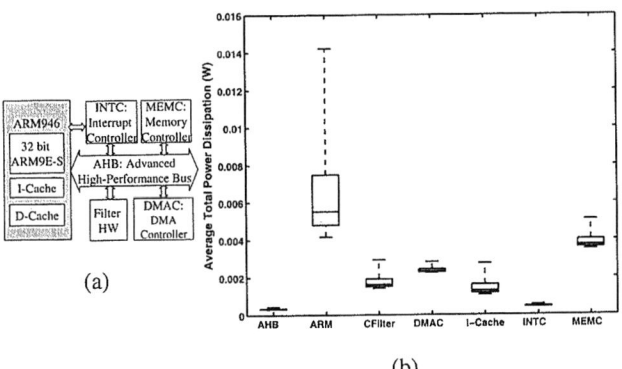

(a)

(b)

Figure 1: Case study in chip-to-chip power variations for an example SoC: (a) system-level block diagram; (b) inter-quartile range of power variations for system components

3. CONSIDERING VARIATIONS IN POWER ANALYSIS

The goal of variation-aware power analysis is to generate a power *distribution* rather than a deterministic estimate. Moreover, this distribution must be estimated accurately and efficiently. There are two main approaches currently used for system-level power estimation: (i) enhancing system-level simulation with power models for various system components, and (ii) simple spreadsheet analysis based on rough metrics such as total gate count, switching activity factors, *etc.* We consider simple extensions of these approaches to consider variations, and evaluate their merits and drawbacks.

Direct Monte Carlo simulation: In this method, full-system simulation-based power estimation is iteratively performed in order to generate a power consumption distribution for the SoC. The power models for system components are initially constructed based on nominal (typical) values of process parameters. For each simulation, the power models are refined by randomly generating a sample point from a pre-defined distribution of process parameters (*e.g.,* transistor channel lengths), and calculating a variation factor for dynamic and leakage power. The power models are assumed to be sensitive to the power-states (active, idle, sleep, *etc.*) of the respective components, since the power variability will be different in different states. When simulation-based system-level power analysis is performed, functionality and power-state transitions are simulated in an integrated manner, and total power consumption is determined. This is repeated for a desired number of sample points in order to obtain a distribution of total power consumption for the SoC under process variations.

The larger the sample size, the higher is the degree of confidence in the accuracy of the computed power distribution. For the ex-

ample SoC shown in Figure 1(a), we calculated that in order to be 95% confident that the estimated mean of the power distribution differs from the actual mean by no more than 5%, at least 607 simulations would be required, which is a computationally challenging task since each simulation may require anywhere from several minutes to hours, depending on the length of the simulation trace. In summary, direct Monte Carlo simulation can be accurate, but is too time consuming to use for architectural exploration.

Spreadsheet-based analysis: In this method, simple equations are used to analytically calculate the distribution of leakage power, given SoC characteristics such as the gate count and activity profile, and a distribution of process parameters. It is possible to analytically relate variations in transistor leakage to variations in process parameters such as channel length and oxide thickness [5, 6]. The SoC's characteristics, such as gate count and activity factor, are then used to estimate the average number of leaking transistors, and compute distributions for leakage power and total SoC power.

Figure 2 compares the distribution obtained through such spreadsheet-based analysis to one obtained via direct Monte Carlo simulations under a given test bench. Clearly, there is a notable discrepancy between the two distributions. This discrepancy arises because, for a specific workload, the spreadsheet-based approach fails to consider the extent to which different system components contribute to power variations. For example, components with long idle periods may power down, and may contribute only slightly to leakage power variations. Also, for components that are mostly active, the larger contribution of dynamic power may overwhelm the variations in leakage power. Clearly, the spreadsheet-based approach, while simple and computationally efficient, cannot capture workload characteristics and component power-states. As illustrated earlier, these factors significantly influence the impact of variations on power.

4. PROPOSED METHODOLOGY

The proposed methodology for considering process variations during system-level power analysis is illustrated in Figure 3. It has three main phases. In the first phase, *leakage power modeling* is performed to obtain leakage distributions for all the power-states of the SoC components, while taking into account circuit and process characteristics. In phase 2, conventional *system simulation and power analysis* is performed to obtain a set of dynamic power and power-state traces for all components. These distributions and the power-state traces are used as inputs in phase 3, namely *Monte Carlo Analysis*. In this phase, the parameter space is sampled, leakage power (LUT) is computed for all power-states and stored in a lookup-table (LUT).

This LUT, along with the traces obtained in the first phase, are used by an efficient *trace analysis* step to determine the average power consumption (considering both dynamic and leakage power) for the current sample. This step is repeated for all sample points. The larger the size of the sample space, the lower is the sampling

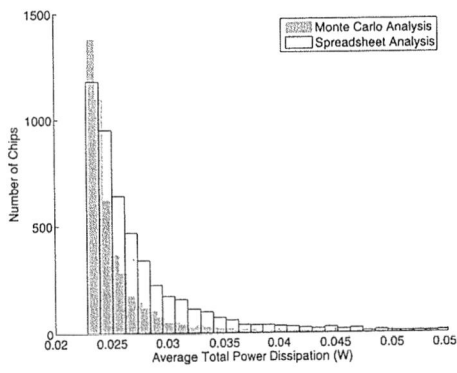

Figure 2: Comparing distributions obtained through the two analysis methods

343

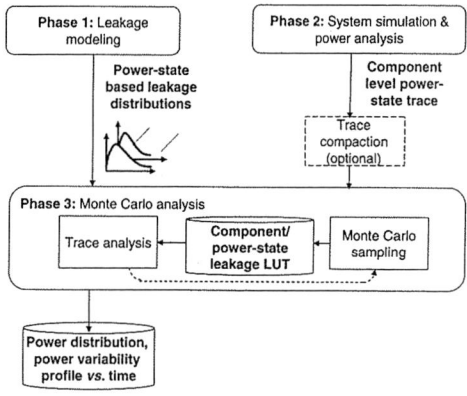

Figure 3: Variation-aware system-level power analysis

error associated with the resulting power distribution. Note that, unlike the direct Monte Carlo approach described earlier, time consuming simulations are not part of the sampling loop in the proposed methodology. Therefore, very large numbers of samples can be analyzed efficiently with the proposed approach, leading to low sampling error. The power traces can be optionally processed by a *trace compaction* step prior to Monte Carlo simulations for even faster analysis. The output of the methodology is a distribution of system-level power, and optionally, a power variability profile versus time. We next describe each phase in detail.

4.1 Leakage Power Modeling

In this phase, process, circuit, and system-level characteristics are analyzed to develop variation-aware leakage power models for each SoC component. As illustrated in Section 2, for accurate incorporation of variations into system-level power analysis, it is important to consider the dependence of leakage on component power-states. A procedure to accomplish this is shown in Figure 4.

In **Step 1**, for each component, a set of power-states (*e.g.*, active, idle, sleep, deep-sleep, *etc.*) are identified. The rationale for identifying power-states is that the leakage power can be distinctly different for different power-states and depends on the number of transistors that are powered on in a particular power-state.

In **Step 2**, for each power-state of each component, circuit-level parameters are extracted for subsequent use by leakage power models. For each power-state, the total device width associated with N-type and P-type devices that are powered on is estimated. Typically at this stage of design, detailed physical implementations are not available. Therefore, high-level estimation techniques based on gate-counts (similar to the ones used in [14]) are used to estimate the total "active" device width associated with each power-state.

In **Step 3**, statistical leakage power models that consider process variations are calibrated using low-level simulation data. Most empirical leakage power models are of the form $I_{leakage} = \alpha e^{f(L,T_{ox})}$ [5, 6], capturing the exponential dependence of leakage currents on process parameters such as channel length (L) and gate-oxide (T_{ox}). Here, α depends on the gate characteristics and is directly proportional to the device width and f is a polynomial function of the process parameters and hence, is a random variable under process variations. Assuming process parameters are normally distributed, leakage current follows a log-normal distribution. The leakage of a circuit is given by the sum of the correlated log-normal

leakages of its constituent gates, which is approximated with another log-normal distribution [5, 6].

The leakage current of the circuit (I_{ckt}) is sum of the leakages in the N-type transistors ($I_{n_{ckt}}$) and the P-type transistors ($I_{p_{ckt}}$). The corresponding distribution parameters are given by $\mu(I_{ckt}) = \mu(I_{n_{ckt}}) + \mu(I_{p_{ckt}})$, and $\sigma^2(I_{ckt}) = \sigma^2(I_{n_{ckt}}) + \sigma^2(I_{p_{ckt}}) + 2 * COV(I_{n_{ckt}}, I_{p_{ckt}})$. Modeling the circuit as a sea of identical gates and ignoring spatial variations within the circuit, we can write $I_{n_{ckt}} = \sum I_{n_i} = K_n * I_n$, and $I_{p_{ckt}} = \sum I_{p_i} = K_p * I_p$, where, K_n and K_p depend on circuit level characteristics identified in **Step 2**, and I_n and I_p represent the leakage current of unit width N-type and P-type devices, respectively. The distribution parameters are computed using $\mu(I_{n_{ckt}}) = K_n * \mu(I_n)$, $\sigma(I_{n_{ckt}}) = K_n * \sigma(I_n)$, $\mu(I_{p_{ckt}}) = K_p * \mu(I_p)$, and $\sigma(I_{p_{ckt}}) = K_p * \sigma(I_p)$. The parameters $\mu(I_n)$, $\mu(I_p)$, $\sigma(I_n)$ $\sigma(I_p)$ and $COV(I_n, I_p)$ are determined through SPICE simulations of accurate MOSFET models [15], where the transistor parameters are varied through Monte Carlo sampling. Similar data could also be obtained through detailed measurements of transistor off current for a set of test chips. The result of this step is a set of random variables with specified distribution parameters that model leakage power for each sub-circuit.

4.2 System Simulation and Power Analysis

In this phase (Phase 2 of Figure 3), simulation-based system-level power estimation is performed for the target SoC architecture. The inputs consist of a system-level test bench that models typical operating conditions, an architectural model of the target system, and a set of power models that track the dynamic power of each SoC component at the cycle level. The output is a set of cycle-level traces of dynamic power consumption over time for each component. In addition, a trace of the power-states for each component over time is also generated.

Trace Compaction: In this optional step, the generated traces are compacted, enabling more efficient analysis at the expense of temporal detail. The original cycle-level trace contains fields for the cycle number, dynamic power, and the power-state. High cycle-level profiling accuracy can be achieved with this trace. Trace compaction *preserving temporal information* involves collapsing consecutive cycles, in which the power-state of the component remains the same, into a single trace entry that contains the total number of collapsed cycles, average dynamic power for those cycles, and the power-state. Since temporal ordering of power-states and dynamic power is preserved, albeit at a coarser granularity, power and variability profiles *vs.* time can still be generated. If trace compaction is performed *without preserving temporal information*, more significant compaction can be achieved. The trace is reduced to a distribution of power-states (the number of clock cycles spent by each component in each state), and the corresponding dynamic power estimates that are averaged over all occurrences of each state. Using this compact representation, the analysis is extremely fast but is incapable of estimating profiles of power, or power variability *vs.* time. These schemes provide the flexibility of trading analysis efficiency for temporal resolution, based on user requirements.

4.3 Monte Carlo Analysis

In phase 3 of the methodology (Figure 3), the following two steps are executed iteratively for a fixed number of samples, to generate system-level power distributions and power variability profiles.

Monte Carlo Leakage Sampling: In this step, Monte Carlo sampling is performed on the distributions generated in phase 1 to obtain leakage power estimates for each SoC component in each of its power-states. If we are modeling only inter-chip variations, a single sample is used to determine leakage power estimates for all the components (this can be easily extended to model intra-chip variations as well). Leakage estimates are stored in a LUT that is indexed by component name and power-state.

Trace Analysis: The trace analysis procedure makes a single pass over the trace. For each trace entry, it obtains the component's leakage power from the LUT based on the power-state specified

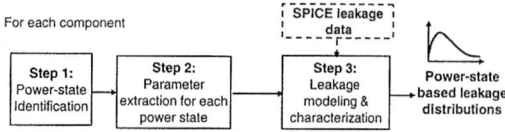

Figure 4: Power-state based leakage power modeling

344

in the trace entry. The result is added to the dynamic power consumption (available in the trace). When cycle-level traces are used, this step generates a system-level power trace at the cycle level. If trace compaction preserving temporal information is used, power profiles are generated at a granularity determined by the frequency at which components undergo power-state transitions. In both the cases, average power dissipation of the system for the entire trace is calculated. When temporal information is not preserved, the procedure simply computes the average power associated with each power-state by adding the average dynamic power in that state to the corresponding leakage sample obtained from the LUT. A weighted average is then computed, using the probabilities of occurrence of the power-states as weights, to generate a power sample. By iterating the leakage sampling and trace analysis steps, we generate a distribution of total system power consumption.

5. EXPERIMENTAL RESULTS

In this section, we describe our experimental set up, and present results that compare the proposed method with the simple extensions of existing techniques described in Section 3.

Experimental Setup: For our experiments, we modeled the SoC described in Section 2 using cycle-accurate SystemC [16] models for all the hardware components, an instruction-set simulator for the CPU, and transaction-level models for the on-chip bus. All the components were enhanced with dynamic power models [13] and leakage power models (Section 4). The models were calibrated using data collected from Monte Carlo HSPICE simulations using 90 nm BSIM3 models [17]. We considered inter-die channel length variations with $3\sigma/\mu$ of 30%. Trace analysis was implemented using the C-MEX utility in MATLAB [18].

Efficiency Comparison: Table 1 presents the time required to generate the power distribution for the example SoC (column 2) and the resulting speed up (column 3), using different analysis methods. It can be seen that our method is about 2 orders of magnitude faster than direct Monte Carlo simulation. Efficiency improving techniques such as trace compaction enable even higher speedups of 3-5 orders of magnitude compared to direct Monte Carlo simulation, resulting in efficiency comparable to spreadsheet-based analysis.

Accuracy Comparison: Table 1 also presents the estimates of the "power yield", *i.e.*, the fraction of manufactured chips that meet a specified power budget during normal operation (columns 4-6), as estimated by different methods. Our approach results in no accuracy loss with respect to Monte Carlo simulation (rows 2-5). Rows 6 and 7 present the power yield estimates obtained using spreadsheet-based analysis. In the pessimistic approach (row 6) all transistors are assumed to be leaking all the time. Therefore, the SoC leakage distribution ($\mu = 6.36$ mW, $\sigma = 12.35$ mW) is directly added to the dynamic power estimate. In the optimistic approach (row 7) the idle components are assumed to be in deep-sleep state, thereby consuming negligible leakage power in those periods. It can be seen that the spreadsheet-based approach provides very loose quantitative bounds on the actual power yield. The proposed methodology accurately captures the times spent by each SoC component in each of its power-states, enabling a more accurate consideration of their contributions to the total system power.

Power Variability Profile: Figure 5 shows the temporal variation in the system-level power and its spread due to process variations, as generated by our methodology. The three waveforms from bottom to top, represent the 25th, 50th and 75th percentiles of the power distribution respectively. This information can be used to identify intervals during which the system's power variations are high, and the corresponding power-state combinations of SoC components. Such information could be used to modify the system architecture, or to design appropriate power management schemes.

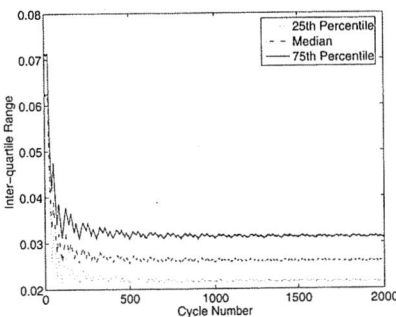

Figure 5: Power variability profile versus time

6. CONCLUSION

In this work, we demonstrated that efficient trace analysis, coupled with power-state based leakage modeling and Monte Carlo sampling provides an effective means of analyzing the impact of variations on system-level power consumption. Potential extensions of this work include the incorporation of dynamically varying parameters such as temperature and voltage, and further studies on application of such tools to variation-tolerant system design.

7. REFERENCES

[1] S. Borkar, T. Karnik, S. Narendra, J. Tschanz, A. Keshavarzi, and V. De, "Parameter variations and impact on circuits and microarchitecture," in *Proc. DAC*, pp. 338–342, 2003.

[2] L. W. Liebmann, "Layout impact of resolution enhancement techniques: Impediment or opportunity?," in *Proc. ISPD*, pp. 110–117, Apr. 2003.

[3] J. W. Tschanz et. al., "Adaptive Body Bias for Reducing Impacts of Die-to-Die and Within-Die Parameter Variations on Microprocessor Frequency and Leakage," *IEEE JSSC*, vol. 37, pp. 1396–1402, Nov. 2002.

[4] J. A. G. Jess, K. Kalafala, S. R. Naidu, R. H. J. M. Otten, and C. Visweswariah, "Statistical timing for parametric yield prediction of digital integrated circuits," in *Proc. DAC*, pp. 932–937, 2003.

[5] R. Rao, A. Srivastava, D. Blaauw, and D. Sylvester, "Statistical estimation of leakage current considering inter- and intra-die process variation," in *Proc. ISLPED*, pp. 88–89, 2003.

[6] H. Chang and S. S. Sapatnekar, "Full-chip analysis of leakage power under process variations, including spatial correlations," in *Proc. DAC*, pp. 523–528, 2005.

[7] D. Marculescu and E. Talpes, "Energy awareness and uncertainty in microarchitecture-level design," *IEEE Micro*, vol. 25, no. 5, pp. 64–76, 2005.

[8] N. Azizi, M. M. Khellah, V. De, and F. N. Najm, "Variations-aware low-power design with voltage scaling," in *Proc. DAC*, pp. 529–534, 2005.

[9] H. Wang, M. Miranda, A. Papanikolaou, F. Catthoor, and W. Dehaene, "Variable tapered pareto buffer design and implementation allowing run-time configuration for low-power embedded SRAMs," *IEEE Trans. VLSI Systems*, vol. 13, pp. 1127–1135, Oct. 2005.

[10] http://www.arm.com/products/CPUs/ARM946ES.html.

[11] "AMBA 2.0 Specification." http://www.arm.com/armtech/AMBA.

[12] "Cell Based IC CB-90 L/M/H Type Features/Basic Specifications, NEC Electronics." http://www.necel.com/cbic/en/cb90/cb90.html.

[13] N.Bansal, K.Lahiri, A.Raghunathan, and S.T.Chakradhar, "Power Monitors: A framework for system-level power estimation using heterogeneous power models," in *Proc. Int. Conf. VLSI Design*, pp. 579–585, 2005.

[14] "BACPAC - Berkeley Advanced Chip Performance Calculator." http://www.eecs.umich.edu/~dennis/bacpac.

[15] "Industry standard deep submicron SPICE MOS device model for circuit designs." http://www-device.eecs.berkeley.edu/~bsim3.

[16] "The Open SystemC initiative." http://www.systemc.org.

[17] "BPTM - Berkeley Predictive Technology Models." http://www-device.eecs.berkeley.edu/~ptm.

[18] "MATLAB - High-level technical computing environment." http://www.mathworks.com/products/matlab.

Table 1: Experimental Results

Analysis Method	Efficiency Computation time		Accuracy Predicted power yield (%)		
	Time taken	Speed up	27.5 mW	30 mW	40 mW
Monte Carlo	2936m	1	80.52	87.78	95.44
Trace-based (no compaction)	14.49m	203			
Trace-based (Trace compaction with temporal information)	52s	3387	80.52	87.78	95.44
Trace-based (Trace compaction without temporal information)	<1s	>10⁵			
Spreadsheet-Pessimistic	<1s	>10⁵	64.78	76.56	92.12
Spreadsheet-Optimistic			85.56	91.66	98.18

Synchronization-Driven Dynamic Speed Scaling for MPSoCs

Mirko Loghi
Politecnico di Torino
Torino, Italy
mirko.loghi@polito.it

Massimo Poncino
Politecnico di Torino
Torino, Italy
massimo.poncino@polito.it

Luca Benini
Università di Bologna
Bologna, Italy
lbenini@deis.unibo.it

ABSTRACT

Equalizing the ratios between workloads and speeds of processing elements provides the optimal speed allocation. Based on that principle, this work describes a dynamic speed setting policy for multi-processor systems-on-chip (MPSoCs) that relies on the estimation of processor idle times specifically due to the synchronization work. The policy provides two advantages: first, it does not rely on any assumption about the communication pattern of the application executed by the system. Second, it is purely architectural; it automatically detects changes in the system workload and sets processors speeds accordingly by means of a custom hardware block.

Results on a parallel MPEG video decoding application show an EDP saving above 55%, averaged over several datasets, corresponding to an energy saving above 50%, and a corresponding penalty in performance below 8%.

Categories and Subject Descriptors: C.1.2 [Computer Systems Organization] : Multiprocessors; J.6 [Computer Applications]: COMPUTER-AIDED ENGINEERING

General Terms: Algorithm, Design, Experimentation.

Keywords: Power Optimization, MPSoC, Dynamic Voltage/Frequency Scaling.

1. INTRODUCTION

The power crisis in current digital integrated systems is fueling an architectural paradigm shift toward multi-core architectures: Single-chip multi-core engines, that have first become widespread in embedded computing, are now making deep inroads in general purpose computing [1].

Energy-efficient multi-core design is getting an increasing level of technology support. The concept of "voltage islands" [2] has gathered momentum, and many recently announced SoC architectures feature tens of power domains [3]. Multi-core architectures with independently controllable supply voltage and clock frequency enable an unprecedented level of control on the performance vs. power/energy tradeoff. Voltage and speed setting allows to virtually eliminate all wasteful mismatches created by non-uniform workloads allocated to cooperating cores.

Permission to make digital or hard copies of all or part of this work for personal or classroom use is granted without fee provided that copies are not made or distributed for profit or commercial advantage and that copies bear this notice and the full citation on the first page. To copy otherwise, to republish, to post on servers or to redistribute to lists, requires prior specific permission and/or a fee.
ISLPED'06, October 4–6, 2006, Tegernsee, Germany.
Copyright 2006 ACM 1-59593-462-6/06/0010 ...$5.00.

A tough challenge in this area is the accurate and timely detection of workload mismatches at execution time. In many realistic use cases, applications cannot be accurately pre-characterized, and efficient run-time mechanisms are required to detect when one or more cores are over-clocked with respect to their workload. To address this challenge, several techniques have been proposed that require modification in the applications ([6]–[8]), or impose a particular inter-processor communication style that helps making speed mismatches more easily detectable [9, 10, 11, 12].

This paper moves towards the development of minimally-invasive detection of workload mismatches. Instead of requiring additional efforts from the programmer's side, we augment the hardware architecture with self-monitoring and autonomous control capabilities. We *monitor synchronization* to detect non-constructive idleness, which is a direct manifestation of non-optimal speed setting. We infer the idleness by observing the memory accesses performed by processors, and use such an information to drive a dynamic speed setting policy.

Results on a MPEG decoder show that our approach can save more than 50% of the energy spent with respect to a system running at the maximum speed, also improving by 25% a static assignment of speeds based on the off-line monitoring of the system workload.

2. PREVIOUS WORK

The problem of voltage/frequency selection is a quite mature research topic: many techniques have been proposed in the literature, requiring various levels of support by either hardware or software (a survey of DVS/DFS techniques can be found in [4, 5]). All these schemes, however, target single-processors systems.

In the multi-processor system domain, conversely, the spectrum of solution is much more limited. Most approaches have focused on a task-based model of the system, in which system behavior consists of a set of tasks with well-defined execution times and deadlines. The availability of an RTOS is usually assumed, and the issue of speed selection is embedded into the task scheduling problem with real-times constraints ([6]–[8]).

The approaches followed by [11, 12] introduce a control-theoretic perspective of the speed assignment problem, which is solved by modeling the system as a queue network. Characterizing embedded applications in terms of service and arrival rates tends to be difficult, however; these approaches thus tend to be more suitable for an analytical exploration of workload allocation policies than for detailed frequency assignments.

In [9], authors concentrate on the balance between computation and communication, concurrently determining an optimal speed assignment for both communication and computation tasks. This work is explicitly targeted for network processors, for which communication power is significant.

Our scheme shares with [9] the assumption that tasks are statically mapped to processing elements: in this scenario, scheduling of the tasks is immaterial, and only speed assignment is relevant. Our solution differs from this approach in two main aspects: First, we do not make *any relevant assumptions about the application running on the system*, including its data-flow or communication patterns; second, our method is *purely architectural*, in the sense that it automatically detects, through a custom hardware block, changes in the system workload, and sets processors speeds accordingly.

3. SYNCHRONIZATION-DRIVEN SPEED SCALING

The proposed speed scaling scheme is based on the idea of *relating the speed assignment of a given processor to the amount of time spent waiting for synchronization*.

In this work we consider applications that rely on a shared-memory paradigm, and execute on a multi-core platform explicitly based on shared-memory. We therefore assume that synchronization is based on shared variables. Our "low-level" synchronization primitive is therefore based on the polling of some shared variables.

We then assume that the system consists of n cores numbered $1, 2, \ldots, n$. Each core i can be set to a speed S_i, which can take values from a set of m possible speeds $S_k, k = 1, ..., m$. The i-th core executes a *workload* W_i, measured in terms of the amount of work (e.g., instructions) to be performed; its actual unit is irrelevant since our formulation relies on the ratio between workload and speed W_i/S_i, which yields a measure of time.

Finally, we denote by T_i and T_{S_i}, respectively, the time spent by processor i executing its workload and busy waiting for some synchronization variable to become available.

3.1 Energy, Workload and Synchronization Time

The time spent for synchronization by one processor represents a form of non-constructive idleness; unlike idleness due to other factors (e.g., I/O waiting), this idleness is caused by the *interference* of other processors, making thus its nature (and thus the issue of speed setting) quite different from the single-processor case.

Intuitively, time spent during synchronization wastes energy, since a core busy-waits without performing useful computation. In order to support this intuition, we ran an experiment to establish a relation between the time spent for synchronization and energy, consisting of an exhaustive exploration on a synthetic application.

The latter implements a prime number search on a shared vector, implementing a parallel version of the Eratosthenes' algorithm. By allocating slices of the vector with different sizes to cores, it is possible to split the total amount of computational work as desired, thus allowing explorations with different workload allocations.

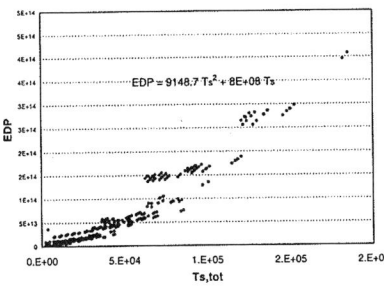

Figure 1: Relation Between EDP vs. Total Synchronization Time.

Figure 1 shows the energy-delay product versus the total synchronization time $T_{S,tot} = \sum_{i=1}^{n} T_{S_i}$ (values of T_{S_i} have been measured through simulation on the platform described in Section 4). The plot clearly supports the intuition and shows how execution time and energy consumption are strongly impacted by the amount of time spent in synchronization. Using these observations, we derive a speed setting policy driven by the idle synchronization time. Le us consider an arbitrary time window TW; over this window, the time spent by the application running on the system will consists, in general, of three distinct components: the time for executing "useful" instructions, the time spent for synchronizing with other processors, and the time used to wait for the bus to access its private memory. In formula: $TW = T_i + T_{S_i} + T_{BW_i}$. Without a significant loss in accuracy, we can neglect T_{BW_i}, since caches will filter out most of the memory accesses.

Reformulating, and by exposing processor speed, we get:

$$TW = \frac{W_i}{S_i} + T_{S_i} \qquad (1)$$

Notice that TW is an invariant, that is, it is the same for all cores. Based on the results of Figure 1, the objective is to reduce T_{S_i} to 0; this can be achieved by either slowing down some processors, or by speeding up some others. Consider for instance the situation depicted in Figure 2, where processor P_2 waits, for a time T_{S_2}, processor P_1 to produce an event (e.g., unlock of a semaphore). This

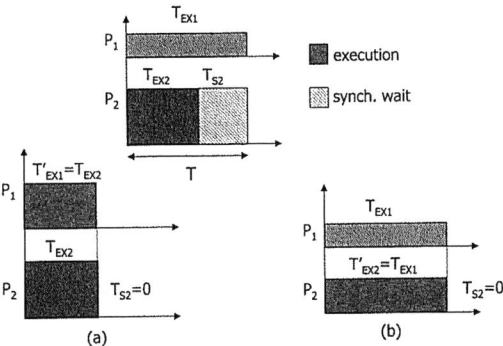

Figure 2: Alternative Speed Assignments: Speeding Up P_1 (a) or Slowing Down P_2 (b).

time can be reduced to 0 by speeding up P_1 of a quantity T_{S_2}/T (a), or by slowing down P_2 of the same quantity. The two alternatives differ in the execution time, which is correlated to the speed of the two processors. While from the point of view of EDP they are both identical (because $T_{S_2} = 0$), we should tend to privilege the scheme where the overall execution time is smaller.

The example suggests a possible qualitative criterion for speed setting: *If a processor P_i waits, then slow it down if it is running at the maximum speed, otherwise speed up the other cores.*

3.2 Speed Scaling Algorithm

We consider time to be split into a set of disjoint time windows of the same size TW; the speeds of the cores are determined upon expiration of each time window, based on the quantities measured in the previous window. This discretization relies on the assumption that the *workloads are slowly changing with respect to the observation window*, which is quite reasonable if TW is much smaller than the total execution time.

Computation of the optimal speeds requires the estimation of the incoming workloads; from Equation 1, a workload can be espressed as $W_i = (TW - T_{S_i}) \cdot S_i$. At the end of a time window, all the three quantities on the right hand side of this equation are known (TW

347

and S_i), or can be measured (T_{S_i}). This allows to compute W_i for the just expired time slot; under the assumption that workload will not change abruptly in the next slot, we use W_i as an estimate of the incoming workload in the next slot.

In order to achieve the optimal speed assignment ($T_{S_i} \equiv 0, \forall i$), we must enforce that all workloads are executed in the same time, that is, all the ratios W_i/S_i are identical:

$$\frac{W_1}{S_1} = \frac{W_2}{S_2} = \ldots = \frac{W_n}{S_n} \qquad (2)$$

This condition implies that all processors will execute their workloads without synchronization idle time (e.g., as in Figure 2).

Formula 2 is a system of $n-1$ equations with n unknowns. In fact, the time to which each W_i/S_i evaluates to depends on the speed assignments and it is not defined a priori.

Since we aim at obtaining the same performance of a full speed system, still with a reduced energy consumption, we solve Equation 2 by *forcing the processor with the maximum workload to run at maximum speed*. All the other speeds will thus be automatically derived based on Equation 2.

More precisely, let j be the index corresponding to the processor with maximum workload during the k-th time window:

$$j = \text{index}(\max_{1,\ldots,n} W_i)$$

The speed setting for the $k+1$-th time window is thus defined as

$$\begin{cases} S_{j,opt} = S_{max} \\ S_{i,opt} = \frac{W_i}{W_j} \cdot S_{max} \end{cases} \qquad (3)$$

where S_{max} is the maximum available speed in the system.

Notice how this simple speed setting criterion is able to deal naturally with the most critical difficulty for a DVS scheme, that is, the problem of workloads that are variable over time.

4. EXPERIMENTAL RESULTS

To verify the effectiveness of the proposed policy, we first run a set of parametric benchmarks that expose different workload allocations and different kinds of parallelism. We have then applied it to a real-life embedded application, (an MPEG video decoder), for which some parametric exploration has also been carried out.

In both experiments, we compared our policy against (i) a full-speed system (all cores running at the maximum speed), and (ii) one static speed assignment. Since the only way to determine the optimal static assignment would be by exhaustive exploration of all possible frequency assignments, we determined a "reasonably good" assignment, obtained by extracting some execution information from the application code, and by a further exploration of other points around this possible optimal value. This procedure is supposed to emulate the choice of a designer that has to select a static assignment for a target application.

4.1 The Multiprocessor Platform

We implemented our policy by augmenting a virtual multiprocessor platform ([13]), consisting of a configurable number of processing elements, with their private memories, a shared memory, some hardware device for basic interprocessor synchronization (a hardware interrupt module, and a hardware device that provides test-and-set features), a shared-bus interconnect, and a clock divider that feeds the cores.

We augmented the platform by adding a speed setting device based on the above scheme. The device estimates the T_{S_i} and updates the processors' frequencies accordingly. Synchronization is evaluated by monitoring accesses performed by each core. We assume that

a processor is performing a busy wait when it repeatedly accesses the same address with no accesses to other locations between them. Accesses must be detected by inspecting the interface between each core and its cache controller, since many accesses are filtered by the data cache.

4.2 Application Kernels

- **Eratost**: A prime number search on a shared vector, implementing a parallel version of the Eratosthenes' algorithm. Each processor scans a disjoint portion of the vector, and cores synchronize using barriers.

- **Mergesort**: A parallel version of the mergesort algorithm; it sorts a pool of randomly generated vectors stored in shared memory. The application is parallelized in a barrier-based fashion for the sorting phase, and in a pipeline style for the final merging phase. Therefore, when the merging processor is working on the i-th vector, the other ones can sort the $(i+1)$-th. The workload allocation among cores is not uniform and it varies during the execution.

- **DES**: A DES algorithm parallelized in a master-workers-slave fashion. The master core sends the stream to two workers that encode it and send the encoded data to the slave core. The resulting workload is not uniform among cores (workers bear the heavier computational load) but it is static, since the amount of computation does not depend on data streams.

- **FFT**: a variant from the *SPLASH-2* [15] suite, modified to perform two concurrent FFTs on two distinct vectors of different sizes. Half of processors are in charge to perform the computation of the first vector, while the other cores handle the second vector. The computation is repeated several times and the role of processing elements is changed after each iteration, in order to obtain a non-uniform and non-static workload allocation.

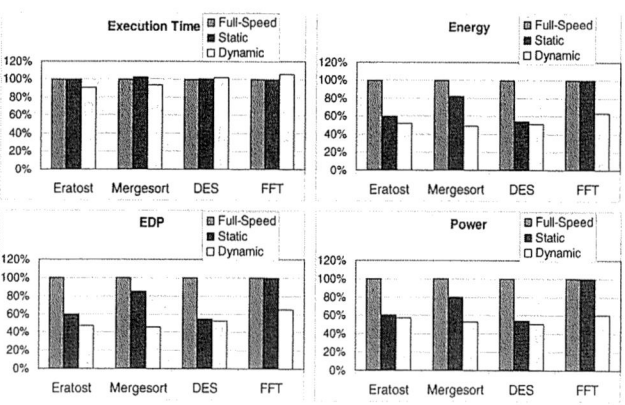

Figure 3: Results for Application Kernels.

Results of those experiments are shown in Figure 3. For each benchmark, we compare execution time, energy consumption, energy-delay product, and power, all normalized to the full-speed case. We can observe that our policy outperforms the static policy whenever the workload is variable in time (as usually happens in real complex applications). Conversely, when the workload is static (as in *DES*), the dynamic policy provides results comparable to the "optimal" static policy.

For the *FFT* application, since the best static allocation is the one with all cores at full speed, no savings can be obtained. Conversely, our dynamic policy, following the workload variations, can achieve appreciable benefits even in such a condition.

Notice also that, in some cases, a good speed setting policy can provide, as a side effect, a small improvement in terms of execution time. This happens because slowing down some processor reduces the traffic on the shared medium, hence allowing faster accesses to cores that execute at higher speeds.

4.3 Case Study

The application used in our case study decodes an MPEG video stream. One core is in charge of decoding the frame headers, and performs the entropic decompression of the data stream (Huffman and RLE decoding). The other 2 cores perform the de-quantization and the inverse-DCT (for I-frames), or they apply the motion compensation (for P- and B-frames). The last processor collects decoded data and sends them to the output buffer. The workload is not static, because different kinds of frames in a MPEG stream require different amounts of computation to be decoded.

Figure 4 shows how our policy outperforms the static policy in decoding a video stream where 20% of frames are I-frames. We reported data for execution time, energy spent, energy-delay product, and average power, all normalized with respect to the full-speed case (as a quantitative measure, the full-speed execution requires 472 million cycles and 293 mJ).

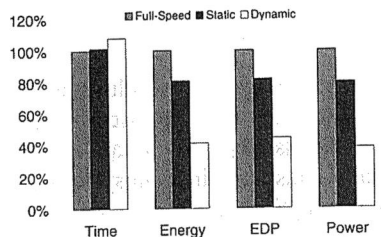

Figure 4: Comparison of Different Policies.

Since the workload behavior depends on the ratio between the various frame types, we used streams with different features to test our policy, varying the (Bframes + Pframes)/Iframes ratio. Figure 5 shows results of this exploration: we can observe how savings and penalties are just marginally affected by the input characteristics (time penalties do not exceed 8%, while the energy saving ranges between 56% and 64% with respect to full speed execution).

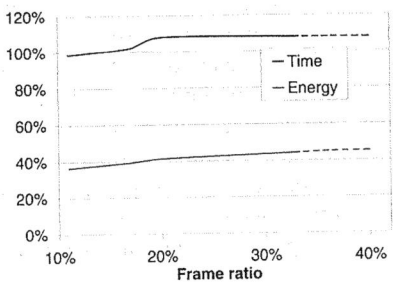

Figure 5: Dependency on the input features.

Last, we explored the dependency of our policy on TW (the size of the sampling window). Figure 6 shows how execution time and energy vary for window sizes between 2 and 64 KCycles: it is evident from the plot how the behavior of our policy is substantially independent of the window size.

5. CONCLUSIONS

We presented a purely architectural solution for the reduction of dynamic energy in MPSoCs, based on the principle of the equalization

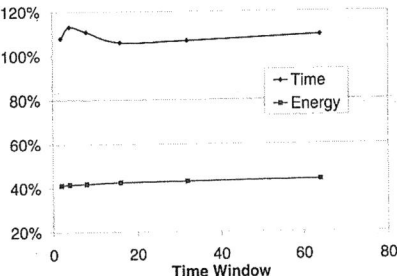

Figure 6: Dependency on the size of the sampling window.

of the normalized workload of the cores. Balancing is driven by the idleness due to synchronization, and provides relevant benefits in terms of energy consumption, with marginal impact on performance. The dynamic speed allocation policy does not require any software support, since it can infer the time spent in synchronization by examining the behavior of the processors on-line.

Results on a real life application (a parallel MPEG decoder) show that our policy outperforms other static policies, and it does not suffer from relevant performance penalties. We reduce energy by more than 40% with respect to a static assignment, and about 60% with respect to the full-speed system. Performance penalties remain below 8% with respect to the system with all cores running at the maximum speed, thus leading to savings in energy-delay product larger than 50% with respect to a full-speed execution.

6. REFERENCES

[1] David Geer, "Industry Trends: Chip Makers Turn to Multicore Processors," *IEEE Computer*, Vol. 38, No. 5, pp. 11-13, May, 2005.

[2] D. Lackey, et al., "Managing Power and Performance for SOC Designs using Voltage Islands," *ICCAD 2002*, pp. 195–202, Nov. 2002.

[3] "A Power Management Scheme Controlling 20 Power Domains for a Single-Chip Mobile Processor," Y. Kanno, et al., *ISSCC'06*, pp 540–541, Feb. 2006.

[4] W. Kim, D. Shin, H. S. Yun, J. Kim, S. L. Min, "Performance comparison of dynamic voltage scaling algorithms for hard real-time systems", *RTAS'02:IEEE Real-Time and Embedded Technology and Applications Symposium*, pp. 219–228, 2002.

[5] B. Zhai, D. Blaauw, D. Sylvester, K. Flautner, "Theoretical and Practical Limits of Dynamic Voltage Scaling", *DAC'04:41st Design Automation Conference*, pp. 868–873, 2004.

[6] P. Yang, et al., "Energy-Aware Runtime Scheduling for Embedded Multiprocessor SOCs," *IEEE Design and Test of Computers*, pp. 46–58, Vol. 18, No. 5, 2001.

[7] D. Zhu, R. Melhem, B. Childers. "Scheduling with Dynamic Voltage/Speed Adjustment Using Slack Reclamation in Multi-Processor Real-Time Systems", *IEEE Transactions on Parallel and Distributed Systems*, Vol. 14, No. 7, pp. 686–700, July 2003.

[8] W. Kwon and T. Kim. "Optimal Voltage Allocation Techniques for Dynamically Variable Voltage Processors". *IEEE Transactions on VLSI*, pp. 125–130, June 2003.

[9] J. Liu, P. Chou, N. Bagherzadeh, "Communication Speed Selection for Embedded Systems with Networked Voltage-Scalable Processors", *CODES'02*, pp. 169–174, May 2002.

[10] Z. Lu, J. Lach, M. Stan, "Reducing Multimedia Decode Power using Feedback Control," *ICCD'03:*, pp. 489-494.

[11] Q. Wu, P. Juang, M. Martonosi, L.-S. Peh and D.W. Clark. "Formal Control Techniques for Power-Performance Management". *IEEE Micro*, Vol. 25, No. 5, Sep./Oct. 2005, pp. 52–62.

[12] A. Alimonda, S. Carta, A. Pisano, A. Acquaviva, "A Control Theoretic Approach to Run-Time Energy Optimization of Pipelined Elaboration in MPSoC," *DATE'06*, pp. 876–877.

[13] MPARM Home Page, *www-micrel.deis.unibo.it/sitonew/research/mparm.html*

[14] A. Bona, V. Zaccaria, R. Zafalon, "System-Level Power Modeling and Simulation of High-End Industrial NoC", *DATE'04*, pp. 318–323, Mar. 04.

[15] J. P. Singh, W.-D. Weber, A. Gupta, "SPLASH: Stanford Parallel Applications for Shared-Memory", *Computer Architecture News*, Vol. 20, No. 1, pages 5–44, March 1992.

Power Phase Variation in a Commercial Server Workload

W. L. Bircher and L. K. John
Laboratory for Computer Architecture
Department of Electrical and Computer Engineering
The University of Texas at Austin
{bircher,ljohn}@ece.utexas.edu

ABSTRACT

Many techniques have been developed for adaptive power management of computing systems. These techniques rely on the presence of varying power phases to detect opportunities for adaptation. However, little information is available regarding the extent of power phases in real systems. This paper illustrates available power phases ranging from 1 millisecond to 1 second using a commercial workload running on enterprise class hardware. Data is obtained using a server instrumented for power measurement at the subsystem level. The analysis shows that chipset, memory and disk subsystems have the most homogenous phase behavior with greater than 71% of samples within phases of 100 milliseconds or shorter. In contrast, CPU and I/O subsystems have much more variation with only 26% of samples within phases of 10 milliseconds or shorter.

Categories and Subject Descriptors

C.4 [**Computer Systems Organization**]: Performance of Systems

General Terms

Measurement, Performance and Experimentation

Keywords

power, program phase, commercial workload characterization, microprocessor, chipset, memory, disk

1. INTRODUCTION

Power consumption is a problem for designers at all levels from transistor architects to system programmers. For many it is a first order design constraint. In recent years, designers have made progress in addressing this problem. At the transistor and microarchitecture levels, these improvements are applied without knowledge of what type of workloads are present. However, at the instruction set level or higher these improvements are applied dynamically in response to workload variation.

This research was supported in part by NSF grant 0429806 and by the IBM, Intel and AMD corporations.

Permission to make digital or hard copies of all or part of this work for personal or classroom use is granted without fee provided that copies are not made or distributed for profit or commercial advantage and that copies bear this notice and the full citation on the first page. To copy otherwise, or republish, to post on servers or to redistribute to lists, requires prior specific permission and/or a fee.
ISLPED '06, October 4–6, 2006, Tegernsee, Germany.
Copyright 2006 ACM 1-59593-462-6/06/0010...$5.00.

Typically, this involves identifying performance independent phases within a program. These phases allow for the exchange of performance for power savings without a loss in perceived performance.

While there has been much research on adaptive power aware architectures at various granularities, there has not been sufficient study on the actual phase granularities of real programs. Many past studies used simulations. Some used performance counter based data to extrapolate power in order to identify phases. In this paper, we study phases based on actual measurement. We measure power samples for several periodicities to study the granularities of phases that exist in commercial servers.

Using 10 KHz instrumentation, we illustrate power phases beyond typical coarse-grain phases used in servers running commercial workloads [1]. By quantifying how much power is typically consumed in a subsystem *and* how long the power consumption is stable enough to justify application of power adaptations, effective adaptations can be selected for a workload. This paper considers power phase durations of 1 ms, 10 ms, 100 ms and 1000 ms. Phases in this range are applicable to more fine grain adaptations such as dynamic voltage scaling, throttling, or other microarchitectural approaches.

This paper makes two primary contributions. The first is a measurement framework for the fine-grain study of subsystem power consumption. By simultaneously measuring the power of multiple subsystems, it is possible to observe complex interactions between subsystems without the need for simulation. Second, using this framework we characterize power phase behavior of a commercial workload. Unlike previous studies, the characterization includes power-phase duration and amplitude distribution, which can be used to predict the amount of detectable phase behavior in a workload.

2. METHODOLOGY
2.1 Power Sampling

For this study, we utilize an existing measurement framework from a previous processor power study [2] and extend it to provide additional functionality required for subsystem level study. One of the most significant differences between the studies of processor level versus subsystem level is the requirement for simultaneously sampling multiple power domains. To fulfill this requirement we chose the IBM x440 server. The x440 is 4-way SMP server utilizing 2GHz Xeon processors. Typical server components are used such as dual channel memory and high-speed SCSI disks. By choosing this server, instrumentation is greatly simplified due to the presence of current sensing resistors on the major subsystem power domains. Five power domains are considered: CPU, chipset, memory, I/O, and disk.

In order to limit the loss of power in the sense resistors and to prevent excessive drops in regulated supply voltage, the server system designer used a particularly small resistance. Even at maximum power consumption, the corresponding voltage drop is in the tens of millivolts. In order to improve noise immunity and sampling resolution we designed a custom circuit board to amplify the observed signals to levels more appropriate for our measurement environment. The entire measurement environment is shown in Figure 1.

2.2 Database Configuration

We chose the dbt-2 transaction processing workload from Open Source Development Labs [3] as a representative commercial workload. This workload imitates the TPC-C benchmark. It represents a wholesale parts supplier accepting orders and distributing parts to various sales districts. Dbt-2 dictates the warehouse/client configuration to maintain similarity with TPC-C. Within these requirements, we discovered that disk space is the primary bottleneck of our x440. Therefore, the 28 GBytes of available space on a dedicated disk yielded a 160-warehouse workload.

2.3 Phase Classification

Classification of power phases is presented in three ways: amplitude distribution, duration strata and combined amplitude and duration distribution. For our purposes, amplitude distribution is defined as a probability distribution of sampling power at a particular amplitude (Watts). Duration is the length of a phase in milliseconds.

Amplitude results are presented as probability distributions of all power samples. The samples are stratified into groups with a range equal to one tenth of the difference between maximum and minimum sampled value. The shape of the distribution can be used to direct power management policies. For example, multimodal power amplitude distributions suggest multiple distinct power phases. In contrast, normally distributed power consumption, suggests a simpler phase behavior. For the purpose of dynamic phase detection and adaptation, the widely distributed (large standard deviation) and/or multimodal distributed (multiple peaks) offer the best opportunities, due to the presence of multiple distinct behaviors. Very narrowly distributed power behavior indicates highly homogeneous power consumption and consequently, little opportunity to detect power phases. Finally, the location of the distribution center provides a single, representative power consumption value for the subsystem.

Presented power phase duration results do not have as fine a granularity (4 levels) as the amplitude results (10 levels). Rather than performing an exhaustive search of all possible phase durations, we instead selected four groups: 1ms, 10ms, 100ms and 1000ms. These groups are intended to cover the range of useful, but underutilized durations. Phase durations much greater than 1s are fairly well known and utilized in the server environment [4][1]. No phases less than 1 milliseconds are considered due to the overhead of current adaptation mechanisms.

Our mechanism for defining a phase is similar to the phase comparison metric used by Lau [5]. In order for a series of samples to be considered a phase they must have a coefficient of variation (CoV) less than the limit specified for the experiment. Our results show that a CoV of 0.05 yields representative phases that differ from the sampled data by 3.2% on average.

Phase groupings by duration are inclusive of all phases greater than or equal to their duration size, yet smaller than the next larger duration group. For example, all phases with durations from 10 ms to 99 ms are placed in the 10ms group. In addition, phases are mutually exclusive. Grouping in the largest possible duration is preferred. For example, though a 100 ms phase is composed of 10, 10 ms phases, the 10 ms phases are not placed in the 10 ms group. This approach favors identifying the maximum number of long duration phases, since long phases give the best opportunity for amortizing the cost of identification and power adaptation.

Combined power phase amplitude and duration results are presented as a three dimensional distribution. The horizontal x-y plane contains power and duration values while the vertical z-axis contains the probability of a sample at a particular amplitude-duration combination. By considering the distribution shape, conclusions can be drawn regarding power phase behavior.

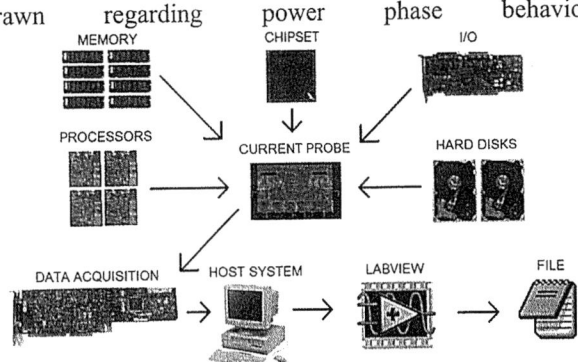

Figure 1 Measurement Environment

3. POWER ANALYSIS

3.1 Amplitude

Figure 2 shows the average power breakdown for the various subsystems. Not surprisingly, CPUs are the dominant power users. However, unlike distributed, scientific [6] and mobile, productivity workloads such as [7], I/O and disk power are significant. While differences in average subsystem power are large at 138% for disk compared to CPU, the variations within an individual subsystem are even greater. A comparison of subsystem power-amplitude distributions is made in Figure 3. Note that the CPU distribution is truncated at 60 Watts to prevent obscuring results from the other subsystems. A small number of phases (6.5%) exist above 60 Watts and extending to 163 Watts.

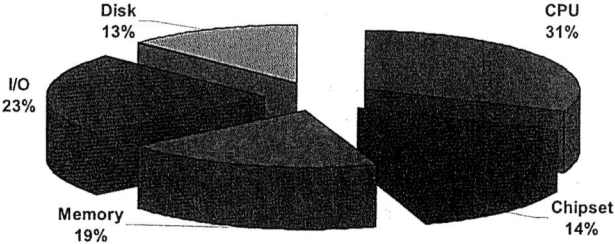

Figure 2 Server Average Power Consumption

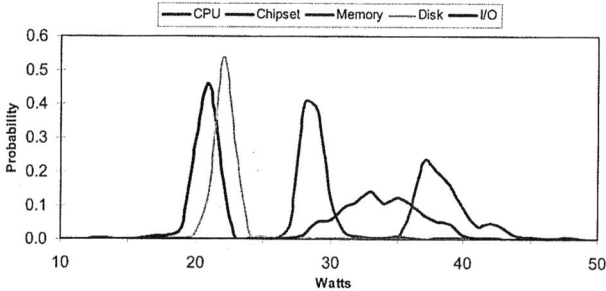

Figure 3 Subsystem Amplitude Distributions

These distributions suggest that there are significant opportunities for phase-based power savings for CPU and I/O. These subsystems have wider and/or multimodal distributions. The larger variations in power consumption provide greater opportunity to use runtime detection techniques such as [8][9]. In contrast, chipset, memory and disk have very homogeneous behavior suggesting nearly constant power consumption and less opportunity for phase detection.

3.2 Duration

The presence of power variation is not sufficient to motivate the application of power adaptation. Due to the overhead of detection and transition, adapting for short duration phases may not be worthwhile. Table 1 presents the percentage of samples that are classifiable as phases with durations of 1 ms, 10ms, 100ms and 1000ms. These results assume a group of samples can be defined as phases of multiple durations. As described in Section 3.3, a 100 ms phase would be made up of 10-10ms phases. The effect of narrow chipset, memory and disk distributions is evident in their high rates of classification. For these, at least half of all samples can be classified as 1000 ms phases. In

contrast, CPU and I/O have no 1000 ms phases and considerably fewer phases classified at finer granularities. These results can be used to plan power management strategies for a particular workload. For example, by noting that the I/O subsystem has few phases longer than 1 ms, the designer would be required to use very low latency adaptations to avoid performance loss. In contrast, chipset and memory subsystems have a large percentage of classifiable samples at phase lengths greater than 1ms. It should be noted that this is not a guarantee of detectable phase behavior. By also considering that most of the chipset and memory samples are very close to the average (standard deviations of 0.9 Watts and 1.4 Watts respectively) there may be insufficient variation for runtime phase detection.

Table 1 Percent of Samples Classifiable at CoV = 0.05

Duration (ms)	CPU	Chipset	Memory	I/O	Disk
1	61.6	88.3	97.7	98.4	92.1
10	25.5	78.0	91.2	32.0	91.1
100	6.00	63.2	78.6	18.5	91.3
1000	0.00	64.4	50.0	0.00	99.7

3.3 Amplitude - Duration

The following phase duration classification can be used to direct selection of power saving techniques. The phase duration can select the power management type based on similar transition times. The power level and frequency work in opposition to each other as a policy control. For example, though a particular phase may only occur 5% of the time, since it is such a high power case it would be valuable to reduce its power. This is similar to the case of the CPU in which 4% of power samples were over 2X the average sample value and 2% were over 3X. At the other extreme, a phase may consume very low power, but since it occurs frequently, it would be valuable to address.

Simultaneously considering the distribution of amplitude and duration, the subsystem power consumption can be divided into two groups: high homogeneity and low homogeneity. The homogeneity of a distribution is defined as the amount of variation with respect to phase amplitude and duration. The narrow distribution of Section 3.3 is typical of high homogeneity, while the wide and multimodal distributions are considered typical of low homogeneity.

3.3.1 High Homogeneity

While chipset power samples are mostly symmetric about the average (Figure 4), memory power has small numbers (2.7%) of phases above the average (Figure 5). If these rare cases could be eliminated, worst-case power consumption could be reduced by 25% from 40 Watts to 30 Watts. Depending on the level of resultant performance loss, the reduction in cooling and power cost could be worthwhile. In contrast, the disk subsystem (Figure 6) has few significant fluctuations to provide an opportunity for phase detection.

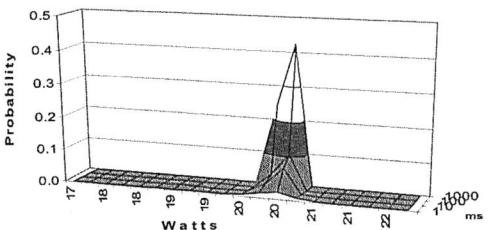

Figure 4 Chipset Amplitude-Duration Distribution

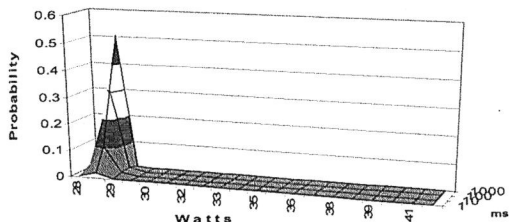

Figure 5 Memory Amplitude-Duration Distribution

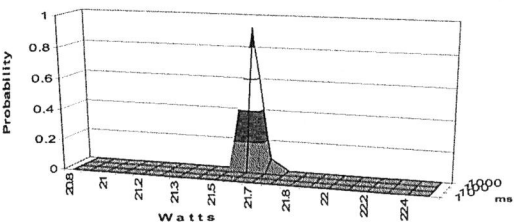

Figure 6 Disk Amplitude-Duration Distribution

3.3.2 Low Homogeneity

CPU and I/O present two distinct versions of low homogeneity. The CPU distribution in Figure 7 is characterized by vast swings in amplitude. Excluding the large variations, most samples were tightly grouped around 40 Watts. Like the memory power distribution (and to a greater extent) worst-case power consumption is a significant issue for the CPU subsystem. Reducing power supply and cooling capacity to only accommodate the common 40-Watt case, could significantly reduce cost.

Finally, in Figure 8, the I/O power distribution contains significant phase detection opportunities. Centered about the average, most samples are in phases with durations from 1 ms to 100 ms. Across this range, the distribution is nearly constant with 18%-20% of samples in each phase duration group. However, moving symmetrically away from the average, two 1 ms local maximums occur at 31 Watts and 36 Watts.

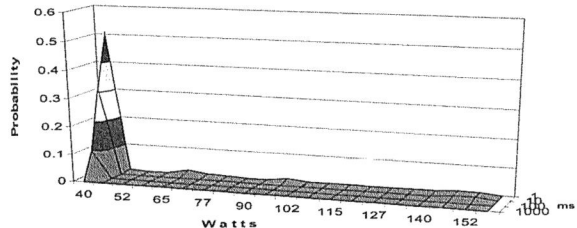

Figure 7 CPU Amplitude-Duration Distribution

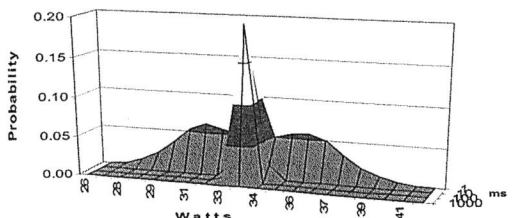

Figure 8 I/O Amplitude-Duration Distribution

4. CONCLUSION

In this paper, we have presented a framework for measuring power at a fine grain. The fine grain sampling illustrates distinct power phases ranging in duration from milliseconds to seconds and amplitude variations from 30% to 300%. Further, we suggest that CPU and I/O subsystems have a greater potential for phase detection compared to chipset, memory and disk, due to their larger phase variation.

5. REFERENCES

[1] Yiyu Chen, Amitayu Das, Wubi Qin, Anand Sivasubramaniam, Qian Wang and Natarajan Gautam, Managing Server Energy and Operational Costs in Hosting Centers. ACM SIGMETRICS, pp 303-314, June 2005.

[2] W. L. Bircher, M. Valluri, J. Law, L. K. John: Runtime identification of microprocessor energy saving opportunities. International Symposium on Low Power Electronics and Design, pp 275-280, August 2005.

[3] Open Source Development Lab, Database Test 2, www.osdl.org/lab_activities/kernel_testing/osdl_database_te st_suite/osdl_dbt-2/, February 2006.

[4] K. Rajamani and C. Lefurgy. On Evaluating Request-Distribution Schemes for Saving Energy in Server Clusters. Proceedings of the IEEE International Symposium on Performance Analysis of Systems and Software, pp 111-122, March 2003.

[5] Jeremy Lau, Stefan Schoenmackers and Brad Calder. Structures for Phase Classification. IEEE International Symposium on Performance Analysis of Systems and Software, pp 57-67, March 2004.

[6] Xizhou Feng, Rong Ge, Kirk W. Cameron: Power and Energy Profiling of Scientific Applications on Distributed Systems. International Parallel & Distributed Processing Symposium, pp 34-50, April 2005.

[7] Aqeel Mahesri and Vibhore Vardhan, Power Consumption Breakdown on a Modern Laptop, Workshop on Power Aware Computing Systems, 37th International Symposium on Microarchitecture, December 2004.

[8] Canturk Isci and Margaret Martonosi, Phase Characterization for Power: Evaluating Control-Flow-Based and Event-Counter-Based Techniques. 12th International Symposium on High-Performance Computer Architecture, pp 122-133, February 2006.

[9] A. Dhodapkar and J. Smith. Comparing program phase detection techniques. 36th International Symp. on Microarchitecture, pages 217-228, December 2003.

[10] Pat Bohrer, Elmootazbellah N. Elnozahy, Tom Keller, Michael Kistler, Charles Lefurgy, Chandler McDowell, and Ram Rajamony, The Case For Power Management in Web Servers. IBM Research, Austin TX 78758, USA.

Reducing Power through Compiler-Directed Barrier Synchronization Elimination*

Mahmut Kandemir   Seung Woo Son
Dept. of Computer Science and Engineering
The Pennsylvania State University
University Park, PA 16802
{kandemir,sson}@cse.psu.edu

ABSTRACT

Interprocessor synchronization, while extremely important for ensuring execution correctness, can be very costly in terms of both power and performance overheads. Unfortunately, many parallelizing compilers are very conservative in inserting barrier synchronizations at the end of each and every parallel loop. This can lead to significant power consumption in chip multiprocessor based execution environments. This paper proposes a compiler-directed approach for eliminating such synchronization calls between neighboring parallel loops. It achieves its goal by partitioning loop iterations across processors such that each processor executes iterations from both the loops that access the same set of array elements. We implemented the proposed approach using an experimental compilation framework and made experiments with ten SPEC benchmark codes. Our experiments clearly show that the proposed compiler-directed approach is very effective and reduces energy overheads due to synchronizations by about 75.5%, and this corresponds to around 5.48% saving on average in overall energy consumption.

Categories and Subject Descriptors: D.3.4 [Programming Languages]: Processors–*Compilers, Optimization*
General Terms: Algorithms, Design, Experimentation, Performance
Keywords: Barrier Elimination, Compiler, Low Power

1. INTRODUCTION AND MOTIVATION

Synchronization is an important concept in parallel computing to ensure data integrity and eliminate timing related errors. However, excessive synchronization among parallel processors can lead to both performance and power overheads which can easily offset the potential benefits expected from parallel execution. Minimizing synchronization is particularly important in array-dominated applications, e.g., those from the domain of embedded signal processing, that exhibit high degrees of parallelism. While an experienced user can insert a minimum number of synchronization calls in the application code to reduce the overheads, this task is generally very difficult to handle and automated compiler's help can be very useful.

Unfortunately, many parallelizing compilers are conservative in their insertion of synchronization calls in application codes. A typical situation is depicted in Figure 1, where the compiler inserts a barrier synchronization between two parallel loops to ensure that all processors are synchronized before they start to execute their

---

*This work is supported in part by NSF career award #0093082 and a grant from GSRC.

Permission to make digital or hard copies of all or part of this work for personal or classroom use is granted without fee provided that copies are not made or distributed for profit or commercial advantage and that copies bear this notice and the full citation on the first page. To copy otherwise, to republish, to post on servers or to redistribute to lists, requires prior specific permission and/or a fee.
ISLPED'06, October 4–6, 2006, Tegernsee, Germany.
Copyright 2006 ACM 1-59593-462-6/06/0010 ...$5.00.

```
for i:= 1, n              forall i:= LB_k, UB_k
  ...                       ...
endfor                    endforall
for j:= 1, n      ⟹      ⟨barrier_synch⟩
  ...                     forall j:= LB'_k, UB'_k
endfor                      ...
                          endforall
```

Figure 1: Left: A code fragment. Right: Parallel code to be executed by processor k. $LB_k(LB'_k)$ and $UB_k(UB'_k)$ denote the lower and upper bounds, respectively, for the portions of iteration spaces assigned to processor k.

```
for i:= 1, n          forall i:= 1, n/4        forall i:= 1, n/4
  A(i) = B(i)+1         A(i) = B(i) + 1          A(i) = B(i) + 1
endfor                endforall                endforall
for j:= 1, n          ⟨barrier_synch⟩          forall j:= n-n/4+1, n
  C(n-j+1) =           forall j:= 1, n/4         C(n-j+1) =
       A(n-j+1)/2.0      C(n-j+1) =                   A(n-j+1)/2.0
endfor                      A(n-j+1)/2.0        endforall
                       endforall
       (a)                    (b)                     (c)
```

Figure 2: (a) Original code. (b) Straightforward iteration distribution for a processor. (c) Code generated by our approach.

portions from the second loop. This synchronization is important since it is possible that an array element written (updated) in the first loop by a processor might be read within the second loop by another processor. If these two processors do not get synchronized at the barrier, the second processor can read the old values of the said array element and this leads to wrong computation in most cases.

Such synchronizations inserted into parallel applications are problematic from the power perspective in three ways. First, the processors that come to synchronization points earlier than the last processor wait for the latter, and during this waiting period, they consume dynamic energy. In fact, two recent papers [8, 9] have already documented that the energy consumption in such periods can be significant (around 15% on average) and proposed methods to reduce this energy. Second, since these synchronizations increase execution time of the application (as the slowest processor determines the overall latency), they contribute to extra leakage consumption. Third, the synchronization calls themselves consume some energy as the instructions to implement the synchronization have to be brought from memory to the instruction cache. These three energy-related overheads form a strong motivation for investigating ways of eliminating synchronization calls if it is possible to do so without affecting program correctness.

This paper is a step towards that direction. Specifically, it proposes a compiler-directed approach for eliminating such synchronization calls between parallel loops. It achieves its goal by partitioning loop iterations across processors such that each processor executes iterations from the neighboring loops that access the same set of array elements. Figure 2 illustrates the proposed approach using a simple example. The original sequential code and its parallel version without using our approach are shown in Figures 2(a) and (b), respectively. We assume that there are four processors in the underlying parallel architecture and the code in (b) is for the first one. Figure 2(c) on the other hand shows the code generated

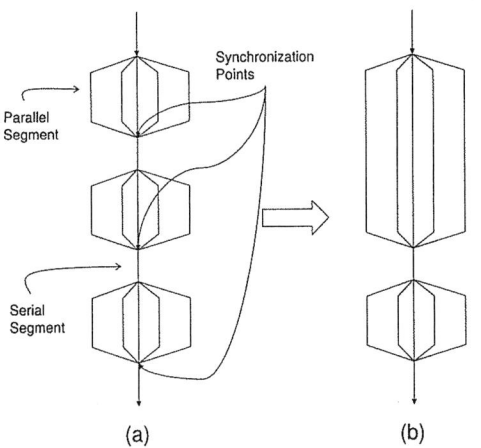

Figure 3: Fork-join style of parallelization. (a) Conservative synchronization. (b) Eliminating the synchronization point between the two parallel loops.

using our proposed approach. In this code, no synchronization is needed between the parallel loops since the array element updated by a processor in the first loop is read in the second loop by the same processor. In other words, each processor can proceed independently of the others, which is not the case in Figure 2(b). We also show in this paper how our approach can convert barrier synchronizations to point-to-point synchronizations in case it is not possible to eliminate the former completely.

To implement the proposed approach, we use the Presburger arithmetic [13], which is the first-order theory of the natural numbers with addition. More specifically, we represent the set of array elements accessed by a processor and the set of iterations assigned to a processor using Presburger sets and ensure that each processor uses the data that it has updated in the previous loop nest. To enumerate the Presburger sets we build, we employ the Omega Library, a polyhedral tool from the University of Maryland [5].

We implemented the proposed approach using an experimental compilation framework and made experiments with nine SPEC benchmark codes. Our experiments clearly show that the proposed compiler-directed approach is very effective and reduces energy overheads due to synchronizations by about 75.5%, and this corresponds to around 5.48% saving on average in overall energy consumption.

The remainder of this paper is organized as follows. Section 2 introduces the fork-join style of parallelization model, which is assumed as the base case in this paper. Section 3 explains the mathematical framework behind our compiler-directed approach. Section 4 discusses the experimental results we collected and Section 5 presents our concluding remarks.

2. EXECUTION MODEL

We adopt a shared memory based parallelization model depicted in Figure 3(a), which is also referred to as the *fork-join parallelism*. In this model, the application is divided into sequential and parallel segments. The sequential segments are executed by a master thread, whereas the parallel segments are executed (in parallel) using all threads, including the master. During the executions of the parallel segments, processors can also engage in communication, depending on interprocessor data sharing. In current practice, at the end of each parallel segment, the processors synchronize (typically using a barrier as explained above) so that they continue together with the rest of the application without worrying about dependencies. The goal behind the approach proposed in this work is to eliminate some of these synchronizations and reduce energy consumption. In a pictorial form, if the barrier synchronization between the first two loop nests (parallel segments) shown in Figure 3(a) can be eliminated, we want to obtain the execution profile shown in Figure 3(b), given the one in Figure 3(a).

In this work, the target set of applications considered is array-intensive codes frequently used in embedded signal processing. Most of these codes are constructed using a series of loop nests, each operating on a subset of the arrays (of signals) declared in the program code. We assume that the array references in the application code are affine functions of enclosing loop indices and loop-invariant constants. We handle non-affine references conservatively. It is important to note that, in this work, we do not propose a new loop parallelization strategy; instead, we assume that the set of loops whose iterations will be executed in parallel are decided by a prior pass (phase) during compilation. What our approach does is to decide the set of iterations that will be assigned to each processor in the chip multiprocessor architecture. In other words, given the set of parallel loops, our approach performs the *iteration-to-processor mapping* in a fashion that the number of synchronization calls in the application is minimized. We also assume that codes to be optimized are first translated to the array-SSA form, i.e., in a loop nest, each element of a left-hand-side array can be updated at most once. In our implementation, we use the approach in [6] for this purpose.

3. SYNCHRONIZATION ELIMINATION

The key observation behind our barrier synchronization elimination scheme is that, if a processor updates a data element in a loop nest, it should be the only one that accesses the same element in the next loop nest. Clearly, this requirement may not be possible to satisfy in every case. However, in cases where one can satisfy it, one can reduce the number of synchronizations and the corresponding power overheads they bring. Let Λ^j be the set of iterations from a loop nest j that that will be executed in parallel. As noted earlier, we assume in this work that this set is determined by a previous compiler pass. Assuming that P is the total number of processors[1], our goal is to decide a partitioning of Λ^j into P sets, namely Λ_1^j, $\Lambda_2^j, \cdots \Lambda_P^j$, such that:

$$\bigcup_{i=1}^{P} \Lambda_i^j = \Lambda^j \quad \text{and} \quad \bigcap_{i=1}^{P} \Lambda_i^j = \emptyset.$$

Let $\vec{\lambda}$ be an iteration that belongs to Λ^j and $\vec{\lambda'}$ be an iteration that belongs to $\Lambda^{j'}$[2], the nest that comes after j. If there is a left-hand-side reference R within j and a right-hand-side reference R' within j' such that $R(\vec{\lambda}) = R'(\vec{\lambda'}) = \vec{d}$, where $\vec{d}$ is an array element, then we need to ensure that $\vec{\lambda} \in \Lambda_i^j$, on $\vec{\lambda'} \in \Lambda_i^{j'}$, where $1 \leq i \leq P$, and there should not be a $\vec{\lambda''} \in \Lambda_{i'}^{j'}$ such that $\vec{d} = R''(\vec{\lambda''})$ where $i' \neq i$, for a right-hand-side reference R'' in j'. That is, the loop iterations that access the same array element $\vec{d}$ have to be mapped to the *same* processor. It needs to be noted at this point that we apply a similar logic when we move to the nest (j'') that follows j'. That is, if the same element $\vec{d}$ is accessed by both j and j'', the loop iterations that access the said element, $\vec{\lambda}$ and $\vec{\lambda''}$, are assigned to the same processor i if it is possible to do so, and no other processor should access $\vec{d}$ in j''.

In our framework, this condition is captured using the Presburger sets and manipulated using a polyhedral tool called the Omega Library [5]. The Omega Library is a set of C++ classes for manipulating integer tuple relations and sets. It has routines for creating, manipulating, and enumerating Presburger sets. Clearly, working at an individual array element granularity will not be a scalable

---

[1]Our approach can be used in any parallel architecture. In this paper, however, we assume a chip multiprocessor.

[2]We use vectors to denote loop iterations. A loop iteration vector, $\vec{\lambda}$, contains the values of loop indices from the outermost loop in the nest to the innermost one. Similarly, array elements are represented by vectors; each such vector, $\vec{d}$, represents the index of the corresponding array element

```
Compute $\Lambda_i^1$ for all $1 \le i \le P$.
Let $R_1, R_2, \ldots, R_q$ be the left-hand-side references
  in the first loop nest.
Compute $\Theta_i^1(1), \Theta_i^1(2), \ldots, \Theta_i^1(q)$ for all $1 \le i \le P$.
$\Theta_i^1 = \Theta_i^1(1) \cup \Theta_i^1(2) \cup \ldots \cup \Theta_i^1(q)$.
for $j = 2$, number of nests
    Compute $\Lambda_i^j$ using $\Theta_i^{j-1}$.
    Compute $\Theta_i^j$.
    if $(\Theta_i^j \cap \Theta_{i'}^j \ne \emptyset)$ for any $i \ne i'$ then
        insert point-to-point synchronization between $i$ and $i'$.
```

Figure 4: Our synchronization elimination algorithm.

approach, and therefore, we prefer to work with sets. More specifically, let us assume that j is the first loop nest that assigns new values to elements of an array and $j+1, j+2, \cdots, j+s$ are the loop nests that use the elements of the same array. We assume that the iterations of j are distributed across processors using any existing scheme, and after this distribution, processor i accesses the following set of elements (in executing j) from the array in question:

$$\Theta_i^j = \{\vec{d} \mid \exists R, \vec{\lambda} \text{ such that } \vec{\lambda} \in \Lambda_i^j \text{ and } R(\vec{\lambda}) = \vec{d}\}.$$

Here, R is an array reference that appears in loop nest j. We now focus on loop nest $j+1$ and our job is to identify the set of iterations that need to be mapped to processor i so that the barrier synchronization between j and $j+1$ can be eliminated. We can use the following Presburger set for this purpose:

$$\Lambda_i^{j+1} = \{\vec{\lambda} \mid \exists R, \vec{d} \text{ such that } \vec{d} \in \Theta_i^j \text{ and } R(\vec{\lambda}) = \vec{d}\}.$$

Note that similar sets, $\Lambda_i^{j+2}, \cdots, \Lambda_i^{j+s}$, can be defined for nests $j+2, \cdots, j+s$ as well. If there is no element $\vec{d}$ such that $R(\lambda_i^{j+1}) = R'(\lambda_{i'}^{j+1})$ for a given processor pair (i, i'), we do not need synchronization between processors i and i'. If this condition is satisfied for all processor pairs, then we do not need any synchronization between loop j and j'.

We now focus on the scenario where two (or more) arrays are updated in nest j. In this case, let us use $\Theta_i^j(1)$ and $\Theta_i^j(2)$ to refer to the data elements accessed within loop nest j by processor i from these two arrays. We can now define $\Theta_i^j = \Theta_i^j(1) \cup \Theta_i^j(2)$ and proceed to compute Λ_i^{j+1} as explained above. It needs to be noted that, there are cases where our approach may not be able to eliminate the inter-processor synchronization between loop nests j and $j+1$ completely. This occurs when Λ_i^{j+1} and $\Lambda_{i'}^{j+1}$, where $i \ne i'$, access a common data element, i.e., when $\Theta_i^{j+1} \cap \Theta_{i'}^{j+1}$ is not empty. While in this case it is not possible to get rid of the synchronization requirement, our approach can still be used for *converting barrier synchronization to point-to-point synchronization*. More specifically, we identify all processor pairs i and i' such that $\Theta_i^{j+1} \cap \Theta_{i'}^{j+1} \ne \emptyset$ and insert point-to-point synchronization calls between them. For a given processor i, the set of processors with which point-to-point communications are needed between nests j and $j+1$ can be captured by the following Presburger set:

$$\begin{aligned} PP_{i,(j,j+1)} = \quad &\{i' \mid \exists R, R', \vec{\lambda}, \vec{\lambda}', \vec{d} \text{ such that } \vec{\lambda} \in \Lambda_i^{j+1} \\ &\text{and } \vec{\lambda}' \in \Lambda_{i'}^{j+1} \text{ and } R(\vec{\lambda}) = R'(\vec{\lambda}')= \vec{d}\}. \end{aligned}$$

The important point here is that such sets can be converted, using the Omega Library or similar polyhedral tools, to loop nests that enumerate the elements in them and these loop nests can be executed – at compile time – to determine, for each processor, the set of processors to synchronize with.

Handling Load Imbalance: The compiler-based approach explained so far is oriented towards eliminating barrier synchronizations if it is possible to do so or convert them to point-to-point synchronizations when complete elimination is not possible. However, such a synchronization-centric view can also lead to load imbalances across the processors. The load imbalance can offset the benefits coming from eliminating interprocessor synchronization. Our

Table 1: Default simulation parameters.

Simulation Parameter	Default Values
L1 Capacity, Line Size, Associativity	32KB (per processor), 32 bytes, 2-way
L1 Latency	2 cycles
L2 Capacity, Line Size, Associativity	3MB (shared), 64 bytes, 4-way
L2 Latency	10 cycles
Off-Chip Memory Latency	150 cycles
Bus Arbitration Delay	5 cycles
Replacement Policy	Strict LRU
Processor Count	8

Table 2: Benchmarks.

Benchmark	Exec Cycles (M)	Total Energy (mJ)	Barrier Energy (mJ)
swim	147,533	124.0	33.1
mgrid	524,928	401.8	75.4
applu	1,312,018	973.5	220.2
galgel	218,634	196.5	32.7
apsi	622,055	476.3	128.6
gafort	807,501	601.8	105.4
fma3d	724,386	589.3	124.7
art	673,491	416.1	96.3
ammp	905,716	720.8	136.9

approach detects such load imbalances when they occur and fixes them. It is built upon the observation which indicates that, when a severe load imbalance occurs, usually one or more processors are totally idle (this was always the case in all applications we tested). To see whether there will be a load imbalance across processors during execution of loop nest j, our approach checks whether Λ_i^j is $\emptyset$ for any processor i ($1 \le i \le P$). If this is the case, it adopts the following load balancing strategy. Assuming that N is the total number of iterations in the nest to be distributed (say j), we divide these iterations across all P processors as evenly as possible, trying to satisfy the following two conditions, assuming that Λ_i^j contains the set of iterations that would be assigned to processor i by our approach explained so far and μ_i^j is the set of iterations to be executed by i after load balancing:

- For any $\vec{i} \in \mu_i^j$, either $\vec{i} \in \Lambda_i^j$ or $\Lambda_i^j = \emptyset$, and
- For any processor pair (i, i'), the set of data elements accessed by Λ_i^{j-1} and that accessed by $\mu_{i'}^j$ do not have any common data elements.

The important point here is that the first condition can always be satisfied and helps us ensure that, after load balancing, any iteration assigned to a processor should be from the set of iterations (if any) it has been originally assigned by our approach. The second condition on the other hand tries to ensure that, in reassigning the iterations (for load balancing purposes), we do not create any cross-processor dependences. Since this condition may not be satisfied for all processor pairs, we may need to insert synchronization. However, by being careful in redistribution, we can keep the amount of synchronization at minimum, which can in turn allow us employ the point-to-point synchronizations instead of the barrier synchronization. A sketch of our synchronization elimination algorithm (without load balancing part) is given in Figure 4.

4. EXPERIMENTAL EVALUATION

We used SUIF [3] to analyze codes and the Omega Library [5] to build and enumerate the Presburger sets. To implement our approach, we used the SIMICS simulation infrastructure [10], with the default parameters shown in Table 1. SIMICS is a full-system simulator capable of running target hardware at high-performance speeds. We extended the baseline SIMICS tool with accurate timing and energy models (based on Wattch [1]). The chip multiprocessor considered in this work is of the shared memory based type, where a certain number of CPUs share the memory address space. We assume that each CPU has its own private L1 data and instruction caches that are accessible to it without the need of going across the shared interconnects. Several proposed chip multiprocessor designs from industry and academia already employ such

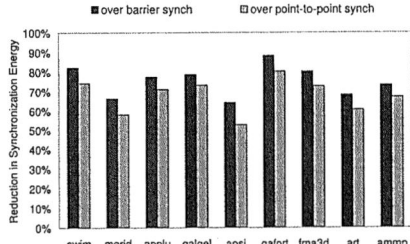
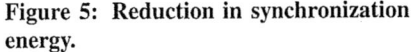
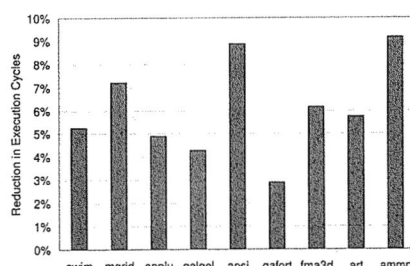
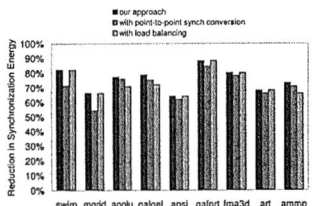

Figure 5: Reduction in synchronization energy.

Figure 6: Reduction in execution cycles.

Figure 7: Effects of converting some barriers to point-to-point synchronizations and load balancing.

private L1 based configurations [2, 7, 12]. We assume that the on-chip interconnect is a bus but our approach can work with NoC type of communication fabric as well. We also employ the MOESI protocol for keeping the cache coherent, though its choice is orthogonal to the main focus of this paper. We also assume the existence of a shared L2 cache. The applications used in this study from the SPEC [14] benchmark suite and are listed in Table 2, along with their important characteristics. The total sizes of the data processed by these applications range from 2.6 MB to 754.3 MB. The second column of this table gives the execution cycles. The last two columns show the total energy consumption and energy consumption due to barrier synchronizations at the end of parallel loops.

Figure 5 gives the reductions in synchronization energy over-heads as a result of applying our strategy. The first bar gives the reduction provided by our approach with respect to the energy numbers shown in the last column of Table 2. On an average, our approach brings around 75.5%. The second bar for an application on the other hand gives the percentage energy reductions with respect to a scenario where a barrier synchronization in the original parallel code has been replaced by a point-to-point synchronization when it is possible to do so (i.e., the base case against which we compare our approach uses point-to-point synchronizations where possible). The average savings in this case is about 67.9%, still a very large value. Since our approach is successful under both barrier synchronization and point-to-point synchronization cases, in the remaining part of this section, we present results with respect to the barrier synchronization alone. While these savings in synchronization energy are significant, one may be more interested in savings in overall energy consumption (i.e., over the values listed under the third column of Table 2). We found that the savings in total energy consumption range from 2.04% to 9.63%, averaging on 5.48%.

Figure 6 gives the savings in execution cycles as a result of the proposed approach. These savings, which range between 2.88% and 9.16%, are mainly due to evening out the imbalances among the execution times of the processors across the loop nests. As an example, consider three loop nests that are executed successively. It is possible that the latest processors in the first and second loop nests are different, and by eliminating the synchronization points between the nests, our approach helps to balance execution times of different processors.

Recall that so far in the experimental evaluation our approach has been used to eliminate barrier synchronizations (with no conversion to point-to-point synchronizations) and without any load balancing. Our last set of results quantify the effect of converting some barriers to point-to-point synchronizations and load balancing, and are presented in Figure 7. We see that converting the unelimated barrier synchronizations to the point-to-point synchronizations bring significant benefits in two applications (due to nearest neighbor communication). On the other hand, the load balancing helps in three applications (applu, galgel, and ammp) which experience load imbalance after our approach. While not presented here in detail, in these three applications, load balancing improved execution cycles over our baseline approach by about 5.7%.

To summarize, our experimental evaluation shows that compiler-directed synchronization elimination is very effective in reducing the energy overheads due to barrier synchronizations.

5. CONCLUSIONS

Increasing employment of chip multiprocessors require hardware designers and software writers to look for techniques that reduce execution cycles and power consumption of the applications mapped to these architectures. The main contribution of this paper is a compiler-driven barrier synchronization elimination algorithm. This algorithm eliminates most of the barrier synchronization calls inserted between parallel loops by being careful in assigning loop iterations to processors. We implemented the proposed approach using the SUIF compilation framework and the Omega Library, and performed experiments with benchmarks from the SPEC benchmark suite. Our experiments clearly show that the proposed compiler-directed approach is very effective and reduces energy overheads due to synchronizations by about 75.5%, and this corresponds to around 5.48% saving on average in overall energy consumption.

6. REFERENCES

[1] D. Brooks, V. Tiwari, and M. Martonosi. Wattch: A Framework for Architectural-Level Power Analysis and Optimizations. In *Proc. of the 27th Annual International Symposium on Computer Architecture*, pages 83–94, 2000.

[2] J. D. Davis, J. Laudon., K. Olukotun. Maximizing CMP Throughput with Mediocre Cores. In *Proc. of the 14th International Conference on Parallel Architectures and Compilation Techniques*, pages 51–62, 2005.

[3] M. W. Hall, J. M. Anderson, S. P. Amarasinghe, B. R. Murphy, S.-W. Liao, E. Bugnion, and M. S. Lam. Maximizing Multiprocessor Performance with the SUIF Compiler. *IEEE Computer*, vol. 29, no. 12, pp. 84–89, December 1996.

[4] M. T. Kandemir, P. Banerjee, A. N. Choudhary, J. Ramanujam, and N. Shenoy. A Global Communication Optimization Technique based on Data-Flow Analysis and Linear Algebra. *ACM Transactions on Programming Languages and Systems*, 21(6):1251–1297, 1999.

[5] W. Kelly, V. Maslov, W. Pugh, E. Rosser, T. Shpeisman, and D. Wonnacott. The Omega Library Interface Guide. *Technical Report CS-TR-3445*, CS Dept., University of Maryland, College Park, March 1995.

[6] K. Knobe and V. Sarkar. Array SSA Form and Its Use in Parallelization. In *Proc. of the 25th ACM SIGPLAN-SIGACT Symposium on Principles of Programming Languages*, pages 107–120, 1998.

[7] P. Kongetira, K. Aingaran, and K. Olukotun. Niagara: A 32-Way Multithreaded SPARC Processor. *IEEE MICRO*, 25(2):21–29, 2005.

[8] J. Li, J. F. Martínez, and M. C. Huang. The Thrifty Barrier: Energy-Aware Synchronization in Shared-Memory Multiprocessors. In *Proc. of the 10th International Conference on High-Performance Computer Architecture*, pages 14–23, 2004.

[9] C. Liu, A. Sivasubramaniam, M. Kandemir, and M. J. Irwin. Exploiting Barriers to Optimize Power Consumption of CMPs, In *19th IEEE International Parallel and Distributed Processing Symposium*, April 2005.

[10] P. S. Magnusson, M. Christensson, J. Eskilson, D. Forsgren, G. Hållberg, J. Högberg, F. Larsson, A. Moestedt, and B. Werner. Simics: A Full System Simulation Platform. *Computer*, 35(2):50–58, 2002.

[11] M. F. P. O'Boyle, L. Kervella, and F. Bodin. Synchronization Minimization in a SPMD Execution Model. *Journal of Parallel and Distributed Computing*, 29(2):196–210, 1995.

[12] K. Olukotun, B. A. Nayfeh, L. Hammond, K. Wilson, and K.-Y. Chang. The Case for a Single-Chip Multiprocessor. In *Proc. of the 7th International Symposium on Architectural Support for Parallel Languages and Operating Systems*, pages 2–11, 1996.

[13] W. Pugh and D. Wonnacott. An Exact Method for Analysis of Value-based Array Data Dependences. In *Proc. of the 6th International Workshop on Languages and Compilers for Parallel Computing*, pages 546–566, London, UK, 1994. Springer-Verlag.

[14] SPEC, Specfp 2000. http://www.specbench.org/cpu2000/CFP2000/, 2000.

Minimizing Energy Consumption of Banked Memories Using Data Recomputation [*]

H. Koc
Department of EECS
Syracuse University
hkoc@ecs.syr.edu

O. Ozturk, M. Kandemir,
S.H.K. Narayanan
The Pennsylvania State
University
{ozturk, kandemir,
snarayan}@cse.psu.edu

E. Ercanli
Department of EECS
Syracuse University
eercanli@ecs.syr.edu

ABSTRACT

Banking has been identified as one of the effective methods using which memory energy can be reduced. We propose a novel approach that improves the energy effectiveness of a banked memory architecture by performing extra computations if doing so makes it unnecessary to reactivate a bank which is in the low-power operating mode. More specifically, when an access to a bank, which is in the low-power mode, is to be made, our approach first checks whether the data required from that bank can be recomputed by using the data that are currently stored in already active banks. If this is the case, we do not turn on the bank in question, and instead, recalculate the value of the requested data using the values of the data stored in the active banks. Given the fact that the contribution of the leakage consumption to overall energy budget keeps increasing, the proposed approach has the potential of being even more attractive in the future. Our experimental results collected so far clearly show that this recomputation based approach can reduce energy consumption significantly.

Categories and Subject Descriptors: B.3 [Memory Structures]: Miscellaneous

General Terms: Experimentation, Management, Algorithms.

Keywords: Energy, memory bank, multiple operating modes.

1. INTRODUCTION

Banking has been identified as one of the effective methods using which memory energy can be reduced. Two factors contribute to this: reduced effective capacitance brought by banking as compared to a single monolithic memory and existence of multiple low-power operating modes that work with banked memories. As a result, prior research has shown that banking can reduce memory energy for different types of applications in the context of both embedded and high end computing platforms. Several studies have also proposed techniques - in hardware and/or software - that help increase the effectiveness of low-power operating modes.

---

[*] This work is supported in part by NSF career award #0093082 and a grant from GSRC.

Permission to make digital or hard copies of all or part of this work for personal or classroom use is granted without fee provided that copies are not made or distributed for profit or commercial advantage and that copies bear this notice and the full citation on the first page. To copy otherwise, to republish, to post on servers or to redistribute to lists, requires prior specific permission and/or a fee.
ISLPED'06, October 4–6, 2006, Tegernsee, Germany.
Copyright 2006 ACM 1-59593-462-6/06/0010 ...$5.00.

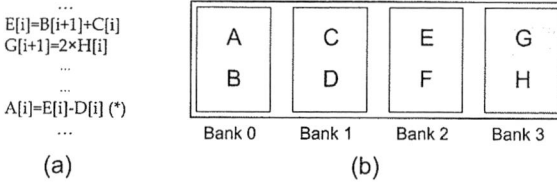

Figure 1: An example scenario with four banks. a) Sample code fragment, b) Data mapping into arrays

This paper builds upon this prior work and proposes a novel approach that further improves energy effectiveness of a banked memory architecture by performing *extra computations* if doing so makes it unnecessary to reactivate a bank which is in the low-power operating mode. More specifically, when an access to a bank which is placed into a low-power mode is to be made, our approach first checks whether the data required from that bank can be recomputed by using the data that are currently stored in already active banks. If this is the case, we do not turn on the bank in question, and instead, recalculate the value of the data using the values of the data stored in the active banks. Given the fact that the contribution of leakage consumption to overall energy budget keeps increasing, the proposed approach has the potential of being even more attractive in the future.

Figure 1 illustrates the proposed approach using an example. Consider the code fragment shown in Figure 1(a), which appears in the body of a loop whose index is i. Let us assume that the memory system has four banks and, at a particular moment in execution, only two of these banks (Bank 0 and Bank 1) are active, whereas the remaining two (Bank 2 and Bank 3) are put in the low-power operating mode. Figure 1(b) depicts how the data manipulated by the code fragment in Figure 1(a) are mapped to our banks. Assuming now that the execution is currently about the touch the statement marked using "*", the first data element to be accessed is $E[i]$, which is stored in Bank 2. Now, we have two options: either transition Bank 2 to the active (fully operational) mode and access $E[i]$ from there or recompute the value of $E[i]$ using arrays $B[i+1]$ and $C[i]$, which are stored in Bank 0 and Bank 1, respectively. Our approach chooses the second option if it is beneficial to do so from the energy perspective.

We implemented the proposed approach and performed experiments with six array-intensive benchmark programs. Our experimental results collected so far clearly show that this recomputation based approach can reduce energy consumption by about 12.8% when averaged over all codes in our experimental suite.

The rest of this paper is organized as follows. Section 2 introduces the banked memory architecture we consider in this work

Power Mode	Dynamic Energy/Access (nJ)	Leakage Energy/Bit (fJ)	Resynch. Cost (cycles)
Active	0.294	1.369	0
Power-down	-	0.218	2

Table 1: Energy consumptions and resynchronization costs for our operating modes, given for a bank size of 64KB.

and discusses the related work on exploiting low-power operating modes for memory energy saving. Section 3 presents our compiler algorithm that implements selective recomputation for energy saving and explains how it operates using an example. Section 4 presents the data we collected from our experiments. Section 5 concludes the paper and briefly discusses the planned future work.

2. BANKED MEMORY AND RELATED WORK

Our architectural model is based on a multi-bank memory system, in which SRAM banks can be placed into low-power modes independently. More specifically, each memory bank can be in one of two operating modes: *active* or *power-down* (also called *drowsy*) at any point during execution. Read/write requests are serviced only in the active mode.

Low-power operating modes are typically implemented by disabling certain parts of the memory chip. In addition to this, the different implementation techniques such as gated-V_{dd} [12], ABB-MTCMOS [10], and dynamic voltage scaling [6] have been utilized to enable low-power operating modes in SRAM circuits. Consequently, each low-power mode can have a different energy consumption and a different resynchronization cost (i.e., the penalty of reactivation) from the rest. This reactivation penalty is mainly due to bringing a low-powered memory bank back to the active mode. Table 1 shows energy consumption and resynchronization costs used in our evaluation. The energy consumed during transition from power-down to active mode is 25.6pJ. One must perform tradeoff analysis between energy savings and performance penalty when choosing a low-power mode for an idle bank. The time between successive accesses to the same memory bank is the main factor in selecting the most suitable operating mode.

Power management of banked memories have been investigated from the different angles including hardware, OS, and compiler. Delaluz et al [3] investigated software and hardware techniques to exploit the memory mode control capabilities. Using memory access patterns in embedded systems, [1] proposed an algorithm to partition on-chip SRAM into multi-banks that can be accessed independently. Flautner et al [6] used a simple technique to put cold cache lines into a state preserving low power mode to improve the leakage power consumption. Kim et al [8] extended this technique for instruction caches. Fan et al [4] presented memory controller policies for memory architectures with low-power operating modes. The impact of classical loop optimizations on energy consumption of banked memories has been evaluated in [7]. Farrahi et al [5] discussed how a sleep mode can be exploited for memory partitions. The impact of loop optimizations (loop splitting and loop distribution) and array placement strategies on a banked off-chip memory architecture are presented in [2]. Lyuh and Kim [9] used a compiler directed approach to determine the operating modes of memory banks after scheduling the memory access operations. Panda [11] addressed the problem of incorporating the application-specific customization of memory bank configuration into behavioral synthesis. In comparison, the work described in this paper shows that it is possible to further increase energy savings of a banked memory by using recomputation, on top of the

energy savings coming from exploiting the available low-power operating modes.

3. DETAILS OF OUR APPROACH

3.1 Energy Evaluation Model

In this section, we present the energy model used in this work. Please note that we tried to keep our model simple and at the high level for the sake of clarity. Total energy consumption E_{total} of an SRAM-based memory system can be modeled, at a high level of abstraction, as summation of dynamic, $E_{dynamic}$, (during access) and static, $E_{leakage}$, (during access and idle periods) components, in addition to the energy spent during transitions, E_{trans}, between operating modes (comprising both dynamic and static energy components).

The first component of the equation, $E_{dynamic}$ is directly proportional to the number of accesses, N_access, to the memory. Then, we multiply the number of accesses with the required energy per memory access, DE_access, to determine dynamic energy consumption, $E_{dynamic}$. In this model, the energies consumed during memory reads and that during writes are assumed to be equal.

$$E_{dynamic} = N\_access \times DE\_access$$

On the other hand, the leakage energy consumption, $E_{leakage}$, is proportional to the capacity of the memory system, M_size, and the number of cycles, N_cycles, required for the execution of the application. Since our memory system has two different operating modes, we need to consider the percentage of the execution time the memory is in the active mode, P_active, and the percentage of the time during which the memory is in the power-down mode $(1 - P\_active)$. Finally, we need to take into account the leakage energy per bit for each power mode. In our case, they are the leakage energy per bit for the active mode, LE_active, and the leakage energy per bit for the power-down mode, LE_down. Please note that this model could easily be modified to capture the systems with more than one low-power operating modes. The energy equation for leakage is shown below:

$$\begin{aligned} E_{leakage} \ = \ & M\_size \times N\_cycles \times ((P\_active \times LE\_active) \\ & + \ (1 - P\_active) \times LE\_down) \end{aligned}$$

Transition energy, E_{trans}, is calculated by multiplying the number of transitions from power-down mode to active mode, N_trans, with the energy consumed per transition, E_trans. In addition to this, the number of cycles required for transition, T_cycles, should be accounted for when computing the leakage energy consumption (they are captured in N_cycles).

$$E_{trans} = N\_trans \times E\_trans$$

Note that since the memory model used in this paper has multiple banks, the models for $E_{leakage}$ and E_{trans} should be employed for each bank.

Let us now discuss the benefits and overheads of the recomputation considering the energy model. Recomputation brings extra computations into consideration. This increases the number of memory accesses by N_access_extra (number of extra memory accesses) and the execution time by N_cycles_extra (number of extra execution cycles). Consequently, the overall energy overhead, E_over, is calculated as follows:

$$E\_over = N\_access\_extra \times DE\_access$$
$$+ N\_cycles\_extra \times LE\_active$$

On the other hand, recomputation prevents the banks switching back and forth between the active and power-down modes. This, in turn, reduces the number of cycles by N_access_less (the number of cycle reduced); the number of transitions from the power-down to active mode by N_trans_less (the reduced number of transitions from the power-down to active mode); and significantly increases the percentage of the execution time in which a bank is in the power-down mode by P_gain. As a result, the energy improvement can be determined as follows:

$$E\_gain = N\_trans\_less \times E\_trans + N\_access\_less \times$$
$$DE\_access + P\_gain \times M\_size \times (LE\_active - LE\_down)$$

Based on the above-mentioned formulation, our approach can save energy if $E\_gain > E\_over$.

3.2 Recomputation

This section briefly discusses our recomputation-based approach and presents the algorithm used. Our approach targets embedded applications consisting of loops that manipulate array elements. Particularly, we assume that all scalar variables are stored in registers and hence do not play a part in our analysis. Let code C consist of a series of L loops in which a total of M arrays are manipulated. Let the loops be denoted by $L_1, L_2, ..., L_N$, where a loop can be nested at an arbitrary depth. For example, consider the loop L_i to have a nesting depth of k. The lth loop in this nest is denoted by L_{i_l} and the lower and upper bounds of this loop are given by $L_{i_{l_S}}$ and $L_{i_{l_E}}$, respectively. Finally, an array reference, as part of an expression, in this loop nest is denoted by $A_{i_l}^R(a_1, a_2...a_l)$ and $A_{i_l}^L(a_1, a_2...a_l)$ for the right hand side and the left hand side, respectively, of the expression. The subscript i_l denotes the loop number (i) and the depth of the nest l. Let us consider a memory architecture which has N banks and let the arrays of the given code be partitioned among these N banks.

Recomputation in our work is defined as the computation of a previously computed value for an array reference instead of looking up the computed value that is stored in a memory bank. This process is performed in three main steps. First, possible recomputation opportunities are found in the given application code. Then, suitable arrays are selected for recomputation based on their performance cost. Total cost of all recomputations must be less than the total allowable Cost-Overhead $\mathcal{T_C}$ that is the maximum number of extra execution cycles allowed. Finally, the code is restructured by replacing the selected array references with the expression that computes its value.

Algorithm 1 gives a sketch of our approach. The input to this algorithm is the original code, the mapping of arrays to the memory banks, and $\mathcal{T_C}$. The output is the restructured code. First, the set of selected elements, $\mathcal{S}$, is set to ϕ. Each element of this set is a tuple, formed by the array reference, $A_{i_l}^L(a_1, a_2...a_l)$, that has been selected for recomputation and the set of array references, $\mathcal{L}$, that compute the values of the same data space as the array reference. After that, in lines [5-18], the set $\mathcal{S}$ is calculated. To calculate it; first, for each array reference in the input code, $A_{i_l}^L(a_1, a_2...a_l)$, the set $\mathcal{L}$ is determined. Then, if the data space accessed by the reference is a subset of that accessed by $\mathcal{L}$ and if the cost is not prohibitive, and finally, if the banks that the reference $A_{i_l}^L(a_1, a_2...a_l)$ accesses are OFF and those that $\mathcal{L}$ accesses are ON, the new tuple $\{(A_{i_l}^L(a_1, a_2...a_l), \mathcal{L}\}$ is added to $\mathcal{S}$. Otherwise, the bank accessed by $A_{i_l}^L(a_1, a_2...a_l)$ is considered to be ON. Finally, in the

---

Algorithm 1 *Recomputation()*

1: Input : Source code, array mapping to banks, and $\mathcal{T_C}$
2: Output : Restructured code
3: $\mathcal{S} = \phi$
4: Initialize all banks as OFF
5: **for** all array references $A_{i_l}^L(a_1, a_2...a_l)$ **do**
6:   $\mathcal{L} = \phi$
7:   **for** all $j < i$ and $\exists A_{i_l}^R(a_1, a_2...a_l)$ **do**
8:     $\mathcal{L} = \mathcal{L} \bigcup A_{i_l}^R(a_1, a_2...a_l)$
9:   **end for**
10:   **if** $\mathcal{D_S}(A_{i_l}^L(a_1, a_2...a_l)) \subset \mathcal{D_S}(\mathcal{L})$ and $Bank(A_{i_l}^L(a_1, a_2...a_l)) = OFF$ and $Bank(\mathcal{L}) = ON$ **then**
11:     **if** $\mathcal{T_C} > 0$ **then**
12:       $\mathcal{S} = \mathcal{S} \bigcup \{A_{i_l}^R(a_1, a_2...a_l), \mathcal{L}\}$
13:       $\mathcal{T_C} = \mathcal{T_C} - \mathcal{C}(T(A_{i_l}^R(a_1, a_2...a_l)), N(A_{i_l}^R(a_1, a_2...a_l)))$
14:     **end if**
15:   **else**
16:     $Bank(A_{i_l}^L(a_1, a_2...a_l)) = ON$
17:   **end if**
18: **end for**
19: **for** each element $\{A_{i_l}^R(a_1, a_2...a_l), \mathcal{L}\}$ of $\mathcal{S}$ **do**
20:   **for** each element $\mathcal{L}_j \in \mathcal{L}$ **do**
21:     Calculate the loops bounds $L_{i_l s}$ and $L_{i_l e}$ for
22:     $\mathcal{D_S}(A_{i_l}^R(a_1, a_2...a_l)) \bigcap \mathcal{D_S}(\mathcal{L}_j)$
23:     Restructure code using $L_{i_l s}$ and $L_{i_l e}$
24:   **end for**
25: **end for**

---

lines [19-25], the code for the each selected array reference is generated. First, the loop bounds are calculated and then, these bounds are used to generate the output code.

4. EXPERIMENTAL EVALUATION

This section presents the experimental evaluation of our proposed approach. We used six array-intensive benchmarks from the Spec and Perfect Club benchmark suites. Each benchmark is written in C language and comprises the manipulations of one- to three-dimensional data arrays. The data set sizes of these benchmarks range from 170KB to 256KB. In our experiments, we used a single level of SRAM with four data banks, each 64KB, as our memory architecture. Execution time is a significant factor during the calculation of leakage energy consumption. Execution time of each application with/without recomputation is evaluated using an UltraSPARC-III based system with 750 MHz of clock frequency. In our experiments with baseline configuration, we found that the benchmarks incurred around 7% performance degradation on the average when recomputation is introduced.

Our first set of results is given in Table 2, under the performance bound of 20% (i.e., we allow up to 20% increase in original execution cycles). The first column of this table gives the name of the benchmark. The next two columns are the energy consumptions without and with the proposed recomputation based approach. The last column indicates the percentage energy improvements. The average energy improvement obtained through the recomputation based approach is about 12.8%. As can be seen in Table 2, the improvement is small for the first benchmark *bmcm*. This is because the execution time increases almost 20% when recomputation is introduced. This, in turn, increases not only dynamic energy but also leakage.

In our next set of experiments we kept the total memory size (capacity) fixed at 256KB and changed the number of banks. Recall that our experiments shown in Table 2 assume that 4 data banks are available. Figure 2 shows the normalized energy savings with the different bank counts. In this case, we focus only on benchmarks *wss* and *tomcatv*. As can be seen from the figure, an increase in the

Benchmark	Energy w/o Recomp. (mJ)	Energy w/ Recomp. (mJ)	Improvement (%)
bmcm	212	210	0.6
eflux	786	625	20.5
mxm	133	100	24.4
tomcatv	1240	1120	9.8
interp	3050	2830	7.4
wss	1980	1700	13.8

Table 2: Energy consumptions with and without recomputation for different benchmarks.

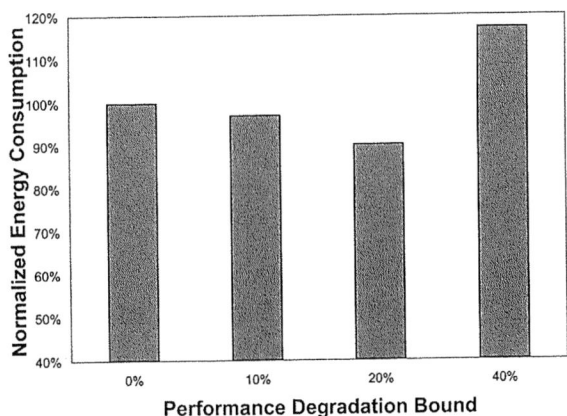

Figure 3: Normalized energy consumptions with different recomputation scenarios for *tomcatv*.

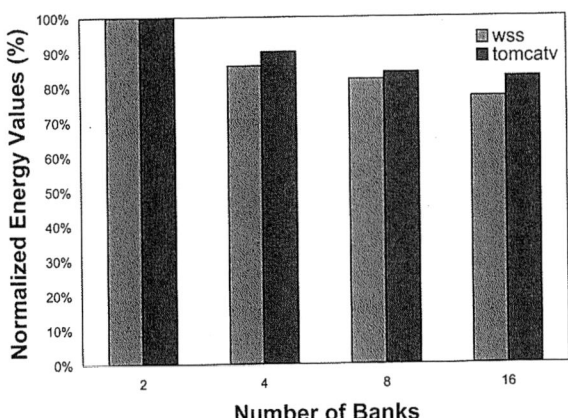

Figure 2: Normalized energy consumptions with different number of banks for *wss* and *tomcatv*.

number of banks improves the normalized energy savings archived by our recomputation based multibank memory system. This is due to two main reasons: a) our approach provides an opportunity to power down a bank, which cannot be powered down in the original case due to lifetime conflicts between the data arrays that reside in the bank, and b) our approach makes it possible to keep a bank in the power down mode for a longer period of time. As expected, there is no energy improvement for the benchmarks when only two data banks are available.

Next, we studied the tradeoff between performance and energy. In other words, we evaluated energy consumptions under different recomputation scenarios. Figure 3 plots the normalized energy consumption values for each scenario for the benchmark *tomcatv*. The first bar with 0% performance degradation represents the case with no recomputation. Each bar after that introduces extra recomputation over the previous one (i.e., it increases the tolerable increase in the execution cycles). As can be observed from the figure, when the amount of recomputation is increased, it is possible to achieve better energy savings (the cases with 10% and 20% performance degradation bound). However, the case with 40% performance degradation bound has a negative effect in energy consumption. The main reason for this is the increase in leakage energy consumption due to excessive execution time overhead.

5. CONCLUSIONS AND FUTURE WORK

Low-power operating modes have been employed to improve the energy consumption of banked memory architectures. Reduction in energy consumption is achieved by putting inactive memory banks into a low-power operating mode. In this paper, we propose a novel energy saving scheme based on recomputation to further increase the energy savings that could be obtained from in multi-bank memory systems. Basically, when an access to a powered-down bank is to be made, our approach first checks if it is possible to obtain the

required data by recomputing the data placed in the active banks, instead of reactivating the powered-down bank and fetching the data from it. We performed experiments with several benchmarks and the results collected show the effectiveness of the proposed recomputation based approach. Our future work involves extending this idea for multiple low-power operating modes with compiler-directed recomputation decisions.

6. REFERENCES

[1] L. Benini, A. Macii, and M. Poncino. A recursive algorithm for low-power memory partitioning. In *Proceedings of the International Symposium on Low Power Electronics and Design*, pages 78–83, New York, NY, USA, 2000. ACM Press.

[2] V. Delaluz, M. Kandemir, N. Vijaykrishnan, and M. J. Irwin. Energy-oriented compiler optimizations for partitioned memory architectures. In *Proceedings of the International Conference on Compilers, Architecture, and Synthesis for Embedded Systems*, pages 138–147, 2000.

[3] V. Delaluz, M. Kandemir, N. Vijaykrishnan, A. Sivasubramaniam, and M. J. Irwin. Dram energy management using sofware and hardware directed power mode control. In *Proceedings of the International Symposium on High-Performance Computer Architecture*, page 159, 2001.

[4] X. Fan, C. Ellis, and A. Lebeck. Memory controller policies for dram power management. In *Proceedings of the International Symposium on Low Power Electronics and Design*, pages 129–134, 2001.

[5] A. Farrahi, G. Tellez, and M. Sarrafzadeh. Exploiting sleep mode for memory partitions and other applications. In *Proceedings of the VLSI Design*, pages 271–287, 1998.

[6] K. Flautner, N. S. Kim, S. Martin, D. Blaauw, and T. Mudge. Drowsy caches: simple techniques for reducing leakage power. In *Proceedings of the Annual International Symposium on Computer Architecture*, pages 148–157, Washington, DC, USA, 2002.

[7] M. Kandemir, I. Kolcu, and I. Kadayif. Influence of loop optimizations on energy consumption of multi-bank memory systems. In *Proceedings of the International Conference on Compiler Construction*, pages 276–292, 2002.

[8] N. S. Kim, K. Flautner, D. Blaauw, and T. Mudge. Drowsy instruction caches: leakage power reduction using dynamic voltage scaling and cache sub-bank prediction. In *Proceedings of the Annual International Symposium on Microarchitecture*, pages 219–230, Los Alamitos, CA, USA, 2002.

[9] C.-G. Lyuh and T. Kim. Memory access scheduling and binding considering energy minimization in multi-bank memory systems. In *Proceedings of the Annual Conference on Design Automation*, pages 81–86, New York, NY, USA, 2004.

[10] K. Nii, H. Makino, Y. Tujihashi, C. Morishima, Y. Hayakawa, H. Nunogami, T. Arakawa, and H. Hamano. A low power SRAM using auto-backgate-controlled MT-CMOS. In *Proceedings of the International Symposium on Low Power Electronics and Design*, pages 293–298, New York, NY, USA, 1998.

[11] P. R. Panda. Memory bank customization and assignment in behavioral synthesis. In *Proceedings of the International Conference on Computer-Aided Design*, pages 477–481, 1999.

[12] M. Powell, S.-H. Yang, B. Falsafi, K. Roy, and T. N. Vijaykumar. Gated-Vdd: a circuit technique to reduce leakage in deep-submicron cache memories. In *Proceedings of the International Symposium on Low Power Electronics and Design*, pages 90–95, New York, NY, USA, 2000.

Design Contest Presentations

Chair: Barry Pangrle
Chair's Affiliation

D1.1 **500-MS/s 5-b ADC in 65-nm CMOS With Split Capacitor Array**
Brian P. Ginsburg and Anantha P. Chandrakasan
Massachusetts Institute of Technology

D1.2 **Power Reduction with Dynamic Sampling and All-Digital I/Q-Mismatch Calibration for a MB-OFDM UWB Baseband Transceiver**
Jui-Yuan Yu, Ching-Che Chung, Hsuan-Yu Liu, and Chen-Yi Lee
National Chiao-Tung University

D1.3 **AmbiMax: Autonomous Energy Harvesting Platform for Multi-Supply Wireless**
Chulsung Park and Pai H. Chou
University of California, Irvine

D1.4 **A Continuous-Time Programmable Digital FIR Filter**
Yee William Li, Kenneth L. Shepard, and Yannis P. Tsividis
Columbia University

D1.5 **Design and Implementation of Power-Aware Mobile 3D Graphics**
Chanmin Park, Hyunhee Kim, and Jihong Kim
Seoul National University

Energy Optimality and Variability in Subthreshold Design

Scott Hanson   Bo Zhai   David Blaauw   Dennis Sylvester   Andres Bryant   Xinlin Wang

University of Michigan
{hansons,bzhai,blaauw,dmcs}@umich.edu

IBM T.J. Watson Research Center
{bryanta,xinlinw}@us.ibm.com

ABSTRACT

Recent progress in the development of subthreshold circuit design techniques has created the opportunity for dramatic energy reductions in many applications. However, energy efficiency comes at the price of timing and energy variability due to process variations. We explore energy optimality in the subthreshold regime, discuss variability in this region, and highlight the energy and variability characteristics of a real subthreshold design.

Categories and Subject Descriptors:

B.7.1 Integrated Circuits - Types and Design Styles

General Terms: Design

Keywords: Subthreshold circuits, variability, ultra-low energy

1. INTRODUCTION

The low voltage design community has grown explosively during the past few years. New research has built a theoretical foundation for subthreshold ($V_{dd} < V_{th}$) design strategies and has begun to address the more critical problems of memory design and process variability. Subthreshold designs, which can improve energy efficiency by several orders of magnitude, will play an important role in the continuing drive toward mobile, pervasive computing. Applications like environmental monitoring, biomedical sensing, and supply chain management will be bolstered by complex subthreshold logic. High performance servers may also receive significant benefit from the subthreshold design community through low power, massively parallel systems of subthreshold processors. In this paper, we address three key points. In the first section, we show how subthreshold operation can be used to achieve energy optimality. We then highlight one of the primary problems facing subthreshold circuit designers: variability. In the final section, we relate the topics of the first two sections to measurements of a real subthreshold design.

2. MINIMIZING ENERGY

Though power has received more attention than energy in optimizing high performance and embedded processors, energy is a more suitable metric for mobile applications. Both theory [1] and hardware measurements [2][3] have shown that the energy consumed by a processor per operation is typically minimized in the subthreshold regime, where the supply voltage (V_{dd}) is smaller than the device threshold voltage (V_{th}). As V_{dd} is reduced toward V_{th}, the transistor drive current evolves from drift-dominated strong-inversion current to diffusion-dominated

Permission to make digital or hard copies of all or part of this work for personal or classroom use is granted without fee provided that copies are not made or distributed for profit or commercial advantage and that copies bear this notice and the full citation on the first page. To copy otherwise, or republish, to post on servers or to redistribute to lists, requires prior specific permission and/or a fee.
ISLPED '06, October 4–6, 2006, Tegernsee, Germany.
Copyright 2006 ACM 1-59593-462-6/06/0010...$5.00.

weak-inversion current. Though the magnitude of the weak-inversion current is small, it is large enough to charge and discharge nodes in the manner required by digital logic.

It is well known that digital logic may function correctly at extremely low voltages, but it is worthwhile to consider whether there are energy benefits to low voltage operation. The energy consumed by an inverter chain is described by Equation 1 and plotted in Figure 1 for a 130nm technology. Energy is composed of two components: dynamic energy and leakage energy. Short-circuit energy is commonly ignored but may be accounted for through appropriate adjustment of the activity factor.

$$E = E_{dyn} + E_{leak} = \frac{1}{2} \cdot C_s \cdot V_{dd}^2 \cdot \alpha + I_{leak} \cdot V_{dd} \cdot t_p \quad \textbf{(EQ 1)}$$

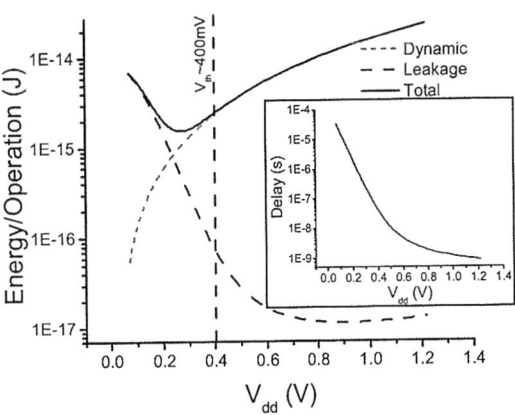

Figure 1: Delay and energy consumption for a chain of 50 inverters with activity factor of 0.2. For this circuit, energy reaches a minimum at V_{dd}=266 mV (130nm technology)

Figure 1 also shows how the delay of an inverter chain increases dramatically as V_{dd} scales into the subthreshold regime. For many applications, the performance penalty paid for subthreshold operation is tolerable. However, the implications of increased delay extend beyond performance concerns. As Figure 1 shows, the leakage energy of the inverter chain under test increases significantly as voltage reduces. Though leakage *power* ($I_{leak}V_{dd}$) reduces with supply voltage, delay increases exponentially and forces leakage *energy* ($I_{leak}V_{dd}t_p$) to increase. Consequently, total energy reaches a minimum value at a voltage called V_{min}. The expression for V_{min} was first solved numerically in [1] as:

$$V_{min} = \left[1.587 \cdot \ln\left(\eta \cdot \frac{n}{\alpha} \right) - 2.355 \right] \cdot m \cdot v_T \quad \textbf{(EQ 2)}$$

Most notably absent from Equation 2 is V_{th}. Assuming that the circuit is operating safely in the subthreshold regime, V_{min} does not depend on V_{th}. Equation 2 also highlights the fact that V_{min} is closely tied to the dynamic to leakage ratio through the n and α terms. Circuits with longer paths (larger n) tend to consume more leakage and tend to have larger values of V_{min}. Similarly, circuits with lower switching activity tend to be more leakage dominated and have larger values of V_{min}. These sensitivities are important when thinking about V_{dd} selection for a larger system. Caches, for example, will achieve optimality at much higher V_{dd} than general logic due to differences in switching activity.

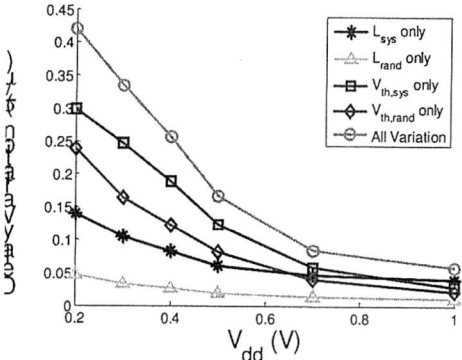

Figure 2: Delay variability (σ/μ) as a function of supply voltage (65 nm technology)

3. ADDRESSING VARIABILITY

The last section showed that energy/operation may be, in theory, minimized by operating in the subthreshold regime. In practice, achieving energy optimality is not as simple as reducing V_{dd} to V_{min}. Process-induced variability leads to problems with both functionality and energy efficiency.

We can take a simplistic but accurate view of variability by assuming that there are four types of variation: systematic V_{th} variation, random V_{th} variation, systematic gate length variation, and random gate length variation. There is also some component of threshold and gate length variation that varies from region to region on the chip, but this can be safely grouped with global systematic variation for our simple discussion.

Figure 2 shows how the delay variation of a chain of 10 inverters changes as V_{dd} scales in a 65 nm technology. Since subthreshold current is exponentially dependent on V_{th}, variation in V_{th} becomes more problematic at subthreshold voltages. Conversely, subthreshold current is inversely proportional to gate length and has a relatively weak exponential dependence through DIBL-induced V_{th} variations. V_{th} variations are therefore the most important concern for subthreshold designers.

3.1 Timing Variability

The increased sensitivity to threshold voltage fluctuations in the subthreshold regime leads to dramatic variations in gate delay, which may result in both late-mode and early-mode timing failures. Late-mode failures occur when a circuit path delay exceeds the clock period and may be fixed by increasing the clock period. Monte Carlo simulations show that the clock period for a 10-inverter chain in a 65 nm technology must increase by 10% at V_{dd}=1V and an astonishing 230% at V_{dd}=300mV to eliminate late-mode errors introduced by variability. The performance and energy implications of addressing late-mode failures are clearly undesirable. Early-mode failures can occur when excessive clock skew allows data to be latched one clock cycle early at a receiving latch. Monte Carlo simulations suggest that clock skew (in terms of FO4 delay) can increase by more than 10X as voltage is scaled from 1V to 300mV. Early-mode failures must be fixed by adding delay elements to short paths or by designing variation-tolerant clock distribution networks.

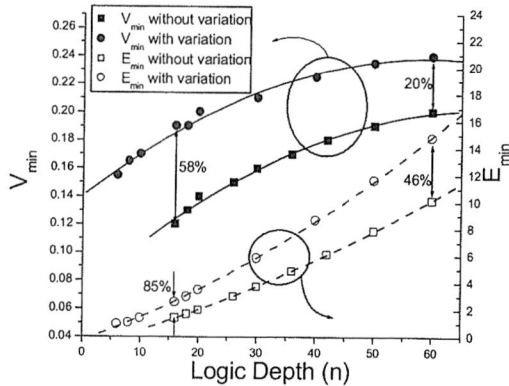

Figure 3: Variation in worst-case ($\mu+3\sigma$) V_{min} and E_{min} for an inverter chain of length n gates. The relative increases in V_{min} and E_{min} are less severe at large n (130 nm technology).

3.2 Energy Variability

For most subthreshold designs, energy will be the most important metric. It is therefore very important to understand how V_{min}, identified in Section 2, is affected by variability. While dynamic energy remains relatively constant with variability, worst-case delay and worst-case leakage energy increase dramatically [4]. Assuming a fixed frequency and supply voltage across all chips, process variation leads to a lower operating frequency and consequently, a dramatic increase in leakage. Figure 3 shows the variation in the worst-case V_{min} and energy for an inverter chain of length n in a 130 nm technology.

3.3 Mitigating Variability

Though variability is one of the primary problems in the subthreshold regime, careful design will help alleviate its effects. We begin by considering device-level optimizations. The FinFET promises to reduce threshold variation (δV_{th}) induced by random dopant fluctuations (RDF) as well as δV_{th} induced by short channel effects (SCEs). With its double gate and thin body, the FinFET can achieve excellent SCE control down to very short channels in the absence of channel doping [5]. Figures 4(a-c) show FIELDAY simulation results comparing the SCEs of a planar poly-gate PDSOI FET, a poly-gate FinFET with body doping, and a mid-gap work-function metal gate FinFET without body doping. Even without body doping, the FinFET V_{th}, drain barrier lowering, and subthreshold slope dependence on gate length are much weaker than those observed for planar PDSOI FETs. Thus, the FinFET enables the elimination of RDF-induced δV_{th} while still reducing SCE-induced δV_{th}. However, the elimination of body doping will require the development of metal gates of varying work-functions to set thresholds to their optimum values. In Figure 4(a), the mid-gap metal gate FinFET has a V_{th} that may be too high for some applications. A quarter-gap metal gate would reduce V_{th} to a more reasonable level. An alternative approach would be to use a split-gate FinFET, where one gate is active and the other gate is used to adjust V_{th}. With this "back-gate" approach, FETs with multiple thresholds can coexist in the same circuit, and thresholds can be adjusted dynamically to optimize power-performance tradeoffs [5].

In addition to novel device design, there are several well-studied circuit techniques that can reduce variability significantly. Systematic variations can be addressed globally with techniques like dynamic voltage scaling (DVS) and adaptive body biasing (ABB), though both techniques incur area and complexity overheads. Random variations are somewhat more difficult to address. Figure 3 shows that relative energy (E_{min}) and V_{min} variations can be reduced dramatically by increasing path lengths in a circuit. Recent research has also shown that timing and energy fluctuations due to random variation can be reduced by using larger gate sizes [4][6]. Both longer paths and larger gates attempt to "average out" variability over larger gate area. When using these techniques, designers must be careful to weigh the variability benefits against the accompanying energy penalties.

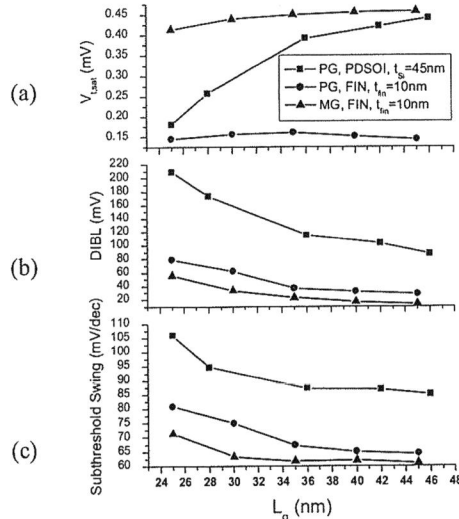

Figure 4: (a) $V_{th,sat}$ (b) DIBL and (c) subthreshold swing characteristics for a poly-gate (PG) partially-depleted SOI (PDSOI) device, a poly-gate FINFET (FIN) with body doping, and a metal-gate (MG) FINFET without body doping

Successful subthreshold operation will undoubtedly require the development of error-tolerant architectures. Error correction codes (ECC) and memory redundancy have been studied extensively, but it will be important to create whole architectures that can dynamically detect and correct errors. Without such techniques, designers will need to incorporate large margins in design parameters to avoid timing errors and will need to resort to expensive statistical techniques like Monte Carlo simulation.

4. A 2.6 PJ SUBTHRESHOLD PROCESSOR

A number of subthreshold circuits have been successfully demonstrated recently [2][3][7]. In this section, we focus on the design described in [2] to put the topics from Sections 2 and 3 into the context of a real design. The chip under test is a subthreshold sensor network processor with an 8-bit CISC architecture and a 2-kbit memory. The processor, fabricated in a 130nm technology, occupies an area of 85,022 μm^2.

Figure 5(a) shows the energy consumption of the processor for a typical instruction stream. The total energy consumption clearly reaches the minimum predicted by Equation 2. The energy minimum of 2.6 pJ per instruction occurs at V_{dd}=400 mV, which is significantly higher than the minimum energy voltage observed for the inverter chain in Figure 1. Recall that V_{min} tends to be higher for circuits with lower switching activity. The memory, which has a very low switching activity compared to typical logic, accounts for 65% of the total transistor area and is largely responsible for the higher V_{min}.

Figure 5(b) shows measured energy distributions for 26 chips under different frequency and voltage selection schemes. As expected, the worst-case energy is minimized by allowing each chip to operate at its own V_{min} and frequency. V_{min} varies by 80 mV across the measured chips with worst-case energy variation ($3\sigma/\mu$) of 16.99%. These fluctuations in V_{min} and energy are in agreement with the trends discussed in the last section but are not nearly as large as one might expect when looking at the variability characteristics of a single gate (where $3\sigma/\mu$ for delay can be greater than 200% in the subthreshold regime).

Allowing each chip to operate at the optimal supply voltage and frequency requires adaptive voltage and frequency tuning and significant design overhead. This design overhead can be minimized to some extent by fixing the supply voltage at a mean voltage but allowing frequency to vary. As shown in Figure 5(b), the energy penalty for choosing this scheme is small. The energy minimum is shallow, so the choice of supply voltage can deviate from V_{min} without a significant penalty. However, fixing both V_{dd} and frequency (also shown in Figure 5(b)) imposes a significant energy penalty and is only desirable when the overheads of adaptive voltage and frequency tuning are large.

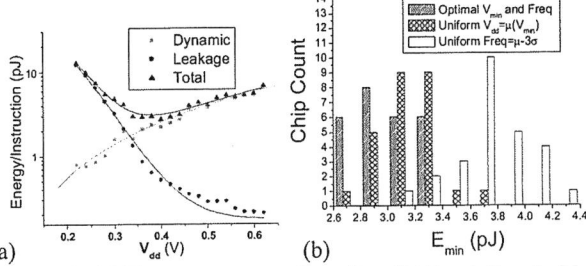

Figure 5: (a) Energy consumption of an 8-bit subthreshold processor (130 nm technology) (b) Energy distributions for 26 measured subthreshold processors using different supply voltage and frequency selection schemes (130 nm technology)

5. CONCLUSIONS

In this paper, we used both simulated and measured hardware to demonstrate that energy efficiency can be maximized by operating in the subthreshold regime. One of the primary problems that must be addressed by subthreshold designers is the management of variability. Threshold variability, in particular, becomes problematic at low voltage. Existing device, circuit and architectural techniques offer some reprieve to this problem, but further innovation will be necessary to ensure that subthreshold design gets widespread acceptance.

6. REFERENCES

[1] B. Zhai, D. Blaauw, D. Sylvester, K. Flautner, "Theoretical and Practical Limits of Dynamic Voltage Scaling," *DAC*, 2004, pp. 868-873.
[2] B. Zhai, et al., "A 2.60pJ/Inst Subthreshold Sensor Processor for Optimal Energy Efficiency," *VLSI Circuits Symposium*, 2006.
[3] B. Calhoun, A. Wang, A. Chandrakasan, "Device sizing for minimum energy operation in subthreshold circuits," *CICC*, 2004, pp. 95-98.
[4] B. Zhai, S. Hanson, D. Blaauw, D. Sylvester, "Analysis and Mitigation of Variability in Subthreshold Design," *ISLPED*, 2005, pp. 20-25.
[5] W. Haensch, et al., " Silicon CMOS devices beyond scaling," *IBM J. Res. & Dev. 50*,No. 4, pp. 339-362.
[6] J. Kwong, A. Chandrakasan, "Variation-Driven Device Sizing for Minimum Energy Sub-threshold Circuits," *ISLPED*, 2006.
[7] C. Kim, H. Soeleman, K. Roy, "Ultra-Low-Power DLMS Adaptive Filter for Hearing Aid Applications," *IEEE Trans on VLSI Systems 11*, No. 6, pp. 1058-1067.

Sub-Threshold Design: The Challenges of Minimizing Circuit Energy

B. H. Calhoun[1], A. Wang[2], N. Verma[3], and A. Chandrakasan[3]
[1]University of Virginia, [2]Texas Instruments, [3]Massachussetts Institute of Technology
bcalhoun@virginia.edu, aliwang@ti.com, {nverma,anantha}@mtl.mit.edu

ABSTRACT

In this paper, we identify the key challenges that oppose sub-threshold circuit design and describe fabricated chips that verify techniques for overcoming the challenges.

Categories and Subject Descriptors

B.7.1 [ICs]: Types and Design Styles

General Terms

Performance, Design, Reliability

Keywords

Sub-threshold digital circuits, low voltage memory, dynamic voltage scaling, process variations, sub-threshold logic

1. INTRODUCTION

Sub-threshold operation for digital circuits first was shown as the means to minimizing CMOS V_{DD} in 1972 [1]. Analog sub-threshold circuits subsequently received a lot of attention for low power applications (e.g. [2][3]). Interest in digital sub-threshold was revived in the late 1990s [4], and a multiplier was demonstrated operating in sub-V_T at 0.475V that used body bias to balance p/n currents [5]. A sub-V_T ring oscillator also employed body biasing and functioned at 80mV [6].

The primary motivation for using sub-V_T circuits is to reduce energy. Analysis of energy contours in [7] demonstrated that minimum energy operation occurs in the sub-threshold region. Once $V_{DD}<V_T$, delay increases exponentially with additional voltage scaling. Leakage current integrates over the longer delay until leakage energy per operation exceeds the active energy and causes the minimum point. Models capture this effect and illustrate the impact of various parameters in [8][9].

The potential for minimizing energy at the cost of speed degradation defines the set of applications for which sub-threshold circuits are well-suited. First, energy-constrained applications such as wireless sensor nodes, RFID tags, or implants are dominated by the need to minimize energy consumption. Speed is a secondary consideration for this class of applications, so sub-V_T circuits offer a good solution. Secondly, many burst-mode applications require high performance for brief time periods between extended sections of low performance operation. Sub-threshold circuits can minimize energy for computations executed during the low performance slots. Finally, the parallelism inherent in many signal processing and communications circuits can be exploited to scale voltages into sub-V_T, providing a low energy solution for throughput-centric applications (e.g. [10]).

This paper describes the key challenges that confront sub-threshold circuit designers and presents chips that overcome the challenges.

2. Sub-Threshold Logic: FFT Processor

Static CMOS gates continue to function in sub-V_T, but some challenges make logic design more difficult. First, CMOS processes are designed with strong-inversion operation in mind, so the ratio of drive current in sub-V_T is frequently imbalanced relative to the case where pMOS and nMOS are symmetrical. The shaded region in Figure 1 shows the operational range for a ring oscillator in 0.18μm CMOS at the worst-case corners. V_{DD} is minimized when the p/n sizing ratio is 12, which indicates that the process is imbalanced such that p/n current is 1/12 relative to the symmetric case. This unfriendly sort of technology imbalance can aggravate process variations and even require different circuit designs for different imbalance scenarios. In addition, the low V_{DD} results in a reduced I_{on}/I_{off} ratio that can reduce robustness, especially for circuits with parallel leakage paths [11].

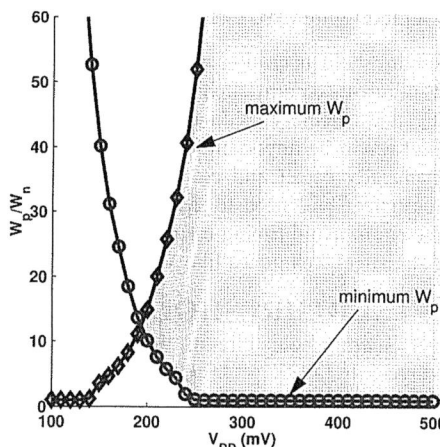

Figure 1: Minimum achievable voltage for 10%-90% output swing for 0.18μm ring oscillator at worst case process corners (simulation).

A 0.18μm CMOS FFT processor uses circuits that account for these challenges: static CMOS logic is used for robustness, gates with parallel leakage paths are redesigned, large stacks are avoided to improve I_{on}/I_{off}, and a register-file memory uses logic-based structures. The chip is fully functional for 128, 256, 512, and 1024 FFT lengths (8-bit and 16-bit precision) at V_{DD} from 180mV to 900mV [11]. Figure 2 shows the measured energy consumption for 8-bit and 16-bit processing as a function of voltage. 8-bit processing has a lower activity factor and thus has lower switching energy. However, because the leakage energy is the same for both 8-bit and 16-bit processing, the minimum

---

Permission to make digital or hard copies of all or part of this work for personal or classroom use is granted without fee provided that copies are not made or distributed for profit or commercial advantage and that copies bear this notice and the full citation on the first page. To copy otherwise, or republish, to post on servers or to redistribute to lists, requires prior specific permission and/or a fee.
ISLPED'06, October 4–6, 2006, Tegernsee, Germany.
Copyright 2006 ACM 1-59593-462-6/06/0010...$5.00.

energy point increases to 400mV from 350mV. At the 16-bit optimum, the chip runs at 10 kHz and consumes 155nJ/FFT, which is 350X more energy efficient than a typical low-power microprocessor and 8X more energy efficient than a standard ASIC implementation [11].

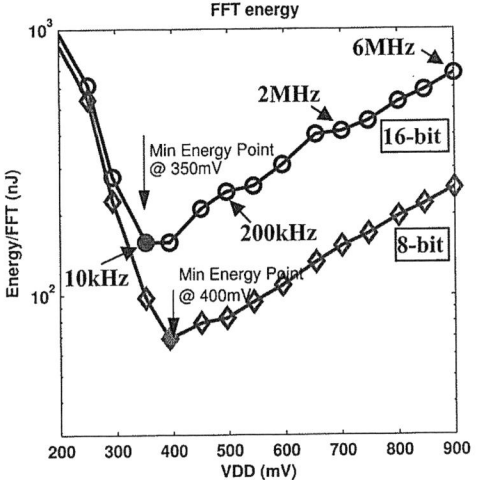

Figure 2: Measured energy per 8- and 16-bit FFT vs. V_{DD}.

3. Scaling Performance: Ultra-DVS

Burst mode applications cannot exclusively utilize sub-threshold operation because they require periodic high speed functionality. Traditional dynamic voltage scaling (DVS) could be extended to include sub-threshold operation, but the overhead of providing the necessary voltages can be large. Adjustable DC-DC converters tend to have limited efficiency over broad voltage ranges, and they take 100s of micro-seconds to switch. An alternative implementation method called local voltage dithering (LVD) offers a reduced overhead means for implementing ultra-DVS (UDVS) down to the sub-threshold region. LVD uses power switches to select from among two or more V_{DD} supplies at the local block level [12]. Figure 3 shows an example system that has 3 V_{DD}s. As the required rate (normalized frequency) for processing incoming data changes, each block spends a different fraction of its operating time at different voltage levels. The averaging effect of this dithering produces an energy consumption profile that nears the optimal (e.g. infinite voltage levels) profile.

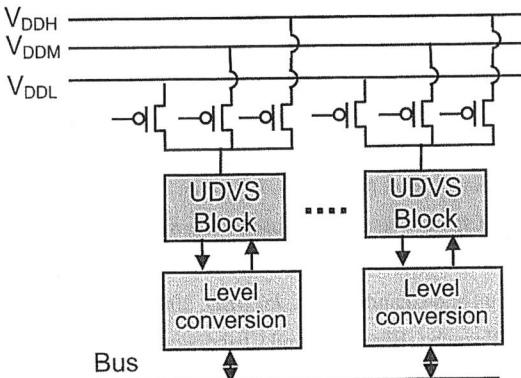

Figure 3: Example UDVS system using LVD and three V_{DD}s.

A 90nm CMOS test chip uses LVD to implement UDVS for 32-bit Kogge-Stone adders [12]. Measurements from the chip show that high rate (e.g. >0.1) dithering can occur in 1 cycle due to the local granularity of the headers. Figure 4 shows an example energy profile for a UDVS system using energy measurements from the test chip. For high rates, the blocks dither between the top two supplies (1.1V and 0.8V in the figure) to achieve near-optimal energy consumption. When performance requirements relax for low rate operation, the blocks can hop to the V_{DD} that gives minimum energy operation (330mV for the 90nm adder block) to achieve 9X savings in energy consumption.

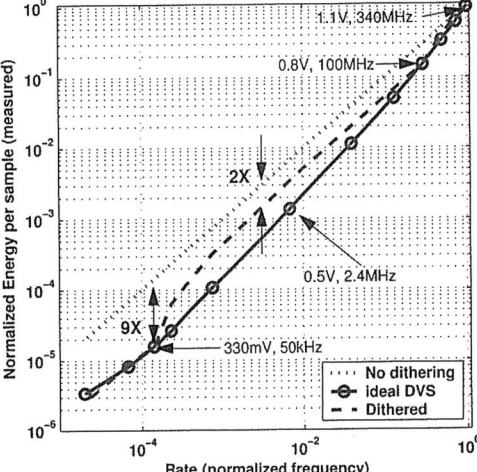

Figure 4: Energy profile based on 90nm chip measurements for example 3-V_{DD} system.

4. Sub-Threshold SRAM

SRAM is an important component of many ICs, and it can contribute a large fraction of the active and leakage power consumption. It is important to have sub-V_T compatible SRAMs for sub-V_T systems. However, the nature of SRAM circuits makes them a melting pot of all of the major sub-V_T challenges.

Random variation fundamentally affects the geometry and threshold voltage of CMOS devices and is increasingly prominent in scaled technologies. The large array nature of SRAM implies that extreme tails of the distributions limit yield. The problem is exacerbated in sub-V_T, where device strength depends exponentially on threshold voltage, and, in the presence of variation, relative strengths cannot be guaranteed by sizing. As a result, the widely used 6T SRAM cell, which relies on ratioed operation and is used to maintain density, fails to operate in sub-V_T. Figure 5a,b show the read/hold and write static nose margins [13] respectively for a typical 6T cell and for the 3σ case. At reduced voltages, read margin is negative and write margin is positive, indicating failure for both operations.

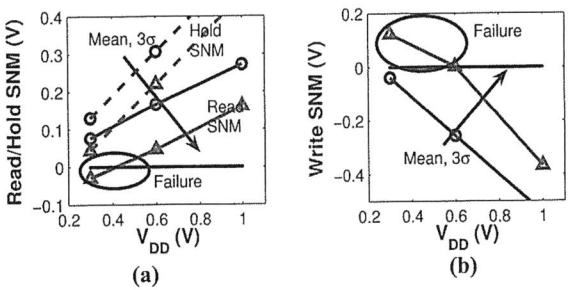

Figure 5: Simulated SNM for (a) read/hold and (b) write.

The increased impact of variation on device strength in sub-V_T also has a limiting effect on SRAM performance and integration. SRAM cell read current, I_{RD}, decreases exponentially in sub-V_T, but the speed is ultimately set by the weakest cell in the array. Figure 6a plots I_{RD} for cells on the weak side of the distribution normalized to the mean (i.e. $I_{RD}/\mu(I_{RD})$). The limiting effect of cell strength variation is amplified in sub-V_T where cells can be over an order of magnitude weaker than the mean.

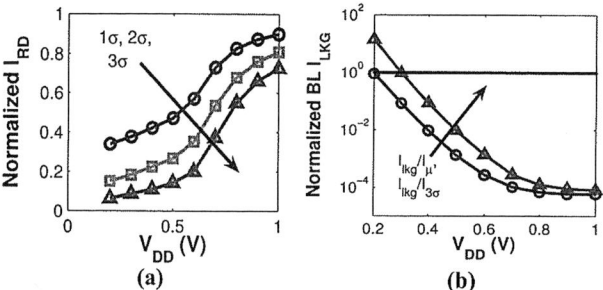

Figure 6: Effect of cell variation on (a) worst case read current and (b) bit-line leakage.

Parallel leakage also limits voltage scaling for SRAM. In conventional 6T SRAM, a stored "1" is read dynamically from a precharged bit-line. However, the reduced I_{on}/I_{off} ratio in sub-V_T is lowered even more due to the unaccessed cells sharing the bit-line, which results in a degraded logic level. Sub-V_T bit-line leakage is less problematic at high voltages where the discharge time of an accessed cell is much faster than that of the aggregate unaccessed cells. However, where variation extends the required discharge time, bit-line leakage severely limits the number of cells that can be integrated onto a column. Figure 6b shows the leakage current of 127 unaccessed cells normalized to the drive current of a single accessed cell weakened by variation. Values greater than unity, which occur in sub-V_T, imply that drive current is indistinguishable from leakage, making reliable read accesses impossible.

Numerous techniques have been reported to mitigate the low-voltage SRAM problems described above. For instance, reduced bit-line precharge voltages and negative word-line bias for unaccessed cells have been used to increase the read SNM. Similarly, increased word-line bias and negative bit-line voltages have been used to improve the write SNM. While these approaches can improve the situation for sub-V_T SRAM, approaches that address the problems more fundamentally provide a better solution for robust operation in sub-threshold.

A 65nm test chip implements a 256kb memory that overcomes the problems and provides functionality in the sub-threshold region to below 400mV [14]. The SRAM uses a 10T bit-cell, shown in Figure 7. M7-M10 form a read buffer that isolates the internal storage nodes, Q and QB, so that a read upset is not possible. This eliminates the read SNM problem of Figure 5a, and stability is instead limited by the hold SNM. Measurements from the test chip show that the cell can hold data correctly below 250mV. Write operations in Figure 5b fail since the access devices in a 6T bit-cell are too weak to over-power the internal cell feedback, which is made worse by process imbalance that makes pMOS sub-threshold current higher than nMOS by an order of magnitude. Robust write in the new 10T cell is performed by weakening the feedback structure by floating VV_{DD}. Finally,

bit-line leakage on RBL is minimized by unconditionally raising the voltage of QBB for unaccessed cells. This relies on either the active pull-up current through M9, or the ratio of its leakage current to that of M10's. In either case, M8's V_{GS} becomes negative, resulting in vanishingly small sub-threshold leakage current to the bit-line. This structure allows 256 bit-cells to be integrated per column.

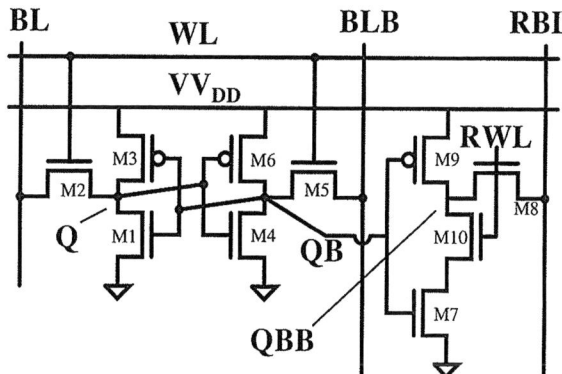

Figure 7: Schematic of 10T sub-threshold bit-cell [14].

5. Conclusions

Numerous problems increase the challenge of designing robust sub-threshold circuits. Some time-testing design practices, such as ratioed write in SRAM, become unreliable due to the exponential dependence of sub-threshold drive current on parameters with large process variations. We have presented an overview of the types of circuits and architectures that overcome these problems and produce working designs. Functional implementations of a sub-threshold FFT processor [11], an energy-scalable UDVS test chip [12], and a sub-threshold SRAM [14] attest that robust sub-threshold systems can practically offer minimum energy operation.

6. ACKNOWLEDGEMENTS

We acknowledge DARPA and Texas Instruments for funding.

7. REFERENCES

[1] Swanson and Meindl, *JSSC*, 1972.

[2] Vittoz and Fellrath, *JSSC*, 1977.

[3] Mead, Addison-Wesley, 1989.

[4] Soeleman and Roy, *ISLPED*, 1999.

[5] Paul, Soeleman, and Roy, *ESSCIRC*, 2001.

[6] Deen, Kazemeini, and Naseh, *ICCDCS*, 2002.

[7] Wang, Chandrakasan, and Kosonocky, *SVLSI*, 2002.

[8] Zhai, Blaauw, Sylvester, and Flautner, *DAC*, 2004.

[9] Calhoun and Chandrakasan, *ISLPED*, 2004.

[10] Sze, Blazquez, Bhardwaj, and Chandrakasan, *ICASSP*, 2006.

[11] Wang and Chandrakasan, *ISSCC*, 2004.

[12] Calhoun and Chandrakasan, *ISSCC*, 2005.

[13] Seevinck, List, and Lohstroh, *JSSC*, 1987.

[14] Calhoun and Chandrakasan, *ISSCC*, 2006.

Design and Power Management of Energy Harvesting Embedded Systems

Vijay Raghunathan
NEC Labs America
Princeton, NJ 08540
vijay@nec-labs.com

Pai H. Chou
University of California
Irvine, CA 92697
phchou@uci.edu

ABSTRACT

Harvesting energy from the environment is a desirable and increasingly important capability in several emerging applications of embedded systems such as sensor networks, biomedical implants, etc. While energy harvesting has the potential to enable near-perpetual system operation, designing an efficient energy harvesting system that actually realizes this potential requires an in-depth understanding of several complex tradeoffs. These tradeoffs arise due to the interaction of numerous factors such as the characteristics of the harvesting transducers, chemistry and capacity of the batteries used (if any), power supply requirements and power management features of the embedded system, application behavior, etc. This paper surveys the various issues and tradeoffs involved in designing and operating energy harvesting embedded systems. System design techniques are described that target high conversion and storage efficiency by extracting the most energy from the environment and making it maximally available for consumption. Harvesting aware power management techniques are also described, which reconcile the very different spatio-temporal characteristics of energy availability and energy usage within a system and across a network.

Categories and Subject Descriptors

C.3 [**Special-Purpose and Application-Based Systems**]: Real-time and embedded systems

General Terms

Management, Measurement, Performance, Design

Keywords

Energy harvesting, power management, wireless sensors, solar power

1. INTRODUCTION

Energy harvesting embedded systems (EHES) have received growing attention in recent years. Miniaturization and wireless or logging capabilities open up brand new applications by enabling a complete system to be mounted on or implanted inside many more

---

Permission to make digital or hard copies of all or part of this work for personal or classroom use is granted without fee provided that copies are not made or distributed for profit or commercial advantage and that copies bear this notice and the full citation on the first page. To copy otherwise, to republish, to post on servers or to redistribute to lists, requires prior specific permission and/or a fee.
ISLPED'06, October 4–6, 2006, Tegernsee, Germany.
Copyright 2006 ACM 1-59593-462-6/06/0010 ...$5.00.

---

objects than ever before. Although some applications such as automobiles provide their own infrastructure for supplying power, many other targets such as trees being monitored at a remote location do not readily supply electrical power. Batteries with limited capacities will eventually drain long before the service life of the system. Although radioisotopes and other radioactive materials can supply steady power for decades and are currently used in exit signs and other applications, their radioactive nature poses additional user concerns that prevent them from wider adoption [1]. As a result, for the great majority of embedded applications, designers will have to choose from more conventional energy sources that will scale well with economy.

Energy harvesting itself is not new, but what is new is how to build efficient energy harvesting capabilities into modern embedded systems while satisfying all their constraints. For instance, windmills and hydroelectric generators have been in use for a long time, and solar panels have been powering satellites and space stations for decades. It is possible to miniaturize these power sources and use them to power embedded systems. However, straightforward implementations will usually result in low efficiency.

Efficiency can be divided into several parts: *conversion efficiency* from one form of energy to another (*e.g.*, from light to electricity), *transfer efficiency* from the source to the supply, *buffering efficiency* once it has been harvested, and *consumption efficiency* in terms of the amount of useful work given the harvestable energy. Although research work has been proposed to optimize the efficiency at each of these levels, it is crucial to consider these techniques together in the context of the entire system, or else the gain at one level may come at the price of efficiency loss or high overhead elsewhere, rendering the system much less efficient than before.

1.1 System classification

One way to classify EHESs is based on the energy source of their application in relation to the system. They can be divided into *environmentally embedded* and *wearable*. Environmentally embedded systems are those that are embedded into an environment, which can be a building, a habitat, a greenhouse, or many other environments. In some such environments, abundant energy is available for harvesting, including sunlight [2, 3, 4, 5], wind [6], etc [7], A special case of the environment is a wearable [8, 9] or implantable system, where the subject is a person or an animal. In such systems, the source of energy can be from the subject itself, possibly in addition to the environment in which the subject operates.

In other environments, if the device is buried underground or inside walls, then very little energy can be harvested. In these cases, energy harvesting is difficult, but *wireless energy transfer* can be performed. For instance, inductive charging can be used to receive energy from electromagnetic emission [10]. Although related, this paper will consider wireless energy transfer to be a separate topic as it requires additional coordination with the energy source.

1.2 Mechanisms for Energy Harvesting

The form of energy that can be harvested include mechanical, thermal, photovoltaic, electromagnetic, biological, and chemical. Mechanical energy is possibly the most prevalent and are found in the forms of wind [6], limb movement [8], strain [11], ambient vibration [12, 13, 14, 15, 16], car wheel rotation, and many more. Heat differential can also be used to generate electricity [17]. The most well known source is light. Each given form of energy can be harvested by a different class of generator that performs conversion to electricity. To the system, the key differences are the output power level (current, voltage), AC vs. DC, the dynamic range, and the impedance model. For instance, windmills [6], magnetic coil generators [8], piezoelectric generators [11, 12, 13, 14, 15, 16], and magnetic induction [10] output AC power, whereas thermal [17] and photovoltaic [2, 4, 3, 5] power sources output DC power. Since most digital systems run on DC power, the default option is to rectify the current. An emerging alternative is to design self-timed circuits that will run directly on rectified AC power with minimal conversion loss [18]. Before such technologies become available in all components, it might not be possible to adopt them in more conventional designs. Even for DC power sources, it is often necessary to convert the DC power to a different voltage. In both cases, AC-DC or DC-DC conversion will incur additional loss.

The goal of this paper is to survey techniques that can be readily adopted in conventional systems that harvest energy. We divide the discussion into system design issues and power management issues. The former covers system functionality that must be designed-in in order to support efficient energy harvesting. The latter covers power management policies that specifically optimize for energy harvesting.

2. SYSTEM DESIGN ISSUES

The issues at the system level can be divided into voltage and current of the supply, form of energy storage, maximum power point tracking, and use of multiple power sources. It is important to consider the cost and operating range (V, I) associated with each level when trying to improve its efficiency, or else it can be counterproductive.

2.1 Voltage and Current

The very first consideration in energy harvesting is the voltage. Without high enough voltage, it is difficult or impossible to either power the system directly or to charge an energy storage device. Ideally, the circuit would adapt its power consumption and performance by tracking the available power without additional conversion. Asynchronous circuits can do this, but most commercially available components use synchronous designs. Moreover, systems such as wireless sensor nodes contain analog and RF components, which are sensitive to noise in the power supply. Therefore, it is necessary to perform voltage conversion to a known, controllable level, and then consume regulated power.

Voltage regulators are used to bridge the gap between the supply and the consumer. Linear regulators output clean, stable power, which are required by analog/RF components. However, they have lower conversion efficiency, incur a voltage drop, and dissipates more heat. On the other hand, switching regulators have much higher efficiency and are commonly used for digital subsystems, which are more immune to noise.

Switching regulators are further divided into buck, boost, and buck-boost regulators. Buck regulators perform voltage step-down

conversion and are efficient, but the input voltage must be higher than the output or else it does not work properly. Boost regulators perform voltage step-up conversion but are less efficient. Buck-boost regulators, as the name implies, can work as a buck or a boost depending on if the input voltage is higher or lower than the output.

In the case of power supply design, there is almost always an energy storage device such as a battery or a supercapacitor. It acts as a consumer to the harvesting device when charging, and acts as a supply to the system it is powering. Thus, regulators may need to be inserted on both ends. The overall conversion efficiency is no longer just a ratio of output / input power; it must also take the *operating range* of all stages into account. Specifically, even though a buck regulator has higher *power efficiency*, it does not necessarily have a higher *energy efficiency* when regulating solar output to charge a battery. This is because by performing step-down conversion, it stops charging the battery when the solar output drops below the battery voltage [16]. A similar issue occurs in EHESs that use a voltage comparator and prioritize drawing current from the energy harvesting source when its voltage is higher than the battery or capacitor. When it is lower, the power below the useful threshold level is simply discarded [3, 2], causing the system to draw energy from the storage. In both cases, the power conversion efficiency may be higher while within the operating range, but the overall energy conversion efficiency may be lower. This is known as the *(internal) power fragmentation* problem.

A few solutions are possible. One simple way is to use a boost regulator to raise the voltage above the threshold, which then makes the power usable again. For instance, the MIT shoes harvest from a high-impedance piezoelectric generator, which outputs AC. To power 5V digital devices, the authors propose a circuit that includes a full-wave rectifier bridge, uses a boost (step-up) regulator to raise the voltage to charge the capacitor to up to 12.6V, above which it turns on the 5V regulator. However, boost regulators are usually much less efficient than buck (step-down) regulators or DC-DC converters. The large gap between the input and output voltage may be another source of inefficiency. An interesting problem is to choose an alternative capacitor configuration (*e.g.*, parallel vs. series composition, difference capacitance values) that will decrease this voltage gap for higher regulator efficiency.

2.2 Maximum Power Point Tracking

Maximum power point tracking (MPPT) refers to drawing power from an energy harvesting source at a level that maximizes the power output. For DC sources such as solar panels, the maximum power point (MPP) is a voltage-current combination that maximize the power output under a given sunlight condition, and a given temperature[1]. The power is maximized when the supply and the load are impedance matched [4]. For AC sources such as piezoelectric, on the other hand, the MPP is actually related to the resonant frequency of the device in addition to the magnitude of the physical oscillation. Although MPPT is not strictly required for energy harvesting to work, the efficiency loss can be tremendous that 65% to 90% of the available power may simply be discarded.

2.2.1 Measurement Method for MPPT

MPPT requires that the input intensity be known, so that the MPP in terms of voltage and current can be determined. The input intensity can be determined by measurement either *before* or *after* conversion to electricity.

For instance, the MPP for solar panels is determined mainly by the light intensity and secondarily by temperature. One may perform direct measurement of sunlight before conversion by includ-

---

[1] The MPP is only weakly dependent on temperature.

ing a light sensor and possibly a temperature sensor to provide the readings, so that the MPP can be looked up or computed. While straightforward, the light sensor covers a much smaller area than the solar panel and might not yield a representative reading, if the dust or shadow on the panel does not cover the light sensor in the same proportion. MPPT applies to other energy sources as well, including windmills and even fuel cells. For a windmill, a rotation speed sensor can be used.

An alternative is to sense the input intensity after conversion. This means measuring the voltage and current from the solar panel. One may measure either the open-circuit voltage or short-circuit current. Both require the load to be temporarily disconnected from the supply during its measurement. This is possible if there is another energy storage device such as either a battery or a capacitor to continue powering the system while the measurement is being taken. This method can better track the entire area of exposure, although discrete sampling assumes the input level does not alter rapidly.

2.2.2 MPPT for AC

For AC generators that rely on vibration, the MPP depends on not only the amplitude but also frequency of the vibration [8]. Ottman *et al.* [16] show that the power is maximized if the rectifier voltage is maintained at 1/2 open circuit voltage, and the duty cycle at approximately the square-root:

$$V_{rect} = \frac{I_P}{2\omega C_P} \qquad (1)$$

$$D_{opt} \approx \sqrt{\frac{4\omega L C_p f_s}{\pi}} \qquad (2)$$

Applying their tracking method to control a DC-DC converter has been shown to result in 400% improvement in efficiency. However, the controller is implemented with a DSP which may consume nontrivial power, making this difficult to apply to μW-level energy harvesting.

2.2.3 Software vs. Hardware MPPT Controller

The control for MPPT can be implemented either in hardware or software. A hardware implementation usually means a special designed circuit. It entails taking the output of a sensor, usually before conversion to electricity, and control a DC-DC converter or a programmable regulator, all without additional computation. On the other hand, a software implementation entails sampling the voltage level, usually after conversion to electricity, either performing a table look-up or running a DSP algorithm, and then controlling the power circuitry accordingly. The former is usually called *autonomous* in that MPPT can be performed with low overhead and be part of the power subsystem in a modular way, without the involvement from the microcontroller or DSP. The latter requires the MPPT to be scheduled as part of the software, and it is acceptable if it is scheduled as a low duty cycle task. Also, software-based schemes are more suitable for higher power energy harvesting systems, since the minimum power requirement for processors are usually significantly higher. Other downsides include more complex software, the use of precious I/O pins for control, and inability to operate if the DSP or MCU itself is powered down. Hardware implementations are almost mandatory for very low-power MPPT, or if the input power level can change rapidly. However, most hardware implementations are very simple in order to keep the overhead low, but they tend to track the MPP with a hysteresis band [6].

The MPPT overhead must be considered for the entire system, or else MMPT may be counterproductive. Software MPPT usually is counterproductive for microwatt-level energy harvesting. For ex-

ample, MPPT enables one piezoelectric generator to improve its power output by 400%, from 16.43mW to 70.42mW after 18.87mW efficiency loss due to the DC-DC converter. However, the TMS320C31 DSP that runs the MPPT algorithm consumes 80mA of current, which is already higher than the whole energy harvesting source. The voltage at $67.3V_{oc}$ is also very high for harvesting 30.66mW [19]. Therefore, it can actually be a loss in this case, unless the overhead can be amortized over a large enough array of these harvesting devices.

2.3 Power Defragmentation

One problem with harvesting energy from environmental sources is the wide dynamic range of power. Even with MPPT, the available power may be so low that it is below the useful threshold. To solve this problem, one may wish to harvest energy from multiple sources. For instance, one might include both a solar panel and a wind generator in a heterogeneous power harvesting system. This will harvest more energy than relying just on one single source. However, it is difficult to "add" power by simply composing heterogeneous sources such as a battery and a windmill in series or in parallel. Depending on the target system to be powered, each source might not be sufficiently powerful to power the entire system. In this case, the power must be discarded, even if it is available at a non-trivial level. This is known as the *(external) power fragmentation* problem.

To address the power fragmentation problem, power matching switches have been proposed [20]. The idea is to divide up the system into subsystems that can be powered separately. The supply-to-power are related by a one-to-many mapping. This way, instead of discarding the power when it is below the threshold of the system, now it is possible to continue utilizing power until the subsystem threshold, which is much lower.

2.4 Energy Storage Devices

Many energy harvesting embedded systems need energy storage because they need to continue operation even when there is no energy to harvest (*e.g.*, at night for a solar-powered system). By default, rechargeable batteries are used for longer term storage (several days to weeks). Batteries have the problem of non-ideal effects including aging and rate capacity effects.

More recently, alternative energy storage technologies have become available, including supercapacitors and fuel cells. Supercapacitors, also called ultracapacitors, are high-value (*i.e.*, from mF to hundreds of farads), portable sized (10s of cubic-cm) capacitors. They are commonly used for buffering transient energy (from minutes to hours). For instance, electric and gasoline-electric hybrid vehicles use supercapacitors to store energy in regenerative brakes, and they are starting to be used in EHESs as well [4, 5, 21, 22, 10, 6, 3]. Supercapacitors do not have the aging and rate-capacity problems. They do have limited energy capacity and higher leakage, but these are less of a problem if energy is regularly replenished. Another often cited problem is the linear discharge curve and inaccessible charge below the target operating range, but this can be addressed easily with the buck-boost regulator described in Section 2.1.

Even though capacitor charging and MPPT are active areas of research, the combination of MPPT using supercapacitors as an energy storage device poses new challenges. A capacitor of hundreds of farads that is attached to the power rails appears as a short circuit on a cold start and throughout much of the charge cycle. The capacitor will still charge but at a very inefficient level. One attempt to solve this problem is to insert a voltage regulator between the supply and the capacitor. However, this will not work because the regulator as a feedback control circuitry see the low output voltage and attempts to drive up the voltage, which also results in effectively a short circuit. Other converters either require AC [23] or incur high overhead [24]. Instead of a feedback mechanism, a feed-forward PFM (pulse frequency modulation) regulator [4] has been proposed to address these problems.

3. POWER MANAGEMENT ISSUES

Complementary to the problem of designing a harvesting device that efficiently extracts, stores, and transfers power to the load is the issue of adapting the embedded system's power management policy to be aware of the operating conditions and state of the energy harvesting device, a process termed as *harvesting aware power management*. This section illustrates how harvesting aware power management improves upon conventional battery-based power management and surveys recent work in designing such harvesting aware power management methodologies.

As mentioned in Section 2, there are several non-idealities (*e.g.*, power fragmentation, inefficiency of energy storage elements, *etc.*) that manifest in energy harvesting systems. We use the following example to illustrate the potential benefits of harvesting aware power management.

EXAMPLE 1. *Consider the task of routing data in a simple sensor network where two route options exist from the data source to the sink, one of which uses node A and the other node B. Nodes A and B receive the same amount of solar energy per day, E_s, but due to obstacles such as trees, node A receives all of its energy in the morning, whereas node B receives all of its energy in the afternoon. Both nodes begin with the same residual battery energy, E_b, and the battery round trip efficiency is η. A node uses energy E_r for one hour of routing activity, and the daily workload consists of an hour of routing activity in the morning and another hour in the afternoon. We compare two routing schemes, $\mathcal{H}$, which explicitly uses information about the solar energy availability pattern, and $\mathcal{B}$, which operates based on residual battery levels alone and is representative of state-of-the-art power aware routing schemes. On the first morning, $\mathcal{H}$ chooses node A to route data (since it knows that node A receives solar energy in the morning) while $\mathcal{B}$ may pick either node, as each has the same battery level. Say that it chooses node B. At noon, in the system running $\mathcal{H}$, node A has energy^2 $E_b + (E_s - E_r)\eta$, and B has energy E_b. In the system running $\mathcal{B}$, node A has $E_b + E_s\eta$ and B has $E_b - E_r$. In the afternoon, algorithm $\mathcal{H}$ will choose node B (since it is aware that node B receives solar energy in the afternoon), and the residual battery energy at the end of the day is $E_b + (E_s - E_r)\eta$ at A and $E_b + (E_s - E_r)\eta$ at B. Algorithm $\mathcal{B}$ will instead choose node A due to its higher battery level, resulting in battery levels of $E_b + E_s\eta - E_r$ at A and $E_b + E_s\eta - E_r$ at B. At each node, the nodes following algorithm $\mathcal{H}$ have a higher energy, ΔE, given by:*

$$\Delta E = E_b + (E_s - E_r)\eta - (E_b + E_s\eta - E_r)$$
$$= E_r(1 - \eta) \qquad (3)$$

Note that, at the end of the day, both nodes in each system have equal energy. Hence the process may repeat on the next day, increasing the energy gap between $\mathcal{H}$ and $\mathcal{B}$.

The above example shows that modifying the power management policy to be harvesting aware can provide improved energy

---

^{2}Since the energy for routing is supplied from the solar panel and only the remainder is stored in the battery. It is assumed that $E_s \geq E_r$, although a similar reasoning can be followed when $E_s < E_r$.

efficiency compared to a conventional system-level power management scheme that operates without any knowledge of the spatio-temporal characteristics of the environmental energy source.

3.1 Energy Neutrality in Harvesting Systems

A significant difference between battery-powered systems and energy harvesting systems from the perspective of power management is that conventional energy optimization metrics might not be suitable in an energy harvesting scenario. For instance, a commonly used objective in battery-powered sensor networks is to maximize network lifetime under a total energy constraint. Clearly, this changes if energy harvesting is allowed since the amount of energy available itself depends on the time duration for which the system operates. Instead, a more relevant design objective might be to operate in an *energy neutral* mode, consuming only as much energy as harvested. Such a mode of operation raises the possibility of indefinitely long lifetime, limited only by the hardware longevity. Note that reducing the power consumption below the level needed for energy neutrality will not increase system lifetime any further.

Achieving energy neutrality in a harvesting system depends on several factors such as the average power generated by the harvesting device, the capacity of the energy storage device, *etc.*, which are influenced by design choices made by the system architect. In order to help designers of harvesting systems make the right design decisions, a systematic framework is needed that can capture and analyze the various energy-neutrality related requirements quantitatively. In [25], the authors develop such a framework to better understand the energy neutral mode of operation.

3.1.1 Modeling environmental energy sources

The first step in analyzing a harvesting system's energy neutrality is to analytically model the power generated by the harvesting transducer. However, this is a non-trivial task due to the inherent temporal variability present in environmental energy sources. In [25], the authors introduce the following model for characterizing time-varying environmental energy sources.

DEFINITION 3.1. *A non-negative, continuous, and bounded function $P(t)$ is said to be a $(\rho, \sigma_1, \sigma_2)$ function if and only if, for any positive, finite real numbers τ and T, the following is satisfied:*

$$\rho T - \sigma_2 \leq \int_{\tau}^{\tau+T} P(t)dt \leq \rho T + \sigma_1 \qquad (4)$$

For instance, let $P_s(t)$ denote the power output by the transducer at time t. Intuitively, if $P_s(t)$ is a $(\rho_1, \sigma_1, \sigma_2)$ function, then the average rate at which energy is available from the transducer is ρ_1, and the burstiness caused by temporal variations is bounded by σ_1 and σ_2. Note that this model can also be used to model power consumers. The power consumption profile of the load, $P_c(t)$, may be modeled as a (ρ_2, σ_3) function, where the parameters ρ_2 and σ_3 are used in a constraint similar to the upper bound inequality of Equation (4) to place a upper limit on the power consumption, while no constraint is placed on the minimum power consumption.

3.1.2 Analyzing Energy Neutrality Requirements

The harvesting theory as presented in [25] considers two non-idealities associated with energy storage, namely the round trip efficiency and self-discharge. Round trip efficiency is related to the energy loss that occurs when energy is stored into an energy storage element such as a battery and retrieved later. Self-discharge is the energy loss due to leakage paths in the energy storage element. Based on the energy production, consumption, and energy storage models discussed above, the condition for energy neutrality, the following theorem provides a sufficient condition for guaranteeing energy neutrality of the system [25]

THEOREM 3.2. *Consider a harvesting system in which the energy production profile is characterized by a* $(\rho_1, \sigma_1, \sigma_2)$ *function, the load is characterized by a* (ρ_2, σ_3) *function, and the energy buffer is characterized by parameters* η *for storage efficiency, and* ρ_{leak} *for leakage power. The following conditions are sufficient for the system to achieve energy neutrality:*

$$\rho_2 \leq \eta\rho_1 - \rho_{leak} \tag{5}$$
$$B_0 \geq \eta\sigma_2 + \sigma_3 \tag{6}$$
$$B \geq B_0 \tag{7}$$

where B denotes the capacity of the energy buffer and B_0 *is the initial energy stored in the buffer.*

This theorem has two important design implications. First, it characterizes the sustainable performance level that may be supported in energy neutral mode. This is significant for system design both in hardware and software. At the hardware level, if the sustainable power consumption supported is too low, changes may be made to increase the harvested energy (*e.g.*, by using a larger solar panel). In software, this level will help determine the appropriate power scaling required based on the relationship between energy consumption and system performance. Second, it specifies the minimum capacity of the energy storage element required to achieve energy neutrality for given burstiness bounds on energy production and consumption. This size can be directly used while designing the system. Using a higher capacity battery will not yield an increase in sustainable performance or lifetime.

3.2 Node Level Power Management

While the harvesting theory presented above enables us to design the embedded system in order to guarantee energy neutrality, equally crucial are techniques that adapt the performance and power consumption of the embedded system at runtime in response to the spatial and temporal variations in harvested energy. The goal of these techniques is to maximize system performance while not violating the energy neutrality requirement.

In [26], the authors present an algorithm for harvesting-aware duty cycling of wireless sensor nodes. The authors choose to use duty-cycling between active and low power modes for the purpose of performance/power scaling since most sensor nodes provide at least one low power mode in which the power consumption is negligible. More sophisticated performance/power scaling methods, such as dynamic voltage scaling, may be used when available. Their algorithm consists of three steps, namely (a) learning the harvested energy profile at run-time, (b) adapting the power consumption level to match the harvested energy, and (c) fine tuning the power scaling algorithm to account for battery non-idealities. The last step is important because it helps minimize the energy loss due to battery inefficiency.

A significant challenge with harvesting-aware power management is that determining the optimal duty cycle for a node at a given point in time requires information about the harvested energy availability in the future. The authors overcome this by learning the daily energy generation profile for the harvesting device and use this information to predict the energy availability for the near future. The authors argue that, on a typical day, the energy generation is expected to be similar to the energy generation at the same time on previous days. Based on this, they use an Exponentially Weighted Moving-Average (EWMA) filter based prediction model. The method is designed to exploit the diurnal cycle in solar energy but at the same time adapt to weather and seasonal variations. The predicted energy generated for each 30 minute slot is calculated as a weighted average of the energy received in the same time slot during previous days. The weights are exponential, resulting in decaying weights for older data.

Using a user defined system utility function that is a function of the duty cycle, the authors then solve a lightweight optimization problem that determines duty cycles for each time slot in order to maximize the utility of the system over the course of the entire day. As a post processing step, they account for any errors in the predicted energy profile by adjusting the duty cycle as the day progresses using an approach similar to slack redistribution techniques during power-aware task scheduling [27]. The authors demonstrate that their algorithm utilizes environmental energy more efficiently (up to 58%) compared to duty cycling techniques that are not harvesting aware.

3.3 Network Level Power Management

The technique presented above adapts the performance at a single node in response to the temporal variations in harvested energy. Next, we consider a distributed network of harvesting enabled nodes and discuss how the network as a whole can be power managed to address the spatial variations in harvesting opportunity across the different nodes. The authors of [7] addressed this problem in the context of a data routing application presented a harvesting aware routing scheme which attempts to align the allocation of routing energy consumption with the harvested energy profiles across the network.

Energy aware routing protocols in sensor networks typically use battery energy based routing cost metrics [28]. The objective is to choose data routes appropriately such that the routing load is distributed uniformly across the network to leverage the total battery resource for maximizing lifetime and preventing individual nodes from being over-used and hence, running out of battery quickly. In an energy harvesting network, battery awareness is not sufficient to select the best routes, as we saw in Example 1. In [7], the authors propose a harvesting enhanced cost metric for data routing. The authors assume that each node can learn the expected rate of energy harvesting, ρ_i, at node i. They propose an enhanced routing cost metric that considers both the harvesting potential of a node as well as its residual battery level (B_i). They define an energy potential, E_i, at node i as follows:

$$E_i = w \cdot \rho_i + (1 - w) \cdot B_i \tag{8}$$

where w is a weight parameter, $0 \leq w \leq 1$ and is typically set close to 1. A low value of w may be relevant in networks powered predominantly by batteries with a small harvesting opportunity. The authors use the inverse of the energy potential at a node as the communication cost for all wireless links into that node. In other words, in the directed graph representing each sensor node as a vertex, v_i and each wireless link between two nodes i and j as an edge e_{ij}, they associate the following cost for each edge into node i:

$$c_{ki}(e_{ki}) = \frac{1}{E_i} \quad \forall k \in \{k | e_{ki} \in E_{comm}\} \tag{9}$$

where E_{comm} represents the set of edges across which radio communication is feasible for the deployed network topology and the radio hardware used. Thus, the authors derive a graph representation of the wireless network with each link represented as a weighted directed edge. They use a distributed Bellman-Ford algorithm to compute minimum cost routes between a given source and destination. They show that such a harvesting-aware data routing scheme significantly increases the energy scalability of the network.

4. CONCLUSIONS

Energy harvesting in embedded systems is representing a fruitful area of research as made possible by the convergence of low-power designs, miniaturization, and advances in materials and mechanical devices. The power consumption has been reduced to the same level as the harvesting devices are capable of outputting. The potentially perpetual operation of these EHESs are just starting to enable a brand new class of applications. However, it seems unlikely that existing systems can automatically operate efficiently by just adding an energy harvesting module. We believe the entire system must be optimized in a holistic way from the design of the architecture to power management at the application and networking levels.

5. REFERENCES

[1] Amit Lal, Rajesh Duggirala, and Hui Li. Pervasive power: A radioisotope-powered piezoelectric generator. *IEEE Pervasive Computing*, 4(1):53–61, January–March 2005.

[2] V. Raghunathan, A. Kansal, J. Hsu, J. Friedman, and M. Srivastava. Design considerations for solar energy harvesting wireless embedded systems. In *Proc. 4th Int. Conf. on Information Processing in Sensor Networks*, pages 457–462, Apr. 2005.

[3] X. Jiang, J. Polastre, and D. Culler. Perpetual environmentally powered sensor networks. In *Proc. 4th Int. Conf. on Information Processing in Sensor Networks*, pages 463–468, Apr. 2005.

[4] Farhan Simjee and Pai H. Chou. Everlast: Long-life, supercapacitor-operated wireless sensor node. In *Proc. ISLPED*, October 2006.

[5] Henri Dubois-Ferrière, Laurent Fabre, Roger Meier, and Pierre Metrailler. TinyNode: a comprehensive platform for wireless sensor network applications. In *IPSN '06: Proceedings of the fifth international conference on Information processing in sensor networks*, pages 358–365, New York, NY, USA, 2006. ACM Press.

[6] Chulsung Park and Pai H. Chou. AmbiMax: Efficient, autonomous energy harvesting system for multiple-supply wireless sensor nodes. In *to appear, Proc. Third Annual IEEE Communications Society Conference on Sensor, Mesh, and Ad Hoc Communications and Networks (SECON)*, 2006.

[7] Aman Kansal and Mani B. Srivastava. An environmental energy harvesting framework for sensor networks. *Proceedings of the 2003 international symposium on Low power electronics and design*, pages 481–486, 2003.

[8] John Kymissis, Clyde Kendall, Joseph A. Paradiso, and Neil Gershenfeld. Parasitic power harvesting in shoes. In *Proc. of the Second IEEE International Symposium on Wearable Computing ISWC*, pages 132–139, 1998.

[9] Joseph A. Paradiso and Thad Starner. Energy scavenging for mobile and wireless electronics. *IEEE Pervasive Computing*, 4(1):18–27, January–March 2005.

[10] Olivier Chevalerias, Terence O'Donnell, Daithi Power, Neil O'Donovan, Gerald Duffy, Gary Grant, and Sean Cian O'Mathuna. Inductive telemetry of multiple sensor modules. *IEEE Pervasive Computing*, 4(1):46–52, January–March 2005.

[11] David L. Churchill, Michael J. Hamel, Christopher P. Townsend, and Steven W. Arms. Strain energy harvesting for wireless sensor networks. *Proceedings of SPIE*, 5055:319, 2003.

[12] S. Kim, W.W. Clark, and Q.M. Wang. Piezoelectric energy harvesting using diaphragm structure. *Proceedings of SPIE*, 5055:307, 2003.

[13] Henry A. Sodano, Daniel J. Inman, and Gyuhae Park. A review of power harvesting from vibration using piezoelectric materials. *The Shock and Vibration Digest*, 36(3):197–205, 2004.

[14] Sunghwan Kim. *Low power energy harvesting with piezoelectric generators*. PhD thesis, University of Pittsburgh, 2002.

[15] Shad Roundy, Eli S. Leland, Jessy Baaker, Eric Carleton, Elizabeth Reilly, Elaine Lai, Brian Otis, Jan M. Rabaey, Paul K. Wright, and V. Sundararajan. Improving power output for vibration-based energy scavengers. *IEEE Pervasive Computing*, 4(1):28–36, January–March 2005.

[16] Geffrey K. Ottman, Heath F. Hofmann, Archin C. Bhatt, and George A. Lesieutre. Adaptive piezoelectric energy harvesting circuit for wireless remote power supply. *IEEE Transactions on Power Electronics*, 17(5):669–676, 2002.

[17] Ingo Stark. Invited talk: Thermal energy harvesting with Thermo Life. In *International Workshop on Wearable and Implantable Body Sensor Networks (BSN'06)*, pages 19–22, Los Alamitos, CA, USA, 2006. IEEE Computer Society.

[18] J. Siebert, J. Collier, and R. Amirtharajah. Self-timed circuits for harvesting AC power supplies. In *Proc. International Symposium on Low Power Electronics and Design*, 2005.

[19] Geffrey K. Ottman, Heath F. Hofmann, and George A. Lesieutre. Optimized piezoelectric energy harvesting circuit using step-down converter in discontinuous conduction mode. *IEEE Transactions on Power Electronics*, 18(2):696–703, March 2003.

[20] Chulsung Park and Pai H. Chou. Power Utility Maximization for Multi-Supply Systems by a Load-Matching Switch. In *ISLPED'04*, pages 168–173, August 2004.

[21] P. Enjeti, J.W. Howze, and L. Palma. An approach to improve battery run-time in mobile applications with supercapacitors. In *IEEE 34th Annual Power Electronics Specialists Conference*, 2003.

[22] T.A. Smith, J.P. Mars, and G.A. Turner. Using supercapacitors to improve battery performance. In *Power Electronics Specialists Conference*, 2002.

[23] H. Pollock. High efficiency, high frequency power supplies for capacitor and battery charging. In *IEE Colloquium on Power Electronics for Demanding Applications*, pages 901 – 910, Apr. 1999.

[24] R.M. Nelms and J.E. Schatz. A capacitor charging power supply utilizing a ward converter. *IEEE Trans. Ind. Electronics*, 39:421–428, Oct. 1992.

[25] Aman Kansal, Dunny Potter, and Mani B. Srivastava. Performance aware tasking for environmentally powered sensor networks. In *SIGMETRICS*, pages 223–234, 2004.

[26] J. Hsu, S. Zahedi, A. Kansal, V. Raghunathan, and M. Srivastava. Adaptive duty cycling for energy harvesting systems. In *Proc. ACMInternational Symposium on Low Power Electronics and Design*, 2006.

[27] Niraj K. Jha. Low power system scheduling and synthesis. In *ICCAD*, pages 259–263, 2001.

[28] Rahul C. Shah and Jan M. Rabaey. Energy aware routing for low energy ad hoc sensor networks. In *Proc. IEEE Wireless Communications and Networking Conference (WCNC)*, pages 350–355, 2002.

Panel

Flexibility and Low Power;
A Contradiction in Terms?

Can Configurable or Re-Configurable Computing Offer Solutions?

Moderator
Peter Wintermayr
Editor in Chief Markt und Technik

Panelists:

Reiner Hartenstein
Univ. Kaiserslautern

Heinrich Meyr
RWTH Aachen University and
Chief Scientific Officer
CoWare

Steve Leibson
Tensilica

Both configurable computing paradigms as well as re-configurable computing paradigms have gained significant impact within the last few years. Both paradigms have shown to be effective when power consumption is a major design constraint even though the philosophies behind are quite different: configurable approaches aim to adapt an embedded processor to an application through, for example, an extensible instruction set plus other parameters that are determined during design time. They come in two basic flavors:

a) starting with a fixed core that is extended by the system designer or,

b) designing the instruction set from scratch for a specific application.

Re-configurable approaches on the other side gain most of their benefits through run-time re-configuration. A high degree of parallelism is needed to overcome the physical deficiencies of re-configurable fabrics (e.g. FPGAs), though.

The panel will discuss advantages and disadvantages of these paradigms with respect to low power.

Copyright is held by the author/owner(s).
ISLPED '06, October 4–6, 2006, Tegernsee, Germany.
ACM 1-59593-462-6/06/0010.

Efficient Scan-Based BIST Scheme for Low Power Testing of VLSI Chips

Malav Shah
Dhirubhai Ambani Institute of Information and Communication Technology (DA-IICT),
Gandhinagar 382 007, Gujarat, INDIA
Contact No: +91 9886731033

shah_malav@da-iict.org

ABSTRACT

It is seen that power dissipation during test mode is quite high compared to that during the functional mode of operation of a digital circuit. This may lead to damage of certain chips only because they are tested, leading to unnecessary loss of yield. This paper presents a simple yet efficient low power scheme for scan-based BIST. It reduces test length and switching-activity in CUTs reducing power dissipation during test mode without compromising fault coverage. Experiments conducted on ISCAS89 benchmark circuits demonstrate that proposed scheme gives better fault coverage with a large reduction in transitions reducing power dissipation during testing.

Categories and Subject Descriptors

B.8.1 [**Performance and Reliability**]: Reliability, Testing, and Fault-Tolerance.

General Terms

Design, Reliability.

Keywords

Scan, test-per-scan, test-per-clock, partial scan, switching activity, test length.

1. INTRODUCTION

Scan-based BIST integrates scan design and built-in self-test methodologies. Using scan, the memory elements are connected into a shift register and thus their values can be controlled or observed during the test mode by shifting the desired values into the register or shifting out the content of the register. BIST is a technique that makes the circuit test itself without using automatic test equipment. It includes on-chip test pattern generator (TPG) [for e.g., Linear Feedback Shift Register (LFSR)] and response analyzer [for e.g., Multiple Input Signature Register (MISR)].

Permission to make digital or hard copies of all or part of this work for personal or classroom use is granted without fee provided that copies are not made or distributed for profit or commercial advantage and that copies bear this notice and the full citation on the first page. To copy otherwise, or republish, to post on servers or to redistribute to lists, requires prior specific permission and/or a fee.
ISLPED '06, October 4–6, 2006, Tegernsee, Germany.
Copyright 2006 ACM 1- 59593-462-6/06/0010…$5.00.

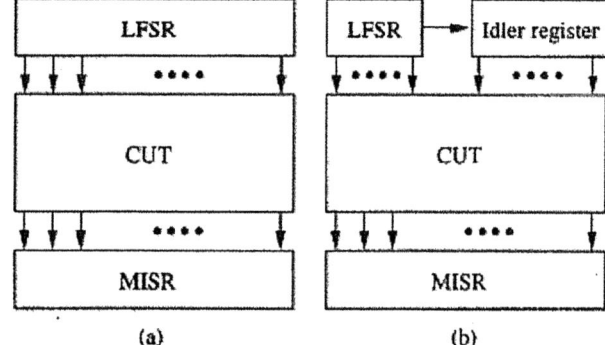

Figure 1. Test-per-clock BIST architecture.

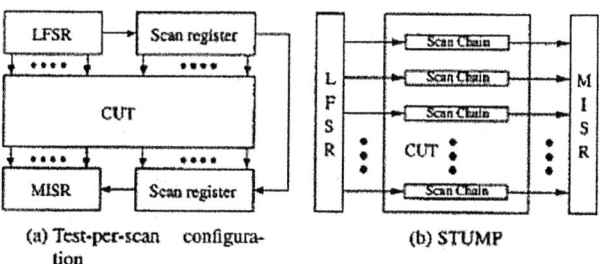

(a) Test-per-scan configuration

(b) STUMP

Figure 2. Test-per-scan BIST architecture.

The LFSR is commonly used as a TPG in low overhead BIST schemes. The correlation between consecutive random patterns generated by an LFSR is low but a significant correlation exists between consecutive patterns at the primary inputs (PIs) during the normal operation of circuit. For DFT, the situation is even worse. Here, during test mode, it breaks all the secondary input (SI) and secondary output (SO) to apply any desired value at state FFs i.e. breaks a sequential circuit into a combinational block and state elements (scan elements/D FFs). This destroys the correlation that typically exists between successive states of FSM leading to even lesser correlation. Due to all these, the switching activity in a circuit can be significantly higher during BIST than that during its normal operation, which can cause very destructive consequences like excessive heat dissipation, electro-migration and simultaneous switching noise, i.e. power / ground noise. Hence, excessive switching leads to the failure of some good chips only because they are tested, leading to unnecessary loss of yield. Thus, reducing the heat dissipated during test application is becoming a very important objective these days, particularly for the large VLSI systems.

2. BACKGROUND

BIST schemes can be classified into (a) *test-per-clock* and (b) *test-per-scan*, according to the way in which test patterns are applied to the CUT. In *test-per-clock* BIST (Fig.1), the outputs of a test pattern generator are directly connected to the inputs of a CUT and a new test pattern is applied to the inputs of the CUT at every clock. Also, the response to a test pattern applied to the CUT is loaded into a response analyzer at every test cycle. BILBO and circular BIST are the examples of *test-per-clock* BIST.

In *test-per-scan* BIST (Fig.2), a test pattern is applied to a CUT every $m+1$ clock cycles, where m is the number of flip-flops in a scan chain, since the patterns are applied via the scan chain. The response to a test pattern applied to the CUT is captured into the scan chain and scanned out during next m scan cycles and loaded into a response analyzer; the next test pattern is scanned in concurrently. When a CUT has long scan chains, most switching activity in the CUT will occur during scan operation. STUMPS (Fig.2) is a *test-per-scan* BIST architecture.

In this paper, a test scheme is proposed for scan-based BIST that can reduce SA in CUTs as well as test application time during BIST, does not degrate CUT performance and can be implemented with little area overhead. Section 3 briefly mentions the some related work that has been done in this area. Section 4 discusses the existing Low Transition LFSR (LT-LFSR) based test-per-scan scheme targeted for low heat dissipation during test by reducing the number of transitions at the cost of reduced fault coverage. Section 5 describes a *combined approach* using both test-per-scan and test-per-clock application schemes and then presents proposed BIST capability built on top of a partial scan circuit adding above LT-LFSR as the TPG and MISR for signature analysis. This is expected to take optimal advantage of the *combined approach*. Section 6 gives experimental results conducted on different ISCAS89 benchmark circuits which shows that proposed scheme gives quite better fault coverage (almost comparable to conventional LFSR) that too, with a large reduction in   test lengths and transitions (and hence heat dissipation during testing).

3. RELATED WORK

Minimizing heat dissipation during testing along with better test efficiency and performance has become an important goal of BIST design. The existing techniques include optimal scan chain partitioning [2], weighted random BIST [3], mixed mode approach or hybrid BIST [5], low energy BIST [6], and techniques with multiple capture cycles [8]. Most of these techniques require either involving complex test generation, higher area overhead or complex routing resulting in high routing overhead as well.

4. OVERVIEW OF LT-RTPG DESIGN

The LT-RTPG (Fig. 3) reduces switching activity during BIST by reducing transitions at scan flip-flops during scan shift operations.

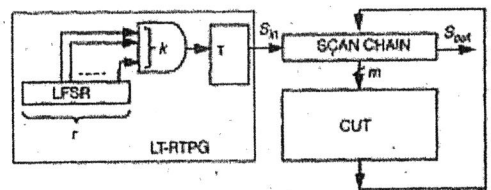

Figure 3. LT-RTPG and overall test-per-scan BIST.

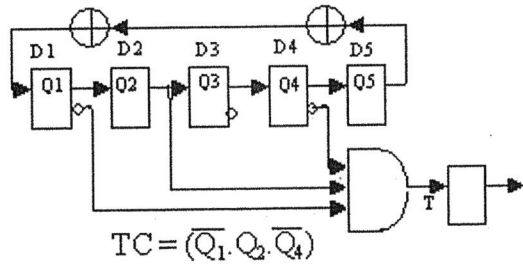

Figure 4. An example tap configuration.

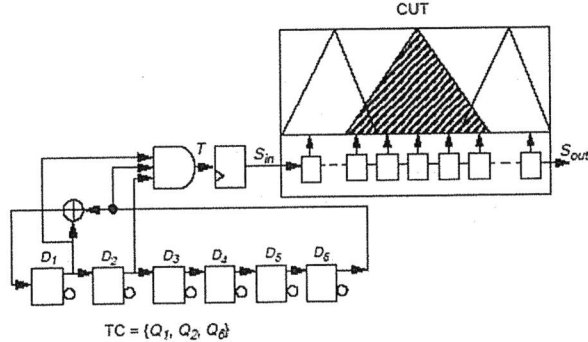

Figure 5. A CUT with span S=5.

An example implementation of the LT-RTPG is shown in fig. 4. It is comprised of an r-stage LFSR, a k-input AND gate, and a T flip-flop. Hence, it can be implemented with very little hardware. Each of k inputs of the AND gate is connected to either normal or inverting outputs of the r LFSR stages. Since a T flip-flop holds previous values until the input of the T flip-flop is assigned a 1, the same value $v \in \{0,1\}$ is repeatedly scanned into the scan chain until the value at the output of the AND gate becomes 1. Hence, if large k is used i.e., the AND gate has many inputs, neighbouring scan flip-flops are assigned identical values in most test patterns and the scan chain input have fewer transitions during scan shift operations. Since most switching activity during scan BIST occurs during scan shift operations, the LT-RTPG can reduce switching activity during overall scan testing. Various properties have been studied and proved and a detailed methodology for its design is presented [9]. Here, some of the main results have been mentioned.

4.1 Analysis of LT-RTPG

The combinational part of a sequential circuit can be viewed a collection of (overlapping) *output cones*, where an output cone Ω_j is composed of all the logic and inputs (primary and state) that form the transitive fanin of its *j*-th output.

377

A pair of inputs are said to be *compatible* if there exits no circuit cone to which they both belong. For faults such as stuck-at, any correlation between the values applied to a pair of compatible inputs does not reduce fault coverage for any given test length.

Consider a full scan circuit with a single scan chain. Let the *span S_j of cone Ω_j* be the *distance* between the first and the last flip-flops in the scan chain as shown in Fig. 5, whose outputs drive the state inputs of cone. If the α-th and β-th flip-flops of the scan chain drive the first and the last flip-flops of Ω_j, then $S_j = \beta - \alpha + 1$. The *span S of the circuit* is defined as the maximum of spans of all its cones. For the above type of faults, it is sufficient to apply all possible patterns to each set of S consecutive flip-flops of the scan chain to guarantee coverage of all faults.

4.2 Methodology to design efficient LT-RTPG

The number of transitions that occur at state inputs during scan shifting is determined by only the number of taps of the LT-RTPG *(k = |TC|)* and independent of the tap configuration of the LT-RTPG. Test sequences generated by LT-RTPG's with more taps cause fewer transitions at state inputs. However, since consecutive state inputs are assigned identical values more frequently, test sequences generated by LT-RTPG's with more taps typically attain lower fault coverage than those generated by LT-RTPG's with fewer taps.

Hence, if high fault coverage is desired, then an LT-RTPG with k=2 should be used to obtain a less reduction in heat dissipation compared to the case when k=3. Here, k =2 is used since mixed-mode testing targeting hard faults to compensate for the reduction in fault coverage is not implemented. In order to design an LT-RTPG that can achieve high fault coverage, three key facts should be considered.

Firstly, the tap configuration of the LT-RTPG should be carefully selected. A tap configuration that has the longest forbidden sequence, *i.e.,* a tap configuration of the form *TC = {Qi, Qi+a}* or *TC = {Qi', Qi+a')*, should be selected.

Secondly, the separation between the first and last tapped stage of the LFSR within the LT-RTPG should be greater than or equal to *S' (S' = S − 1)*. If the separation between the first and last tapped stages is less than *S'*, the LT-RTPG cannot generate all possible 2^S distinct patterns at S consecutive stages of the scan chain, even if an LFSR with infinite stages is used.

Finally, the LFSR that is used in an LT-RTPG to generate random sequences should have at least 2S' stages to generate all possible 2^S distinct patterns at any *S* consecutive stages of the scan chain.

4.3 LT-RTPG Results and Comparison

The above design has been implemented and results have been obtained for different ISCAS89 benchmark circuits using HOPE fault simulator. For all benchmark circuits, single scan chain has been assumed here for convenience though the scheme is perfectly valid for multiple scan chains as well and have been successfully worked out. Also, single stuck-at fault model is used. Number of transitions have been calculated using C program.

Table I shows comparison of fault coverage between LT-RTPG and conventional LFSR BIST. Here, PI is number of primary inputs, SI is number of secondary inputs (F/Fs) present in circuit.

Table 1. Fault Coverage

Circuit	SI	% FC LFSR	% FC LT-RTPG	% Reduction
s298	14	100	96.48	3.52
s344	15	100	99.73	0.27
s349	15	99.23	99.12	0.11
s386	6	98.93	97.13	1.89
s953	29	98.90	96.16	2.27
s1423	74	98.78	97.65	1.14
s1512	57	93.48	92.12	1.45
s3271	116	99.03	98.32	0.72
s3330	132	92.68	91.54	1.23
s5378	179	96.80	95.79	1.04

Table 2. Number of Transitions (NT)

Circuit	NT LFSR	NT LT-RTPG	% Reduction
s298	1	.72	28
s344	1	.69	31
s349	1	.73	27
s386	1	.70	30
s953	1	.65	35
S1423	1	.67	33
S1512	1	.76	24
S3271	1	.64	36
S3330	1	.69	31
S5378	1	.65	35

Table 3. Signature Analysis

Ckt	PO	SI	MISR Stage	Signature	Active Fault	Faulty Signature
s27	1	3	4	1010	G8/$_1$	1100
s298	6	14	7	1101010	G50/$_0$	0010011
s386	4	6	5	11101	G25/$_1$	11001

As is expected, there is a reduction in fault coverage which is the disadvantage of LT-RTPG BIST (this is almost overcome by employing combined approach discussed in next section). But, Table II depicts the clear advantage of this scheme for which it is used. It clearly shows very high reduction in the number of transitions (NT) that varies from 11-35% which is quite good. Table III depicts the signature analysis of this complete BIST scheme. It is shown only for some smaller circuits requiring less number of MISR stages since for others the MISR stages and hence length of signature is quite large (>10 bits) to represent here. Also, fault activation is demonstrated by randomly

378

activating a fault at some internal node of the circuit (e.g. G50/0 indicates node G50 stuck-at-0) which results into a faulty signature.

5. PROPOSED SCHEME

5.1 A Combined Approach

It is seen that in the above LT-RTPG, due to assigning of identical values to nearby inputs to reduce transitions degrades the uniformity of distribution. Hence, faults that require neighboring inputs to be assigned opposite values for detection may escape most test patterns. This is what results into the degradation in fault coverage.

Now, in any scan-based BIST, the entire sequential circuit is, in a way, broken down into the combinational part and the flip-flops or the memory elements forming scan chains. LT-RTPG was a true *test-per-scan* BIST where during the test (scan-shift) mode, the scan in data is fed to the scan chain and no inputs are changed at the primary inputs and hence no activity takes place in the combinational part of the circuit during the test mode. But, there are all possible chances of several faults being detected into the combinational part and their effects being captured at some of the primary outputs if the combinational part were to be active during test mode.

This is the observation that forms the basis of a combined approach which combines both conventional *test-per-scan* and *test-per-clock* BIST schemes.

Here, while the scan inputs are scanned into the scan chains from the output of T flip-flop of LT-RTPG during the test mode, simultaneously, primary inputs are also supplied with changing inputs taken from the various different tappings of the LFSR stages. This results into active combinational part during test mode as well and hence making detectable faults propagate to one of the primary outputs. Thus, this results into achieving a desired FC faster with a similar hardware overhead comparing to conventional *test-per-scan* scheme. The price paid here, is the increase in the switching activity during the test mode to some extent compared to LT-RTPG test-per-scan. But when compared with conventional LFSR *test-per-scan*, it certainly results into lesser number of overall switching activity as is demonstrated by the results in the next section.

5.2 Partial Scan Incorporation

It has been shown [10] that scanning a flip-flop may actually deteriorate the observability of its data input. In the above *combined approach*, the values at the primary outputs are observed in *every clock cycle* but the values appear at the data input of a scan flip-flop are captured only *once in every m+1 clock cycles*. Therefore, the observability of the data input of a scan flip-flop is much smaller than that of a primary output, especially when the scan chain is long. On the other hand, a non-scan flip-flop will capture the response at its data input in every clock cycle. Once a fault effect is captured, if the observability of the non-scan flip-flop is reasonably high, the captured fault effect may then be propagated to the primary outputs so that faults can be detected earlier. As a result, it may be beneficial not to scan a flip-flop with reasonably high observability through the primary outputs to increase the overall random testability of the circuit.

Because resolving the lack of observability of a scan flip-flop is of major concern, it is easier to address this problem by removing flip-flops from the scan chains of a full-scan circuit and then evaluate the testability impact due to the removal of the flip-flops. Removing flip-flops from the scan chains affects the observability as well as the controllability of the circuit. Attempting to increase the observability by unscanning a flip-flop may not justify the loss of its controllability. Therefore, a cost function which reflects both controllability and observability of the circuit is needed. Here, the cost functions and the testability defined in [10] are used. Also, the non-scan flip-flop identification algorithm of [10] is implemented.

After computing the testability, those flip-flops that satisfy the following two conditions are considered as candidates for removing from the scan chains.

1) They will *not* create global feedback loops if not scanned (a global feedback loop is a loop containing at least two non-scan flip-flops).

2) Their corresponding PSIs have observabilities greater than a given observability threshold.

Note that unscanning those flip-flops, which create global feedback loops, could drastically reduce the testability of the circuit. Therefore, it is preferable to exclude them from further consideration.

6. EXPERIMENTAL RESULTS

The final results, different analysis and comparisons for this efficient BIST design have been presented in this section.

Experimental results have been obtained for different ISCAS89 benchmark circuits using HOPE fault simulator. As above, here also, for all benchmark circuits, single scan chain has been assumed here for convenience though the scheme is perfectly valid for multiple scan chains as well and have been successfully worked out. Also, single stuck-at fault model is used. Number of transitions have been calculated using C program.

Here, analysis has been done for conventional LFSR based test-per-scan BIST, existing LT-RTPG based test-per-scan BIST, modified LT-RTPG based BIST combining both *test-per-scan* and *test-per-clock* BIST schemes (which is quoted as "combined approach" in all the results) and finally, an efficient BIST scheme built on top of a partial scan circuit using a combined approach adding above LT-RTPG as the TPG and MISR as signature analyzer (quoted as "Partial Scan" in all the results).

Table IV shows the fault coverage comparision for same number of test cycles. It is seen that LT-RTPG test-per-scan (as also shown in Table I) has a degradation in fault coverage compared to conventional LFSR scheme but the combined approach gives about 0-4 % increase in fault coverage which is quite significant otherwise it is very difficult to increase fault coverage even by one percent beyond 95% FC (that sometimes even doubles the number of cycles required to increase by just 1%). This is further improved significantly for most of the circuits by incorporating partial scan as shown by the last column. Table V shows the test length comparison of different circuits for similar fault coverage.

Table 4. Fault Coverage Comparison *for same number of test cycles*

Circuit	# of FFs	% FC Conventional LFSR	% FC LT-RTPG (full scan)		Partial Scan	
			Test-per-scan	Combined Approach	# of scan FFs	% FC
s298	14	100	96.48	97.40	10	98.01
s344	15	100	99.73	99.73	12	99.72
s349	15	99.23	99.12	99.56	09	99.68
s386	6	98.93	97.13	99.01	06	99.21
s953	29	98.90	96.16	98.42	07	98.64
s1423	74	98.78	97.65	99.21	72	99.43
s1512	57	93.48	92.12	94.23	41	95.24
s3271	116	99.03	98.32	99.34	92	99.32
s3330	132	92.68	91.54	95.42	104	97.74
s5378	179	96.80	95.79	98.46	174	98.12

Table 5. Test Length Comparison *for same fault coverage*

Circuit	# of FFs	% FC Conventional LFSR	% FC LT-RTPG (full scan)		Partial Scan	
			Test-per-scan	Combined Approach	# of scan FFs	% FC
s298	14	100	96.48	97.40	10	98.01
s344	15	100	99.73	99.73	12	99.72
s349	15	99.23	99.12	99.56	09	99.68
s386	6	98.93	97.13	99.01	06	99.21
s953	29	98.90	96.16	98.42	07	98.64
s1423	74	98.78	97.65	99.21	72	99.43
s1512	57	93.48	92.12	94.23	41	95.24
s3271	116	99.03	98.32	99.34	92	99.32
s3330	132	92.68	91.54	95.42	104	97.74
s5378	179	96.80	95.79	98.46	174	98.12

It shows a considerable reduction between 17-50% for the combined approach compared to test-per-scan alone (both using LT-RTPG as TPG). Once again, this is also improved further for all circuits giving about 18-79% reduction in the test length which ultimately reduces the test application time. Table VI shows the disadvantage of this combined approach over the LT-RTPG test-per-scan scheme producing comparatively higher number of transitions which is the price paid for better performance in terms of fault coverage and test length and hence application time. But, upon comparing the second and the last column, it is quite clear that partial scan does not compromise with the increase in number of transitions and has almost similar reduction in the number of transitions which is very satisfactory.

Also, it is to be noted that Table VI shows percentage reduction in the number of transitions compared to conventional LFSR and hence, afterall, all these have an advantage in terms of number of transitions which ultimately results into low heat dissipation.

7. CONCLUSION

In this paper, it is shown that the combined test-per-scan and test-per-clock scheme when used with LT-RTPG as the TPG, there is a slight increase in the number of transitions which seems to contradict the main objective of low dissipation during testing of VLSI chips. But, this increase, to a large extent, is compensated by the reduced test lengths and hence lesser test application time. Also, area overhead is reduced since now primary inputs (PIs) are not a part of scan chain and hence extra scan flip-flops are not required. Moreover, comparatively lesser transitions takes place into the scan chain since the length of scan chain is reduced due to exclusion of PIs from being a part of scan chain.

Furthermore, it is shown that there is considerable improvement in the test quality in all respects with the incorporation of parital scan into the circuits. This is because partial scan incorporation takes optimal advantage of the combination of both the approaches, i.e. the combined test-per-scan and test-per-clock.

Table 6. Number of Transitions Comparison *for same fault coverage*

Circuit	% Reduction in no. of Transitions (NT) compared to conventional LFSR test-per-scan		
	LT-RTPG (full scan)		Partial Scan
	Test-per-scan	Combined Approach	
s298	28	22	26
s344	31	27	29
s349	27	21	29
s386	30	22	28
s953	35	27	31
s1423	33	30	32
s1512	24	21	21
s3271	36	27	30
s3330	31	21	34
s5378	35	24	35

This gives quite reasonable and satisfactory fault coverage, test lengths with less test application time and lesser number of transitions implying reduced heat dissipation during testing.

Thus, partial scan circuit with a combined test-per-scan and test-per-clock approach using LT-RTPG as a TPG makes a simple yet efficient low power scan-based BIST test scheme.

Decrease in fault coverage due to use of LT-RTPG can be reduced even further by employing scan chain ordering such that compatible state inputs are placed in consecutive positions of scan chains. In such cases, fault coverage does not decrease even if adjacent inputs are always assigned identical values, as is the case with LT-RTPG. Here, power dissipation will not be affected by scan chain ordering.

8. ACKNOWLEDGEMENTS

I would like to thank Dr. Dipankar Nagchoudhuri for continuously helping out with this work which is a part of my thesis work under his guidance. I am also grateful to Dr. Vishwani Agrawal and Dr. C.P.Ravikumar for giving valuable suggestions and sharing some of their helpful slides with me. Finally, I would like to thank Dr. Dong Xiang of Tsinghua University, Beijing, China for his valuable online help and suggestions.

9. REFERENCES

[1] V. Agrawal and M. Bushnell, *Essentials of Electronic Testing For Digital, Memory and Mixed-Signal VLSI Circuits,* Kluwer Academic Publishers, 2000.

[2] O. Sinanoglu and A. Orailoglu, "A Novel Scan Architecture for Power Efficient, Rapid Test," *IEEE/ACM International Conference on Computer Aided Design,* 2002.

[3] S. Wang, "Low Hardware Overhead Scan based 3-Weight Weighted Random BIST," *IEEE Proc. International Test Conference,* 2001.

[4] N. Touba and E. McCluskey, "Altering a Pseudo-Random Bit Sequence for Scan-Based BIST," *IEEE Proc. International Test Conference,* 1996.

[5] S. Wang, "Generation of Low Power Dissipation and High Fault Coverage Patterns for Scan-Based BIST," *IEEE Proc. International Test Conference,* 2002.

[6] B. Bhattacharya, S. Sheth and S. Zhang, "Low-Energy BIST Design for Scan-based Logic Circuits," *Proc. Of VLSI Design,* 2003.

[7] S. Hellebrand, S. Tarnik and J. Rajski, "Generation of Vector Patterns through Reseeding of Multiple-Polynomial LFSR," *IEEE Proc. International Test Conference,* 1992.

[8] H. Tsai, K. Cheng and S. Bhawmik, "On Improving test quality of Scan-based BIST," *IEEE Trans. Computer-Aided Design Integr. Circuits Systems,* 2000.

[9] S. Wang and S. Gupta, "LT-RTPG: A New Test-Per-Scan BIST TRG for Low Heat Dissipation," *IEEE Proc. International Test Conference,* 1999.

[10] H. Tsai, S. Bhawmik, K. Cheng, " An Almost Full-Scan BIST Solution – Higher Fault Coverage and Shorter Test Application Time," *IEEE Proc. International Test Conference,* 1998.

Modeling and Analysis of Leakage Induced Damping Effect in Low Voltage LSIs

Jie Gu, John Keane, Chris Kim

University of Minnesota, 200 Union Street S.E., Minneapolis, MN, 55455
Phone: (612) 625-1029, jiegu@umn.edu

ABSTRACT

Although there has been extensive research on controlling leakage power, the fact that leaky transistors can act as a damping element for supply noise has been long ignored or unnoticed in the design community. This paper investigates the leakage induced damping effect that helps suppress the supply noise. By developing physics-based impedance models for active and leakage currents, we show that leakage, particularly gate tunneling leakage, provides more damping than strong-inversion current. Simulations were performed in a 32nm CMOS technology to validate our models under PVT variations and to explore the voltage dependent behavior of this phenomenon. Design example utilizing leakage induced damping such as decap assignment is discussed with results showing 15.6% saving in decap area.

Categories and Subject Descriptors

B.7.2 [Hardware]: Integrated Circuits – Design Aides.

General Terms

Management, Performance, Design,

Keywords

Damping effect, subthreshold leakage, gate leakage, supply noise

1. INTRODUCTION

With aggressive CMOS scaling, the power density of microprocessors is increasing exponentially towards that of a nuclear reactor [1]. Power supply noise which includes IR and Ldi/dt components has become a major design concern due to the ever-increasing current density, higher switching speeds and reduced operating voltages with technology scaling. Uncontrolled supply noise threatens circuits in nanoscale technologies by causing problems such as timing violations, reduced noise margin in storage circuits, substrate noise coupled into analog devices, and reliability issues due to hot carrier injection.

One challenging task in designing robust power supply networks is the damping of supply noise, particularly at the resonant frequency determined by the bonding wire inductance and on-chip capacitance [2]. Resonant supply noise can be excited even by weak sub-harmonics of the clock signal [3]. Without sufficient damping, this noise can result in a severe degradation of circuit

Permission to make digital or hard copies of all or part of this work for personal or classroom use is granted without fee provided that copies are not made or distributed for profit or commercial advantage and that copies bear this notice and the full citation on the first page. To copy otherwise, or republish, to post on servers or to redistribute to lists, requires prior specific permission and/or a fee.
ISLPED '06, October 4–6, 2006, Tegernsee, Germany.
Copyright 2006 ACM 1-59593-462-6/06/0010...$5.00.

performance due to the large impedance of the supply network at the resonant frequency.

A common method for suppressing resonant noise is to lower the ac impedance of the supply network by adding a large amount of decoupling capacitors (decaps) [4]. However, this approach has limitations such as the significant increase in gate leakage, which has already been reported to account for 15% to 20% of the total power in current microprocessors [5]. Another commonly used method to suppress resonant noise is adding passive or active resistors in series or parallel to the supply and ground networks to provide more damping [6-7]. The drawback of adding resistors in series with the power line is the increase in IR droop, while adding resistors between V_{dd} and Gnd will introduce more static current. Circuit techniques have also been proposed to deal with the supply noise issue. Ang demonstrated a switched decoupling capacitor circuit to boost the effectiveness of decaps when resonant noise is excited [8]. Rahal-Arabi proposed a clock/data compensation scheme for resonant noise tolerance where extra timing margin is obtained by matching the clock delay with the circuit delay [9]. This scheme relies on the fact that for a sudden current spike, the supply voltage starts to oscillate at the resonant frequency and lasts a few clock cycles.

Leakage power consumption, including both subthreshold leakage and gate leakage, has become the major barrier for the continued scaling of CMOS devices. According to the International Technology Roadmap for Semiconductors (ITRS), leakage power consumption will take up as much as 50% of the total power consumption in deeply scaled CMOS technologies [10]. However, a commonly ignored fact is that a device conducting leakage current between V_{dd} and Gnd acts as a resistor (linear or nonlinear) that can help damp the supply noise. As will be shown in the paper, ignoring this effect can lead to a pessimistic power supply network design resulting in wasted area and power.

In this work, we study, model, and verify the damping effect induced by on-chip current components including both active and leakage current. To the best of our knowledge, the only previous publication that examined this phenomenon is [11]. However, no device level modeling was provided, and the simulation was only performed on a voltage-controlled resistor instead of real CMOS transistors. Here we have studied and examined this damping phenomenon in great detail. Simulations in this paper are based on a 32nm Predictive Technology Model with a supply voltage of 0.9V [12]. The contributions of this paper are as follows:

- For the first time, simple yet accurate damping models for active current, subthreshold leakage, and gate leakage are provided and verified by simulations.
- The voltage dependent behaviors of leakage induced damping effects are explored under Process-Voltage-Temperature (PVT) variations.

- Design issues such as decap assignment and supply mesh planning are discussed and simulation results are provided to support our conclusions.

The rest of this paper is organized as follows. In section 2, a power supply network model is introduced with the various resistance components. In section 3, physical models of damping resistance for various current components are derived. Section 4 shows simulations on damping effects under PVT variations. Section 5 discusses the associated circuit design issues. Finally, conclusions are drawn in section 6.

2. POWER SUPPLY MODELING AND RESONANT DAMPING

Fig.1 shows the *RLC* model of a power supply mesh with the supply noise spectrum. The peak response at f_{res} in Fig. 1(b) indicates the resonant frequency of the supply network, which is equal to $1/(2\pi\sqrt{LC})$, where L is the inductance of the bonding wire and C is the on-chip capacitance. Although on-chip power grids also introduce parasitic inductance which accounts for the small peaks at higher frequencies, the noise at f_{res} is a magnitude higher than the others. As a result, much of the design effort is put into suppressing the supply noise at the resonant frequency f_{res}.

A popular method used to reduce the resonant noise is to provide sufficient damping by adding more resistance to supply network. Fig.1 (b) shows the noise level before and after adding a damping resistor to the power lines. The resonant noise is shown to be greatly reduced with the additional damping. As mentioned, the drawback of adding more damping is increased *IR* droop and additional power consumption.

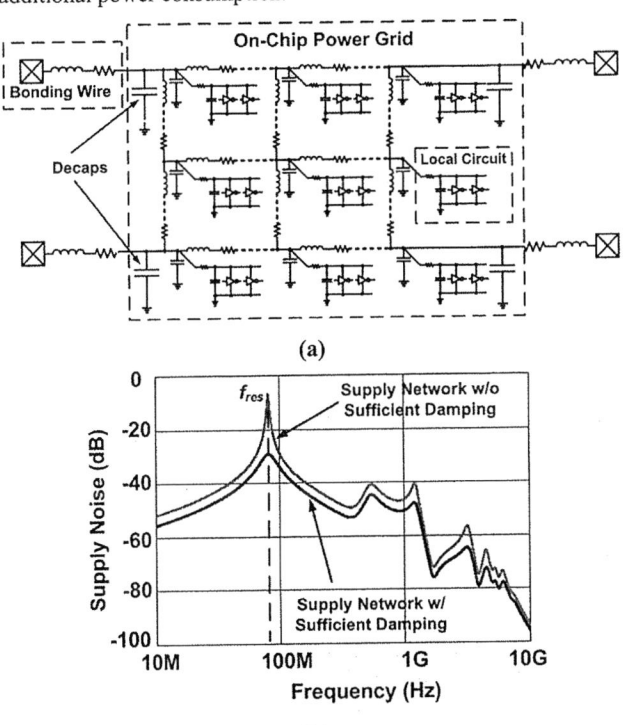

(a)

(b)

Fig. 1: (a) Model of power supply mesh; (b) Simulated frequency spectrum of supply noise from the model in (a).

Fig. 2(a) shows a simplified supply network model used in this paper. Fig. 2(b) shows the corresponding impedance models including bond wire inductance L, on-chip capacitance C, supply network resistance R_{wire-p}, and circuit resistance $R_{circuit}$. Note that for simplicity of calculation, the series connected wire resistance R_{wire} in Fig. 2(a) can be converted into a parallel resistor using equation $R_{wire-p}=R_{wire}\cdot(1+Q^2)$ where Q is the Q-factor given by

$$Q=\frac{1}{R_{wire}}\cdot\sqrt{\frac{L}{C}}.$$ The circuit resistor $R_{circuit}$ comes from the

switching and non-switching devices which can have active current I_{on}, subthreshold leakage I_{sub}, or gate leakage I_{gate}. The resistance of each current component is equivalent to the inverse of the slope of the I-V curves in Fig. 2(b) at the nominal operating point. Inspecting the I-V curves of each current component, we find that the leakage current is highly nonlinear to supply voltage. Thus, the equivalent resistance value of leakage current cannot be treated as a constant. Furthermore, the fact that the gate leakage rises most rapidly as supply voltage increases, indicates substantial damping induced by gate leakage. The above observations tell us that the existence of $R_{circuit}$ can no longer be ignored and a correct model of each resistance component will be important for designing efficient on-chip power supply networks.

Simplified Power Supply Network

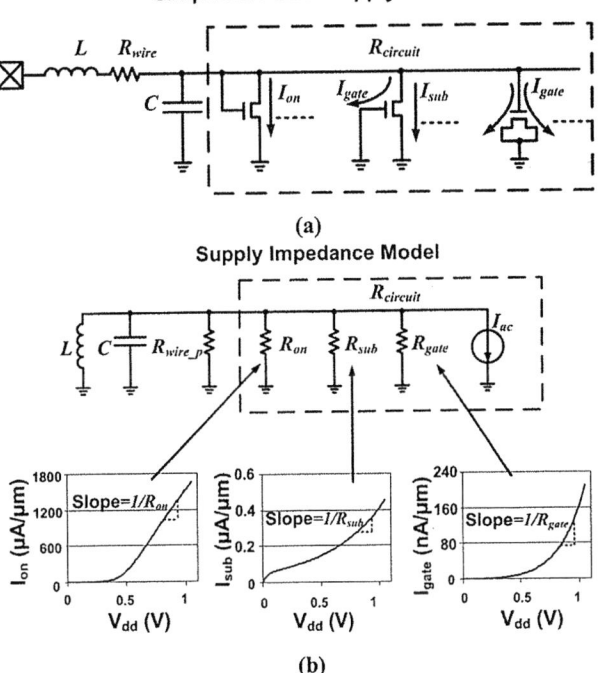

(a)

Supply Impedance Model

(b)

Fig. 2: (a) Simplified power supply model used in this paper; (b) Supply impedance model including the resistance of the circuit itself. Each current component has different damping effects due to the difference in equivalent resistance.

Given the impedance model in Fig. 2(b), the power supply noise at a particular frequency can be calculated as the multiplication of operating current at that frequency and the corresponding supply network impedance. At the resonant frequency, the impedance becomes $R_{wire-p}//R_{circuit}$ since the inductance and capacitance completely cancel each other out. The smaller this resistance is, the more damping is provided and the less supply noise we will have. Because of the small value of R_{wire} (or large corresponding R_{wire-p}), the damping provided from the power network wiring is not sufficient. This means that additional damping has to be added to suppress the resonant noise. $R_{circuit}$ that comes from the existing circuits can actually provide this additional damping requirement. The goal of this paper is to develop a proper model for $R_{circuit}$ and investigate its variation under different conditions. It is also important to mention that the damping effect provided by R_{wire-p} and $R_{circuit}$ are effective for Ldi/dt noise at all frequencies as shown in Fig. 1(b) rather than for only that at f_{res}. However, since the supply noise is most pronounced at the resonant frequency, we will focus on the damping effect at f_{res} in this paper.

3. MODELING THE DAMPING EFFECT FOR ACTIVE AND LEAKAGE CURRENT

In this section, we discuss the modeling of equivalent resistance and conductance for various on-chip current components, including active current I_{on}, subthreshold leakage I_{sub} and gate leakage I_{gate}. Because supply noise is typically controlled below 15% of V_{dd}, the damping conductance will be calculated based on a small-signal analysis; i.e. the dI/dV_{dd} value will be derived at the nominal V_{dd} point, where I can be one of the following: I_{on}, I_{sub} and I_{gate}. The equivalent damping resistance is then simply the inverse of the calculated conductance.

3.1 Damping Model for Active Current

Active current can be modeled as:

$$
\begin{aligned}
I_{on} &= \mu_e \cdot C_{ox} \cdot \frac{W}{L} \cdot (V_{gs} - V_{th})^\alpha \\
&\approx \mu_e \cdot C_{ox} \cdot \frac{W}{L} \cdot (V_{dd} - (V_{th0} - \lambda V_{dd}))
\end{aligned}
\tag{1}
$$

using the alpha-power law current equation from [13], with $\alpha \approx 1$. In this equation, V_{th0} is the zero-biased threshold voltage, and λ is the Drain Induced Barrier Lowering (DIBL) coefficient. From (1), we can calculate the damping conductance as:

$$
\begin{aligned}
dI_{on}/dV_{dd} &= \mu_e \cdot C_{ox} \cdot \frac{W}{L} \cdot (1 + \lambda) \\
&= \frac{\overline{I_{on}}}{\overline{V_{dd}} - (V_{th0} - \lambda \overline{V_{dd}})} \cdot (1 + \lambda) \\
&= \frac{(1 + \lambda)}{(1 + \lambda)\overline{V_{dd}} - V_{th0}} \cdot \overline{I_{on}}
\end{aligned}
\tag{2}
$$

where $\overline{V_{dd}}$ is the nominal supply voltage and $\overline{I_{on}}$ is the nominal active current at $\overline{V_{dd}}$. A parameter denoted with a bar in this paper refers to the dc bias value, i.e. the operating point for small signal analysis. The V_{gs} ($=V_{dd}$) term in the current equation and DIBL contributes to the damping conductance. Equation (2) illustrates that the damping conductance of the active current is a constant determined only by its operating points $\overline{V_{dd}}$ and $\overline{I_{on}}$.

3.2 Damping Model for Subthreshold Leakage Current

Subthreshold leakage current can be described as:

$$
\begin{aligned}
I_{sub} &= \mu_e \cdot C_{ox} \cdot \frac{W}{L} \cdot \left(\frac{kT}{q}\right)^2 \cdot (m-1) \cdot e^{q(0-V_{th})/mkT} \\
&= \mu_e \cdot C_{ox} \cdot \frac{W}{L} \cdot \left(\frac{kT}{q}\right)^2 \cdot (m-1) \cdot e^{-q(V_{th0} - \lambda V_{dd})/mkT}
\end{aligned}
\tag{3}
$$

where m is the body effect coefficient and kT/q is the thermal voltage. The derivative of I_{sub} is then calculated to be:

$$
\begin{aligned}
dI_{sub}/dV_{dd} &= \frac{q\lambda}{mkT} \cdot I_{sub} \\
&\approx \frac{q\lambda}{mkT} \cdot \left(1 + \frac{q\lambda}{mkT} \cdot \Delta V_{dd}\right) \cdot \overline{I_{sub}}
\end{aligned}
\tag{4}
$$

where $\overline{I_{sub}}$ is the nominal leakage current at $\overline{V_{dd}}$ and ΔV_{dd} is the power supply noise. The finite damping conductance for subthreshold leakage mainly comes from the DIBL effect. Unlike active current, dI_{sub}/dV_{dd} has to be modeled as a function of supply noise ΔV_{dd} because of the nonlinear exponential relationship between I_{sub} and V_{dd}. The linearization of I_{sub} with respect to ΔV_{dd} in (4) is performed using the Taylor expansion around the nominal supply voltage $\overline{V_{dd}}$. Similar to the case for active current, equation (4) shows that the damping conductance from subthreshold leakage is proportional to the dc leakage current $\overline{I_{sub}}$. However, the damping conductance of subthreshold leakage is a linear function of ΔV_{dd}, so its value changes with supply voltage. This voltage dependent resistance behavior provides additional damping performance.

3.3 Damping Model for Gate Leakage Current

The gate leakage current can be expressed as:

$$
I_{gate} = W \cdot L \cdot A \cdot \frac{V_{dd}^2}{T_{ox}^2} \cdot \exp\left(\frac{-B\left(1 - \left(1 - \frac{V_{dd}}{\phi_{ox}}\right)^{3/2}\right)}{\frac{V_{dd}}{T_{ox}}}\right)
\tag{5}
$$

where T_{ox} is the oxide thickness and ϕ_{ox} is the barrier height for tunneling electrons or holes [14]. A and B are given as:

$$
A = \frac{q^3}{16\pi^2 \eta \phi_{ox}} \quad, \quad B = \frac{4\sqrt{2m^*}\phi_{ox}^{3/2}}{3\eta q}.
$$

where m^* is the effective mass of tunneling electrons or holes and $\hbar$ is the reduced form of Plank's constant.

By performing a first order Taylor expansion on the derivative of I_{gate} and ignoring higher order terms, we find:

$$
dI_{gate}/dV_{dd} = \left(\frac{2}{\overline{V_{dd}}} + C + \left(\frac{2}{\overline{V_{dd}}^2} + \frac{4}{\overline{V_{dd}}} \cdot C\right) \cdot \Delta V_{dd}\right) \cdot \overline{I_{gate}}
\tag{6}
$$

where $C = \left(1 - \sqrt{1 - \frac{\overline{V_{dd}}}{\phi_{ox}}} \cdot \left(1 + \frac{\overline{V_{dd}}}{2\phi_{ox}}\right)\right) \cdot \frac{T_{ox} \cdot B}{\overline{V_{dd}}^2}$.

Although (6) requires a more involved calculation, the behavior of damping from gate leakage is similar to subthreshold leakage, in the sense that it is proportional to the dc bias current, and is dependent on the supply noise ΔV_{dd}.

3.4 Damping Effect Comparison for I_{on}, I_{sub}, and I_{gate}

Fig. 3 shows the current and damping conductance of each component (I_{on}, I_{sub}, I_{gate}) as a function of supply voltage. For the purpose of comparison, results are shown when the three current values are set to be the same at the nominal condition (V_{dd}=0.9V). This figure shows that for an equal amount of current, gate leakage provides a much stronger damping effect than the other two current components. Our derivation models the effective conductance of subthreshold and gate leakage as linear function of V_{dd} as opposed to a constant value derived for the on-current. This is verified in Fig. 3(b) where the damping conductance increases almost linearly with supply voltage for both subthreshold and gate leakage.

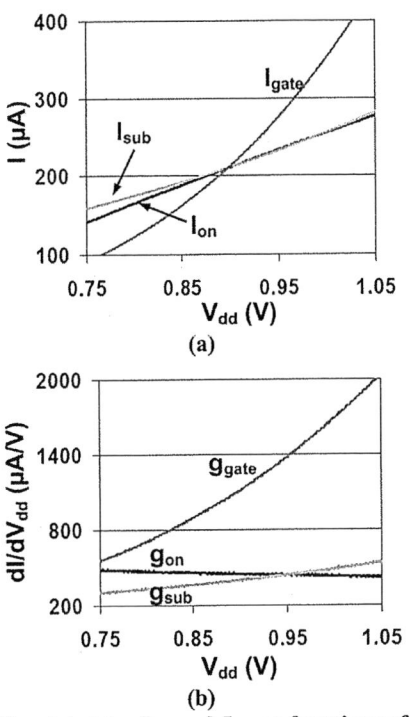

Fig. 3: (a) Simulated I_{on}, I_{sub} and I_{gate} as functions of V_{dd}; (b) Corresponding conductance of each current component.

Table 1 summaries the effective damping conductance for each current component. The coefficients calculated using the developed models are compared with the coefficients obtained from Hspice simulation. Our damping conductance models using physical parameters show a good fit with simulation results, which indicate that the developed model is capable of accurately capturing the behavior of the current induced damping effects. Small discrepancies between our model and the simulation exist due to the following reasons: (a) the α parameter in equation (1) is actually between 1 and 1.2 which leads to a smaller I_{on} conductance value predicted by the model; (b) the parasitic source/drain resistance in MOSFET device accounts for the slight drop on g_{on} in Fig. 3(b) but is not modeled in our simplified current equation; (c) the modeling of gate tunneling leakage for an ultra-thin oxide device involves a more complex mathematical representation and therefore an error exists when using the simple equation in (5) to obtain the solution. However, it is important to realize that the purpose of this modeling is to explore the physical principle of the leakage induced damping effect and to provide a simple closed-form equation for damping estimation. Hence the advantage of this model is that for a given leakage current consumption, the damping resistance can be easily obtained from

table 1. For improved accuracy, extensive simulations can be performed at the cost of computation time.

Table 1. Comparison of the current induced damping conductance from derived model and Hspice simulation.

	Derived Model	Hspice Simulation
$g_{on}=1/R_{on}$	$1.80 \cdot \overline{I_{on}}$	$2.13 \cdot \overline{I_{on}}$
$g_{sub}=1/R_{sub}$	$(1.80+3.24\,\Delta V_{dd}) \cdot \overline{I_{sub}}$	$(1.92+3.55\,\Delta V_{dd}) \cdot \overline{I_{sub}}$
$g_{gate}=1/R_{gate}$	$(5.51+17.1\,\Delta V_{dd}) \cdot \overline{I_{gate}}$	$(5.22+19.6\,\Delta V_{dd}) \cdot \overline{I_{gate}}$

Below we evaluate the effect of voltage dependent conductance. At resonant frequency f_{res}, the supply noise is calculated as:

$$\Delta V_{dd} = I_{ac} \cdot R = I_{ac}/(g_0 + A \cdot \Delta V_{dd}) \qquad (7)$$

where I_{ac} is the exciting current at frequency f_{res}. g_0 and A are the constant term and linear coefficient in our resistive model in table 1; i.e. $g = g_0 + A \cdot \Delta V_{dd}$. Solving (7), we find the supply noise as:

$$\Delta V_{dd} = \frac{-g_0 + \sqrt{g_0^2 + 4AI_{ac}}}{2A} \qquad (8)$$

By using this expression, for a ΔV_{dd} of 0.1V, the voltage dependent term $A \cdot \Delta V_{dd}$ for subthreshold leakage provides an additional 14.5% reduction in power supply noise compared with a constant conductance g_0. Similarly, the voltage dependent term in gate leakage provides an extra 22% reduction in supply noise. In general, a larger A and larger noise amplitude make the voltage dependent resistance more beneficial for supply noise damping.

Fig. 4 shows the simulated transient damping performance at resonant frequency for different current components. Note that for subthreshold leakage test, a small positive gate biased is applied to amplify the subthreshold leakage so that the gate leakage becomes negligible in proportion. The results in Fig.4 are consistent with our resistive model given in table 1. Each current component provides significant damping effect with gate leakage being the strongest.

4. LEAKAGE INDUCED DAMPING EFFECT UNDER PVT VARIATION

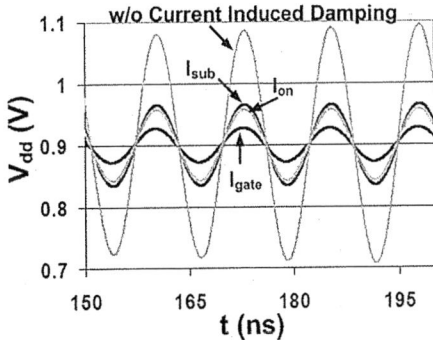

Fig. 4: Supply resonant noise damped by different current components. Gate leakage provides largest damping effect for the same amount of current.

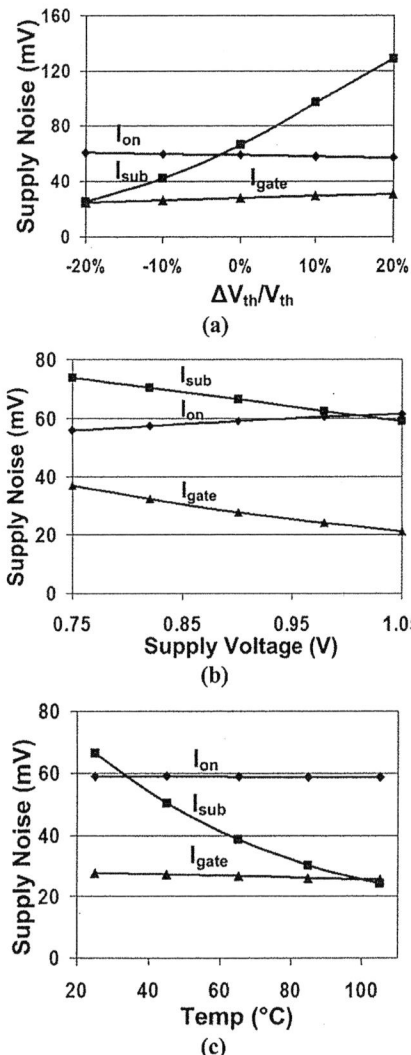

Fig. 5: Simulated damping effects under PVT variation. (a) Threshold voltage variation; (b) Supply voltage variation; (c) Temperature variation. Note that larger damping effect leads to smaller supply noise.

PVT variation has become an increasingly critical issue in deeply scaled LSI circuits. Due to the fact that different leakage components have different behaviors under PVT variation, it is necessary to evaluate the leakage induced damping effects under various operating conditions.

Fig. 5 compares the resonant supply noise for each damping current component while varying the PVT parameters. As before, all current components have equal current amplitude at the nominal condition (V_{th} at typical corner, 0.9V supply voltage, and room temperature). Fig. 5(a) shows the impact of threshold voltage variation on damping effects of each current component. The figure shows that the damping effect of subthreshold leakage varies the most under V_{th} variation because of its exponential dependency. This observation indicates that transistors in a fast corner die (lower V_{th}) will introduce a larger damping effect, which in turn can compensate for the increased IR and Ldi/dt droop due to larger transient currents. Fig. 5(b) shows the impact of supply voltage variations on damping effects. Both

subthreshold leakage and gate leakage show an increased damping performance as V_{dd} rises while damping from active current is slightly reduced. This observation is consistent with Fig. 3(b) and can be explained by the models developed in the previous section. Fig. 5(c) shows the temperature dependency of the damping effects from each current component. With the increase in temperature, the damping from subthreshold leakage increases dramatically, resulting in a significantly reduced noise level as seen in the figure. By inspecting the subthreshold damping model in equation (4), we find that the damping coefficient actually decreases with temperature. However, the dc leakage current $\overline{I_{sub}}$ increases exponentially with temperature, and therefore the overall damping performance by subthreshold leakage increases. This positive temperature dependency can potentially compensate the slow-down of circuits at high temperatures.

Fig. 6 shows the comparison of transient noise waveforms at 25°C and 110°C. We used an ISCAS85 benchmark circuit (C3540, 8-bit ALU) clocked at 640MHz for generating the supply noise. Six additional circuit blocks of the same kind are placed in idle mode to provide the subthreshold and gate leakage. To excite the resonant noise at 80MHz (=640MHz/8), a large capacitive load is driven by a divided clock that is 8 times slower than the system clock. Fig. 6 confirms the increased damping effect at a higher temperature due to the increase in subthreshold leakage as explained in Fig. 5(c).

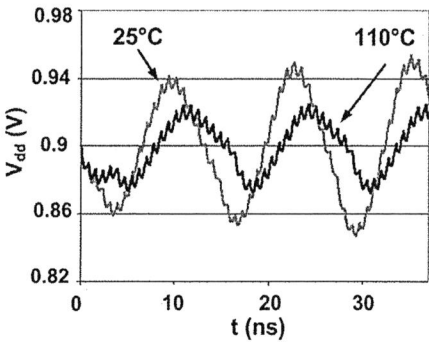

Fig. 6: Simulated noise waveforms at 25°C and 110°C for ISCAS benchmark circuits.

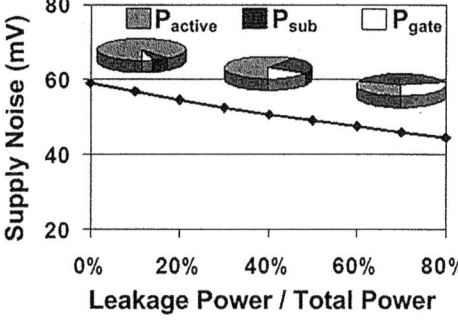

Fig. 7: Resonant supply noise versus leakage power ratio.

The overall damping effect in a realistic VLSI system depends on the ratio between leakage power and active power. For the same current amplitude, leakage currents have larger damping coefficients than active current as shown in table 1. Fig. 7 displays how the resonant supply noise changes when varying the leakage versus active current ratio. For simplicity, we assumed a fixed ratio (60%:40%) between the subthreshold and gate leakage. The figure shows increased damping and less supply noise in leakier chips. For example, a chip with 70% leakage power has 19% less resonant noise compared to a chip with 10% leakage power, due to the leakage induced damping effect.

5. DESIGN CONSIDERATIONS WITH LEAKAGE INDUCED DAMPING EFFECT

The phenomenon of leakage induced damping has normally been ignored or unnoticed in previous IC designs. However, as leakage becomes the dominant component of total power dissipation, this effect can no longer be ignored. In fact, since leakage induced damping helps to suppress the power supply noise, a design which does not consider this phenomenon can be too pessimistic. The issues associated with leakage induced damping effects include (1) design of power grid with proper resistive damping (2) decap assignment. The first issue is obvious because considering the damping effect discussed in this paper can alleviate the amount of effort of putting sufficient damping in power grid. Leakage induced damping effect also plays an important role in decap assignment. Detailed discussion with an example is given below.

In order to meet the supply noise constraints, the most common solution was to add more on-chip decaps [5]. Adding decaps is inefficient because one has to pay a huge amount of area overhead to bring down the Q factor of the RLC network which is given as $Q = \frac{1}{R} \cdot \sqrt{\frac{L}{C}}$. This is especially the case when the supply network resistance R is small. Fig. 8 shows simulation results on a medium-scale circuit with the parameters given in table 2. Decap of the original circuit is approximately 80nF causing the resonant supply noise to reach 15% of the nominal supply voltage. However, to meet a supply noise target of 8%, an additional 80nF of decap has to be added if the leakage induced damping effect is not considered. Our simulation results show that when leakage induced damping is considered, only 55nF of decap needs to be added, which results in a total decap saving of 15.6%.

Table 2. Simulation parameters for decaps assignments.

Technology	32nm CMOS	# of min. sized transistors	8.4M
P_{total}	0.4W	I_{sub}	120mA
I_{gate}	80mA	f_{res}	80MHz
R_{wire}	2mΩ	L	0.1nH
Decap w/o Leak. Damp.	160nF	Decap w/ Leak. Damp.	135nF

6. CONCLUSIONS

Leakage in modern VLSI circuits is becoming comparable to active power. A fact that has generally gone unnoticed is that on-chip leakage current can provide damping for power supply noise, especially resonant noise, which can severely degrade the circuit performance once excited. This paper studies the phenomenon of leakage induced damping in low voltage ICs where leakage and noise are the two most important design constraints. Models for the damping effect induced by the on-chip active current, subthreshold leakage, and gate leakage are developed and verified with Hspice simulations in a 32nm CMOS technology. The

impact of PVT variations on damping effect was also investigated to gain a better understanding of the damping behavior. Simulation shows that under normal conditions gate leakage provides more damping than the other two components. Damping effect of subthreshold leakage varies the most with PVT variations, and can compensate for the performance loss at high temperatures and high transient noise situations. Finally, the associated design issues are discussed. By considering the leakage induced damping effect, decap saving of 15.6% can be achieved which prevents over-design of power supply networks.

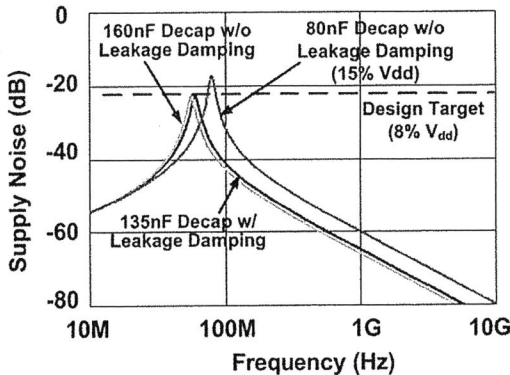

Fig. 8: Frequency diagram of supply noise w/ and w/o considering the leakage damping effects for decap assignment.

7. REFERENCES

[1] P. Gelsinger, Keynote talk at Intel Developer Forum, Feb. 2002

[2] B. Garben, etc. "Frequency Dependencies of Power Noise", *IEEE Trans. On Adv. Packaging*, vol. 25, no. 2, pp. 166-173, May 2002.

[3] E. Hailu, D. Boerstler, K. Miki, etc. "A Circuit for Reducing Large Transient Current Effects on Processor Power Grids", *Intl. Solid-State Circuits Conf.*, pp. 548-549, February 2006.

[4] N. Na, J. Choi, M. Swaminathan, J. P. Libous, etc. "Modeling and Simulation of Core Switching Noise for ASICs", *IEEE Trans On Advanced Packaging*, vol. 25, no. 1, pp. 4-11, February 2002.

[5] M. Gowan, L. Biro and D. Jackson, "Power Considerations in the Design of the Alpha 21264 Microprocessor", *Design Automation Conference*, pp. 726–731, June 1998.

[6] G. Ji, T. R. Arabi and G. Taylor, "Design and Validation of a Power Supply Noise Reduction Technique", *IEEE Trans On Advanced Packaging*, vol. 28, no. 3, pp. 445-448, August 2005.

[7] P. Larsson, "Resonance and Damping in CMOS Circuits with On-Chip Decoupling Capacitance", *IEEE Trans. On Circuits and Systems — I: Fundamental Theory and Applications*, vol. 45, no. 8, pp. 849-858, August 1998.

[8] M. Ang, R. Salem and A. Taylor, "An On-Chip Voltage Regulator using Switched Decoupling Capacitors", *Intl. Solid-State Circuits Conf.*, pp. 438-439, February 2000.

[9] T. Rahal-Arabi, G. Taylor, J. Barkatullah, etc. "Enhancing Microprocessor Immunity to Power Supply Noise with Clock/Data Compensation", *Symp. On VLSI Circuits*, pp. 16-19, June 2005.

[10] International Technology Roadmap for Semiconductors, online: http://public.itrs.net/ .

[11] B. Garben, etc. "Influence of Damping and Voltage Dependent Leakage Resistance on Mid-Frequency Power Noise", *IEEE Workshop on Sign. Propagation on Interconn.*, pp. 45-48, May 2004.

[12] Predictive Technology Model, online: http://www.eas.asu.edu/~ptm/ .

[13] T. Sakurai, etc. "Alpha-Power Law MOSFET Model and its Applications to CMOS Inverter Delay and Other Formulas", IEEE J. of Solid-State Circuits, vol. 25, no. 2, pp. 584-594, April 1990.

[14] S. Mukhopadhyay, C. Neau, etc. "Gate leakage Reduction for Scaled Devices Using Transistor Stacking", IEEE Trans. On VLSI Systems, vol. 11, no. 4, pp. 716-730, August 2003.

Dithering Skip Modulator with a Novel Load Sensor for Ultra-wide-load High-Efficiency DC-DC Converters

Hong-Wei Huang[*], Hsin-Hsin Ho[+], Ke-Horng Chen[+], and Sy-Yen Kuo[*], *Fellow, IEEE*

[*] Graduate Institute of Electronics Engineering, National Taiwan University, Taipei, Taiwan

[+] Department of Electrical and Control Engineering, National Chiao Tung University, Hsinchu, Taiwan

d94943014@ntu.edu.tw, mnop543@yahoo.com.tw, khchen@cn.nctu.edu.tw, sykuo@cc.ee.ntu.edu.tw

ABSTRACT

Dithering skip mode with a novel load sensor for DC-DC converters is proposed to maintain a high efficiency over a wide load range. Due to the efficiency drop of the transition from the pulse-width modulation (PWM) to pulse-frequency modulation (PFM), a novel dithering skip modulation (DSM) is introduced to smooth the efficiency curve. Importantly, DSM mode can dynamically skip the number of gate driving pulses, which is inverse proportional to load current. Besides, a novel proposed load sensor can automatically select the optimum modulation method from these three modulation methods without an external selection pin. Simulation results shows DSM can maintain the efficiency of converters as high as about 89% over a wide load current range from 3mA to 500mA.

Categories and Subject Descriptors:

B.7.1 [Integrated Circuits]: Types and Design Styles

General Terms: Design, Performance

Keywords: Switching converter, load sensing circuit, delay-line chain, and dithering skip modulation.

1. INTRODUCTION

As we know, dynamic (or adaptive) voltage scaling (DVS) technique is widely used as one of the most effective means for achieving energy-efficiency design [1]. Generally speaking, power consumption has become the most important issue in portable battery-powered applications and high-performance desktop and server applications. The attractive salient features of DVS systems trigger the design of fast and adaptive-output-voltage DC-DC converters with high efficiency over a wide load range.

A popular technique to improve the efficiency over a wide load range is the hybrid mode, which is composed of pulse-width modulation (PWM) and pulse-frequency modulation (PFM) [2-3]. Hybrid mode achieves a high efficiency for the load current region A and B in Figure 1. However, there exists an efficiency dropping region C in Figure 1. It means that the efficiency curve is not smooth at the transition between PWM mode and PFM mode. It is a matter of efficiency and current load range for hybrid-mode modulation technique. The hybrid-mode modulation can maintain a high efficiency by closing the two peak efficiency values to reduce the efficiency drop at the sacrifice of load range. Therefore, a dithering

Permission to make digital or hard copies of all or part of this work for personal or classroom use is granted without fee provided that copies are not made or distributed for profit or commercial advantage and that copies bear this notice and the full citation on the first page. To copy otherwise, or republish, to post on servers or to redistribute to lists, requires prior specific permission and/or a fee.
ISLPED '06, October 4–6, 2006, Tegernsee, Germany.
Copyright 2006 ACM 1-59593-462-6/06/0010...$5.00.

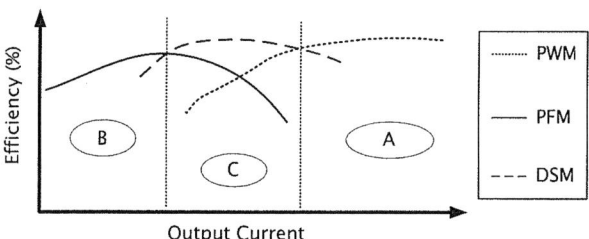

Figure 1. Three control modes converter that efficiency as functions of output current.

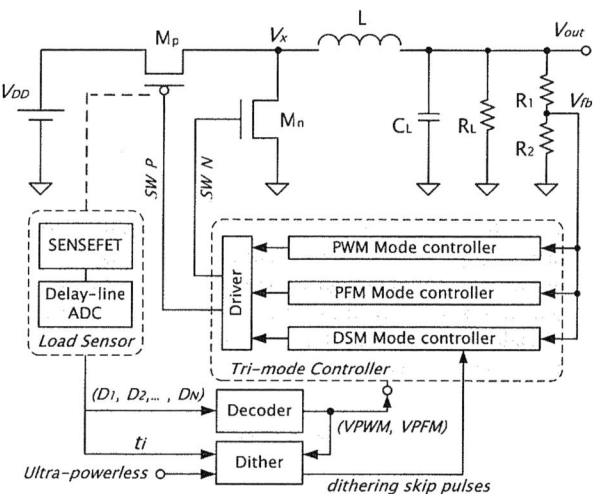

Figure 2. Block diagram of the tri-mode buck converter.

skip modulator is proposed to raise the efficiency between PWM and PFM curves in Figure 1. In other words, the efficiency drop between PWM and PFM modes can be raised by the novel DSM mode. Besides, a novel load sensor is also proposed for DSM in order to dynamically switch among these three modes, which are PWM, PFM, and DSM modes. Furthermore, compared with PSM mode and burst mode [4-6], DSM mode uses the dithering technique to reduce the output ripple [7]. Due to the insertion of DSM mode, a wide load range and high efficiency can be achieved without sacrificing the load range of conventional hybrid mode. Besides, the improved result is expected as the smooth efficiency curve from PWM mode curve to DSM mode curve and further extending to PFM mode without increasing the output voltage ripple.

The architecture of DC-DC buck converter with DSM mode is illustrated in Section II. The implementation of tri-mode control is described in Section III. Simulation results are shown in Section IV and we make the conclusions in Section V.

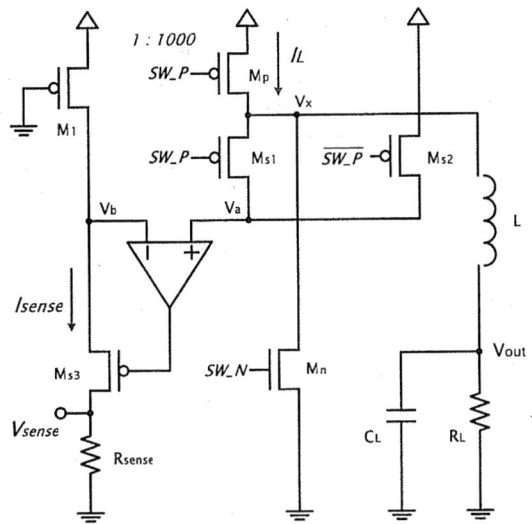

Figure 3. Current sensing technique with SENSEFET topology [8].

Table 1. Controller mode table.

	VPWM	VPFM
PWM mode	1	0
DSM mode	0	0
PFM mode	0	1

2. ARCHITECTURE OF A BUCK DC-DC CONVERTER WITH DSM MODE

In Figure 2, the buck DC-DC converter is modulated by a tri-mode controller, which is composed of PWM, PFM, and DSM modes. Besides, the load sensor estimates the load condition and sends the digital decision code ($D_1, D_2,..., D_N$) to decoder in order to dynamically select an optimum modulator among these three modulators. Compared with the prior design [2], the buck DC-DC converter does not need an external pin to decide the optimum modulator because of the novel load sensor.

In. Figure 3, a formal current sensing technique called SENSEFET topology is proposed by prior literature [8]. During the sensing period, the P-type power MOSFET is turned on by setting signal SW_P low and the sensing current I_{sense} is equal to a thousandth of inductor current I_L. By the sensing resistor R_{sense}, the sensing current can be transferred to sensing voltage V_{sense}. Thus, the peak value of sensing voltage can stand for the load condition of the output. However, even though V_{sense} is direct proportional to the load current, it is difficult for simple comparators to decide the switching points of three controllers because the variation of V_{sense} is too small.

Thus, a novel load sensor is proposed to determine the load condition, which is composed of current sensing circuit with SENSEFET topology and a current-mode delay-line A/D converter. At first, sensing voltage V_{sense} is sampled and held to get a peak voltage by a simple sample-and-hold circuit. The V-I converter converts the peak value of V_{sense} to a current signal for driving the delay-line A/D converter. Owing to the temperature independent characteristic of the V-I converter [9], the current signal I is independent of temperature variation. It means that the delay-line A/D converter is also independent of temperature variation. Hence, the output of delay-line A/D converter is decoded by the mode decoder to generate two mode-selected signals $VPWM$ and $VPFM$.

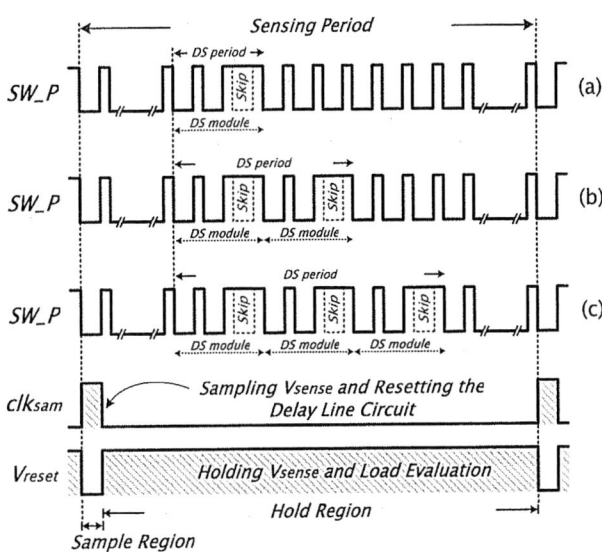

Figure 4. Timing diagrams for dithering skip modulator.

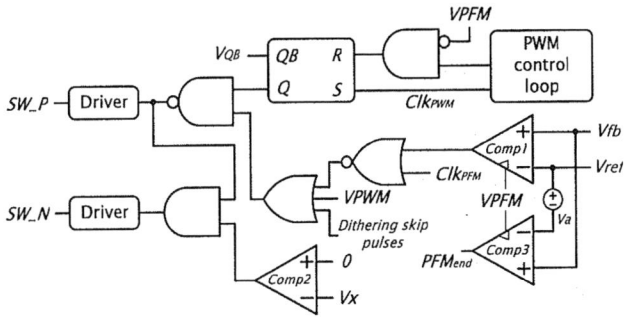

Figure 5. Tri-mode controller is composed of PWM, PFM, and DSM modes.

These two bits can select one optimum modulator from the three modulators. Table 1 shows the codes for determine the operation mode.

The concept of DSM mode is illustrated in Figure 4. The decreasing load current increases the size of *DS period* as shown in Figure 4 (a) to (c). The latter section will prove the size of *DS period* is inversely proportional to the load current. Thus, a *DS period* gradually contains more *DS modules* when the load current continuously decreases. In order to reduce the output voltage ripple, the dithering skip technique is implemented by the *DS module*. The function of a *DS module* is to make the DC-DC converter skip one switching pulse among three continuous switching cycles. Certainly, much power can be retrenched by reducing the switching consumption of power MOSFET because of the gradual decrease of load current.

3. IMPLEMENTATION OF TRI-MODE CONTROL TOPOLOGY

3.1 Tri-Mode Controller

According to the operation codes in Table 1, the tri-mode controller composed of three modulators is shown in Figure 5. Depending on the load condition, the tri-mode controller selects one optimum modulator from these three modulators to generate the switching signals SW_P and SW_N for P-type power MOSFET M_p and N-type power MOSFET M_n, respectively. For the PWM mode, the conventional current-mode with feedback control is adopted. The comparator named as 'comp1' is used for PFM mode to determine whether the output voltage is equal to the desired output voltage level or not. If $VPFM$ signal is equal to "0", the comparators 'comp1' and 'comp3' are turned off and their output values are set to "1". The other comparator 'comp2' is utilized to turn off N-type power MOSFET M_n to prevent negative inductor current.

The digital decision code ($VPWM$, $VPFM$) decides one optimum modulator from the three modulators for the tri-mode controller. If the code ($VPWM$, $VPFM$) is equal to (1, 0), the converter is switched to PWM mode. Similarly, the converter is switched to DSM mode when the code ($VPWM$, $VPFM$) is equal to (0, 0). In the meanwhile the dithering skip circuit is ready to skip some pulses of PWM mode to save much power consumption. Owing to the dithering technique, the output voltage ripple can be smaller than that of PSM mode or burst mode [4-6]. Certainly, the power consumption can be reduced by the skipping pulses and the skipping pulses do not dramatically affect the output voltage because of dithering technique.

As we know, if the load condition continues to reduce, the efficiency of DSM mode will be worse than PFM mode. Thus, the load sensor will send a code ($VPWM$, $VPFM$) equal to (0, 1) and the tri-mode controller is switched to the PFM mode. The high efficiency can be obtained again by this simple modulation when the load condition is light. The load sensor is turned off to save much power consumption in PFM mode. However, it is a problem for coming back to DSM or PWM modes when the load condition varies from light load to heavy load. Owing to PFM mode is operated in the discontinuous conduction mode (DCM), the peak value of V_{sense} can not be indicative of the average load current even if the load sensor is not turned off. Thus, the delay-line A/D converter will be reset and adopt the comparator named as 'comp3' to end the PFM mode when the load condition is heavy. A lower threshold voltage is set to determine whether the PFM mode can supply the output loading or not. This lower threshold voltage is designed as a voltage level with an offset voltage V_a that is lower than the reference voltage V_{ref}. Furthermore, the PFM mode is switched to PWM mode not DSM mode in order to supply much energy to reduce the output voltage ripple.

3.2 Current-Mode Delay-Line A/D Converter with Temperature Independent Characteristic

Resort to a high-resolution A/D converter is not economic for the implementation of the load sensor. The reason is that a high-resolution consumes much power and chip area for energy-efficiency of DC-DC converters. Thus, a simple delay-line A/D converter is a better solution for DC-DC converters. The principle of the voltage-mode delay-line A/D converter [10] is based on the propagation delay of a logic gate is inversely proportional to the gate supply voltage. Owing to the load condition can be represented by the peak value of V_{sense} voltage, a sample-and-hold circuit is used to hold this peak value. Therefore, the peak value of V_{sense} voltage can be used as the supply voltage of delay-line chain to convert the voltage level to a digital word. This digital word represents the loading condition of DC-DC converters and varies with the peak value of V_{sense} voltage during load variation.

However, the temperature dependence is an encumbrance for the accuracy of the voltage-mode delay-line A/D converter proposed in [10]. As we know, the increasing temperature makes the mobility

and threshold voltage all decrease. Thus, the driving current of each delay cell of the voltage-mode delay-line chain will be affected seriously to make the conversion digital word varying with temperature. Besides, the peak value of V_{sense} voltage level is needed to be converted to the voltage level of the succeeding logic circuit. In the meanwhile an extra voltage level shifter is needed for the conversion. Higher sensing speed needs more driving current consumption for the voltage level shifter. In other words, the voltage-mode delay-line A/D converter is not suitable for load sensor. In this paper, a current-mode delay-line A/D converter is adopted to improve the temperature dependent drawback as shown in Figure 6.

The proposed current-mode delay-line A/D converter is composed of a sample-and-hold circuit, a V-I converter, a current-mode delay-line chain, and a register. During a sensing period, there are two regions to complete the generation of dithering skip pulses. At the beginning of the sensing period, a sampling clock clk_{sam} starts the first region, which is the sample region. In order to avoid unnecessary mode switching and save the power consumption, the sensing clock clk_{sam} is periodically generated for several times of switching cycles.

The sampling clock clk_{sam} is inverted to generate the active-low resetting pulse V_{reset}. Furthermore, the load sensor is only working in PWM and DSM modes not in PFM mode. Thus, the logic function of V_{reset} in the sample region is written as:

$$V_{reset} = \overline{(VPFM + Clk_{sam})} \qquad (1)$$

An active-low resetting pulse V_{reset} resets the delay-line circuit. At the same time, the capacitor C_1 samples the peak value of V_{sense} voltage. At the negative falling edge of the sampling clock clk_{sam}, the hold region starts to evaluate the load condition by triggering the current-mode delay-line chain. In the meanwhile V-I converter converts the peak V_{sense} voltage to a current signal as a temperature independent current source for driving the delay-line chain.

In Figure 6, if $g_{m2}R_s$ is greater than one, the voltages V_{sense} can be converted to I_{M2} written as:

$$I_{M2} = \frac{V_{sense} + V_{SG1}}{R_s} = \left(\frac{I_L}{1000} \times \frac{R_{sense}}{R_s}\right) + \frac{V_{SG1}}{R_s} \qquad (2)$$

Similarly, I_{M4} is generated by connecting the gate of M_3 to ground for eliminating the extra signal V_{SG1} from equation (2). Subtract I_{M4} from I_{M2} and the output current of the V-I converter I_{M7} is written as equation (3). Owing to the temperature independence of R_{sense}/R_s, the V-I converter provides a temperature independent current to the delay-line chain.

$$I_{M7} = I_{M2} - I_{M4} = \frac{I_L}{1000} \times \frac{R_{sense}}{R_s}$$
$$\text{where } I_{M4} = \frac{V_{SG3}}{R_s} \qquad (3)$$

In Figure 6, each delay cell contributes delay time T_d and the driving current I_D are written as:

$$T_d = \frac{2C_{tot}V_{DD}}{I_D}$$
$$\text{where } C_{tot} = \left(\frac{5}{2}\right)C_{ox}\left(W_pL_p + W_nL_n\right) \qquad (4)$$

In equation (4), C_{tot} is the equivalent output capacitance of each

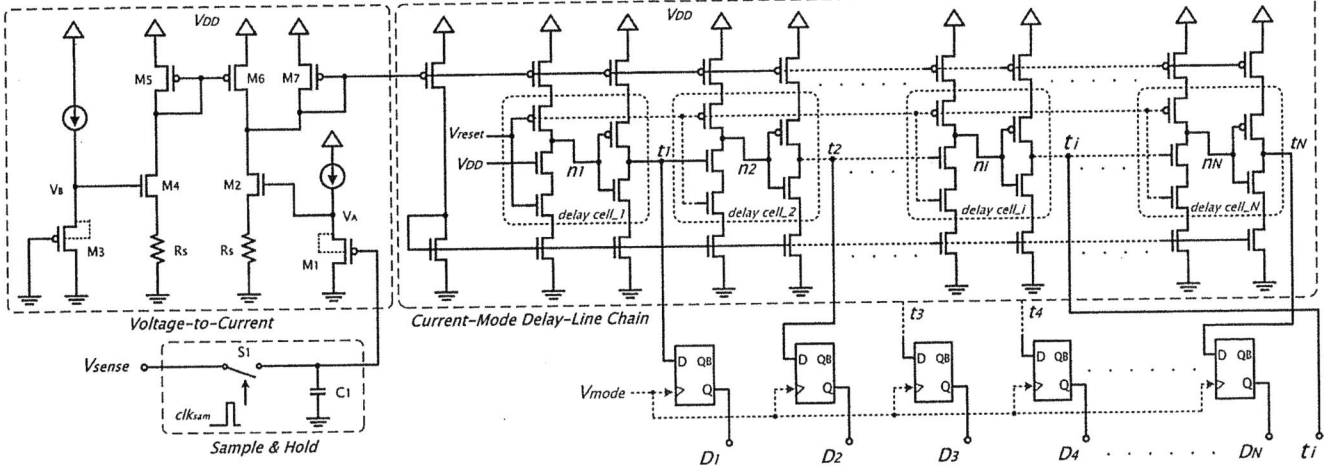

Figure 6. Schematic of the current-mode delay-line A/D converter.

delay cell. $L_{n(p)}$ and $W_{n(p)}$ are the channel length and width for N(P)-type MOSFETs of each delay cell, respectively. Larger load current raises the peak V_{sense} voltage by the SENSEFET circuit [8]. At the same time, the temperature independent current source is also raised by V-I converter to decrease the propagation delay time T_d of each delay cell in the delay-line chain. The relationship between delay time T_d and inductor current I_L is shown as the following equation. Clearly, the propagation delay time T_d is inverse proportional to load current without temperature dependent characteristic.

$$T_d = \frac{2C_{tot}V_{DD}}{I_D} = 2C_{tot}V_{DD} \cdot \frac{1000}{I_L} \cdot \frac{R_s}{R_{sense}} = \frac{k}{I_L}$$

$$\text{where } k = 2000C_{tot}V_{DD} \cdot \frac{R_s}{R_{sense}} \qquad (5)$$

An example of the timing diagram generated by the delay-line chain is shown in Figure 7. A periodic pulse V_{mode} triggers the register in the load sensor to latch the outputs of the delay-line chain. In order to increase the total delay time in DSM mode, extra delay cells can be added to the delay-line chain in Figure 6. In other words, the signals ($t_1 \sim t_n$) does not need to select from consecutive delay cells. Figure 8 demonstrates the linear conversion of analog load current to the output digital word. Furthermore, *DS period* in Figure 4 is determined by two signals, V_{mode} and t_i. A fixed positive rising edge of *DS period* starts the operation of the dithering skip code at the positive rising edge of V_{mode}. The negative falling edge of *DS period* synchronized with one of the outputs of the delay-line chain t_i, which ends the operation of the dithering skip code. The choice of t_i is a trade-off between the efficiency and output voltage ripple in DSM mode. Output load current determines the value of t_i and the length of *DS period* because the value of t_i is inversely proportional to the load current.

A digital word (D_1, D_2,..., D_N) stands for current load condition. The succeeding decoder can use this digital word to switch the tri-mode controller to an optimum operation mode as shown in Figure 2. The following subsection will discuss the decoder and the operation of dithering skip code.

3.3 Decoder and Dithering Skip Code Generator

A simple example shows the implementation of the mode decoder in Figure 9. Three digital mode bits are selected from the

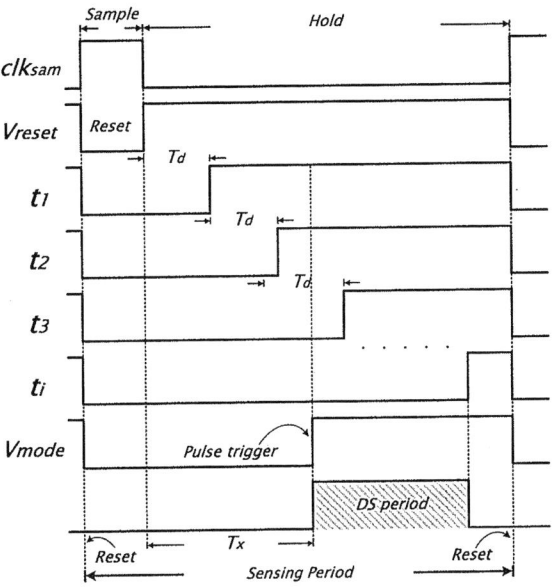

Figure 7. The timing waveforms of delay-line chain.

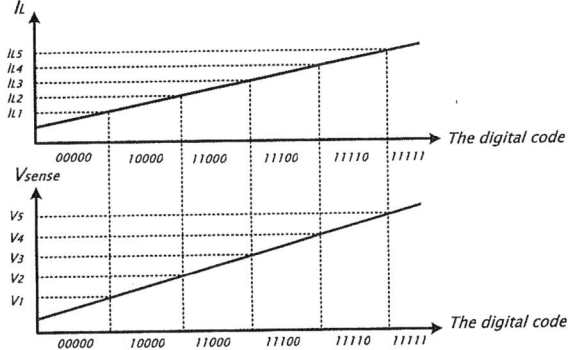

Figure 8. Load current and sensing voltage V_{sense} versus the digital word.

digital word (D_1, D_2,..., D_N). The selection of digital mode bits defines the sizes of three mode region. Thus, the selection rule is determined by the trade-off between efficiency and output voltage

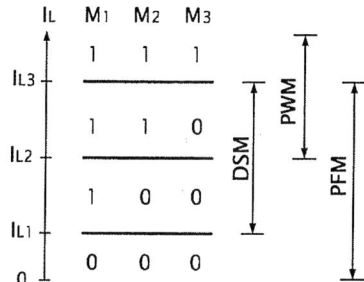

Figure 9. Diagram of the mode decoder.

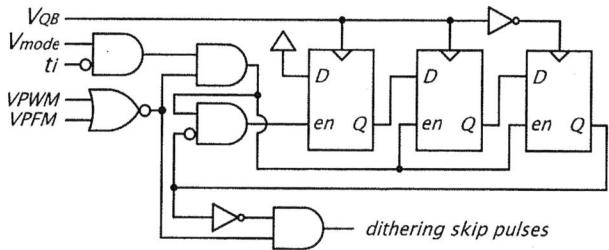

Figure 10. Dither skip pulses generator.

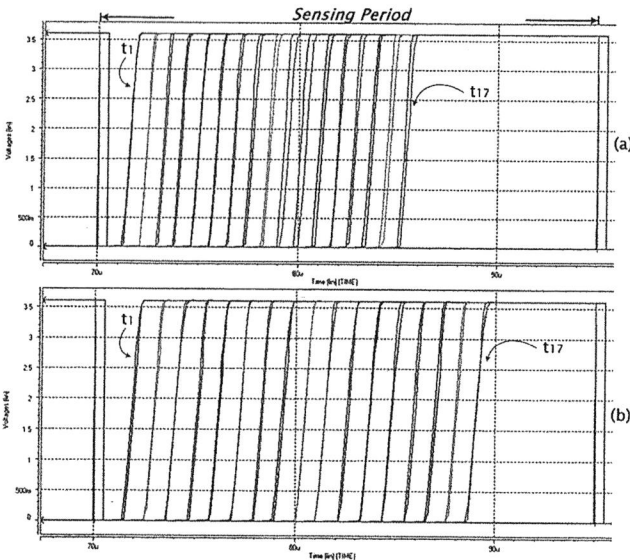

Figure 11. Waveforms of delay-line chain with temperature variation (-40°C~140°C). (a) Load current = 120 mA. (b) Load current = 40 mA.

ripple according to the applications. In Figure 9, the Boolean function of mode signals *VPWM* and *VPFM* are written as:

$$VPWM_i = M_1 M_2 M_3 + M_2 (PWM)_{i-1} \qquad (6)$$

$$VPFM_i = PFM_{end}\left(\overline{M_1}\ \overline{M_2}\ \overline{M_3} + (PFM)_{i-1}\right) \qquad (7)$$

VPWM$_i$ and *VPFM$_i$* are the current situation of load current, and *VPWM$_{i-1}$* and *VPFM$_{i-1}$* are the previous situation of load current. It is worth noting that a hysteretic region is needed between PWM and DSM mode to avoid unnecessary mode switching. Therefore, if PWM mode changes to DSM mode, the load current must be lower than I_{L2}. Similarly, load current must be higher than I_{L3} to make control mode coming back to PWM mode. Furthermore, a sudden heavy load current switches the operation mode back to PWM mode not DSM mode to prevent the output voltage from dropping too lower when the current operation is PFM mode.

Figure 10 shows the generation of the dithering skip pulse when DC-DC converter operates in DSM mode. It generates many *DS modules* according the size of the *DS period* in Figure 4. Each *DS module* contains three pulses, which is composed of two ON signals and one OFF signal by a switching signal V_{QB} that is from SR latch of tri-mode controller. If the ultra-powerless is turned on, each *DS module* is composed of one ON signal and two OFF signals to get many skipping pulses.

4. SIMULATION RESULTS

Dithering skip mode with a novel load sensor for DC-DC converters are simulated by TSMC 0.35μm process. Specifications of DC-DC buck converter are listed in Table 2 and the component values are shown in Table 3.

The output waveforms of delay-line circuit are shown in Figure 11 (a) and (b) when temperature is from -40°C to 140°C. Waveforms represent the output of delay-line chain which is composed of seventeen delay cells, and the DC-DC converter operates in 120 mA and 40 mA, respectively. Obviously, the

thermal effect is almost eliminated by current-mode delay-line A/D converter. Then, no matter what the temperature is, the proposed DC-DC converter can work all the time.

During DSM mode, the simulation results are shown in Figure 12 (a)-(c). In Figure 12 (a), the output ripple of converter is 31.8mV and the number of *DS module* is one. With the gradual decrease of load current, the number of *DS module* increase to two and three for Figure 12 (b) and (c), and output ripple of them are 30.6 mV and 18.6 mV, respectively. Therefore, much power can be retrenched by reducing the switching consumption of power MOSFET when load current reduces gradually.

The efficiency of proposed DC-DC buck converter is shown in Figure 13. As illustrated in Figure 9, the (I_{L3}, I_{L2}, I_{L1}) is set to (120 mA, 80 mA, 40 mA) in this designed. With the insertion of DSM mode, the efficiency between PWM curve and PFM curve can be improved. Thus, the efficiency is above 89% achieved over a wide load current range from 3mA to 500mA.

5. CONCLUSIONS

We propose a novel load sensor to dynamically switch among the three modulation modes including PWM, DSM, and PFM mode. With the automatic switching, the DC-DC converter can get high conversion efficiency for DVS systems. Simulation data indicate that the efficiency of the buck converter is about 92-94.5% in DSM mode for the load current from 40mA to 120mA. With the insertion of DSM mode, the efficiency between PWM curve and PFM curve can be improved significantly.

The concept of DSM mode is to decrease the switching conduction loss. According to the load current, DSM mode will dynamic increase the *DS period* to skip the switching pulse and get better efficiency. Furthermore, the power consumption can be reduced by the skipping pulses and the skipping pulses do not dramatically affect the output voltage because of dithering technique. As a result, this tri-mode buck converter can have high efficiency over a wide load range.

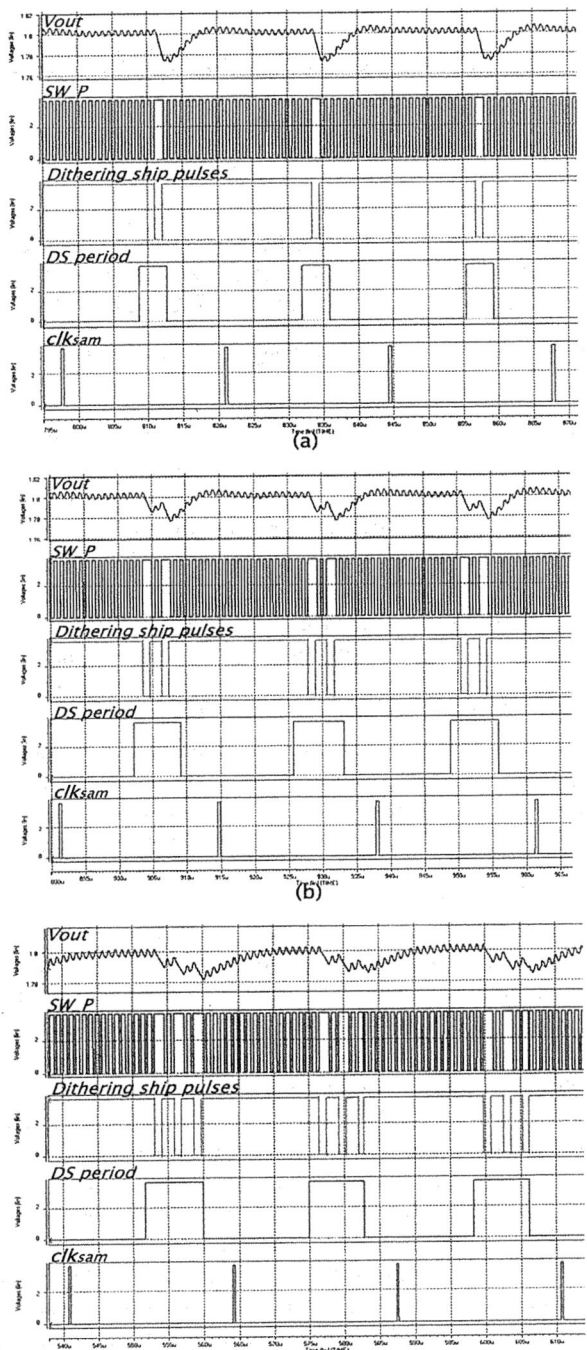

Figure 12. Waveforms of proposed DC-DC buck converter. (a) Load current = 120 mA. (b) Load current = 80 mA. (c) Load current = 40 mA.

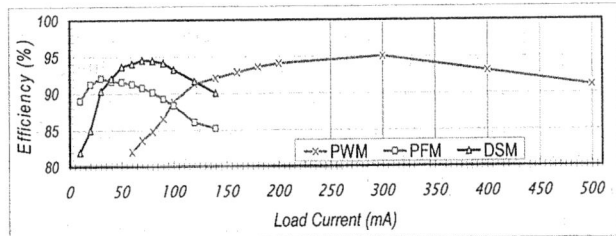

Figure 13. Measured PWM, DSM and PFM mode converter efficiency as V_{DD} = 3.6v, V_{out} = 1.8v.

Table 2. Specifications of buck converter.

Technology	TSMC 0.35-μm process
Input Voltage	3.6 V
Output Voltage	1.8 V
Switching frequency	1 MHz
Output Current Range	3 mA – 500 mA
Pulse-Width Modulation Mode	
Load Region	500 mA ~ 80mA
Output Ripple Voltage	< 10 mV
Dithering skip Modulation Mode	
Load Region	120 mA ~ 40mA
Output Ripple Voltage	< 32 mV
Pulse-Frequency Modulation Mode	
Load Region	120 mA ~ 3mA

Table 3. Component values.

L	4.7 μH	R_{sense}	1 MΩ
C_L	4.7 μF	R_s	500 kΩ

4. REFERENCES

[1] George Patounakis, Yee Willian Li, and Kenneth L. Shepard, "A Fully Integrated On-Chip DC-DC Conversion and Power Management System," *IEEE J. Solid-State Circuits*, vol. 39, no.3, pp. 443-451, March 2004.

[2] Jinwen Xiao, Angel V. Peterchev, Jianhui Zhang, and Seth R. Sanders, "A 4-uA Quiescent-Current Dual-Mode Digitally Controlled Buck Converter IC for Cellular Phone Applications," *IEEE J. Solid-State Circuits*, vol. 39, no.12, pp. 2342-2348, Dec. 2004.

[3] Biranchinath Sahu and Gabriel A. Rincon-Mora, "A high-efficiency, dual-mode, dynamic, buck-boost power supply IC for portable applications," *VLSI Design 2005*, pp. 858-861, Jan. 2005

[4] Paolo Sandri, Maria Rosa Borghi, and Luca Rigazio, "DC-to-DC Converter Function in a Pulse-Skipping Mode with Low Power Consumption and PWM Inhibit," US Patent 5,745,352, Apr. 28, 1998.

[5] Luo Ping, Li Zhaoji, Xiong Fugui, and Chenguangju; "Fuzzy pulse skip modulation mode in DC-DC converter," *Power Electronics Congress* pp. 87-91, Oct. 2004.

[6] Joey Martin Esteves and Randy Guy Flatness, "Adjustable Minimum Peak Inductor Current Level For Burst Mode In Current-Mode DC/DC Regulators," US Patent 6,724,174, Apr. 20, 2004.

[7] Angel V. Peterchev, "Digital Pulse-Width Modulation Control in Power Electronic Circuits: Theory and Applications," Ph.D. thesis, University of California, Berkeley, 2005.

[8] Hassan Pooya Forghani-zadeh, Gabriel A. Rincon-Mora, "Current-sensing techniques for DC-DC converters," *MWSCAS-2002* vol.2, pp. II-577 - II-580, Aug. 2002.

[9] Cheung Fai Lee and Philip K. T. Mok, "A Monolithic Current-Mode CMOS DC-DC Converter with On-Chip Current-Sensing Technique," *IEEE J. Solid-State Circuits*, vol. 39, no. 1, pp. 3-14, Jan. 2004.

[10] Benjamin J. patella, Aleksandar Prodic, Art Zirger, and Dragan Maksimovic, "High-frequency digital PWM controller IC for DC-DC converters," *IEEE Trans. Power Electronics*, vol. 18, no.1, pp. 438-446, Jan. 2003.

Adaptive On-Chip Power Supply with Robust One-Cycle Control Technique

Dongsheng Ma
Integrated System Design Laboratory
The University of Arizona
Tucson, AZ 85721, USA

ma@ece.arizona.edu

Janet Wang
Digital VLSI Design Laboratory
The University of Arizona
Tucson, AZ 85721, USA

wml@ece.arizona.edu

Pablo Vozqua
Integrated System Design Laboratory
The University of Arizona
Tucson, AZ 85721, USA

pablo@ece.arizona.edu

ABSTRACT

In this paper, an integrated adaptive output switching converter is proposed. The design uses a one-cycle control for fast line regulation and a single outer loop for load regulation and fine tuning. A switched-capacitor integrator is introduced in the one-cycle control to obtain positive integration with a single positive power supply, allowing a standard low-cost CMOS fabrication process. To improve the efficiency, a dynamic loss control technique is presented. HSPICE Simulations are performed with transistor models from TSMC 0.35μm N-well CMOS fabrication process. With a supply voltage of 3 V, a voltage ripple of less than 15 mV is measured. The maximum efficiency is 92% with a load power of 475 mW. The converter exhibits a tracking speed of 2 μs/V for both start-up and reference voltage transitions. The recovery time for a 20% load change is approximately 2.6μs. The overall design exhibits superior system robustness to battery perturbations, noise and process variations.

Categories and Subject Descriptors

B1, B7, B8.

General Terms

Design, Performance

Keywords

Switching converter, adaptive output, robust design, one-cycle control

1. INTRODUCTION

In the age of portable electronics, one of the most critical design considerations is the battery run-time. It is desirable for devices such as cell phones, PDAs and laptops to have low power consumption in order to maximize the run-time of the battery. Since the majority of these devices have different power consumption levels during distinct

Permission to make digital or hard copies of all or part of this work for personal or classroom use is granted without fee provided that copies are not made or distributed for profit or commercial advantage and that copies bear this notice and the full citation on the first page. To copy otherwise, or republish, to post on servers or to redistribute to lists, requires prior specific permission and/or a fee.
ISLPED'06, October 4–6, 2006, Tegernsee, Germany.
Copyright 2006 ACM 1-59593-462-6/06/0010...$5.00.

states (i.e. idle, active, sleep mode), it is very desirable to have a variable power supply that regulates the output power adaptively according to real-time system power and speed demands. In fact, dynamic power management with a variable power supply has been proved to be one of most effective methods for low power operations in VLSI systems [1-3]. This poses a challenge on the design of such a power supply. The supply must have good line regulation to maintain a constant voltage level during perturbations to the battery. It should also have a tight load regulation to be insensitive to load variations. In addition, fast transient response is expected to minimize switching losses and latency when switching between the desired output voltage levels. Also of interest is the efficiency of the converter, which varies with different output power levels.

This paper proposes solutions to the above obstacles in designing an adaptive on-chip power converter. The control method employed here is inspired from a nonlinear control called one-cycle control, first introduced in [4] for discrete power factor correctors. It makes use of feed-forward control to achieve fast line regulation. In contrast with a feedback control, the feed-forward control will make corrections every switching cycle without dependence on the previous cycles, thus providing an unconditionally stable and robust design. In addition, a second control loop connected at the output of the converter incorporates the output voltage information to the one-cycle control algorithm. This modification leads to much improved load regulation. To enhance the efficiency of the converter, by reducing power losses in the power stage, a dynamic loss control technique is proposed to adaptively adjust the sizes of the power transistors.

The paper is organized into five sections. Following this introduction, **Section 2** introduces the system architecture of the proposed converter and describes the functionality of the one-cycle control with the outer loop. In **Section 3** the implementation of the design is explained along with the operation principles of the circuit elements. **Section 4** presents the results from transistor-level simulations, using the models from a 0.35μm standard CMOS fabrication process. The results demonstrate the performance of the proposed converter. Finally, in **Section 5**, a recap of the design techniques and benefits will conclude this work.

2. SYSTEM CONTROL & ARCHITECTURE

In this design, the one-cycle control method is employed for fast line regulation. This approach makes use of a feed-forward control methodology, since the voltage is sensed at node V_X instead of at the output node V_O. The DC-DC converter and the feed-forward one-cycle controller are shown in Fig. 1. For improved load regulation a second outer loop will be incorporated into the one-cycle control. The modified architecture is shown in Fig. 2.

2.1 Original One-Cycle Control & Architecture

Fig. 1 shows a buck DC-DC switching converter with the original one-cycle control. When the power transistor $\mathbf{M_p}$ is turned on, the diode is reverse-biased, and V_X will be equal to the supply voltage V_g, if the voltage drop across M_p is negligible, which is usually true. When the power transistor M_p is off, the diode will be forward biased, grounding V_X and discharging the current in the inductor to the filtering capacitor and load resistor.

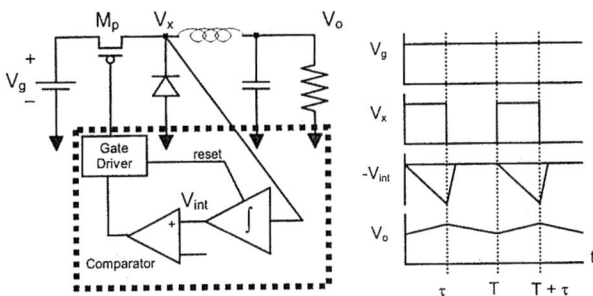

Figure 1. Block diagram of one-cycle control converter with main waveforms

Integrating V_X to get the average value during one cycle results in the following equation:

$$\overline{V_X} = \frac{1}{T} \int_0^\tau V_g \ dt \ = \ V_g \frac{\tau}{T} \cdot \tag{1}$$

Here T is the switching cycle of the converter and τ is the on-time of M_p. Also, the input/output voltage relation of a buck converter, operating in CCM (Continuous Conduction Mode), is given by

$$V_O = D V_g \tag{2}$$

Equating (1) and (2) suggests that the duty cycle D is equal to the integration constant τ / T in (1). Thus the output voltage of the converter can be indirectly controlled by averaging V_X at every switching cycle.

For the controller part, at the beginning of each switching cycle, a clock will turn on M_p and V_X will be integrated. The output of the integrator V_{int} will then be compared to a constant reference voltage set by the user. Once V_{int} reaches V_{ref} the gate driver will turn off M_p and

reset the integrator, which will resume integrating a zero until the beginning of the next clock cycle.

The integration of V_X at each cycle allows for next cycle corrections to perturbations in the power supply. This fast response provides the one-cycle control with good line regulation. It also eases the design stress on the battery. Because the energy delivered to V_o is measured by taking average of V_x, the output V_o of the power converter is robust to any noise, process variations and battery discharging.

Typical waveforms for the critical voltage nodes are presented in Fig. 1. Note that V_{int} and V_{ref} are both negative values if a traditional inverting configuration integrator is used [4]. This leads to implementation difficulties in low-cost single-well CMOS process. For example, in a N-well CMOS process, the lowest possible voltage nodes should be tied to the ground. No negative voltage is allowed on-chip to avoid leakage and latch-up.

2.2 The Proposed Design

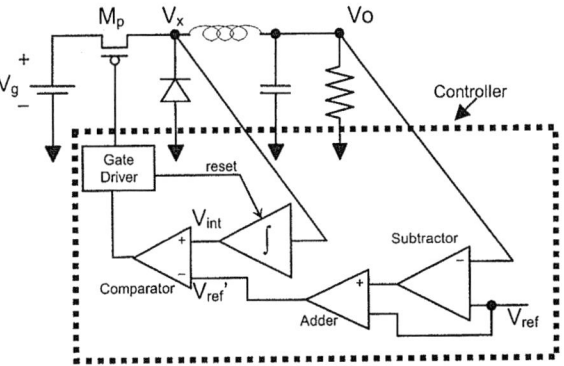

Figure 2. Block diagram of the proposed converter

Although the one-cycle control allows for a fast transient response, through its tight line regulation, it suffers from poor load regulation. No information on the output voltage V_O is collected in the one-cycle control algorithm. Hence, load variations are not easily detected and/or corrected. In addition, the parasitic resistance of the power devices introduces integration errors and thus degrades the regulation accuracy at V_o [5]. To account for such variations, a second loop is proposed for fine tuning, as depicted in Fig. 2. This outer loop detects the output voltage and incorporates it into the one-cycle control using a negative feedback configuration. The load regulation, through the outer loop, provides a robust design that regulates load changes effectively, as well as corrects possible process variations in the fabrication of the integrated controller. Because the loop is only activated during voltage fine-turning period, it will not affect the transient response and stability of the converter noticeably.

395

3. CIRCUIT IMPLEMENTATION & THEORY

3.1 SC non-inverting integrator

In this design, the integrator in the one-cycle control loop is implemented by a switched-capacitor circuit. This technique offers some advantages over methods employed in [4-6]. **First,** a traditional RC integrator with the inverting configuration produces a negative integration voltage at the output [4, 6], requiring the use of bipolar supply rails and a more expensive fabrication process. A DC level shifting technique was proposed in [5] to eliminate the need for a bipolar supply, which would allow circuit implementation with a standard digital CMOS process. This approach, however, increases the number of components and thus the silicon area. **Second,** the system bandwidth is reduced due to the added components that must propagate the signal. This affect will limit the converter switching frequency. **Lastly,** in both the above approaches RC components are used for the integration constants. This leads to the matching issue in the fabrication process, as values for distinct components are difficult to match. The resistive elements in CMOS processes can have variations of up to 20%.

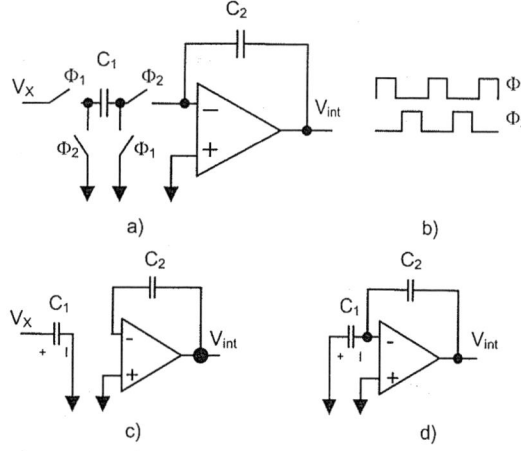

Figure 3. a) non-inverting SC integrator, b) non-overlapping clocks signals, c) the integrator during Φ_1, and d) the integrator during Φ_2.

As a result, Fig. 3 illustrates the proposed switched-capacitor (SC) integrator. It consists of only one op-amp, two capacitors and four CMOS switches. Because the integration time constant is only determined by the ratio of the capacitors C_1 and C_2, the integration error can be controlled with 0.05% in a standard CMOS process [7]. The operation of the circuit is explained as follows. Φ_1 and Φ_2 are a pair of complimentary, non-overlapping clock signals with a clock frequency of f_s. The switches are driven by these clocks to charge and discharge capacitor C_1. According to SC circuit theory, by adjusting the value of C_1

and f_s, an equivalent resistance can be obtained between V_x and the inverting input of the op-amp, with a value of

$$ R_{eq} = \frac{1}{f_s C_1}. \tag{3} $$

The operation of the integrator can be divided into two phases Φ_1 and Φ_2, as shown in Figs. 3c and 3d. During Φ_1, C_1 is charged to V_x and the total charge stored will be equal to $Q = C_1 V_x$. During Φ_2, C_1 is discharged with its positive terminal grounded. Thus, the charge transferred into the inverting terminal of the op-amp is negative. As a result, an equivalent positive charge appears at the output of the integrator V_{int}, which can be expressed as

$$ V_{int} = \frac{1}{R_{eq}C_2} \int_0^\tau V_x \ dt = f_s \frac{C_1}{C_2} \int_0^\tau V_x \ dt. \tag{4} $$

This non-inverting integration allows the use of only one positive supply rail for the op-amp and a standard CMOS process for fabrication.

3.2 Fine-tuning outer loop design

Due to the missing information at V_o, original one-cycle control suffers from poor load regulation, as discussed in Section 2.2. Approaches have been proposed to solve this problem. One such approach involves a direct sensing of the current in the inductor. Such sensing circuits, however, introduce an increase in the number of components and could cause unwanted harmonics in the system. Another solution in [6] is to integrate the inductor voltage V_L instead of V_x, allowing for information at the output to be taken into account. This design again requires bipolar power supplies, making the integration difficult to achieve in a standard CMOS process. In [5] the output voltage is compared to an upper bound and lower bound (fixed voltages). If the output voltage falls out of this bounded range, the error difference voltage is added to the reference voltage. Although this approach makes the CMOS integration possible, the circuit, which involves two loops for the upper and lower bounds, requires many circuit components. In this paper, the number of components for the outer loop load error correction is greatly reduced. Instead of eight op-amps, 24 resistors, and 2 switches in [5], the proposed circuit only needs one op-amp and 2 resistors as shown in Fig. 4.

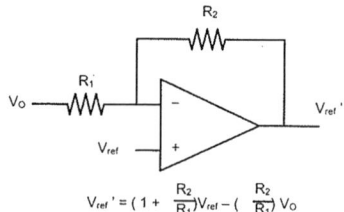

Figure 4. Outer loop for load error correction

The operation of the circuit can be explained with reference to Fig. 2. The output voltage V_O is subtracted from the reference voltage V_{ref} to determine the variation, ΔV, which is then added back into V_{ref} to produce a new voltage reference V_{ref}', which is equal to

$$V_{ref}' = 2V_{ref} - V_O. \tag{5}$$

By choosing the same resistance for R_1 and R_2, the circuit in Fig. 4 can easily achieve the function in (5). V_{ref}' will then be compared with V_{int} and determine the new duty ratio for the converter. Thus, if the output voltage drops below the desired reference voltage V_{ref}, the outer loop will increase the value V_{ref}' and the output voltage will increase accordingly. A similar error correction mechanism holds when the output voltage is higher than the desired reference voltage V_{ref}.

3.3 Dynamic loss control technique

In a traditional power converter with fixed load and output voltage, the size of each power transistor can be determined by minimizing the sum of the conduction and switching losses. However, for an adaptive-output converter, both the load and the output voltage are variable. It is desirable to dynamically adjust the sizes of the power transistors to have the minimum power loss in different working modes.

There are mainly two components, P_{sw} and P_{cond}, contributing to the power losses of the power transistors. Here, P_{sw} is the power loss due to the charging and discharging the gate capacitance of the transistor, while P_{cond} is the power conduction loss due to the turn-on resistance of the transistor. The total power loss due to the two factors can be expressed in terms of circuit parameters of interests, including the length (L) and width (W) of the transistor, the load current (I_O), and the supply voltage (Vdd), as shown in the following equations.

$$P_{cond} = I_O^2 \frac{L}{\mu_n C_{ox}(V_{dd} - V_T)W} \tag{7}$$

$$P_{sw} = WLC_{ox}V_{dd}^2 f_s \tag{8}$$

$$P_{loss} \approx P_{cond} + P_{sw} \tag{9}$$

Note that the total loss is inversely proportional to W in P_{cond}, but linearly proportional to W in P_{sw}. Hence, a minimum P_{loss} can be obtained by optimizing W. This method is very similar to the technique proposed in [8].

Fig. 5 shows the circuit schematic used for dynamic power loss control. The entire power range is divided into four levels in this design. A higher-level power matches a higher-level voltage and a faster operation speed. Accordingly, the regulated output voltage range can be divided into four levels from V_1 to V_4 as shown in the

figure. During the real-time operation, the actual output voltage V_o will be compared with these pre-defined voltages to determine the size of the power transistor at the instant. For example, if V_o is lower than V_1, a small power transistor with a width of W_1 will be used in the converter to reduce the switching power P_{sw}. When V_o is higher than V_4, a large power transistor with a width of ($W_1+W_2+W_3+W_4$) will be employed to reduce the conduction power loss P_{cond}.

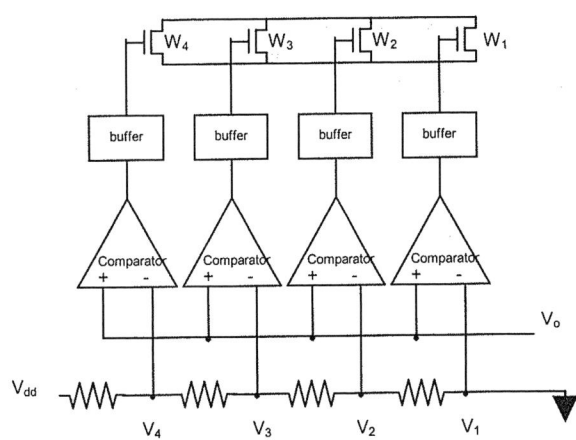

Figure 5. Dynamic loss control on a power transistor

4. SIMULATION RESULTS

The proposed converter was designed and simulated with TSMC 0.35 μm N-well CMOS process, using Hspui for Windows 2001.4.2v1.1 HSPICE software. The resulting simulation waveforms were viewed with AvanWaves 2001.4.2 software. The layout of the design has been sent for fabrication. Fig. 6 shows the layout photo of the converter. With fully on-chip power transistors, the overall active die area is 0.552 mm^2.

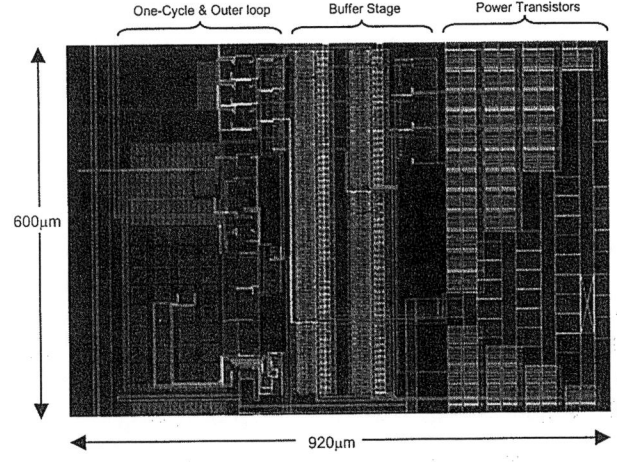

Figure 6. Layout of the proposed converter

4.1 Switched-capacitor integrator

Simulation result on the integration of V_x in the steady state, using the switched-capacitor circuit described earlier, is shown in Fig. **7**. Here, the reference voltage V_{ref} is set to be 1.8 V and input supply voltage V_g is 3 V. The integration follows a stair-step pattern and contains 20 sampling levels during each switching cycle. The output of the integration determines the duty cycle of the converter, as stated in Sect. 2. In Fig. 7, the duty cycle can be measured as 0.6 (=120ns/200ns), which verifies our theoretical design target – an output voltage of 1.8V.

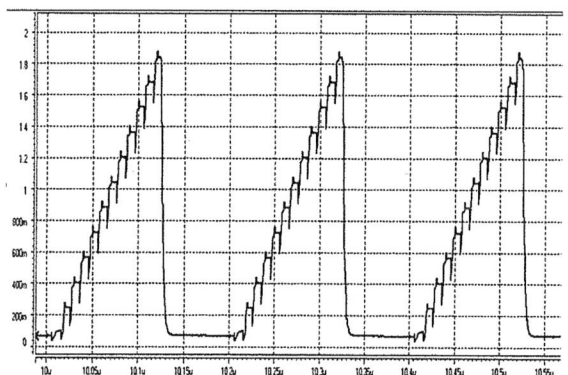

Figure 7. Switched-capacitor integration of Vx

4.2 Steady-State Measurement

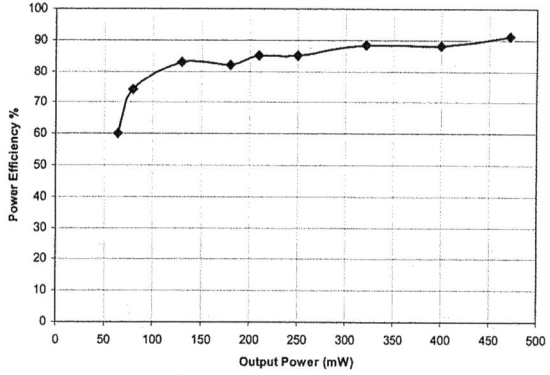

Fig. 8 Efficiency of the proposed converter

The efficiency of the converter, with the dynamic loss control, is plotted in Fig. 8. With a full power load of 475 mW, the converter reaches its maximum efficiency of 92%. The overall power range spans from 70 mW to 475 mW. Due to the effectiveness of dynamic loss control, the efficiency stays above 60 % in the entire range.

Fig. 9 shows the steady-state output voltage regulated at 1.8 V and 0.8 V, respectively. The peak-to-peak output ripple voltage for V_{ref} = 1.8 V is measured to be less than 15 mV. Likewise, for V_{ref} = 0.8 V the voltage ripple is less than

10 mV. Different from the steady-state output voltage in traditional analog switching converters, the peak and valley voltages vary with time due to the nonlinear regulation nature of the outer loop design. In addition, a slight offset to the desired reference is also observed. This offset is around 24 mV to 26 mV in the two measured cases, primarily due to the discrete nature of the switched-capacitor integrator. As shown in Fig. 7 the output contains discrete levels and is thus subject to quantization errors. Also contributing to this offset is the finite gain of the op-amp used in the outer error correction loop. This error can be overcome by pre-measurement calibration.

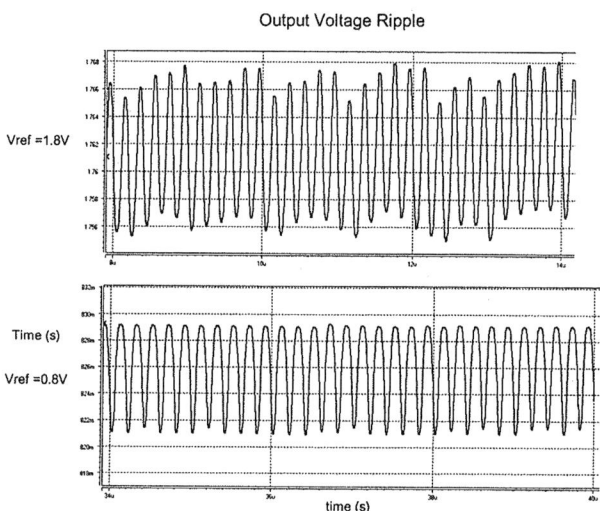

Figure 9. Output voltage regulated at 1.8 V (top) and 0.8 V (bottom), respectively.

4.3 Transient Response

As was mentioned earlier, one of the advantages of the one-cycle control is its fast transient response. In this section, all the transient response cases are measured, including start-up, output voltage tracking and load transient responses.

First, the start-up transient and output voltage tracking are measured in Fig. 10 and Fig. 11 respectively. The converter is initially relaxed. Then it is charged up to the desired level of 1.8 V. Again, the slight offset is due to the factors discussed in Sect. 4.2. The converter settles in approximate 2μs with a voltage reference of 1.8 V. Note that this start-up time is measured from initial conditions to 90% of the desired voltage level.

A similar approach was taken to measure the transient time from one voltage level to another. Here the voltage reference was changed from 1.8V to 0.8V and the transient time was calculated as the time it takes the voltage to fall from 1.8V to within 10% of the final settling value. The measured transient time is slightly less than 2μs and we define a tracking speed of 2μs/V. This performance is much faster than most of the state-of-the-arts adaptive switching converter designs compared in [5].

Load regulation was measured by increasing the load current by 20% of its nominal value of 180 mA, at the moment of 10 μs in Fig. 12. Here when the voltage V_O recovers, the output current I_O will increase to accommodate the load change. The measured recovery time is approximately 2.6μs.

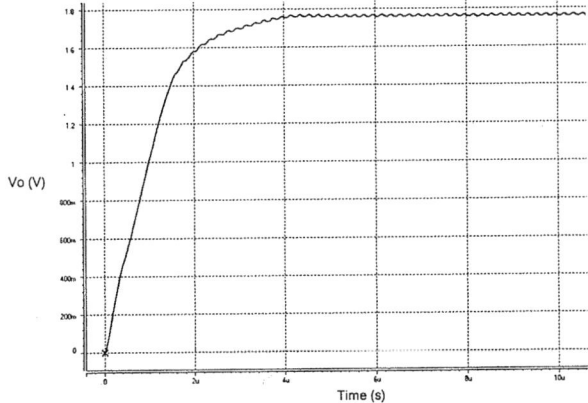

Figure 10. Start-up transient response of the converter

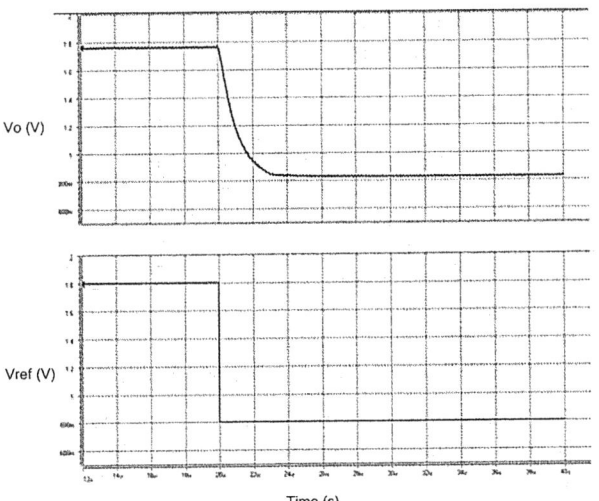

Figure 11. Transient response to the reference voltage

5. CONCLUSIONS

In this paper, a new integrated one-cycle controlled buck switching converter is presented. In order to allow using standard low-cost CMOS process and to achieve fast transient response, a non-inverting switched-capacitor integrator is employed in the controller. Tight load regulation was achieved through the use of an additional outer loop to incorporate the output voltage information into the one-cycle control. Simulation results show a fast

transient response, in comparison to current state of the art designs, and a very desirable output ripple. These values were achieved with a converter frequency of 5MHz and a sampling frequency (used for the switched-cap integrator) of 100MHz. These characteristics are very desirable in systems that use voltage scheduling in order to minimize power losses. Furthermore, the design techniques can be applied to other DC-DC or AC-DC converter designs.

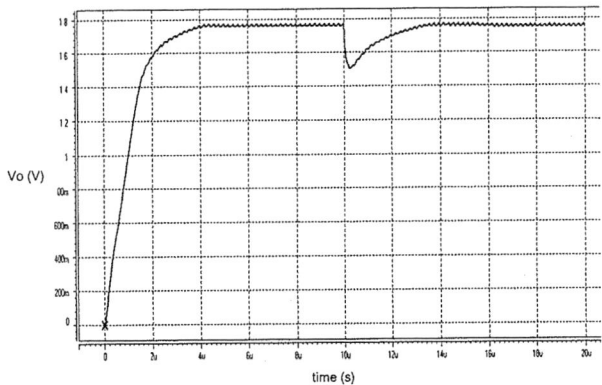

Figure 12. Transient response to load variation

6. REFERENCES

[1] J. Goodman et al, "An Energy/Security Scalable Encryption Processor Using an Embedded Variable Voltage DC/DC Converter," *IEEE J. of Solid-State Circuits*, Vol. 33, November 1998, pp 1799-1809.

[2] F. Ichida et al, "Variable Supply Voltage Scheme with 95% Efficiency DC-DC Converter for MPEG-4 Codec," *IEE Int. Symp. Low Power Electronics*, 1999, pp 54-59.

[3] T. D. Burd et al, "A Dynamic Voltage Scaled Microprocessor System," *IEEE J. of Solid-State Circuits*, Vol. 35, Nov. 2000, pp 1571-1580.

[4] K. Smedley, S. Cuk, "One-Cycle Control of Switching Converters," *IEEE Trans. Power Electronics,* Vol. 10, Nov. 1995, pp 625-633.

[5] D. Ma, Wing-Hung Ki, and Chi-Ying Tsui, "An Integrated One-Cycle Control Buck Converter With Adaptive Output and Dual Loops for Output Error Correction," *IEEE J. of Solid-State Circuits*, Vol. 39, No. 1, Jan. 2004, pp 140-149.

[6] Pallab Midya, Philip T. Krein, and Matthew F. Greuel, "Sensorless Current Mode Control- An Observer-Based Technique for DC-DC Converters," *IEEE Transactions on Power Electronics*, Vol. 16, No. 4, July 2001, pp 522-526.

[7] Dabrowski, Adam, "Multirate and Multiphase Switched-Capacitor Circuits" Chapman and Hall, London 1997.

[8] D. Ma et al, "Design and Optimization on Dynamic Power System for Self-Powered Integrated Wireless Sensing Nodes," *ISLPED '05*, San Diego, CA USA, August 8-10, 2005, pp 303-306.

Robust Multiple-Phase Switched-Capacitor DC-DC Converter with Digital Interleaving Regulation Scheme

Dongsheng Ma
Integrated System Design Laboratory
The University of Arizona
Tucson, AZ 85721, USA

ma@ece.arizona.edu

ABSTRACT

An integrated switched-capacitor (SC) DC-DC converter with a digital interleaving regulation scheme is presented. By interleaving the newly-structured charge pump (CP) cells in multiple phases, the input current ripple and output voltage ripple are reduced significantly. The converter exhibits excellent robustness, even when one of the CP cells fails to operate. A fully digital controller is employed with a hysteretic control algorithm. It features dead-beat system stability and fast transient response. Hspice post-layout simulation shows that, with a 1.5 V input power supply, the SC converter accurately provides an adjustable regulated power output in a range of 1.6 to 2.7 V. The maximum output ripple is 40 mV when a full load of 0.54 W is supplied. Transient response of 1.8 μs is observed when the load current switches from half- to full-load (from 100 to 200 mA).

Categories and Subject Descriptors
B1, B7, B8.

General Terms: Design, Performance

Keywords
Switched-Capacitor DC-DC Converter, Interleaving Regulation

1. INTRODUCTION

As the feature size of MOSFET transistor scales down with Moore's Law, it is predicted that the power supply voltage in digital VLSI system will drop below 1 V for low power operation in the near future. However, for many high-precision analog and mixed-signal circuits, higher supply voltages are still desirable [1, 2]. Voltage boosting technique thus becomes very critical. Switch mode and switched-capacitor (SC) DC-DC power converters are considered as the best candidates for on-chip voltage boosting. However, switch mode power converter is usually not favorable in low-noise designs, due to large EMI and coupling noise caused by bulky inductive components. We thus explore low

---

Permission to make digital or hard copies of all or part of this work for personal or classroom use is granted without fee provided that copies are not made or distributed for profit or commercial advantage and that copies bear this notice and the full citation on the first page. To copy otherwise, or republish, to post on servers or to redistribute to lists, requires prior specific permission and/or a fee.
ISLPED'06, October 4–6, 2006, Tegernsee, Germany.
Copyright 2006 ACM 1-59593-462-6/06/0010...$5.00.

---

ripple (noise) high performance design solutions from SC power converters.

A power stage, usually referred as a charge pump, is one most critical part in a SC DC-DC converter. Nowadays, traditional pipe-line type charge pumps, such as Dickson charge pump [3], are rarely used in low-voltage VLSI systems. Due to large threshold voltage drops of power diodes or power transistors, power efficiency of these charge pumps is inherently low. In recent years, cross-coupled voltage doubler becomes more popular, because the voltage drop in the power stage is mainly caused by much lower drain-to-source voltage instead of transistor's threshold voltage. The circuit can thus achieve higher efficiency with a low supply voltage [4, 5]. However, this circuit also has a few serious design drawbacks.

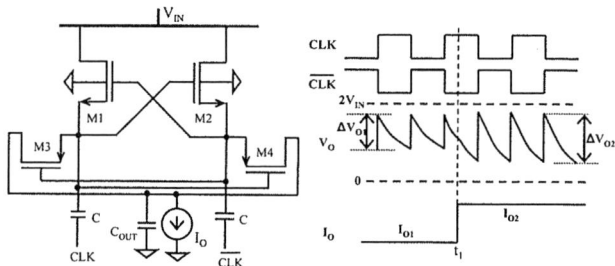

Fig. 1 Schemtic of a CMOS voltage doubler with timing diagram.

Fig. 1 illustrates the circuit schematic of the voltage doubler and its timing diagram. *CLK* and $\overline{CLK}$ are two complementary non-overlapping clocks, determining the switching actions among M_1-M_4. Because the charging time of the pumping capacitor C is usually designed much smaller than the discharging time of the output capacitor C_{OUT}, its output voltage V_O can thus be plotted as in the figure. V_O keeps decreasing during each half clock cycle, since the pumping capacitor connected to V_O would not be recharged until the next half clock cycle begins. If a load change from I_{O1} to I_{O2} occurs at a moment of t_1, a large voltage ripple will be observed at V_O, since the circuit cannot response to this change until the current half clock cycle expires. This affects the transient response and leads to large ripple voltage. One possible solution is to increase the clock frequency f_{CLK} of the charge pump. However, this also increases the switching power loss and thus degrades the efficiency. In addition, because M_1 and M_2 are required to be turned on respectively in two non-overlapping phases, the input current of power supply V_{IN} is discontinuous with large ripple. This current ripple cause large switching noise, which can be coupled into the entire system through the power supply. Recently, several control schemes have been proposed to reduce the input current ripple [6] or output voltage ripple [7]. However, these require an extra large power switch cascaded with the input or output power stage, which increased silicon area and design complexity.

In this paper, we propose a new SC power converter with a fully digital interleaving controller to overcome the aforementioned drawbacks. The rest of the paper is organized as follows. Section 2 introduces the power stage architecture and interleaving control scheme of the proposed converter. In Section 3, a digital hysteretic controller is designed, which exhibits much improved noise immunity compared to its analog counterparts. In addition, this controller features a very fast transient response. Post-layout simulation results are presented in Section 4. Finally, a conclusion is made in Section 5 to summarize the work.

2. POWER STAGE DESIGN WITH INTERLEAVING REGULATION SCHEME

2.1 Prior Arts of Interleaving Regulation Scheme

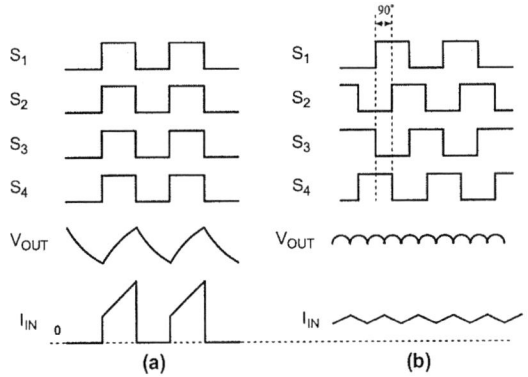

Fig. 2 Switch mode power converter with interleaving regulation

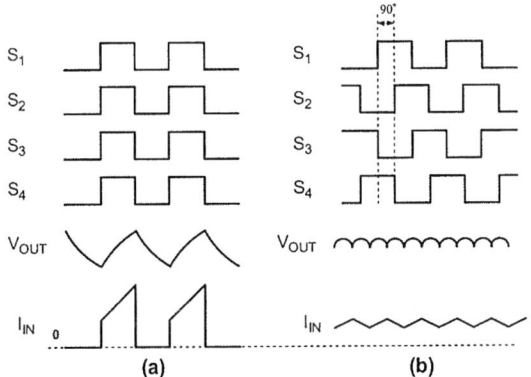

Fig. 3 Timing diagrams of the converter in Fig. 2: (a) without interleaving scheme, and (b) with interleaving scheme

The original interleaving regulation scheme was first introduced in switch mode DC-DC converter designs [8, 9]. Figs. 2 and 3 depict the simplified circuit and control timing diagram, respectively. In Fig. 2, N buck switch mode sub-converters are connected in parallel, sharing the same power input and output V_{IN} and V_{OUT}. If the same control signals on the power switches S_1 to S_N are applied, the entire circuit can be regarded as a traditional buck converter. The operation of the converter can be described with the waveforms in Fig. 3a, where we take $N=4$. Since all the power switches of S_1 to S_4 are turned on and off simultaneously, the sub-converters reach the peak current and voltage at the same time, which results in large current ripple at the input current I_{IN} and large voltage ripple at the output voltage V_{OUT}. These ripple noises are coupled into substrate and power supply, affecting system accuracy and stability. Fig. 3b shows another scenario where the sub-converters operate in an interleaving way. Peak currents and voltages are avoided to occur simultaneously, leading to obvious reductions on current and voltage ripples. Especially for the input current I_{IN}, it is discontinuous in Fig. 3a, but turns into a continuous one in Fig. 3b.

In general, the interleaving operation on N cells yields a reduction in peak ripple magnitude by a factor of N. It also extends the system bandwidth by N times [7]. Another advantage of the interleaving operation is fault tolerance. For example, if one sub-converter in Fig. 2 fails to operation due to any malfunctions or failures, the other N-1 sub-converters will compensate the loss. Although the ripples will increase as a result, the converter still has better performances than non-interleaving case. System robustness is much improved.

2.2 The Proposed Charge Pump Cells with Interleaving Regulation Scheme

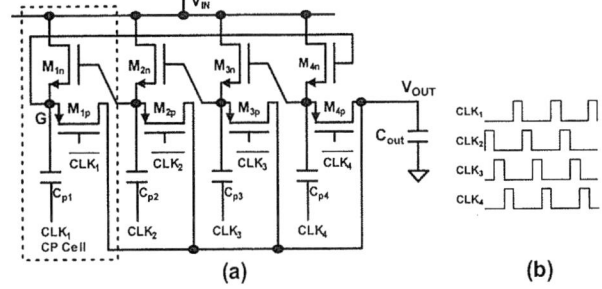

Fig. 4 (a) Schematic of a four-phase interleaved CP in a loop connection, and (b) clock signals

The difficulty of applying interleaving scheme into SC power converter lies in the fact that the charging and discharging mechanisms in the pumping capacitors C_{pi} as shown in Fig. 4a are more difficult to control than those in the inductors. In addition, not every interleaving scheme can effectively reduce the ripple noise addressed in Section 1. An instinct implementation of interleaved charge pump is to connect the four CP cells in series as a close loop, as shown in Fig. 4a. The clock signals of the circuit are shown in Fig. 4b, represented by CLK_i s, i = 1...4, respectively. Each clock signal has 90°phase difference to its adjacent cells. $\overline{CLK_i}$ s, (i = 1...4) are complementary signals of CLK_i , but swing between 0 and $2V_{IN}$, instead of 0 and V_{IN}. The voltage level of $2V_{IN}$ guarantees that PMOS transistors can be completely turned off when the corresponding pumping capacitor clock signal is *LOW*. This requires another voltage boosting circuit to generate $2V_{IN}$. Moreover, the input current is discontinuous since the clock signals are non-overlapping. Hence, the input current ripple is not improved. Last, when one clock signal is *HIGH*, only one of the other three pumping capacitors is charged. For example, when CLK_1 is *HIGH*, the node G will be $2V_{dd}$. M_{4n} is thus turned on to charge C_{p4} to V_{IN}, while C_{p2} and C_{p3} have to wait to be charged in next clock cycle. This slows down the transient response.

To overcome the above drawbacks, we propose a new charge pump architecture shown in Fig. 5a. The clock signals are shown in Fig. 5b. From the circuit connection and clock waveform, it is easy to identify that this is in fact a parallel connection of two cross-coupled voltage doublers with 90° phase difference. By introducing 90° phase overlapping between neighboring CP cells, the input current becomes continuous and has low ripples. All the gate clock signals of the PMOS and NMOS transistor are generated internally, with no extra clock signals needed. At any instant when two clock signals are *HIGH*, the pumping capacitors associated with the other two complementary clocks are charged to V_{IN}. For example, when CLK_1 and CLK_4 are *HIGH*, the nodes 1 and 4 become *HIGH*. The transistors M_{3n} and M_{2n} are thus turned on and the pumping capacitors C_{p3} and C_{p2} are charged to V_{IN}. This ensures a faster

transient response than the previous design. Here, because the gate clock signals are generated internally, it eliminates the necessity of extra digital circuits and buffers, which simplifies the design. Hence, the new architecture overcomes all the drawbacks in the circuit of Fig. 4a.

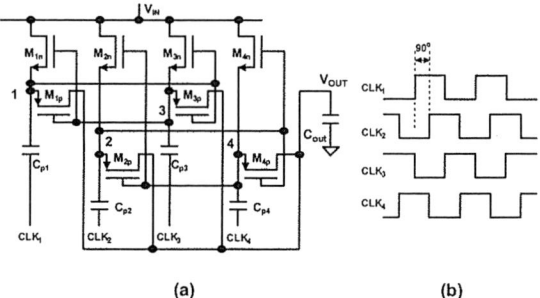

Fig. 5 (a) The proposed four-phase charge pump with two cross-coupled connection, and (b) clock signals

3. DIGITAL HYSTERETIC CONTROL & CLOSE-LOOP SYSTEM DESIGN

It is well known that digital circuits usually have larger noise margin than analog ones. It is desirable if the controller in this design is implemented in digital manner. In addition, currently signal processor has been commonly used as core circuit in many mixed-signal circuit systems. A digital controller of a power converter can be implemented as part of the processor. No extra silicon area is needed, while more advanced control scheme can be achieved by the software in the processor. The circuit block diagram of the entire proposed SC DC-DC converter is shown in Fig. 6. In general, it includes the proposed four-phase charge pump in Section 2, a digital hysteretic controller and a reference clock generator. We will discuss the core circuits and close-loop system design in due course.

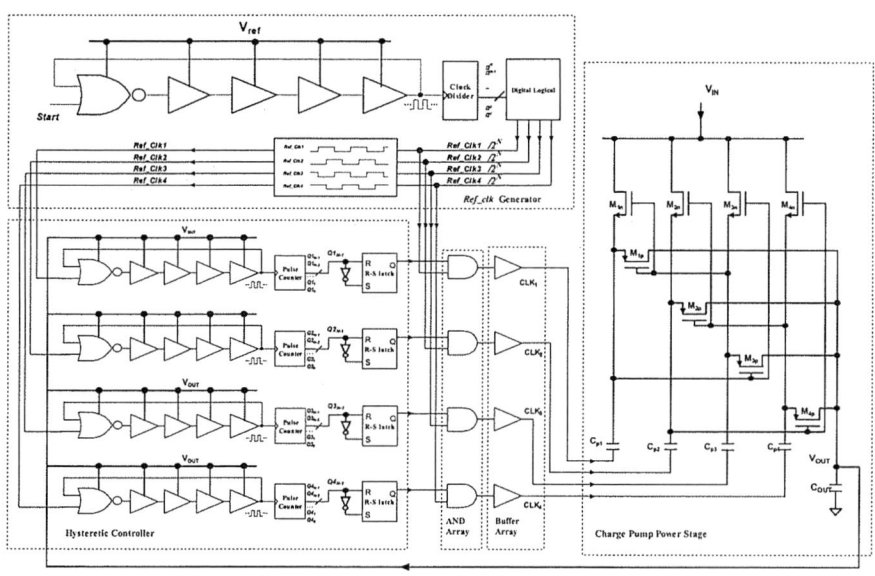

Fig. 6 The proposed system block diagram

3.1 Ring Oscillator Based A/D Converter

The output signal of the charge pump is an analog voltage. In order to implement the digital control, an analog to digital (A/D) converter is required to convert analog output voltage into digital signals. Traditional A/D converter is not preferred, because it occupies too much silicon area, consumes large power and is very sensitive to noise. Recently, ring-oscillator and delay-line based A/D converter has been reported [10]. Compared with traditional design, it is more area- and power-efficient. Since both of them choose digital logic gates as building blocks, it has larger noise margin and is more robust than analog A/D converters.

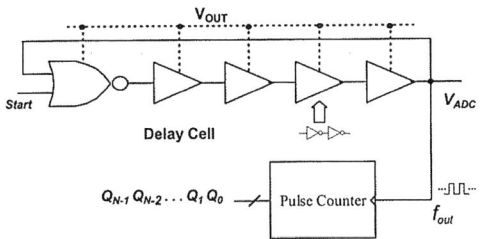

Fig. 7 Ring Oscillator A/D converter

Compared to the delay-line based design, ring oscillator based A/D converter is even more area efficient because the delay elements can be re-used even within a single switching clock cycle. We propose a new ring-oscillator based A/D converter as shown in Fig. 7. The circuit consists of one NOR gate, four delay cells and one pulse counter. Each delay cell simply includes two inverters. The pulse counter is an asynchronous positive edge triggered N-bit counter. Note that the NOR gate and the delay cells are powered by V_{OUT}, which is the output of the SC DC-DC converter. When the *Start* signal is *HIGH*, the loop will keep in a static state and the outputs of delays cells keep *Low*. Otherwise, the loop oscillates and a series of pulse are generated at V_{ADC} with an oscillating frequency of f_{out}. By examining $Q_{N-1}...Q_0$ at the output of the counter, the voltage V_{OUT} is presented in form of f_{out}, as in the following equation:

$$f_{out} = \frac{\beta(V_{OUT} - V_{th})^2}{2k \cdot n_{stages} \cdot C_{load} V_{OUT}},$$

where k and β are process parameters, n_{stages} is the number of stages in the ring oscillator and C_{load} is the load capacitor for one delay cell.

3.2 Sensor Circuit

To detect the regulation error between the output V_{OUT} and the desired reference voltage, two identical ring oscillator A/D converters are used as a sensor, depicted in Fig. 8. The upper ring oscillator, powered by the constant reference voltage V_{ref}, generates a reference clock signal with a frequency of f_{ref}. This signal passes through a clock divider and generates another clock signal Ref_clk with a frequency equal to $f_{ref}/2^N$, where N is the number of bits in the clock divider. Ref_clk is then used as the *Start* signal of

the second ring oscillator powered by the power converter's output voltage V_{OUT}. When Ref_clk is *LOW*, the lower side oscillator sensor generates a clock signal with a frequency of f_{out}, which drives the pulse counter to start counting the number of pulses in a fixed time period. The number of pulses is displayed as the counter output as $(N-1)$-bit binary signals $Q_{N-1}...Q_0$.

- If $V_{OUT} > V_{ref}$, $Q_{N-1}...Q_0 > $ '10 ... 0'
- If $V_{OUT} = V_{ref}$, $Q_{N-1}...Q_0 = $ '10 ... 0'
- If $V_{OUT} < V_{ref}$, $Q_{N-1}...Q_0 < $ '10 ... 0'

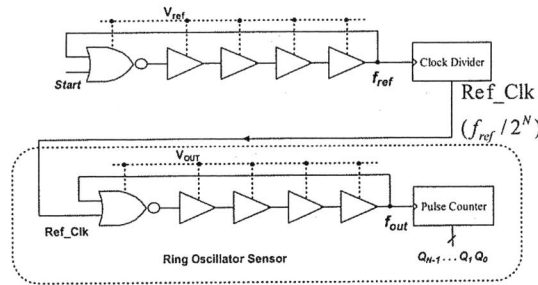

Fig. 8 Ring oscillator sensor circuit

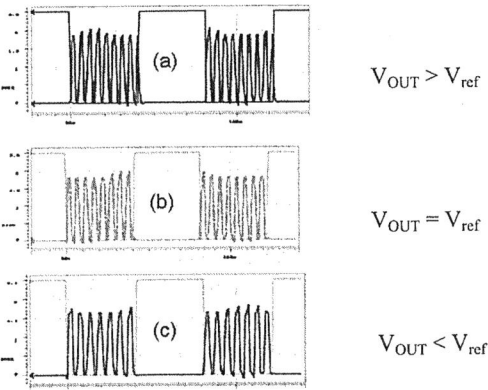

Fig. 9 Simulated results of the sensor, when (a) $V_{OUT} > V_{ref}$, (b) $V_{OUT} = V_{ref}$, and (c) $V_{OUT} < V_{ref}$

Based on the value of $Q_{N-1}...Q_0$, the range of V_{OUT} can be determined and then the controller can use this information to regulate V_{OUT} to V_{ref}. Fig. 9 demonstrates simulation results with $N = 4$. In Fig. 9a, when the reference clock is *LOW*, there are 9 pulses during that interval, so $Q_{N-1}...Q_0 = $ '1001', which is larger than '1000' and agrees well with the statement. Similar analysis can be done when $V_{OUT} = V_{ref}$ and $V_{OUT} < V_{ref}$, where '1000' and '0111' pulses are found from Figs. 9b and 9c, respectively.

3.3 The Closed-Loop System Design

Hysteretic control used to be employed in switch mode DC-DC converter to regulate the output voltage. Its non-linear mechanism makes the close loop system unconditionally dead-beat stable since the control only depends on the current-state comparison between the actual output voltage and one of the hysteretic bounds, performed by a hysteretic comparator. It never relies on the information in past switching cycles. Its drawback occurs when there is a load change. Large voltage ripples are observed at the output, because the inductor current cannot be promptly changed to adapt to the load change. Charge pump does not suffer from this due to the absence of inductive elements. When the output voltage is higher than the reference, the PMOS switches are turned off and the output voltage drops immediately. When the output voltage is lower than the reference, the PMOS switches are turned on. The charge stored on the pumping capacitor is transferred to the output. The output voltage thus rises immediately. Hence, hysteretic control can be applied here with a fast transient response.

The overall system architecture employing hysteretic closed-loop control has been shown in Fig. 6. The reference voltage V_{ref} powers a ring oscillator and generates four Ref_clk_is, $i = 1...4$, through digital logic gates. These four clocks then are used as the *Start* signals of four identical sensors powered by the output voltage V_{out}. Recall that in section 3.2, if $V_{out} > V_{ref}$, Q_{N-1} will be **1** and if $V_{out} < V_{ref}$, Q_{N-1} will be **0**. Thus, the sensors' outputs, Qi_{N-1} s, $i = 1...4$, can indicate whether the output voltage is larger or smaller than V_{ref}. In addition, Q_{N-1} and $\overline{Q_{N-1}}$ are used as the inputs to a *HIGH* effective R-S latch with Q_{N-1} connected to R and $\overline{Q_{N-1}}$ connected to S. The output of the R-S latch is used to determine the instants of transferring energy to the output of the converter. Here the hysteretic band is set as 1 LSB of the ring-oscillator A/D converter.

4. SIMULATION RESULTS

The proposed SC power converter is designed, layout and simulated using a TSMC 0.35 μm CMOS N-well process. The design has been sent out for fabrication. Hspice post-layout simulations show that the converter successfully provides a well-regulated output at a range of 1.6 to 2.7 V, with a supply voltage of 1.5 V. The nominal output voltage is 2.5 V with the pumping capacitors of 1 μF and output capacitor of 2.2 μF. The maximum efficiency is 81%. The following simulation results are performed under two different situations: normal and fault-tolerance operation.

4.1 Normal Operation Simulations

Fig. 10 shows the output ripple voltage, when the converter powers a 12.5-Ω load. The peak to peak voltage is around 40 mV, which can be further improved by increasing A/D converter's resolution. In this design, the A/D converter resolution is 20 mV with $N=7$. Note that the steady state waveform here is different from the periodic waveform from an analog controller, where the regulation is continuous and an accurate duty ratio can be obtained. While in digital control, due to finite-resolution on duty ratio, the ripple voltage will not be periodic. We add/substract one error-correction short pulse after several switching clocks to make the average duty ratio equal to the ideal case. Hence, a few

'unusual' ripples are observed in Fig. 10, which is beneficial to the stability of the converter. In Fig. 11, the load transient response is simulated. We introduce a step change in the load current between 100 mA and 200 mA, to observe the dynamic behavior. The top trace in Fig. 11a shows that V_{out} has no obvious variations to load change. Figs. 11b and 11c show the close-ups when the load current steps up and down, respectively. It takes the converter less than 2 us to recover with only 22 mV variation. Fig. 12 shows the simulated reference clock signals Ref_clk_i and the pumping capacitor signals, during the load change. It illustrates that, when the load becomes heavy, more *High* appears in pumping capacitor signals, thus more energy is transferred to the output capacitor through the pumping capacitor.

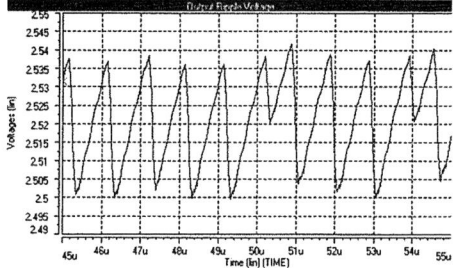

Fig. 10 Output ripple voltage with 12.5 Ω load

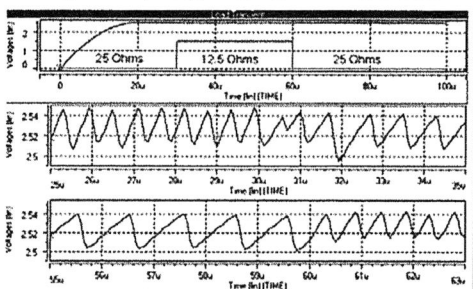

Fig. 11 Load transient at V_{out}: (top) the overall view, (middle) a close-up when the load current steps up, (bottom) a close-up when the load current steps down.

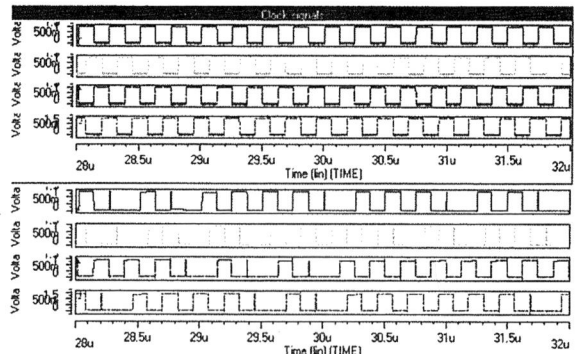

Fig. 12 Clock signals during load transient: (top) clock signals, (bottom) pumping capacitor signals.

4.2 Fault-Tolerance Simulations

We make one CP cell or one control loop fail to examine the converter's robustness. Without losing genericness, we assume that one device/circuit failure makes

$Q3_{N-1}$ always *LOW* and CLK_3 thus keeps having a 50% duty cycle. With this failure, Fig. 13 shows that the peak-to-peak ripple at V_{OUT} is still less than 40 mV. The load transient simulation is shown in Fig. 14. With the same simulation conditions used in normal operation simulations, the converter demonstrates a prompt transient response performance of 3.2 µs. Because only three CP cells are pumping the energy to the output, the response time is a little slower than the regular case. However, compared to most existing designs, it is still very competitive. The clock signals for pump capacitor are shown in Fig. 15. Due to the failure, *CLK3* always has 50 % duty cycle. But the other CP cells compensate the failure and the output voltage can still be regulated well.

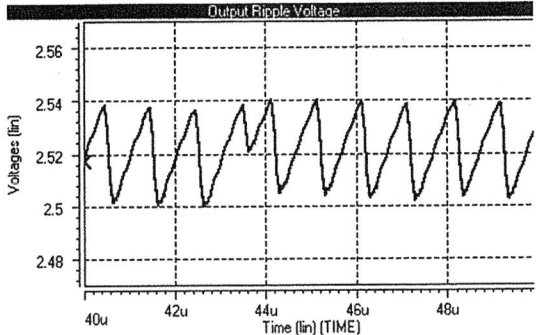

Fig. 13 Output ripple voltage in fault-tolerance simulation

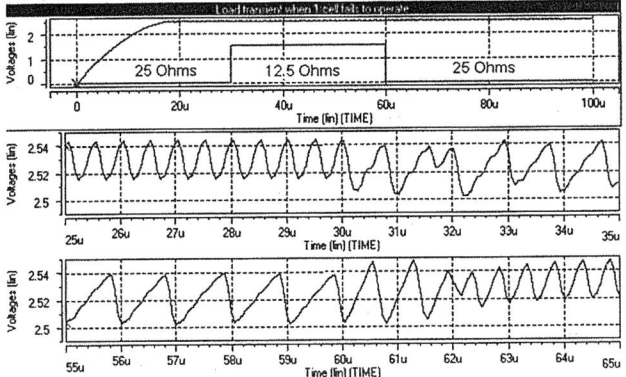

Fig. 14 Load transient response at V_{out} in fault-tolerance simulation: (top) the overall view, (middle) a close-up when the load current steps up, (bottom) a close-up when the load current steps down.

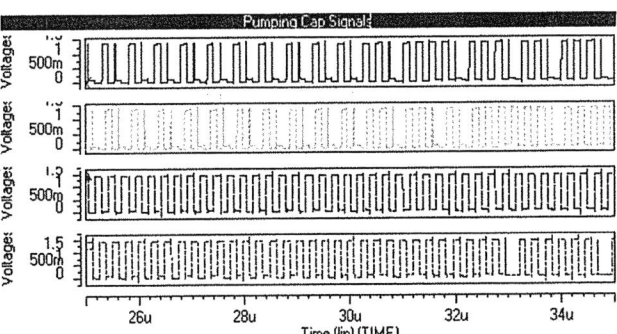

Fig. 15 Pumping capacitor clock signals in fault-tolerance simulation.

5. CONCLUSIONS

In this paper, we propose a new integrated SC DC-DC converter. With the proposed interleaving operation, both input current ripple and output voltage ripple in the converter are significantly reduced. A fully digital control makes the system more robust to noise and even device failure. Hysteretic control is adopted in the controller to achieve fast transient response. Simulation results show that the interleaving structure enhances the system robustness.

6. REFERENCES

[1] M. Nagata, "Limitations, innovations, and challenges of circuits and devices into a half micrometer and beyond", *IEEE J. of Solid-State Circuits*, Apr. 1992, pp. 465-472.

[2] R. Gonzalez, et al, "Supply and threshold voltage scaling for low power CMOS", *IEEE J. of Solid-State Circuits*, Vol. 32, Aug. 1997, pp. 1210-1216.

[3] J. F. Dickson, "On-chip high-voltage generation in MNOS integrated circuits using an improved voltage multiplier technique", *IEEE J. of Solid-State Circuits*, June 1976, pp. 374-378.

[4] J. A. Starzyk, et al, "A DC-DC charge pump design based on voltage doubles", *IEEE Trans. on Circuits & Systems*, March 2001, pp.350-359.

[5] Y. Moisiadis, et al, "A CMOS charge pump for low voltage operation", *IEEE ISCAS*, May 00, pp. 577-580.

[6] S. H. Nork, "Charge Pump DC/DC Converters with Reduced Input Noise" U.S. Patent 6,411,531, 2002.

[7] T. L. Botker, H. Zhang, "Charge Pump Having Very Low Voltage Ripple", U.S. Patent 6,661,683, 2003.

[8] D. J. Perreault, J. G. Kassakian, "Distributed interleaving of paralleled power converters", *IEEE Trans. on Circuits & Systems-I*, Aug. 1997, pp. 728-734.

[9] T. Saito, et al., "Interleaved Buck Converters Based on Winner-Take-All Switching", *IEEE Trans on Circuits & Systems I*, Volume 52, Aug. 2005, pp. 1666-1672.

[10] J. Kim, M. Horowitz, "An Efficient Digital Sliding Controller for Adaptive Power-Supply Regulation", *IEEE Symp. on VLSI Circuits*, June 2001, pp. 133-136.

A Low Power Viterbi Decoder Implementation using Scarce State Transition and Path Pruning Scheme for High Throughput Wireless Applications

Jie Jin and Chi-Ying Tsui

Dept. of Electrical and Computer Engineering, The Hong Kong University of Science and Technology,
Clear Water Bay, Kowloon, Hong Kong
{eetsui, eejinjie}@ee.ust.hk

ABSTRACT

This paper presents a low power Viterbi decoder design based on Scarce State Transition (SST). We propose an approach which seamlessly integrates the path pruning techniques with the SST decoding to reduce the average add-compare-select (ACS) computation. The scheme has very low overhead and is practical for implementation. We also propose an uneven-partitioned memory architecture for the survivor memory unit to reduce the memory access power during the trace back operation. The proposed decoder is implemented in SMIC 0.18μm CMOS process. Simulation results show that significant power consumption reduction can be achieved for high throughput wireless systems such as MB-OFDM Ultrawide-band applications.

Categories and Subject Descriptors

B.7.1 [**Integrated Circuits**]: Types and Design Styles –
Algorithms implemented in hardware.

General Terms: Design.

Keywords

Convolutional code, low power, Viterbi algorithm.

1. INTRODUCTION

Convolutional codes are widely used in modern digital wireless communication systems such as IEEE 802.11, IEEE 802.16 and MultiBand OFDM UWB systems. The Viterbi algorithm (VA) [1] is an optimal solution for decoding the convolutional codes. Because of the highly regular computation and storage operation, VLSI architecture for Viterbi decoder has been widely deployed for the channel decoder for high speed wireless systems. However, as the trend of appealing for higher data rate continues in the wireless applications, the power consumption of the Viterbi decoder, which could account for as much as one third of the power consumption of the baseband processing [2], becomes one of the most critical issues for designing Viterbi decoder.

A conventional Viterbi decoder contains three main units: 1) a branch metric unit (BMU) which calculates the branch metrics; 2)

Permission to make digital or hard copies of all or part of this work for personal or classroom use is granted without fee provided that copies are not made or distributed for profit or commercial advantage and that copies bear this notice and the full citation on the first page. To copy otherwise, or republish, to post on servers or to redistribute to lists, requires prior specific permission and/or a fee.
ISLPED '06, October 4–6, 2006, Tegernsee, Germany.
Copyright 2006 ACM 1-59593-462-6/06/0010...$5.00.

an add–compare–select unit (ACSU) which recursively accumulates the branch metrics as the path metrics (PM), makes decision to select the most likely state transitions and generates the corresponding decision bits and 3) a survivor memory unit (SMU), which stores the decision bits and generates the decoded output. Among these three units, the ACSU and SMU consume most power of the decoder.

To meet the high throughput requirement of the modern communication systems, e.g. 480Mbps for UWB system, the fully parallel architecture is commonly used for implementing the Viterbi decoder. In the ASCU, 2^{K-1} ACS computation units are used and operate in parallel, where K is the constraint length of the convolutional code. Since many ACS units are running at a high clock frequency and hence the ACSU consumes a lot of power. The SMU consumes more than half of the power of the conventional Viterbi decoder because of the large number of memory accesses and hence is another high power consumption unit [2]. There are two methods in the implementation of SMU [3], i.e. register exchange (RE) and trace back (TB). In general, RE has the advantage of high speed, low latency and simple control. However it consumes more power than the TB mechanism since it needs to move the data among the memories in every cycle. Therefore TB mechanism is the most commonly used implementation for the SMU. In [3] a *k*-pointer algorithm was proposed for the efficient implementation of the TB-based SMU design. Here, the SMU is divided into several memory blocks. Simultaneous traceback and decode operations are carried out in order to provide enough bandwidth for the SMU decode operation. In this architecture, several memory read operations are required in order to decode one bit. Thus the amount of the memory access operation is large.

Numerous methods have been proposed to reduce the power consumption of the Viterbi decoder by exploring the different aspects of the system characteristics. Limited search algorithms [2], [4] and [5] has been proposed to reduce the number of average ACS computation and the path storage required by VA. One of the examples is the T-algorithm [6] which is basically a breadth-first decoding algorithms. Instead of computing and keeping all the 2^{K-1} states in each stage as in the traditional VA, some paths are purged according to certain criterion. Specifically, at each decoding stage, only some of the most likely paths with the accumulative path metric satisfying a certain pre-set threshold from the best path metric are kept. Large amount of the ACS computation can be reduced, while the degradation in performance is small. However, it is a challenge to implement a parallel high-throughput T-algorithm as we need to do serial sorting/comparison operation in order to search for the best path metric at each stage. This limits the decoder to achieve a high throughput [10].

To realize a low power Viterbi decoder, we want to use the T-algorithm which has the superior algorithmic performance in reducing the average number of ACS and the survivor paths. At the same time, we want to eliminate the serial sorting operation to find the best path metric in order to achieve a high throughput and reduce the power overhead of the searching. In this work, we proposed a novel architecture that implements the T-algorithm based on the Scarce-state-transition (SST) decoder structure. SST [9] was first introduced to reduce the switching activities of the Viterbi decoder. However, it cannot reduce the average number of the ACS calculation. By seamlessly integrating the T-algorithm and the SST together, we can reduce the complexity without the need of finding the best path metric at each decoding stage. Also the overhead of the implementation is very small, making the implementation very practical. Moreover, by exploiting the probability distribution of the maximum likelihood states in the SST decoder, we propose an uneven-partitioned memory architecture for the trace-back unit of the SMU to reduce the power consumption due to the memory access of the SMU. We applied our proposed technique on the design of a Viterbi decoder targeting for the MBOA-OFDM UWB [11] system, which has a very high throughput requirement. Experimental results show that both the power of the ACSU and the SMU are reduced significantly.

The rest of the paper is organized as follows. Section II gives the background of the T-algorithm and the SST decoding scheme. The proposed path pruning scheme is presented in section III. The architecture for the proposed scheme is then described in section IV. In section V, experimental results on the implementation of the proposed Viterbi decoder targeting UWB applications is presented. Conclusions are drawn in section VI.

2. BACKGROUD

2.1 T-algorithm

The T-algorithm is similar to VA except that the number of the survivor paths is not constant. Unlike the traditional VA which retains all the 2^{K-1} states, only some of the most-likely paths are kept at every trellis stage in the T-algorithm. The detail of the T-algorithm is described as follows.

Every surviving path at the trellis stage $l-1$ is expanded and its successors at stage l are kept if their corresponding path metric values are smaller or equal to $d_m + T$, where T is a preset pruning threshold decided by user and d_m is the smallest path metric of all the survivor states at stage $l-1$. There are different variations of the scheme. In [7], the number of the states or survivor paths stored is restricted to a maximum number N_{max} set by user, which is less than 2^{K-1}. Among the N_{max} states, only the states with cumulative path metric satisfying the path threshold restriction are kept. By using a proper threshold T, significant amount of the paths are pruned, while the BER performance is maintained. The corresponding ACS computation for the pruned paths are saved, thus the computation complexity is reduced. The detail discussion of the choice of the value of T on the performance and the number of pruned path can be found in [5] [6].

One of the requirements of the T-algorithm is a serial comparison operation for searching the best metric in each decoding stage. This limit the T-algorithm to be used for high throughput applications. In the worst case, there are 2^{K-1} states and it requires 2^{K-1} comparisons to find the best path metric. For low throughput applications, the comparisons can be done in multiple cycles. Most of the previous architectures for T-algorithms are proposed for the low throughput applications [7] [8]. However, in high throughput applications, the

fully parallel ACS units are implemented and the ACS computation for each stage is completed in one clock cycle. Thus, the comparison must be computed in one cycle and the hardware and power overhead to find the best metric in one cycle is great, especially when the number of the states is large. The work in [10] tried to solve the problem by relieving the requirement for the comparison in one cycle to v cycles, where v is the latency of the comparison operation. The best path metric is estimated with errors and then is corrected every v cycles. In this work, we propose an alternative way to address this issue by totally eliminating the requirement of finding the best path metric. Instead the best path metric is approximated by the default best path metrics of the SST decoder. In the following section, we will give a brief introduction on the SST decoding scheme.

2.2 SST decoding algorithm

The scarce state transition (SST) Viterbi decoder [9] was proposed to minimize the switching activity of the decoded bits and also reduce the truncation length. Fig. 1 shows a block diagram of a rate 1/3 SST decoder.

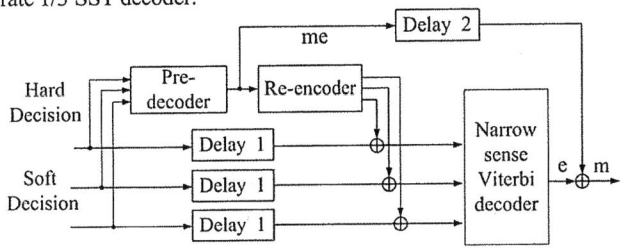

Fig. 1 Block diagram of a SST Viterbi decoder with R = 1/3

The received data is first pre-decoded by a simple pre-decoder, which performs the inverse of the encoder. The pre-decoded signal which contains the information sequence and channel errors is re-encoded and ex-or with the original received data before input to the Viterbi decoder. The input is thus mainly the error of the sequence. Then the Viterbi decoder is used to correct the errors of the information sequence. At last, the final decoded sequence is obtained by adding the decoded output of the Viterbi decoder with the pro-decoded sequence using modulo-2 addition. The SST decoder has the following properties: When the channel errors are small, most of decoded output bits of the Viterbi decoder are zero. Thus the switching activity of the SST decoder is much smaller than that of the conventional Viterbi decoder. This is true for most of the practical SNR ranges for a typical communication system. Most of the time the survivor path i.e. the decoded sequence will pass through the zero state and the zero state most likely has the smallest path metric. Thus the probability distribution of the maximum likelihood states is no longer equal to that of the original VA. By exploring this new state probability distribution of the SST decoding, we propose a new path pruning scheme to facilitate the implementation of the T-algorithm for high throughput applications in the following section.

3. PATH PRUNING SCHEME BASED ON SST VITERBI DECODER

With the SST scheme, the zero state is most likely to be the best state. Most of the time, the cumulative path metric d_0 of the zero state equals to the best path metric d_m at high SNR. Thus we can use d_0 instead of d_m as the basis for the path pruning. The complex sorting or comparing operation to find d_m per trellis stage is eliminated. As the value d_0 is computed in the normal ACS calculation, there is no overhead in obtaining the estimated best path metric. The proposed scheme is expressed in the following:

Let s, k represent two states in the trellis diagram and s is the predecessor state of k. Let bm_s^k denote the branch metric of the state transition s to k. The path metric of the state k at stage l is donated as $d_k(l)$ and the path metric of the zero state at stage $l-1$ is donated as $d_0(l-1)$. When calculating the path metric at stage l, only the paths meet the following threshold condition are kept:

$$d_k(l) = d_s(l-1) + bm_s^k \leq d_0(l-1) + T \qquad (1)$$

The path metric $d_0(l-1)$ is used instead of $d_0(l)$ so that the decision to determine whether a path at stage l should be kept or not can be made without waiting for $d_0(l)$ to be computed.

Since in the ACS computation, only the difference between two candidate path metrics will affect the results, thus we can subtract $d_0(l-1) + T$ from all the path metrics. Equation (1) can be expressed as

$$d_s(l-1) + bm_s^k - \left(d_0(l-1) + T \right) \leq 0 \qquad (2)$$

Let $q = bm_s^k - \left(d_0(l-1) + T \right)$ denote the new branch metric, then (2) is expressed as

$$d_s(l-1) + q \leq 0 \qquad (3)$$

The left hand side of (3) is now the new path metric. By checking the sign of the path metric, we can determine whether this path should be kept or not instead of comparing it with the threshold in (1) like the T-algorithm. By using the above transformation, the overhead for the pruning scheme is kept to minimum since the number of the branch metrics is usually very small. It can be seen that the predecessor of the zero state is most likely from the zero state also. Thus, $d_0(l)$ most likely equals to $d_0(l-1) + q = bm_0^0 - T$. If bm_0^0 is subtracted from all the path metrics, the path metric of the zeros state most likely will be $-T$, which is a constant and the switching activities of the zero state is reduced. The new branch metric is now expressed as:

$$q' = bm_s^k - \left(d_0(l-1) + T + bm_0^0 \right) \qquad (4)$$

and it is computed by subtracting $d_0(l-1) + T + bm_0^0$ from the original branch metric which can be easily implemented by modifying the conventional BMU. Comparing with the structures proposed in [7] [8] and [10] for T-algorithm, the sorting or comparison units are eliminated.

The zero state may be deviated from the maximum likelihood states and the number of the survivor paths kept for the proposed scheme can be larger than that of the traditional T-algorithm for the same threshold value T. This will result in a lower saving in ACS reduction. Thus a different threshold value T should be set. By setting a proper threshold value T, significant amount of the ACS computation can be saved, while the BER performance of the VA is maintained. The effect of the different values of T on the BER performance and computation reduction is shown in Fig. 2 and Fig. 3. The simulation is performed for the transmission mode of the data rate 160Mbps of the UWB system using the CM1 channel environments [12]. The design SNR for the data rate of 160Mbps is around 8dB.

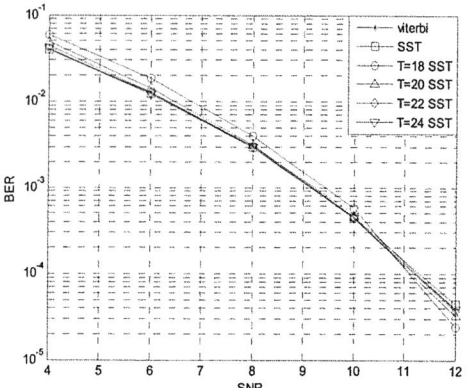

Fig. 2 BER performance for CM1 with data rate=160Mbps

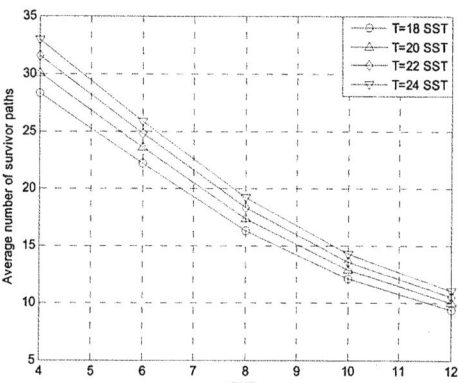

Fig. 3 Average number of survivor paths for CM1 with data rate=160Mbps

Fig. 2 shows that the proposed scheme can achieve the similar performance of the traditional VA and the original SST, if $T \geq 22$. Fig.3 shows that around 45%~80% of the paths can be pruned without affecting the performance. For channels with higher SNR values, the proposed scheme works better i.e. the average number of pruned path increases. However, a large amount of reduction in the computation does not guarantee to have a significant power saving. Proper hardware should be designed to transform the reduced computation at the algorithm level to the reduced switching activities in the hardware. To realize the power consumption reduction, a Viterbi decoder for the MBOA-OFDM based Ultra-wide-band (UWB) system is presented.

4. PROPOSED VLSI ARCHITECTURE

Fig. 4 shows the block diagram of the proposed architecture. 5-bit soft metrics are input to the decoder. The BMU is modified to support the calculation of the new branch metric as stated in equation (4). The ACSU contains 64 parallel modified ACS units which support the path pruning and also the stopping of the ACS calculation when the path is pruned. Traceback mechanism is used for the SMU. We employed the 3-pointer even algorithm [3] which is also adopted in [13] for low power consumption. As the zero state is most likely the maximum likelihood state, the traceback operation always begins with the zero state.

4.1 ACS design

The structure of an ACS unit in the ACSU is shown in Fig. 5. An additional signal S is used to indicate whether the path is pruned or not. S is determined by the sign (i.e. MSB) of the calculated path metric $d_0(l)$ as discussed in the previous section. S is also used to gate the path metric register to reduce the switching activity of the ACS if the path is pruned. In addition it is used to mask the input to the adder and the comparator of the ACS if the path is pruned. We cannot just gate the branch metric as it is used in other ACS units. Also we cannot completely disable the whole ACS unit as there are cases that only one path is pruned while the other path is active. To reduce the switching activity, we use the S signal together with an AND gate to mask the input signals to the adders and the comparator instead of using complicated clock gating control. For the pruned paths, most of the time the input to the adder and the comparator will be zero and there is not switching activity. To eliminate glitches, we make sure that the S signal to the AND gate is always the earliest input. The area overhead for the additional hardware is small. The cell area for the conventional ACSU and the proposed ACSU for a 0.18μm CMOS process are given in Table I.

4.2 SMU

The 3-pointer even algorithm proposed in [3] is used in the implementation of the SMU. However, the large number of memory accesses during the traceback and decoding stage and the wide memory word width lead to large power consumption in the SMU. In general, in order to generate the decoded output at the required truncation length L, more read operations are required than write operations in the TB.

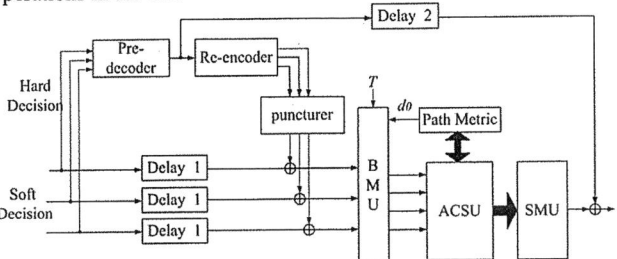

Fig. 4 Block diagram of proposed decoder

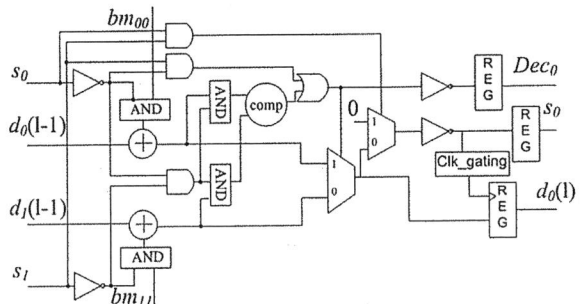

Fig. 5 The proposed ACS structure

TABLE I
AREA COMPARISON OF THE ACS UNIT

Designation	Cell Area (mm²)
Conventional Viterbi	0.25
SST-Path-thresholding	0.28

TABLE II
TOTAL MEMORY AREA WITH DIFFERENT BIT WIDTH

Memory partitioning configuration[1]	32* (64*2)	16* (64*4)	4* (64*16)	2* (64*32)	1* (64*64)
Area (mm²)	1.43	0.82	0.37	0.29	0.25

[1](No. * # of entry * bitwidth)

For example, for the 3-pointer even algorithm used in [13], 6 banks of memories are used. To decode a single bit, one write and three read access of the memory of 64 bits wide are required. One way to reduce the power consumption of the SMU is to reduce the power consumption for the read operation.

For the traceback read operation, it is inefficient to read all the 64 decision bits at each stage. Instead we should just read out the required bit and not access the other bits. The power consumption for the read operation will be greatly reduced. However in order to do this we need to partition the memory into many smaller units which can be addressed and enabled separately. The area overhead for having a large number of small memories can be very large. Table II gives the cell area for different memory partition configurations with different bit width for a 64x64 memory unit. The embedded memories are generated by Artisan memory generator with 0.18μm technology. We can see that having large number of small memory partitions to facilitate the low power reading has significant area overhead. E.g. the area is increased by almost 5 times if we partition the memory into 32 blocks of 2-bit wide memory. Therefore we want to have a small number of partitions and at the same time can reduce the power of the read access. Here we propose an uneven-partitioned memory architecture for the SMU based on the maximum likelihood state probability distribution of the SST scheme.

In the SST decoding, the Viterbi decoder is used to decode the errors of the information sequence. When the channel errors are small, the decoded bits are most likely to be zero. Thus the zero state is most of the time the maximum likelihood state, while for conventional VA, the maximum likelihood state is evenly distributed across all states. Therefore the probabilities of the states being the maximum likelihood state are no longer equal to that of the conventional VA. Fig. 6 and 7 show the probability distribution of all the 64 states for the UWB systems. The data was obtained for a data rate of 160Mbps at SNR=8dB and 10dB with the CM1 channel environments. It shows that the zero state has highest probability to be the maximum likelihood state. As SNR increases, the channel errors are fewer and the probability of the zero state increases. From Figures 6 and 7, we can also see that besides the zero states, the probabilities of the other states are also not evenly distributed. There are groups of states that have higher probabilities. These are the states that are more directly reachable from the zero state (e.g. state 32) or directly transit to the zero state (e.g. state 1).

Based on this uneven state probability distribution, we propose an uneven-partitioned memory architecture for the SMU. The path of the decoded sequence most likely passes through the states with high probability. Therefore the decision bits of the states with higher probability are more likely to be read. We store the decision bits of the states with higher probability into memory with smaller bit-width and the decision bits of the states with lower probability into another memory with large bit-width. The read operation thus will most of the time access the smaller memory and the overall power consumption of the read operation is reduced compared with that of reading all the 64 bits out in every cycle. The number of the partitioned memory should be small in order to reduce the area overhead. Table III gives the estimated power consumption of the traceback read operation for several different partitioning configurations, assuming the power consump- tion of the read access is proportional to the bit-width of the memory. We chose the configuration 1, 15, 48 as it provides the best area and power tradeoff. The one bit memory partition is used to store the decision bits of the zero state and is implemented in registers instead of embedded SRAM. The 15-bit memory is used to store the decision bits of the fifteen states with the next highest probabilities besides the zero state.

The block diagram of the proposed memory unit is shown in Fig. 8. A simple memory enable signal generator is used to enable the read access of the two embedded SRAMs. The average read access rate of the different memories for the UWB system with a data rate 160Mbps under various SNRs is shown in Fig. 9. It can be seen that most the time, the memory for the zero state is read and the corresponding access rate increases as the SNR increases. For the low SNR range, significant amount of power can still be saved. The total access rate for the 15-bit memory and 48-bit memory two is up to 60%.

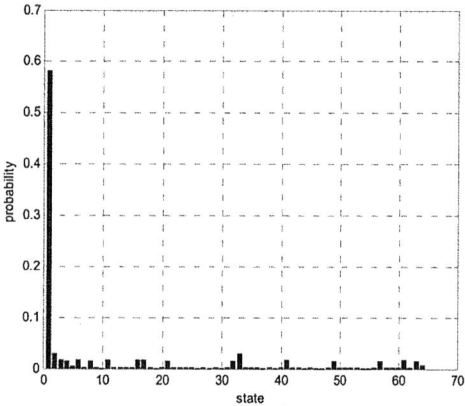

Fig. 6 Distribution probability of the maximum likelihood states at SNR=8dB for UWB system

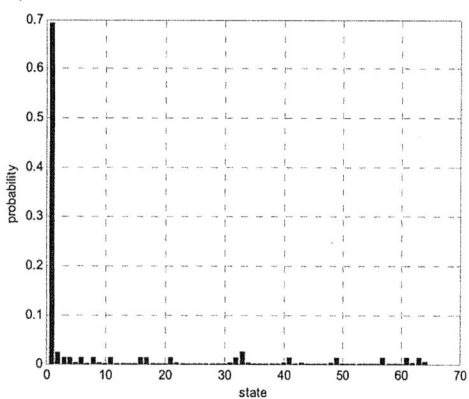

Fig. 7 Distribution probability of the maximum likelihood states at SNR=10dB for UWB system

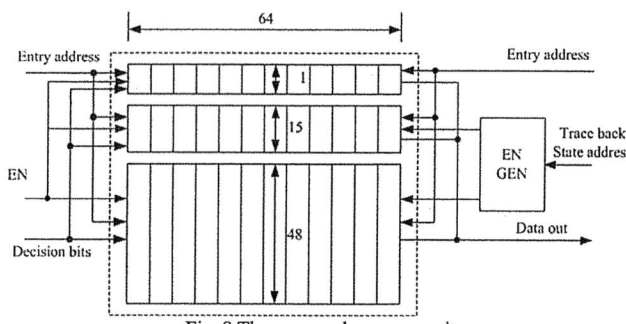

Fig. 8 The proposed memory unit

TABLE III
ESTIMATED POWER CONSUMPTIOIN FOR TRACEBACK READ OPERATION

Memory	Normalized Power for reading			
	6db	8db	10db	12db
64	1	1	1	1
1,63	0.56	0.42	0.31	0.24
1,31,32	0.28	0.21	0.16	0.12
1,15,48	0.26	0.18	0.13	0.09
1,15,16,32	0.17	0.13	0.09	0.07

5. Experimental Results

We implemented the Viterbi decoder for a MBOA-OFDM based Ultra-wide -band system [9] using our method. We also compared the power consumption with the conventional Viterbi decoder. For the UWB system, a convolutional code with constraint length 7 is used. The generator polynomials are 133_8, 165_8 and 171_8, respectively. We simulated the performance of the system using the CM1 channel environment [12] with 100 channel realizations. The received symbols are quantized to 5-bit soft metric. The Viterbi decoders of the VA, SST and the SST-path thresholding scheme were implemented in VHDL and then synthesized with Synopsys (Design Compiler) using the Artisan's SMIC 0.18μm standard cell library.

The embedded SRAM is generated by the Artisan's memory generator. The power consumption was simulated using synopsys VCS-MX and power compiler. One frame of the data generated by the UWB system under different SNRs is used to simulate the power consumption. The supply voltage is 1.8V and the clock frequency is 200MHz.

Tables IV and V summarize the power consumption of different part of the Viterbi decoder. Table IV shows the power of the computational parts of the decoder (i.e. the BMU, ACSU and the additional logic for SST decoding). Table V shows the power consumption of the traceback and decoding of the SMU. From table IV, we can see that 30%~76% reduction in power consumption is obtained for the computational parts over the traditional design for different SNR values. For the read access during TB and decoding, the power consumption can be saved by as much as 80% when uneven-partitioned memory is used. Table VI
summaries the overall cell area comparison of different schemes. With a small area overhead, the power consumption of a high-throughput parallel Viterbi decoder can be reduced significantly.

6. Conclusion

A low power Viterbi decoder based on SST and path pruning was proposed to reduce the ACS power consumption. By exploiting the characteristic of the maximum likelihood state probability distribution of the SST decoder, we proposed an uneven-partitioned memory architecture for the survivor memory unit. With these two schemes, the proposed decoder can achieve a significant power reduction.

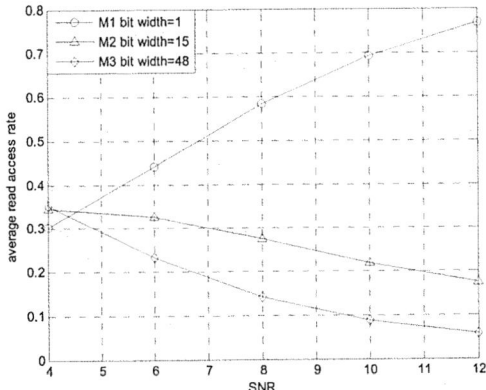

Fig. 9 The read access rate of memories for data rate 160Mbps for UWB system

TABLE IV
POWER ESTIMATION RESULTS (IN mW) OF THE COMPUTATIONAL PARTS FOR DIFFERENT DECDOERS

Designation	Power under different SNRs				
	4dB	6 dB	8 dB	10 dB	12 dB
Conventional Viterbi	78.84	79.18	78.68	78.88	78.63
SST	80.58	80.48	77.18	75.10	72.25
SST-Path Thresholding	64.44	58.88	37.78	27.15	17.27

TABLE V
POWER ESTIMATION RESULTS (IN mW) OF SMU

SMU	4db	6db	8db	10db	12db
Conventional Viterbi	194.34	194.34	194.34	194.34	194.34
SST-Path Thresholding	75.90	57.09	40.47	29.13	22.17

TABLE VI
CELL AREA OF DIFFERENT DECODERS

Designation	Area (mm^2)
Conventional Viterbi	1.06
SST	1.10
SST-Path-thresholding	1.26

References

[1] A. J. Viterbi, "Error bounds for convolutional codes and asymptotically optimum decoding algorithm," *IEEE Trans. Inf. Theory*, vol. IT-13, no.2, pp. 260–269, Apr. 1967.

[2] R. Henning and C. Chakrabarti, "An Approach for Adaptively Approximating the Viterbi Algorithm to Reduce Power Consumption While Decoding Convolutional Codes," *IEEE Trans. Signal processing,* vol. 52. May 2004.

[3] G. Feygin and P. Gulak, "Architectural tradeoffs for survivor sequence memory management in Viterbi decoders," *IEEE Trans. Commun.*, vol. 41, no. 3, pp. 425–429, Mar. 1993.

[4] J. B. Anderson, "Limited search trellis decoding of convolutional codes," *IEEE Trans. Inf. Theory*, vol. 35, pp. 944–955, Sep. 1989.

[5] R. Chan and David Haccoun, "Adaptive Viterbi Decoding of Convolutional Codes over Memoryless Channels," *IEEE Trans. Communications,* vol. 45, Nov. 1997.

[6] S. J. Simmons, "Breadth-first trellis decoding with adaptive effort," *IEEE Trans. Commun.*, vol. 38, pp. 3–12, Jan. 1990.

[7] R. Tessier and S. Swaminathan and R. Ramaswamy, D. Goeckel and W. Burleson, "A Reconfigurable, Power-Efficient Adatpive Viterbi Decode -r, " *IEEE Trans. VLSI Systems,* vol. 13, Apr. 2005.

[8] Man Guo, M. O. Ahmad, M. N. S. Swamy, and Chunyan Wang, "FPGA design and implementation of a low-power systolic array -based adaptive Viterbi decoder," *IEEE Trans. Circuits and Systems I,* vol. 52. Feb. 2005.

[9] S. Kubota, S. Kato and T. Ishitani, "Novel Viterbi Decoder VLSI Implementation and its Performance," *IEEE Trans. Communications,* vol. 41, Aug. 1993.

[10] Fei Sun and Tong Zhang, "Parallel High-Throughput Limited Search Trellis Decoder VLSI Design," *IEEE Trans. VLSI Systems,* vol. 13, Sept. 2005.

[11] Anuj Batra et al., Texas Instruments et al., Multi-band OFDM Physical Layer Specification, Release 1.0, Apr. 27, 2005.

[12] J. Foerster, Ed., "Channel modeling sub-committee report final," IEEE802.15-02/490.

[13] C. C. Lin, Y.H. Shih, H. C. Chang, and C. Y. Lee, "Design of a Power-Reduction Viterbi Decoder for WLAN Applications," *IEEE Trans. Circuits and Systems I,* vol.52, June, 2005.

SmartSaver: Turning Flash Drive into a Disk Energy Saver for Mobile Computers

Feng Chen
The Ohio State University
Columbus, OH 43210, USA
fchen@cse.ohio-state.edu

Song Jiang
Wayne State University
Detroit, MI 48202, USA
sjiang@ece.eng.wayne.edu

Xiaodong Zhang
The Ohio State University
Columbus, OH 43210, USA
zhang@cse.ohio-state.edu

ABSTRACT

In a mobile computer the hard disk consumes a considerable amount of energy. Existing dynamic power management policies usually take conservative approaches to save disk energy, and disk energy consumption remains a serious issue. Meanwhile, the flash drive is becoming a must-have portable storage device for almost every laptop user on travel. In this paper, we propose to make another highly desired use of the flash drive — saving disk energy. This is achieved by using the flash drive as a standby buffer for caching and prefetching disk data. Our design significantly extends disk idle times with careful and deliberate consideration of the particular characteristics of the flash drive. Trace-driven simulations show that up to 41% of disk energy can be saved with a relatively small amount of data written to the flash drive.

Categories and Subject Descriptors

D.4.2 [**Storage Management**]: Secondary Storage

General Terms

Design, Experimentation, Performance

Keywords

Hard disk, flash drive, energy saving, mobile computer

1. INTRODUCTION

As one of the major energy consumers in a mobile computer [9], the hard disk accounts for a considerable amount of energy consumption. Constrained by its mechanical nature, hardware support for disk energy conservation has not changed much over the years. Existing *Dynamic Power Management* policies are still using a simple timeout strategy to save disk energy: Once disk is idle for a specific period (timeout threshold), it is spun down to save energy. Upon arrival of a request, the disk is spun up to service the request. This strategy is also the basis of most existing disk energy-saving schemes for mobile computers [3, 4, 6, 13, 15].

Recently, with a rapid technology improvement, the flash drive has quickly taken the place of the floppy disk as a convenient portable storage device. Attracted by its compact size

Permission to make digital or hard copies of all or part of this work for personal or classroom use is granted without fee provided that copies are not made or distributed for profit or commercial advantage and that copies bear this notice and the full citation on the first page. To copy otherwise, to republish, to post on servers or to redistribute to lists, requires prior specific permission and/or a fee.
ISLPED'06, October 4–6, 2006, Tegernsee, Germany.
Copyright 2006 ACM 1-59593-462-6/06/0010 ...$5.00.

and low price, almost every mobile computer user now carries a flash drive on travel. Different from the hard disk, the flash drive is made of solid-state chips without any mechanical components, such as disk platters, which consume a considerable amount of energy. In addition, as a non-volatile storage device, the flash drive does not need power to maintain its data as main memory does. Compared with the hard disk, the energy consumption of the flash drive is almost negligible (its standby energy consumption is only around 1% of disk standby power consumption [14]). Unfortunately, in current mobile systems the flash drive is only used for file transfer or temporary storage, while its low energy consumption advantage has not yet been well exploited. In this paper we present a novel and practical scheme to utilize the low-price and ubiquitous flash drive to achieve a highly desired goal — saving disk energy.

This idea is challenged by two particular characteristics of the flash drive. First, the bandwidth of the flash drive is usually much lower than the peak bandwidth of the hard disk. For example, the Transcend TS1GJF2A Flash drive has a read bandwidth of 12MB/sec and write bandwidth of 8MB/sec [14], while the 4200RPM Hitachi-DK23DA hard disk can achieve a bandwidth of 35MB/sec. Second, the flash drive has a limited number of erasure (rewrite) cycles. Typically, a flash memory cell could wear out with over 100,000 overwrites. These two characteristics of the flash drive must be carefully considered in the scheme design to effectively use it for saving disk energy.

In this paper we present a comprehensive disk energy-saving scheme, called *SmartSaver*. This scheme uses the flash drive as a standby buffer for caching and prefetching disk data to service requests without disturbing disk. Although main memory can also be used as a buffer, it is undesired to use main memory for disk energy saving, as main memory itself is a big energy consumer [11].

Compared with existing disk energy-saving schemes, *Smart-Saver* has the following merits: (1) It effectively exploits the low power consumption feature of the flash drive to save disk energy. (2) It carefully takes the flash drive's particular characteristics into consideration. (3) It is designed to be widely applicable in various operating systems with minimal changes to existing OS kernels. (4) Our experiments show that it can achieve significant disk energy saving by effectively extending disk idle times.

2. RELATED WORK

Existing disk energy-saving schemes for mobile computers can be classified into three groups. The first group of work focuses on the selection of timeout threshold, which could be a fixed time period, such as 2-5 seconds [4], or be adaptively adjusted at runtime [3, 6]. These schemes passively monitor disk I/O operations without extending disk idle time, which greatly limits their energy saving potential. The second group of work customizes system or application software for saving disk energy [13,15]. The proposed scheme in [13] uses aggressive prefetching to create bursty disk accesses for long disk idle periods. While the scheme can extend disk idle time, significant changes are required to modify existing buffer cache management policies in OS kernels. Furthermore, as the buffer cache management policy is a performance-critical component in a kernel, practitioners have to be very cautious about potential performance degradation when they orient the component towards energy saving. In contrast, our scheme can effectively extend disk idle time with minimal changes to existing kernel policies. The third group of work proposes hardware designs to dynamically change disk rotational speed on the fly so that disk energy can be saved by reducing disk speed [5]. Our solution is complementary to this hardware mechanism, while such speed-adjustable disks are not yet available in the mainstream commercial market.

Using the flash memory for disk energy saving was discussed in an earlier work [10]. However, the work only provides a preliminary algorithm and many important design issues are missing, including file prefetching, balanced flash space allocation among cached and prefetched blocks, and consideration of the characteristics of the flash memory. Some of their design choices are inappropriate from today's point of view. For example, they write every missed block into flash memory when it is read from disk. Today's applications, such as movie player, tend to access a large volume of disk data. If all the data have to go through the flash memory, its low write bandwidth and limited erasure cycles would pose a serious problem. Another work [1] proposes to redirect write-back traffic to the flash drive when the disk is spun down so that the number of costly disk spin-up/downs could be reduced. However, this approach simply uses the flash drive as a write-only cache and it has no caching or prefetching mechanisms. Some other work also mentions the low power consumption characteristics of flash memory [12,16]. However, no further detailed design considering the particular characteristics of the flash drive for disk energy saving was proposed. In this paper we provide a comprehensive flash memory management scheme that addresses these issues.

3. ISSUES AND CHALLENGES

A straightforward approach [10] to use the flash drive for disk energy saving is to simply use the flash drive as a new layer between the main memory and the hard disk in the storage hierarchy. This so-called 'cache all' policy simply caches every byte that is read from disk or evicted from memory into the flash drive, and uses the LRU replacement algorithm to free space once the flash drive is full.

After carefully examining this approach, here we summarize its several critical weaknesses, which also serve as technical bases for us to design an efficient flash-drive-based disk energy-saving scheme.

1. **Flash drive does not fit well in the storage hierarchy.** Since the write bandwidth of the flash drive is usually far less than that of the hard disk, simply caching all data that are transferred between the disk and memory in the flash drive can easily make the flash drive a bottleneck in the storage hierarchy.

2. **One-time access data should not be cached at all**. The 'cache all' approach inevitably stores one-time access data in the flash drive, which is a waste of its space and its precious erasure cycles.

3. **What to cache is critical**. Energy saving can be maximized by caching two types of data in the flash drive to effectively extend disk idle times: (1) very frequently reused data, and (2) data that move slowly from/to the disk, such as multimedia data with think time between disk reads and widely scattered data in the disk with long seek times to access. The 'cache all' approach pays no special attentions to these two types.

4. **What to replace is also critical.** Following the same rule mentioned above, we'd better not replace these two types of data when the flash drive is full. However, a standard LRU replacement algorithm could easily replace the second type of data that are infrequently used.

To address these issues, we must hold two principles in the design of our scheme. (1) A flash drive is not simply a 'smaller disk' or a 'larger memory'. Its low write bandwidth and limited erasure cycles need to be carefully considered. (2) Our scheme needs to balance the benefit (how much energy is saved) and the cost (how much flash memory space is demanded). In other words, it should identify and store the most valuable data in the flash drive for energy saving.

4. THE DESIGN OF SMARTSAVER

Typically, the hard disk has four power-consumption states — *active*, *idle*, *standby*, and *sleep*. To save energy, a disk is spun down to the standby state if it is idle for a specific period, and it can be spun up later to the active state for servicing a request. We call the time period between the disk spin-up and its consecutive spin-down a *busy period*, and the time period between the disk spin-down and its consecutive spin-up a *quiet period*. Since spinning up/down disk consumes substantial energy, the disk has to stay in the standby state for a sufficiently long period to compensate the energy overhead. The minimum interval to pay off the overhead is referred to as *break-even* time. Obviously, the longer disk is idle, the more energy can be saved.

To effectively extend disk idle periods, *SmartSaver* uses the flash drive as a buffer to store disk data, which can be used to service requests without disturbing the disk. The available flash drive space managed by *SmartSaver* plays three roles: (1) A caching area for holding data that is likely to be reused; (2) A prefetching area for storing data preloaded from the disk; and (3) A writeback area for temporarily storing the dirty blocks flushed from memory. Accordingly, *SmartSaver* partitions available flash drive space into three areas for caching, prefetching, and writeback, respectively.

4.1 Energy Saving Rate

The goal of caching is to avoid a future busy period by holding disk data that are likely to be reused in the flash drive. The energy that could be saved by avoiding a busy period may vary greatly. For example, *gzip* can compress a 20MB file in a few seconds, while *glimpse* may need a few minutes to build an index for the same file. Obviously, avoiding the busy period of *glimpse* can keep disk idle longer and save more energy than avoiding the busy period of *gzip*, though they have the same flash space cost.

To quantitatively measure the energy-saving potential of avoiding a busy period, we introduce a metric, *Energy Saving Rate (ESR)*. ESR is the energy that could be saved if a busy period was avoided over the amount of data accessed during the busy period. The amount of saved energy equals to the energy spent in the busy period (the sum of disk active energy, disk idle energy, and the energy overhead of spinning up/down disk) subtracted by the disk standby energy spent in a quiet period of the same length. ESR describes the amount of saved energy each cached block could contribute if its busy period was avoided.

4.2 Caching

It is important to manage all the blocks accessed during one busy period as a whole. If we cache only a part of the accessed data blocks, the disk still has to be spun up for accessing the remaining blocks uncached in the flash drive, thus the effort of avoiding a busy period is foiled. To this end,

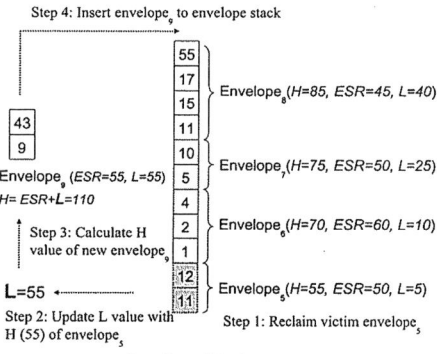

Envelope Stack

Figure 1: An example of envelope stack with four envelopes, among which $envelope_5$ with the smallest H value (55) is at the stack bottom and $envelope_8$ with the largest H value (85) is at the stack top. When a new envelope $envelope_9$ is to be inserted into the full envelope stack, the algorithm performs four steps as shown in the figure.

SmartSaver uses a data structure, called *envelope*, to record the metadata of all accessed data blocks during a busy period. Each envelope is associated with a busy period. When a block is requested from the disk during a busy period, its metadata is recorded in its associated envelope. The blocks in an envelope are organized in the LRU order. When the busy period is completed, its ESR value can be calculated. If we decide to cache the envelope, all the data blocks in it are written into the flash drive to maintain the integrity of the busy period. Cached envelopes are placed in a queue, called *envelope stack*, as shown in Figure 1.

When the caching area is full, we need to identify and reclaim the most 'valueless' blocks in terms of energy saving. A block's value for energy saving is determined not only by how much energy could be saved, which can be presented by ESR, but also by how likely it is to be accessed. The challenge is to evaluate and compare the blocks' value for energy saving simultaneously using these two orthogonal metrics.

To address the challenge, we adopt an algorithm similar to *GreedyDual-Size* [2], which is originally designed for web-caching. In the algorithm when an envelope e is inserted into the envelope stack or accessed in the stack, it is given an H value, where $H(e) = ESR(e) + L$. In the formula $ESR(e)$ is its corresponding busy period's ESR value, L is a global inflation value and is set to the H value of the most recently reclaimed envelope. L is initialized to 0. Envelopes in the stack are sorted in ascending order of their H values from the bottom to the top. When the caching area is full and a replacement is needed, the envelope at the stack bottom is selected, and the blocks in the envelope are reclaimed one by one in the LRU order. Once all blocks in an envelope are reclaimed, the envelope is destroyed, and L is updated using the H value of the destroyed envelope. Thus, the L value keeps being inflated. The most recently accessed block always uses the up-to-date L value, which represents its locality, and its ESR value, which represents its energy-saving potential, to calculate its H value. The H value determines the block's position in the stack and its timing to be reclaimed. Therefore, the block with the smallest H value is the most 'valueless' block and is located at the stack bottom for reclamation.

4.3 Prefetching

Prefetching is used for disk energy saving through *condensing* the sequential disk accesses at the beginning of an access sequence, which is called *stream*. After an initial prefetching of data into the flash drive, future requests to the data can be satisfied from the flash drive without accessing data on the disk. In other words, prefetching makes the *evenly* distributed disk accesses more *bursty*. In this section, we first explain how to identify the sequential disk accesses and then describe the prefetching mechanism itself.

Effective prefetching in *SmartSaver* requires an accurate identification of sequential accesses of files. In order to avoid intrusive changes to existing buffer cache management in OS kernels, *SmartSaver* does not conduct its own sequential access pattern detection. Instead, it relies on the kernel to provide hints of file access patterns to conduct its prefetching. In Linux, the file-level prefetching mechanism uses two readahead windows to detect access patterns, and the sizes of readahead windows are dynamically adjusted according to current accesses. *SmartSaver* monitors the readahead window sizes and uses them as the hints for its prefetching decisions. When the windows are enlarged to their full sizes, which indicates the kernel concludes that the file is being accessed sequentially, *SmartSaver* initiates a prefetching stream for it. When the windows are shrunk, which indicates a change of file access patterns, *SmartSaver* terminates the associated prefetching stream. If there exist multiple prefetching streams, we consider them as one aggregate stream to create a common disk idle period. When the disk is spun up or a stream uses up its prefetched blocks, all the streams are refilled.

Considering the low write bandwidth and limited erasure cycles of the flash drive, we need to reduce the amount of data written to the flash drive by avoiding inefficient prefetching. In a mobile computing environment, many applications are rate-based applications [13], which hold a steady rate of data consumption, such as *mplayer* and *xmms*. For such applications, *Prefetching Efficiency* for energy saving (the percentage of energy that can be saved through prefetching) is affected by two factors: data consumption rate and prefetching time (the maximum interval during that requests can be serviced with prefetched data in the flash drive). To guarantee efficient prefetching, we set some rules as follows: First, if the data consumption rate of a stream is not lower than the flash write bandwidth or the data consumption rate is so large that we can not use available space in the prefetching area to keep the disk idle longer than the disk break-even time, *SmartSaver* does not prefetch data for the stream. Second, *SmartSaver* sets the lower bound of prefetching time to the disk break-even time, and the upper bound to 300 seconds, as the prefetching time that goes too large brings diminishing benefits.

In a multi-task system, other concurrent disk events may break a long disk idle period created through prefetching. For example, we prefetch 4 minutes of movie clips in an attempt to keep the disk idle for that long period of time. However, if a virus-scanning application touches the disk every 10 seconds during that period, the expected 4 minutes of idle period cannot be realized. To coordinate prefetching with other disk events, we classify disk events that spin up a standby disk into two types: (1) *Bumps*, which are disk spin-up events that take place when streams use up their prefetched data and need to be refilled. A bump can be avoided by increasing prefetching time. (2) *Holes*, which are the other disk spin-up events waking up a standby disk, such as cold misses. If one bump takes place, which indicates that the disk could be kept idle longer if the prefetching time was increased, we then double the current prefetching time so that the prefetching becomes more aggressive. When a hole appears, which indicates that the current prefetching has been overshot, we reduce the prefetching time to the length of the most recent interval between two holes to avoid overshooting the expected disk idle period time.

4.4 Writeback

In most OS kernels, dirty blocks in main memory are periodically flushed to the disk to avoid losing data in volatile memory. For example, Linux writes dirty blocks older than 30 seconds back to the disk every 5 seconds. Such periodical disk accesses conflict with disk energy-saving efforts. Linux laptop mode recommends to accommodate a large ratio of dirty pages in memory and reduce the frequency of writing back of dirty pages. However, this solution increases the risk of losing data in volatile memory. *SmartSaver* solves this problem by redirecting the writeback traffic to the flash drive, a non-volatile buffer, to avoid breaking a long disk idle period.

In *SmartSaver*, the writeback area serves as a destaging buffer for temporarily holding dirty blocks. When a disk is in the standby state, dirty blocks are written to the writeback area to avoid spinning up the disk. Otherwise, dirty blocks are directly written to the hard disk. When the disk is back to the active state and over 90% of the writeback area is used, the dirty blocks in the writeback area are flushed to the disk. When the writeback area is overflowed, the disk is spun up to write back all the dirty blocks. In this way, *SmartSaver* achieves both energy saving and data safety purposes.

4.5 Balancing the three areas

As the flash drive is partitioned into the caching, prefetching, and writeback areas, *SmartSaver* employs a balancing mechanism to dynamically adjust the sizes of these three areas to optimize overall energy-saving efficiency. The principle for the mechanism is that the area that can save more energy with the same amount of additional buffer space should be given more buffers. *SmartSaver* monitors accesses to each area and periodically evaluates the amount of energy saved due to the addition of another N blocks. Each time N blocks are reclaimed from the least 'productive' area, where adding N more blocks can achieve less increased energy saving than adding them to other areas, and allocated to the most 'productive' one.

5. PERFORMANCE EVALUATION

5.1 Simulation

We wrote a trace-driven simulator to evaluate our disk energy-saving scheme. It simulates the management of three storage devices: main memory, hard disk, and flash drive. The simulator emulates the policies used for Linux buffer cache management, including the 2Q-like memory page replacement algorithm, the two-window readahead policy that prefetches up to 32 pages, and the I/O request clustering mechanism for grouping consecutive blocks in multiple requests into a large request.

The disk simulated in our experiment is the Hitachi-DK23DA hard disk [7]. It has a 30GB capacity, 4200 RPM and 35MB/sec peak bandwidth. Its average seek time is 13.0 ms, and its average rotation time is 7.1ms. Its energy consumption parameters are listed in Table 1. The timeout threshold for disk spin-down is set as 20 seconds, the default value for Linux laptop mode.

P_{active}	Active Power	2.00 W
P_{idle}	Idle Power	1.60 W
$P_{standby}$	Standby Power	0.15 W
E_{spinup}	Spin up Energy	5.00 J
$E_{spindown}$	Spin down Energy	2.94 J
T_{spinup}	Spin up Time	1.60 sec
$T_{spindown}$	Spin down Time	2.30 sec

Table 1: The energy consumption parameters for the Hitachi-DK23DA hard disk.

The simulated flash drive is the Transcend TS1GJF2A flash drive [14] with a 1GB capacity. Its read and write bandwidths are 12MB/sec and 8MB/sec, respectively. Its maximum active power consumption, 0.37W, is conservatively adopted as both read and write power consumptions in our simulations. Its sleep power consumption is 0.60mW.

To measure the energy consumed in the disk, we need to evaluate the period length of each disk state. The period length of disk active state can be broken down to three components: transfer time, rotation time, and seek time. Among them, the transfer time is the amount of requested data divided by disk bandwidth. As the rotation time is variable, we use a random value between 0 and 14.2ms (the maximum rotation delay). The seek time is determined by the seek distance between two consecutive requests. To estimate the seek time, we trace the latency of real disk accesses with different seek distances and plot a seek profile as done in [8]. We can then estimate seek times for various seek distances using the seek profile curve.

In the experiments, besides *SmartSaver* we also simulate the disk energy-saving scheme presented in [10], which is denoted as *Baseline*, and the original Linux laptop mode without a flash drive, which is denoted as *Linux*.

5.2 Traces

We modified the *strace* utility in Linux to collect traces to drive our simulator. The modified *strace* can intercept system calls related to file operations, such as open(), close(), read(), write(), and lseek(). For each system call, we collected the following information: PID, file descriptor, inode number, offset, size, type, timestamp, and duration. Blocks of the traced files are sequentially mapped to a simulated disk with a small random distance between files to simulate an actual layout of files on disk.

Eight traces of seven applications that are typically used in a mobile computing environment were collected, as listed in Table 2. Besides the five single-application traces, *thunderbird*, *scp-r*, *make*, *grep*, and *xmms*, we also collected traces *scp-mplayer* and *ftp-mplayer*, where two applications ran concurrently as representatives of multi-stream cases.

Name	Description	# of files	Size
thunderbird	an email client tool	283	188.0
scp-r	a remote copy tool	12669	191.0
make	a Linux kernel make tool	2579	72.5
grep	a text search tool	1332	50.4
xmms	a mp3 player	116	47.9
ftp-mplayer	a ftp client tool & a movie player	91	364.7
scp-mplayer	a remote copy tool & a movie player	91	364.7

Table 2: Trace descriptions. Sizes are in units of MBs. Both *scp-mplayer* and *ftp-mplayer* concurrently run two programs to access two separate sets of identical files. *Scp* transfers data in 8MB/sec, and *ftp* transfers data in 20KB/sec.

We synthesized two typical mobile computing scenarios by concatenating individual application traces one by one. The first experiment simulates a programming scenario, where a user is programming, searching codes, listening to music, and performing remote file transfer. The second experiment simulates a networking scenario, where a user is checking and searching emails, performing remote file transfer, using FTP service, and watching a movie. They are referred to as *programming* and *networking*, respectively.

5.3 Case 1: Programming

The *programming* scenario is composed of eight stages. Each stage is a replay of one trace, *make*, *grep*, *xmms*, *make*, *grep*, *scp-r*, *make*, and *grep*, in that order. This scenario consists of both non-sequential accesses on a large number of small files (*make*, *grep*, and *scp-r*), and relatively long sequential accesses on large files (*xmms*).

Since disk idle time is critical for energy saving, we plot the Cumulative Distribution Function (CDF) curves of disk idle times for three schemes, as shown in Figure 2. The vertical line in the figure is the disk break-even time. Because disk energy could be saved only when the disk is idle longer than

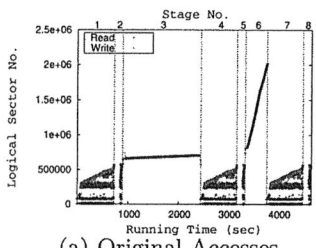

(a) Original Accesses

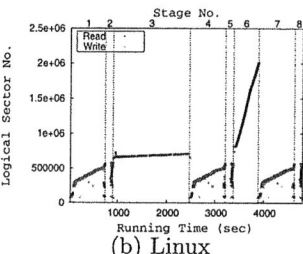

(b) Linux

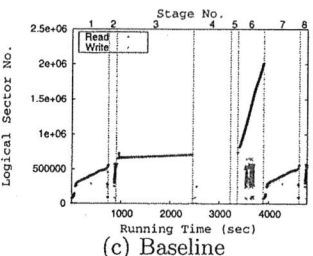

(c) Baseline

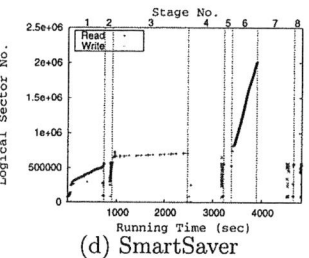

(d) SmartSaver

Figure 3: Case 1 *programming*: Disk I/O accesses in original application, *Linux*, *Baseline*, and *SmartSaver* schemes. All experiments are configured with a 64MB memory and a 128MB flash drive.

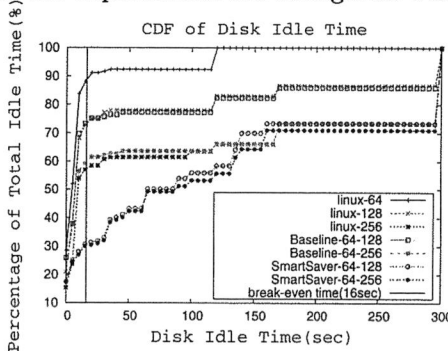

Figure 2: The CDF curves of disk idle times for *programming*. Linux-64, Linux-128, and Linux-256 are the original *Linux* scheme with a 64MB, 128MB, and 256MB main memory, respectively. Baseline-64-128 and Baseline-64-256 are the *Baseline* scheme with a 64MB memory and a 128MB or 256MB flash drive, respectively. SmartSaver-64-128 and SmartSaver-64-256 are the *SmartSaver* scheme with a 64MB memory and a 128MB or 256MB flash drive, respectively. The vertical line is the disk break-even time, 16 seconds.

Scheme	Mem	Flash	Energy(J)	Flash R/W
Linux	64	N/A	7325.0	N/A
Linux	128	N/A	6317.3	N/A
Linux	256	N/A	5231.6	N/A
Baseline	64	128	6365.7	64/493
Baseline	64	256	5312.9	104/442
SmartSaver	64	128	4318.0	96/263
SmartSaver	64	256	4046.2	110/316

Table 3: The energy consumption and the Read/Write volume in the flash drive for *programming*. The memory sizes, flash drive sizes, and flash R/W volumes are in units of MBs.

the break-even time, we call the percentage of the sum of idle periods that are shorter than the break-even time over the total disk idle time *Unusable Idleness Percentage* (UIP). The smaller an UIP is, the more energy could be saved.

Figure 2 shows that *Linux* with a 64MB memory has an UIP as large as 89.1%. This percentage is reduced to 73.3% and 57.2% with a 128MB and 256MB memory, respectively. This is because the working sets of *make* and *grep* are more likely to be held in a larger memory to avoid disk accesses. In a 64MB memory system, *Baseline* with a 128MB flash drive has an UIP of 73.9%, which is almost identical to that for *Linux* with a 128MB memory. This is because *Baseline* manages the flash drive as a larger memory to hold all data blocks transferred between memory and disk. In contrast, *SmartSaver* with a 128MB flash drive has an UIP of only 30.8%. Correspondingly, *SmartSaver* consumes only 4318.0J disk energy, compared with 6365.7J for *Baseline* and 7325.0J for *Linux*, as shown in Table 3.

SmartSaver attempts to save the limited erasure cycles of the flash drive. With a 128MB flash drive *SmartSaver* writes only 263MB data to the flash drive, which is 46.6% less than 493MB for *Baseline*. Meanwhile, with a 128MB flash drive *SmartSaver* reads 96MB data from flash drive, while *Baseline* reads only 64MB data. The read/write ratios clearly show that *SmartSaver* uses the flash drive more efficiently. This is a much desired advantage considering limited erasure cycles of the flash drive. Using existing wear-levelling techniques, such as JFFS2, *SmartSaver* can distribute a small amount of overwrites evenly on the flash drive. For *programming*, if we use a 1GB flash drive and configure *SmartSaver* with 128MB flash space for energy saving, one cell could be overwritten once approximately every 4.5 hours.

For a detailed analysis of disk accesses at each stage, we plot the disk I/O activities with a setting of a 64MB memory and a 128MB flash drive in Figure 3. As shown in this figure, there are disk accesses at all stages for *Linux*, because *Linux* cannot accommodate the working set in memory. *Baseline* performs better with a flash drive. For example, at stage 4 and stage 5 the disk keeps idle for 822 seconds, as the working set of *make* and *grep* is held in the flash drive. However, since *Baseline* caches every byte transferred between memory and disk, at stage 6 *scp-r* flushes the data of *make* out from the flash drive, which causes the disk to be spun up to serve *make* at the next stage. In contrast, *SmartSaver* protects the data of *make* from being replaced by the data of *scp-r* and the disk remains idle at stage 7. This figure also shows that most writeback requests are absorbed by the flash drive and reshaped as a few bursts, which helps maintain a long disk idle period.

5.4 Case 2: Networking

The *networking* scenario includes three stages, *thunderbird*, *scp-mplayer*, and *ftp-mplayer*. Except *thunderbird*, which accesses one or multiple email files, each sequentially and non-continuously, *scp*, *mplayer*, and *ftp* sequentially access files.

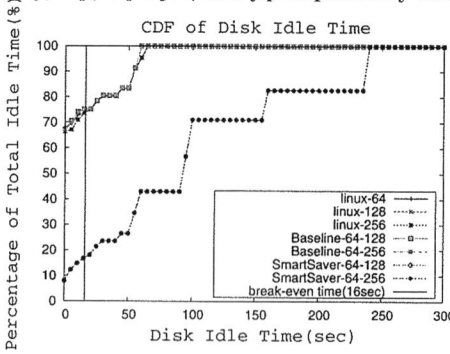

Figure 4: The CDF curves of disk idle times for *networking*.

Figure 4 shows the CDF curves of disk idle times. Both *Linux* and *Baseline* have an UIP of around 73%, and no improvements can be achieved with additional memory or flash

416

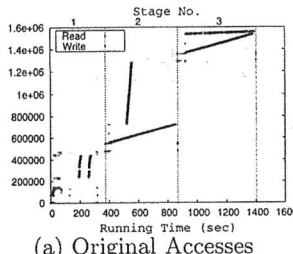 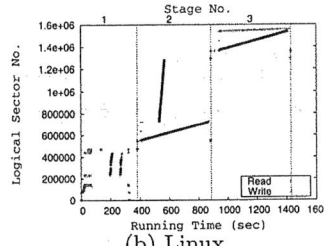 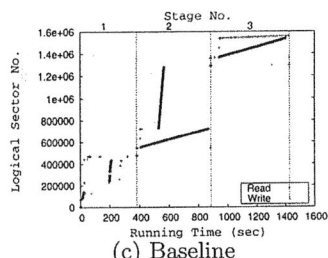 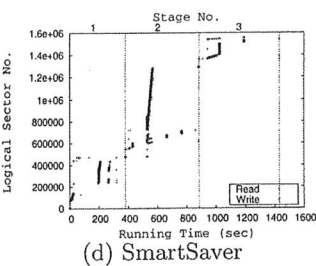

| (a) Original Accesses | (b) Linux | (c) Baseline | (d) SmartSaver |

Figure 5: Case 2 *networking*: Disk I/O accesses in original application, *Linux*, *Baseline*, and *SmartSaver* schemes. All experiments are configured with a 64MB memory and a 128MB flash drive.

Scheme	Mem	Flash	Energy(J)	Flash R/W
Linux	64	N/A	2076.4	N/A
Linux	128	N/A	2052.5	N/A
Linux	256	N/A	2052.1	N/A
Baseline	64	128	2103.2	26/574
Baseline	64	256	2102.2	26/552
SmartSaver	64	128	1207.5	210/443
SmartSaver	64	256	1208.0	210/447

Table 4: The energy consumption and Read/Write volume in the flash drive for *networking*. **The memory sizes, flash drive sizes, and flash R/W volumes are in units of MBs.**

drive space, because the disk activities in this scenario are dominated by one-time accesses. In contrast, *SmartSaver* significantly extends disk idle time through prefetching. With a 128MB flash drive, it reduced the UIP to 16.7%. As shown in Table 4, *Linux* with a 64MB memory consumes 2076.4J, while *Baseline* with a 64MB memory and a 128MB flash drive consumes 2103.2J, which is even more than that for *Linux*. This is because *Baseline* wastes energy on caching the data that are never to be reused in the flash drive. With the same setting, *SmartSaver* consumes only 1207.5J, which is 41.8% less than that for *Linux*. We also observe that *SmartSaver* is not sensitive to the size of the flash drive. This is because the upper-bound of prefetching time limits the consumption of flash drive space by avoiding too aggressive prefetching. As in *programming*, *SmartSaver* with a 128MB flash drive writes 22.8% less data to the flash drive (443MB) than *Baseline* (574MB), but reads 707.6% more data (210MB) than *Baseline* (26MB).

Figure 5 plots the disk I/O activities for *networking*. Both *Linux* and *Baseline* cannot avoid disk busy periods in each stage. In contrast, *SmartSaver* reshapes most evenly distributed disk accesses into several bursts of disk accesses through prefetching. Figure 5(d) shows that during the overlapped 36 seconds of *scp* and *mplayer* at stage 2, there exist disk accesses. This is because *SmartSaver* does not prefetch for *scp*, whose data consumption rate is 8MB/sec, the flash write bandwidth. The prefetching for *mplayer* is not carried out either, because *scp* keeps the disk busy and prefetching for *mplayer* is meaningless at that time. After *scp* completes, the prefetching time for *mplayer* quickly increases to the upper bound. At stage 3, the prefetching is conducted for both *ftp*, whose consumption rate is only 20KB/sec, and *mplayer*.

6. CONCLUSION

In this paper, we identify some critical issues involved in the flash memory management to save disk energy. We provide a comprehensive set of solutions to maintain an effective use of the flash drive for three goals: (1) accommodating the unique properties of the flash memory; (2) minimizing OS kernel changes; and (3) significantly improving disk energy saving. Trace driven simulations for typical mobile computing scenarios demonstrate that our scheme can save up to 41% disk energy compared with existing policies used in Linux.

7. ACKNOWLEDGMENT

We thank anonymous reviewers for their comments and suggestions. We appreciate William L. Bynum for reading the paper and his comments. Some technical discussions with Xiaoning Ding are also helpful. This work is partially supported by the National Science Foundation under grants CNS-0098055 and CNS-0405909.

8. REFERENCES

[1] T. Bisson and S. Brandt. Reducing energy consumption with a non-volatile storage cache. In *Proc. of International Workshop on Software Support for Portable Storage*, San Francisco, CA, March 2005.

[2] P. Cao and S. Irani. Cost-aware www proxy caching algorithms. In *Proc. of USENIX'97*, Dec. 1997.

[3] F. Douglis, P. Krishnan, and B. Bershad. Adaptive disk spin-down policies for mobile computers. In *Computing Systems*, volume 8(4), pages 381–413, 1995.

[4] F. Douglis, P. Krishnan, and B. Marsh. Thwarting the power-hungry disk. In *Proc. of USENIX'94*, Jan. 1994.

[5] S. Gurumurthi, A. Sivasubramaniam, M. Kandemir, and H. Franke. Drpm: Dynamic speed control for power management in server class disks. In *Proc. of ISCA'03*, 2003.

[6] D. P. Helmbold, D. D. E. Long, and B. Sherrod. A dynamic disk spin-down technique for mobile computing. In *Proc. of the 2nd Annual International Conference on Mobile Computing and Networking*, 1996.

[7] HITACHI. http://www.hitachigst.com/tech/techlib.nsf/products/DK23DA_Series.

[8] H. Huang, W. Hung, and K. G. Shin. Fs2: dynamic data replication in free disk space for improving disk performance and energy consumption. In *Proc. of SOSP'05*, Brighton, U.K., Oct. 2005.

[9] Intel. Intel mobile platform vision guide for 2003.

[10] B. Marsh, F. Douglis, and P. Krishnan. Flash memory file caching for mobile computers. In *Proc. of the 27th Hawaii Conference on Systems Science*, Hawaii, 1994.

[11] MICRON. http://download.micron.com/pdf/technotes/TN4603.pdf.

[12] E. B. Nightingale and J. Flinn. Energy-efficiency and storage flexibility in the blue file system. In *Proc. of OSDI'04*, SF, CA, Dec. 2004.

[13] A. E. Papathanasiou and M. L. Scott. Energy efficient prefetching and caching. In *Proc. of USENIX'04*, 2004.

[14] TRANSCEND. http://www.transcend.com.tw/Support/DLCenter/Datasheet/TS1GJF2A.pdf.

[15] A. Weissel, B. Beutel, and F. Bellosa. Cooperative io - a novel io semantics for energy-aware applications. In *Proc. of OSDI'02*, Dec. 2002.

[16] F. Zheng, N. Garg, S. Sobti, C. Zhang, R. E. Joseph, A. Krishnamurty, and R. Y. Wang. Considering the energy consumption of mobile storage alternatives. In *Proc. of MASOCTS'03*, Oct. 2003.

An Energy-Efficient Virtual Memory System with Flash Memory as the Secondary Storage

Hung-Wei Tseng, Han-Lin Li, Chia-Lin Yang
Department of Computer Science and Information Engineering,
National Taiwan University, Taipei, Taiwan, R.O.C
{r92022,b90124,yangc}@csie.ntu.edu.tw

ABSTRACT

The traditional virtual memory system is designed for decades assuming a magnetic disk as the secondary storage. Recently, flash memory becomes a popular storage alternative for many portable devices with the continuing improvements on its capacity, reliability and much lower power consumption than mechanical hard drives. The NAND flash memory is organized with blocks, and each block contains a set of pages. The characteristics of flash memory are quite different from a magnetic disk. Therefore, in this paper, we revisit virtual memory system design considering limitations imposed by flash memory. In particular, we study the effects of the subpaging technique and storage cache management. In the traditional virtual memory system, a full page is written back to the secondary storage on a page fault. We found that this could result in unnecessary writes thereby wasting energy. The subpaging technique that partitions a page into subunits, and only dirty subpages are written to flash memory is beneficial to the energy efficiency. For the storage cache management, unlike traditional disk cache management, care needs to be taken to guarantee that the flash pages of a main memory page are replaced from the cache in sequence. Experimental results show that the average energy reduction of combined subpaging and caching techniques is 35.6%.

Categories and Subject Descriptors

B.3.2 [**Memory Structures**]: Design Styles—Virtual memory; D.4.2 [**Operating Systems**]: Storage Management—Secondary storage, Virtual memory

General Terms

Algorithms, Measurement, Performance, Design

Keywords

Page replacement, Virtual memory, NAND flash memory, Embedded systems, Embedded storages

---

Permission to make digital or hard copies of all or part of this work for personal or classroom use is granted without fee provided that copies are not made or distributed for profit or commercial advantage and that copies bear this notice and the full citation on the first page. To copy otherwise, to republish, to post on servers or to redistribute to lists, requires prior specific permission and/or a fee.
ISLPED'06, October 4–6, 2006, Tegernsee, Germany.
Copyright 2006 ACM 1-59593-462-6/06/0010 ...$5.00.

1. INTRODUCTION

Modern operation system often adopts the virtual memory approach to allow the physical memory shared among multiple tasks. The traditional virtual memory system is designed for decades assuming a magnetic disk as the secondary storage. When a page fault occurs, the replaced page is written into the disk. To improve the disk access latency, most modern hard disks contain an integrated cache to buffer recently accessed data. Recently, flash memory becomes a popular storage alternative for many portable devices with the continuing improvements on its capacity, reliability and much lower power consumption than mechanical hard drives. The characteristics of flash memory are quite different from a magnetic disk. Therefore, to achieve energy efficiency of a flash-based storage system, in this paper, we revisit virtual memory system design considering limitations imposed by flash memory.

NAND flash memory is commonly used for data storage due to its lower cost and higher density compared with NOR flash. The NAND flash memory is organized with blocks, and each block contains a set of pages. The typical block size and page size are 16KB and 512B, respectively. There are three types of operations in flash memory: read, write and erase. Data on a flash are read/written in an unit of one page. Due to the write-once feature, a page cannot be overwritten. Therefore, flash memory performs out-place updates. That is, data is written to a free page, and the old page is invalidated. Those with invalid data are called dead pages. After a certain number of writes, free space on flash memory would be low. Flash memory must reclaim dead pages through erase operations. Such reclaiming process is called garbage collection. Erasing is done in an unit of one block. Since a block consists of multiple pages, live pages of the victim block must be copied to free pages before the block is erased. Since flash memory has limited number of erase operations, frequent garbage collection does not only degrade performance, it also shortens the life time of flash memory.

From the above discussion, we know that writes are problematic. Frequent writes result in high flash utilization that would in turn trigger more garbage collection thereby causing more energy consumption. Furthermore, writes also consume more power than reads. Table 1 shows the energy consumption for read, write and erase from [9]. Therefore, to reduce energy of the flash storage system, we need to reduce writes to flash memory. In the traditional virtual memory system, the full victim page is written back to the disk when a page miss occurs. While this is OK for a disk,

operation	latency	energy consumption
page read	47.2 us	679 nJ
page write	533 us	7.66 uJ
block erase	3 ms	43.2 uJ

Table 1: NAND flash characteristics

application	dirty ratio	application	dirty ratio
kword	89.73%	kword+juk	69.41%
mozilla	48.49%	mozilla+juk	40.90%
kspread	88.61%	kspread+juk	72.40%
openoffice	66.59%	openoffice+juk	59.31%
gqview	98.86%	gqview+juk	97.62%

Table 2: Dirty ratio of workloads[1]

it is not energy-efficient for flash memory. A 4K virtual memory page results in 8 writes to flash memory assuming a 512B flash memory page. We observed that a victim page often contains unmodified data. Therefore, writing a full page results in unnecessary writes to flash memory. In this paper, we study the effect of subpaging on the energy efficiency of flash memory. The subpaging technique partitions a virtual memory page in the granularity of flash page size. Only dirty subpages are written to flash memory on a page fault. Experimental results show that the subpaging technique can achieve average 20% energy saving for the workload tested in this paper. The second issue we investigate in this paper is the storage cache management. Unlike traditional disk cache management which caches all reads/writes, only writes are cached in SRAM here. To increase the cache utilization, we investigate two approaches for the cache management. In the first approach, every write is cached and the replacement policy is based on both the access time and frequency factors (TF policy). The second approach is to identify frequently written data and only those data are stored in the cache. One problem particular to the flash cache management is that higher cache hit rate does not necessarily result in more energy savings. In addition to cache hit rate, keeping all sugpages of a virtual memory page is critical. One key principle to reduce garbage collection overhead is to allocate data accessed close in time (i.e., locality) to the same flash block so a victim block with small amount of live pages could be found when the garbage collection is triggered [1, 13]. On a virtual memory system, a page fault results in series of flash writes. We refer to this locality as "intra-page locality". It is critical to preserve intra-page locality when writing back data from the storage cache to flash memory. The experimental results show that the TF policy with page locality gathering can achieve 19.5% energy reduction with a 1M cache. The average energy reduction of combined subpaging and caching techniques is 35.6%.

The rest of the paper is organized as follows. Section 2 describes the details of the proposed energy efficient flash storage system. Section 3 describes our experimental methodology. Section 4 presents and discusses the experimental results. Section 5 discusses related work. Finally, Section 6 concludes this paper.

2. ENERGY-EFFICIENT FLASH MEMORY STORAGE SYSTEM

In this section, we describe the proposed energy-aware virtual memory system with flash memory as the secondary storage. To reduce unnecessary writes to flash memory, we divide a main memory page into a set of subpages, and only dirty subpages are written into the flash memory when a page fault occurs. An SRAM is used to cache frequently written data, which is called HotCache. To avoid accidentally power break, we can add an additional small rechargeable battery with energy enough to write back all HotCache content.

2.1 Subpaging

In the traditional virtual memory system, a full victim page is written back to the disk when a page miss occurs. A typical flash page size is 512B or 2KB, while a main memory page size could be 4KB, 2MB, or 4MB. With a 4KB virtual memory page and 512B flash page, each page fault incurs 8 writes to flash memory. For the applications tested in this paper, we find that the victim page often contains a significant amount of unmodified data. Table 2 shows the ratio of dirty blocks in a victim page assuming a 512B block, and 4K virtual memory page. Therefore, writing a full victim page to flash memory is not energy efficient.

The subpaging technique divides a virtual memory page into a set of subpages. The subpaging technique was previously proposed to reduce the transferring size and latency of remote memory in a networked system [3]. To tailor the subpaging technique for flash memory, we divide a page in the granularity of flash page size. Each subpage is associated with a dirty bit. On a page fault, only dirty subpages are written into flash memory. Park et al. [6] proposed a new replacement policy CFLRU (Clean First Least Recently Used) to reduce writes to flash memory by keeping dirty pages in memory as long as possible. Although this method reduces the energy consumption of flash memory effectively, it could incur more page faults. In contrast, the subpaging technique reduces writes to flash memory without increasing page faults.

2.2 HotCache

To reduce writes to flash memory, we propose to keep frequent writes to an SRAM, which is referred to as HotCache in this paper. HotCache is organized as a fully-associative cache with the HotCache block size equal to the page size of flash memory.

The HotCache management policy affects the performance of the HotCache. eNVy [13] proposed to use an SRAM as a write buffer. That is, every write request is cached and the FIFO (First In First Out) policy is adopted for replacement. In this paper, we investigate three new policies for the Hot-Cache management. Below we detail these three policies.
Time-Frequency (TF)
In the Time-Frequency policy, every write request is cached in HotCache. The replacement is based on the following weight function:

$$timestamp \times write\_counts$$

The HotCache block with the smallest weight is selected as the victim when a replacement occurs. This policy considers both the time and frequency factors. The advantage of this TF policy over the traditional LRU is that it prevents a

---
[1]The dirty ratio is defined as
$$\frac{the\ number\ of\ dirty\ 512B\ block\ in\ a\ dirty\ memory\ page}{the\ number\ of\ 512B\ blocks\ in\ a\ main\ memory\ page}$$

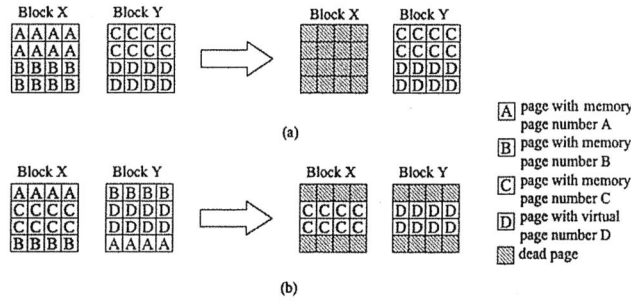

Figure 1: An example that TF breaks intra-page locality

Application	Memory footprint	write/total inst.	Scenario
kword	39.2 MB	15.7%	edits several lines and reform the columns in a document
mozilla	42.1 MB	12.9%	browses web sites including Yahoo, Google, Gmail, Amazon, and etc.
kspread	28.4 MB	11.7%	calculates sum, average, min, max of numerical data and sorts them.
openoffice	74.2 MB	14.3%	plays about twenty slides.
gqview	52.9 MB	32.5%	views some images
juk	19.9 MB	13.8%	plays a list of mp3 files summing up to 1 hour

Table 3: Workloads used in our experiments and their characteristics

hot page from being replaced by a recently accessed cold page[2].

Time-Frequency-Locality (TFL)

One important data allocation principle in flash memory is to cluster hot data such that we can easily find a victim block with small amount of live pages when garbage collection is required. In the virtual memory system, flash pages in one main memory page are always written back to back. This attribute is called "intra-page locality" in this paper. Consider two data allocation methods shown in Figure 1. Assume a main memory page contains 8 flash pages. In Figure 1, two flash blocks contain four main memory pages A, B, C, and D. In Figure 1(a), flash pages in one main memory page are allocated in one block, while in Figure 1(b), they are scattered in two blocks. Assume after a period of time, memory page A and B are swapped out. Therefore, for the data allocation in Figure 1(a), block X contains only dead flash pages, while for the data allocation in Figure 1(b), both block X and Y contains dead and live pages.

With the TF policy described above, the flash pages of a virtual memory page is not guaranteed to be allocated in the same block since they may not be replaced out of the HotCache in sequence. Therefore, to avoid destroying intra-page locality, we enhance the TF policy by forcing all the pages of the same virtual memory page to be replaced in sequence. The virtual page number of the victim block is recorded and a counter is used to keep track of how many HotCache blocks in the same main memory pages

---

[2]Hot (cold) pages are those frequently (rarely) accessed.

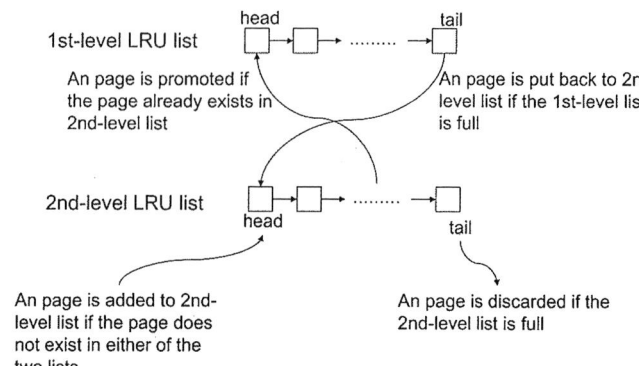

Figure 2: The operation of two-level LRU list mechanism

device	operation	energy consumption
SRAM (512 KB) [11]	access energy	1.82 nJ
SRAM (1 MB) [11]	access energy	3.02 nJ
NAND (page size = 512B) [9]	page read	679 nJ
	page write	7.66 uJ
	block erase	43.2 uJ
NAND (page size = 2KB) [10]	page read	2.36 uJ
	page write	14.5 uJ
	block erase	54.0 uJ

Table 4: Simulation parameters

have been replaced. A HotCache block with the smallest $timestamp \times write\_counts$ and has the same recorded virtual page number is chosen as the victim HotCache block. Once the counter reaches zero, a HotCache block with smallest $timestamp \times write\_counts$ is selected as the victim. Its main memory page number is then recorded and the counter is reset. Note that since the virtual memory page number of a HotCache block can be obtained directly from the cache tag[3], we do not need to record the page number of each HotCache block. The enhanced TF policy is called TF-Locality (TFL).

Two-Level LRU (2L)

Different from the TF and TFL policies, the two-level LRU policy observes a page for a period of time to determine whether this page should be allocated in HotCache. Figure 2 shows the operation of the two-level LRU list mechanism.

We use the first-level LRU list records the pages considered as hot data, and the second-level list records the pages considered as the candidates to be hot data. The difference is that we allocate hot data, which pages are recorded in first level list, to HotCache, and cold data are written to flash memory.

The length of the first-level LRU is the number of HotCache blocks. That is, every HotCache block has a corresponding entry in the first level LRU list. Note that the 2L policy does not destroy the intra-page locality since it considers only the time factor for replacement. Therefore, flash pages in the same virtual page are guaranteed to be replaced back to back.

---

[3]$virtual\ page\ number = cache\ tag \times flash\ page\ size \div main\ memory\ page\ size$

Application	512KB					1MB				
	FIFO	LRU	2L	TF	TFL	FIFO	LRU	2L	TF	TFL
kword	0.33%	0.35%	2.24%	1.67%	1.67%	1.24%	1.37%	2.24%	3.87%	3.87%
mozilla	0.05%	0.06%	8.29%	7.43%	7.39%	1.11%	1.10%	15.01%	16.18%	16.13%
kspread	0.09%	0.09%	1.09%	2.02%	2.02%	5.64%	6.60%	2.48%	8.03%	8.03%
openoffice	0.40%	0.62%	5.77%	6.15%	6.13%	3.80%	4.50%	10.43%	11.32%	11.29%
gqview	0.00%	0.00%	0.05%	0.10%	0.10%	0.03%	0.12%	0.05%	0.47%	0.47%
kword+juk	0.68%	1.23%	8.87%	10.29%	10.24%	6.20%	7.52%	16.94%	17.56%	17.55%
mozilla+juk	0.05%	0.06%	6.07%	6.19%	6.15%	0.55%	0.79%	11.28%	12.33%	12.30%
kspread+juk	1.92%	1.81%	7.71%	8.62%	8.63%	9.79%	10.69%	14.35%	15.53%	15.49%
openoffice+juk	0.57%	0.67%	7.00%	7.48%	7.44%	4.95%	6.34%	13.37%	14.32%	14.29%
gqview+juk	0.00%	0.01%	0.11%	0.21%	0.21%	0.05%	0.10%	0.11%	0.41%	0.41%
average	0.41%	0.49%	4.72%	5.02%	5.00%	3.34%	3.91%	8.62%	10.00%	9.98%
average normalized energy	0.933	0.941	0.904	0.910	0.895	0.856	0.863	0.833	0.814	0.805

Table 5: Write hit rates of HotCache mechanisms (HotCache size: 512KB/1MB)

application	page size		application	page size	
	512B	2KB		512B	2KB
kword	12.3%	3.7%	kword+juk	27.9%	10.8%
mozilla	42.9%	14.7%	mozilla+juk	50.8%	19.1%
kspread	10.8%	6.5%	kspread+juk	26.8%	10.3%
openoffice	18.0%	5.9%	openoffice+juk	38.1%	20.4%
gqview	1.8%	2.3%	gqview+juk	1.6%	3.3%
overall average				23.1%	9.7%

Table 6: Reduced writes by subpaging

3. EXPERIMENTAL METHODOLOGY

We adopt a trace-driven simulation in this paper. Our simulator contains a main memory paging system and flash storage with an SRAM. We use Valgrind [5] on an x86-linux machine to collect memory traces. The applications we tested in this study are listed in Table 3 : kword, a word processor; mozilla, a web browser; kspread, a spreadsheet application; openoffice, a popular office suite similar to Microsoft office; gqview, a image viewer. We also create multi-programming workloads by running juk, an MP3 jukebox program, with applications listed in Table 3. The OS adopts the round-robin scheduling policy.

Our baseline model contains 16MB main memory, and the main memory page size is 4KB. The virtual memory is managed using the LRU policy. For flash memory, we assume a 16KB block with 512B page size or a 128KB block with 2KB page size. We adopt the cost-benefit policy [4] as our garbage collection policy. We assume the initial utilization of flash memory is 97%. The SRAM sizes of the HotCache considered in this paper are 512KB and 1MB.

The access energy of SRAM and flash memory assumed in our experiments are listed in Table 4. The SRAM access energy is obtained using the CACTI [11] assuming 0.13um technology. The CACTI is an integrated cache access time, power and area model, and has been widely used in studies on cache architecture [2, 12]. The energy consumption of flash memory are based on the data sheet of Samsung K9F1208R0B [9] (page size = 512B) and K9K2G08X0A [10] (page size = 2KB) NAND flash.

4. SIMULATION RESULTS

4.1 Subpaging

Figure 3 shows the energy consumption of the subpaging

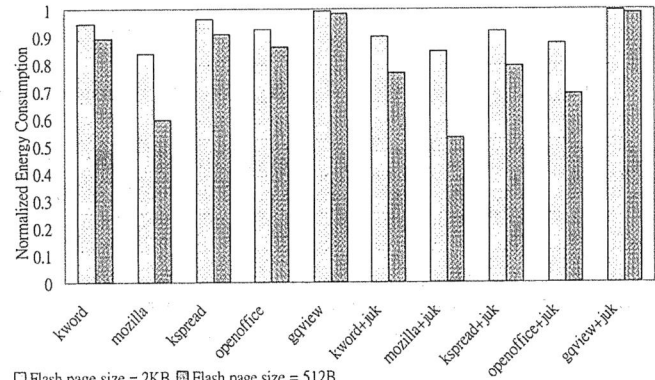

Figure 3: Relative energy consumption of subpaging technique in 512B and 2KB flash page size

technique normalized to the baseline storage system (without subpaging) assuming both the 512B and 2KB flash page size. We also show write reduction rates in Table 6. We can see that with a 512B flash page, the subpaging could reduce up to 50.8% writes and 47.1% of flash energy. Since there are only few writes in gqview, the subpaging technique only reduces 1.8% of writes. With a 2KB flash page, the subpaging is less effective since a main memory page only contains two flash pages. The effects of subpaging depends on the dirty ratio which is defined in Section 2.1. The lower the dirty ratio is, the higher energy reduction we expect to see by adopting subpaging. In single-programming workloads, mozilla and openoffice show most significant energy reduction via subpaging since the dirty ratio of these two applications are much lower than others. Multiprogramming workloads show lower dirty ratio than single-programming ones since there are more contention for the memory resource. With a 512B flash page, the subpaging achieves about 24.7% energy saving on the average for multiprogramming workloads, and 15.2% for single-programming workloads.

4.2 HotCache

In this section, we evaluate the HotCache hit rates and energy efficiency of various caching policies discussed in Section 2.2. We also show two commonly used policies: FIFO

421

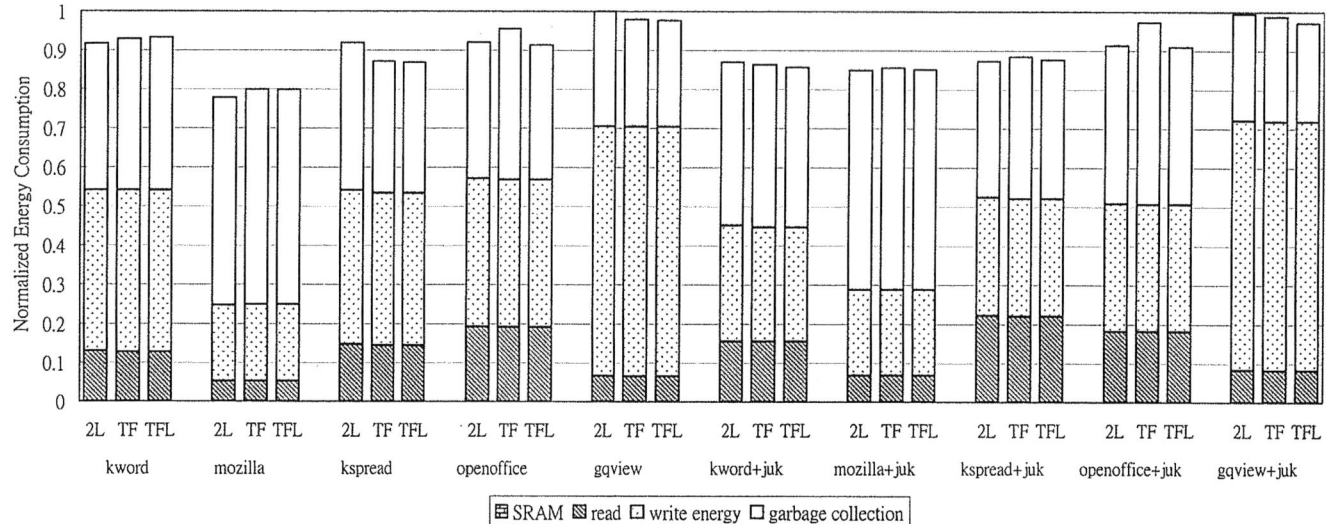

Figure 4: The average energy consumption under different HotCache schemes (HotCache size: 512K)

	FIFO	LRU	2L	TF	TFL
number of garbage collections	0.96	0.98	0.90	0.97	0.89
garbage collection overhead	0.96	0.99	0.96	1.05	0.96

Table 7: Garbage collection frequency and copying overhead of openoffice+juk under different Hot-Cache schemes (HotCache size: 512KB)

and LRU. Table 5 list the write hit rates and energy reductions of a 512KB and 1MB HotCache, respectively.

From Table 5, we can see that the TF policy has the highest HotCache hit rate, 5.02% for a 512 KB cache and 10% for a 1 MB cache. The FIFO policy used in eNVy [13] has much lower hit rate compared with the TF policy, 0.41% for a 512B cache and 3.34% for a 1 MB cache. With a 1MB HotCache, the TFL policy could reduce about 19.5% of flash energy on the average.

One problem particular to the cache management is that higher cache hit rate does not necessarily result in more energy savings. For example, openoffice+juk with a 512KB cache, the TF policy has higher cache hit rate than 2L as shown in Table 5 (7.48% vs. 7%), but it achieves less energy savings than 2L as shown in Figure 4 (2.73% vs. 8.63%). The cause of this abnormal behavior is that the TF policy destroys intra-page locality. This results in higher garbage collection overhead. Figure 4 shows the energy normalized to the baseline architecture (without the HotCache) for various cache management policies. We break down the energy consumption of the HotCache scheme into four components: garbage collection, write, read and accessing HotCache. We can see that the TF policy has higher garbage collection energy than TFL in some applications. For openoffice+juk, the TF policy incurs about 6.08% more garbage collection energy than 2L. Table 7 shows the normalized garbage collection frequency and average copying overhead incurred by each garbage collection for openoffice+juk. We can see that with the TF policy, the garbage collection frequency is lower than the baseline, but the copying overhead is increased by 5%, while other policies are beneficial for re-

ducing the garbage collection frequency and copying overheads.

From Figure 4, we can see that SRAM accessing energy is negligible. Reducing writes to flash memory does not only reduce energy consumed in writes, but it also actually reduces even more garbage collection energy consumption.

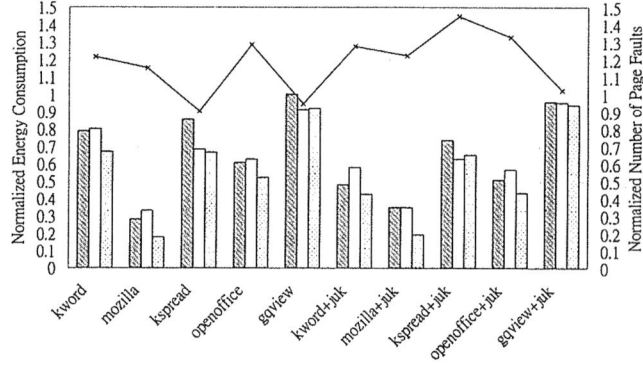

Figure 5: Relative energy consumption of applying sram cache, subpaging technique and CFLRU

4.3 HotCache and Subpaging

Figure 5 shows the combined effect of HotCache and Subpaging. In this set of results, we assume 1MB HotCache managed in the TFL policy, and 512KB flash pages. We also compare our scheme with the work - CFLRU (Clean First Least Recently Used) [6] which proposed a new virtual memory replacement policy to reduce writes to flash memory. Since the CFLRU could incur more page faults, we plot the number of page faults of the CFLRU normalized to the baseline architecture in Figure 5.

The experimental results show that the energy reduction of adopting a 1MB HotCache and subpaging together ranges from 8.7% to 66.8%. The CFLRU is also quite effective in

422

saving flash memory energy except for applications that have many writes, e.g., gqview, since the CFLRU is not able to find clean pages to replace in this case. The main problem of the CFLRU is its impact on performance. From Figure 5, we can see that the CFLRU increases the number of page faults significantly for several applications, such as kspread+juk (44.7%). In contrast, our scheme achieves energy savings without causing adverse effect on performance.

Since our scheme is orthogonal to the CFLRU, the CFLRU could be used alone with the HotCache and subpaging. So we also show the combined effects of the CFLRU and our scheme in Figure 5. The results show that using the CFLRU on top of our scheme could further reduce the flash memory energy by 15.8% at most in mozilla+juk. Therefore, for applications that could trade performance for energy savings, we could use the CFLRU and HotCache/subpaging together to achieve higher energy savings.

5. RELATED WORK

Previous works combine SRAM with flash memory mainly for performance consideration. Wu, et al. [13] propose the eNVy system which adopting SRAM as write buffers to allow better response time of flash memory and reduce invalidations from write merges. Park, et al. [7] propose a NAND XIP architecture applying an SRAM cache with prefetching to reduce the access latency to the same level with NOR flash. In the area of virtual memory system, Park et al. [6] proposed a new replacement policy CFLRU (Clean First Least Recently Used) to reduce writes to flash memory by keeping dirty pages in memory as long as possible. The subpaging technique [3] is first used to reduce transfer latency in a networked system. This paper is the first to study the effect of subpaging for energy savings of flash memory.

Other works on flash memory look at how to reduce garbage collection overheads, and increase I/O access parallelism. Rosenblum, et al. [8] propose a cost-benefit garbage collection policy using value-based heuristic for log-structured file systems. Chang et al. [1] propose the adaptive stripping architecture to exploit the I/O parallelism from multi-banked flash memory system by using dynamic bank assignment policy rather than static policies.

6. CONCLUSIONS

In this paper, we study the design issue of an energy-efficient virtual memory system with flash memory as the secondary storage. In particular, we look at the subpaging technique and storage cache management. We find that the subpaging technique reduces about 20% of flash memory energy on the average, and 24.7% for multi-programming workloads. For the storage buffer management, we find that higher buffer hit rate does not necessarily lead to higher flash energy savings. The intra-page locality needs to be preserved when writing data from the storage buffer to flash memory. The TF-Locality (TFL) policy can achieve about 19.5% of energy saving. Joint use of the subpaging and TFL storage buffer can reduce 35.6% of flash memory energy on average.

7. ACKNOWLEDGEMENT

This work is supported in part by research grants from ROC Industrial Technology Research Institute 95-S-B43,

National Science Council NSC 95-2215-E-002-021 and NSC 95-2752-E-002-008-PAE.

8. REFERENCES

[1] L.-P. Chang and T.-W. Kuo. An adaptive striping architecture for flash memory storage systems of embedded systems. In *Proceedings of The 8th IEEE Real-Time and Embedded Technology and Applications Symposium*, pages 24–27, September 2002.

[2] M. Huang, J. Renau, S.-M. Yoo, and J. Torrellas. The design of DEETM: a framework for dynamic energy efficiency and temperature management. *Journal of Instruction-Level Parallelism*, 3, 2002.

[3] H. A. Jamrozik, M. J. Feeley, G. M. Voelker, J. Evans, A. R. Karlin, H. M. Levy, and M. K. Vernon. Reducing network latency using subpages in a global memory environment. In *Proceedings of the 7th ACM Conference on Architectural Support for Programming Languages and Operating Systems*, pages 258–267, 1996.

[4] A. Kawaguchi, S. Nishioka, and H. Motoda. A flash-memory based file system. In *Proceedings of the 1995 USENIX Technical Conference*, pages 155–164, January 1995.

[5] N. Nethercote and J. Seward. Valgrind: A program supervision framework. *Electronic Notes in Theoretical Computer Science*, 89(2), 2003.

[6] C. Park, J.-U. Kang, S.-Y. Park, and J.-S. Kim. Energy-aware demand paging on NAND flash-based embedded storages. In *Proceedings of the IEEE/ACM International Symposium on Low Power Electronics and Design*, Auguest 2004.

[7] C. Park, J. Seo, S. Bae, H. Kim, S. Kim, and B. Kim. A low-cost memory architecture with NAND XIP for mobile embedded systems. In *Proceedings of the 1st IEEE/ACM/IFIP international conference on Hardware/software codesign and system synthesis*, pages 138–143, 2003.

[8] M. Rosenblum and J. Ousterhout. The design and implementation of a log-structured file system. In *Proceedings of the 13th Symposium on Operating System Principles*, pages 1–15, October 1991.

[9] Samsung Electronics CO.,LTD. *Datasheet of Samsung K9F1208R0B NAND flash*, 2004.

[10] Samsung Electronics CO.,LTD. *Datasheet of Samsung K9K2G08X0A NAND flash*, 2006.

[11] P. Shivakumar and N. P. Jouppi. CACTI 3.0: An integrated cache timing, power and area model. *Technical report, Compaq Computer Corporation*, August 2001.

[12] S. Steinke, L. Wehmeyer, B. Lee, and P. Marwedel. Assigning program and data objects to scratchpad for energy reduction. In *Proceedings of the 2002 Design, Automation and Test in Europe Conference and Exhibition*, pages 409–417, March 2002.

[13] M. Wu and W. Zwaenepoel. eNVy: A non-volatile, main memory storage system. In *Proceedings of International Conference on Architectural Support for Programming Languages and Operating Systems*, pages 86–97, October 1994.

Maximizing the Lifetime of Embedded Systems Powered by Fuel Cell-Battery Hybrids*

Jianli Zhuo[1], Chaitali Chakrabarti[1], Naehyuck Chang[2], Sarma Vrudhula[3]
[1]Dept. of Electrical Engineering, Arizona State University, Tempe, AZ, 85287, U.S.
[2]School of CSE, Seoul National University, Seoul, Korea
[3]Dept. of CSE, Arizona State University, Tempe, AZ, 85287, U.S.
jianli@asu.edu, chaitali@asu.edu, naehyuck@snu.edu.kr, vrudhula@asu.edu

ABSTRACT

Fuel cells are a viable alternative power source for portable applications. They have higher energy density than traditional Li-ion batteries and can achieve longer lifetime for the same weight or volume. However, because of their limited power density, they can not track fluctuations in the load current fast. A hybrid power source, that consists of a fuel cell and a Li-ion battery, has the advantages of long lifetime and good load following capabilities. In this work, we consider the problem of extending the lifetime of a fuel-cell based hybrid source that is used to provide power to a DVFS processor. We propose a new algorithm that is built on top of an energy based optimization framework. The algorithm simultaneously adjusts the fuel flow rate (at the producer end), and judiciously scales the load current (at the consumer end) to minimize the energy loss of the hybrid system. Simulations on randomly generated task sets demonstrate the superiority of this algorithm with respect to an algorithm that does not allow adjustment of the fuel flow rate.

Categories and Subject Descriptors

D.4.1 [**Operating Systems**]: Process Management—*Scheduling*

General Terms

Algorithms

Keywords

Fuel cell, Battery, Hybrid systems, DVFS system, Task scaling

1. INTRODUCTION

Fuel cells are clean power sources that have attracted a great deal of attention in recent years. Fuel cells have a distinct advantage over batteries in that they have very high energy density compared to batteries. They are thus expected to generate power longer (4 to 10x) than a battery package of the same size and weight [1, 2].

---

*This research was funded in part by the NSF grant (CSR-EHS 05059540), the Consortium for Embedded Systems, ASU, and LG Yonam Foundation.

Permission to make digital or hard copies of all or part of this work for personal or classroom use is granted without fee provided that copies are not made or distributed for profit or commercial advantage and that copies bear this notice and the full citation on the first page. To copy otherwise, to republish, to post on servers or to redistribute to lists, requires prior specific permission and/or a fee.
ISLPED'06, October 4–6, 2006, Tegernsee, Germany.
Copyright 2006 ACM 1-59593-462-6/06/0010 ...$5.00.

Compared to other alternative power source such as solar powers [3, 4, 5], fuel cells are easily controllable, have a stable power output and so have been widely used in automobiles, power plants, and more recently, portable applications.

Fuel cells have limited load following capacity, ie, the power range is limited and the response is slow. Thus it is widely accepted that a hybrid power source (fuel cells plus secondary energy storage) is much more efficient than a fuel cell alone power source. A hybrid power system consists of a primary power source with high energy density and a secondary power source with high power density, and improves both the power capacity and the response time. The secondary power source could be a battery or an ultra capacitor or a combination [6].

Most of the prior work on fuel cell control has been in the context of hybrid automobiles. These include fuel cell models that capture the chemical kinetics, and mechanisms to control fuel cell operations [7, 8, 9]. Control techniques to coordinate the operations of the fuel cell, battery and super capacitor to achieve longer battery lifetime and higher system efficiency have also been proposed [10, 11]. The superiority of hybrid fuel cell/battery system over pure fuel cell systems has been demonstrated in terms of system efficiency, dynamic control requirement, etc in [7, 8]. Unfortunately, techniques developed for hybrid automobiles cannot be easily adapted for portable embedded systems. This is because while automotive systems are reactive, in an embedded system, the tasks can be executed in any manner as long as they meet a specified performance level. Thus for an embedded system powered by fuel cell-battery hybrids, techniques that jointly optimize the control parameters at the producer end (fuel cell and battery) and the consumer end (embedded processor) have to be developed.

In this paper, we describe a procedure to enhance the lifetime of a DVFS based embedded system powered by a hybrid source. The hybrid power source is built with a PEM (Proton Exchange Membrane) fuel cell that works at room temperature and a Li-ion battery. The proposed procedure to enhance the fuel cell lifetime is built on top of an energy based optimization framework. Maximizing the fuel cell lifetime is equivalent to minimizing the overall energy loss, defined as the sum of the energy consumed by the load and the energy that is lost through the bleeder bypass when the battery is fully charged. The algorithm minimizes the energy loss by scaling the power of the DVFS based embedded system subject to deadline and battery constraints on the one hand and adjusting the fuel cell output power on the other hand. We first show how the energy based optimization framework can be used to set the fuel cell power output and DVFS setting for a single task, and then we describe the procedure for multiple tasks in a static (off-line) scheduling environment. Basically, we first determine the scaling factor that minimizes the energy consumption of the embedded system, and then determine the fuel cell power that minimizes the wasted energy. This is an extension of our prior work presented in [12], where the fuel cell output power was assumed to be fixed and the scaling factor of the DVFS based embedded system was the only control knob.

The rest of the paper is organized as follows. Section 2 describes the fuel cell/battery characteristics and provides an overview of the hybrid system. The fuel cell efficient algorithm which jointly controls the DVFS scaling factor and the fuel cell current is explained in Section 3. Section 4 presents the performance of the proposed algorithm for randomly generated task set. The paper is concluded in section 5.

2. OVERVIEW OF THE HYBRID SYSTEMS

2.1 Fuel cell/battery characteristics

The I-V-P characteristic curves of the fuel cell are shown in Figure 1. They can be explained by electron transfer phenomenon that occurs at very low currents, and mass transport phenomenon that occurs at high currents. From the curves, we see that as the current density increases, the voltage drops, and that the power first increases then drops. The operating point of the fuel cell is typically set around $P=\frac{2}{3}$ maxpower, and not P=maxpower because of stability considerations.

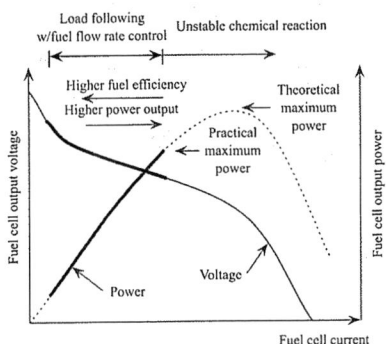

Figure 1: I-V-P curves for room temperature fuel cell

The chemical to electrical efficiency of a single fuel cell is proportional to its output voltage [13]. However, the efficiency of a fuel cell system is not only impacted by the fuel cell stack, but also impacted by other factors such as fuel utilization, fuel transform efficiency and power conditioning efficiency. Figure 2 shows the fuel cell efficiency curve of the room temperature fuel cell considered in our work. We can see that the system efficiency is 44-46% in the power range 5W to 18W. So in our work, we assume that the fuel cell system efficiency is a constant.

The fuel cell power output can be varied in a certain range (*load following* region) by controlling the fuel input rate. The load following capability of the fuel cell is quite slow, as indicated in [2]. When the membrane is wet, it takes the fuel cell around 10 sec to generate the full current (corresponding to 2/3 of the max power). Assuming that the output increases linearly, it takes about 1 sec to adjust the fuel cell output current by 10%, which is quite long for embedded applications.

A traditional battery (such as a Li-ion battery), on the other hand, has superior load following capabilities and can respond to the current change immediately. However, Li-ion batteries have much lower energy density (< 200Whr/kg) compared to the fuel cell (2000 Whr/kg) [14]. A hybrid power source has the advantage of provid-

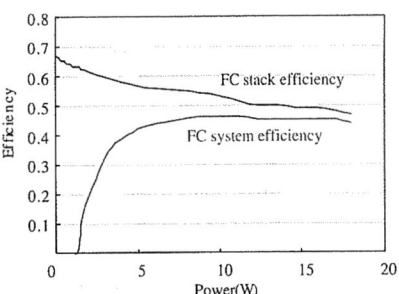

Figure 2: Fuel cell stack efficiency and fuel cell system efficiency of the fuel cell used in our work (measured values)

ing the power density of the Li-ion battery and the energy density of the fuel cell. Figure 3 shows how, in order to ensure system safety, a fuel cell only system may be over-designed if the load current variance is large (max load current is 3X of the average load current). This would result in increase in total size and weight of the fuel cell. Since the Li-ion battery can supply the additional current when necessary, a fuel cell in conjunction with the battery is more efficient with respect to power/energy density and size/weight.

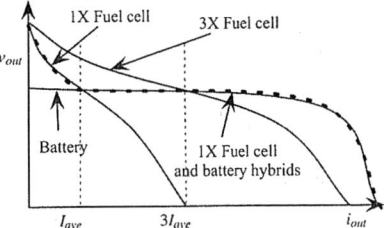

Figure 3: Advantage of using hybrid power source

2.2 Fuel cell hybrid system

The hybrid system under consideration is shown in Figure 4. The hybrid power source consists of a sodium borohydride ($NaBH_4$) PEM (Proton Exchange Membrane) fuel cell system that works at room temperature and a Li-ion battery. The fuel cell system has an output power P_F. The Li-ion battery has a fixed capacity and is capable of providing multiple levels of output power. The charging and discharging of the battery is controlled by the charge management system (CMS). At the consumer end is a dynamic voltage scalable (DVFS) processor based embedded system whose load power, P_L, is a function of several variables such as the processor supply voltage, the power management state of various peripherals, and the type of energy management policy used. When $P_F < P_L$, the Li-ion battery provides $P_L - P_F$; when $P_F > P_L$, the Li-ion battery is charged. If the battery gets fully charged and $P_F > P_L$, then the excess power is dissipated through the bleeder circuit.

The hybrid control system accepts the state of the embedded system as input, and then determines the action of the power source. Specifically, the controller can adjust the fuel cell output power by adjusting the fuel rate flow (expressed in milliliter/second) as well as the air flow. At the consumer end, the load power can be adjusted by task scheduling, processor frequency/voltage scaling, memory frequency scaling, etc. In this paper, we show how a combination of the dynamic frequency/voltage scaling at the consumer end and fuel cell power adjustment at the producer end can help achieve a longer lifetime for the hybrid power system.

The optimization metric in embedded systems powered by fuel

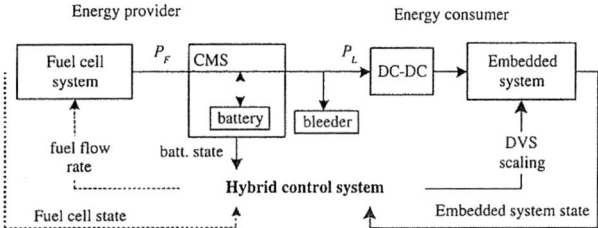

Figure 4: The schematic view of the hybrid system

cell/battery hybrids is lifetime of the fuel cell. The lifetime is related to the fuel consumption of the fuel cell. We assume that a given volume (or weight) of fuel can provide a certain amount of energy. If the fuel energy consumption is less, as a result of lower energy consumed at the consumer end, the lifetime of the fuel cell is longer. So the objective of extending the lifetime of the fuel cell is transformed to minimizing the fuel energy consumption during task execution.

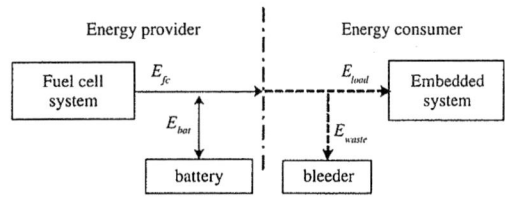

Figure 5: The energy flow of the hybrid system

The energy flow of the hybrid system is shown in Figure 5. At the energy provider end, E_{fc} is the electrical energy provided by the fuel cell and E_{bat} is the energy provided by the battery. The battery works as an energy buffer. Assume that the initial state of charge of the battery is B^{ini} and final state of charge is B^{end}. Then the energy provided by the battery is $E_{bat} = B^{ini} - B^{end}$ (if $E_{bat} < 0$, then the battery buffer stores energy). At the energy consumer end, E_{load} is the energy consumed by the embedded system and E_{waste} is the energy wasted through the bleeder. Recall that when the battery is fully charged but $P_F > P_L$, the excess power is dissipated through the bleeder circuit. The energy flow is summarized by the following equation.

$$E_{fc} + E_{bat} = E_{load} + E_{waste} \qquad (1)$$

Our objective is to minimize the fuel energy consumption given by $E_{fc} + E_{bat}$. This is equivalent to minimizing $E_{load} + E_{waste}$, according to Equation (1).

3. FUEL CELL EFFICIENT SCALING

3.1 Definitions

We begin with the notations that have been used in the rest of this paper. The hybrid power source is characterized by the fuel cell power, P_F that has a value in the range $[P_F^{min}, P_F^{max}]$, and B^{max}, the energy capacity of the Li-ion battery. For task T_k, the fuel cell power is $P_{F,k}$, the battery stored energy at the beginning of the execution is B_k^{ini} and at the end of the execution is B_k^{end}.

Let s_k be the frequency scaling factor while executing task T_k. The task execution time is then $s_k \times \tau_k$, where τ_k is the worst case execution time at the highest frequency (corresponding to $s_k = 1$).

The total power consumption of the DVFS processor is given by

$$P = C \cdot V_{dd}^2 \cdot f + P_{on} + V_{dd} \cdot I_{static} \qquad (2)$$

where the first term is the dynamic power, the second term is the intrinsic power, and the last term is the static power [15, 16]. If we assume that for scaling factor s_k, the voltage scales by s_k, and both P_{on} and I_{static} are constant, then the total power consumption of task T_k, $P_k(s_k)$ is given by

$$P_k(s_k) = \alpha_1 \cdot P_k(1) \cdot s_k^{-3} + \alpha_2 \cdot P_k(1) + (1 - \alpha_1 - \alpha_2) \cdot P_k(1) \cdot s_k^{-1} \qquad (3)$$

where $P_k(1)$ is the total power consumption at $s_k = 1$, α_1 and α_2 are the ratios of dynamic power to total power and intrinsic power to the total power at $s_k = 1$. In this paper, we assume that $\alpha_1 + \alpha_2 = 0.8$. This implies that the static power is 20% of the total power.

The load power P_L seen by the power source is actually the DC-DC inlet power. If we assume a constant DC-DC efficiency [17], then the load power equation for task T_k is given by

$$P_{L,k}(s_k) = \alpha_1 \cdot P_{L,k}(1) \cdot s_k^{-3} + \alpha_2 \cdot P_{L,k}(1) + (1 - \alpha_1 - \alpha_2) \cdot P_{L,k}(1) \cdot s_k^{-1} \qquad (4)$$

The metric to measure the performance of difference policies is the total energy consumption of the system, E_{total}, which includes the load (embedded system) energy consumption E_{load} and the wasted energy through the bleeder, E_{waste}. The scaling factor which minimizes E_{total} is called as the optimal scaling factor, s^{opt}. The scaling factor which only considers minimizing E_{load} is called s^{load}. Table 1 provides a comprehensive list of all parameters.

Table 1: Definition of fuel cell system parameters

$P_{F,k}$	the output power of the fuel cell.
P_F^{max}	the maximum power in the load following range.
P_F^{min}	the minimum power in the load following range.
B^{max}	the charge capacity of the battery
T_k	the $k-th$ task in the task profile
s_k	voltage / frequency scaling factor of T_k, $s_k \geq 1$
τ_k	the worst case execution time of T_k
$P_{L,k}(s_k)$	the load power when task T_k is scaled by s_k
α_1	the ratio of the dynamic power to the total power when $s_k=1$
α_2	the ratio of the intrinsic power to the total power when $s_k=1$; $\alpha_1 + \alpha_2 = 0.8$ in this chapter
B_k^{ini}	the battery charge value when task T_k starts
B_k^{end}	the battery charge value when task T_k finishes
E_{load}	energy consumed by the load (DC-DC inlet)
E_{waste}	energy wasted through bleeder bypass
E_{total}	the total energy consumption, given by $E_{load} + E_{waste}$
s_k^{opt}	the scaling factor which minimizes E_{total}
s_k^{load}	the scaling factor which minimizes E_{load}

To simplify the analysis, we make the following assumptions: (1) the battery charge management system has 100% efficiency; (2) we only do inter-task scaling, so the load power and the fuel cell output power do not change during the execution of a single task; (3) the task sequence is determined apriori, we only consider voltage/frequency scaling of the tasks.

3.2 Motivational example

Consider a DVFS system whose frequency can be scaled from 1 to 2.5 in steps of 0.1. The fuel cell power can vary in [5, 15]W. The battery capacity is $B^{max} = 1500$J and the initial state of the battery is half of B^{max}. The task configuration is given in Table 2.

We first assume that we do not use the load following capability, and the fuel cell output power is fixed at $P_F = 15$W. By applying Algorithm fc_scale [12], we obtain the scaling profile and the energy metrics as shown in Figure 6(a).

Now if we control both the fuel cell output power and the load power, we get significant energy savings. In the profile shown in Figure 6(b), the fuel cell power, $P_{F,1}$ is set to 8.225 W and the load power, $P_{L,1}$ is scaled to 5.1W for task T_1, and $P_{F,2}$ is set to 5W and $P_{L,2}$ is scaled to 4.99W for task T_2. The values of all three energy metrics are significantly smaller for this profile. For instance, E_{load} reduces from 5229J to 3907J and E_{total} reduces from 5730J to 3924J. Thus the lifetime of the fuel cell can be extended significantly if the fuel cell has load following capabilities.

Table 2: Task parameters of the illustrative example

T_k	τ_k	$P_L(1)$	$\alpha_1 : \alpha_2$	D
T_1	2min	15W	4:1	15min
T_2	5min	12W	3:1	15min

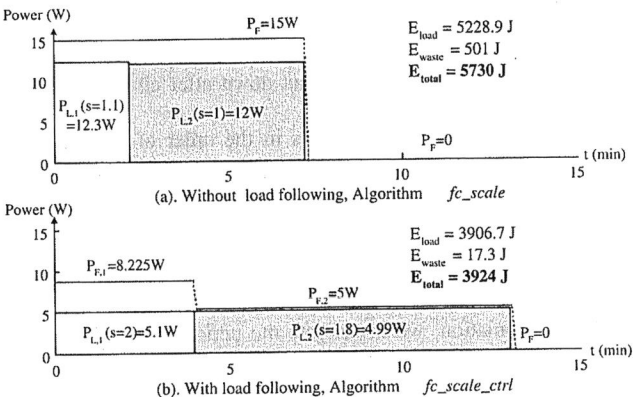

(a). Without load following, Algorithm fc_scale

(b). With load following, Algorithm fc_scale_ctrl

Figure 6: Motivational example: (a) no load following, (b) with load following.

3.3 Algorithm

The input to the task scaling algorithm is a sequence of tasks, along with their specifications (deadlines, WCETs, current), B^{max} and the state of the battery, B_k^{ini}. Each task is scaled such that the total energy consumption E_{total} is minimized subject to the deadline constraint and the battery constraint.

3.3.1 Determining $P_{F,k}$ and s_k^{opt} for a single task

To minimize E_{total} for a single task execution, the two parameters that need to be determined are the fuel cell system output $P_{F,k}$ and the task scaling factor s_k^{opt}.

The task duration is $s_k \times \tau_k$, and total energy consumption is $E_{total}(P_{F,k}, s_k) = E_{load}(s_k) + E_{waste}(P_{F,k}, s_k)$. If we can minimize both E_{load} and E_{waste}, then the total energy loss E_{total} is minimized.

The load power consumption E_{load}, wasted energy E_{waste}, and the battery constraint are as follows:

$$E_{load}(s_k) = P_{L,k}(s_k) \times s_k \times \tau_k \quad (5)$$

$$E_{waste}(P_{F,k}, s_k) = \max\left(0, \left(P_{F,k} - P_{L,k}(s_k)\right)s_k\tau_k + B_k^{ini} - B^{max}\right) \quad (6)$$

$$Battery\ constraint : \left(P_{L,k}(s_k) - P_{F,k}\right) \times s_k \times \tau_k \le B_k^{ini} \quad (7)$$

First, since E_{load} is independent of the fuel cell current $P_{F,k}$, we minimize E_{load}. Recall that for every task, there exists a scaling factor s_k which minimizes $E_{load}(s_k)$. We call this scaling factor s_k^{load}. Since $P_{L,k}(s_k) = \left(\alpha_1 \cdot s_k^{-3} + \alpha_2 + (1 - \alpha_1 - \alpha_2) \cdot s_k^{-1}\right) \times P_{L,k}(1)$, $s_k^{load} = \sqrt[3]{\frac{2 \times \alpha_1}{\alpha_2}}$ (if there is no deadline constraints). If s_k^{max} is the maximum scaling factor determined by the deadline constraint, then the scaling factor s_k^{load} has to be bounded by s_k^{max}.

Since $E_{total} = E_{load} + E_{waste}$, and E_{load} is minimized by scaling factor s_k^{load}, $E_{load}(P_{F,k}, s_k^{load})$ can be minimized if the fuel cell output power $P_{F,k}$ is set to a value that makes $E_{waste}(P_{F,k}, s_k^{load}) = 0$. This means, $P_{F,k}$ should satisfy $\frac{B^{max} - B_k^{ini}}{\tau_k} \ge \left(P_{F,k} - P_{L,k}(s_k^{load})\right) \times s_k^{load}$ or $P_{F,k} \le P_{L,k}(s_k^{load}) + \frac{B^{max} - B_k^{ini}}{s_k^{load} \times \tau_k}$. The value of $P_{F,k}$ should also satisfy the charge constraint in Equation (7). So the range of $P_{F,k}$ is

$$\Phi_k = \left[P_{L,k}(s_k^{load}) - \frac{B_k^{ini}}{s_k^{load} \times \tau_k}, P_{L,k}(s_k^{load}) + \frac{B^{max} - B_k^{ini}}{s_k^{load} \times \tau_k} \right] \quad (8)$$

Now assume the load following range limits of the fuel cell under consideration is $\Omega = [P_F^{min}, P_F^{max}]$. If Φ_k overlaps Ω, that is $\Psi_k \bigcap \Omega \ne \emptyset$, then $P_{F,k}$ could be set to any value in the overlapped region $\Psi_k \bigcap \Omega$. In our algorithm, we choose the largest value in $\Psi_k \bigcap \Omega$ because then the battery can get charged more and it could benefit the next task T_{k+1} (the charge constraint of T_{k+1} will be more relaxed if B_{k+1}^{ini} is larger). In the case when $\Psi_k \bigcap \Omega = \emptyset$, then we set $P_{F,k}$ to either P_F^{min} or P_F^{max} (set to the value which is closer to the set Ψ_k), and then re-calculate the optimal scaling factor s_k^{opt} by minimizing the function $E_{total}(P_{F,k}, s_k) = E_{load}(s_k) + E_{waste}(P_{F,k}, s_k)$ under battery constraint and deadline constraint.

3.3.2 Determining $P_{F,k}$ and s_k^{load} for a sequence of tasks

Assume that there are n tasks $(T_1, T_2, ..., T_n)$; all tasks arrive at time 0 and share the same deadline D. The total energy consumption E_{total} during task execution is

$$E_{total}(P_{F,1}, .. P_{F,n}; s_1, .., s_n) = \sum_{k=1}^{n} E_{total}(P_{F,k}, s_k)$$

$$= \sum_{k=1}^{n} E_{load}(s_k) + \sum_{k=1}^{n} E_{waste}(P_{F,k}, s_k) \quad (9)$$

Our objective is to find the scaling factors $(s_1^{load}, s_2^{load}, ..., s_n^{load})$ which minimize the load energy consumption $\sum_{k=1}^{n} E_{load}(s_k)$ under the deadline constraint, where s_k^{load} is a specific value of s_k.

$$E_{load}(s_1, s_2, .., s_n) = \sum_{k=1}^{n} \left(P_{L,k}(s_k) \times s_k \times \tau_k\right) \quad (10)$$

$$Deadline\ constraint : \sum_{k=1}^{n} (s_k \times \tau_k) \le D \quad (11)$$

These values can be determined analytically by the Lagrange multiplier approach for the given load power model. Since the objective function $E_{load}(s_1, s_2, .., s_n)$ is a sum of convex functions $E_{load}(s_k)$, and the deadline constraint is also a convex function, we can introduce a Lagrange multiplier λ and construct a function $f(s_1, s_2, .., s_n)$ as follows

$$f(s_1, .., s_n) = E_{load}(s_1, .., s_n) - \lambda\left((\sum_{k=1}^{n} s_k\tau_k) + x^2 - D\right)$$

$$= \sum_{k=1}^{n} \left(\left(\alpha_1 \cdot s_k^{-3} + \alpha_2 + (1 - \alpha_1 - \alpha_2) \cdot s_k^{-1}\right) \times P_{L,k}(1) \times s_k\tau_k\right)$$

$$- \lambda \times \left(\sum_{k=1}^{n} (s_k \times \tau_k) + x^2 - D\right) \quad (12)$$

where we include x^2 in the function because the deadline constraint is not a equation.

We calculate the partial derivative of $f(s_1, s_2, .., s_n)$ w.r.t. to each

s_k, x and λ, and set the partial differential functions to value zero.

$$\frac{\partial f}{\partial s_k} = \left((\alpha_2 - 2\alpha_1 s_k^{-3}) P_{L,k}(1) - \lambda \right) \tau_k = 0, \forall k \leq n \quad (13)$$

$$\frac{\partial f}{\partial \lambda} = \sum_{k=1}^{n} (s_k \times \tau_k) - D = 0 \quad (14)$$

$$\frac{\partial f}{\partial x} = 2 \times x \times \lambda = 0 \quad (15)$$

The solution of Equation (15) can be either $\lambda = 0$ or $x = 0$. If $\lambda = 0$, then the the simultaneous equations in Equation (13) have an unconstrained solution: $s_k = \sqrt[3]{\frac{2 \times \alpha_1}{\alpha_2}} = s_k^{load}$.

If $\lambda \neq 0$ and $x = 0$, the solution $(s_1, s_2, ..s_n, \lambda), x$ for the $n + 1$ functions in Equation (13) and Equation (14) exists, but it is difficult to find analytically. However, Equation (13) tells us that the value of $(s_1, s_2, .., s_n)$ which minimizes the objective function should satisfy the condition $\left(P_{L,k}(s_k) \times s_k \right)'_{s_k} = \left(P_{L,j}(s_j) \times s_j \right)'_{s_j} \leq 0, \forall j, k \leq n$. If $\alpha_1 : \alpha_2$ is identical for all tasks, then s_k^{load} is also identical for all tasks and the value can be obtained by evenly distributing the slack to all tasks. When $\alpha_1 : \alpha_2$ is different from each other, we can find these values numerically.

After we determine s_k^{load}, we find the value of $P_{F,k}$ which minimizes wasted charge and satisfies charge constraint for each individual task. Note that it may be necessary to re-calculate s_k^{opt} if $\Psi_k \cap \Omega = \emptyset$.

3.3.3 Algorithm description

The proposed fuel cell efficient task scaling algorithm consists of three main steps. The details are shown in Algorithm 1.

Step1: Calculate the scaling factor s_k^{opt} of the task sequence based on unlimited load following range (line 1).

Step2: Determine $P_{F,k}$ based on the load following range (lines 5-9).

Step3: Re-determine s_k^{opt} to minimize the total energy consumption (line 10).

Algorithm 1 Task Scaling Algorithm fc_scale_ctrl

1: **Determine** $(s_1^{opt}, s_2^{opt}, .., s_n^{opt})$ **by assuming unlimited load following range.**
2: k=1, flag=0;
3: **WHILE** $k \leq n$ **DO**
4: **Input:** task T_k with τ_k, s_k^{load}, **battery with** B^{max} and B_k^{ini}.
5: $P_{low} = P_{L,k}(s_k^{opt}) - \frac{B_k^{ini}}{s_k^{opt} \times \tau_k}$;
6: $P_{high} = P_{L,k}(s_k^{opt}) + \frac{B^{max} - B_k^{ini}}{s_k^{opt} \times \tau_k}$;
7: **if** $P_{low} > P_F^{max}$ **then** $P_{F,k} = P_F^{max}$, flag=1 **end if**;
8: **if** $P_{high} < P_F^{min}$ **then** $P_{F,k} = P_F^{min}$, flag=1 **end if**;
9: **if** flag==0 **then** $P_{F,k} = \min(P_{high}, P_F^{max})$;
10: **else** determine s_k^{opt} by minimizing $E_{chm}(s_k)$, and adjust s_j^{opt} for $k < j \leq n$ if s_k^{opt} changes;
11: **end if**
12: execute task T_k by s_k^{opt} till it finishes;
13: $B_k^{end} = \min(B^{max}, B_k^{ini} + (P_{F,k} - P_{L,k}(s_k^{opt})) \times s_k^{opt} \times \tau_k)$;
14: **if** $B_k^{end} < 0$ **then** return FAILURE;
15: **else** $B_{k+1}^{ini} = B_k^{end}$, k=k+1, flag=0; **end if**
16: **END WHILE**

---

4. SIMULATION RESULTS

In this section, we compare the performance of the proposed algorithm fc_scale_ctrl with the algorithm fc_scale [12] which is proposed for minimizing the total energy consumption of a hybrid system with fixed fuel cell output power P_F.

4.1 Experimental setting

The DVS system supports CPU scaling factor from 1 to 2.5 with steps of 0.1. In addition, we vary the value of $P_L(1)$ and the ratio $\alpha_1 : \alpha_2$ for each task.

The task set used in the experiments is a task graph where the task order is already scheduled beforehand, and all tasks share the same deadline. Here each task sequence consists of 50 to 100 tasks. Each task has τ_k from 1 to 2 min. The task density is $\mu = \sum_{k=1}^{n} \frac{\tau_k}{D}$, where D is the deadline. μ varies from 0.3 to 0.7. For each task density value, we run 100 task sequences, and then get the average values for all the feasible cases.

The battery capacity is set to $B^{max} = 15$KJ, the initial value is assumed to be half of the capacity. The fuel cell load following range is [5,15]W. When we apply Algorithm fc_scale for no load following case, we set a constant fuel cell output power $P_F = 15$W. We assume that the fuel cell is shut down after all the tasks are completed.

Since the task execution time is in the order of minutes and the battery capacity is quite large, we assume that the additional charge/discharge because of the delay in adjusting $P_{F,k}$ (≈ 1 sec for 10% change) is handled by the battery.

4.2 Experiment 1: Embedded system w/o load power variance

In this simulation, we investigate the performance of the proposed algorithm for an embedded system where the load power $P_L(1)$ is the same for all tasks. Assume that $P_{L,k}(1) = 15$W. The power ratio $\alpha_1 : \alpha_2$ is different from task to task, and randomly (based on the uniform distribution) chosen from the range [3, 7].

Table 3 shows the total energy consumption E_{total}, load energy consumption E_{load}, wasted energy E_{waste} for the different algorithms. The energy consumption values in Table 3 show that the total energy consumption, E_{total}, of Algorithm fc_scale_ctrl is much lower than fc_scale.

Table 3: Experiment 1: Performance comparison when $P_{L,k}(1) = 15$W, and $\alpha_1 : \alpha_2$ varies randomly between 3 to 7

Item	μ	0.7	0.5	0.3
E_{total}(KJ)	fc_scale	99.60	100.16	100.42
	fc_scale_ctrl	**75.09**	**72.17**	**73.86**
E_{load}(KJ)	fc_scale	99.57	100.13	100.40
	fc_scale_ctrl	75.09	67.67	67.48
E_{waste}(KJ)	fc_scale	0.03	0.03	0.02
	fc_scale_ctrl	0	4.49	6.38

4.3 Experiment 2: Embedded system with load current variation

Next, we vary the load power, $P_{L,k}(1)$ for each task. The load power is randomly chosen based on the uniform distribution in the range [10,15]W. The power ratio $\alpha_1 : \alpha_2$ is still randomly chosen from the range [3, 7]. Table 4 compares the energy performance of the different algorithms.

In this case, the difference between Algorithm fc_scale_ctrl and fc_scale is larger than when the load power is constant (as in Experiment 1). Recall that in Algorithm fc_scale, $P_F = 15$W which is based on the highest load power value 15W. When the load current variance is quite high (a realistic scenario since different tasks may use different resources) and the fuel cell power has been set to a high value, then many a time E_{waste} is high since the battery is not big enough to hold all the excess charge. In such scenarios, changing the fuel cell current P_F reduces E_{waste} and E_{total} as shown in Table 4.

Table 4: Experiment 2: Performance comparison when $P_{L,k}(1)$ varies within [10,15]W and $\alpha_1 : \alpha_2$ varies between 3 to 7

Item	μ	0.7	0.5	0.3
E_{total}(KJ)	fc_scale	94.751	95.482	95.813
	fc_scale_ctrl	**59.981**	**65.786**	**67.872**
E_{load}(KJ)	fc_scale	79.926	80.212	80.370
	fc_scale_ctrl	59.952	54.120	54.007
E_{waste}(KJ)	fc_scale	14.825	15.271	15.443
	fc_scale_ctrl	0.029	11.666	13.865

4.4 Experiment 3: Configuration of the hybrid power source

For a given embedded system, if we know the application domain and the corresponding load power variance, then it is possible to come up with a set of candidate configurations which satisfy the energy performance and then choose the configuration that satisfies the weight/size requirement.

To illustrate this, consider the set of tasks with utilization $\mu = 0.7$ in Experiment 1 that have load power value $P_L(1)$ which is 15W, and $\alpha_1 : \alpha_2$ which varies randomly between 3 to 7. First, assume that we have already chosen the battery with $B^{max} = 64$kJ and that the initial charge B^{ini} is half of B^{max}. We vary the fuel cell load following range increasing the value of PF^{max} from 5W in steps of 1W, and the value of P_F^{min} is set to $P_F^{max}/3$. Figure 7(a) plots the system performance, E_{total} (averaged over 100 random task sets) as a function of P_F^{max}. The infeasible region corresponds to the case when P_F^{max} is too small so that the tasks cannot complete with sufficient power or the state of charge of the battery falls below a certain value (it is set to half of B^{ini} in this experiment) at the end of the task profile. Figure 7(a) shows that $P_F^{max} = 6$W is sufficient for this particular system, and that increasing P_F^{max} does not change the system performance. Now if we have a smaller battery, say, $B^{max} = 16$kJ, then P_F^{max} should be chosen to a value not less than 8W, as shown in Figure 7(b).

Since the feasible regions in Figure 7(a) and (b) have the same energy performance, we can choose any candidate configuration from the feasible region without any performance degradation. The final choice will depend on the availability of the components as well as the weight and size requirements.

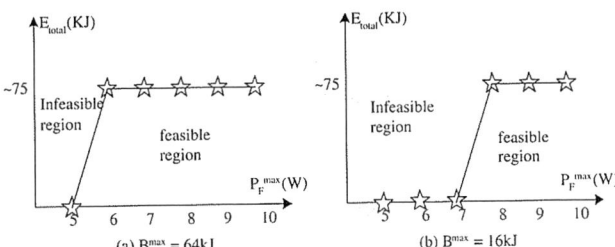

Figure 7: Different configurations: (a) when $B^{max} = 64$kJ, $P_F^{max} = 6$W is sufficient, (b) when $B^{max} = 16$kJ, $P_F^{max} = 8$W is sufficient.

5. CONCLUSION AND FUTURE WORK

In this paper, we presented a procedure to increase the lifetime of fuel cell/battery hybrids. The procedure, built on top of an energy based optimization framework, simultaneously adjusts the fuel flow rate (at the producer end), and judiciously scales the load current (at the consumer end) to minimize the energy loss of the hybrid sys-

tem. Simulations on randomly generated task sets demonstrated the superior performance of this algorithm compared to the algorithm that does not allow adjustment of the fuel flow rate.

In deriving the procedure, we assumed that the fuel cell power efficiency is constant. In the near future, we will take a more detailed look at the fuel cell system efficiency as a function of the output power. Also, to account for a more realistic scenario, we will include the delay of the fuel cell load following. This can typically be handled by proper off-line scheduling. But for very short task durations, we may experience uncertainty that is not trivial to handle by off-line methods.

6. ACKNOWLEDGEMENT

We sincerely thank Dr. Don Gervasio and Sonja Tasic (*Flexible Display Center, ASU*), and Kyungsoo Lee (*School of Computer Science and Engineering, SNU*) for help with the fuel cell setup and measurement.

7. REFERENCES

[1] D. Gervasio, S. Tasic, and F. Zenhausern, "A room temperature micro-hydrogen-generator," *Journal of Power Sources*, vol. 149, pp. 15–21, 2005.

[2] D. Gervasio, "Fuel-cell system for hand-carried portable power," *International Fuel Cell R&D Forum*, Nov. 2005.

[3] P. H. Chou and D. Li, "Maximizing efficiency of solar-powered systems by load matching," in *ISLPED'04*, August 2004.

[4] V. Raghunathan, A. Kansal, J. Hsu, J. Friedman, and M. B Srivastava, "Design considerations for solar energy harvesting wireless embedded systems," in *IPSN'05*, April 2005.

[5] X. Jiang, J. Polastre, and D. Culler, "Perpetual environmentally powered sensor networks," in *IPSN'05*, April 2005.

[6] J. P. Zheng, T. R. Jow, and M. S. Ding, "Hybrid power sources for pulsed current applications," *IEEE Trans. Aerospace and Electronic Sys.*, vol. 37, no. 1, pp. 288–292, 2001.

[7] Y. Guezennec, Ta-Young Choi, G. Paganelli, and G. Rizzoni, "Supervisory control of fuel cell vehicles and its link to overall system efficiency and low-level control requirements," in *Proc. American Control Conf. (ACC)*, 2003, vol. 3, pp. 2055–2061.

[8] A. Vahidi, A. Stefanopoulou, and Huei Peng, "Model predictive control for starvation prevention in a hybrid fuel cell system," in *Proc. American Control Conf. (ACC)*, 2004, vol. 1, pp. 834–839 vol.1.

[9] A. Nasiri, V. S. Rimmalapudi, A. Emadi, D. J. Chmielewski, and S. Al-Hallaj, "Active control of a hybrid fuel cell-battery system," in *Proc. Intl'l Power Electronics and Motion Control Conf. (IPEMC)*, 2004, vol. 2, pp. 491–496.

[10] Z. Jiang, L. Gao, and R. A. Dougal, "Flexible multiobjective control of power converter in active hybrid fule cell/battery power sources," *IEEE Transactions on Power Electronics*, vol. 20, no. 1, pp. 244–253, Jan 2005.

[11] M. J. Gielniak and Z. J. Shen, "Power management strategy based on game theory for fuel cell hybrid electric vehicles," in *Proc. Vehicular Technology Conference (VTC)*, 2004, vol. 6, pp. 4422–4426.

[12] J. Zhuo, C. Chakrabarti, N. Chang, and S.Vrudhula, "Extending the lifetime of fuel cell based hybrid systems," in *43rd DAC*, July 2006.

[13] J. Larminie and A. Dicks, *Fuel Cell Systems Explained*, John Wiley & Sons, LTD, 2000.

[14] C. K. Dyer, "Fuel cells and portable electronics," in *Symposium on VLSI circuits (digest of technical papers)*, June 2004, pp. 124–127.

[15] R. Jejurikar, C. Pereira, and R. Gupta, "Leakage aware dynamic voltage scaling for real-time embedded systems," in *41st DAC*, June 2004, pp. 275–280.

[16] J. Zhuo and C. Chakrabarti, "System-level energy-efficient dynamic task scheduling," in *42nd DAC*, June 2005, pp. 628–631.

[17] Y. Choi, N. Chang, and T. Kim, "Dc-dc converter-aware power management for battery-operated embedded systems," in *42nd DAC*, June 2005, pp. 895–900.

Author Index

Agrawal, V. D. 232
Albonesi, D. H. 55
Alhajj, T. 330
Amelifard, B. 334
Amirtharajah, R. 20
Anis, M. 119
Balkan, D. 37
Benini, L. 162, 346
Bero, B. 131
Bircher, W. L. 350
Blaauw, D. 338, 363
Bleakley, C. J. 95
Boni, A. 280
Bowman, K. 79
Brown, R. 73
Bryant, A. 363
Burns, D. 191
Bushnell, M. L. 214
Cai, L. 186
Calhoun, B. H. 366
Cao, N. 85
Casas-Sanchez, M. 95
Castro, F. 268
Chakrabarti, C. 101, 292, 424
Chakraborty, A. 162
Chandra, S. 342
Chandrakasan, A. P. 8, 366
Chang, I. J. 14
Chang, N. 292, 424
Chao, C.-F. 89
Chaparro, P. 49
Chaver, D. 268
Chen, A. 20
Chen, F. 412
Chen, K.-H. 388
Chen, Q. 2
Cheng, W.-C. 89
Cheung, W. T. 226
Cho, S. 147
Choi, K. 310
Chou, P. H. 197, 369
Dabiri, F. 207
De, V. 79
Dennington, B. 213

Deogun, H. S. 73
Dey, S. 342
Donald, J. 304
Duraisami, K. 162
Ehmen, G. 220
Eom, H. 127
Ercanli, E. 358
Facen, A. 280
Fallah, F. 334
Fong, X. 2
Fujiwara, H. 61
Garg, A. 268
Ghoneima, M. 79
Ghose, K. 37, 244
González, A. 49
González, J. 49
Gu, J. 67, 382
Guilar, N. 20
Guo, S. 322
Hamoui, A. A. 330
Hanson, S. 338, 363
Harlow III, J. E. 174
Hartenstein, R. 375
He, L. 156, 168, 326
Heidel, D. 85
Helms, D. 220
Henzler, S. 286
Heydari, P. 274
Hill, E. L. 31
Ho, H.-H. 388
Holzhauer, D. 191
Hsu, C.-F. 89
Hsu, J. 180
Hu, F. 232
Hu, Y. 168
Huang, H.-W. 388
Huang, M. 268
Ismail, Y. 79
Itoh, K. 123
Jacob, B. 25
Jafari, R. 207
Jaffari, J. 119
Jiang, S. 412
Jin, J. 406

Jin, L. 147
John, L. K. 350
Joseph, R. 262
Kandemir, M. 354
Kandemir, M. 358
Kansal, A. 180
Karnik, T. 156, 326
Katsavounidis, I. X. 107
Kawaguchi, H. 61
Keane, J. 127, 382
Khellah, M. 79
Kil, J. 67
Kim, J.-J. 14
Kim, C. H. 67
Kim, T.-H. 127
Kim, C. 127, 382
Kim, Y. 310
King, K.-J. 143
Kleeburg, T. 20
Koc, H. 358
Koeppe, S. 286
Kondo, M. 43
Koziri, M. G. 107
Kuo, S.-Y. 388
Kutter, C. 1
Kwong, J. 8
Lahiri, K. 342
Lee, H. 101, 322
Lee, W. 316
Leibson, S. 375
Li, H.-L. 418
Lilja, D. J. 298
Lin, C.-H. 143
Lin, Y. 168
Lipasti, M. H. 31
Liu, B. 135
Liu, Y. 238
Loghi, M. 346
Long, C. 326
Lu, P.-F. 85
Lu, Y.-H. 186
Luo, R. 238
Ma, D. 394, 400
Macii, A. 162

Macii, E. ... 162	Ranganathan, N. ... 174	Tschanz, J. ... 79
Magklis, G. ... 49	Raghunathan, V. ... 168, 180, 369	Tseng, H.-W. ... 418
Mahalingam, V. ... 174	Rao, R. ... 292	Tsui, C.-Y. ... 406
Martonosi, M. ... 304	Raychowdhury, A. ... 2	Varatkar, G. V. ... 113
Massey, T. ... 207	Reddy, S. ... 326	Veneris, A. ... 250
Meng, K. ... 262	Reimer, A. ... 151	Verma, N. ... 366
Meyr, H. ... 375	Rizo-Morente, J. ... 95	Vozqua, P. ... 394
Miyakoshi, J. ... 61	Robertazzi, R. ... 85	Vrudhula, S. ... 292, 424
Morita, Y. ... 61	Rodriguez, S. ... 25	Wang, A. ... 366
Moshovos, A. ... 250	Roy, K. ... 2, 14	Wang, H. ... 238
Mudge, T. ... 101	Sachdev, M. ... 256	Wang, J. ... 394
Murachi, Y. ... 61	Safi, E. ... 250	Wang, L. ... 139
Nahapetan, A. ... 207	Sapatnekar, S. S. ... 298	Wang, S. ... 139
Nakamura, H. ... 43	Sarrafzadeh, M. ... 207	Wang, X. ... 363
Narayanan, S. H. K. ... 358	Sasaki, H. ... 43	Wang, Y. ... 238
Nassif, S. R. ... 203	Sathanur, A. ... 162	Wintermayr, P. ... 375
Nebel, W. ... 151, 220	Schulz, A. ... 151	Woltgens, P. ... 85
Nii, K. ... 61	Sekiguchi, T. ... 123	Wong, N. ... 226
Nookala, V. ... 298	Senger, R. ... 73	Wu, Q. ... 155, 191
Nowka, K. ... 73	Shah, M. ... 376	Yang, C.-L. ... 143, 418
Nyathi, J. ... 131	Shameli, A. ... 274	Yang, H. ... 238
Ozturk, O. ... 358	Shanbhag, N. R. ... 113	Ye, Y. ... 135
Paek, Y. ... 310	Sharifkhani, M. ... 256	Yoshimoto, M. ... 61
Pamarti, S. ... 326	Sharkey, J. ... 37	Yu, B. ... 214
Pangrle, B. ... 362	Shi, Y. ... 156	Yu, H. ... 156
Park, I. ... 310	Sigal, L. ... 85	Zahedi, S. ... 180
Patel, K. ... 316	Simjee, F. ... 197	Zeng, H. ... 244
Pedarm, M. ... 334, 316	Sithambaram, P. ... 162	Zhai, B. ... 363
Piñuel, L. ... 268	Son, S. W. ... 354	Zhang, J. ... 135
Poncino, M. ... 162, 346	Srivastava, M. ... 180	Zhang, X. ... 412
Ponomarev, D. ... 37	Stamoulis, G. I. ... 107	Zhu, YK. ... 55
Prieto, M. ... 268	Sylvester, D. ... 73, 338, 363	Zhuo, J. ... 424
Qiu, Q. ... 191	Taherzadeh-Sani, M. ... 330	
Raghunathan, A. ... 342	Takemura, R. ... 123	

IEEE Catalog Number: 06TH8925
ISBN: 1-59593-462-6

9781595934628